Lecture Notes in Computer Science 5687

Commenced Publication in 1973
Founding and Former Series Editors:
Gerhard Goos, Juris Hartmanis, and Jan van Leeuwen

T0189833

Irit Dinur   Klaus Jansen
Joseph Naor   José Rolim (Eds.)

Approximation, Randomization, and Combinatorial Optimization

Algorithms and Techniques

12th International Workshop, APPROX 2009
and 13th International Workshop, RANDOM 2009
Berkeley, CA, USA, August 21-23, 2009
Proceedings

Volume Editors

Irit Dinur
The Weizmann Institute of Science
Dept. of Applied Math and Computer Science
Rehovot, 76100, Israel
E-mail: irit.dinur@weizmann.ac.il

Klaus Jansen
University of Kiel
Institute for Computer Science and Applied Mathematics
Olshausenstr. 40, 24098 Kiel, Germany
E-mail: kj@informatik.uni-kiel.de

Joseph (Seffi) Naor
Technion, Computer Science Department
Haifa, 32000, Israel
E-mail: naor@cs.technion.ac.il

José Rolim
University of Geneva
Centre Universitaire d'Informatique
Battelle Bat. A, 7 rte de Drize, 1227 Carouge, Switzerland
E-mail: jose.rolim@unige.ch

Library of Congress Control Number: 2009932147

CR Subject Classification (1998): F.2, G.2, G.1, G.3, G.4, I.1, I.2.2

LNCS Sublibrary: SL 1 – Theoretical Computer Science and General Issues

ISSN   0302-9743
ISBN-10   3-642-03684-8 Springer Berlin Heidelberg New York
ISBN-13   978-3-642-03684-2 Springer Berlin Heidelberg New York

springer.com

© Springer-Verlag Berlin Heidelberg 2009
Printed in Germany

Typesetting: Camera-ready by author, data conversion by Scientific Publishing Services, Chennai, India
Printed on acid-free paper   SPIN: 12734290   06/3180   5 4 3 2 1 0

Preface

This volume contains the papers presented at the 12th International Workshop on Approximation Algorithms for Combinatorial Optimization Problems (APPROX 2009) and the 13th International Workshop on Randomization and Computation (RANDOM 2009), which took place concurrently at the HP Auditorium in UC Berkeley, USA, during August 21–23, 2009. APPROX focuses on algorithmic and complexity issues surrounding the development of efficient approximate solutions to computationally difficult problems, and was the 12th in the series after Aalborg (1998), Berkeley (1999), Saarbrücken (2000), Berkeley (2001), Rome (2002), Princeton (2003), Cambridge (2004), Berkeley (2005), Barcelona (2006), Princeton (2007), and Boston (2008). RANDOM is concerned with applications of randomness to computational and combinatorial problems, and was the 13th workshop in the series following Bologna (1997), Barcelona (1998), Berkeley (1999), Geneva (2000), Berkeley (2001), Harvard (2002), Princeton (2003), Cambridge (2004), Berkeley (2005), Barcelona (2006), Princeton (2007), and Boston (2008).

Topics of interest for APPROX and RANDOM are: design and analysis of approximation algorithms, hardness of approximation, small space algorithms, sub-linear time algorithms, streaming algorithms, embeddings and metric space methods, mathematical programming methods, combinatorial problems in graphs and networks, game theory, markets, and economic applications, geometric problems, packing, covering, scheduling, approximate learning, design and analysis of online algorithms, randomized complexity theory, pseudorandomness and derandomization, random combinatorial structures, random walks/Markov chains, expander graphs and randomness extractors, probabilistic proof systems, error-correcting codes, average-case analysis, property testing, computational learning theory, and other applications of approximation and randomness.

The volume contains 25 contributed papers, selected by the APPROX Program Committee out of 56 submissions, and 28 contributed papers, selected by the RANDOM Program Committee out of 57 submissions.

We would like to thank all of the authors who submitted papers and the members of the Program Committees:

APPROX 2009

Nikhil Bansal	IBM T. J. Watson Research Center
Ziv Bar-Yossef	Google
Artur Czumaj	University of Warwick
Michel Goemans	MIT
Sudipto Guha	University of Pennsylvania
Magnus Halldorsson	Reykjavik University
Dorit Hochbaum	University of California, Berkeley

Elias Koutsoupias   University of Athens
Robert Krauthgamer   Weizmann Institute of Science
Ravi Kumar   Yahoo! Research
Lap Chi Lau   Chinese University of Hong Kong
Joseph (Seffi) Naor   Technion - Israel Institute of Technology
                           (Chair)
Tim Roughgarden   Stanford University
Bruce Shepherd   McGill University
Tami Tamir   The Interdisciplinary Center, Herzliya

RANDOM 2009

Irit Dinur   Weizmann Institute of Science (Chair)
Vitaly Feldman   IBM Almaden Research Center
Parikshit Gopalan   Microsoft Research
Danny Gutfreund   MIT
Prahladh Harsha   University of Texas, Austin
Avinatan Hassidim   MIT
Russel Impagliazzo   University of California, San Diego
Mark Jerrum   University of Edinburgh
Tali Kaufman   MIT
Subhash Khot   New York University
J. Radhakrishnan   Tata Institute of Fundamental Research
Dana Randall   Georgia Institute of Technology
Michael Saks   Rutgers University
Adi Shraibman   Weizmann Institute of Science
Emanuele Viola   Northeastern University

We would also like to thank the external subreferees: Dimitris Achlioptas, Adi Akavia, Andris Ambainis, Alexandr Andoni, Eli Ben-Sasson, Nayantara Bhatnagar, Arnab Bhattacharya Andrej Bogdanov, Niv Buchbinder, Arkadev Chattopadhyay, Bernard Chazelle, Kai-Min Chung, Amin Coja-Oghlan, Artur Czumaj, Amit Deshpande, Robert Elsasser, Joseph Emerson, Funda Ergun, Vitaly Feldman, Elena Grigorescu, Moritz Hardt, Jason Hartline, Tom Hayes, Elad Hazan, Rahul Jain, T.S. Jayram, Adam Kalai, Swastik Kopparty, Robert Krauthgamer, Oded Lachish, Homin Lee, Troy Lee, Yury Lifshits, Shachar Lovett, Aleksander Madry, Arie Matsliah, Or Meir, Manor Mendel, Sarah Miracle, Michael Mitzenmacher, Michael Molloy, Dana Moshkovitz, Elchanan Mossel, Hariharan Narayanan, Jelani Nelson, Marc Noy, Ryan O'Donnell, Krzysztof Onak, Amanda Pascoe, Seth Pettie, Prasad Raghavendra, Sofya Raskhodnikova, Ran Raz, Omer Reingold, Atri Rudra, Alex Samorodnitsky, Shubhangi Saraf, Nitin Saxena, Pranab Sen, Rocco Servedio, C. Seshadhri, Devavrat Shah, Asaf Shapira, Mohit Singh, Sasha Sodin, Daniel Spielman, Srikanth Srinivasan, Salil Vadhan, Rakesh Venkat, Elad Verbin, Eric Vigoda, Danny Vilenchik, Andrew Wan, Enav Weinreb, Udi Wieder, and Yi Wu.

We gratefully acknowledge the support from the Deptartment of Computer Science at the Technion in Israel, the Deptartment of Computer Science and Applied Mathematics of the Weizmann Institute in Israel, the Institute of Computer Science of the Christian-Albrechts-Universität zu Kiel and the Department of Computer Science of the University of Geneva.

The invited talk this year was dedicated to the memory of Rajeev Motwani from Stanford University who died in tragic circumstances on June 5, 2009. Throughout his career Rajeev made fundamental contributions to many areas of computer science including foundations, search and retrieval, databases, privacy, robotics, and more. Some of the most striking contributions were algorithmic in nature, in many branches of the field. The talk was given by Prabhakar Raghavan who is the head of Yahoo! Research and was a close friend and collaborator of Rajeev.

Finally, many thanks to Parvaneh Karimi-Massouleh for editing the proceedings.

August 2009   Irit Dinur
                                                               Klaus Jansen
                                                         Joseph (Seffi) Naor
                                                            José D.P. Rolim

Table of Contents

Contributed Talks of APPROX

Contributed Talks of RANDOM

Erratum

Approximation Algorithms and Hardness Results for Packing Element-Disjoint Steiner Trees in Planar Graphs

Ashkan Aazami[1], Joseph Cheriyan[1], and Krishnam Raju Jampani[2]

[1] Dept. of Comb. & Opt., U. Waterloo, Waterloo ON Canada N2L 3G1
aaazami@uwaterloo.ca, jcheriyan@uwaterloo.ca
[2] Dept. of Comp. Sci., U. Waterloo, Waterloo ON Canada N2L 3G1
krjampani@uwaterloo.ca

Abstract. We study the problem of packing element-disjoint Steiner trees in graphs. We are given a graph and a designated subset of terminal nodes, and the goal is to find a maximum cardinality set of element-disjoint trees such that each tree contains every terminal node. An *element* means a non-terminal node or an edge. (Thus, each non-terminal node and each edge must be in at most one of the trees.) We show that the problem is APX-hard when there are only three terminal nodes, thus answering an open question.

Our main focus is on the special case when the graph is planar. We show that the problem of finding two element-disjoint Steiner trees in a planar graph is NP-hard. We design an algorithm for planar graphs that achieves an approximation guarantee close to 2. In fact, given a planar graph that is k element-connected on the terminals (k is an upper bound on the number of element-disjoint Steiner trees), the algorithm returns $\lfloor \frac{k}{2} \rfloor - 1$ element-disjoint Steiner trees. Using this algorithm, we get an approximation algorithm for the edge-disjoint version of the problem on planar graphs that improves on the previous approximation guarantees. We also show that the natural LP relaxation of the planar problem has an integrality ratio approaching 2.

1 Introduction

In the STEINER TREE PACKING problem we are given an (undirected) graph $G = (V, E)$ and a subset of nodes $R \subseteq V$; each node in R is called a terminal node, and each node in $V - R$ is called a *Steiner* node or a non-terminal node. A Steiner node or an edge is called an *element*. A tree that contains all terminal nodes in R is called an R-Steiner tree (or Steiner tree, for short). The goal is to find a set of element-disjoint R-Steiner trees of maximum cardinality; that is, find as many R-Steiner trees as possible such that each Steiner node and each edge is in at most one of the trees. Our main focus is on approximation algorithms and hardness results for this problem. There is a closely related problem that we

I. Dinur et al. (Eds.): APPROX and RANDOM 2009, LNCS 5687, pp. 1–14, 2009.

call the EDGE-DISJOINT STEINER TREE PACKING problem; here, the goal is to find a set of edge-disjoint R-Steiner trees of maximum cardinality; that is, find as many R-Steiner trees as possible such that each edge is in at most one of the trees.

1.1 Previous Literature

Consider the special case of the STEINER TREE PACKING problem where all of the nodes are terminal nodes (i.e., $R = V$). Then the problem is the same as finding a maximum-cardinality set of edge-disjoint spanning trees. Tutte [26] and Nash-Williams [22] independently proved the following min-max theorem for this special case: An undirected graph G has k edge-disjoint spanning trees if and only if for any partition $\mathcal{P}$ of V into $|\mathcal{P}|$ non-empty subsets we have $e(\mathcal{P}) \geq k(|\mathcal{P}| - 1)$, where $e(\mathcal{P})$ is the number of edges in G with end-nodes in different sets of $\mathcal{P}$. Frank, Kiraly and Kriesell [7] extended this result to hypergraphs via the notion of partition-connectivity (see Section 2 for details): A hypergraph $\mathcal{H}$ decomposes into k hyperedge-disjoint partition-connected hypergraphs if and only if the partition-connectivity of $\mathcal{H}$ is at least k.

We say that the set of terminals R is k-element connected if there exist k element-disjoint paths between every pair of nodes in R; that is, for any two nodes $s, t \in R$, there exist k paths between s and t such that each element occurs in at most one of these k paths. Similarly, we say that the set of terminals R is k-edge connected if there exist k edge-disjoint paths between every pair of nodes in R. We use n to denote the number of nodes in the input graph. Also, we call the terminal nodes *black* nodes, and the non-terminal nodes *white* nodes. An edge between two white nodes is called a *white* edge.

Kaski [16] proved that the problem of finding two edge-disjoint Steiner trees is NP-hard, and also showed that the EDGE-DISJOINT STEINER TREE PACKING problem is NP-hard even with 7 terminals. The problem was proved to be APX-hard even with 4 terminals in [3]. Jain, Mahdian and Salavatipour [15] presented an approximation algorithm with a guarantee of $O(|R|)$. Later, Lau [19,20], using the result of Frank et al. [7], proved that if the terminals are $24k$-edge connected, then there exist k edge-disjoint Steiner trees, and he gave an approximation algorithm with a guarantee of 24.

Cheriyan and Salavatipour [4] studied the element-disjoint STEINER TREE PACKING problem; they observed that the problem is hard to approximate within a factor of $\Omega(\log n)$, and they designed a randomized approximation algorithm with a guarantee of $O(\log n)$. Subsequently, Calinescu, Chekuri and Vondrak [1] designed a simpler algorithm with a similar approximation guarantee, and also, they derandomized their algorithm.

To the best of our knowledge, the systematic study of problems of this type was started by Grötschel et al., see [10,8,11,9,12]. They were motivated by applications in VLSI circuit design, see [12,21]. They focused on a generalization of the EDGE-DISJOINT STEINER TREE PACKING problem, where we are given a list of terminal sets, $R_1, R_2, R_3, \ldots, R_q$ and the goal is to find edge-disjoint

Steiner trees $T_1, T_2, T_3, \ldots, T_q$ such that T_i contains (and connects) all the terminal nodes in R_i, for $i = 1, \ldots, q$. Their problem is quite different from the problems of interest to us, and is more general; for example, their problem contains the EDGE-DISJOINT PATHS problem as a special case, namely, the special case where each terminal set R_i has size two. Consequently, any hardness result that applies to the EDGE-DISJOINT PATHS problem applies also to the generalization of the EDGE-DISJOINT STEINER TREE PACKING problem of Grötschel et al., but those hardness results may not apply to the problems of interest to us. Further results and applications of the generalized problem are discussed by Wagner [27] and Korte et al., [17], also see Naves and Sebő [23], but note that the NP-hardness results in [17] (where different Steiner trees have different terminal sets) do not apply to the problems of interest to us. The main focus of the work on the generalized problem was to obtain computational procedures for finding an optimal solution, based on mathematical programming. Some algorithmic results on the generalized problem are presented by Wagner [27], but those results are "disjoint" from our results.

The generalized problem has other well-known applications including multicasting in wireless networks [6], and broadcasting large data streams, such as videos, over the Internet [15].

Chekuri and Korula [2] recently obtained some related results, including a 5-approximation algorithm for the element-disjoint Steiner Forest Packing problem on planar graphs. The two papers are independent of each other.

1.2 Results in This Paper

Our focus is on approximation algorithms and hardness results for the element-disjoint STEINER TREE PACKING problem on planar graphs. We call this the PLANAR STEINER TREE PACKING problem. Our main results are as follows:

- In Section 2, we present an approximation algorithm with a guarantee of (almost) 2 for the PLANAR STEINER TREE PACKING problem; more precisely, given a planar graph and a set of terminal nodes R such that R is k-element connected, our algorithm finds at least $\max(1, \lfloor \frac{k}{2} \rfloor - 1)$ element-disjoint Steiner trees; here, k is a positive integer. Based on this, we get an approximation algorithm with a guarantee of (almost) 4 for the edge-disjoint version of the problem on planar graphs. To the best of our knowledge, this improves on the known approximation guarantees for the EDGE-DISJOINT STEINER TREE PACKING problem on planar graphs. The planarity of the graph is used at only one point in our analysis, and there we use the upperbound on the number of edges in a planar bipartite simple graph. Our methods extend to larger classes of graphs, namely, graphs that exclude a fixed minor, to give approximation guarantees that depend on the order of the forbidden minor.

  We conjecture that a planar graph that is k-element connected on the terminals has at least $\lfloor \frac{k}{2} \rfloor$ element-disjoint Steiner trees.

- In Section 3, we prove that the STEINER TREE PACKING problem is APX-hard even with three terminals (i.e., $|R| = 3$). This answers an open question in the literature, see Floréen, et al. [6, Page 119].

  Then, we show that the problem of finding two element-disjoint Steiner trees in a planar graph is NP-hard. An immediate implication is that one cannot improve on the approximation guarantee of 2 for the PLANAR STEINER TREE PACKING problem without further assumptions.

- In Section 4, we show that even on planar graphs the standard LP (linear programming) relaxation of the element-disjoint STEINER TREE PACKING problem has an integrality ratio $\geq 2 - \frac{2}{|R|} - \epsilon$, where the additive term ϵ is a function of $|R|$ and the element-connectivity of the terminals, k, and for fixed $|R|$, $\epsilon \to 0$ as $k \to \infty$. Our approximation guarantee of (almost) 2 for planar graphs (mentioned above) implies that the integrality ratio on planar graphs approaches 2 as $k \to \infty$.

  The significance of our lower bound on the integrality ratio comes from the fact that the optimal value of this LP relaxation gives the best upper bound known (as far as we know) on the maximum number of element-disjoint Steiner trees. Thus, for planar graphs, our result shows that the approximation guarantee of 2 cannot be improved by any algorithm or analysis that relies on an upper bound that is dominated by the LP bound.

  Moreover, we modify our construction to get a similar lower bound on the integrality ratio for the edge-disjoint version of the problem on planar graphs.

2   Approximation Algorithms

2.1   Element-Disjoint Steiner Trees

We present an approximation algorithm for packing element-disjoint Steiner trees in planar graphs that achieves an approximation guarantee close to 2 (details below). Our method consists of two steps. First, we transform to a planar bipartite graph, while preserving the terminals and their element-connectivity. Then, we view the bipartite graph as a hypergraph, and apply a method of Frank et al. [7] to decompose the set of hyperedges $\mathcal{E}$ into a number of disjoint sets $\mathcal{E}_1, \mathcal{E}_2, \ldots$ such that each set $\mathcal{E}_i$ induces a Steiner tree of our bipartite graph. Each of these "bipartite" Steiner trees transforms back to a Steiner tree of the original graph. The planarity of the graph is used at only one point in our analysis, and there we use the upperbound on the number of edges in a planar bipartite simple graph. Our methods extend to larger classes of graphs, namely, graphs that exclude a fixed minor, to give approximation guarantees that depend on the order of the forbidden minor.

The following theorem is the main result of this section.

Theorem 1. *Let $G = (V, E)$ be an undirected planar graph, let $R \subseteq V$ be the set of terminals, and assume that R is k-element connected. Then there are at least $\lfloor \frac{k}{2} \rfloor - 1$ element-disjoint Steiner trees in G. Moreover, there is an algorithm with a running time of $O(|V|^{4.5})$ that finds at least $\lfloor \frac{k}{2} \rfloor - 1$ element-disjoint Steiner trees in G.*

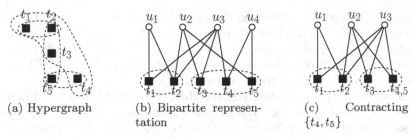

(a) Hypergraph   (b) Bipartite represen-   (c) Contracting
                        tation $\{t_4, t_5\}$

Fig. 1. A Hypergraph and its bipartite representation

We define the BIPARTITE STEINER TREE PACKING problem to be a subproblem of the element-disjoint STEINER TREE PACKING problem such that the graph is bipartite, all terminal nodes are in one part of the bipartition, and all Steiner nodes are in the other part. Consider a planar instance of the element-disjoint STEINER TREE PACKING problem, i.e., the associated graph is planar. We can transform it into a planar instance of BIPARTITE STEINER TREE PACKING by using the following theorem. The theorem is due to Hind and Oellermann, see [14], and a short proof is given in [4].

Theorem 2. *[14] Consider a graph $G = (V, E)$ that has a set of terminals R such that R is k-element connected. There is a polynomial-time algorithm that repeatedly deletes or contracts white edges to obtain a bipartite graph G' from G such that R stays k-element connected, and moreover, R forms one part of the bipartition of G'.*

A hypergraph is a pair $\mathcal{H} = (V, \mathcal{E})$ where V is the node-set of $\mathcal{H}$ and $\mathcal{E}$ is a collection of non-empty subsets of V. A subset $Z \in \mathcal{E}$ is called a *hyperedge* of $\mathcal{H}$. Given a partition $\mathcal{P} = \{V_1, \ldots, V_t\}$ of V into non-empty subsets, a hyperedge $Z \in \mathcal{E}$ is called a *crossing* hyperedge if it intersects at least two subsets of $\mathcal{P}$ and otherwise it is called an *internal* hyperedge. We use $|\mathcal{P}|$ to denote the number of sets V_i in $\mathcal{P}$, and we denote the number of crossing hyperedges corresponding to the partition $\mathcal{P}$ by $e_{\mathcal{H}}(\mathcal{P})$ (or simply, by $e(\mathcal{P})$). Given a hypergraph $\mathcal{H} = (V, \mathcal{E})$, we associate a bipartite graph $G_{\mathcal{H}} = (V, U; E)$ to $\mathcal{H}$ as follows. Corresponding to each hyperedge $Z \in \mathcal{E}$ we have a node $u_Z \in U$. A node $v \in V$ is adjacent to $u_Z \in U$ if $v \in Z$; note that the degree of u_Z in $G_{\mathcal{H}}$ is the size of Z.

Consider the hypergraph $\mathcal{H}$ shown in Figure 1(a). The node-set of $\mathcal{H}$ is $V = \{t_1, t_2, t_3, t_4, t_5\}$, and the hyperedges of $\mathcal{H}$ are $Z_1 = \{t_1, t_2\}$, $Z_2 = \{t_2, t_3, t_5\}$, $Z_3 = \{t_1, t_2, t_3, t_4\}$ and $Z_4 = \{t_4, t_5\}$. Figure 1(b) shows the bipartite graph, $G = (V, U; E)$, associated with $\mathcal{H}$. Consider the partition $\mathcal{P} = \{\{t_1, t_2\}, \{t_3, t_4, t_5\}\}$ of V; this partition is shown in dashed lines in Figure 1(b). The hyperedges Z_2, Z_3 are crossing hyperedges w.r.t. (with respect to) $\mathcal{P}$, and hyperedges Z_1, Z_4 are internal hyperedges w.r.t. $\mathcal{P}$. Thus, $e(\mathcal{P}) = 2$, since there are two crossing hyperedges in $\mathcal{P}$. Given a partition $\mathcal{P}$, a useful operation is to *contract* an internal hyperedge: we identify all nodes in Z into a single node and remove Z from the hypergraph. For example, Figure 1(c) shows the bipartite representation of the hypergraph obtained by contracting the internal hyperedge $Z_4 = \{t_4, t_5\}$. If we further contract $Z_1 = \{t_1, t_2\}$ we get a copy of $K_{2,3}$. If we contract some internal hyperedges (w.r.t. $\mathcal{P}$) of $\mathcal{H}$, then we obtain a "shrunk" hypergraph $\mathcal{H}'$ and a partition $\mathcal{P}'$ of $V(\mathcal{H}')$; note that the crossing hyperedges of $\mathcal{H}$ (w.r.t. $\mathcal{P}$) are the same as the crossing hyperedges of $\mathcal{H}'$ (w.r.t. $\mathcal{P}'$).

Let $G = (R, U; E)$ be an instance of the BIPARTITE STEINER TREE PACKING problem, where R is the set of terminal nodes and U is the set of Steiner nodes. We associate a hypergraph $\mathcal{H}_G = (R, \mathcal{E})$ to G as follows. The terminal nodes of G are the nodes in $\mathcal{H}_G$, and corresponding to each Steiner node $u \in U$ we have a hyperedge Z_u that contains the set of neighbors of u in G. Also, given any hypergraph $\mathcal{H}$, we may view its associated graph $G_{\mathcal{H}}$ as an instance of the BIPARTITE STEINER TREE PACKING problem.

A hypergraph $\mathcal{H}$ is k-partition connected if $e_{\mathcal{H}}(\mathcal{P}) \geq k(|\mathcal{P}| - 1)$ for every partition $\mathcal{P}$ of V. A 1-partition connected hypergraph is simply called partition-connected. If a hypergraph $\mathcal{H}$ is partition-connected, then it is easy to see that the associated bipartite graph $G_{\mathcal{H}}$ is connected, and so it contains a Steiner tree (with terminal set $V(\mathcal{H})$). But the converse does not hold: for a connected instance of BIPARTITE STEINER TREE PACKING, the associated hypergraph may not be partition-connected. Frank et al. [7] proved the following generalization of the Tutte–Nash-Williams theorem.

Theorem 3 (Theorem 2.8 in [7]). *A hypergraph $\mathcal{H} = (V, \mathcal{E})$ is k-partition connected if and only if $\mathcal{E}$ partitions into k subsets $\mathcal{E}_1, \ldots, \mathcal{E}_k$ such that each of the sub-hypergraphs $\mathcal{H}_i = (V, \mathcal{E}_i)$ is partition-connected.*

Therefore, we can obtain ℓ element-disjoint Steiner trees in G if $\mathcal{H}_G$ is ℓ-partition connected. Now we prove the following lemma that completes the proof of Theorem 1.

Lemma 1. *Let $G = (R, U; E)$ be a bipartite planar graph such that R is k-element connected. Then the hypergraph $\mathcal{H}_G = (R, \mathcal{E})$ associated with G is $\lfloor \frac{k-2}{2} \rfloor$-partition connected.*

Proof. We may assume that G is connected. Consider the hypergraph $\mathcal{H}$ and define the *fractional* partition-connectivity, λ^*, as follows:

$$\lambda^* = \min_{\mathcal{P}} \frac{e(\mathcal{P})}{|\mathcal{P}| - 1}, \tag{1}$$

where the minimum is over all partitions $\mathcal{P}$ of R with $|\mathcal{P}| \geq 2$. Let λ denote the partition-connectivity of $\mathcal{H}$. It follows from the definition of partition-connectivity that $\lambda = \lfloor \lambda^* \rfloor$. Let $\mathcal{P}^* = \{X_1, X_2, \ldots, X_\ell\}$ be a partition that achieves the minimum ratio λ^*. In the rest of the proof, except where mentioned otherwise, crossing hyperedges and internal hyperedges are w.r.t. $\mathcal{P}^*$.

Consider the Steiner nodes of G that correspond to the internal hyperedges. We contract all the edges of G that are incident to these Steiner nodes, and we call the resulting graph G'. In more detail, consider each internal hyperedge $Z_u \in \mathcal{E}$ and contract all edges in G adjacent to the Steiner node u corresponding to hyperedge Z_u. We may ignore all parallel edges in G' formed by these edge contractions.

Claim. The obtained graph G' is a bipartite planar graph and has the following properties: 1) All of the remaining Steiner nodes in G' correspond to crossing hyperedges in $\mathcal{H}$, and they form one part of the bipartition 2) The other part of the bipartition has $|\mathcal{P}^*|$ nodes, and each node has degree at least k.

Proving this completes the lemma. This follows because G' has at least $k|\mathcal{P}^*|$ edges and at most $2(e(\mathcal{P}^*) + |\mathcal{P}^*|) - 4$ edges since it is a bipartite planar graph. Hence, we have

$$k|\mathcal{P}^*| \leq 2(e(\mathcal{P}^*) + |\mathcal{P}^*|) - 4 \Longrightarrow e(\mathcal{P}^*) \geq \frac{(k-2)|\mathcal{P}^*|}{2} + 2 \Longrightarrow \lambda^* > \frac{k-2}{2}.$$

Proof of the above claim: Consider a set $X_i \in \mathcal{P}^*$ of size at least 2 and arbitrarily partition it into two non-empty sets X_i' and X_i'', and let $\mathcal{P}'$ be the obtained partition. Since $\mathcal{P}^*$ is the minimum ratio partition, we have $\lambda' = \frac{e(\mathcal{P}')}{|\mathcal{P}'|-1} \geq \lambda^*$. Hence, $e(\mathcal{P}') \geq \lambda^*(|\mathcal{P}'| - 1) > \lambda^*(|\mathcal{P}^*| - 1) = e(\mathcal{P}^*)$. Hence, there exists a hyperedge that is crossing w.r.t. $\mathcal{P}'$ but is not crossing w.r.t. $\mathcal{P}^*$; that is, one of the internal hyperedges w.r.t. $\mathcal{P}^*$ intersects both X_i' and X_i''. This reasoning applies to each set $X_i \in \mathcal{P}^*$ and to each 2-partition X_i', X_i'' of X_i; hence, for each $X_i \in \mathcal{P}^*$, the subgraph of G induced by X_i and the Steiner nodes corresponding to the hyperedges internal to X_i is connected. Thus, contracting all edges in G adjacent to the Steiner nodes corresponding to the internal hyperedges (w.r.t. $\mathcal{P}^*$) will shrink each set X_i of $\mathcal{P}^*$ into a single node. The obtained graph G' is planar, and it is easy to see that it is bipartite with all the Steiner nodes corresponding to the crossing hyperedges (w.r.t $\mathcal{P}^*$) in one part of the partition and all of the "contracted" nodes in the other part. Now we prove that the degree of each contracted node is at least k using the fact that the terminals are k-element connected in G. To see this, consider a shrunk node v_i corresponding to a subset $X_i \in \mathcal{P}^*$, and assume that it has less than k neighbors in G'. Let Y' be the set of neighbors of v_i, so $|Y'| < k$. Note that Y' separates v_i from any other contracted node v_j in G', i.e., v_i and v_j are in different connected components of $G' \setminus Y'$. Now focus on the original hypergraph $\mathcal{H}$ and note that Y' (viewed as a subset of $\mathcal{E}(\mathcal{H})$) contains all hyperedges that intersect both X_i and $R \setminus X_i$; thus, in the original graph G, we see that Y' (viewed as a subset of U) separates X_i from the rest of the terminals, because Y' contains all Steiner nodes that are adjacent to both X_i and $R \setminus X_i$. This is a contradiction because the terminals are k-element connected in G; that is, for any set of white nodes Y whose deletion separates a pair of terminals, we must have $|Y| \geq k$. This shows that each contracted node has degree at least k in G'.   □

We have planar examples showing that the analysis in Lemma 1 is tight, but we are omitting these examples here.

Running time of the above algorithm: The algorithm has two steps. In the first step, we reduce the given graph $G = (V, E)$ to an instance G' of the BIPARTITE STEINER TREE PACKING problem using Theorem 2. In the second step, using results of Frank et al. [7] and Edmonds [5], we decompose the associated hypergraph $\mathcal{H}$ of G' into the maximum number of partition-connected sub-hypergraphs. The running time of the first step is $O(kn^2 |R|)$, and the second step can be implemented using the Matriod intersection algorithm in time $O(n^{4.5})$. Hence, the total running time of our algorithm on planar graphs is $O(n^{4.5})$.

2.2 Edge-Disjoint Steiner Trees

The above result extends to the packing of edge-disjoint Steiner trees in planar graphs, to give the following result. The proof is given in the full paper.

Theorem 4. *Let $G = (V, E)$ be an undirected planar graphs, let $R \subseteq V$ be the set of terminals, and assume that R is k-edge connected. Then there are at least $\lfloor \frac{k}{4} \rfloor - 1$ edge-disjoint Steiner trees in G. Moreover, there is an algorithm with a running time of $O(|V|^{4.5})$ that finds at least $\lfloor \frac{k}{4} \rfloor - 1$ edge-disjoint Steiner trees in G.*

2.3 Element-Disjoint Steiner Trees in H-Minor-Free Graphs

It is known that an H-minor-free graph G has at most $c_H \cdot |V(G)|$ edges [18,25], where $c_H = \frac{c}{2} |V(H)| \sqrt{\log_2 |V(H)|}$ for some constant $c \leq 324$. Our analysis for planar graphs extends to the H-minor-free graphs to give the following result.

Theorem 5. *Let H be a fixed graph. Let $G = (V, E)$ be an undirected graph that has no H minor, let $R \subseteq V$ be the set of terminals, and assume that R is k-element connected. Then there are at least $\left\lfloor \frac{k}{c_H} \right\rfloor - 1$ element-disjoint Steiner trees in G. Moreover, there is an algorithm with a running time of $O(n^{4.5} + k |R| c_H^2 n^2)$ that finds this number of element-disjoint Steiner trees in G.*

3 Hardness Results

This section has two main results. In the first subsection, we show that the edge-disjoint STEINER TREE PACKING problem with 3 terminal nodes is APX-hard; then we extend this to prove APX-hardness for the element-disjoint STEINER TREE PACKING problem on three terminal nodes. This settles an open question in the literature, see [6, Page 119 second column]. In the second subsection, we show that the problem of finding two element-disjoint Steiner trees in a planar graph is NP-hard.

3.1 APX-Hardness for General Graphs with 3 Terminal Nodes

In this subsection, we prove that the edge-disjoint STEINER TREE PACKING problem with 3 terminals is APX-hard. Our result is obtained by a reduction from the INTEGER2COMMODITY problem that is known to be APX-hard

[13, Corollary 4.1]. We also show that the element-disjoint STEINER TREE PACK-
ING problem with 3 terminals is APX-hard, by using a simple reduction from the
edge-disjoint version.

Theorem 6. *The edge-disjoint* STEINER TREE PACKING *problem with* 3 *terminals is* APX-*hard.*

We prove our result by a reduction from the INTEGER2COMMODITY problem,
which is as follows: We are given an undirected graph $G = (V, E)$ and distinct
nodes $x_1, y_1, x_2, y_2 \in V$; the goal is to find a maximum-size collection of edge-
disjoint paths, each joining either x_1 to y_1 or x_2 to y_2.

Theorem 7 ([13, Corollary 4.1]). *The* INTEGER2COMMODITY *problem is*
APX-*hard.*

In the hardness construction in the proof of the above theorem (see [13, Section 4.1.1]), the nodes x_1 and y_1 both have degree d_1, and the node y_2 has
degree d_2; so there are at most d_i edge-disjoint paths between x_i and y_i for each
$i \in \{1, 2\}$. In the "yes" instances of the problem the objective value is $d_1 + d_2$,
whereas in the "no" instance the objective value is at most $(d_1 + d_2)(1 - \epsilon)$, for
some $\epsilon > 0$. We denote this instance of the INTEGER2COMMODITY problem by
$\mathcal{I} = (G; x_1, y_1, d_1; x_2, y_2, d_2)$.

Reduction
Let $\mathcal{I} = (G; x_1, y_1, d_1; x_2, y_2, d_2)$ be an instance of the INTEGER2COMMODITY
problem.

1. Start from a copy of G and add 3 terminal nodes $\{t, t_1, t_2\}$ and two non-
   terminal nodes s_1, s_2 to G.
2. Add d_1 parallel edges from s_1 to each of t, t_2, x_1, and similarly we add d_2
   parallel edges from s_2 to each of t, t_1, x_2.
3. Finally, we add d_1 parallel edges from y_1 to t_1, and d_2 parallel edges from
   y_2 to t_2.
4. Let H be the obtained graph, and let $R = \{t, t_1, t_2\}$ (see Figure 2 for an
   illustration).

For the analysis of the above construction and the reduction to the element-
disjoint version of the problem refer to the full paper.

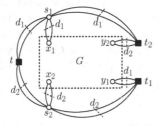

Fig. 2. Hardness construction

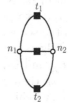

Fig. 3. Basic Gadget

3.2 NP-Hardness of Packing 2 Element-Disjoint Steiner Trees in Planar Graphs

Our NP-hardness proof is based on two previous results, namely, Kaski's proof [16] that the problem of finding two edge-disjoint Steiner trees in general graphs is NP-hard, and Plesník's proof [24] that the Hamiltonian cycle problem in planar digraphs with degree bound two is NP-hard.

A *Basic Gadget* or BG is a complete bipartite graph with 3-terminals and 2-Steiner nodes (see Figure 3). In any planar embedding of this graph, the outer face consists of two terminals and the two Steiner nodes. If H is a BG we use $H(t_1)$ and $H(t_2)$ to denote its terminals on the outer face and $H(n_1)$ and $H(n_2)$ to denote its Steiner nodes (also on the outer face). Note that any solution to the (planar) 2-element disjoint trees problem on H, contains $H(n_1)$ and $H(n_2)$ in different trees.

Reduction: Let $\mathcal{I} = Q_1 \wedge Q_2 \wedge \cdots Q_m$, be an instance of NAE-3SAT where clause $Q_j = P_{j_1} \vee P_{j_2} \vee P_{j_3}$, with literals $P_{j_k} \in \{x_1, \bar{x}_1, x_2, \bar{x}_2, \cdots, x_n, \bar{x}_n\}$. Given $\mathcal{I}$, we describe how to create an instance $G = (U \cup R, E)$ of the planar 2-element disjoint Steiner trees problem, such that $\mathcal{I}$ has two complimentary satisfying assignments if and only if G has two element-disjoint Steiner trees. We first describe the construction of a partial planar graph G_p along with its embedding, which would aid us in constructing G. We define G_p and its embedding as follows. (See Figure 4(a) for an example).

1. We add a sequence of "clause" BG's $C_1, C_2, \cdots, C_{3m}$ such that adjacent BG's share their outer terminals: i.e. for $i \in \{1, \cdots, 3m - 1\}$, $C_i(t_2) = C_{i+1}(t_1)$ and all nodes $C_i(n_2)$ are on the same side in the embedding (see Figure 4(a)).
2. For each clause Q_i, we add a terminal q_i and connect it to Steiner nodes $C_{3i-2}(n_1)$, $C_{3i-1}(n_1)$ and $C_{3i}(n_1)$.
3. We add a sequence of "literal" BG's $L_1, L_2, \cdots, L_{2n}$ such that adjacent BG's share their outer terminals: i.e for $i \in \{1, \cdots, 2n - 1\}$, $L_i(t_2) = L_{i+1}(t_1)$ and all nodes $L_i(n_1)$ are on the same side in the embedding.
4. For each pair of BG's L_{2i-1} and L_{2i} (where $1 \in \{1, \cdots, n\}$), we add a new terminal v_i and connect it to $L_{2i-1}(n_2)$ and $L_{2i}(n_2)$.
5. We add the edges $(C_1(t_1), L_1(t_1))$ and $(C_{3m}(t_2), C_{2n}(t_2))$.
6. Finally, we add certain constraints between the Steiner nodes of G_p called *switching lines*. A switching line (s_1, s_2) between Steiner nodes s_1 and s_2 ensures the two nodes are in different Steiner trees in any solution to G. Later in the section, we give a procedure to replace the switching lines with certain gadgets that "implement" them. For each clause $Q_i = P_{i_1} \vee P_{i_2} \vee P_{i_3}$, let L_{j_1}, L_{j_2} and L_{j_3} be the literal BG's corresponding to P_{i_1}, P_{i_2} and P_{i_3}. (e.g., If $P_{i_1} = x_k$ then $j_1 = 2k - 1$ and if $P_{i_1} = \bar{x}_k$, then $j_1 = 2k$). We add the following switching lines to G_p:

$$(C_{3i-2}(n_2), L_{j_1}(n_1)), (C_{3i-1}(n_2), L_{j_2}(n_1)), (C_{3i}(n_2), L_{j_3}(n_1)).$$

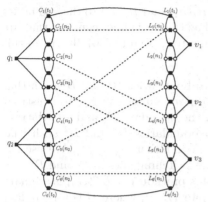 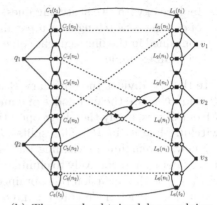

(a) Graph G_p for the instance $\mathcal{I} =$ (x₁ ∨ x̄₂ ∨ x₃) ∧ (x₁ ∨ x₂ ∨ x̄₃)

(b) The graph obtained by applying the Uncross operation on the switching line $(C_5(n_2), L_3(n_1))$

Fig. 4. Planar construction

7. **Embedding of G_p:** Let H be the graph G_p without the switching lines. We embed H in the plane such that the clause BG's are aligned vertically to the left, the literal BG's are aligned vertically to the right and the cycle $B_p = C_1(t_1), C_1(n_2), C_2(t_1), C_2(n_2), \cdots, C_{3m}(t_1), C_{3m}(n_2), C_{3m}(t_2), L_{2n}(t_2), L_{2n}(n_1), L_{2n}(t_1), L_{2n-1}(n_1), L_{2n-1}(t_1), \cdots, L_1(n_1), L_1(t_1)$ forms an (internal) face of H (see Figure 4(a)). We refer to B_p as the *boundary* of G_p. We define a boundary as a cycle whose interior contains no nodes or edges but may contain switching lines. Now, we represent each switching line of G_p with a straight (dashed) line joining its end nodes. Note that these line segments would all be present inside (the embedding of) B_p. Also the line segments may cross each other. But without loss of generality, we assume that no three switching lines cross at the same point.

We now describe the procedure for obtaining G from G_p. Given a boundary B and a switching line e in B, the following operation replaces e with a subgraph $\hat{S}_e$ and adjusts the interior of B, splitting it into two boundaries.

Uncross (B, e)

1. If e doesn't cross any other switching line, then define $\hat{S}_e$ to be a path of length two connecting the nodes of e and having a (new) terminal in the middle. Delete e and embed $\hat{S}_e$ along the straight line corresponding to e.
2. Otherwise let e crosses switching lines $e_1, e_2, \cdots, e_k$. In this case, define $\hat{S}_e$ as a sequence of BG gadgets $R_1, R_2, \cdots, R_k$, such that adjacent BG's share an outer terminal (i.e., $R_i(t_2) = R_{i+1}(t_1)$ for $i \in \{1, \cdots, k-1\}$). Now connect the terminal nodes $R_1(t_1)$ and $R_k(t_2)$ to the end nodes of e.

   Delete e and embed $\hat{S}_e$ such that all terminals in $\hat{S}_e$ lie along the straight line corresponding to e and for each R_i the Steiner nodes $R_i(n_1)$ and $R_i(n_2)$ lie along the straight line corresponding to e_i. Also, for each i ($\in \{1, \cdots, k\}$), replace e_i with two switching lines, from $R_i(n_1)$ and $R_i(n_2)$ to the end nodes of e_i. These switching lines are embedded as (disjoint) line segments that are contained in the line segment corresponding to e_i. Figure 4(b) illustrates this with an example.

Note that $\hat{S}_e$ divides the boundary B into two boundaries B^1 and B^2 such that B^1 and B^2 share the end nodes of e and the outer terminals of $\hat{S}_e$. To construct G from G_p, we apply the above operation on the boundary B_p and an arbitrary switching line e_p in B_p. This splits B_p into boundaries B_p^1 and B_p^2 such that any new or remaining switching line is present in either B_p^1 or B_p^2. We use the above operation recursively to eliminate all the switching lines in B_p^1 and B_p^2. Let $SLT(G_p)$ be the recursion tree obtained by this procedure. $SLT(G_p)$ is a binary tree in which each node is represented by a pair (B, e), where B is a boundary and e is a switching line in B. A node (B', e') is a child of (B, e) if boundary B' is one of the two boundaries obtained by applying the uncross operation at (B, e). If (B, e) is a leaf node then B doesn't contain any switching lines inside it, and we define e to be empty. We assign an integer number called the level to each pair (B, e) in the above construction. The level of the pair (B_p, e_p) is defined to be 0. If (B', e') is a child of a pair (B, e) at level i, then we define the level of (B', e') to be $i + 1$. Let h denote the maximum level over all pairs; i.e., h is the height of the recursion tree $SLT(G_p)$. In the construction of G, we assign a level to each of the following objects: terminal nodes, Steiner nodes, boundary faces, and switching lines. The objects in G_p (before applying any uncross operation) are defined to be at level 0. When we apply the uncross operation to a pair at level i in $SLT(G_p)$, we define the level of new objects (i.e., terminal nodes, Steiner nodes, switching lines, and the two new boundary faces) to be $i + 1$.

Our NP-hardness result follows from the next theorem (proved in the full paper).

Theorem 8. *The instance $\mathcal{I}$ of NAE-3SAT is satisfiable if and only if G has two element-disjoint Steiner trees.*

4   Integrality Ratio for Packing Steiner Trees in Planar Graphs

In this section, we show that the following "standard" linear programming relaxation of the element-disjoint STEINER TREE PACKING problem has integrality ratio approaching 2, even on planar graphs. For notational convenience, we assume there are no edges between terminals, by subdividing edges if needed.

$$\text{(LP-element)} \quad z_{LP}(G) = \max \sum_{T \in \mathcal{T}} x_T$$

$$\text{subject to} \sum_{T \in \mathcal{T}: v \in T} x_T \le 1 \qquad \forall v \in V \setminus R$$

$$x_T \ge 0 \qquad \forall T \in \mathcal{T}$$

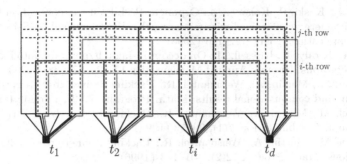

Fig. 5. Integrality ratio example for the element-disjoint problem

Construction: Start from a $2k \times 2kd$ grid and subdivide the alternate edges of the last row of the grid. Now add d terminal nodes $R = \{t_1, \ldots, t_d\}$ to the outer face of the grid. Next connect each terminal node t_i to k consecutive subdivided nodes (see Figure 5 for an illustration). Let G be the obtained graph. First, we prove that G has at most $\frac{kd}{2(d-1)} + d - 2$ element-disjoint Steiner trees. Next, we claim that LP-element has optimal value of $z_{LP} = k$. We give a sketch of our construction; the details can be found in the full paper. We construct k pairs of half-integral Steiner trees. Each pair is obtained from two consecutive rows by connecting each terminal to these two rows using two consecutive columns. Figure 5 shows two pairs of half-integral Steiner trees and shows how they cross each other. It is easy to check that these $2k$ Steiner trees form a feasible solution to LP-element. Thus, we have $z_{LP} \geq k$. This gives us the following theorem. The above construction extends to give the same lower bound on the integrality ratio for the *planar edge-disjoint* STEINER TREE PACKING problem.

Theorem 9. *The LP relaxation of the element-disjoint* STEINER TREE PACK-ING *problem has an integrality ratio* $\geq 2 - \frac{2}{|R|} - \epsilon$ *even on planar graphs, where the additive term ϵ is a function of k and $|R|$ and for fixed $|R|$, $\epsilon \to 0$ as $k \to \infty$ (here, k denotes the element-connectivity of the terminals).*

References

1. Calinescu, G., Chekuri, C., Vondrak, J.: Disjoint bases in a polymatroid. To appear in Random Structures and Algorithms
2. Chekuri, C., Korula, N.: A Graph Reduction Step Preserving Element-Connectivity and Applications. In: Albers, S., et al. (eds.) ICALP 2009. LNCS, vol. 5555, pp. 254–265. Springer, Heidelberg (2009)
3. Cheriyan, J., Salavatipour, M.R.: Hardness and approximation results for packing Steiner trees. Algorithmica 45(1), 21–43 (2006)
4. Cheriyan, J., Salavatipour, M.R.: Packing element-disjoint Steiner trees. ACM Transactions on Algorithms 3(4) (2007)
5. Edmonds, J.: Minimum partition of a matroid into independent subsets. Journal of Research National Bureau of Standards Section B 69, 67–72 (1965)

6. Floréen, P., Kaski, P., Kohonen, J., Orponen, P.: Lifetime maximization for multi-casting in energy-constrained wireless networks. IEEE Journal on Selected Areas in Communications 23(1), 117–126 (2005)
7. Frank, A., Király, T., Kriesell, M.: On decomposing a hypergraph into k connected sub-hypergraphs. Discrete Applied Mathematics 131(2), 373–383 (2003)
8. Grötschel, M., Martin, A., Weismantel, R.: Packing Steiner trees: a cutting plane algorithm and computational results. Math. Program. 72(2), 125–145 (1996)
9. Grötschel, M., Martin, A., Weismantel, R.: Packing Steiner trees: further facets. European J. Combinatorics 17(1), 39–52 (1996)
10. Grötschel, M., Martin, A., Weismantel, R.: Packing Steiner trees: polyhedral investigations. Math. Prog. A 72(2), 101–123 (1996)
11. Grotschel, M., Martin, A., Weismantel, R.: Packing Steiner trees: Separation algorithms. SIAM J. Discret. Math. 9(2), 233–257 (1996)
12. Grötschel, M., Martin, A., Weismantel, R.: The Steiner tree packing problem in VLSI design. Math. Program. 78(2), 265–281 (1997)
13. Guruswami, V., Khanna, S., Rajaraman, R., Shepherd, B., Yannakakis, M.: Near-optimal hardness results and approximation algorithms for edge-disjoint paths and related problems. J. Comput. Syst. Sci. 67(3), 473–496 (2003)
14. Hind, H.R., Oellermann, O.: Menger-type results for three or more vertices. Congressus Numerantium 113, 179–204 (1996)
15. Jain, K., Mahdian, M., Salavatipour, M.R.: Packing Steiner trees. In: SODA, pp. 266–274 (2003)
16. Kaski, P.: Packing Steiner trees with identical terminal sets. Inf. Process. Lett. 91(1), 1–5 (2004)
17. Korte, B., Prömel, H.J., Steger, A.: Steiner trees in VLSI-layout. In: Korte, B., Lovász, L., Prmel, H.J., Schrijver, A. (eds.) Paths, Flows, and VLSI-Layout, pp. 185–214. Springer, Berlin (1990)
18. Kostochka, A.V.: Lower bound of the Hadwiger number of graphs by their average degree. Combinatorica 4(4), 307–316 (1984)
19. Lau, L.C.: On approximate min-max theorems for graph connectivity problems. PhD thesis, University of Toronto (2006)
20. Lau, L.C.: An approximate max-Steiner-tree-packing min-Steiner-cut theorem. Combinatorica 27(1), 71–90 (2007)
21. Martin, A., Weismantel, R.: Packing paths and Steiner trees: Routing of electronic circuits. CWI Quarterly 6, 185–204 (1993)
22. Nash-Williams, C.S.J.A.: Edge-disjoint spanning trees of finite graphs. Journal of the London Mathemathical Society 36, 445–450 (1961)
23. Naves, G., Sebő, A.: Multiflow feasibility: an annotated tableau. In: Research Trends in Combinatorial Optimization, pp. 261–283. Springer, Heidelberg (2008)
24. Plesník, J.: The NP-completeness of the Hamiltonian cycle problem in planar digraphs with degree bound two. Inf. Process. Lett. 8(4), 199–201 (1979)
25. Thomason, A.: An extremal function for contractions of graphs. Math. Proc. Cambridge Philos. Soc. 95, 261–265 (1984)
26. Tutte, W.T.: On the problem of decomposing a graph into n connected factors. Journal of the London Mathemathical Society 36, 221–230 (1961)
27. Wagner, D.: Simple algorithms for Steiner trees and paths packing problems in planar graphs. CWI Quarterly 6(3), 219–240 (1993)

Adaptive Sampling for k-Means Clustering

Ankit Aggarwal[1], Amit Deshpande[2], and Ravi Kannan[2]

[1] IIT Delhi
zenithankit@gmail.com
[2] Microsoft Research India
{amitdesh,kannan}@microsoft.edu

Abstract. We show that *adaptively* sampled $O(k)$ centers give a constant factor bi-criteria approximation for the k-means problem, with a constant probability. Moreover, these $O(k)$ centers contain a subset of k centers which give a constant factor approximation, and can be found using LP-based techniques of Jain and Vazirani [JV01] and Charikar et al. [CGTS02]. Both these algorithms run in effectively $O(nkd)$ time and extend the $O(\log k)$-approximation achieved by the k-means++ algorithm of Arthur and Vassilvitskii [AV07].

1   Introduction

k-means is a popular objective function used for clustering problems in computer vision, machine learning and computational geometry. The k-means clustering problem on given n data points asks for a set of k centers that minimizes the sum of squared distances between each point and its nearest center. To write it formally, the k-means problem asks: Given a set $X \subseteq \mathbb{R}^d$ of n data points and an integer $k > 0$, find a set $C \subseteq \mathbb{R}^d$ of k centers that minimizes the following potential function.

$$\phi(C) = \sum_{x \in X} \min_{c \in C} \|x - c\|^2$$

We denote by $\phi_A(C) = \sum_{x \in A} \min_{c \in C} \|x - c\|^2$ the contribution of points in a subset $A \subseteq X$. Let C_{OPT} be the set of optimal k centers. In the optimal solution, each point of X is assigned to its nearest center in C_{OPT}. This induces a natural partition on X as $A_1 \cup A_2 \cup \cdots \cup A_k$ into disjoint subsets.

There is a variant of the k-means problem known as the *discrete k-means problem* where the centers have to be points from X itself. Note that the optima of the k-means problem and its discrete variant are within constant factors of each other. There are other variants where the objective is to minimize the sum of p-th powers of distances instead of squares (for $p \geq 1$), or to be more precise,

$$\left(\sum_{x \in X} \min_{c \in C} \|x - c\|^p \right)^{1/p}.$$

The $p = 1$ case is known as the *k-median problem* and the $p = \infty$ case is known as the *k-center problem*. Moreover, one can also ask the discrete k-means problem over arbitrary metric spaces instead of $\mathbb{R}^d$.

I. Dinur et al. (Eds.): APPROX and RANDOM 2009, LNCS 5687, pp. 15–28, 2009.

1.1   Previous Work

It is NP-hard to solve the k-means problem exactly, even for $k = 2$ [ADHP09], [Das08, KNV08] and even in the plane [MNV09]. Constant factor approximation algorithms are known based on linear programming techniques used for facility location problems but their running time is super-linear in n [JV01]. Kanugo et al. [KMN$^+$04] give a $(9 + \epsilon)$-approximation via local search but in running time $O(n^3\epsilon^{-d})$ that has exponential dependence on d. There are polynomial time approximation schemes with running time linear in n and d but exponential or worse in k [dlVKKR03, HPM04, KSS04, Mat00, Che09]. Such a dependence on k may well be unavoidable, as shown in the case of the discrete k-median problem [GI03].

On the other hand, the most popular algorithm for the k-means problem is a simple iterative-refinement heuristic due to Lloyd [Llo82]: start with k arbitrary (or random) centers, compute the clusters defined by them, define the means of these clusters as the new centers, re-compute clusters and repeat. Lloyd's method is fast in practice but is guaranteed to converge only to a local optimum. In theory, the worst-case running time of Lloyd's heuristic is exponential even in the plane [Vat09]; however, a plausible explanation for its popularity could be its polynomial smoothed complexity [AMR09].

In attempts to bridge this gap between theory and practice, several randomized algorithms have been proposed based on the idea of sampling a subset of points as centers to get a constant factor approximation in time effectively $O(nkd)$. These centers could then be used to initialize the Lloyd's method. Mettu and Plaxton [MP02] and Ostrovsky et al. [ORSS06] give constant factor approximations but their results do not work unconditionally for all data sets.

The most relevant to our paper is a randomized algorithm called k-means++ due to Arthur and Vassilvitskii [AV07]. They propose a simple *adaptive* sampling scheme (they call it as D^2 sampling): in each step, pick a point with probability proportional to its current cost (i.e, its squared distance to the nearest center picked so far) and add it as a new center. This is similar to a greedy 2-approximation algorithm for the k-center problem that picks a point with the maximum cost in each step [Gon85]. Arthur and Vassilvitskii show that *adaptively* sampled k centers give, in expectation, an $O(\log k)$-approximation for the k-means problem. This also means, by Markov inequality, that we get an $O(\log k)$-approximation with a constant probability.

Similar sampling schemes have appeared in the literature on clustering of data streams [GMM$^+$03, COP03] and online facility location [Mey01]. However, these sampling schemes are not as simple and their analysis is quite different.

Arthur and Vassilvitskii's analysis of their $O(\log k)$-approximation relies heavily on a non-trivial induction argument (Lemma 3.3 of [AV07]). Reverse engineering the same argument, they show a lower bound example where adaptively sampled k centers give $\Omega(\log k)$-approximation, in expectation. However, their lower bound is misleading in the sense that even though the expected error for adaptive sampling on this example is high, it gives an $O(1)$-approximation with high probability. The starting point for our work was the following question: Do adaptively sampled k centers *always* give a *constant* factor approximation, with a *constant* probability?

1.2   Our Results

In Section 2, we extend the results of Arthur and Vassilvitskii to show that adaptively sampled $O(k)$ centers give a constant factor bi-criteria approximation for the k-means problem, with a constant probability. This probability of success can be boosted to arbitrary $(1-\delta)$ by repeating the algorithm $O\left(\log(1/\delta)\right)$ times and taking the best solution.

In Section 3, we show that our adaptively picked $O(k)$ centers contain a subset of k centers that gives a constant factor approximation for the k-means problem, and this k-subset can be found by solving a weighted k-means problem on $O(k)$ points using the LP-based techniques of Jain and Vazirani [JV01] and Charikar et al. [CGTS02]. This gives us a randomized $O(1)$-approximation for the k-means problem with running time effectively $O(nkd)$.

Our proof techniques bypass the inductive argument of [AV07] and are general enough so as to be applicable in a wide range of other problems, such as facility location, where *adaptive* sampling could be useful.

In Appendix 4, we give a simpler proof of Arthur and Vassilvitskii's $\Omega(\log k)$ lower bound on the expected error of adaptively picked k centers to explain why their lower bound is misleading.

2   Bi-criteria Approximation by Adaptive Sampling

For a given set of centers, the current cost that each point pays in the k-means objective is its squared distance to the nearest center. In each step of *adaptive* sampling, we pick a point with probability proportional to its current cost and make it a new center. In this section, we show that adaptively sampling $O(k)$ points from the given data set itself gives a constant factor approximation for the k-means problem, with a constant probability.

Bi-criteria approximation by adaptive sampling
```
Input: a set X ⊆ ℝ^d of n points and k > 0.
Output: a set S ⊆ X of size t = ⌈16(k + √k)⌉.
Initialize S₀ = ∅.
For i = 1 to t do:
```

1. Pick a point x from the following distribution:
    $\Pr(\text{picking } x) \propto \phi_{\{x\}}(S_{i-1}) = \min_{c \in S_{i-1}} \|x - c\|^2$.
    (Note: For $i = 1$ step, the distribution is uniform.)
2. $S_i \leftarrow S_{i-1} \cup \{x\}$.
3. $i \leftarrow i + 1$.

Return $S \leftarrow S_t$.

Theorem 1. *Let $S \subseteq X$ be the subset of $t = \lceil 16(k + \sqrt{k}) \rceil = O(k)$ points picked by the sampling algorithm given above. Then*

$$\phi(S) \leq 20\phi(C_{OPT}),$$

with probability at least 0.03. (This probability could be boosted to $1 - \delta$ by repeating the algorithm $\log(1/\delta)$ times and picking the best of the subsets.) The running time of our algorithm is $O(nkd)$.

To prove correctness of our algorithm, we first analyze one step. Let S_{i-1} be the set of points obtained after the $(i-1)$-th step of our algorithm. In step i, we define

$$\mathsf{Good}_i = \{A_j \ : \ \phi_{A_j}(S_{i-1}) \leq 10\phi_{A_j}(C_{OPT})\}$$
$$\mathsf{Bad}_i = \{A_1, A_2, \ldots, A_k\} \setminus \mathsf{Good}_i$$

Observe that at each step we pick a point with probability proportional to its cost at the current step. We first show that at each step, either we are already within a small constant factor of the optimum or we pick a point from Bad_i with high probability.

Lemma 1. *In the i-th step of our algorithm, either $\phi(S_{i-1}) \leq 20\phi(C_{OPT})$ or else the probability of picking a point from some cluster in Bad_i is $\geq 1/2$.*

Proof. Suppose $\phi(S_{i-1}) > 20\phi(C_{OPT})$. Then the probability of picking x from some cluster in Bad_i is equal to

$$\Pr\left(x \in A_j \text{ from some } A_j \in \mathsf{Bad}_i\right) = \frac{\sum_{A_j \in \mathsf{Bad}_i} \phi_{A_j}(S_{i-1})}{\phi(S_{i-1})}$$

$$= 1 - \frac{\sum_{A_j \in \mathsf{Good}_i} \phi_{A_j}(S_{i-1})}{\phi(S_{i-1})}$$

$$\geq 1 - \frac{10\sum_{A_j \in \mathsf{Good}_i} \phi_{A_j}(C_{OPT})}{20\phi(C_{OPT})}$$

$$\geq 1 - 1/2$$

$$= 1/2.$$

Note that once a cluster becomes good at some stage then it continues to remain good, i.e. $\mathsf{Good}_i \subseteq \mathsf{Good}_{i+1}$. Good clusters are those clusters that are being covered well enough by the centers we have chosen so far. We analyze a bad cluster and show how the algorithm makes it good.

Here is an important fact about the mean of a point set that we will use throughout the analysis. It can be thought of as an analog of the parallel axis theorem about moment of inertia from elementary physics.

Proposition 1. *Let μ be the mean of a set of points $A \subseteq \mathbb{R}^d$ and let $y \in \mathbb{R}^d$ be any point. Then*

$$\sum_{x \in A} \|x - y\|^2 = \sum_{x \in A} \|x - \mu\|^2 + |A| \, \|y - \mu\|^2 .$$

Proof. Folklore. See Lemma 2.1.

Consider a cluster $A \in \mathsf{Bad}_i$. Let μ be the center of A in C_{OPT} and let $|A| = m$. (We drop the subscript j in A_j for the sake of simplicity.) Define $r = \sqrt{\phi_A(C_{OPT})/m}$, the root-mean-square optimal cost for points in A. Furthermore, let y be the point closest to μ in S_{i-1} and $d = \|\mu - y\|$. Observe that since $A \in \mathsf{Bad}_i$,

$$
\begin{aligned}
10\phi_A(C_{OPT}) &\leq \phi_A(S_{i-1}) && \text{because } A \in \mathsf{Bad}_i \\
&= \sum_{x \in A} \min_{c \in S_{i-1}} \|x - c\|^2 \\
&\leq \sum_{x \in A} \|x - y\|^2 \\
&= \phi_A(C_{OPT}) + m \|\mu - y\|^2 && \text{by Proposition 1} \\
&= \phi_A(C_{OPT}) + md^2
\end{aligned}
$$

Therefore,

$$
d \geq \sqrt{\frac{9\phi_A(C_{OPT})}{m}} = 3r.
$$

Define $B(\alpha) = \{x \in A \ : \ \|x - \mu\| \leq \alpha r\}$, where $0 \leq \alpha \leq 3 \leq d/r$. This is the set of points from A which are close to the center. The set $B(\alpha)$ is a good set to sample points from because any point $b \in B(\alpha)$ makes A a good cluster as shown below.

Lemma 2. *Let A be any cluster defined by C_{OPT} and let $b \in B(\alpha)$, for $0 \leq \alpha \leq 3$. Then*

$$
\phi_A(S_{i-1} \cup \{b\}) \leq 10\phi_A(C_{OPT}).
$$

Proof

$$
\begin{aligned}
\phi_A(S_{i-1} \cup \{b\}) &= \sum_{x \in A} \min_{c \in S_{i-1}\cup\{b\}} \|x - c\|^2 \\
&\leq \sum_{x \in A} \|x - b\|^2 \\
&= \phi_A(C_{OPT}) + m \|\mu - b\|^2 && \text{by Proposition 1} \\
&\leq \phi_A(C_{OPT}) + m(\alpha r)^2 \\
&= (1 + \alpha^2)\phi_A(C_{OPT}) \\
&\leq 10\phi_A(C_{OPT}) && \text{since } \alpha \leq 3.
\end{aligned}
$$

Now we show that $B(\alpha)$ contains a large fraction of points in A.

Lemma 3

$$
|B(\alpha)| \geq m \left(1 - \frac{1}{\alpha^2}\right), \qquad \text{for } 1 \leq \alpha \leq 3.
$$

Proof

$$\phi_A(C_{OPT}) \geq \phi_{A \setminus B(\alpha)}(C_{OPT})$$

$$= \sum_{x \in A \setminus B(\alpha)} \min_{c \in C_{OPT}} \|x - c\|^2$$

$$= \sum_{x \in A \setminus B(\alpha)} \|x - \mu\|^2$$

$$\geq |A \setminus B(\alpha)| (\alpha r)^2$$

$$= \left(1 - \frac{|B(\alpha)|}{m}\right) m(\alpha r)^2$$

$$= \left(1 - \frac{|B(\alpha)|}{m}\right) \alpha^2 \phi_A(C_{OPT}),$$

which implies that

$$|B(\alpha)| \geq m \left(1 - \frac{1}{\alpha^2}\right).$$

The following lemma states that the cost of $B(\alpha)$ is a substantial fraction of the cost of A with respect to the current S_{i-1} and thus also lower bounds the probability of the next point being chosen from $B(\alpha)$ given that it belongs to A.

Lemma 4

$$\Pr\left(x \in B(\alpha) \mid x \in A \text{ and } A \in \mathsf{Bad}_i\right) = \frac{\phi_{B(\alpha)}(S_{i-1})}{\phi_A(S_{i-1})} \geq \frac{(3 - \alpha)^2}{10} \left(1 - \frac{1}{\alpha^2}\right).$$

Proof. To prove the above lemma, we obtain an upper bound on $\phi_A(S_{i-1})$ and a lower bound on $\phi_{B(\alpha)}(S_{i-1})$ as follows.

$$\phi_A(S_{i-1}) = \sum_{x \in A} \min_{c \in S_{i-1}} \|x - c\|^2$$

$$\leq \sum_{x \in A} \|x - y\|^2$$

$$= \phi_A(C_{OPT}) + m \|\mu - y\|^2 \qquad \text{by Proposition 1}$$

$$= m(r^2 + d^2).$$

Observe that $\alpha r \leq d$ and $d = \|\mu - y\| \min_{c \in S_{i-1}} \|\mu - c\|$. For any $b \in B(\alpha)$ and any $c \in S_{i-1}$, we have

$$\|b - c\| \geq \|\mu - c\| - \|b - \mu\| \geq d - r\alpha \qquad \text{by triangle inequality.}$$

Thus, $\min_{c \in S_{i-1}} \|b - c\| \geq d - r\alpha$. Using this, we lower bound $\phi_{B(\alpha)}(S_{i-1})$ as follows.

$$\phi_{B(\alpha)}(S_{i-1}) = \sum_{b \in B(\alpha)} \min_{c \in S_{i-1}} \|b - c\|^2$$

$$\geq |B(\alpha)| (d - \alpha r)^2$$

$$\geq m \left(1 - \frac{1}{\alpha^2}\right) (d - \alpha r)^2 \qquad \text{from Lemma 3.}$$

Putting these together we get

$$\Pr\left(x \in B(\alpha) \mid x \in A \text{ and } A \in \mathsf{Bad}_i\right) = \frac{\phi_{B(\alpha)}(S_{i-1})}{\phi_A(S_{i-1})} \geq \frac{(1 - 1/\alpha^2)(d - \alpha r)^2}{r^2 + d^2}.$$

Observe that $(d - \alpha r)^2/(r^2 + d^2)$ is an increasing function of d for $d \geq 3r \geq \alpha r$. Therefore,

$$\Pr\left(x \in B(\alpha) \mid x \in A \text{ and } A \in \mathsf{Bad}_i\right) \geq \left(1 - \frac{1}{\alpha^2}\right) \frac{(3 - \alpha)^2}{10}.$$

Lemma 5. *Suppose the point x picked by our algorithm in the i-th step is from $A \in \mathsf{Bad}_i$ and $S_i = S_{i-1} \cup \{x\}$. Then*

$$\Pr\left(\phi_A(S_i) \leq 10\phi_A(C_{OPT}) \mid x \in A \text{ and } A \in \mathsf{Bad}_i\right) \geq 0.126.$$

Proof. Immediately follows from Lemma 2 and Lemma 4 using $\alpha = 1.44225$ (by numerically maximizing the expression in α).

We want to show that in each step, with high probability, we pick a bad cluster A and make it good. Our proof uses the following well known facts about super-martingales.

Definition 1. *A sequence of real valued random variables $J_0, J_1, \ldots, J_t$ is called a super-martingale if for every $i > 1$, $\mathsf{E}[J_i \mid J_0, \ldots, J_{i-1}] \leq J_{i-1}$.*

Super-martingales have the following concentration bound.

Theorem 2. *(Azuma-Hoeffding inequality) If $J_0, J_1, \ldots, J_t$ is a super-martingale with $J_{i+1} - J_i \leq 1$, then $\Pr\left(J_t \geq J_0 + \delta\right) \leq \exp(-\delta^2/2t)$.*

Proof. (Proof of Theorem 1) By Lemma 1 and Lemma 5, we have

$\Pr\left(|\mathsf{Bad}_{i+1}| < |\mathsf{Bad}_i|\right)$

$= \Pr\left(x \in A \text{ for some } A \in \mathsf{Bad}_i\right) \Pr\left(\phi_A(S_i) \leq 10\phi_A(C_{OPT}) \mid x \in A \text{ and } A \in \mathsf{Bad}_i\right)$

$\geq \dfrac{1}{2} \cdot 0.126$

$= 0.063.$

For each step define an indicator variable X_i as follows.

$$X_i = \begin{cases} 1 & \text{if } |\mathsf{Bad}_{i+1}| = |\mathsf{Bad}_i| \\ 0 & \text{otherwise.} \end{cases}$$

Thus, $\Pr(X_i = 0) \geq p = 0.063$ and $\mathsf{E}[X_i] \leq 1 - p$. Further, we define

$$J_i = \sum_{1 \leq j \leq i} (X_j - (1-p)).$$

Then $J_{i+1} - J_i \leq 1$ and

$$
\begin{aligned}
\mathsf{E}[J_i \mid J_0, \ldots, J_{i-1}] &= \mathsf{E}[J_{i-1} + X_i - (1-p) \mid J_0, \ldots, J_{i-1}] \\
&= J_{i-1} + \mathsf{E}[X_i \mid J_0, \ldots, J_{i-1}] - (1-p) \\
&\leq J_{i-1},
\end{aligned}
$$

which means that $J_1, J_2, \ldots, J_t$ is a super-martingale. So using Theorem 2 we get the following bound.

$$\Pr(J_t \geq J_0 + \delta) \leq \exp(-\delta^2/2t),$$

which means

$$\Pr\left(\sum_{i=1}^{t}(1 - X_i) \geq pt - \delta\right) \geq 1 - \exp(-\delta^2/2t).$$

Choosing $t = (k + \sqrt{k})/p \leq 16(k + \sqrt{k})$ and $\delta = \sqrt{k}$, we obtain

$$\Pr\left(\sum_{i=1}^{(k+\sqrt{k})/p}(1 - X_i) \geq k\right) \geq 1 - \exp\left(\frac{-pk}{2(k + \sqrt{k})}\right)$$

$$\geq 1 - \exp(-p/4).$$

Therefore,

$$\Pr\left(\text{there are no bad clusters after } (k + \sqrt{k})/p \text{ steps}\right) \geq 1 - \exp(-p/4) \geq 0.03,$$

or equivalently

$$\Pr(\phi(S) \leq 10\phi(C_{OPT})) \geq 0.03.$$

There is nothing special about the approximation factor 20 in the proof above. One could start with any factor more than 4 and repeat the same proof. The higher the approximation factor, the better are the bounds on the probability and the number of centers picked. We get the following result as a straightforward generalization.

Theorem 3. *Our bi-criteria algorithm, when run for $t = O(k/\epsilon \cdot \log(1/\epsilon))$ steps, gives a $(4 + \epsilon)$-approximation for the k-means problem, with a constant probability.*

3   Picking a k-Subset of S

If we use our bi-criteria solution S to cluster X, then every $x \in X$ is assigned to its closest point in S. This induces a natural partition of $X = X_1 \cup X_2 \cup \cdots \cup X_t$ into t disjoint subsets. Let $|X_i| = n_i$ and μ_i be the mean of points in X_i. Then for all i,

$$\phi_{X_i}(\{\mu_i\}) \leq \phi_{X_i}(S)$$

Weighted k-means clustering: Given a set $X \subseteq \mathbb{R}^d$ and weights w_i for each point $x_i \in X$, find a set $C \subseteq \mathbb{R}^d$ of k centers that minimizes the following potential function.

$$\phi'(C) = \sum_{x_i \in X} \min_{c \in C} w_i \|x_i - c\|^2 .$$

We denote by $\phi'_A(C) = \sum_{x_i \in A} \min_{c \in C} w_i \|x_i - c\|^2$ the contribution of points in a subset $A \subseteq X$.

Using the bi-criteria solution S, we define a weighted k-means problem with points $X' = \{\mu_i \ : \ 1 \leq i \leq t\}$ and weights n_i assigned to point μ_i, respectively. Let C'_{OPT} denote the optimal solution for this weighted k-means problem.

Lemma 6

$$\phi'(C'_{OPT}) \leq 2\phi(C_{OPT}) + 2\phi(S).$$

Proof. By triangle inequality, for any $x \in X$ we have

$$\min_{c \in C_{OPT}} \|\mu_i - c\| \leq \|\mu_i - x\| + \min_{c \in C_{OPT}} \|x - c\| .$$

Therefore,

$$\min_{c \in C_{OPT}} \|\mu_i - c\|^2 \leq 2 \|\mu_i - x\|^2 + 2 \min_{c \in C_{OPT}} \|x - c\|^2$$

Summing over all $x \in X_i$,

$$\min_{c \in C_{OPT}} n_i \|\mu_i - c\|^2 \leq \sum_{x \in X_i} 2 \|\mu_i - x\|^2 + 2 \min_{c \in C_{OPT}} \|x - c\|^2)$$

$$\leq 2\phi_{X_i}(S) + 2\phi_{X_i}(C_{OPT}).$$

Thus,

$$\phi'(C'_{OPT}) \leq \phi'(C_{OPT})$$

$$= \sum_{1 \leq i \leq t} \min_{c \in C_{OPT}} n_i \|\mu_i - c\|^2$$

$$\leq \sum_{1 \leq i \leq t} 2\phi_{X_i}(S) + 2\phi_{X_i}(C_{OPT})$$

$$= 2\phi(S) + 2\phi(C_{OPT})$$

Theorem 4. *Let C be an β-approximation to the weighted k-means problem, i.e., $\phi'(C) \leq \beta\phi'(C'_{OPT})$. Then,*

$$\phi(C) \leq (2\beta + 1)\phi(S) + 2\beta\phi(C_{OPT}).$$

Proof. In the solution C, let μ_i be assigned to the center $c_j \in C$.

$$\sum_{x \in X_i} \min_{c \in C} \|x - c\|^2 \leq \sum_{x \in X_i} \|x - c_j\|^2$$

$$= \sum_{x \in X_i} \|x - \mu_i\|^2 + n_i \|\mu_i - c_j\|^2 \quad \text{by Proposition 1}$$

$$\leq \phi_{X_i}(S) + n_i \min_{c \in C} \|\mu_i - c\|^2.$$

Therefore,

$$\phi(C) = \sum_{x \in X} \min_{c \in C} \|x - c\|^2$$

$$= \sum_{1 \leq i \leq t} \sum_{x \in X_i} \min_{c \in C} \|x - c\|^2$$

$$\leq \sum_{1 \leq i \leq t} \phi_{X_i}(S) + n_i \min_{c \in C} \|\mu_i - c\|^2$$

$$= \phi(S) + \phi'(C)$$

$$\leq \phi(S) + \beta\phi'(C'_{OPT})$$

$$\leq (2\beta + 1)\phi(S) + 2\beta\phi(C_{OPT}).$$

Note that Theorem 4 implies that a constant factor approximation to the weighted k-means problem constructed from our bi-criteria solution S is also a constant factor approximation to our original k-means problem. The advantage is that the weighted k-means problem is defined only on $O(k)$ points instead of n points. Interestingly, previous works on k-means clustering and a closely related problem of k-median clustering ([JV01],[CGTS02]) generalize to weighted k-means problem as well. This is because [CGTS02] solves the weighted k-median problem and the solution generalizes to distances where even a weak triangle inequality is satisfied. In case of squared Euclidean distance, for example,

$$\|x - z\|^2 \leq 2 \left(\|x - y\|^2 + \|y - z\|^2 \right).$$

We omit the details as the proofs are essentially the same as in [CGTS02]. These are LP-based algorithms and since the number of variables in our weighted k-means instance is $O(k)$ the overall running time of our sampling coupled with the LP-based algorithm for the resulting weighted k-means problem has running time $O(nkd + \text{poly}(k, \log n))$, which is effectively $O(nkd)$.

4   Simplified Lower Bound

Arthur and Vassilvitskii [AV06] prove that adaptive sampling for the k-means clustering gives an $O(\log k)$ approximation, in expectation. They also show an example where adaptive sampling gives expected error at least $\Omega(\log k)$ times the optimum. Both these proofs are based on a tricky inductive argument.

In this note, we give a simplified proof of their lower bound. The example for lower bound is the same. Consider n points where they are grouped into k sets $S_1, S_2, \ldots, S_k$ of size n/k each. The points in each S_i form vertices of a regular simplex and the centers of these simplices $S_1, S_2, \ldots, S_k$ form vertices of a larger regular simplex. The smaller simplices live in different dimensions so that

$$\|x - y\| = \begin{cases} \delta & \text{if } x, y \in S_i \text{ for the same } i \\ \Delta & \text{if } x \in S_i \text{ and } y \in S_j \text{ for } i \neq j \end{cases}$$

The optimal k-means clustering uses centers of these regular simplices $S_1, S_2, \ldots, S_k$ and has error

$$\text{OPT} = \frac{n-k}{2}\delta^2.$$

The probability that adaptive sampling picks all k centers from different S_i's is

$$\Pr\left(\text{adaptive sampling covers all } S_1, S_2, \ldots, S_k\right)$$

$$= \prod_{i=1}^{k-1}\left(1 - \frac{i\left(\frac{n}{k} - 1\right)\delta^2}{\frac{n}{k}(k-i)\Delta^2 + i\left(\frac{n}{k} - 1\right)\delta^2}\right)$$

$$\geq \prod_{i=1}^{k-1}\left(1 - \frac{i(n-k)\delta^2}{n(k-i)\Delta^2}\right)$$

$$\geq 1 - \sum_{i=1}^{k-1}\frac{i(n-k)\delta^2}{n(k-i)\Delta^2} \qquad \text{by Weierstrass product inequality}$$

$$= 1 - \frac{n-k}{n}\frac{\delta^2}{\Delta^2}\sum_{i=1}^{k-1}\frac{i}{k-i}$$

$$\geq 1 - \frac{\delta^2}{\Delta^2}\sum_{i=1}^{k-1}\frac{k-i}{i}$$

$$\geq 1 - \frac{\delta^2}{\Delta^2}k\left(\sum_{i=1}^{k-1}\frac{1}{i} - 1\right)$$

$$\geq 1 - \frac{\delta^2}{\Delta^2}k\log k.$$

In fact, we will fix n, k, δ and use $\Delta \gg n, k, \delta$.

$$\Pr\left(\text{adaptive sampling covers all } S_1, S_2, \ldots, S_k\right)$$

$$= \prod_{i=1}^{k-1}\left(1 - \frac{i\left(\frac{n}{k} - 1\right)\delta^2}{\frac{n}{k}(k-i)\Delta^2 + i\left(\frac{n}{k} - 1\right)\delta^2}\right)$$

$$\leq \prod_{i=1}^{k-1} \left(1 - \frac{i(n-k)\delta^2}{2n(k-i)\Delta^2}\right)$$

$$\leq 1 - \frac{1}{2}\sum_{i=1}^{k-1} \frac{i(n-k)\delta^2}{2n(k-i)\Delta^2} \qquad \text{for } \Delta \gg k\delta$$

$$= 1 - \frac{n-k}{2n}\frac{\delta^2}{\Delta^2}\sum_{i=1}^{k-1} \frac{i}{k-i}$$

$$\leq 1 - \frac{\delta^2}{4\Delta^2}\sum_{i=1}^{k-1} \frac{k-i}{i} \qquad \text{for } n \gg k$$

$$\leq 1 - \frac{\delta^2}{8\Delta^2}k\left(\sum_{i=1}^{k-1} \frac{1}{i} - 1\right)$$

$$= 1 - \frac{\delta^2}{8\Delta^2}k\log k.$$

Thus

$$\Pr\left(\text{adaptive sampling covers all } S_1, S_2, \ldots, S_k\right) = 1 - \Theta\left(\frac{\delta^2}{\Delta^2}k\log k\right).$$

If our adaptive sampling covers all $S_1, S_2, \ldots, S_k$ then it's error is

$$\text{Err}_{\text{no miss}} = (n-k)\delta^2,$$

whereas even if we miss (i.e., do not cover) one of the S_i's the error is at least

$$\text{Err}_{\text{some miss}} \geq \frac{n}{k}\Delta^2.$$

So the expected error for adaptive sampling is given by

$$\mathsf{E}\left[\text{Err}\right] \geq \left(1 - \Theta\left(\frac{\delta^2}{\Delta^2}k\log k\right)\right)\text{Err}_{\text{no miss}} + \Theta\left(\frac{\delta^2}{\Delta^2}k\log k\right)\text{Err}_{\text{some miss}}$$

$$\geq \left(1 - \Theta\left(\frac{\delta^2}{\Delta^2}k\log k\right)\right)(n-k)\delta^2 + \Theta\left(\frac{\delta^2}{\Delta^2}k\log k\right)\frac{n}{k}\Delta^2$$

$$\geq (n-k)\delta^2 + \frac{1}{\Delta^2} \cdot \text{some term} + \Theta(\log k)n\delta^2$$

$$= \Omega(\log k)\frac{n-k}{2}\delta^2 \qquad \text{using } n \gg k \text{ and } \Delta \to \infty$$

$$= \Omega(\log k)\text{OPT}.$$

Notice that even though the expected error is $\Omega(\log k)\text{OPT}$, we get a constant factor approximation when the adaptive sampling covers all $S_1, S_2, \ldots, S_k$, which happens with a high probability.

5   Conclusion

We present a *simple* bi-criteria constant factor approximation algorithm for the k-means problem using *adaptive* sampling. Our proof techniques can be generalized to prove similar results for other variants of the k-means problem such as the k-median problem, or more generally, the ℓ_p version where we want to minimize the sum of p-th powers of distances rather than squares. This follows because of the weak triangle inequalities satisfied by the p-th powers of Euclidean distances, which gives us a weak form of the parallel axis theorem (i.e., Proposition 1). For the ℓ_p version, we get a similar bi-criteria algorithm where the number of centers picked by the algorithm is $O(k)$, where the constant depends exponentially on p.

Arthur and Vassilvitskii [AV07] show that adaptively sampled k centers give an $O(\log k)$-approximation for the k-means problem, in expectation (and hence also with a constant probability, by Markov inequality). In this paper, we show that adaptively sampled $O(k)$ centers give an $O(1)$-approximation for the k-means problem, with a constant probability. Looking at the lower bound example example (see Appendix 4) it is tempting to conjecture that adaptively sampled k centers give an $O(1)$-approximation for the k-means problem, with a constant probability. It would be nice to settle this conjecture.

Acknowledgements. The second author would like to thank Kasturi Varadarajan for several helpful discussions and Jaikumar Radhakrishnan for suggesting the analogy of Proposition 1 with the parallel axis theorem in elementary physics.

References

[ADHP09]   Aloise, D., Deshpande, A., Hansen, P., Popat, P.: NP-hardness of Euclidean sum-of-squares clustering. Machine Learning 75(2), 245–248 (2009)

[AMR09]   Arthur, D., Manthey, B., Röglin, H.: k-means has polynomial smoothed complexity (2009), http://arxiv.org/abs/0904.1113

[AV06]   Arthur, D., Vassilvitskii, S.: How slow is the k-means method?. In: Annual Symposium on Computational Geometry (SOCG) (2006)

[AV07]   Arthur, D., Vassilvitskii, S.: k-means++: The advantages of careful seeding. In: ACM-SIAM Symposium on Discrete Algorithms (SODA) (2007)

[CGTS02]   Charikar, M., Guha, S., Tardos, M., Shmoys, D.: A constant factor approximation for the k-median problem. Journal of Computer and System Sciences (2002)

[Che09]   Chen, K.: On coresets for k-median and k-means clustering in metric and euclidean spaces and their applications. Submitted to SIAM Journal on Computing (SICOMP) (2009)

[COP03]   Charikar, M., O'Callaghan, L., Panigrahy, R.: Better streaming algorithms for clustering problems. In: ACM Symposium on Theory of Computing (STOC), pp. 30–39 (2003)

[Das08]   Dasgupta, S.: The hardness of k-means clustering, Tech. Report CS2008-0916, UC San Diego (2008) [dlVKKR03] de la Vega, F., Karpinski, M., Kenyon, C., Rabani, Y.: Approximation schemes for clustering problems. In: ACM Symposium on Theory of Computing (STOC), pp. 50–58. ACM Press, New York (2003)

[GI03] Guruswami, V., Indyk, P.: Embeddings and non-approximability of geometric problems. In: ACM-SIAM Symposium on Discrete Algorithms (SODA) (2003)

[GMM⁺03] Guha, S., Meyerson, A., Mishra, N., Motwani, R., O'Callaghan, L.: Clustering data streams: Theory and practice. IEEE Transactions on Knowledge and Data Engineering 15(3), 515–528 (2003)

[Gon85] Gonzalez, T.: Clustering to minimize the maximum intercluster distance. Theoretical Computer Science 38, 293–306 (1985)

[HPM04] Har-Peled, S., Mazumdar, S.: On core-sets for k-means and k-median clustering. In: ACM Symposium on Theory of Computing (STOC), pp. 291–300 (2004)

[JV01] Jain, K., Vazirani, V.: Approximation algorithms for metric facility loca- tion and k-median problems using the primal-dual schema and Lagrangian relaxation. Journal of ACM 48, 274–296 (2001)

[KMN⁺04] Kanugo, T., Mount, D., Netanyahu, N., Piatko, C., Silverman, R., Wu, A.: A local search approximation algorithm for k-means clustering. Computational Geometry 28(2-3), 89–112 (2004)

[KNV08] Kanade, G., Nimbhorkar, P., Varadarajan, K.: On the NP-hardness of the 2-means problem (unpublished manuscript) (2008)

[KSS04] Kumar, A., Sabharwal, Y., Sen, S.: A simple linear time $(1 + \epsilon)$- approximation algorithm for k-means clustering in any dimensions. In: IEEE Symposium on Foundations of Computer Science (FOCS), pp. 454–462 (2004)

[Llo82] Lloyd, S.: Least squares quantization in pcm. IEEE Transactions on Information Theory 28(2), 129–136 (1982)

[Mat00] Matoušek, J.: On approximate geometric k-clustering. Discrete and Computational Geometry 24(1), 61–84 (2000)

[Mey01] Meyerson, A.: Online facility location. In: IEEE Symposium on Foundations of Computer Science (FOCS) (2001)

[MNV09] Mahajan, M., Nimbhorkar, P., Varadarajan, K.: The planar k-means problem is NP-hard. In: Das, S., Uehara, R. (eds.) WALCOM 2009. LNCS, vol. 5431, pp. 274–285. Springer, Heidelberg (2009)

[MP02] Mettu, R., Plaxton, C.: Optimal time bounds for approximate clustering. Machine Learning, 344–351 (2002)

[ORSS06] Ostrovsky, R., Rabani, Y., Schulman, L., Swamy, C.: The effectiveness of Lloyd-type methods for the k-means problem. In: IEEE Symposium on Foundations of Computer Science (FOCS), pp. 165–176 (2006)

[Vat09] Vattani, A.: k-means requires exponentially many iterations even in the plane. In: Annual Symposium on Computational Geometry (SOCG) (2009)

Approximations for Aligned Coloring and Spillage Minimization in Interval and Chordal Graphs

Douglas E. Carroll[1,*], Adam Meyerson[2], and Brian Tagiku[2]

[1] Raytheon Company
decarroll@raytheon.com
[2] University of California, Los Angeles
{awm,btagiku}@cs.ucla.edu

Abstract. We consider the problem of aligned coloring of interval and chordal graphs. These problems have substantial applications to register allocation in compilers and have recently been proven NP-Hard. We provide the first constant approximations: a $\frac{4}{3}$-approximation for interval graphs and a $\frac{3}{2}$-approximation for chordal graphs. We extend our techniques to the problem of minimizing spillage in these graph types.

1 Introduction

One of the most complex and time-consuming aspects of a compiler is the register allocation process where variables are assigned to registers. The implementation of the register allocator is of upmost importance as it has a substantial impact on the efficiency of code generated [10]. The seminal works of Chaitin *et al.* [7, 8] established a connection between register allocation and graph coloring. Since these results, a number of heuristic techniques based on splitting of live ranges and hierarchical divide and conquer have been proposed [4, 5, 6, 9, 13]. However, provably good approaches have eluded researchers, primarily because graph coloring in the general case is NP-Hard and also difficult to approximate [15].

Despite these hardness results, graph coloring is more tractable on special classes of graphs. Two examples are the class of *chordal graphs* and its subclass of *interval graphs*, for which coloring can be computed optimally in linear time [12]. These graphs arise in many applications and, fortunately, many real programs do in fact correspond to chordal and interval graphs [14, 17]. Thus, algorithmic results for these classes are meaningful to register allocation and a broad range of other applications.

Modern work in register allocation [1, 2, 3, 16, 18, 20] now focuses on *heterogeneous* register architectures where variables are either single-word (requiring one register) or double-word (requiring two registers). Double-word variables must be stored in adjacent registers starting on an even address (*i.e.* the registers must

---

* Douglas Carroll's contribution to this research was conducted while he was a Ph.D. student at UCLA.

I. Dinur et al. (Eds.): APPROX and RANDOM 2009, LNCS 5687, pp. 29–41, 2009.

be *aligned*). Indeed, most modern architectures are heterogeneous which stresses the importance of this model. The corresponding coloring problem is called the *aligned* coloring problem, introduced by Lee, Palsberg, and Pereira [16] who prove it to be NP-Hard even when restricted to chordal or interval graphs. We give the first approximation results for aligned coloring: a $\frac{4}{3}$-approximation for interval graphs and a $\frac{3}{2}$-approximation for chordal graphs.

We also consider the problem of minimizing spillage both in the aligned and unaligned case. Here, we are given a specific number of colors (or registers) and asked to color as many graph nodes as possible subject to this constraint. This problem relates directly to register allocation since the number of colors (registers) is a fixed property of the machine architecture. The unaligned problem is known to be NP-Hard even for chordal graphs [21]. We show that it is hard to approximate to $\Omega(\log n)$ based on a reduction from set cover [11] and give a matching $O(\log n)$-approximation. If we are permitted to run in time exponential in the number of colors, there is an exact algorithm for chordal graphs with running time $O(nm(\tau + 1)^c)$ for c colors, n nodes, m edges, and tree-width τ. We show that if we are willing to accept a $1 + \epsilon$ approximation, we can improve this running time to $O(nm(c + \frac{c}{\epsilon})^c)$ (independent of the tree-width).

For the aligned version over chordal graphs, we can apply dynamic programming to get a $1 + \epsilon$ approximation in running time $O(nm(c + 1)^{c+c/\epsilon})$. If we are willing to accept a bicriteria approximation in which we use extra colors to obtain at most the spillage needed by optimum using c colors, then we can give bounds of $(\frac{3}{2}, \log n)$ and $(\frac{4}{3}, 1)$ for chordal and interval graphs respectively, where the first factor is on the number of colors and the second is on the spillage.

2   Preliminaries and Notation

Our work will be restricted to the class of chordal and interval graphs. A graph is *chordal* if every cycle of four or more vertices has a chord (an edge joining two non-adjacent vertices in the cycles). A graph is *interval* if its vertices correspond to intervals of the real line and two vertices are adjacent if and only if their corresponding intervals intersect. For convenience, we will sometimes express an interval graph using a collection of intervals rather than show the graph representation.

We consider a generalization of graph coloring in which certain vertices require *two* colors instead of just one. Thus, graphs $G = (V, E)$ will have vertex weights $w : V \rightarrow \{1, 2\}$ indicating the number of colors a vertex requires. We will call these *1,2-vertex-weighted graphs*. We extend the notion of *clique number* $\omega(G)$ to be the maximum total weight of any clique of G. We represent colors numerically starting with color 0 and any vertex requiring two colors also requires that the colors are $2i$ and $2i + 1$ for integer i. Formally:

Definition 1 (Aligned c-Coloring). *An* aligned c-coloring *for a 1,2-vertex-weighted graph $G = (V, E)$ with weights w is a mapping $\phi : V \rightarrow \{0, \ldots, c - 1\}$ such that for all $(u, v) \in E$ we have $\phi(u) \neq \phi(v)$. Additionally, if $w(u) = 2$, then $\phi(u)$ must be even and for all $(u, v) \in E$ we must have $\phi(u) + 1 \neq \phi(v)$. This*

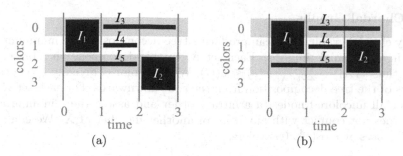

Fig. 1. Colorings of a collection of intervals of time. Thick intervals have weight 2 while thin intervals have weight 1. (a) A proper aligned coloring. (b) A minimum unaligned coloring.

corresponds to assigning any vertex u with $w(u) = 2$ to two consecutive colors starting from an even value.

At times we will say weight-1 vertex v *blocks* color $2i$ (and accordingly, $2i$ is *blocked* by v) if $\phi(v) = 2i$ or $\phi(v) = 2i + 1$. Figure 1 gives examples of aligned and unaligned colorings of a collection of intervals of time (thus, corresponding to an interval graph). Note that Figure 1 also shows that an aligned coloring may require more than $\omega(G)$ colors.

We use $a\chi(G)$ to denote the aligned chromatic index of G (*i.e.*, the smallest c for which G is aligned c-colorable). Note that we can easily lower bound $a\chi(G)$ using the clique number of G:

$$\omega(G) \le a\chi(G). \tag{1}$$

For any integer $k \ge 0$, we let $a\chi(k)$ denote the maximum aligned chromatic index over all 1,2-vertex-weighted graphs G with $\omega(G) = k$.

3   The Aligned Coloring Problem

Problem 1 (Aligned Coloring Problem). *Given 1,2-vertex-weighted graph $G = (V, E)$, find an aligned c-coloring ϕ of G where c is minimal.*

In this section we give approximations for the aligned coloring problem. Since the problem is hard to approximate for general graphs [15], we restrict our attention to either *chordal graphs* or *interval graphs*. Although standard minimum coloring can be solved in linear time for either class [12], the aligned coloring problem remains NP-Complete [16]. It is straightforward to produce a 2-approximation for the problem by simply splitting the graph into two subgraphs based on vertex weights and then coloring each using a distinct set of colors. We will provide the first approximation algorithms which improve upon this factor, giving a $\frac{3}{2}$-approximation for chordal graphs and a $\frac{4}{3}$-approximation for interval graphs.

3.1 Chordal Graphs

For any chordal graph G, we can produce a tree decomposition of minimum tree width in polynomial time [12]. Each set X_i in this decomposition represents a clique, thus $w(X_i) = \sum_{v \in X_i} w(v) \leq \omega(G)$. We will color G greedily by considering sets of the tree decomposition from the root downwards. For each set X, we consider all uncolored nodes in arbitrary order and assign the minimum color which does not conflict with any color of another member of X. We can show that this uses at most $\frac{3}{2}\omega(G)$ colors.

Theorem 1. *The above algorithm successfully produces an aligned coloring using at most $\frac{3}{2}\omega(G)$ colors. Thus it is a $\frac{3}{2}$-approximation to the problem of aligned coloring of a chordal graph.*

Proof. First we show that a valid aligned coloring is produced. Consider any edge $(u, v) \in E$. Let X be the first tree decomposition set we consider which contains both $u, v \in X$. Since we consider sets from the root downwards and since we know that u, v were not both in the parent of X, one of u, v must be appearing for the first time. Thus, we can assign a color such that there are no conflicts.

Now we bound the number of colors. Consider any node u with $w(u) = 1$. Let X be the tree decomposition set in which we first encounter u. Since $w(X) \leq \omega(G)$, at least one of colors 0 through $\omega(G) - 1$ must be available. Thus, all weight-1 vertices are assigned a color at most $\omega(G) - 1$.

Now consider a node u with $w(u) = 2$ and let X be the tree decomposition set in which we first encounter u. Suppose that there are x_1 weight-1 nodes and x_2 weight-2 nodes (excluding u) in X. Then $w(X - u) = x_1 + 2x_2 \leq \omega(G) - 2$. When we assign a color to u, suppose the color assigned were at least $\frac{3}{2}\omega(G)$. It would follow that since all nodes with $w(v) = 1$ are assigned colors at most $\omega(G)$, we must have at least $x_2 \geq \frac{1}{4}\omega(G)$ as otherwise we could have selected a smaller color. On the other hand, each node in X can block at most one even color, so we must have $x_1 + x_2 \geq \frac{3}{4}\omega(G)$. Combining these two inequalities gives a contradiction. $\square$

3.2 Chordal Graph Lower Bound

There do in fact exist aligned chordal graph coloring problems where we need $\frac{3}{2}\omega(G)$ colors. This does not necessarily imply hardness of approximation, but it does indicate that we will need to use a different lower bound on optimum if we are to improve upon our $\frac{3}{2}$-approximation.

Fix $n > 0$. We build a 1,2-vertex weighted graph G_n with $\omega(G_n) = 4n$ as follows: G_n will have a clique of $4n$ weight-1 vertices. Call this clique K_1 and let V_1 be its vertices. For each $S \subseteq V_1$ with $|S| = 2n$, we add a distinct clique K_S of n weight-2 vertices and all possible edges between vertices in S and K_S. It is easy to check that G_n is chordal and that $\omega(G_n) = 4n$.

Theorem 2. *For each $n > 0$, there exists no aligned coloring of G_n using less than $\frac{3}{2}\omega(G_n) = 6n$ colors.*

Proof. Assume, by way of contradiction, that we have an aligned coloring of G_n using at most $6n - 1$ colors. Consider V_1. These vertices form a clique so they must all be given unique colors. Then at least $2n$ even colors are blocked by vertices in V_1. Let $X \subset V_1$ be a set of $2n$ vertices that each block a distinct even color. Then notice that there are $n - 1$ even colors that are not blocked by X. But then K_X does not have enough even colors for its n weight-2 vertices. This contradicts the existence of our coloring and proves the claim.   □

3.3   Interval Graphs

We will show that every 1,2-vertex-weighted interval graph G has an aligned coloring using $\frac{4}{3}\omega(G)$ colors. Our algorithm works by first decomposing G into multiple smaller graphs, then approximately coloring each of the smaller graphs with disjoint sets of colors. We perform this decomposition as described below:

Theorem 3. *For every integer $\alpha \geq 1$, every 1,2-vertex-weighted interval graph G with $\omega(G) > \alpha + 1$ can be decomposed (in polynomial-time) into 1,2-vertex-weighted interval graphs H and G' such that $\omega(H) \leq \alpha+1$ and $\omega(G') = \omega(G)-\alpha$.*

Proof. Order the vertices of G so that for each vertex v all its neighbors appearing before it in the ordering form a clique (*i.e.* in a reverse perfect vertex elimination scheme). Such an ordering can be computed in linear time [19]. We partition the vertices of G into V_H and V' and let $H = G[V_H]$ and $G' = G[V']$ (where $G[X]$ is the subgraph of G induced by vertex set X). This is done by taking vertices one at a time in order. We add the next vertex to V_H if it can be added without increasing $\omega(H)$ beyond $\alpha + 1$. Otherwise, we add it to V'.

It is clear that $\omega(H)$ will lie between α and $\alpha+1$. Suppose a vertex v is added to V' resulting in $\omega(G[V']) > \omega(G) - \alpha$. Let $j = w(v)$. Since v was not added to V_H, there must be set of vertices of total weight $\alpha+2-j$ adjacent to v. Similarly, there must be a set of vertices of total weight $\omega(G) - \alpha + 1 - j$ adjacent to v in V'. By our ordering, all of these neighbors of v must form a clique. Thus, we've found a clique of total weight $\omega(G) + 3 - j > \omega(G)$. This contradicts the definition of $\omega(G)$.   □

We can now recursively apply our theorem to obtain a decomposition of G into $\left\lceil \frac{\omega(G)}{\alpha} \right\rceil$ interval graphs of clique index bounded by $\alpha + 1$. Thus, if we can find aligned c-colorings for all H with $\omega(H) = \alpha + 1$ then we can color each graph in the decomposition using a disjoint set of colors and obtain an approximate solution.

Corollary 1. *Suppose for some α, we have a polynomial-time algorithm which can color any 1,2-vertex-weighted interval graph H with $\omega(H) \leq \alpha + 1$ using at most c colors. Then we have a polynomial-time $\frac{c}{\alpha}$-approximation algorithm for any 1,2-vertex-weighted interval graph.*

We remark that there is a small additive term in our approximation, induced by the ceiling in our division. This can be eliminated by providing a solution with approximation factor $\frac{c}{\alpha}$ whenever $\omega(H) \leq \alpha$.

We now show how to reduce all interval graphs to a restricted form. This allows us to enforce a problem structure that can be later exploited. We will assume that the corresponding intervals have integer start and end points and are closed towards $-\infty$ and open towards ∞. Moreover, we assume that all intervals have non-negative endpoints and we let B denote the latest endpoint of any interval. Thus, the entire instance is bounded within $[0, B]$. We say interval u and interval v are *tightly compatible* if the startpoint of one is the endpoint of another. We will enforce the following restrictions:

1. **Uniform:** For $t \in [0, B]$ the total weight of intervals containing t is $\omega(G)$.
2. **United:** No two tightly compatible intervals have the same weight.
3. **Unique:** No two weight-1 intervals have identical endpoints.
4. **Nested:** If two weight-2 intervals intersect, one is contained in the other.
5. **Staged:** No two weight-2 intervals share a common endpoint.

We will show that given any interval graph G, we can produce a new interval graph G' of equal clique index and satisfying the above properties such that a coloring of G' can be used to efficiently compute a coloring of G (with equal number of colors). We will then describe a polynomial-time algorithm to compute a coloring of G' using at most $\frac{4}{3}\omega(G')$ colors.

Theorem 4. *Every 1,2-vertex-weighted interval graph G can be transformed into a 1,2-vertex-weighted interval graph G' (in polynomial time) with $\omega(G') = \omega(G)$ such that all five of the above properties hold. Any aligned c-coloring of G' can be used to construct an aligned c-coloring of G (in polynomial time).*

Proof. Any interval graph G can be made uniform by adding an appropriate amount of weight-1 intervals during the deficient times. Since this is an extension of G, any aligned c-coloring directly gives us an aligned c-coloring of G.

We can make this graph united and unique by merging any tightly compatible weight-1 intervals, then merging any identical weight-1 intervals into a weight-2 interval, then merging any tightly compatible weight-2 intervals. While we may have $a\chi(G') \geq a\chi(G)$, an aligned c-coloring of G can be recovered from an aligned c-covering of G' by recording which intervals were merged and assigning colors appropriately.

A united interval graph can be made nested by taking any non-nested, intersecting weight-2 intervals u, v and replacing them with weight-2 intervals $u \cap v$ and $u \cup v$. We repeat until G' is nested. Again, an aligned c-coloring of G can be recovered from an aligned c-coloring of G' if we remember how intervals were spliced; we take the color pairs of $u \cap v$ and of $u \cup v$ and exchange their roles after the splice point. Figure 2 illustrates this process.

We can make an interval graph staged by arbitrarily perturbing one of the intervals and two tightly-compatible weight-1 intervals (selected arbitrarily as well). $\square$

We define a *stage* to be a maximal interval of time during which the number of weight-2 intervals is constant. An i-stage will be a stage during which i weight-2

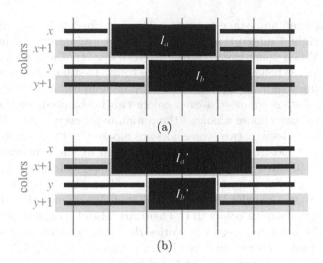

Fig. 2. Splicing colors. (a) Coloring of a non-nested instance. (b) Coloring of a corresponding nested instance.

intervals are active. Once all the requirements are enforced, there are a number of nice properties we can exploit. First, as time progresses, stages can only increase or decrease by one level. Second, intervals only start and end when the instance is changing stages. Third, when going from an i-stage to an $(i+1)$-stage, exactly one weight-2 interval begins and two weight-1 intervals end. Lastly, when going from an $(i+1)$-stage to an i-stage, exactly two weight-1 intervals begin and one weight-2 interval ends.

We will show that $a\chi(10) = 12$ for instances satisfying the five properties described (and thus for all instances) and that such colorings can be found in polynomial-time. Our coloring algorithm will color an instance by its stages.

Theorem 5. $a\chi(10) = 12$. *Moreover, we can construct an aligned 12-coloring in polynomial time.*

Proof. We start by removing all weight-2 intervals active during a 1-stage (and thus bordering a 0-stage) and give them color 10 (and 11). The remainder of the instance never has more than total weight 8 of active intervals except for the times where no weight-2 intervals are active but 10 weight-1 intervals are active. Call these times (-1)-stages, thus the remainder of our instance now has stages of level -1 through level 4. Notice that exactly two weight-1 intervals start at the beginning of a (-1)-stage and exactly two end when the (-1)-stage concludes. We will show that we need only 10 additional colors for the remainder of this instance.

Starting from the first stage, greedily assign colors to the active weight-2 intervals, then to the active weight-1 intervals. This will use no more than 10 colors since the maximum weighted overlap during the first stage is at most 10. We proceed by coloring from stage to stage.

When going from an i-stage to an $(i-1)$-stage (for $i > 0$) we simply assign the starting weight-1 intervals the colors that the terminating weight-2 interval was assigned. When going from a (-1)-stage to a 0-stage, there are no incoming intervals to color. When going from a 0-stage to a (-1)-stage, we can simply give the two incoming weight-1 intervals the two remaining colors. When going from a 3-stage to a 4-stage at most 3 even colors can be blocked, so the incoming weight-2 interval can choose amongst the remaining 2 even colors. When going from a 2-stage to a 3-stage, two even colors are blocked by the continuing weight-2 intervals and at most two even colors are blocked by the continuing weight-1 intervals. This still leaves one open even color for the incoming weight-2 interval.

The interesting cases occur when we go from a 1-stage to a 2-stage and from a 0-stage to a 1-stage. Consider the former case first. We assume that the weight-2 from the 1-stage occupies colors 0/1. There are also 4 weight-1 intervals that conflict with the incoming weight-2. Notice that if we cannot color the weight-2, it must be that each weight-1 is blocking colors 2/3, 4/5, 6/7, 8/9. Thus, WLOG, we assume that these weight-1s are in colors 2, 4, 6, and 8. Then one of the following cases must hold:

1. All weight-2 intervals seen so far are colored 0/1.
2. Let t be the ending time of the latest weight-2 that isn't colored 0/1 (without loss of generality, we'll assume its colored 2/3). At least one of the colors 4 through 9 either have a weight-1 interval starting at t OR are unoccupied between t and $t+1$.
3. None of the colors 4 through 9 have a (weight-1) interval starting at t and all are occupied between t and $t+1$.

In the first case, where no weight-2's have been colored 2 through 9, it is clear that we can simply swap colors 3 and 4, then color the incoming weight-2 colors 4/5. Now consider the second case. Let the "culprit" color be the color that either contains a weight-1 starting at t or is unoccupied at t. If the culprit color is either 4, 6 or 8, then we can swap that color with color 3 from time t onwards. This allows us to assign the incoming weight-2 to the culprit color and its partner. If the culprit color is 5, 7 or 9, then we can swap that color with color 2 from time t onwards. This allows us to color the incoming weight-2 colors 2/3.

Let us consider the last case. Since just prior to time t, we had a weight-2, this means that the stage immediately following time t cannot be a (-1)-stage. In particular, there are 4 colors at time t that either have no assigned active interval or have a weight-1 starting at t. Since case 2 doesn't hold, it follows that these four colors are 0,1,2 and 3. Moreover, since no two weight-2 intervals share endpoints (by nestedness), color pair 0/1 cannot have an active weight-2 between times $t-1$ and t. Thus, we can swap the contents of colors 0/1 with those of colors 2/3 from time 0 up until t. The result is a coloring in which no weight-2 appears in colors 3-10 after time $t-1$. Thus, one of case 1, case 2 or case 3 with t strictly earlier must hold. If case 3 holds, we can repeat this process only finitely many times (at most B times) until case 1 or case 2 must hold.

Finally, let us check that we can color when going from a 0-stage to a 1-stage. Note that there must be 6 weight-1 intervals intersecting the incoming weight-2

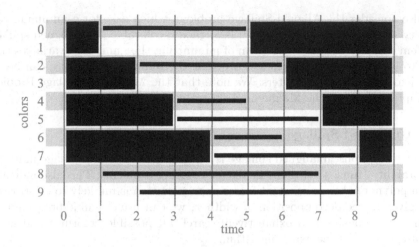

Fig. 3. Example where $a\chi(G) = \omega(G) + 2$

interval. Collectively, these must block all color pairs. Thus, we assume that two of these are in colors 0/1 and the others are colored 2,4,6,8. However, from this point on, we can use the same argument as in the proof of going from a 1-stage to a 2-stage (this proof did not rely on a weight-2 being in 0/1).

Thus, we can modify our coloring to accomodate all stages in the instance. This proves that $a\chi(10) \leq 12$. Figure 3 gives a tight example showing $a\chi(8) \geq 10$ (which can be extended to show $a\chi(10) \geq 12$ by simply adding intervals).

To color, we do the following for each interval. We retrieve the available colors and their last appearing weight-2 interval and swap some color assignments in linear time. We repeat this process until we can place the interval. However, since there are a linear number of timesteps, we can color a sub-instance in quadratic time. Since we need only record the current color assignment, a linear amount of space suffices.   □

Theorems 3, 4 and 5 give us a $\frac{4}{3}$-approximation for aligned coloring of general interval graphs. This algorithm decomposes the graph into subgraphs and colors each subgraph independently. Let n_i be the size of subgraph i. Since we can color each subgraph in time $O(n_i^2)$ and space $O(n_i)$, it follows that our algorithm has time complexity $O(n^2)$ and space complexity $O(n)$.

4   Minimum Spillage

We now consider the Minimum Spillage Problem defined as follows:

Definition 2 (Minimum Spillage Problem). *We are given a graph $G = (V, E)$ and a number of colors c. We wish to select a minimum cardinality set $V' \subseteq V$ and a c-coloring ϕ of the subgraph $G[V - V']$ of G induced by $V - V'$.*

We also consider the Aligned Spillage problem, which asks for a minimum cardinality set V' such that $G[V - V']$ is aligned c-colorable. The aligned spillage problem corresponds to the problem of minimizing the number of variables that need to be "spilled" into memory such that the remaining variables can be assigned to the available registers. We note that the Minimum Spillage Problem is complementary to the Maximum c-Colorable Subgraph problem.

4.1   Unaligned Spillage

For general graphs, unaligned spillage is NP-Complete since 0 spillage indicates c-colorability (thus, solving the unaligned coloring problem). This also indicates that a polynomial-time constant-factor approximation is unlikely to exist. However, given a tree decomposition of width τ, we can use dynamic programming on this decomposition to exhaustively search all possible colorings and get a $O(nm(c+1)^{\tau+1})$-time exact algorithm.

Unaligned spillage on *chordal graphs* is known to be NP-Hard [21]. However, the reduction of [21] is not approximation-preserving and thus does not give a bound on hardness of approximation. We offer an alternative reduction which establishes that unaligned spillage cannot be approximated to within a logarithmic factor.

Theorem 6. *The unaligned spillage problem is NP-complete even for chordal graphs. There is no $o(\log |V|)$-approximation unless $NP \subseteq DPTIME(n^{O(\log n)})$.*

Proof. This can be shown via a simple approximation-preserving reduction from set cover. Suppose we would like to solve a set cover instance with n elements and m sets. We create a vertex for each of the m sets, and connect all of these vertices into a clique. For each element x we create $m + 1 - \delta(x)$ vertices, where $\delta(x)$ is the number of sets containing x. We connect all of these vertices into a clique. Additionally, we connect each of these vertices to the vertices representing sets containing x. It is simple to check that this graph is chordal.

We now ask to color the graph with $c = m$ colors. Suppose that we can compute a partial coloring in polynomial time, which approximately minimizes the number of spilled vertices. We observe that for any vertex representing an element, that vertex and its neighbors form a clique of size exactly $m + 1$, from which it follows that for every element, either one of the vertices representing that element is spilled, or one of the vertices representing a set containing that element is spilled. If a vertex representing an element is spilled, we can pick a set containing that element and swap its color for the spilled vertex. This does not increase the number of spilled vertices, and also cannot violate the validity of the coloring. We conclude that we can find a partial coloring with the same number of spilled vertices, where only set vertices are spilled, and for every element one of the sets containing that element has a spilled vertex. Thus the spilled vertices imply a set cover. Conversely, we can spill the vertices of any set cover and get a valid coloring (color other vertices greedily). This approximation-preserving reduction from set cover combined with [11] implies the hardness results given.   □

On a chordal graph, we can obtain a $O(\log |V|)$-approximation by reducing to a set multicover instance. Our elements are supernodes X_i of a tree decomposition and our sets are $\{X_i \mid v \in X_i\}$ for each $v \in V$. We now wish to find a minimum collection of sets so that each element X_i is covered $|X_i| - c$ times. Alternately, we can use dynamic programming to obtain a $O\left(cnm(\tau + 1)^c\right)$-time exact algorithm [21].

We can remove the dynamic program runtime's dependence on τ by doing the following: Fix $\alpha > 1$. For each X_i in the tree decomposition with $|X_i| \geq \alpha c$, we will spill all of X_i, then remove every vertex in X_i from the tree decomposition. We repeat this process until all supernodes of the tree decomposition have weight less than αc. At this point, we can use dynamic programming to find the best coloring in $O(nm(\alpha c + 1)^c)$-time.

Theorem 7. *The dynamic programming algorithm yields a $\frac{\alpha}{\alpha-1}$-approximation.*

Proof. Order the supernodes of the tree decomposition $X_1, \ldots, X_p$. Let $I = \{i_1, i_2, \ldots, i_q\}$ be the set of indices corresponding to the X_i that were declared spilled. Let $N_k = \bigcup_{j<k} X_{i_j}$ and N_k^* be the set of nodes of N_k that were spilled by the optimum solution. We would like to show that $|N_q^*| \geq \frac{\alpha-1}{\alpha}|N_q|$. We prove this by induction:

Clearly,

$$|N_1^*| \geq |X_1| - c \geq |X_1|\left(1 - \frac{1}{\alpha}\right) = |N_1|\left(1 - \frac{1}{\alpha}\right)$$

where the second inequality follows from the fact that $|X_1| \geq \alpha c$.

Now, assume that $|N_{j-1}^*| \geq \frac{\alpha-1}{\alpha}|N_{j-1}|$. We know that since X_j was declared spilled we have $|X_j - N_{j-1}| \geq \alpha c$ which gives $c \leq \frac{|X_j - N_{j-1}|}{\alpha}$. By definition of N_j, we also have $N_j - N_{j-1} = X_j - N_{j-1}$. Then notice:

$$|N_j^* - N_{j-1}^*| \geq |X_j - N_{j-1}^*| - c \geq |N_j - N_{j-1}|\left(1 - \frac{1}{\alpha}\right)$$

Summing this with our inductive hypothesis gives us $|N_j^*| \geq |N_j|\left(1 - \frac{1}{\alpha}\right)$.   □

We note that unaligned spillage on interval graphs is polynomial-time solvable by a standard greedy removal of intervals to get a clique index of c [21]. We simply find the smallest time t during which more than c intervals are active, remove the interval with latest endpoint and repeat the process.

4.2   Aligned Spillage

The dynamic programming algorithm for chordal graphs is faster than for general graphs because we do not actually need to track the colorings, only the set of spilled nodes. This will not work for aligned spillage, because in an aligned chordal graph the number of colors needed can be greater than the clique index. However, we can still apply the general spillage algorithm in time

$O(nm(c + 1)^{\tau+1})$. We can make use of the technique of theorem 7 to obtain a $\frac{\alpha}{\alpha-1}$-approximation in time $O(nm(c + 1)^{\alpha c+1})$, an improvement if $\tau \gg c$.

Alternately, we can produce a bicriterion approximation by first reducing the clique number to c by spilling the minimum number of nodes (for an interval graph) or $\log |V|$ times the minimum number of nodes (for a chordal graph), then apply our approximation algorithms for aligned coloring. This yields $(\frac{4}{3}, 1)$ and $(\frac{3}{2}, \log |V|)$ approximations for interval and chordal graphs respectively, where an (α, β) approximation implies that we color with αc colors while spilling at most β times the number of nodes which optimum would have to spill in order to color with c colors.

Acknowledgements. We would like to thank Professor Jens Palsberg and Dr. Fernando Magno Quintão Pereira for introducing us to the Aligned Coloring problem and for many useful discussions about register allocation.

References

1. Ahn, M., Lee, J., Paek, Y.: Optimistic coalescing for heterogeneous register architectures. SIGPLAN Not. 42(7), 93–102 (2007)
2. Briggs, P.: Register allocation via graph coloring. Technical Report TR92-183, Rice University (1998)
3. Briggs, P., Cooper, K., Torczon, L.: Coloring register pairs. Letters on Programming Languages 1(1), 3–13 (1992)
4. Briggs, P., Cooper, K.D., Kennedy, K., Torczon, L.: Coloring heuristics for register allocation. SIGPLAN Not. 39(4), 283–294 (2004)
5. Briggs, P., Cooper, K.D., Torczon, L.: Improvements to graph coloring register allocation. ACM Trans. Program. Lang. Syst. 16(3), 428–455 (1994)
6. Callahan, D., Koblenz, B.: Register allocation via hierarchical graph coloring. SIGPLAN Not. 26(6), 192–203 (1991)
7. Chaitin, G.J.: Register allocation and spilling via graph coloring. SIGPLAN Notices 17, 98–105 (1982)
8. Chaitin, G.J., Auslander, M.A., Chandra, A.K., Cocke, J., Hopkins, M.E., Markstein, P.W.: Register allocation via coloring. Computer Languages 6, 47–57 (1981)
9. Chow, F., Hennessy, J.: Register allocation by priority-based coloring. SIGPLAN Not. 19(6), 222–232 (1984)
10. Cooper, K., Torczon, L.: Engineering a Compiler. Morgan Kaufmann, San Francisco (2003)
11. Feige, U.: A threshold of $\ln n$ for approximating set cover. Journal of the ACM 45(4), 634–652 (1998)
12. Gavril, F.: Algorithms for minimum coloring, maximum clique, minimum covering by cliques, and maximum independent set of a chordal graph. Siam Journal of Computing 1(2), 180–187 (1972)
13. Gupta, R., Soffa, M.L., Steele, T.: Register allocation via clique separators. In: PLDI 1989: Proceedings of the ACM SIGPLAN 1989 Conference on Programming language design and implementation, pp. 264–274 (1989)
14. Hack, S., Goos, G.: Optimal register allocation for SSA-form programs in polynomial time. Information Processing Letters 98(4), 150–155 (2006)

15. Khot, S.: Improved inapproximability results for maxclique, chromatic number and approximate graph coloring. In: Proceedings of the 42nd IEEE Symposium on Foundations of Computer Science (FOCS), pp. 600–609 (2001)
16. Lee, J.K., Palsberg, J., Pereira, F.M.Q.: Aliased register allocation for straight-line programs is NP-complete. In: Arge, L., Cachin, C., Jurdziński, T., Tarlecki, A. (eds.) ICALP 2007. LNCS, vol. 4596, pp. 680–691. Springer, Heidelberg (2007)
17. Pereira, F.M.Q., Palsberg, J.: Register allocation via coloring of chordal graphs. In: Proceedings of the Asian Symposium on Programming Lanugages and Systems (APLAS), Tsukuba, Japan, pp. 315–329 (2005)
18. Pereira, F.M.Q., Palsberg, J.: Register allocation by puzzle solving. In: Proceedings of the 2008 ACM SIGPLAN conference on Programming language design and implementation, pp. 216–226 (2008)
19. Rose, D.J., Tarjan, R.E.: Algorithmic aspects of vertex elimination. In: STOC 1975: Proceedings of seventh annual ACM symposium on Theory of computing, pp. 245–254 (1975)
20. Smith, M.D., Ramsey, N., Holloway, G.: A generalized algorithm for graph-coloring register allocation. In: Proceedings of the ACM SIGPLAN 2004 conference on Programming language design and implementation, pp. 277–288 (2004)
21. Yannakakis, M., Gavril, F.: The maximum k-colorable subgraph problem for chordal graphs. Information Processing Letters 24(2), 133–137 (1987)

Unsplittable Flow in Paths and Trees and Column-Restricted Packing Integer Programs

Chandra Chekuri*, Alina Ene, and Nitish Korula**

Dept. of Computer Science, University of Illinois, Urbana, IL 61801
{chekuri,ene1,nkorula2}@illinois.edu

Abstract. We consider the unsplittable flow problem (UFP) and the closely related column-restricted packing integer programs (CPIPs). In UFP we are given an edge-capacitated graph $G = (V, E)$ and k request pairs $R_1, \ldots, R_k$, where each R_i consists of a source-destination pair (s_i, t_i), a demand d_i and a weight w_i. The goal is to find a maximum weight subset of requests that can be routed unsplittably in G. Most previous work on UFP has focused on the *no-bottleneck* case in which the maximum demand of the requests is at most the smallest edge capacity. Inspired by the recent work of Bansal *et al.* [3] on UFP on a path without the above assumption, we consider UFP on paths as well as trees. We give a simple $O(\log n)$ approximation for UFP on trees when all weights are identical; this yields an $O(\log^2 n)$ approximation for the weighted case. These are the first non-trivial approximations for UFP on trees. We develop an LP relaxation for UFP on paths that has an integrality gap of $O(\log^2 n)$; previously there was no relaxation with $o(n)$ gap. We also consider UFP in general graphs and CPIPs without the no-bottleneck assumption and obtain new and useful results.

1 Introduction

In the Unsplittable Flow Problem (hereafter, UFP), the input is a graph $G(V, E)$ (directed or undirected; in this paper, we chiefly focus on the latter case) with a capacity c_e on each edge $e \in E$, and a set $\mathcal{R} = \{R_1, R_2, \ldots R_k\}$ of *requests*. Each request R_i consists of a pair of vertices (s_i, t_i), a demand d_i, and a weight/profit w_i. To route a request R_i is to send d_i units of flow along a *single* path (hence the name *unsplittable flow*) in G from s_i to t_i. The goal is to find a maximum-profit set of requests that can be simultaneously routed without violating the capacity constraints; that is, the total flow on an edge e should be at most c_e. A special case of UFP when $d_i = 1$ for all i and $c_e = 1$ for all e is the classical maximum edge-disjoint path problem (MEDP). MEDP has been extensively studied, and its approximability in directed graphs is better understood — the best approximation ratio known is is $O(\min\{\sqrt{m}, n^{2/3} \log^{1/3} n)\})$ [20,33], while it is NP-Hard to approximate to within a factor better than $n^{1/2-\varepsilon}$ [18];

---

* Partially supported by NSF grants CCF 07-28782 and CNS-0721899.
** Partially supported by NSF grant CCF 07-28782.

I. Dinur et al. (Eds.): APPROX and RANDOM 2009, LNCS 5687, pp. 42–55, 2009.

here n and m are the number of vertices and edges respectively in the input graph. For undirected graphs there is a large gap between the known upper and lower bounds on the approximation ratio: there is an $O(\sqrt{n})$-approximation [13] while the best known hardness factor is $\Omega(\log^{\frac{1}{2}-\varepsilon} n)$ under the assumption that $NP \not\subseteq ZPTIME(n^{O(\text{polylog}(n))})$ [1]. Thus UFP is difficult in general graphs even without the packing constraints imposed by varying demand values; one could ask if UFP is harder to approximate than MEDP. Most of the work on UFP has been on two special cases. One is the uniform capacity UFP (UCUFP) in which $c_e = C$ for all e and the other is UFP with the no-bottleneck assumption (UFP-NBA) where one assumes that $\max_i d_i \leq \min_e c_e$. Note that UCUFP is a special case of UFP-NBA. Kolliopoulos and Stein [22] showed, via grouping and scaling techniques, that certain linear programming based approximation algorithms for MEDP can be extended with only an extra constant factor loss to UFP-NBA. This reduction holds even when one considers restricted families of instances, say, those induced by planar graphs. See [15] for a precise definition of when the reduction applies. In [7,30], a different randomized rounding approach was used for UFP-NBA.

In this paper we are primarily interested in UFP instances that *do not* necessarily satisfy the no-bottleneck assumption. UCUFP and UFP-NBA have many applications and are of interest in themselves. However, the general UFP, due to algorithmic difficulties, has received less attention. One can extend some results for MEDP and UFP-NBA to UFP by separately considering requests that are within say a factor of 2 of each other; this geometric grouping incurs an additional factor of $\log d_{\max}/d_{\min}$ in the approximation ratio, which could be as large as a factor of n [18]. Azar and Regev showed that UFP in directed graphs is $\Omega(n^{1-\varepsilon})$-hard unless $P = NP$; note that the hardness for UFP-NBA is $\Omega(n^{1/2-\varepsilon})$ [18]. Chakrabarti et al. [12] observed that the natural LP relaxation has $\Omega(n)$ integrality gap even when G is a path. In contrast, the integrality gap for the path is $O(1)$ for UFP-NBA [12,15]. One could argue that the integrality gap of the natural LP has been the main bottleneck in addressing UFP.

This paper is inspired by the recent work of Bansal et al. [3] who gave an $O(\log n)$ approximation for UFP on a path. Interestingly, this was the first nontrivial approximation for this problem; previously there was a quasi-polynomial time approximation scheme [4], provided the capacities and demands are quasi-polynomially bounded in n. We note that UFP even on a single edge is NP-Hard, since it is equivalent to the knapsack problem. UFP on a path has received considerable attention, not only as an interesting special case of UFP, but also as a problem that has direct applications to resource allocation where one can view the path as modeling the availability of a resource over time. See [5,6,9,12,4,3] for previous work related to UFP on a path. The algorithm in [3] is combinatorial and bypasses the $\Omega(n)$ lower bound on the integrality gap of the natural LP. An open problem raised in [3] is whether UFP on trees also has a poly-logarithmic approximation. The difficulty of UFP on paths and trees is not because of routing (there is a unique path between any two nodes) but entirely due to the difficulty of choosing the subset to route. We note that this subset selection problem is easy on a path if $d_i = 1$ for all i (the natural LP is integral since the incidence matrix is totally unimodular) while this special case is already NP-Hard (and APX-Hard to approximate) on capacitated trees [17]. A constant factor approximation is known for UFP-NBA on trees [15]. We prove the following theorem, answering positively the question raised in [3].

Theorem 1. *There is an $O(\log n)$ approximation for UFP on n-vertex trees when all weights are equal. There is an $O(\log n \cdot \min\{\log n, \log k\})$ approximation for arbitrary non-negative weights.*

We borrow a crucial high-level idea from [3] of decomposing the given instance into one in which the demands all intersect. We, however, deviate from their approach of using dynamic programming for "large" demands which does not (seem to) generalize from paths to trees; our algorithm for trees is significantly simpler than the complex dynamic programming for the path used by [3]. We show that for the unit-weight case, a greedy algorithm is a 2-approximation if all requests go through a common vertex in the tree. This insight into the performance of the greedy algorithm allows us to develop a new linear programming relaxation for paths.

Theorem 2. *There is a linear programming relaxation for UFP on the path that has an integrality gap of $O(\log n \cdot \min\{\log n, \log k\})$ and there is a polynomial time algorithm that obtains a feasible $O(1)$-approximate solution to the relaxation.*

The separation oracle for the exponential-sized relaxation we develop is non-trivial. The integrality gap of the relaxation may very well be $O(1)$; resolving this is an interesting open problem. We underscore the novelty of our relaxation by showing that some reasonable approaches to strengthening the natural relaxation fail to improve the gap. In particular we show that the relaxation obtained after applying t rounds of the Sherali-Adams lift-and-project scheme [29] to the natural relaxation has a gap of $\Omega(n/t)$.

Column-Restricted Packing Integer Programs: UFP on paths and trees are special cases of column-restricted packing integer programming problems (CPIP). A packing integer program (PIP) is an optimization problem of the form $\max\{wx \mid Ax \le b, x \in \{0,1\}^n\}$ where A is a non-negative matrix; we use (A, w, b) to define a PIP. A CPIP has the additional restriction that all the non-zero entries in each column of A are identical. It is easy to write UFP on a tree as a CPIP (see Section 5 for formal details). The common coefficient of each column is the "demand" of that column. UFP on general graphs can also be related to CPIPs by using the path formulation and additional constraints [22]. A 0-1 PIP is one in which all entries of A are in $\{0,1\}$; note that it is also a CPIP. 0-1 PIPs capture the maximum independent set problem (MIS) as a special case and the strong inapproximability results for MIS [19] imply that no $n^{1-\varepsilon}$-approximation is possible for 0-1 PIPs unless $P = NP$; here n is the number of columns of A. However, an interesting question is the following. Suppose a 0-1 PIP has a small integrality gap because A has some structural properties. For example, if A is totally unimodular, then the integrality gap is

1. What can be said about a CPIP that is derived from A? In other words, one is asking how the "demand version" of a CPIP is related to its "unit-demand" version (see [22,28,15]). A CPIP satisfies the no-bottleneck assumption (NBA) if $\max_{i,j} A_{ij} \leq \min_i b_i$. Kolliopoulos and Stein [22] showed that for CPIPs that satisfy the NBA, one can relate the integrality gap of a CPIP to the gap of its underlying 0-1 PIP; there is only an extra constant factor. These ideas are what allows one to relate UFP-NBA to MEDP.

As with UFP, we are interested in this paper in CPIPs where we do not make the NBA assumption. As above, one could ask whether the integrality gap for the demand version of CPIP can be related to its unit-demand version. (Here, we refer to the "natural" relaxation in which one simply relaxes the integrality constraints.) However, the gap example for UFP on the path shows that unlike the no-bottleneck case, such a relationship is not possible. The unit-demand version of UFP on the path has integrality gap 1 while the demand-version has a gap of $\Omega(n)$. It is therefore natural to look for an intermediate case. In particular, suppose we have a CPIP (A, w, b) such that $\max_j A_{ij} \leq (1 - \delta)b_i$ for each i; this corresponds to the assumption that each demand is at most $(1 - \delta)$ times the *bottleneck* capacity for that demand. We call such a CPIP a δ-bounded CPIP. We informally state below a result that we obtain; the formal statement can be found in Section 5.

The integrality gap of a δ-bounded CPIP is at most $O(\log(1/\delta)/\delta^3)$ times the integrality gap of its unit-demand version.

The proof of the above is not difficult and is based on the grouping and scaling ideas of [22] with an additional trick. However, this has not been observed or stated before and the corollary below was not known previously.

Corollary 1. *For each fixed $\delta > 0$, there is an $O(\log(1/\delta)/\delta^3)$ approximation for UFP on paths and trees if the demand of each request is at most $(1-\delta)$ times the capacity of the edges on the unique path of the request.*

One class of CPIPs that have been studied before are those in which the maximum number of non-zero entries in any column is at most L. Baveja and Srinivasan [7] showed that the integrality gap of such CPIPs is $O(L)$ if A satisfies the no-bottleneck assumption. In recent and independent work, Pritchard [25] considered PIPs that have at most L non-zero entries per column, calling them L-column-sparse PIPs, and gave an $O(2^L L^2)$ approximation for them. We follow his notation, but obtain a tighter bound by restricting our attention to L-sparse CPIPs.

Theorem 3. *There is an $O(L)$-approximation for L-sparse CPIPs via the natural LP relaxation, even without the no-bottleneck assumption. If w is the all 1's vector then a simple greedy algorithm gives an L-approximation to the integral optimum (not necessarily with respect to the LP optimum).*

As corollaries we obtain the following results. We refer to UFP in which the paths for the routed requests have to contain at most L edges as L-bounded-UFP.

Corollary 2. *There is an $O(L)$-approximation for L-bounded-UFP in directed graphs.*

The demand-matching problem considered by Shepherd and Vetta [28] is an instance of a 2-bounded CPIP and therefore we have.

Corollary 3. *There is an $O(1)$-approximation for the demand-matching problem. Moreover, there is a 2-approximation for the cardinality version.*

Note that [28] gives a 3.264 approximation for general graphs and a 3-approximation for the cardinality version, both with respect to the LP optimum. Our $O(L)$ bound for L-bounded CPIPs has a larger constant factor since it does not take the structure of the particular problem into account, however the algorithm is quite simple. On the other hand, the greedy 2-approximation for the cardinality case was not noticed in [28].

Due to space constraints, we omit most of the proofs. A full version of the paper will be available on the authors' websites.

Other Related Work and Discussion: UFP and MEDP are extensively studied and we refer the reader to [1,13,14,16,20,21] for various pointers on approximation algorithms and hardness results. Schrijver [27] discusses known results on exact algorithms in great detail. We focus on UFP on paths and trees and have already pointed to the relevant literature. We mention some results on UFP for the special case when $w_i = d_i$. Kolman and Schiedeler [24] considered this special case in directed graphs and obtained an $O(\sqrt{m})$-approximation. Kolman [23] extended the results in [33] for UCUFP to this special case. We note that the $\Omega(n)$ integrality gap for the path [12] does not hold if $w_i = d_i$. In a technical sense, one can reduce a UFP instance with $w_i = d_i$ to an instance in which the ratio d_{max}/d_{min} is polynomially bounded. Two approximation techniques for UFP-NBA are greedy algorithms [20,22,2] and randomized rounding of the multi-commodity flow based LP relaxation [30,7,9,12]. These methods when dealing with UFP-NBA classify demands as "large" ($d_i \geq d_{max}/2$) and "small". Large demands can be reduced to uniform demands and handled by MEDP algorithms (since $d_{max} \leq \min_e c_e$) and small demands behave well for randomized rounding. This classification does not apply for UFP. A simple observation we make is that if we are interested in the cardinality problem then it is natural to consider the greedy algorithm that gives preference to smaller demands; under various conditions this gives a provably good algorithm. Another insight is that the randomized rounding algorithm followed by alteration [31,12] has good behaviour if we sort the demands in decreasing order of their size — this observation was made in [12] but its implication for general UFP was not noticed. Finally, the modification of the grouping and scaling ideas to handle δ-bounded demands is again simple but has not been noticed before. Moreover, for UFP on paths and trees one obtains constant factor algorithms for any fixed δ. We remark that this result is not possible to derive from the randomized rounding and alteration approach for paths (or trees) because the alteration approach needs to insert requests based on left end point to take advantage of the path structure while one needs the requests to be sorted in decreasing demand value order to handle the fact that we cannot separate small and large demands any more.

Strengthening LP relaxations by adding valid inequalities is a standard methodology in mathematical programming. There are various generic as well as problem specific approaches known. The knapsack problem plays an important role since each linear constraint in a relaxation can be thought of inducing a separate knapsack constraint. Knapsack cover inequalities [10] have been found to be very useful in reducing the integrality gap of covering problems [10,25]. However, it is only recently that Bienstock [8], answering a question of Van Vyve and Wolsey [32], developed an explicit system of inequalities for the knapsack packing problem (the standard maximization problem) that yields an approximation scheme. Wolsey (as reported in [15]) raises the question of how multiple knapsack constraints implied by the different linear constraints of a relaxation interact since that is what ultimately determines the strength of the relaxation. UFP on a path is perhaps a good test case for examining this question. The $\Omega(n)$ gap example shows the need to consider multiple constraints simultaneously — we hope that our formulation and its analysis is a step forward in tackling other problems.

2   UFP on Trees

Recall that each request R_i consists of a pair of vertices s_i, t_i, a demand d_i and a profit/weight w_i, and if selected, the entire d_i units of demand for this request must be sent along a single path. When the input graph is a tree, there is a *unique* path between each s_i and t_i. For such instances, we refer to this unique path P_i as being the request path for R_i.

The following flow-based LP relaxation is natural for UFP on trees: Here, x_i indicates whether flow is routed from s_i to t_i.

$$\max \quad \sum_{i=1}^{k} w_i x_i \quad \text{s.t.}$$
$$\sum_{i:\ P_i \ni e} d_i x_i \ \leq\ c_e \qquad\qquad (\forall e \in E(G))$$
$$x_i \ \in\ [0,1] \qquad\qquad (\forall i \in \{1,\ldots,k\})$$

This relaxation has an $O(1)$ integrality gap for UFP-NBA on trees [15]. Unfortunately, without NBA, the gap can be as large as $\Omega(n)$ even when the input graph is a path, as shown in [12] (see Section 3). No relaxations with gap $o(n)$ were previously known, even for UFP on paths. The difficulty appeared to lie in dealing with requests for which the demands are very close to the capacity constraints; we confirm this intuition by proving Corollary 1 in Section 5: For UFP on trees, if each $d_i \leq (1 - \delta) \min_{e \in P_i} c_e$, the natural LP relaxation has an integrality gap of $O(\text{poly}(1/\delta))$. In Section 4, we show how to handle large demands for UFP on paths by giving a new relaxation with an integrality gap of $O(\log n \cdot \min\{\log n, \log k\})$.

In this section, we give a simple combinatorial algorithm that achieves an $O(\log n \cdot \min\{\log n, \log k\})$-approximation for UFP on trees. We first obtain an

$O(\log n)$ approximation for unit-profit instances of UFP on trees with n vertices. To do this, we note that if all the request paths must pass through a common vertex, a simple greedy algorithm achieves a 2-approximation.

Lemma 1. *Consider unit profit instances of UFP on trees, for which there exists a vertex v such that all request paths pass through v. There exists a 2-approximation algorithm for such instances.*

Proof Sketch: We order the requests in increasing order according to their demands. We consider the requests in this order and, if adding the current request maintains feasibility, we add the request to our set. □

Lemma 2. *There exists an $O(\log n)$-approximation algorithm for unit profit instances of UFP on trees.*

Proof Sketch: It is well known that any n-vertex tree T has a vertex v, called a *center*, such that each component of $T \backslash v$ has at most $n/2$ vertices. If many request paths pass through the center v, we use Lemma 1 and are done; if not, most paths are entirely contained in the subtrees (each of size at most $n/2$) obtained after deleting v from T, and we can recurse. □

Theorem 1 now follows from Lemma 2 and Lemma 3 below, which is proved using standard profit-scaling.

Lemma 3. *Suppose there exists an r-approximation algorithm for unit profit instances of UFP on a given graph. Then there exists an $O(r \min\{\log n, \log k\})$-approximation algorithm for arbitrary instances of UFP on the graph, where k is the number of requests.*

3   LP Relaxations for UFP on Paths

The following Linear Programming relaxation is natural for UFP on paths. There is a variable x_i for each request R_i to indicate whether it is selected, and the constraints enforce that the total demand of selected requests on each edge is at most its capacity.

Standard LP $\max \sum_i w_i x_i$

$$\sum_{i:\, e \in P_i} d_i x_i \leq c_e \qquad (\forall e \in E(G))$$
$$x_i \in [0, 1] \qquad (\forall i \in \{1, \ldots, k\})$$

It is shown in [12] that the integrality gap of this LP relaxation is $\Theta(\log \frac{d_{\max}}{d_{\min}})$ where $d_{\max}$ and $d_{\min}$ are $\max_i d_i$ and $\min_i d_i$ respectively. Unfortunately, this gap can be as bad as $\Omega(n)$, as shown in the following example from [12]: the input path has n edges with edge i having capacity 2^i; request R_i is for 2^i units of capacity on edges i through n, and has profit 1. (See Fig. 1.) An integral solution can only route a single request, for a profit of 1; however, setting $x_i = 1/2$ for

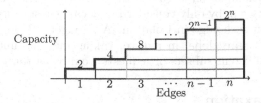

Fig. 1. An instance of UFP on paths with large integrality gap

each i is a feasible fractional solution to the LP, for a total profit of $n/2$. We refer to this instance as the *canonical integrality gap example*.

Though an $O(\log n)$-approximation algorithm for UFP on paths was given in [3], no LP with an integrality gap of $o(n)$ was known for this problem, and obtaining such an LP has been an interesting open question. One could attempt to write a configuration LP for the problem, or to consider strengthening the natural LP, for instance, via the Sherali-Adams hierarchy of relaxations. We remark that these relaxations also have feasible fractional solutions of profit $\Omega(n)$ for the canonical integrality gap example. For both of the relaxations below, we use $\mathcal{R}_e$ to denote the set of requests passing through edge e.

A Configuration LP: In the configuration LP below, there is a variable $x_{S,e}$ for each set $S \subseteq \mathcal{R}_e$ if the total demand d_S of the requests in S is at most the capacity c_e. Though this LP has an exponential number of variables, we can separate over its dual, which has a polynomial number of variables and constraints that are essentially equivalent to the knapsack problem (with polynomially bounded profits, since we assume that the profits of the original instance are integers in $\{1, \ldots, k^2\}$). However, the integrality gap of the configuration LP is also $n/2$, as shown by the canonical example; set $x_i = 1/2$ for each i, and for the jth edge e_j, set $x_{\{R_j\},e_j} = 1/2$, and $x_{S_j,e_j} = 1/2$, where $S_j = \{1, \ldots, j-1\}$. (On edge e_1, set $x_{\emptyset,e_1} = 1/2$.)

Config LP $\max \sum_i w_i x_i$

$$\sum_{S:\, S \subseteq \mathcal{R}_e} x_{S,e} = 1 \qquad\qquad (\forall e \in E(G))$$
$$x_i \le \sum_{S:\, S \subseteq \mathcal{R}_e} x_{S,e} \quad (\forall i \in \{1, \ldots, k\}, e \in P_i)$$
$$x_{S,e} \ge 0 \qquad\qquad (\forall e \in E(G), S \subseteq \mathcal{R}_e, d_S \le c_e)$$

The Sherali-Adams hierarchy for the **Standard LP***:* For a zero-one programming problem, let P denote the feasible integer polytope, and P_0 denote the convex polytope of an LP relaxation for P. The Sherali-Adams Hierarchy [29] is a sequence $P_0, P_1, P_2 \ldots P_n = P$ of (successively tighter) relaxations of P. We refer the reader to [29] for a more complete description of the Sherali-Adams Hierarchy; here, we simply note that the integrality gap of P_t is $\Omega(n/t)$.

Theorem 4. *After applying t rounds of the Sherali-Adams hierarchy to the relaxation* **Standard LP***, the integrality gap of the LP obtained is $\Omega(n/t)$.*

The two preceding examples show that it is difficult to write an LP relaxation with small integrality gap by only considering "local" constraints, which bound the capacity used on each edge in isolation. A stronger LP needs to introduce constraints that are more global in nature, taking into account that different edges may prevent different subsets of requests from being routed.

4 A New Relaxation

We now describe a Linear Programming Relaxation for the UFP on paths with an $O(\log^2 n)$ integrality gap. Corollary 1 implies that **Standard LP** has small integrality gap if the demand of each request is small compared to the capacity constraints; recall that in the canonical example with integrality gap $n/2$, every request, if routed, uses the *entire* capacity of the leftmost edge on its path. In the new LP relaxation, we keep the previous constraints to handle "small" requests, and introduce new rank constraints to deal with "big" requests.

For each request R_i, let the *bottleneck* for R_i be the edge in P_i with least capacity. (If multiple edges have the same minimum capacity, let the bottleneck be the leftmost edge.) Let $\mathcal{S} \subseteq \mathcal{R}$ be the set of all requests R such that the demand of R is smaller than $(3/4) \cdot c(e)$ where e is the bottleneck edge for R. Let $\mathcal{B} = \mathcal{R} \backslash \mathcal{S}$ denote the remaining ("big") requests, and let $\mathcal{B}_e$ denote the set of requests R in $\mathcal{B}$ such that the path for R passes through edge e. For each request R_i, we have a variable x_i denoting whether this request is selected or not. For each set $B \subseteq \mathcal{B}$ of big requests, let $f(B)$ denote the maximum *number* of requests in B that can be simultaneously routed without violating the capacity constraints. For each set B of "big" requests that pass through a common edge, we introduce a *rank* constraint which requires that the total extent to which requests in B are selected by the LP must be at most the number of requests in B that can be routed integrally.

UFP-LP $\max \sum_i w_i x_i$

$$\sum_{i:\, e \in P_i} d_i x_i \le c_e \quad (\forall e \in E(G)) \qquad \text{[capacity constraints]}$$
$$\sum_{R_i \in B} x_i \le f(B) \ (\forall e \in E(G), B \subseteq \mathcal{B}_e) \quad \text{[rank constraints]}$$
$$x_i \in [0,1] \ (\forall i \in \{1, \ldots, k\})$$

The new constraints enforce a small integrality gap; we prove Theorem 5 in Section 4.2. The upper bound on the integrality gap is not known to be tight; the integrality gap could be $O(\log n)$ or even $O(1)$.

Theorem 5. *The   LP   relaxation   **UFP-LP**   has   integrality   gap $O(\log n \cdot \min\{\log n, \log k\})$ for instances of UFP on paths, where n is the length of the path and k is the number of requests.*

An interesting question is obtaining a separation oracle for **UFP-LP**, which has an an exponential number of constraints. We describe an algorithm SEPARATION ORACLE and prove the following theorem in Section 4.1 below; together, Theorems 5 and 6 imply Theorem 2.

Theorem 6. *Let $x \in [0,1]^n$ and suppose there exists a set $B \subseteq \mathcal{B}_e$ such that $\sum_{R_i \in B} x_i > 18f(B)$. Then the algorithm* SEPARATION ORACLE(e) *returns a violated constraint.*

An approximate separation oracle such as the one guaranteed by Theorem 6 can be used to find an approximate solution to **UFP-LP**; this follows for a large class of packing and covering problems (see [11]); we omit details.

One can write a relaxation similar to **UFP-LP** for UFP on trees; though it has small integrality gap, we do not know a separation oracle as in Theorem 6.

4.1   A Separation Oracle

We now describe an approximate separation oracle for **UFP-LP**. We can obviously check in polynomial time whether there exists a capacity constraint that is violated (and return such a constraint if one exists). Therefore we may assume that all the capacity constraints are satisfied and hence we can safely ignore the requests in $\mathcal{S}$. We give an algorithm to detect a violated rank constraint at edge e if some rank constraint at e is violated by a factor of at least 18. We first introduce some notation:

We define $x(S) = \sum_{R_i \in S} x_i$. Let $\mathcal{B}_{\text{left}}(e) \subseteq \mathcal{B}_e$ be the set of requests R_i such that the bottleneck for R_i is to the *left* of edge e, and $\mathcal{B}_{\text{right}}(e)$ be the set of requests R_i with bottleneck to the right of e. (If the bottleneck for $R_i \in \mathcal{B}_e$ is edge e, R_i can be added to either $\mathcal{B}_{\text{left}}(e)$ or $\mathcal{B}_{\text{right}}(e)$.)

Let $left(e)$ denote the set of edges to the left of e, together with edge e, and let $right(e)$ be the set of edges to the right of e (again including e). For requests R_i, R_j both in $\mathcal{B}_{\text{left}}(e)$ (respectively, both in $\mathcal{B}_{\text{right}}(e)$) we say that R_i blocks R_j if there is an edge $e' \in left(e)$ (respectively, $right(e)$) such that $d_i + d_j > c_{e'}$ and both P_i and P_j pass through e'.

SEPARATION ORACLE(EDGE e):

> for each request $R_i \in \mathcal{B}_{\text{left}}(e)$
>> let $S = \{R_i\} \cup \{R_j | R_j \in \mathcal{B}_{\text{left}}(e), d_j > d_i, R_i \text{ blocks } R_j\}$
>> if $x(S) > 1$
>>> return S $\langle\!\langle f(S) = 1 \text{ by construction}\rangle\!\rangle$
>
> for each request $R_i \in \mathcal{B}_{\text{right}}(e)$
>> let $S = \{R_i\} \cup \{R_j | R_j \in \mathcal{B}_{\text{right}}(e), d_j > d_i, R_i \text{ blocks } R_j\}$
>> if $x(S) > 1$
>>> return S $\langle\!\langle f(S) = 1 \text{ by construction}\rangle\!\rangle$

Lemma 4. *For any set S returned by* SEPARATION ORACLE, $f(S) = 1$.

Thus, if this algorithm returns a set $S \subseteq \mathcal{B}_e$, the constraint corresponding to S and e is violated, as $x(S) > 1$. To prove Theorem 6, it remains only to show that if constraints are sufficiently violated, the algorithm will always return some set S corresponding to a violated constraint. The proof is somewhat involved, though the outline is simple: Given a set $B \subseteq \mathcal{B}_e$ such that $x(B) > 18f(B)$, we show the existence of a set $S \subseteq B$ with "simpler" structure, such that $f(S) = 1$ and $x(S) > 1$; that is, S is a violated set. The structure of S is such that the algorithm SEPARATION ORACLE(e) can find it.

Given an edge e and a set $S' \subseteq \mathcal{B}_e$, we say that S' is *feasible on the left* (respectively, *on the right*), if all requests in S' can be routed simultaneously without exceeding the capacity of any edge in $left(e)$ (respectively, $right(e)$). For any set $S \subseteq \mathcal{B}_e$, let $f_\ell(S)$ denote the maximum size subset of S that is feasible on the left and $f_r(S)$ denote the maximum size subset of S that is feasible on the right. (Equivalently, $f_\ell(S)$ is $f(S)$ in the instance obtained by truncating all requests at the right endpoint of e.)

Lemma 5. *If there exists a set $B \subseteq \mathcal{B}_e$ such that $x(B) > \alpha f(B)$, there exists a set $B' \subseteq \mathcal{B}_{left}(e)$ such that $x(B') > \frac{\alpha}{2} f_\ell(B')$ or a set $B'' \subseteq \mathcal{B}_{right}(e)$ such that $x(B'') > \frac{\alpha}{2} f_r(B'')$.*

By symmetry, we assume w.l.o.g. that there exists $B' \subseteq \mathcal{B}_{left}(e)$ such that $x(B') > (\alpha/2) f_\ell(B')$. For brevity, we complete the proof of Theorem 6 by only stating the remaining lemmas for the "left" side.

Lemma 6. *If there exists a set $S' \subseteq \mathcal{B}_{left}(e)$ such that $f_\ell(S') = 1$ and $x(S') > 1$, the algorithm SEPARATION ORACLE returns such a set.*

The previous two lemmas show that: (a) If there is a constraint violated by a factor α, there is one violated by a factor of $\alpha/2$ either "on the left" or "on the right", and (b) If there is a constraint corresponding to set S' violated on the left (or on the right) such that $f_\ell(S')$ (or $f_r(S')$) $= 1$, the algorithm detects it. To complete the proof of Theorem 6, our final lemma shows that if there is a constraint violated on the left by a large factor, there is a violated constraint corresponding to a set S' such that $f_\ell(S') = 1$.

Lemma 7. *If there exists a set $B \subseteq \mathcal{B}_{left}(e)$ such that $x(B) > \beta f_\ell(B)$ for some $\beta > 9$, there exists a set $S' \subseteq B$ such that $f_\ell(S') = 1$ and $x(S') > 1$.*

4.2 Bounding the Integrality Gap

Given an fractional solution to the LP of profit OPT, we show how to round it to obtain an integral solution of comparable profit. For any set S of requests, we define profit(S) as $\sum_{R_i \in S} w_i x_i$. We round "small" and "big" jobs separately; note that one of profit($\mathcal{S}$) or profit($\mathcal{B}$) is at least OPT/2. If profit($\mathcal{S}$) $\geq$ OPT/2 then one obtains from Corollary 1 that there is an integral solution of value $\Omega(\text{OPT})$; recall that for each request $R_i \in \mathcal{S}$, we have $d_i \leq (3/4) \min_{e \in P_i} c_e$.

The difficulty in bounding the integrality gap of LPs has been in dealing with the "big" requests. However, the new rank constraints allow one to overcome this. The proof essentially follows the combinatorial algorithm from Section 2. We apply the same arguments as in Section 2, now with respect to the LP solution instead of an integral optimum solution, to reduce the problem to an intersecting instance with unit weights; this loses an $O(\log n \min\{\log n, \log k\})$ factor. For an intersecting instance with unit weights, the rank constraints trivially show that the LP optimum is equal to the integral optimum. We omit further details. This completes the proof of Theorem 5.

5 UFP and Column-Restricted Packing Integer Programs

In this section, we consider a class of packing problems, so-called *Column-Restricted* Packing Integer Programs (hereafter, CPIPs), introduced by Kolliopoulos and Stein [22]. Let A be an arbitrary $m \times n$ $\{0, 1\}$ matrix, and d be an n-element vector with d_j denoting the jth entry in d. Let $A[d]$ denote the matrix obtained by multiplying every entry of column j in A by d_j. A CPIP is a problem of the form $\max wx$, subject to $A[d]x \leq b, x \in \{0,1\}^n$, for some integer vectors w, d, b.[1] (Intuitively, a CPIP is a 0-1 packing program in which all non-zero coefficients of a variable x_j are the same.) It is easy to see that the natural LP for UFP in paths and trees is a CPIP.

CPIPs were studied in [22,15], and it was shown that the integrality gap of a CPIP with $\max_j d_j \leq \min_i b_i$ is at most a constant factor more than the integrality gap of the corresponding "unit-demand" version; we explain this more formally below, using the notation introduced by [15].

Let P be a convex body in $[0,1]^n$ and $w \in \mathbf{R}^n$ be an objective vector; for any choice of P, w, we obtain a maximization problem $\max\{wx : x \in P\}$. Let γ denote the fractional optimum value of this program, and γ^* denote the optimum *integral* value, which is given by $\max wx$ over all integer vectors $x \in P$. The *integrality gap* of P is γ/γ^*, the ratio between the value of the optimal fractional and integral solutions. A class $\mathcal{P}$ of integer programs is given by problems induced by pairs P, w as above; the integrality gap for a class of problems $\mathcal{P}$ is the supremum of integrality gaps for each problem in $\mathcal{P}$.

We say that a collection of vectors $W \subseteq \mathbf{Z}^n$ is *closed* if for each $w \in W$, replacing any entry w_i with 0 gives a vector $w' \in W$. Subsequently, for each $m \times n$ matrix A and closed collection of vectors W in Z^n, we use $\mathcal{P}(A, W)$ to denote the class of problems of the form $\max\{wx : Ax \leq b, x \in [0,1]^n\}$, where $w \in W$ and b is a vector in $\mathbf{Z}_+^m$. We let $\mathcal{P}^{dem}(A, W)$ denote the class of problems of the form $\max\{wx : A[d]x \leq b, x \in [0,1]^n\}$ where $w \in W, b \in \mathbf{Z}_+^m, d \in \mathbf{Z}_+^n$. Finally, we use $\mathcal{P}_{nba}^{dem}(A, W)$ to denote the class of problems of the same form that satisfy $\max_j d_j \leq \min_i b_i$. For UFP, the condition $\max_j d_j \leq \min_i b_i$ corresponds to the no-nottleneck assumption.

Using techniques introduced in [22], the following theorem was proved in [15]:

Theorem 7 ([15]). *Let A be a $\{0, 1\}$ matrix and W be a closed collection of vectors. If the integrality gap for the collection of problems $\mathcal{P}(A, W)$ is at most Γ, then the integrality gap for the collection of problems $\mathcal{P}_{nba}^{dem}(A, W)$ is at most $11.542\Gamma \leq 12\Gamma$.*

The above theorem is used in [15] to give an $O(1)$-approximation for UFP-NBA on trees. Unfortunately, the analogous theorem is not true for $\mathcal{P}^{dem}(A, W)$, as

---

[1] If vectors w, d, b are rational, we can scale them as necessary.

shown by the canonical integrality gap example for the UFP linear program **Standard LP**. In this section, we note that if there exists $\delta < 1$ such that for each i, we have $\max_j A_{ij}d_j \leq (1 - \delta)b_i$, we *can* obtain an analogous theorem, with integrality gap depending on δ. More precisely, let $\mathcal{P}_\delta^{dem}(A, W)$ denote the class of problems of the form $\max\{wx : A[d]x \leq b, x \in [0, 1]^n\}$ where $w \in W, b \in \mathbf{Z}_+^m, d \in \mathbf{Z}_+^n$, and $\forall i$, $\max_j A_{ij}d_j \leq (1 - \delta)b_i$.

Theorem 8. *Let A be a $\{0, 1\}$ matrix and W be a closed collection of vectors. If the integrality gap for $\mathcal{P}(A, W)$ is at most Γ, the integrality gap for $\mathcal{P}_\delta^{dem}(A, W)$ is at most $O(\frac{\log(1/\delta)}{\delta^3} \cdot \Gamma)$.*

Thus, we obtain Corollary 1 as a special case of Theorem 8.

6   Concluding Remarks

Is there an $O(1)$-approximation for UFP on paths, and more generally on trees? Is the integrality gap of **UFP-LP** $O(1)$? Is there an LP relaxation for UFP on trees with poly-logarithmic integrality gap?

We recently obtained an $O(L^2)$-approximation ratio and integrality gap bound for L-sparse PIPs using the iterated rounding idea of Pritchard [25]; this improves his bound of $O(2^L L^2)$. Can the bound be improved to $O(L)$, matching the lower bound on the integrality gap?

References

1. Andrews, M., Chuzhoy, J., Khanna, S., Zhang, L.: Hardness of the undirected edge-disjoint paths problem with congestion. In: Proc. of IEEE FOCS, pp. 226–241 (2005)
2. Azar, Y., Regev, O.: Combinatorial algorithms for the unsplittable flow problem. Algorithmica 441(1), 49–66 (2006)
3. Bansal, N., Friggstad, Z., Khandekar, R., Salavatipour, M.R.: A logarithmic approximation for unsplittable flow on line graphs. In: Proc. of ACM-SIAM SODA, pp. 702–709 (2009)
4. Bansal, N., Chakrabarti, A., Epstein, A., Schieber, B.: A Quasi-PTAS for unsplittable flow on line graphs. In: Proc. of ACM STOC, pp. 721–729 (2006)
5. Bar-Noy, A., Bar-Yehuda, R., Freund, A., Naor, J., Schieber, B.: A unified approach to approximating resource allocation and scheduling. JACM 48(5), 1069–1090 (2001)
6. Bar-Noy, A., Guha, S., Naor, J., Schieber, B.: Approximating the Throughput of Multiple Machines in Real-Time Scheduling. SICOMP 31(2), 331–352 (2001)
7. Baveja, A., Srinivasan, A.: Approximation algorithms for disjoint paths and related routing and packing problems. Math. Oper. Res. 25(2), 255–280 (2000)
8. Bienstock, D.: Approximate formulations for 0-1 knapsack sets. Oper. Res. Lett. 36(3), 317–320 (2008)
9. Calinescu, G., Chakrabarti, A., Karloff, H., Rabani, Y.: Improved approximation algorithms for resource allocation. In: Proc. of IPCO, pp. 439–456 (2001)
10. Carr, R.D., Fleischer, L., Leung, V.J., Phillips, C.A.: Strengthening integrality gaps for capacitated network design and covering problems. In: Proc. of ACM-SIAM SODA, pp. 106–115 (2000)

11. Carr, R., Vempala, S.: Randomized meta-rounding. Random Structures and Algorithms 20(3), 343–352 (2002)
12. Chakrabarti, A., Chekuri, C., Gupta, A., Kumar, A.: Approximation Algorithms for the Unsplittable Flow Problem. Algorithmica 47(1), 53–78 (2007)
13. Chekuri, C., Khanna, S., Shepherd, F.B.: An $O(\sqrt{n})$ Approximation and Integrality Gap for Disjoint Paths and Unsplittable Flow. Theory of Computing 2, 137–146 (2006)
14. Chekuri, C., Khanna, S., Shepherd, F.B.: Edge-Disjoint Paths in Planar Graphs with Constant Congestion. In: Proc. of ACM STOC, pp. 757–766 (2006)
15. Chekuri, C., Mydlarz, M., Shepherd, F.B.: Multicommodity Demand Flow in a Tree and Packing Integer Programs. ACM Trans. on Algorithms 3(3) (2007)
16. Chuzhoy, J., Guruswami, V., Khanna, S., Talwar, K.: Hardness of Routing with Congestion in Directed Graphs. In: Proc. of ACM STOC, pp. 165–178 (2007)
17. Garg, N., Vazirani, V.V., Yannakakis, M.: Primal-dual approximation algorithms for integral flow and multicut in trees. Algorithmica 18(1), 3–20 (1997)
18. Guruswami, V., Khanna, S., Shepherd, B., Rajaraman, R., Yannakakis, M.: Near-Optimal Hardness Results and Approximation Algorithms for Edge-Disjoint Paths and Related Problems. J. of Computer and System Sciences 67(3), 473–496 (2003)
19. Håstad, J.: Clique is Hard to Approximate within $n^{1-\varepsilon}$. Acta Mathematica 182, 105–142 (1999)
20. Kleinberg, J.M.: Approximation algorithms for disjoint paths problems. PhD thesis, MIT EECS (1996)
21. Kolliopoulos, S.G.: Edge-disjoint Paths and Unsplittable Flow. In: Gonzalez, T.F. (ed.) Handbook of Approximation Algorithms and Metaheuristics. Chapman & Hall/CRC, Boca Raton (2007)
22. Kolliopoulos, S.G., Stein, C.: Approximating disjoint-path problems using greedy algorithms and Packing Integer Programs. Math. Prog. A (99), 63–87 (2004)
23. Kolman, P.: A Note on the Greedy Algorithm for the Unsplittable Flow Problem. Information Processing Letters 88(3), 101–105 (2003)
24. Kolman., P., Scheideler, C.: Improved bounds for the unsplittable flow problem. J. of Algorithms 61(1), 20–44 (2006)
25. Pritchard, D.: Approximability of Sparse Integer Programs. In: Proc. of ESA (to appear, 2009); arXiv.org preprint, arXiv:0904.0859v1, http://arxiv.org/abs/0904.0859
26. Raghavan, P.: Probabilistic Construction of Deterministic Algorithms: Approximating Packing Integer Programs. JCSS 37(2), 130–143 (1988)
27. Schrijver, A.: Combinatorial Optimization: Polyhedra and Efficiency. Springer, Heidelberg (2003)
28. Shepherd, B., Vetta, A.: The Demand Matching Problem. Math. of Operations Research 32(3), 563–578 (2007)
29. Sherali, H., Adams, W.P.: A Hierarchy of Relaxations and Convex Hull Characterizations for Mixed-Integer Zero-One Programming Problems. Discrete Applied Mathematics and Combinatorial Operations Research and Computer Science 52 (1994)
30. Srinivasan, A.: Improved approximations for edge-disjoint paths, unsplittable flow, and related routing problems. In: Proc. of IEEE FOCS, pp. 416–425 (1997)
31. Srinivasan, A.: New Approaches to Covering and Packing Problems. In: Proc. of ACM-SIAM SODA, pp. 567–576 (2001)
32. Van Vyve, M., Wolsey, L.A.: Approximate Extended Formulations. Math. Prog. 105, 501–522 (2006)
33. Varadarajan, K., Venkataraman, G.: Graph Decomposition and a Greedy Algorithm for Edge-disjoint Paths. In: Proc. of ACM-SIAM SODA, pp. 379–380 (2004)

Truthful Mechanisms via Greedy Iterative Packing[*]

Chandra Chekuri[1] and Iftah Gamzu[2]

[1] Department of Computer Science, University of Illinois, Urbana, IL 61801, USA
chekuri@cs.illinois.edu
[2] Blavatnik School of Computer Science, Tel-Aviv University, Tel-Aviv 69978, Israel
iftgam@tau.ac.il

Abstract. An important research thread in algorithmic game theory studies the design of efficient truthful mechanisms that approximate the optimal social welfare. A fundamental question is whether an α-approximation algorithm translates into an α-approximate truthful mechanism. It is well-known that plugging an α-approximation algorithm into the VCG technique may not yield a truthful mechanism. Thus, it is natural to investigate properties of approximation algorithms that enable their use in truthful mechanisms.

The main contribution of this paper is to identify a useful and natural property of approximation algorithms, which we call loser-independence; this property is applicable in the single-minded and single-parameter settings. Intuitively, a loser-independent algorithm does not change its outcome when the bid of a losing agent increases, unless that agent becomes a winner. We demonstrate that loser-independent algorithms can be employed as sub-procedures in a greedy iterative packing approach while preserving monotonicity. A greedy iterative approach provides a good approximation in the context of maximizing a non-decreasing submodular function subject to independence constraints. Our framework gives rise to truthful approximation mechanisms for various problems. Notably, some problems arise in online mechanism design.

1 Introduction

Algorithmic aspects of mechanism design have become an important area of research in recent years. A central research theme focuses on the design of efficient mechanisms for algorithmic problems in strategic settings. These mechanisms must take into account both standard computational efficiency considerations and strategic behavior of the participants. The latter goal commonly correlates with the development of truthful mechanisms, namely, mechanisms that are robust against manipulation by the participants. The primary technique of mechanism design, i.e., VCG mechanisms [17, 25, 41], is known to be truthful for optimizing social welfare. Unfortunately, implementing VCG is

---

[*] Proofs and details omitted from this extended abstract appear in the full version of this paper. The first author was partially supported by NSF grants CCF-0728782 and CNS-0721899. The second author was supported by the Binational Science Foundation, by the Israel Science Foundation, by the European Commission under the Integrated Project QAP funded by the IST directorate as Contract Number 015848, and by a European Research Council (ERC) Starting Grant.

I. Dinur et al. (Eds.): APPROX and RANDOM 2009, LNCS 5687, pp. 56–69, 2009.
© Springer-Verlag Berlin Heidelberg 2009

computationally intractable in many (even simple) settings of interest since the underlying optimization problem that needs to solved is NP-Hard. An important research thread in algorithmic game theory, starting with the work of Nisan and Ronen [37], focussed on designing efficient truthful mechanisms that approximate the optimal social welfare. A fundamental question is whether an α-approximation algorithm translates into an α-approximate truthful mechanism. It is well-known that plugging an α-approximation algorithm into the VCG mechanism may not yield a truthful mechanism [31, 38]. Thus, it is natural to investigate properties of approximation algorithms that enable their use in truthful mechanisms.

The problem of combinatorial auctions has gained the status of the paradigmatic problem in the field of algorithmic mechanism design. For a detailed overview, see [11]. In the context of single-minded agents, Lehmann, O'Callaghan and Shoham [31] established that an approximation algorithm can support a truthful mechanism if it satisfies a monotonicity property. Consequently, monotone approximation algorithms and techniques have been developed for various combinatorial optimization problems that underlie special cases of combinatorial auctions such as multi-unit auctions [12, 34]. One interesting set of techniques, devised by Mu'alem and Nisan [34], enables one to combine approximation algorithms while preserving monotonicity. In particular, they identified a special case of monotonicity, which they name bitonicity, and demonstrated that bitonic algorithms may be combined via the "max" operation.

1.1 Our Results

The main contribution of this paper is to identify a useful and natural property of approximation algorithms, which we name *loser-independence*. Intuitively, a loser-independent algorithm does not change its outcome when the bid of a losing agent increases, unless that agent becomes a winner. We demonstrate that loser-independent algorithms can be employed as sub-procedures in a greedy iterative packing approach while preserving monotonicity. A greedy iterative approach provides good approximation in the context of maximizing a non-decreasing submodular function subject to independence constraints such as matroid constraints [22, 24, 35]. There are various interesting problems that can be cast as special instances of this family (see, e.g., [13, 42]). We note that our loser-independence property is somewhat orthogonal to the notion of *composability* presented by Aggarwal and Hartline [1]. Intuitively, a composable algorithm does not change its outcome when the bid of a winning agent varies above its critical winning bid. Moreover, combining our property with the composability property yields, in the current setting, the *stability* condition suggested by Dobzinski and Sundararajan [19]. This condition states that if the bid of an agent changes but its allocation stays the same then the allocations to all other agents also do not change.

Our framework gives rise to efficient truthful approximation mechanisms for several problems. Notably, some of these problems arise in online mechanism design. We view the framework and the identification of the loser-independence property as the key contribution, and hence, we focus on those rather than the improvements for specific problems. We illustrate the applicability of the framework by briefly outlining two representative results that we derive.

An offline setting. A truthful $(2 + \epsilon)$-approximate mechanism for the multiple knapsack problem (MKP) among single-minded agents. This result improves and generalizes a 6-approximation mechanism for a special case of MKP among single-parameter agents [4]. In addition, we show that an almost identical mechanism attains an approximation ratio of $2 + \epsilon$ for the generalized assignment problem (GAP) among single-parameter agents. This is the first non-trivial approximate truthful mechanism for this problem when the number of knapsacks is part of the input; a monotone PTAS exists for this problem when the number of knapsacks is a fixed constant [12].

An online setting. A truthful 2-competitive mechanism for the online problem of dynamic auction with expiring items. This mechanism is essentially identical to the mechanism devised by Hajiaghayi et al. [27]. Furthermore, we achieve a truthful $(2+\epsilon)$-competitive mechanism for the generalization of the problem in which the underlying auction in each time-slot is a multi-item auction among single-minded agents rather than a single-item auction.

1.2 Related Work

It is widely known that many common techniques that are broadly used by approximation algorithms cannot be used in a strategic setting since they violate certain monotonicity properties which are imperative for truthfulness. Correspondingly, recent years have seen an ever-growing line of work addressing the development of monotone algorithmic alternatives. Mu'alem and Nisan [34] seem to have been the first to pay attention to this issue. They presented sufficient conditions for composing monotone algorithms via two basic operators, namely "max" and "if-then-else". Briest, Krysta and Vöcking [12] devised a general approach to transform a pseudo-polynomial algorithm into a monotone FPTAS, and demonstrated that primal-dual greedy algorithms may be used to derive truthful mechanisms. Lavi and Swamy [30] designed a general technique to convert approximation algorithms in packing domains to randomized approximation mechanisms that are truthful in *expectation*. Babaioff, Lavi and Pavlov [8] presented a method that translates any given algorithm to a truthful mechanism in single parameter domains. However, their method degrades the performance guarantee of the resulting mechanism by a factor of $O(\log \rho)$, where ρ denotes the ratio between the largest and smallest valuations. Recently, Azar and Gamzu [4] presented a monotone partition framework for approximating packing integer programs.

Focusing on the previously-mentioned representative problems from a purely algorithmic point of view, MKP is known to admit a PTAS by the work of Chekuri and Khanna [14], while GAP is known to be approximable within a factor that is slightly better than $e/(e-1) \approx 1.582$ by the work of Feige and Vondrák [21]. The dynamic auction with expiring items problem is equivalent to online scheduling of unit-length jobs on a single machine to maximize weighted throughput. The best known deterministic online algorithm for this problem has a competitive ratio of about 1.828 [20] (see also [32]), while it is known that no deterministic online algorithm can achieve a competitive ratio better than $\phi \approx 1.618$ [2, 16, 26]. Turning to the randomized setting, the best online algorithm attains a ratio of $e/(e-1)$ [9, 15], while it is known that no online algorithm can attain a ratio better than 1.25 [16]. This problem can be solved optimally in the offline setting.

2   The General Setting

In this section, we study the truthfulness properties of an iterative packing approach for a general class of maximizing assignment problems with packing constraints. We illustrate our ideas by restricting attention to the separable assignment problem [23]. In Section 4, we discuss our approach in the context of maximizing a non-decreasing submodular function subject to independence constraints. Note that the separable assignment problem is an instance of maximizing a non-decreasing submodular function over a partition matroid [13, 23].

An instance of the single-parameter variant of the *separable assignment problem* consists of a collection B of m bins and a set U of n items. Each bin $j \in B$ has a separable independence system $I_j \subseteq 2^U$, representing the subsets of items that may be packed in that bin[1]. Each item $i \in U$ has a positive value v_i, which is gained by assigning the item to one of the bins. The objective is to find a maximum value subset of items $S \subseteq U$, along with an assignment of these items to the bins, so that all the items in S can be simultaneously placed in their designated bins, while preserving the constraints induced by the independence systems. In particular, the set of items assigned to bin j, namely, S_j, must satisfy $S_j \in I_j$. In the game theoretic version of this problem there are n strategic single parameter agents, each of which controls an item, and may be untruthful about its value.

We consider the following *iterative packing approach* for approximately solving the mentioned problem: assume the existence of an α-approximation oracle for the single bin sub-problem, and build a solution by iteratively packing each of the bins (without backtracking) using the oracle. More precisely, the single bin sub-problem corresponding to bin j is to find a maximum value subset S_j that satisfies $S_j \in I_j$, and the iterative packing approach utilizes the approximation oracle to generate a packing $S_1 \subseteq U$ for the first bin, then it is used to generate a packing $S_2 \subseteq U \setminus S_1$ for the second bin, and so on. In what follows, we investigate the truthfulness properties of the iterative packing approach. Specifically, we focus on the approximation oracle, and establish a sufficient condition which guarantees that the iterative approach will lead to a monotone algorithm, and hence, a truthful mechanism.

2.1   Preliminaries

We introduce some notation and terminology that will be used throughout the paper, and describe a characterization that links monotone algorithms with truthful mechanisms. The reader is encouraged to refer to [11, 36] for a more comprehensive overview of the underlying concepts.

We will mainly concentrate on two types of agents: single-parameter and single-minded. *Single-parameter* agents have private data that consists of a single number, namely, their value. *Single-minded* agents [12, 31] have private data which consists of

---

[1] An *independence system* $I \subseteq 2^U$ is a family of subsets that is downward closed, that is, $A \in I$ and $B \subseteq A$ implies that $B \in I$. Note that the packing constraints are implicit from the independence systems which guarantee that if some subset of items is feasible for a bin then any subset of it is also feasible.

a pair (o, v), where o is an object that the agent is interested in and v is the valuation of the agent for attaining o. Remark that the interpretation of the object o depends on the problem at hand. For example, an object may represent a bandwidth demand of an agent (as in network routing), and it might stand for a set of items that an agent wants (as in combinatorial auctions). The valuation function of an agent whose data is (o, v) is a step function with respect to o. Specifically, if the agent obtains the object o or any object that extends it then its valuation is v; otherwise, its valuation is 0. We use the notation $\tilde{o} \leq o$ to indicate that object o *extends* object $\tilde{o}$. Again, the interpretation of the term extension depends on the problem at hand. For instance, if the object $\tilde{o}$ represents bandwidth demand then the object o extends it if it represents a higher bandwidth demand, and if the object $\tilde{o}$ stands for a set of items then the object o extends it if it stands for a superset of the items.

We now present the notion of monotonicity for single-minded agents, and then turn to describe a characterization that reduces the goal of designing truthful mechanisms to that of designing monotone algorithms. Note that similar definitions can be made for single-parameter agents by refining the monotonicity property and the characterization theorem. Specifically, both of them need to be defined only with respect to the value of every agent, and the objects-related terms need to be cast off. We say that an agent is a *selected* if it is assigned the object o or any object that extends it.

Definition 1. An algorithm $\mathcal{A}$ is said to be *monotone* with respect to the bid of an agent if it satisfies the following property: if algorithm $\mathcal{A}$ selects the agent when its bid is (o, v) then it selects the agent when its bid is $(\tilde{o}, \tilde{v})$, where $\tilde{o} \leq o$ and $\tilde{v} \geq v$, and the bids of all the other agents are fixed.

Theorem 2. ([12]) *If algorithm $\mathcal{A}$ is monotone with respect to the bid of every agent then there exists a corresponding truthful mechanism which can be efficiently computed using algorithm $\mathcal{A}$.*

2.2 A Motivating Example

Let us consider the single-parameter variant of MKP. This problem is a special case of the separable assignment problem, where the single bin sub-problem is the *knapsack problem*. Specifically, an instance of the *multiple knapsack problem* (MKP) consists of a collection B of m bins, and a set U of n items. Each bin $j \in B$ has a capacity W_j, and each item $i \in U$ is characterized by a pair (w_i, v_i), where w_i is the size of the item and v_i is its positive value. The goal is to select a maximum value subset of items $S \subseteq U$, along with an assignment of these items to the bins, so that all the items in S can be simultaneously placed in their designated bins while preserving the capacities of the bins. Note that the single bin sub-problem that corresponds to bin j is to find a maximum value subset of items whose overall size does not exceed W_j.

We focus on algorithm MaxGreedy, formally described below, which approximately solves the knapsack problem. This algorithm initially computes two assignments: one based on a greedy approach with respect to the values of the items, and another based on a greedy approach with respect to the *profit density* ratio of the items, that is, a value to size ratio. Then, it returns the assignment having maximum value. This algorithm was considered by Mu'alem and Nisan [34], who proved that it is monotone with respect to the value, and that it achieves 2-approximation. Due to its monotone properties, it may seem natural to use this algorithm as the single bin approximation oracle in the iterative approach attending to MKP. Unfortunately, as the following theorem states, the resulting iterative algorithm fails to be monotone.

Algorithm 1. MaxGreedy

Input: A set of items U, and the capacity of the bin W
Output: A set of items S to be assigned to the bin

1: $U_1 \leftarrow U, U_2 \leftarrow U, S_1 \leftarrow \emptyset, S_2 \leftarrow \emptyset$
2: **while** $U_1 \neq \emptyset$ **do**
3:     **remove** the item i that has a maximum value from U_1
4:         **if** $\sum_{\ell \in S_1} w_\ell + w_i \leq W$ **then add** i to S_1
5: **end while**

6: **while** $U_2 \neq \emptyset$ **do**
7:     **remove** the item i that has a maximum profit density from U_2
8:         **if** $\sum_{\ell \in S_2} w_\ell + w_i \leq W$ **then add** i to S_2
9: **end while**

10: **return** the maximum value allocation between S_1 and S_2

Theorem 3. *The iterative packing approach that employs algorithm* MaxGreedy *as the single bin approximation oracle is not monotone.*

It is worth noting that algorithm MaxGreedy is not only monotone, but also bitonic with respect to the value [34]. Informally, an algorithm is *bitonic* if its outcome value as a function of the value of any single agent i has the pattern that it does not increase as long as agent i is not selected, and it does not decrease as long as agent i is selected. This implies that both monotonicity and bitonicity of the single bin oracle are not sufficient to ensure the monotonicity of the corresponding iterative packing approach.

2.3 A Sufficient Condition

In the following, we establish a sufficient condition for the single bin approximation oracle. This condition guarantees that the iterative packing approach, which employs the oracle as the single bin sub-procedure, will satisfy monotonicity. We present the condition for single-minded agents, and an analogous condition for single-parameter agents can be derived in a similar manner.

We briefly motivate the sufficient condition by using the algorithm and the MKP instance described in the previous subsection. Recall that monotonicity guarantees that if a selected agent improves its bid then it continues to be selected. In particular, the monotonicity of algorithm MaxGreedy implies that if an agent that is selected for the first bin increases its value then it continues to be selected for that bin. However, the key difficulty appears when we consider an agent that was selected in a later bin, say the second one. In this case, when that agent increases its value then (from the perspective of the first bin) it is like a non-selected agent increases its value. Consequently, no guarantees can be made with respect to the assignment generated for the first bin. As a result, the agent that increased its value may compete against a different set of agents for the second bin, and may not be selected. One way to deal with this difficulty is to restrict the set of algorithms that may be employed by the iterative packing approach to those that are loser-independent, as formally defined below. Intuitively, a loser-independent algorithm does not disturb an assignment when the value of a losing agent increases, unless that agent becomes a winner. Note that this requirement is satisfied by an optimal (1-approximation) algorithm.

Definition 4. An algorithm $\mathcal{A}$ is said to be *loser-independent* with respect to the bid of an agent if it satisfies the following property: if algorithm $\mathcal{A}$ generates the solution a in which the agent is not selected when its bid is (o, v) then algorithm $\mathcal{A}$ either generates the same solution a or selects the agent when its bid is $(\tilde{o}, \tilde{v})$, where $\tilde{o} \leq o$ and $\tilde{v} \geq v$, and the bids of all the other agents are fixed.

Theorem 5. *If algorithm $\mathcal{A}$ is loser-independent and monotone with respect to the bid of every agent then the iterative packing approach, which employs it as the single bin oracle, is monotone with respect to the bid of every agent.*

2.4   Applications

The domain of problems for which the mentioned characterization is useful is broad. Essentially, one may take any single-minded or single-parameter version of a social welfare maximization packing problem π, and design a new "multiple" variant of this problem via the separable assignment problem paradigm. In what follows, we demonstrate the applicability of the characterization by utilizing it in the context of several well-known or highly-motivated problems. Note that the characterization reduces the task of designing a monotone algorithm for the "multiple" variant of π to that of designing a loser-independent and monotone algorithm for π.

We begin by pointing out two known approximation properties of the iterative packing approach. We will utilize these properties when analyzing the approximation ratio of the iterative approach for problems under consideration. The first property states that given an α-approximation oracle for the single bin sub-problem of the separable assignment problem, the iterative packing approach has an approximation ratio of at most $\alpha + 1$ [13, 22, 24]. The second property pertains to the special case of the separable assignment problem in which all the separable independence systems are identical, that is, $I_1 = \cdots = I_m$. In this case, it is known that given an α-approximation oracle for the single bin sub-problem, the iterative packing approach achieves approximation ratio of at most $e^{1/\alpha}/(e^{1/\alpha} - 1)$ [24, 35]. Note that these properties are additional motivation for our use of the iterative packing approach as they show that the approximation ratio of the iterative approach for the "multiple" variant problem degrades by constants with respect to that of the single bin oracle.

The multiple knapsack and generalized assignment problems. In the following, we consider the single-minded variant of MKP, and the single-parameter variant of the GAP. Both problems are special cases of the separable assignment problem, where the single bin sub-problem is the knapsack problem. The single-minded variant of MKP generalizes the single-parameter variant presented in Subsection 2.2 by allowing each agent i to be dishonest about the pair (w_i, v_i), that is, it may by untruthful about both the size and the value of the corresponding item. The single-parameter variant of the *generalized assignment problem* (GAP) extends the single-parameter variant of MKP by characterizing each item i with a pair (w_i, v_i), where w_i is a *vector* of length m that represents the size that item i occupies in each of the bins, and v_i is its positive value. Note that the private data of agent i consists only of the value v_i, while the vector w_i is public knowledge.

We demonstrate the utility of the new condition via these problems. We begin by designing a relatively simple 2-approximate algorithm for the knapsack problem that may be utilized as the single bin oracle. It is instructive to measure our algorithm against algorithm MaxGreedy as our algorithm is loser-independent. Algorithm HalfGreedy, formally described below, begins by computing two assignments: one which consists only of the maximum value item, and another that is based on a greedy approach with respect to the profit density ratio. Note that profit density greedy approach in our algorithm has three key differences from the profit density greedy approach applied by algorithm MaxGreedy. The first is that it only considers *small* items, namely, items whose size is no more than half the capacity of the bin; the second is that it stops adding items to the assignment once their overall size is at least half of the capacity of the bin; and the third is that it defines the value of the assignment, marked by V_2, to be the overall value of the items that fill exactly half of the capacity of the bin (unless the overall size of all small items is less than that). Particularly, in the former case, only a portion of the value of the last item included in the assignment is taken into account. This portion corresponds to the portion of the size of the item contained before the half-way mark of the bin. Note that V_2 is a lower bound on the overall value of the items in the profit density greedy assignment. Then, the algorithm returns the assignment corresponding to a greater assignment value.[2]

Theorem 6. *Algorithm HalfGreedy achieves 2-approximation and maintains monotonicity and loser-independence with respect to the bid of every agent.*

Theorem 6, Theorem 5 and the previously mentioned approximation properties imply the following corollary.

Corollary 7. *There is a truthful 3-approximation mechanism for MKP among single-minded agents. This mechanism attains an approximation ratio of $e^{1/2}/(e^{1/2} - 1) \approx 2.541$ when bin capacities are identical. Moreover, there is a truthful 3-approximation mechanism for GAP among single-parameter agents.*

As we are interested in better performance guarantees, we turn to study the monotone FPTAS for the knapsack problem, developed by Briest, Krysta and Vöcking [12]. We prove that this algorithm is loser-independent, and thus, may be employed by an iterative packing approach.

---

[2] There are high-level similarities between algorithm HalfGreedy and algorithm AK of [1]; however, the properties for which the algorithms designed for are quite different. We thank Tim Roughgarden for pointing out [1].

Algorithm 2. HalfGreedy

 Input: A set of items U, and the capacity of the bin W
 Output: A set of items S to be assigned to the bin

1: $U_1 \leftarrow U, U_2 \leftarrow \{i \in U : w_i \le W/2\}, S_1 \leftarrow \emptyset, S_2 \leftarrow \emptyset$
2: **let** S_1 be the singleton set that consists of the maximum value item of U_1
3: **let** V_1 be the value of the single item in S_1
4: **while** ($\sum_{\ell \in S_2} w_\ell < W/2$ and $U_2 \ne \emptyset$) **do**
5: **remove** the item i that has a maximum profit density from U_2
6: **add** i to S_2
7: **end while**
8: **let** $i_1, \ldots, i_k$ be the items selected to S_2 according to their inspection order.
9: **let** $V_2 = \sum_{\ell=1}^{k-1} v_{i_\ell} + v_{i_k}/w_{i_k} \cdot \min\{w_{i_k}, W/2 - \sum_{\ell=1}^{k-1} w_{i_\ell}\}$
10: **if** $V_1 \ge V_2$ **then return** S_1 **else return** S_2

Theorem 8. *There is an FPTAS for the knapsack problem that maintains monotonicity and loser-independence with respect to the bid of every agent.*

Corollary 9. *There is a truthful $(2 + \epsilon)$-approximation mechanism for MKP among single-minded agents. This mechanism attains an approximation ratio of $e/(e - 1) + \epsilon \approx 1.582$ when bin capacities are identical. In addition, there is a truthful $(2 + \epsilon)$-approximation mechanism for GAP among single-parameter agents.*

Additional applications. In what follows, we briefly list several additional packing problems whose "multiple" variant can be solved by exploiting the characterization. In particular, we identify the corresponding loser-independent algorithms.

Combinatorial auctions. The *multi-unit* combinatorial auction problem (see, e.g., [3, 10]) is a natural generalization of the celebrated combinatorial auction problem in which each good has several copies. One may interpret the "multiple" variant of this problem as adding *group constraints* to the basic problem. Specifically, in this variant, each good is associated with a group, and goods from different groups cannot be used to form a bundle satisfying an agent. One can demonstrate that the algorithm for single-minded combinatorial auction [31], and the algorithms for the single-minded multi-unit version [5, 12] maintain loser-independence.

The *single value* combinatorial auction problem is a special case of the combinatorial auction problem in which the valuation of each agent is represented by a single value. Particularly, each (multi-minded) agent is interested in several different bundles, but obtains the same value from any non-zero outcome. It is clear that our characterization is not applicable since the agents are multi-minded. Still, if the agents are *known*, that is, all their data besides their values is publicly known, then one can establish that the characterization is still suitable. Essentially, this follows from the observation that truthfulness in known agents setting reduces to value monotonicity. Similarly to before, one may interpret the "multiple" variant of this problem as adding group constraints to the basic problem, and may prove that the algorithm for single value combinatorial auction among known multi-minded agents [7] maintains loser-independence.

Advertisement space auctions. The theme of selling advertisement space on a newspaper page can be modelled by packing convex figures in a plane. One may interpret the "multiple" variant of this problem as increasing the advertisement space to several pages, and may demonstrate that the algorithms presented in [6] maintain loser-independence.

Network routing. The task of routing in networks is commonly modelled using the unsplittable flow problem. One may interpret the "multiple" variant of this problem as adding wavelength constraints to the basic problem. These constraints prevent serving requests across different wavelengths. One can verify that the algorithms presented in [5, 12] maintain loser-independence.

3 The Online Setting

In this section, we extend our results for an online environment in which agents arrive and leave dynamically over time and there is uncertainty about the set of decisions to be made in the future. We illustrate our ideas by considering the online version of the separable assignment problem. In this variant, bins are aligned with discrete time slots, and items arrive and depart dynamically. In particular, an item is not known prior to its arrival and cannot be assigned after its departure. The goal is to generate a maximum value assignment of items to bins in an online fashion. Specifically, any assigned item must be packed in a bin that corresponds to a time slot between its arrival and departure times. In the game theoretic version of this problem, each agent controls an item, and may be untruthful about its value, arrival time, and departure time. In adherence with previous results in an online setting [27, 40], we assume *no early-arrival* and *no late-departure* misreports. That is, agents cannot report an arrival time earlier than their true arrival time or a departure time later than their true departure time. Note that Lavi and Nisan [29] considered the special case of separable assignment problem in which any bin can only accommodate a single item, and proved that it is impossible to attain bounded competitive ratio without restricting the misreports.

Our approach to solve this online variant is identical to before. Namely, we assume the existence of an α-approximation oracle for the single bin sub-problem, and build a solution by iteratively employing it to generate an assignment for each of the bins. Notice that a single bin oracle optimizes with respect to a current state of agents and does not take into account the global system-wide view, and hence, if bins are considered according to their time order then the iterative approach constitute an online algorithm.

3.1 A Sufficient Condition

In what follows, we reformulate the sufficient condition for the single bin approximation oracle, exhibited in Subsection 2.3, for online environments. We begin by presenting revised definitions of single-minded agents and monotonicity for an online setting. Remark that the forthcoming definitions can be refined for single-parameter agents in a similar manner to before. Additionally, we encourage the reader to refer to [39] for a more detailed overview of online mechanisms.

The private data of single-minded agent in an online setting consists of a quadruple (o, v, a, d), where o is an object that the agent is interested in, v is the valuation of the agent for attaining o, and a and d are the arrival and departure times of the agent, respectively.

Definition 10. An online algorithm $\mathcal{A}$ is said to be *monotone* with respect to the bid of an agent if it satisfies the following property: if algorithm $\mathcal{A}$ selects the agent when its bid is (o, v, a, d) then algorithm $\mathcal{A}$ selects the agent when its bid is $(\tilde{o}, \tilde{v}, \tilde{a}, \tilde{d})$, where $\tilde{o} \leq o$, $\tilde{v} \geq v$, $\tilde{a} \leq a$ and $\tilde{d} \geq d$, and the bids of all the other agents are fixed.

Theorem 11. ([39]) *If online algorithm $\mathcal{A}$ is monotone with respect to the bid of every agent then there exists a corresponding truthful mechanism which can be computed using algorithm $\mathcal{A}$.*

We are ready to prove that an online iterative packing approach, which employs a monotone and loser-independent oracle as the single bin sub-procedure, satisfies monotonicity. Note that the monotonicity and loser-independence of the oracle are with respect to the non-temporal part of the bid, that is, the object-value pair (o, v).

Theorem 12. *If algorithm $\mathcal{A}$ is loser-independent and monotone with respect to the non-temporal bid of every agent then the online iterative packing approach, which employs it as the single bin oracle, is monotone with respect to the bid of every agent.*

3.2 Applications

Similarly to the offline setting, the domain of problems for which the mentioned characterization is useful is broad. Basically, one may take any single-minded or single-parameter version of a social welfare maximization packing problem, and design an online variant of this problem via the online separable assignment problem paradigm. Several straightforward examples are the problems presented in Subsection 2.4. Note that the online iterative packing approach achieves a competitive ratio of at most $\alpha + 1$, assuming an α-approximation oracle for the single bin sub-problem. This claim can be established by using nearly identical arguments to the ones used to prove the corresponding offline claim.

An additional interesting application is the problem of *dynamic auction with expiring items*. This problem is a special case of the online separable assignment problem, where the single bin sub-problem is a *single-item auction*. An instance of this problem consists of a collection of unit-capacity bins, each associated with a distinct time-slot. An additional ingredient of the input is an online sequence of unit-size items, each of which is characterized by a triple (v, a, d), where v is its positive value, a is its arrival time, and d is its departure time. The objective is to generate a maximum value assignment of items to bins in an online fashion. In particular, this assignment should place at most one item in each bin, and each assigned item must be placed in a bin that corresponds to a time slot between its arrival and departure times. Focusing on the single bin sub-problem, one can notice that it admits a trivial optimal algorithm which places the most valuable item in a bin. As previously mentioned, any optimal algorithm is monotone and loser-independent. Hence, the characterization and the claimed approximation property imply the following corollary.

Corollary 13. *There is a truthful 2-competitive mechanism for dynamic auction with expiring items among single-parameter agents.*

Interestingly, this simple online iterative packing algorithm is identical to the algorithm presented by Hajiaghayi et al. [27], and it is best possible [18, 27]. Specifically, no deterministic truthful mechanism can obtain a competitive ratio better than 2. A natural generalization of this problem can be obtained by replacing the single bin sub-problem of single-item auction with *multi-item auction*. We refer to this problem as *dynamic auction with expiring multi-items*. It is well-known that the combinatorial optimization problem that underlie multi-unit auction among single-minded agents is the knapsack problem. In correspondence with previous results, we yield the following corollary.

Corollary 14. *There is a truthful $(2 + \epsilon)$-competitive mechanism for dynamic auction with expiring multi-items among single-minded agents.*

4 Additional Applications via Submodular Function Maximization

As mentioned before, the greedy iterative approach provides good approximation in the broad context of maximizing a non-decreasing submodular function subject to independence constraints. More formally, let $f : 2^N \to \mathcal{R}^+$ be a non-decreasing submodular function on a finite ground set N, and let $(N, \mathcal{I})$ be an independence family. In other words, $\mathcal{I} \subseteq 2^N$ is a family of subsets that is downward closed, that is, $A \in \mathcal{I}$ and $B \subseteq A$ imply $B \in \mathcal{I}$. The optimization problem is then $\max_{S \in \mathcal{I}} f(S)$. Interesting independence families are matroids, intersection of a small number k of matroids, and somewhat more general notions such as k-independence and k-extendible systems (see [13, 28, 33]). The greedy approach is then simple; start with an empty set, and incrementally build a solution by greedily adding an element that (approximately) improves the current solution the most while maintaining its independence. It is known that the greedy approach gives a $(k\alpha + 1)$-approximation for the above problem if there is an α-approximation for picking the element that most improves the current solution [22] (see [13, 24] for recent and more easily available proofs).

When the underlying optimization problem of mechanism design, in particular the winner determination problem, can be cast as a special case of submodular function maximization subject to independence constraints, one may be able to use the greedy approach. In this case, if the (approximation) algorithm employed by the greedy incremental step is monotone and loser-independent then one can show that the overall greedy approach is monotone. We remark that the greedy approach here is somewhat different from the one presented in Section 2 for separable assignment problems; for the latter case, we employed a *local greedy* approach which considers the bins according to an *arbitrary* ordering and packs each bin with the approximate best solution. However, a global greedy approach would have considered all empty bins in each step and then pack the bin that most improves the solution. We note that the local greedy approach works for partition matroids [22], and is essential for the applications in online settings. Still, more general independence constraints requires the global greedy approach. As we remarked, loser-independence is still applicable. We give a concrete application to illustrate it.

Consider MKP (or GAP), and suppose we add a constraint that at most $m' < m$ of the bins can be used in the packing. The resulting optimization problem becomes a submodular function maximization problem subject to a laminar matroid constraint (as observed in [13]). In this setting, the global greedy approach needs to pick in each step the best bin to pack by trying all remaining bins. One can easily extend Theorem 5 to this setting, and prove that if the single bin algorithm is monotone and loser-independent then the greedy approach is monotone. We hope that additional applications to mechanism design problems will be found by using the above high-level approach.

Acknowledgments. The authors thank Yossi Azar, Jason Hartline, Tim Roughgarden and Jan Vondrák for useful discussions and comments on topics related to this paper.

References

1. Aggarwal, G., Hartline, J.D.: Knapsack auctions. In: 17th SODA, pp. 1083–1092 (2006)
2. Andelman, N., Mansour, Y., Zhu, A.: Competitive queueing policies for qos switches. In: 14th SODA, pp. 761–770 (2003)
3. Archer, A., Papadimitriou, C.H., Talwar, K., Tardos, É.: An approximate truthful mechanism for combinatorial auctions with single parameter agents. In: 14th SODA, pp. 205–214 (2003)
4. Azar, Y., Gamzu, I.: Truthful unification framework for packing integer programs with choices. In: Aceto, L., Damgård, I., Goldberg, L.A., Halldórsson, M.M., Ingólfsdóttir, A., Walukiewicz, I. (eds.) ICALP 2008, Part I. LNCS, vol. 5125, pp. 833–844. Springer, Heidelberg (2008)
5. Azar, Y., Gamzu, I., Gutner, S.: Truthful unsplittable flow for large capacity networks. In: 19th SPAA, pp. 320–329 (2007)
6. Babaioff, M., Blumrosen, L.: Computationally-feasible truthful auctions for convex bundles. In: Jansen, K., Khanna, S., Rolim, J.D.P., Ron, D. (eds.) RANDOM 2004 and APPROX 2004. LNCS, vol. 3122, pp. 27–38. Springer, Heidelberg (2004)
7. Babaioff, M., Lavi, R., Pavlov, E.: Mechanism design for single-value domains. In: 20th AAAI, pp. 241–247 (2005)
8. Babaioff, M., Lavi, R., Pavlov, E.: Single-value combinatorial auctions and implementation in undominated strategies. In: 17th SODA, pp. 1054–1063 (2006)
9. Bartal, Y., Chin, F.Y.L., Chrobak, M., Fung, S.P.Y., Jawor, W., Lavi, R., Sgall, J., Tichý, T.: Online competitive algorithms for maximizing weighted throughput of unit jobs. In: Diekert, V., Habib, M. (eds.) STACS 2004. LNCS, vol. 2996, pp. 187–198. Springer, Heidelberg (2004)
10. Bartal, Y., Gonen, R., Nisan, N.: Incentive compatible multi unit combinatorial auctions. In: 9th TARK, pp. 72–87 (2003)
11. Blumrosen, L., Nisan, N.: Combinatorial auctions. In: Nisan, N., Roughgarden, T., Tardos, E., Vazirani, V. (eds.) Algorithmic Game Theory, ch. 11. Cambridge University Press, Cambridge (2007)
12. Briest, P., Krysta, P., Vöcking, B.: Approximation techniques for utilitarian mechanism design. In: 37th STOC, pp. 39–48 (2005)
13. Calinescu, G., Chekuri, C., Pál, M., Vondrák, J.: Maximizing a submodular set function subject to a matroid constraint (Extended abstract). In: Fischetti, M., Williamson, D.P. (eds.) IPCO 2007. LNCS, vol. 4513, pp. 182–196. Springer, Heidelberg (2007)
14. Chekuri, C., Khanna, S.: A polynomial time approximation scheme for the multiple knapsack problem. SICOMP 35(3), 713–728 (2005)
15. Chin, F.Y.L., Chrobak, M., Fung, S.P.Y., Jawor, W., Sgall, J., Tichý, T.: Online competitive algorithms for maximizing weighted throughput of unit jobs. J. Discrete Algorithms 4(2), 255–276 (2006)
16. Chin, F.Y.L., Fung, S.P.Y.: Online scheduling with partial job values: Does timesharing or randomization help? Algorithmica 37(3), 149–164 (2003)

17. Clarke, E.H.: Multipart pricing of public goods. Public Choice 8, 17–33 (1971)
18. Cole, R., Dobzinski, S., Fleischer, L.K.: Prompt mechanisms for online auctions. In: Monien, B., Schroeder, U.-P. (eds.) SAGT 2008. LNCS, vol. 4997, pp. 170–181. Springer, Heidelberg (2008)
19. Dobzinski, S., Sundararajan, M.: On characterizations of truthful mechanisms for combinatorial auctions and scheduling. In: 9th EC, pp. 38–47 (2008)
20. Englert, M., Westermann, M.: Considering suppressed packets improves buffer management in qos switches. In: 18th SODA, pp. 209–218 (2007)
21. Feige, U., Vondrák, J.: Approximation algorithms for allocation problems: Improving the factor of 1 - 1/e. In: 47th FOCS, pp. 667–676 (2006)
22. Fisher, M.L., Nemhauser, G.L., Wolsey, L.A.: An analysis of approximations for maximizing submodular set functions ii. Mathematical Programming Study 8, 73–87 (1978)
23. Fleischer, L., Goemans, M.X., Mirrokni, V.S., Sviridenko, M.: Tight approximation algorithms for maximum general assignment problems. In: 17th SODA, pp. 611–620 (2006)
24. Goundan, P.R., Schulz, A.S.: Revisiting the greedy approach to submodular set function maximization (manuscript) (2007)
25. Groves, T.: Incentives in teams. Econemetrica 41(4), 617–631 (1973)
26. Hajek, B.: On the competitiveness of online scheduling of unit-length packets with hard deadlines in slotted time. In: Conference on Information Sciences and Systems, pp. 434–438 (2001)
27. Hajiaghayi, M.T., Kleinberg, R.D., Mahdian, M., Parkes, D.C.: Online auctions with reusable goods. In: 6th EC, pp. 165–174 (2005)
28. Jenkyns, T.A.: The efficiency of the "greedy" algorithm. In: 7th South Eastern Conference on Combinatorics, Graph Theory and Computing, pp. 341–350 (1976)
29. Lavi, R., Nisan, N.: Online ascending auctions for gradually expiring items. In: 16th SODA, pp. 1146–1155 (2005)
30. Lavi, R., Swamy, C.: Truthful and near-optimal mechanism design via linear programming. In: 46th FOCS, pp. 595–604 (2005)
31. Lehmann, D.J., O'Callaghan, L., Shoham, Y.: Truth revelation in approximately efficient combinatorial auctions. JACM 49(5), 577–602 (2002)
32. Li, F., Sethuraman, J., Stein, C.: Better online buffer management. In: 18th SODA, pp. 199–208 (2007)
33. Mestre, J.: Greedy in approximation algorithms. In: Azar, Y., Erlebach, T. (eds.) ESA 2006. LNCS, vol. 4168, pp. 528–539. Springer, Heidelberg (2006)
34. Mu'alem, A., Nisan, N.: Truthful approximation mechanisms for restricted combinatorial auctions. GEB 64, 612–631 (2008)
35. Nemhauser, G.L., Wolsey, L.A., Fisher, M.L.: An analysis of approximations for maximizing submodular set functions i. Mathematical Programming 14, 265–294 (1978)
36. Nisan, N.: Introduction to mechanism design (for computer scientists). In: Nisan, N., Roughgarden, T., Tardos, E., Vazirani, V. (eds.) Algorithmic Game Theory, ch. 9. Cambridge University Press, Cambridge (2007)
37. Nisan, N., Ronen, A.: Algorithmic mechanism design. GEB 35, 166–196 (2001)
38. Nisan, N., Ronen, A.: Computationally feasible vcg mechanisms. JAIR 29, 19–47 (2007)
39. Parkes, D.C.: Online mechanisms. In: Nisan, N., Roughgarden, T., Tardos, E., Vazirani, V. (eds.) Algorithmic Game Theory, ch. 16. Cambridge University Press, Cambridge (2007)
40. Porter, R.: Mechanism design for online real-time scheduling. In: 5th EC, pp. 61–70 (2004)
41. Vickery, W.: Counterspeculation, auctions and competitive sealed tender. Journal of Finance 16, 8–37 (1961)
42. Vondrák, J.: Optimal approximation for the submodular welfare problem in the value oracle model. In: 40th STOC, pp. 67–74 (2008)

Resource Minimization Job Scheduling

Julia Chuzhoy[1] and Paolo Codenotti[2]

[1] Toyota Technological Institute, Chicago, IL 60637
Supported in part by NSF CAREER award CCF-0844872
cjulia@tti-c.org
[2] Department of Computer Science, University of Chicago, Chicago, IL 60637
paoloc@cs.uchicago.edu

Abstract. Given a set J of jobs, where each job j is associated with release date r_j, deadline d_j and processing time p_j, our goal is to schedule all jobs using the minimum possible number of machines. Scheduling a job j requires selecting an interval of length p_j between its release date and deadline, and assigning it to a machine, with the restriction that each machine executes at most one job at any given time. This is one of the basic settings in the resource-minimization job scheduling, and the classical randomized rounding technique of Raghavan and Thompson provides an $O(\log n / \log \log n)$-approximation for it. This result has been recently improved to an $O(\sqrt{\log n})$-approximation, and moreover an efficient algorithm for scheduling all jobs on $O((\mathsf{OPT})^2)$ machines has been shown. We build on this prior work to obtain a constant factor approximation algorithm for the problem.

1 Introduction

In one of the basic scheduling frameworks, the input consists of a set J of jobs, and each job $j \in J$ is associated with a subset $\mathcal{I}(j)$ of time intervals, during which it can be executed. The sets $\mathcal{I}(j)$ of intervals can either be given explicitly (in this case we say we have a *discrete* input), or implicitly by specifying the release date r_j, the deadline d_j and the processing time p_j of each job (*continuous* input). In the latter case, $\mathcal{I}(j)$ is the set of all time intervals of length p_j contained in the time window $[r_j, d_j]$. A schedule of a subset $J' \subseteq J$ of jobs assigns each job $j \in J'$ to one of the time intervals $I \in \mathcal{I}(j)$, during which j is executed. In addition to selecting a time interval, each job is also assigned to a machine, with the restriction that all jobs assigned to a single machine must be executed on non-overlapping time intervals.

In this paper we focus on the Machine Minimization problem, where the goal is to schedule all the jobs, while minimizing the total number of machines used. We refer to the discrete and the continuous versions of the problem as Discrete and Continuous Machine Minimization, respectively. Both versions admit an $O(\log n / \log \log n)$-approximation via the Randomized LP-Rounding technique of Raghavan and Thompson [8], and this is the best currently known approximation for Discrete Machine Minimization. Chuzhoy and Naor [7] have shown

I. Dinur et al. (Eds.): APPROX and RANDOM 2009, LNCS 5687, pp. 70–83, 2009.
© Springer-Verlag Berlin Heidelberg 2009

that the discrete version is $\Omega(\log\log n)$-hard to approximate. Better approximation algorithms are known for Continuous Machine Minimization: an $O(\sqrt{\log n})$-approximation algorithm was shown by Chuzhoy et. al. [6], who also obtain better performance guarantees when the optimal solution cost is small. Specifically, they give an efficient algorithm for scheduling all jobs on $O(k^2)$ machines, where k is the number of machines used by the optimal solution. In this paper we improve their result by showing a constant factor approximation algorithm for Continuous Machine Minimization. Combined with the lower bound of [6], our result proves a separation between the discrete and the continuous versions of Machine Minimization.

Related Work. A problem that can be seen as dual to Machine Minimization is Throughput Maximization, where the goal is to maximize the number of jobs scheduled on a single machine. This problem has an $\left(\frac{e}{e-1}+\epsilon\right)$-approximation for any constant ϵ, in both the discrete and the continuous settings [5]. The discrete version is MAX-SNP hard even when each job has only two intervals [9] (i.e., $|\mathcal{I}(j)| = 2$ for all j), while no hardness of approximation results are known for the continuous version. In the more general weighted setting of Throughput Maximization, each job j is associated with weight w_j, and the goal is to maximize the total weight of scheduled jobs. The best current approximation factor for this problem is 2 for both the discrete and the continuous versions [2].

A natural generalization of Throughput Maximization is the Resource Allocation problem, where each job j is also associated with height (or bandwidth) h_j. The goal is again to maximize the total weight of scheduled jobs, but now the jobs are allowed to overlap in time, as long as the total height of all jobs executed at each time point does not exceed 1. For the weighted variant of this problem, Bar-Noy et. al. [3] show a factor 5-approximation, while the unweighted version can be approximated up to factor $(2e - 1)/(e - 1) + \epsilon$ for any constant ϵ [5]. For the special case of Resource Allocation where each job has exactly one time interval (i.e., $|\mathcal{I}(j)| = 1$ for all j), Calinescu et. al. [4] show a factor $(2+\epsilon)$-approximation for any ϵ, and Bansal et. al. [1] give a Quasi-PTAS.

Our Results and Techniques. We show a constant factor approximation algorithm for Continuous Machine Minimization. Our algorithm builds on the work of Chuzhoy et. al. [6]. Since the basic linear programming relaxation for the problem is known to have an $\Omega(\log n/\log\log n)$ integrality gap, [6] design a stronger recursive linear programming relaxation for the problem. The solution of this LP involves dynamic programming, where each entry of the dynamic programming table is computed by solving the LP relaxation on the corresponding sub-instance. Using the LP solution, [6] then partition the input set J of jobs into $k = \lceil OPT \rceil$ subsets, $J^1, \ldots, J^k$. They show that each subset J^i can be scheduled on $O(k_i)$ machines, where k_i is the total number of machines used to schedule all jobs in J^i by the fractional solution. Since in the worst case k_i can be as large as k for all i, they eventually use $O(k^2)$ machines to schedule all jobs.

We perform a similar partition of jobs into subsets. One of our main ideas is to define, for each job class J^i, a function $f_i(t)$, whose value is the total fractional weight of intervals of jobs in J^i containing time point t. We then find a schedule for each job class J^i, with at most $O(\lceil f_i(t) \rceil)$ jobs being scheduled at each time point t. The algorithm for finding the schedule itself is similar to that of [6], but more work is needed to adapt their algorithm to this new setting.

2   Preliminaries

In the Continuous Machine Minimization problem the input consists of a set J of jobs, and each job $j \in J$ is associated with a release date r_j, a deadline d_j and a processing time p_j. The goal is to schedule all jobs, while minimizing the number of machines used. In order to schedule a job j, we need to choose a time interval $I \subseteq [r_j, d_j]$ of length p_j during which job j will be executed, and to assign the job to one of the machines. The chosen intervals of jobs assigned to any particular machine must be non-overlapping.

We denote by $\mathcal{I}(j)$ the set of all time intervals of job j, so $\mathcal{I}(j)$ contains all intervals of length p_j contained in the time window $[r_j, d_j]$. For convenience we will assume that these intervals are open. If $I \in \mathcal{I}(j)$, then we say that interval I *belongs* to job j. Notice that $|\mathcal{I}(j)|$ may be exponential in the input length. Given any solution, if interval I is chosen for job j, we say that j is scheduled on interval I, and for each $t \in I$ we say that j is scheduled at time t. We denote by $\mathcal{T}$ the smallest time interval containing all the input job intervals, and denote by OPT both the optimal solution and its cost. We refer to the time interval $[r_j, d_j]$ as the *time window* of job j. We will use the following simple observation.

Claim. Let S be a set of intervals containing exactly one interval $I \in \mathcal{I}(j)$ for each job $j \in J$. Moreover, assume that for each $t \in \mathcal{T}$, the total number of intervals in S containing t is at most k. Then all jobs in J can be scheduled on k machines, and moreover, given S, such a schedule can be found efficiently.

Proof. Consider the interval graph defined by set S. The size of the maximum clique in this graph is at most k, and therefore it can be efficiently colored by k colors. Each color will correspond to a distinct machine.   □

Our goal is therefore to select a time interval $I \in \mathcal{I}(j)$ for each job j, while minimizing the maximum number of jobs scheduled at any time point t.

The Linear Programming Relaxation. We now describe the linear programming relaxation of [6], which is also used by our approximation algorithm. We start with the following basic linear programming relaxation for the problem. For each job $j \in J$, for each interval $I \in \mathcal{I}(j)$, we have an indicator variable $x(I, j)$ for scheduling job j on interval I. We require that each job is scheduled on at least one interval, and that the total number of jobs scheduled at each time point $t \in T$ is at most z, the value of the objective function.

$$\text{(LP1)} \quad \min \quad z$$
$$\text{s.t.} \quad \sum_{I \in \mathcal{I}(j)} x(I, j) = 1 \quad \forall j \in J$$
$$\sum_{j \in J} \sum_{\substack{I \in \mathcal{I}(j): \\ t \in I}} x(I, j) \le z \; \forall t \in T$$
$$x(I, j) \ge 0 \qquad \forall j \in J, \forall I \in \mathcal{I}(j)$$

It is well-known however that the integrality gap of (LP1) is $\Omega\left(\frac{\log n}{\log \log n}\right)$ (e.g. see [6]). To overcome this barrier, Chuzhoy et. al. [6] propose a stronger relaxation for the problem. Consider first the special case where the optimal solution uses only one machine, that is, $\mathsf{OPT} = 1$. Let $I \in \mathcal{I}(j)$ be some job interval, and suppose there is another job $j' \ne j$, whose entire time window $[r_{j'}, d_{j'}]$ is contained in I. Then interval I is called *forbidden interval* for job j. Since $\mathsf{OPT} = 1$, job j cannot be scheduled on interval I. Therefore, we can add the valid constraint $x(I, j) = 0$ to the LP for all jobs j and intervals I, where I is a forbidden interval for job j. Chuzhoy et. al. show an LP-rounding algorithm for this stronger LP relaxation that schedules all jobs on a constant number of machines for this special case of the problem.

When the optimal solution uses more than one machine, constraints of the form $x(I, j) = 0$, where I is a forbidden interval for job j, are no longer valid. Instead, [6] define a function $m(T)$ for each time interval $T \subseteq \mathcal{T}$, whose intuitive meaning is as follows. Let $J(T)$ be the set of jobs whose time window is completely contained in T. Then $m(T)$ is the minimum number of machines needed to schedule jobs in $J(T)$. Formally, $m(T) = \lceil z \rceil$, where z is the optimal solution of the following linear program:

$$\text{(LP(T))} \quad \min \quad z$$
$$\text{s.t.} \quad \sum_{I \in \mathcal{I}(j)} x(I, j) = 1 \qquad \forall j \in J(T)$$
$$\sum_{j \in J(T)} \sum_{\substack{I \in \mathcal{I}(j): \\ t \in I}} x(I, j) \le z \qquad \forall t \in T \tag{1}$$
$$\sum_{j \in J(T)} \sum_{\substack{I \in \mathcal{I}(j): \\ T' \subseteq I}} x(I, j) \le z - m(T') \; \forall T' \subseteq T \tag{2}$$
$$x(I, j) \ge 0 \qquad \forall j \in J(T), \forall I \in \mathcal{I}(j)$$

Observe that for integral solutions, where $x(I, j) \in \{0, 1\}$ for all $j \in J, I \in \mathcal{I}(j)$, the value $m(T)$ is precisely the number of machines needed to schedule all jobs in $J(T)$. Constraint (2) requires that for each time interval $T' \subseteq T$, the total number of jobs scheduled on intervals containing T' is at most $m(T) - m(T')$. This is a valid constraint, since at least $m(T')$ machines are needed to schedule all jobs in $J(T')$. Therefore, $\lceil \mathsf{OPT}(\mathcal{T}) \rceil \le \mathsf{OPT}$. Notice that the number of constraints in $LP(T)$ may be exponential in the input size. This difficulty is overcome in [6] as follows. First they define, for each job $j \in J$ a new discrete subset $\mathcal{I}'(j)$ of time intervals, with $|\mathcal{I}'(j)| = \text{poly}(n)$. Sets $\mathcal{I}'(j)$ of intervals for $j \in J$ define a new instance of Discrete Machine Minimization, whose optimal solution cost is at most $3\,\text{OPT}$. Moreover, any solution for the new instance implies a feasible solution for the original instance of the same cost. Next they define the set $D \subseteq \mathcal{T}$ of time points, consisting of all release dates and deadlines of jobs in J, and all endpoints of intervals in $\{\mathcal{I}'(j)\}_{j \in J}$. Clearly, the size of D is polynomially bounded. Finally they modify $LP(\mathcal{T})$, so that Constraint (1) is only defined for $t \in D$ and Constraint (2) is only applied to time intervals T with both endpoints in D. The new LP relaxation can be solved in polynomial time and its solution cost is denoted by OPT'. We are guaranteed that $\lceil \text{OPT}' \rceil \leq 3\,\text{OPT}$. Moreover, any feasible solution to the new LP implies a feasible solution to the original LP. From now on we will denote by x this near-optimal fractional solution, and by $\text{OPT}'(\mathcal{T})$ its value, $\lceil \text{OPT}'(\mathcal{T}) \rceil \leq 3\,\text{OPT}$. For each job $j \in J$, let $\mathcal{I}^*(j) \subseteq \mathcal{I}(j)$ be the subset of intervals I for which $x(I, j) > 0$. For any interval $I \in \mathcal{I}^*(j)$, we call $x(I, j)$ the *LP-weight* of I.

3   The Algorithm

Our algorithm starts by defining a recursive partition of the time line into blocks. This recursive partition in turn defines a partition of the jobs into *job classes* $J^1, J^2, \ldots$ Our algorithm then defines, for each job class J^i, a function $f_i : \mathcal{T} \to \mathbb{R}$, where $f_i(t)$ is the summation of values $x(I, j)$ over all jobs $j \in J^i$ and intervals $I \in \mathcal{I}(j)$ containing t. We then consider each of the job classes J^i separately, and show an efficient algorithm for scheduling jobs in J^i so that at most $O(\lceil f_i(t) \rceil)$ jobs of J^i are executed at each time point $t \in \mathcal{T}$.

3.1   Partition into Blocks and Job Classes

Let T be any time interval, and let $\mathcal{B}$ be any set of disjoint sub-intervals of T. Then we say that $\mathcal{B}$ defines a partition of T into blocks, and each interval $B \in \mathcal{B}$ is referred to as a *block*. Notice that we do not require that the union of the intervals in $\mathcal{B}$ is T.

Let $k = \lceil m(\mathcal{T}) \rceil$ be the cost of the near-optimal fractional solution. We define a recursive partition of the time interval $\mathcal{T}$ into blocks. We use a partitioning sub-routine, that receives as input a time interval T and a set $J(T)$ of jobs whose time windows are contained in T. The output of the procedure is a partition $\mathcal{B}$ of T into blocks. This partition in turn defines a partition of the set $J(T)$ of jobs, as follows. For each $B \in \mathcal{B}$, we have a set $J_B \subseteq J(T)$ of jobs whose time window is contained in B, so $J_B = \{j \in J(T) \mid [r_j, d_j] \subseteq B\}$. Let $J'' = \cup_{B \in \mathcal{B}} J_B$, and let $J' = J(T) \setminus J''$. Notice that $J' \dot\cup (\bigcup_{B \in \mathcal{B}} J_B)$ is indeed a partition of $J(T)$, and that for each $j \in J'$, r_j and d_j lie in distinct blocks. The partitioning procedure will also guarantee the following properties: (i) For each job $j \in J'$, each interval $I \in \mathcal{I}^*(j)$ has a non-empty intersection with at most two blocks; and (ii) For each $B \in \mathcal{B}$, there is a job $j \in J'$ and a job interval $I \in \mathcal{I}^*(j)$, with $B \subseteq I$.

A partitioning procedure with the above properties is provided in [6]. For the sake of completeness we briefly sketch it here. Let $T = [L, R]$. We start with $t = L$ and $\mathcal{B} = \emptyset$. Given a current time point t, the next block $B = (\ell, r)$ is defined as follows. If there is any job $j \in J(T)$ with a time interval $I \in \mathcal{I}^*(j)$ containing t, we set the left endpoint of our block to be $\ell = t$. Otherwise, we set it to be the first (i.e., the leftmost) time point t for which such a job and such an interval exist. To define the right endpoint of the block, we consider the set S of all job intervals with non-zero LP-weight containing ℓ, so $S = \{I \mid \ell \in I \text{ and } \exists j \in J(T) : I \in \mathcal{I}^*(j)\}$. Among all intervals in S, let I^* be the interval with rightmost right endpoint. We then set r to be the right endpoint of I^*. Block $B = (\ell, r)$ is then added to $\mathcal{B}$, we set $t = r$ and continue.

We are now ready to describe our recursive partitioning procedure. We have k iterations. Iteration h, for $1 \leq h \leq k$, produces a partition $\mathcal{B}^h$ of T into blocks, refining the partition $\mathcal{B}^{h-1}$. Additionally, we produce a partition of the set J of jobs into k classes $J^1, \ldots, J^k$. In the first iteration, we apply the partitioning procedure to time interval T and the set J of jobs. We set $\mathcal{B}^1$ to be the partition into blocks produced by the procedure. We denote the corresponding partition of the jobs as follows: $J^1 = J'$, and for all $B \in \mathcal{B}^1$, we denote J_B by J_B^1. In general, to obtain partition $\mathcal{B}^h$, we run the partitioning algorithm on each of the blocks $B \in \mathcal{B}^{h-1}$, together with the associated subset J_B^{h-1} of jobs. For each block $B \in \mathcal{B}^{h-1}$, we denote by $\mathcal{B}_B$ the new block partition and by $J_B^{h-1} = (J'_B, J''_B)$ the new job partition computed by the partitioning procedure. We then set $\mathcal{B}^h = \bigcup_{B \in \mathcal{B}^{h-1}} \mathcal{B}_B$, $J^h = \bigcup_{B \in \mathcal{B}^{h-1}} J'_B$, and for each block $B' \in \mathcal{B}^h$, let $J_{B'}^h$ denote the subset of jobs in J^{h-1}, whose time windows are contained in B'. This finishes the description of the recursive partitioning procedure. An important property, established in the next claim, is that every job is assigned to one of the k classes $J^1, \ldots, J^k$. Due to lack of space the proof is omitted.

Claim. $J = J^1 \cup \cdots \cup J^k$.

We have thus obtained a recursive partition $\mathcal{B}^1, \ldots, \mathcal{B}^k$ of T into blocks, and a partition $J = \bigcup_{h=1}^{k} J^h$ of jobs into classes. For simplicity we denote $\mathcal{B}^0 = \{T\}$.

The algorithm of [6] can now be described as follows. Consider the set J^h of jobs, for $1 \leq h \leq k$, together with the partition $\mathcal{B}^{h-1}$ of T into blocks. Recall that for each block $B \in \mathcal{B}^{h-1}$, J_B^{h-1} is the subset of jobs whose time windows are contained in B, and $J^h \subseteq \bigcup_{B \in \mathcal{B}^{h-1}} J_B^{h-1}$. Consider now some block $B \in \mathcal{B}^{h-1}$ and the corresponding subset $\tilde{J} = J^h \cap J_B^{h-1}$. Let $\mathcal{B}' = \mathcal{B}_B$ be the partition of B into blocks returned by the partitioning procedure when computing $\mathcal{B}^h$. This partition has the property that each interval $I \in \mathcal{I}^*(j)$ of each job $j \in \tilde{J}$ has a non-empty intersection with at most two blocks in $\mathcal{B}'$, and furthermore for each $j \in \tilde{J}$, the window of j is not contained in any single block $B \in \mathcal{B}'$. These two properties are used in [6] to extend a simpler algorithm for the special case where $\mathsf{OPT} = 1$ to the more general setting, where an arbitrary number of machines is used. In particular, if OPT_h is the fractional number of machines used to schedule jobs in J^h (i.e., OPT_h is the maximum value, over time points t, of $\sum_{j \in J^h} \sum_{I \in \mathcal{I}(j): t \in I} x(I, j)$), then all jobs in J^h can be efficiently scheduled on $O(\lceil OPT_h \rceil)$ machines. In the worst case, OPT_h can be as large as OPT for all $h : 1 \leq h \leq k$, and so overall $O(k^2)$ machines are used in the algorithm of [6].

In this paper, we refine this algorithm and its analysis as follows. For each $h : 1 \leq h \leq k$, we define a function $f_h : \mathcal{T} \to \mathbb{R}$, where $f_h(t)$ is the total fractional weight of intervals containing t that belong to jobs in J^h. Clearly, for all t, $\sum_h f_h(t) \leq k$. We then consider each one of the job classes J^h separately. For each job class J^h we find a schedule for jobs in J^h, such that for each time point $t \in \mathcal{T}$, at most $O(\lceil f_h(t) \rceil)$ jobs are scheduled on intervals containing t. The algorithm for scheduling jobs in J^h and its analysis are similar to those in [6]. We partition all jobs in J^h into a constant number of subsets, according to the way the fractional weight is distributed on their intervals. We then schedule each one of the subsets separately. The analysis is similar to that of [6], but does not follow immediately from their work. In particular, more care is needed in the analysis of the subsets of jobs j that have substantial LP-weight on intervals lying inside blocks to which r_j or d_j belong.

We now proceed to describe our algorithm more formally. For each job class $J^h : 1 \leq h \leq k$, let $f_h : \mathcal{T} \to \mathcal{R}$ be defined as follows. For each $t \in \mathcal{T}$, $f_h(t) = \sum_{j \in J^h} \sum_{\substack{I \in \mathcal{I}(j): \\ t \in I}} x(I, j)$. Our goal is to prove the following theorem:

Theorem 1. *For each job class $J^h : 1 \leq h \leq k$, we can efficiently schedule jobs in J^h so that, for each time point $t \in \mathcal{T}$, at most $O(\lceil f_h(t) \rceil)$ jobs are scheduled on intervals containing t.*

We prove the theorem in the next section. We show here that a constant factor approximation algorithm for Continuous Machine Minimization follows from Theorem 1. For each time point $t \in \mathcal{T}$, the total number of jobs scheduled on intervals containing point t is at most $\sum_h O(\lceil f_h(t) \rceil)$. Since $\sum_h f_h(t) \leq k$, $\sum_{h=1}^k \lceil f_h(t) \rceil \leq 2k$, and so the solution cost is $O(k)$.

3.2   Proof of Theorem 1

Consider a job class J^h and the block partition $\mathcal{B}^{h-1}$. For each block $B \in \mathcal{B}^{h-1}$, let $J_B^* = J_B^{h-1} \cap J^h$ be the set of jobs whose windows are contained in B, and so $J^h = \bigcup_{B \in \mathcal{B}^{h-1}} J_B^*$. Clearly, for blocks $B \neq B'$, the windows of jobs in J_B^* and $J_{B'}^*$ are completely disjoint, and therefore they can be considered separately. From now on we focus on scheduling jobs in J_B^* inside a specific block $B \in \mathcal{B}^{h-1}$. For simplicity, we denote $J^* = J_B^*$, and $\mathcal{B}^*$ is the partition of B into blocks obtained when computing $\mathcal{B}^h$. Recall that we have the following properties: (i) For each job $j \in J^*$, r_j and d_j lie in distinct blocks of $\mathcal{B}^*$; and (ii) For each job $j \in J^*$, each interval $I \in \mathcal{I}^*(j)$ has a non-empty intersection with at most two blocks.

For each $t \in B$, let $g(t) = \lceil f_h(t) \rceil$. Observe that $g(t)$ is a step function. Our goal is to schedule all jobs in J^h so that, for each $t \in B$, at most $O(g(t))$ jobs are scheduled on intervals containing t. The rest of the algorithm consists of three steps. In the first step, we partition the area "below" the function $g(t)$ into a set $\mathcal{R}$ of rectangles of height 1. In the second step we assign each job interval

$I \in \mathcal{I}^*(j)$ for $j \in J^*$ to one of the rectangles $R \in \mathcal{R}$, such that the total LP-weight of intervals assigned to R at each time point $t \in R$ is at most 5. In the third step, we partition all jobs in J^* into 7 types, and find a schedule for each one of the types separately. The assignment of job intervals to rectangles found in Step 2 will help us find the final schedule.

Step 1: Defining Rectangles. A rectangle R is defined by a time interval $W(R)$, and we think of R as the interval $W(R)$ of height 1. We say that time point t belongs to R iff $t \in W(R)$ and we say that interval I is contained in R iff $I \subseteq W(R)$. We denote by ℓ_R and r_R the left and the right endpoints of $W(R)$ respectively. We find a nested set $\mathcal{R}$ of rectangles, such that for each $t \in T$, the total number of rectangles containing t is exactly $g(t)$.

To compute the set $\mathcal{R}$ of rectangles, we maintain a function $g' : B \to \mathbb{Z}$. Initially $g'(t) = g(t)$ for all t and $\mathcal{R} = \emptyset$. While there is a time point $t \in B$ with $g'(t) > 0$, we perform the following: Let I be the longest consecutive sub-interval of B with $g'(t) \geq 1$ for all $t \in I$. We add a rectangle R of height 1 with $W(R) = I$ to $\mathcal{R}$ and decrease the value $g'(t)$ for all $t \in I$ by 1. Consider the final set $\mathcal{R}$ of rectangles. For each $t \in B$, let $\mathcal{R}(t) \subseteq \mathcal{R}$ be the subset of rectangles containing the point t. Then for each $t \in B$, $|\mathcal{R}(t)| = g(t)$. Furthermore, it is easy to see that $\mathcal{R}$ is a nested set of rectangles, and for every pair $R, R' \in \mathcal{R}$ of rectangles with non-empty intersection, either $W(R) \subseteq W(R')$ or $W(R') \subseteq W(R)$ holds. Notice also that a rectangle $R \in \mathcal{R}$ may contain several blocks or be contained in a block. Its endpoints also do not necessarily coincide with block boundaries.

Step 2: Assigning Job Intervals to Rectangles. We start by partitioning the set $\mathcal{R}$ of rectangles into k layers as follows. The first layer L_1 contains all rectangles $R \in \mathcal{R}$ that are not contained in any other rectangle in $\mathcal{R}$. In general layer L_z contains all rectangles $R \in \mathcal{R} \setminus (L_1 \cup \cdots L_{z-1})$ that are not contained in any other rectangle in $\mathcal{R} \setminus (L_1 \cup \cdots L_{z-1})$ (if we have identical rectangles then at most one of them is added to each layer, breaking ties arbitrarily). Since $\mathcal{R}$ is a nested set of rectangles, each $R \in \mathcal{R}$ belongs to one of the layers $L_1, \ldots, L_k$, and the rectangles in each layer are disjoint.

Let $\mathcal{I} = \{I \in \mathcal{I}^*(j) \mid j \in J^*\}$ be the set of all intervals of jobs in J^* with non-zero weight. For $I \in \mathcal{I}$, we say that I belongs to layer z_I iff z_I is the largest index, for which there is a rectangle $R \in L_z$ containing I. If I belongs to layer L_{z_I}, then for each layer $L_{z'}$, $1 \leq z' \leq z_I$, there is a *unique* rectangle $R(I, z') \in L_{z'}$ containing I. Let $\mathcal{I}_z \subseteq \mathcal{I}$ be the set of intervals belonging to layer z. Then $\mathcal{I} = \bigcup_{z=1}^k \mathcal{I}_z$.

We process intervals in $\mathcal{I}_1, \ldots, \mathcal{I}_k$ in this order, while intervals belonging to the same layer are processed in non-increasing order of their lengths, breaking ties arbitrarily. Let $I \in \mathcal{I}_z$ be some interval, and assume that $I \in \mathcal{I}^*(j)$. Consider the rectangles $R(I, 1), \ldots, R(I, z_I)$. For each $z' : 1 \leq z' \leq z_I$, we say that I is *feasible* for $R(I, z')$ iff, for each time point $t \in I$, the total LP-weight of intervals currently assigned to R that contain t is at most $5 - x(I, j)$. We select any rectangle $R(I, z')$, $1 \leq z' \leq z_I$, for which I is feasible and assign I to $R(I, z')$. In order to show that this procedure succeeds, it is enough to prove the following:

Claim. When interval I is processed, there is at least one rectangle $R(I, z')$, with $1 \le z' \le z_I$, for which I is feasible.

Proof. Assume otherwise. Let $I' \in \mathcal{I}$ be any interval that has already been processed. It is easy to see that $I' \not\subseteq I$: If I' and I belong to the same layer, then the length of I' should be greater than or equal to the length of I, so $I' \not\subseteq I$. If I' belongs to some layer z and I belongs to layer $z_I > z$, then by the definition of layers it is impossible that $I' \subseteq I$ (since then any rectangle containing I would also contain I'). Therefore, any job interval that has already been processed and overlaps with I must contain either the right or the left endpoint of I. Let ℓ and r denote the left and the right endpoints of I, respectively.

Let R be any rectangle in $\{R(I, 1), \ldots, R(I, z_I)\}$. Let $w_\ell(R)$ denote the total LP-weight of job intervals assigned to R that contain ℓ, and define $w_r(R)$ similarly for r. Since I cannot be assigned to R, $w_\ell(R) + w_r(R) > 4$. Therefore, either $\sum_{z=1}^{z_I} w_\ell(R(I, z)) > 2z_I$ or $\sum_{z=1}^{z_I} w_r(R(I, z)) > 2z_I$. Assume w.l.o.g. that it is the former. So we have a set S of job intervals belonging to layers $1, \ldots, z_I$, all containing point ℓ, whose total LP-weight is greater than $2z_I$. Let t_1, t_2 be the time points closest to ℓ on left and right respectively, such that $g(t_i) < z_I + 1$ for $i \in \{1, 2\}$. Then there is a layer-$(z_I + 1)$ rectangle $R \in \mathcal{R}$ with $W(R) = [t_1, t_2]$. Let I' be any interval in S. Since I' belongs to one of the layers $1, \ldots, z_I$, it is not contained in $W(R)$, and so either $t_1 \in I'$ or $t_2 \in I'$. Therefore, either the total LP-weight of intervals I' in S containing t_1 is more than z_I, or the total LP-weight of intervals I' in S containing t_2 is more than z_I. But this contradicts the fact that $g(t_i) < z_I + 1$. $\qquad\square$

Step 3: Scheduling the Jobs Given a rectangle $R \in \mathcal{R}$, let $\mathcal{I}(R) \subseteq \mathcal{I}$ be the set of job intervals assigned to R. For simplicity from now on we denote J^* by J and the block partition $\mathcal{B}^*$ by $\mathcal{B}$. As before, for each time point t, $\mathcal{R}(t) \subseteq \mathcal{R}$ denotes the set of rectangles containing t. We partition the jobs into 7 types $Q_1, \ldots, Q_7$. We then schedule each of the types separately. Each job $j \in J$ will be scheduled on one of its time intervals $I \in \mathcal{I}(j)$. If $I \in \mathcal{I}(R)$, then we say that j is scheduled *inside* R. Given a subset S of jobs scheduled inside a rectangle R, we say that the schedule *uses* α *machines* iff for each time point $t \in R$, the total number of jobs of S scheduled on intervals in $\mathcal{I}(R)$ containing t is at most α. We will ensure that for each job type Q_i, for each rectangle $R \in \mathcal{R}$, all jobs of Q_i scheduled inside R use a constant number of machines. Since $|\mathcal{R}(t)| = g(t)$ for all $t \in B$, overall we obtain a schedule where the number of jobs scheduled at time t is at most $O(g(t))$ for all $t \in B$, as desired. We start with a high level overview. The set Q_1 contains jobs with a large LP-weight on intervals intersecting block boundaries. The set Q_2 contains all jobs with large LP-weight on intervals I whose length is more than half the length of $R(I)$. These two job types are taken care of similarly to type 1 and 2 jobs in [6]. The sets Q_3 and Q_5 contain jobs j with large LP-weight on intervals belonging to rectangles that contain d_j. These sets corresponds to jobs of type 3 in [6]. However, in our more general setting, we need to consider many different rectangles contained in a block simultaneously, and so these job types require more care and the algorithm and its analysis are more complex. Job types 4 and 6 are similar to types 3 and 5, except that we use release dates instead of deadlines. Finally, type 7 contains all remaining jobs, and we treat them similarly to jobs of type 5 in [6]. We now proceed to define the partition of jobs into 7 types, and show how to schedule jobs of each type.

Type 1. Let P be the set of time points that serve as endpoints of blocks in $\mathcal{B}$. We say that $I \in \mathcal{I}$ is a type-1 interval, and denote $I \in \mathcal{I}_1$, iff it contains a point in P. We define the set of jobs of type 1: $Q_1 = \left\{ j \in J \mid \sum_{I \in \mathcal{I}(j) \cap \mathcal{I}_1} x(I, j) \geq 1/7 \right\}$. These jobs are treated similarly to type-1 jobs in [6], via a simple max flow computation. We omit the details due to lack of space.

We will now focus on the set $\mathcal{I}' = \mathcal{I} \setminus \mathcal{I}_1$ of intervals that do not cross block boundaries. We can now refine our definition of rectangles to intersections of blocks and rectangles. More formally, for each $R \in \mathcal{R}$, the partition $\mathcal{B}$ of B into blocks also defines a partition of R into a collection $\mathcal{C}(R, \mathcal{B})$ of rectangles. We then define a new set $\mathcal{R}' = \bigcup_{R \in \mathcal{R}} \mathcal{C}(R, \mathcal{B})$ of rectangles. The set $\mathcal{I}(R')$ of intervals assigned to $R' \in \mathcal{C}(R, \mathcal{B})$ is the set of intervals in $\mathcal{I}(R)$ that are contained in R'. We will schedule the remaining jobs inside the rectangles of $\mathcal{R}'$, such that the schedule inside each $R \in \mathcal{R}'$ uses a constant number of machines. Recall that for each $R, R' \in \mathcal{R}$, if $R \cap R' \neq \emptyset$, then either $W(R) \subseteq W(R')$ or $W(R') \subseteq W(R)$. It is easy to see that the same property holds for rectangles in $\mathcal{R}'$.

Type 2. An interval $I \in \mathcal{I}'$ is called *large* iff the length of the rectangle $R \in \mathcal{R}'$, where $I \in \mathcal{I}(R)$, is at most twice the length of I. Let $\mathcal{I}_2$ denote the set of all large intervals. We define $Q_2 = \left\{ j \in J \setminus Q_1 \mid \sum_{I \in \mathcal{I}(j) \cap \mathcal{I}_2} x(I, j) \geq 1/7 \right\}$. These jobs are scheduled similarly to type-2 jobs in [6], using a simple max-flow computation. We omit details due to lack of space.

Type 3. Consider an interval $I \in \mathcal{I}(j)$ for some job $j \in J \setminus (Q_1 \cup Q_2)$, and assume that $I \in \mathcal{I}(R)$ for $R \in \mathcal{R}'$. We say that I is *deadline large* iff $d_j \in R$ and $p_j > \frac{1}{2}(d_j - \ell_R)$. Let $\mathcal{I}_3$ be the set of all deadline large intervals. We define the set Q_3 of jobs of type 3 as follows: $Q_3 = \left\{ j \in J \setminus (Q_1 \cup Q_2) \mid \sum_{I \in \mathcal{I}(j) \cap \mathcal{I}_3} x(I, j) \geq 1/7 \right\}$.

For each job $j \in Q_3$, define the interval $\Gamma_j = (d_j - p_j, d_j)$. Notice that Γ_j is the right-most interval in $\mathcal{I}(j)$. We simply schedule each job $j \in Q_3$ on interval Γ_j.

Claim. The total number of jobs of Q_3 scheduled at any time t is at most $O(g(t))$.

Proof. For each job $j \in Q_3$, for each rectangle $R \in \mathcal{R}'$, with $\mathcal{I}(j) \cap \mathcal{I}(R) \cap \mathcal{I}_3 \neq \emptyset$, we define a fractional value $x''_R(\Gamma_j, j)$. We will ensure that for each $j \in Q_3$, $\sum_{R \in \mathcal{R}'} x''_R(\Gamma_j, j) = 1$, and for each rectangle $R \in \mathcal{R}'$, for each $t \in \mathcal{R}'$, $\sum_{j:t \in \Gamma_j} x''_R(\Gamma_j, j) \leq 70$. Since for each point t, $|\mathcal{R}(t)| = g(t)$, the claim follows.

Consider now some fixed rectangle $R \in \mathcal{R}'$. We change the fractional schedule of intervals inside R in two steps. In the first step, for each $j \in Q_3$, we set $x'(I, j) = x(I, j) / \sum_{I \in \mathcal{I}_3 \cap \mathcal{I}(j)} x(I, j)$ for each $I \in \mathcal{I}(j) \cap \mathcal{I}(R) \cap \mathcal{I}_3$. By the definition of jobs of type 3, we now have that

$$\forall t \in R \qquad \sum_{j \in Q_3} \sum_{\substack{I \in \mathcal{I}(j) \cap \mathcal{I}(R): \\ t \in I}} x'(I, j) \le 35 \qquad\qquad (3)$$

Next, for each job $j \in Q_3$ with $\Gamma_j \subseteq R$, we set $x''_R(\Gamma_j, j) = \sum_{I \in \mathcal{I}(R)} x'(I, j)$. Notice that since $j \in Q_3$, $\sum_{R \in \mathcal{R}'} x''_R(\Gamma_j, j) = 1$. It is now enough to prove that for each time point $t \in R$, $\sum_{j \in Q_3 : t \in \Gamma_j} x''_R(\Gamma_j, j) \le 70$.

Assume otherwise. Let t be some time point, such that $\sum_{j \in Q_3 : t \in \Gamma_j} x''_R(\Gamma_j, j) > 70$. Let S_t be the set of jobs $j \in Q_3$ with $t \in \Gamma_j$ and $x''_R(\Gamma_j, j) > 0$, and let $j' \in S_t$ be the job with smallest processing time. Consider the time point $t' = d_{j'} - p_{j'}$. We claim that for each $j \in S_t$, for each interval $I \in \mathcal{I}(j) \cap \mathcal{I}(R)$, either $t' \in I$ or $t \in I$. If this is true then we have that either $\sum_{j \in Q_3} \sum_{\substack{I \in \mathcal{I}(j) \cap \mathcal{I}(R): \\ t \in I}} x'(I, j) > 35$ or $\sum_{j \in Q_3} \sum_{\substack{I \in \mathcal{I}(j) \cap \mathcal{I}(R): \\ t' \in I}} x'(I, j) > 35$, contradicting (3).

Consider some job $j \in S_t$ and assume for contradiction that there is some time interval $I \in \mathcal{I}(j) \cap \mathcal{I}(R)$ that contains neither t nor t'. Then I must lie completely to the left of t' and hence to the left of $\Gamma_{j'}$. But since $p_j \ge p_{j'}$, we have that $t' - \ell_R \ge p_j \ge p_{j'}$, and so $d_{j'} - \ell_R \ge 2p_{j'}$, contradicting the fact that $j' \in S_t$. $\qquad \square$

Type 4. Same as type 3, but for release date instead of deadline. Is treated similarly to Type 3. The set of type 4 jobs is denoted by Q_4.

Type 5. Consider some interval $I \in \mathcal{I}(j)$ for $j \in J \setminus (Q_1 \cup \cdots \cup Q_4)$, and assume that $I \in \mathcal{I}(R)$ for $R \in \mathcal{R}'$. We say that I is of type 5 ($I \in \mathcal{I}_5$) iff $d_j \in R$ and $I \notin \mathcal{I}_3$ (so $d_j - \ell_R \ge 2p_j$). We define the set Q_5 of jobs of type 5 as follows:

$$Q_5 = \left\{ j \in J \setminus (Q_1 \cup \cdots \cup Q_4) \mid \sum_{I \in \mathcal{I}(j) \cap \mathcal{I}_5} x(I, j) \ge 1/7 \right\}.$$

For a job $j \in Q_5$ and a rectangle $R \in \mathcal{R}'$, we say that R is *admissible* for j iff $d_j \in R$ and $d_j - \ell_r \ge 2p_j$. We say that an interval $I \in \mathcal{I}(j)$ is admissible for j iff $I \in \mathcal{I}_5$. Notice that if $j \in Q_5$ then the sum of values $x(I, j)$ where I is admissible for j is at least $1/7$. Let $R \in \mathcal{R}'$ be any rectangle, and let $S \subseteq Q_5$ be any subset of jobs of type 5. We say that set S is *feasible* for R iff R is admissible for each $j \in S$, and, for each time point $t \in R$, $\sum_{j \in S : d_j \le t} p_j < 70(t - \ell_R)$. We now proceed as follows. First we show that if S is feasible for R, then we can schedule all jobs of S inside R on at most 140 machines. After that we show how to assign all jobs of Q_5 to rectangles such that each rectangle is assigned a feasible subset. We start with the following lemma.

Lemma 1. *If $S \subseteq Q_5$ is a feasible subset of jobs for R then all jobs in S can be scheduled inside R on at most 140 machines.*

Proof. We will schedule all jobs of S on 140 machines inside the time interval $W(R)$. We scan all 140 machines simultaneously from left to right starting from time point ℓ_R. Whenever any machine becomes idle, we schedule on it the job with earliest deadline among all available jobs of S. It is easy to see that all jobs are scheduled: Assume otherwise, and let j be the first job that we are unable to schedule. Consider the time point $t = d_j - p_j$. All the machines are occupied at time t, and they only contain jobs whose deadline is before d_j. Therefore, $\sum_{j' \in S : d_{j'} < d_j} p_{j'} \geq 140(t - \ell_R)$. But since $d_j - \ell_R \geq 2p_j$, we have that $t - \ell_R = d_j - p_j - \ell_R \geq \frac{1}{2}(d_j - \ell_R)$, and so $\sum_{j' \in S : d_{j'} < d_j} p_{j'} \geq 70(d_j - \ell_R)$, contradicting the fact that S is feasible for R. □

We now show how to assign jobs of Q_5 to rectangles, such that each rectangle is assigned a feasible subset. Consider some block $B' \in \mathcal{B}$. Let $\mathcal{R}(B') \subseteq \mathcal{R}'$ be the set of rectangles contained in B', and let $H(B') \subseteq Q_5$ be the subset of jobs of type 5 whose deadline is inside B'. We will assign jobs in $H(B')$ to rectangles in $\mathcal{R}(B')$. Recall the partition of the set $\mathcal{R}(B')$ of rectangles into layers. Layer i, denoted by L_i, consists of all rectangles that are not contained in any other rectangle of $\mathcal{R}(B') \setminus (L_1 \cup \cdots \cup L_{i-1})$ (if we have identical rectangles then at most one of them is assigned to each layer and we break the ties arbitrarily). Consider some job $j \in H(B')$. Let $z(j)$ be the maximum index i, such that some rectangle $R \in L_i$ is admissible for j. Then for each $z : 1 \leq z \leq z(j)$, there is a unique layer-z rectangle $R_z(j)$ that is admissible for j.

We will assign a subset $A(R)$ of jobs to each rectangle $R \in \mathcal{R}(B')$. We start with $A(R) = \emptyset$ for all R. We process jobs of $H(B')$ in non-decreasing order of their deadlines. When job j is processed, it is assigned to $R_z(j)$, where z is the maximum index, $1 \leq z \leq z(j)$, such that $A(R) \cup \{j\}$ is feasible for R. It now only remains to prove is that every job j can be assigned to a rectangle. The next lemma will finish the analysis of the algorithm for type-5 jobs.

Lemma 2. *For each job $j \in H(B')$, when j is processed, there is a rectangle $R_z(j)$, $1 \leq z \leq z(j)$, such that j can be assigned to $R_z(j)$.*

Proof. Assume otherwise, and let j be the first job that cannot be assigned to any such rectangle. We now proceed as follows. We construct a subset $\tilde{\mathcal{R}} \subseteq \mathcal{R}(B')$ of rectangles, and for each $R \in \tilde{\mathcal{R}}$ we define a time point $t_R \in R$. For each $R \in \tilde{\mathcal{R}}$, we define a subset $\tilde{J}(R) \subseteq A(R)$ of jobs whose deadline is before t_R and show that the total processing time of jobs in $\tilde{J}(R)$ is more than $35(t_R - \ell_R)$. On the other hand, we ensure that for each $j \in \bigcup_{R \in \tilde{\mathcal{R}}} \tilde{J}(R)$, for each admissible interval I for j, if $I \in \mathcal{I}(R)$, then $R \in \tilde{\mathcal{R}}$ and $I \subseteq [\ell_R, t_R]$. This leads to a contradiction, since for each $j \in \tilde{J}$, at least $1/7$ of the LP weight is on admissible intervals, and all such intervals are contained in the intervals $[\ell_R, t_R]$ for $R \in \tilde{\mathcal{R}}$. On the other hand, for each rectangle $R \in \tilde{\mathcal{R}}$, for each time point $t \in R$, the total LP-weight of intervals of R containing t is at most 5.

Let $R \in \mathcal{R}(B')$ be any rectangle, and let $t \in R$. We say that R is *overpacked for* t iff $\sum_{j' \in A(R) : d_{j'} \leq t} p_{j'} > 35(t - \ell_R)$. We process the rectangles layer-by-layer. At the beginning, we set $\tilde{\mathcal{R}} = \emptyset$ and $\tilde{J} = \emptyset$. In the first iteration, we consider the rectangles of layer L_1. Let $R = R_1(j)$. We add R to $\tilde{\mathcal{R}}$ and set $t_R = d_j$. Note that since j could not be assigned to R, rectangle R must be overpacked for t_R. We add to $\tilde{J}$ all jobs in $A(R) \cup \{j\}$.

In iteration i, we consider rectangles $R \in L_i$. Consider the set $Y(R)$ of jobs j' for which $z(j') \geq i$ and $R_i(j') = R$. If $\tilde{J} \cap Y(R)$ is non-empty, we add R to $\tilde{\mathcal{R}}$, and set t_R to be the maximum deadline of any job $j' \in \tilde{J} \cap Y(R)$. Notice that since j' was not assigned to R, rectangle R is overpacked for t_R. Let $\tilde{J}(R)$ be the set of all jobs $j'' \in A(R)$ with $d_{j''} \le t_R$. We add jobs in $\tilde{J}(R)$ to $\tilde{J}$.

Consider the final set $\tilde{\mathcal{R}}$ of rectangles and the set $\tilde{J}$ of jobs. Clearly, the set $\tilde{J}$ of jobs is the disjoint union of sets $\tilde{J}(R)$ for $R \in \tilde{\mathcal{R}}$. Recall that $\tilde{J}(R)$ contains all jobs $j' \in A(R)$ with $d_{j'} \le t_R$. Since each rectangle $R \in \tilde{\mathcal{R}}$ is overpacked for t_R, we have that $\sum_{j \in \tilde{J}} p_j > 35 \sum_{R \in \tilde{\mathcal{R}}} (t_R - \ell_R)$. On the other hand, the next claim shows that for each job $j \in \tilde{J}$, for each admissible interval I of j, if $I \in \mathcal{I}(R)$, then $R \in \tilde{\mathcal{R}}$ and I lies to the left of t_R.

Claim 2. Let $j \in \tilde{J}$, let I be any admissible interval for j, and assume that $I \in \mathcal{I}(R)$. Then $R \in \tilde{\mathcal{R}}$, and $I \subseteq [\ell_R, t_R]$.

We now obtain a contradiction as follows. We have shown that $\sum_{j \in \tilde{J}} p_j > 35 \sum_{R \in \tilde{\mathcal{R}}} (t_R - \ell_R)$. On the other hand, for each job $j \in \tilde{J}$, at least $1/7$ LP-weight lies on admissible intervals. Since al these admissible intervals are contained inside intervals $[\ell_R, t_R]$ for $R \in \tilde{\mathcal{R}}$, we have that $\sum_{R \in \tilde{\mathcal{R}}} \sum_j \sum_{\substack{I \in \mathcal{I}(j) \cap \mathcal{I}(R): \\ I \subseteq [\ell_R, t_R]}} x(I, j) \ge \frac{1}{7} \sum_{j \in \tilde{J}} p_j > 5 \sum_{R \in \tilde{\mathcal{R}}} (t_R - \ell_R)$. This contradicts the fact that for every rectangle $R \in \mathcal{R}'$, for each $t \in R$, $\sum_j \sum_{I \in \mathcal{I}(j) \cap \mathcal{I}(R): t \in I} x(I, j) \le 5$. It now only remains to prove Claim 2.

Proof (Of Claim 2). Consider some job $j' \in \tilde{J}$, and suppose it was added to $\tilde{J}$ in iteration i. Let I be any admissible interval of j'. Then there must be an index $z : 1 \le z \le z(j')$ such that $I \in \mathcal{I}(R_z(j))$. We now consider three cases. First, if $z = i$, then let $R = R_i(j')$. Then, since j' was added to $\tilde{J}$ in iteration i, $j' \in \tilde{J}(R)$ and so $I \subseteq [\ell_R, t_R]$. Clearly, $R \in \tilde{\mathcal{R}}$. Assume now that $z > i$ and let $R = R_z(j')$. Then $j' \in \tilde{J}$ in iteration z, and so when R was considered, $j' \in Y(R) \cap \tilde{J}$. So R has been added to $\tilde{\mathcal{R}}$ and t_R has been set to be at least $d_{j'}$. Finally, assume that $z < i$. Let $R = R_i(j')$ and $R' = R_z(j')$. Then $R \subseteq R'$. It is then enough to prove the following claim:

Claim. Let $R \in L_i$ and $R' \in L_{i-1}$, with $R \subseteq R'$. Assume that $R \in \tilde{R}$. Then $R' \in \tilde{R}$, and moreover $t_{R'} \ge t_R$.

Proof. Consider the iteration i when R was added to $\tilde{\mathcal{R}}$, and let $j'' \in Y(R)$ be the job which determined t_R, so $t_R = d_{j''}$. Two cases are possible. If $j'' \in A(R')$, then j'' has been added to $\tilde{J}$ in iteration $i - 1$ when R' was processed. So $R' \in \tilde{R}$ and $t_{R'} \ge d_{j''} = t_R$. Otherwise, j'' was in $\tilde{J}$ when R' was processed. Since $R \subseteq R'$ and R is admissible for j'', so is R'. Therefore, $j'' \in Y(R') \cap \tilde{J}$ and so $R' \in \tilde{R}$ and $t_{R'} \ge d_{j''} = t_R$.   □ □ □

Type 6. Like type 5, but for release date.

Type 7. All other jobs. The algorithm for these jobs is the same as the one used in [6], substituting rectangles for blocks. We omit details due to lack of space.

References

1. Bansal, N., Chakrabarti, A., Epstein, A., Schieber, B.: A quasi-ptas for unsplittable flow on line graphs. In: STOC 2006: Proceedings of the thirty-eighth annual ACM Symposium on Theory of Computing, pp. 721–729. ACM Press, New York (2006)
2. Bar-Noy, A., Guha, S., Naor, J., Schieber, B.: Approximating the throughput of multiple machines in real-time scheduling. In: Proceedings of the 31st Annual ACM Symposium on Theory of Computing, pp. 622–631 (1999)
3. Bar-Noy, A., Bar-Yehuda, R., Freund, A., Naor, J(S.), Schieber, B.: A unified approach to approximating resource allocation and scheduling. J. ACM 48(5), 1069–1090 (2001)
4. Calinescu, G., Chakrabarti, A., Karloff, H., Rabani, Y.: Improved approximation algorithms for resource allocation. In: Cook, W.J., Schulz, A.S. (eds.) IPCO 2002. LNCS, vol. 2337, pp. 401–414. Springer, Heidelberg (2002)
5. Chuzhoy, J., Ostrovsky, R., Rabani, Y.: Approximation algorithms for the job interval selection problem and related scheduling problems. Math. Oper. Res. 31(4), 730–738 (2006)
6. Chuzhoy, J., Guha, S., Khanna, S., Naor, J.: Machine minimization for scheduling jobs with interval constraints. In: FOCS, pp. 81–90 (2004)
7. Chuzhoy, J., Naor, J(S.): New hardness results for congestion minimization and machine scheduling. J. ACM 53(5), 707–721 (2006)
8. Raghavan, P., Thompson, C.D.: Randomized rounding: A technique for provably good algorithms and algorithmic proofs. Combinatorica 7 (1987)
9. Spieksma, F.C.R.: On the approximability of an interval scheduling problem. Journal of Scheduling 2, 215–227 (1999)

The Power of Preemption
on Unrelated Machines and Applications to
Scheduling Orders*

José R. Correa[1], Martin Skutella[2], and José Verschae[2]

[1] Departamento de Ingeniería Industrial, Universidad de Chile, Santiago, Chile
joser.correa@gmail.com
[2] Institute of Mathematics, TU Berlin, Germany
{skutella,verschae}@math.tu-berlin.de

Abstract. Scheduling jobs on unrelated parallel machines so as to minimize the makespan is one of the basic, well-studied problems in the area of machine scheduling. In the first part of the paper we prove that the power of preemption, i.e., the ratio between the makespan of an optimal nonpreemptive and an optimal preemptive schedule, is exactly 4. This result is a definite answer to an important basic open problem in scheduling. The proof of the lower bound is based on a clever iterative construction while the rounding technique we use to prove the upper bound is an adaptation of Shmoys and Tardos' rounding for the generalized assignment problem. In the second part of the paper we apply this adaptation to the more general setting in which orders, consisting of several jobs, have to be processed on unrelated parallel machines so as to minimize the sum of weighted completion times of the orders. We obtain the first constant factor approximation algorithms for the preemptive and nonpreemptive case, improving and extending a recent result by Leung et. al.

1 Introduction

Problem description and basic results. Consider the classical scheduling problem of minimizing the makespan on unrelated parallel machines. In this problem we are given a set of jobs $J = \{1, \ldots, n\}$ and a set of machines $M = \{1, \ldots, m\}$ to process the jobs. Each job $j \in J$ has associated processing times p_{ij}, denoting the amount of time that it takes to process job j on machine i. Every job has to be scheduled on exactly one machine without interruption and each machine can schedule at most one job at a time. The objective is to find a schedule minimizing the point in time at which the last job is completed, i.e., minimizing

---

* This work was partially supported by Berlin Mathematical School, by DFG research center MATHEON in Berlin, by CONICYT, through grants FONDECYT 1060035 and Anillo en Redes ACT08. The authors thank Nikhil Bansal for stimulating discussions on the material in Section 2.

I. Dinur et al. (Eds.): APPROX and RANDOM 2009, LNCS 5687, pp. 84–97, 2009.

$C_{\max} := \max_{j \in J} C_j$, where C_j is the completion time of job j. In the standard three-field scheduling notation (see, e.g., Lawler et al. [14]) this problem is denoted by $R||C_{\max}$.

In a seminal work, Lenstra, Shmoys and Tardos [16] give a 2-approximation algorithm for $R||C_{\max}$, and show that the problem is NP-hard to approximate within a factor better than $3/2$. On the other hand, Lawler and Labetoulle [13] show that the preemptive version of this problem, denoted $R|pmtn|C_{\max}$, where jobs can be interrupted and resumed later on the same or a different machine, can be formulated as a linear program and thus be solved in polynomial time.

Power of preemption. The power of preemption is the worst-case ratio between the makespan of an optimal preemptive and an optimal nonpreemptive solution. This ratio has been studied in the literature for various scheduling problems [4,21,22]. One contribution of this work is to prove that this ratio is exactly 4 for the considered problem on unrelated machines. The proof consists of two steps — proving an upper and a lower bound of 4. For the upper bound, we consider an optimal solution to the linear programming formulation of Lawler and Labetoulle [13] for $R|pmtn|C_{\max}$, and round it to obtain an assignment of jobs to machines in which the makespan is increased at most by a factor of 4. The rounding consists in setting to zero all variables whose corresponding processing time is too large compared to the makespan, and then amplifying the remaining values so that a feasible fractional assignment is maintained. Then, the technique of Shmoys and Tardos [23] is applied to obtain a nonpreemptive solution. The proof of the lower bound is based on a clever recursive construction, where in each iteration the gap of the instance is increased.

Scheduling orders of jobs. In the second part of the paper, we apply the rounding technique used for the previous result to a more general setting. Consider the natural scheduling problem where clients place orders, consisting of several products, to a manufacturer owning m unrelated parallel machines. Each product has a machine dependent processing requirement. The manufacturer has to find an assignment of products to machines (and a schedule within each machine) so as to give the best possible service to his clients.

More precisely, we are given a set of machines $M = \{1, \ldots, m\}$, a set of jobs $J = \{1, \ldots, n\}$ (as before) and a set of orders $O \subseteq 2^J$, such that $\bigcup_{L \in O} L = J$. Each job $j \in J$ takes p_{ij} units of time to be processed in machine $i \in M$, and each order L has a weight factor w_L depending on how important it is for the manufacturer and the client. Also, job j is associated with a release date r_{ij}, so it can only start being processed on machine i by time r_{ij}. An order $L \in O$ is completed once all its jobs have been processed. Therefore, if C_j denotes the time at which job j is completed, $C_L = \max\{C_j : j \in L\}$ denotes the completion time of order L. The goal of the manufacturer is to find a nonpreemptive schedule on the m available machines so as to minimize the sum of weighted completion times of orders, i.e., $\min \sum_{L \in O} w_L C_L$. Let us remark that in this general framework we are not restricted to the case where the orders are disjoint, and therefore one job may contribute to the completion time of more than one order.

We adopt the standard three-field scheduling notation by denoting this problem $R|r_{ij}|\sum w_L C_L$, or $R||\sum w_L C_L$ in case all release dates are zero. When the processing times p_{ij} do not depend on the machine, we replace "R" with "P". Also, when we impose the additional constraint that orders are disjoint subsets of jobs we will add *part* in the second field of the notation.

Relation to other scheduling problems. It is easy to see that this setting generalizes several classical machine scheduling problems. In particular our problem becomes $R||C_{\max}$ when the total number of orders is one. Thus, it follows from [16] that $R||\sum w_L C_L$ cannot be approximated within a factor better than $3/2$, unless $P = NP$. On the other hand, if orders are singletons our problem becomes $R||\sum w_j C_j$. In this setting each job $j \in J$ is associated with a processing time p_{ij} and a weight w_j, and the goal is to find a schedule of the jobs so as to minimize the sum of weighted completion times. In other words, if C_j denotes the completion time of job j in a given schedule, the goal is to minimize $\sum_{j=1}^{n} w_j C_j$. As in the makespan case, this problem was shown to be APX-hard [12] and therefore there is no PTAS, unless $P = NP$. Using randomized rounding techniques based on a linear relaxation, Schulz and Skutella [22] proposed an approximation algorithm for this problem with performance guarantee $3/2 + \varepsilon$ in the case without release dates, and $2 + \varepsilon$ in the more general case. Later, Skutella [25] slightly improved this result by using randomized rounding over a convex cuadratic relaxation, obtaining approximation algorithms with performance guarantee $3/2$ and 2, respectively.

However, for the more general setting $R|r_{ij}|\sum w_L C_L$, there is no constant factor approximation known. The best known result, due to Leung, Li, Pinedo, and Zhang [18], is an approximation algorithm for the special case of related machines without release dates, denoted $Q||\sum w_L C_L$, where $p_{ij} = p_i/s_i$ and s_i is the speed of machine i. The performance ratio of their algorithm is $1 + \rho(m - 1)/(\rho + m - 1)$, where ρ is the ratio of the speed of the fastest machine to that of the slowest machine. In general this guarantee is not constant and can be as bad as $m/2$.

Identical parallel machines. For the special case of identical parallel machines, our problem $P||\sum w_L C_L$ also generalizes $P||C_{\max}$ and $P||\sum w_j C_j$. These two problems are well known to be NP-hard, even for the case of only two machines, since the well-known PARTITION problem can be reduced to them. For the makespan objective, Graham [9] showed that a simple list scheduling algorithm yields a 2-approximation algorithm. Furthermore, Hochbaum and Shmoys [11] present a PTAS for the problem. On the other hand, for the sum of weighted completion times objective, a sequence of approximations algorithms had been proposed until Skutella and Woeginger [24] found a PTAS (see also [1]).

On the even more restricted setting of a single machine, the two previously mentioned problems $1||C_{\max}$ and $1||\sum w_j C_j$ can be easily solved, the first one by any feasible solution with no idle time, and the second one by applying *Smith's rule* [26]. However, our problem $1||\sum w_L C_L$ is NP-hard, as it is equivalent to $1|prec|\sum w_j C_j$. In the latter problem, there is a partial order $\preceq$ over the jobs, meaning that job j must be processed before job k if $j \preceq k$.

Lemma 1. *The approximability thresholds of $1|prec|\sum w_j C_j$ and $1||\sum w_L C_L$ coincide.*

Due to space restrictions, the proof of the lemma is omitted. The scheduling problem $1|prec|\sum w_j C_j$ has attracted much attention since the sixties. Lenstra and Rinnooy Kan [15] showed that this problem is strongly NP-hard even with unit weights. On the other hand, several 2-approximation algorithms have been proposed [10,6,5,20]. Furthermore, the results in [2,7] imply that $1|prec|\sum w_j C_j$ is a special case of *vertex cover*. However, hardness of approximation results were unknown until recently Ambühl, Mastrolilli and Svensson [3] proved that there is no PTAS unless NP-hard problems can be solved in randomized subexponential time. In particular, the same result holds for $P||\sum w_L C_L$. Nonetheless, the reduction used in the previous lemma does not work on the more restrictive case where orders are disjoint, $P|part|\sum w_L C_L$, and thus the question whether there is a PTAS for this latter problem remains open. However, we were able to develope a PTAS for the special cases in which either the orders are of constant size, or there is a constant number of orders, or there is a constant number of machines. Due to space restrictions this result is left for the full version of the paper (see [27] for details).

Our Contribution. Our tight result on the power of preemption for unrelated parallel machine scheduling with makespan objective have already been outlined above. In addition to the result stated in Lemma 1, we present the first constant factor approximation algorithm for the general problem $R|r_{ij}|\sum w_L C_L$ and its preemptive variant $R|r_{ij}, pmtn|\sum w_L C_L$. This is achieved by considering the interval indexed linear programs proposed by Dyer and Wolsey [8] and Hall et al. [10], and then applying essentially the same rounding technique that is used to prove the upper bound on the power of preemption. This approximation result improves upon the previously mentioned result of Leung, Li, Pinedo, and Zhang [18] for the special case $Q||\sum w_L C_L$.

2   A Simple Rounding Technique

We start by showing that the power of preemption for $R||C_{\max}$ is at most 4. As shown by Lawler and Labetoulle [13], we can obtain the optimal value of the preemptive version of this problem by solving the following linear program, whose variables x_{ij} denote the fraction of job j that is processed on machine i, and C the makespan of the solution: [LL] minimize C such that $\sum_{i \in M} x_{ij} = 1$ for all $j \in J$, $\sum_{j \in J} p_{ij} x_{ij} \leq C$ for all $i \in M$, $\sum_{i \in M} p_{ij} x_{ij} \leq C$ for all $j \in J$ and $x_{ij} \geq 0$ for all i, j.

Let x_{ij} and C be any feasible solution to [LL]. To round this fractional solution we proceed in two steps: First, we eliminate fractional variables whose corresponding processing time is too large; Then, we use the rounding technique developed by Shmoys and Tardos [23] for the general assignment problem. In the general assignment problem, we are given m machines and n jobs with machine dependant processing times p_{ij}. We also consider a cost of assigning job j to machine i, denoted by c_{ij}. Given a total budget B and makespan C, the question is to decide whether there exists a schedule with total cost at most B and makespan at most C. The main result of [23] is subsumed in the next theorem.

Theorem 1 (Shmoys and Tardos [23]). *Given a nonnegative fractional solution to the following system of equations:*

$$\sum_{j \in J} \sum_{i \in M} c_{ij} x_{ij} \leq B, \tag{1}$$

$$\sum_{i \in M} x_{ij} = 1, \qquad \text{for all } j \in J, \tag{2}$$

there exists an integral solution $\hat{x}_{ij} \in \{0, 1\}$ satisfying (1),(2), and also,

$$x_{ij} = 0 \implies \hat{x}_{ij} = 0 \qquad \text{for all } i \in M, j \in J, \tag{3}$$

$$\sum_{j \in J} p_{ij} \hat{x}_{ij} \leq \sum_{j \in J} p_{ij} x_{ij} + \max\{p_{ij} : x_{ij} > 0\} \qquad \text{for all } i \in M. \tag{4}$$

Furthermore, such integral solution can be found in polynomial time.

To proceed with our rounding, let $\beta > 1$ be a fixed parameter that we will specify later. We first define a modified solution x'_{ij} as follows:

$$x'_{ij} = \begin{cases} 0 & \text{if } p_{ij} > \beta C, \\ \frac{x_{ij}}{X_j} & \text{else,} \end{cases} \quad \text{where } X_j = \sum_{i : p_{ij} \leq \beta C} x_{ij} \text{ for all } j \in J.$$

Note that,

$$1 - X_j = \sum_{i : p_{ij} > \beta C} x_{ij} < \sum_{i : p_{ij} > \beta C} x_{ij} \frac{p_{ij}}{\beta C} \leq 1/\beta,$$

where the last inequality follows from [LL]. Therefore, x'_{ij} satisfies that $x'_{ij} \leq x_{ij}\beta/(\beta - 1)$ for all $j \in J$ and $i \in M$, and thus $\sum_{j \in J} x'_{ij} p_{ij} \leq C\beta/(\beta - 1)$ for all $i \in M$. Also, note that by construction $\sum_{i \in M} x'_{ij} = 1$ for all $j \in J$, and $x'_{ij} = 0$ if $p_{ij} > \beta C$. Then, we can apply Theorem 1 to x'_{ij} (for $c_{ij} = 0$), to obtain a feasible integral solution $\hat{x}_{ij}$ to [LL], and thus a feasible solution to $R||C_{\max}$, such that for all $i \in M$,

$$\sum_{j \in J} \hat{x}_{ij} p_{ij} \leq \sum_{j \in J} x'_{ij} p_{ij} + \max\{p_{ij} : x_{ij} > 0\} \leq \frac{\beta}{\beta - 1} C + \beta C = \frac{\beta^2}{\beta - 1} C.$$

Therefore, by optimally choosing $\beta = 2$, the makespan of the rounded solution is at most $\beta^2/(\beta - 1) = 4$ times larger than the makespan of the fractional solution.

Power of Preemption for $R||C_{\max}$

We now give a family of instances showing that the integrality gap of [LL] is arbitrarily close to 4. Surprisingly, this implies that the rounding technique showed in the last section is best possible. Note that this is equivalent to saying that the optimal nonpreemptive schedule is within a factor of 4, and no better than 4, of the optimal preemptive schedule.

Let us fix $\beta \in [2, 4)$, and $\varepsilon > 0$ such that $1/\varepsilon \in \mathbb{N}$. We now construct an instance $I = I(\beta, \varepsilon)$ such that its optimal nonpreemptive makespan is at most $(1 + \varepsilon)C$, and that any nonpreemptive solution of I has makespan at least βC. The construction is done iteratively, maintaining at each iteration a preemptive schedule of makespan $(1 + \varepsilon)C$, and where the makespan of any nonpreemptive solution is increased at each step. Due to the equivalence between [LL] and $R|pmtn|C_{\max}$ we can use assignment variables to denote preemptive schedules.

Base Case. We begin by constructing an instance I_0, which will later be our first iteration. To this end consider a set of $1/\varepsilon$ jobs $J_0 = \{j(0; 1), j(0; 2), \ldots, j(0; 1/\varepsilon)\}$ and a set of $1/\varepsilon + 1$ machines $M_0 = \{i(1), i(0; 1), \ldots, i(0; 1/\varepsilon)\}$. Every job $j(0; \ell)$ can only be processed in machine $i(0; \ell)$, where it takes βC units of time to process, and in machine $i(1)$, where it takes a very short time. More precisely, for all $\ell = 1, \ldots, 1/\varepsilon$ we define,

$$p_{i(0;\ell)j(0;\ell)} := \beta C \text{ and } p_{i(1)j(0;\ell)} := \varepsilon C \frac{\beta}{\beta - 1}.$$

The rest of the processing times are defined as infinity. Note that a feasible fractional assignment is given by setting $x_{i(0;\ell)j(0;\ell)} = 1/\beta$, $x_{i(1)j(0;\ell)} := f_0 := (\beta - 1)/\beta$ and setting to zero all other variables. The makespan of this fractional solution is exactly $(1 + \varepsilon)C$. Indeed, the load of each machine $i \in M_0$ is exactly C, and the load associated to each job in J_0 equals $C + \varepsilon C$. Furthermore, any nonpreemptive solution with makespan less than βC must process all jobs $j(0; \ell)$ in $i(1)$. This yields a makespan of $C/f_0 = \beta C/(\beta - 1)$. Therefore, the makespan of any nonpreemptive solution is $\min\{\beta C, C/f_0\}$. If β is chosen as 2, the makespan of any nonpreemptive solution must be at least $2C$, and therefore the gap of the instance tends to 2 when ε tend to zero.

Iterative Procedure. To increase the integrality gap we proceed iteratively as follows. Starting from instance I_0, which will be the base case, we show how to construct instance I_1. An analogous procedure can be used to construct instance I_{n+1} from instance I_n.

Begin by making $1/\varepsilon$ copies of instance I_0, I_0^ℓ for $\ell = 1, \ldots, 1/\varepsilon$, and denote the set of jobs and machines of I_0^ℓ as J_0^ℓ and M_0^ℓ respectively. We impose that jobs in J_0^ℓ can only be processed on machines in M_0^ℓ by setting $p_{ij} = \infty$, for all $j \in J_0^\ell$ and $i \in M_0^k$ such that $k \neq \ell$. Also, denote as $i(1; \ell)$ the copy of machine $i(1)$ belonging to M_0^l. Consider a new job $j(1)$ for which $p_{i(1;\ell)j(1)} = C\beta - C/f_0$ for all $\ell = 1, \ldots, 1/\varepsilon$ (and ∞ otherwise), and define $x_{i(1;\ell)j(1)} = \varepsilon C/p_{i(1;\ell)j(1)}$. This way, the load of each machine $i(1; \ell)$ in the fractional solution is $(1 + \varepsilon)C$, and the load corresponding to job $j(1)$ is exactly C. Nevertheless, depending on the value of β, job $j(1)$ may not be completely assigned. A simple calculation shows that for $\beta = (3 + \sqrt{5})/2$, job $j(1)$ is completely assigned in the fractional

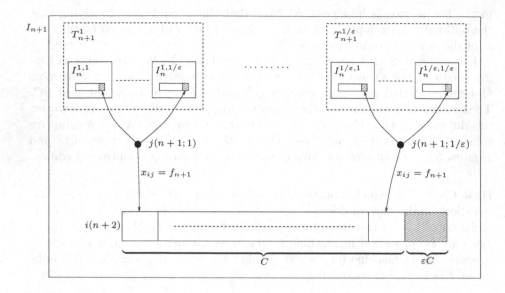

Fig. 1. Construction of instance $I_{n+1}(\beta)$

assignment. Furthermore, as justified before, in any nonpreemptive schedule of makespan less than βC, all jobs in J_0^l must be processed in machine $i(1; \ell)$. Since also job $j(1)$ must be processed in some machine $i(1; \ell)$, the load of that machine must be $\sum_{j \in J_0^l} p_{i(1;\ell)j} + p_{i(1;\ell)j(1)} = C/f_0 + C(\beta - 1/f_0) = \beta C$. Then, the gap of the instance already constructed converges to $\beta = (3 + \sqrt{5})/2 \approx 2.618$ when ε tend to 0, thus improving the gap of 2 shown before.

On the other hand, for $\beta > (3 + \sqrt{5})/2$ (as we would like) there will be some fraction of job $j(1)$, $f_1 := 1 - \sum_{\ell=1}^{1/\varepsilon} x_{i(1;\ell)j(1)} = ((\beta - 1)f_0 - 1)/(\beta f_0 - 1)$ that must be processed elsewhere. To overcome this, we do as follows. Let us denote the instance consisting of jobs $\bigcup_{\ell=1}^{1/\varepsilon} J_0^l$ and machines $\bigcup_{\ell=1}^{1/\varepsilon} M_0^\ell$ as T_1, and construct $1/\varepsilon$ copies of instance T_1, T_1^k for $k = 1, \ldots, 1/\varepsilon$. Define the processing times of jobs in T_1^ℓ to infinity in all machines of T_1^k, for all $k \neq \ell$, so that jobs of T_1^ℓ can only be processed in machines of T_1^ℓ. Also, consider $1/\varepsilon$ copies of job $j(1)$, and denote them by $j(1; k)$ for $k = 1, \ldots, 1/\varepsilon$. As shown before, we can assign a fraction $1 - f_1$ of each job $j(1; k)$ to machines of T_1^k. To assign the remaining fraction f_1, we add an extra machine $i(2)$, with $p_{i(2)j(1;\ell)} := \varepsilon C/f_1$ (and ∞ for all other jobs), so that the fraction f_1 of each job $j(1; \ell)$ takes exactly εC to process in $i(2)$. Then, defining $x_{i(2)j(1;\ell)} = f_1$, the total load of each job $j(1; \ell)$ equals $(1 + \varepsilon)C$, while the load of machine $i(2)$ is exactly C. Let us denote the instance we have constructed so far as I_1.

Following an analogous procedure to the one just described, we can construct a sequence of instances and fractional assignments (see Figure 1). Each instance I_n satisfies the following properties:

(i) The fraction of each job $j(n; 1), \ldots, j(n, 1/\varepsilon)$ assigned to machine $i(n+1)$ is given by $f_n = ((\beta-1)f_{n-1}-1)/(\beta f_{n-1}-1)$.

(ii) Job $j(n+1)$ (or any of its copies) has processing time equal to $C(\beta-1/f_n)$ on each machine $i(n; \ell)$.

(iii) In any nonpreemptive solution of makespan less than βC, every job $j(n+1; \ell)$ must be processed in machine $i(n+2)$. Therefore the makespan of any nonpreemptive solution is at least $\min\{\beta C, C/f_{n+1}\}$.

(iv) The makespan of the fractional solution constructed is $(1+\varepsilon)C$. In particular the load of machine $i(n+2)$ is C, and therefore a fraction of a job which takes less than εC can still be processed in this machine without increasing the makespan.

To finish the construction procedure, notice that if there is some n^* such that $f_{(n^*-1)} \leq 1/(\beta-1)$, then there is no need to construct the whole instance I_{n^*}, but rather instance T_{n^*} suffices. Indeed, if this is the case job $j(n^*)$ can be totally assigned to machines $i(n^*; \ell)$ on the fractional solution, by defining $x_{i(n^*;\ell)j(n^*)} = \varepsilon$ for all $\ell = 1, \ldots, 1/\varepsilon$. This yields a valid assignment since $\sum_{\ell=1}^{1/\varepsilon} p_{i(n^*;\ell)j(n^*)} x_{i(n^*;\ell)j(n^*)} = C(\beta-1/f_{n^*-1}) \leq C$. Also, by Property (iii), any nonpreemtpive solution of makespan less than βC assigns a load of C/f_{n^*-1} to any machine $i(n^*; \ell)$. Furthermore, job $j(n^*)$ must be processed in some machine $i(n^*; \ell)$, which will have a makespan of $C/f_{n^*-1} + (\beta C - C/f_{n^*-1}) = \beta C$. With this we have sketched the proof of the following lemma.

Lemma 2. *If the procedure finishes, then it returns an instance with a gap of at least $\beta/(1+\varepsilon)$.*

Then, we just need to show that the construction terminates, i.e., that $f_{n^*-1} \leq 1/(\beta-1)$ for some n^*. For that, notice the following.

Lemma 3. *For each $\beta \in [2, 4)$, if $f_n > 1/\beta$, then $f_{n+1} \leq f_n$.*

Lemma 4. *The procedure finishes.*

Proof. If the procedure does not finish, then $f_n > 1/(\beta-1) > 1/\beta$ for all $n \in \mathbb{N}$. Then Lemma (3) implies that $\{f_n\}_{n \in \mathbb{N}}$ is a decreasing sequence. Therefore f_n must converge to some real number $L \geq 1/(\beta-1)$. Thus, Property (i) implies that $L = ((\beta-1)L-1)/(\beta L-1)$, and therefore L is a real root of equation $-\beta x^2 + \beta x - 1$ which is a contradiction if $\beta \in [2, 4)$. $\square$

Theorem 2. *The integrality gap of relaxation [LL] is 4.*

3   A $(4+\varepsilon)$-Approximation for $R|r_{ij}, pmtn| \sum w_L C_L$

In this section we adapt the rounding technique discussed in the previous chapter to derive a $(4+\varepsilon)$-approximation algorithm for the preemptive version of $R|r_{ij}| \sum w_L C_L$. Our algorithm is based on a time-indexed linear program, whose variables correspond to the fraction of each job processed at each time in each machine. This kind of linear relaxation was originally introduced by Dyer and Wolsey [8] for $1|r_j|\sum w_jC_j$, and was extended by Schulz and Skutella [22], who used it to obtain a $(3/2 + \varepsilon)$-approximation and a $(2 + \varepsilon)$-approximation for $R||\sum w_jC_j$ and $R|r_j|\sum w_jC_j$ respectively.

Let us consider a time horizon T, large enough so it upper bounds the greatest completion time of any reasonable schedule, for instance $T = \max_{i\in M, k\in J}\{r_{ik} + \sum_{j\in J} p_{ij}\}$. We divide the time horizon into exponentially-growing time intervals, so that there is only polynomially many of them. For that, let ε be a fixed parameter, and let q be the first integer such as $(1 + \varepsilon)^{q-1} \geq T$. Then, we consider the intervals $[0,1], (1, (1+\varepsilon)], ((1+\varepsilon), (1+\varepsilon)^2], \ldots, ((1+\varepsilon)^{q-2}, (1+\varepsilon)^{q-1}]$. To simplify the notation, let us define $\tau_0 = 0$, and $\tau_\ell = (1 + \varepsilon)^{\ell-1}$, for each $\ell = 1\ldots q$. With this, the ℓ-th interval corresponds to $(\tau_{\ell-1}, \tau_\ell]$. In what follows we will assume, without loss of generality, that all processing times are positive integers.

Given any preemptive schedule, let $y_{ij\ell}$ the fraction of job j that is processed in machine i in the ℓ-th interval. Then, $p_{ij}y_{ij\ell}$ is the amount of time that job j is processed in machine i in the ℓ-th interval. With this interpretation is easy to see that the following linear program is a relaxation of $R|r_{ij}, pmtn|\sum w_L C_L$:

$$[\text{DW}] \quad \min \sum_{L\in O} w_L C_L$$

$$\sum_{i\in M}\sum_{\ell=1}^{q} y_{ij\ell} = 1 \qquad \text{for all } j \in J, \quad (5)$$

$$\sum_{j\in J} p_{ij}y_{ij\ell} \leq \tau_\ell - \tau_{\ell-1} \quad \text{for all } \ell = 1,\ldots,q \text{ and } i \in M, \quad (6)$$

$$\sum_{i\in M} p_{ij}y_{ij\ell} \leq \tau_\ell - \tau_{\ell-1} \quad \text{for all } \ell = 1,\ldots,q \text{ and } j \in J, \quad (7)$$

$$\sum_{i\in M}\left(y_{ij1} + \sum_{\ell=2}^{q} \tau_{\ell-1}y_{ij\ell}\right) \leq C_L \qquad \text{for all } L \in O \text{ and } j \in L, \quad (8)$$

$$y_{ij\ell} = 0 \qquad \qquad \text{for all } j,i,\ell : r_{ij} > \tau_\ell, \quad (9)$$

$$y_{ij\ell} \geq 0 \qquad \qquad \text{for all } i,j,\ell. \quad (10)$$

Let $y^*_{ij\ell}$ and C^*_L be the optimal solution of [DW]. Using the same ideas as in Section 2, we round this solution by taking to zero all variables $y^*_{ij\ell}$ having a co-efficient that is too large in (8), and then rescale to obtain a feasible assignment. Then, we use the result in [13], to construct a feasible preemptive schedule inside each interval. More precisely, let $j \in J$, and $L = \arg\min\{C^*_{L'}|j \in L' \in O\}$. For each parameter $\beta > 1$, we define:

$$y'_{ij\ell} = \begin{cases} 0 & \text{if } \tau_{\ell-1} > \beta C^*_L, \\ \dfrac{y^*_{ij\ell}}{Y_j} & \text{else,} \end{cases} \quad \text{where } Y_j = \sum_{i\in M}\sum_{\ell:\, \tau_{\ell-1}\leq \beta\cdot C^*_L} y^*_{ij\ell}. \quad (11)$$

Lemma 5. *The modified solution y' obtained by applying Equation (11) to y^*, satisfies Equation (5). Furthermore, $y'_{ij\ell} = 0$ if $\tau_{\ell-1} > \beta C^*_L$, for all $L \in O$ and $j \in L$, and y' satisfies equations (6) and (7) when their righthand sides are amplified by a factor of $\beta/(\beta - 1)$.*

The proof of the lemma follows the ideas of the rounding in Section 2. Note that since y' only satisfy equations (6) and (7) when their righthand side are amplified, the amount of load assign to each interval may not fit in the available space. Thus, we will have to increase the size of every interval in a factor $\beta/(\beta-1)$. Furthermore, the variables $y'_{ij\ell}$ only assign jobs to intervals that start before βC^*_L in case $j \in L$, allowing us to easily bound the cost of the solution. With the latter observations, we are ready to describe the algorithm.

ALGORITHM: GREEDY PREEMPTIVE LP

1. Solve [DW] to optimality and call the solution y^* and $(C^*_L)_{L \in O}$.
2. Define $y'_{ij\ell}$ using Equation (11).
3. Construct a preemptive schedule S as follows.
   (a) For each $\ell = 1, \ldots, q$, define $x_{ij} = y'_{ij\ell}$ and $C_\ell = (\tau_\ell - \tau_{\ell-1})\beta/(\beta - 1)$, and apply the algorithm by Lawler and Labetoulle [13] to this fractional solution, to obtain a preemptive schedule (i.e., no job is processed in parallel by two machines) of makespan C_ℓ. Call the preemptive schedule obtained S_ℓ.
   (b) For each job $j \in J$ that is processed by schedule S_ℓ at time $t \in [0, C_\ell]$ in machine $i \in M$, make schedule S process j in machine i at time $\tau_{\ell-1}\beta/(\beta - 1) + t$.

Theorem 3. ALGORITHM: GREEDY PREEMPTIVE LP *yields a feasible schedule where the completion time of each order $L \in O$ is less than $C^*_L(1+\varepsilon)\beta^2/(\beta-1)$. Moreover, for $\beta = 2$, the algorithm is a $(4+\varepsilon)$-approximation for the preemptive version of $R|r_{ij}| \sum w_L C_L$.*

4   A Constant Factor Approximation for $R|r_{ij}| \sum w_L C_L$

In this section we propose the first constant factor approximation algorithm for the nonpreemptive version of the problem just described, $R|r_{ij}| \sum w_L C_L$, improving the results in [18]. Our algorithm consists on applying the rounding shown in Section 2 to an adaptation of the interval-index linear programming relaxation developed by Hall, Schulz, Shmoys and Wein [10].

Let us consider a large enough time horizon T as in last section. We divide the time horizon into exponentially-growing time intervals, so that there is only polynomially many. For that, let $\alpha > 1$ be a parameter which will determine later and let q be the first integer such as $\alpha^{q-1} \geq T$. With this, consider the intervals $[1, 1], (1, \alpha], (\alpha, \alpha^2], \ldots, (\alpha^{q-2}, \alpha^{q-1}]$.

To simplify the notation, let us define $\tau_0 = 1$ and $\tau_\ell = \alpha^{\ell-1}$ for each $\ell = 1, \ldots, q$. With this, the ℓ-th interval corresponds to $(\tau_{\ell-1}, \tau_\ell]$. Note that, for technical reasons, these definitions slightly differ from the ones on the previous section.

To model the scheduling problem we consider the variables $y_{ij\ell}$, indicating whether job j is finished in the machine i and in the ℓ-th interval. These variables allow us to write the following linear program based on that in [10], which is a relaxation of the scheduling problem even when integrality constraints are imposed,

$$[\text{HSSW}] \qquad \min \sum_{L \in O} w_L C_L$$

$$\sum_{i \in M} \sum_{\ell=1}^{q} y_{ij\ell} = 1 \qquad\qquad \text{for all } j \in J, \qquad (12)$$

$$\sum_{s=1}^{\ell} \sum_{j \in J} p_{ij} y_{ijs} \le \tau_\ell \qquad \text{for all } i \in M \text{ and } \ell = 1, \ldots, q, \qquad (13)$$

$$\sum_{i \in M} \sum_{\ell=1}^{q} \tau_{\ell-1} y_{ij\ell} \le C_L \qquad \text{for all } L \in O \text{ and } j \in L, \qquad (14)$$

$$y_{ij\ell} = 0 \qquad \text{for all } i, \ell, j : p_{ij} + r_{ij} > \tau_\ell, \qquad (15)$$

$$y_{ij\ell} \ge 0 \qquad\qquad \text{for all } i, j, \ell. \qquad (16)$$

It is clear that [HSSW] is a relaxation of our problem. Indeed, (12) guarantees that each job finishes in some time interval. The left hand side of (13) corresponds to the total load processed on machine i and interval $[0, \tau_\ell]$, and therefore the inequality is valid. The sum in inequality (14) corresponds exactly to $\tau_{\ell-1}$, where ℓ is the interval where job j finishes, so that is at most C_j, and therefore it is upper bounded by C_L if $j \in L$. Also, it is clear that (15) must hold since no job j can finish processing on machine i before $p_{ij} + r_{ij}$.

Let $(y_{ij\ell}^*)_{ij\ell}$ and $(C_L^*)_L$ be an optimal solution to [HSSW]. To obtain a feasible schedule we need to round such solution into an integral one. To this end, Hall et. al. [10] used Shmoys and Tardos' result given in Theorem 1. If in [HSSW] all orders are singleton (as in Hall et al's situation), (14) becomes an equality so that one can use Theorem 1 to round a fractional solution to an integral solution of smaller total cost and such that the righthand side of equation (13) is increased to $\tau_\ell + \max\{p_{ij} : y_{ij\ell} > 0\} \le 2\tau_\ell$, where the last inequality follows from (15). This can be used to derive a constant factor approximation algorithm for the problem. In our setting however, it is not possible to apply Theorem 1 directly, due to the nonlinearity of the objective function. To overcome this difficulty, consider $j \in J$ and $L = \arg\min\{C_{L'}^* | j \in L' \in O\}$, and apply (11) to y^*, thus obtaining a new fractional assignment y'. With this we obtain a solution in which job j is never assigned to an interval starting after βC_L^*. Moreover, the following lemma holds.

Lemma 6. *The modified solution* $y'_{ij\ell} \geq 0$ *satisfies* (12), (15), *and:*

$$\sum_{s=1}^{\ell} \sum_{j \in J} p_{ij} y'_{ijs} \leq \frac{\beta}{\beta - 1} \tau_{\ell} \qquad \qquad \text{for all } i \in M, \quad (17)$$

$$y'_{ij\ell} = 0 \qquad \qquad \text{if } \tau_{\ell-1} > \beta C_L^*, \text{ for all } i, j, \ell, L : j \in L. \quad (18)$$

With the previous lemma on hand we are in position to apply Theorem 1 by interpreting a machine-interval pair (i, ℓ) on [HSSW] as a virtual machine on the theorem. We thus obtain a rounded solution $\hat{y}_{ij\ell} \in \{0, 1\}$ satisfying (12), (15), (18) and

$$\sum_{j \in J} p_{ij} \hat{y}_{ij\ell} \leq \sum_{j \in J} p_{ij} y'_{ij\ell} + \max_{j \in J} \{p_{ij} : y'_{ij\ell} > 0\} \leq \sum_{j \in J} p_{ij} y'_{ij\ell} + \tau_{\ell}, \quad (19)$$

where the first inequality follows from (4) and the second follows since y' satisfies (15).

To obtain a feasible schedule we do as follows. Define $J_{i\ell} = \{j \in J : \hat{y}_{ij\ell} = 1\}$, and greedily schedule in each machine i all jobs in $\bigcup_{\ell=1}^{q} J_{i\ell}$, starting from those in J_{i1} until we reach J_{iq} (with an arbitrary order inside each set $J_{i\ell}$), respecting the release dates. Let us call the algorithm just described GREEDY-LP.

For simplicity, we only show that GREEDY-LP is a constant factor approximation algorithm for the case in which all release dates are zero. The case with nontrivial release dates follows from a similar argument.

Theorem 4. *Procedure* GREEDY-LP *is a (27/2)-approximation algorithm for* $R||\sum w_L C_L$.

Proof. Let us fix a machine i and take a job $j \in L$ such that $\hat{y}_{ij\ell} = 1$, so that $j \in J_{i\ell}$. Clearly, C_j, the completion time of job j in algorithm GREEDY-LP, is at most the total processing time of jobs in $\bigcup_{k=1}^{\ell} J_{ik}$. Then,

$$C_j \leq \sum_{s=1}^{\ell} \sum_{k \in J} p_{ik} \hat{y}_{iks} \leq \sum_{s=1}^{\ell} \left(\sum_{k \in J} p_{ik} y'_{iks} + \tau_s \right) \leq \frac{\beta}{\beta - 1} \tau_{\ell} + \sum_{s=1}^{\ell} \tau_s$$

$$\leq \left(\frac{\beta \alpha}{\beta - 1} + \frac{\alpha^2}{\alpha - 1} \right) \tau_{\ell-1} \leq \beta \alpha \left(\frac{\beta}{\beta - 1} + \frac{\alpha}{\alpha - 1} \right) C_L^*.$$

The second inequality follows from (19), the third from (17), and the fourth follows from the definition of τ_k. The last inequality follows since, by condition (3), $\hat{y}_{ij\ell} = 1$ implies $y'_{ij\ell} > 0$, so that by (18) we have $\tau_{\ell-1} \leq \beta C_L^*$. Optimizing over the approximation factor, the best possible guarantee given by this method is attained at $\alpha = \beta = 3/2$, and thus we conclude that $C_j \leq 27/2 \cdot C_L^*$ for all $L \in O$ and $j \in L$. $\square$

Theorem 5. GREEDY-LP *is a (27/2)-approximation for* $R|r_{ij}|\sum w_L C_L$.

5 Further Results

Beyond the results shown in this paper, we have also considered the problem $P|part| \sum w_L C_L$, where no job can simultaneously belong to more than one order. Following Afrati et al. [1], we were able to develop a PTAS for some restricted versions of this problem, namely, when the number of jobs in each order is constant, the number of machines is constant, or the number of orders is constant. Thus, our algorithm generalizes the known PTAS's in [1,11,24]. The main extra difficulty compared to the case in [1], is that we might have orders that are processed through a long period of time, and their costs are only realized when they are completed. To overcome this issue, and thus be able to apply the dynamic programming ideas of Afrati et al., we simplify the instance and prove that there is a near-optimal solution in which every order is fully processed in a restricted time span. This requires some careful enumeration plus the introduction of artificial release dates. Due to space restrictions this result is left for the full version of this paper (see [27] for details).

References

1. Afrati, F., Bampis, E., Chekuri, C., Karger, D., Kenyon, C., Khanna, S., Milis, I., Queyranne, M., Skutella, M., Stein, C., Sviridenko, M.: Approximation schemes for minimizing average weighted completion time with release dates. In: Proceedings of the 40th Annual IEEE Symposium on Foundations of Computer Science (FOCS), pp. 32–43 (1999)
2. Ambühl, C., Mastrolilli, M.: Single Machine Precedence Constrained Scheduling is a Vertex Cover Problem. In: Azar, Y., Erlebach, T. (eds.) ESA 2006. LNCS, vol. 4168, pp. 28–39. Springer, Heidelberg (2006)
3. Ambühl, C., Mastrolilli, M., Svensson, O.: Inapproximability Results for Sparsest Cut, Optimal Linear Arrangement, and Precedence Constrained Scheduling. In: Proceedings of the 48th Annual IEEE Symposium on Foundations of Computer Science (FOCS), pp. 329–337 (2007)
4. Canetti, R., Irani, S.: Bounding the Power of Preemption in Randomized Scheduling. SIAM J. Computing 27, 993–1015 (1998)
5. Chekuri, C., Motwani, R.: Precedence constrained scheduling to minimize sum of weighted completion times on a single machine. Discrete Applied Mathematics 98, 29–38 (1999)
6. Chudak, F., Hochbaum, D.S.: A half-integral linear programming relaxation for scheduling precedence-constrained jobs on a single machine. Oper. Res. Let. 25, 199–204 (1999)
7. Correa, J.R., Schulz, A.S.: Single Machine Scheduling with Precedence Constraints. Math. Oper. Res. 30, 1005–1021 (2005)
8. Dyer, M.E., Wolsey, L.A.: Formulating the single machine sequencing problem with release dates as a mixed integer program. Discrete Applied Mathematics 26, 255–270 (1999)
9. Graham, R.L.: Bounds for certain multiprocessing anomalies. Bell Systems Technical Journal 45, 1563–1581 (1966)
10. Hall, L.A., Schulz, A.S., Shmoys, D.B., Wein, J.: Scheduling to minimize average completion time: off-line and on-line approximation algorithms. Math. Oper. Res. 22, 513–544 (1997)

11. Hochbaum, D., Shmoys, D.: Using dual approximation algorithm for scheduling problems: Theoretical and practical results. J. ACM 34, 144–162 (1987)
12. Hoogeveen, H., Schuurman, P., Woeginger, G.J.: Non-approximability results for scheduling problems with minsum criteria. INFORMS J. Computing 13, 157–168 (2001)
13. Lawler, E.L., Labetoulle, J.: On Preemptive Scheduling of Unrelated Parallel Processors by Linear Programming. J. ACM 25, 612–619 (1978)
14. Lawler, E.L., Lenstra, J.K., Rinnooy Kan, A.H.G., Shmoys, D.B.: Sequencing and scheduling: Algorithms and complexity. In: Graves, S.C., Rinnooy Kan, A.H.G., Zipkin, P.H. (eds.) Logistics of Production and Inventory, Handbooks in Oper. Res. and Management Science, vol. 4, pp. 445–522. North-Holland, Amsterdam (1993)
15. Lenstra, J.K., Rinnooy Kan, A.H.G.: Complexity of scheduling under precedence constrains. Operations Research 26, 22–35 (1978)
16. Lenstra, J.K., Shmoys, D.B., Tardos, E.: Approximation algorithms for scheduling unrelated parallel machines. Mathematical Programming 46, 259–271 (1990)
17. Leung, J., Li, H., Pinedo, M.: Approximation algorithm for minimizing total weighted completion time of orders on identical parallel machines. Naval Research Logistics 53, 243–260 (2006)
18. Leung, J., Li, H., Pinedo, M., Zhang, J.: Minimizing Total Weighted Completion Time when Scheduling Orders in a Flexible Environment with Uniform Machines. Information Processing Letters 103, 119–129 (2007)
19. Leung, J., Li, H., Pinedo, M.: Scheduling orders for multiple product types to minimize total weighted completion time. Discrete Applied Mathematics 155, 945–970 (2007)
20. Margot, F., Queyranne, M., Wang, Y.: Decompositions, network flows, and a precedence constrained single machine scheduling problem. Operations Research 51, 981–992 (2003)
21. Shachnai, H., Tamir, T.: Multiprocessor Scheduling with Machine Allotment and Parallelism Constraints. Algorithmica 32, 651–678 (2002)
22. Schulz, A., Skutella, M.: Scheduling unrelated machines by randomized rounding. SIAM J. Discrete Math. 15, 450–469 (2002)
23. Shmoys, D.B., Tardos, E.: An approximation algorithm for the generalized assignment problem. Mathematical Programming 62, 561–574 (1993)
24. Skutella, M., Woeginger, G.J.: Minimizing the total weighted completion time on identical parallel machines. Math. Oper. Res. 25, 63–75 (2000)
25. Skutella, M.: Convex quadratic and semidefinite programming relaxations in scheduling. J. ACM 48, 206–242 (2001)
26. Smith, W.E.: Various optimizers for single-stage production. Naval Research Logics Quarterly 3, 59–66 (1956)
27. Verschae, J.: Approximation algorithms for scheduling orders on parallel machines. Mathematical engineering thesis. Universidad de Chile, Santiago, Chile (2008)

New Hardness Results for Diophantine Approximation

Friedrich Eisenbrand* and Thomas Rothvoß

Institute of Mathematics
EPFL, Lausanne, Switzerland
{friedrich.eisenbrand,thomas.rothvoss}@epfl.ch

Abstract. We revisit *simultaneous Diophantine approximation*, a classical problem from the geometry of numbers which has many applications in algorithms and complexity. The input to the decision version of this problem consists of a rational vector $\alpha \in \mathbb{Q}^n$, an error bound ε and a denominator bound $N \in \mathbb{N}_+$. One has to decide whether there exists an integer, called the *denominator* Q with $1 \le Q \le N$ such that the distance of each number $Q \cdot \alpha_i$ to its nearest integer is bounded by ε. Lagarias has shown that this problem is NP-complete and optimization versions have been shown to be hard to approximate within a factor $n^{c/\log\log n}$ for some constant $c > 0$. We strengthen the existing hardness results and show that the optimization problem of finding the *smallest denominator* $Q \in \mathbb{N}_+$ such that the distances of $Q \cdot \alpha_i$ to the nearest integer are bounded by ε is hard to approximate within a factor 2^n unless P = NP.

We then outline two further applications of this strengthening: We show that a directed version of Diophantine approximation is also hard to approximate. Furthermore we prove that the *mixing set* problem with arbitrary capacities is NP-hard. This solves an open problem raised by Conforti, Di Summa and Wolsey.

1 Introduction

Diophantine approximation is one of the fundamental topics in mathematics. Roughly speaking, the objective is to replace a number or a vector, by another number or vector which is very close to the original, but less complex in terms of fractionality. A famous example is the Gregorian calendar, which approximates a solar year with its leap year rule.

Since the invention of the LLL algorithm [15], simultaneous Diophantine approximation has been a very important object of study also in computer science. One powerful result, for example, is the one of Frank and Tardos [7] who provided an algorithm based on Diophantine approximation and the LLL algorithm

* Supported by the Deutsche Forschungsgemeinschaft (DFG) within Priority Programme 1307 "Algorithm Engineering" and by the Swiss National Science Foundation (SNSF).

I. Dinur et al. (Eds.): APPROX and RANDOM 2009, LNCS 5687, pp. 98–110, 2009.
© Springer-Verlag Berlin Heidelberg 2009

which, among other things, shows that a combinatorial 0/1-optimization problem is polynomial if and only if it is strongly polynomial.

Let us denote the distance of a real number $x \in \mathbb{R}$ to its nearest integer by $\{x\} = \min\{|x - z| : z \in \mathbb{Z}\}$ and the distance of a vector $v \in \mathbb{R}^n$ to its nearest integer vector w.r.t. the infinity norm ℓ_∞ by $\{\{v\}\} = \min\{\|v - z\|_\infty : z \in \mathbb{Z}^n\}$.

Lagarias [14] has shown that it is NP-complete to decide whether there exists an integer $Q \in \{1, \ldots, N\}$ with $\{\{Q \cdot \alpha\}\} \le \varepsilon$, given $\alpha \in \mathbb{Q}^n$, $N \in \mathbb{N}_+$ and $\varepsilon > 0$. The *best approximation error* δ_N of a vector $\alpha \in \mathbb{Q}^n$ with denominator bound $N \in \mathbb{N}_+$ is defined as $\delta_N = \min\{\{\{Q \cdot \alpha\}\} : Q \in \{1, \ldots, N\}\}$. Lagarias [14] showed also that the existence of a polynomial algorithm, which computes on input $\alpha \in \mathbb{Q}^n$ and $N \in \mathbb{N}_+$ a number $Q \in \{1, \ldots, 2^{n/2} \cdot N\}$ with $\{\{Q \cdot \alpha\}\} \le \delta_N$ implies NP = co-NP.

Lagarias' reduction was then sharpened to an inapproximability result by Rössner and Seifert [21] and Chen and Meng [1] to the extent that, given $\alpha \in \mathbb{Q}^n$ and N as above, it is NP-hard to compute a $Q \in \{1, \ldots, \lfloor n^{c/\log\log n} \rfloor N\}$ with $\{\{Q \cdot \alpha\}\} \le n^{c/\log\log n} \delta_N$ where $c > 0$ is a constant. We revisit the reduction technique of Lagarias [14] and its sharpening by Rössner and Seifert [21] to obtain the following theorem.

Theorem 1. *There exists a constant $c > 0$ and a polynomial time transformation which maps an instance C of SAT to an instance $\alpha \in \mathbb{Q}^n$, $N \in \mathbb{N}_+$, $\varepsilon \in \mathbb{Q}_+$ of simultaneous Diophantine approximation such that the following holds.*

  i) If C is satisfiable, then there is a $Q \in \{\lceil N/2 \rceil, \ldots, N\}$ with $\{\{Q \cdot \alpha\}\} \le \varepsilon$.
  ii) If C is not satisfiable, then one has $\{\{Q \cdot \alpha\}\} \ge n^{c/\log\log n} \cdot \varepsilon$ for each $Q \in \{1, \ldots, 2^n \cdot N\}$.
 iii) The error bound ε satisfies $\varepsilon \le 1/(2^{2n})$.

The crucial differences between our result and the result in [21] are as follows. In case i), there exists a good Q which is at least $\lceil N/2 \rceil$ whereas the result in [21] guarantees only a good Q in the interval $\{1, \ldots, N\}$. In case ii) each Q which is bounded by $2^n \cdot N$ is violating the distance bound by $n^{c/\log\log n}$, whereas the reduction of [21] together with the result of [1] guarantees this violation only for $Q \in \{1, \ldots, \lfloor n^{c/\log\log n} \rfloor \cdot N\}$. These differences facilitate the application of our hardness result to other problems from the geometry of numbers and integer programming. We describe three such applications in this paper.

Applications

One immediate consequence of Theorem 1 is that the *best denominator problem*

$$\min\{Q \in \mathbb{N}_+ : \{\{Q \cdot \alpha\}\} \le \varepsilon\}$$

cannot be approximated within a factor of 2^n unless P = NP, see Corollary 1. Furthermore, it follows that the existence of a polynomial algorithm, which computes on input $\alpha \in \mathbb{Q}^n$, $N \in \mathbb{N}_+$ a number $Q \in \{1, \ldots, 2^n \cdot N\}$ with $\{\{Q \cdot \alpha\}\} \le \delta_N$ implies P = NP improving the result of Lagarias [14] mentioned above to the extent of replacing the factor $2^{n/2}$ and the assumption NP $\neq$ co-NP by 2^n and P $\neq$ NP respectively, see Corollary 2.

We then provide a strong inapproximability result for *directed Diophantine approximation*, where the distance to the nearest integer vector which is greater than or equal to $Q \cdot \alpha$ has to be small. Directed Diophantine approximation was for example considered by Henk and Weismantel [12] in the context of an integer programming problem and an optimization version of directed Diophantine approximation was shown to be hard to approximate within a constant factor by the authors of this paper [6].

Finally we apply our results to solve an open problem raised by Conforti, Di Summa and Wolsey [3] concerning the complexity of a linear optimization problem over a *mixing set with arbitrary capacities*, a type of integer program which frequently appears in production planning.

2   A Strengthening of the Lagarias, Rössner-Seifert Reduction

The goal of this section is to prove Theorem 1. To do this, we rely on several results from the literature. Our starting point is a similar result for the *shortest integer relation* problem. Here, one is given a vector $a \in \mathbb{Z}^n$ and the goal is to find a nonzero integral solution $x \in \mathbb{Z}^n$ of the equation $a^T x = 0$ of minimum infinity norm. By modifying a reduction from *Super-Sat* to shortest vector in the infinity norm by Dinur [5], Chen and Meng [1] showed that there exists a reduction from SAT to shortest integer relation with the property that if C is satisfiable, then the optimum value of the shortest integer relation problem is one and if C is unsatisfiable, then the optimum value of the shortest integer relation problem is at least $n^{c/\log\log n}$ for some constant $c > 0$. This can be extended to the following result which we prove in the appendix. The only difference to the stated result above is the presence of condition c).

Lemma 1. *There exists a constant $c > 0$ and a polynomial time algorithm, which maps a SAT-formula C to an instance $a \in \mathbb{Z}^n$ of shortest integer relation with the following properties:*

a) *If C is satisfiable, then $\min\{\|x\|_\infty : a^T x = 0, x \in \mathbb{Z}^n - 0\} = 1$.*
b) *If C is not satisfiable, then $\min\{\|x\|_\infty : a^T x = 0, x \in \mathbb{Z}^n - 0\} \geq n^{c/\log\log n}$.*
c) *There exists an optimum solution x of $\min\{\|x\|_\infty : a^T x = 0, x \in \mathbb{Z}^n - 0\}$ with $x_1 \geq 1$.*

We proceed from Lemma 1 to show the existence of a reduction from SAT to simultaneous Diophantine approximation with properties i), ii) and iii). For this, by Lemma 1 it is enough to provide a reduction from a shortest integer relation problem $\min\{\|x\|_\infty : a^T x = 0, x \in \mathbb{Z}^n - 0\}$ with the property that there exists an optimum solution x with $x_1 \geq 1$ to an instance of simultaneous Diophantine approximation $\alpha_0, \ldots, \alpha_n, \varepsilon, N$ such that the following assertions hold.

I) If the optimum value of the shortest integer relation problem is one, then
there exists a $Q \in \{\lceil N/2 \rceil, \ldots, N\}$ with $\{\{Q \cdot \alpha\}\} \le \varepsilon$.

II) For each $\rho \in \{1, \ldots, n\}$ the following statement is true: If the optimum
value of the shortest integer relation problem is larger than ρ, then $\{\{Q \cdot \alpha\}\} > \rho \cdot \varepsilon$ for each $Q \in \{1, \ldots, 2^n \cdot N\}$.

III) The error bound ε satisfies $\varepsilon \le 1/(2^{2n})$.

The rest of the proof of Theorem 1 follows closely the proof of Lagarias [14] and
the one of Rössner and Seifert [21]. Let $\min\{\|x\|_\infty : a^T x = 0, x \in \mathbb{Z}^n - 0\}$ be
the instance of shortest integer relation. One can efficiently find different primes
$p, q_1, \ldots, q_n$ as well as natural numbers R and T in polynomial time, such that

1. $n \cdot \sum_{j=1}^{n} |a_j| < p^R < q_1^T < q_2^T < \ldots < q_n^T < (1 + \frac{1}{n}) \cdot q_1^T$
2. p and all q_i are co-prime to all a_j
3. $q_1^T > 2^{2n} \cdot p^R$
4. The values of $T, R, p, q_1, \ldots, q_n$ are bounded by a polynomial in the input
length of a.

A proof of this claim with weaker bounds is presented in [14,21]. The crucial
difference to the results in these papers is the bound 3), which before stated
that p^R times a polynomial in the input encoding is at most q_1^T. Here we have
the exponential factor 2^{2n} instead. The full proof is in the Appendix.

The following system of congruences appears already in [16] and is also crucial
in the reductions presented in [14,21].

$$r_j \equiv_{p^R} a_j \tag{1}$$
$$r_j \equiv_{q_i^T} 0 \; \forall i \ne j \tag{2}$$
$$r_j \not\equiv_{q_j} 0 \tag{3}$$

For each j, this is a system of congruences with co-prime moduli and thus, the
Chinese remainder theorem (see, e.g. [18]) guarantees that there exists a solution
r_j for each $j = 1, \ldots, n$.

Lemma 2. *The systems*

$$\sum_{j=1}^{n} x_j a_j = 0 \quad and \quad \sum_{j=1}^{n} x_j r_j \equiv_{p^R} 0 \tag{4}$$

have the same set of integral solutions $x \in \mathbb{Z}^n$ with $\|x\|_\infty \le n$.

Proof. Since $a_j \equiv_{p^R} r_j$, each solution $x \in \mathbb{Z}^n$ of the equation on the left is also a
solution of the congruence equation on the right. If $x \in \mathbb{Z}^n$ is a solution for the
congruence on the right, then $\sum_{j=1}^{n} a_j x_j \equiv_{p^R} 0$. Assume furthermore $\|x\|_\infty \le n$.
If we can infer that the absolute value of $\sum_{j=1}^{n} a_j x_j$ is strictly less than p^R, then
$\sum_{j=1}^{n} a_j x_j = 0$ follows. But

$$|\sum_{j=1}^{n} x_j a_j| \le n \cdot \sum_{j=1}^{n} |a_j| < p^R$$

by the choice of the prime numbers. $\qquad\square$

We now provide the construction of the instance $\alpha_0, \ldots, \alpha_n, \varepsilon, N$ of the simultaneous Diophantine approximation problem for our reduction. By $r_j^{-1} \in \mathbb{Z}$ we denote the unique integer in $\{1, \ldots, q_j^T - 1\}$ with $r_j \cdot r_j^{-1} \equiv_{q_j^T} 1$. This must exist since $r_j \not\equiv_{q_j} 0$ implies that r_j is a unit in the ring $\mathbb{Z}_{q_j^T}$. The instance is

$$\alpha_0 = \frac{1}{p^R}$$

$$\alpha_j = \frac{r_j^{-1}}{q_j^T}, \quad j = 1, \ldots, n$$

$$N = \sum_{j=1}^n r_j$$

$$\varepsilon = \frac{1}{q_1^T}.$$

The bound iii) on ε follows from $q_1^T > 2^{2n} \cdot p^R$. Let $x \in \mathbb{Z}^n$ be a solution of the shortest integer relation problem with $\|x\|_\infty \leq 1$. Consider the integer $Q = \sum_{j=1}^n r_j \cdot x_j$ whose absolute value is bounded by $N = \sum_{j=1}^n r_j$. What is the distance of $Q \cdot \alpha$ to the nearest integer vector in the infinity norm?

Since $\sum_{j=1}^n r_j \cdot x_j \equiv_{p^R} \sum_{j=1}^n a_j \cdot x_j = 0$ it follows that p^R divides $\sum_{j=1}^n r_j \cdot x_j$ which means that $\{Q\alpha_0\} = 0$. For $i \geq 1$ one has $r_i^{-1} \cdot \sum_{j=1}^n r_j \cdot x_j \equiv_{q_i^T} x_i$ (since $r_j \equiv_{q_i^T} 0$ for $i \neq j$) and since $x_i \in \{0, \pm 1\}$ one has $\{Q \cdot \alpha_i\} \leq 1/q_i^T \leq 1/q_1^T = \varepsilon$. In other words, Q is an integer whose absolute value is bounded by N which satisfies $\{\{Q \cdot \alpha\}\} \leq \varepsilon$. This is almost condition I), except that $Q \in \{\lceil N/2 \rceil, \ldots, N\}$ might not be satisfied.

To achieve this additional bound on Q we use the fact that there exists an optimal solution of the shortest integer relation problem which satisfies $x_1 \geq 1$ and we choose r_1 significantly larger than the other r_j. Consider again the system of congruences (1-3). Let $B = p^R \prod_{j=1}^n q_j^T$ and let $0 \leq r_j' \leq B/q_j^T$ be a solution to (1) and (2). If $r_j' \not\equiv_{q_j^T} 0$, then $r_j = r_j'$ otherwise $r_j = r_j' + B/q_j^T$. Thus each r_j is bounded by $0 \leq r_j \leq 2 \cdot B/q_j^T$. We choose r_1 however considerably larger, namely $r_1 = r_1' + 12nB/q_1^T$ or $r_1' + (12n+1)B/q_1^T$. In this way we have $r_1 \geq 6n \cdot r_j$. By choosing the r_j in this way, we obtain the following lemma.

Lemma 3. *If* $\min\{\|x\|_\infty : a^T x = 0, x \in \mathbb{Z}^n - 0\} = 1$, *then there exists a* $Q \in \{\lceil N/2 \rceil, \ldots, N\}$ *such that* $\{\{Q \cdot \alpha\}\} \leq \varepsilon$.

Proof. By our assumption, there exists an optimum solution $x \in \mathbb{Z}^n$ of the shortest integer relation problem with $x_1 = 1$. Let Q, as in the discussion above, be $Q = \sum_{j=1}^n r_j x_j$. We have already seen that $\{\{Q \cdot \alpha\}\} \leq \varepsilon$ holds and clearly $Q \leq \sum_{j=1}^n r_j = N$. On the other hand $x_1 \geq 1$, $\|x\|_\infty = 1$ and $r_1 \geq 6nr_j$ for each $j = 2, \ldots, n$ implies $Q \geq N/2$. $\square$

The next lemma provides condition II.

Lemma 4. *Let ρ be any number in $\{1,\ldots,n\}$ and suppose there exists a $Q \in \{1,\ldots,2^n N\}$ with $\{\{Q \cdot \alpha\}\} \leq \rho \cdot \varepsilon$. Then, the optimum value of the shortest integer relation problem is at most ρ.*

Proof. We construct a solution x of the shortest integer relation instance: Let x_j be the smallest integer in absolute value with

$$Q r_j^{-1} \equiv_{q_j^T} x_j.$$

We need to show three things, namely

$$\|x\|_\infty \leq \rho, \quad x \neq 0 \quad \text{and} \quad a^T x = 0. \tag{5}$$

The first assertion of (5) follows from the fact that $q_1^T < q_j^T < (1 + 1/\rho) \cdot q_1^T$ which implies the strict inequality in

$$\left| \frac{x_j}{q_j^T} \right| = \left\{ \frac{Q r_j^{-1}}{q_j^T} \right\} \leq \rho \cdot \varepsilon = \frac{\rho}{q_1^T} < \frac{\rho + 1}{q_j^T}.$$

Observe that Q is a multiple of p^R. If this was not the case, then

$$\{Q \alpha_0\} = \left\{ \frac{Q}{p^R} \right\} \geq \frac{1}{p^R} > \frac{\rho}{q_1^T} = \rho \cdot \varepsilon,$$

since $q_1^T > 2^{2 \cdot n} p^R$ and $\rho \leq n$. We next show that $Q = \sum_{i=1}^n x_i r_i$. This implies directly that $x \neq 0$, since $Q \geq 1$. Furthermore $Q \equiv_{p^R} 0$ and Lemma 2 imply together with $\|x\|_\infty \leq \rho$ that $a^T x = 0$ and (5) is proved.

Multiplying the equation $Q \cdot r_j^{-1} \equiv_{q_j^T} x_j$ with r_j yields $Q \equiv_{q_j^T} r_j x_j$. Let $D = \prod_{j=1}^n q_j^T$. We have $Q \equiv_{q_i^T} r_i x_i$ and $0 \equiv_{q_i^T} r_j x_j$ for $j \neq i$ and thus $Q \equiv_{q_i^T} \sum_{j=1}^n r_j x_j$. Since the moduli q_i^T are co-prime, this implies that $Q \equiv_D \sum_{j=1}^n r_j x_j$. We are done with the proof, once we have shown that $Q < D/2$ and $|\sum_{j=1}^n x_j r_j| < D/2$, since then both values must coincide if they are congruent to each other modulo D.

We first bound the value of $|\sum_{j=1}^n x_j r_j|$. This is at most $\rho \cdot \sum_{j=1}^n r_j \leq n \cdot N$. Applying the bound $r_j \leq 13 \cdot np^R D/q_1^T$ and $q_1^T > 2^{2n} \cdot p^R$ we can bound N by

$$N \leq 13 \cdot n^2 \cdot D/2^{2n}.$$

Consequently

$$|\sum_{j=1}^n x_j r_j| \leq 13 \cdot n^3 \cdot D/2^{2n}$$

which is smaller than $D/2$ for n sufficiently large. Finally Q is bounded by $2^n N$ which is also bounded by $D/2$ for n large enough. The claim follows. $\qquad\square$

This proves Theorem 1.

2.1   Hardness of the Best Denominator

We now discuss the hardness of the *best denominator problem*. The input to this problem is $\alpha_1, \ldots, \alpha_n, \varepsilon \in \mathbb{Q}$ and the task is to find a smallest $Q \in \mathbb{N}_+$ with $\{\{Q \cdot \alpha\}\} \leq \varepsilon$. The following corollary is an immediate consequence of Theorem 1.

Corollary 1. *If $P \neq NP$, then there does not exist a polynomial time approximation algorithm for the* best denominator problem *with an approximation factor 2^n.*

Furthermore we can strengthen the result of Lagarias [14] which states that, if there exists a polynomial time algorithm which, on input $\alpha \in \mathbb{Q}^n$ and $N \in \mathbb{N}_+$ computes a $Q \in \{1, \ldots, 2^{n/2}N\}$ with $\{\{Q \cdot \alpha\}\} \leq \delta_N$, then NP = co-NP. Recall that $\delta_N = \min\{\{\{Q \cdot \alpha\}\}: Q \in \{1, \ldots, N\}\}$. The strengthening is as follows.

Corollary 2. *If there exists a polynomial time algorithm which computes on input $\alpha \in \mathbb{Q}^n$ and $N \in \mathbb{N}_+$ a $Q \in \{1, \ldots, 2^n \cdot N\}$ with $\{\{Q \cdot \alpha\}\} \leq \delta_N$, then $P = NP$.*

Proof. Consider an instance α, N, ε which stems from the reduction of a SAT-formula C as in Theorem 1 and suppose that there exists an algorithm which computes in polynomial time a $Q \in \{1, \ldots, 2^n N\}$ with $\{\{Q \cdot \alpha\}\} \leq \delta_N$. If $\{\{Q \cdot \alpha\}\} \leq \varepsilon$, then C is satisfiable. Otherwise, C is unsatisfiable. This implies the assertion. $\qquad\square$

3   Directed Diophantine Approximation

In this section we consider a variant of the classical Diophantine approximation problem, in which we measure the distance of the vector $Q \cdot \alpha$ to the nearest integer vector which is in each component greater or equal than $Q \cdot \alpha$. We use the notation $\{x\}^\uparrow$ for the distance of the real number $x \in \mathbb{R}$ to the nearest integer which is greater or equal to x, $\{x\}^\uparrow = \min\{z - x: z \in \mathbb{Z}, z \geq x\}$. For a vector $\alpha \in \mathbb{R}^n$ we denote its distance to the nearest integer greater or equal to α by $\{\{\alpha\}^\uparrow\}$, in other words

$$\{\{\alpha\}^\uparrow\} = \min\{\|x - \alpha\|_\infty : x \in \mathbb{Z}^n, x \geq \alpha\}.$$

An instance of directed Diophantine approximation consists of $\alpha_1, \ldots, \alpha_n, \varepsilon, N$ with $\alpha_i \in \mathbb{Q}$, $\varepsilon \in \mathbb{Q}$ and $N \in \mathbb{N}_+$. The goal of this section is to show the following theorem.

Theorem 2. *There is a constant $c > 0$ and a polynomial time transformation which maps a SAT instance C to an instance $\alpha_0, \ldots, \alpha_n, \varepsilon, N$ of directed Diophantine approximation such that the following conditions hold.*

i') If C is satisfiable, then there exists a $Q \in \{\lceil N/2 \rceil, \ldots, N\}$ with $\{\{Q \cdot \alpha\}^\uparrow\} \leq \varepsilon$.

ii') If C is unsatisfiable, then for each $Q \in \{1, \ldots, \lfloor n^{c/\log\log n}\rfloor N\}$ one has
$\{\{Q \cdot \alpha\}^\uparrow\} > 2^n \varepsilon$.

iii') The error bound ε satisfies $\varepsilon \leq 3/2^n$.

Proof. For the proof of this theorem, we rely on Theorem 1. Let $\alpha_1, \ldots, \alpha_n, \varepsilon, N$ be a simultaneous Diophantine approximation instance which results from the transformation from SAT. From this, we construct an instance of directed Diophantine approximation $\alpha'_1, \ldots, \alpha'_{2n}, N, \varepsilon'$ with

$$\alpha'_i = \alpha_i - \delta \quad i = 1, \ldots, n$$
$$\alpha'_{i+n} = -\alpha_i - \delta \quad i = 1, \ldots, n$$
$$\varepsilon' = 3\varepsilon,$$

where $\delta = 2\varepsilon/N$.

Suppose that there exists a $Q \in \{\lceil N/2\rceil, \ldots, N\}$ with $\{\{Q\alpha\}\} \leq \varepsilon$ and let z_i be the nearest integer to $Q \cdot \alpha_i$. Since $Q \cdot \delta \geq \varepsilon$ it follows that $Q(\alpha_i + \delta) \geq z_i$ and thus that the distance of $Q(\alpha_i+\delta)$ to $\lfloor Q(\alpha_i+\delta)\rfloor$ is bounded by $|Q(\alpha_i+\delta) - z_i| \leq |Q\alpha_i - z_i| + |Q\delta| \leq 3\varepsilon$. This means that $\{Q(-\alpha_i-\delta)\}^\uparrow \leq 3\varepsilon$. Similarly, $Q(\alpha_i-\delta) \leq z_i$ and thus $\{Q(\alpha_i - \delta)\}^\uparrow$ is bounded by $|Q(\alpha_i - \delta) - z_i| \leq |Q\alpha_i - z_i| + |Q\delta| \leq 3\varepsilon$. This implies property i').

Next let $\rho \in \{1, \ldots, n\}$ and suppose that there exists a $Q \in \{1, \ldots, \rho N\}$ with $\{\{Q\alpha'\}^\uparrow\} \leq 2^n \varepsilon'$. We show that this implies that $\{\{Q\alpha\}\} \leq 2\rho\varepsilon$ which in turn shows that property ii') holds.

For each $i \in \{1, \ldots, n\}$ there exists an integer z_i which lies between $Q(\alpha_i - \delta)$ and $Q(\alpha_i + \delta)$, since otherwise one of the values $\{Q(\alpha_i - \delta)\}^\uparrow$ or $\{Q(-\alpha_i-\delta)\}^\uparrow$ is at least $1/2$. But $\{\{Q \cdot \alpha'\}^\uparrow\} \leq 2^n \varepsilon' = 2^n 3\varepsilon < 1/2$, a contradiction. Then $Q(\alpha_i - \delta) \leq z_i \leq Q(\alpha_i + \delta)$ implies

$$|Q\alpha_i - z_i| \leq Q\delta \leq 2\rho\varepsilon. \qquad \square$$

4   Hardness of Mixing Set

In recent integer programming approaches for production planning the study of simple integer programs which are part of more sophisticated models has become very successful in practice, see, e.g. [19]. One of these simple integer programs is the so-called *mixing set* [9,2]. The constraint system of a mixing set problem is of the form

$$
\begin{aligned}
s + a_i y_i &\geq b_i \quad i = 1, \ldots, n, \\
s &\geq 0 \\
y_i &\in \mathbb{Z} \quad i = 1, \ldots, n, \\
s &\in \mathbb{R}.
\end{aligned}
\tag{6}
$$

where $a_i, b_i \in \mathbb{Q}$. Optimizing a linear function over this mixed integer set can be done in polynomial time if all a_i are equal to one [9,17] or if a_{i+1}/a_i is an integer for each $i = 1, \ldots, n-1$ [22], see also [3,4] for subsequent simpler approaches.

Conforti et al. [3] pose the problem, whether one can optimize a linear function over the set of mixed-integer vectors defined by (6) also in the general case, to which they refer as the case with *arbitrary capacities*, in polynomial time. In this section, we apply our results on directed Diophantine approximation to show that this problem is NP-hard.

Suppose we have an instance of the directed Diophantine approximation problem α, N, ε, where we are supposed to round down to the nearest integer vector. By using the notation $\{x\}^{\downarrow} = \min\{x - z : z \leq x, z \in \mathbb{Z}\}$ for $x \in \mathbb{R}$ and $\{\{v\}^{\downarrow}\} = \min\{\|v - z\|_{\infty} : z \in \mathbb{Z}^n, z \leq v\}$ and the observation that $\{x\}^{\downarrow} = \{-x\}^{\uparrow}$ it follows that Theorem 2 is also true if the rounding up operation is replaced by rounding down. We next formulate an integer program to compute a Q which yields a good approximation by rounding down and satisfies the denominator bound $Q \in \{1, \ldots, N\}$.

$$\min \sum_{i=1}^n Q(\alpha_i - y_i)$$
$$Q - 1/\alpha_i \cdot y_i \geq 0 \quad i = 1, \ldots, n$$
$$Q \geq 1$$
$$Q \leq N$$
$$Q, y_1, \ldots, y_n \in \mathbb{Z}.$$

The goal is to transform this integer program into a linear optimization problem over a mixing set. Consider the following mixing set.

$$Q - 1/\alpha_i \cdot y_i \geq 0 \quad i = 1, \ldots, n$$
$$Q + 0 \cdot y_0 \geq 1$$
$$Q - y_{-1} \geq 0 \tag{7}$$
$$Q \in \mathbb{R}$$
$$y_{-1}, y_0, y_1, \ldots, y_n \in \mathbb{Z}.$$

We now argue that, if the linear optimization problem over this mixing set can be done in polynomial time, then P = NP.

Suppose that the linear optimization problem can be solved in polynomial time. Then, we can also solve the linear optimization problem over the non-empty face of the convex hull of the solutions which is induced by the inequality $Q - y_{i-1} \geq 0$, see, e.g., [8]. This enforces Q to be an integer. Next consider the following objective function

$$\min \sum_{i=1}^n Q(\alpha_i - y_i) + (2^{n-1}\varepsilon/N)(Q - N). \tag{8}$$

The sum on the left is measuring the distance of $Q \cdot \alpha$ to its nearest integer vector from below in the ℓ_1-norm. The term on the right stems from the removal of the constraint $Q \leq N$, which would not be allowed in a system defining a mixing set. In fact, we thereby follow a Lagrangian relaxation approach, which is common in approximation algorithms, see e.g. [20], in order to show a hardness result.

Theorem 3. *Optimizing a linear function over a mixing set is NP-hard.*

Proof. Let $\alpha_1, \ldots, \alpha_n, N, \varepsilon$ be an instance of directed Diophantine approximation with rounding down, which stems from a transformation from SAT, as in Theorem 2 and suppose that one can solve the linear optimization problem with objective function (8) over the convex hull of the mixing set. Then we can also optimize this over the face induced by $Q - y_{-1} \geq 0$. This merely means that we can find a pure integer optimum solution over the mixing set (7).

Our instance $\alpha_1, \ldots, \alpha_n, N, \varepsilon$ has the following property. If the originating SAT formula is satisfiable, then there exists a $Q \in \{\lceil N/2 \rceil, \ldots, N\}$ with $\{\{Q \cdot \alpha\}^{\downarrow}\} \leq \varepsilon$ and if not, then there does not exist a $Q \in \{1, \ldots, \lfloor n^{c/\log\log n} \rfloor N\}$ with $\{\{Q \cdot \alpha\}^{\downarrow}\} \leq 2^n \varepsilon$.

In the case where the SAT formula is satisfiable, let $Q \in \{\lceil N/2 \rceil, \ldots, N\}$ with $\{\{Q \cdot \alpha\}^{\downarrow}\} \leq \varepsilon$. The objective function value of this Q with the appropriate y_i yields an objective function value bounded by $n \cdot \varepsilon$.

Suppose now that the SAT formula is not satisfiable and consider a solution Q with appropriate y_i of the mixing set problem. If $Q \in \{1, \ldots, \lfloor n^{c/\log\log n} \rfloor N\}$, then the objective function is at least

$$2^n \varepsilon - 2^{n-1} \varepsilon = 2^{n-1} \varepsilon.$$

If Q is larger than $\lfloor n^{c/\log\log n} \rfloor N$, then the objective function value is at least

$$2^{n-1} \varepsilon (\lfloor n^{c/\log\log n} \rfloor - 1).$$

Thus problem of optimizing a linear function over a mixing set with arbitrary capacities is NP-hard. □

References

1. Chen, W., Meng, J.: An improved lower bound for approximating shortest integer relation in l_∞ norm (SIR$_\infty$). Information Processing Letters 101(4), 174–179 (2007)
2. Conforti, M., Di Summa, M., Wolsey, L.A.: The mixing set with flows. SIAM Journal on Discrete Mathematics 21(2), 396–407 (2007) (electronic)
3. Conforti, M., Summa, M.D., Wolsey, L.A.: The mixing set with divisible capacities. In: Lodi, A., Panconesi, A., Rinaldi, G. (eds.) IPCO 2008. LNCS, vol. 5035, pp. 435–449. Springer, Heidelberg (2008)
4. Conforti, M., Zambelli, G.: The mixing set with divisible capacities: a simple approach (manuscript)
5. Dinur, I.: Approximating SVP$_\infty$ to within almost-polynomial factors is NP-hard. Theoretical Computer Science 285(1), 55–71 (2000); Algorithms and complexity, Rome (2000)
6. Eisenbrand, F., Rothvoß, T.: Static-priority realtime-scheduling: Response time computation is NP-hard. In: RTSS 2008 (2008)
7. Frank, A., Tardos, É.: An application of simultaneous Diophantine approximation in combinatorial optimization. Combinatorica 7, 49–65 (1987)
8. Grötschel, M., Lovász, L., Schrijver, A.: Geometric algorithms and combinatorial optimization, 2nd edn. Algorithms and Combinatorics, vol. 2. Springer, Berlin (1993)

9. Günlük, O., Pochet, Y.: Mixing mixed-integer inequalities. Mathematical Programming. A Publication of the Mathematical Programming Society 90(3, Ser. A), 429–457 (2001)
10. Heath-Brown, D.R.: The number of primes in a short interval. Journal für die Reine und Angewandte Mathematik 389, 22–63 (1988)
11. Heath-Brown, D.R., Iwaniec, H.: On the difference between consecutive primes. American Mathematical Society. Bulletin. New Series 1(5), 758–760 (1979)
12. Henk, M., Weismantel, R.: Diophantine approximations and integer points of cones. Combinatorica 22(3), 401–407 (2002)
13. Kannan, R.: Polynomial-time aggregation of integer programming problems. Journal of the Association for Computing Machinery 30(1), 133–145 (1983)
14. Lagarias, J.C.: The computational complexity of simultaneous Diophantine approximation problems. SIAM Journal on Computing 14(1), 196–209 (1985)
15. Lenstra, A.K., Lenstra Jr., H.W., Lovász, L.: Factoring polynomials with rational coefficients. Mathematische Annalen 261(4), 515–534 (1982)
16. Manders, K.L., Adleman, L.: NP-complete decision problems for binary quadratics. Journal of Computer and System Sciences 16(2), 168–184 (1978)
17. Miller, A.J., Wolsey, L.A.: Tight formulations for some simple mixed integer programs and convex objective integer programs. Mathematical Programming 98(1-3, Ser. B), 73–88 (2003); Integer programming, Pittsburgh, PA (2002)
18. Niven, I., Zuckerman, H.S., Montgomery, H.L.: An introduction to the theory of numbers, 5th edn. John Wiley & Sons Inc., New York (1991)
19. Pochet, Y., Wolsey, L.A.: Production planning by mixed integer programming. Springer Series in Operations Research and Financial Engineering. Springer, New York (2006)
20. Ravi, R., Goemans, M.X.: The constrained minimum spanning tree problem (extended abstract). In: Karlsson, R., Lingas, A. (eds.) SWAT 1996. LNCS, vol. 1097, pp. 66–75. Springer, Heidelberg (1996)
21. Rössner, C., Seifert, J.P.: Approximating good simultaneous Diophantine approximations is almost NP-hard. In: Penczek, W., Szałas, A. (eds.) MFCS 1996. LNCS, vol. 1113, pp. 494–505. Springer, Heidelberg (1996)
22. Zhao, M., de Farias Jr., I.R.: The mixing-MIR set with divisible capacities. Mathematical Programming. A Publication of the Mathematical Programming Society 115(1, Ser. A), 73–103 (2008)

Appendix

Shortest Integer Relation

By modifying a reduction from *Super-Sat* to shortest vector in the infinity norm by Dinur [5], Chen and Meng [1] showed that there exists a reduction from SAT to shortest integer relation with the property that if C is satisfiable, then the optimum value of the shortest integer relation problem is one and if C is unsatisfiable, then the optimum value of the shortest integer relation problem is at least $n^{c/\log\log n}$ for some constant $c > 0$. Here, we show that this can be extended such that there exists an optimum solution of shortest integer relation, whose first component is nonzero, thus give a proof of Lemma 1.

Let $\min\{\|x\|_\infty : a^T x = 0, x \in \mathbb{Z}^n - 0\}$ be an instance of a shortest integer relation problem. Consider the matrix

$$A = \begin{pmatrix} 0 & a^T & \mathbf{0}^T & \dots & \mathbf{0}^T \\ 0 & \mathbf{0}^T & a^T & \dots & \mathbf{0}^T \\ \vdots & \vdots & \vdots & \ddots & \vdots \\ 0 & \mathbf{0}^T & \mathbf{0}^T & \dots & a^T \\ -1 & e_1^T & e_2^T & \dots & e_n^T \end{pmatrix} \in \mathbb{Z}^{(n+1)\times(n^2+1)}$$

containing n copies of a^T on a shifted diagonal and having $(-1, e_1^T, e_2^T, \dots, e_n^T)$ as last row, where e_i is the i-th n-dimensional unit column vector. The rest is filled by zeros.

Clearly, the optimization problems $\min\{\|x\|_\infty : a^T x = 0, x \in \mathbb{Z}^n - 0\}$ and $\min\{\|x\|_\infty : Ax = 0, x \in \mathbb{Z}^{n^2+1} - 0\}$ are equivalent and the second optimization problem has the property that there is always an optimum solution with nonzero first entry. Kannan [13] provided an algorithm replacing a system $Ax = 0$ by one equation $a'^T x = 0$ in polynomial time such that the sets $\{x \in \mathbb{Z}^{n^2+1} : Ax = 0, \|x\|_\infty \leq \mu\}$ and $\{x \in \mathbb{Z}^{n^2+1} : a'x = 0, \|x\|_\infty \leq \mu\}$ are identical. His algorithm is polynomial in the encoding length of A and μ. Choosing $\mu = n$ is enough for our purposes so that Kannan's algorithm yields the desired shortest integer relation instance $\min\{\|x\|_\infty : a'^T x = 0, x \in \mathbb{Z}^{n^2+1} - 0\}$.

Computing Dense Primes

In the reduction from shortest integer relation to simultaneous Diophantine approximation (Sect. 2) we rely on the fact that one can efficiently compute prime numbers $p, q_1, \dots, q_n$ and integers R and T with

1. $n \cdot \sum_{j=1}^n |a_j| < p^R < q_1^T < q_2^T < \dots < q_n^T < (1 + \frac{1}{n}) \cdot q_1^T$,
2. p and all q_i are co-prime to all a_j,
3. $q_1^T > 2^{2n} \cdot p^R$,
4. the values of $T, R, p, q_1, \dots, q_n$ are bounded by a polynomial in the input length of a.

The algorithm which we now present is almost identical, up to better bounds, to the one proposed by Lagarias [14] and uses two deep results from number theory. The first one is the *prime number theorem*, which states that $\pi(n) \approx n/\log n$, see, e.g. [18]. The second result is the following theorem by Heath-Brown and Iwaniec [10,11].

Theorem 4. *For each $\delta > 11/20$, there exists a constant c_δ such the interval $[z, z + z^\delta]$ contains a prime for each $z > c_\delta$.*

Let m be the binary encoding length of a. The number of different primes which divide a component of a is bounded by m. We can compute the first $m + 1$ prime numbers with the sieve of Eratosthenes. Here the prime number theorem is used, since we run the sieve on the first $O(m \log m)$ natural numbers. Out of these primes we choose one which is co-prime to all components of a. This is the prime p from above. Next, we compute the smallest integers R and T such that $p^R > n \cdot \sum_{j=1}^{n} |a_j|$ and $2^T > 2^{2n} p^R$. The values of R and T are bounded by a polynomial in m.

Next, the result of Heath-Brown and Iwaniec comes into play. Let $\delta = 3/5$ and consider the sequence

$$z_i = T^{20} + i \cdot (2T)^{12}, \quad \text{for } i = 0, \ldots, T^2 - 1.$$

Each interval $[z_i, z_i + z_i^{3/5}]$ contains a prime number, since may may assume $T > c_\delta$. The number $z_i^{3/5}$ can be bounded by

$$
\begin{aligned}
z_i^{3/5} &= \left(T^{20} + i(2T)^{12}\right)^{3/5} \\
&< \left(T^{20} + T^2(2T)^{12}\right)^{3/5} \\
&\leq (2T)^{12}.
\end{aligned}
$$

From this it follows that $z_i + z_i^{3/5} < z_{i+1}$, which implies that the interval $[T^{20}, T^{20} + T^2(2T)^{12}]$ contains T^2 prime numbers. Since $T^2(2T)^{12} < T^{15}$ for T large enough, we infer that the interval

$$[T^{20}, T^{20} + T^{15}]$$

contains T^2 primes. If we denote the largest and smallest prime in this interval by $p_{\max}$ and $p_{\min}$ respectively, then $p_{\max}/p_{\min} \leq 1 + (1/T)^5$ and consequently

$$(p_{\max}/p_{\min})^T \leq (1 + (1/T)^5)^T \leq e^{1/T^4} \leq 1 + 2/T^4 \leq 1 + \frac{1}{n}.$$

Here, we used the inequality $1 + x \leq e^x$ and $e^x \leq 1 + 2x$ for $x \in [0, 1]$.

By choosing T larger than $m + n + 1$, we may obtain prime numbers $q_1 < \ldots < q_n$ from the interval $[T^{20}, T^{20} + T^{15}]$, which are co-prime to p and each a_j and hence satisfy the conditions (1–4).

PASS Approximation
A Framework for Analyzing and Designing Heuristics

Uriel Feige[1,*], Nicole Immorlica[2,*], Vahab S. Mirrokni[3,*],
and Hamid Nazerzadeh[4,*]

[1] Weizmann Institute, Rehovot, Israel
uriel.feige@weizmann.ac.il
[2] Northwestern University, Chicago, IL
nickle@eecs.northwestern.edu
[3] Google Research, New York, NY
mirrokni@theory.csail.mit.edu
[4] Stanford University, Stanford, CA
hamidnz@stanford.edu

Abstract. We introduce a new framework for designing and analyzing algorithms. Our framework applies best to problems that are inapproximable according to the standard worst-case analysis. We circumvent such negative results by designing guarantees for classes of instances, parameterized according to properties of the optimal solution. We also make sure that our parameterized approximation, called *PArametrized by the Signature of the Solution (PASS)* approximation, is the best possible. We show how to apply our framework to problems with additive and submodular objective functions such as the capacitated maximum facility location problems. We consider two types of algorithms for these problems. For greedy algorithms, our framework provides a justification for preferring a certain natural greedy rule over some alternative greedy rules that have been used in similar contexts. For LP-based algorithms, we show that the natural LP relaxation for these problems is not optimal in our framework. We design a new LP relaxation and show that this LP relaxation coupled with a new randomized rounding technique is optimal in our framework.

In passing, we note that our results strictly improve over previous results of Kleinberg, Papadimitriou and Raghavan [JACM 2004] concerning the approximation ratio of the greedy algorithm.

1 Introduction

Many important optimization problems in practice are inapproximable in theory. Practitioners deal with inapproximability issues by designing heuristics that, while provably bad on some instances, appear to perform well in practice. But for theoreticians, designing a formal framework to help guide algorithmic development for inapproximable problems has proved largely elusive.

---

* Work performed in part at Microsoft Research.

I. Dinur et al. (Eds.): APPROX and RANDOM 2009, LNCS 5687, pp. 111–124, 2009.
© Springer-Verlag Berlin Heidelberg 2009

In this paper, we present a new framework, called *PArametrized by the Signature of the Solution (PASS) approximations*. Our framework attempts to categorize instances according to how "easy" or "hard" they are, and design guarantees for all instances simultaneously with a single algorithm (the offered guarantee depends on the class of the instance and will degrade to arbitrarily bad factors for inapproximable problems, but in a controlled way). We show how this framework can be applied to a general class of optimization problem, including *capacitated maximum facility location*, that can be described as maximizing a non-decreasing submodular revenue function minus a linear cost function. We then show how the new framework affects the choice of algorithms. Two standard approaches for handling such problems are via greedy and LP-based algorithms. We study a natural greedy algorithm and prove that it is an optimal PASS approximation whereas other greedy algorithms that give optimal worst-case approximations are not. For LP-based algorithms, we show that a natural LP relaxation cannot be used to design an optimal PASS approximation. Instead, we provide a different LP relaxation and an associated rounding technique that is optimal. Our new LP relaxation is unconventional in the sense that instead of providing an upper bound on the optimal solution (this is a maximization problem), it provides a lower bound.

The current paper outlines the theory of PASS approximations. We describe the general technique and how to apply this technique to a wide range of theoretical problems using both greedy and LP-based algorithms. In a companion paper [3], we apply the notion of approximation developed here to a specific problem (banner advertising) of practical significance. This problem is a special case of the broad class of problems studied in this paper.

The rest of the paper is organized as follows. After defining the problems, in Section 2, we describe the theory of PASS approximation, and compare it with previous approaches proposed to deal with the hardness of approximation. The summary of results is given in Section 3. The greedy and an LP-based algorithms for these problems are presented respectively in Sections 4 and 5.

Problem Studied

In this paper, we mainly focus on the maximum facility location problem [1,2]:

Maximum Facility Location (MFL). *A set $\mathcal{F}$ of m facilities is given. For every facility i, there is an opening cost of c_i. There is also a set $\mathcal{J}$ of n clients. The revenue of connecting client j to facility i is $u_{ij} \geq 0$ (this may be interpreted as a client revenue minus a connection cost). Every client can connect to at most one open facility (or none). The goal in MFL is to open some facilities and connect clients to them so as to maximize the total revenue from the connected clients minus the total cost of the opened facilities.*

For comparison with some previous work [6], we shall discuss also the following problem that [6] call the variable catalog segmentation problem.

Catalog Segmentation Problem. A company has a collection of products and a collection of potential clients. Clients have various levels of interest associated with each type of product. The company wishes to produce several types of catalogs, each type containing a subset of the products (the number of products in a catalog may be limited by considerations such as weight), and mail to every potential client at most one catalog (presumably, of a type that would be of interest to the client). Assuming that producing a catalog-type has unit cost, and that for each type i and client j there is a expected revenue of u_{ij} from mailing a catalog of type i to client j, which catalogs should the company produce in order to maximize its expected profit (expected benefit minus production cost)? If all potential types of catalogs can be listed beforehand and all values u_{ij} are known, then this is a special case of MFL, with the catalogs serving as facilities. (In [6] it is assumed that all types of catalogs cost the same to produce, and we follow this assumption in our presentation. More generally, we may associate a cost c_i for producing the catalog of type i, and then the problem becomes equivalent to MFL.).

Most of our results apply to a general class of maximizing submodular set functions, called *submodular maximum facility location*, that can be described as maximizing a non-decreasing submodular revenue function minus a linear cost function.

Submodular Maximum Facility Location (SMFL). Consider a set N of n facilities and a set function $f : 2^N \to R^+$. For any subset $S \subset N$, $f(S) = R(S) - c(S)$, where R is a non-negative non-decreasing submodular set function corresponding to the revenue, and $c(S) = \sum_{i \in S} c_i$ is a linear cost function. As a result, set function f is a non-monotone submodular function and the goal is to find a subset S that maximizes $f(S)$.[1] We assume a value oracle for the revenue function R and a description for the cost c (this is of polynomial size) are given.

MFL is a special case of SMFL. Moreover, one can show (details omitted) that capacitated maximum facility location (CMFL), in which every facility has a capacity that limits the number of clients that it can serve, is also a special case of SMFL. Other examples include a variety of optimization problems such as set buying, catalog segmentation [6], banner ad allocation problem with guaranteed delivery [3], maximizing influence in social networks [5,8], and optimal sensor installation for outbreak detection [7].

2   The Theory of PASS Approximation

In this section we describe the notion of PASS approximation for the maximum facility location problem; in Section 4.2 we show how this notion extends to submodular maximum facility location. First note that for these problems the value of the objective function may be negative for some feasible solutions. As is often the case with objective functions that may be negative, the MFL problems are NP-hard to approximate within any constant factor (see for example Theorem 1). Therefore, researchers have attempted to present other types of

---

[1] Note that function f can be possibly negative and therefore the result of Feige et al. [4] does not apply.

performance guarantees. As we discuss in the next section, most of previous attempts suffer in that they do not prove guarantees with regard to the real optimum value on every instance. Nonetheless, there are large classes of interesting instances in which the approximation ratio can be much better than the worst-case guarantees, for example if the cost of opening facilities is far from the revenue one can get from the open facilities. Our goal is to get a better understanding of the approximation ratio, exposing classes of input instances for which a constant approximation ratio is possible.

Let us first describe an attempt that fails to resolve our concerns.

Relatively small costs. Based on the intuition of the previous paragraph, for $0 < \alpha < 1$, let us call an instance α-bounded if for every facility, the cost of opening the facility is at most α times the revenue one gets by connecting all clients to the facility. Is it the case that when α is sufficiently small there is a constant approximation for MFL for α-bounded instances? The answer is negative. The proof involves starting from a hard to approximate instance of MFL, and adding an additional client that provides revenue $\max(c_i/\alpha)$ regardless of which facility services it. This forces the instance to be α-bounded, while increasing the value of an optimal solution by only $\max(c_i/\alpha)$. An appropriate choice of parameters leads to the desired hardness result.

We now discuss a performance measure introduced by Kleinberg et al [6]. Unlike the notion of an α-bounded instance discussed previously, the idea is to use the notion of α-boundedness not with respect to the input instance, but rather with respect to its optimal solution. Namely, call a solution α-bounded if the total cost of opening the facilities in this solution is at most an α-fraction of the total revenue derived from all clients in the solution.[2] In [6], it is shown that for the catalog segmentation problem, whenever α is bounded away from 1, the approximation ratio of a natural greedy algorithm is a constant (that tends to 1 as α tends to 0). An exact statement of this result of [6] appears in Theorem 4. Our notion of performance guarantee can be viewed as a generalization of the notion used in [6]. A more detailed comparison between our work and that of [6] will appear in Section 4.

Consider an arbitrary MFL instance I and an arbitrary feasible solution S. For each facility i open in S, let c_i be its opening cost, and let $r_i = \sum u_{ij}$ (where the sum is taken over clients j connected to facility i in S) be the total revenue derived from clients connected in S to facility i.[3] Let α_i denote the ratio c_i/r_i. Intuitively, a value of α_i close to 0 indicates that opening the facility i was a favorable decision, because the revenue r_i that resulted from this opening came at relatively little cost. A value of α_i close to 1 indicates that the opening of facility i may have been questionable, as most of the revenue r_i is offset by the cost c_i. On a global scale, the total revenue of S is $R(S) = \sum_{i \in S} r_i$, the total cost

---

[2] Technically, in [6] a different parameter μ is considered, which in our terminology is $\mu = \frac{1}{\alpha} - 1$. It is straightforward to translate results expressed in terms of μ to results expressed in terms of α and vice versa.

[3] A better notation might be to write $r_i(S)$ instead of r_i, but we use r_i for brevity.

is $C(S) = \sum_{i \in S} c_i$, and the value of solution S is $V(S) = R(S) - C(S)$. Similar to the local values α_i, we shall use α to denote an aggregate value $\alpha = C(S)/R(S)$.

Definition 1. *Given an instance I of MFL and a feasible solution S, and using notation as above.*

- *The* expanded signature *of S is the collection $\{(q_i, \alpha_i)\}$, where i ranges over all facilities open in S, $q_i = r_i/R(S)$, and $\alpha_i = c_i/r_i$.*
- *The* signature $sig(S)$ *of S is the collection $\{(q_i, \alpha_i)\}$ obtained from the expanded signature by unifying components that share the same value of α_i. Namely, in the signature i no longer refers to a specific facility, all α_i are distinct, r_i denotes the total revenue that comes from open facilities which share the same α_i value (namely $r_i = \sum_{\text{facilities } i':\alpha'_i = \alpha_i} r_{i'}$), and q_i denotes the fraction of revenue that comes from open facilities which share the same α_i value (namely, $q_i = r_i/R(S)$).*

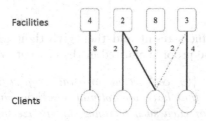

Fig. 1. An instance of the facility location problem: the costs of the facilities and the revenue from the clients (on the edges). An optimal solution is depicted by solid lines. The expanded signature is $\{(\frac{1}{2}, \frac{1}{2}), (\frac{1}{4}, \frac{1}{2}), (\frac{1}{4}, \frac{3}{4})\}$, the signature is $\{(\frac{3}{4}, \frac{1}{2}), (\frac{1}{4}, \frac{3}{4})\}$, and the summary signature is $\frac{9}{16}$.

Note that for every signature $\sum q_i = 1$, and that if all open facilities in S have the same value α_i then the signature is $(1, \alpha)$, in which case we abbreviate it to α. When open facilities have different values of α_i we may view $\alpha = C(S)/R(S)$ as a parameter that to some extent summarizes the signature, even though it does not have the same distinguishing power among solutions as the signature does. Using α as a *summary signature* will be convenient when we compare our results against previous results of [6]. Also, it is important to distinguish between the *expanded signature* and the *signature* in order to be able to talk about asymptotics in the hardness results as for any fixed expanded signature there are a fixed number of facilities. For the positive results, the notions of expanded signature and signature are interchangeable by changing the index of summation, and the reader may find it easier to interpret the positive results using the expanded signature.

In our framework of *PASS approximation*, we express the approximation ratios of algorithms as a function of the signature. Observe that an instance may have multiple different signatures (one for each feasible solution). Our approximation ratios will apply to all of them (and hence to the best of them). Nevertheless, the reader may find it convenient to think of the signature of an instance as that of (one of) its optimal solution(s). Given any feasible solution (e.g., an optimal one) with signature S, for every index i solution S generates a value of $r_i(1 - \alpha_i)$ from facilities with α value equal to α_i. Our algorithms may open facilities different than those opened by S, but our accounting method will show that our algorithms recover value at least $\hat{v}_i = r_i(1 - \alpha_i - \alpha_i \ln \frac{1}{\alpha_i})$ in exchange to the value generated by S from index i. This parameter $\hat{v}_i$ is therefore called the *recoverable value*, and, as we will prove, it is the optimal recoverable value (i.e., it is NP-hard to recover more). Note that $0 \le \hat{v}_i \le r_i$, with $\hat{v}_i = 0$ when $\alpha_i = 1$ (i.e., we can't recover any value from facilities whose cost equals their revenue) and $\hat{v}_i = r_i$ when $\alpha_i = 0$ (i.e., we can recover all the revenue from facilities with zero cost). To simplify the presentation in this paper, and with no significant effect on the results, we pretend that quantities such as $\ln x$ can be computed exactly in polynomial time for every x.

3   Our Results

We first present a hardness result and then give tight greedy and LP-based algorithms for MFL. The proof is omitted in the current version.

Theorem 1. *Let $sig = \{(q_i, \alpha_i)\}$ be an arbitrary signature, and consider the class of MFL instances that have an optimal solution with signature sig. For simplicity of notation, for each such instance, normalize the costs and revenues such that the revenue of the optimal solution having signature sig is 1, and hence its value is $1 - \sum q_i \alpha_i$. Then on this class of instances, for every $\epsilon > 0$, it is NP-hard to find a solution of value $\sum \hat{v}_i + \epsilon$ where $\hat{v}_i = q_i(1 - \alpha_i - \alpha_i \ln \frac{1}{\alpha_i})$.*

Corollary 1. *For any $\epsilon > 0$ and $\alpha = C(S)/R(S)$, for any optimal solution S, it is NP-hard to approximate MFL within a ratio better than $\frac{1-\alpha-\alpha \ln \frac{1}{\alpha}}{1-\alpha} + \epsilon$.*

We show that there are algorithms with approximation ratios that match the hardness results, and moreover do so for wider classes of problems. The first class of algorithms that we consider is that of greedy algorithms. We shall distinguish between two types of greedy algorithms depending on whether it is greedy with respect to margin or to rate. Only one of these versions is optimal in our framework.

Theorem 2. *Let I be an arbitrary instance of MFL, let S be an arbitrary feasible solution and let $\{(q_i, \alpha_i)\}$ be the signature of S. For simplicity of notation, normalize the costs and revenues in I such that the revenue of S is 1, and hence its value is $1 - \sum q_i \alpha_i$. Then the greedy-rate algorithm produces a solution of value at least $\sum \hat{v}_i$ where $\hat{v}_i = q_i(1 - \alpha_i - \alpha_i \ln \frac{1}{\alpha_i})$.*

Corollary 2. *The greedy-rate algorithm approximates MFL within a ratio of at least $\frac{1-\alpha-\alpha \ln \frac{1}{\alpha}}{1-\alpha}$, where $\alpha = C(S)/R(S)$ for any optimal solution S.*

Remark: As shown in Section 4.2, the result above holds for SMFL problem, under the appropriate definition of signature. We also remark that Corollaries 1 and 2 are each stronger than previous results proved in [6]. These issues will be discussed in Section 4.

The next class of algorithms that we consider is based on linear programming. It is also possible to show that the natural linear programming relaxation does not result in approximation ratios that match the hardness results of Theorem 1 (the details are omitted in this version). Hence, we introduce a new linear program, called the *recoverable value LP*, whose objective is to maximize the (fractional) recoverable value rather than the (fractional) true value.[4] We then show that the LP can be rounded to give a feasible solution of value not lower than the recoverable value of the LP.

Theorem 3. *The recoverable value LP for MFL can be solved in polynomial time. For every input instance and feasible solution S with signature $\{(q_i, \alpha_i)\}$, the LP has a solution of value at least as high as $\sum \hat{v}_i$, where $\hat{v}_i = r_i(1 - \alpha_i - \alpha_i \ln \frac{1}{\alpha_i})$. Any solution of the LP can be rounded in random polynomial time to give a feasible solution of expected value at least as high as the value of the objective function in the LP solution.*

3.1   Why Use Our Notion of PASS Approximation?

In this section we present arguments in favor of our notion of PASS approximation. The point that we will try to make is that performance measures guide the design of algorithms, and our performance measure appears to us to be a very good guide. We assume in the discussion below that the true goal is to maximize revenue minus cost, and compare various approaches that can be used in order to circumvent the inapproximability results for this measure. Recall that our approach of PASS approximation is to express the approximation ratio not as a function of the size of the input instance, but as of its signature.

As mentioned, MFL and SMFL are NP-hard to approximate. Therefore, researchers have attempted to present other types of performance guarantees. One existing theoretical approach for coping with inapproximable problems is to change the objective function in a way that preserves the spirit of the original problem. For example, one might consider the complement of the objective (e.g., vertex cover as opposed to independent set), or bicriteria approximations (e.g., bisection in graphs). In [2,1] the approach taken was to measure the quality of a solution on a shifted scale which is always nonnegative. This is equivalent to changing the objective function by adding to it a sufficiently large constant that ensures that all solutions have nonnegative value. As an example, consider algorithms for MFL based on linear programming. In [1] a combination of a linear program and rounding technique is designed. They show that the approximation ratio of $2(\sqrt{2} - 1)$ that they obtain is best possible (matches the integrality

---

[4] We note this relaxation is not a relaxation in the usual sense, because the value of the objective function of the LP is a lower bound on the value of an optimal solution, rather than an upper bound.

gap), but with respect to a shifted scale of the objective function. As we do not claim the same about our linear programming approach, then clearly there are instances in which the algorithm of [1] is better than ours. Likewise, there are instances on which our LP plus rounding gives better results (because we are optimal with respect to the structural approximation measure, whereas [1] are not). Hence it appears as if the results are incomparable. Nevertheless, we would like to convince the reader that even though the result of [1] is interesting mathematically, it does not really provide the kind of algorithmic insights that are relevant to the original problem. To obtain an approximation ratio of $2(\sqrt{2} - 1)$ with respect to the shifted scale, it is safe to open every facility with probability at least $1 - 2(\sqrt{2} - 1)$ (and at most 1), regardless of the cost of the facility, and regardless of whether any client wants to connect to the facility. This is a simple (and obviously counterproductive) rule of thumb that comes out of the shifted scale performance measure, and in fact the algorithm of [1] follows it. We view this as evidence that the shifted scale performance measure is not a good guide in the design of algorithms (with respect to the original objective function).

One can also show that other approaches designed to analyze heuristics, including optimizing with respect to a budget constraint, average-case and smoothed analysis, do not provide a desirable performance measure for the MFL problem. Details are omitted.

In comparison with [6], note that the approximation ratio in [6] is expressed as a function of one parameter that we refer to as the summary signature α. What is the advantage of presenting the more complicated signature $\{(q_i, \alpha_i)\}$? We see two advantages (beyond the obvious advantage of always providing a performance guarantee that is at least as good as that provided by the summary signature). One is *prescriptive*: the design of our LP is a natural consequence of our signature, valuing each star according to its own recoverable value. It would have been very difficult to design and analyze it without having at least implicitly a notion similar to the detailed signature. The other advantage is *conceptual*: our signature enjoys closure properties that the summary signature does not have. Given two disjoint instances of SMFL, the detailed signature becomes simply the union of the original detailed signatures, and the output guarantee (approximation ratio times value of optimal solution) is simply the sum of output guarantees of the two instances. For the summary signature, this is not true.

LP-based vs. greedy algorithms. The performance guarantees that we prove for the greedy algorithm and the LP-based algorithm are the same, and for the greedy algorithm we prove this performance guarantee for a wider range of instances (SMFL rather than just MFL). So what is the point of having an LP-based algorithm? There are several reasons to do this.

Most importantly, there is a conceptual difference between the use of PASS approximation framework for our greedy versus LP-based algorithms. For our greedy algorithm, the theory of structural approximation is *descriptive*. It describes the approximation ratios of existing algorithms, and may guide us in the choice of the greedy rule to use. For the LP-based approach, however, the theory of PASS approximation is not only descriptive, but also *prescriptive*. It guides us in the design of new algorithms. The definitions of the signature and recoverable value define for us the linear program and the rounding technique. While in our examples, the greedy algorithms happens to be tight, the LP-based approach may still be of value for other problems precisely because of it's prescriptive nature – *it is designed to produce tight algorithms*. For this reason, we consider the LP-based approach to be a significant contribution of our paper.

Another reason is so as to diversify our algorithmic toolbox. Even though the current paper is concerned with a class of problems for which the LP approach does not seem to offer significant advantages over the greedy approach, this need not be the case for other classes of problems. The development of a methodology of how to use linear programming relaxations in the context of structural approximation (which turns out to be different than the way linear programming relaxations are typically used in "classical" approximation) is anticipated to lead to rewards in future work.

4   A Greedy Approach

One standard approach for the MFL problems are greedy algorithms. In this section, we describe two plausible greedy algorithms for SMFL, and prove that one of them is optimal with regards to the PASS approximation. Given a set S of facilities, let $C(S)$ denote the total cost of facilities in S, let $R(S)$ denote the revenue of the optimum assignment given that the open facilities are those in S, and let $V(S) = R(S) - C(S)$ denote the total value of S. Given a facility i, let $M(i|S)$ denote the marginal revenue of i with respect to S. Namely, $M(i|S) = R(S \cup \{i\}) - R(S)$. If $i \in S$ then $M(i|S) = 0$.

The greedy algorithms construct a solution iteratively by selecting facilities that maximize some function of the marginal revenue $M(i|S)$. Given a partial solution (set of open facilities) S, the *greedy-rate* algorithm opens the facility i which maximizes the *rate* of increase in value, i.e., $\frac{M(i|S)-c_i}{M(i|S)}$, provided that this rate is positive. The *greedy-margin* algorithm simply opens the facility with the largest marginal value, i.e., $M(i|S) - c_i$, provided that this value is positive.

The greedy step can be implemented in polynomial time for the special cases of SMFL mentioned in Section 1. (For example, for CMFL, implementing the greedy step involves computing the optimal assignment of clients to the open facilities subject to the capacity constraints. This can be solved in polynomial time via an algorithm for the so called B-matching problem in bipartite graphs.) However, in general, the greedy step for SMFL might be NP-hard.

4.1   Comparison to KPR

The greedy-margin algorithm was studied by Kleinberg, Papadimitriou, and Raghavan [6] for the catalogue segmentation problem (and generalizations which maintain the property of uniform-cost facilities).[5] They proved the following theorem (Theorems 2.3 and 2.4 in [6]):

Theorem 4. [Kleinberg, Papadimitriou and Raghavan.] *For the catalogue segmentation problem, the greedy-margin algorithm achieves an approximation ratio of at least $1 + \alpha - 2\sqrt{\alpha}$, where $\alpha = C(S)/R(S)$ for any optimal solution S. There are instances on which the approximation ratio of the algorithm is no better than $1 - \alpha$.*

In this section, we will improve upon this result by generalizing the analysis to accommodate non-uniform facility costs and providing improved approximation guarantees (Theorem 2) together with a matching NP-hardness result (Theorem 1). In doing so, we must be careful with our choice of greedy algorithm. We have defined two natural greedy algorithms – the greedy-rate and the greedy-margin algorithm – and in fact in uniform-cost settings such as that of [6] these two algorithms coincide as the rate of a facility is monotone in its marginal revenue. But for non-uniform facility costs, as the following simple example illustrates, the greedy-margin algorithm gives very poor results.

Example. There are n clients and $n + 1$ facilities. Facility i, $1 \le i \le n$, has cost 1, revenue 2 for client i, and 0 revenue for all other clients; Facility $n+1$ has cost $n - 2$ and revenue 1 per client. The optimal solution will open the first n facilities for a value of n, whereas the greedy-margin rule will open only facility $n + 1$ for a value of 2.

By contrast, for the greedy-rate algorithm, the approximation ratio that we prove is strictly better than that proved in [6]. See Figure 2 for a detailed comparison of our bounds with those of [6].

4.2   Approximation as a Function of α

In this section, we first prove Corollary 2 which gives an approximation factor for the greedy rate algorithm as a function of the summary signature α. Later in this section, we give an approximation factor as the function of the signature, proving Theorem 2.

The following simple observation (appearing in [6]) is of key importance, and hence we state it as a lemma.

Lemma 1. *Let i be a facility and let S and T be sets of facilities. Then the marginal revenue of facility i with respect to S is at least as large as the loss in marginal revenue of T when facility i is added to S.*

$$M(i|S) \ge M(T|S) - M(T|S \cup \{i\})$$

---

[5] The greedy algorithm specified prior to Theorem 2.3 in [6] does not specify a rule of which facility to open next, as long as its marginal revenue is larger than its cost. However, the proof of Theorem 2.4 in [6] is based on the use of a greedy-margin rule, without stating this explicitly.

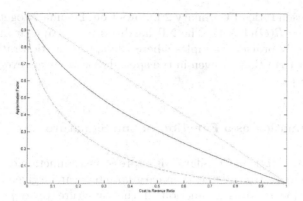

Fig. 2. The solid line is the approximation ratio of the greedy-rate algorithm plotted for $\alpha \in [0,1]$, as proved in Corollary 2. For every value of α improving over this approximation ratio is NP-hard, as proved in Theorem 1. The dashed and dotted lines depict the lower and upper bounds proved in [6].

Proof. By the fact that the revenue function is nondecreasing, we have $R(S \cup \{i\} \cup T) \geq R(S \cup T)$. Breaking each revenue to a sum of marginal revenues we have $R(S) + M(i|S) + M(T|S \cup \{i\}) \geq R(S) + M(T|S)$. Canceling the $R(S)$ and subtracting $M(T|S \cup \{i\})$ from both sides, the lemma is proved.

We now proceed to prove our improved bounds. As stated, our analysis applies to any problem with a nondecreasing submodular revenue function and linear cost function.

Lemma 2. *The value of the greedy-rate algorithm is at least $R(O)(1-\alpha-\alpha\ln\frac{1}{\alpha})$ where O is an optimal solution, and $\alpha = C(O)/R(O)$.*

Proof. We analyze the value of greedy-rate up to the first point in time in which its total revenue meets or exceeds $R(O)-C(O)$. Let $\mu(x)$ denote the rate at which the value obtained by the greedy-rate algorithm increases when it has already made a revenue of x. Observe that at a point when greedy has already made a revenue of $x < R(O) - C(O)$, the marginal revenue of O is at least $R(O) - x$ (by Lemma 1). By submodularity of the revenue function, at this point there must be at least one facility of O with rate $\frac{R(O)-x-C(O)}{R(O)-x}$. The rate at which the value increases at each point in time is at least as high as the rate one would get by choosing the highest rate among the facilities of O at the same time. Therefore, $\mu(x) \geq \frac{R(O)-x-C(O)}{R(O)-x}$. As the total value of greedy-rate, $V(G)$, is the integral of the rate of increase of the value, we have:

$$V(G) = \int \mu(x)dx \geq \int_0^{R(O)-C(O)} \frac{R(O) - x - C(O)}{R(O) - x} dx$$

$$= R(O) \int_0^{1-\alpha} \frac{1 - x/\alpha - \alpha}{1 - x/\alpha} dx$$

$$= R(O)(1 - \alpha - \alpha \ln\frac{1}{\alpha})$$

The approximation ratio of Corollary 2 follows from Lemma 2 together with the fact that $V(O) = R(O)(1 - \alpha)$. The NP-hardness results appear in Theorem 1, and they naturally provide examples where the approximation ratio of greedy-rate is no better than claimed even in the special case of linear revenue functions and uniform costs.

4.3   Approximation as a Function of the Signature

The approximation ratio in Corollary 2 is expressed as a function of the summary signature α, whereas stronger performance guarantees can be given by expressing the approximation ratio as a function of the signature $\{q_i, \alpha_i\}$, as stated in Theorem 2.

In fact, we prove Theorem 2 for the general submodular facility location problems. To do so, we should extend the notion of a signature to a solution for submodular maximum facility location. The difficulty is that even though the cost of every open facility is well defined, its revenue is not. Hence we refine the notion of a solution to be represented not as a set of open facilities, but as an ordered set (a tuple). Namely, the open facilities are given (after renaming) in some order $1, 2, \ldots$ (even though this order is irrelevant to the actual value of the solution). Thereafter, a refined parameter α_i' in the expanded signature is defined relative to the marginal revenue of facility i with respect to this order. That is, $\alpha_i' = c_i/M_i$, where here M_i is shorthand notation for $M(i|\{1, \ldots, i-1\})$. Likewise, we define $q_i' = M_i/\sum M_j$. Using this notation, we can now strengthen Lemma 2. The proofs of the lemmas are omitted.

Lemma 3. *Let S be an arbitrary (ordered) solution for submodular maximum facility location with expanded signature $\{(q_i', \alpha_i')\}$ and total revenue normalized to 1. Then the value of the greedy-rate algorithm is at least $\sum_{i \in S} q_i'(1 - \alpha_i' - \alpha_i' \ln \frac{1}{\alpha_i'})$.*

To motivate the following lemma, observe that Lemma 3 by itself does not capture the notion of PASS approximation that we have for the special case of MFL. For a given set of open facilities in MFL, the optimal choice of allocation of clients might not correspond to revenues r_i per facility that are equal to M_i for any ordering of facilities. For example, if there are two facilities and two clients, where client i has revenue 2 if connected to facility i and revenue 1 if connected to the other facility, then in the optimal solution $r_1 = r_2 = 2$, whereas for any ordering $M_1 = 3$ and $M_2 = 1$.

Lemma 4. *Let S be an arbitrary solution for maximum facility location with expanded signature $\{(q_i, \alpha_i)\}$. Then there is an ordering of the facilities of S giving $\sum_{i \in S} q_i'(1 - \alpha_i' - \alpha_i' \ln \frac{1}{\alpha_i'}) \geq \sum_{i \in S} q_i(1 - \alpha_i - \alpha_i \ln \frac{1}{\alpha_i})$.*

The combination of Lemmas 3 and 4 imply Theorem 2 (and also the generalization of Theorem 2 to SMFL).

5 A Linear Programming Approach

In this section, we develop an LP-based approach for MFL. It is based on an interplay between the notions of the true value of a solution and the recoverable value of the solution. Recall that the recoverable value (see definition in Section 2), which in general is lower than the true value, represents our approximation goal in the sense that we wish to find a solution of true value at least equal to that of the recoverable value of the best integral solution. First we introduce a new LP relaxation for the general problem called the *recoverable value relaxation*. This LP captures the natural constraints for the MFL problem, but has an objective function describing the recoverable value of the solution rather than the true value. Hence the LP provides a fractional solution that maximizes the recoverable value, and we denote this value by $\hat{V}_f$. We round this fractional solution to an integral one of (expected) true value at least $\hat{V}_f$, thus meeting our approximation goal. Moreover, we can solve the recoverable value LP in polynomial time.

While our approach is general, we have been unable to analyze it for CMFL or more general variants, and leave this as an open question.

5.1 An LP Relaxation

Recall that in the MFL problem, each facility $i \in \mathcal{F}$ has an opening cost of c_i and each client $j \in \mathcal{J}$ has a revenue u_{ij} for being connected to facility i. We call pair (i, T) of a facility i and a subset T of clients connected to it a *star*. Let x_{iT} be an indicator variable of star (i, T), i.e., that facility i is opened and connected to clients $j \in T$. The revenue of connecting the clients in T to facility i is $r_{iT} = \sum_{j \in T} u_{ij}$. For every star (i, T) we associate a *recoverable value* which is $\hat{v}_{iT} = r_{iT}(1 - \alpha_{iT} - \alpha_{iT} \ln \frac{1}{\alpha_{iT}})$, where $\alpha_{iT} = \frac{c_i}{r_{iT}}$. Then the optimal fractional recoverable value is described by the following LP, called the *recoverable value LP relaxation*.

$$\text{maximize} \quad \sum \hat{v}_{iT} x_{iT} \tag{1}$$

$$\text{subject to} \quad \sum_{i, T : j \in T} x_{iT} \leq 1 \qquad j \in \mathcal{J}$$

$$\sum_{T \subseteq \mathcal{J}} x_{iT} \leq 1 \qquad i \in \mathcal{F}$$

$$x_{iT} \geq 0 \qquad i \in \mathcal{F}, T \subseteq \mathcal{J}$$

The first inequality guarantees that each client contributes revenue to at most one facility and the second inequality guarantees that each facility is opened at most once. Every integral solution satisfies these constraints. Hence the value of the LP is at least as large as the recoverable value of the best integer solution (the one maximizing the recoverable value).

Let $\hat{V}_f$ be the optimal fractional recoverable value, namely, the optimal value to the above LP. Let V_f be the fractional true value associated with this solution, namely $\sum(r_{iT} - c_i)x_{iT}$. Typically, LP-relaxations provide upper bounds for maximization problems. In contrast, it is not in general true that V_f provides an upper bound on the true value of the best integer solution. Instead, as Lemma 5 will show, $\hat{V}_f$ provides a lower bound.

Our LP has exponentially many variables; however, we can solve it using the ellipsoid method. We solve the separation oracle of the dual linear program using a greedy algorithm. This algorithm exploits concavity of the recoverable values and some other structural properties of the dual.

Our randomized rounding procedure is composed of two steps: The first step considers facilities independently. Facilities of 0-cost are always opened. For the remaining facilities, $\alpha_{iT} > 0$. For each such facility i and each star (i, T) let $\beta_{iT} = x_{iT} \ln \frac{1}{\alpha_{iT}}$. Let $\beta_i = \sum_T \beta_{iT}$. We open facility i with probability $\min[\beta_i, 1]$. The first step might open several facilities with overlapping sets of clients. In the second step, we assign any over-demanded client j to the facility to which it contributes the maximum revenue. The following lemma yields Theorem 3. For the lack of space, the proof is omitted here.

Lemma 5. *Consider an optimal fractional solution of LP (1), with fractional recoverable value $\hat{V}_f$. For the MFL problem, our randomized rounding technique achieves an integral solution of expected (true) value at least $\hat{V}_f$.*

Acknowledgements

The work of the first author was supported in part by The Israel Science Foundation (grant No. 873/08).

References

1. Ageev, A., Sviridenko, M.: An 0.828-approximation algorithm for uncapacitated facility location problem. Discrete Applied Mathematics 93, 289–296 (1999)
2. Cornuejols, G., Nemhauser, G., Wolsey, L.: Locations of bank accounts to optimize float: an analytic study of exact and approximate algorithms. Management Science 23, 789–810 (1977)
3. Feige, U., Immorlica, N., Mirrokni, V., Nazerzadeh, H.: A combinatorial allocation mechanism with penalties for banner advertising. Proceedings of the 17th International World Wide Web Conference (WWW) (2008)
4. Feige, U., Mirrokni, V., Vondrak, J.: Maximizing non-monotone submodular functions. In: Proceedings of the forty-eight annual IEEE symposium on Foundations of Computer Science (FOCS), pp. 461–471 (2007)
5. Kempe, D., Kleinberg, J., Tardos, É.: Maximizing the spread of influence through a social network. In: KDD 2003: Proceedings of the ninth ACM SIGKDD international conference on Knowledge discovery and data mining, pp. 137–146. ACM, New York (2003)
6. Kleinberg, J.M., Papadimitriou, C.H., Raghavan, P.: Segmentation problems. J. ACM 51(2), 263–280 (2004)
7. Leskovec, J., Krause, A., Guestrin, C., Faloutsos, C., VanBriesen, J., Glance, N.S.: Cost-effective outbreak detection in networks. In: KDD, pp. 420–429 (2007)
8. Mossel, E., Roch, S.: On the submodularity of influence in social networks. In: Proceedings of ACM STOC, pp. 128–134 (2007)

Optimal Sherali-Adams Gaps from Pairwise Independence

Konstantinos Georgiou[1,*], Avner Magen[1,*], and Madhur Tulsiani[2,**]

[1] Department of Computer Science, University of Toronto
{cgeorg,avner}@cs.toronto.edu
[2] Computer Science Division, University of California, Berkeley
madhurt@cs.berkeley.edu

Abstract. This work considers the problem of approximating fixed predicate constraint satisfaction problems (MAX k-CSP(P)). We show that if the set of assignments accepted by P contains the support of a balanced pairwise independent distribution over the domain of the inputs, then such a problem on n variables cannot be approximated better than the trivial (random) approximation, even using $\Omega(n)$ levels of the Sherali-Adams LP hierarchy.

It was recently shown [3] that under the Unique Game Conjecture, CSPs with predicates with this condition cannot be approximated better than the trivial approximation. Our results can be viewed as an unconditional analogue of this result in the restricted computational model defined by the Sherali-Adams hierarchy. We also introduce a new generalization of techniques to define consistent "local distributions" over partial assignments to variables in the problem, which is often the crux of proving lower bounds for such hierarchies.

1 Introduction

A constraint satisfaction problem (CSP) consists of a set of constraints that seek a universal solution. In the maximization version (MAX-CSP) one tries to maximize the number of constraints that can be simultaneously satisfied. The most standard family of CSPs arise from Boolean predicates P with bounded support k. In their generality, the predicates are defined over an alphabet $\{0, 1, \ldots, q - 1\} = [q]$ and they can be thought as functions $P : [q]^k \to \{0, 1\}$. A constraint is defined by the predicate P applied to a k-tuple of literals $(x_1 + b_1 \bmod q, \ldots, x_k + b_k \bmod q)$, where $b_i \in [q]$, and is said to be satisfied by some assignment on $(x_1, \ldots, x_k)$ if the predicate evaluates to 1. Given some predicate P, an instance of the MAX k-CSP(P) problem is a collection of constraints as above and the objective is to maximize the number of constraints that can be satisfied simultaneously. As a special case, we can obtain all well studied MAX-CSP problems,

---

* Funded in part by NSERC.
** Supported by the NSF grants CCF-0515231 and CCF-0729137 and by US-Israel BSF grant 2006060.

I. Dinur et al. (Eds.): APPROX and RANDOM 2009, LNCS 5687, pp. 125–139, 2009.

e.g. MAX k-SAT, MAX-CUT etc. When the predicate to be used in different constraints is not fixed we simply refer to the problem as MAX k-CSP.

The MAX k-CSP problem is NP-hard for $k \geq 2$, and a lot of effort has been devoted in determining the true inapproximability of the problem. In general, the inapproximability of the MAX k-CSP depends on the size of alphabet over which literals are valued. For the case of Boolean variables, Samorodnitsky and Trevisan [19] proved that the problem is hard to approximate better than a factor of $2^{2\sqrt{k}}/2^k$, which was improved to $2^{\sqrt{2k}}/2^k$ by Engebresten and Holmerin [9]. Later Samorodnitsky and Trevisan [20] showed that it is Unique-Games-hard to approximate the same problem with factor better than $2^{\lceil \log k+1 \rceil}/2^k$. For the more general case of q-ary variables (MAX k-CSP$_q$), Guruswami and Raghavendra [13] showed a hardness ratio of q^2k/q^k when q is a prime.

In a very general result which captures all the above ones, Austrin and Mossel [3] showed that if $P : [q]^k \to \{0,1\}$ is a predicate such that the set of accepted inputs $P^{-1}(1)$ contains the support of a balanced pairwise independent distribution μ on $[q]^k$, then MAX k-CSP(P) is UG-hard to approximate better than a factor of $|P^{-1}(1)|/q^k$. Considering that a random assignment satisfies $|P^{-1}(1)|/q^k$ fraction of all the constraints, this is the strongest result one can get for a predicate P. Using appropriate choices for the predicate P, this then implies hardness ratios of $kq^2(1 + o(1))/q^k$ for general $q \geq 2$, $q(q-1)k/q^k$ when q is a prime power, and $(k + O(k^{0.525}))/2^k$ for $q = 2$.

We study the inapproximability of such a predicate P (which we call promising) in the hierarchy of linear programs defined by Sherali and Adams. In particular, we show an unconditional analogue of the result of Austrin and Mossel in this hierarchy.

Hierarchies of Linear and Semidefinite Programs. A standard approach in approximating NP-hard problems, and therefore MAX k-CSP, is to formulate the problem as a 0-1 integer program and then relax the integrality condition to get a linear (or semidefinite) program which can be solved efficiently. The quality of such an approach is intimately related to the *integrality gap* of the relaxation, namely, the ratio between the optimum of the relaxation and that of the integer program.

Several methods (or procedures) were developed in order to obtain tightenings of relaxations in a systematic manner. These procedures give a sequence or a *hierarchy* of increasingly tighter relaxations of the starting program. The commonly studied ones include the hierarchies defined by Lovász-Schrijver [16], Sherali-Adams [24], and Lasserre [14] (see [15] for a comparison). Stronger relaxations in the sequence are referred to as higher *levels* of the hierarchy. It is known for all these hierarchies that for a starting program with n variables, the program at level n has integrality gap 1, and that it is possible to optimize over the program at the rth level in time $n^{O(r)}$.

Many known linear (semidefinite) programs can be captured by constant many levels of the Sherali-Adams (Lasserre) hierarchy. Fernández de la Vega and Kenyon-Mathieu [11] have provided a PTAS for Max Cut in dense graphs using Sherali-Adams. In [17] it is shown how to get a Sherali-Adams based PTAS for Vertex-Cover and Max-Independent-Set in minor-free graphs, while recently Mathieu and Sinclair [18] showed that the integrality gap for the matching poly-tope is asymptotically $1 + 1/r$, and Bateni, Charikar and Guruswami [4] that the integrality gap for a natural LP formulation of the MaxMin allocation problem has integrality gap at most $n^{1/r}$, both after r many Sherali-Adams tightenings. Chlamtac [7] and Chlamtac and Singh [8] gave an approximation algorithm for Max-Independent-Set in hypergraphs based on the Lasserre hierarchy, with the performance depending on the number of levels.

Lower bounds in these hierarchies amount to showing that the integrality gap remains large even after many levels of the hierarchy. Integrality gaps for $\Omega(n)$ levels can be seen as unconditional lower bounds (as they rule out even exponential time algorithms obtained by the hierarchy) in a restricted (but still fairly interesting) model of computation. Considerable effort was invested in proving lower bounds (see [2,26,25,23,5,10,1,22,12,11]). For CSPs in particular, strong lower bounds ($\Omega(n)$ levels) were proved recently for the Lasserre hierarchy (which is the strongest) by [21] and [27], who showed a factor 2 integrality gap for MAX k-XOR and factor $2^k/2k$ integrality gap for MAX k-CSP respectively.

Our Result and Techniques. Both the results in the Lasserre hierarchy (and previous analogues in the Lovász-Schrijver hierarchy) seemed to be heavily rely-ing on the structure of the predicate for which the integrality gap was proven, as being some system of linear equations. It was not clear if the techniques could be extended using only the fact that the predicate is promising (which is a much weaker condition). In this paper, we try to explore this issue, proving $\Omega(n)$ level gaps for the (admittedly weaker) Sherali-Adams hierarchy.

Theorem 1. *Let* $P : [q]^k \to \{0,1\}$ *be predicate such that* $P^{-1}(1)$ *contains the support of a balanced pairwise independent distribution* μ. *Then for every con-stant* $\zeta > 0$, *there exist* $c = c(q, k, \zeta)$ *such that for large enough* n, *the integrality gap of* MAX k-CSP(P) *for the tightening obtained by* cn *levels of the Sherali-Adams hierarchy applied to the standard LP[1] is at least* $\dfrac{q^k}{|P^{-1}(1)|} - \zeta$.

We note that $\Omega(n^\delta)$-level gaps for these predicates can also be deduced via reductions from the recent result of [6] who obtained $\Omega(n^\delta)$-level gaps for Unique Games, where $\delta \to 0$ as $\zeta \to 0$.

A first step in achieving our result is to reduce the problem of a level-t gap to a question about family of distributions over assignments associated with sets of variables of size at most t. These distributions should be (a) supported only on satisfying (partial) assignments and (b) should be consistent among themselves, in the sense that for $S_1 \subseteq S_2$ which are subsets of variables, the distributions over S_1 and S_2 should be equal on S_1. The second requirement guarantees that the obtained solution is indeed feasible, while the first implies that the solution achieves objective value that corresponds to satisfying *all* the constraints of the instance.

---

[1] See the resulting LP in section 2.3.

The second step is to come up with these distributions! We explain why the simple method of picking a uniform distribution (or a reweighting of it according to the pairwise independent distribution that is supported by P) over the satisfying assignments cannot work. Instead we introduce the notion of "advice sets". These are sets on which it is "safe" to define such simple distributions. The actual distribution for a set S we use is then the one induced on S by a simple distribution defined on the advice-set of S. Getting such advice sets heavily relies on notions of expansion of the constraints graph. In doing so, we use the fact that random instances have inherently good expansion properties. At the same time, such instances are highly unsatisfiable, ensuring that the resulting integrality gap is large.

Arguing that it is indeed "safe" to use simple distributions over the advice sets relies on the fact that the predicate P in question is promising, namely $P^{-1}(1)$ contains the support of a balanced pairwise independent distribution. We find it interesting and somewhat curious that the condition of pairwise independence comes up in this context for a reason very different than in the case of UG-hardness. Here, it represents the limit to which the expansion properties of a random CSP instance can be pushed to define such distributions.

2   Preliminaries and Notation

2.1   Constraint Satisfaction Problems

For an instance Φ of MAX k-CSP$_q$, we denote the variables by $\{x_1, \ldots, x_n\}$, their domain $\{0, \ldots, q-1\}$ by $[q]$ and the constraints by $C_1, \ldots, C_m$. Each constraint is a function of the form $C_i : [q]^{T_i} \to \{0,1\}$ depending only on the values of the variables in the ordered tuple T_i with $|T_i| \leq k$.

For a set of variables $S \subseteq [n]$, we denote by $[q]^S$ the set of all mappings from the set S to $[q]$. In context of variables, these mappings can be understood as partial assignments to a given subset of variables. For $\alpha \in [q]^S$, we denote its projection to $S' \subseteq S$ as $\alpha(S')$. Also, for $\alpha_1 \in [q]^{S_1}, \alpha_2 \in [q]^{S_2}$ such that $S_1 \cap S_2 = \emptyset$, we denote by $\alpha_1 \circ \alpha_2$ the assignment over $S_1 \cup S_2$ defined by α_1 and α_2.

We shall prove results for constraint satisfaction problems where every constraint is specified by the same Boolean predicate $P : [q]^k \to \{0,1\}$. We denote the set of assignments which the predicate evaluates to 1 by $P^{-1}(1)$. A CSP instance for such a problem is a collection of constraints of the form of P applied to k-tuples of *literals*. For a variable x with domain $[q]$, we take a literal to be $(x + a) \bmod q$ for any $a \in [q]$. More formally,

Definition 1. *For a given $P : [q]^k \to \{0,1\}$, an instance Φ of MAX k-CSP$_q(P)$ is a set of constraints $C_1, \ldots, C_m$ where each constraint C_i is over a k-tuple of variables $T_i = \{x_{i_1}, \ldots, x_{i_k}\}$ and is of the form $P(x_{i_1} + a_{i_1}, \ldots, x_{i_k} + a_{i_k})$ for some $a_{i_1}, \ldots, a_{i_k} \in [q]$. We denote the maximum number of constraints that can be simultaneously satisfied by $\mathsf{OPT}(\Phi)$.*

2.2 Expanding CSP Instances

For an instance Φ of MAX k-CSP$_q$, define its constraint graph G_Φ, as the following bipartite graph from L to R. The left hand side L consists of a vertex for each constraint C_i. The right hand side R consists of a vertex for every variable x_j. There is an edge between a constraint-vertex i and a variable-vertex j, whenever variable x_j appears in constraint C_i. When it is clear from the context, we will abbreviate G_Φ by G.

For $C_i \in L$ we denote by $\Gamma(C_i) \subseteq R$ the neighbors $\Gamma(C_i)$ of C_i in R. For a set of constraints $\mathcal{C} \subseteq L$, $\Gamma(\mathcal{C})$ denotes $\cup_{C_i \in \mathcal{C}} \Gamma(C_i)$. For $S \subseteq R$, we call a constraint $C_i \in L$, S-dominated if $\Gamma(C_i) \subseteq S$. We denote by $G|_{-S}$ the bipartite subgraph of G that we get after removing S and all S-dominated constraints. Finally, we also denote by $\mathcal{C}(S)$ the set of all S-dominated constraints.

Our result relies on set of constraints that are well expanding. We make this notion formal below.

Definition 2. *Consider a bipartite graph $G = (V, E)$ with partition L, R. The* **boundary expansion** *of $X \subset L$ is the value $|\partial X|/|X|$, where $\partial X = \{u \in R : |\Gamma(u) \cap X| = 1\}$. G is (r, e) **boundary expanding** if the boundary expansion for all subsets of L of size at most r is at least e.*

2.3 The Sherali-Adams Hierarchy

Below we present a relaxation for the MAX k-CSP$_q$ problem as it is obtained by applying a level-t Sherali-Adams tightening of the standard LP formulation of some instance Φ of MAX k-CSP$_q$. A well known fact states that the level-n Sherali-Adams tightening provides a perfect formulation, i.e. the integrality gap is 1 (see [24] or [15] for a proof).

The intuition behind the level-t Sherali-Adams tightening is the following. Note that an integer solution to the problem can be given by a single mapping $\alpha_0 \in [q]^{[n]}$, which is an assignment to all the variables. Using this, we can define 0/1 variables $X_{(S,\alpha)}$ for each $S \subseteq [n]$ such that $|S| \leq t$ and $\alpha \in [q]^S$. The intended solution is $X_{(S,\alpha)} = 1$ if $\alpha_0(S) = \alpha$ and 0 otherwise. We introduce $X_{(\emptyset,\emptyset)}$ which is intended to be 1. By relaxing the integrality constraint on the variables, we obtain the level-t Sherali-Adams LP tightening.

Level-t (for $t \geq k$) Sherali-Adams LP tightening for a MAX k-CSP$_q$ instance Φ

maximize $\displaystyle\sum_{i=1}^{m} \sum_{\alpha \in [q]^{T_i}} C_i(\alpha) \cdot X_{(T_i, \alpha)}$

subject to $\displaystyle\sum_{j \in [q]} X_{(S \cup \{i\}, \alpha \circ j)} = X_{(S,\alpha)}$ $\forall S$ s.t. $|S| < t,\ \forall i \notin S, \alpha \in [q]^S$

$X_{(S,\alpha)} \geq 0$ $\forall S$ s.t. $|S| \leq t,\ \forall \alpha \in [q]^S$

$X_{(\emptyset,\emptyset)} = 1$

For an LP formulation of MAX k-CSP$_q$, and for a given instance Φ of the problem, we denote by FRAC(Φ) the LP (fractional) optimum, and by OPT(Φ) the integral optimum. For the particular instance Φ, the integrality gap is then defined as FRAC(Φ)/OPT(Φ). The integrality gap of the LP formulation is the supremum of integrality gaps over all instances.

Next we give a sufficient condition for the existence of a solution to the level-t Sherali-Adams LP tightening for a MAX k-CSP$_q$ instance Φ.

Lemma 1. *Consider a family of distributions* $\{\mathcal{D}(S)\}_{S \subseteq [n] : |S| \leq t}$, *where each* $\mathcal{D}(S)$ *is defined over* $[q]^S$. *If for every* $S \subseteq T \subseteq [n]$ *with* $|T| \leq t$, *the distributions* $\mathcal{D}(S), \mathcal{D}(T)$ *are equal on* S, *then*

$$X_{(S,\alpha)} = \Pr_{\mathcal{D}(S)}[\alpha]$$

satisfy the above level-t Sherali-Adams tightening.

Proof. Consider some $S \subseteq [n]$, $|S| < t$, and some $i \notin S$. Note that the distributions $\mathcal{D}(S), \mathcal{D}(S \cup \{i\})$ are equal on S, and therefore we have

$$\sum_{j \in [q]} X_{(S \cup \{i\}, \alpha \circ j)} = \sum_{j \in [q]} \Pr_{\beta \sim \mathcal{D}(S \cup \{i\})}[\beta = \alpha \circ j]$$

$$= \sum_{j \in [q]} \Pr_{\beta \sim \mathcal{D}(S \cup \{i\})}[(\beta(i) = j) \wedge (\beta(S) = \alpha)]$$

$$= \Pr_{\beta \sim \mathcal{D}(S \cup \{i\})}[\beta(S) = \alpha]$$

$$= \Pr_{\beta' \sim \mathcal{D}(S)}[\beta' = \alpha]$$

$$= X_{(S,\alpha)}.$$

The same argument also shows that if $S = \emptyset$, then $X_{(\emptyset, \emptyset)} = 1$. Finally, it is clear that all linear variables are assigned non negative values completing the lemma.

2.4 Pairwise Independence and Approximation Resistant Predicates

We say that a distribution μ over variables $x_1, \ldots, x_k$, is a balanced pairwise independent distribution over $[q]^k$, if we have

$$\forall j \in [q]. \forall i. \Pr_\mu[x_i = j] = \frac{1}{q} \quad \text{and} \quad \forall j_1, j_2 \in [q]. \forall i_1 \neq i_2. \Pr_\mu[(x_{i_1} = j_1) \wedge (x_{i_2} = j_2)] = \frac{1}{q^2}.$$

A predicate P is called approximation resistant if it is hard to approximate the MAX k-CSP$_q(P)$ problem better than using a random assignment. Assuming the Unique Games Conjecture, Austrin and Mossel [3] show that a predicate is approximation resistant if it is possible to define a balanced pairwise independent distribution μ such that P is always 1 on the support of μ.

Definition 3. *A predicate* $P : [q]^k \to \{0, 1\}$ *is called* **promising**, *if there exist a distribution supported over a subset of* $P^{-1}(1)$ *that is pairwise independent and balanced. If* μ *is such a distribution we say that* P *is promising supported by* μ.

3 Towards Defining Consistent Distributions

To construct valid solutions for the Sherali-Adams LP tightening, we need to define distributions over every set S of bounded size as is required by Lemma 1. Since we will deal with promising predicates supported by some distribution μ, in order to satisfy consistency between distributions we will heavily rely on the fact that μ is a balanced pairwise independent distribution.

Consider for simplicity that μ is uniform over $P^{-1}(1)$ (the intuition for the general case is not significantly different). It is instructive to think of $q = 2$ and the predicate P being k-XOR, $k \geq 3$. Observe that the uniform distribution over $P^{-1}(1)$ is pairwise independent and balanced. A first attempt would be to define for every S, the distribution $\mathcal{D}(S)$ as the uniform distribution over all consistent assignments of S. We argue that such distributions are in general problematic. This follows from the fact that satisfying assignments are not always extendible. Indeed, consider two constraints $C_{i_1}, C_{i_2} \in L$ that share a common variable $j \in R$. Set $S_2 = T_{i_1} \cup T_{i_2}$, and $S_1 = S_2 \setminus \{j\}$. Assuming that the support of no other constraint is contained in S_2, we get that distribution $\mathcal{D}(S_1)$ maps any variable in S_1 to $\{0, 1\}$ with probability $1/2$ independently, but some of these assignments are not even extendible to S_2 meaning that $\mathcal{D}(S_2)$ will assign them with probability zero.

Thus, to define $\mathcal{D}(S)$, we cannot simply sample assignments satisfying all constraints in $\mathcal{C}(S)$ with probabilities given by μ. In fact the above example shows that any attempt to blindly assign a set S with a distribution that is supported on all satisfying assignments for S is bound to fail. At the same time it seems hard to reason about a distribution that uses a totally different concept. To overcome this obstacle, we take a two step approach:

1. For a set S we define a superset $\overline{S}$ such that $\overline{S}$ is "global enough" to contain sufficient information, while it also is "local enough" so that $\mathcal{C}(\overline{S})$ is not too large. We require the property of such sets that if we remove $\overline{S}$ and $\mathcal{C}(\overline{S})$, then the remaining graph $G|_{-\overline{S}}$ still has good expansion. We deal with this in Section 3.1.
2. The distribution $\mathcal{D}(S)$ is going to be the uniform distribution over satisfying assignments in $\overline{S}$. In the case that μ is not uniform over $P^{-1}(1)$, we give a natural generalization to the above uniformity. We show how to define distributions, which we denote by $\mathcal{P}_\mu(S)$, such that for $S_1 \subseteq S_2$, the distributions are guaranteed to be consistent if $G|_{-S_1}$ *has good expansion*. This appears in Section 3.2.

We then combine the two techniques and define $\mathcal{D}(S)$ according to $\mathcal{P}_\mu(\overline{S})$. This is done in section 4.

3.1 Finding Advice-Sets

We now give an algorithm below to obtain a superset $\overline{S}$ for a given set S, which we call the advice-set of S. It is inspired by the "expansion correction" procedure in [5].

Algorithm Advice

The input is an (r, e_1) boundary expanding bipartite graph $G = (L, R, E)$, some $e_2 \in (0, e_1)$, and some $S \subseteq R$, $|S| < (e_1 - e_2)r$, with some order $S = \{x_1, \ldots, x_t\}$.

Initially set $\overline{S} \leftarrow \emptyset$ and $\xi \leftarrow r$
For $j = 1, \ldots, |S|$ **do**
    $M_j \leftarrow \emptyset$
    $\overline{S} \leftarrow \overline{S} \cup \{x_j\}$
    If $G|_{-\overline{S}}$ is not (ξ, e_2) boundary expanding **then**
        Find a maximal $M_j \subset L$ in $G|_{-\overline{S}}$, such that $|M_j| \leq \xi$ in $G|_{-\overline{S}}$ and
$|\partial M_j| \leq e_2 |M_j|$
        $\overline{S} \leftarrow \overline{S} \cup \partial M_j$
        $\xi \leftarrow \xi - |M_j|$
Return $\overline{S}$

Theorem 2. *Algorithm Advice, with internal parameters e_1, e_2, r, returns $\overline{S} \subseteq R$ such that (a) $G|_{-\overline{S}}$ is (ξ_S, e_2) boundary expanding, (b) $\xi_S \geq r - \frac{|S|}{e_1 - e_2}$, and (c) $|\overline{S}| \leq \frac{e_1 |S|}{e_1 - e_2}$.*

Proof. Suppose that the loop terminates with $\xi = \xi_S$. Then $\sum_{j=1}^{t} |M_j| = r - \xi_S$. Since G is (r, e_1) boundary expanding, the set $M = \cup_{j=1}^{t} M_j$ has initially at least $e_1(r - \xi_S)$ boundary neighbors in G. During the execution of the while loop, each set M_j has at most $e_2 |M_j|$ boundary neighbors in $G|_{-\overline{S}}$. Therefore, at the end of the procedure M has at most $e_2(r - \xi_S)$ boundary neighbors in $G|_{-S}$. It follows that $|S| + e_2(r - \xi_S) \geq e_1(r - \xi_S)$, which implies (b).

From the bound size of S we know that $\xi_S > 0$. In particular, ξ remains positive throughout the execution of the while loop. Next we identify a loop invariant: $G|_{-\overline{S}}$ is (ξ, e_2) boundary expanding.

Indeed, note that the input graph G is (ξ, e_1) boundary expanding. At step j consider the set $\overline{S} \cup \{x_j\}$, and suppose that $G_{-(\overline{S} \cup \{x_j\})}$ is not (ξ, e_2) boundary expanding. We find maximal M_j, $|M_j| \leq \xi$, such that $|\partial M_j| \leq e_2 |M_j|$. We claim that $G_{-(\overline{S} \cup \{x_j\} \cup \partial M_j)}$ is $(\xi - |M_j|, e_2)$ boundary expanding (recall that since ξ remains positive, $|M_j| < \xi$). Now consider the contrary. Then, there must be $M' \subset L$ such that $|M'| \leq \xi - |M_j|$ and such that $|\partial M'| \leq e_2 |M'|$. Consider then $M_j \cup M'$ and note that $|M_j \cup M'| \leq \xi$. More importantly $|\partial(M_j \cup M')| \leq e_2 |M_j \cup M'|$, and therefore we contradict the maximality of M_j; (a) follows.

Finally note that $\overline{S}$ consists of S union the boundary neighbors of all M_j. From the arguments above, the number of those neighbors does not exceed $e_2(r - \xi_S)$ and hence $|\overline{S}| \leq |S| + e_2(r - \xi_S) \leq |S| + \frac{e_2 |S|}{e_1 - e_2} = \frac{e_1 |S|}{e_1 - e_2}$, which proves (c).

3.2 Defining the Distributions $\mathcal{P}_\mu(S)$

We now define for every set S, a distribution $\mathcal{P}_\mu(S)$ such that for any $\alpha \in [q]^S$, $\Pr_{\mathcal{P}_\mu(S)}[\alpha] > 0$ only if α satisfies all the constraints in $\mathcal{C}(S)$. For a constraint C_i

with set of inputs T_i, defined as $C_i(x_{i_1}, \ldots, x_{i_k}) \equiv P(x_{i_1} + a_{i_1}, \ldots, x_{i_k} + a_{i_k})$, let $\mu_i : [q]^{T_i} \to [0, 1]$ denote the distribution

$$\mu_i(x_{i_1}, \ldots, x_{i_k}) = \mu(x_{i_1} + a_{i_1}, \ldots, x_{i_k} + a_{i_k})$$

so that the support of μ_i is contained in $C_i^{-1}(1)$. We then define the distribution $\mathcal{P}_\mu(S)$ by picking each assignment $\alpha \in [q]^S$ with probability proportional to $\prod_{C_i \in \mathcal{C}(S)} \mu_i(\alpha(T_i))$. Formally,

$$\mathrm{Pr}_{\mathcal{P}_\mu(S)}[\alpha] = \frac{1}{Z_S} \cdot \prod_{C_i \in \mathcal{C}(S)} \mu_i(\alpha(T_i)) \tag{1}$$

where $\alpha(T_i)$ is the restriction of α to T_i and Z_S is a normalization factor given by

$$Z_S = \sum_{\alpha \in [q]^S} \prod_{C_i \in \mathcal{C}(S)} \mu_i(\alpha(T_i)).$$

To understand the distribution, it is easier to think of the special case when μ is just the uniform distribution on $P^{-1}(1)$ (like in the case of MAX k-XOR). Then $\mathcal{P}_\mu(S)$ is simply the uniform distribution on assignments satisfying all the constraints in $\mathcal{C}(S)$. When μ is not uniform, then the probabilities are weighted by the product of the values $\mu_i(\alpha(T_i))$ for all the constraints[2]. However, we still have the property that if $\mathrm{Pr}_{\mathcal{P}_\mu(S)}[\alpha] > 0$, then α satisfies all the constraints in $\mathcal{C}(S)$.

In order for the distribution $\mathcal{P}_\mu(S)$ to be well defined, we need to ensure that $Z_S > 0$. The following lemma shows how to calculate Z_S if G is sufficiently expanding, and simultaneously proves that if $S_1 \subseteq S_2$, and if $G|_{-S_1}$ is sufficiently expanding, then $\mathcal{P}_\mu(S_1)$ is consistent with $\mathcal{P}_\mu(S_2)$ over S_1.

Lemma 2. *Let Φ be a MAX k-CSP(P) instance as above and $S_1 \subseteq S_2$ be two sets of variables such that both G and $G|_{-S_1}$ are $(r, k-2-\delta)$ boundary expanding for some $\delta \in (0, 1)$ and $|\mathcal{C}(S_2)| \leq r$. Then $Z_{S_2} = q^{|S_2|}/q^{k|\mathcal{C}(S_2)|}$, and for any $\alpha_1 \in [q]^{S_1}$*

$$\sum_{\substack{\alpha_2 \in [q]^{S_2} \\ \alpha_2(S_1) = \alpha_1}} \mathrm{Pr}_{\mathcal{P}_\mu(S_2)}[\alpha_2] = \mathrm{Pr}_{\mathcal{P}_\mu(S_1)}[\alpha_1].$$

Proof. Let $\mathcal{C} = \mathcal{C}(S_2) \backslash \mathcal{C}(S_1)$ be given by the set of t many constraints $C_{i_1}, \ldots, C_{i_t}$ with each C_{i_j} being on the set of variables T_{i_j}. Some of these variables may be fixed by α_1. Also, any α_2 consistent with α_1 can be written as $\alpha_1 \circ \alpha$ for some $\alpha \in [q]^{S_2 \backslash S_1}$. Below, we express these probabilities in terms the product of μ on the constraints in $\mathcal{C}(S_2) \setminus \mathcal{C}(S_1)$.

---

[2] Note however that $\mathcal{P}_\mu(S)$ is not a product distribution because different constraints in $\mathcal{C}(S)$ may share variables.

Note that the equations below are still correct even if we haven't shown $Z_{S_2} > 0$ (in that case both sides are 0). In fact, replacing S_1 by $\emptyset$ in the same calculation will give the value of Z_{S_2}.

$$
Z_{S_2} \cdot \sum_{\substack{\alpha_2 \in [q]^{S_2} \\ \alpha_2(S_1) = \alpha_1}} \mathrm{Pr}_{\mathcal{P}_\mu(S_2)}[\alpha_2] = \sum_{\alpha \in [q]^{S_2 \setminus S_1}} \prod_{C_i \in \mathcal{C}(S_2)} \mu_i((\alpha_1 \circ \alpha)(T_i))
$$

$$
= \left(\prod_{C_i \in \mathcal{C}(S_1)} \mu_i(\alpha_1(T_i)) \right) \sum_{\alpha \in [q]^{S_2 \setminus S_1}} \prod_{j=1}^{t} \mu_{i_j}((\alpha_1 \circ \alpha)(T_{i_j}))
$$

$$
= (Z_{S_1} \cdot \mathrm{Pr}_{\mathcal{P}_\mu(S_1)}[\alpha_1]) \sum_{\alpha \in [q]^{S_2 \setminus S_1}} \prod_{j=1}^{t} \mu_{i_j}((\alpha_1 \circ \alpha)(T_{i_j}))
$$

$$
= (Z_{S_1} \cdot \mathrm{Pr}_{\mathcal{P}_\mu(S_1)}[\alpha_1]) \cdot q^{|S_2 \setminus S_1|} \mathop{\mathbb{E}}_{\alpha \in [q]^{S_2 \setminus S_1}} \left[\prod_{j=1}^{t} \mu_{i_j}((\alpha_1 \circ \alpha)(T_{i_j})) \right]
$$

The following claim, whose proof can be found in the Appendix, lets us calculate this expectation conveniently using the expansion of $G|_{-S_1}$.

Claim. Let $\mathcal{C}$ be as above. Then there exists an ordering $C_{i'_1}, \ldots, C_{i'_t}$ of constraints in $\mathcal{C}$ and a partition of $S_2 \setminus S_1$ into sets of variables $F_1, \ldots, F_t$ such that for all j, $F_j \subseteq T_{i'_j}$, $|F_j| \geq k - 2$, and

$$
\forall j \ F_j \cap \left(\cup_{l > j} T_{i'_l} \right) = \emptyset.
$$

Using this decomposition, the expectation above can be split as

$$
\mathop{\mathbb{E}}_{\alpha \in [q]^{S_2 \setminus S_1}} \left[\prod_{j=1}^{t} \mu_{i_j}(\alpha_1 \circ \alpha(T_{i_j})) \right] = \mathop{\mathbb{E}}_{\beta_t \in [q]^{F_t}} \left[\mu_{i'_t} \cdots \mathop{\mathbb{E}}_{\beta_2 \in [q]^{F_2}} \left[\mu_{i'_2} \mathop{\mathbb{E}}_{\beta_1 \in [q]^{F_1}} \left[\mu_{i'_1} \right] \right] \cdots \right]
$$

where the input to each $\mu_{i'_j}$ depends on α_1 and $\beta_j, \ldots, \beta_t$ but not on $\beta_1, \ldots, \beta_{j-1}$.

We now reduce the expression from right to left. Since F_1 contains at least $k - 2$ variables and $\mu_{i'_1}$ is a balanced pairwise independent distribution,

$$
\mathop{\mathbb{E}}_{\beta_1 \in [q]^{F_1}} \left[\mu_{i'_1} \right] = \frac{1}{q^{|F_1|}} \cdot \mathrm{Pr}_\mu[(\alpha_1 \circ \beta_2 \ldots \circ \beta_t)(T_{i'_1} \setminus F_1)] = \frac{1}{q^k}
$$

irrespective of the values assigned by $\alpha_1 \circ \beta_2 \circ \ldots \circ \beta_t$ to the remaining (at most 2) variables in $T_{i'_1} \setminus F_1$. Continuing in this fashion from right to left, we get that

$$
\mathop{\mathbb{E}}_{\alpha \in [q]^{S_2 \setminus S_1}} \left[\prod_{j=1}^{t} \mu_{i_j}((\alpha_1 \circ \alpha)(T_{i_j})) \right] = \left(\frac{1}{q^k} \right)^t = \left(\frac{1}{q^k} \right)^{|\mathcal{C}(S_2) \setminus \mathcal{C}(S_1)|}
$$

Hence, we get that

$$
Z_{S_2} \cdot \sum_{\substack{\alpha_2 \in [q]^{S_2} \\ \alpha_2(S_1) = \alpha_1}} \mathrm{Pr}_{\mathcal{P}_\mu(S_2)}[\alpha_2] = \left(Z_{S_1} \cdot \frac{q^{|S_2 \setminus S_1|}}{q^{k|\mathcal{C}(S_2) \setminus \mathcal{C}(S_1)|}} \right) \mathrm{Pr}_{\mathcal{P}_\mu(S_1)}[\alpha_1]. \quad (2)
$$

Summing over all $\alpha_1 \in [q]^{S_1}$ on both sides gives

$$Z_{S_2} = Z_{S_1} \cdot \frac{q^{|S_2 \setminus S_1|}}{q^{k|\mathcal{C}(S_2) \setminus \mathcal{C}(S_1)|}}.$$

Since we know that G is $(r, k - 2 - \delta)$ boundary expanding, we can replace S_1 by $\emptyset$ in the above equation to obtain $Z_{S_2} = q^{|S_2|}/q^{k|\mathcal{C}(S_2)|}$ as claimed. Also note that since $\mathcal{C}(S_1) \subseteq \mathcal{C}(S_2)$, $Z_{S_2} > 0$ implies $Z_{S_1} > 0$. Hence, using equation (2) we get

$$\sum_{\substack{\alpha_2 \in [q]^{S_2} \\ \alpha_2(S_1) = \alpha_1}} \text{Pr}_{\mathcal{P}_\mu(S_2)}[\alpha_2] = \text{Pr}_{\mathcal{P}_\mu(S_1)}[\alpha_1]$$

which proves the lemma.

4   Constructing the Integrality Gap

We now show how to construct integrality gaps using the ideas in the previous section. For a given promising predicate P, our integrality gap instance will be random instance Φ of the MAX k-CSP$_q(P)$ problem. To generate a random instance with m constraints, for every constraint C_i, we randomly select a k-tuple of distinct variables $T_i = \{x_{i_1}, \ldots, x_{i_k}\}$ and $a_{i_1}, \ldots, a_{i_k} \in [q]$, and put $C_i \equiv P(x_{i_1} + a_{i_1}, \ldots, x_{i_k} + a_{i_k})$. It is well known and used in various works on integrality gaps and proof complexity (e.g. [5], [1], [22] and [21]), that random instances of CSPs are both highly unsatisfiable and highly expanding. We capture the properties we need in the lemma below (for a proof see e.g. [27]).

Lemma 3. *Let $\epsilon, \delta > 0$ and a predicate $P : [q]^k \to \{0,1\}$ be given. Then there exist $\gamma = O(q^k \log q/\epsilon^2)$, $\eta = \Omega((1/\gamma)^{10/\delta})$ and $N \in \mathbb{N}$, such that if $n \geq N$ and Φ is a random instance of MAX k-CSP(P) with $m = \gamma n$ constraints, then with probability $1 - o(1)$*

1. $\text{OPT}(\Phi) \leq \frac{|P^{-1}(1)|}{q^k}(1 + \epsilon) \cdot m$.
2. For any set $\mathcal{C}$ of constraints with $|\mathcal{C}| \leq \eta n$, we have $|\partial(\mathcal{C})| \geq (k - 2 - \delta)|\mathcal{C}|$.

Let Φ be an instance of MAX k-CSP$_q$ on n variables for which G_Φ is $(\eta n, k - 2 - \delta)$ boundary expanding for some $\delta < 1/2$, as in Lemma 3. For such a Φ, we now define the distributions $\mathcal{D}(S)$.

For a set S of size at most $t = \eta \delta n/4k$, let $\overline{S}$ be subset of variables output by the algorithm Advice when run with input S and parameters $r = \eta n, e_1 = (k - 2 - \delta), e_2 = (k - 2 - 2\delta)$ on the graph G_Φ. Theorem 2 shows that

$$|\overline{S}| \leq (k - 2 - \delta)|S|/\delta \leq \eta n/4.$$

We then use (1) to define the distribution $\mathcal{D}(S)$ for sets S of size at most $\delta \eta n/4k$ as

$$\text{Pr}_{\mathcal{D}(S)}[\alpha] = \sum_{\substack{\beta \in [q]^{\overline{S}} \\ \beta(S) = \alpha}} \text{Pr}_{\mathcal{P}_\mu(\overline{S})}[\beta].$$

Using the properties of the distributions $\mathcal{P}_\mu(\overline{S})$, we can now prove that the distributions $\mathcal{D}(S)$ are consistent.

Claim. Let the distributions $\mathcal{D}(S)$ be defined as above. Then for any two sets $S_1 \subseteq S_2 \subseteq [n]$ with $|S_2| \leq t = \eta\delta n/4k$, the distributions $\mathcal{D}(S_1), \mathcal{D}(S_2)$ are equal on S_1.

Proof. The distributions $\mathcal{D}(S_1), \mathcal{D}(S_2)$ are defined according to $\mathcal{P}_\mu(\overline{S}_1)$ and $\mathcal{P}_\mu(\overline{S}_2)$ respectively. To prove the claim, we show that $\mathcal{P}_\mu(\overline{S}_1)$ and $\mathcal{P}_\mu(\overline{S}_2)$ are equal to the distribution $\mathcal{P}_\mu(\overline{S}_1 \cup \overline{S}_2)$ on $\overline{S}_1, \overline{S}_2$ respectively (note that it need not be the case that $\overline{S}_1 \subseteq \overline{S}_2$).

Let $S_3 = \overline{S}_1 \cup \overline{S}_2$. Since $|\overline{S}_1|, |\overline{S}_2| \leq \eta n/4$, we have $|S_3| \leq \eta n/2$ and hence $|\mathcal{C}(S_3)| \leq \eta n/2$. Also, by Theorem 2, we know that both $G|_{-\overline{S}_1}$ and $G|_{-\overline{S}_2}$ are $(2\eta n/3, k-2-2\delta)$ boundary expanding. Thus, using Lemma 2 for the pairs $(\overline{S}_1, S_3)$ and $(\overline{S}_2, S_3)$, we get that

$$
\begin{aligned}
\Pr_{\mathcal{D}(S_1)}[\alpha_1] &= \sum_{\substack{\beta_1 \in [q]^{\overline{S}_1} \\ \beta_1(S_1)=\alpha_1}} \Pr_{\mathcal{P}_\mu(\overline{S}_1)}[\beta_1] \\
&= \sum_{\substack{\beta_3 \in [q]^{S_3} \\ \beta_3(S_1)=\alpha_1}} \Pr_{\mathcal{P}_\mu(S_3)}[\beta_3] \\
&= \sum_{\substack{\beta_2 \in [q]^{\overline{S}_2} \\ \beta_2(S_1)=\alpha_1}} \Pr_{\mathcal{P}_\mu(\overline{S}_2)}[\beta_2] \\
&= \sum_{\substack{\alpha_2 \in [q]^{S_2} \\ \alpha_2(S_1)=\alpha_1}} \Pr_{\mathcal{D}(S_2)}[\alpha_2]
\end{aligned}
$$

which shows that $\mathcal{D}(S_1)$ and $\mathcal{D}(S_2)$ are equal on S_1.

It is now easy to prove the main result.

Theorem 3. *Let $P : [q]^k \to \{0,1\}$ be a promising predicate. Then for every constant $\zeta > 0$, there exist $c = c(q, k, \zeta)$, such that for large enough n, the integrality gap of MAX k-CSP(P) for the tightening obtained by cn levels of the Sherali-Adams hierarchy is at least $\dfrac{q^k}{|P^{-1}(1)|} - \zeta$.*

Proof. We take $\epsilon = \zeta/q^k, \delta = 1/4$ and consider a random instance Φ of MAX k-CSP(P) with $m = \gamma n$ as given by Lemma 3. Thus, $\text{OPT}(\Phi) \leq \frac{|P^{-1}(1)|}{q^k}(1+\epsilon) \cdot m$.

On the other hand, by Claim 4 we can define distributions $\mathcal{D}(S)$ over every set of at most $\delta\eta n/4k$ variables such that for $S_1 \subseteq S_2$, $\mathcal{D}(S_1)$ and $\mathcal{D}(S_2)$ are consistent over S_1. By Lemma 1 this gives a feasible solution to the LP obtained by $\delta\eta n/4k$

levels. Also, by definition of $\mathcal{D}(S)$, we have that $\Pr_{\mathcal{D}(S)}[\alpha] > 0$ only if α satisfies all constraints in $\mathcal{C}(S)$. Hence, the value of $\mathsf{FRAC}(\Phi)$ is given by

$$\sum_{i=1}^{m} \sum_{\alpha \in [q]^{T_i}} C_i(\alpha) X_{(T_i,\alpha)} = \sum_{i=1}^{m} \sum_{\alpha \in [q]^{T_i}} C_i(\alpha) \Pr_{\mathcal{D}(T_i)}[\alpha] = \sum_{i=1}^{m} \sum_{\alpha \in [q]^{T_i}} \Pr_{\mathcal{D}(T_i)}[\alpha] = m.$$

Thus, the integrality gap after $\delta \eta n / 4k$ levels is at least

$$\frac{\mathsf{FRAC}(\Phi)}{\mathsf{OPT}(\Phi)} = \frac{q^k}{|P^{-1}(1)|(1+\epsilon)} \geq \frac{q^k}{|P^{-1}(1)|} - \zeta.$$

References

1. Alekhnovich, M., Arora, S., Tourlakis, I.: Towards strong nonapproximability results in the Lovász-Schrijver hierarchy. In: Proceedings of the 37th Annual ACM Symposium on Theory of Computing, Baltimore, MD, USA, May 22-24, 2005, ACM Press, New York (2005)
2. Arora, S., Bollobás, B., Lovász, L., Tourlakis, I.: Proving integrality gaps without knowing the linear program. Theory of Computing 2(2), 19–51 (2006)
3. Austrin, P., Mossel, E.: Approximation resistant predicates from pairwise independence. In: IEEE Conference on Computational Complexity, pp. 249–258. IEEE Computer Society Press, Los Alamitos (2008)
4. Bateni, M.H., Charikar, M., Guruswami, V.: MaxMin allocation via degree lower-bounded arborescences. In: STOC 2009. ACM Press, New York (2009)
5. Buresh-Oppenheim, J., Galesi, N., Hoory, S., Magen, A., Pitassi, T.: Rank bounds and integrality gaps for cutting planes procedures. Theory of Computing 2(4), 65–90 (2006)
6. Charikar, M., Makarychev, K., Makarychev, Y.: Integrality gaps for Sherali-Adams relaxations. In: STOC 2009. ACM Press, New York (2009)
7. Chlamtac, E.: Approximation algorithms using hierarchies of semidefinite programming relaxations. In: FOCS: IEEE Symposium on Foundations of Computer Science (FOCS), pp. 691–701 (2007)
8. Chlamtac, E., Singh, G.: Improved approximation guarantees through higher levels of SDP hierarchies. In: Goel, A., Jansen, K., Rolim, J.D.P., Rubinfeld, R. (eds.) APPROX and RANDOM 2008. LNCS, vol. 5171, pp. 49–62. Springer, Heidelberg (2008)
9. Engebretsen, L., Holmerin, J.: More efficient queries in pCPs for NP and improved approximation hardness of maximum CSP. In: Diekert, V., Durand, B. (eds.) STACS 2005. LNCS, vol. 3404, pp. 194–205. Springer, Heidelberg (2005)
10. Feige, U., Krauthgamer, R.: The probable value of the Lovász-Schrijver relaxations for maximum independent set. SICOMP: SIAM Journal on Computing 32(2), 345–370 (2003)
11. de la Vega, W.F., Kenyon-Mathieu, C.: Linear programming relaxations of maxcut. In: SODA 2007: Proceedings of the eighteenth annual ACM-SIAM symposium on Discrete algorithms, Philadelphia, PA, USA, pp. 53–61. Society for Industrial and Applied Mathematics (2007)

12. Georgiou, K., Magen, A., Pitassi, T., Tourlakis, I.: Integrality gaps of $2 - o(1)$ for Vertex Cover SDPs in the Lovász-Schrijver hierarchy. In: Proceedings of the 47th IEEE Symposium on Foundations of Computer Science, pp. 702–712 (2007)
13. Guruswami, V., Raghavendra, P.: Constraint satisfaction over a non-boolean domain: Approximation algorithms and unique-games hardness. In: Goel, A., Jansen, K., Rolim, J.D.P., Rubinfeld, R. (eds.) APPROX and RANDOM 2008. LNCS, vol. 5171, pp. 77–90. Springer, Heidelberg (2008)
14. Lasserre, J.B.: An explicit exact SDP relaxation for nonlinear 0-1 programs. In: Aardal, K., Gerards, B. (eds.) IPCO 2001. LNCS, vol. 2081, pp. 293–303. Springer, Heidelberg (2001)
15. Laurent, M.: A comparison of the Sherali-Adams, Lovász-Schrijver, and Lasserre relaxations for 0-1 programming. Math. Oper. Res. 28(3), 470–496 (2003)
16. Lovász, L., Schrijver, A.: Cones of matrices and set-functions and 0-1 optimization. SIAM Journal on Optimization 1(2), 166–190 (1991)
17. Magen, A., Moharrami, M.: Sherali-Adams based polynomial approximation schemes for NP-hard problems on planar and minor-free graphs (manuscript) (2008)
18. Mathieu, C., Sinclair, A.: Sherali-Adams relaxations of the matching polytope. In: STOC 2009. ACM Press, New York (2009)
19. Samorodnitsky, A., Trevisan, L.: A PCP characterization of NP with optimal amortized query complexity. In: Proceedings of the 32nd Annual ACM Symposium on Theory of Computing, STOC 2000, Portland, Oregon,, May 21-23, 2000, pp. 191–199. ACM Press, New York (2000)
20. Samorodnitsky, A., Trevisan, L.: Gowers uniformity, influence of variables, and PCPs. In: STOC 2006, pp. 11–20 (2006)
21. Schoenebeck, G.: Linear level lasserre lower bounds for certain k-CSPs. In: FOCS, pp. 593–602. IEEE Computer Society Press, Los Alamitos (2008)
22. Schoenebeck, G., Trevisan, L., Tulsiani, M.: A linear round lower bound for Lovász-Schrijver SDP relaxations of vertex cover. In: IEEE Conference on Computational Complexity, pp. 205–216. IEEE Computer Society Press, Los Alamitos (2007)
23. Schoenebeck, G., Trevisan, L., Tulsiani, M.: Tight integrality gaps for Lovász-Schrijver LP relaxations of vertex cover and max cut. In: Proceedings of the 39th Annual ACM Symposium on Theory of Computing, San Diego, California, USA, June 11-13, 2007. ACM Press, New York (2007)
24. Sherali, H.D., Adams, W.P.: A hierarchy of relaxations between the continuous and convex hull representations for zero-one programming problems. SIAM J. Discrete Math. 3(3), 411–430 (1990)
25. Tourlakis, I.: Towards optimal integrality gaps for hypergraph vertex cover in the Lovász-Schrijver hierarchy. In: Chekuri, C., Jansen, K., Rolim, J.D.P., Trevisan, L. (eds.) APPROX 2005 and RANDOM 2005. LNCS, vol. 3624, pp. 233–244. Springer, Heidelberg (2005)
26. Tourlakis, I.: New lower bounds for vertex cover in the Lovász-Schrijver hierarchy. In: Proceedings of the 21st IEEE Conference on Computational Complexity, pp. 170–182. IEEE Computer Society Press, Los Alamitos (2006)
27. Tulsiani, M.: CSP gaps and reductions in the Lasserre hierarchy. In: STOC 2009. ACM Press, New York (2009)

Appendix

Proof. (of Claim 3.2) We build the sets F_j inductively using the fact that $G|_{-S_1}$ is $(r, k-2-\delta)$ boundary expanding.

Start with the set of constraints $\mathcal{C}_1 = \mathcal{C}$. Since $|\mathcal{C}_1| = |\mathcal{C}(S_2) \setminus \mathcal{C}(S_1)| \leq r$, this gives that $|\partial(\mathcal{C}_1) \setminus S_1| \geq (k-2-\delta)|\mathcal{C}_1|$. Hence, there exists $C_{i_j} \in \mathcal{C}_1$ such that $|T_{i_j} \cap (\partial(\mathcal{C}_1) \setminus S_1)| \geq k-2$. Let $T_{i_j} \cap (\partial(\mathcal{C}_1) \setminus S_1) = F_1$ and $i'_1 = i_j$. We then take $\mathcal{C}_2 = \mathcal{C}_1 \setminus \{C_{i'_1}\}$ and continue in the same way.

Since at every step, we have $F_j \subseteq \partial(\mathcal{C}_j) \setminus S_1$, and for all $l > j$ $\mathcal{C}_l \subseteq \mathcal{C}_j$, F_j shares no variables with $\Gamma(\mathcal{C}_l)$ for $l > j$. Hence, we get $F_j \cap \left(\cup_{l>j} T_{i'_l} \right) = \emptyset$ as claimed.

An Approximation Scheme for Terrain Guarding

Matt Gibson, Gaurav Kanade, Erik Krohn, and Kasturi Varadarajan*

Department of Computer Science
University of Iowa
Iowa City, IA 52242-1419, USA
{mrgibson,gkanade,eakrohn,kvaradar}@cs.uiowa.edu

Abstract. We obtain a polynomial time approximation scheme for the terrain guarding problem improving upon several recent constant factor approximations. Our algorithm is a local search algorithm inspired by the recent results of Chan and Har-Peled [2] and Mustafa and Ray [15]. Our key contribution is to show the existence of a planar graph that appropriately relates the local and global optimum.

1 Introduction

A *1.5D terrain* is a polygonal chain in the plane that is x-monotone, that is, any vertical line intersects the chain at most once. A terrain T consists of a set of m vertices $\{v_1, v_2, \ldots, v_m\}$. The vertices are ordered in increasing order with respect to their x-coordinates. There is an edge connecting v_i with v_{i+1} for all $i = 1, 2, \ldots, m - 1$. For any two points $a, b \in T$, we say that a sees b if the line segment $\overline{ab}$ lies entirely above or on the terrain.

In this paper, we consider the *discrete terrain guarding problem* in which we are given a terrain T and finite sets $X, G \subseteq T$. For a set $G' \subseteq G$, we say that G' covers/sees/guards X if every point in X can be seen by at least one point in G'. The goal of the problem is to find a minimum cardinality subset of G that covers X.

The motivation for guarding terrains comes from placing street lights or security sensors along roads, as well as constructing line-of-sight networks for radio broadcasting and other communication networks [1].

A closely related problem is the *art gallery problem*. Again the goal is to find a minimum cardinality guarding set, but in this setting we must guard a simple polygon. The basic version of this problem, *vertex guarding*, requires guards to be placed at the vertices of the polygon. Another version, *point guarding*, allows guards to be placed anywhere inside the polygon.

The art gallery problem was shown to be NP-complete by Lee and Lin [14] and was later shown to be APX-hard by Eidenbenz [7]. This means that there is an $\epsilon > 0$ such that no polynomial time algorithm can compute a guarding set whose cardinality is within a $(1 + \epsilon)$ factor of the cardinality of an optimal guarding set, unless P = NP. Ghosh gives an $O(\log n)$-approximation algorithm

---

* Partially supported by NSF CAREER award CCR 0237431.

I. Dinur et al. (Eds.): APPROX and RANDOM 2009, LNCS 5687, pp. 140–148, 2009.
© Springer-Verlag Berlin Heidelberg 2009

for vertex guarding an n-vertex simple polygon [11]. The point guarding problem seems to be much more difficult as not as much is known about it [5]. A constant factor approximation is given by Nilsson for the special case of the problem when the polygon is x-monotone [16]. Based on his result, Nilsson gives an $O(OPT^2)$-approximation algorithm for rectilinear polygons.

Previous Work on Terrain Guarding. Chen et al. [3] claimed that terrain guarding is NP-hard, but the proof was never completed formally [12]. Most of the past research has gone into developing approximation algorithms. The first constant factor approximation was a combinatorial algorithm given by Ben-Moshe et al. [1]. Clarkson and Varadarajan [4] also give a constant factor approximation based on rounding a linear programming relaxation. King gave a simple combinatorial 4-approximation which was later determined to actually be a 5-approximation [12]. Recently, Elbassioni et al. [8] gave a 4-approximation that also works for the weighted case.

All of the approximation algorithms use the following "order claim":

Claim. Let a, b, c, d be four points on the terrain in increasing order according to x-coordinate. If a sees c and b sees d then a sees d.

Natural attempts at constructing NP-hardness reductions for the terrain guarding problem do not work because of the order claim. Recently Krohn and King [13] were able to get around the order claim and prove NP-hardness.

Our Contribution. We give a polynomial time algorithm that returns a guard cover whose cardinality is at most $(1+\epsilon) \cdot OPT$ for any $\epsilon > 0$. Here, OPT denotes the cardinality of an optimal guard cover. Thus we obtain the first PTAS for the problem improving upon several recent constant factor approximations. Given the hardness result [13], this settles the computational complexity of the problem.

The inspiration for our work comes from the recent results of Chan and Har-Peled [2] and Mustafa and Ray [15]. Chan and Har-Peled show that a local search algorithm actually yields a PTAS for the maximum independent set problem given a collection of disks. Unlike a previous PTAS for the problem [9], their analysis does not use packing arguments and thus also applies to "pseudo-disks". Mustafa and Ray consider several geometric hitting set and set cover problems and describe local search algorithms that yield PTASs. For instance, in a rather surprising result they obtain a PTAS for the problem of covering a set of points by the smallest number of a given set of disks. Both papers use separator theorems for planar graphs. In particular, they show that there exists a planar graph that relates the locally optimal solution returned by the local search and the global optimal solution. The separator theorem is then used to show that the locally optimal solution is not too much worse than the global optimum.

Our PTAS for the terrain guarding problem is also based on local search. Our key contribution is to show the existence of an appropriate planar graph even for the terrain guarding context. Having shown this, the rest of the analysis is very similar to that of Mustafa and Ray [15].

2   Guarding Terrains via Local Search

Recall that our input is a polygonal terrain, a set X of points on the terrain that need to be guarded, a set G of possible guard locations, and a parameter $0 < \epsilon < 1$. For purposes of exposition, we will initially assume that $X \cap G = \emptyset$. We later show how this assumption can be removed. We describe a polynomial time algorithm that returns a subset $Q \subseteq G$ that sees X, so that $|Q|$ is at most a factor $(1 + \epsilon)$ times the size of the smallest subset of G that sees X. Let n denote the input size – the number of vertices in the terrain, plus $|G|$, plus $|X|$.

We say that a subset of G that sees X is b-locally optimal if one cannot obtain a smaller set of guards that sees X by deleting at most b guards from it and inserting at most $b - 1$ guards.

Our algorithm simply returns a b-locally optimal solution for $b = \frac{\alpha}{\epsilon^2}$, where α is a suitably large constant, by performing local search. We start with some arbitrary $Q \subseteq G$ that covers X. For every subset $S \subseteq Q$ of size at most b, we see if there exists a subset $T \subseteq G \setminus Q$ of size at most $|S| - 1$ such that $(Q \setminus S) \cup T$ guards X. If so, we set $Q \leftarrow (Q \setminus S) \cup T$. Every such exchange decreases the size of Q by at least one, and as such can happen at most n times. Since there are $\binom{n}{b}$ subsets S to consider, the running time is bounded by $n^{O(b)}$.

2.1   Approximation Analysis

Let R' denote the optimal cover for X, and B' the set of guards output by our local search algorithm on termination. We show that $|B' \setminus R'| \leq (1+\epsilon)|R' \setminus B'|$, and thus $|B'| \leq (1+\epsilon)|R'|$. Let $R \equiv R' \setminus B'$, $B \equiv B' \setminus R'$, and abusing notation, let X denote the set after removing all points seen by $R' \cap B'$. So now both R and B cover X and we wish to show that $|B| \leq (1+\epsilon)|R|$. We will refer to points in B as blue points and points in R as red points.

The following lemma is our main contribution; it shows that the *locality condition* of Mustafa and Ray [15] is satisfied.

Lemma 1. *There exists a planar graph $\mathcal{G} = (V \equiv R \cup B, E)$ with the property that for each $x \in X$, there is an edge (r, b) in $\mathcal{G}$ between guards $r \in R$ and $b \in B$ that both see x.*

Before giving the proof, we show how the lemma implies that $|B| \leq (1 + \epsilon)|R|$; this is similar to [2,15]. We need the following partition theorem on planar graphs due to Frederickson [10]. For $U \subseteq V$, let $\Gamma(U)$ denote the set of neighbors in $\mathcal{G}$ of vertices in U with U excluded. Let $\mu = |V|$.

Lemma 2. *For any parameter $1 \leq r \leq \mu$, we can find a set $S \subseteq V$ of size at most $c_1\mu/\sqrt{r}$ and a partition of $V \setminus S$ into μ/r sets $V_1, V_2, \ldots, V_{\mu/r}$, satisfying (i) $|V_i| \leq c_2 r$, (ii) $|\Gamma(V_i)| \leq c_3\sqrt{r}$, and (iii) $(V_i \cup \Gamma(V_i)) \cap V_j = \emptyset$ for $i \neq j$. Here, c_1, c_2, and c_3 are absolute positive constants.*

Let us apply the lemma with $r \equiv b/(c_2+c_3)$. We have $|V_i \cup \Gamma(V_i)| \leq c_2 r + c_3\sqrt{r} \leq b$. Thus, letting $R_i = R \cap V_i$ and $B_i = B \cap V_i$, we must have $|B_i| \leq |R_i| + |\Gamma(V_i)|$.

For otherwise, the local search can replace B_i by $R_i \cup \Gamma(V_i)$ and obtain a smaller set that still covers X (Lemma 1), a contradiction. Thus

$$|B| \le |S| + \sum_i |B_i| \le |S| + \sum_i |R_i| + \sum_i |\Gamma(V_i)| \le |R| + c\frac{\mu}{\sqrt{r}}$$

$$\le |R| + c'\frac{|R| + |B|}{\sqrt{b}},$$

where c and c' are positive constants. With b a large enough constant times $1/\epsilon^2$, this implies that $|B| \le (1 + \epsilon)|R|$.

2.2 Proof of Lemma 1

We begin with some notation. For points a and b on the terrain, we say $a \le b$ to mean that the x-coordinate $a.x$ of a is at most $b.x$. We use the notation of intervals that this implies – for instance, $[a, b]$ denotes all points c on the terrain so that $a \le c \le b$.

We now prove Lemma 1. Let us first construct the planar graph $\mathcal{G}$. For each $x \in X$, let $\lambda(x)$ denote the leftmost point that sees x among points in $R \cup B$ to the left of x, assuming such a point does exist. Similarly, let $\rho(x)$ denote the rightmost point that sees x among points in $R \cup B$ to the right of x, assuming such a point does exist. Note that at least one of $\lambda(x)$ or $\rho(x)$ does exist.

Let A_1 denote the set of segments $\overline{\lambda(x)x}$, for $x \in X$. Because of the order claim, these segments do not cross. For each $v \in R \cup B$, shoot a vertical ray up from v; if this ray hits some segment in A_1, let $\overline{\lambda(y)y}$ denote the first such segment hit; we add the edge $(v, \lambda(y))$ to a set E_1 if v and $\lambda(y)$ are of opposite colors.

Now, the edges in $A_1 \cup E_1$ can be embedded above the terrain in a non-crossing way. To see this, let A_1 be embedded as the original straight line segments. To embed an edge of the form $(v, \lambda(y)) \in E_1$ as above, we travel straight up from v till we hit $\overline{\lambda(y)y}$, and then slide along the segment $\overline{\lambda(y)y}$ to reach $\lambda(y)$. See Figure 1. A more formal argument that $A_1 \cup E_1$ can be so embedded is given in the appendix.

Let A_2 denote the set of segments $\overline{x\rho(x)}$, for $x \in X$. Again, these segments do not cross. For each $v \in R \cup B$, shoot a vertical ray up from v; if this ray hits some segment in A_2, let $\overline{y\rho(y)}$ denote the first such segment hit; we add the edge $(v, \rho(y))$ to a set E_2 if v and $\rho(y)$ are of opposite colors.

The edges in $A_2 \cup E_2$ can also be embedded above the terrain in a non-crossing way. We "flip" the embedding of $A_1 \cup E_1$ to obtain a non-crossing embedding below the terrain; see Figure 2. This gives us a planar embedding of $A_1 \cup E_1 \cup A_2 \cup E_2$.

Finally, for each $x \in X$, we add the edge $(\lambda(x), \rho(x))$ to a set E_3 if $\lambda(x)$ and $\rho(x)$ are of opposite colors. Our graph $\mathcal{G}$ consists of the edge set $E_1 \cup E_2 \cup E_3$. This is a planar graph; just embed E_1 and E_2 as above, and for each $(\lambda(x), \rho(x)) \in E_3$, embed it using the embedding of the segments $\overline{\lambda(x)x}$ and $\overline{x\rho(x)}$.

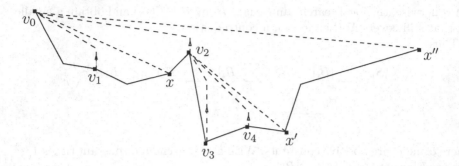

Fig. 1. The embedding of $A_1 \cup E_1$, with $X = \{x, x', x''\}$, and $R \cup B = \{v_0, v_1, v_2, v_3, v_4\}$. Segments in A_1 are shown in dashed lines, and the edges in E_1 are embedded as dashed curves with arrows. Note that $v_0 = \lambda(x) = \lambda(x'')$, and $v_2 = \lambda(x')$.

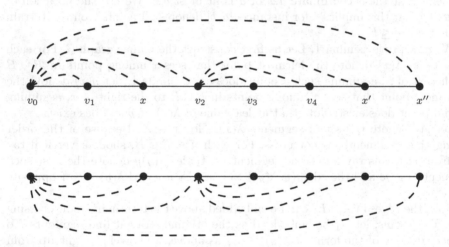

Fig. 2. A combinatorial embedding of $A_1 \cup E_1$ from Figure 1, and flipping it so that $A_1 \cup E_1$ is now embedded below the terrain. Note that only the edges in $A_1 \cup E_1$ are being flipped to make room for $A_2 \cup E_2$; the vertex set $R \cup B \cup X$ retains its embedding.

Now we need to show that for each $x \in X$, there are points $r \in R$ and $b \in B$ that see x, and $(r, b) \in E_1 \cup E_2 \cup E_3$. Fix an $x \in X$. If $\lambda(x)$ and $\rho(x)$ are of opposite colors, then $(\lambda(x), \rho(x)) \in E_3$, and we are done. Otherwise, it must be the case that there are red and blue points to the left of x that see x, or that there are red and blue points to the right of x that see x.

Let us assume that the first case holds (there are red and blue points to the left of x that see x), and that $\lambda(x)$ is red. The other situations are symmetric. Let b be the leftmost blue point that sees x; it must be that $b \in (\lambda(x), x)$. Thus the ray shot up from b hits $\overline{\lambda(x)x}$; let $\overline{\lambda(y)y}$ be the first segment in A_1 that it hits. See Figure 3. Because segments in A_1 don't cross, it must be that $\lambda(y) \in [\lambda(x), b)$

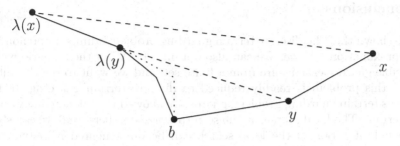

Fig. 3. Here, $\lambda(y)$ sees y and b sees x. By the order claim, $\lambda(y)$ sees x.

and $y \in (b, x)$. The order claim (applied to $\lambda(y)$, b, y, and x) implies that $\lambda(y)$ sees x. Now $\lambda(y)$ cannot be blue, otherwise b is not the leftmost blue point that sees x. Thus $\lambda(y) \in R$, $b \in B$, both $\lambda(y)$ and b see x and $(b, \lambda(y)) \in E_1$. This completes the proof.

2.3   Relaxing the Disjointness Assumption

For ease of exposition, we have so far assumed that the set of possible guard locations has an empty intersection with the set of points to be guarded. We now relax that assumption. The algorithm remains unchanged, and we indicate how the analysis is modified. For each $x \in X$, the point $\lambda(x)$ (resp. $\rho(x)$) denotes the leftmost (resp. rightmost) point that sees x among points in $R \cup B$ *strictly* to the left (resp. right) of x, assuming such a point does exist. With this understanding, the construction of the planar graph $\mathcal{G}$ proceeds with the sets A_1, E_1, A_2, and E_2 defined exactly as above.

There is a change in the construction of E_3. For each $x \in X$ that is not in $R \cup B$, we proceed as before and add the edge $(\lambda(x), \rho(x))$ to E_3 if $\lambda(x)$ and $\rho(x)$ are of opposite colors. For $x \in X$ that is also in $R \cup B$, we add the edge $(\lambda(x), x)$ to E_3 if $\lambda(x)$ exists and $\lambda(x)$ and x are of opposite colors; we also add the edge $(x, \rho(x))$ to E_3 if $\rho(x)$ exists and $\rho(x)$ and x are of opposite colors.

This completes the construction of the graph, which is readily seen to be planar. To show Lemma 1, we need to argue that for each $x \in X$, there are points $r \in R$ and $b \in B$ that see x, and $(r, b) \in E_1 \cup E_2 \cup E_3$. If $x \notin R \cup B$ then the argument is exactly as before. Without loss of generality, assume that $x \in R$. Since B sees X, there must be a point in B that sees x. Assume that such a blue point lies to the left of x. The other case is symmetric. In this case, it must be that $\lambda(x)$ exists. If $\lambda(x) \in B$, then we are done since we added $(\lambda(x), x)$ to E_3. Therefore assume that $\lambda(x) \in R$. Let b be the leftmost blue point that sees x. Now we can use the reasoning in the last paragraph of the previous section to show that there is an edge $(b, u) \in E_1$ such that both b and u see x, b is blue, and u is red.

3   Conclusions

We have shown that the discrete terrain guarding problem admits a polynomial-time approximation scheme. We can also obtain a PTAS in the scenario where the possible guard locations are from a finite set, and we want to see the entire terrain – this problem is readily reduced to discrete terrain guarding. In the continuous terrain guarding problem, guards are allowed to be located anywhere on the terrain. The local search can be seen to work even here, and we can show that a single iteration of the local search can be implemented in polynomial time along the lines of Section 4 of [6]. There is one issue that remains, however, and this is to bound the number of bits needed to represent the guards maintained by the local search. We are currently investigating how this can be handled.

Acknowledgements. We thank the anonymous reviewers for useful feedback.

References

1. Ben-Moshe, B., Katz, M.J., Mitchell, J.S.B.: A constant-factor approximation algorithm for optimal terrain guarding. In: SODA, pp. 515–524 (2005)
2. Chan, T., Har-Peled, S.: Approximation algorithms for maximum independent set of pseudo-disks. In: Symposium on Computational Geometry (to appear, 2009)
3. Chen, D.Z., Estivill-Castro, V., Urrutia, J.: Optimal guarding of polygons and monotone chains (extended abstract) (1996)
4. Clarkson, K.L., Varadarajan, K.: Improved approximation algorithms for geometric set cover. In: SCG 2005: Proceedings of the twenty-first annual symposium on Computational geometry, pp. 135–141. ACM Press, New York (2005)
5. Deshpande, A., Kim, T., Demaine, E.D., Sarma, S.E.: A pseudopolynomial time $o(\log^2 n)$-approximation algorithm for art gallery problems. In: Dehne, F., Sack, J.-R., Zeh, N. (eds.) WADS 2007. LNCS, vol. 4619, pp. 163–174. Springer, Heidelberg (2007)
6. Efrat, A., Har-Peled, S.: Guarding galleries and terrains. Information Processing Letters 100(6), 238–245 (2006)
7. Eidenbenz, S.: Inapproximability results for guarding polygons without holes. LNCS, pp. 427–436. Springer, Heidelberg (1998)
8. Elbassioni, K., Krohn, E., Matijevic, D., Mestre, J., Severdija, D.: Improved approximations for guarding 1.5-dimensional terrains. In: Albers, S., Marion, J.-Y. (eds.) 26th International Symposium on Theoretical Aspects of Computer Science (STACS 2009), Dagstuhl, Germany, Schloss Dagstuhl - Leibniz-Zentrum fuer Informatik, Germany (2009)
9. Erlebach, T., Jansen, K., Seidel, E.: Polynomial-time approximation schemes for geometric intersection graphs. SIAM Journal on Computing 34(6), 1302–1323 (2005)
10. Frederickson, G.N.: Fast algorithms for shortest paths in planar graphs, with applications. SIAM J. Comput. 16(6), 1004–1022 (1987)
11. Ghosh, S.: Approximation algorithms for art gallery problems. In: Proc. Canadian Information Processing Society Congress (1987)

12. King, J.: A 4-approximation algorithm for guarding 1.5-dimensional terrains. In: Correa, J.R., Hevia, A., Kiwi, M. (eds.) LATIN 2006. LNCS, vol. 3887, pp. 629–640. Springer, Heidelberg (2006)
13. Krohn, E., King, J.: The complexity of guarding terrains (manuscript, 2009)
14. Lee, D., Lin, A.: Computational complexity of art gallery problems. IEEE Transactions on Information Theory 32(2), 276–282 (1986)
15. Mustafa, N.H., Ray, S.: Ptas for geometric hitting set problems via local search. In: Symposium on Computational Geometry (to Appear, 2009)
16. Nilsson, B.J.: Approximate guarding of monotone and rectilinear polygons. In: Caires, L., Italiano, G.F., Monteiro, L., Palamidessi, C., Yung, M. (eds.) ICALP 2005. LNCS, vol. 3580, pp. 1362–1373. Springer, Heidelberg (2005)

A   Appendix

We present here a more formal argument for a piece of Lemma 1 that shows that the graph $(X \cup R \cup B, A_1 \cup E_1)$ has a planar embedding with $X \cup R \cup B$ on a horizontal line and $A_1 \cup E_1$ drawn above the line. We now think of A_1 as a set of combinatorial edges rather than as line segments. Let us say that two edges in $A_1 \cup E_1$ *cross* if all four end points are distinct, and can be ordered as $a < b < c < d$ with the edges being (a, c) and (b, d). Notice that crossing in this sense is determined entirely by the ordering of the endpoints, and makes no reference to a drawing of the edges. We first argue that no two edges in $A_1 \cup E_1$ cross.

1. Let $(\lambda(x), x)$ and $(\lambda(y), y)$ be any two edges in A_1. They do not cross, for if $\lambda(x) < \lambda(y) < x < y$, then by the order claim, y sees $\lambda(x)$, contradicting the definition of $\lambda(y)$.

2. We argue that an edge in A_1 and an edge in E_1 do not cross. Let us recall how edges in E_1 are defined. Let us say that an edge $(\lambda(x), x)$ is *above* $v \in R \cup B$ if $\lambda(x) < v < x$. We look at all edges in A_1 that are above v, and if this set is non-empty we find the "innermost" such edge $(\lambda(y), y)$. We add the edge $(\lambda(y), v)$ to E_1 if $\lambda(y)$ and v are of opposite colors. Let us say that $(\lambda(y), y) \in A_1$ *defines* $(\lambda(y), v)$ in this case.

   Now suppose $(\lambda(y), v)$ crosses some $(\lambda(x), x) \in A_1$. Now if $\lambda(x) < \lambda(y) < x < v$, then $(\lambda(x), x)$ and $(\lambda(y), y)$ cross, a contradiction. On the other hand, if $\lambda(y) < \lambda(x) < v < x$, there are two cases: either $x \leq y$, in which case the contradiction is that $(\lambda(y), y)$ is not the innermost edge in A_1 that is above v; or $x > y$, in which case the contradiction is that $(\lambda(y), y)$ and $(\lambda(x), x)$, which are both edges in A_1, cross.

3. We now show that no two edges in E_1 cross. Consider two such edges $(\lambda(y_1), v_1)$ and $(\lambda(y_2), v_2)$, defined by $(\lambda(y_1), y_1)$ and $(\lambda(y_2), y_2)$ respectively. These edges do not cross, for if $\lambda(y_1) < \lambda(y_2) < v_1 < v_2$, then $(\lambda(y_1), v_1)$ crosses $(\lambda(y_2), y_2)$, which contradicts the fact that an edge in E_1 and an edge in A_1 do not cross.

Thus, no two edges in $A_1 \cup E_1$ cross. From this, it follows that the required embedding exists. For instance, embed the vertices $X \cup B \cup R$ as distinct points in the correct order on the lower half of the unit circle, and draw the edges in $A_1 \cup E_1$ as straight line segments. Using convexity arguments, this can be seen to be a planar embedding. Now we "bend" the lower half of the unit circle into a horizontal segment, allowing the drawing of edges in $A_1 \cup E_1$ to now become curved.

Scheduling with Outliers

Anupam Gupta[1,*], Ravishankar Krishnaswamy[1,*],
Amit Kumar[2,**], and Danny Segev[3,*]

[1] Department of Computer Science, Carnegie Mellon University
[2] Department of Computer Science and Engineering, Indian Institute of Technology
New Delhi, India, 110016
[3] Sloan School of Management, Massachusetts Institute of Technology

Abstract. In classical scheduling problems, we are given jobs and machines, and have to schedule all the jobs to minimize some objective function. What if each job has a specified profit, and we are no longer required to process all jobs? Instead, we can schedule any subset of jobs whose total profit is at least a (hard) target profit requirement, while still trying to approximately minimize the objective function.

We refer to this class of problems as *scheduling with outliers*. This model was initiated by Charikar and Khuller (SODA '06) for minimum max-response time in broadcast scheduling. In this paper, we consider three other well-studied scheduling objectives: the generalized assignment problem, average weighted completion time, and average flow time, for which LP-based approximation algorithms are provided. Our main results are:

- For the *minimum average flow time* problem on identical machines, we give an LP-based logarithmic approximation algorithm for the unit profits case, and complement this result by presenting a matching integrality gap.
- For the *average weighted completion time* problem on unrelated machines, we give a constant-factor approximation. The algorithm is based on randomized rounding of the time-indexed LP relaxation strengthened by knapsack-cover inequalities.
- For the *generalized assignment problem* with outliers, we outline a simple reduction to GAP without outliers to obtain an algorithm whose makespan is within 3 times the optimum makespan, and whose cost is at most $(1 + \epsilon)$ times the optimal cost.

1 Introduction

In classical scheduling problems, we are given jobs and machines, and have to schedule all jobs to minimize some objective function. *What if we are given a (hard) profit constraint, and merely want to schedule a "profitable" subset of jobs?*

---

* Supported in part by NSF awards CCF-0448095 and CCF-0729022, and an Alfred P. Sloan Fellowship.
** Work partly done at MPI, Saarbrücken, Germany.

I. Dinur et al. (Eds.): APPROX and RANDOM 2009, LNCS 5687, pp. 149–162, 2009.

In this paper, we consider three widely studied scheduling objectives—makespan, weighted average completion time, and average flow-time—and give approximation algorithms for these objectives in the model of scheduling with outliers.

Formally, the *scheduling with outliers* model is defined as follows: given an instance of some classical scheduling problem, imagine each job j also comes with a certain *profit* π_j. Given a target profit Π, the goal is now to pick a subset of jobs S whose total profit $\sum_{j \in S} \pi_j$ is at least Π, and to schedule them to minimize the underlying objective function. (Equivalently, we could define the "budget" $B = \sum_j \pi_j - \Pi$, and discard a subset of "outlier" jobs whose total profit is at most B.) Note that this model introduces two different sources of computational difficulty: on one hand, the task of choosing a set of jobs to achieve the profit threshold captures the knapsack problem; on the other hand, the underlying scheduling problem may itself be an intractable problem.

The goal of picking some subset of jobs to process as efficiently as possible, so that we attain a minimum level of profit or "happiness", is a natural one. In fact, various problems of scheduling with job rejections have been studied previously: a common approach, studied by Bartal et al. [3], has been to study "prize-collecting" scheduling problems (see, e.g., [9,2,10,17]), where we attempt to minimize the scheduling objective *plus the total profit of unscheduled jobs*. One drawback of this prize-collecting approach is that we lose fine-grained control on the individual quantities—the scheduling cost, and the lost profit—since we naïvely sum up these two essentially incomparable quantities. In fact, this makes our model (with a hard target constraint) interesting also from a technical standpoint: while we can reduce the prize-collecting problem to the target profit problem by guessing the lost profit in the optimal prize-collecting solution, reductions in the opposite direction are known only for a handful of problems with very restrictive structure (see Section 1.2 for a discussion).

To the best of our knowledge, the model we investigate was introduced by Charikar and Khuller [5], who considered the problem of minimizing the maximum response time in the context of broadcast scheduling; one of our results is to resolve an open problem from their paper. Scheduling problems with outliers were also implicitly raised in the context of model-based optimization with budgeted probes: Guha and Munagala [15] gave an LP-based algorithm for completion-time scheduling with outliers which violated budgets by a constant factor—we resolve an open problem from their paper by avoiding any violation of the budgets.

1.1 Our Results

GAP and Makespan. As a warm-up, we study the Generalized Assignment Problem, a generalization of makespan minimization on unrelated machines, in Section 2. For this problem, we give a simple reduction to the non-outlier version to obtain a solution approximating the makespan and cost by factors of 3 and $(1 + \epsilon)$, respectively. Recall that the best known non-outlier guarantee is a 2-approximation [27] without violating the cost; however, it is easy to show that, in the presence of outliers, the $(1 + \epsilon)$ loss in cost is unavoidable unless $P = NP$.

Average Completion Time. We then consider in Section 3 the problem of minimizing the sum of weighted completion times on unrelated machines with release dates, and propose a randomized $O(1)$-approximation algorithm. Note that the best non-outlier upper bound for $R|r_j| \sum_j w_j C_j$ is a 2-approximation due to Skutella [28]; this problem is also known to be APX-hard [20]. Our approach is based on approximately solving the time-indexed LP relaxation of Schulz and Skutella [25], strengthened with *knapsack-cover* inequalities, followed by randomized rounding. We improve on this result to obtain an FPTAS for *unweighted* sum of completion times on a constant number of machines.

Average Flow Time. This is the technical heart of the paper, where the problem is to minimize the average (preemptive) flow time on identical machines, $P|r_j, pmtn, \text{outliers}| \sum_j F_j$. Our main result is an $O(\log P)$-approximation algorithm for the case of unit-profit jobs, where P is the ratio between the largest and smallest processing times. This comes close to matching the best known bound of $O(\log \min\{P, n/m\})$ for the non-outlier version due to Leonardi and Raz [21]. However, this problem seems to be much harder with outliers, as we obtain the same approximation guarantee even for a single machine, in contrast to the non-outlier single-machine case, which can be solved optimally. We demonstrate that our analysis is tight, as the LP relaxation used is shown to have an $\Omega(\log P)$ integrality gap.

Due to space limitations, some proofs and technical details are omitted from this extended abstract. We refer the reader to the full version of this paper (available online at http://arxiv.org/abs/0906.2020), in which all missing information is provided.

1.2   Related Work

Scheduling with Rejections. As mentioned above, previous papers on this topic considered the "prize-collecting" version which minimizes the scheduling objective plus the total profit of unscheduled jobs. Existing techniques do not seem to extend to scheduling with outliers, in which we have a strict budget constraint on the total penalty of rejected jobs. Bartal et al. [3] considered offline and online makespan minimization and gave best-possible algorithms for both cases. Makespan minimization with preemptions was investigated in [17,26]. Epstein et al. [10] examined scheduling unit-length jobs. Engels et al. [9] studied the prize-collecting version of weighted completion-time minimization (on single or parallel machines), and gave PTASs or constant-factor approximations; they also proposed a general framework for designing algorithms for such problems.

Outlier Versions of Other Problems. Also called *partial-covering* problems, these have been widely studied: e.g., the k-MST problem [11], the k-center and facility location problem [6], k-median with outliers [7], partial vertex cover (e.g., [23] and references therein) and k-multicut [14,22]. Chudak et al. [8] distilled the ideas of Jain and Vazirani [18] on converting "Lagrange-multiplier preserving" algorithms for prize-collecting Steiner tree into one for k-MST;

Könemann et al. [19] gave a general framework to convert prize-collecting algorithms into algorithms for outlier versions (see also [24]). However, in the context of this paper, it is not clear how to make prize-collecting scheduling algorithms to also be Langrange-multiplier preserving, or whether the above-mentioned framework is applicable in scheduling-related scenarios.

2   GAP and Makespan

As a warm-up, we consider the generalized assignment problem, which is an extension of minimizing makespan on unrelated machines with outliers. Formally, an instance $\mathcal{I}$ has m machines and n jobs. Each job j has a processing time of p_{ij} on machine i, an assignment cost of c_{ij}, and a profit of π_j. Given a profit requirement Π, cost bound C and makespan bound T, the goal is to compute a feasible schedule satisfying these requirements. Of course, since the problem is NP-hard, we look at finding solutions where we may slightly violate the cost and makespan bounds, but not the (hard) profit requirement. We now show how to reduce this problem to its non-outlier version, while incurring small additional losses in the approximation guarantees.

Theorem 1. *Given an instance $\mathcal{I}$ of GAP with outliers, there is a polynomial-time algorithm to compute an assignment with cost at most $(1+\epsilon)C$ and makespan at most $3T$.*

Proof. Given the instance $\mathcal{I}$, construct the following instance $\mathcal{I}'$ of the standard GAP, where there are no profits or outliers. There are $m+1$ machines: machines $1, \ldots, m$ are identical to those in $\mathcal{I}$, while machine $m+1$ is a "virtual profit machine". We have n jobs, where job j has a processing time of p_{ij} and an assignment cost of c_{ij} when scheduled on machine i (for $1 \leq i \leq m$). If job j is scheduled on the virtual machine $m+1$, it incurs a processing time of π_j and cost zero, i.e., $p_{m+1,j} = \pi_j$ and $c_{m+1,j} = 0$. For this instance, we set a cost bound of C, makespan bound of T for machines $1, \ldots, m$, and a makespan bound of $\sum_{j=1}^{n} \pi_j - \Pi$ for the virtual profit machine. Note that any feasible solution for $\mathcal{I}$ is also feasible for $\mathcal{I}'$, with the outliers being scheduled on the virtual profit machine, since the total profit of the outliers is at most $\sum_{j=1}^{n} \pi_j - \Pi$.

We can now use the algorithm of Shmoys and Tardos [27] which guarantees an assignment S for the GAP instance $\mathcal{I}'$ with the following properties: (a) The cost of S is at most C; (b) the makespan induced by S on machine i (for $1 \leq i \leq m$) is at most $T + \min\{\max_j p_{ij}, T\}$; and (c) The makespan of S on the virtual machine $m+1$ is at most $(\sum_{i=1}^{n} \pi_j - \Pi) + \max_j \pi_j$.

Note that the assignment S is *almost* feasible for the outlier problem $\mathcal{I}$, since the makespan on any real machine is at most $T + \max_j p_{ij}$, and the assignment cost is at most C. However, the profit of scheduled jobs is only guaranteed to be at least $\Pi - \max_j \pi_j$, instead of Π. This shortcoming is easy to fix: we choose a job j' assigned by S to the virtual machine which has the largest profit, and schedule j' on the machine where it has the least processing time. Now the modified assignment has cost at most $C + \max_{ij'} c_{ij'}$, makespan at most

$T + 2\min\{\max_j p_{ij}, T\}$, and the total profit of scheduled jobs is at least Π. (We assume that any job j where $\min_i p_{ij} > T$ has already been discarded.) This is almost what we want, apart from the cost guarantee. To this end, suppose we "guess" the $\lceil 1/\epsilon \rceil$ most expensive assignments in OPT, in time $O((mn)^{1/\epsilon})$, and hence we can focus only on the jobs having $c_{ij} \leq \epsilon C$ for all possible remaining assignments. Now the assignment cost is at most $C + \max_{ij} c_{ij} \leq (1 + \epsilon)C$, and the makespan is at most $3T$. $\qquad\qquad\square$

In fact, the $(1 + \epsilon)$ loss in cost is inevitable since we can reduce the knapsack problem to the single machine makespan minimization with outliers problem. As for the makespan guarantee, the 3/2-hardness of Lenstra et al. [20] carries over.

3 Weighted Sum of Completion Times

We now turn our attention to average completion time with outliers. The main result of this section is a constant factor approximation for this problem. Not surprisingly, the integrality gap of standard LP relaxations is unbounded[1], and hence we strengthen the time-indexed formulation with knapsack-cover inequalities [4,29]. We show that a randomized rounding scheme similar to that of Schulz and Skutella [25] gives us the claimed guarantees on the objective function. In the full version of this paper, we also give an FPTAS for the single-machine case of unweighted sum of completion times.

3.1 $O(1)$ Approximation for Weighted Sum of Completion Times

We have a collection of m machines and n jobs, where each job j is associated with a profit π_j, a weight w_j, and a release date r_j. When job j is scheduled on machine i, it incurs a processing time of p_{ij}. Given a parameter $\Pi > 0$, the objective is to identify a set of jobs S and a feasible schedule such that $\sum_{j \in S} \pi_j \geq \Pi$ and such that $\sum_{j \in S} w_j C_j$ is minimized. Here, C_j denotes the completion time of job j.

A Time Indexed LP Relaxation. For the non-outlier version, in which all jobs have to be scheduled, Schulz and Skutella [25] gave a constant factor approximation by making use of a time-indexed LP. We first describe a natural extension of their linear program to the outlier case, while also *strengthening* it.

minimize $\sum_{j=1}^{n} w_j C_j$

subject to (1) $\quad C_j = \sum_{i=1}^{m} \sum_{t=0}^{T} \left(\frac{x_{ijt}}{p_{ij}} \left(t + \frac{1}{2} \right) + \frac{x_{ijt}}{2} \right) \qquad \forall j$

$\qquad\qquad$ (2) $\quad y_j = \sum_{i=1}^{m} \sum_{t=0}^{T} \frac{x_{ijt}}{p_{ij}} \qquad\qquad\qquad\quad \forall j$

$\qquad\qquad$ (3) $\quad \sum_{j=1}^{n} x_{ijt} \leq 1 \qquad\qquad\qquad\qquad\quad \forall i, t$

$\qquad\qquad$ (4) $\quad \sum_{j \notin \mathcal{A}} \pi_j^{\mathcal{A}} y_j \geq \Pi - \Pi(\mathcal{A}) \qquad\quad \forall \mathcal{A} : \Pi(\mathcal{A}) < \Pi$

$\qquad\qquad$ (5) $\quad x_{ijt} = 0 \qquad\qquad\qquad\qquad\qquad \forall i, j, t : t < r_j$

$\qquad\qquad$ (6) $\quad x_{ijt} \geq 0, 0 \leq y_j \leq 1 \qquad\qquad\quad \forall i, j, t$

---

[1] Implicit in the work of Guha and Munagala [15] is an algorithm which violates the profit requirement by a constant factor. They also comment on the integrality gap, and pose the problem of avoiding this violation.

In this formulation, the variable x_{ijt} stands for the fractional amount of time machine i spends on processing job j in the time interval $[t, t+1)$; note that the LP schedule may be preemptive. The variable C_j, defined by constraint (1), is a measure for the completion time of job j. In any integral solution, where job j is scheduled from t to $t + p_{ij}$ on a single machine i, it is not difficult to verify that C_j evaluates to $t + p_{ij}$. The variable y_j, defined by constraint (2), is the fraction of job j being scheduled. Constraint (3) ensures that machine i spends at most one unit of processing time in $[t, t+1)$.

We first observe that replacing the set of constraints (4) by a single inequality, $\sum_{j=1}^n \pi_j y_j \geq \Pi$, would result in an unbounded integrality gap. Consider a single job of profit M, and $\Pi = 1$; the LP can schedule a $1/M$ fraction of the job, incurring a cost which is only $1/M$ times the optimum. We therefore add in the family of constraints (4), known as the *knapsack-cover (KC) inequalities*. Let $\mathcal{A}$ be any set of jobs, and let $\Pi(\mathcal{A}) = \sum_{j \in \mathcal{A}} \pi_j$ be the sum of profits over all jobs in $\mathcal{A}$. Then, $[\Pi - \Pi(\mathcal{A})]^+$ is the profit that needs to be collected by jobs not in $\mathcal{A}$ when all jobs in $\mathcal{A}$ are scheduled. Further, if $\mathcal{A}$ does not fully satisfy the profit requirement, any job $j \notin \mathcal{A}$ has a marginal contribution of at most $\pi_j^{\mathcal{A}} = \min\{\pi_j, \Pi - \Pi(\mathcal{A})\}$. Therefore, for every set $\mathcal{A}$ such that $\Pi(\mathcal{A}) < \Pi$, we add a constraint of the form $\sum_{j \notin \mathcal{A}} \pi_j^{\mathcal{A}} y_j \geq \Pi - \Pi(\mathcal{A})$. Note that there are exponentially many such constraints, and hence we cannot naively solve this LP.

"Solving" the LP. We will not look to find an optimal solution to the above LP; for our purposes, it suffices to compute a solution vector $(\widehat{x}, \widehat{y}, \widehat{C})$ satisfying the following:

(a) Constraints (1)-(3) and (5)-(6) are satisfied.
(b) Constraint (4) is satisfied for the single set $\{j : \widehat{y}_j \geq 1/2\}$.
(c) $\sum_{j=1}^n w_j \widehat{C}_j \leq 2 \cdot \mathsf{Opt}$, where Opt is the cost of an optimal integral solution.

We compute this solution vector by first guessing Opt up to a multiplicative factor of 2 (call the guess $\widetilde{\mathsf{Opt}}$), and add the explicit constraint $\sum_{j=1}^n w_j C_j \leq \widetilde{\mathsf{Opt}}$. Then, we solve the LP using the ellipsoid algorithm. For the separation oracle, in each iteration, we check if the current solution satisfies properties (a)–(c) above. If none of these properties is violated, we are done; otherwise, we have a violated constraint. We now present our rounding algorithm (in Algorithm 1).

3.2 Analysis

We show that the expected weighted sum of completion times is $O(1) \cdot \mathsf{Opt}$, and also that with constant probability, the total profit obtained is at least Π.

Lemma 1. *The expected weighted sum of completion times is at most* $16 \cdot \mathsf{Opt}$.

Proof. Let C_j^R be a random variable, standing for the completion time of job j; if this job has not been scheduled, we set $C_j^R = 0$. Since $\sum_{j=1}^n w_j \widehat{C}_j \leq 2 \cdot \mathsf{Opt}$, it is sufficient to prove that $\mathrm{E}[C_j^R] \leq 8\widehat{C}_j$ for every j. To this end, note that

$$\mathrm{E}\left[C_j^R\right] = \sum_{i=1}^m \sum_{t=0}^T \Pr[\tau_j = (i,t)] \cdot \mathrm{E}\left[C_j^R \,\middle|\, \tau_j = (i,t)\right] \leq \sum_{i=1}^m \sum_{t=0}^T \frac{2\widehat{x}_{ijt}}{p_{ij}} \cdot \mathrm{E}\left[C_j^R \,\middle|\, \tau_j = (i,t)\right],$$

Algorithm 1. Weighted Sum of Completion Times

1: Given a solution $(\widehat{x}, \widehat{y}, \widehat{C})$ satisfying properties (a)–(c), let $\mathcal{A}^* = \{j : \widehat{y}_j \geq 1/2\}$.
2: For each job j, do the following steps

    2a: If $j \in \mathcal{A}^*$, for each (i,t) pair, set $l_{ijt} = \widehat{x}_{ijt}/(p_{ij}\widehat{y}_j)$. Note that for such jobs $j \in \mathcal{A}^*$, we have $\sum_{i=1}^m \sum_{t=0}^T l_{ijt} = 1$ from constraint (2) of the LP.
    2b: If $j \notin \mathcal{A}^*$, set $l_{ijt} = 2\widehat{x}_{ijt}/p_{ij}$. In this case, note that $\sum_{i=1}^m \sum_{t=0}^T l_{ijt} = 2\widehat{y}_j$.
    2c: Partition the interval $[0,1]$ in the following way: assign each (i,t) pair a sub-interval I_{it} of $[0,1]$ of length l_{ijt} such that these sub-intervals are pairwise disjoint. Then choose a uniformly random number $r \in [0,1]$ and set τ_j to be the (i,t) pair such that $r \in I_{it}$. If there is no such pair, leave j unmarked.

3: For each machine i, consider the jobs such that $\tau_j = (i, *)$; order them in increasing order of their marked times; schedule them as early as possible (subject to the release dates) in this order.

---

where the last inequality holds since $\Pr[\tau_j = (i,t)] = l_{ijt} \leq 2\widehat{x}_{ijt}/p_{ij}$, regardless of whether $j \in \mathcal{A}^*$ or not. Now let us upper bound $\mathrm{E}[C_j^R | \tau_j = (i,t)]$. The total time for which job j must wait before being processed on machine i can be split in the worst case into: (a) the idle time on this machine before j is processed, and (b) the total processing time of other jobs marked (i, t') with $t' \leq t$. If job j has been marked (i,t), the idle time on machine i before j is processed is at most t. Also, the total expected processing time mentioned in item (b) is at most

$$\sum_{k \neq j} p_{ik} \sum_{t'=0}^{t} \Pr\left[\tau_k = (i,t') \,|\, \tau_j = (i,t)\right] = \sum_{k \neq j} p_{ik} \sum_{t'=0}^{t} \Pr\left[\tau_k = (i,t')\right]$$

$$\leq \sum_{k \neq j} p_{ik} \sum_{t'=0}^{t} \frac{2\widehat{x}_{ikt'}}{p_{ik}} = 2 \sum_{t'=0}^{t} \sum_{k \neq j} \widehat{x}_{ikt'} \leq 2(t+1),$$

where the last inequality follows from constraint (3). Combining these observations and constraint (1), we have

$$\mathrm{E}\left[C_j^R\right] \leq 2 \sum_{i=1}^{m} \sum_{t=0}^{T} \frac{\widehat{x}_{ijt}}{p_{ij}} \left(t + 2(t+1) + p_{ij}\right) \leq 8 \sum_{i=1}^{m} \sum_{t=0}^{T} \left(\frac{\widehat{x}_{ijt}}{p_{ij}}\left(t + \frac{1}{2}\right) + \frac{\widehat{x}_{ijt}}{2}\right) = 8\widehat{C}_j.$$

$\square$

Lemma 2. *With probability at least $1/5$, the resulting schedule meets the profit requirement.*

Proof. Clearly, when the jobs in $\mathcal{A}^*$ collectively satisfy the profit requirement, we are done since the algorithm picks every job in $\mathcal{A}^*$. In the opposite case, consider the Knapsack Cover inequality for $\mathcal{A}^*$, stating that $\sum_{j \notin \mathcal{A}^*} \pi_j^{\mathcal{A}^*} \widehat{y}_j \geq \Pi - \Pi(\mathcal{A}^*)$. The total profit collected from these jobs can be lower bounded by $Z = \sum_{j \notin \mathcal{A}^*} \pi_j^{\mathcal{A}^*} Z_j$, where each Z_j is a random variable indicating whether job j is picked or not.

Since our rounding algorithm picks all jobs in $\mathcal{A}^*$, the profit requirement is met if Z is at least $\Pi - \Pi(\mathcal{A}^*)$. To provide an upper bound on the probability that Z falls below $\Pi - \Pi(\mathcal{A}^*)$, notice that by the way the algorithm marks jobs in Step 2, we have that each job not in $\mathcal{A}^*$ is marked with probability $2\widehat{y}_j$, *independently* of other jobs. Therefore,

$$\mathrm{E}\left[Z\right] = \mathrm{E}\left[\sum_{j \notin \mathcal{A}^*} \pi_j^{\mathcal{A}^*} Z_j\right] = 2 \sum_{j \notin \mathcal{A}^*} \pi_j^{\mathcal{A}^*} \widehat{y}_j \geq 2(\Pi - \Pi(\mathcal{A}^*)).$$

Consequently, if we define $\alpha_j = \pi_j^{\mathcal{A}^*}/(\Pi - \Pi(\mathcal{A}^*))$, then

$$\Pr\left[Z \leq \Pi - \Pi(\mathcal{A}^*)\right] = \Pr\left[\sum_{j \notin \mathcal{A}^*} \frac{\pi_j^{\mathcal{A}^*}}{\Pi - \Pi(\mathcal{A}^*)} Z_j \leq 1\right]$$

$$\leq \Pr\left[\sum_{j \notin \mathcal{A}^*} \alpha_j Z_j \leq \frac{1}{2} \cdot \mathrm{E}[\sum_{j \notin \mathcal{A}^*} \alpha_j Z_j]\right] \leq \exp\left(-\frac{1}{8} \cdot \mathrm{E}\left[\sum_{j \notin \mathcal{A}^*} \alpha_j Z_j\right]\right) \leq e^{-1/4} < \frac{4}{5},$$

where the first and third inequalities hold since $\mathrm{E}[\sum_{j \notin \mathcal{A}^*} \alpha_j Z_j] \geq 2$, and the second inequality follows from bounding the lower tail of the sum of independent $[0,1]$ random variables (see, e.g., [1, Thm. 3.5]).   □

The above two lemmas combine to give the following theorem.

Theorem 2. *There is an $O(1)$-approximation algorithm for minimizing weighted completion times on unrelated machines with outliers.*

While the LP formulation as stated has exponentially many time intervals of length 1, we can make our algorithm polynomial in the input size (with a small loss in the approximation guarantee) by considering geometrically increasing sizes for the time intervals (see, for instance, [16]).

4   Average Flow Time on Identical Machines

Finally, we consider the problem of minimizing the average (preemptive) flow time on identical machines ($P|r_j, pmtn, \text{outliers}| \sum F_j$) with *unit profits*. We present an LP rounding algorithm that produces a preemptive non-migratory schedule[2] whose flow time is within $O(\log P)$ of optimal, where P is the ratio between the largest and smallest processing times.

This is the technical heart of our paper; in sharp contrast to the problems in Sections 2 and 3, it is not clear how to easily modify existing algorithms for this problem to handle its outliers version. While we use the same LP as in previous papers, our rounding algorithm has to substantially extend previous non-outlier algorithms. Since our algorithms are somewhat involved, we first

---

[2] That is, no job is scheduled on multiple machines.

present an algorithm for the *single machine* case. In the full version of this paper, we show how to combine our single machine algorithm along with ideas drawn from [12] to obtain an $O(\log P)$ approximation for the more general case of identical machines.

For the remainder of this section, consider the following setup: we are given a single machine and a collection of n jobs, where each job j has a release date $r_j \in \mathbb{Z}$ and a processing time $p_j \in \mathbb{Z}$. Given a parameter $\Pi > 0$, we want to identify a set of jobs S and a preemptive schedule minimizing $\sum_{j \in S} F_j$ subject to $|S| \geq \Pi$. Here, $F_j = C_j - r_j$ is the flow time of job j.

4.1   The Flow-Time LP Relaxation

Our LP relaxation is a natural outlier extension of the one used in earlier flow-time algorithms [12,13]. We first describe what the variables and constraints correspond to: (i) f_j is the fractional flow time of job j; (ii) x_{jt} is the fraction of job j scheduled in the time interval $[t, t+1)$; and (iii) y_j is the fraction of job j scheduled. Constraint (1) keeps track of the flow time of each job, while constraints (2), (3), and (4) make sure that the solution is feasible with respect to the profit constraint. Notice that in constraint (1), we use the quantity $\widetilde{p}_j$, which denotes the processing time p_j rounded up to the next power of 2, instead of p_j. This modification is present only in constraint (1) which dictates the LP cost, and not in constraint (2) which measures the extent to which each job is scheduled. The quantity T is a guess for the time at which the optimal solution completes processing jobs (in fact, any upper bound of it would suffice). Our algorithm has a running time which is polynomial in T and n. We also assume that a parameter $k^* \in \mathbb{Z}$ was guessed in advance, such that the optimal solution only schedules jobs with $p_j \leq 2^{k^*}$.

$$\text{minimize } \sum_{j=1}^{n} f_j$$

$$\text{subject to (1)} \quad f_j = \sum_{t=0}^{T} \left(\frac{x_{jt}}{\widetilde{p}_j} \left(t + \frac{1}{2} - r_j \right) + \frac{x_{jt}}{2} \right) \qquad \forall j$$

$$(2) \quad p_j y_j = \sum_{t=0}^{T} x_{jt} \qquad \forall j$$

$$(3) \quad \sum_{j=1}^{n} x_{jt} \leq 1 \qquad \forall t$$

$$(4) \quad \sum_{j=1}^{h} y_j \geq \Pi$$

$$(5) \quad x_{jt} = 0 \qquad \forall j, t : t < r_j$$

$$(6) \quad x_{jt} \geq 0, \, 0 \leq y_j \leq 1 \qquad \forall j, t$$

Lemma 3 (Relaxation). $\mathsf{Opt}(\mathsf{LP}) \leq \mathsf{Opt}$, *where* Opt *denotes the optimal sum of flow times.*

Theorem 3 (Integrality Gap). *There are instances in which* $\mathsf{Opt} = \Omega(\log P) \cdot \mathsf{Opt}(\mathsf{LP})$, *where* P *is the ratio between the largest and smallest processing times.*

Our gap instance is on a single machine, for which the shortest remaining processing time policy (SRPT) is known to be optimal in the non-outlier case. However, our results eventually show that this is as bad as it gets — we establish an upper bound of $O(\log P)$ on the integrality gap even for identical machines.

4.2   The Rounding Algorithm: Game Plan and Some Hurdles

Before we present our algorithm in detail, let us give a high-level picture and indicate some of the complicating factors over earlier work. Previous LP-based rounding techniques [12,13] relied on the fact that if we rearrange the jobs of length roughly 2^k — call such jobs "class-k" jobs — among the time slots they occupy in the fractional solution, the objective function does not change much. These algorithms then use such rearrangements to make the schedule feasible (no job simultaneously scheduled on two machines) and even non-migratory across machines. We are currently considering the single machine case, so these issues are irrelevant for the time being; however, we need to handle jobs that are fractionally picked by the LP. In particular, we need to swap "mass" between jobs to pick an integral number of jobs to schedule, and it is this step which increases the LP cost even in the single machine case. Note that we essentially care only about the y_j value for each job j, which indicates the extent to which this job is scheduled — if we could make the y_j's integral without altering the objective by much, we would be done!

However, naïve approaches to make the y_j's integral may have bad approximation guarantees. E.g., consider taking two consecutive fractional jobs j and j' with similar processing times[3] and scheduling more of the first one over the second. If the second job j' has *even slightly smaller* processing time than j has, we would run out of space trying to schedule an equal fraction of j over j', and this loss may hurt us in the (hard) profit requirement. In such a case, we could try to schedule j' over j, observing that the later job j' would not advance too much in time, since j and j' were consecutive in that class and have similar processing times. The eventual hope is that, given a small violation of the release dates, we may be able to shift the entire schedule by a bit and regain feasibility. However, this strategy could lead to arbitrarily bad approximations: we could keep fractionally growing a job j until (say) 2/3 of it is scheduled, only to meet a job j' subsequently that also has 2/3 of it scheduled, but j' has smaller processing time and therefore needs to be scheduled over j. In this case, j would shrink to 1/3, and then would start growing again. Repeated occurrences of these events might cause the flow time for j to be very high.

Indeed, trying to avoid such situations leads us to our algorithm, where we look at a *window* of jobs and select an appropriate one to schedule, rather than greedily running a swapping process. To analyze our algorithm, we charge the total increase in the fractional flow time to the *fractional makespan* of the LP solution, and show that each class of jobs charges the fractional makespan at most twice.

4.3   Notation and Preliminaries

We partition the collection of jobs into *classes*, with jobs in class C_k having $p_j \in (2^{k-1}, 2^k]$. Notice that $\widetilde{p}_j = 2^k$ for every $j \in C_k$, and the class of interest with highest index is C_{k^*}. Given a fractional solution (x, y, f), we say that job j is *fully*

---

[3] Observe that jobs with similar processing times have similar contributions to the objective, except for the release date component.

scheduled if $y_j = 1$, and *dropped* if $y_j = 0$; in both cases, j is *integrally scheduled*. Let flow$(x, y, f) = \sum_{j=1}^{n} f_j$ be the fractional cost; note that this is *not the same* as the actual flow time given by this solution, but rather an approximation. Let $\mathcal{P}(x, y, f) = \sum_{j=1}^{n} \sum_{t=0}^{T} x_{jt}$ be the total fractional processing time. Since each job j gets x_{jt} amount of processing time in $[t, t+1)$, the cost of (x, y, f) remains unchanged if all jobs are processed during the first part $[t, t + \sum_{j=1}^{n} x_{jt})$ of this unit interval; we therefore refer to $[t + \sum_{j=1}^{n} x_{jt}, t + 1)$ as the *free time interval* in $[t, t+1)$.

We say that an LP solution (x, y, f) is *non-alternating* across each class if the fractional schedule does not alternate between two jobs of the same class. Formally, the schedule is non-alternating if for class k and any two class-k jobs j and j', if $y_j, y_{j'} > 0$ and $r_j < r_{j'}$ (or $r_j = r_{j'}$ and $j < j'$), then for any times t, t' such that $x_{jt} > 0$ and $x_{j't'} > 0$, it holds that $t \leq t'$. We call a solution *packed* if there is no free time between the release date of a job, and the last time it is scheduled by the LP solution.

Lemma 4. *There is an optimal LP solution (x^*, y^*, f^*) that is non-alternating and packed.*

4.4   The Rounding Algorithm

We assume that an optimal LP solution, non-alternating and packed, has already been computed. At a high level, the rounding algorithm proceeds in two stages.

- In Stage I, for each k, we completely schedule almost as many class-k jobs as the LP does fractionally (up to an additive two jobs). The main challenge, as sketched above, is to do this with only a small change in the fractional flow time and in the processing time of these jobs.
- In Stage II, we add in at most two class-k jobs to compensate for the loss of jobs in Stage I. Since we add only two jobs per class, we can show that the additional flow time can be controlled.

Flow-Time Rounding: Stage I. Recall that we want to convert the non-alternating and packed optimal solution (x^*, y^*, f^*) into a new solution (x', y', f') where at least $\lfloor \sum_{j \in C_k} y_j^* \rfloor - 1$ class-k jobs are *completely scheduled*. The algorithm operates on the classes $1, \ldots, k^*$ one by one. For each class, it performs a *swapping phase* where mass is shifted between jobs in this class (potentially violating release dates), followed by a *shifting phase* to handle all the release-date violations.

Swapping Phase for Class-k. Given the non-alternating and packed solution (x^*, y^*, f^*), we execute the swapping phase given in Algorithm 2.

Shifting Phase for Class-k. After the above swapping phase for class-k jobs, we perform a *shifting phase* to handle any violated release dates. Specifically, consider the collection of time intervals occupied either by class-k jobs or by free time. By the process given above, this collection remains fixed over the execution of the swapping phase. We now shift all class-k jobs to the right by 2^{k+1} within these intervals. Of course, we need to prove that this takes care of all release date violations.

Algorithm 2. Class-k Swapping

1: Set $(x', y', f') := (x^*, y^*, f^*)$. Repeat steps 2-5 until $\lfloor \sum_{j \in C_k} y_j^* \rfloor - 1$ class-k jobs are completely scheduled in (x', y', f').

2: Advance all class-k jobs as much as possible without violating release dates within the time intervals that are either free or are occupied by class-k jobs. Jobs already violating their release dates are not advanced any further.

3: Let j_1 be the first fractionally scheduled job in the current LP solution (x', y', f'). Let j_{q+1} be the first class-k job scheduled after j_1 which has processing time $p_{j_{q+1}} < p_{j_1}$, and say the class-k jobs that are scheduled between j_1 and j_{q+1} are $j_2, \ldots, j_q$. Note that all these jobs must have processing times greater than p_{j_1}. Also, let free denote the total free time between j_1 and j_{q+1} in the current schedule.

4: If $\sum_{k=2}^{q} y_k' + \text{free}/p_{j_1} \geq 1 - y_{j_1}'$, we know that j_1 can be *completely* scheduled over the jobs $j_2, \ldots, j_q$ and the free time; for $k = 2$ to q, do the following

   - If there is some free time (of total length, say, L) between j_{k-1} and j_k, schedule a fraction $\Delta = \min(1 - y_{j_1}', L/p_{j_1})$ of j_1 in the free time, and delete a fraction Δ from class-k jobs at the rear end of the schedule. Update (x', y', f').
   - Schedule a fraction $\Delta = \min(1 - y_{j_1}', y_{j_k}')$ of j_1 over a fraction Δ of job j_k (possibly creating some free space). Update (x', y', f').
   - If $k = q$ and there is some free time (of total length, say, L) between j_q and j_{q+1}, schedule a fraction $\Delta = \min(1 - y_{j_1}', L/p_{j_1})$ of j_1 in the free time, and delete a fraction Δ from class-k jobs at the rear end of the schedule. Update (x', y', f').

5: Else if $\sum_{k=2}^{q} y_k' + \text{free}/p_{j_1} < 1 - y_{j_1}'$, do the following

   - delete a total fraction of $\min(\sum_{k=1}^{q} y_k', y_{j_{q+1}}')$ from a prefix of jobs $j_1, \ldots, j_q$, and *advance* the current fractional schedule of the job j_{q+1} to occupy the space created. Update the solution (x', y', f'). Note that it may or may not have been possible to schedule j_1 in the space fractionally occupied by jobs $j_2, \ldots, j_q$ and free time in this interval; for accounting reasons we do the same in both cases.

Lemma 5. *The following properties are satisfied at the end of Stage I:*

(i) $\mathcal{P}(x', y', f') \leq 2\mathcal{P}(x^*, y^*, f^*)$.

(ii) $\text{flow}(x', y', f') \leq 4 \cdot \text{flow}(x^*, y^*, f^*) + 6k^*\mathcal{P}(x^*, y^*, f^*)$.

(iii) *The total flow time over all fully scheduled jobs is at most* $2 \cdot \text{flow}(x', y', F') + k^*\mathcal{P}(x', y', F')$.

The analysis proceeds by a delicate charging argument, where the basic idea is the following. In Step 4 of the algorithm, suppose that a Δ fraction of a job j_1 is being scheduled over a Δ fraction of a job j_k: we will *charge* every point in the interval (r_{j_1}, r_{j_k}) by an amount of Δ. In the case when a Δ fraction of j_1 is being scheduled over an interval of free time beginning at t, we will then charge every point in the interval (r_{j_1}, t) by the fraction Δ. We then proceed by showing that $\text{flow}(x', y', f') - \text{flow}(x^*, y^*, f^*)$ is not too much more than the total charge accumulated by the interval $[0, T]$, where T is the last time at which the LP scheduled some fractional job. To complete the proof, we argue that the total charge accumulated is $O(\log P)\mathcal{P}(x^*, y^*, f^*)$. In the full version, we restate Stage I in a slightly different way, where we also define a charging scheme associated with each step, and give the complete proof of Lemma 5.

Flow-Time Rounding: Stage II. The fractional solution (x', y', F') may not be feasible, since we have only scheduled $\lfloor \sum_{j \in C_k} y_j^* \rfloor - 1$ jobs from class k. Hence, for each class k, arbitrarily pick the minimum number of non-fully-scheduled jobs to bring this number to $\lceil \sum_{j \in C_k} y_j^* \rceil$ (at most two per class). These jobs are preemptively scheduled as soon as possible after their release date. Since at most two jobs per class are added, the flow time does not change much.

Lemma 6. *The total flow time of all added jobs is at most $k^*(P(x', y', F') + 2^{k^*+2})$.*

Proof. For a class k, we may have to complete two additional jobs. When we schedule an extra job as soon as possible, it waits only for jobs that were fully scheduled during stage II or for jobs that were added in previous iterations of the current stage. Therefore, its flow time can be at most $P(x', y', F') + 2\sum_{k=1}^{k^*} 2^k$, and therefore the total flow time of added jobs is at most $k^*(P(x', y', F') + 2^{k^*+2})$. □

We now point out that since f_j is lower bounded by $\sum_t x_{jt}^*/2$, we have $P(x^*, y^*, f^*) \le 2 \cdot \text{Opt}$. Therefore, Lemmas 5 and 6 in conjunction with the inequalities $P(x', y', F') \le 2P(x^*, y^*, f^*) \le 4 \cdot \text{Opt}$ and $k^* \le \log P + 1$, prove the following result for minimizing flow time on a single machine.

Theorem 4. *The problem of minimizing flow time on a single machine with unit profits can be approximated within a factor of $O(\log P)$.*

References

1. Mayr, E.W., Prömel, H.J., Steger, A. (eds.): Dagstuhl Seminar 1997. LNCS, vol. 1367. Springer, Heidelberg (1998)
2. Bansal, N., Blum, A., Chawla, S., Dhamdhere, K.: Scheduling for flow-time with admission control. In: Di Battista, G., Zwick, U. (eds.) ESA 2003. LNCS, vol. 2832, pp. 43–54. Springer, Heidelberg (2003)
3. Bartal, Y., Leonardi, S., Marchetti-Spaccamela, A., Sgall, J., Stougie, L.: Multi-processor scheduling with rejection. In: SODA 1996, pp. 95–103 (1996)
4. Carr, R.D., Fleischer, L., Leung, V.J., Phillips, C.A.: Strengthening integrality gaps for capacitated network design and covering problems. In: SODA 2000, pp. 106–115 (2000)
5. Charikar, M., Khuller, S.: A robust maximum completion time measure for scheduling. In: SODA 2006, pp. 324–333 (2006)
6. Charikar, M., Khuller, S., Mount, D.M., Narasimhan, G.: Algorithms for facility location problems with outliers. In: SODA 2001, pp. 642–651 (2001)
7. Chen, K.: A constant factor approximation algorithm for k-median clustering with outliers. In: SODA 2008, pp. 826–835 (2008)
8. Chudak, F.A., Roughgarden, T., Williamson, D.P.: Approximate k-msts and k-steiner trees via the primal-dual method and lagrangean relaxation. Mathematical Programming 100(2), 411–421 (2004)

9. Engels, D.W., Karger, D.R., Kolliopoulos, S.G., Sengupta, S., Uma, R.N., Wein, J.: Techniques for scheduling with rejection. In: Bilardi, G., Pietracaprina, A., Italiano, G.F., Pucci, G. (eds.) ESA 1998. LNCS, vol. 1461, pp. 175–191. Springer, Heidelberg (1998)
10. Epstein, L., Noga, J., Woeginger, G.J.: On-line scheduling of unit time jobs with rejection: minimizing the total completion time. Operations Research Letters 30(6), 415–420 (2002)
11. Garg, N.: Saving an epsilon: a 2-approximation for the k-MST problem in graphs. In: STOC 2005, pp. 396–402 (2005)
12. Garg, N., Kumar, A.: Better algorithms for minimizing average flow-time on related machines. In: Bugliesi, M., Preneel, B., Sassone, V., Wegener, I. (eds.) ICALP 2006. LNCS, vol. 4051, pp. 181–190. Springer, Heidelberg (2006)
13. Garg, N., Kumar, A.: Minimizing average flow-time: Upper and lower bounds. In: FOCS 2007, pp. 603–613 (2007)
14. Golovin, D., Nagarajan, V., Singh, M.: Approximating the k-multicut problem. In: SODA 2006, pp. 621–630 (2006)
15. Guha, S., Munagala, K.: Model-driven optimization using adaptive probes. In: SODA 2007, pp. 308–317 (2007)
16. Hall, L.A., Shmoys, D.B., Wein, J.: Scheduling to minimize average completion time: off-line and on-line algorithms. In: SODA 1996, pp. 142–151 (1996)
17. Hoogeveen, H., Skutella, M., Woeginger, G.J.: Preemptive scheduling with rejection. Mathematical Programming, Ser. B 94(2-3), 361–374 (2003)
18. Jain, K., Vazirani, V.V.: Approximation algorithms for metric facility location and k-median problems using the primal-dual schema and Lagrangian relaxation. Journal of the ACM 48(2), 274–296 (2001)
19. Könemann, J., Parekh, O., Segev, D.: A unified approach to approximating partial covering problems. In: Azar, Y., Erlebach, T. (eds.) ESA 2006. LNCS, vol. 4168, pp. 468–479. Springer, Heidelberg (2006)
20. Lenstra, J.K., Shmoys, D.B., Tardos, É.: Approximation algorithms for scheduling unrelated parallel machines. Mathematical Programming 46, 259–271 (1990)
21. Leonardi, S., Raz, D.: Approximating total flow time on parallel machines. In: STOC 1997, pp. 110–119 (1997)
22. Levin, A., Segev, D.: Partial multicuts in trees. Theoretical Computer Science 369(1-3), 384–395 (2006)
23. Mestre, J.: A primal-dual approximation algorithm for partial vertex cover: making educated guesses. In: Chekuri, C., Jansen, K., Rolim, J.D.P., Trevisan, L. (eds.) APPROX 2005 and RANDOM 2005. LNCS, vol. 3624, pp. 182–191. Springer, Heidelberg (2005)
24. Mestre, J.: Lagrangian relaxation and partial cover (extended abstract). In: STACS 2008, pp. 539–550 (2008)
25. Schulz, A.S., Skutella, M.: Scheduling unrelated machines by randomized rounding. SIAM Journal on Discrete Mathematics 15(4), 450–469 (2002)
26. Seiden, S.S.: Preemptive multiprocessor scheduling with rejection. Theoretical Computer Science 262(1), 437–458 (2001)
27. Shmoys, D.B., Tardos, É.: An approximation algorithm for the generalized assignment problem. Mathematical Programming 62, 461–474 (1993)
28. Skutella, M.: Convex quadratic and semidefinite programming relaxations in scheduling. Journal of the ACM 48(2), 206–242 (2001)
29. Wolsey, L.A.: Faces for a linear inequality in 0-1 variables. Mathematical Programming 8, 165–178 (1975)

Improved Inapproximability Results for Maximum k-Colorable Subgraph

Venkatesan Guruswami* and Ali Kemal Sinop

Computer Science Department,
School of Computer Science,
Carnegie Mellon University,
Pittsburgh, PA
{venkatg,asinop}@cs.cmu.edu

Abstract. We study the maximization version of the fundamental graph coloring problem. Here the goal is to color the vertices of a k-colorable graph with k colors so that a maximum fraction of edges are properly colored (i.e. their endpoints receive different colors). A random k-coloring properly colors an expected fraction $1 - \frac{1}{k}$ of edges. We prove that given a graph promised to be k-colorable, it is NP-hard to find a k-coloring that properly colors more than a fraction $\approx 1 - \frac{1}{33k}$ of edges. Previously, only a hardness factor of $1 - O\left(\frac{1}{k^2}\right)$ was known. Our result pins down the correct asymptotic dependence of the approximation factor on k. Along the way, we prove that approximating the Maximum 3-colorable subgraph problem within a factor greater than $\frac{32}{33}$ is NP-hard.

Using semidefinite programming, it is known that one can do better than a random coloring and properly color a fraction $1 - \frac{1}{k} + \frac{2\ln k}{k^2}$ of edges in polynomial time. We show that, assuming the 2-to-1 conjecture, it is hard to properly color (using k colors) more than a fraction $1 - \frac{1}{k} + O\left(\frac{\ln k}{k^2}\right)$ of edges of a k-colorable graph.

1 Introduction

1.1 Problem Statement

A graph $G = (V, E)$ is said to be k-colorable for some positive integer k if there exists a k-coloring $\chi : V \to \{1, 2, \ldots, k\}$ such that for all edges $(u, v) \in E$, $\chi(u) \neq \chi(v)$. For $k \geq 3$, finding a k-coloring of a k-colorable graph is a classic NP-hard problem. The problem of coloring a graph with the fewest number of colors has been extensively studied. In this paper, our focus is on hardness results for the following maximization version of graph coloring: Given a k-colorable graph (for some fixed constant $k \geq 3$), find a k-coloring that maximizes the fraction of properly colored edge. (We say an edge is properly colored under a coloring if its endpoints receive distinct colors.) Note that for $k = 2$ the problem is trivial — one can find a proper 2-coloring in polynomial time when the graph is bipartite (2-colorable).

* Research supported in part by NSF CCF 0835814 and a Packard Fellowship.

I. Dinur et al. (Eds.): APPROX and RANDOM 2009, LNCS 5687, pp. 163–176, 2009.
© Springer-Verlag Berlin Heidelberg 2009

We will call this problem Max k-Colorable Subgraph. The problem is equivalent to partitioning the vertices into k parts so that a maximum number of edges are cut. This problem is more popularly referred to as Max k-Cut in the literature; however, in the Max k-Cut problem the input is an arbitrary graph that need not be k-colorable. To highlight this difference that our focus is on the case when the input graph is k-colorable, we use Max k-Colorable Subgraph to refer to this variant. We stress that we will use this convention throughout the paper: Max k-Colorable Subgraph **always** *refers to the "perfect completeness" case, when the input graph is k-colorable.*[1] Since our focus is on hardness results, we note that this restriction only makes our results stronger.

A factor $\alpha = \alpha_k$ approximation algorithm for Max k-Colorable Subgraph is an efficient algorithm that given as input a k-colorable graph outputs a k-coloring that properly colors at least a fraction α of the edges. We say that Max k-Colorable Subgraph is NP-hard to approximate within a factor β if no factor β approximation algorithm exists for the problem unless P = NP. The goal is to determine the approximation threshold of Max k-Colorable Subgraph: the largest α as a function of k for which a factor α approximation algorithm for Max k-Colorable Subgraph exists.

1.2   Previous Results

The algorithm which simply picks a random k-coloring, without even looking at the graph, properly colors an expected fraction $1 - 1/k$ of edges. Frieze and Jerrum [1] used semidefinite programming to give a polynomial time factor $1 - 1/k + 2 \ln k/k^2$ approximation algorithm for Max k-Cut, which in particular means the algorithm will color at least this fraction of edges in a k-colorable graph. This remains the best known approximation guarantee for Max k-Colorable Subgraph to date. Khot, Kindler, Mossel, and O'Donnell [2] showed that obtaining an approximation factor of $1 - 1/k + 2 \ln k/k^2 + \Omega(\ln \ln k/k^2)$ for Max k-Cut is Unique Games-hard, thus showing that the Frieze-Jerrum algorithm is essentially the best possible. However, due to the "imperfect completeness" inherent to the Unique Games conjecture, this hardness result does *not* hold for Max k-Colorable Subgraph when the input is required to be k-colorable.

For Max k-Colorable Subgraph, the best hardness known prior to our work was a factor $1 - \Theta(1/k^2)$. This is obtained by combining an inapproximability result for Max 3-Colorable Subgraph due to Petrank [3] with a reduction from Papadimitriou and Yannakakis [4]. It is a natural question whether is an efficient algorithm that could properly color a fraction $1 - 1/k^{1+\varepsilon}$ of edges given a k-colorable graph for some absolute constant $\varepsilon > 0$. The existing hardness results do not rule out the possibility of such an algorithm.

For Max k-Cut, a better hardness factor was shown by Kann, Khanna, Lagergren, and Panconesi [5] — for some absolute constants $\beta > \alpha > 0$, they showed that it is NP-hard to distinguish graphs that have a k-cut in which a fraction $(1-\alpha/k)$ of the edges cross the cut from graphs whose Max k-cut value is at most

---

[1] While a little non-standard, this makes our terminology more crisp, as we can avoid repeating the fact that the hardness holds for k-colorable graphs in our statements.

a fraction $(1 - \beta/k)$ of edges. Since MaxCut is easy when the graph is 2-colorable, this reduction does not yield any hardness for Max k-Colorable Subgraph.

1.3 Our Results

Petrank [3] showed the existence of a $\gamma_0 > 0$ such that it is NP-hard to find a 3-coloring that properly colors more than a fraction $(1 - \gamma_0)$ of the edges of a 3-colorable graph. The value of γ_0 in [3] was left unspecified and would be very small if calculated. The reduction in [3] was rather complicated, involving expander graphs and starting from the weak hardness bounds for bounded occurrence satisfiability. We prove that the NP-hardness holds with $\gamma_0 = \frac{1}{33}$. In other words, it is NP-hard to obtain an approximation ratio bigger than $\frac{32}{33}$ for Max 3-Colorable Subgraph. The reduction is from the constraint satisfaction problem corresponding to the adaptive 3-query PCP with perfect completeness from [6].

By a reduction from Max 3-Colorable Subgraph, we prove that for every $k \geqslant 3$, the Max k-Colorable Subgraph is NP-hard to approximate within a factor greater than $\approx 1 - \frac{1}{33k}$ (Theorem 2). This identifies the correct asymptotic dependence on k of the best possible approximation factor for Max k-Colorable Subgraph. The reduction is similar to the one in [5], though some crucial changes have to be made in the construction and some new difficulties overcome in the soundness analysis when reducing from Max 3-Colorable Subgraph instead of MaxCut.

In the quest for pinning down the *exact* approximability of Max k-Colorable Subgraph, we prove the following *conditional* result. Assuming the so-called 2-to-1 conjecture, it is hard to approximate Max k-Colorable Subgraph within a factor $1 - \frac{1}{k} + O\left(\frac{\ln k}{k^2}\right)$. In other words, the Frieze-Jerrum algorithm is optimal up to lower order terms in the approximation ratio *even for instances of* Max k-Cut *where the graph is k-colorable*.

Unlike the Unique Games Conjecture (UGC), the 2-to-1 conjecture allows perfect completeness, i.e., the hardness holds even for instances where an assignment satisfying *all* constraints exists. The 2-to-1 conjecture was used by Dinur, Mossel, and Regev [7] to prove that for every constant c, it is NP-hard to color a 4-colorable graph with c colors. We analyze a similar reduction for the k-coloring case when the objective is to maximize the fraction of edges that are properly colored by a k-coloring. Our analysis uses some of the machinery developed in [7], which in turn extends the invariance principle of [8]. The hardness factor we obtain depends on the spectral gap of a certain $k^2 \times k^2$ stochastic matrix.

Remark 1. In general it is far from clear which Unique Games-hardness results can be extended to hold with perfect completeness by assuming, say, the 2-to-1 (or some related) conjecture. In this vein, we also mention the result of O'Donnell and Wu [9] who showed a tight hardness for approximating satisfiable constraint satisfaction problems on 3 Boolean variables assuming the d-to-1 conjecture for any fixed d. While the UGC assumption has led to a nearly complete understanding of the approximability of constraint satisfaction problems [10], the approximability of *satisfiable* constraint satisfaction problems remains a mystery to understand in any generality.

Remark 2. It has been shown by Crescenzi, Silvestri and Trevisan [11] that any hardness result for weighted instances of Max k-Cut carries over to unweighted instances assuming the total edge weight is polynomially bounded. In fact, their reduction preserves k-colorability, so an inapproximability result for the weighted Max k-Colorable Subgraph problem also holds for the unweighted version. Therefore all our hardness results hold for the unweighted Max k-Colorable Subgraph problem.

2   Unconditional Hardness Results for Max k-Colorable Subgraph

We will first prove a hardness result for Max 3-Colorable Subgraph, and then reduce this problem to Max k-Colorable Subgraph.

2.1   Inapproximability Result for Max 3-Colorable Subgraph

Petrank [3] showed that Max 3-Colorable Subgraph is NP-hard to approximate within a factor of $(1 - \gamma_0)$ for some constant $\gamma_0 > 0$. This constant γ_0 is presumably very small, since the reduction starts from bounded occurrence satisfiability (for which only weak inapproximability results are known) and uses expander graphs. We prove a much better inapproximability factor below, via a simpler proof.

Theorem 1 (Max 3-Colorable Subgraph **Hardness**). *The* Max 3-Colorable Subgraph *problem is NP-hard to approximate within a factor of* $\frac{32}{33} + \varepsilon$ *for any constant* $\varepsilon > 0$.

Proof. For the proof of this theorem, we will use reduce from a hard to approximate constraint satisfaction problem (CSP) underlying the adaptive 3-query PCP given in [6]. This PCP has perfect completeness and soundness $1/2 + \varepsilon$ for any desired constant ε (which is the best possible for 3-query PCPs).

We first state the properties of the CSP. An instance of the CSP will have variables partitioned into three parts $\mathcal{X}, \mathcal{Y}$ and $\mathcal{Z}$. Each constraint will be of the form $(x_i \vee (Y_j = z_k)) \wedge (\overline{x_i} \vee (Y_j = z_l))$, where $x_i \in \mathcal{X}$, $z_k, z_l \in \mathcal{Z}$ are variables (unnegated) and Y_j is a literal ($Y_j \in \{y_j, \overline{y_j}\}$ for some variable $y_j \in \mathcal{Y}$). For YES instances of the CSP, there will be a Boolean assignment that satisfies **all** the constraints. For No instances, every assignment to the variables will satisfy at most a fraction $(1/2 + \varepsilon)$ of the constraints.

Remark 3. We remark the condition that the instance is tripartite, and that the variables in $\mathcal{Z}$ never appear negated are not explicit in [6]. But these can be ensured by an easy modification to the PCP construction in [6]. The PCP in [6] has a bipartite structure: the proof is partitioned into two parts called the A-tables and B-tables, and each test consists of probing one bit $A(f)$ from an A table and 3 bits $B(g), B(g_1), B(g_2)$ from the B table, and checking $(A(f) \vee (B(g) = B(g_1)) \wedge (\overline{A(f)} \vee (B(g) = B(g_2)))$. Further these tables are *folded* which is

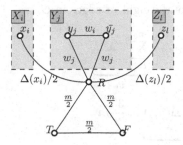

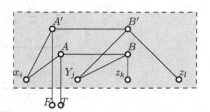

Fig. 1. Global gadget for truth value assignments. Blocks X_i, Y_j and Z_l are replicated for all vertices in $\mathcal{X}$, $\mathcal{Y}$ and $\mathcal{Z}$. Edge weights are shown next to each edge.

Fig. 2. Local gadget for each constraint of the form $(x_i \vee Y_j = z_k) \wedge (\overline{x_i} \vee Y_j = z_l)$. All edges have unit weight. Labels A, A', B, B' refer to the local nodes in each gadget.

a technical condition that corresponds to the occurrence of negations in the CSP world. If the queries at locations g_1 and g_2 are made in a parallel C-table, and even if the C-table is not folded (though the A and B tables need to be folded), one can verify that the analysis of the PCP construction still goes through. This then translates to a CSP with the properties claimed above.

Let $\mathcal{I}$ be an instance of such a CSP with m constraints of the above form on variables $\mathcal{V} = \mathcal{X} \cup \mathcal{Y} \cup \mathcal{Z}$. Let $\mathcal{X} = \{x_1, x_2, \ldots, x_{n_1}\}$, $\mathcal{Y} = \{y_1, y_2, \ldots, y_{n_2}\}$ and $\mathcal{Z} = \{z_1, z_2, \ldots, z_{n_3}\}$. From the instance $\mathcal{I}$ we create a graph G for the Max 3-Colorable Subgraph problem as follows. There is a node x_i for each variable $x_i \in \mathcal{X}$, a node z_l for each $z_l \in \mathcal{Z}$, and a pair of nodes $\{y_j, \overline{y_j}\}$ for the two literals corresponding to each $y_j \in \mathcal{Y}$. There are also three global nodes $\{R, T, F\}$ representing boolean values which are connected in a triangle with edge weights $m/2$ (see Fig. 1).

For each constraint of the CSP, we place the local gadget specific to that constraint shown in Figure 2. Note that there are 10 edges of unit weight in this gadget. The nodes $y_j, \overline{y_j}$ are connected to node R by a triangle whose edge weights equal $w_j = \frac{\Delta(y_j) + \Delta(\overline{y_j})}{2}$. Here $\Delta(X)$ denotes the total number of edges going from node X *into* all the local gadgets. The nodes x_i and z_l connected to R with an edge of weight $\Delta(x_i)/2$ and $\Delta(z_l)/2$ respectively. The proofs of the following (simple) lemmas will appear in the full version.

Lemma 1 (Completeness). *Given an assignment of variables $\sigma : \mathcal{V} \to \{0, 1\}$ which satisfies at least c of the constraints, we can construct a 3-coloring of G with at most $m - c$ improperly colored edges (each of weight 1).*

Lemma 2 (Soundness). *Given a 3-coloring of G, χ, such that the total weight of edges that are not properly colored by χ is at most $\tau < m/2$, we can construct an assignment $\sigma' : \mathcal{V} \to \{0, 1\}$ to the variables of the CSP instance that satisfies at least $m - \tau$ constraints.*

Returning to the proof of Theorem 1, the total weight of edges in G is

$$10m + \frac{3m}{2} + \underbrace{\sum_{i=1}^{n_1} \frac{\Delta(x_i)}{2} + \sum_{j=1}^{n_2} 3w_j + \sum_{l=1}^{n_3} \frac{\Delta(z_l)}{2}}_{m \qquad\qquad\qquad\qquad m}$$

$$= \frac{27}{2}m + \underbrace{\frac{3}{2}\sum_{j=1}^{n_2}(\Delta(y_i) + \Delta(\overline{y_j}))}_{2m} = \frac{33}{2}m .$$

By the completeness lemma, YES instances of the CSP are mapped to graphs G that are 3-colorable. By the soundness lemma, NO instances of the CSP are mapped to graphs G such that every 3-coloring miscolors at least a fraction $\frac{(1/2-\varepsilon)}{33/2} = \frac{1-2\varepsilon}{33}$ of the total weight of edges. Since $\varepsilon > 0$ is an arbitrary constant, the proof of Theorem 1 is complete.[2]

2.2   Max k-Colorable Subgraph Hardness

Theorem 2. *For every integer $k \geqslant 3$ and every $\varepsilon > 0$, it is NP-hard to approximate* Max k-Colorable Subgraph *within a factor of* $1 - \frac{1}{33(k+c_k)+c_k} + \varepsilon$ *where* $c_k = k \mod 3 \leqslant 2$.

Proof. We will reduce Max 3-Colorable Subgraph to Max k-Colorable Subgraph and then apply Theorem 1. Throughout the proof, we will assume k is divisible by 3. At the end, we will cover the remaining cases also. The reduction is inspired by the reduction from MaxCut to Max k-Cut given by Kann *et al.* [5] (see Remark 4). Some modifications to the reduction are needed when we reduce from Max 3-Colorable Subgraph, and the analysis has to handle some new difficulties. The details of the reduction and its analysis follow.

Let $G = (V, E)$ be an instance of Max 3-Colorable Subgraph. By Theorem 1, it is NP-hard to tell if G is 3-colorable or every 3-colors miscolors a fraction $\frac{1}{33} - \varepsilon$ of edges. We will construct a graph H such that H is k-colorable when G is 3-colorable, and a k-coloring which miscolors at most a fraction μ of the total weight of edges of H implies a 3-coloring of G with at most a fraction μk of miscolored edges. Combined with Theorem 1, this gives us the claimed hardness of Max k-Colorable Subgraph.

Let $K'_{k/3}$ denote the complete graph with loops on $k/3$ vertices. Let G' be the tensor product graph between $K_{k/3}$ and G, $G' = K'_{k/3} \otimes G$ as defined by Weichsel [12]. Identify each node in G' with $(u, i), u \in V(G), i \in \{1, 2, \ldots, k/3\}$. The edges of G' are $((u, i), (v, i'))$ for $(u, v) \in E$ and any $i, i' \in \{1, \ldots, k/3\}$. Next we make 3 copies of G', and identify the nodes with $(u, i, j), (u, i) \in V(G'), j \in \{1, 2, 3\}$, then put edges between all nodes of the form (u, i, j) and (u, i', j') if

---

[2] Our reduction produced a graph with edge weights, but by Remark 2, the same inapproximability factor holds for unweighted graphs as well.

either $i \neq i'$ or $j \neq j'$ with weight $\frac{2}{3}d_u$, where d_u is degree of node u. The total weight of edges in this new construction H equals

$$\sum_{u \in V} \left(\binom{k}{2} \frac{2}{3} d_u + \frac{3}{2} \left(\frac{k}{3} \right)^2 d_u \right) \leqslant k^2 m \, .$$

Lemma 3. *If G is 3-colorable, then H is k-colorable.*

Proof. Let $\chi_G : V(G) \to \{1,2,3\}$ be a 3-coloring of G. Consider the following coloring function for H, $\chi_H : V(H) \to \{1,2,\ldots,k\}$. For node (u,i,j), let $\chi_H((u,i,j)) = \pi^j(\chi_G(u)) + 3(i-1)$. Here π is the permutation $\begin{pmatrix} 1\,2\,3 \\ 2\,3\,1 \end{pmatrix}$, and $\pi^j(x) = \underbrace{\pi(\ldots(\pi(x)))}_{\text{j times}}$. Equivalently $\pi(x) = x \mod 3 + 1$.

Consider edges of the form $\{(u,i,j),(v,i',j)\}$. If $i \neq i'$, then colors of the endpoints are different. Else we have $\chi((u,i,j)) - \chi((v,i,j)) \equiv \chi(u) - \chi(v) \not\equiv 0 \mod 3$. For edges of the form $\{(u,i,j),(u,i',j')\}$, if $i \neq i'$, clearly edge is satisfied. When $i = i', j \neq j'$, $\chi((u,i,j)) - \chi((u,i,j')) \equiv \pi^j(u) - \pi^{j'}(u) \equiv j - j' \not\equiv 0 \mod 3$.

Lemma 4. *If H has a k-coloring that properly colors a set of edges with at least a fraction $(1 - \mu)$ of the total weight, then G has a 3-coloring which colors at least a fraction $(1 - \mu k)$ of its edges properly.*

Proof. Let χ_H be the coloring of H, $S_u^j = \{\chi_H((u,i,j)) \mid 1 \leqslant i \leqslant k/3\}$ and $S_u = \bigcup_j S_u^j$. Denote the total weight of uncut edges in this solution as

$$C^{total} = \sum_{u \in V(G)} \frac{2}{3} d_u C_u^{within} + C^{between}, \tag{1}$$

where C_u^{within} and $C^{between}$ denotes the number of improperly colored edges within the copies of node u and between copies of different vertices $u, v \in V(G)$ respectively. We have the following relations:

$$C^{between} = \sum_{j=1}^{3} \sum_{uv \in E(G)} \sum_{1 \leqslant i \leqslant i' \leqslant k/3} 1_{\chi_H((u,i,j))=\chi_H((v,i',j))} \tag{2}$$
$$\geqslant \sum_{j=1}^{3} \sum_{uv \in E(G)} |S_u^j \cap S_v^j|$$

$$\begin{aligned}
C_u^{within} &= \sum_{c \in S_u} \binom{|\chi_H^{-1}(c) \cap B_u|}{2} & (B_u = \{(u,i,j)|\forall i,j\}) \\
&= \sum_{c \in S_u} \frac{|B_{u,c}|^2}{2} - \frac{k}{2} & (B_{u,c} = B_u \cap \chi_H^{-1}(c)) \\
&\geqslant \frac{1}{2|S_u|} \left(\sum_{c \in S_u} |B_{u,c}| \right)^2 - \frac{k}{2} & \text{(Cauchy-Schwarz)} \\
&\geqslant \frac{k}{2} \left(\frac{k}{|S_u|} - 1 \right) \geqslant \frac{k}{2} \frac{|S_u|}{|S_u|} \geqslant \frac{|S_u|}{2}
\end{aligned} \tag{3}$$

Now we will find a (random) 3-coloring χ_G for G. Pick c from $\{1,2,\ldots,k\}$ uniformly at random. If $c \notin S_u$, select $\chi_G(u)$ uniformly at random from $\{1,2,3\}$. If $c \in S_u$, set $\chi_G(u) = j$ if j is the smallest index for which $c \in S^j(u)$. With this coloring $\chi_G(u)$, the probability that an edge $(u,v) \in E(G)$ will be improperly colored is:

$$\Pr\left[\chi_G(u) = \chi_G(v)\right] \leqslant \sum_{j=1}^{3} \Pr_c\left[c \in S_u^j \cap S_v^j\right] + \frac{1}{3}\Pr_c\left[c \in \overline{S_u}, c \in S_v\right]$$

$$+ \frac{1}{3}\Pr_c\left[c \in S_u, c \in \overline{S_v}\right] + \frac{1}{3}\Pr_c\left[c \in \overline{S_u}, c \in \overline{S_v}\right]$$

$$\leqslant \sum_{j=1}^{3} \frac{|S_u^j \cap S_v^j|}{k} + \frac{|\overline{S_u}|}{3k} + \frac{|\overline{S_v}|}{3k}$$

We can thus bound the expected number of miscolored edges in the coloring χ_G as follows.

$$\mathbb{E}\left[\sum_{(u,v)\in E(G)} 1_{\chi_G(u)=\chi_G(v)}\right] \leqslant \sum_{uv \in E}\left[\left(\sum_{j=1}^{3}\frac{|S_u^j \cap S_v^j|}{k}\right) + \frac{|\overline{S_u}|}{3k} + \frac{|\overline{S_v}|}{3k}\right]$$

$$\leqslant \frac{1}{k}\left(C^{between} + \sum_{u \in V(G)}\frac{d_u}{3}|\overline{S_u}|\right) \quad \text{(using (2))}$$

$$\leqslant \frac{1}{k}\left(C^{between} + \sum_{u \in V(G)}\frac{2d_u}{3}C_u^{within}\right) = \frac{C^{total}}{k}$$

This implies that there exists a 3-coloring of G for which the number of improperly colored edges in G is at most $\frac{C^{total}}{k}$. Therefore if H has a k-coloring which improperly colors at most a total weight $\mu k^2 m$ of edges, then there is a 3-coloring of G which colors improperly at most a fraction $\frac{\mu k^2 m}{km} = \mu k$ of its edges.

This completes the proof of Theorem 2 when k is divisible by 3. The other cases are easily handled by adding $k \mod 3$ extra nodes connected to all vertices by edges of suitable weight. Due to space considerations, the details will appear in the full version.

Remark 4 (Comparison to [5]). The reduction of Kann *et al* [5] converts an instance G of MaxCut to the instance $G' = K'_{k/2} \otimes G$ of Max k-Cut. Edge weights are picked so that the optimal k-cut of G' will give a set S_u of $k/2$ different colors to all vertices in each $k/2$ clique (u,i), $1 \leqslant i \leqslant k/2$. This enables converting a k-cut of G' into a cut of G based on whether a random color falls in S_u or not. In the 3-coloring case, we make 3 copies of G' in an attempt to enforce three "translates" of S_u, and use those to define a 3-coloring from a k-coloring. But we cannot ensure that each k-clique is properly colored, so these translates might overlap and a more careful soundness analysis is needed.

3 Conditional Hardness Results for Max k-Colorable Subgraph

We will first review the (exact) 2-to-1 Conjecture, and then construct a noise operator, which allows us to preserve k-colorability. Then we will bound the stability of coloring functions with respect to this noise operator. In the last section, we will give a PCP verifier which concludes the hardness result.

3.1 Preliminaries

We begin by reviewing some definitions and d-to-1 conjecture.

Definition 1. *An instance of a bipartite Label Cover problem represented as $\mathcal{L} = (U, V, E, W, R_U, R_V, \Pi)$ consists of a weighted bipartite graph over node sets U and V with edges $e = (u, v) \in E$ of non-negative real weight $w_e \in W$. R_U and R_V are integers with $1 \leqslant R_U \leqslant R_V$. Π is a collection of projection functions for each edge: $\Pi = \{\pi_{vu} : \{1, \ldots, R_V\} \to \{1, \ldots, R_U\} | u \in U, v \in V\}$. A labeling ℓ is a mapping $\ell : U \to \{1, \ldots, R_U\}$, $\ell : V \to \{1, \ldots, R_V\}$. An edge $e = (u, v)$ is satisfied by labeling ℓ if $\pi_e(\ell(v)) = \ell(u)$. We define the value of a labeling as sum of weights of edges satisfied by this labeling normalized by the total weight. $\mathrm{Opt}(\mathcal{L})$ is the maximum value over any labeling.*

Definition 2. *A projection $\pi : \{1, \ldots, R_V\} \to \{1, \ldots, R_U\}$ is called d-to-1 if for each $i \in \{1, \ldots, R_U\}$, $|\pi^{-1}(i)| \leqslant d$. It is called exactly d-to-1 if $|\pi^{-1}(i)| = d$ for each $i \in \{1, 2, \ldots, R_U\}$.*

Definition 3. *A bipartite Label-Cover instance $\mathcal{L}$ is called d-to-1 Label-Cover if all projection functions, $\pi \in \Pi$ are d-to-1.*

Conjecture 1 (d-to-1 Conjecture [13]). For any $\gamma > 0$, there exists a d-to-1 Label-Cover instance $\mathcal{L}$ with $R_V = R(\gamma)$ and $R_U \leqslant dR_V$ many labels such that it is NP-hard to decide between two cases, $\mathrm{Opt}(\mathcal{L}) = 1$ or $\mathrm{Opt}(\mathcal{L}) \geqslant \gamma$. Note that although the original conjecture involves d-to-1 projection functions, we will assume that it also holds for *exactly* d-to-1 functions (so $R_U = dR_V$), which is the case in [7].

Using the reductions from [7], it is possible to show that the above conjecture still holds given that the graph $(U \cup V, E)$ is left-regular and unweighted, i.e., $w_e = 1$ for all $e \in E$.

3.2 Noise Operators

For a positive integer M, we will denote by $[M]$ the set $\{0, 1, \ldots, M - 1\}$. We will identify elements of $[M^2]$ with $[M] \times [M]$ in the obvious way, with the pair $(a, b) \in [M]^2$ corresponding $a + Mb \in [M^2]$.

Definition 4. *A Markov operator T is a linear operator which maps probability measures to other probability measures. In a finite discrete setting, it is defined by a stochastic matrix whose (x, y)'th entry $T(x \to y)$ is the probability of* *transitioning from x to y. Such an operator is called symmetric if $T(x \to y) = T(y \to x) = T(x \leftrightarrow y)$.*

Definition 5. *Given $\rho \in [-1, 1]$, the Beckner noise operator, T_ρ on $[q]$ is defined by as $T_\rho(x \to x) = \frac{1}{q} + \left(1 - \frac{1}{q}\right) \rho$ and $T_\rho(x \to y) = \frac{1}{q}(1 - \rho)$ for any $x \neq y$.*

Observation 1. *All eigenvalues of the operator T_ρ are given by $1 = \lambda_0(T_\rho) \geqslant \lambda_1(T_\rho) = \ldots = \lambda_{q-1}(T_\rho) = \rho$. Any orthonormal basis $\alpha_0, \alpha_1, \ldots, \alpha_{q-1}$ with α_0 being constant vector, is also a basis for T_ρ.*

Lemma 5. *For an integer $q \geqslant 6$, there exists a symmetric Markov operator T on $[q]^2$ whose diagonal entries are all 0 and with eigenvalues $1 = \lambda_0 \geqslant \lambda_1 \geqslant \ldots \geqslant \lambda_{q^2-1}$ such that the spectral radius $\rho(T) = \max\{|\lambda_1|, |\lambda_{q^2-1}|\}$ is at most $\frac{4}{q-1}$.*

Proof. Consider the symmetric Markov operator T on $[q]^2$ such that, for $x = (x_1, x_2), y = (y_1, y_2) \in [q]^2$,

$$T(x \leftrightarrow y) = \begin{cases} \alpha & \text{if } \{x_1, x_2\} \cap \{y_1, y_2\} = \emptyset \text{ and } x_1 \neq x_2, y_1 \neq y_2, \\ \beta & \text{if } x_1 \notin \{y_1, y_2\} \text{ and } x_1 = x_2, y_1 \neq y_2, \\ \beta & \text{if } y_1 \notin \{x_1, x_2\} \text{ and } x_1 \neq x_2, y_1 = y_2, \\ 0 & \text{else,} \end{cases}$$

where $\alpha = \frac{1}{(q-1)(q-3)}$ and $\beta = \frac{1}{(q-1)(q-2)}$. It is clear that T is symmetric and doubly stochastic.

To bound the spectral radius of T, we will bound the second largest eigenvalue $\lambda_1(T^2)$ of T^2. Notice that T^2 is also a symmetric Markov operator. Moreover $\lambda_i(T^2) = \lambda_i^2(T)$, therefore $\lambda_1(T^2) \geqslant \max(\lambda_1^2(T), \lambda_{q^2-1}^2(T)) \geqslant \rho(T)^2$.

Notice that $T^2(x \leftrightarrow y) > 0$ for all pairs $x, y \in [q]^2$. Consider the variational characterization of $1 - \lambda_1(T^2)$ [14]:

$$\min_\psi \frac{\sum_{x,y} (\psi(x) - \psi(y))^2 \pi(x) T^2(x \leftrightarrow y)}{\sum_{x,y} (\psi(x) - \psi(y))^2 \pi(x) \pi(y)}$$

$$\geqslant \min_{\psi, x, y} \frac{\pi(x)(\psi(x) - \psi(y))^2 T^2(x \leftrightarrow y)}{(\psi(x) - \psi(y))^2 \pi(x) \pi(y)} = \min_{x,y} q^2 T^2(x \leftrightarrow y)$$

For any two pairs $(x_1, x_2), (y_1, y_2) \in [q]^2$, let $l = |[q] \setminus \{x_1, x_2, y_1, y_2\}|$. Then we have

$$T^2((x_1, x_2) \leftrightarrow (y_1, y_2)) = \begin{cases} l(l-1)\beta^2 \geqslant (q-3)^2 \beta^2 & \text{if } x_1 = x_2 \text{ and } y_1 = y_2, \\ l(l-1)\alpha\beta \geqslant (q-4)^2 \alpha\beta & \text{if } x_1 \neq x_2 \text{ and } y_1 = y_2, \\ l(l-1)\alpha\beta \geqslant (q-4)^2 \alpha\beta & \text{if } x_1 = x_2 \text{ and } y_1 \neq y_2, \\ l(l-1)\alpha^2 + l\beta^2 \geqslant (q-4)\left[(q-5)\alpha^2 + \beta^2\right] \\ \qquad\qquad\qquad\qquad \text{if } x_1 \neq x_2 \text{ and } y_1 \neq y_2. \end{cases}$$

$$\geqslant \frac{(q-5)(q-4)}{(q-3)^2(q-2)(q-1)}$$

So $\rho(T) \leqslant \sqrt{\lambda_1(T^2)} \leqslant \sqrt{1 - \frac{(q-5)(q-4)q^2}{(q-3)^2(q-2)(q-1)}} \leqslant \frac{3}{q} + \frac{8}{q^2} \leqslant \frac{4}{q-1}$ for $q \geqslant 6$.

3.3 q-Ary Functions, Influences, Noise Stability

We define inner product on space of functions from $[q]^N$ to $\mathbb{R}$ as $\langle f, g \rangle = \mathbb{E}_{x \sim [q]^N}[f(x)g(x)]$. Here $x \sim \mathcal{D}$ denotes sampling from distribution $\mathcal{D}$ and $\mathcal{D} = [q]^N$ denotes the uniform distribution on $[q]^N$.

Given a symmetric Markov operator T and $x = (x_1, \ldots, x_N) \in [q]^N$, let $T^{\otimes N} x$ denote the product distribution on $[q]^N$ whose i^{th} entry y_i is distributed according to $T(x_i \leftrightarrow y_i)$. Therefore $T^{\otimes N} f(x) = \mathbb{E}_{y \sim T^{\otimes N} x}[f(y)]$.

Definition 6. *Let $\alpha_0, \alpha_1, \ldots, \alpha_{q-1}$ be an orthonormal basis of $\mathbb{R}^q$ such that α_0 is all constant vector. For $x \in [q]^N$, we define $\alpha_x \in \mathbb{R}^{q^N}$ as*

$$\alpha_x = \alpha_{x_1} \otimes \ldots \otimes \alpha_{x_N}.$$

Definition 7 (Fourier coefficients). *For a function $f : [q]^N \to \mathbb{R}$, define $\hat{f}(\alpha_x) = \langle f, \alpha_x \rangle$.*

Definition 8. *Let $f : [q]^N \to \mathbb{R}$ be a function. The* influence *of ith variable on f, $\mathsf{Inf}_i(f)$ is defined by*

$$\mathsf{Inf}_i(f) = \mathbb{E}\left[\mathrm{Var}\left[f(x)|x_1, \ldots, x_{i-1}, x_{i+1}, \ldots, x_N\right]\right]$$

where $x_1, \ldots, x_N$ are uniformly distributed. Equivalently,

$$\mathsf{Inf}_i(f) = \sum_{x : x_i \neq 0} \hat{f}^2(\alpha_x).$$

Definition 9. *Let $f : [q]^N \to \mathbb{R}$ be a function. The* low-level influence *of ith variable of f is defined by*

$$\mathsf{Inf}_i^{\leq t}(f) = \sum_{x : x_i \neq 0, \ |x| \leq t} \hat{f}^2(\alpha_x).$$

Observation 2. *For any function f,*

$$\sum_i \mathsf{Inf}_i^{\leq t}(f) = \sum_{x : |x| \leq t} \hat{f}^2(\alpha_x)|x| \leq t \sum_x \hat{f}^2(\alpha_x) = t\|f\|_2^2.$$

If $f : [q]^N \to [0, 1]$, then $\|f\|_2^2 \leq 1$, so $\sum_i \mathsf{Inf}_i^{\leq t}(f) \leq t$.

Definition 10 (Noise stability). *Let f be a function from $[q]^N$ to $\mathbb{R}$, and let $-1 \leq \rho \leq 1$. Define the* noise stability *of f at ρ as*

$$\mathbb{S}_\rho(f) = \langle f, T_\rho^{\otimes n} f \rangle = \sum_x \rho^{|x|} \hat{f}_i^2(\alpha_x)$$

where T_ρ is the Beckner operator as in Definition 5.

A natural way to think about a q-coloring function is as a collection of q-indicator variables summing to 1 at every point. To make this formal:

Definition 11. *Define the unit q-simplex as* $\Delta_q = \{(x_1, \ldots, x_q) \in \mathbb{R}^q \mid \sum x_i = 1, x_i \geqslant 0\}$.

Observation 3. *For positive integers Q, q and any function $f = (f_1, \ldots, f_q)$:* $[Q]^N \to \Delta_q$, $\sum_i \mathsf{Inf}_i^{\leqslant t}(f) = \sum_i \sum_j \mathsf{Inf}_i^{\leqslant t}(f_j) \leqslant t \sum_j \|f_j\|^2 \leqslant t$.

We want to prove a lower bound on the stability of q-ary functions with noise operators T. The following proposition is generalization of Proposition 11.4 in [2] to general symmetric Markov operators T with small spectral radii. The proof is also very similar, so it is left out and will appear in the full version.

Proposition 1. *For integers $Q, q \geqslant 3$, and a symmetric Markov operator T on $[Q]$ with spectral radius $\rho(T) \leqslant \frac{c}{q-1}$, for some $c > 0$, there is a small enough $\delta = \delta(q) > 0$ and $t = t(q) > 0$ such that for any function $f = (f_1, \ldots, f_q)$: $[Q]^N \to \Delta_q$ with $\mathsf{Inf}_i^{\leqslant t}(f) \leqslant \delta$, for all i, satisfies*

$$\sum_{j=1}^q \langle f_j, T^{\otimes N} f_j \rangle \geqslant 1/q - 2c \ln q/q^2 - C \ln \ln q/q^2$$

for some universal constant $C < \infty$.

Definition 12 (Moving between domains). *For any $x = (x_1, \ldots, x_{2N}) \in [q]^{2N}$, denote $\overline{x} \in [q^2]^N$ as*

$$\overline{x} = ((x_1, x_2), \ldots, (x_{2N-1}, x_{2N})) .$$

Similarly for $y = (y_1, \ldots, y_N) \in [q^2]^N$, denote $\underline{y} \in [q]^{2N}$ as

$$\underline{y} = (y_{1,1}, y_{1,2}, \ldots, y_{N,1}, y_{N,2}),$$

where $y_i = y_{i,1} + y_{i,2}q$ such that $y_{i,1}, y_{i,2} \in [q]$. For a function f on $[q]^{2N}$, define $\overline{f}$ on $[q^2]^N$ as $\overline{f}(y) = f(\underline{y})$.

The relationship between influences of variables for functions f and $\overline{f}$ are given by the following claim (Claim 2.7 in [7]).

Claim. For any function $f : [q]^{2N} \to \mathbb{R}$, $i \in \{1, \ldots, N\}$ and any $t \geqslant 1$, $\mathsf{Inf}_i^{\leqslant t}(\overline{f}) \leqslant \mathsf{Inf}_{2i-1}^{\leqslant 2t}(f) + \mathsf{Inf}_{2i}^{\leqslant 2t}(f)$.

3.4   PCP Verifier for Max k-Colorable Subgraph

This verifier uses ideas similar to the Max k-Cut verifier given in [2] and the 4-coloring hardness reduction in [7]. Let $\mathcal{L} = (U, V, E, R, 2R, \Pi)$ be a 2-to-1 bipartite, unweighted and left regular Label-Cover instance as in Conjecture 1. Assume the proof is given as the Long Code over $[k]^{2R}$ of the label of every vertex $v \in V$. Below for a permutation σ on $\{1, \ldots, n\}$ and a vector $x \in \mathbb{R}^n$, $x \circ \sigma$ denotes $(x_{\sigma(1)}, x_{\sigma(2)}, \cdots, x_{\sigma(n)})$. For a function f on $\mathbb{R}^n$, $f \circ \sigma$ is defined as $f \circ \sigma(x) = f(x \circ \sigma)$.

- Pick u uniformly at random from U, $u \sim U$.
- Pick v, v' uniformly at random from u's neighbors. Let π, π' be the associated projection functions, $\chi_v, \chi_{v'}$ be the (supposed) Long Codes for the labels of v, v' respectively.
- Let T be the Markov operator on $[k]^2$ given in Lemma 5. Pick $x \sim [k^2]^R$ and $y \sim T^{\otimes R}x$. Let $\sigma_v, \sigma_{v'}$ be two permutations of $\{1, \ldots, 2R\}$ such that $\pi(\sigma_v^{-1}(2i-1)) = \pi(\sigma_v^{-1}(2i)) = \pi'(\sigma_{v'}^{-1}(2i-1)) = \pi'(\sigma_{v'}^{-1}(2i))$ (both π and π' are exactly 2-to-1, so such permutations exist).
- Accept iff $\chi_v \circ \sigma_v(\underline{x})$ and $\chi_{v'} \circ \sigma_{v'}(\underline{y})$ are different.

The proofs of the following two lemmas are very similar to the ones in [2], and they are left out for space considerations.

Lemma 6 (Completeness). *If the original* 2-to-1 *Label-Cover instance $\mathcal{L}$ has a labeling which satisfies all constraints, then there is a proof which makes the above verifier always accept.*

Lemma 7 (Soundness). *There is a constant C such that, if the above verifier passes with probability exceeding $1 - 1/k + O(\ln k/k^2)$, then there is a labeling of $\mathcal{L}$ which satisfies $\gamma' = \gamma'(k)$ fraction of the constraints independent of label set size R.*

Note that our PCP verifier makes "k-coloring" tests. By the standard conversion from PCP verifiers to CSP hardness, and Remark 2 about conversion to unweighted graphs with the same inapproximability factor, we conclude the main result of this section by combining Lemmas 6 and 7.

Theorem 3. *For any constant $k \geqslant 3$, assuming* 2-to-1 *Conjecture, it is NP-hard to approximate* Max k-Colorable Subgraph *within a factor of $1 - 1/k + O(\ln k/k^2)$.*

References

1. Frieze, A.M., Jerrum, M.: Improved approximation algorithms for max k-cut and max bisection. Algorithmica 18(1), 67–81 (1997)
2. Khot, S., Kindler, G., Mossel, E., O'Donnell, R.: Optimal inapproximability results for MAX-CUT and other 2-variable CSPs? SIAM J. Comput. 37(1), 319–357 (2007)
3. Petrank, E.: The hardness of approximation: Gap location. Computational Complexity 4, 133–157 (1994)
4. Papadimitriou, C.H., Yannakakis, M.: Optimization, approximation, and complexity classes. J. Comput. Syst. Sci. 43(3), 425–440 (1991)
5. Kann, V., Khanna, S., Lagergren, J., Panconesi, A.: On the hardness of approximating max k-cut and its dual. Chicago J. Theor. Comput. Sci. 1997 (1997)
6. Guruswami, V., Lewin, D., Sudan, M., Trevisan, L.: A tight characterization of NP with 3 query PCPs. In: Proceedings of the 39th Annual IEEE Symposium on Foundations of Computer Science, pp. 8–17 (1998)
7. Dinur, I., Mossel, E., Regev, O.: Conditional hardness for approximate coloring. In: Proceedings of the 38th Annual ACM Symposium on Theory of Computing, pp. 344–353 (2006)

 8. Mossel, E., O'Donnell, R., Oleszkiewicz, K.: Noise stability of functions with low influences: invariance and optimality. In: Proceedings of the 46th Annual IEEE Symposium on Foundations of Computer Science, pp. 21–30 (2005)
 9. O'Donnell, R., Wu, Y.: Conditional hardness for satisfiable CSPs. In: Proceedings of the 41st Annual ACM Symposium on Theory of Computing (to appear, 2009)
10. Raghavendra, P.: Optimal algorithms and inapproximability results for every CSP? In: Proceedings of the 40th ACM Symposium on Theory of Computing, pp. 245–254 (2008)
11. Crescenzi, P., Silvestri, R., Trevisan, L.: On weighted vs unweighted versions of combinatorial optimization problems. Inf. Comput. 167(1), 10–26 (2001)
12. Weichsel, P.M.: The kronecker product of graphs. Proceedings of the American Mathematical Society 13(1), 47–52 (1962)
13. Khot, S.: On the power of unique 2-prover 1-round games. In: Proceedings of the 34th Annual ACM Symposium on Theory of Computing, pp. 767–775 (2002)
14. Sinclair, A.: Improved bounds for mixing rates of markov chains and multicommodity flow. Combinatorics, Probability and Computing 1, 351–370 (1992)

Improved Absolute Approximation Ratios for Two-Dimensional Packing Problems

Rolf Harren and Rob van Stee*

Max-Planck-Institut für Informatik (MPII),
Campus E1 4, 66123 Saarbrücken, Germany
{rharren,vanstee}@mpi-inf.mpg.de

Abstract. We consider the two-dimensional bin packing and strip packing problem, where a list of rectangles has to be packed into a minimal number of rectangular bins or a strip of minimal height, respectively. All packings have to be non-overlapping and orthogonal, i.e., axis-parallel. Our algorithm for strip packing has an absolute approximation ratio of 1.9396 and is the first algorithm to break the approximation ratio of 2 which was established more than a decade ago. Moreover, we present a polynomial time approximation scheme ($\mathcal{PTAS}$) for strip packing where rotations by 90 degrees are permitted and an algorithm for two-dimensional bin packing with an absolute worst-case ratio of 2, which is optimal provided $\mathcal{P} \neq \mathcal{NP}$.

Keywords: two-dimensional bin packing, strip packing, rectangle packing, approximation algorithm, absolute worst-case ratio.

1 Introduction

In the two-dimensional bin packing problem, a list $I = \{r_1, \ldots, r_n\}$ of rectangles of width $w_i \leq 1$ and height $h_i \leq 1$ is given. An unlimited supply of equally-sized, rectangular bins is available to pack all items from I such that no two items overlap and all items are packed axis-parallel into the bins. The goal is to minimize the number of bins used. We assume that the bins have unit size, which can be achieved by scaling the items appropriately. For the strip packing problem, the given items have to be packed into a strip of unit width and minimal height.

Both problems have many applications, for instance in stock-cutting or scheduling on partitionable resources. In many applications, rotations are not allowed because of the pattern of the cloth or the grain of the wood. This is the main case that we consider in this paper. Note that the assumption of unit-sized bins is a restriction in the case where rotations are permitted.

Most of the previous work on two-dimensional packing problems has focused on the *asymptotic* approximation ratio, i.e., the behavior of the algorithm on instances with large optimal value. The asymptotic approximation ratio is defined

---

* Research supported by German Research Foundation (DFG).

I. Dinur et al. (Eds.): APPROX and RANDOM 2009, LNCS 5687, pp. 177–189, 2009.
© Springer-Verlag Berlin Heidelberg 2009

as follows. Let $\text{ALG}(I)$ be the value, i.e., the height of the strip or the number of bins, of a packing produced by algorithm ALG on input I. Denote the optimal algorithm by OPT. The asymptotic approximation ratio of packing algorithm ALG is defined to be

$$\limsup_{n \to \infty} \sup_I \left\{ \frac{\text{ALG}(I)}{\text{OPT}(I)} \ \middle| \ \text{OPT}(I) = n \right\}.$$

Kenyon & Rémila [12] and Jansen & van Stee [10] gave asymptotic fully polyno-mial approximation schemes ($\mathcal{FPTAS}$'s) for strip packing without rotations and with rotations, respectively. The additive constant was recently improved from $\mathcal{O}(1/\varepsilon^2)$ to 1 by Jansen & Solis-Oba [9] at the cost of a higher running time.

Caprara [5] was the first to present an algorithm with an asymptotic approx-imation ratio less than 2 for two-dimensional bin packing. Indeed, he considered 2-stage packing, in which the items must first be packed into shelves that are then packed into bins, and showed that the asymptotic worst case ratio between two-dimensional bin packing and 2-stage packing is $T_\infty = 1.691\ldots$. Therefore the asymptotic $\mathcal{FPTAS}$ for 2-stage packing by Caprara, Lodi & Monaci [6] achieves an asymptotic approximation guarantee arbitrarily close to T_∞.

Recently, Bansal, Caprara & Sviridenko [2] presented a general framework to improve subset oblivious algorithms and obtained asymptotic approximation guarantees arbitrarily close to $1.525\ldots$ for packing with rotations of 90 degrees or without rotations. These are the currently best-known asymptotic approximation ratios for general two-dimensional bin packing problems. For packing squares into square bins, Bansal, Correa, Kenyon & Sviridenko [4] gave an asymptotic $\mathcal{PTAS}$. On the other hand, the same paper showed the $\mathcal{APX}$-hardness of two-dimensional bin packing without rotations, thus no asymptotic $\mathcal{PTAS}$ exists unless $\mathcal{P} = \mathcal{NP}$. Chlebík & Chlebíková [7] were the first to give explicit lower bounds of $1 + 1/3792$ and $1 + 1/2196$ on the asymptotic approximability of rectangle packing with and without rotations, respectively.

It should be noted that for the positive results for bin packing mentioned above, the approximation ratio only gets close to the stated value for very large inputs. In particular, the 1.525-approximation by Bansal et al. [2] has an additive constant which is not made explicit in the paper but which the authors believe is extremely large [1]. Thus, for any reasonable input, the actual (absolute) approximation ratio of their algorithm is much larger than 1.525, and it therefore makes sense to consider alternative algorithms and in particular, an alternative performance measure.

In the current paper, we consider the absolute approximation ratio. This is defined simply as $\sup_I \text{ALG}(I)/\text{OPT}(I)$, where the supremum is taken over all inputs. Proving a bound on the absolute approximation gives us a performance guarantee for all inputs, not just for (very) large ones.

Steinberg [15] and Schiermeyer [14] presented absolute 2-approximation al-gorithms for strip packing. Especially Steinberg's algorithm has been used in many subsequent bin packing and strip packing papers as subroutines. Since one-dimensional bin packing is a natural subproblem of strip packing, there ex-ists no $(3/2 - \varepsilon)$-approximation for any $\varepsilon > 0$. Jansen & Solis-Oba [9] showed an absolute $\mathcal{PTAS}$ for strip packing with rotations on instances with optimal height at least 1.

For the bin packing problem, Zhang [17] presented an absolute 3-approximation algorithm. For the special case of packing squares into bins, van Stee [16] showed that an absolute 2-approximation is possible. Moreover, Harren & van Stee [8] gave an absolute 2-approximation for bin packing with rotations. They also showed that the algorithm Hybrid First Fit has an absolute approximation ratio of 3 for packing without rotations, as conjectured by Zhang [17].

Our contribution. We present an approximation algorithm for strip packing with an absolute approximation ratio of 1.9396. Although Schiermeyer [14] already expected in his work in 1994 that this bound can be reduced below 2, this is the first improvement on the absolute approximability of strip packing since Schiermeyer's work. For strip packing with rotations we show that an (absolute) $\mathcal{PTAS}$ can easily be derived using a result by Bansal, Caprara & Sviridenko [3]. This improves upon the restricted $\mathcal{PTAS}$ from [9].

Moreover, we present an approximation algorithm for two-dimensional bin packing with an absolute approximation ratio of 2. As Leung et al. [13] showed that it is strongly $\mathcal{NP}$-complete to decide whether a set of squares can be packed into a given square, this is best possible unless $\mathcal{P} = \mathcal{NP}$.

2   Important Tools and Preparations

Let $I = \{r_1, \ldots, r_n\}$ be the set of given rectangles, where $r_i = (w_i, h_i)$. For $\delta \leq 1/2$, let $W_\delta = \{r_i \mid w_i > 1 - \delta\}$ be the set of so-called δ-*wide* items and let $H_\delta = \{r_i \mid h_i > 1-\delta\}$ be the set of δ-*high* items. To simplify the presentation, we denote the $1/2$-wide items as *wide* items and the $1/2$-high items as *high* items. Let W and H be the sets of wide and high items, respectively. The set of *small* items, i.e., items r_i with $w_i \leq 1/2$ and $h_i \leq 1/2$, is denoted by S. Finally, we call items that are wide and high at the same time *big*.

For a set T of items, let $\mathcal{A}(T) = \sum_{i \in T} w_i h_i$ be the total area and let $h(T) = \sum_{r_i \in T} h_i$ and $w(T) = \sum_{r_i \in T} w_i$ be the total height and total width, respectively. Finally, let $w_{\max}(T) = \max_{r_i \in T} w_i$ and $h_{\max}(T) = \max_{r_i \in T} h_i$.

Steinberg [15] proved the following theorem for his algorithm that we use as a subroutine.

Theorem 1 (Steinberg's algorithm). *If the following inequalities hold,*

$$w_{\max}(T) \leq a, \quad h_{\max}(T) \leq b, \qquad \qquad \text{and}$$
$$2\mathcal{A}(T) \leq ab - (2w_{\max}(T) - a)_+ (2h_{\max}(T) - b)_+$$

where $x_+ = \max(x, 0)$, then it is possible to pack all items from T into $R = (a, b)$ in time $\mathcal{O}((n \log^2 n)/\log \log n)$.

Bansal, Caprara & Sviridenko [3] considered the two-dimensional knapsack problem in which each item $r_i \in I$ has an associated profit p_i and the goal is to maximize the total profit that is packed into a unit-sized bin. Using a very technical

Structural Lemma they derived an algorithm that we call BCS algorithm in this paper. We use the following corollary of their analysis for the case where we want to maximize the total packed area, i.e., $p_i = w_i h_i$ for all items $r_i \in I$. Let $\text{OPT}_{(a,b)}(T)$ denote the maximum area of items from T that can be packed into the rectangle (a, b), where individual items in T do not necessarily fit in (a, b).

Corollary 1. *For any fixed $\varepsilon > 0$, the BCS algorithm returns a packing of $I' \subseteq I$ in a rectangle of width $a \leq 1$ and height $b \leq 1$ such that $\mathcal{A}(I') \geq \text{OPT}_{(a,b)}(I) - \varepsilon$.*

3   Strip Packing

An important link between strip packing and two-dimensional bin packing is the interpretation of a strip of height 1 as a bin of unit size. This link is especially crucial as handling instances that fit into one bin turns out to be a major challenge for bin packing. Moreover, strip packing can essentially be reduced to the packing of instances with optimal value at most 1 as the following lemma shows.

Lemma 1. *Let $0 < \varepsilon < 1/4$. If there exists a polynomial-time algorithm for strip packing that packs any instance I with optimal value at most 1 into a strip of height h, then there also exists a polynomial-time algorithm for strip packing with absolute approximation ratio at most $h + \varepsilon$.*

Proof. Let ALG be the algorithm that packs any instance I with optimal value at most 1 into a strip of height h and assume that $h \leq 2$ by otherwise applying Steinberg's algorithm. Let ε' be the maximal value with $\varepsilon' \leq \varepsilon/(4h)$ such that $1/\varepsilon'$ is integer. We guess the optimal value approximately and apply ALG on an appropriately scaled instance. To do this, we first apply Steinberg's algorithm on I to get a packing into height $h' \leq 2\,\text{OPT}(I)$. We split the interval $J = [h'/2, h']$ into $1/\varepsilon'$ subintervals $J_i = [(1 + \varepsilon'(i - 1))h'/2, (1 + \varepsilon'i)h'/2]$ for $i = 1, \ldots, 1/\varepsilon'$. Then we iterate over $i = 1, \ldots, 1/\varepsilon'$, scale the heights of all items by $2/((1 + \varepsilon'i)h')$ and apply the algorithm ALG on the scaled instance I'. Convert the packing to a packing of the unscaled instance I and finally output the minimal packing that was derived. We eventually consider $i^* \in \{1, \ldots, 1/\varepsilon'\}$ with $\text{OPT}(I) \in J_{i^*}$. Then we have

$$1 - 2\varepsilon' < 1 - \frac{\varepsilon'h}{1 + \varepsilon'i^*h} = \frac{1 + \varepsilon'(i^* - 1)h}{1 + \varepsilon'i^*h} \leq \text{OPT}(I') \leq \frac{1 + \varepsilon'i^*h}{1 + \varepsilon'i^*h} = 1$$

and thus

$$\frac{\text{ALG}(I)}{\text{OPT}(I)} = \frac{\text{ALG}(I')}{\text{OPT}(I')} < \frac{h}{1 - 2\varepsilon'} = h + \frac{2\varepsilon'h}{1 - 2\varepsilon'} \leq h + 4\varepsilon'h \leq h + \varepsilon. \qquad \square$$

Thus we concentrate on approximating instances that fit into a strip of height 1 and therefore assume $\text{OPT}(I) \leq 1$ for the remainder of this section. The overall approach for our algorithm for strip packing consists of two parts. First, we use the BCS algorithm to pack instances where the total height of the δ-wide items is small relative to δ into a strip of height $2 - x$ for some positive value x. Second, we derive an area guarantee for instances that could not be packed in the previous step and use this guarantee to successfully pack the instance into a strip of height $2 - x$.

Finally, we will show that x can be chosen as large as $(1 - \ln 2)/(3 + 3 \ln 2) - \varepsilon$ and with Lemma 1 we get the following theorem for any $0 < \varepsilon < 10^{-5}/2$.

Theorem 2. *There exists a polynomial-time approximation algorithm for strip packing with absolute approximation ratio*

$$2 - x + \varepsilon = \frac{5 + 7 \ln 2}{3 + 3 \ln 2} + 2\varepsilon < 1.9396.$$

Assume that we have a fixed $x \in [0, 1/6 - 5/3\,\varepsilon)$ and $0 < \varepsilon \leq 10^{-5}/2$.

3.1 Small Total Height of the δ-Wide Items

In the following we describe an important subroutine that is used by our algorithms for strip and bin packing. We consider the case that the total height of the δ-wide items is small relative to some δ, i.e.,

$$h(W_\delta) \leq \frac{\delta(1 - x) - 2x - \varepsilon}{1 + 2\delta} =: f(\delta)$$

for some $\delta \in ((2x + \varepsilon)/(1 - x), 1/2]$ (the lower bound is required as otherwise $f(\delta) < 0$). We want to derive a packing of I into two bins such that only a height of $1 - x$ is used in the second bin. For strip packing this directly gives a height of $2 - x$ by putting the second bin on top of the first. And for bin packing we get a feasible solution for all $x \geq 0$.

Let $\gamma := f(\delta) + x = (\delta(1 + x) - x - \varepsilon)/(1 + 2\delta) < 1/2$. In the first step, we show that a packing of almost all items into a unit bin and with a special structure exists. This special structure consists of a part of width $w(H_\gamma)$ for the γ-high items and a part of width $1 - w(H_\gamma)$ for the other items. The following lemma shows that almost all other items can be packed.

Lemma 2. *We have* $\mathrm{OPT}_{(1 - w(H_\gamma), 1)}(I \setminus H_\gamma) \geq \mathcal{A}(I \setminus H_\gamma) - 2\gamma$.

Proof. Consider an optimal packing of I into a bin. Remove all items that are completely contained in the top or bottom γ-margin. After this step there is no item directly above or below any item of $H_\gamma = \{r_i \mid h_i > 1 - \gamma\}$. Thus we can cut the remaining packing at the left and right side of any item from H_γ. These cuts partition the packing into parts which can be swapped without losing any further items. Move all items of H_γ to the left of the bin and move all other parts of the packing to the right. The total area of the removed items is at most 2γ and thus a total area of at least $\mathcal{A}(I \setminus H_\gamma) - 2\gamma$ fits into the rectangle of size $(1 - w(H_\gamma), 1)$ to the right of H_γ. $\square$

In the second step, we actually derive a feasible packing that is based on the structure described above (see Figure 1(a)). First, pack H_γ into a stack of width

$w(H_\gamma)$ at the left side of the first bin. Note that $w(H_\gamma) \leq 1$. This leaves an empty space of width $1 - w(H_\gamma)$ and height 1 at the right. We therefore apply the BCS algorithm on $I \setminus H_\gamma$ and a rectangle of size $(1 - w(H_\gamma), 1)$ using an accuracy of ε. Lemma 2 and Corollary 1 yield that at least a total area of $\mathcal{A}(I \setminus H_\gamma) - 2\gamma - \varepsilon$ is packed by the algorithm.

Let T be the set of remaining items with $\mathcal{A}(T) \leq 2\gamma + \varepsilon$. Pack the remaining δ-wide items, i.e., the items of $T \cap W_\delta$, in a stack at the bottom of the second bin. The total area of the remaining items $T \setminus W_\delta$ is

$$\mathcal{A}(T \setminus W_\delta) \leq \mathcal{A}(T) - (1 - \delta)h(T \cap W_\delta) \leq 2\gamma + \varepsilon - (1 - \delta)h(T \cap W_\delta).$$

We pack these items into the free rectangle of size (a, b) with $a = 1$ and $b = 1 - h(T \cap W_\delta) - x$ above the stack of $T \cap W_\delta$ in the second bin. A short calculation shows that Steinberg's algorithm is applicable.

So far we assumed the knowledge of $\delta \in ((2x + \varepsilon)/(1 - x), 1/2]$ for which $h(W_\delta) \leq f(\delta)$. It is easy to see that this value can be computed by calculating $h(W_\delta)$ for $\delta = 1 - w_i$ for all $r_i = (w_i, h_i)$ with $w_i > 1/2$. As $h(W_\delta)$ changes only for these values of δ, we will necessarily find a suitable δ if one exists. We therefore have the following lemma.

Lemma 3. *For any fixed $\varepsilon > 0$, there exists a polynomial-time algorithm that, given an instance I with $\mathrm{OPT}(I) = 1$ and $h(W_\delta) \leq f(\delta)$ for some $\delta \in ((2x + \varepsilon)/(1 - x), 1/2]$, returns a packing of I into two bins such that only a height of $1 - x$ is used in the second bin.*

3.2   Using an Area Guarantee for the Wide Items

In this section we describe how to use a guarantee on the total area of the wide items for the instances that cannot be packed into a strip of height $2 - x$ by Lemma 3. Consider a strip with the lower left corner at the origin of a cartesian coordinate system and consider the stack of wide items ordered by non-increasing width and aligned with the lower right corner of the strip. If there exists a $\delta \in ((2x + \varepsilon)/(1 - x), 1/2]$ such that $h(W_\delta) \leq f(\delta)$ then we use the algorithm of Lemma 3 to pack the instance into a strip of height $2 - x$ (see Figure 1(a)). Otherwise the stack of wide items exceeds the function $f(\delta)$ for *all* $\delta \in ((2x + \varepsilon)/(1 - x), 1/2]$ (see Figure 1(b)). Then we have

$$\mathcal{A}(W) > \int_{\frac{2x+\varepsilon}{1-x}}^{1/2} \frac{\delta(1 - x) - 2x - \varepsilon}{1 + 2\delta} d\delta + \frac{h(W)}{2} > \xi(x) + \frac{h(W)}{2} \tag{1}$$

for $\xi(x) := \frac{1}{4}(1 - \ln 2) - \frac{1}{4}x(1 + 3\ln 2) - \frac{1}{2}\varepsilon \ln 2$ (this function corresponds to a lower bound of the area in darker shade below $f(\delta)$ in Figure 1(b))—see full version for the calculation.

We use this lower bound for the area of W to derive a packing into a strip of height $2 - x$. Assume that $2 - h(W) - x \geq 1$. Stack the wide items in the

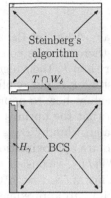
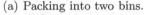

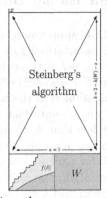

(a) Packing into two bins.

(b) Using the area guarantee of $\xi(x) + h(W)/2$ to pack into height $2 - x$.

Fig. 1. Main cases for strip packing

bottom of the strip and use Steinberg's algorithm to pack $I \setminus W$ above this stack into a rectangle of size (a, b) with $a = 1$ and $b = 2 - h(W) - x$. Then we have $h_{\max}(I \setminus W) \leq 1 \leq b$, $w_{\max}(I \setminus W) \leq 1/2$ and for

$$2\mathcal{A}(I \setminus W) \leq 2 - 2\xi(x) - h(W) \leq 2 - h(W) - x$$
$$= ab = ab - (2w_{\max} - a)_+ (2h_{\max} - b)_+$$

we require $x \leq 2\,\xi(x)$. This is satisfied for

$$x \leq \frac{1 - \ln 2}{3 + 3\ln 2} - \varepsilon \leq \frac{1 - \ln 2 - 2\varepsilon \ln 2}{3 + 3\ln 2}.$$

We give a simple algorithm that also has the requirement $x \leq 2\,\xi(x)$ for the other case in the full version. Thus we can choose $x = (1 - \ln 2)/(3 + 3\ln 2) - \varepsilon$ and together with Lemmas 1 and 3 we proved Theorem 2.

Strip packing with rotations. The BCS algorithm also works when rotations by 90 degrees are permitted. Thus by Corollary 1 we can pack a total area of $A \geq \mathcal{A}(I) - \varepsilon$ into a strip of height 1. Rotating the remaining items $r_i \in T$ such that $w_i \geq h_i$ allows us to pack items with $w_i > 1/2$ into a stack of height at most 2ε. For all other items T' we have $w_{\max}(T') \leq 1/2$, $h_{\max}(T') \leq \sqrt{\varepsilon}$ and $\mathcal{A}(T') \leq \varepsilon$. Thus Steinberg's algorithm allows us to pack T' into a height of $\max(h_{\max}(T'), 2\mathcal{A}(T')) \leq \sqrt{\varepsilon}$ above this stack. With Lemma 1 and by appropriately scaling ε we get the following result.

Theorem 3. *There exists a polynomial time approximation scheme for strip packing with rotations.*

4   Two-Dimensional Bin Packing

As the asymptotic approximation ratio of the algorithm by Bansal, Caprara & Sviridenko [2] is arbitrarily close to $1.525\ldots$, there exists a constant k such that for any instance I with optimal value larger than k, their algorithm gives a solution of value at most $2\,\mathrm{OPT}(I)$. This constant k is not explicitly known as we already mentioned in the introduction. We show how to approximate the problem within an absolute factor of 2, provided that the optimal value of the given instance is less than k. Combined with the algorithm by Bansal et al., this proves the existence of an algorithm with an absolute approximation ratio of 2.

Our approach for packing instances I with $\mathrm{OPT}(I) < k$ consists of two parts. First, we give an algorithm that is able to pack instances I with $\mathrm{OPT}(I) = 1$ in two bins in Section 4.1 and second, we show how to approximate instances with $1 < \mathrm{OPT}(I) < k$ within a factor of 2 in Section 4.2. This at first glance surprising distinction is due to the inherent difficulty of packing wide and high items together into a single bin. In the case $\mathrm{OPT}(I) = 1$ we cannot ensure a separation of the wide and high items into easily feasible sets whereas for $\mathrm{OPT}(I) > 1$ this is possible in many cases.

The approach to solve instances with optimal value greater than some constant k with an asypmotic algorithm is similar to the 2-approximation for two-dimensional bin packing with rotations in [8] but the methods we use here to handle the instances with smaller optimal value are much more involved. The reason for this is that we cannot use rotations to avoid the necessity to combine wide and high items in a bin. Our approach for solving instances I with $1 < \mathrm{OPT}(I) < k$ is comparable to the main algorithm in [8] as it is also based on an enumeration of the large items. However, a new ingredient in this paper is a separation of the wide and high items after this enumeration. Another crucial novelty in our algorithm is the use of the BCS algorithm to ensure a good area guarantee for at least one bin. In total we show the following theorem.

Theorem 4. *There exists a polynomial-time approximation algorithm for two-dimensional bin packing with absolute approximation ratio 2.*

4.1   Packing Instances That Fit into One Bin

Throughout this section we assume that the given instance I can be packed into a single bin, i.e., $\mathrm{OPT}(I) = 1$. At first glance it seems surprising that packing such an instance into two bins is difficult. However, we need to carefully analyse different cases to be able to give a polynomial-time algorithm that solves this problem.

Let $\varepsilon := 1/52$. In a first step we consider instances I that satisfy the requirements of Lemma 3 for $x = 0$, i.e., we have $h(W_\delta) \leq f(\delta) = (\delta - \varepsilon)/(1 + 2\delta)$ for some $\delta \in (\varepsilon, 1/2]$. Obviously, we can apply Lemma 3 to the high items instead of the wide items as well. We get the following Lemma from Inequality (1) for $\xi = 0.075 < \xi(0)$.

Lemma 4. *For any input which cannot be packed in two bins by the methods of Lemma 3, we have*

$$\mathcal{A}(W \cup H) \geq 2\xi + \frac{w(H) + h(W)}{2}.$$

It is crucial for our work that we get this additional area guarantee of $2\xi = 0.15$ on top of the trivial guarantee of $w(H)/2 + h(W)/2$ here. We use this area guarantee to give different methods to pack the input, depending on the total height of the wide items. To do this, we assume that we have $h(W) \geq w(H)$ by otherwise rotating the whole instance and apply different methods for $w(H) > 1/2$ and $w(H) \leq 1/2$. In all cases we are able to pack the input into at most two bins. Before we show how to solve both cases above we need the following lemma that allows us to pack *all* wide items and high items of almost half of their total width (see full version for a proof of this lemma).

Lemma 5. *For any fixed $\varepsilon > 0$, there exists a polynomial-time algorithm that, given sets W and H of wide and high items with $\mathrm{OPT}(W \cup H) = 1$, returns a packing of $W \cup H'$ into a bin with $H' \subseteq H$ and $w(H') > w(H)/2 - \varepsilon$.*

With these preparations, the following lemma is easy to show.

Lemma 6. *Let $\varepsilon > 0$ and let I be an instance with $\mathrm{OPT}(I) = 1$, $h(W) \geq w(H) > 1/2$, and $h(W_\delta) > f(\delta)$ and $w(H_\delta) > f(\delta)$ for all $\delta \in (\varepsilon, 1/2]$. There exists a polynomial-time algorithm that returns a packing of I into two bins.*

Proof. Use Lemma 5 to pack $W \cup H'$ with $H' \subseteq H$ and $w(H') > w(H)/2 - \varepsilon$ in the first bin. Build a stack of the remaining high items $H \setminus H'$ and align it with the left side of the second bin. The width of this stack is $w(H \setminus H') < w(H)/2 + \varepsilon$. Note that $w(H \setminus H') \leq 1/2$, as otherwise $h(W) \geq w(H) \geq 1 - 2\varepsilon$ and $\mathcal{A}(W \cup H) \geq 2\xi + (w(H) + h(W))/2 \geq 2\xi + 1 - 2\varepsilon > 1$ (by Lemma 4) which is a contradiction to $\mathrm{OPT}(I) = 1$. Pack the remaining items T with Steinberg's algorithm in the free rectangle of size (a, b) with $a = 1 - w(H \setminus H')$ and $b = 1$ next to the stack of $H \setminus H'$. This is possible since $w_{\max}(T) \leq 1/2 \leq 1 - w(H \setminus H')$, $h_{\max}(T) \leq 1/2$ and with Lemma 4 we have (see full version)

$$2\mathcal{A}(T) \leq 2\left(1 - 2\xi - \frac{w(H) + h(W)}{2}\right) < 1 - w(H \setminus H')$$

$$= ab - (2w_{\max} - a)_+(2h_{\max} - b)_+. \qquad \square$$

In the following we assume that $w(H) \leq 1/2$ as otherwise we could pack the instance into two bins with the algorithms of Lemma 3 or Lemma 6. Furthermore, we still have our initial assumption $h(W) \geq w(H)$. Using Steinberg's algorithm it is straightforward to prove the following lemma.

Lemma 7. *Any set $T = \{r_1, \ldots, r_m\}$ where $r_i = (w_i, h_i)$ with $w_i \leq 1/2$, $h_i \leq 1 - h(W)$ for $i = 1, \ldots, m$ and total area $\mathcal{A}(T) \leq 1/2 - h(W)/2$ can be packed together with W.*

Obviously, Lemma 7 can also be formulated such that we pack the high items together with a set of small items of total area at most $1/2 - w(H)/2$ (in this case we do not need a condition like $h_i \leq 1 - h(W)$, as $w(H) \leq 1/2$ and thus all remaining items fit into the free rectangle next to the stack of H). This suggests partitioning the small items into sets with these area bounds in order to pack them with the wide and high items. This is possible in all but two special cases, which we deal with separately. In the end, we find the following lemma. Details are in the full version.

Lemma 8. *Let $\varepsilon > 0$ and let I be an instance with $\mathrm{OPT}(I) = 1$, $w(H) \leq 1/2$, and $h(W_\delta) > f(\delta)$ and $w(H_\delta) > f(\delta)$ for all $\delta \in (\varepsilon, 1/2]$. There exists a polynomial-time algorithm that returns a packing of I into two bins.*

This concludes our algorithm for instances I with $\mathrm{OPT}(I) = 1$ as the Lemmas 3, 6 and 8 cover all the cases.

4.2   Packing Instances That Fit into a Constant Number of Bins

In the following we give a brief description of our algorithm that packs the instances I with $2 \leq \mathrm{OPT}(I) < k$ into $2\,\mathrm{OPT}(I)$ bins.

Let $\varepsilon := 1/(20k^3 + 2)$. Let $L = \{r_i \mid w_i h_i > \varepsilon\}$ be the set of *large* items and let $T = \{r_i \mid w_i h_i \leq \varepsilon\}$ be the set of *tiny* items. As defined in Section 2 we refer to items as wide (W), high (H), small (S) and big, according to their side lengths. Note that the terms *large* and *tiny* refer to the area of the items whereas *big*, *wide*, *high* and *small* refer to their widths and heights. Also note that, e.g., an item can be tiny and high, or wide and big at the same time.

We guess $\ell = \mathrm{OPT}(I) < k$ and open 2ℓ bins that we denote by $B_1, \ldots, B_\ell$ and $C_1, \ldots, C_\ell$. By *guessing* we mean that we iterate over all possible values for ℓ and apply the remainder of this algorithm on every value. As there are only a constant number of values, this is possible in polynomial time. We assume that we know the correct value of ℓ as we eventually consider this value in an iteration. For the ease of presentation, we also denote the sets of items that are associated with the bins by $B_1, \ldots, B_\ell$ and $C_1, \ldots, C_\ell$. We will ensure that the set of items that is associated with a bin is feasible and a packing is known or can be computed in polynomial time. To do this we use the following corollary from Theorem 1 for some of these sets.

Corollary 2 (Jansen & Zhang [11]). *If the total area of a set T of items is at most $1/2$ and there are no wide items (except a possible big item) then the items in T can be packed into a bin.*

Obviously, this corollary also holds for the case that there are no high items (except a possible big item). This corollary is an improvement upon Theorem 1 if there is a big item in T as in this case Theorem 1 would give a worse area bound.

Let I_i^* be the set of items in the i-th bin in an optimal solution. We assume w.l.o.g. that $\mathcal{A}(I_i^*) \geq \mathcal{A}(I_j^*)$ for $i < j$. Then we have

$$\mathcal{A}(I) = \mathcal{A}(I_1^*) + \cdots + \mathcal{A}(I_\ell^*) \leq \ell \cdot \mathcal{A}(I_1^*). \tag{2}$$

In a first step, we guess the assignment of the large items to bins. Using this assignment and the BCS algorithm we pack a total area of at least $\mathcal{A}(I_1^*) - \varepsilon$ into B_1 and keep C_1 empty. This step has the purpose of providing a good area bound for the first bin and leaving a free bin for later use. We ensure that the large items that are assigned to B_1 are actually packed. For all other bins we reserve B_i for the wide and small items (except the big items) and C_i for the high and big items for $i = 2, \ldots, \ell$. This separation enables us to use Steinberg's algorithm (Corollary 2) to pack up to half of the bins' area. In detail, the first part of the algorithm works as follows.

1. Guess $L_i = I_i^* \cap L$ for $i = 1, \ldots, \ell$.
2. Apply the BCS algorithm on $L_1 \cup T$ while ensuring that L_1 is actually packed (see full version for the details). Assign the output to bin B_1 and keep an empty bin C_1.
3. For $i = 2, \ldots, \ell$, assign the wide and small items of L_i to B_i (omitting big items) and assign the high and big items of L_i to C_i. That is, $B_i = L_i \setminus H$ and $C_i = L_i \cap H$.
4. For $i = 2, \ldots, \ell$, greedily add tiny wide items from $T \cap W$ by non-increasing order of width to B_i as long as $\mathcal{A}(B_i) \leq 1/2$ and greedily add tiny high items from $T \cap H$ by non-increasing order of height to C_i as long as $w(C_i) \leq 1$.

Corollary 2 shows that using Steinberg's algorithm the bins $B_2, \ldots, B_\ell$ can be packed as there are no wide items and the total area is at most $1/2$. The bins $C_2, \ldots, C_\ell$ can be packed with a simple stack as they contain only high items of total width at most 1. Observe that in Step 4 we only add to a new bin B_i if the previous bins contain items of total area at least $1/2 - \varepsilon$ and we only add to a new bin C_i if the previous bins contain items of total width at least $1 - 2\varepsilon$ (as the width of the tiny high items is at most 2ε) and thus of total area at least $1/2 \cdot (1 - 2\varepsilon) = 1/2 - \varepsilon$. After the application of this first part of the algorithm, some tiny items $T' \subseteq T$ might remain unpacked. Note that if $\mathcal{A}(B_\ell) < 1/2 - \varepsilon$, then there are no wide items in T' and if $\mathcal{A}(C_\ell) < 1/2 - \varepsilon$ then there are no high items in T' (as these items would have been packed in Step 4).

In the full version we prove that

$$\mathcal{A}(B_1) \geq \mathcal{A}(I_1^*) - \varepsilon. \tag{3}$$

We distinguish different cases to continue the packing according to the filling of the last bins B_ℓ and C_ℓ.

Exemplarily assume that $\mathcal{A}(B_\ell) < 1/2 - \varepsilon$ and $\mathcal{A}(C_\ell) < 1/2 - \varepsilon$. In this case T' does not contain any wide or high items as these items would have been packed to B_ℓ or C_ℓ. Greedily add items from T' into all bins except B_1 as long as the bins contain items of total area at most $1/2$. This process packs all remaining items as otherwise we had a packed area of at least $\mathcal{A}(I_1^*) - \varepsilon + (2\ell - 1)(1/2 - \varepsilon) > \ell\mathcal{A}(I_1^*)$ by Inequality (3) (see full version for the calculation) which is a contradiction to Inequality (2).

All other cases are more complex and we refer to the full version of the paper for a detailed description. In total we showed the following lemma which concludes our presentation of the 2-approximation algorithm for two-dimensional bin packing.

Lemma 9. *There exists a polynomial-time algorithm that, given an instances I with $1 < \mathrm{OPT}(I) < k$, returns a packing in $2\,\mathrm{OPT}(I)$ bins.*

Acknowledgement

The authors would like to thank Reto Spöhel for inspiring discussions.

References

1. Bansal, N.: Personal communication (2008)
2. Bansal, N., Caprara, A., Sviridenko, M.: Improved approximation algorithms for multidimensional bin packing problems. In: FOCS: Proc. 47th IEEE Symposium on Foundations of Computer Science, pp. 697–708 (2006)
3. Bansal, N., Caprara, A., Sviridenko, M.: A structural lemma in 2-dimensional packing, and its implications on approximability, IBM Research Division, RC24468, W0801-070 (2008),
   http://domino.research.ibm.com/library/cyberdig.nsf/index.html
4. Bansal, N., Correa, J.R., Kenyon, C., Sviridenko, M.: Bin packing in multiple dimensions - inapproximability results and approximation schemes. Mathematics of Operations Research 31(1), 31–49 (2006)
5. Caprara, A.: Packing d-dimensional bins in d stages. Mathematics of Operations Research 33(1), 203–215 (2008)
6. Caprara, A., Lodi, A., Monaci, M.: Fast approximation schemes for two-stage, two-dimensional bin packing. Mathematics of Operations Research 30(1), 150–172 (2005)
7. Chlebík, M., Chlebíková, J.: Inapproximability results for orthogonal rectangle packing problems with rotations. In: CIAC: Proc. 6th Conference on Algorithms and Complexity, pp. 199–210 (2006)
8. Harren, R., van Stee, R.: Absolute approximation ratios for packing rectangles into bins. Journal of Scheduling (to appear, 2009)
9. Jansen, K., Solis-Oba, R.: New approximability results for 2-dimensional packing problems. In: MFCS: Proc. 32nd International Symposium on Mathematical Foundations of Computer Science, pp. 103–114 (2007)
10. Jansen, K., van Stee, R.: On strip packing with rotations. In: STOC: Proc. 37th ACM Symposium on Theory of Computing, pp. 755–761 (2005)
11. Jansen, K., Zhang, G.: Maximizing the total profit of rectangles packed into a rectangle. Algorithmica 47(3), 323–342 (2007)
12. Kenyon, C., Rémila, E.: A near optimal solution to a two-dimensional cutting stock problem. Mathematics of Operations Research 25(4), 645–656 (2000)
13. Leung, J.Y.-T., Tam, T.W., Wong, C.S., Young, G.H., Chin, F.Y.: Packing squares into a square. Journal of Parallel and Distributed Computing 10(3), 271–275 (1990)

14. Schiermeyer, I.: Reverse-fit: A 2-optimal algorithm for packing rectangles. In: van Leeuwen, J. (ed.) ESA 1994. LNCS, vol. 855, pp. 290–299. Springer, Heidelberg (1994)
15. Steinberg, A.: A strip-packing algorithm with absolute performance bound 2. SIAM Journal on Computing 26(2), 401–409 (1997)
16. van Stee, R.: An approximation algorithm for square packing. Operations Research Letters 32(6), 535–539 (2004)
17. Zhang, G.: A 3-approximation algorithm for two-dimensional bin packing. Operations Research Letters 33(2), 121–126 (2005)

On the Optimality of Gluing over Scales*

Alex Jaffe, James R. Lee, and Mohammad Moharrami

University of Washington, Seattle WA 98105, USA

Abstract. We show that for every $\alpha > 0$, there exist n-point metric spaces (X, d) where every "scale" admits a Euclidean embedding with distortion at most α, but the whole space requires distortion at least $\Omega(\sqrt{\alpha \log n})$. This shows that the scale-gluing lemma [Lee, SODA 2005] is tight, and disproves a conjecture stated there. This matching upper bound was known to be tight at both endpoints, i.e. when $\alpha = \Theta(1)$ and $\alpha = \Theta(\log n)$, but nowhere in between.

More specifically, we exhibit n-point spaces with doubling constant λ requiring Euclidean distortion $\Omega(\sqrt{\log \lambda \log n})$, which also shows that the technique of "measured descent" [Krauthgamer, et. al., *Geometric and Functional Analysis*] is optimal. We extend this to L_p spaces with $p > 1$, where one requires distortion at least $\Omega((\log n)^{1/q}(\log \lambda)^{1-1/q})$ when $q = \max\{p, 2\}$, a result which is tight for every $p > 1$.

1   Introduction

Suppose one is given a collection of mappings from some finite metric space (X, d) into a Euclidean space, each of which reflects the geometry at some "scale" of X. Is there a non-trivial way of gluing these mappings together to form a global mapping which reflects the entire geometry of X? The answers to such questions have played a fundamental role in the best-known approximation algorithms for Sparsest Cut [6,9,4,1] and Graph Bandwidth [15,6,10], and have found applications in approximate multi-commodity max-flow/min-cut theorems in graphs [15,6]. In the present paper, we show that the approaches of [6] and [9] are optimal, disproving a conjecture stated in [9].

Let (X, d) be an n-point metric space, and suppose that for every $k \in \mathbb{Z}$, we are given a non-expansive mapping $\phi_k : X \to L_2$ which satisfies the following. For every $x, y \in X$ with $d(x, y) \geq 2^k$, we have

$$\|\phi_k(x) - \phi_k(y)\| \geq \frac{2^k}{\alpha}.$$

The Gluing Lemma of [9] (generalizing the approach of [6]) shows that the existence of such a collection $\{\phi_k\}$ yields a Euclidean embedding of (X, d) with distortion $O(\sqrt{\alpha \log n})$. (See Section 1.1 for the relevant definitions on embeddings and distortion.) This is known to be tight when $\alpha = \Theta(1)$ [14] and also when $\alpha = \Theta(\log n)$ [11,2], but nowhere in between. In fact, in [9], the second

---

* Research partially supported by NSF CCF-0644037.

I. Dinur et al. (Eds.): APPROX and RANDOM 2009, LNCS 5687, pp. 190–201, 2009.

named author conjectured that one could achieve $O(\alpha + \sqrt{\log n})$ (this is indeed stronger, since one can always construct $\{\phi_k\}$ with $\alpha = O(\log n)$).

In the present paper, we give a family of examples which shows that the $\sqrt{\alpha \log n}$ bound is tight for any dependence $\alpha(n) = O(\log n)$. In fact, we show more. Let $\lambda(X)$ denote the *doubling constant* of X, i.e. the smallest number λ so that every open ball in X can be covered by λ balls of half the radius. In [6], using the method of "measure descent," the authors show that (X, d) admits a Euclidean embedding with distortion $O(\sqrt{\log \lambda(X) \log n})$. (This is a special case of the Gluing Lemma since one can always find $\{\phi_k\}$ with $\alpha = O(\log \lambda(X))$ [5]). Again, this bound was known to be tight for $\lambda(X) = \Theta(1)$ [7,8,5] and $\lambda(X) = n^{\Theta(1)}$ [11,2], but nowhere in between. We provide the matching lower bound for any dependence of $\lambda(X)$ on n. We also generalize our method to give tight lower bounds on L_p distortion for every fixed $p > 1$.

Construction and Analysis. In some sense, our lower bound examples are an interpolation between the multi-scale method of [14] and [7], and the expander Poincaré inequalities of [11,2,12]. We start with a vertex-transitive expander graph G on m nodes. If D is the diameter of G, then we create $D + 1$ copies $G^1, G^2, \ldots,$ G^{D+1} of G where $u \in G^i$ is connected to $v \in G^{i+1}$ if (u, v) is an edge in G, or if $u = v$. We then connect a vertex s to every node in G^1 and a vertex t to every node in G^{D+1} by edges of length D. This yields the graph $\overrightarrow{G}$ described in Section 2.2.

In Section 3, we show that whenever there is a non-contracting embedding f of $\overrightarrow{G}$ into L_2, the following holds. If $\gamma = \frac{\|f(s)-f(t)\|}{d_{\overrightarrow{G}}(s,t)}$, then some edge of $\overrightarrow{G}$ gets stretched by at least $\sqrt{\gamma^2 + \Omega(\log m)^2}$, i.e. there is a "stretch increase." This is proved by combining the uniform convexity of L_2 (i.e. the Pythagorean theorem), with the well-known contraction property of expander graphs mapped into Hilbert space. To convert the "average" nature of this contraction to information about a specific edge, we symmetrize the embedding over all automorphisms of G (which was chosen to be vertex-transitive).

To exploit this stretch increase recursively, we construct a graph $\overrightarrow{G}^{\oslash k}$ inductively as follows: $\overrightarrow{G}^{\oslash k}$ is formed by replacing every edge of $\overrightarrow{G}^{\oslash k-1}$ by a copy of $\overrightarrow{G}$ (see Section 2.1 for the formal definitions). Now a simple induction shows that in a non-contracting embedding of $\overrightarrow{G}^{\oslash k}$, there must be an edge stretched by at least $\Omega(\sqrt{k} \log m)$. In Section 3.1, a similar argument is made for L_p distortion, for $p > 1$, but here we have to argue about "quadrilaterals" instead of "triangles" (in order to apply the uniform convexity inequality in L_p), and it requires slightly more effort to find a good quadrilateral.

Finally, we observe that if $\widetilde{G}$ is the graph formed by adding two tails of length $3D$ hanging off s and t in $\overrightarrow{G}$, then (following the analysis of [7,8]), one has $\log \lambda(\widetilde{G}^{\oslash k}) \lesssim \log m$. The same lower bound analysis also works for $\widetilde{G}^{\oslash k}$, so since $n = |V(\widetilde{G}^{\oslash k})| = 2^{\Theta(k \log m)}$, the lower bound is

$$\sqrt{k} \log m \approx \sqrt{\log m \log n} \gtrsim \sqrt{\log \lambda(\widetilde{G}^{\oslash k}) \log n},$$

completing the proof.

1.1 Preliminaries

For a graph G, we will use $V(G), E(G)$ to denote the sets of vertices and edges of G, respectively. Sometimes we will equip G with a non-negative length function $\mathsf{len} : E(G) \to \mathbb{R}_+$, and we let d_{len} denote the shortest-path (semi-)metric on G. We refer to the pair (G, len) as a *metric graph*, and often len will be implicit, in which case we use d_G to denote the path metric. We use $\mathsf{Aut}(G)$ to denote the group of automorphisms of G.

Given two expressions E and E' (possibly depending on a number of parameters), we write $E = O(E')$ to mean that $E \le CE'$ for some constant $C > 0$ which is independent of the parameters. Similarly, $E = \Omega(E')$ implies that $E \ge CE'$ for some $C > 0$. We also write $E \lesssim E'$ as a synonym for $E = O(E')$. Finally, we write $E \approx E'$ to denote the conjunction of $E \lesssim E'$ and $E \gtrsim E'$.

Embeddings and Distortion. If $(X, d_X), (Y, d_Y)$ are metric spaces, and $f : X \to Y$, then we write

$$\|f\|_{\mathrm{Lip}} = \sup_{x \ne y \in X} \frac{d_Y(f(x), f(y))}{d_X(x, y)}.$$

If f is injective, then the *distortion of f* is defined by $\mathrm{dist}(f) = \|f\|_{\mathrm{Lip}} \cdot \|f^{-1}\|_{\mathrm{Lip}}$. A map with distortion D will sometimes be referred to as *D-bi-lipschitz*. If $d_Y(f(x), f(y)) \le d_X(x, y)$ for every $x, y \in X$, we say that f is *non-expansive*. If $d_Y(f(x), f(y)) \ge d_X(x, y)$ for every $x, y \in X$, we say that f is *non-contracting*. For a metric space X, we use $c_p(X)$ to denote the least distortion required to embed X into some L_p space.

Finally, for $x \in X$, $r \in \mathbb{R}_+$, we define the open ball $B(x, r) = \{y \in X : d(x, y) < r\}$. Recall that the *doubling constant* of a metric space (X, d) is the infimum over all values λ such that every ball in X can be covered by λ balls of half the radius. We use $\lambda(X, d)$ to denote this value.

We now state the main theorem of the paper.

Theorem 1. *For any positive nondecreasing function $\lambda(n)$, there exists a family of n-vertex metric graphs $\widetilde{G}^{\oslash k}$ such that $\lambda(\widetilde{G}^{\oslash k}) \lesssim \lambda(n)$, and for every fixed $p > 1$,*

$$c_p(\widetilde{G}^{\oslash k}) \gtrsim (\log n)^{1/q} (\log \lambda(n))^{1-1/q},$$

where $q = \max\{p, 2\}$.

2 Metric Construction

2.1 $\oslash$-Products

An *s-t graph* G is a graph which has two distinguished vertices $s, t \in V(G)$. For an s-t graph, we use $s(G)$ and $t(G)$ to denote the vertices labeled s and t, respectively. We define the length of an s-t graph G as $\mathsf{len}(G) = d_{\mathsf{len}}(s, t)$.

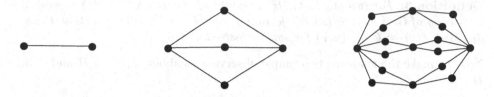

Fig. 1. A single edge H, $H \oslash K_{2,3}$, and $H \oslash K_{2,3} \oslash K_{2,2}$

Definition 1 (Composition of s-t graphs). *Given two s-t graphs H and G, define $H \oslash G$ to be the s-t graph obtained by replacing each edge $(u, v) \in E(H)$ by a copy of G (see Figure 1). Formally,*

- *$V(H \oslash G) = V(H) \cup (E(H) \times (V(G) \setminus \{s(G), t(G)\}))$.*
- *For every edge $e = (u, v) \in E(H)$, there are $|E(G)|$ edges,*

$$\left\{\left((e, v_1), (e, v_2)\right) \mid (v_1, v_2) \in E(G) \text{ and } v_1, v_2 \notin \{s(G), t(G)\}\right\} \cup$$

$$\left\{\left(u, (e, w)\right) \mid (s(G), w) \in E(G)\right\} \cup \left\{\left((e, w), v\right) \mid (w, t(G)) \in E(G)\right\}$$

- *$s(H \oslash G) = s(H)$ and $t(H \oslash G) = t(H)$.*

If H and G are equipped with length functions $\mathrm{len}_H, \mathrm{len}_G$, respectively, we define $\mathrm{len} = \mathrm{len}_{H \oslash G}$ as follows. Using the preceding notation, for every edge $e = (u, v) \in E(H)$,

$$\mathrm{len}\left((e, v_1), (e, v_2)\right) = \frac{\mathrm{len}_H(e)}{d_{\mathrm{len}_G}(s(G), t(G))} \mathrm{len}_G(v_1, v_2)$$

$$\mathrm{len}\left(u, (e, w)\right) = \frac{\mathrm{len}_H(e)}{d_{\mathrm{len}_G}(s(G), t(G))} \mathrm{len}_G(s(G), w)$$

$$\mathrm{len}\left((e, w), v\right) = \frac{\mathrm{len}_H(e)}{d_{\mathrm{len}_G}(s(G), t(G))} \mathrm{len}_G(w, t(G)).$$

This choice implies that $H \oslash G$ contains an isometric copy of $(V(H), d_{\mathrm{len}_H})$.

Observe that there is some ambiguity in the definition above, as there are two ways to substitute an edge of H with a copy of G, thus we assume that there exists some arbitrary orientation of the edges of H. However, for our purposes the graph G will be symmetric, and thus the orientations are irrelevant.

Definition 2 (Recursive composition). *For an s-t graph G and a number $k \in \mathbb{N}$, we define $G^{\oslash k}$ inductively by letting $G^{\oslash 0}$ be a single edge of unit length, and setting $G^{\oslash k} = G^{\oslash k-1} \oslash G$.*

The following result is straightforward.

Lemma 1 (Associativity of $\oslash$). *For any three graphs A, B, C, we have $(A \oslash B) \oslash C = A \oslash (B \oslash C)$, both graph-theoretically and as metric spaces.*

Definition 3. *For two graphs G, H, a subset of vertices $X \subseteq V(H)$ is said to be a* copy *of G if there exists a bijection $f : V(G) \to X$ with distortion 1, i.e. $d_H(f(u), f(v)) = C \cdot d_G(u, v)$ for some constant $C > 0$.*

Now we make the following two simple observations about copies of H and G in $H \oslash G$.

Observation 2. *The graph $H \oslash G$ contains $|E(H)|$ distinguished copies of the graph G, one copy corresponding to each edge in H.*

Observation 3. *The subset of vertices $V(H) \subseteq V(H \oslash G)$ form an isometric copy of H.*

2.2   A Stretched Version of G

Let $G = (V, E)$ be an unweighted graph, and put $D = \text{diam}(G)$. We define a metric s-t graph $\overrightarrow{G}$ which has $D + 1$ layers isomorphic to G, with edges between the layers, and a pair of endpoints s, t. Formally,

$$V(\overrightarrow{G}) = \{s, t\} \cup \{v^{(i)} : v \in V, i \in [D + 1]\}$$
$$E(\overrightarrow{G}) = \{(s, v^{(1)}), (v^{(D+1)}, t) : v \in V\}$$
$$\cup \left\{(u^{(i)}, v^{(i+1)}), (u^{(j)}, v^{(j)}) : (u, v) \in E, i \in [D], j \in [D + 1]\right\}$$
$$\cup \{(v^{(i)}, v^{(i+1)}) : v \in V, i \in [D]\}.$$

We put $\text{len}(s, v^{(1)}) = \text{len}(v^{(D+1)}, t) = D$ for $v \in V$, $\text{len}(u^{(i)}, v^{(i+1)}) = \text{len}(u^{(j)}, v^{(j)}) = 1$ for $(u, v) \in E$, $i \in [D]$, $j \in [D + 1]$ and $\text{len}(v^{(i)}, v^{(i+1)}) = 1$ for $v \in V, i \in [D]$. We refer to edges of the form $(u^{(i)}, v^{(i)})$ as *vertical edges*. All other edges are called *horizontal edges*. In particular, there are $D + 1$ copies $G^{(1)}, \ldots, G^{(D+1)}$ of G in $\overrightarrow{G}$ which are isometric to G itself, and their edges are all vertical.

A Doubling Version, Following Laakso. Let $\overrightarrow{G}$ be a stretched graph as in Section 2.2, with $D = \text{diam}(G)$, and let $s' = s(\overrightarrow{G}), t' = t(\overrightarrow{G})$. Consider a new metric s-t graph $\widetilde{G}$, which has two new vertices s, t and two new edges $(s, s'), (t', t)$ with $\text{len}(s, s') = \text{len}(t', t) = 3D$.

Claim. For any graph G with $|V(G)| = m$, and any $k \in \mathbb{N}$, we have $\log \lambda(\widetilde{G}^{\oslash k}) \lesssim \log m$.

The proof of the claim is similar to [7,8], and follows from the following three results.

We define $\text{tri}(G) = \max_{v \in V(G)}(d_{\text{len}}(s, v) + d_{\text{len}}(v, t))$. For any graph G, we have $\text{len}(\widetilde{G}) = d(s, t) = 9D$, and it is not hard to verify that $\text{tri}(\widetilde{G}^{\oslash k}) \leq \text{len}(\widetilde{G}^{\oslash k})(1 + \frac{1}{9D-1})$. For convenience, let G_0 be the top-level copy of $\widetilde{G}$ in $\widetilde{G}^{\oslash k}$, and H be the graph $\widetilde{G}^{\oslash k-1}$. Then for any $e \in E(G_0)$, we refer to the copy of H along edge e as H_e.

Observation 4. *If* $r > \frac{\mathrm{tri}(\widetilde{G}^{\oslash k})}{3}$, *then the ball* $B(x, r)$ *in* $\widetilde{G}^{\oslash k}$ *may be covered by at most* $|V(\widetilde{G})|$ *balls of radius* $r/2$.

Proof. For any $e \in E(G_0)$, we have $r > \frac{\mathrm{len}(e)}{\mathrm{len}(H)} \mathrm{tri}(H)$, so every point in H_e is less than $r/2$ from an endpoint of e. Thus all of $\widetilde{G}^{\oslash k}$ is covered by placing balls of radius $\frac{\mathrm{tri}(\widetilde{G}^{\oslash k})}{6}$ around each vertex of $\widetilde{G}$.

Lemma 2. *If* $s \in B(x, r)$, *then one can cover the ball* $B(x, r)$ *in* $\widetilde{G}^{\oslash k}$ *with at most* $|E(\widetilde{G})||V(\widetilde{G})|$ *balls of radius* $r/2$.

Proof. First consider the case in which $r > \frac{\mathrm{len}(\widetilde{G}^{\oslash k})}{6}$. Then for any edge e in $\widetilde{G}^{\oslash k}$, we have $r > \frac{\mathrm{len}(e)}{\mathrm{len}(H)} \cdot \frac{\mathrm{tri}(H)}{3}$. Thus by Observation 4, we may cover H_e by $|V(\widetilde{G})|$ balls of radius $r/2$. This gives a covering of all of $\widetilde{G}^{\oslash k}$ by at most $|E(\widetilde{G})||V(\widetilde{G})|$ balls of radius $r/2$.

Otherwise, assume $\frac{\mathrm{len}(\widetilde{G}^{\oslash k})}{6} \geq r$. Since $s \in B(x, r)$, but $2r \leq \frac{\mathrm{len}(\widetilde{G}^{\oslash k})}{3}$, the ball must be completely contained inside $H_{(s, s')}$. By induction, we can find a sufficient cover of this smaller graph.

Lemma 3. *We can cover any ball* $B(x, r)$ *in* $\widetilde{G}^{\oslash k}$ *with at most* $2|V(\widetilde{G})||E(\widetilde{G})|^2$ *balls of radius* $r/2$.

Proof. We prove this lemma using induction. For $\widetilde{G}^{\oslash 0}$, the claim holds trivially. Next, if any H_e contains all of $B(x, r)$, then by induction we are done. Otherwise, for each H_e containing x, $B(x, r)$ contains an endpoint of e. Then by Lemma 2, we may cover H_e by at most $|E(\widetilde{G})||V(\widetilde{G})|$ balls of radius $r/2$. For all other edges $e' = (u, v)$, $x \notin H_{e'}$, so we have:

$$V(H_{e'}) \cap B(x, r) \subseteq B(v, \max(0, r - d(x, v))) \cup B(u, \max(0, r - d(x, u))).$$

Thus, using Lemma 2 on both of the above balls, we may cover $V(H_{e'}) \cap B(x, r)$ by at most $2|E(\widetilde{G})||V(\widetilde{G})|$ balls of radius $r/2$. Hence, in total, we need at most $2|V(\widetilde{G})||E(\widetilde{G})|^2$ balls of radius $r/2$ to cover all of $B(x, r)$.

Proof (Proof of Claim 2.2). First note that $|V(\widetilde{G})| = m(D + 1) + 2 \lesssim m^2$. By Lemma 3, we have

$$\lambda(\widetilde{G}^{\oslash k}) \leq 2|V(\widetilde{G})||E(\widetilde{G})|^2 \leq 2|V(\widetilde{G})|^5 \lesssim m^{10}.$$

Hence $\log \lambda(\widetilde{G}^{\oslash k}) \lesssim \log m$.

3   Lower Bound

For any $\pi \in \mathsf{Aut}(G)$, we define a corresponding automorphism $\tilde{\pi}$ of $\widetilde{G}$ by $\tilde{\pi}(s) = s$, $\tilde{\pi}(t) = t$, $\tilde{\pi}(s') = s'$, $\tilde{\pi}(t') = t'$, and $\tilde{\pi}(v^{(i)}) = \pi(v)^{(i)}$ for $v \in V, i \in [D + 1]$.

Lemma 4. *Let G be a vertex transitive graph. Let $f : V(\widetilde{G}) \to L_2$ be an injective mapping and define $\bar{f} : V(\widetilde{G}) \to L_2$ by*

$$\bar{f}(x) = \frac{1}{\sqrt{|\mathsf{Aut}(G)|}} \left(f(\widetilde{\pi}x) \right)_{\pi \in \mathsf{Aut}(G)}.$$

Let β be such that for every $i \in [D+1]$ there exists a vertical edge $(u^{(i)}, v^{(i)})$ with $\|\bar{f}(u^{(i)}) - \bar{f}(v^{(i)})\| \geq \beta$. Then there exists a horizontal edge $(x, y) \in E(\widetilde{G})$ such that

$$\frac{\|\bar{f}(x) - \bar{f}(y)\|^2}{d_{\widetilde{G}}(x, y)^2} \geq \frac{\|\bar{f}(s) - \bar{f}(t)\|^2}{d_{\widetilde{G}}(s, t)^2} + \frac{\beta^2}{36} \tag{1}$$

Proof. Let $D = \operatorname{diam}(G)$. We first observe four facts about $\bar{f}$.

(F1) $\|\bar{f}(s) - \bar{f}(t)\| = \|\bar{f}(s) - \bar{f}(t)\|$
(F2) For all $u, v \in V$,

$$\|\bar{f}(s) - \bar{f}(v^{(1)})\| = \|\bar{f}(s) - \bar{f}(u^{(1)})\|,$$
$$\|\bar{f}(t) - \bar{f}(v^{(D+1)})\| = \|\bar{f}(t) - \bar{f}(u^{(D+1)})\|.$$

(F3) For every $u, v \in V$, $i \in [D]$,

$$\|\bar{f}(v^{(i)}) - \bar{f}(v^{(i+1)})\| = \|\bar{f}(u^{(i)}) - \bar{f}(u^{(i+1)})\|.$$

(F4) For every pair of vertices $u, v \in V$ and $i \in [D+1]$,

$$\langle \bar{f}(s) - \bar{f}(t), \bar{f}(u^{(i)}) - \bar{f}(v^{(i)}) \rangle = 0.$$

Let $z = \frac{\bar{f}(s) - \bar{f}(t)}{\|\bar{f}(s) - \bar{f}(t)\|}$. Fix some $r \in V$ and let $\rho_0 = |\langle z, \bar{f}(s) - \bar{f}(r^{(1)}) \rangle|$, $\rho_i = |\langle z, \bar{f}(r^{(i)}) - \bar{f}(r^{(i+1)}) \rangle|$ for $i = 1, 2, \ldots, D$ and $\rho_{D+1} = |\langle z, \bar{f}(t) - \bar{f}(r^{(D+1)}) \rangle|$. Note that, by (F2) and (F3) above, the values $\{\rho_i\}$ do not depend on the representative $r \in V$. In this case, we have

$$\sum_{i=0}^{D+1} \rho_i \geq \|\bar{f}(s) - \bar{f}(t)\| = 9\gamma D, \tag{2}$$

where we put $\gamma = \frac{\|\bar{f}(s) - \bar{f}(t)\|}{d_{\widetilde{G}}(s, t)}$. Note that $\gamma > 0$ since f is injective.

Recalling that $d_{\widetilde{G}}(s, t) = 9D$ and $d_{\widetilde{G}}(s, r^{(1)}) = 4D$, observe that if $\rho_0^2 \geq \left(1 + \frac{\beta^2}{36\gamma^2} \right) (4\gamma D)^2$, then

$$\max \left(\frac{\|\bar{f}(s) - \bar{f}(s')\|^2}{d_{\widetilde{G}}(s, s')^2}, \frac{\|\bar{f}(s') - \bar{f}(r^{(1)})\|^2}{d_{\widetilde{G}}(s', r^{(1)})^2} \right) \geq \gamma^2 + \frac{\beta^2}{36},$$

verifying (1). The symmetric argument holds for ρ_{D+1}, thus we may assume that

$$\rho_0, \rho_{D+1} \leq 4\gamma D \sqrt{1 + \frac{\beta^2}{36\gamma^2}} \leq 4\gamma D \left(1 + \frac{\beta^2}{72\gamma^2} \right).$$

In this case, by (2), there must exist an index $j \in [D]$ such that

$$\rho_j \geq \left(1 - \frac{8\beta^2}{72\gamma^2}\right)\gamma = \left(1 - \frac{\beta^2}{9\gamma^2}\right)\gamma.$$

Now, consider a vertical edge $(u^{(j+1)}, v^{(j+1)})$ with $\|\bar{f}(u^{(j)}) - \bar{f}(v^{(j)})\| \geq \beta$, and $u' = \bar{f}(u^{(j)}) + \rho_j z$. From (F4), we have

$$\max(\|\bar{f}(u^{(j)}) - \bar{f}(u^{(j+1)})\|^2, \|\bar{f}(u^{(j)}) - \bar{f}(v^{(j+1)})\|^2) =$$
$$\|\bar{f}(u^{(j)}) - u'\|^2 + \max(\|u' - \bar{f}(v^{(j+1)})\|^2, \|u' - \bar{f}(u^{(j+1)}\|^2))$$
$$\geq \rho_j^2 + \frac{\beta^2}{4}$$
$$\geq \left(1 - \frac{2\beta^2}{9\gamma^2}\right)\gamma^2 + \frac{\beta^2}{4}$$
$$\geq \gamma^2 + \frac{\beta^2}{36},$$

again verifying (1) for one of the two edges $(u^{(j)}, v^{(j+1)})$ or $(u^{(j)}, u^{(j+1)})$.

The following lemma is well-known, and follows from the variational characterization of eigenvalues (see, e.g. [13, Ch. 15]).

Lemma 5. *If $G = (V, E)$ is a d-regular graph with second Laplacian eigenvalue $\mu_2(G)$, then for any mapping $f : V \to L_2$, we have*

$$\mathbb{E}_{x,y \in V} \|f(x) - f(y)\|^2 \lesssim \frac{d}{\mu_2(G)} \mathbb{E}_{(x,y) \in E} \|f(x) - f(y)\|^2 \qquad (3)$$

The next lemma shows that when we use an expander graph, we get a significant increase in stretch for edges of $\widetilde{G}$.

Lemma 6. *Let $G = (V, E)$ be a d-regular vertex-transitive graph with $m = |V|$ and $\mu_2 = \mu_2(G)$. If $f : V(\widetilde{G}) \to L_2$ is any non-contractive mapping, then there exists a horizontal edge $(x, y) \in E(\widetilde{G})$ with*

$$\frac{\|f(x) - f(y)\|^2}{d_{\widetilde{G}}(x, y)^2} \geq \frac{\|f(s) - f(t)\|^2}{d_{\widetilde{G}}(s, t)^2} + \Omega\left(\frac{\mu_2}{d}(\log_d m)^2\right). \qquad (4)$$

Proof. We need only prove the existence of an $(x, y) \in E(\widetilde{G})$ such that (4) is satisfied for $\bar{f}$ (as defined in Lemma 4), as this implies it is also satisfied for f (possibly for some other edge (x, y)).

Consider any layer $G^{(i)}$ in $\widetilde{G}$, for $i \in [D + 1]$. Applying (3) and using the fact that f is non-contracting, we have

$$\mathbb{E}_{(u,v) \in E} \|\bar{f}(u^{(i)}) - \bar{f}(v^{(i)})\|^2 = \mathbb{E}_{(u,v) \in E} \|f(u^{(i)}) - f(v^{(i)})\|^2$$
$$\gtrsim \frac{\mu_2}{d} \mathbb{E}_{u,v \in V} \|f(u^{(i)}) - f(v^{(i)})\|^2$$
$$\geq \frac{\mu_2}{d} \mathbb{E}_{u,v \in V} d_G(u, v)^2$$
$$\gtrsim \frac{\mu_2}{d}(\log_d m)^2.$$

In particular, in every layer $i \in [D + 1]$, at least one vertical edge $(u^{(i)}, v^{(i)})$ has $\|\bar{f}(u^{(i)}) - \bar{f}(v^{(i)})\| \gtrsim \sqrt{\frac{\mu_2}{d}} \log_d m$. Therefore the desired result follows from Lemma 4.

We now to come our main theorem.

Theorem 5. *If $G = (V, E)$ is a d-regular, m-vertex, vertex-transitive graph with $\mu_2 = \mu_2(G)$, then*

$$c_2(\widetilde{G}^{\oslash k}) \gtrsim \sqrt{\frac{\mu_2 k}{d}} \log_d m.$$

Proof. Let $f : V(\widetilde{G}^{\oslash k}) \to L_2$ be any non-contracting embedding. The theorem follows almost immediately by induction: Consider the top level copy of $\widetilde{G}$ in $\widetilde{G}^{\oslash k}$, and call it G_0. Let $(x, y) \in E(G_0)$ be the horizontal edge for which $\|f(x) - f(y)\|$ is longest. Clearly this edge spans a copy of $\widetilde{G}^{\oslash k-1}$, which we call G_1. By induction and an application of Lemma 6, there exists a (universal) constant $c > 0$ and an edge $(u, v) \in E(G_1)$ such that

$$\frac{\|f(u) - f(v)\|^2}{d_{\widetilde{G}^{\oslash k}}(u, v)^2} \geq \frac{c\mu_2(k-1)}{d} (\log_d m)^2 + \frac{\|f(x) - f(y)\|^2}{d_{\widetilde{G}^{\oslash k}}(x, y)^2}$$

$$\geq \frac{c\mu_2(k-1)}{d} (\log_d m)^2 + \frac{c\mu_2}{d} (\log_d m)^2 + \frac{\|f(s) - f(t)\|^2}{d_{\widetilde{G}^{\oslash k}}(s, t)^2},$$

completing the proof.

Corollary 1. *If $G = (V, E)$ is an $O(1)$-regular m-vertex, vertex-transitive graph with $\mu_2 = \Omega(1)$, then*

$$c_2(\widetilde{G}^{\oslash k}) \gtrsim \sqrt{k} \log m \approx \sqrt{\log m \log N},$$

where $N = |V(\widetilde{G}^{\oslash k})| = 2^{\Theta(k \log m)}$.

3.1   Extension to Other L_p Spaces

Our previous lower bound dealt only with L_2. We now prove the following.

Theorem 6. *If $G = (V, E)$ is an $O(1)$-regular m-vertex, vertex-transitive graph with $\mu_2 = \Omega(1)$, for any $p > 1$, there exists a constant $C(p)$ such that*

$$c_p(\widetilde{G}^{\oslash k}) \gtrsim C(p)k^{1/q} \log m \approx C(p)(\log m)^{1-1/q}(\log N)^{1/q}$$

were $N = |V(\widetilde{G}^{\oslash k})|$ and $q = \max\{p, 2\}$.

The only changes required are to Lemma 5 and Lemma 4 (which uses orthogonality). The first can be replaced by Matoušek's [12] Poincaré inequality: If $G = (V, E)$ is an $O(1)$-regular expander graph with $\mu_2 = \Omega(1)$, then for any $p \in [1, \infty)$ and $f : V \to L_p$,

$$\mathbb{E}_{x,y \in V} \|f(x) - f(y)\|_p^p \leq O(2p)^p \, \mathbb{E}_{(x,y) \in E} \|f(x) - f(y)\|_p^p.$$

Generalizing Lemma 4 is more involved.

Lemma 7. *Let G be a vertex transitive graph, and suppose $p > 1$. If $q = \max\{p, 2\}$, then there exists a constant $K(p) > 0$ such that the following holds. Let $f : V(\widetilde{G}) \to L_p$ be an injective mapping and define $\bar{f} : V(\widetilde{G}) \to L_p$ by*

$$\bar{f}(x) = \frac{1}{|\mathrm{Aut}(G)|^{1/p}} \left(f(\widetilde{\pi}x) \right)_{\pi \in \mathrm{Aut}(G)}.$$

Suppose that β is such that for every $i \in [D+1]$, there exists a vertical edge $(u^{(i)}, v^{(i)})$ which satisfies $\|\bar{f}(u^{(i)}) - \bar{f}(v^{(i)})\|_p \geq \beta$. Then there exists a horizontal edge $(x, y) \in E(\widetilde{G})$ such that

$$\frac{\|\bar{f}(x) - \bar{f}(y)\|_p^q}{d_{\widetilde{G}}(x, y)^q} \geq \frac{\|f(s) - f(t)\|_p^q}{d_{\widetilde{G}}(s, t)^q} + K(p)\beta^q. \tag{5}$$

Proof. Let $D = \mathrm{diam}(G)$. For simplicity, we assume that D is even in what follows. We first observe three facts about $\bar{f}$.

(F1) $\|\bar{f}(s) - \bar{f}(t)\|_p = \|f(s) - f(t)\|_p$
(F2) For all $u, v \in V$,

$$\|\bar{f}(s) - \bar{f}(v^{(1)})\|_p = \|\bar{f}(s) - \bar{f}(u^{(1)})\|_p,$$
$$\|\bar{f}(t) - \bar{f}(v^{(D+1)})\|_p = \|\bar{f}(t) - \bar{f}(u^{(D+1)})\|_p.$$

(F3) For every $u, v \in V$, $i \in [D]$,

$$\|\bar{f}(v^{(i)}) - \bar{f}(v^{(i+1)})\|_p = \|\bar{f}(u^{(i)}) - \bar{f}(u^{(i+1)})\|_p.$$

Fix some $r \in V$ and let $\rho_0 = \|\bar{f}(s) - \bar{f}(r^{(1)})\|_p$, $\rho_i = \|\bar{f}(r^{(2i-1)}) - \bar{f}(r^{(2i+1)})\|_p$ for $i = 1, \ldots, D/2$, $\rho_{D/2+1} = \|\bar{f}(t) - \bar{f}(r^{(D+1)})\|_p$. Also let $\rho_{i,1} = \|\bar{f}(r^{(2i-1)}) - \bar{f}(r^{(2i)})\|_p$ and $\rho_{i,2} = \|\bar{f}(r^{(2i)}) - \bar{f}(r^{(2i+1)})\|_p$ for $i = 1, \ldots, D/2$.

Note that, by (F2) and (F3) above, the values $\{\rho_i\}$ do not depend on the representative $r \in V$. In this case, we have

$$\sum_{i=0}^{D/2+1} \rho_i \geq \|\bar{f}(s) - \bar{f}(t)\|_p = 9\gamma D, \tag{6}$$

where we put $\gamma = \frac{\|f(s) - f(t)\|_p}{d_{\widetilde{G}}(s, t)}$. Note that $\gamma > 0$ since f is injective.

Let $\delta = \delta(p)$ be a constant to be chosen shortly. Recalling that $d_{\widetilde{G}}(s, t) = 9D$ and $d_{\widetilde{G}}(s, r^{(1)}) = 4D$, observe that if $\rho_0^q \geq \left(1 + \delta \frac{\beta^q}{\gamma^q}\right)(4\gamma D)^q$, then

$$\max\left(\frac{\|\bar{f}(s) - \bar{f}(s'))\|_p^q}{d_{\widetilde{G}}(s, s')^q}, \frac{\|\bar{f}(s') - \bar{f}(r^{(1)})\|_p^q}{d_{\widetilde{G}}(s', r^{(1)})^q}\right) \geq \gamma^q + \delta\beta^q,$$

verifying (5). The symmetric argument holds for $\rho_{D/2+1}$, thus we may assume that

$$\rho_0, \rho_{D/2+1} \leq 4\gamma D \left(1 + \delta\frac{\beta^q}{\gamma^q}\right)^{1/q} \leq 4\gamma D \left(1 + \delta\frac{\beta^q}{\gamma^q}\right).$$

Similarly, we may assume that $\rho_{i,1}, \rho_{i,2} \leq \gamma \left(1 + \delta\frac{\beta^q}{\gamma^q}\right)^{1/q}$ for every $i \in [D/2]$.

In this case, by (6), there must exist an index $j \in \{1, 2, \ldots, D/2\}$ such that

$$\rho_j \geq \left(1 - 8\delta\frac{\beta^q}{\gamma^q}\right) 2\gamma.$$

Now, consider a vertical edge $(u^{(2j)}, v^{(2j)})$ with $\|f(u^{(2j)}) - f(v^{(2j)})\|_p \geq \beta$. Also consider the vertices $v^{(2j-1)}$ and $v^{(2j+1)}$. We now replace the use of orthogonality ((F4) in Lemma 4) with the following well-known 4-point inequalities in L_p spaces (see [3, App. A]). If $1 < p \leq 2$, then for every $u, v, w, x \in L_p$,

$$\|u - w\|_p^2 + (p-1)\|x - v\|_p^2 \leq \|u - v\|_p^2 + \|v - w\|_p^2 + \|x - w\|_p^2 + \|u - x\|_p^2.$$

On the other hand, if $p \geq 2$, then for every $u, v, w, x \in L_p$,

$$\|u - w\|_p^p + \|x - v\|_p^p \leq 2^{p-2} \left(\|u - v\|_p^p + \|v - w\|_p^p + \|x - w\|_p^p + \|u - x\|_p^p\right).$$

We apply one of these two inequalities with $x = f(u^{(2j)}), v = f(v^{(2j)}), u = f(v^{(2j-1)}), w = f(v^{(2j+1)})$. In the case $p \geq 2$, we conclude that

$$\|f(u^{(2j)}) - f(v^{(2j-1)})\|_p^p + \|f(u^{(2j)}) - f(v^{(2j+1)})\|_p^p \geq 2^{-p+2}\rho_j^p + 2^{-q+2}\beta^p - \rho_{j,1}^p - \rho_{j,2}^p$$
$$\geq 2\gamma^p + 2^{-p+2}\beta^p - 34\delta p\beta^p.$$

Thus choosing $\delta = \frac{2^{1-p}}{34p}$ yields the desired result for one of $(u^{(2j)}, v^{(2j-1)})$ or $(u^{(2j)}, v^{(2j+1)})$.

In the case $1 \leq p \leq 2$, we conclude that

$$\|f(u^{(2j)}) - f(v^{(2j-1)})\|_p^2 + \|f(u^{(2j)}) - f(v^{(2j+1)})\|_p^2 \geq \rho_j^2 + (p-1)\beta^2 - \rho_{j,1}^2 - \rho_{j,2}^2.$$

A similar choice of δ again yields the desired result.

References

1. Arora, S., Lee, J.R., Naor, A.: Euclidean distortion and the sparsest cut [extended abstract]. In: STOC 2005: Proceedings of the 37th Annual ACM Symposium on Theory of Computing, pp. 553–562. ACM Press, New York (2005)
2. Aumann, Y., Rabani, Y.: An O(logk) approximate min-cut max-flow theorem and approximation algorithm. SIAM J. Comput. 27(1), 291–301 (1998)
3. Benyamini, Y., Lindenstrauss, J.: Geometric nonlinear functional analysis. vol. 1. American Mathematical Society Colloquium Publications, vol. 48. American Mathematical Society, Providence (2000)

4. Chawla, S., Gupta, A., Räcke, H.: An improved approximation to sparsest cut. In: Proceedings of the 16th Annual ACM-SIAM Symposium on Discrete Algorithms, Vancouver. ACM Press, New York (2005)
5. Gupta, A., Krauthgamer, R., Lee, J.R.: Bounded geometries, fractals, and low-distortion embeddings. In: 44th Symposium on Foundations of Computer Science, pp. 534–543 (2003)
6. Krauthgamer, R., Lee, J.R., Mendel, M., Naor, A.: Measured descent: A new embedding method for finite metrics. Geom. Funct. Anal. 15(4), 839–858 (2005)
7. Laakso, T.J.: Plane with A_∞-weighted metric not bi-Lipschitz embeddable to $\mathbb{R}^N$. Bull. London Math. Soc. 34(6), 667–676 (2002)
8. Lang, U., Plaut, C.: Bilipschitz embeddings of metric spaces into space forms. Geom. Dedicata 87(1-3), 285–307 (2001)
9. Lee, J.R.: On distance scales, embeddings, and efficient relaxations of the cut cone. In: SODA 2005: Proceedings of the sixteenth annual ACM-SIAM symposium on Discrete algorithms, Philadelphia, PA, USA, pp. 92–101. Society for Industrial and Applied Mathematics (2005)
10. Lee, J.R.: Volume distortion for subsets of Euclidean spaces. Discrete Comput. Geom. 41(4), 590–615 (2009)
11. Linial, N., London, E., Rabinovich, Y.: The geometry of graphs and some of its algorithmic applications. Combinatorica 15(2), 215–245 (1995)
12. Matoušek, J.: On embedding expanders into l_p spaces. Israel J. Math. 102, 189–197 (1997)
13. Matoušek, J.: Lectures on discrete geometry. Graduate Texts in Mathematics, vol. 212. Springer, New York (2002)
14. Newman, I., Rabinovich, Y.: A lower bound on the distortion of embedding planar metrics into Euclidean space. Discrete Comput. Geom. 29(1), 77–81 (2003)
15. Rao, S.: Small distortion and volume preserving embeddings for planar and Euclidean metrics. In: Proceedings of the 15th Annual Symposium on Computational Geometry, pp. 300–306. ACM Press, New York (1999)

On Hardness of Pricing Items for Single-Minded Bidders

Rohit Khandekar, Tracy Kimbrel, Konstantin Makarychev,
and Maxim Sviridenko

IBM T.J. Watson research center, Yorktown Heights, NY 10598, USA
{rohitk,kimbrel,konstantin,sviri}@us.ibm.com

Abstract. We consider the following *item pricing* problem which has received much attention recently. A seller has an infinite numbers of copies of n items. There are m buyers, each with a budget and an intention to buy a fixed subset of items. Given prices on the items, each buyer buys his subset of items, at the given prices, provided the total price of the subset is at most his budget. The objective of the seller is to determine the prices such that her total profit is maximized.

In this paper, we focus on the case where the buyers are interested in subsets of size at most two. This special case is known to be APX-hard (Guruswami et al [1]). The best known approximation algorithm, by Balcan and Blum, gives a 4-approximation [2]. We show that there is indeed a gap of 4 for the combinatorial upper bound used in their analysis. We further show that a natural linear programming relaxation of this problem has an integrality gap of 4, even in this special case. Then we prove that the problem is NP-hard to approximate within a factor of 2 assuming the Unique Games Conjecture; and it is unconditionally NP-hard to approximate within a factor 17/16. Finally, we extend the APX-hardness of the problem to the special case in which the graph formed by items as vertices and buyers as edges is *bipartite*.

We hope that our techniques will be helpful for obtaining stronger hardness of approximation bounds for this problem.

1   Introduction

Many pricing questions in the IT industry stem from a specific cost structure: high fixed cost of production, but near-zero or zero variable cost of production. This cost structure characterizes a class of technology products which are collectively termed *digital goods*. Put differently, the cost of producing the first unit of a digital good is very high, but the cost of producing each additional unit is virtually zero. For instance, Microsoft spends hundreds of millions of dollars on developing each version of its Windows operating system. Once this first copy of the OS has been developed, however, it can be replicated at no cost. Other examples of digital goods are pay-per-view television programs, downloadable audio files, etc.

I. Dinur et al. (Eds.): APPROX and RANDOM 2009, LNCS 5687, pp. 202–216, 2009.

In this paper, we consider a problem of pricing digital goods that has received a lot of attention in the computer science community recently. Consider a *monopolistic* market with a single seller who has n digital goods to sell. Since the variable cost of production is near-zero, we assume that the seller has infinite copies of each good. Suppose that there are m buyers, each buyer i associated with a fixed budget $b_i > 0$, which is the maximum amount of money he is willing to spend. Each buyer is interested in buying some bundles of digital goods. For example, a buyer may be interested in buying an operating system together with an anti-virus software; but he may not be interested in buying them separately.

We further focus on the case where each buyer is interested in exactly one subset of goods. This setting is often referred to as a market with *single-minded* buyers. While this assumption may seem unnatural, it turns out that even this special case is computationally hard for the optimization problem we consider. The seller, who is assumed to know the demand and budget information, is then posed with the following problem of pricing goods. The seller must set a price $p_j \geq 0$ for each good j — she is not allowed to price the same item differently for different buyers. For a subset S of goods, let $p(S) = \sum_{j \in S} p_j$ denote the total price of goods in S. Once the prices are fixed, each buyer i buys his subset S_i of items if its total price is at most his budget, i.e., $p(S_i) \leq b_i$. If a buyer i satisfies this condition, he pays $p(S_i)$ to the seller. If on the other hand, this condition is not satisfied, buyer i buys nothing and pays nothing to the seller. In such a model, a natural objective for the seller is to price the items so as to maximize the total profit generated, i.e., to find prices $\{p_j\}$ so as to maximize $\sum_{i:p(S_i) \leq b_i} p(S_i)$.

1.1 Related Work

The problem of profit-maximizing pricing of goods in unlimited supply was introduced by Goldberg, Hartline, Karlin, Saks, and Wright [3]. In their setting, the buyers were interested in single goods and hence the optimization problem was trivial, and they focused on designing truthful mechanisms to maximize profit. There has been a lot of subsequent work on this and related models — below, we briefly survey only those results that are directly relevant to the problem we consider.

Guruswami, Hartline, Karlin, Kempe, Kenyon, and McSherry [1] considered the problem of profit maximization in a variety of settings, including single-minded bidders. They showed a logarithmic approximation guarantee and APX-hardness for the profit maximization problem. For single-minded bidders, a polylogarithmic hardness result was obtained by Demaine, Feige, Hajiaghayi, and Salavatipour [4]. The problem of the single-minded bidder case, where the size of the bundles demanded by the buyers was at most k, was considered by Briest and Krysta [5] who gave an $O(k^2)$ approximation for the problem, and was improved by Balcan and Blum [2] to $O(k)$. For the special case of $k = 2$, they obtain a 4-approximation algorithm.

The case of $k = 2$ (also called as the *graph pricing* problem) can be thought of as the following graph problem with goods as vertices and buyers as edges.

Consider an undirected graph on n vertices and m edges. There may be parallel edges and loops. Each edge e has a budget $b_e \geq 0$. Given prices $p_v \geq 0$ on the vertices v, an edge $e = (u, v)$ is *satisfied* if $p_u + p_v \leq b_e$. The goal is to set the prices to maximize the total profit generated: $\sum_{e=(u,v)\in E: p_u + p_v \leq b_e} (p_u + p_v)$. The 4-approximation algorithm of Balcan and Blum [2] for this case first reduces the problem to the case where G is a bipartite graph by losing a factor of 2 in the approximation. It then gives a 2-approximation on the bipartite graphs. Recently, Krauthgamer, Mehta, and Rudra [6] focused on the case $k = 2$ with further restriction that the budgets b_e are same for all the edges; but the graph may have self-loops. In such a case, they gave an LP-rounding algorithm that yields an approximation of $\frac{6+\sqrt{2}}{5+\sqrt{2}} \approx 1.15$. They also showed a matching integrality gap for these instances.

If we assume that the goods that are being sold are the edges of a graph and that buyers are purchasing paths in this graph, we can interpret this as the problem of pricing network connections, street segments (therefore termed the tollbooth problem [1]), or other types of transportation links (e.g., railway or flight connections). If the underlying graph is a path itself, then this problem is called the highway problem [1]. Interestingly, even this very restricted variant turns out to be intriguingly complex [5,7,1]. Hartline and Koltun [8] have presented a near-linear-time FPTAS for the practically relevant case that the number of goods for sale is a fixed constant.

1.2   Our Results and Techniques

In this paper, we focus on the graph pricing problem described above. The bundles of the buyers have at most two goods each, i.e., $k = 2$.

We first prove that the problem is hard to approximate within a factor of 2 assuming the Unique Games Conjecture, and within a factor of 17/16 assuming $P \neq NP$. To this end, we introduce a new problem which we call the Restricted Maximum Acyclic Subgraph problem: we are given a directed graph and our goal is to arrange its vertices on the real line so as to maximize the number of forward edges. However, unlike the Maximum Acyclic Subgraph problem, we can place every vertex v only in a specified set of positions S_v (see Section 2 for details). We show that the Graph Pricing problem is at least as hard to approximate as the Restricted Maximum Acyclic Subgraph problem (in Section 3). This immediately gives us a lower bound of 2, since Maximum Acyclic Subgraph is a special case of Restricted Maximum Acyclic Subgraph, and Maximum Acyclic Subgraph as was recently shown by Guruswami, Manokaran, Raghavendra [9], is hard to approximate within a factor of 2 assuming the Unique Games Conjecture.

MAX DICUT on directed acyclic subgraphs is also a special case of the Restricted Maximum Acyclic Subgraph problem. We can show that MAX DICUT on directed acyclic subgraphs is at least as hard to approximate as MAX CUT. (We omit the proof from this extended abstract.) This gives us an unconditional NP-hardness of 17/16. The inapproximability of MAX CUT was established by Håstad [10].

Then we initiate a study of several algorithmic approaches that might improve the approximation guarantee. Note the following trivial upper bound on the value of the optimal solution: allow each node v to collect its maximum profit $R(v)$ from the incident edges assuming that all its neighbors are priced at 0. The overall upper bound on the optimum solution is then $\sum_v R(v)$. This observation was used by Balcan and Blum [2] in their approximation algorithm that computes a solution of value at least $\frac{1}{4} \sum_v R(v)$, and thus gives a 4-approximation. It was not known however whether this analysis of the algorithm could be improved. We show that this upper bound indeed has a gap of 4. Therefore new upper bounds are required to get a better approximation factor.

A natural linear programming relaxation (LP) gives such an upper bound. This linear program can be thought of as a generalization of the one used by Krauthgamer, Mehta and Rudra [6] to the case of arbitrary budgets. Unfortunately, it turns out that this LP also has an integrality gap of 4. The proof again uses our reductions from Restricted Maximum Acyclic Subgraph and MAX DI-CUT on directed acyclic subgraphs. We take a directed acyclic graph $G = (V, A)$ in which every directed cut contains at most a $(1/4 + o(1))$ fraction of all edges. (A family of such graphs was recently constructed by Alon, Bollobàs, Gyàrfàs, Lehel, and Scott [11].) We show how to transform G to an instance of the Graph Pricing problem whose solutions correspond to directed cuts in G. Therefore, every combinatorial solution to this instance has value at most $(1/4 + o(1))|A|$. Meanwhile, there is an LP solution that collects a profit 1 from every edge, and thus has value $|A|$. We describe this transformation and its analysis in Section 3.2.

Finally, we analyze the bipartite case. Note that if we improved the algorithm for bipartite graphs, we would get an improvement over the 4-approximation of Balcan and Blum for general graphs. In particular, if we could solve the problem for bipartite graphs exactly, we would get a 2-approximation for general graphs. Unlike the general case of the graph pricing problem, the bipartite case was not even known to be NP-hard. We show that it is in fact APX-hard by a reduction from MAX CUT.

2   Preliminaries

Let us fix some notation. An instance of the Graph Pricing problem $\Pi = (G, b)$ is a pair consisting of a graph $G = (V, E)$ and a set of budgets $\{b_e\}$, $e \in E$. Throughout the paper we assume that the budgets are positive integers and that the graph does not have parallel edges or self-loops. A solution of the problem is an arbitrary assignment of prices to the vertices, i.e., a set of nonnegative real numbers $\{p_v\}_{v \in V}$. The profit of the solution is

$$\text{profit}_\Pi(p) = \sum_{e=(u,v) \in E} \begin{cases} p_u + p_v, & \text{if } p_u + p_v \leq b_e; \\ 0, & \text{otherwise.} \end{cases}$$

We denote the profit of the optimal solution by $OPT_\Pi = \max_{p_v \in \mathbb{R}^+ \cup \{0\}} \text{profit}_\Pi(p)$.

In the proof we consider a more general version of the Graph Pricing problem, in which the graph may have parallel edges and edges are weighted. We denote the weight of an edge e by w_e. We define the profit of a solution $\{\tilde{p}_v\}_{v \in V}$ of the generalized problem $\tilde{\Pi} = (\tilde{G}, \tilde{b})$ as

$$\text{profit}_{\tilde{\Pi}}(\tilde{p}) = \sum_{u,v} \sum_{e \in E(u,v)} w_e \cdot \begin{cases} (\tilde{p}_u + \tilde{p}_v), & \text{if } \tilde{p}_u + \tilde{p}_v \leq \tilde{b}_e; \\ 0, & \text{otherwise}; \end{cases}$$

here $E(u,v)$ denotes the set of edges going from u to v. We shall show that the Generalized Graph Pricing problem, even if we allow budgets and weights to be exponential in the number of vertices, is not harder than the standard Graph Pricing problem.

The Generalized Graph Pricing problem is a special case of the general constraint satisfaction problem with constraints depending on two variables (MAX 2GCSP). In our case, the variables are vertices; the constraints or payoff functions are functions

$$f_{\tilde{b}_e}(\tilde{p}_u, \tilde{p}_v) = \begin{cases} \tilde{p}_u + \tilde{p}_v, & \text{if } \tilde{p}_u + \tilde{p}_v \leq \tilde{b}_e; \\ 0, & \text{otherwise}. \end{cases}$$

Strictly speaking, prices can be arbitrary nonnegative real numbers, and thus the domain is infinite. However, if all budgets are positive integers in the range from 1 to B, then the prices in the optimal solution are semi-integral numbers in the range from 0 to B.

Lemma 1. *Consider an instance $\Pi = (G, b)$ of the Generalized Graph Pricing problem. Suppose that the budgets $\{b_e\}$ are integers in the range from 1 to B, then prices in one of the optimal solutions are semi-integral numbers in the range from 0 to B.*

Proof (sketch). Consider an arbitrary optimal solution $\{p_v\}_{v \in V}$. Let E' be the set of satisfied edges: $E' = \cup_{u,v}\{e \in E(u,v) : p_u + p_v \leq b_e\}$. Then $\{p_v\}_{v \in V}$ is a solution of the LP: maximize $\sum_{u,v} \sum_{e \in E' \cap E(u,v)} p_u + p_v$ subject to $p_u + p_v \leq b_e$ for all u, v, and $e \in E' \cap E(u,v)$. The LP is semi-integral and thus either all the p_v's are semi-integral numbers or another solution with the same objective value is semi-integral.

Since we consider only problem instances with integral budgets, we shall assume that all prices are semi-integral. Then the domain size equals $2B + 1$. Note that we could reduce the domain size even further to $O(\log_{(1+\varepsilon)} B) = O(\log(B)/\varepsilon)$ by rounding prices down to powers of $(1 + \varepsilon)$. This reduces the profit of the solution, but by no more than a factor of $(1 + \varepsilon)$.

We now show how to transform an arbitrary Generalized Graph Pricing instance $\tilde{\Pi} = (\tilde{G}, \tilde{b})$ to an unweighted Graph Pricing instance $\Pi = (G, b)$ without parallel edges. We use a relatively standard probabilistic construction that works for arbitrary constraint satisfaction problems. Without loss of generality we assume that the maximum weight is 1.

Input: an instance of Generalized Graph Pricing problem $\tilde{\Pi} = (\tilde{G} = (\tilde{V}, \tilde{E}), \tilde{b})$; a positive ε

Output: an unweighted instance of the Graph Pricing problem $\Pi = (G = (V, E), b)$

1. Let m be the total number of edges in the graph $\tilde{G}$; let w be the minimum (non-zero) edge weight.
2. Set $N = \lceil m/(w\varepsilon) \rceil^4$.
3. For every vertex v of the graph $\tilde{G}$, create N new vertices $v_1, \ldots, v_N$ in the graph G.
4. For every edge e between vertices u and v add an unweighted edge between u_i and v_j with probability $\alpha_e = \varepsilon w_e/m$. Set the budget of the new edge to be $b_{(u_i,v_j)} = \tilde{b}_e$. We call this edge a copy of e.
5. If an edge (u_i, v_j) is a copy of e and e' ($e \neq e'$) then remove (u_i, v_j) from G.

Lemma 2. *Consider an instance of the Generalized Graph Pricing problem $\tilde{\Pi} = (\tilde{G} = (\tilde{V}, \tilde{E}), \tilde{b})$ and an instance of the of the Graph Pricing problem $\Pi = (G = (V, E), b)$ obtained via the reduction above. Let $\gamma = \varepsilon N^2/m$. Then G is an unweighted graph without parallel edges; and with probability $1 - e^{-N}$,*

$$\frac{OPT_\Pi}{\gamma \, OPT_{\tilde{\Pi}}} = 1 + O(\varepsilon).$$

Proof. Consider an edge e between two vertices u and v in $\tilde{G}$. We add a copy of e between u_i and v_j at step 4 with probability α_e. The probability that we remove the edge at the last step is less than $\alpha_e \times m\alpha_e \leq \varepsilon$. Thus the probability β_e that the obtained graph G has the edge (u_i, v_j) is between $(1 - \varepsilon)\alpha_e$ and α_e.

Let $\mathcal{V}^u$ and $\mathcal{V}^v$ be arbitrary subsets of $\{u_i : 1 \leq i \leq N\}$ and $\{v_j : 1 \leq j \leq N\}$ respectively. Denote by $E_e(\mathcal{V}^u, \mathcal{V}^v)$ the set of copies of e going from $\mathcal{V}^u$ to $\mathcal{V}^v$. The expected size of $E_e(\mathcal{V}^u, \mathcal{V}^v)$ is $\beta_e|\mathcal{V}^u||\mathcal{V}^v|$. By a Bernstein or Chernoff type inequality,

$$\Pr\left(\left||E_e(\mathcal{V}^u, \mathcal{V}^v)| - \beta_e|\mathcal{V}^u| \cdot |\mathcal{V}^v|\right| \leq 4N^{3/2}\right) \leq 2e^{-\frac{16N^3}{2(\beta_e|\mathcal{V}^u|\cdot|\mathcal{V}^v|+N^{3/2}/3)}} \leq e^{-4N}.$$

The number of ways we can choose sets $\mathcal{V}^u$ and $\mathcal{V}^v$ is 2^{2N}. Thus, by the union bound, with probability at least $1 - e^{-2N}$, for all $\mathcal{V}^u \subset \{u_i : 1 \leq i \leq N\}$ and $\mathcal{V}^v \subset \{v_j : 1 \leq j \leq N\}$,

$$\left||E_e(\mathcal{V}^u, \mathcal{V}^v)| - \beta_e|\mathcal{V}^u| \cdot |\mathcal{V}^v|\right| \leq 4N^{3/2}.$$

Moreover, since the number of edges m is less than e^N, with probability $1 - e^{-N} > 0$, for all $u, v, e \in E(\mathcal{V}^u, \mathcal{V}^v)$, $\mathcal{V}^u \subset \{u_i : 1 \leq i \leq N\}$ and $\mathcal{V}^v \subset \{v_j : 1 \leq j \leq N\}$,

$$\left||E_e(\mathcal{V}^u, \mathcal{V}^v)| - \beta_e|\mathcal{V}^u| \cdot |\mathcal{V}^v|\right| \geq 4N^{3/2}. \tag{1}$$

We fix one of the random instances satisfying this condition. Given an arbitrary semi-integral solution p_{v_i} of the problem Π, we define a probabilistic solution of the original problem $\tilde{\Pi}$ as follows: for every vertex v pick a random i from 1 to

N and set $\tilde{p}_v = p_{v_i}$. For all v and all semi-integral q, let $\mathcal{V}_q^v = \{v_i : \tilde{p}_{v_i} = q\}$. The probability that we assign price q to u and s to v equals $|\mathcal{V}_q^u \times \mathcal{V}_s^v|/N^2$. Thus the expected profit of $\tilde{p}$ equals

$$\mathbb{E}\left[\text{profit}_{\tilde{\Pi}}(\tilde{p})\right] = \sum_{u,v} \sum_{e \in E(u,v)} \sum_{q,s} \frac{|\mathcal{V}_q^u||\mathcal{V}_s^v|}{N^2} \times w_e f_{b_e}(q,s).$$

The profit of p equals $\text{profit}_\Pi(p) = \sum_{u,v} \sum_{e \in E(u,v)} \sum_{q,s} |E_e(\mathcal{V}_q^u, \mathcal{V}_s^v)| \times f_{b_e}(q,s)$. Thus,

$$
\begin{aligned}
&\text{profit}_\Pi(p) - \gamma \cdot \mathbb{E}\left[\text{profit}_{\tilde{\Pi}}(\tilde{p})\right] \\
&= \sum_{u,v} \sum_{e \in E(u,v)} \sum_{q,s} (|E_e(\mathcal{V}_q^u, \mathcal{V}_s^v)| - \alpha_e |\mathcal{V}_q^u||\mathcal{V}_s^v|) \times f_{b_e}(q,s) \\
&\leq \sum_{u,v} \sum_{e \in E(u,v)} \sum_{q,s} (|E_e(\mathcal{V}_q^u, \mathcal{V}_s^v)| - \beta_e |\mathcal{V}_q^u||\mathcal{V}_s^v|) \times f_{b_e}(q,s) \quad \text{(from (1))} \\
&\leq m \times 4N^{3/2} \times \max_e b_e \leq \varepsilon \times \gamma w \max_e b_e.
\end{aligned}
$$

Since $OPT_\Pi \geq w \max_e b_e$, we have $OPT_\Pi \geq \gamma\, OPT_{\tilde{\Pi}}(1 + O(\varepsilon))$.

Similarly, given a solution $\{p_v\}_{v \in V}$ of the problem $\tilde{\Pi}$, we define a solution of Π as $\tilde{p}_{v_i} = p_v$. Then

$$
\begin{aligned}
&(1 - \varepsilon)\frac{\varepsilon N^2}{m} \cdot \text{profit}_{\tilde{\Pi}}(\tilde{p}) - \text{profit}_\Pi(p) \\
&\leq \sum_{u,v} \sum_{e \in E(u,v)} \sum_{q,s} ||E_e(\mathcal{V}_q^u, \mathcal{V}_s^v)| - \beta_e |\mathcal{V}_q^u||\mathcal{V}_s^v|| \times f_{b_e}(q,s) \\
&\leq m \times 4N^{3/2} \times \max_e b_e \leq \varepsilon \times \gamma w \max_e b_e.
\end{aligned}
$$

Thus, $(1 - \varepsilon)OPT_{\tilde{\Pi}} \geq \gamma\, OPT_\Pi(1 + O(\varepsilon))$.

Corollary 1. *Fix a positive integer B. Suppose that it is NP-hard to approximate the Generalized Graph Pricing problem within a factor of ρ if all budgets are bounded by B. Then for every positive ε, it is NP-hard to approximate the Graph Pricing problem within a factor $(1 - O(\varepsilon))\rho$.*

Proof. Consider an instance $\tilde{\Pi} = (\tilde{G}, \tilde{b})$ of the Generalized Graph Pricing problem with budgets bounded by B. Let m be the number of edges in the graph $\tilde{G}$. Rescale all weights so that the maximum weight equals 1. Remove all edges with weight less than $\varepsilon m/B$. This decreases OPT_Π by at most ε. We now transform the instance $\tilde{\Pi}$ to Π using the reduction from Lemma 2. By Lemma 2, $OPT_{\tilde{\Pi}} = (1 + O(\varepsilon))OPT_\Pi/\gamma$. Thus it is NP-hard to approximate the Graph Pricing problem within a factor $(1 - O(\varepsilon))\rho$.

Theorem 1. *Suppose that it is weakly NP-hard to approximate the Generalized Graph Pricing problem within a factor of ρ (i.e. it is NP-hard to approximate the problem within a factor of ρ when the budgets and weights can be exponentially* *large in the problem size). Then, assuming the Unique Games Conjecture, for every positive ε, it is NP-hard to approximate the Graph Pricing problem within a factor $(1 - O(\varepsilon)\rho)$.*

Proof. We show that there exists a finite set of budgets $\mathcal{B}$ such that if we require all budgets to be from the set $\mathcal{B}$, then the Graph Pricing problem is NP-hard to approximate within a factor of $(1 - O(\varepsilon))\rho$. This is an easy corollary from the recent result of Raghavendra [12]. Raghavendra showed that, assuming the Unique Games Conjecture, the best approximation ratio we can achieve for every 2GCSP problem Λ is at least the integrality gap of the problem Λ (up to any positive constant ε). The problem Λ is defined by a finite set of possible payoff functions and their finite domain.

As mentioned above, we may assume that prices take values in a domain of size $O(\log(B)/\varepsilon)$. Write the standard assignment SDP relaxation for the Generalized Graph Pricing problem (see e.g. Raghavendra [12] SDP (I)). This SDP can be solved in polynomial time. Thus its integrality gap is $(1 - O(\varepsilon))\rho$. Fix an integrality gap example with gap $(1 - O(\varepsilon))\rho$. Let $\mathcal{B} = \{1, \ldots, B\}$ be the set containing all budgets from this example. We now consider MAX 2GSP with the set of payoff functions $\{f_b\}_{b \in \mathcal{B}}$ and domain $\{0, 1/2, 1, \ldots, B\}$. Its integrality gap is at least $(1 - O(\varepsilon))\rho$. Thus by Raghavendra's theorem [12], it is NP-hard to approximate this MAX 2GCSP problem within a factor $(1 - O(1))\rho$. However, this MAX 2GCSP problem is just the Generalized Graph Pricing problem with budgets bounded by the constant B.

3   Reduction from Maximum Acyclic Subgraph

We introduce a new problem, which we call Restricted Maximum Acyclic Subgraph. We are given a graph $G = (V, A)$ and a collection of disjoint label sets $S_v \subset \mathbb{N}$ for all vertices v. The goal is to assign a label l_v from the set $S_v \cup \{0\}$ to every vertex v so as to maximize the number of arcs $(u, v) \in A$ for which $l_u < l_v$. The value of a solution is the number of such arcs. We denote the value of the solution $\{l_v\}_{v \in V}$ by $\text{value}_{(G,S)}(l)$; we denote the value of the optimal solution by $OPT_{(G,S)}$.

We now reduce the Restricted Maximum Acyclic Subgraph problem to the Generalized Graph Pricing problem. Given an arbitrary Restricted Maximum Acyclic Subgraph instance $G = (V, A)$, $\{S_v\}_v$ we construct an instance of the Generalized Graph Pricing problem $\Pi = (H, b)$ as follows. The vertices of the graph $H = (V, E)$ are the vertices of the graph G. The edges are triples $(u, v)_l$, where $(u, v) \in A$ and $l \in S_v$. The edge $(u, v)_l$ goes from u to v, has weight M^{-l} and budget $M^l(1 + 1/M)$, where M is a sufficiently large number we specify later. It is convenient to think that the edges are directed; whenever we write $(u, v)_l$ we mean that $(u, v) \in A$. The profit of a solution $\{p_v\}_{v \in V}$ equals

$$\text{profit}_\Pi(p) = \sum_{\substack{(u,v)_l \in E \\ p_u + p_v \leq M^l(1 + 1/M)}} M^{-l}(p_u + p_v).$$

We define the principal profit of the solution as

$$\sum_{\substack{(u,v)_l \in E \\ p_u + p_v \leq M^l(1+1/M)}} M^{-l} p_v;$$

and a principal profit of an edge $(u,v)_l$ as $M^{-l} p_v$. We say that a solution $\{p_v\}_{v \in V}$ is *canonical* if $p_v \in \{0\} \cup \{M^l : l \in S_v\}$ for all v. Every solution $\{l_v\}_{v \in V}$ of the Restricted Maximum Acyclic Subgraph problem corresponds to a canonical solution of the Generalized Graph Pricing problem:

$$p_v = \begin{cases} M^{l_v}, & \text{if } l_v \neq 0; \\ 0, & \text{otherwise.} \end{cases}$$

The principal profit of this solution satisfies

$$\sum_{\substack{(u,v)_l \in E \\ p_u + p_v \leq M^l(1+1/M)}} M^{-l} p_v \geq \sum_{\substack{(u,v) \in A \\ p_u + p_v \leq M^{l_v}(1+1/M)}} M^{-l_v} \cdot M^{l_v}$$

$$= \sum_{(u,v) \in A} \begin{cases} 1, & \text{if } l_v > l_u; \\ 0, & \text{otherwise;} \end{cases} = \text{value}_{(G,S)}(l).$$

Thus $\text{profit}_\Pi(p) \geq \text{value}_{(G,S)}(l)$; and $OPT_\Pi \geq OPT_{(G,S)}$. We now show that OPT_Π cannot be much bigger than $OPT_{(G,S)}$. First, we show that the principal profit of every solution almost equals the total profit.

Lemma 3. *The profit of an arbitrary solution $\{p_v\}_{v \in V}$ of the Generalized Graph Pricing problem $\Pi = (H, b)$ defined above is bounded as follows:*

$$\text{profit}_\Pi(p) \equiv \sum_{\substack{(u,v)_l \in E \\ p_u + p_v \leq M^l(1+1/M)}} M^{-l}(p_u + p_v) \leq \sum_{\substack{(u,v)_l \in E \\ p_u + p_v \leq M^l(1+1/M)}} M^{-l} p_v + 2n.$$

Proof. We need to show that $\sum_{(u,v)_l \in E : p_u + p_v \leq M^l(1+1/M)} M^{-l} p_u \leq 2n$. Fix a vertex u. All its outgoing edges have distinct weights and all weights are powers of M. Thus the sequence $M^{-l} p_u$ (where $(u,v)_l \in E$; $p_u \leq M^l(1+1/M)$) is a subsequence of a geometric progression with the largest term at most $(1+1/M)$. Hence

$$\sum_{\substack{v:(u,v)_l \in E \\ p_u + p_v \leq M^l(1+1/M)}} M^{-l} p_u \leq (1+1/M) \sum_{l=0}^{\infty} M^{-l} \leq 2.$$

We now show how every Generalized Graph Pricing solution can be transformed into a canonical solution.

Lemma 4. *For every solution $\{p_v\}_v$ of the problem Π defined above there exists a canonical solution $\{p'_v\}_v$ with the principal profit*

$$\sum_{\substack{(u,v)_l \in E \\ p'_u + p'_v \le M^l(1+1/M)}} M^{-l} p'_v \ge \sum_{\substack{(u,v)_l \in E \\ p_u + p_v \le M^l(1+1/M)}} M^{-l}(p_u + p_v) - (m/M + 2n).$$

Proof. Define

$$p'_v = \begin{cases} M^l, & \text{if } M^{l-1}(1+1/M) < p_v \le M^l(1+1/M) \text{ for some } l \in S_v \\ 0, & \text{otherwise} \end{cases}$$

We compare the principal profit of $\{p'_v\}_{v \in V}$ with the principal profit of $\{p_v\}_{v \in V}$. Consider an edge $(u,v)_l$ with a nonnegative contribution to the profit of $\{p_v\}_{v \in V}$. Then $p_u + p_v \le M^l(1+1/M)$ and both $p'_u, p'_v \le M^l$. Moreover, since $l \in S_v$ and thus $l \notin S_u$ (the sets S_u and S_v are disjoint), $p'_u \le M^{l-1}$. Therefore $p'_u + p'_v \le M^l(1+1/M)$. If $p_v > M^{l-1}(1+1/M)$, then $p'_v = M^l$; and the principal profit of the edge is $M^{-l}p'_v = 1$. If $p_v \le M^{l-1}(1+1/M)$, then $M^{-l}p_v < 2/M$. Hence, the difference $M^{-l}p_v - M^{-l}p'_v$ is always less than $2/M$. We get

$$\sum_{\substack{(u,v)_l \in E \\ p_u + p_v \le M^l(1+1/M)}} M^{-l}p_v \le \sum_{\substack{(u,v)_l \in E \\ p'_u + p'_v \le M^l(1+1/M)}} M^{-l}p'_v + 2m/M;$$

and by Lemma 3,

$$\sum_{\substack{(u,v)_l \in E \\ p_u + p_v \le M^l(1+1/M)}} M^{-l}(p_u + p_v) \le \sum_{\substack{(u,v)_l \in E \\ p'_u + p'_v \le M^l(1+1/M)}} M^{-l}(p'_u + p'_v) + m/M + 2n.$$

Theorem 2. *Consider an instance $(G = (V, A), S)$ of the Restricted Maximum Acyclic Subgraph problem. Let $\Pi = (H = (V, E), b)$ be the instance of the Graph Pricing problem obtained through the reduction described above. Then*

$$OPT^{RMAS}_{(G,S)} + 2m/M + 2n \ge OPT_\Pi \ge OPT^{RMAS}_{(G,S)}. \tag{2}$$

Proof. We have already proved that $OPT_\Pi \ge OPT^{RMAS}_{(G,S)}$. Thus we only need to prove the first inequality. Consider an arbitrary solution $\{p_v\}_{v \in V}$. By Lemma 4 there exists a canonical solution $\{p'_v\}_v$ with the principal profit

$$\sum_{\substack{(u,v)_l \in E \\ p'_u + p'_v \le M^l(1+1/M)}} M^{-l}p'_v \ge \sum_{\substack{(u,v)_l \in E \\ p_u + p_v \le M^l(1+1/M)}} M^{-l}(p_u + p_v) - (m/M + 2n).$$

Set labels l_u as follows: $l_u = \log_M p_u$ if $p_u \neq 0$; and $l_u = 0$ otherwise. If the principal profit of an edge $(u,v)_l$ is greater than $1/M$ then $p_v = M^l$ and $p_u \leq M^{l-1}$. Thus $l_u \leq l_v$ and the arc (u,v) contributes 1 to the value of solution. We get

$$OPT_{(H,S)}^{RMAS} + m/M \geq \sum_{\substack{(u,v)_l \in E \\ p'_u + p'_v \leq M^l(1+1/M)}} M^{-l} p'_v$$

$$\geq \sum_{\substack{(u,v)_l \in E \\ p_u + p_v \leq M^l(1+1/M)}} M^{-l}(p_u + p_v) - (m/M + 2n).$$

Hence $OPT_{(H,S)}^{RMAS} + 2m/M + 2n \geq OPT_G$.

3.1   UG Hardness

Theorem 3. *Assuming the Unique Games Conjecture, it is NP-hard to approximate the Graph Pricing problem within a factor $2 - \varepsilon$, for every positive ε.*

Proof. Guruswami, Manokaran, and Raghavendra [9] showed that it is NP-hard to approximate the Maximum Acyclic Subgraph problem within a factor of $2-\varepsilon$. Observe that the Maximum Acyclic Subgraph problem is a special case the Restricted Maximum Acyclic Subgraph problem, where sets S_v are chosen so that any ordering of vertices is possible. Hence, by Theorem 2 we can transform any graph G to an instance of the Generalized Graph Pricing problem Π satisfying

$$OPT_G^{MAS} + 2m/M + 2n \geq OPT_\Pi \geq OPT_G^{MAS},$$

where OPT_G^{MAS} denotes the size of maximum acyclic subgraph in G. Assume for a moment that $2m/M + 2n \leq \varepsilon\, OPT_G^{MAS}$. Then $(1 + \varepsilon)OPT_G^{MAS} \geq OPT_\Pi \geq OPT_G^{MAS}$; and thus the Generalized Graph Pricing problem is NP-hard to approximate within a factor of $2 - O(\varepsilon)$. Theorem 1 implies that the Graph Pricing problem is then also NP-hard to approximate within a factor of $2 - O(\varepsilon)$.

We now take care of the term $2m/M + 2n$. We replace every vertex v in G by $K = \lceil 1/\varepsilon \rceil$ new vertices $v_1, \ldots, v_K$ and every edge (u,v) with K^2 edges (u_i, v_j). The number of vertices in the graph increases K times; the number of edges and the size of the maximum acyclic subgraph increases exactly K^2 times. (Since vertices $v_1, \ldots v_K$ have exactly the same neighbors they can be arranged consecutively in the optimal solution.) Pick $M = mK^2$. Then $2m/M + 2n \leq 2 + \varepsilon OPT_G^{MAS}$.

3.2 LP Integrality Gap

We study the following LP relaxation:

$$\max \sum_{(u,v)\in E} \sum_{\substack{q,s \\ q+s\leq b_{(u,v)}}} (q+s)y_{uv}(q,s),$$

$$\text{subject to} \quad \sum_q x_u(q) = 1 \text{ for all } u$$

$$\sum_s y_{uv}(q,s) = x_u(q) \text{ for all } u,v,q$$

$$y_{uv}(q,s) = y_{vu}(q,s) \text{ for all } u,v,q,s$$

$$0 \leq x_u(q) \leq 1 \text{ for all } u,q$$

$$0 \leq y_{uv}(q,s) \leq 1 \text{ for all } u,v,q,s$$

In the *intended* integral solution, each $x_u(q)$ is the indicator variable of the event "the vertex u has price q," i.e., $x_u(q) = 1$, if $p_u = q$; $y_{uv}(q,s) = 1$, if u has price s, v has price q; and is equal to 0, otherwise. It is easy to see that in the intended integral solution all the constraints are satisfied. As before we assume that budgets $\{b_e\}_{e\in V}$ are integral and indexes q,s take semi-integral values in the range 0 to $\max_e b_e$. Note that the LP upper bound on the optimal solution is stronger than the combinatorial upper bound of Balcan and Blum (see the introduction). Indeed, for all u, we have

$$\sum_v \sum_{q,s:q+s\leq b_{(u,v)}} y_{uv}(q,s)\times q \leq \sum_q x_u(q) \sum_{v:q\leq b_{(u,v)}} q \leq \max_q \sum_v f_{(u,v)}(q,0) = R(u).$$

Our LP integrality gap example is based on the construction of Alon, Bollobàs, Gyàrfàs, Lehel, and Scott [11].

Theorem 4 (Alon et al. [11]). *There exists a directed acyclic graph G having m edges and $n = o(m)$ vertices, such that every directed cut of G contains at most $(1/4 + o(1))m$ edges.*

Theorem 5. *The integrality gap of the LP is $(4 - \varepsilon)$, for every positive ε.*

Proof. Let $G = (V, A)$ be the graph of Alon et al. [11]. We order the vertices of G in the reverse topological order. For every $v \in V$, let $o_v \in \{1, \ldots, n\}$ be the position of the vertex v in the ordering. Then if $(u, v) \in A$, $o_u > o_v$. Fix an integer parameter T. Construct an instance of the Restricted Maximum Acyclic Subgraph problem on graph G. Set $S_v = \{o_v \times T, o_v \times T + 1, \ldots, o_v \times T + T - 1\}$. For every edge $(u, v) \in A$, valid assignments of labels l_u and l_v that satisfy the inequality $l_u < l_v$ are $l_u = 0$; $l_v \in S_v$. Thus the value of any solution $\{l_v\}_{v\in V}$ equals the size of the directed cut between the sets $\{u : l_u = 0\}$ and $\{v : l_v \in S_v\}$. Therefore, the optimal value of the solution is at most $(1/4+o(1))m$. We transform (G, S) to an instance of the Generalized Graph Pricing problem (using Theorem 2) and then to an unweighted instance of the Graph Pricing problem (using Lemma 2). The profit of the optimal solution of the obtained problem $\Pi = (H = (V_H, E_H), b)$ is at most $(1/4 + O(\varepsilon))m \times \gamma N^2$ (if we choose M to be sufficiently large).

We now describe an LP solution of value $(1 - 1/T)m \times \gamma N^2$. Recall that the vertices of H are pairs in $V \times \{1, \ldots, N\}$ denoted v_i. The set of edges is a random subset of triples $(u_i, v_j)_l$, where $(u, v) \in A$, $l \in S_v$. The budget of $(u_i, v_j)_l$ is M^l. The probability that the edge $(u_i, v_j)_l$ is present in the graph is $\alpha'_{(u_i, v_j)_l} = \gamma/(M^l N^2)(1 - O(\varepsilon))$. We choose edges, so that the graph does not have parallel edges.

Set LP variables $x_{v_i}(M^l) = 1/T$, and $x_v(0) = 1/T$ for all vertices v_i and $l \in S_v$. Note that S_v contains exactly $T - 1$ elements, thus $x_{v_i}(0) + \sum_l x_{v_i}(M^l) = 1$. For every edge $(u, v)_l$ set $y_{u_i v_j}(0, M^l) = 1/T$. Set all other $y_{u_i v_j}(s, q)$ arbitrary to satisfy the LP constraints (e.g. $y_{uv}(M^l, 0) = 1/T$; $y_{uv}(M^{l'}, M^{l'}) = 1/T$ for $l' \neq l$).

If an edge $(u_i, v_j)_l$ is present in the graph, then its contribution to the LP objective function is at least $(0 + M^l) \times y_{u_i v_j}(0, M^l) = M^l/T$. Thus for every $(u, v) \in A$, the expected contribution of all edges $(u_i, v_j)_l$ is at least

$$\sum_{l \in S_v} \sum_{1 \leq i, j \leq N} \frac{(1 - O(\varepsilon))\gamma}{M^l N^2} \frac{M^l}{T} = (T-1) \cdot N^2 \cdot \frac{(1 - O(\varepsilon)) \cdot \gamma}{N^2 \cdot T} = (1 + O(\varepsilon)) \cdot \frac{(T-1) \cdot \gamma}{T}.$$

We have proved that for every positive ε, there exists a graph with the cost of the optimal solution at most $\gamma m/4 \times (1 + O(\varepsilon))$ and the cost of the LP at least $\gamma m \times (1 - O(\varepsilon))$. Hence the integrality gap is $4 - O(\varepsilon)$.

Remark 1. A similar construction shows that the problem is (unconditionally) at least as hard as MAX CUT, which as was shown by Håstad [10] cannot be approximated better than within a factor of $17/16$ (unless $P = NP$).

4   Hardness of the Bipartite Case

Balcan and Blum achieve a 4-approximation by reducing the general problem to the bipartite case, and they note that any improvement over the trivial 2-approximation for the bipartite case would immediately improve the 4-approximation for the general case. Here we show that the bipartite case is APX-hard, which to the best of our knowledge was previously unknown.

Theorem 6. *The Graph Pricing Problem in the bipartite case is APX-hard.*

Proof. We reduce the APX-hard problem MAX CUT to graph pricing in a bipartite graph. Let $G = (V, E)$ be a graph. For each vertex $u \in V$, we will construct a vertex u in our bipartite graph $G' = (V_1, V_2, E')$. For convenience we will refer to these corresponding nodes using the same names. All the original nodes in G will be on the same side of G', say V_1. For each edge $(u, v) \in E$ we construct the gadget shown in Figure 1. The proof of the following claim is omitted from this extended abstract.

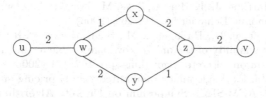

Fig. 1. Bipartite pricing gadget. Note that despite the layout, u and v belong to the same side of the bipartition.

Claim. If u and v both charge 0 or both charge 1, the maximum profit that can be gained from the gadget is 8, whereas profit 9 can be obtained if one charges 0 and the other charges 1.

Unfortunately, profit greater than 8 can be extracted from our gadget via fractional charges at u and/or v. To ensure that each node corresponding to a node in the original graph charges either 1 or ε for some small ε, we use the gadget displayed in Figure 2.

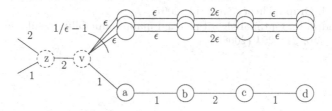

Fig. 2. Price-enforcing gadget: node v charges 1 or ε

Claim. For any given solution, a solution in which node v charges either 1 or ε and whose value is at least that of the given solution, can be found in polynomial time.

It should be clear now that (neglecting at most 2ε per edge), from a solution of value $24m+C$ to our bipartite graph pricing instance, we can recover a cut of size C to the original MAX CUT instance, where m is the number of edges. We can make the error insignificant by appropriate choice of ε. Thus the APX-hardness of MAX CUT implies the APX-hardness of the bipartite graph pricing.

References

1. Guruswami, V., Hartline, J., Karlin, A., Kempe, D., Kenyon, C., McSherry, F.: On profit-maximizing envy-free pricing. In: Proceedings, ACM-SIAM Symposium on Discrete Algorithms, pp. 1164–1173 (2005)
2. Balcan, M.-F., Blum, A.: Approximation algorithms and online mechanisms for item pricing. Theory of Computing 3, 179–195 (2007)

3. Goldberg, A., Hartline, J., Karlin, A., Saks, M., Wright, A.: Competitive auctions. Games and Economic Behavior 55, 242–269 (2006)
4. Demaine, E.D., Feige, U., Hajiaghayi, M., Salavatipour, M.R.: Combination can be hard: Approximability of the unique coverage problem. In: Proceedings, ACM-SIAM Symposium on Discrete Algorithms, pp. 162–171 (2006)
5. Briest, P., Krysta, P.: Single-minded unlimited supply pricing on sparse instances. In: Proceedings, ACM-SIAM Symposium on Discrete Algorithms, pp. 1093–1102 (2006)
6. Krauthgamer, R., Mehta, A., Rudra, A.: Pricing commodities, or how to sell when buyers have restricted valuations. In: Proceedings, Workshop on Approximation and Online Algorithms (2007)
7. Elbassioni, K., Sitters, R., Zhang, Y.: A quasi-ptas for profit-maximizing pricing on line graphs. In: Proceedings, European Symposium on Algorithms, pp. 451–462 (2007)
8. Hartline, J., Koltun, V.: Near-optimal pricing in near-linear time. In: Proceedings, Workshop on Algorithms and Data Structures, pp. 422–431 (2005)
9. Guruswami, V., Manokaran, R., Raghavendra, P.: Beating the random ordering is hard: Inapproximability of maximum acyclic subgraph. In: FOCS, pp. 573–582 (2008)
10. Håstad, J.: Some optimal inapproximability results. J. ACM 48, 798–859 (2001)
11. Alon, N., Bollobàs, B., Gyàrfàs, L.A.J., Scott, A.: Maximum directed cuts in acyclic digraphs. J. Graph Theory 55, 1–13 (2007)
12. Raghavendra, P.: Optimal algorithms and inapproximability results for every csp? In: STOC, pp. 245–254 (2008)

Real-Time Message Routing and Scheduling*

Ronald Koch, Britta Peis, Martin Skutella, and Andreas Wiese

TU Berlin, Institut für Mathematik, Straße des 17. Juni 136, 10623 Berlin, Germany
www.math.tu-berlin.de/coga/

Abstract. Exchanging messages between nodes of a network (e.g., embedded computers) is a fundamental issue in real-time systems involving critical routing and scheduling decisions. In order for messages to arrive on time, one has to determine a suitable (short) origin-destination path for each message and resolve conflicts between messages whose paths share a communication link of the network. We provide efficient routing strategies yielding origin-destination paths of bounded dilation and congestion. In particular, we can give good a priori guarantees on the time required to send a given set of messages which, under certain reasonable conditions, implies that all messages can be scheduled to reach their destination on time. Our algorithm uses a path-based LP-relaxation and iterative rounding. Finally, for message routing along a directed path (which is already $\mathcal{NP}$-hard), we identify a natural class of instances for which a simple scheduling heuristic yields provably optimal solutions.

1 Introduction

In a distributed real-time system, processes residing at different nodes of the network communicate by passing messages. One of the most challenging and important tasks for the design of a distributed system is the problem of sending a given set of messages through the network from the respective origin- to the destination nodes on time.

The message routing problem. To model the problem we represent the communication network by a (directed or undirected) graph $G = (V, E)$, whose edges correspond to the communication links of the network. In the *message routing problem*, each message $M_i = (s_i, t_i, d_i)$ of a given set of messages $\{M_i\}_{i \in I}$ consists of d_i packets of unit size that have to be sent from the origin node $s_i \in V$ to the destination node $t_i \in V$ within a certain time horizon $T > 0$. Usual constraints are (see e.g., [1,2], or [3, Chapter 37]):

(i) it takes one time unit to send a packet on any edge $e \in E$,
(ii) at most one packet can traverse an edge per time unit,
(iii) a message has to be completely received by a node before the node can start to transmit it to any other node.

---

* This work was partially supported by Berlin Mathematical School, by DFG research center MATHEON and by the DFG Focus Program 1307 within the project "Algorithm Engineering for Real-time Scheduling and Routing".

I. Dinur et al. (Eds.): APPROX and RANDOM 2009, LNCS 5687, pp. 217–230, 2009.

The last constraint is due to integrity checks performed by each node and implies that each message M_i has to be sent along a unique path P_i from its origin to its destination node.

Example 1. Consider the problem illustrated in Figure 1 where three messages need to be routed through a grid graph within a time horizon of twelve time units. Suppose we decide to send each message along the (unique) shortest path. Then, after three time steps there is a conflict between the second packet of message 1 and the first packet of message 2 that both want to traverse edge e in time step four. No matter which message is assigned a higher priority, we need at least 13 time steps to send all message from their sources to their destinations. On the other hand, if we choose the longer path $\{a, b, c, d, f, g\}$ for message 1, all messages can be sent within twelve time units since all paths are edge-disjoint.

Store-and-forward packet routing. In the special case where each message consists of only one packet, message routing reduces to *store-and-forward packet routing*, a fundamental routing problem in interconnection networks (see, e.g., Leighton's survey [4]). Store-and-forward packet routing can be formulated as an *integral dynamic multicommodity flow problem* with unit capacities and unit transit times on the edges. While this problem is known to be $\mathcal{NP}$-hard [5], store-and-forward packet routing can be solved efficiently by calculating a maximum flow over time in case all packets share the same origin and destination. In contrast, the message routing problem turns out to be $\mathcal{NP}$-hard even in the special case where all messages have the same origin and destination [1]. Thus, message routing is considerably harder than packet routing.

We would like to mention that the possibility of storing packets is crucial in the message routing model we consider, since packets need to wait at intermediate nodes for the entire message. Therefore, our problem considerably differs from the well-studied *direct routing problem* in which the packets are not allowed to be stored at intermediate nodes on the way to their destination.

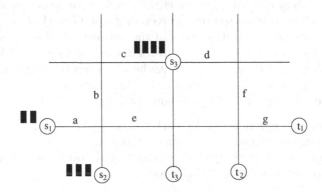

Fig. 1. Message routing problem with three messages and time horizon twelve. The messages consist of two, three, and four packets, respectively.

Routing and scheduling. A natural approach for solving the message routing problem is the following two-stage strategy. In the first stage (the *routing stage*), determine the set of paths $\{P_i\}_{i \in I}$. Then, in the second stage (the *scheduling stage*), resolve conflicts between messages sharing an edge. Of course, in order to determine good solutions, the paths chosen in the routing stage must feature certain desirable properties that guarantee the existence of good solutions to the second stage scheduling problem.

Congestion and dilation. If the paths $\{P_i\}_{i \in I}$ are given, we immediately obtain two trivial lower bounds on the minimum amount of time needed to send all messages, which we call the *makespan* of the problem. The first one is the *congestion*

$$C = \max_{e \in E} \sum_{i \in I : e \in P_i} d_i,$$

i.e., the maximum number of packets that have to traverse a single edge. The second one is the *dilation*

$$D = \max_{i \in I}(d_i|P_i|),$$

i.e., the maximum time necessary to send a message without any delays from its origin to its destination. As usual, $|P_i|$ denotes the number of edges in path P_i. As we will see in the following, C and D not only provide lower bounds on the makespan but also good upper bounds in terms of C and D can be determined.

A related job shop scheduling problem. Given paths $\{P_i\}_{i \in I}$, it remains to declare priorities on the messages whenever two packets of different messages meet at an intermediate node and want to use the same outgoing edge. However, this is exactly an instance of the well-studied *acyclic preemptive job shop scheduling problem*. Every edge corresponds to a machine and a message is a job that has to be consecutively processed on the machines corresponding to the edges on its path. In shop scheduling, the processing requirement of a job is usually machine-dependent. In our case, however, we have the special property that the processing requirement of a job/message is identical (namely equal to the size of the message) on each machine/edge on its path.

It is well-known that even this special case of acyclic preemptive job shop scheduling is $\mathcal{NP}$-hard and even $\mathcal{NP}$-hard to approximate[1] with performance guarantee $\frac{5}{4} - \epsilon$ for any $\epsilon > 0$ [6]. On the positive side, Feige and Scheideler [7] prove the existence of a schedule with makespan $\mathcal{O}(C + D \log \log d_{\max})$ for the preemptive job shop scheduling problem in general by using the non-constructive General Lovász Local Lemma (LLL). (Here, $d_{\max}$ denotes the maximum operation-length, resp. message-size.) An algorithmic version of the General LLL can be found in [8]. In the special case where all operation lengths

---

[1] An *α-approximation algorithm* for an optimization problem is a polynomial-time algorithm which computes a solution whose value is at most a factor α away from the optimum. The number α is called the performance guarantee of the algorithm.

are identical, Leighton et al. [9,10] even establish an efficient randomized algorithm which computes a schedule with makespan $\mathcal{O}(C + D)$. Busch, Magdon-Ismail, and Mavronicolas [11] prove that intermediate storage of packets can be avoided at the cost of an additional poly-logarithmic factor in the makespan.

Desirable properties of paths. We now return to our discussion of the two-stage approach to message routing discussed above. As a consequence of the scheduling results mentioned in the previous paragraph, a promising approach is to determine a set of paths in the routing stage such that $C + D$ is relatively small. The first constant-factor approximation for the special case of store-and-forward packet routing, established by Srinivasan and Teo [12], is also based on this idea. Basically, Srinivasan and Teo establish a constant-factor approximation for the problem to find paths minimizing $C + D$. Combining this result with the $\mathcal{O}(C + D)$-schedule for acyclic job shop scheduling with constant operation lengths (proved in [9,10]), they obtain a constant-factor approximation for packet routing. A similar idea has been used by Fleischer and Skutella [13] in the general context of dynamic network flow problems.

Our contributions. In Section 2 we describe an algorithm that, given a set of messages $\{M_i\}_{i \in I}$ on a communication network, and a desired dilation Δ, finds a set of paths of dilation at most Δ and congestion smaller than $C^*(\Delta) + \Delta$, where $C^*(\Delta)$ denotes the congestion of an optimal fractional solution with dilation at most Δ. The dilation Δ that is given to the algorithm as an input can be chosen arbitrarily (e.g., $\Delta = T/2$). Of course, the smaller the dilation Δ is, the larger is the optimal congestion $C^*(\Delta)$. In practice it is thus reasonable to try several values of $\Delta \leq T$ in order to find a good tradeoff between dilation and congestion. In theory, one can, for example, use binary search in order to determine Δ such that $\Delta + C^*(\Delta)$ or $\Delta + (C^*(\Delta) + \Delta)$ (or some other function of Δ and $C^*(\Delta)$) is minimal.

Although our algorithm can be applied for arbitrary message lengths, it even improves upon the performance guarantee of [12] for the special case of store-and-forward packet routing by a multiplicative factor of two. The main difference between our approach and the approach in [12] is our use of a path-based linear programming formulation which turns out to be efficiently solvable as the corresponding separation problem is a special case of the *length-bounded shortest path problem*. (The latter can be solved with a modification of Dijkstra's algorithm). Given an optimal solution to the linear program, we apply iterative rounding to turn the fractional solution into an integral one, and guarantee that the congestion is not increased by more than Δ.

Our path-finding algorithm works for arbitrary directed or undirected graphs. Combined with either approximation algorithms for the acyclic job shop scheduling problem, or with suitable priority heuristics, it therefore returns solutions for the message routing problem in general. In many situations in practice, however, the communication graphs are very simple. It therefore makes sense to consider the problem on special graph classes. In Section 3 we consider the message routing problem on directed paths (which is already $\mathcal{NP}$-hard [1]), and show that the *Farthest-Destination-First Algorithm* works optimally on a directed path P in case the messages are not nested, i.e., in case

$$s_i <_P s_j \quad \Longrightarrow \quad t_i \leq_P t_j \quad \forall i, j \in I.$$

2   Routing with Small Congestion and Dilation

Note that any set of edge-disjoint paths $\{P_i\}_{i \in I}$, where the length of each path P_i is bounded by $\frac{T}{d_i}$, forms a solution to the message routing problem: all messages can be sent directly without any delay from their origin to their destination nodes where they arrive before time T. Of course, such length-bounded edge-disjoint paths do not necessarily exist (it is $\mathcal{NP}$-hard to decide whether they do exist or not [14]). However, some delays are allowed if the path-lengths do not meet the upper bounds $(\frac{T}{d_i})_{i \in I}$. Thus, we restrict to shorter paths on which we minimize the congestion.

Given a suitable value $\Delta \leq T$ (which can, for example, be determined by binary search), we define for each $i \in I$ the set of paths

$$\mathcal{P}_i := \left\{ s_i, t_i\text{-paths in } G \text{ of length at most } \frac{\Delta}{d_i} \right\}$$

and $\mathcal{P} := \bigcup_{i \in I} \mathcal{P}_i$. Among $\mathcal{P}$, we are looking for a set of representatives $\{P_i\}_{i \in I}$ with minimal congestion. That is, we are interested in an optimal integral solution to the following linear program

$$
\begin{aligned}
\min \; & C \\
\text{s.t.} \; & \sum_{P \in \mathcal{P}_i} x_P \geq 1 & \forall i \in I, \\
& \sum_{i \in I} \sum_{P \in \mathcal{P}_i : e \in P} d_i x_P \leq C & \forall e \in E, \\
& x_P \geq 0 & \forall P \in \mathcal{P}.
\end{aligned}
$$

Note that the paths in the support of any feasible integral solution $\hat{x} \in \{0, 1\}^{|\mathcal{P}|}$ of the linear program above with objective value $\hat{C}$ yield a set of representatives $\{P_i\}_{i \in I}$ with dilation at most Δ and congestion $\hat{C}$: the first set of constraints ensures that at least one path is found for each message, while the second set of constraints guarantees that the total number of packets traversing a single edge does not exceed $\hat{C}$.

2.1   Optimal Fractional Solutions

To find a good integral solution to the linear program above, we first determine an optimal fractional solution x^* with objective value C^*, and then, in a second step, round x^* to an integral solution $\hat{x} \in \{0,1\}^{|\mathcal{P}|}$ whose congestion is at most $C^* + \Delta$. At first sight, it seems to be impossible to find an optimal fractional solution in polynomial time, since the number of variables is in general exponential in the size of the underlying network G. However, if we consider the dual linear program, we get

$$\max \sum_{i \in I} z_i$$

$$\text{s.t.} \sum_{e \in E} y_e \leq 1$$

$$\sum_{e \in P} y_e \geq \frac{z_i}{d_i} \qquad \forall P \in \mathcal{P}_i, i \in I$$

$$y_e, z_i \geq 0 \qquad \forall e \in E, i \in I.$$

The corresponding separation problem can be formulated as a *length-bounded shortest path problem*: find a shortest s_i, t_i-path with respect to the edge costs y_e among those paths containing at most $\frac{\Delta}{d_i}$ edges. In contrast to the general length-bounded shortest path problem with arbitrary edge lengths (which is known to be $\mathcal{NP}$-hard [14]), this problem can be solved efficiently with a modification of Dijkstra's algorithm (*sketch:* in each iteration of Dijkstra's algorithm determine a shortest path among those with at most $1, 2, \ldots$ edges). Thus, by the equivalence of optimization and separation [15], an optimal fractional solution to the dual and thus also to the primal linear program can be found in polynomial time. (I.e., we do not need to consider all path-variables in the LP. Instead, we iteratively solve the LP for small subsets of variables, where in each step a variable corresponding to the shortest length-bounded path is added to the LP in case the reduced costs are negative.) In practice, column generation seems to be the most suitable technique to actually solve the primal linear programming problem.

2.2   Iterative Rounding

Given the upper bound Δ on the dilation of paths and an optimal fractional solution x^* with objective value C^* to the corresponding linear program, we now describe how to round the fractional solution to an integral one while increasing congestion at most by Δ.

In the rounding algorithm described below, we iteratively solve a linear programming relaxation and fix a path P_i for message i as soon as the corresponding variable x_{P_i} attains value 1. In the following, F is the set of those messages i for which a path P_i has already been fixed. Initially, F is empty. The messages in F are removed from I such that I only contains the messages for which a path remains to be fixed. In each step of the algorithm, we thus solve the following linear program (LP):

$$\min \ C$$

$$\sum_{P \in \mathcal{P}_i} x_P \geq 1 \qquad\qquad\qquad \forall i \in I \qquad\qquad (1)$$

$$\sum_{i \in I} \sum_{P \in \mathcal{P}_i : e \in P} d_i x_P \leq C - \sum_{i \in F : e \in P_i} d_i \qquad\qquad \forall e \in E \qquad\qquad (2)$$

$$x_P \geq 0 \qquad\qquad\qquad \forall P \in \mathcal{P}.$$

The basic idea of the algorithm is as follows: in each iteration, we fix the integral variables and drop at least one of the constraints, before we solve the (LP) again. That is, in each iteration, whenever there is an index i with $x_P^* = 1$ for some $P \in \mathcal{P}_i$, we move index i from I to F. Moreover, we remove all paths not in the support of x^* from $\mathcal{P}$. After fixing the integral variables, we can easily find a constraint of type (2) which can be dropped from the updated (LP): the reason is that even if all remaining variables are rounded up to 1, the right-hand side of the inequality is not violated by more than Δ (see Theorem 1).

Algorithm 1 (Iterative Rounding Algorithm)
1. *Initialize: $F \leftarrow \emptyset$;*
2. *Compute a basic optimum solution x^* to (LP);*
3. *For $i \in I$, let $\mathcal{P}_i \leftarrow \{P \in \mathcal{P}_i \mid x_P^* > 0\}$;*
4. *WHILE $\exists i \in I$ and $P_i \in \mathcal{P}_i$ with $x^*(P_i) = 1$ DO*
   - *Set $I \leftarrow I \setminus \{i\}$;*
   - *Set $F \leftarrow F \cup \{i\}$;*
5. *Set $\mathcal{P} \leftarrow \bigcup_{i \in I} \mathcal{P}_i$;*
6. *WHILE $\mathcal{P} \neq \emptyset$ DO*
   - *Drop a constraint of type (2) with*

$$\sum_{i \in I} \sum_{P \in \mathcal{P}_i : e \in P} d_i < C^* - \sum_{i \in F : e \in P_i} d_i + \Delta;$$

   - *GoTo step 2;*

Note that in a single iteration of our algorithm, we do not round fractional variables explicitly, but simply fix the integral variables. The "rounding" is thus done by solving in each iteration the modified linear program corresponding to the remaining fractional variables. It remains to show that the algorithm is well-defined, i.e., we need to show the following: in case the set $\mathcal{P}$ of non-integral components is non-empty, we can find an edge $e \in E$ such that the congestion cannot be violated by more than Δ, even if all non-integral components are rounded up to one.

Theorem 1. *If x^* is a basic optimum solution to (LP) with $0 < x_P^* < 1$ for all $P \in \mathcal{P}$, then there exists a constraint of type (2) such that for the corresponding edge $e \in E$ holds*

$$\sum_{i \in I} \sum_{P \in \mathcal{P}_i : e \in P} d_i < C^* - \sum_{i \in F : e \in P_i} d_i + \Delta.$$

The theorem can be derived from a more general result shown in [16], stating that any fractional solution x^* of a linear equality system $Ax = b$ can be rounded to an integral vector $\hat{x}$ satisfying $A\hat{x} < b + \Delta$, whenever the sum of positive entries in each column of matrix A is bounded from above by Δ, and the sum of negative entries in each column is bounded from below by $-\Delta$. However, the proof turns out to be much simpler for our special inequality system:

Proof. Let $n = |\mathcal{P}|$. Since x^* is a basic feasible solution, there exist linearly independent tight constraints $\mathcal{T}_1$ and $\mathcal{T}_2$ of type (1) and (2), respectively, such that

$$n = |\mathcal{T}_1| + |\mathcal{T}_2|.$$

Observe that for each constraint $j \in \mathcal{T}_1$ we have

$$\Delta \sum_{P \in \mathcal{P}_j} x_P^* = \Delta. \tag{3}$$

Suppose by contradiction that for each e corresponding to a constraint in $\mathcal{T}_2$, we have

$$\sum_{i \in I} \sum_{P \in \mathcal{P}_i : e \in P} d_i \geq C^* - \sum_{i \in F : e \in P_i} d_i + \Delta. \tag{4}$$

Since

$$\sum_{i \in I} \sum_{P \in \mathcal{P}_i : e \in P} d_i x_P^* = C^* - \sum_{i \in F : e \in P_i} d_i$$

holds by the tightness of the constraint, equation (4) turns out to be equivalent to

$$\sum_{i \in I} \sum_{P \in \mathcal{P}_i : e \in P} d_i (1 - x_P^*) \geq \Delta. \tag{5}$$

Summing up the inequalities of type (3) and (5) for all constraints in $\mathcal{T}_1$ and $\mathcal{T}_2$, we get

$$n\Delta \leq \sum_{j \in \mathcal{T}_1} \Delta \sum_{P \in \mathcal{P}_j} x_P^* + \sum_{e \in \mathcal{T}_2} \sum_{i \in I} \sum_{P \in \mathcal{P}_i : e \in P} d_i (1 - x_P^*)$$

$$= \sum_{i \in I} \sum_{P \in \mathcal{P}_i} \left(\chi_i^{\mathcal{T}_1} \Delta x_P^* + \sum_{e \in \mathcal{T}_2 \cap P} d_i (1 - x_P^*) \right)$$

$$\leq \sum_{i \in I} \sum_{P \in \mathcal{P}_i} \left(\Delta x_P^* + \Delta (1 - x_P^*) \right) = n\Delta,$$

where $\chi_i^{\mathcal{T}_1} \in \{0, 1\}$ is an indicator variable with $\chi_i^{\mathcal{T}_1} = 1$ iff $i \in \mathcal{T}_1$. Since $0 < x_P^* < 1$ for all $P \in \mathcal{P}$, the second inequality in the derivation above is an equality only if for all $i \in I$ and all paths $P \in \mathcal{P}_i$ the following two conditions are satisfied.

1. $\chi_i^{\mathcal{T}_1} = 1$, and
2. $\sum_{e \in \mathcal{T}_2 \cap P} d_i = \Delta$.

If we now consider each column of (LP) separately, add the column's entries corresponding to constraints of type (2) and subtract the column's entries corresponding to constraints of type (1), we achieve a result of 0 in each column. This demonstrates that T_1 and T_2 must be linearly dependent constraints. A contradiction!   □

Thus, after at most $|E|$ iterations, the algorithm terminates with an integral vector $\hat{x} \in \{0,1\}^{|\mathcal{P}|}$, whose support contains a path P_i for each message $i \in I$. It is guaranteed that each path P_i does not contain more than $\frac{\Delta}{d_i}$ edges, and that the congestion of the paths violates the congestion of the optimal fractional solution by at most Δ.

Corollary 1. *Given Δ, the rounding algorithm determines a set of paths $\{P_i\}_{i \in I}$ with dilation $\leq \Delta$ and congestion $\leq C^* + \Delta$, where C^* is the minimum possible congestion of fractional paths with dilation Δ.*

2.3   Individual Deadlines

In a more general model of the message routing problem, each message M_i is additionally equipped with a certain deadline $D_i > 0$, denoting the latest point in time when the message must be received by the destination node t_i. We want to emphasize that our algorithm might as well be applied in this more general setting: we simply restrict the path lengths with respect to the deadlines. That is, instead of choosing a value Δ which is not greater than the overall time horizon T, we choose a factor $q \in (0,1]$ and consider for each message M_i the collection of paths

$$\mathcal{P}_i := \left\{ s_i, t_i\text{-paths in } G \text{ of length at most } q\frac{D_i}{d_i} \right\}.$$

This guarantees a dilation of at most

$$\Delta = \max_{i \in I} qD_i$$

and a congestion of at most $C^* + \Delta$.

2.4   Arbitrary Travel Times

The algorithm above can also be applied in a further extension of the message routing problem, where travel times $\tau(e) \in \mathbb{N}_{>0}$ are associated with all edges $e \in E$. Here $\tau(e)$ denotes the time it takes for one packet to traverse e. Thus, a message of size d_i completely traverses edge e in $\tau(e) + d_i - 1$ time units. Further, if message $i \in I$ is to be sent along path P_i, it takes at least

$$\tau^i(P_i) := \sum_{e \in P_i} (d_i + \tau(e) - 1)$$

time steps before the message is completely received by its destination node t_i. These observations show that the dilation for a given set of paths $\{P_i\}_{i \in I}$ in this more general model becomes

$$D := \max_{i \in I} \tau^i(P_i),$$

while the congestion $C = \max_{e \in E} \sum_{i \in I : e \in P_i} d_i$ remains unchanged. Note that we can adopt our algorithm to handle travel times by defining for a given value $\Delta \leq T$ the collections of paths

$$\mathcal{P}_i := \{s_i, t_i\text{-paths with } \tau^i(P) \leq \Delta\} \quad \forall i \in I.$$

However, with arbitrary travel times, the corresponding separation problem to our linear relaxation (LP) is the *general* length-bounded shortest path problem. While this problem is $\mathcal{NP}$-hard, it can be solved approximately in the following sense: for any $\epsilon > 0$, one can find in time polynomial in the size of the network G and $\frac{1}{\epsilon}$ an s_i, t_i-path P with $\tau^i(P) \leq (1 + \epsilon)\Delta$ whose cost is bounded from above by the cost of a shortest path in $\mathcal{P}^i$ [17,18,19]. As before, the fractional solution (which is now a $(1+\epsilon)$-approximation to the optimal one) can be turned into an integral solution with the rounding algorithm described above, since the inequality $\sum_{e \in P} d_i \leq \Delta$ still holds for each path $P \in \mathcal{P}_i$ and $i \in I$. Thus, we achieve the following result.

Corollary 2. *Even if each edge $e \in E$ is equipped with a travel time $\tau(e) \in \mathbb{N}_{>0}$, a slight modification of the algorithm above returns a set of paths whose dilation is bounded by $(1+\epsilon)\Delta$ and whose congestion differs from the optimal congestion by an additive factor of at most $(1 + \epsilon)\Delta$. Here, $\epsilon > 0$ can be chosen arbitrarily small.*

Given the set of paths $\{P_i\}_{i \in I}$ with congestion C and dilation D, the remaining problem of determining priority rules in order to minimize the makespan, can again be formulated as an acyclic job shop scheduling problem: to incorporate the travel times, we simply define for each message $i \in I$ and each edge $e \in E$ with $e \in P_i$ an additional machine e^i. After job i has been executed on machine e for d_i time steps, it needs to be processed on machine e^i for $\tau(e) - 1$ time steps, before it can proceed to the next machine corresponding to the successive edge of e in P_i.

   Note that processing times in the resulting acyclic job shop scheduling problem depend on both, the job and the machine. However, as already mentioned in the introduction, schedules of length $\mathcal{O}(C + D \log \log \ell_{\max})$ can be found for this more general problem. (In our model, $\ell_{\max}$ denotes the maximum of all travel times and message sizes).

3   Message Routing on Paths

In this section we consider instances of the message routing problem where the underlying network is a directed path. Since the path taken by any message is unique on such instances, no routing decisions but only scheduling decisions have to be taken. That is, an algorithm for the message routing problem must only resolve conflicts if two messages want to traverse the same edge at the same point in time. This can be done by assigning priorities to the messages such that a message with higher priority is sent first. More precisely, even if a message is currently being sent while a message with a strictly higher priority arrives, the latter message is sent instantaneously. Thus, an interruption of the message of lower priority occurs.

The following example illustrates that a wrong choice of a priority rule can lead to arbitrarily bad schedules.

Example 2. Suppose n messages $\{M_i\}_{i=0}^{n-1}$ start at the same origin node and need to be sent along a directed path. Each message M_i consists of $d_i = 2^i$ packets and needs to traverse 2^{n-i} edges before it reaches its destination. First we consider a schedule, where messages with farther destination get a higher priority. In order to send message i we wait at the origin until the first $i - 1$ messages are sent and then traverse the path without any additional delay. Thus message i arrives at its destination at time $\sum_{k=0}^{i-1} 2^k + 2^i \cdot 2^{n-i} \le 2^{n+1}$. Therefore the optimal makespan is at most 2^{n+1}.

In contrast, we next consider a schedule where messages with farther destination are assigned lower priorities. Then the makespan is determined by the completion time of the smallest message 0. Furthermore, any message i is sent without additional delay on its last $2^{n-i} - 2^{n-i-1} + 1$ edges and each message smaller than i is sent immediately after i on these edges. Thus each messages i adds at least $(2^{n-i} - 2^{n-i-1})2^i = 2^n - 2^{n-1}$ time units to the completion time of message 0. Thus the makespan of this schedule is at least $n(2^n - 2^{n-1}) = \frac{n}{4}2^{n+1}$. This shows that the gap to the optimal makespan can grow linearly in the number of messages.

In this example the *Farthest-Destination-First Algorithm* (FDFA for short) leads to an optimal schedule. FDFA assigns a higher priority to messages which have a farther destination according to the order of the underlying path. In case of ties, messages with a later origin node get higher priority. If both origin and destination of two messages coincide, ties are broken arbitrarily.

FDFA seems to be a good choice for the message routing problem on directed paths in general. But, since the problem is known to be $\mathcal{NP}$-hard [1], there surely exist examples where FDFA is not optimal:

Example 3. Consider a directed path consisting of four edges and three messages 1, 2, and 3. Message 1 must be sent from the first to the last edge and has size 1, whereas messages 2 and 3 must be sent from the second to the third edge and have both size $1 + \epsilon$ for small enough $\epsilon > 0$ (see Figure 2).

Then the optimum solution has a makespan of $4 + 2\epsilon$ and the solution of FDFA has a makespan of $5 + 2\epsilon$. Thus the performance guarantee of FDFA cannot be better than $\frac{5}{4}$.

In this section, we identify a large class of problems where FDFA is guaranteed to be optimal. But before, let us introduce some notation. For a message routing

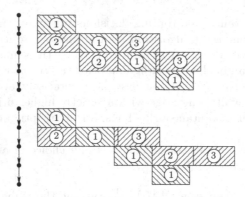

Fig. 2. Schedules of Example 3 showing that the approximation ratio of FDFA is not less than $\frac{5}{4}$. The optimum schedule is illustrated above the FDFA-schedule.

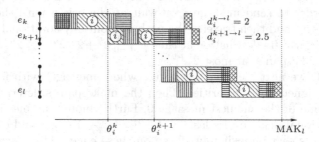

Fig. 3. Setting of Lemma 1

instance the underlying directed path P is given by node set $V(P) := \{v_1, \ldots, v_n\}$ and edge set $E(P) := \{e_k := (v_k, v_{k+1}) \mid k = 1, \ldots, n-1\}$. We say that a message *experiences additional delay* or *is additionally delayed* on edge e in a given schedule, if the starting time of i on e is strictly greater than the end time of i on the predecessor edge. The *makespan* on an edge e is the earliest point in time when all its messages have been sent through e. A time interval where no message traverses a particular edge is called *idle time*. (The infinitely long time interval after the makespan of an edge is not called idle time).

We show that the Farthest-Destination-First algorithm computes an optimum solution on non-nested instances. For this we need improved bounds on the minimum makespan combining dilation and congestion.

Lemma 1. *Consider an arbitrary feasible schedule. Let $e_k, e_l \in E(P)$ with $k \leq l$ be two edges of P and $i \in I$ be a message which must pass these edges. Let θ_i^k be the time when i has completely traversed e_k and let $d_i^{k \to l}$ be the total amount of messages passing e_k and e_l and traversing e_k after time θ_i^k (see Figure 3). Then a lower bound on the makespan occurring on e_l is $\theta_i^k + d_i^{k \to l} + d_i(l - k)$.*

Proof. The proof is illustrated in Figure 3. We prove this by induction over $l - k$. If $l - k = 0$ then the statement is of course true. Let MAK_l be the makespan of edge l. By the induction hypothesis we know for given k and l with $l - k \geq 1$ that

$$\text{MAK}_l \geq \theta_i^{k+1} + d_i^{k+1\to l} + d_i(l - k - 1). \tag{6}$$

Further let Δ be the total size of messages passing e_k and e_l, traversing e_k after time θ_i^k and e_{k+1} before time θ_i^{k+1}. Then we get:

$$\theta_i^{k+1} \geq \theta_i^k + d_i + \Delta \tag{7}$$
$$\Delta \geq d_i^{k\to l} - d_i^{k+1\to l} \tag{8}$$

Combining these inequalities leads to

$$\text{MAK}_l \geq \theta_i^k + d_i^{k\to l} + d_i(l - k). \tag{9}$$

This completes the proof. □

Note that the bounds in the previous lemma depend on the considered schedule. The following corollary states a lower bounds on the minimum makespan on a particular edge over all feasible schedules.

Corollary 3. *Let $e_k, e_l \in E(P)$ with $k \leq l$ be two edges of P and $i \in I$ be a message which must pass both of these edges. Let $d^{k\to l}$ be the total size of messages passing e_k and e_l. Then a lower bound on the minimum makespan on edge e_l is $d^{k\to l} + d_i(l - k)$.*

Proof. Given an arbitrary schedule and a message i passing e_k and e_l we know $d^{k\to l} \leq \theta_i^k + d_i^{k\to l}$. Since $d^{k\to l}$ and d_i are independent of the considered schedule, the corollary follows directly from Lemma 1. □

Next we show that FDFA computes a schedule minimizing the makespan if the underlying instance does not contain nested messages. Let $<_P$ be the topological order of P. Recall that two messages $i_1, i_2 \in I$ are nested if one is strictly contained in the other one, i.e., if $s_{i_1} <_P s_{i_2} \leq_P t_{i_2} <_P t_{i_1}$ or vice versa.

Theorem 2. *Consider an instance of the message routing problem where no two messages are nested. Then FDFA computes a schedule which minimizes the makespan on each edge simultaneously.*

Proof. Due to the lack of space, we refer for the proof to the full version of the paper. □

Acknowledgements. The authors are much indebted to Fritz Eisenbrand for helpful discussions on the topic of this paper.

References

1. Leung, J.Y.T., Tam, T.W., Wong, C.S., Young, G.H.: Routing messages with release time and deadline constraints. Journal of Parallel and Distributed Computing 31, 65–76 (1995)
2. Leung, J.Y.T., Tam, T.W., Wong, C.S., Young, G.H.: Online routing of real-time messages. Journal of Parallel and Distributed Computing 34, 211–217 (1996)

3. Leung, J.Y.T.: Handbook of Scheduling: Algorithms, Models and Performance Analysis. Chapman and Hall/CRC, Boca Raton (2004)
4. Leighton, F.: Methods for message routing in parallel machines. In: Proceedings of the ACM Symposium on the Theory of Computing, pp. 77–96 (1992)
5. Hall, A., Hippler, S., Skutella, M.: Multicommodity flows over time: Efficient algorithms and complexity. Theoretical Computer Science 379, 387–404 (2007)
6. Williamson, D.P., Hall, L.A., Hoogeveen, J.A., Hurkens, C.A.J., Lenstra, J.K., Sevast'janov, A.V., Shmoys, D.B.: Short shop schedules. Operations Research 45, 288–294 (1997)
7. Feige, U., Scheideler, C.: Improved bounds for acyclic job shop scheduling. Combinatorica 3(22), 361–399 (2002)
8. Czumaj, A., Scheideler, C.: A new algorithmic approach to the general Lovász Local Lemma with applications to scheduling and satisfiability problems. In: Proc. of the thirty-second annual ACM symposium on Theory of computing, Portland, USA, pp. 38–47 (2000)
9. Leigthon, F.T., Maggs, B.M., Rao, S.B.: Packet routing and job shop scheduling in O(congestion+dilation) steps. Combinatorica 14, 167–186 (1994)
10. Leigthon, F.T., Maggs, B.M., Richa, A.: Fast algorithms for finding O(congestion+dilation) packet routing schedules. Combinatorica 19, 375–401 (1999)
11. Busch, C., Magdon-Ismail, M., Mavronicolas, M.: Universal bufferless packet switching. SIAM Journal on Computing 37, 1139–1162 (2007)
12. Srinivasan, A., Teo, C.-P.: A constant-factor approximation algorithm for packet routing and balancing local vs. global criteria. SIAM Journal on Computing 30(6), 2051–2068 (2001)
13. Fleischer, L., Skutella, M.: Quickest flows over time. SIAM Journal on Computing 36, 1600–1630 (2007)
14. Garey, M.R., Johnson, D.S.: Computers and Intractability. W. H. Freeman and Co., New York (1979)
15. Grtschel, M., Lovsz, L., Schrijver, A.: The ellipsoid method and its consequences in combinatorial optimization. Combinatorica 1(2), 169–197 (1981)
16. Karp, R.M., Leigthon, F.T., Rivest, R.L., Thompson, C.D., Vazirani, U.V., Vazirani, V.V.: Global wire routing in two-dimensional arrays. Algorithmica 2, 113–129 (1987)
17. Hassin, R.: Approximation schemes for the restricted shortest path problem. Math. Oper. Research 17, 36–42 (1992)
18. Lorenz, D.H., Raz, D.: A simple efficient approximation scheme for the restricted shortest path problem. Oper. Res. Letters 28, 213–219 (2001)
19. Phillips, C.A.: The network inhibition problem. In: Proc. of the 25th Annual ACM Symposium on the Theory of Computing, San Diego, CA, pp. 776–785 (1993)

Approximating Some Network Design Problems with Node Costs

Guy Kortsarz[1] and Zeev Nutov[2]

[1] Rutgers University, Camden
guyk@camden.rutgers.edu
[2] The Open University of Israel
nutov@openu.ac.il

Abstract. We study several multi-criteria undirected network design problems with node costs and lengths with all problems related to the node costs Multicommodity Buy at Bulk (MBB) problem in which we are given a graph $G = (V, E)$, demands $\{d_{st} : s, t \in V\}$, and a family $\{c_v : v \in V\}$ of subadditive cost functions. For every $s, t \in V$ we seek to send d_{st} flow units from s to t on a single path, so that $\sum_v c_v(f_v)$ is minimized, where f_v the total amount of flow through v. In the Multicommodity Cost-Distance (MCD) problem we are also given lengths $\{\ell(v) : v \in V\}$, and seek a subgraph H of G that minimizes $c(H) + \sum_{s,t \in V} d_{st} \cdot \ell_H(s, t)$, where $\ell_H(s, t)$ is the minimum ℓ-length of an st-path in H. The approximation for these two problems is equivalent up to a factor arbitrarily close to 2. We give an $O(\log^3 n)$-approximation algorithm for both problems for the case of demands polynomial in n. The previously best known approximation ratio for these problems was $O(\log^4 n)$ [Chekuri et al., FOCS 2006] and [Chekuri et al., SODA 2007]. This technique seems quite robust and was already used in order to improve the ratio of Buy-at-bulk with protection (Antonakopoulos et al FOCS 2007) from $\log^3 h$ to $\log^2 h$. See [3].

We also consider the Maximum Covering Tree (MaxCT) problem which is closely related to MBB: given a graph $G = (V, E)$, costs $\{c(v) : v \in V\}$, profits $\{p(v) : v \in V\}$, and a bound C, find a subtree T of G with $c(T) \leq C$ and $p(T)$ maximum. The best known approximation algorithm for MaxCT [Moss and Rabani, STOC 2001] computes a tree T with $c(T) \leq 2C$ and $p(T) = \Omega(\text{opt}/\log n)$. We provide the first nontrivial lower bound and in fact provide a *bicriteria* lower bound on approximating this problem (which is stronger than the usual lower bound) by showing that the problem admits no better than $\Omega(1/(\log \log n))$ approximation assuming NP $\not\subseteq$ Quasi(P) *even if the algorithm is allowed to violate the budget by any universal constant* ρ. This disproves a conjecture of [Moss and Rabani, STOC 2001].

Another related to MBB problem is the Shallow Light Steiner Tree (SLST) problem, in which we are given a graph $G = (V, E)$, costs $\{c(v) : v \in V\}$, lengths $\{\ell(v) : v \in V\}$, a set $U \subseteq V$ of terminals, and a bound L. The goal is to find a subtree T of G containing U with $\text{diam}_\ell(T) \leq L$ and $c(T)$ minimum. We give an algorithm that computes

I. Dinur et al. (Eds.): APPROX and RANDOM 2009, LNCS 5687, pp. 231–243, 2009.
© Springer-Verlag Berlin Heidelberg 2009

a tree T with $c(T) = O(\log^2 n) \cdot \mathsf{opt}$ and $\mathsf{diam}_\ell(T) = O(\log n) \cdot L$. Previously, a polylogarithmic bicriteria approximation was known only for the case of edge costs and edge lengths.

Keywords: Network design, Node costs, Multicommodity Buy at Bulk, Covering tree, Approximation algorithm, Hardness of approximation.

1 Introduction

Network design problems require finding a minimum cost (sub-)network that satisfies prescribed properties, often connectivity requirements. The most fundamental problems are the ones with $0, 1$ connectivity requirements. Classic examples are: Shortest Path, Min-Cost Spanning Tree, Min-Cost Steiner Tree/Forest, Traveling Salesperson, and others. Examples of problems with high connectivity requirements are: Min-Cost k-Flow, Min-Cost k-Edge/Node-Connected Spanning Subgraph, Steiner Network, and others. All these problems also have practical importance in applications.

Two main types of costs are considered in the literature: the edge costs and the node costs. We consider the latter, which is usually more general than the edge costs variants; indeed, for most undirected network design problems there is a very simple reduction that transforms edge costs to node costs, but the inverse is, in general, not true. The study of network design problems with node costs is already well motivated and established from both theoretical as well as practical considerations [5, 6, 8, 10, 12]. For example, in telecommunication networks, expensive equipment such as routers and switches are located at the nodes of the underlying network, and thus it is natural to model some of these problems by assigning costs on the nodes rather than to the edges.

For some previous work on undirected network-design problems with node costs see the work of Klein and Ravi [10], Guha et al. [8], Moss and Rabani [12], and Chekuri et al. [5, 6]. We mostly focus on resolving some open problems posed in these papers.

1.1 Problems Considered

Given a *length function* ℓ on edges/nodes of a graph H, let $\ell_H(s,t)$ denote the ℓ-distance between s, t in H, that is, the minimum ℓ-length of an st-path in H (including the lengths of the endpoints). Let $\mathsf{diam}_\ell(H) = \max_{s,t \in V(H)} \ell_H(s,t)$ be the ℓ-*diameter of* H, that is the maximum ℓ-distance between two nodes in H. We consider the following two related problems on undirected graphs.

Multicommodity Buy at Bulk (MBB)
Instance: A graph $G = (V, E)$, a family $\{c_v : v \in V\}$ of sub-additive monotone non-decreasing cost functions, a set D of pairs from V, and positive demands $\{d_{st} : \{s,t\} \in D\}$.
Objective: Find a set $\{P_{st} : \{s,t\} \in D\}$ of st-paths so that $\sum_{v \in V} c_v(f_v)$ is minimized, where $f_v = \sum\{d_{st} : \{s,t\} \in D, v \in P_{st}\}$.

Multicommodity Cost-Distance (MCD)

Instance: A graph $G = (V, E)$, costs $\{c(v) : v \in V\}$, lengths $\{\ell(v) : v \in V\}$, a set D of pairs from V, and positive integral demands $\{d_{st} : \{s, t\} \in D\}$.

Objective: Find a subgraph H of G that minimizes

$$w(H, D) = c(H) + \sum_{\{s,t\}\in D} d_{st} \cdot \ell_H(s, t) \tag{1}$$

As linear functions are subadditive, MCD is a special case of MBB. The following statement shows that up to a factor arbitrarily close to 2, MCD and MBB are equivalent w.r.t. approximation.

Proposition 1 ([2]). *If there exists a ρ-approximation algorithm for* MCD *then there exists a $(2\rho + \varepsilon)$-approximation algorithm for* MBB *for any $\varepsilon > 0$.*

We consider two other fundamental problems closely related to MBB (see an explanation below):

Maximum Covering Tree (MaxCT)

Instance: A graph $G = (V, E)$, costs $\{c(v) : v \in E\}$, profits $\{p(v) : v \in V\}$, and a bounds C.

Objective: Find a subtree T of G with $c(T) \leq C$ and $p(T)$ maximum.

Shallow-Light Steiner Tree (SLST)

Instance: A graph $G = (V, E)$, costs $\{c(v) : v \in V\}$, lengths $\{\ell(v) : v \in V\}$, a set $U \subseteq V$ of terminals, and a bound L.

Objective: Find a subtree T of G containing U with $\mathrm{diam}_\ell(T) \leq L$ and $c(T)$ minimum.

Each one of the problems MBB and SLST has an "edge version", where the costs/lengths are given on the edges. As was mentioned, the edge version admits an easy approximation ratio preserving reduction to the node version.

1.2   The Unifying Theme of the Problems Considered

A bicriteria approximation algorithm for the following problem was used to derive an $O(\log^4 n)$-approximation algorithm for MBB [9].

k-Buy at Bulk Steiner Tree (k-BBST)

Instance: A graph $G = (V, E)$, costs $\{c(v) : v \in V\}$, lengths $\{\ell(v) : v \in V\}$, a set $U \subseteq V$ of terminals, a root $r \in V - U$, diameter bound L, cost bound C, and an integer k.

Question: Does G has a subtree T containing r and at least k terminals so that $\mathrm{diam}_\ell(T) \leq L$ and $c(T) \leq C$?

Theorem 1 ([9]). *Suppose that there exists a polynomial time algorithm that given a YES-instance of k-BBST finds a tree T containing r with $\Omega(k)$ terminals so that $c(T) = \rho_1 \cdot C$ and $\mathrm{diam}_\ell(T) = \rho_2 \cdot L$. Then* MBB *admits an approximation algorithm with ratio $O(\log n) \cdot \rho_1 + O(\log^3 n) \cdot \rho_2$.*

Theorem 2 ([9]). *There exist a polynomial time algorithm that given a YES-instance of k-BBST finds a a tree T containing at lest $k/8$ terminals so that $c(T) = O(\log^3 n) \cdot C$ and $\mathsf{diam}_\ell(T) = O(\log n) \cdot L$.*

Thus improved algorithm for k-BBST would imply a better approximation algorithm for MBB. It seems hard to improve the bicriteria $O(\log^3 n, \log n)$ approximation for k-BBST given in [9]. Hence we consider *relaxations* of k-BBST, hoping that they may shed light on MBB. Also, MaxCT and SLST are interesting in their own right. The MaxCT problem is similar to k-BBST. For unit terminal costs, setting cost bound k is the same as seeking a tree with k terminals, and maximizing the profit. What makes MaxCT much easier than k-BBST is that MaxCT has no length constrains. In particular, the primal-dual approach of [12] does not seem suitable to handle lengths constrains as well hence does not seem suited to handle k-BBST. The SLST is easier than k-BBST from another point of view. In SLST, given a cost and diameter bounds, a tree that is *both shallow and light* is required. But this is the case $k = |U|$ namely the problem of covering *all* terminals (and not only k as in k-BBST). The difference seems quite significant, and thus k-BBST seems significantly harder to handle than SLST. In summary, one may hope that techniques for MaxCT that are able to find a tree with k terminals and low cost (but cant handle lengths), could somehow be combined with techniques that do produce a tree that is both shallow and light, but work only for $k = |U|$, getting a better approximation for k-BBST.

1.3   Related Work

We survey some results on relevant network design problems with node costs. Klein and Ravi [11] showed that the Node-Weighted Steiner Tree problem is Set-Cover hard, thus it admits no $o(\log n)$ approximation unless P=NP [13]. They also obtained a matching approximation ratio using a greedy merging algorithm. Guha et al. [8] showed $O(\log n)$ integrality gap of a natural LP-relaxation for the problem. The MBB problem is motivated by economies of scale that arise in a number of applications, especially in telecommunication. The problem is studied as the fixed charge network flow problem in operations research. The first approximation algorithm for the problem is by Salman et al. [14]. For the multi-commodity version MBB the first non-trivial result is due to Charikar and Karagiazova [4] who obtained an $O(\log |D| \exp(O(\sqrt{\log n \log \log n})))$-approximation, where $|D|$ is the sum of the demands. In [5] an $O(\log^4 n)$-approximation algorithm is given for the edge costs case, and further generalized to the node costs case in [6]. See [1] for an $\Omega(\log^{1/2-\varepsilon} n)$-hardness result.

The MaxCT problem was introduced in [8] motivated by efficient recovery from power outage. In [8] a pseudo approximation algorithm is presented that returns a subtree T with $c(T) \le 2C$ and $p(T) = \Omega(P/\log^2 n)$, where P is the maximum profit under budget cost C. This was improved in [12] to produce a tree T with $c(T) \le 2C$ and $p(T) = \Omega(P/\log n)$. For a related minimization problem when one seeks to find a minimum cost tree T with $p(T) \ge P$ [12] gives an $O(\ln n)$-approximation algorithm.

1.4 Our Results

The previously best known ratio for MCD/MBB was $O(\log^4 n)$ both for edge costs [5] and node costs [6], and this was also so for polynomial demands. We improve this by using, among other things, a better LP-relaxation for the problem.

Theorem 3. MCD/MBB *with polynomial demands admits an* $O(\log^3 n)$-*approximation algorithm.*

The technique used is quite robust. It was already used in [3] to improve the approximation ratio for Buy-at-bulk with protection (see [3]) from $O(\log^3 h)$ to $O(\log^2 h)$.

Our next result is for the MaxCT problem. In [12] it is conjectured that MaxCT admits an $O(1)$ approximation algorithm (which would have been quite helpful for dealing with k-BBST). We disprove this conjecture. Since the upper bound is a bicriteria upper bound, we give a bicriteria lower bound (which is stronger than the usual lower bound).

Theorem 4. MaxCT *admits no constant approximation algorithm unless* $\mathrm{NP} \subseteq \mathrm{DTIME}(n^{O(\log n)})$ *even if the algorithm is allowed to use a budget of* $\rho \cdot B$ *for any universal constant* ρ. MaxCT *admits no* $o(\log \log n)$ *approximation algorithm unless* $\mathrm{NP} \subseteq \mathrm{DTIME}(n^{\mathrm{polylog}(n)})$ *even if the algorithm is allowed to use* $\rho \cdot B$ *budget for any universal constant* ρ.

Our last result is for the SLST problem. For SLST with *edge costs* and *edge lengths*, the algorithm of [11] computes a tree T with $c(T) = O(\log n) \cdot$ opt and $\mathrm{diam}_\ell(T) = O(\log n) \cdot L$. We consider the more general case of node costs and node lengths.

Theorem 5. SLST *with node costs and lengths admits an approximation algorithm that computes a tree* T *with* $c(T) = O(\log^2 n) \cdot$ opt *and* $\mathrm{diam}_\ell(T) = O(\log n) \cdot L$.

Theorems 3 and 4 are proved in Sections 2 and 3.

2   Improved Algorithm for MBB

In this section we prove Theorem 3. We give an $O(\log^2 n \cdot \log N)$-approximation algorithm for MCD with running time polynomial in N, where N is the sum of the demands plus n. If N is polynomial in n, the running time is polynomial in n, and the approximation ratio is $O(\log^3 n)$. We may assume (by duplicating nodes) that all demands are 1. Then our problem is:

Instance: A graph $G = (V, E)$, costs $\{c(v) : v \in V\}$, lengths $\{\ell(v) : v \in V\}$, and a set D of node pairs.

Objective: Find a subgraph H of G minimizing $w(H, D) = c(H) + \displaystyle\sum_{\{s,t\}\in D} \ell_H(s,t)$.

For the latter problem, we give an $O(\log^2 n \cdot \log |D|)$-approximation algorithm.

2.1   Approximate Greedy Algorithm and Junction Trees

We use a result about the performance of a *Greedy Algorithm* for the following type of problems:

Covering Problem
Instance: A groundset Π and functions ν, w on 2^Π with $\nu(\Pi) = 0$.
Objective: Find $\mathcal{P} \subseteq \Pi$ with $\nu(\mathcal{P}) = \nu(\Pi)$ and with $w(\mathcal{P})$ minimized.

Let $\rho > 1$ and let opt be the optimal solution value for the Covering Problem. The ρ-*Greedy Algorithm* starts with $\mathcal{P} = \emptyset$ and iteratively adds subsets of $\Pi - \mathcal{P}$ to $\mathcal{P}$ one after the other using the following rule. As long as $\nu(\mathcal{P}) > \nu(\Pi)$ it adds to $\mathcal{P}$ a set $\mathcal{R} \subseteq \Pi - \mathcal{P}$ so that

$$\sigma_\mathcal{P}(\mathcal{R}) = \frac{w(\mathcal{R})}{\nu(\mathcal{P}) - \nu(\mathcal{P} + \mathcal{R})} \leq \frac{\rho \cdot \text{opt}}{\nu(\mathcal{P}) - \nu(\Pi)} . \tag{2}$$

The following known statement follows by a standard set-cover analysis, c.f., [10].

Theorem 6. *If ν is decreasing and w is increasing and subadditive, then the ρ-Greedy Algorithm computes a solution $\mathcal{P}$ with $w(\mathcal{P}) \leq \rho \cdot [\ln(\nu(\emptyset) - \nu(\Pi)) + 1] \cdot \text{opt}$.*

In our setting, Π is the family of all st-paths, $\{s, t\} \in D$. For a set $\mathcal{R} \subseteq \Pi$ of paths connecting a set R of pairs in D, let $\nu(\mathcal{R}) = |D| - |R|$ be the number of pairs in D not connected by paths in $\mathcal{R}$, and let $w(\mathcal{R}) = c(\mathcal{R}) + \sum_{\{s,t\} \in R} \ell(P_{st})$, where $c(\mathcal{R})$ denotes the cost of the union of the paths in $\mathcal{R}$, and P_{st} is the shortest st-path in $\mathcal{R}$. Note that $\nu(\Pi) = 0$ and $\nu(\emptyset) = |D|$. We will show how to find such $\mathcal{R}$ satisfying (2) with $\rho = O(\log^2 n)$. W.l.o.g., we may consider the case $\mathcal{P} = \emptyset$. (Otherwise, we consider the residual instance obtained by excluding from D all pairs connected by $\mathcal{P}$ and setting $\mathcal{P} = \emptyset$; it is easy to see that if $\mathcal{R}$ satisfies (2) for the residual instance, then this is also so for the original instance.) Assuming $\mathcal{P} = \emptyset$, (2) can be rewritten as:

$$\sigma(\mathcal{R}) = \frac{c(\mathcal{R})}{|R|} + \frac{\sum_{\{s,t\} \in R} \ell(P_{st})}{|R|} \leq \frac{\rho \cdot \text{opt}}{|D|} . \tag{3}$$

The quantity $\sigma(\mathcal{R})$ in (3) is the *density* of $\mathcal{R}$; it is a sum of "cost-part" $c(\mathcal{R})/|R|$ and the remaining "length-part". The following key statement from [5] shows that with $O(\log n)$ loss in the length part of the density, we may restrict ourselves to very specific $\mathcal{R}$, as given in the following definition; in [5] it is stated for edge-costs, but the generalization to node-costs is immediate.

Definition 1. *A tree T with a designated node r is a* junction tree *for a subset $R \subseteq D$ of node pairs in T if the unique paths in T between the pairs in R all contain r.*

Lemma 1 ([5], The Junction Tree Lemma). *Let H^* be an optimal solution to an MCD instance with $\{0, 1\}$ demands. Let $C = c(H^*)$ and let $L = \sum_{\{s,t\} \in D} \ell_{H^*}(s, t)$. Then there exists a junction tree T for a subset $R \subseteq Q$ of pairs, so that $\text{diam}_\ell(T) = O(\log n) \cdot L/|D|$ and $c(T)/|R| = O(C/|D|)$.*

If we could find a pair T, R as in Lemma 1 in polynomial time, then we would obtain an $O(\log|D| \cdot \log n)$-approximation algorithm, by Theorem 6. In [5] it is shown how to find such a pair that satisfies (3) with $\rho = O(\log^3 n)$. We will show how to find such a pair with $\rho = O(\log^2 n)$.

Theorem 7. *There exists a polynomial time algorithm that given an instance of* MCD *with* $\{0, 1\}$ *demands computes a set* R *of paths connecting a subset* $R \subseteq D$ *of pairs satisfying* (3) *with* $\rho = O(\log^2 n)$.

Motivated by Lemma 1, the following LP was used in [5, 6]. Guess the common node r of the paths in R of the junction tree T. Let U be the union of pairs in D. Relax the integrality constraints by allowing "fractional" nodes and paths. For $v \in V$, x_v is the "fraction of v" taken into the solution. For $u \in U$, y_u is the total amount of flow v delivers to r. In the LP, we require $y_s = y_t$ for every $\{s, t\} \in D$, so $y_s = y_t$ amount of flow is delivered from s to t via r. For $u \in U$ let Π_u be the set of all ur-paths in Π, and thus $\Pi = \cup_{u \in U} \Pi_u$. For $P \in \Pi$, f_P is the amount of flow through P. Dividing all variables by $|R|$ (note that this does not affect the objective value), gives the following LP:

$$(\text{LP1}) \min \ \sum_{v \in V} c(v) \cdot x_v + \sum_{P \in \Pi} \ell(P) \cdot f_P$$

$$
\begin{array}{ll}
\text{s.t.} \quad \sum_{u \in U} y_u = 1 & \\
\sum_{\{P \in \Pi_u | v \in P\}} f_P \leq x_v & v \in V, \ u \in U \\
\sum_{P \in \Pi_u} f_P \geq y_u & u \in U \\
y_s - y_t = 0 & \{s, t\} \in D \\
x_v, f_P, y_u \geq 0 & v \in V, \ P \in \Pi, \ u \in U
\end{array}
$$

2.2   The LP Used

Let $A \cdot \log n \cdot L/|D|$ be the bound on the lengths of the paths in R guaranteed by Lemma 1. We use almost the same LP as (LP1), except that we seek to minimize the cost only, and restrict ourselves to paths of length at most $A \cdot \log n \cdot L/|D|$, which reflects better the statement in Lemma 1. For $\Pi' \subseteq \Pi$ let $\tilde{\Pi}' = \{P \in \Pi' : \ell(P) \leq A \cdot \log n \cdot L/|D|\}$. Again recall that y_u is the flow delivered from u to r. The LP we use is:

$$(\text{LP2}) \min \ \sum_{v \in V} c(v) \cdot x_v$$

$$
\begin{array}{ll}
\text{s.t.} \quad \sum_{u \in U} y_u = 1 & \\
\sum_{\{P \in \tilde{\Pi}_u | v \in P\}} f_P \leq x_v & v \in V, \ u \in U \\
\sum_{P \in \tilde{\Pi}_u} f_P \geq y_u & u \in U \\
y_s - y_t = 0 & \{s, t\} \in D \\
x_v, f_P, y_u \geq 0 & v \in V, \ P \in \tilde{\Pi}, \ u \in U
\end{array}
$$

Although the number of variables in (LP2) might be exponential, any basic feasible solution to (LP2) has $O(N^2)$ non-zero variables.

Lemma 2. (LP2) *can be solved in polynomial time.*

By Lemma 1 there exists a solution to (LP2) of value $O(C/|D|)$. Indeed, let $T, R, \mathcal{R}$ be as in Lemma 1; in particular, $c(T)/|R| = O(C/|D|)$. For $u \in T$ let P_u be the unique ur-path in T. Define a feasible solution for (LP2) as follows: $x_v = 1/|R|$ for every $v \in T$, $y_u = f_{P_u} = 1/|R|$ for every u that belongs to some pair in R, and x_u, y_u, f_P are zero otherwise. It easy to see that this solution is feasible for (LP2), and its value (cost) is $c(T)/|R| = O(C/|D|)$.

2.3   Proof of Theorem 3

We now proceed similarly to [5, 6]. We may assume that $\max\{1/y_u : u \in U\}$ is polynomial in n, see [6]. Partition U into $O(\log n)$ sets $U_j = \{u \in U : 1/2^{j+1} \le y_u \le 1/2^j\}$. There is some U_j that delivers $\Omega(1/\ln n)$ flow units to r. Focus on that U_j. Clearly, $|U_j| = \Theta(2^j)/\log n$. Setting $x'_v = \min\{\Theta(2^j) \cdot x_v, 1\}$ for all $v \in V$ and $f'_P = \min\{\Theta(2^j) \cdot f_P, 1\}$ for all $P \in \Pi$, gives a feasible solution for the following LP that requires from every node in U_j to deliver a flow unit to r.

$$\text{(LP3) min} \quad \sum_{v \in V} c(v) \cdot x'_v + \sum_{P \in \Pi} \ell(P) \cdot f'_P$$

$$\text{s.t.} \sum_{\{P \in \Pi_u | v \in P\}} f'_P \le x'_v \qquad v \in V, \ u \in U_j$$
$$\sum_{P \in \Pi_u} f'_P \ge 1 \qquad u \in U_j$$
$$x'_v, f'_P \ge 0 \qquad v \in V, \ P \in \Pi$$

We bound the value of the above solution x', f' for (LP3). Since we have $\sum_{v \in V} c(v) x_v = O(C/|D|)$,

$$\sum_{v \in V} c(v) x'_v = O(2^j) \cdot C/|D| \ .$$

We later see that, since $|U_j| = \Theta(2^j/\log n)$, an extra $\log n$ factor is invoked in the cost-density part of our solution; if, e.g., $|U_j| = 2^j$ would hold, this $\log n$ factor would have been saved. Our main point is that the length-part of the density does not depend on the size of U_j. We show this as follows. All paths used in (LP2) are of length $O(\log n \cdot L/|D|)$. First, assure that $\sum_{P \in \tilde{\Pi}_u} f'_P$ is not too large. For any $u \in U_j$ the fractional values of $\{f'_P : P \in \Pi_u\}$ only affect u, namely, if $u \ne u'$ then $\tilde{\Pi}_u \cap \tilde{\Pi}_{u'} = \emptyset$. Therefore, if $\sum_{P \in \tilde{\Pi}_u} f_P >> 1$, we may assure that the sum is at most $3/2$ as follows. If a single path carries at least $1/2$ a unit of flow then (scaling values by only 2) this path can be used as the solution for u. Else, any minimal collection of paths delivering at least one unit of flow, delivers at most $3/2$ units of flow to r. Hence the contribution of a single node u to the fractional length-part is

$$O(\log n \cdot L/|D|) \sum_{P \in \tilde{\Pi}_u} f'_P = O(\log n \cdot L/|D|) \ .$$

Over all terminals, the contribution is $O(|U_j| \cdot \log n \cdot L/|D|)$. Now, use the main theorem of [6]:

Theorem 8 ([6]). *There exists a polynomial time algorithm that finds an integral solution to* (LP3) *of value* $O(\log n)$ *times the optimal fractional value of* (LP3).

Hence we can find in polynomial time a tree T containing r and U_j with $c(T) = O(\log n \cdot 2^j \cdot C/|D|)$ and $\sum_{u \in U_j} \ell_T(u, r) = O(|U_j| \cdot \log^2 n \cdot L/|D|)$.

Note that if the tree contains i terminals then it contains $i/2$ pairs. This is due to the constraint $y_s = y_t$. Since the tree spans $\Theta(2^j/\log n)$ pairs, its cost-part density is $O(\log^2 n) \cdot C/|D|$. Clearly, the length-part density is $O(\log^2 n) \cdot L/|D|$. This finishes the proof of Theorem 7, and thus also the proof of Theorem 3 is complete.

3   A Lower Bound for MaxCT

Here we prove the following statement that implies Theorem 4. We first prove a *non-bicriteria* lower bound.

Theorem 9. MaxCT *admits no better than c-approximation algorithm, unless* $\mathrm{NP} \subseteq \mathrm{DTIME}(n^{O(c \cdot \ln c \cdot \exp(5c))})$.

Clearly, this implies that MaxCT admits no constant approximation algorithm unless P=NP. Also, the problem admits no $B \log \log n$-ratio approximation for some universal constant B unless $\mathrm{NP} \subseteq \mathrm{DTIME}(n^{\mathrm{polylog}\,n})$.

Remark: The size of the instance produced is $s = n^{O(c \cdot \ln c \cdot \exp(9c))}$ and thus $c = \Theta(\log \log s)$. Therefore, it is not possible to get a stronger hardness than $\log \log n$ unless we get a better gap in terms of c.

3.1   The Gap of Set-Cover

The Set-Cover problem is as follows. Given a collection A of sets on a groundset B, find a minimum size subcollection $A' \subseteq A$ so that the union of the sets in A' is B. We consider the decision version, and present the problem in terms of the incidence bipartite graph $H = (A + B, E)$ of A and B, where $ab \in E$ if the set $a \in A$ contains the element $b \in B$. For $A' \subseteq A$ the set of elements covered by A' is the set $\Gamma(A') = \{b \in B : ab \in E \text{ for some } a \in A'\}$ of *neighbors* of A in H. Let opt denote the optimum solution value for an instance of Set-Cover at hand.

Set-Cover (decision version)
Instance: A bipartite graph $H = (A + B, E)$.
Question: Does there exists $A' \subseteq A$ with $|A'| = \mathrm{opt}$ and $|\Gamma_H(A')| = B$?

Theorem 10 ([7]). *For any* NPC *language* I *with* $|I| = n$ *there exists an* $O(n^{O(\log \log n)})$ *time reduction from* I *to an instance of* Set-Cover *so that:*

- *For a* YES-*instance there exists* $A' \subseteq A$ *with* $|A'| = \mathrm{opt}$ *so that* $\Gamma(A') = B$.
- *For a* NO-*instance* $|A'| \geq \mathrm{opt} \cdot \ln |B|$ *for any* $A' \subseteq A$ *with* $\Gamma(A') = B$.

Corollary 1. *Unless* $\mathrm{NP} \subseteq \mathrm{DTIME}(n^{O(\log \log n)})$, Set-Cover *admits no polynomial time algorithm that for some* $1 \leq \alpha < \ln |B|$ *finds* $A' \subseteq A$ *with* $|A'| \leq \alpha \cdot \mathrm{opt}$ *and* $|\Gamma(A')| \geq (1 - 1/e^{\alpha+1})|B|$.

Proof. Suppose that we can find in polynomial time $A' \subseteq A$ with $|A'| = \alpha \cdot \text{opt}$ and $\Gamma(A') \geq (1 - \beta)|B|$, $\beta \leq 1$. For the residual instance of Set-Cover, we still need to cover $\beta|B|$ nodes in B. We can find a cover of size $\text{opt} \cdot [1 + \ln(b|B|)]$ of the remaining nodes using the Greedy Algorithm. So, we can find a cover of size $\text{opt} \cdot [\alpha + 1 + \ln(\beta|B|)]$ of all B. But this cannot be smaller than $\text{opt} \cdot \ln|B|$, by Theorem 10. So, we get that $\alpha + 1 + \ln(\beta|B|) \geq \ln|B|$. This gives $\beta \geq 1/e^{\alpha+1}$.

3.2   The Reduction

Define a sequence of graphs $G^1, G^2, \ldots$ by induction (see Fig. 1). To obtain G^1, take H, add a root r, and connect r to every node in A. Let $A_1 = A$ and $B_1 = B$. To obtain G^i from G^{i-1}, $i \geq 2$, take G^1 and $|B|$ copies of G^{i-1}, each corresponding to a node in B_1, and for every copy identify its root with the node corresponding to it in B_1. As the construction resembles a tree, we borrow some terms from the terminology of trees. A copy of H has *level i* if its A sets have distance $2i - 1$ to the root r. The copies of H at level i are ordered arbitrarily. A typical copy of H at level i is denoted by $H_{ij} = (A_{ij}, B_{ij}, E_{ij})$ with i the level of the copy and j the *index* of the copy. This means that the A_{ij} sets are at distance $2i - 1$ from the root and the index j is the order statistic of the copy inside level i. Let $A^i = \bigcup_j A_{ij}$ and $B^i = \bigcup_j B_{ij}$.

An H_{ij} is an *ancestor* of a terminal y if y belongs to the subgraph rooted by some $v \in B_{ij}$; such v is called *the elements ancestor* of y in level i and is denoted $ans^i(y)$. Note that $ans^i(y)$ is *unique*.

The *terminals* of G^h are $\bigcup_j B_{h,j}$, and each of them has profit 1; other nodes have profit 0. The cost of every node in A_{ij} is $1/|B|^{i-1}$ (so the nodes in $A_1 = A_{11}$ have cost 1), and the cost of any other node is 0. The cost bound is $C = h \cdot \text{opt}$. The number h of levels in the construction is defined as:

$$h = 4\frac{c \cdot \exp(4c + 1)}{2c - 1} \ln c . \qquad (4)$$

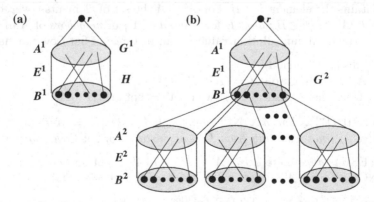

Fig. 1. (a) The graph G^1. (b) The graph G^2; if instead of copies of G^1 we "attach" to nodes in B_1 roots of the copies of G^{i-1}, then we obtain G^i.

Fact 11. *The size (and the construction time) of the construction is* $n^{O(h)}$, *where* $n = \max\{|A|, |B|\}$.

3.3   Analysis

While increasing the level by 1, the number Set-Cover instances grows up by $|B|$ but the node costs go down by $|B|$. Hence the total cost of every level i is $|A|$, and the total cost of G is $h \cdot |A|$. We may assume that any solution T to the obtained instance of MaxCT contains r. Otherwise, we may add the shortest path from r to T; the cost added is negligible in our context.

Lemma 3 (The YES-instance). *The obtained* MaxCT *instance* G, C *admits a feasible solution* T *that contains all terminals.*

Proof. Consider the graph T induced in G by r and all the copies of $A'_{ij} \cup B_{ij}$ so that $|A'_{ij}| = \mathsf{opt}$ and A'_{ij} covers B_{ij}. This graph contains all terminals. Since every A'_{ij} covers B_{ij}, T is connected. The cost of all copies of A'_{ij} at any level i is opt. Summing over all levels gives total cost $c(T) = h \cdot \mathsf{opt} = C$, as claimed.

We now deal with the MaxCT instance derived from a NO-instance. Fix a feasible solution T for MaxCT. Intuitively, T has an average cost of opt to spend on every level i. Averaging over all (A_{ij}, B_{ij}, E_{ij}) copies, $|T \cap A_{ij}|$ should be about opt for every i, j. In such a case the total cost would be $\mathsf{opt} \cdot h$.

Definition 2. *Level* i *in* G *is* cheap *(w.r.t.* T*) if* $|T \cap A^i| < 2\mathsf{opt}$ *and is* expensive *otherwise. A copy* H_{ij} *in a cheap level* i *is called* expensive *if* $|T \cap A_{ij}| \geq 4 \cdot c \cdot \mathsf{opt}$.

Lemma 4 (The NO-instance). *If* MaxCT *derived from a NO-instance then* T *contains less than* $1/c$ *fraction of the terminals.*

Proof. Let us say that a node v is *active* if it belongs to T; else v is *lost*. Initiate all terminals to be active. We gradually prove that some of them are actually lost, and at the end we will show that at most $1/c$ fraction of them can be active. At each level we are going to have some *already lost elements* B_{ij} for several different j. This means that an element ancestor of those B_{ij} was proven to be lost. This indicates that all their terminals descendants are lost (because every terminal ℓ has a unique ancestor $ans^i(\ell)$). The rest will be active elements.

We only consider terminals lost at cheap levels, ignoring those that may get lost in expensive levels. Let i be a cheap level. and let R_i be the number of terminals still declared "active" after we go via level i. Let $j > i$ be the next cheap level. We divide the leaves with respect to level j into *automatically active*, and *unsure*. The automatically active leaves are descendents of heavy H_{ij} copies.

Note that at most $1/2c$ of the $(A_{j,k}, B_{jk}, E_{jk})$ copies at level j may be heavy (because the total cost invested on active copies is still at most 2opt). Thus by symmetry at most $1/2c$ fraction of R_i leaves may become automatically active. The number of unsure leaves is at least $|R_i|(1 - 1/2c)$.

Those remaining $(1 - 1/2c)R_i \; H_{ij}$ copies are cheap and satisfy $|T \cap A_{jk}| \leq 4c \cdot \text{opt}$. By Claim 1, at least $\exp(-4c - 1)$ fraction of the the elements in the cheap copies at level j become lost. This means that at least

$$\left(1 - \frac{1}{2c}\right) \cdot \exp(-4c - 1)$$

fraction of the previously active terminals are lost at level j. This follows by symmetry (every element has the same number of leaf descendants) and because distinct elements have disjoint collection of descendants. Hence, at every cheap level the active terminals decrease by a factor of at least

$$1 - \frac{2c - 1}{2c} \cdot \exp(-4c - 1).$$

Because the total budget bound is C we get that at most half of the levels are expensive, so at least $h/2$ levels are cheap. Thus the fraction of active terminals remaining at the end is at most

$$\left(1 - \frac{2c - 1}{2c} \cdot \exp(4c + 1)\right)^{h/2} < 1/c.$$

The last inequality follows by the choice of h in (4). Thus T contains at most $1/c$ of the terminals, which concludes the proof of the lemma.

Theorem 4 directly follows from Lemma 3 and Lemma 4.

References

1. Andrews, M.: Hardness of buy-at-bulk network design. In: Proc. FOCS, pp. 115–124 (2004)
2. Andrews, M., Zhang, L.: Approximation algorithms for access network design. Algorithmica 34(2), 197–215 (2002)
3. Antonakopoulos, S., Chekuri, C., Shepherd, F.B., Zhang, L.: Buy-at-bulk network design with protection. In: FOCS, pp. 634–644 (2007)
4. Charikar, M., Karagiozova, A.: On non-uniform multicommodity buy-at-bulk network design. In: Proc. STOC, pp. 176–182 (2005)
5. Chekuri, C., Hajiaghayi, M.T., Kortsarz, G., Salavatipour, M.R.: Approximation algorithms for non-uniform buy-at-bulk network design. In: Proc. FOCS, pp. 677–686 (2006)
6. Chekuri, C., Hajiaghayi, M.T., Kortsarz, G., Salavatipour, M.R.: Approximation algorithms for node-weighted buy-at-bulk network design. In: SODA, pp. 1265–1274 (2007)
7. Feige, U.: A threshold of for approximating set cover. J. ACM 45, 634–652 (1998)
8. Guha, S., Moss, A., Naor, J.S., Schieber, B.: Efficient recovery from power outage. In: Proc. STOC, pp. 574–582 (1999)

9. Hajiaghayi, M.T., Kortsarz, G., Salavatipour, M.R.: Approximating buy-at-bulk and shallow-light k-steiner trees. In: Díaz, J., Jansen, K., Rolim, J.D.P., Zwick, U. (eds.) APPROX 2006. LNCS, vol. 4110, pp. 152–163. Springer, Heidelberg (2006)

10. Klein, C., Ravi, R.: A nearly best-possible approximation algorithm for node-weighted steiner trees. Journal of Algorithms 19(1), 104–115 (1995)

11. Marathe, M.V., Ravi, R., Sundaram, R., Ravi, S.S., Rosenkrantz, D.J., Hunt III., H.B.: Bicriteria network design problems. J. Algorithms 28(1), 142–171 (1998)

12. Moss, A., Rabani, Y.: Approximation algorithms for constrained node weighted Steiner tree problems. In: Proc. STOC, pp. 373–382 (2001)

13. Raz, R., Safra, S.: A sub-constant error-probability low-degree test, and a sub-constant error-probability PCP characterization of NP. In: Proc. STOC, pp. 475–484 (1997)

14. Salman, F.S., Cheriyan, J., Ravi, R., Subramanian, S.: Approximating the single-sink link-installation problem in network design. SIAM J. on Optimization 11(3), 595–610 (2000)

Submodular Maximization over Multiple Matroids via Generalized Exchange Properties

Jon Lee[1], Maxim Sviridenko[1], and Jan Vondrák[2],*

[1] IBM T.J. Watson Research Center
{jonlee,sviri}@us.ibm.com
[2] IBM Almaden Research Center
jvondrak@gmail.com

Introduction

In this paper, we consider the problem of maximizing a non-negative submodular function f, defined on a (finite) ground set N, subject to matroid constraints. A function $f : 2^N \to \mathbb{R}$ is *submodular* if for all $S, T \subseteq N$, $f(S \cup T) + f(S \cap T) \leq f(S) + f(T)$. Furthermore, all submodular functions that we deal with are assumed to be non-negative. Throughout, we assume that our submodular function f is given by a *value oracle*; i.e., for a given set $S \subseteq N$, an algorithm can query an oracle to find the value $f(S)$. Without loss of generality, we take the ground set N to be $[n] = \{1, 2, \ldots, n\}$.

We assume some familiarity with matroids [26] and associated algorithmics [28]. Briefly, a matroid $\mathcal{M}$ is an ordered pair $(N, \mathcal{I})$, where N is the ground set of $\mathcal{M}$ and $\mathcal{I}$ is the set of independent sets of $\mathcal{M}$. For a given matroid $\mathcal{M}$, the associated *matroid constraint* is $S \in \mathcal{I}(\mathcal{M})$. In our usage, we deal with k matroids $\mathcal{M}_i = (N, \mathcal{I}_i)$, $i = 1, \ldots, k$, on the common ground set N. We assume that each matroid is given by an *independence oracle*, answering whether $S \in \mathcal{I}_i$ or not. It is no coincidence that we use N for the ground set of our submodular function f as well as for the ground set of our matroids $\mathcal{M}_i = (N, \mathcal{I}_i)$, $i = 1, \ldots, k$. Indeed, our optimization problem is

$$\max \left\{ f(S) \ : \ S \in \cap_{i=1}^{k} \mathcal{I}_i \right\}.$$

Where necessary, we make some use of other standard matroid notation. For a matroid $\mathcal{M} = (N, \mathcal{I})$, we denote its rank function by $r_{\mathcal{M}}$ and its dual by $\mathcal{M}^*$. A base of $\mathcal{M}$ is a maximal independent set $J \in \mathcal{I}$, having cardinality $r_{\mathcal{M}}(N)$. For a set $S \subset N$, we let $\mathcal{M} \backslash S$, $\mathcal{M}/S$, and $\mathcal{M}|S$ denote deletion of S, contraction of S, and restriction to S, respectively.

Previous Results. Optimization of submodular functions is a central topic in combinatorial optimization [22, 28]. While submodular minimization is polynomially solvable [18, 29], maximization variants are usually NP-hard because they include either Max Cut, variants of facility location, and set coverage problems.

---

* This work was done while the last author was at Princeton University.

I. Dinur et al. (Eds.): APPROX and RANDOM 2009, LNCS 5687, pp. 244–257, 2009.

A classical technique for submodular maximization is the greedy algorithm. The greedy algorithm was first applied to a wide range of submodular maximization problems in the late-70's and early-80's [8, 9, 10, 14, 15, 19, 23, 24]. The most relevant result for our purposes is the proof that the greedy algorithm gives a $1/(k+1)$-approximation for the problem of maximizing a monotone submodular function subject to k matroid constraints [24]. Due to a simple reduction, this problem also encapsulates the problem of maximizing a linear function subject to $k + 1$ matroid constraints.[1] Thus we get a $1/k$-approximation for maximizing a linear function subject to k matroid constraints, $k \geq 3$ (this result appeared first in [15]). Until recently, the greedy algorithm had the best established performance guarantee for these problems under general matroid constraints.

Recently, improved results have been achieved using the *multilinear extension* of a submodular function and *pipage rounding* [1, 5, 6, 30]. In particular, Vondrák [30] designed the continuous greedy algorithm which achieves a $(1 - 1/e)$-approximation for our problem with $k = 1$, i.e. monotone submodular maximization subject to a single matroid constraint (see also [6]). This result is optimal in the oracle model even for the case of a uniform matroid constraint [25], and also optimal unless $P = NP$ for the special case of maximum coverage[11].

Another algorithmic technique that has been used for submodular maximization is local search. Cornuéjols et al. [9] show that a local-search algorithm achieves a constant-factor approximation guarantee for the maximum uncapacitated facility-location problem which is a special case of submodular maximization. Analogously, Nemhauser et al. [23] show a similar result for the problem of maximizing a monotone submodular function subject to a single cardinality constraint (i.e. a uniform matroid constraint). We remark that local search in this case is known to yield only a $1/2$-approximation, i.e. it performs worse than the greedy algorithm [23].

The maximum k-dimensional matching problem is a problem of maximizing a linear function subject to k special partition matroid constraints. Improved algorithms for maximum k-dimensional matching have been designed using local search. The best known approximation factors are $2/(k + \varepsilon)$ in the unweighted case (i.e., 0/1 weights), and $2/(k+1+\varepsilon)$ for a general linear function, even in the more general cases of weighted set packing [17] and independent set problems in $(k + 1)$-claw free graphs [3]. The latter result was obtained after a series of improvements over the basic local-search algorithm [2, 3, 7].

However, general matroid constraints seem to complicate the matter. Prior to this paper, the best approximation for the problem of maximum independent set in the intersection of $k \geq 3$ matroids was $1/k$ (for a recent discussion see [27]). On the hardness side, it is known that unless $P = NP$, there is no approximation better than $O(\log k/k)$ for k-dimensional matching [16], and hence neither for the intersection of k general matroids. The $1/(k + 1)$-approximation for submodular maximization subject to k matroids [24] can be improved in the case when all k

---

[1] Given a problem $\max\{w(S) : S \in \bigcap_{i=0}^{k} \mathcal{I}_i\}$ where $w(S)$ is linear, we can equivalently consider the problem $\max\{f(S) : S \in \bigcap_{i=1}^{k} \mathcal{I}_i\}$, where $f(S) = \max\{w(I) : I \subseteq S, I \in \mathcal{I}_0\}$, the *weighted rank function* of $\mathcal{M}_0$, is known to be monotone submodular.

constraints correspond to partition matroids. For any fixed $k \geq 2$ and $\varepsilon > 0$, a simple local-search algorithm gives a $1/(k + \varepsilon)$-approximation for this variant of the problem [21]. The analysis strongly uses the properties of partition matroids. It is based on relatively simple exchange properties of partition matroids that do not hold in general.

Local-search algorithms were also designed for non-monotone submodular maximization. The best approximation guarantee known for unconstrained submodular maximization is $2/5 - \varepsilon$ [12]. For the problem of non-monotone submodular maximization subject to k matroid constraints, the best known approximation is $1/(k + 2 + 1/k + \varepsilon)$ (for any constant $k \geq 1$ and $\varepsilon > 0$) [21].

Our Results and Techniques. In this paper we analyze a natural local-search algorithm: Given a feasible solution, i.e. a set S that is independent in each of the k matroids, our local-search algorithm tries to add at most p elements and delete at most kp elements from S. If there is a local move that generates a feasible solution and improves the objective value, our algorithm repeats the local-search procedure with that new solution, until no improvement is possible. Our main result is that for $k \geq 2$, every locally-optimal feasible solution S satisfies the inequality

$$(k + 1/p) \cdot f(S) \geq f(S \cup C) + (k - 1 + 1/p) \cdot f(S \cap C),$$

for every feasible solution C. We also provide an approximate variant of the local-search procedure that finds an approximate locally-optimal solution in polynomial time, while losing a factor of $1 + \varepsilon$ on the left-hand side of the above inequality (Lemma 11). Therefore, for any fixed $k \geq 2$ and $\varepsilon > 0$, we obtain a polynomial-time algorithm with approximation guarantee $1/(k + \varepsilon)$ for the problem of maximizing a monotone non-decreasing submodular function subject to k matroid constraints. This algorithm gives a $1/(k - 1 + \varepsilon)$-approximation in the case when the objective function is linear. These results are tight for our local search algorithm, which follows from [2].

We also obtain an approximation algorithm for non-monotone submodular functions. In this case, one round of local search is not enough, but applying the local search iteratively, as in [21], one can obtain an approximation algorithm with performance guarantee of $1/(k + 1 + 1/(k - 1) + \varepsilon)$.

The main technical contributions of this paper are two new exchange properties for matroids. One is a generalization of the classical Rota Exchange Property (Lemma 8) and another is an exchange property for the intersection of two matroids (Lemma 5), which generalizes an exchange property based on augmenting paths which was used in [21] for partition matroids. We believe that both properties and their proofs are interesting in their own right.

In §1, we establish some useful properties of submodular functions. In §2, we establish our exchange properties for matroids. In §3, we describe and analyze our local-search algorithm.

1   Some Useful Properties of Submodular Functions

Lemma 1. *Let f be a submodular function on N. Let $S, C \subseteq N$ and let $\{T_l\}_{l=1}^t$ be a collection of subsets of $C \setminus S$ such that each element of $C \setminus S$ appears in exactly k of these subsets. Then*

$$\sum_{l=1}^t [f(S \cup T_l) - f(S)] \geq k \, (f(S \cup C) - f(S)).$$

Proof. Let $s = |S|$ and $c = |C \cup S|$. We will use the notation $[n]$ to denote the set $\{1, \ldots, n\}$ (by convention $[0] = \emptyset$). Without loss of generality, we can assume that $S = \{1, 2, \ldots, s\}$ and that $C \setminus S = \{s+1, s+2, \ldots, c\}$. Then for any $T \subseteq C \setminus S$, by submodularity: $f(S \cup T) - f(S) \geq \sum_{p \in T} [f([p]) - f([p-1])]$. Summing up over all sets T_l, we get

$$\sum_{l=1}^t [f(S \cup T_l) - f(S)] \geq \sum_{l=1}^t \sum_{p \in T_l} [f([p]) - f([p-1])]$$
$$= k \sum_{p=s+1}^c [f([p]) - f([p-1])] = k [f(S \cup C) - f(S)].$$

The first equality follows from the fact that each element in $\{s+1, \ldots, c\}$ appears in exactly k sets T_l, and the second equality follows from a telescoping summation. □

Lemma 2. *Let f be a submodular function on N. Let $S' \subseteq S \subseteq N$, and let $\{T_l\}_{l=1}^t$ be a collection of subsets of $S \setminus S'$ such that each element of $S \setminus S'$ appears in exactly k of these subsets. Then*

$$\sum_{l=1}^t (f(S) - f(S \setminus T_l)) \leq k \, (f(S) - f(S')).$$

Proof. Let $s = |S|$ and $c = |S'|$. Without loss of generality, we can assume that $S' = \{1, 2, \ldots, c\} = [c] \subseteq \{1, 2, \ldots, s\} = [s] = S$. For any $T \subseteq S$, $f(S) - f(S \setminus T) \leq \sum_{p \in T} (f([p]) - f([p-1]))$ by submodularity. Using this we obtain

$$\sum_{l=1}^t (f(S) - f(S \setminus T_l)) \leq \sum_{l=1}^t \sum_{p \in T_l} (f([p]) - f([p-1]))$$
$$= k \sum_{i=c+1}^s (f([i]) - f([i-1])) = k \, (f(S) - f(S')).$$

The first equality follows from $S \setminus C = \{c+1, \ldots, s\}$ and the fact that each element of $S \setminus C$ appears in exactly k of the sets $\{T_l\}_{l=1}^t$. The last equality is due to a telescoping summation. □

2   New Exchange Properties of Matroids

2.1   Intersection of Two Matroids

An exchange digraph is a well-known construct for devising efficient algorithms for exact maximization of linear functions over the intersection of two matroids (for example, see [28]). We are interested in submodular maximization, k matroids and approximation algorithms; nevertheless, we are able to make use of such exchange digraphs, once we establish some new properties of them.

Let $\mathcal{M}_l = (N, \mathcal{I}_l)$, $l = 1, 2$, be two matroids on ground set N. For $I \in \mathcal{I}_1 \cap \mathcal{I}_2$, we define two digraphs $D_{\mathcal{M}_1}(I)$ and $D_{\mathcal{M}_2}(I)$ on node set N as follows:

- For each $i \in I, j \in N \setminus I$ with $I - i + j \in \mathcal{I}_1$, we have an arc (i, j) of $D_{\mathcal{M}_1}(I)$;
- For each $i \in I, j \in N \setminus I$ with $I - i + j \in \mathcal{I}_2$, we have an arc (j, i) of $D_{\mathcal{M}_2}(I)$.

The arcs in $D_{\mathcal{M}_l}(I)$, $l = 1, 2$, encode valid swaps in $\mathcal{M}_l$.

In what follows, we assume that I is our current solution and J is the optimal solution. We also assume that $|I| = |J|$. If not, we extend I or J by dummy elements so that we maintain independence in both matroids (more details later). When we refer to a matching (or perfect matching) in $D_{\mathcal{M}_l}(I)$ for $l = 1, 2$ we mean a matching in an undirected graph where the arcs of the graph $D_{\mathcal{M}_l}(I)$ are treated as undirected edges. We use two known lemmas from matroid theory.

Lemma 3 ([28, Corollary 39.12a]). *If $|I| = |J|$ and $I, J \in \mathcal{I}_l$ ($l = 1$ or 2), then $D_{\mathcal{M}_l}(I)$ contains a perfect matching between $I \setminus J$ and $J \setminus I$.*

Lemma 4 ([28, Theorem 39.13]). *Let $|I| = |J|$, $I \in \mathcal{I}_l$, and assume that $D_{\mathcal{M}_l}(I)$ has a unique perfect matching between $I \setminus J$ and $J \setminus I$. Then $J \in \mathcal{I}_l$.*

Next, we define a digraph $D_{\mathcal{M}_1, \mathcal{M}_2}(I)$ on node set N as the union of $D_{\mathcal{M}_1}(I)$ and $D_{\mathcal{M}_2}(I)$. A dicycle in $D_{\mathcal{M}_1, \mathcal{M}_2}(I)$ corresponds to a chain of feasible swaps. However, observe that it is not necessarily the case that the entire cycle gives a valid exchange in both matroids.

If $|I| = |J|$ and $I, J \in \mathcal{I}_1 \cap \mathcal{I}_2$, this means we have two perfect matchings on $I \triangle J$ which together form a collection of dicycles in $D_{\mathcal{M}_1, \mathcal{M}_2}(I)$. However, only the uniqueness of a perfect matching assures us that we can legally perform the exchange. This motivates the following definition.

Definition 1. *We call a dicycle C in $D_{\mathcal{M}_1, \mathcal{M}_2}(I)$ irreducible if $C \cap D_{\mathcal{M}_1}(I)$ is the unique perfect matching in $D_{\mathcal{M}_1}(I)$ and $C \cap D_{\mathcal{M}_2}(I)$ is the unique perfect matching in $D_{\mathcal{M}_2}(I)$ on their vertex set $V(C)$. Otherwise, we call C reducible.*

The following, which is our main technical lemma, allows us to consider only irreducible cycles. The proof follows the ideas of matroid intersection (see [28, Lemma 41.5α]). This lemma holds trivially for partition matroids with $s = 0$.

Lemma 5. *Let $\mathcal{M}_l = (N, \mathcal{I}_1)$, $l = 1, 2$, be matroids on ground set N. Suppose that $I, J \in \mathcal{I}_1 \cap \mathcal{I}_2$ and $|I| = |J|$. Then there is $s \geq 0$ and a collection of irreducible dicycles $\{C_1, \ldots, C_m\}$ (allowing repetition) in $D_{\mathcal{M}_1, \mathcal{M}_2}(I)$, using only elements of $I \triangle J$, so that each element of $I \triangle J$ appears in exactly 2^s of the dicycles.*

Proof. Consider $D_{\mathcal{M}_1, \mathcal{M}_2}(I) = D_{\mathcal{M}_1}(I) \cup D_{\mathcal{M}_2}(I)$. By Lemma 3, there is a perfect matching between $I \setminus J$ and $J \setminus I$, both in $D_{\mathcal{M}_1}(I)$ and $D_{\mathcal{M}_2}(I)$. We denote these two perfect matchings by M_1, M_2. The union $M_1 \cup M_2$ forms a subgraph of out-degree 1 and in-degree 1 on $I \triangle J$. Therefore, it decomposes into a collection of dicycles $C_1, \ldots, C_m$. If they are all irreducible, we are done and $s = 0$.

If C_i is not irreducible, it means that either $M_1' = C_i \cap D_{\mathcal{M}_1}(I)$ or $M_2' = C_i \cap D_{\mathcal{M}_2}(I)$ is not a unique perfect matching on $V(C_i)$. Let us assume, without loss of generality, that there is another perfect matching M_1'' in $D_{\mathcal{M}_1}(I)$. We consider the disjoint union $M_1' + M_1'' + M_2' + M_2'$, duplicating arcs where necessary.

This is a subgraph of out-degree 2 and in-degree 2 on $V(C_i)$, which decomposes into dicycles $C_{i1}, \ldots, C_{it}$, covering each vertex of C_i exactly twice:

$$V(C_{i1}) + V(C_{i2}) + \ldots + V(C_{it}) = 2V(C_i).$$

Because $M_1' \neq M_1''$, we have a chord of C_i in M_1'', and we can choose the first dicycle so that it does not cover all of $V(C_i)$. So we can assume that we have $t \geq 3$ dicycles, and at most one of them covers all of $V(C_i)$. If there is such a dicycle among $C_{i1}, \ldots, C_{it}$, we remove it and duplicate the remaining dicycles. Either way, we get a collection of dicycles $C_{i1}, \ldots, C_{it'}$ such that each of them is shorter than C_i and together they cover each vertex of C_i exactly twice.

We repeat this procedure for each reducible dicycle C_i. For irreducible dicycles C_i, we just duplicate C_i to obtain $C_{i1} = C_{i2} = C_i$. This completes one stage of our procedure. After the completion of the first stage, we have a collection of dicycles $\{C_{ij}\}$ covering each vertex in $I \Delta J$ exactly twice.

As long as there exists a reducible dicycle in our current collection of dicycles, we perform another stage of our procedure. This means decomposing all reducible dicycles and duplicating all irreducible dicycles. In each stage, we double the number of dicycles covering each element of $I \Delta J$. To see that this cannot be repeated indefinitely, observe that every stage decreases the size of the longest reducible dicycle. All dicycles of length 2 are irreducible, and therefore the procedure terminates after a finite number of stages s. Then, all cycles are irreducible and together they cover each element of $I \Delta J$ exactly 2^s times. $\square$

We remark that of course the procedure in the proof of Lemma 5 is very inefficient, but it is not part of our algorithm — it is only used for this proof.

Next, we extend this Lemma 5 to sets I, J of different size, which forces us to deal with dipaths as well as dicycles.

Definition 2. *We call a dipath or dicycle A feasible in $D_{\mathcal{M}_1, \mathcal{M}_2}(I)$, if*

- $I \Delta V(A) \in \mathcal{I}_1 \cap \mathcal{I}_2$, *and*
- *For any sub-dipath $A' \subset A$ such that each endpoint of A' is either an endpoint of A or an element of I, we also have $I \Delta V(A') \in \mathcal{I}_1 \cap \mathcal{I}_2$.*

First, we establish that irreducible dicycles are feasible.

Lemma 6. *Any irreducible dicycle in $D_{\mathcal{M}_1, \mathcal{M}_2}(I)$ is also feasible in $D_{\mathcal{M}_1, \mathcal{M}_2}(I)$.*

Proof. An irreducible dicycle C consists of two matchings $M_1 \cup M_2$, which are the unique perfect matchings on $V(C)$, in $D_{\mathcal{M}_1}(I)$ and $D_{\mathcal{M}_2}(I)$ respectively. Therefore, we have $I \Delta V(C) \in \mathcal{I}_1 \cap \mathcal{I}_2$ by Lemma 4.

Consider any sub-dipath $A' \subset C$ whose endpoints are in I. (C has no endpoints, so the other case in Definition 2 does not apply.) This means that A' has even length. Suppose that $a_1 \in V(A')$ is the endpoint incident to an edge in $M_1 \cap A'$ and $a_2 \in V(A')$ is the other endpoint, incident to an edge in $M_2 \cap A'$. Note that any subset of M_1 or M_2 is again a unique perfect matching on its respective vertex set, because otherwise we could produce a different perfect matching on $V(C)$. We can view $I \Delta V(A')$ in two possible ways:

- $I \Delta V(A') = (I - a_1)\Delta(V(A') - a_1)$; because $V(A') - a_1$ has a unique perfect matching $M_2 \cap A'$ in $D_{\mathcal{M}_2}(I)$, this shows that $I \Delta V(A') \in \mathcal{I}_2$.
- $I \Delta V(A') = (I - a_2)\Delta(V(A') - a_2)$; because $V(A') - a_2$ has a unique perfect matching $M_1 \cap A'$ in $D_{\mathcal{M}_1}(I)$, this shows that $I \Delta V(A') \in \mathcal{I}_1$. □

Finally, we establish the following property of possible exchanges between arbitrary solutions I, J (not necessarily of the same size).

Lemma 7. *Let $\mathcal{M}_1 = (N, \mathcal{I}_1)$ and $\mathcal{M}_2 = (N, \mathcal{I}_2)$ be two matroids and let $I, J \in \mathcal{I}_1 \cap \mathcal{I}_2$. Then there is $s \geq 0$ and a collection of dipaths/dicycles $\{A_1, \ldots, A_m\}$ (possibly with repetition), feasible in $D_{\mathcal{M}_1, \mathcal{M}_2}(I)$, using only elements of $I \Delta J$, so that each element of $I \Delta J$ appears in exactly 2^s dipaths/dicycles A_i.*

Proof. If $|I| = |J|$, we are done by Lemmas 5 and 6. If $|I| \neq |J|$, we extend the matroids by new "dummy elements" E, independent of everything else (in both matroids), and add them to I or J, to obtain sets of equal size $|\tilde{I}| = |\tilde{J}|$. We denote the extended matroids by $\tilde{\mathcal{M}}_1 = (N \cup E, \tilde{\mathcal{I}}_1)$, $\tilde{\mathcal{M}}_2 = (N \cup E, \tilde{\mathcal{I}}_2)$. We consider the graph $D_{\tilde{\mathcal{M}}_1, \tilde{\mathcal{M}}_2}(\tilde{I})$. Observe that the dummy elements do not affect independence among other elements, so the graphs $D_{\mathcal{M}_1, \mathcal{M}_2}(I)$ and $D_{\tilde{\mathcal{M}}_1, \tilde{\mathcal{M}}_2}(\tilde{I})$ are identical on $I \cup J$.

Applying Lemma 5 to $\tilde{I}, \tilde{J}$, we obtain a collection of irreducible dicycles $\{C_1, \ldots, C_m\}$ on $\tilde{I} \Delta \tilde{J}$ such that each element appears in exactly 2^s dicycles. Let $A_i = C_i \setminus E$. Obviously, the sets $V(A_i)$ cover $I \Delta J$ exactly 2^s times. We claim that each A_i is either a feasible dicycle, a feasible dipath, or a collection of feasible dipaths (in the original digraph $D_{\mathcal{M}_1, \mathcal{M}_2}(I)$).

First, assume that $C_i \cap E = \emptyset$. Then $A_i = C_i$ is an irreducible cycle in $D_{\mathcal{M}_1, \mathcal{M}_2}(I)$ (the dummy elements are irrelevant). By Lemma 6, we know that $A_i = C_i$ is a feasible dicycle.

Next, assume that $C_i \cap E \neq \emptyset$. C_i is still a feasible dicycle, but in the extended digraph $D_{\tilde{\mathcal{M}}_1, \tilde{\mathcal{M}}_2}(\tilde{I})$. We remove the dummy elements from C_i to obtain $A_i = C_i \setminus E$, a dipath or a collection of dipaths. Consider any sub-dipath A' of A_i, possibly $A' = A_i$, satisfying the assumptions of Definition 2. A_i does not contain any dummy elements. If both endpoints of A' are in I, it follows from the feasibility of C_i that $\tilde{I} \Delta V(A') \in \tilde{\mathcal{I}}_1 \cap \tilde{\mathcal{I}}_2$, and hence $I \Delta V(A') = (\tilde{I} \Delta V(A')) \setminus E \in \mathcal{I}_1 \cap \mathcal{I}_2$.

If an endpoint of A' is outside of I, then it must be an endpoint of A_i. This means that it has a dummy neighbor in $\tilde{I} \cap C_i$ that we deleted. (Note that this case can occur only if we added dummy elements to I, i.e. $|I| < |J|$.) In that case, extend the path to A'', by adding the dummy neighbor(s) at either end. We obtain a dipath from $\tilde{I}$ to $\tilde{I}$. By the feasibility of C_i, we have $\tilde{I} \Delta V(A'') \in \tilde{\mathcal{I}}_1 \cap \tilde{\mathcal{I}}_2$, and therefore $I \Delta V(A') = (\tilde{I} \Delta V(A'')) \setminus E \in \mathcal{I}_1 \cap \mathcal{I}_2$. □

2.2   A Generalized Rota-Exchange Property

Next, we establish a very useful property for a pair of bases of one matroid.

Lemma 8. *Let $\mathcal{M} = (N, \mathcal{I})$ be a matroid and A, B bases in $\mathcal{M}$. Let $A_1, \ldots, A_m$ be subsets of A such that each element of A appears in exactly q of them. Then* *there are sets $B_1, \ldots, B_m \subseteq B$ such that each element of B appears in exactly q of them, and for each i, $A_i \cup (B \setminus B_i) \in \mathcal{I}$.*

Remark 1. A very special case of Lemma 8, namely when $m = 2$ and $q = 1$, attracted significant interest when it was conjectured by G.-C. Rota and proved in [4, 13, 31]; see [28, (39.58)].

Proof. We can assume for convenience that A and B are disjoint (otherwise we can make $\{B_i\}$ equal to $\{A_i\}$ on the intersection $A \cap B$ and continue with a matroid where $A \cap B$ is contracted).

For each i, we define a matroid $\mathcal{N}_i = (\mathcal{M}/A_i)|B$, where we contract A_i and restrict to B. In other words, $S \subseteq B$ is independent in $\mathcal{N}_i$ exactly when $A_i \cup S \in \mathcal{I}$. The rank function of $\mathcal{N}_i$ is

$$r_{\mathcal{N}_i}(S) = r_{\mathcal{M}}(A_i \cup S) - r_{\mathcal{M}}(A_i) = r_{\mathcal{M}/A_i}(S).$$

Let $\mathcal{N}_i^*$ be the dual matroid to $\mathcal{N}_i$. Recall that the ground set is now B. By definition, $T \subseteq B$ is a spanning set in $\mathcal{N}_i^*$ if and only if $B \setminus T$ is independent in $\mathcal{N}_i$, i.e. if $A_i \cup (B \setminus T) \in \mathcal{I}$. The bases of $\mathcal{N}_i^*$ are minimal such sets T; these are the candidate sets for B_i, which can be exchanged for A_i. The rank function of the dual matroid $\mathcal{N}_i^*$ is (by [28, (Theorem 39.3)])

$$r_{\mathcal{N}_i^*}(T) = |T| - r_{\mathcal{N}_i}(B) + r_{\mathcal{N}_i}(B \setminus T) = |T| - r_{\mathcal{M}}(A_i \cup B) + r_{\mathcal{M}}(A_i \cup (B \setminus T))$$
$$= |T| - |B| + r_{\mathcal{M}}(A_i \cup (B \setminus T)) = r_{\mathcal{M}/(B \setminus T)}(A_i).$$

Observe that the rank of $\mathcal{N}_i^*$ is $r_{\mathcal{N}_i^*}(B) = |A_i|$.

Now, we consider a new ground set $\hat{B} = B \times [q]$. We view the elements $\{(i, j) : j \in [q]\}$ as parallel copies of i. For $T \subseteq \hat{B}$, we define its *projection* to B as

$$\pi(T) = \{i \in B \mid \exists j \in [q] \text{ with } (i, j) \in T\}.$$

A natural extension of $\mathcal{N}_i^*$ to $\hat{B}$ is a matroid $\hat{\mathcal{N}}_i^*$ where a set T is independent if $\pi(T)$ is independent in $\mathcal{N}_i^*$. The rank function of $\hat{\mathcal{N}}_i^*$ is

$$r_{\hat{\mathcal{N}}_i^*}(T) = r_{\mathcal{N}_i^*}(\pi(T)) = r_{\mathcal{M}/(B \setminus \pi(T))}(A_i). \tag{1}$$

The question now is whether $\hat{B}$ can be partitioned into $B_1', \ldots, B_m'$ so that B_i' is a base in $\hat{\mathcal{N}}_i^*$. If this is true, then we are done, because each $B_i = \pi(B_i')$ would be a base of $\mathcal{N}_i^*$ and each element of B would appear in q sets B_i. To prove this, consider the union of our matroids, $\hat{\mathcal{N}}^* := \hat{\mathcal{N}}_1^* \vee \hat{\mathcal{N}}_2^* \vee \ldots \vee \hat{\mathcal{N}}_m^*$. By the matroid union theorem ([28, (Corollary 42.1a)]), this matroid has rank function

$$r_{\hat{\mathcal{N}}^*}(\hat{B}) = \min_{T \subseteq \hat{B}} \left(|\hat{B} \setminus T| + \sum_{i=1}^m r_{\hat{\mathcal{N}}_i^*}(T) \right).$$

We claim that for any $T \subseteq \hat{B}$,

$$\sum_{i=1}^m r_{\hat{\mathcal{N}}_i^*}(T) = \sum_{i=1}^m r_{\mathcal{M}/(B \setminus \pi(T))}(A_i) \geq q \cdot r_{\mathcal{M}/(B \setminus \pi(T))}(A) = q|\pi(T)|.$$

The first equality follows from our rank formula (1). The inequality follows from Lemma 1 applied to the submodular function $r_{\mathcal{M}/(B \setminus \pi(T))}$, with $S = \emptyset$ and $C = A$. The last equality holds because both A and B are bases of $\mathcal{M}$ and the rank of the matroid $\mathcal{M}/(B \setminus \pi(T))$ is $|\pi(T)|$. We also have $|T| \leq q|\pi(T)|$, hence $\sum_{i=1}^{m} r_{\hat{\mathcal{N}}_i^*}(T) \geq q|\pi(T)| \geq |T|$ for any $T \subseteq \hat{B}$. Therefore the rank of $\hat{\mathcal{N}}^*$ is

$$ r_{\mathcal{N}^*}(\hat{B}) = \min_{T \subseteq \hat{B}} \left(|\hat{B} \setminus T| + \sum_{i=1}^{m} r_{\hat{\mathcal{N}}_i^*}(T) \right) = |\hat{B}| . $$

This means that $\hat{B}$ can be partitioned into sets $B_1', \ldots, B_m'$, where B_i' is independent in $\hat{\mathcal{N}}_i^*$. However, the ranks of $\hat{B}$ in the $\hat{\mathcal{N}}_i^*$ sum up to $\sum_{i=1}^{m} r_{\hat{\mathcal{N}}_i^*}(\hat{B}) = \sum_{i=1}^{m} |A_i| = |\hat{B}|$, so this implies that each B_i' is a base of $\hat{\mathcal{N}}_i^*$. Then, each $B_i = \pi(B_i')$ is a base of $\mathcal{N}_i^*$, and these are the sets demanded by the lemma. $\square$

Finally, we give a version of Lemma 8 where the two sets need not be bases.

Lemma 9. *Let $\mathcal{M} = (N, \mathcal{I})$ be a matroid and $I, J \in \mathcal{I}$. Let $I_1, \ldots, I_m$ be subsets of I such that each element of I appears in at most q of them. Then there are sets $J_1, \ldots, J_m \subseteq J$ such that each element of J appears in at most q of them, and for each i, $I_i \cup (J \setminus J_i) \in \mathcal{I}$.*

Proof. We reduce this statement to Lemma 8. Let A, B be bases such that $I \subseteq A$ and $J \subseteq B$. Let q_e be the number of appearances of an element $e \in I$ in the subsets $I_1, \ldots, I_m$ and let $q' = \max_{e \in I} q_e$. Obviously, $q' \leq q$. We extend I_i arbitrarily to A_i, $I_i \subseteq A_i \subseteq A$, so that each element of A appears in exactly q' of them. By Lemma 8, there are sets $B_i \subseteq B$ such that each element of B appears in exactly q' of them, and $A_i \cup (B \setminus B_i) \in \mathcal{I}$ for each i. We define $J_i = J \cap B_i$. Then, each element of J appears in at most $q' \leq q$ sets J_i, and

$$ I_i \cup (J \setminus J_i) \subseteq A_i \cup (B \setminus B_i) \in \mathcal{I}. \qquad \square $$

3   Local-Search Algorithm

At each iteration of our local-search algorithm, given a current feasible solution $S \in \cap_{j=1}^{k} \mathcal{I}_j$, our algorithm seeks an improved solution by looking at a polynomial number of options to change S. If the algorithm finds a better solution, it moves to the next iteration, otherwise the algorithm stops. Specifically, given a current solution $S \in \cap_{j=1}^{k} \mathcal{I}_j$, the local moves that we consider are:

p-exchange Operation: If there is $S' \subseteq N$ and $S' \in \cap_{j=1}^{k} \mathcal{I}_j$ such that **(i)** $|S' \setminus S| \leq p$, $|S \setminus S'| \leq kp$, and **(ii)** $f(S') > f(S)$, then $S \leftarrow S'$.

The p-exchange operation for $S' \subseteq S$ is called a *delete operation*. Our main result is the following lower bound on the value of the locally-optimal solution.

Lemma 10. *For every $k \geq 2$ and every $C \in \cap_{j=1}^{k} \mathcal{I}_j$, a locally-optimal solution S under p-exchanges, satisfies*

$$ (k + 1/p) \cdot f(S) \geq f(S \cup C) + (k - 1 + 1/p) \cdot f(S \cap C). $$

Proof. Our proof is based on the new exchange properties of matroids: Lemmas 7 and 9. By applying Lemma 7 to the independent sets C and S in matroids $\mathcal{M}_1$ and $\mathcal{M}_2$, we obtain a collection of dipaths/dicycles $\{A_1, \ldots, A_m\}$ (possibly with repetition), feasible in $D_{\mathcal{M}_1, \mathcal{M}_2}(S)$, using only elements of $C\Delta S$, so that each element of $C\Delta S$ appears in exactly 2^s paths/cycles A_i.

We would like to define the sets of vertices corresponding to the exchanges in our local-search algorithm, based on the sets of vertices in paths/cycles $\{A_1, \ldots, A_m\}$. The problem is that these paths/cycles can be much longer than the maximal cardinality of a set allowable in a p-exchange operation. To handle this, we index vertices of the set of $C \setminus S$ in each path/cycle A_i for $i = 1, \ldots, m$, in such a way that vertices along any path or cycle are numbered consecutively. The vertices of $S \setminus C$ remain unlabeled. Because one vertex appears in 2^s paths/cycles, it might get different labels corresponding to different appearances of that vertex. So one vertex could have up to 2^s different labels.

We also define $p + 1$ copies of the index sets $\{A_1, \ldots, A_m\}$. For each copy $q = 0, \ldots, p$ of labeled $\{A_1, \ldots, A_m\}$, we throw away appearances of vertices from $C \setminus S$ that were labeled by q modulo $p+1$ from each A_i. By throwing away some appearances of the vertices, we are changing our set of paths in each copy of the original sets $\{A_1, \ldots, A_m\}$. Let $\{A_{q1}, \ldots, A_{qm_q}\}$ be the resulting collection of paths for $q = 0, \ldots, p$. Now each path A_{qi} contains at most p vertices from $C \setminus S$ and at most $p + 1$ vertices from $S \setminus C$.

Because our original collection of paths/cycles was feasible in $D_{\mathcal{M}_1, \mathcal{M}_2}(S)$ (see definition 2), each of the paths in the new collections correspond to feasible exchanges for matroids $\mathcal{M}_1$ and $\mathcal{M}_2$, i.e. $S\Delta V(A_{qi}) \in \mathcal{I}_1 \cap \mathcal{I}_2$. Consider now the collection of paths $\{A_{qi} | q = 0, \ldots, p, i = 1, \ldots, m_q\}$. By construction, each element of the set $S \setminus C$ appears in exactly $(p+1)2^s$ paths, and each element of $C \setminus S$ appears in exactly $p2^s$ paths, because each vertex has $2^s(p+1)$ appearances in total, and each appearance is thrown away in exactly one out of $p+1$ copies of the original sets $\{A_1, \ldots, A_m\}$. Let $L_{qi} = S \cap V(A_{qi})$ denote the set of vertices in the path A_{qi} belonging to the locally-optimal solution S, and let $W_{qi} = C \cap V(A_{qi})$ denote the set of vertices in the path A_{qi} belonging to the set C.

For each matroid $\mathcal{M}_i$ for $i = 3, \ldots, k$, independent sets $S \in \mathcal{I}_i$ and $C \in \mathcal{I}_i$, and collection of sets $\{W_{qi} \mid q = 0, \ldots, p; i = 1, \ldots, m_q\}$ (note that some of these sets might be empty), we apply Lemma 9. For convenience, we re-index the collection of sets $\{W_{qi} \mid q = 0, \ldots, p, i = 1, \ldots, m_q\}$. Let $W_1, \ldots, W_t$ be that collection, after re-indexing, for $t = \sum_{q=0}^{p} m_q$. By Lemma 9, for each $i = 3, \ldots, k$ there exist a collection of sets $X'_{1i}, \ldots, X'_{ti}$ such that $W_j \cup (S \setminus X'_{ji}) \in \mathcal{I}_i$. Moreover, each element of S appears in at most $p2^s$ of the sets from collection $X'_{1i}, \ldots, X'_{ti}$.

We consider the set of p-exchanges that correspond to adding the elements of the set W_j to the set S and removing the set of elements $\Lambda_j = L_j \cup (\cup_{i=3}^{k} X'_{ji})$ for $j = 1, \ldots, t$. Note that, $|\Lambda_j| \leq (p+1) + (k-2)p = (k-1)p+1 \leq kp$. By Lemmas 7 and 9, the sets $W_j \cup (S \setminus \Lambda_j)$ are independent in each of the matroids $\mathcal{M}_1, \ldots, \mathcal{M}_k$. By the fact that S is a locally-optimal solution, we have

$$f(S) \geq f\left((S \setminus \Lambda_j) \cup W_j\right), \qquad \forall j = 1, \ldots, t. \qquad (2)$$

Using inequalities (2) together with submodularity for $j = 1, \ldots, t$, we have

$$f(S \cup W_j) - f(S) \leq f((S \setminus \Lambda_j) \cup W_j) - f(S \setminus \Lambda_j) \leq f(S) - f(S \setminus \Lambda_j). \quad (3)$$

Moreover, we know that each element of the set $C \setminus S$ appears in exactly $p2^s$ sets W_j , and each element $e \in S \setminus C$ appears in $n_e \leq (p+1)2^s + (k-2)p2^s$ sets Λ_j .

Consider the sum of t inequalities (3), and add $(p+1)2^s + (k-2)p2^s - n_e$ inequalities

$$f(S) \geq f(S \setminus \{e\}) \quad (4)$$

for each element $e \in S \setminus C$. These inequalities correspond to the delete operations. We obtain

$$\sum_{j=1}^{t} [f(S \cup W_j) - f(S)] \leq \sum_{j=1}^{t} [f(S) - f(S \setminus \Lambda_j)] +$$

$$\sum_{e \in S \setminus C} ((p+1)2^s + (k-2)p2^s - n_e) [f(S \setminus \{e\}) - f(S)]. \quad (5)$$

Applying Lemma 2 to the right-hand side of the inequality (5) and Lemma 1 to the left-hand side of the inequality (5), we have

$$p2^s [f(S \cup C) - f(S)] \leq ((p+1)2^s + (k-2)p2^s) [f(S) - f(S \cap C)],$$

which is equivalent to

$$(k + 1/p) \cdot f(S) \geq f(S \cup C) + (k - 1 + 1/p) \cdot f(S \cap C).$$

The result follows.   □

Simple consequences of Lemma 10 are bounds on the value of a locally-optimal solution when the submodular function f has additional structure.

Corollary 1. *For $k \geq 2$, a locally-optimal solution S, and any $C \in \cap_{j=1}^{k} \mathcal{I}_j$, the following inequalities hold:*

1. *$f(S) \geq f(C)/(k + 1/p)$ if function f is monotone,*
2. *$f(S) \geq f(C)/(k - 1 + 1/p)$ if function f is linear.*

The local-search algorithm defined at the beginning of this section could run for an exponential amount of time before reaching a locally-optimal solution. To ensure polynomial runtime, we follow the standard approach of approximate local search under a suitable (small) parameter $\varepsilon > 0$ as described in Figure 1. The following is a simple extension of Lemma 10.

Lemma 11. *For an approximate locally-optimal solution S and any $C \in \cap_{j=1}^{k} \mathcal{I}_j$,*

$$(1 + \varepsilon)(k + 1/p) \cdot f(S) \geq f(S \cup C) + (k - 1 + 1/p) \cdot f(S \cap C),$$

where $\varepsilon > 0$ is the parameter used in the procedure of Figure 1.

Input: Finite ground set $N := [n]$, value-oracle access to submodular function $f : 2^N \to \mathbb{R}$, and matroids $\mathcal{M} = (N, \mathcal{I}_i)$, for $i \in [k]$.
1. Set $v \leftarrow \arg\max\{f(u) \mid u \in N\}$ and $S \leftarrow \{v\}$.
2. While the following local operation is possible, update S accordingly:

  p-exchange operation. If there is a feasible S' such that
  (i) $|S' \setminus S| \leq p$, $|S \setminus S'| \leq kp$, and
  (ii) $f(S') \geq (1 + \varepsilon/n^4)f(S)$,

  then $S \leftarrow S'$.
Output: S.

Fig. 1. The approximate local-search procedure

Proof. The proof of this lemma is almost identical to the proof of the Lemma 10 — the only difference is that left-hand sides of inequalities (2) and inequalities (4) are multiplied by $1 + \varepsilon/n^4$. Therefore, after following the steps in the proof of Lemma 10, we obtain the inequality:

$$\left(k + 1/p + \varepsilon\lambda/n^4 p 2^s\right) \cdot f(S) \geq f(S \cup C) + (k - 1 + 1/p) \cdot f(S \cap C),$$

where $\lambda = t + \sum_{e \in S \setminus C} [(p+1)2^s + (k-2)p2^s - n_e]$ is the total number of inequalities (2) and (4). because $t \leq |C|p2^s$ we obtain that $\lambda \leq (n + k)p2^s$. Assuming that $n^4 >> n + k$, we obtain the result. □

Lemma 11 implies the following:

Theorem 1. *For any fixed $k \geq 2$ and fixed constant $\delta > 0$, there exists a polynomial $1/(k + \delta)$-approximation algorithm for maximizing a non-negative non-decreasing submodular function subject to k matroid constraints. This bound improves to $1/(k - 1 + \delta)$ for linear functions.*

Remark 2. Combining techniques from this paper with the iterative local-search from [21], we can improve the performance guarantees of the approximation algorithms for maximizing a general (non-monotone) submodular function subject to $k \geq 2$ matroid constraints from $k + 2 + \frac{1}{k} + \delta$ to $k + 1 + \frac{1}{k-1} + \delta$ for any $\delta > 0$.

4 Tightness of Analysis

Next, we demonstrate that our analysis of local search for maximizing monotone submodular functions is tight. By local search, we mean for fixed $p > 0$, adding $\leq p$ elements and removing $\geq kp$ elements at a time. It was known [2] that such an algorithm cannot give better than $1/(k-1+1/p)$-approximation for the weighted k-set packing problem ($k \geq 3$). From the example of [2], the same bound follows also for weighted k-dimensional matching and hence also for the more general problems of maximizing a linear function subject to k matroid constraints, or a monotone submodular function subject to $k - 1$ matroid constraints.

Proposition 1. *For any $k, p \geq 2$, there are instances of maximizing a linear function subject to k partition matroids, where a local optimum with respect to p-exchanges has value $OPT/(k - 1 + 1/p)$.*

Proof. Let $G = (V, E)$ be a k-regular bipartite graph of girth at least $2p + 2$ (see [20] for a much stronger result), with bipartition $V = A \cup B$. We define vertex weights $w_i = 1$ for $i \in A$ and $w_j = k - 1 + 1/p$ for $j \in B$. Being a k-regular bipartite graph, G can be decomposed into k matchings, $E = M_1 \cup M_2 \cup \ldots \cup M_k$. For each M_i, we define a partition matroid $\mathcal{M}_i = (V, \mathcal{I}_i)$ where $S \in \mathcal{I}_i$ iff S contains at most one vertex from each edge in M_i. We maximize $w(S)$ over $S \in \bigcap_{i=1}^{k} \mathcal{I}_i$. Equivalently, we seek a maximum-weight independent set in G.

Clearly, A and B are both feasible solutions. Because $|A| = |B|$, we have $w(A)/w(B) = 1/(k - 1 + 1/p)$. We claim that A is a local optimum. Consider any set obtained by a local move, $A' = (A \setminus K) \cup L$ where $K \subseteq A$, $L \subseteq B$ and $|L| \leq p$. For A' to be independent, $K \cup L$ must contain all edges incident with L. (Otherwise, there is an edge contained in A'.) Also, $K \cup L$ cannot contain any cycle, because every cycle in G has at least $p + 1$ vertices on each side. Therefore, $K \cup L$ induces a forest with $k|L|$ edges. Hence $|K \cup L| \geq k|L| + 1$, i.e. $|K| \geq (k - 1)|L| + 1$. The value of A' is

$$w(A') = w(A) - |K| + (k - 1 + 1/p)|L| \leq w(A) - |K| + (k - 1)|L| + 1 \leq w(A). \qquad \square$$

References

1. Ageev, A., Sviridenko, M.: Pipage rounding: A new method of constructing algorithms with proven performance guarantee. J. Comb. Opt. 8(3), 307–328 (2004)
2. Arkin, E., Hassin, R.: On local search for weighted k-set packing. Math. of Oper. Research 23(3), 640–648 (1998)
3. Berman, P.: A $d/2$ approximation for maximum weight independent set in d-claw free graphs. Nordic J. Comput. 7(3), 178–184 (2000)
4. Brylawski, T.: Some properties of basic families of subsets. Disc. Math. 6, 333–341 (1973)
5. Calinescu, G., Chekuri, C., Pál, M., Vondrák, J.: Maximizing a monotone submodular set function subject to a matroid constraint. In: Fischetti, M., Williamson, D.P. (eds.) IPCO 2007. LNCS, vol. 4513, pp. 182–196. Springer, Heidelberg (2007)
6. Calinescu, G., Chekuri, C., Pál, M., Vondrák, J.: Maximizing a monotone submodular set function subject to a matroid constraint. SIAM J. on Comp. (to appear)
7. Chandra, B., Halldórsson, M.: Greedy local improvement and weighted set packing approximation. J. Algorithms 39(2), 223–240 (2001)
8. Conforti, M., Cornuéjols, G.: Submodular set functions, matroids and the greedy algorithm: Tight worst-case bounds and some generalizations of the Rado-Edmonds theorem. Disc. Appl. Math. 7(3), 251–274 (1984)
9. Cornuéjols, G., Fischer, M., Nemhauser, G.: Location of bank accounts to optimize float: An analytic study of exact and approximation algorithms. Management Sci. 23, 789–810 (1977)

10. Cornuéjols, G., Fischer, M., Nemhauser, G.: On the uncapacitated location problem. Annals of Disc. Math. 1, 163–178 (1977)
11. Feige, U.: A threshold of ln n for approximating set cover. J. of ACM 45, 634–652 (1998)
12. Feige, U., Mirrokni, V., Vondrák, J.: Maximizing non-monotone submodular functions. In: FOCS 2007, pp. 461–471 (2007)
13. Greene, C.: A multiple exchange property for bases. Proc. Amer. Math. Soc. 39, 45–50 (1973)
14. Hausmann, D., Korte, B.: K-greedy algorithms for independence systems. Z. Oper. Res. Ser. A-B 22(5) (1978)
15. Hausmann, D., Korte, B., Jenkyns, T.: Worst case analysis of greedy type algorithms for independence systems. Math. Prog. Study 12, 120–131 (1980)
16. Hazan, E., Safra, S., Schwartz, O.: On the complexity of approximating k-set packing. Computational Complexity 15(1), 20–39 (2006)
17. Hurkens, C., Schrijver, A.: On the size of systems of sets every t of which have an SDR, with an application to the worst-case ratio of heuristics for packing problems. SIAM J. Disc. Math. 2(1), 68–72 (1989)
18. Iwata, S., Fleischer, L., Fujishige, S.: A combinatorial, strongly polynomial-time algorithm for minimizing submodular functions. J. of ACM 48, 761–777 (2001)
19. Jenkyns, T.: The efficacy of the greedy algorithm. Cong. Num. 17, 341–350 (1976)
20. Lazebnik, F., Ustimenko, V., Woldar, A.: A new series of dense graphs of high girth. Bulletin of the AMS 32(1), 73–79 (1995)
21. Lee, J., Mirrokni, V., Nagarajan, V., Sviridenko, M.: Maximizing non-monotone submodular functions under matroid and knapsack constraints. Submitted for publication, preliminary version appeared in STOC 2009 (2009)
22. Lovász, L.: Submodular functions and convexity. In: Bachem, A., et al. (eds.) Mathematical Programmming: The State of the Art, pp. 235–257
23. Nemhauser, G., Wolsey, L., Fisher, M.: An analysis of approximations for maximizing submodular set functions I. Math. Prog. 14, 265–294 (1978)
24. Fisher, M., Nemhauser, G., Wolsey, L.: An analysis of approximations for maximizing submodular set functions II. Math. Prog. Study 8, 73–87 (1978)
25. Nemhauser, G.L., Wolsey, L.: Best algorithms for approximating the maximum of a submodular set function. Math. Oper. Res. 3(3), 177–188 (1978)
26. Oxley, J.: Matroid theory. Oxford University Press, New York (1992)
27. Reichel, J., Skutella, M.: Evolutionary algorithms and matroid optimization problems. In: Proc. of GECCO 2007, pp. 947–954 (2007)
28. Schrijver, A.: Combinatorial Optimization: Polyhedra and Efficiency. Springer, Heidelberg (2003)
29. Schrijver, A.: A combinatorial algorithm minimizing submodular functions in strongly polynomial time. J. Comb. Theory, Ser. B 80, 346–355 (2000)
30. Vondrák, J.: Optimal approximation for the submodular welfare problem in the value oracle model. In: STOC 2008, pp. 67–74 (2008)
31. Woodall, D.: An exchange theorem for bases of matroids. J. Comb. Theory, Ser. B 16, 227–228 (1974)

Robust Algorithms for MAX INDEPENDENT SET on Minor-Free Graphs Based on the Sherali-Adams Hierarchy

Avner Magen[1] and Mohammad Moharrami[2,*]

[1] University of Toronto
[2] University of Washington

Abstract. This work provides a Linear Programming-based Polynomial Time Approximation Scheme (PTAS) for two classical NP-hard problems on graphs when the input graph is guaranteed to be planar, or more generally Minor Free. The algorithm applies a sufficiently large number (some function of $1/\epsilon$ when $1 + \epsilon$ approximation is required) of rounds of the so-called Sherali-Adams Lift-and-Project system. needed to obtain a $(1 + \epsilon)$-approximation, where f is some function that depends only on the graph that should be avoided as a minor. The problem we discuss are the well-studied problems, the MAX INDEPENDENT SET and MIN VERTEX COVER problems. An curious fact we expose is that in the world of minor-free graph, the MIN VERTEX COVER is harder in some sense than the MAX INDEPENDENT SET.

Our main result shows how to get a PTAS for MAX INDEPENDENT SET in the more general "noisy setting" in which input graphs are not assumed to be planar/minor-free, but only close to being so. In this setting we bound integrality gaps by $1 + \epsilon$, which in turn provides a $1 + \epsilon$ approximation of the optimum value; however we don't know how to actually find a solution with this approximation guarantee. While there are known combinatorial algorithms for the non-noisy setting of the above graph problems, we know of no previous approximation algorithms in the noisy setting. Further, we give evidence that current combinatorial techniques will fail to generalize to this noisy setting.

1 Introduction

A common way to handle NP-hard problems is to design approximation algorithms for them. Often, even a good approximation cannot be achieved if one is concerned with the standard worst-case analysis. For example, it is NP-hard not only to solve MAX INDEPENDENT SET but also to approximate it to within factor of $|V|^\delta$ for any $\delta < 1$ unless NP=ZPP [17]. However, we may be able to to compute good approximations for some classes of inputs. Examples for such classes in the context of graph problem could be graphs with bounded degree, sparse graphs, dense graphs, perfect graphs, etc. In some cases a certain restriction on

---

* This work was done while the author was at University of Toronto.

I. Dinur et al. (Eds.): APPROX and RANDOM 2009, LNCS 5687, pp. 258–271, 2009.

the input renders a problem trivial, such as the case of MAX CLIQUE restricted to bounded-degree graphs; in others, such as SPARSEST CUT on bounded degree graphs are still very hard to approximate. More interesting examples are the semidefinite-programming based algorithm for colouring of perfect graphs [16], or the classical Polynomial Time Approximation Scheme (PTAS) by Arora for Euclidean TSP [1].

In this paper we present algorithms based on Linear Programming (LP), which give rise to a PTAS for the problems of MAX INDEPENDENT SET and MIN VERTEX COVER on minor-free graphs, and in particular on planar graphs.

We first explain how Linear Programming approach may lead to a PTAS, namely algorithms that for each $\epsilon > 0$ give approximation of $1 + \epsilon$ and run in time polynomial in the size of the graph and may depend on ϵ. One can think of this as a sequence of algorithms which give approximation factor that approach 1. To come up with such a sequence using an LP, it is natural to consider a sequence of LP formulations rather than a fixed one. Systematic methods that give rise to such sequences are so-called Lift-and-Project methods. Here, the original LP is tightened repeatedly r times (or levels/rounds). When this process is repeated for $r = n$ times, the obtained LP is equivalent to the original Integer Program, and hence solving it will give the exact solution to the original problem, however the running time of such an algorithm will not be polynomial in general. More specifically, starting from a poly-size LP it takes $n^{O(r)}$ to optimize over the level r tightening. In order to obtain a PTAS using the above paradigm, one should show that $\lim \eta_r = 1$ where η_r is the approximation guaranteed by the LP after r rounds of applying the Lift-and-Project operator.

Different variants of Lift and Project methods exist, and in this work we show that the one due to Sherali-Adams satisfies the condition above with respect to some classical graph optimization problems on planar graphs and their generalization to minor-free graphs. To the best of our knowledge there is only one example for PTAS that is obtained by Lift-and-Project systems due to Fernández de la Vega and Kenyon-Mathieu [14] who have provided a PTAS for MAX CUT in dense graphs using the Sherali-Adams hierarchy.

We further consider the setting where the input graphs are noisy, in the sense that they are obtained by applying some bounded number of changes to graphs in the special classes considered above. We show that in this setting the LP-based approach is still effective: we can bound the integrality gap by $1 + \epsilon$ when $O(1/\epsilon)$ rounds of Sherali-Adams are applied. It is important to note that while the integrality gap is well-bounded by our method (whence the method well-approximates the optimal value), we don't know how to translate this guarantee to a rounding procedure or to any other method that will obtain a solution approximating the optimum. There aren't many examples in the literature where a bound on the integrality gap is known but no integral solution is presented to achieve the bound, and we note [13] as one example of such a scenario.

Previous Work: Tree graphs, bipartite graphs, small tree-width , outerplanar and planar graphs all have been well-studied in the context of restrictions on the type of input of NP-hard problems. Specifically for our problem, algorithms for planar graphs were studied by Baker [2] who gave a PTAS with running time $O(f(\epsilon)n \log n)$ for MAX INDEPENDENT SET and MIN VERTEX COVER on planar graphs. For the minor-free case, The work of DeVos et al.[11] opened the way for algorithms in minor-free graph partitioning, as they provided (proof of existence of) a decomposition of the graph to simple parts. Following their work, there were a series of algorithms for minor-free graphs which were mostly nonconstructive[1] such as [15]. However, later in a work of Demaine et al.[12] it was shown that the decomposition can be done in polynomial time which makes those algorithms constructive. We note that our approach is in general inferior to the combinatorial approach in [2] in terms of running time as the time complexity of optimizing in the r-th level of the Sherali-Adams Hierarchy is $n^{O}(r)$ which means that our algorithm run in time $O(n^{\frac{1}{\epsilon}}$.

In contrast to the above work, no algorithms are known for the noisy setting. In fact, in Section 5 we give evidence that current combinatorial approaches or modification of them are bound to fail.

In the context of PTAS which are LP-based not many examples are known, and we mention two here. In [3], Bienstock shows that a Linear Programming of size polynomial in $1/\epsilon$ and in n to approxmate upto $1 + \epsilon$ the knapsack problem on n items. As in our case, this is an LP-based analogue to an existing combinatorial algorithm, the well known PTAS for Knapsack by Lawler [18]. A second example is due to Avis and Unemoto [10] who show that for dense graphs linear programming relaxations of MAX CUT approximate the optimal solution upto $1 + \epsilon$, where the size of the LP is again polynomial in $1/\epsilon$ and in n. Unlike the current work, however, the LPs in these results are not obtained through the lift-and-project method, but rather they are found in a way customized to the problem. (In fact, for the first result above, even the choice of variables to be used is not obvious.)

Techniques: An essential ingredient in our work is a result by Bienstock and Ozbay [9]. Consider a graph G that has tree-width k, and consider the standard LP relaxation of MAX INDEPENDENT PROBLEM on G. It is shown in [9] that the application of the level k Sherali-Adams (SA) operator gives an *exact* solution to the problem. In other words, the relaxed and integral optimal solutions are the same. The graph-theoretic component of our results uses the theorem of DeVos et. al. [11] mentioned above. The theorem shows that for every positive integer j there is a partition of the vertices of a minor-free graph into j parts so that the removal of *any* of them leaves components of tree-width at most $k(j)$, where $k(j)$ depends only on j and on the minor and not on n. In the special case of planar graphs this decomposition theorem is almost straightforward, with $k(j) = j$. Our approach essentially uses the following simple schema: (i) apply the level-$k(j)$ SA operator, where $j \sim 1/\epsilon$. (ii) bound the *integrality gap* obtained by $1 + \epsilon$. This is made possible by separately bounding the contribution of the

---

[1] This means that for every H there is an algorithm for the H-minor-free case, but there was no uniform algorithm, that given H and an H-minor-free graph provides the required approximation.

solution on the different parts relative to the corresponding integral solution. Notice that ensuring small integrality gap gives an approximation of the *value* of the optimum and that in order to provide algorithms that actually supply good approximated *solutions* we need to know the decomposition and round the fractional solution according to this decomposition. We later elaborate on this interesting aspect of our technique.

The rest of the paper is organized as follows. In Section 2 we give the relevant graph theoretical definitions as well as the description of the Sherali-Adams Hierarchy. In Section 3 we deal with the MAX INDEPENDENT SET problem. We first show how to get a PTAS for the simpler case of planar graphs, and then extend to family of minor-free graphs. In Section 4 we deal with the approximation of MIN VERTEX COVER. We show a general lemma that says that under sufficient conditions it is possible to import results about integrality gaps for certain LPs for the problem of MAX INDEPENDENT SET into ones about MIN VERTEX COVER. Last, we consider the case of graphs which are "noisy versions" of planar or minor-free graphs.. We show that unlike combinatorial approaches, our algorithms can extend to this case. More specifically, we show a PTAS for the *value* of the maximum independent set in noisy planar graphs and, more generally, in noisy H-minor-free graphs.

2 Preliminaries

The tree-width of a graph A tree decomposition of the graph G is a pair (T, X) such that

1. T is tree;
2. for every vertex $v \in G$ there is a tree t_v;
3. $X = \{t_v : V(t_v) \subseteq T\}$ such that each t_v is a subtree of T;
4. and for any edge $e(u, v)$ in $E(G)$, we have $t_v \cap t_u \neq \emptyset$.

We say that a graph G has tree-width k if there exists a tree decomposition of G such that the intersection of every $k + 2$ of t_v's is empty.

The Sherali-Adams Hierarchy. Sherali-Adams is a system that given an LP relaxation produces a tightened LP, that will eventually produce a program that is equivalent to the Integer Program describing the problem. More specifically, given a LP relaxation of some $\{0, 1\}$ integer-program on n variables and a parameter r, the Sherali-Adams lifting of the LP in the rth level an LP that is strictly stronger than the original LP and requires $n^{O(r)}$ time to optimize over. When $r = n$, the generated LP is equivalent to the integer program, hence its solution solves the original problem exactly. While this is not essential for the purpose of the current paper, we give below a full description of the system.

For every two disjoint sets of variables I and J such that $|I \cup J| \leq k$, we have a variable $w[I, J]$. This variable represents $\prod_{i \in I} x_i \prod_{j \in J} (1 - x_j)$ in an integer solution, and in particular, an original variable of the LP is associated with $w[\{i\}, \emptyset]$. The system imposes all possible linear conditions on this set of variable that can be derived by (i) the original inequalities of the LP, and (ii) by the relations of the above products amongst themselves. The inequalities of type (i) that we get are derived by every LP inequality For the first type, we obtain the inequality

$$\sum_{j \notin J} a_j w[I \cup \{j\}, J] \geq b \cdot w[I, J]. \tag{1}$$

for every LP inequality $\sum_i a_i x_i \geq b$ and every I, J as above.

For type (ii) the following inequalities are obtained.

$$w[\emptyset, \emptyset] = 1 \tag{2}$$

$$0 \leq w[I \cup \{j\}, J] \leq w[I, J] \text{ for } j \notin (I \cup J) \tag{3}$$

$$0 \leq w[I, J \cup \{j\}] \leq w[I, J] \text{ for } j \notin (I \cup J) \tag{4}$$

$$w[I, J] = w[I \cup \{j\}, J] + w[I, J \cup \{j\}] \tag{5}$$

The obtained linear program "projects back" to the original set of variables, namely considers $w[\{i\}, \emptyset]$. We shall denote by $\text{SA}^{(t)}(G)$ the polytope of all solutions of the t-th level of the Sherali-Adams Hierarchy (this is the extension of the notion of the polytope associated with an LP relaxation).

Noisy Graphs. Consider a class of graphs. Then a *noisy version* of a graph from the class is simply a perturbation applied to it. We adopt a standard notion of distance to quantify this: the distance between two graph is the minimum number of edges or vertices that should be added or removed from one of the graphs to become isomorphic to the other graph. We extend this notion to distance between a graph G and a family of graphs in the standard way, namely as the minimum distance of G over all the graphs in the family. Notice that when the family is monotone, that is closed under edge removal, as is the case with the families we consider, the distance is simply the number of edges needed to be removed from the graph in order for it to be in the family. It is important not to confuse the notion of "noise" here, which is deterministic, with the notion of noise used to describe random perturbation of objects, and the result we supply are stronger than corresponding results in the random model.

3   A PTAS for Max Independent Set

In the MAX INDEPENDENT SET problem the input is a graph and the output is an independent set, namely a set of maximum size of vertices that share no edges. This is a classical NP-hard problem which is notoriously hard to approximate. Let n be the number of vertices, then it is NP-hard to approximate the problem to within factor of $n^{1-\epsilon}$ [17]. In other words, in the worst case setting not much can be done. This motivates looking at special classes of inputs.

3.1   Planar Graph Case

While the MAX INDEPENDENT SET problem is still NP-hard for planar graphs, the problem of approximating the solution is quite a bit different. Indeed, any four colouring of a planar graph gives rise to an independent set of size at least $n/4$, and hence 4-approximation algorithm. The next natural is whether a polynomial time algorithm exists that approximate the optimum to within $1 + \epsilon$ and what is the dependency in ϵ.

The standard Linear Programming relaxation for the problem is:

$$
\begin{array}{ll}
\text{maximize:} & \sum_{v \in G} x_v \\
\text{for } uv \in E(G)) & x_v + x_u \leq 1 \\
\text{for } u \in V(G) & 0 \leq x_u \leq 1
\end{array}
\tag{6}
$$

Notice that this LP is quite weak as the all $1/2$ solution is always a feasible solution. For graphs with sublinear independent sets this LP is therefore quite useless as it is. However, it is not hard to show that for planar graphs the integrality gap of the LP above cannot be larger than 2. Our goal now is to show that by using higher level of the Sherali-Adams hierarchy much better approximations can be obtained.

Let G be the input graph and $\alpha(G)$ be the size of the largest independent set of G. Furthermore, let y be the projection of optimal solution of the level k SA operator applied to LP (6) onto the singleton variables. For a set of vertices S we define $y(S)$ as $\sum_{u \in S} y_u$, and $y'(S)$ as $\sum_{u \in S} y_u - y_u^2$. Abusing notation, when M is a graph, we may write $y(M)$ instead of $y(V(M))$.

Fix an embedding of a planar graph G into the plane. Graph G is m-outerplanar for some $m > 0$. The vertices of the graph can be partitioned into m sets $V_1, V_2 \ldots V_m$, where V_1 is the set of vertices in the boundary of the outerface, V_2 is the set of vertices in the boundary of the outerface after V_1 is removed and so on. Note that, if $u \in V_i$ and $w \in V_j$ are adjacent then $|i - j| \leq 1$.

We now wish to remove some of the V_i from the graph so that (i) the remaining graph is k-outerplanar, and (ii) the weight of the removed set in the optimal SA solution is small. Let

$$
B(i) = \bigcup_{j \equiv i \,(\mathrm{mod}\, k+1)} V_j.
$$

For every value of k this partitions $V(G)$ into $(k+1)$-outerplanar sets. Note that after removing the vertices in $B(i)$, the resulting graph is k-outerplanar. We now consider an index j for which $y'(B(j)) \leq y'(G)/(k+1)$ and denote $B(j)$ by W.

Let G_i be the subgraph of G induced on $V_i = \{v : v \in V_l, ik + j \leq l \leq (i+1)k + j\}$. Notice that every edge or vertex of G appear in one or two of the G_i, and those vertices not in W appear in precisely one of the G_i. A key observation we need is that applying Sherali-Adams on G and then projecting onto V_i (more precisely, projecting onto all subsets of size at most t in V_i) is a solution in $\mathrm{SA}^{(t)}(G_i)$. This follows from the fact that the LP associated with G is stronger than the one associated with the subgraph G_i (on all common variables) and the same extends to the Sherali-Adams hierarchies. Therefore using [9] we can deduce that the projection of y onto the singleton sets in V_i is a convex combination of integral solutions, namely independent sets of G_i.

Let ρ_i be the corresponding distribution of independent sets for G_i and consider the following experiment (or random rounding): pick a set S_i according to ρ_i, independently for each i. We say that a vertex v is *chosen* if it is in S_i whenever $v \in G_i$. (Notice that for $v \notin W$, v belongs to a unique G_i and the condition is simply that $v \in S_i$, but for $v \in W$, v may belong to both G_i and G_j in which case it is chosen only when $v \in S_i \cap S_j$.) Denote by S the set of chosen vertices. We claim that S is an independent set. Indeed, every edge belongs entirely to some G_i, two neighbours in G_i cannot both be in the independent set S_i, and so they cannot both be chosen.

Since the marginals of ρ_i on $v \in G_i$ is y_v, we get that for vertices $v \notin W$

$$\Pr[v \in S] = y_v$$

and for vertices $v \in W$

$$\Pr[v \in S] \geq y_v^2.$$

From the above conditions we can conclude that

$$\mathsf{E}(|S|) \geq \sum_{v \notin W} y_v + \sum_{v \in W} y_v^2 = \sum_v y_v - \sum_{v \in W} (y_v - y_v^2) = y(G) - y'(W)$$

Now, it is easy to see that $y'(G) \leq \frac{3y(G)}{4}$. It is shown in [4] that a k-outerplanar graph has tree-width at most $3k-1$, therefore in the $3k-1$ level of Sherali-Adams y will be integral on any subgraph of tree-width at most k. We can finish off with the required bound

$$IS(G) \geq \mathsf{E}(|S|) \geq y(G) - \frac{1}{k} y'(G) \geq y(G) - \frac{1}{k} \left(\frac{3y(G)}{4} \right) = \left(1 - \frac{3}{4k} \right) y(G)$$

and get

Theorem 1. *Let G be a planar graph. Then $\alpha(G)$ is at least $1 - \frac{3}{4k}$ times the solution of level $3k - 1$ Sherali-Adams operator applied on the standard LP for* Max Independent Set *(LP (6)). Further, the above algorithm gives rise to a rounding procedure that actually finds an independent set that is at least $(1 - \frac{3}{4k})\alpha(G)$*

3.2   Extending to Minor-Free Graphs

Consider a fixed graph H and consider graphs G which are H-minor-free, namely, they don't contain H as a minor[2] Notice that planar graphs are a special case as they do not contain K_5 (or alternatively, $K_{3,3}$ as a minor. As with the case of planar graphs, the special property of a minor-free which is utilized in algorithms

---

[2] A graph G contains H as a minor if H can be obtained from G by applying a sequence of edge/(isolated)vertex removal and edge contraction.

is the fact that it can be decomposed into simple components when some limited part of it is removed. As with the case of planar graph, we would like "simple" to stand for small tree width. A recent theorem due to DeVos et al. gives precisely that.

Theorem 2. *(DeVos et. al [11]) For every graph H and integer $j \geq 1$ there exist constants $k_V = k_V(H, j)$ and $k_E = k_E(H, j)$ such that the vertices of every graph G with no H-minor can be partitioned into $j + 1$ parts such that the union of every j of them has tree-width at most k_V. In addition, the edges of G can be partitioned into $j + 1$ parts such that the union of every j of them has tree-width at most k_E.*

The above theorem is crucial in the algorithm we present. It is worth noting that for the special case of planar graphs we may take k_V to be as small as $O(j)$.

Theorem 3. *For every H and $\epsilon > 0$ there exists a constant $c = c(\epsilon, H)$ such that for every graph G with no H-minor, the integrality gap of the level-c Sherali-Adams operator of LP (6) is at most $1 + \epsilon$.*

Proof. (sketch) Let $c = k_V(H, \lceil 1/\epsilon \rceil)$, we claim that applying level c SA operator is sufficient to derive $1 + \epsilon$ bound on integrality gap. For any subset of vertices we define $y(S) = \sum_{v \in S} y_v$. Using the result from [9]. We know that for any $S \subseteq G$ with tree-width less than or equal to c, we have $y(S) \leq \alpha(S)$. Now if we take the partitioning of vertices into $V_1, \ldots, V_{j+1}$ according to Theorem 2, and remove the partition with minimum $y(V_i)$ from G the rest of the graph must have tree-width at most c, and furthermore we have

$$y(G \setminus V_i) \geq \frac{j}{j+1} y(G),$$

and we bound the integrality gap

$$y(G)/\alpha(G) \leq (1 + 1/j) y(G \setminus V_i)/\alpha(G) = (1 + 1/j) \alpha(G \setminus V_i)/\alpha(G) \leq 1 + 1/j.$$

4 Vertex Cover

A *vertex cover* for a graph G is a subset of the vertices touching all edges. The MIN VERTEX COVER problem is to find a minimal vertex cover for a graph. For a graph G we denote the minimum vertex cover by $\nu(G)$.

The purpose of this section is to show how to get a SA-based PTAS for MIN VERTEX COVER on minor-free graphs from a similar PTAS for MAX INDEPENDENT SET. Generally speaking, MIN VERTEX COVER is easier problem to approximate than its complement, MAX INDEPENDENT SET, and it can be easily approximated by a factor of 2. Notice that an exact algorithm for one problem can be easily converted into an exact algorithm for the other problem. Similarly, the quality of the *additive* approximation to the problems is still the same. It is well known, however, that for the standard measure of approximation namely multiplicative approximation, the approximation quality of the problems may differ dramatically. The most common scenario exhibiting the above difference are graphs with independent sets of size at most $o(n)$ and vertex covers of size at least $n - o(n)$. For the purpose of this section, though, we are interested in understanding the opposite scenario where the size of some vertex covers is $o(n)$; this is since in these such graphs (the compliment of) a $1 + \epsilon$ approximation of MAX INDEPENDENT SET may provide a very poor approximation for MIN VERTEX COVER. Now, there is a standard trick that reduces any instance of MIN VERTEX COVER into one where the optimal solution is of size at least half the graph. This trick simply finds an optimal solution for the standard LP, and removes the vertices who get value 0 in the solution. What we do next avoids the trick. The advantage of having a direct claim about the integrality gap of *any* graph, rather than using it as a subroutine, is that it allows for argument that involves projection of a solution onto smaller subgraphs. Examples of this sort was shown in Section 3, and a more interesting one will be supplied later in Section 5 in the context of noisy graphs.

The LP for MIN VERTEX COVER is formulated below.

$$\boxed{\begin{array}{ll} \text{minimize:} & \sum_{v \in G} x_v \\ \text{for } uv \in E(G)) & x_v + x_u \geq 1 \\ \text{for } u \in V(G) & x_u \geq 0 \end{array}} \tag{7}$$

The idea behind getting a generic statement allowing us to move from MAX INDEPENDENT SET to MIN VERTEX COVER is quite simple. In fact it uses similar reasoning (even if in a more subtle way) to the "standard trick" described above. We split the graph into two parts, one that "behaves integrally" on which no error is incurred, and the other on which the maximum independent set is smaller than the minimum vertex cover, and then combine the two parts. This split is achieved by looking at the optimal solution of the standard LP to MAX INDEPENDENT SET. We start by defining a property of LP relaxations for MAX INDEPENDENT SET.

Downward Property: We say that an LP relaxation for MAX INDEPENDENT SET has the *downward property* if its solution y satisfies that for any $S \subseteq V(G)$, $y(S) \leq (1 + \epsilon)\alpha(G')$, where G' is subgraph of G induced by S.

Lemma 1. *Let y be an optimal solution to an LP relaxation of* MAX INDEPENDENT SET *that has the downward property, then* $|V(G)| - y(G) \geq (1 - \epsilon)\nu(G)$

Proof. Consider the standard LP for MAX INDEPENDENT SET (LP(6)) and denote its solution by z. It is well known that z can be transformed into a half-integral solution. Partition $V(G)$ to S_0, S_1, and $S_{1/2}$ according to the value of z on the vertices. Also, let $S_{\text{int}} = S_0 \cup S_1$, G_{int} be the induced subgraph on S_{int}, and $G_{1/2}$ the induced subgraph on $S_{1/2}$.

We first argue that the restriction of z on S_{int} is the optimal fractional solution of LP(6) on S_{int}. To see that, let w be any fractional solution to LP(6) on S_{int} and let u be the extension of w to S according to z, that is u agrees with w on

S_{int} and with z on $S_{1/2}$. We now show that $(z+u)/2$ is a solution to LP(6) on G: edges inside S_{int} as well as edges inside $S_{1/2}$ are satisfied by both z and u, and so also by $(z+u)/2$; edges between S_0 and $S_{1/2}$ sum to at most $1/2$ in z and at most $3/2$ in u, and so must sum to at most 1 on $(z+u)/2$. Since there are no edges between S_1 and $S_{1/2}$ in G we have that $(z+u)/2$ is a valid solution. Optimality of z implies that $z(S) \geq u(S)$ and hence $z(S_{\text{int}}) \geq w(S_{\text{int}})$. Of course the same holds for any vector which is a solution to a tightening of LP(6) on S_{int}. In particular

$$y(S_{\text{int}}) \leq z(S_{\text{int}}) = |S_1|. \tag{8}$$

The second fact we require is that maximum independent set in $G_{1/2}$ is smaller than the minimum vertex cover of this graph. Since the all-half vector is solution of LP(6) on $G_{1/2}$, it is also a solution of the standard vertex cover relaxation. But then

$$\nu(G_{1/2}) \geq z(S_{1/2}) \geq \alpha(G_{1/2}). \tag{9}$$

With inequalities (8) and (9) we can easily conclude

$$
\begin{aligned}
n - y(G) &= |S_{\text{int}}| - y(S_{\text{int}}) + |S_{1/2}| - y(S_{1/2}) \\
&\geq |S_{\text{int}}| - |S_1| + |S_{1/2}| - (1+\epsilon)\alpha(G_{1/2}) \\
&= |S_0| + \nu(G_{1/2}) - \epsilon\alpha(G_{1/2}) \\
&\geq |S_0| + \nu(G_{1/2}) - \epsilon\nu(G_{1/2}) \\
&\geq \nu(G) - \epsilon\nu(G_{1/2}) \\
&\geq \nu(G) - \epsilon\nu(G)
\end{aligned}
$$

where the second last inequality follows since the union of S_0 and any vertex cover of $G_{1/2}$ is a vertex cover for G.

For any graph G which is H minor-free all its subgraphs are also H minor-free. This fact shows that we satisfy the conditions of Lemma 1. Now if we use Theorem 3, we can immediately get that applying level c SA operator is sufficient to obtain the $n - y(G) \geq (1 - \epsilon)\nu(G)$ inequality. Specifically, we have

Theorem 4. *After applying level k SA operator the above Linear program, we have a approximation of $1 - 1/f(k)$ for* MIN VERTEX COVER.

Any subgraph of a H minor-free graph is also a H minor-free graph and therefore it satisfies the second condition of Lemma 1. Also it is clear that it satisfies the first condition as SA is a tightening of the LP (6). and therefore the approximation on independent set follows the approximation of vertex cover for planar graphs.

5 Main Result: A PTAS for MAX INDEPENDENT SET on Noisy Minor-Free Graphs

Algorithms that makes assumptions about the nature of their input may completely break down when this assumption is not totally met, even if by just a little. Indeed, try to two-colour a graph that is not quite two-colourable, or to approximate Max2SAT for formulas that are almost satisfiable by using an algorithm that solves 2SAT. Perhaps the most obvious example of this sort is MAX-2LIN, the problem of satisfying a maximal number of linear equations. This problem can be solved easily using Gaussian elimination if there is an assignment satisfying all equations but is hard to approximate when this is not the case, even when the system is nearly satisfiable.

Of course a better scenario is when the algorithms are *robust*. Such algorithms are designed to work well on a special class of inputs but even when the input slightly inconsistent with the class (of course, "slightly" should be well defined in some natural way) then the performance (approximation) of the algorithm may only deteriorate in some controlled way.

As was outlined in the Preliminaries, in the context of graphs we say that a graph is close to being Minor Free if by removing a small number of edges the obtained graph is minor-free. With this in mind, we would like to know whether there are good algorithms when the input graph is either minor-free or it can be made minor-free after, say, $o(n)$ edges are removed from.

We first argue that previous combinatorial algorithms, or even other algorithms that work in the same spirit, are non-robust. Notice that all previous algorithms relied on finding a decomposition of the graph into simpler (small tree-width) parts, in a manner which "resembles" a partition. For simplicity we will consider the spacial case of robustness with respect to planar graphs. Had there been robust combinatorial algorithms we would that along the way such algorithms will provide decomposition of the above nature. But then we should also expect such algorithms to perform the simpler task of deleting a few nodes and edges in such graphs so as to make them planar. Two relevant combinatorial problems come in mind, MAXIMUM PLANAR SUBGRAPH and MINIMUM NON-PLANAR DELETION, the first asking to find a planar subgraph of the input graph G with maximum number of edges, and the second is the complementary problem, that is minimizing the number of edges to delete to make G planar. These problems are well studied and was shown to be APX-hard [8,20].

In contrast, the Sherali-Adams based approach uses such decomposition *only in its analysis* and so the algorithmic difficulty in detecting the "wrong edges" disappears. Here is what we can obtain. We jump right away to the general minor-free case, although similar argument will provide an algorithm for the planar case with improved parameters.

Theorem 5. *For every H and ϵ, there exists a constant $r = r(\epsilon, H)$ such after applying level-r SA operator to LP (6) for* MAX INDEPENDENT SET *with input graph G which has distance $d = O(n/|H|\sqrt{log|H|})$ from an H-minor-free graph, the integrality gap is at most*

$$1 + \epsilon + O(d|H|\sqrt{\log |H|}/n).$$

Proof. Let F be an H-minor-free graph that is closest to G. It is easy to verify that (i) $V(F) \subseteq V(G)$ (ii) $E(F) \subseteq E(G)$ and further that $|E(G) - E(F)| \le d$.

Since the removal of every edge can increase the size of the maximum independent set by 1, and since the removal of an isolated vertex will decrease it by 1, it follows that
$$|\alpha(G) - \alpha(F)| \leq d.$$

The next structural statement we need in order to control the behaviour of G compared to that of F is the strength of $\mathrm{SA}^{(t)}(G)$ compared to that of $\mathrm{SA}^{(t)}(F)$. Let y be the optimal solution of $\mathrm{SA}^{(t)}(G)$. Since $E(F) \subseteq E(G)$ we can use the monotonicity argument as in the proof of Theorem refmain to deduce that the restriction of y to F is a valid solution to $\mathrm{SA}^{(t)}(F)$. This allows us to bound $y(F)$ as if it is obtained in $\mathrm{SA}^{(t)}(F)$ and hence we can use Theorem 3, which sys that there exists a constant $r = r(\epsilon, H)$ such that after applying level r Sherali-Adams operator, we get a bound

$$y(F) \leq (1 + \epsilon)\alpha(F). \tag{10}$$

Recall that $y(F)$ is just a projection of the vector y onto F, hence we obviously have

$$y(G) - y(F) \leq d \tag{11}$$

We next argue that there are large independent sets in F. Indeed, recall that the greedy algorithm that repeatedly takes a vertex of lowest degree to the independent set and removes its neighbours, gives an independent set of size $\Omega(n/\delta)$ where δ is the average degree in F. It is known [19] that H-minor-free graphs have on average degree $O(|H|\sqrt{\log |H|})$, hence an independent set of size $\Omega(n/|H|\sqrt{\log |H|})$ is obtained. Since $d = O(n/|H|\sqrt{\log |H|})$ we get that $d = O(\alpha(F))$. We will assume from now on that the hidden constant is such that

$$d \leq \alpha(F)/4 \tag{12}$$

We now combine inequalities 10, 11 and 12 to get obtained the desired bound on the integrality gap of $\mathrm{SA}^{(t)}(G)$.

$$\frac{y(G)}{\alpha(G)} \leq \frac{y(F) + d}{\alpha(F) - d}$$
$$\leq \frac{y(F)}{\alpha(F) - 2d}$$
$$\leq \frac{(1 + \epsilon)\alpha(F)}{\alpha(F)(1 - 2d/\alpha(F))}$$
$$\leq (1 + \epsilon)(1 + 4d/\alpha(F))$$
$$= 1 + \epsilon + O(d|H|\sqrt{\log |H|}/n).$$

When Min Vertex Cover is Harder than Min Independent Set: Is it possible to import the above result to the MIN VERTEX COVER problem a-la Section 4? We give a strong evidence that the answer is negative. The idea is based on two simple facts. First, a graph on d vertices has distance d from the empty graph. Second, the addition of isolated vertices to a graph the optimal value of the vertex cover LP does not change, nd the same holds to the level r SA operator applied on that LP. By a result of Charikar, Makarychev and Makarychev [7] there are graphs on d nodes for which the integrality gap is $2 - o(1)$ even in the r-th level of the Sherali-Adams hierarchy for $r = d^{\Omega(1)}$. Specifically, the fractional solution (in the hierarchy) is roughly $d/2$ while the minimum vertex cover is $d(1 - o(1))$. Now, take a graph G_0 on d vertices as above and add $n - d$ isolated vertices to it. The obtained graph G will have (i) distance d from the empty graph on $n - d$ vertices (which is of course planar), and (ii) an optimal value of roughly $d/2$ in the $d^{\Omega(1)}$-level of the Sherali-Adams hierarchy. Thinking of d and n as asymptotically the same, say $d = n/100$ we get that even linear-level (in number of vertices) of Sherali-Adams has tight integrality-gap for graphs which are d distance away from planar graph, and so for the MIN VERTEX COVER problem, proximity to planarity does not preclude large integrality gaps.

6   Discussion

We have shown how LP-based algorithms "utilize" graph theoretical concepts in a different way compared to their combinatorial counterparts: While the combinatorial algorithms need to find a partition/decomposition of the graph in order to define the execution of the rest of the algorithm, in the Sherali-Adams world the special structure of the graph is used only in the analysis (at least for the problem of approximating the optimal value). This conceptual difference is what allows the Sherali-Admas approach to be successful where the combinatorial approach is limited.

In the introduction we have mentioned the Euclidean TSP result due to Arora[1]. Other works on connectivity problems for Planar/Euclidean case were since investigated, see [5,6]. The underlying principle that is employed in these works is that a discretization of the space can approximate the problem well. The finer the discretization the better the approximation (at the cost of increased running time). Showing that a Sherali-Adams based algorithm leads to similar PTAS would be very interesting. Again, such a result will give rise to a very simple algorithm, "placing all the difficulty" on the analysis.

Acknowledgement. We thank Robi Krauthgamer who suggested to challenge lift and project systems with hard problems on planar graphs.

References

1. Arora, S.: Nearly linear time approximation schemes for euclidean tsp and other geometric problems. In: FOCS 1997: Proceedings of the 38th Annual Symposium on Foundations of Computer Science (FOCS 1997), Washington, DC, USA, p. 554. IEEE Computer Society Press, Los Alamitos (1997)
2. Baker, B.S.: Approximation algorithms for np-complete problems on planar graphs. J. ACM 41(1), 153–180 (1994)

3. Bienstock, D.: Approximate formulations for 0-1 knapsack sets. Oper. Res. Lett. 36(3), 317–320 (2008)
4. Bodlaender, H.L.: A partial k-arboretum of graphs with bounded treewidth. Theor. Comput. Sci. 209(1-2), 1–45 (1998)
5. Borradaile, G., Kenyon-Mathieu, C., Klein, P.N.: A polynomial-time approximation scheme for steiner tree in planar graphs. In: SODA, pp. 1285–1294 (2007)
6. Borradaile, G., Klein, P., Mathieu, C.: A polynomial-time approximation scheme for euclidean steiner forest. In: FOCS (2008)
7. Charikar, M., Makarychev, K., Makarychev, Y.: Integrality gaps for Sherali-Adams relaxations (manuscript) (2007)
8. Călinescu, G., Fernandes, C.G., Finkler, U., Karloff, H.: A better approximation algorithm for finding planar subgraphs. In: SODA 1996: Proceedings of the seventh annual ACM-SIAM symposium on Discrete algorithms, Philadelphia, PA, USA, pp. 16–25 (1996)
9. Bienstock, N.O.D.: Tree-width and the sherali-adams operator. Discrete Optimization 1(1), 13–21 (2004)
10. Avis, J.U.D.: Stronger linear programming relaxations of max-cut. Mathematical Programming 97(3), 451–469 (2003)
11. DeVos, M., Ding, G., Oporowski, B., Sanders, D.P., Reed, B., Seymour, P., Vertigan, D.: Excluding any graph as a minor allows a low tree-width 2-coloring. J. Comb. Theory Ser. B 91(1), 25–41 (2004)
12. Kawarabayashi, K.-i., Demaine, E.D., Hajiaghayi, M.T.: Algorithmic graph minor theory: Decomposition, approximation, and coloring. In: FOCS, pp. 637–646 (2005)
13. Feige, U.: On allocations that maximize fairness. In: SODA 2008: Proceedings of the nineteenth annual ACM-SIAM symposium on Discrete algorithms, San Francisco, California, pp. 287–293 (2008)
14. de la Vega, W.F., Kenyon-Mathieu, C.: Linear programming relaxations of maxcut. In: Proceedings of the 18th ACM-SIAM Symposium on Discrete Algorithms (2007)
15. Grohe, M.: Local tree-width, excluded minors, and approximation algorithms. Combinatorica 23(4), 613–632 (2003)
16. Grötschel, M., Lovász, L., Schrijver, A.: Geometric Algorithms and Combinatorial Optimization. Springer, Heidelberg (1998)
17. Håstad, J.: Some optimal inapproximability results. J. ACM 48(4), 798–859 (2001)
18. Lawler, E.L.: Fast approximation algorithms for knapsack problems. In: FOCS, pp. 206–213 (1977)
19. Thomason, A.: An extremal function for contractions of graphs. Math. Proc. Cambridge Math. Proc. Cambridge Philos. Soc. 95, 261–265 (1984)
20. Yannakakis, M.: Node and edge deletion np-complete problems. In: STOC, pp. 253–264 (1978)

Minimizing Average Shortest Path Distances via Shortcut Edge Addition

Adam Meyerson and Brian Tagiku

University of California, Los Angeles
{awm,btagiku}@cs.ucla.edu

Abstract. We consider adding k *shortcut edges* (*i.e.* edges of small fixed length $\delta \geq 0$) to a graph so as to minimize the weighted average shortest path distance over all pairs of vertices. We explore several variations of the problem and give $O(1)$-approximations for each. We also improve the best known approximation ratio for metric k-median with penalties, as many of our approximations depend upon this bound. We give a $(1 + 2\frac{(p+1)}{\beta(p+1)-1}, \beta)$-approximation with runtime exponential in p. If we set $\beta = 1$ (to be exact on the number of medians), this matches the best current k-median (without penalties) result.

1 Introduction

Multi-core processors have become popular in modern computer architectures because they provide large gains in performance at relatively low cost. In many of these processors the multiple cores are connected as a Network-on-Chip (NoC) as described in [5]. While each individual core may be slower than a state-of-the-art single-core processor, together they form a processor well-suited for largely parallel applications. Moreover, NoC designs avoid tedious power and heat constraints associated with single-core processor design. Instead, the important concern is how to best connect these multiple cores into a single, efficient network.

NoC designs typically use mesh networks since regular topologies are easier to manufacture. However, many pairs of nodes are far apart in mesh graphs. Thus, it becomes necessary to add several long interconnects to decrease average communication latency. While traditional interconnects become inhibitively slow when too long (see [13]), radio-frequency (RF) interconnects, introduced in [8], exhibit much better performance. Unfortunately, RF interconnects require much more area and cannot completely replace traditional interconnects.

Despite this, Chang *et. al.* show how to reap the benefits of RF interconnects without significantly increasing area. They propose in [7,9] a hybrid architecture which uses an underlying mesh topology (using traditional interconnects) with an overlay of a small number of RF interconnects, each of which forms a fast point-to-point connection between otherwise distant nodes. Yet, Chang *et. al.* leave open the question of how to best place these RF interconnects given the traffic profile (between pairs of cores) of a specific application.

We formulate this as a general network design problem which we call the Average Shortest Path Distance Minimization (ASPDM) problem: Given a graph

I. Dinur et al. (Eds.): APPROX and RANDOM 2009, LNCS 5687, pp. 272–285, 2009.
© Springer-Verlag Berlin Heidelberg 2009

with weights on pairs of nodes, find k shortcut edges (of length $\delta \geq 0$) whose addition minimizes the weighted average shortest path distance over all pairs of nodes. We give the following results, where α is the best approximation known for metric k-Median with Penalties:

1. an α-approximation for Single-Source (one-to-all) ASPDM,
2. a 2α-approximation if all pairs have equal weight (Unweighted ASPDM),
3. a $(4\alpha, 2)$-approximation (*i.e.* a 4α-approximation using at most $2k$ edges) for general ASPDM,
4. an α-approximation if paths can use at most one shortcut (1-ASPDM), and
5. an $(\frac{e}{e-1})$-approximation on the improvement in cost for 1-ASPDM.

We show all the above versions to be NP-complete. We also improve the approximation to k-median with penalties by applying local search to $(1+2\frac{p+1}{\beta(p+1)-1}, \beta)$, where an (α, β)-approximation implies that we achieve an α-approximation on cost using at most βk medians. This gives us a smooth tradeoff between allowing additional medians and reducing the cost, and if we require exactly k medians ($\beta = 1$) it gives $\alpha = 3 + \varepsilon$.

Shortcut addition is frequently used in computer networks to obtain small-world topologies. Yet, existing techniques are either heuristic approaches [17,20] or consider specific graphs [22,15,18,21]. Other related problems are the Buy-at-Bulk [4], Rent-or-Buy [12] and Cost-Distance [19] problems which consider purchasing edges in a network. However, unlike these problems, ASPDM places a hard limit on the number of shortcuts. Our results guarantee constant approximations on general graphs despite this hard constraint.

2   Problem Formulation

Let $G = (V, E)$ be an undirected graph with non-negative edge lengths ℓ_e for each $e \in E$ and non-negative weights w_{uv} on each ordered pair of vertices $u, v \in V$. We use d_{uv} to denote the length of the shortest uv-path for vertices $u, v \in V$. The *weighted one-to-all shortest path sum* $D_u(G)$ from vertex u is defined as

$$D_u(G) = \sum_{v \in V} w_{uv} d_{uv}.$$

We then define the *weighted all-pairs shortest-path sum* $D(G)$ to be

$$D(G) = \sum_{u \in V} D_u(G) = \sum_{u \in V} \sum_{v \in V} w_{uv} d_{uv}.$$

Then the *weighted average shortest path distance* $\bar{D}(G)$ over all pairs of vertices is simply $D(G)$ divided by the sum of all the ordered pair weights. Throughout this paper we will be interested in minimizing $\bar{D}(G)$, but it is easy to see that it is equivalent to minimize $D(G)$.

We can now formally define the Average Shortest Path Distance Minimization via Shortcut Edge Addition problem (ASPDM) as follows:

Problem 1 (ASPDM). *Given an undirected graph $G = (V, E)$ with lengths ℓ_e on the edges $e \in E$, weights w_{uv} for each ordered pair of vertices $u, v \in V$, a shortcut edge length $\delta \geq 0$ and an integer k, find a set $F \subseteq V \times V$ of at most k shortcut edges of length δ such that $\bar{D}(G + F)$ is minimized.*

Of course, $F + G$ may be a multi-graph if $F \cap E \neq \emptyset$. In some cases we can consider *directed* shortcuts, but graph G must remain undirected for reasons stated in Section 4. For simplicity of analysis we assume that $\delta = 0$, but all our results extend to arbitrary $\delta \geq 0$.

We consider several variations of ASPDM. The Single-Source ASPDM problem (SS-ASPDM) is the case where the only non-zero weights are on pairs involving a designated source vertex s. Unweighted ASPDM (U-ASPDM) places equal weight on all pairs (which may be the case for general-application NoC designs where weights are unknown). Finally, the 1-Shortcut Edge Restricted ASPDM (1-ASPDM) restricts that each shortest path uses at most one of the added shortcut edges. 1-ASPDM is a suitable model for NoC design since it reduces the complexity of the routing tables that need to be stored in the design and also reduces congestion along these shortcuts.

3   Preliminaries and Initial Observations

In this section, we review k-median with penalties which we use in many of our results below. We will also analyze an algorithm for SS-ASPDM, which is a useful subroutine for more general results.

3.1   Metric k-Median with Penalties

In k-median with penalties, we are given a set of cities and a set of potential facility locations arranged in a metric space. Each city has a demand that needs to be served by a facility. Each city also has a penalty cost, which we can pay to refuse service to the city. If we choose to serve a city, we must pay the distance between the city and its assigned facility for each unit demand. Our job is to find a set of k facilities to open, a set of cities to be served, and an assignment of cities to open facilities such that our total cost is minimized.

Throughout this paper, we use α to denote the ratio of the best approximation algorithm for k-median with penalties. We use this approximation as a subroutine in many of our algorithms. Because of the inapproximability of asymmetric k-median ([2]), our algorithms only apply to undirected graphs. However, most of our algorithms permit directed shortcuts.

3.2   Single Source ASPDM

In this section we consider SS-ASPDM where only the weights w_{sv} may be non-zero for some designated source s and $v \in V$. Thus, we are simply minimizing $D_s(G)$. This model will become useful in analyzing the complexity of our ASPDM variants as well as for obtaining an approximation for U-ASPDM.

Lemma 1. *For every instance of SS-ASPDM, there exists an optimal set F^* such that each edge $e \in F^*$ is incident on s. Moreover, for every $v \in V$, there exists a shortest sv-path that uses at most one edge in F^*.*

Proof. Let F^* be an optimal set of shortcut edges and consider $e = uv \in F^*$. Suppose p_1 is a shortest sx-path that traverses e in the uv direction and p_2 is a shortest sy-path that traverses e in the vu direction. Then the sy-path p_3 that starts at s, follows p_1 until u then follows p_2 never crosses e and can be no longer than p_2 (otherwise there would exist a sx-path shorter than p_1). Thus, e has an implicit orientation such that it is only ever used in the correct direction.

Since e is only used in one direction (say, u to v), then moving u closer to s only improves our cost. Thus, $F^* - uv + sv$ is at least as good a solution. We can do this for all other edges so that F^* contains only edges incident on s. Notice that now since every shortcut edge is incident on s, there is never any incentive to use more than one shortcut in a shortest path. □

Then we need only find k endpoints for our edges that minimize our cost if for each vertex v we pay either its weighted distance to the nearest endpoint or a penalty $w_{sv}d_{sv}$. This is precisely the k-median with penalties problem, thus we have an α-approximation algorithm for SS-ASPDM.

Theorem 1. *There exists a polynomial-time α-approximation algorithm ALG_{SS} for SS-ASPDM.*

Moreover, this α-approximation holds when adding directed shortcuts (to an undirected graph) since each edge $e \in F^*$ is only ever used in a single orientation.

4   Complexity

Consider unweighted (*i.e.* all non-zero weights are equal) SS-ASPDM. We now show that this problem is NP-Hard via reduction from the well-known Set Cover problem (defined in [11]).

Theorem 2. *Unweighted SS-ASPDM is NP-Hard. Further, for directed graphs, unweighted SS-ASPDM is hard to approximate to better than $\Omega(\log |V|)$.*

Proof. Omitted. Here, we give only the construction: Given an instance of set cover with universe U, subset collection $\mathcal{C}$ and integer k, let G have a vertex v_x for every $x \in U$, a vertex v_S for every $S \in \mathcal{C}$, and a vertex s. There is an edge of length 1 from s to each v_S and an edge of length 1 from v_S to each v_x where $x \in S$. Notice that $D_s(G) = |\mathcal{C}| + 2|U|$. We can now solve set cover by asking if there is a set F of k shortcut edges such that $D_s(G + F) \leq |\mathcal{C}| - k + |U|$. □

Unweighted SS-ASPDM is clearly a restriction of SS-ASPDM and ASPDM. By Lemma 1, SS-ASPDM is also a restriction of 1-ASPDM. The above reduction works for U-ASPDM when we replace s with a sufficiently large clique (connected by length-0 edges). Thus, we immediately get that all these problems are NP-Hard.

Corollary 1. *SS-ASPDM, U-ASPDM, 1-ASPDM, ASPDM are all NP-Hard.*

5 Unweighted ASPDM

In this section, we consider U-ASPDM where all pairs have equal weight. We will give an approximation algorithm which uses our SS-ASPDM algorithm ALG_{SS} as a subroutine. To do this, we must first claim that there exists a vertex x that is sufficiently close to all other vertices.

Lemma 2. *There exists an x such that when used as the source ALG_{SS} returns a 2α-approximation.*

Proof. Let F^* be the optimal solution. The average value of $D_v(G+F^*)$ over all v is $\frac{1}{n}D(G+F^*)$. Thus, some vertex x must not exceed the average. Try adding edge set F so as to minimize $D_x(G+F)$. By Theorem 1, we can do this within α of optimal using ALG_{SS}. Since $D_x(G+F^*)$ is no better than optimal,

$$D_x(G+F) \leq \alpha \cdot D_x(G+F^*) \leq \alpha \cdot \frac{1}{n}D(G+F^*). \tag{1}$$

We can also bound $D(G+F)$ in terms of $D_x(G+F)$. Since d is a metric, for each u, v we have $d_{uv} \leq d_{ux} + d_{xv}$. Summing these inequalities over all pairs gives

$$D(G+F) \leq 2nD_x(G+F). \tag{2}$$

Finally, combining Equations 1 and 2 gives the desired result

$$D(G+F) \leq 2nD_x(G+F) \leq 2n\alpha D_x(G+F^*) \leq 2\alpha D(G+F^*). \qquad \square$$

Thus, treating the all-pairs problem as a single-source problem with source vertex x produces a 2α-approximation. However, since finding x requires knowledge of F^*, we must instead try all possible x and take the best solution. We note that while ALG_{SS} works with directed shortcuts, this algorithm does not since edges may need to be used in both directions.

Theorem 3. *There exists a polynomial-time 2α-approximation algorithm for U-ASPDM.*

6 General ASPDM

We now consider the most general version of the problem where each pair can have an arbitrary weight associated with it. For this version, we offer a bicriteria approximation algorithm that breaks the restriction that only k edges be added.

Theorem 4. *There exists a polynomial-time $(4\alpha, 2)$-approximation algorithm for ASPDM. In particular, this algorithm gives at most $2k - 1$ edges yielding cost at most 4α-times the optimum k-edge cost.*

Proof. Let F^* be the optimal set of k edges. Notice that these edges involve $j \leq 2k$ endpoints. Let $\hat{F}$ be a set of $j - 1 \leq 2k - 1$ edges that connect these endpoints as a star. Thus, we can travel between any two endpoints using two shortcuts giving $D(G + \hat{F}) \leq 2D(G + F^*)$.

Since we do not know the set of endpoints used by F^* *a priori*, we try to find a star F over $2k$ points that minimizes $D(G + F)$. We can use $2k$-median with penalties to find this approximate solution F. To do this, we duplicate each vertex u so that the $2k$-median solution can connect u to some vertices and deny connections to others. We duplicate u a total of $2n - 2$ times introducing u_{uv} and u_{vu} for each $v \neq u$, having weights w_{uv} and w_{vu} and penalties $\max\{0, w_{uv}(d_{uv} - 2\delta)\}$ and $\max\{0, w_{vu}(d_{vu} - 2\delta)\}$, respectively. Since all the vertices corresponding to u are co-located, we need only choose one representative as a potential facility location.

For each pair u, v the $2k$-median instance pays for "connecting" u and v through these medians and never pays more than $2w_{uv}(d_{uv} - 2\delta)$. Adding the cost due to traversing shortcuts between these medians shows the optimum $2k$-median solution will have cost less than $2D(G + \hat{F})$. Using an α approximation gives us a cost of:

$$D(G + F) \leq 2\alpha D(G + \hat{F}) \leq 4\alpha D(G + F^*).$$

It follows that F gives a 4α-approximation for this problem.   □

Notice that when $\delta = 0$ we can actually improve this to a $(2\alpha, 2)$-approximation since we have $D(G + \hat{F}) \leq D(G + F^*)$. In this case, we can also deal with directed shortcuts if we connect the $2k$ endpoints as a directed cycle (thus, using exactly $2k$ shortcuts).

7   1-Shortcut Edge Restricted ASPDM

We consider a restriction that each path must use at most one shortcut edge. This allows us to provide improved approximations (in particular removing the increase over k shortcut edges). For real NoC designs, this kind of restriction ensures no pair monopolizes the RF interconnects and permits simplified routing.

7.1   Approximating Total Cost

We first define a metric over pairs of points $V \times V$.

Theorem 5. *If (V, d) is a metric, then so is the space $(V \times V, \hat{d})$ where*

$$\hat{d}(x_1 y_1, x_2 y_2) = \min(d(x_1, x_2) + d(y_2, y_1), d(x_1, y_2) + d(x_2, y_1)).$$

Proof. Omitted.   □

Note that in this space, we can naturally assign weight w_{uv} and penalty $w_{vu} d_{uv}$ to point uv. Moreover, if we select xy as a shortcut edge, then any 1-shortcut edge restricted shortest uv-path using xy has length $\hat{d}(uv, xy)$. Then adding k shortcut edges is equivalent to picking k medians in this pairs-of-points space. Thus, we can use k-median with penalties to obtain an α approximation.

Corollary 2. *There exists a polynomial-time α-approximation algorithm for 1-ASPDM.*

This works for directed shortcuts if we instead use $\hat{d}(x_1y_1, x_2y_2) = d(x_1, x_2) + d(y_2, y_1)$ which explicitly uses shortcuts in the correct direction.

7.2 Approximating Cost Improvement

The previous result guarantees a solution cost of at most $\alpha D(G+F^*)$. However, if $D(G+F^*) \geq \frac{1}{\alpha}D(G)$, then this guarantee can exceed $D(G)$, which even a trivial solution could satisfy! In such cases, it is more meaningful to approximate the optimum amount of improvement. We define $\Delta(G, H) = D(G) - D(H)$. Then we want our solution F to satisfy

$$\Delta(G, G+F) \geq \frac{1}{\zeta}\Delta(G, G+F^*)$$

for some $\zeta \geq 1$. We can obtain such an approximation using linear programming.

We first give an ILP formulation for 1-ASPDM. We use binary variables $x_{xy}, f_{uv}^{st}, g_{uv}^{st}, h_{xy}^{st}$ for each $s, t \in V$, $uv \in E$ and shortcut edge xy whose addition we are considering. If $x_{xy} = 1$ then edge $xy \in F$. Each pair (s, t) is given one unit of flow that needs to travel from s to t. Variable f_{uv}^{st} indicates the amount of (s, t)-flow over edge uv allowed to use a shortcut edge. Similarly, g_{uv}^{st} indicates the amount of (s, t)-flow over edge uv that has already used a shortcut edge. Finally, $h_{xy}^{st} \in \{0, 1\}$ indicates the amount of (s, t)-flow over shortcut edge xy.

Our ILP formulation is as follows:

$$\text{minimize} \quad \sum_{s,t} \left[w_{st} \cdot \sum_{uv \in E} \ell_{uv} \left(f_{uv}^{st} + g_{uv}^{st} \right) \right] \tag{3}$$

$$\text{subject to} \quad \sum_{x,y} x_{xy} = k \tag{4}$$

$$h_{xy}^{st} \leq x_{xy} \qquad\qquad\qquad \forall s, t, x, y \tag{5}$$

$$\sum_{v \in \Gamma(s)} f_{sv}^{st} + \sum_{y} h_{sy}^{st} = 1 \qquad\qquad \forall s, t \tag{6}$$

$$\sum_{u \in \Gamma(w)} f_{uw}^{st} = \sum_{v \in \Gamma(w)} f_{wv}^{st} + \sum_{y} h_{wy}^{st} \qquad \forall s, t, \forall w \neq s, t \tag{7}$$

$$\sum_{u \in \Gamma(w)} g_{uw}^{st} + \sum_{x} h_{xw}^{st} = \sum_{v \in \Gamma(w)} g_{wv}^{st} \qquad \forall s, t, \forall w \neq s, t \tag{8}$$

$$x_{xy}, f_{uv}^{st}, g_{uv}^{st}, h_{xy}^{st} \in \{0, 1\} \qquad\qquad \forall s, t, u, v, x, y \tag{9}$$

where $\Gamma(v)$ are the neighbors of vertex v in graph G. Equation (4) ensures that exactly k edges are selected and Equation (5) ensures that we only use selected shortcuts. Equation (6) enforces that for each pair (s, t), s adds one unit of (s, t)-flow to the graph. Equation (7) and (8) enforce conservation of flow at each vertex other than s, t (this also stipulates that t sink the one unit of (s, t)-flow). Finally, Equation (9) enforces integrality.

Since solving ILPs is NP-complete in general, we relax the integrality constraints by replacing Equation (9) with

$$0 \leq x_{xy}, f_{uv}^{st}, g_{uv}^{st}, h_{xy}^{st} \leq 1.$$

We can now use the solution to this LP as a guide for our edge selection process.

We build F iteratively using the values assigned to each x_{uv} by the optimal LP solution such that $\Pr[(uv) \in F] = x_{uv}$. Arbitrarily order the edges $e_1, e_2, \ldots, e_m$ and set $\hat{x}_{e_i} = x_{e_i}$ for all i and $F_1 = \emptyset$. In the i-th iteration, we add e_i with probability $\hat{x}_{e_i}$ to get $F_{i+1} = F_i \cup \{e_i\}$ or otherwise set $F_{i+1} = F_i$. After doing this, for each $j > i$ we set

$$\hat{x}_{e_j} \leftarrow \hat{x}_{e_j} \cdot \frac{k - |F_{i+1}|}{k - |F_i| - \hat{x}_{e_i}}.$$

We continue this process to get set $F = F_n$ containing at most k shortcut edges.

Lemma 3. *The above process yields a set F of at most k edges such that for each $e_i, 1 \leq i \leq m$, we have $\Pr[e_i \in F] = x_{e_i}$. Moreover, for any $S_i \subseteq \{e_1, e_2, \ldots, e_{i-1}\}$ we have:*

$$\Pr[e_i \in F \mid S_i \cap F = \emptyset] \geq x_{e_i}.$$

Proof. Omitted. □

We can decompose the flow and calculate expected cost to get the following:

Theorem 6. *Let F^* be the optimal set of edges and F the set of edges generated by the process above. Then*

$$Ex[\Delta(G, G + F)] \geq \left(\frac{e-1}{e}\right) \Delta(G, G + F^*).$$

Proof. Fix the pair (s, t) and consider its associated flow in the LP solution. Decompose this flow into simple paths using at most one shortcut. Let $p_1, p_2, \ldots, p_\alpha$ be the paths (in order of non-decreasing length) using exactly one shortcut. Let f_i be the flow over p_i and e_i the shortcut edge used by p_i. We can assume that each path uses a distinct shortcut (we can reroute the flow from one path to the other path otherwise). By LP optimality, none of these paths are longer than d_{st}.

Let q_i be the probability that at least one of paths $p_1, \ldots, p_i$ exist in $G + F$. Then notice

$$
\begin{aligned}
q_i &= 1 - \Pr[\text{none of paths } p_1, \ldots, p_i \text{ exist}] \\
&= 1 - (1 - \Pr[p_1 \text{ exists}]) \cdots (1 - \Pr[p_i \text{ exists} \mid p_1, \ldots, p_{i-1} \text{ don't exist}]) \\
&\geq 1 - (1 - x_{e_1})(1 - x_{e_2}) \cdots (1 - x_{e_i}) \\
&\geq 1 - (1 - f_1)(1 - f_2) \cdots (1 - f_i)
\end{aligned}
$$

where the first inequality follows from Lemma 3 and the second follows from LP-feasibility. Notice this quantity is minimized when all f_js are equal. Let $S_i = \sum_{j=1}^{i} f_j$ and note that since (s,t) has only one unit of demand we have $S_\alpha = 1$. Then since $(1-x)^{1/x} \leq \frac{1}{e}$ and $0 \leq S_i \leq 1$ we have

$$q_i \geq 1 - (1-f_1)(1-f_2)\cdots(1-f_i) \geq 1 - \left(1 - \frac{S_i}{i}\right)^i \geq 1 - \frac{1}{e^{S_i}} \geq \left(1 - \frac{1}{e}\right)S_i.$$

Thus, our expected cost for the (s,t)-pair is precisely

$$\begin{aligned}
\mathrm{Ex[cost]} &= d_{st} - (\ell_{p_2} - \ell_{p_1})q_1 - (\ell_{p_3} - \ell_{p_2})q_2 - \cdots - (d_{st} - \ell_\alpha)q_\alpha \\
&\leq \frac{1}{e}d_{st} + \left(1 - \frac{1}{e}\right)[\ell_{p_1}S_1 + \ell_{p_2}(S_2 - S_1) + \cdots + \ell_{p_\alpha}(S_\alpha - S_{\alpha-1})] \\
&= \frac{1}{e}d_{st} + \left(1 - \frac{1}{e}\right)[\ell_{p_1}f_1 + \ell_{p_2}f_2 + \cdots + \ell_{p_\alpha}f_\alpha]
\end{aligned}$$

Summing this inequality over all (s,t) pairs gives us

$$\mathrm{Ex}[D(G+F)] \leq \left(1 - \frac{1}{e}\right)LP + \left(\frac{1}{e}\right)D(G) \leq \left(1 - \frac{1}{e}\right)D(G+F^*) + \left(\frac{1}{e}\right)D(G)$$

where LP is the cost of the LP solution. Substituting into our definition of $\Delta(G, G+F)$ finishes the proof.   □

This shows we have a $\frac{e}{e-1}$-approximation algorithm on the total amount of improvement. While this algorithm uses randomness to select the shorcut edges, we can easily derandomize the process using conditional expectations. In other words, when considering e_i, we calculate the conditional expected cost given $e_i \notin F$ and given $e_i \in F$. Once this is calculated, we follow the decision that gives us the smallest expected cost.

Corollary 3. *There exists a polynomial-time $\frac{e}{e-1}$-approximation algorithm on the improvement in cost for 1-ASPDM.*

We note that this algorithm works on directed graphs. Additionally, it works if we restrict the possible shortcuts we can add. We also note that the LP used can be rewritten as a much smaller convex program and may be more efficiently solved.

8   Improved k-Median with Penalties Approximation

We now show that for k-median with penalties, we can use βk medians, $\beta \geq 1$, to acheive a cost of at most $1 + 2\frac{p+1}{\beta(p+1)-1}$ times the optimum cost (using k medians). For $\beta = 1$ this improves upon the 4-approximation for k-median with penalties given in [10] and matches the best approximation known for standard k-median given in [3]. This also improves upon the $(1 + \frac{5}{\varepsilon}, 3 + \varepsilon)$-approximation given in [16] for standard k-median. We note that standard k-median is hard to approximate to within $1 + \frac{2}{e}$ as shown in [14]. Our approach extends the local search based approximation algorithm given in [3] by permitting penalties and by creating a smooth bicriteria tradeoff when the algorithm is permitted to use additional medians.

Let C be the set of cities, F the set of potential facility locations and c the metric distance function. City j has demand w_j and penalty cost p_j. Thus, we are searching for a set $S \subseteq F$ of k facilities to open, a set $T \subseteq C$ of cities to serve and an assigment $\sigma : T \rightarrow S$ of cities to facilities to minimize cost

$$cost(S) = serv(S) + deny(S) = \sum_{j \in T} w_j c_{j,\sigma(j)} + \sum_{j \in C-T} p_j.$$

We say city j is *served* by facility i if $\sigma(j) = i$. Otherwise, city j is *denied service*. The neighborhood $\mathcal{N}_S(i)$ of facility i in solution S is the set of cities served by i. We abuse notation and write $\mathcal{N}_S(A)$ to denote the neighborhood of a set A of facilities. It will be convenient to refer to the cost due only to a set X of cities. Here we use $cost_X(S), serv_X(S), deny_X(S)$ to denote the total cost, service cost and denial cost (respectively) due to cities in X.

8.1 The Local Search Algorithm

Given a set of facilities S, we can easily calculate the best T and σ to use by greedily choosing to either assign each city to its closest open facility or to deny it service. Thus, we perform a local search only on the set S. Each iteration we consider all sets $A \subseteq S$ and $B \subseteq F - S$ with $|A| = |B| \leq p$ for some fixed parameter $p \geq 1$. We choose A, B such that $cost(S - A + B)$ is minimized and iterate until no move yields a decrease in cost. We denote swapping the sets A and B by $\langle A, B \rangle$.

8.2 Analysis

We now bound the locality gap of our algorithm:

Theorem 7. *The local search algorithm in Section 8.1 has a locality gap of at most* $1 + 2\frac{p+1}{\beta(p+1)-1}$.

Proof. Let (S, T, σ) be our solution using βk medians and (S^*, T^*, σ^*) be the optimum solution using k medians. We assume for simplicity that all weights are multiples of some $\delta > 0$. Replace each city j with $\frac{w_j}{\delta}$ copies each with weight δ and penalty $\frac{p_j \delta}{w_j}$. S and S^* treat all copies of j as they did j. Clearly, it is enough to analyze this unweighted case.

For a subset $A \subseteq S$, we will say A *captures* $o \in S^*$ if A serves at least half the cities served by both o in the optimum solution and by some facility in our solution. We then define $capture(A)$ to be the set of optimum facilities that A captures. Thus,

$$capture(A) = \{o \in S^* : |\mathcal{N}_S(A) \cap \mathcal{N}_{S^*}(o)| \geq \frac{1}{2}|\mathcal{N}_{S^*}(o) \cap T|\}.$$

A facility $s \in S$ is *bad* if $|capture(s)| \neq \emptyset$ and is *good* otherwise. Note that if $A, B \subseteq S$ are disjoint then so are $capture(A)$ and $capture(B)$.

Suppose S has $r - 1$ bad facilities. Partition S into $A_1, \ldots, A_r$ and S^* into $B_1, \ldots, B_r$ such that for all $i \leq r - 1$ we have $|A_i| = |B_i|$, $B_i = capture(A_i)$ and A_i contains exactly one bad facility. We can build this partition by adding a bad facility to each A_i then adding good facilities until $|A_i| = |capture(A_i)|$. Since each $o \in S^*$ is captured by at most one facility and $capture(A_1) \cap capture(S - A_1) = \emptyset$, we never run out of good facilities.

In fact, we only care about the A_i with $|A_i| \leq p$ (excluding A_r). Without loss of generality, we assume these to be sets $A_1, \ldots, A_b$. Let $x = \sum_{i=1}^{b} |A_i| = \sum_{i=1}^{b} |B_i|$ and note that $x \geq b$ since each A_i is non-empty. Then there are at most $k - x \leq k - b$ optimum facilities total among sets $B_{b+1}, \ldots, B_{r-1}$. Since all these sets have cardinality greater than p and there is one bad facility per A_i, we can upper bound the number of bad facilities by $b + \frac{k-b}{p+1} = \frac{k+pb}{p+1}$.

We let G be the good facilities in $A_{b+1}, \ldots, A_r$ and $a = |G|$. Then since we have βk medians total, we have

$$a \geq \beta k - \frac{k + pb}{p + 1} = \frac{\beta k p + \beta k - k - pb}{p + 1} \tag{10}$$

For each i such that $|A_i| \leq p$, we consider the swap $\langle A_i, B_i \rangle$. We will refer to these swaps as *set swaps*. We also consider all possible single-facility swaps between optimum facilities in B_i and facilities in G. We will call these swaps *bad singleton swaps*. Lastly, we consider all possible single-facility swaps between the remainder of optimum facilities and facilities in G. We will call these swaps *good singleton swaps*. By local optimality, each swap (either set or singleton) $\langle X, Y \rangle$ satisfies

$$cost(S - X + Y) - cost(S) \geq 0. \tag{11}$$

For each facility $o \in S^*$, partition $\mathcal{N}_{S^*}(o)$ into parts $p_X = \mathcal{N}_{S^*}(o) \cap \mathcal{N}_S(X)$ for each considered swap $\langle X, Y \rangle$ above and $p_{deny} = \mathcal{N}_{S^*}(o) - T$. We let $\pi : \mathcal{N}_{S^*}(o) \to \mathcal{N}_{S^*}(o)$ be a bijection such that $p_{deny} = \pi(p_{deny})$ and for each part $p \neq p_{deny}$ having $|p| < \frac{1}{2}|\mathcal{N}_{S^*}(o) \cap T|$ we have $p \cap \pi(p) = \emptyset$. It is easy to check that such a bijection exists.

Now let $\langle X, Y \rangle$ be a set or singleton swap considered above. When we make this swap, we can make sure to assign $\mathcal{N}_{S^*}(Y)$ to Y, but we also need to reassign any other cities served by X. If S^* denies any of these cities, we will also deny them service. Otherwise, we can reassign j to the facility serving $\pi(j)$. Thus, we can bound our change in cost above by:

$$0 \leq cost(S - X + Y) - cost(S) \leq$$
$$\sum_{j \in \mathcal{N}_{S^*}(Y) \cap T} \left[c_{j,\sigma^*(j)} - c_{j,\sigma(j)} \right] +$$
$$\sum_{j \in \mathcal{N}_{S^*}(Y) - T} \left[c_{j,\sigma^*(j)} - p_j \right] + \tag{12}$$
$$\sum_{j \in (\mathcal{N}_S(X) - \mathcal{N}_{S^*}(Y)) \cap T^*} \left[c_{j,\sigma^*(j)} + c_{\sigma^*(j),\pi(j)} + c_{\pi(j),\sigma(\pi(j))} - c_{j,\sigma(j)} \right] +$$
$$\sum_{j \in (\mathcal{N}_S(X) - \mathcal{N}_{S^*}(Y)) - T^*} \left[p_j - c_{j,\sigma(j)} \right].$$

Consider the inequalities corresponding to Equation 12 for each swap considered. We mutiply the inequalities for set, bad singleton and good singleton swaps by

$\gamma = \frac{p+1}{\beta(p+1)-1}$, $\frac{1-\gamma}{a}$ and $\frac{1}{a}$ (respectively) then sum the resulting inequalities. Notice that each $o \in B_i$ is involved in swaps of total weight one. Thus the first two terms of Equation 12 sum to $serv(S^*) - cost_{T^*}(S)$.

Each bad facility s is involved with a set swap of weight γ or is never swapped. Each good facility s is involved in x bad singleton swaps and $k-x$ good singleton swaps for a total weight of

$$x\left(\frac{1-\gamma}{a}\right) + \frac{k-x}{a} = \frac{1}{a}\left(k - x\frac{p+1}{\beta(p+1)-1}\right) \leq \frac{1}{a}\left(k - b\frac{p+1}{\beta(p+1)-1}\right) \leq \gamma$$

Thus, any $j \in T \cap T^*$ is considered in a weighted total of at most γ swaps. Since $\left[c_{j,\sigma^*(j)} + c_{\sigma^*(j),\pi(j)} + c_{\pi(j),\sigma(\pi(j))} - c_{j,\sigma(j)}\right] \geq 0$ by triangle inequality and $\left[p_j - c_{j,\sigma(j)}\right] \geq 0$ we can assume that each j appears exactly γ times (this only increases the right-hand side of Equation 12). Then the third and fourth terms of Equation 12 sum to at most $\gamma\left(2serv_T(S^*) + deny_T(S^*) - serv_{T-T^*}(S)\right)$.

Thus summing Equation 12 over all swaps and rearranging gives

$$serv(S^*) + 2\gamma serv_T(S^*) + \gamma deny_T(S^*) \geq cost_{T^*}(S) + \gamma serv_{T-T^*}(S). \quad (13)$$

Since $deny_{T-T^*}(S) = 0$ as S does not deny service to any member of T, the right-hand side exceeds $cost(S)$. Since $cost(S^*) = serv(S^*) + deny(S^*)$ the left-hand side is no greater than $(1 + 2\gamma)cost(S^*)$. Thus, we have

$$cost(S) \leq \left(1 + 2\frac{p+1}{\beta(p+1)-1}\right)cost(S^*). \qquad \square$$

9   Experiments and Future Work

We have run some experiments comparing the result of our local search-based approximation for 1-ASPDM against the heuristics described in [6] and obtained a 4-5% improvement in both latency and power. We are conducting further experiments to determine whether local search is producing optimum results in practice, and whether a more complex model might lead to even more improvement.

From a theoretical standpoint, we have given constant factor approximations for all versions of ASPDM except the most general one. Whether the general ASPDM problem has a (single criterion) constant approximation remains an open problem. The problem is related to a series of works in the theory of network design literature (for example Rent-or-Buy problems) in much the same way that k-median relates to facility location (instead of summing two types of cost, we have a hard constraint on one type and seek to minimize the other). If we permit *restrictions* on the set of available shortcuts, then approximation hardness results follow from the work of Andrews [1] but we are not aware of any such results for the case where any pair of nodes can be connected via a shortcut edge.

Acknowledgements. We would like to thank Professor Jason Cong and Chunyue Liu for introducing us to the ASPDM problem and for many other useful discussions about RF-interconnects and NoC architectures.

References

1. Andrews, M.: Hardness of buy-at-bulk network design. In: Annual Symposium on Foundations of Computer Science (2004)
2. Archer, A.: Inapproximability of the asymmetric facility location and k-median problems (unpublished manuscript) (2000)
3. Arya, V., Garg, N., Khandekar, R., Meyerson, A., Munagala, K., Pandit, V.: Local search heuristics for k-median and facility location problems. In: Symposium on Theory of Computing (2001)
4. Awerbuch, B., Azar, Y.: Buy-at-bulk network design. In: Proceedings of the 38th Annual Symposium on Foundations of Computer Science, p. 542 (1997)
5. Benini, L., De Micheli, G.: Networks on chips: A new SoC paradigm. Computer 35(1), 70–78 (2002)
6. Chang, M.-C.F., Cong, J., Kaplan, A., Liu, C., Naik, M., Premkumar, J., Reinman, G., Socher, E., Tam, S.-W.: Power reduction of CMP communication networks via RF-interconnects. In: Proceedings of the 41st Annual International Symposium on Microarchitecture (November 2008)
7. Chang, M.-C.F., Cong, J., Kaplan, A., Naik, M., Reinman, G., Socher, E., Tam, S.-W.: CMP network-on-chip overlaid with multi-band RF-interconnect. In: International Symposium on High Performance Computer Architecture (February 2008)
8. Chang, M.-C.F., Roychowdhury, V.P., Zhang, L., Shin, H., Qian, Y.: RF/wireless interconnect for inter- and intra-chip communications. Proceedings of the IEEE 89(4), 456–466 (2001)
9. Chang, M.-C.F., Socher, E., Tam, S.-W., Cong, J., Reinman, G.: RF interconnects for communications on-chip. In: Proceedings of the 2008 international symposium on Physical design, pp. 78–83. ACM Press, New York (2008)
10. Charikar, M., Khuller, S., Mount, D.M., Narasimhan, G.: Algorithms for facility location problems with outliers. In: Symposium on Discrete Algorithms, pp. 642–651 (2001)
11. Garey, M.R., Johnson, D.S.: Computers and Intractability: A Guide to the Theory of NP-Completeness. W.H. Freeman and Company, New York (1979)
12. Gupta, A., Kumar, A., Pal, M., Roughgarden, T.: Approximation via cost-sharing: a simple approximation algorithm for the multicommodity rent-or-buy problem. In: Proceedings 44th Annual IEEE Symposium on Foundations of Computer Science, pp. 606–615 (October 2003)
13. Ho, R., Mai, K.W., Horowitz, M.A.: The future of wires. Proceedings of the IEEE 89(4), 490–504 (2001)
14. Jain, K., Mahdian, M., Saberi, A.: A new greedy approach for facility location problems. In: Proceedings of the thiry-fourth annual ACM symposium on Theory of computing, pp. 731–740 (2002)
15. Kleinberg, J.: The small-world phenomenon: An algorithmic perspective. In: Proceedings of the 32nd ACM Symposium on Theory of Computing, pp. 163–170 (2000)
16. Korupolu, M.R., Plaxton, C.G., Rajaraman, R.: Analysis of a local search heuristic for facility location problems. In: Proceedings of the ninth annual ACM-SIAM symposium on Discrete algorithms (1998)
17. Manfredi, S., di Bernardo, M., Garofalo, F.: Small world effects in networks: An engineering interpretation. In: Proceedings of the 2004 International Symposium on Circuits and Systems, May 2004, vol. 4, IV–820–3 (2004)

18. Martel, C., Nguyen, V.: Analyzing Kleinberg's (and other) small-world models. In: 23rd ACM Symp. on Principles of Distributed Computing, pp. 179–188. ACM Press, New York (2004)
19. Meyerson, A., Munagala, K., Plotkin, S.: Cost-distance: Two metric network design. In: Proceedings 41st Annual Symposium on Foundations of Computer Science, pp. 624–630 (2000)
20. Ogras, U.Y., Marculescu, R.: It's a small world after all: NoC performance optimization via long-range link insertion. IEEE Transactions on Very Large Scale Integration Systems 14(7), 693–706 (2006)
21. Sandberg, O., Clarke, I.: The evolution of navigable small-world networks. Technical report, Chalmers University of Technology (2007)
22. Watts, D.J., Strogatz, S.H.: Collective dynamics of 'small world' networks. Nature 393, 440–442 (1998)

Approximating Node-Connectivity
Augmentation Problems

Zeev Nutov

The Open University of Israel
nutov@openu.ac.il

Abstract. We consider the (undirected) Node Connectivity Augmentation (NCA) problem: given a graph $J = (V, E_J)$ and connectivity requirements $\{r(u, v) : u, v \in V\}$, find a minimum size set I of new edges (any
edge is allowed) so that $J + I$ contains $r(u, v)$ internally disjoint uv-paths,
for all $u, v \in V$. In the Rooted NCA there is $s \in V$ so that $r(u, v) > 0$
implies $u = s$ or $v = s$. For large values of $k = \max_{u,v \in V} r(u, v)$, NCA is
at least as hard to approximate as Label-Cover and thus it is unlikely to
admit a polylogarithmic approximation. Rooted NCA is at least as hard
to approximate as Hitting-Set. The previously best approximation ratios
for the problem were $O(k \ln n)$ for NCA and $O(\ln n)$ for Rooted NCA.
In [Approximating connectivity augmentation problems, SODA 2005]
the author posed the following open question: Does there exist a function $\rho(k)$ so that NCA admits a $\rho(k)$-approximation algorithm? In this
paper we answer this question, by giving an approximation algorithm
with ratios $O(k \ln^2 k)$ for NCA and $O(\ln^2 k)$ for Rooted NCA. This is the
first approximation algorithm with ratio independent of n, and thus is
a constant for any fixed k. Our algorithm is based on the following new
structural result which is of independent interest. If $\mathcal{D}$ is a set of node
pairs in a graph J, then the maximum degree in the hypergraph formed
by the inclusion minimal tight sets separating at least one pair in $\mathcal{D}$ is
$O(\ell^2)$, where ℓ is the maximum connectivity of a pair in $\mathcal{D}$.

1 Introduction

1.1 Problem Definition

Let $\kappa_G(u, v)$ denote the maximum number of internally-disjoint uv-paths in a
graph G. We consider the following fundamental problem in network design:

Node-Connectivity Augmentation (NCA):
Instance: A graph $J = (V, E_J)$, connectivity requirements $\{r(u, v) : u, v \in V\}$.
Objective: Find a minimum size set I of new edges so that $G = J + I$ satisfies

$$\kappa_G(u, v) \geq r(u, v) \quad \text{for all } u, v \in V . \tag{1}$$

We assume that if $uv \in J$ then $uv \notin I$. In general, all graphs are assumed to be
undirected and simple, unless stated otherwise. For an NCA instance at hand,
let opt denote the optimal solution value, let $k = \max_{u,v \in V} r(u, v)$ denote the

I. Dinur et al. (Eds.): APPROX and RANDOM 2009, LNCS 5687, pp. 286–297, 2009.
© Springer-Verlag Berlin Heidelberg 2009

maximum connectivity requirement, and let $n = |V|$. Note that if NCA has a feasible solution $G = J + I$ (G is a simple graph), then $n \geq k + 1$ must hold.

If all the connectivity requirements are "rooted", namely from a specific node s, then we have the following important particular case of NCA:

Rooted NCA:
Instance: A graph $J = (V, E_J)$, a root $s \in V$, and requirements $\{r(v) : v \in V\}$.
Objective: Find a minimum size set I of new edges so that $G = J + I$ satisfies
$$\kappa_G(s, v) \geq r(v) \text{ for all } v \in V.$$

NCA is an extensively studied particular case of the following problem, that recently received a renewed attention:

Survivable Network Design (SND):
Instance: A complete graph on V with edge-costs, and connectivity requirements $\{r(u, v) : u, v \in V\}$.
Objective: Find a minimum cost subgraph G on V that satisfies (1).

NCA is equivalent to SND with $0, 1$-costs, when E_J is the set of edges of cost 0, and any other edge is allowed and has cost 1. The case of $1, \infty$-costs of SND gives the min-size subgraph problems, when we seek a solution using the edges of cost 1 only.

1.2   Our Results

NCA admits an $O(k \ln n)$-approximation [28], and is unlikely to admit a poly-logarithmic approximation even for $\{0, k\}$-requirements [32]. For rooted requirements, an $O(\ln n)$-approximation is known [28], and this is tight [33]. Motivated by results from [21,22,18,19,33,30,32,5,4,26], the author posed in [32] the following question: Does NCA admit a $\rho(k)$-approximation algorithm? Here $\rho(k)$ is a functions that depends on k only. We resolve this question, thus obtaining a constant ratio for any constant k; furthermore, when $\ln^2 k = o(\ln n)$ our ratios are better than the ones in [28].

Theorem 1. NCA *admits the following approximation ratios:*

- $O(k \ln^2 k)$ *for arbitrary requirements (improving $O(k \ln n)$);*
- $O(\ln^2 k)$ *for rooted requirements (improving $O(\ln n)$).*

Here $k = \max_{u, v \in V} r(u, v)$ is the maximum requirement.

As an intermediate problem, we consider NCA instances with $r(u, v) \leq \kappa_J(u, v) + 1$ for all $u, v \in V$. That is, given a set $\mathcal{D}$ of node pairs, we seek to increase the connectivity by 1 between pairs in $\mathcal{D}$, meaning $r(u, v) = \kappa_J(u, v) + 1$ for all $\{u, v\} \in \mathcal{D}$ and $r(u, v) = 0$ otherwise. Formally:

Simple NCA:
Instance: A graph $J = (V, E_J)$ and a set $\mathcal{D}$ of unordered node pairs from V.
Objective: Find a minimum size edge-set I so that $G = J + I$ satisfies
$$\kappa_G(u, v) \geq \kappa_J(u, v) + 1 \quad \text{for all } \{u, v\} \in \mathcal{D} . \tag{2}$$

Given an edge-set or a graph J and disjoint node-sets X, Y let $\delta_J(X, Y)$ denote the set of edges in J that have one endnode in X and the other in Y; let $\delta_J(X) = \delta_J(X, V - X)$. For $S \subseteq V$ let $\Gamma_J(S) = \Gamma(S) = \{v \in V - S : uv \in E_J$ for some $u \in S\}$ denote the set of *neighbors* of S in V.

Definition 1. *Given an instance of* Simple NCA, *we say that* $S \subseteq V$ *is* uv-tight *if* $u \in S$, $v \notin S$, $|\Gamma_J(S)| = \kappa_J(u, v)$, *and either:* $v \in V - (S \cup \Gamma(S))$, *or* $uv \in E_J$ *and* $\delta_J(S, v) = \{uv\}$. S *is* tight *if it is* uv-tight *for some* $\{u, v\} \in \mathcal{D}$. *Let* $\mathcal{C}_J(\mathcal{D})$ *denote the set of inclusion minimal tight sets in* J *w.r.t.* $\mathcal{D}$, *and in the case of rooted requirements let* $\mathcal{C}_J^s(\mathcal{D}) = \{C \in \mathcal{C}_J(\mathcal{D}) : s \notin C\}$.

The proof of Theorem 1 is based on the following theorem, which is of independent interest:

Theorem 2. *Suppose that* $\max\{\kappa_J(u, v) : \{u, v\} \in \mathcal{D}\} \leq \ell$ *for an instance of* Simple NCA. *Then the maximum degree in the hypergraph* $(V, \mathcal{C}_J(\mathcal{D}))$ *is at most* $(4\ell + 1)^2$. *For rooted requirements, the maximum degree in the hypergraph* $(V, \mathcal{C}_J^s(\mathcal{D}))$ *is at most* $2\ell + 1$.

We believe that the result in the theorem reveals a fundamental property which will have further applications, and turn to be useful to design approximation algorithms for various SND problems. Specifically, the approach in this paper was later used by the author in [35] to obtain an $O(k^2)$-approximation for Rooted SND with arbitrary costs, which is currently the best known ratio for the problem.

Theorems 1 and 2 are proved in Sections 2 and 3, respectively. Section 4 concludes with some open problems.

1.3   Previous and Related Work

Variants of SND, and especially of NCA, were vastly studied. See surveys in [27] and [14]. While the edge-connectivity variant of SND – the so called Steiner Network problem – admits a 2-approximation algorithm by the seminal paper of Jain [20], no such algorithm is known for SND. For directed graphs, Dodis and Khanna [9] showed that $\{0, 1\}$-SND – the so called Directed Steiner Forest problem – is at least as hard to approximate as Label-Cover. By extending the construction of [9], Kortsarz, Krauthgamer, and Lee [24] showed a similar hardness result for Undirected $\{0, k\}$-SND; the same hardness is valid even for $\{0, 1\}$-costs, namely, for NCA, see [33]. However, the edge-connectivity variant of NCA – the so called Edge-Connectivity Augmentation problem admits a polynomial time algorithm due to Frank [13].

In general, for small requirement undirected variants of SND are substantially easier to approximate than the directed ones. For example, Undirected Steiner Tree/Forest admits a constant ratio approximation algorithm, while the directed variants are not known to admit even a polylogarithmic approximation ratio. The currently best known approximation lower bound for Directed Steiner Tree is $\Omega(\ln^{2-\varepsilon} n)$ [17], while a long standing best known ratio is $O(|n|^\varepsilon/\varepsilon^3)$ in $O(|n|^{4/\varepsilon} n^{2/\varepsilon})$ time [2]; this gives an $n^\varepsilon/\varepsilon^3$-approximation scheme. In what follows, we survey results for general SND/NCA, Rooted SND/NCA, and the k-Connected Subgraph (k-CS) problem, for both general and $0, 1$-costs; the latter is a famous particular case of SND when $r(u, v) = k$ for all $u, v \in V$. See also surveys in [23] and [27]. We consider the cases of general costs (SND) and of $0, 1$-costs (NCA) separately. The approximability of various SND problems (prior to our work) is summarized in Table 1.

Table 1. Approximation ratios and hardness results for SND problems

Costs	Req.	Approximability	
		Undirected	*Directed*
general	general	$O(\min\{k^3 \ln n, n^2\}$ [8], $k^{\Omega(1)}$ [1]	$O(n^2)$, $\Omega(2^{\log^{1-\varepsilon} n})$ [9]
general	rooted	$O(\min\{k^2, n\})$ [35], $\Omega(\log^2 n)$ [29]	$O(n)$, $\Omega(\log^2 n)$ [17]
general	k-CS	$O\left(\log k \cdot \log \frac{n}{n-k}\right)$ [34]	$O\left(\log k \cdot \log \frac{n}{n-k}\right)$ [34]
metric	general	$O(\log k)$ [6]	$O(n^2)$, $\Omega(2^{\log^{1-\varepsilon} n})$ [9]
metric	rooted	$O(\log k)$ [6]	$O(n)$, $\Omega(\log^2 n)$ [17]
metric	k-CS	$2 + \frac{k-1}{n}$ [25]	$2 + \frac{k}{n}$ [25]
$0, 1$	general	$O(k \ln n)$ [28], $\Omega(2^{\log^{1-\varepsilon} n})$ [32]	$O(k \ln n)$ [28], $\Omega(2^{\log^{1-\varepsilon} n})$ [32]
$0, 1$	rooted	$O(\log n)$ [28], $\Omega(\log n)$ [33]	$O(\log n)$ [28], $\Omega(\log n)$ [33]
$0, 1$	k-CS	$\min\{\text{opt} + k^2/2, 2\text{opt}\}$ [18]	in P [15]

SND–arbitrary costs: Frank and Tardos [16] gave a polynomial time algorithm for the rooted variant with uniform requirements $r(s, v) = k$ for all $v \in V - s$. Ravi and Williamson [36] gave a 3-approximation algorithm for $\{0, 1, 2\}$-SND, and the ratio was improved to 2 by Fleisher et al. [12]. As was mentioned, SND is unlikely to admit a polylogarithmic approximation [24]; a recent improved hardness result of Chakraborty, Chuzhoy, and Khanna [1] shows that SND with requirements in $\{0, k\}$ is $k^{\Omega(1)}$-hard to approximate. Recently, it was shown by Lando and the author [29] that directed SND problems can be reduced to their corresponding undirected variants with large connectivity requirements; one of the consequences of the result of [29] is that the Rooted SND with requirements in $\{0, k\}$ is at least as hard to approximate as the notorious Directed Steiner Tree problem, for $k \geq n/2$. The reduction of [29] does *not* preserves *metric costs*, and indeed, Cheriyan and Vetta [6] showed that (undirected) SND with metric costs admits an $O(\log n)$-approximation algorithm. However, no sublinear approximation algorithm is known for SND with general requirements and costs. Even for the much easier Directed Steiner Forest problem, the best ratio known in terms of n is $O\left(n^{4/5+\varepsilon}\right)$ [11]. Chakraborty, Chuzhoy, and Khanna [1] initiated recently the study of approximation algorithms for SND problems when the parameter k is not too large; they obtained a randomized $O(k^{O(k^2)} \ln^4 n)$-approximation algorithm for Rooted SND. The ratio was improved by Chekuri and Korula [3] to $O(k^{O(k)} \ln n)$. Slightly later, independently, Chuzhoy and Khanna [7], and the author [34], improved the ratio to $O(k^2 \log n)$. Very recently, in another paper [8], Chuzhoy and Khanna gave an $O(k^3 \log n)$-approximation algorithm for SND based on the iterative rounding method. Subsequently, using techniques from this paper, the author developed an $O(k^2)$-approximation algorithm for Rooted SND.

We note that the most famous variant of SND – the k-Connected Subgraph problem, was vastly studied, see [36,5,25,26,10,34] for only a small sample of papers on the topic. The currently best known ratio for directed/undirected k-Connected Subgraph is $O\left(\log k \cdot \log \frac{n}{n-k}\right)$ due to the author [34], see also an $O(\log^2 k)$-approximation algorithm due to Fackharoenphol and Laekhanukit [10].

NCA–0, 1-costs: While most of the "positive" literature on SND problems with general costs is from the recent 2 years, 0, 1-costs NCA problems were extensively studied already in the 90's. For example, the complexity status of k-Connected Subgraph with 0, 1-costs is among the oldest open problems in network design, see [21,22,18,19,33,30] (however, the directed case is solvable in polynomial time [15]). In his seminal papers [21,22], Jordán gave an opt $+ k/2$ approximation for the problem of increasing the connectivity by 1; this approximation resists improvement, and so far it was not established that the problem is in P, nor that it is NP-hard. A simpler and more efficient version of Jordán's algorithm can be found in [30], and a similar result was obtained for the rooted case by the author in [33]. Jordán's algorithm [21,22] was generalized by Jackson and Jordán [18] who gave an algorithm that computes a solution of size roughly opt $+ k^2/2$ for the general k-Connected Subgraph with 0, 1-costs. Another very interesting result of Jackson and Jordán [19] shows that the problem can be solved exactly in time $2^{f(k)}poly(n)$.

For general requirements, NCA admits an $O(k \ln n)$-approximation [28], and is unlikely to admit a polylogarithmic approximation [32]. For rooted requirements an $O(\ln n)$-approximation is known, and for $k = \Omega(n)$ this is tight [33].

2   The Algorithm (Proof of Theorem 1)

Here we prove Theorem 1, which is restated for the convenience of the reader.

Theorem 1. NCA *admits the following approximation ratios:*

- $O(k \ln^2 k)$ *for arbitrary requirements (improving $O(k \ln n)$);*
- $O(\ln^2 k)$ *for rooted requirements (improving $O(\ln n)$).*

Here $k = \max_{u,v \in V} r(u,v)$ is the maximum requirement.

Theorem 1 is proved in several steps, and relies on Theorem 2, which is proved in the next section. We start with the following known fact that is proved using standard flow-cut techniques.

Proposition 1. *The family $\mathcal{C}_J(\mathcal{D})$ can be computed in polynomial time and $|\mathcal{C}_J(\mathcal{D})| \leq |\mathcal{D}| \leq \binom{n}{2}$.*

Proof. It is well known that given $\{u, v\} \in D$, one max-flow computation suffices to find the unique minimal uv-tight set C_{uv} containing u, and the unique minimal vu-tight set C_{vu} containing v. The family $\mathcal{C}_J(\mathcal{D})$ consists from the inclusion minimal members of the family $\{C_{uv} : \{u, v\} \in \mathcal{D}\}$. The statement follows.

We now describe the lower bound on the solution size of Simple NCA that we use.

Definition 2. *A node set $T \subseteq V$ is a $\mathcal{C}$-transversal of a set family $\mathcal{C}$ if T intersects every $C \in \mathcal{C}$. Let $\tau(\mathcal{C})$ be the minimum size of a $\mathcal{C}$-transversal, and let $\tau^*(\mathcal{C})$ be the minimum value of a fractional $\mathcal{C}$-transversal, namely:*

$$\tau^*(\mathcal{C}) = \min\{\sum_{v \in V} x(v) : \sum_{v \in C} x(v) \geq 1 \ \ \forall C \in \mathcal{C}, \ x(v) \geq 0\} \ .$$

Note that $|I| \geq \tau(\mathcal{C}_J(\mathcal{D}))/2 \geq \tau^*(\mathcal{C}_J(\mathcal{D}))/2$ for any feasible solution I for Simple NCA. Indeed, by Menger's Theorem, I is a feasible solution to Simple NCA if, and only if, for any uv-tight set S with $\{u, v\} \in \mathcal{D}$ there is an edge in I from S to $V - (S + \Gamma(S))$ if $uv \notin J$, or from S to $V - (S + \Gamma(S) - v)$ if $uv \in J$. In particular, $\delta_I(C) \geq 1$ must hold for any $C \in \mathcal{C}_J(\mathcal{D})$. Thus the endnodes of the edges in I form a $\mathcal{C}_J(\mathcal{D})$-transversal, so $|I| \geq \tau_J(\mathcal{D})/2$. Note also that in the case of rooted requirements, $\tau^*(\mathcal{C}_J(\mathcal{D})) \geq \tau^*(\mathcal{C}_J^s(\mathcal{D}))$.

Given a hypergraph $(V, \mathcal{C})$, the greedy algorithm of Lovász [31] computes in polynomial time a $\mathcal{C}$-transversal T of size $\leq H(\Delta(\mathcal{C}))\tau^*(\mathcal{C})$, where $\Delta(\mathcal{C})$ is the maximum degree of the hypergraph and $H(k)$ is the kth Harmonic number. Combining with Theorem 2 we deduce the following statement:

Corollary 1. *For Simple NCA there exists a polynomial time algorithm that computes a $\mathcal{C}_J(\mathcal{D})$-transversal T so that $|T| \leq \tau^*(\mathcal{C}_J(\mathcal{D})) \cdot H\left((4\ell + 1)^2\right)$, where $\ell = \max_{\{u,v\} \in \mathcal{D}} \kappa_J(u, v)$; for Rooted NCA, the algorithm computes a $\mathcal{C}_J^s(\mathcal{D})$-transversal T so that $|T| \leq \tau^*(\mathcal{C}_J^s(\mathcal{D})) \cdot H(2\ell + 1)$.*

Now we show how to obtain an augmenting edge set from a given transversal.

Proposition 2. *There exist a polynomial time algorithm that given an instance of Simple NCA and a $\mathcal{C}_J(\mathcal{D})$-transversal T, computes a feasible solution I so that $|I| \leq (\ell + 2)|T|$. In the case of rooted requirements, given a $\mathcal{C}_J^s(\mathcal{D})$-transversal T, the algorithm computes a feasible solution I so that $|I| \leq 2|T|$.*

Proof. Form an edge set I by choosing an arbitrary set U of $\ell + 2$ nodes and connecting every node in T to every node in U, unless there is already an edge between them. Then $|I| \leq (\ell + 2) \cdot |T|$. We claim that I is a feasible solution. Suppose to the contrary that $\kappa_J(u, v) = \kappa_{J+I}(u, v) = \ell' \leq \ell$ for some $\{u, v\} \in \mathcal{D}$. By Menger's Theorem, there exists a partition X, C, Y of V so that X is uv-tight, Y is vu-tight, $\delta_I(X, Y) = \emptyset$ and either: $|C| = \ell'$ and $\delta_J(X, Y) = \emptyset$, or $|C| = \ell' - 1$ and $\delta_J(X, Y) = \{uv\}$. There is $z \in U - C$ so that $z \notin \{u, v\}$ in the case $|C| = \ell' - 1$. As T is a $\mathcal{C}_J(\mathcal{D})$-transversal there are $x \in X \cap T$ and $y \in Y \cap T$. At least one of the edges zx, zy is in $\delta_I(X, Y)$, which gives a contradiction.

Now consider the case of rooted requirements, when T is a $\mathcal{C}_J^s(\mathcal{D})$-transversal. Let $T_0 = \{t \in T : ts \notin E_J\}$ and $I_0 = \{ts : t \in T_0\}$, so $|I_0| = |T_0|$. Let $J' = J + I_0$, and let $\mathcal{D}' = \{\{u, s\} \in \mathcal{D} : \kappa_{J'}(u, s) = \kappa_J(u, s)\}$ consist from those pairs in $\mathcal{D}$ that are not "satisfied" by addition of I_0 to J. Consider an arbitrary us-tight set S in J' with $\{u, s\} \in \mathcal{D}'$. It is not hard to verify that $T' = T - T_0$ is a $\mathcal{C}_{J'}^s(\mathcal{D}')$-transversal, hence there is $t \in S \cap T'$. As $ts \in J'$, we must have $u = t$, by the definition of a tight set. Consequently, $\mathcal{D}' = \{\{t, s\} : t \in T'\}$. Hence to obtain a feasible solution, it would be sufficient to add to I_0 an edge set I' that increases the connectivity (in J or in J') from every $t \in T'$ to s.

We show how to find a set $I(t)$ of at most 2 new edges whose addition increases the ts-connectivity by 1. Let Π be a set of $\kappa_J(t, s)$ pairwise internally disjoint ts-paths (one of these paths is the edge ts). If there is a node a that does not belong to any path in Π, then $I(t) = \{ta, as\} - E_J$. If there is a path of length at least 3 in Π, say $t - a - b - \cdots - s$, then $I(t) = \{tb, as\} - E_J$. Otherwise, all the paths in Π distinct from the edge ts have length 2 and every node belongs to a path in Π. But then $|V| = \kappa_J(t, s) + 1 \le k$, and thus the problem has no feasible solution I so that $J + I$ is a simple graph. Consequently, $I(t)$ as above exists and can be found in polynomial time.

Let $I' = \bigcup_{t \in T_2} I(t)$. Then $I = I_0 + I'$ is a feasible solution and $|I| \le |I_0| + |I'| = |T_0| + 2(|T| - |T_0|) \le 2|T|$. The statement follows.

Remark: If parallel edges are allowed, then in Rooted NCA by connecting every node in T to s, we obtain a feasible solution of size $\le |T|$. However, allowing parallel edges requires changing the definition of tight sets, and we do not know if then the other parts of our proof remain valid.

From Corollary 1 and Proposition 2 we obtain the following result:

Theorem 3. Simple NCA *admits a polynomial time algorithm that computes a solution I so that* $|I| \le (\ell+2)H\left((4\ell + 1)^2\right) \cdot \tau^*(\mathcal{C}_J(\mathcal{D}))$ *for general requirements, and* $|I| \le 2H(2\ell + 1) \cdot \tau^*(\mathcal{C}_J^s(\mathcal{D}))$ *in the case of rooted requirements, where* $\ell = \max_{\{u,v\} \in \mathcal{D}} \kappa_J(u, v)$, *and $H(k)$ denotes the kth harmonic number.*

The following general statement relates approximability of NCA to approximability of Simple NCA.

Proposition 3. *Suppose that* Simple NCA *admits a polynomial time algorithm that computes a solution of size $\le \alpha(\ell) \cdot \tau^*(\mathcal{C}_J(\mathcal{D}))$, where $\alpha(\ell)$ is increasing in ℓ. Then* NCA *admits a polynomial time algorithm that computes a solution of size $\le 2\text{opt} \cdot \sum_{\ell=0}^{k-1} \frac{\alpha(\ell)}{k-\ell} \le 2H(k) \cdot \alpha(k) \cdot \text{opt}$, where opt denotes the optimal solution size for* NCA. *The same is valid for* Rooted NCA.

Proof. Apply the algorithm for Simple NCA as in the proposition sequentially: at iteration $\ell = 0, \ldots, k - 1$ add to J an augmenting edge set I_ℓ that increases the connectivity between pairs in $\mathcal{D}_\ell = \{\{u, v\} : u, v \in V, \kappa_J(u, v) = r(u, v) - k + \ell\}$ by 1. Note that $\kappa_J(u, v) \le \ell$ for $\{u, v\} \in \mathcal{D}_\ell$, thus the algorithm assumed in the proposition can be used to produce a solution I_ℓ to Simple NCA so that

$|I_\ell| \le \alpha(\ell) \cdot \tau_J^*(\mathcal{D}_\ell)$. After iteration ℓ, we have $\kappa_J(u,v) \ge r(u,v) - k + \ell + 1$ for all $u, v \in V$. Consequently, after $k - 1$ iterations, $\kappa_J(u,v) \ge r(u,v)$ holds for all $u, v \in V$. Hence the computed solution for NCA is feasible. We claim that $|I_\ell| \le 2\mathrm{opt} \cdot \frac{\alpha(\ell)}{k-\ell}$, $\ell = 0, \ldots, k - 1$. For that, it is sufficient to show that $\tau_J^*(\mathcal{D}_\ell) \le 2\mathrm{opt}/(k - \ell)$. For any $C \in \mathcal{C}_J(\mathcal{D}_\ell)$, any feasible solution to NCA has at least $k - \ell$ edges with an endnode in C, by Menger's Theorem. Thus

$$\mathrm{opt} \ge \frac{1}{2} \cdot \min\{\sum_{v \in V} x(v) : \sum_{v \in C} x(v) \ge k - \ell \ \forall C \in \mathcal{C}_J(\mathcal{D}_\ell), \ x(v) \ge 0\}$$

$$= \frac{1}{2}(k - \ell) \cdot \tau^*(\mathcal{C}_J(\mathcal{D}_\ell)) \ .$$

Theorem 1 now follows from Theorem 3 and Proposition 3.

3 Maximum Degree of Hypergraph of Minimal Tight Sets (Proof of Theorem 2)

Here we prove Theorem 2, which is restated for the convenience of the reader.

Theorem 2. *Suppose that* $\max\{\kappa_J(u,v) : \{u,v\} \in \mathcal{D}\} \le \ell$ *for an instance of* Simple NCA. *Then the maximum degree in the hypergraph* $(V, \mathcal{C}_J(\mathcal{D}))$ *is at most* $(4\ell + 1)^2$. *For rooted requirements, the maximum degree in the hypergraph* $(V, \mathcal{C}_J^s(\mathcal{D}))$ *is at most* $2\ell + 1$.

Let $S^* = V - (S \cup \Gamma_J(S))$. To avoid considering "mixed" cuts that contain both nodes and edges, we assume that $uv \notin E_J$ for all $\{u,v\} \in \mathcal{D}$. One way to achieve this is to subdivide every edge uv with $\{u,v\} \in \mathcal{D}$ by a new node. After this simple operation, we have that S is uv-tight if $u \in S$, $v \in S^*$, and $|\Gamma_J(S)| = \kappa_J(u,v)$. Also note that by the definition of tight sets and Menger's Theorem $\max\{|\Gamma_J(C)| : C \in \mathcal{C}_J(\mathcal{D})\} \le \max\{\kappa_J(u,v) : \{u,v\} \in \mathcal{D}\} \le \ell$.

The following "sub-modular" and "posi-modular" properties of the function $\Gamma(\cdot) = \Gamma_J(\cdot)$ is well known, see for example [21] and [32].

Proposition 4. *For any* $X, Y \subseteq V$ *the following holds:*

$$|\Gamma(X)| + |\Gamma(Y)| \ge |\Gamma(X \cap Y)| + |\Gamma(X \cup Y)| \tag{3}$$

$$|\Gamma(X)| + |\Gamma(Y)| \ge |\Gamma(X \cap Y^*)| + |\Gamma(Y \cap X^*)| \tag{4}$$

Lemma 1. *Let* X *be* xx'-*tight and let* Y *be* yy'-*tight. If* $\Gamma(X) \cap \{y, y'\} = \emptyset$ *and* $\Gamma(Y) \cap \{x, x'\} = \emptyset$, *then at least one of the sets* $X \cap Y, X \cap Y^*, Y \cap X^*$ *is* xx'-*tight or is* yy'-*tight.*

Proof. W.l.o.g. assume that $\kappa_J(x, x') \ge \kappa_J(y, y')$. We now consider several cases, see Figure 1.

If $x \in X \cap Y$ and $x' \in X^* \cap Y^*$ then (see Figure 1(a)):

$$2\kappa_J(x, x') \ge |\Gamma(X)| + |\Gamma(Y)| \ge |\Gamma(X \cap Y)| + |\Gamma(X \cup Y)|$$
$$\ge \kappa_J(x, x') + \kappa_J(x, x') = 2\kappa_J(x, x') \ .$$

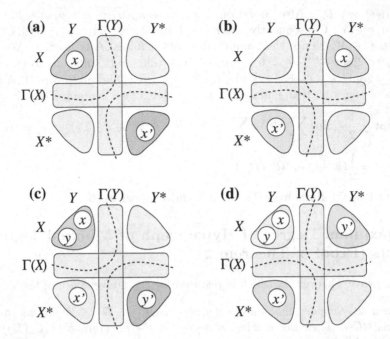

Fig. 1. Illustration to the proof of Lemma 1. Here the sets $X, \Gamma(X), X^*$ are the "rows" and $Y, \Gamma(Y), Y^*$ are the "columns" of a 3×3 "matrix".

Hence equality holds everywhere, so $X \cap Y$ (and also $X \cup Y$) is xx'-tight.

Similarly, if $x \in X \cap Y^*$ and $x' \in X^* \cap Y$ then (see Figure 1(b)):

$$2\kappa_J(x, x') \geq |\Gamma(X)| + |\Gamma(Y)| \geq |\Gamma(X \cap Y^*)| + |\Gamma(X^* \cap Y)|$$
$$\geq \kappa_J(x, x') + \kappa_J(x, x') = 2\kappa_J(x, x') .$$

Hence equality holds everywhere, so both $X \cap Y^*, X^* \cap Y$ are xx'-tight.

The remaining cases are $x, x' \in Y$ or $x, x' \in Y^*$. We consider the case $x, x' \in Y$, and the proof of the case $x, x' \in Y^*$ is similar. If $x, x' \in Y$ then $x \in X \cap Y$ and $x' \in X^* \cap Y$. We have two cases: $y \in Y \cap X$ or $y \in Y \cap X^*$.

If $y \in Y \cap X$ and $y' \in X^* \cap Y^*$ then (see Figure 1(c)):

$$\kappa_J(x, x') + \kappa_J(y, y') = |\Gamma(X)| + |\Gamma(Y)| \geq |\Gamma(X \cap Y)| + |\Gamma(X \cup Y)|$$
$$\geq \kappa_J(x, x') + \kappa_J(y, y') .$$

Hence equality holds everywhere, so $X \cap Y$ is xx'-tight (and $X \cup Y$ is yy'-tight).

If $y \in Y \cap X$ and $y' \in X \cap Y^*$ then (see Figure 1(d)):

$$\kappa_J(x, x') + \kappa_J(y, y') = |\Gamma(X)| + |\Gamma(Y)| \geq |\Gamma(X \cap Y^*)| + |\Gamma(X^* \cap Y)|$$
$$\geq \kappa_J(y, y') + \kappa_J(x, x') .$$

Hence equality holds everywhere, so $X \cap Y^*$ is yy'-tight and $X^* \cap Y$ is xx'-tight.

This concludes the proof of the lemma.

Corollary 2. *Let $C_1, C_2 \in \mathcal{C}_J(\mathcal{D})$ so that C_1 is u_1v_1-tight, C_2 is u_2v_2-tight, and $C_1 \neq C_2$. Then $(u_1, v_1) \neq (u_2, v_2)$. If in addition $C_1 \cap C_2 \neq \emptyset$ then $\Gamma(C_1) \cap \{u_2, v_2\} \neq \emptyset$ or $\Gamma(C_2) \cap \{u_1, v_1\} \neq \emptyset$.*

Proof. The first statement is obvious, as for any $\{u, v\} \in \mathcal{D}$ the minimal uv-tight set is unique. For the second statement, if $\Gamma(C_1) \cap \{u_2, v_2\} = \emptyset$ and $\Gamma(C_2) \cap \{u_1, v_1\} = \emptyset$, then by Lemma 1 at least one of the sets $C_1 \cap C_2, C_1 \cap C_2^*, C_2 \cap C_1^*$ is u_1v_1-tight or is u_2v_2-tight. Since $C_1 \cap C_2 \neq \emptyset$ then this set is strictly contained in C_1 (if it is u_1v_1-tight), or is strictly contained in C_2 (if it is u_1v_1-tight). This contradicts the minimality of one of C_1, C_2.

For $z \in V$ let $\mathcal{C}(z) = \{C \in \mathcal{C}_J(\mathcal{D}) : z \in C\}$ be the set of members in $\mathcal{C}_J(\mathcal{D})$ containing z. Let $q = |\mathcal{C}(z)|$. Construct an auxiliary directed labeled graph $\mathcal{J}(z)$ with labels on the arcs as follows. The node set of $\mathcal{J}(z)$ is $\mathcal{C}(z)$. Add an arc $C'C$ with *label* (u', v') if C' is $u'v'$-tight and $\Gamma(C) \cap \{u', v'\} \neq \emptyset$; from every set of parallel arcs keep only one. Note that $\mathcal{J}(z)$ is a tournament, by Corollary 2.

Lemma 2. *For any $u \in V$, there are at most $2\ell + 1$ arcs that have label (u, v') for some $v' \in V$, and there are at most $2\ell + 1$ arcs that have label (v', u) for some $v' \in V$.*

Proof. Let $u \in V$. We prove there are at most $2\ell + 1$ arcs that have labels (u, v') for some $v' \in V$; the proof for the other case is similar. Consider all the edges with labels of the form (u, v'), say $(u, v_1), \ldots (u, v_t)$, and the corresponding minimal tight sets $C_1, \ldots, C_t$, where C_i is uv_i-tight. We claim that $t \leq 2\ell + 1$. For that, consider the subgraph $\mathcal{J}'$ of $\mathcal{J}(z)$ induced by $C_1, \ldots, C_t$. We have that u belongs to the intersection of the sets C_i for $i = 1, \ldots, t$. Thus for every $i \neq j$ we have $v_i \in \Gamma(C_j)$ or $v_j \in \Gamma(C_i)$, by Corollary 2. As $\mathcal{J}'$ is a tournament, there is a node C in $\mathcal{J}'$ with indegree at least $(t - 1)/2$. Every arc C_iC entering C contributes the node v_i to $\Gamma(C)$; thus $(t - 1)/2 \leq \ell$, since the nodes $v_1, \ldots v_t$ are distinct. This implies $t \leq 2\ell + 1$, as claimed.

Corollary 3. *For any arc with label (u, v) there are at most $4(2\ell + 1)$ arcs with labels (u', v') so that $\{u', v'\} \cap \{u, v\} \neq \emptyset$.*

Proof. If $\{u', v'\} \cap \{u, v\} \neq \emptyset$, then there are 4 cases: $u' = u$, or $v' = v$, or $u' = v$, or $v' = u$. Namely, the label (u', v') belongs to one of the following 4 types: $(u, v'), (u', v), (v, v'), (u', u)$ By Lemma 2, the number of arcs with labels of each one of these types is at most $2\ell + 1$, which implies the statement.

We now finish the proof of Theorem 2. As $\mathcal{J}(z)$ is a tournament, it has a node C of indegree $\geq (q - 1)/2$. Now consider the labels of the arcs entering C in $\mathcal{J}(z)$. By Corollary 3, there are at least $(q - 1)/(16\ell + 8)$ arcs entering C, so that no two arcs have intersecting labels. Each one of these arcs contributes a node to $\Gamma(C)$. Consequently, we must have $(q - 1)/(16\ell + 8) \leq \ell$, which implies $q \leq 8\ell(2\ell + 1) + 1 = (4\ell + 1)^2$. In the case of rooted requirements, the total number of minimal tight sets containing z but not s is at most $2\ell + 1$, by Lemma 2.

The proof of Theorem 2 is complete.

4 Open Problems

- Does SND with arbitrary costs admit a $\rho(k)$-approximation algorithm? The answer is positive for rooted requirements, see [35]. We conjecture the answer is positive for general requirements, motivated also by the results of this paper. As was mentioned, the currently best ratios for SND problems are $O(k^3 \ln n)$ for SND [8], and $O(k^2)$ for Rooted SND [35]. Note that the ratio of [8] for SND depends on n, while in this paper we showed for 0, 1-costs the ratio $O(k \ln^2 k)$ that does not depend on n.
- What versions of SND can be solved exactly and/or well approximated in $t(k)poly(n)$ time? One example of such a problem is k-Connected Subgraph with 0, 1-costs [19].
- Does directed/undirected Simple Rooted SND with requirements in $\{0, k\}$ admit an approximation scheme similar to the one given in [2] for the Directed Steiner Tree problem?

Acknowledgment. I thank an anonymous referee for useful comments.

References

1. Chakraborty, T., Chuzhoy, J., Khanna, S.: Network design for vertex connectivity. In: STOC, pp. 167–176 (2008)
2. Charikar, M., Chekuri, C., Cheung, T., Dai, Z., Goel, A., Guha, S., Li, M.: Approximation algorithms for directed Steiner problems. J. of Algorithms 33, 73–91 (1999)
3. Chekuri, C., Korula, N.: Single-sink network design with vertex-connectivity requirements. In: FSTTCS (2008)
4. Cheriyan, J., Jordán, T., Nutov, Z.: On rooted node-connectivity problems. Algorithmica 30(3), 353–375 (2001)
5. Cheriyan, J., Vempala, S., Vetta, A.: An approximation algorithm for the minimum-cost k-vertex connected subgraph. SIAM Journal on Computing 32(4), 1050–1055 (2002); Preliminary version in STOC 2002
6. Cheriyan, J., Vetta, A.: Approximation algorithms for network design with metric costs. In: STOC, pp. 167–175 (2005)
7. Chuzhoy, J., Khanna, S.: Algorithms for single-source vertex-connectivity. In: FOCS, pp. 105–114 (2008)
8. Chuzhoy, J., Khanna, S.: An $O(k^3 \log n)$-approximation algorithms for vertex-connectivity network design (manuscript, 2009)
9. Dodis, Y., Khanna, S.: Design networks with bounded pairwise distance. In: STOC, pp. 750–759 (2003)
10. Fackharoenphol, J., Laekhanukit, B.: An $O(\log^2 k)$-approximation algorithm for the k-vertex connected subgraph problem. In: STOC, pp. 153–158 (2008)
11. Feldman, M., Kortsarz, G., Nutov, Z.: Improved approximation algorithms for directed steiner forest. In: SODA, pp. 922–931 (2009)
12. Fleischer, L., Jain, K., Williamson, D.P.: Iterative rounding 2-approximation algorithms for minimum-cost vertex connectivity problems. J. Comput. Syst. Sci. 72(5), 838–867 (2006)
13. Frank, A.: Augmenting graphs to meet edge-connectivity requirements. SIAM Journal on Discrete Math. 5(1), 25–53 (1992)

14. Frank, A.: Connectivity augmentation problems in network design. Mathematical Programming: State of the Art, 34–63 (1995)
15. Frank, A., Jordán, T.: Minimal edge-coverings of pairs of sets. J. on Comb. Theory B 65, 73–110 (1995)
16. Frank, A., Tardos, E.: An application of submodular flows. Linear Algebra and its Applications 114/115, 329–348 (1989)
17. Halperin, E., Krauthgamer, R.: Polylogarithmic inapproximability. In: STOC, pp. 585–594 (2003)
18. Jackson, B., Jordán, T.: A near optimal algorithm for vertex connectivity augmentation. In: Lee, D.T., Teng, S.-H. (eds.) ISAAC 2000. LNCS, vol. 1969, pp. 313–325. Springer, Heidelberg (2000)
19. Jackson, B., Jordán, T.: Independence free graphs and vertex connectivity augmentation. J. of Comb. Theory B 94(1), 31–77 (2005)
20. Jain, K.: A factor 2 approximation algorithm for the generalized Steiner network problem. Combinatorica 21(1), 39–60 (2001)
21. Jordán, T.: On the optimal vertex-connectivity augmentation. J. on Comb. Theory B 63, 8–20 (1995)
22. Jordán, T.: A note on the vertex connectivity augmentation. J. on Comb. Theory B 71(2), 294–301 (1997)
23. Khuller, S.: Approximation algorithms for for finding highly connected subgraphs. In: Hochbaum, D.S. (ed.) Approximation Algorithms for NP-hard problems, ch. 6, pp. 236–265. PWS (1995)
24. Kortsarz, G., Krauthgamer, R., Lee, J.R.: Hardness of approximation for vertex-connectivity network design problems. SIAM J. on Computing 33(3), 704–720 (2004)
25. Kortsarz, G., Nutov, Z.: Approximating node connectivity problems via set covers. Algorithmica 37, 75–92 (2003)
26. Kortsarz, G., Nutov, Z.: Approximating k-node connected subgraphs via critical graphs. SIAM J. on Computing 35(1), 247–257 (2005)
27. Kortsarz, G., Nutov, Z.: Approximating minimum-cost connectivity problems. In: Gonzalez, T.F. (ed.) Approximation Algorithms and Metaheuristics, ch. 58. Chapman & Hall/CRC, Boca Raton (2007)
28. Kortsarz, G., Nutov, Z.: Tight approximation algorithm for connectivity augmentation problems. J. Comput. Syst. Sci. 74(5), 662–670 (2008)
29. Lando, Y., Nutov, Z.: Inapproximability of survivable networks. In: Goel, A., Jansen, K., Rolim, J.D.P., Rubinfeld, R. (eds.) APPROX and RANDOM 2008. LNCS, vol. 5171, pp. 146–152. Springer, Heidelberg (2009); To appear in Theoretical Computer Science
30. Liberman, G., Nutov, Z.: On shredders and vertex-connectivity augmentation. Journal of Discrete Algorithms 5(1), 91–101 (2007)
31. Lovász, L.: On the ratio of optimal integral and fractional covers. Discrete Math. 13, 383–390 (1975)
32. Nutov, Z.: Approximating connectivity augmentation problems. In: SODA, pp. 176–185 (2005)
33. Nutov, Z.: Approximating rooted connectivity augmentation problems. Algorithmica 44, 213–231 (2005)
34. Nutov, Z.: An almost $O(\log k)$-approximation for k-connected subgraphs. In: SODA, pp. 912–921 (2009)
35. Nutov, Z.: Approximating minimum cost connectivity problems via uncrossable bifamilies and spider-cover decompositions (manuscript) (2009)
36. Ravi, R., Williamson, D.P.: An approximation algorithm for minimum-cost vertex-connectivity problems. Algorithmica 18, 21–43 (1997)

A 7/9 - Approximation Algorithm for the Maximum Traveling Salesman Problem

Katarzyna Paluch[1,*], Marcin Mucha[2], and Aleksander Mądry[3]

[1] Institute of Computer Science, Wrocław University, Poland
[2] Institute of Informatics, Warsaw University, Poland
[3] CSAIL, MIT, Cambridge, MA, USA

Abstract. We give a deterministic combinatorial 7/9-approximation algorithm for the symmetric maximum traveling salesman problem.

1 Introduction

The traveling salesman problem is one of the most famous and heavily researched problems in computer science. The version we deal with in this paper is the Symmetric Maximum Traveling Salesman Problem, which is defined as follows. For a given complete undirected graph G with nonnegative weights on its edges, we wish to find a tour of the graph of maximum weight. The tour of the graph is a simple cycle that contains each vertex from G. In 1979 Fisher, Nemhauser and Wolsey [8] showed that the *greedy*, the *best neighbour* and the *2-interchange* algorithms have approximation ratio 1/2. In [8] the *2-matching* algorithm is also given, which has a guarantee of $\frac{2}{3}$. In 1994 Kosaraju, Park and Stein [12] presented an improved algorithm having a claimed ratio $\frac{5}{7}$, but the proof contained a flaw and in [2] it was shown to have ratio $\frac{19}{27}$. In the meantime in 1984 Serdyukov [18] presented (in Russian) a simple (to understand) and elegant $\frac{3}{4}$-approximation algorithm. The algorithm is deterministic and runs in $O(n^3)$. Afterwards, Hassin, Rubinstein ([9]) gave a randomized algorithm having *expected* approximation ratio at least $\frac{25(1-\epsilon)}{33-32\epsilon}$ and running in $O(n^2(n+2^{1/\epsilon}))$, where ϵ is an arbitrarly small constant. The first deterministic approximation algorithm with the ratio better than $\frac{3}{4}$ was given in 2005 by Chen, Okamoto, Wang ([4]), which is a $\frac{61}{81}$-approximation and a nontrivial derandomization of the algorithm from [9]. It runs in $O(n^3)$.

Related Work. For the asymmetric version of Max TSP, the best approximation is by Kaplan, Lewenstein, Shafrir, Sviridenko ([11]) and has ratio $\frac{2}{3}$. If additionally in graph G triangle inequality holds, we get two (symmetric and asymmetric) metric versions of the problem. The best approximation bounds for them are $\frac{7}{8}$ ([10]) and $\frac{10}{13}$ ([11]), both of which have been improved by Chen and

---

* Work partly done while the author was in Max Planck Institute for Computer Science, Saarbrücken. Partly supported by MNiSW grant number N N206 1723 33, 2007-2010.

I. Dinur et al. (Eds.): APPROX and RANDOM 2009, LNCS 5687, pp. 298–311, 2009.
© Springer-Verlag Berlin Heidelberg 2009

Nagoya in [5]. The latest improvements are by Kowalik and Mucha ([14] and [15]) and equal, respectively for an asymmetric version $\frac{35}{44}$ and for the symmetric version $\frac{7}{8}$. All four versions of Max TSP are MAX SNP-hard ([6],[7],[16]). A good survey of the maximum TSP is [1].

Our Results. We give an $O(n^3)$ deterministic combinatorial algorithm for the Symmetric Maximum Traveling Salesman problem, with the approximation guarantee equal to $\frac{7}{9}$. To achieve this, we compute the graph described in the following theorem, which is proved in Section 2.

Theorem 1. *Given a complete graph G with nonnegative weights on the edges, we can compute a multisubgraph $H = (V, E_H)$ of $G = (V, E)$ such that H is loopless, 4-regular, each $e \in E_H$ has the same weight as in G, there are at most two edges between a pair of vertices, each connected component has at least 5 vertices and its weight is at least $\frac{35}{18}opt$. (opt denotes the weight of an optimal tour.)*

The combinatorial technique used in this theorem is new and can be used for any optimization problem for which a cycle cover of minimal/maximal weight is a lower/upper bound on the optimal value of the solution. In the proof we exploit the fact that the tour of the graph is a cycle cover of G or in other words a simple perfect 2-matching. Thus a maximum weight cycle cover C of G is an upper bound on *opt*. The tour of the graph in turn is a somewhat special cycle cover, it has some properties we can make use of and the the notion from the matching theory that turns out to be particularly useful is that of an alternating cycle.

Next in the proof of Theorem 2 we show how to extract from H a tour of weight at least $\frac{2}{5} \cdot \frac{35}{18}opt$.

Theorem 2. *If we have a loopless 4-regular graph $H = (V, E_H)$ with nonnegative weights on the edges that can contain at most two edges between a pair of vertices and such that its every connected component has at least 5 vertices, then we can find such a subset E' of its edges that $w(E') \leq 1/5w(H)$ and such that we can 2-path-color the graph $H' = (V, E_H \setminus E')$.*

To 2-path-color the graph means to color its edges into two colors so that no monochromatic cycle arises. The outline of the proof of this theorem is given in Section 3. The whole algorithm runs in time $O(n^3)$, where n denotes the number of vertices in G. The estimation of the approximation ratio is tight. The obstacle to 4/5-approximation is that we are not able to construct an exact gadget for a square. Gadgets for squares are described in Section 2.

For comparison, let us note, that in the case of the Asymmetric Max TSP, which is considered in [11], the authors compute a 2-regular loopless graph G_1 (which is a multisubgraph of G), whose all connected components contain at least 3 vertices and such that its weight is at least $2opt$. Next a tour of weight at least $\frac{1}{3}2opt$ is extracted from G_1. However obtaining graph G_1 in [11] is not combinatorial. It involves using a linear program that is a relaxation of the problem of finding a maximum cycle cover which does not contain 2-cycles. Next scaling up the fractional solution by an appropriate integer D (which is a polynomial in n) to an integral one, which defines a d-regular multigraph, from which a desired graph G_1 is obtained. The running time needed to compute G_1 is $O(n^2 D)$.

2   Upper Bound

Let $G = (V, E)$ be a complete graph with nonnegative weights on the edges, in which we wish to find a traveling salesman tour (a cycle containing all vertices from V) of maximum weight. Let T_{max} denote any such tour and t_{max} its weight.

The weight of the edge $e = (u, v)$ between vertices u and v is denoted by $w(e)$ or $w(u, v)$. By $w(E')$ we denote the weight of the (multi)set of edges $E' \subseteq E$, which is defined as $\sum_{e \in E'} w(e)$. The weight of the graph G is denoted as $w(G) = w(E)$.

One of the natural upper bounds for t_{max} is the weight of a maximum weight cycle cover C of G (C is a cycle cover of G if each vertex of V belongs to exactly one cycle from C). If C contained only cycles of length 5 or more, then by deleting the lightest edge from each cycle and patching them arbitrarily into a tour we would get a solution of weight at least $\frac{4}{5} t_{max}$. C however can of course contain triangles and quadrilaterals. From now on, let C denote a cycle cover of maximum weight and assume that it contains more than one cycle. Further on, we will define the notions of a good cycle cover and alternating weight. They will be strictly connected with C.

We can notice that T_{max} *does not* contain an edge (one or more) from each cycle from C. Since, we aim at a $\frac{7}{9}$-approximation, we will restrict ourselves to bad cycles from C, which are defined as follows. Cycle c of C is said to be **bad** if each edge of c has weight greater than $\frac{2}{9} w(c)$. Let us notice that if a cycle c is bad, then it is a triangle or a quadrilateral. For convenience, let us further on call all quadrilaterals **squares**. We will call a cycle cover C' **good** if for each bad cycle c of C, C' does *not* contain at least one edge from c and if it does not contain a cycle whose vertices all belong to some bad cycle c of C (which means, informally speaking, that C' does not contain cycles that are "subcycles" of the bad cycles from C). Since T_{max} is just one cycle, it is of course good and the weight of a good cycle cover of maximum weight is another upper bound on t_{max}. See Figure 1 for an example of a good cycle cover.

2.1   Approximating a Good Cycle Cover

We will construct graph G' and define a special b-matching B for it, so that B of maximum weight will in a way approximate a good cycle cover of maximum weight in G. (A b-matching is such a generalization of a matching in which every vertex v is required to be matched with $b(v)$ edges.)

Let C' denote a good cycle cover of maximum weight. C and C' are cycle covers or, in other words, simple 2-matchings (2-matchings and their generalizations are desribed, among others, in [17]). Let us look closer at $C \oplus C'$ (i.e. the symmetric difference between sets of edges C and C') and get advantage from the matching theory in order to notice useful properties of a good cycle cover.

First, recall a few notions from matching theory. A path P is **alternating** with respect to a cycle cover C_1 if its edges are alternatingly from C_1 and from $E \setminus C_1$. If an alternating path ends and begins with the same vertex, then it is called an **alternating cycle**. For any two cycle covers C_1 and C_2, $C_1 \oplus C_2$ can be expressed as a set of alternating cycles (with respect to C_1 or C_2).

Since $C' = C \oplus (C \oplus C')$,

$$w(C') = w(C) - w(C \cap (C \oplus C')) + w(C' \cap (C \oplus C')).$$

For convenience, we will also use the notion of **alternating weight** w' and define it for a subset S as $w'(S) = w(S \setminus C) - w(C \cap S)$. Using it we can rephrase the above statement as

$$w(C') = w(C) + w'(C \oplus C'). \qquad 2.1$$

For example in Figure 1 $(C \oplus C_1)$ is an alternating cycle (BE, EG, GB, BC, AC, AB), whose alternating weight amounts to -3. $(C \oplus C')$ is an alternating cycle (BE, EG, GC, CB) whose alternating weight amounts to -5.

If we have an alternating cycle A with respect to C_1, then by **applying** A to C_1 we will mean the operation, whose result is $C_1 \oplus A$.

In view of 2.1 we can look at the task of finding a good cycle cover as at the task of finding a collection A' of alternating cycles with respect to C, such that each bad cycle from C is "touched" (i.e.some edge from a bad cycle c belongs to some alternating cycle from A') by some alternating cycle from A' and the weight of C diminishes in the least possible way as a result of applying A' to C.

In the following fact we describe good cycle covers from the point of view of alternating cycles (with respect to C).

Fact 1. *If C' is a good cycle cover, then if we decompose $C \oplus C'$ into alternating cycles, then for each bad cycle c from C, there exists an alternating cycle K_c containing a subpath $(v_0, v_1, v_2, \ldots, v_k, v_{k+1})$ $(k \in \{2, 4\}$, i.e. a subpath has length 3 or 5) such that vertices $v_1, v_2, \ldots, v_k$ are on c, vertices v_0 and v_{k+1} are not on c and $v_1 \neq v_k$.*

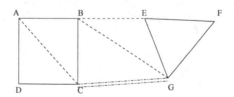

Fig. 1. The weight of the edges drawn with a solid line is 7, with a dashed line 6 and the weight of edge CG is 3. The remaining (not drawn) edges have weight 0. The cycle cover C of maximum weight consists of cycles $ABCD$ and EFG. The cycle cover C_1 consisting of cycles ACD and $BEFG$ is not a good cycle cover as ACD is a subcycle of $ABCD$. A good cycle cover C' consists of cycle $ABEFGCD$.

This fact follows from the definition of a good cycle cover that states that a good cycle cover does not contain cycles that are "subcycles" of cycles from C.

Notice that in Figure 1 C_1 is not a good cycle cover and the alternating cycle $C \oplus C_1$ contains a subpath BC, CA, AB for a square $ABCD$ and it is not such as we desire as it "enters" and "leaves" $ABCD$ with the same vertex B.

We define a graph G' and function b for a b-matching in it as follows.

Definition 1. *The construction of* $G' = (V', E')$:

- *graph G is a subgraph of G',*
- *V' consists of V and also a set S_i of additional vertices for each bad cycle c_i from C: S_i contains a copy v' for each vertex v from c_i and also a set of special vertices T_i. The subgraph of G' induced by S_i is called a gadget U_i corresponding to c_i. If $c_i = (v_1, v_2, v_3)$ is a triangle, then T_i consists of one vertex a_{c_i}. The weight of the edge between v_1' and a_{c_i} is equal to $-w(v_2 v_3)$ and analogously for vertices v_2', v_3'. We set $b(a_{c_i}) = 1$. The description of the gadget for a square is given in Figure 2.*
- *if v_1 is a vertex on some bad cycle c of C, then G' contains edges (v_1', v_2), (v_1', v_2') iff v_2 is not a vertex of the bad cycle c containing v_1. The weight of these edges is the same and equals $w(v_1, v_2)$.*

A b-matching for G' is such that for $v \in V$, we put $b(v) = 2$ and for v' which is a copy of some vertex v, we put $b(v') = 1$.

We define the notion of a **fragment**, that is to denote a possible fragment of an alternating cycle from $C \oplus C'$ contained in a bad cycle. Let $v_1 \neq v_2$ belong to bad cycle c_i from C. Then the fragment connected with v_1, v_2 is any alternating path $(v_1, v_3, v_4, ..., v_k, v_2)$ whose all vertices belong to c_i and such that it begins and

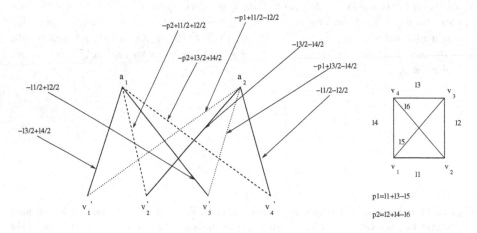

Fig. 2. For a bad cycle $c_i = (v_1, v_2, v_2, v_4)$, T_i contains two additional verices a_{1,c_i}, a_{2,c_i}. We set $b(a_{1,c_i}) = b(a_{2,c_i}) = 1$. The fragment connected with v_1 and v_2 is the edge (v_1, v_2). There are two fragments connected with v_1 and v_3: (v_1, v_2, v_4, v_3) and (v_1, v_4, v_2, v_3).

ends with an edge in C. Thus, if c_i is a triangle and v_1, v_2 are its two different vertices, then the fragment corresponding to them is the edge (v_1, v_2).

A b-matching of G' is defined in such a way that for each good cycle cover C_1 of G, we are able to find a b-matching B of G' that corresponds to it in the sense that alternating cycles $C' \oplus C$ are virtually the same as alternating cycles $B \oplus C$. (These are not quite alternating cycles.) Informally speaking, parts of alternating cycles from $C \oplus C'$ contained in bad cycles correspond in G' to the edges contained in the gadgets and the remaining parts of the alternating cycles are in a way impressed in the graph G.

A b-matching of G' is such that for each bad cycle c_i there are exactly two vertices, say v_1, v_2, such that v'_1, v'_2 will be matched with the edges not contained in the gadget U_i (these edges will be of the form $(v'_1, v_3), (v'_2, v_4)$) and the weight of the edges contained in U_i corresponds to the alternating weight of the fragment connected with v_1 and v_2.

We will say that a b-matching B of G' **lies by an error** $\epsilon \geq 0$ on a bad cycle c_i if for vertices v'_1, v'_2 (such that v_1, v_2 belong to c_i) matched with edges not contained in a gadget U_i, the weight w_i of the edges of B contained in U_i satisfies the following inequality: $w'(f_i) \geq w_i \geq w'(f_i) - \epsilon w(c_i)$, where f_i denotes some fragment connected with v_1, v_2 and $w'(f_i)$ its alternating weight.

We prove

Lemma 1. *Every b-matching B of G' lies on a bad triangle by an error 0 and on a bad square by an error at most $\frac{1}{18}$.*

Proof. We give the proof for a bad square c_i. Suppose that $l_1 + l_3 \leq l_2 + l_4$. Since we have a cycle cover of maximum weight $p_1 + p_2 \geq l_1 + l_3$, which implies that $l_2 + l_4 \geq l_5 + l_6$. If within a gadget vertices that are matched with a_{1,c_i} and a_{2,c_i} are v'_3, v'_4, then the weight of the edges within a gadget is equal to $-l_1$, which is equal to exactly the alternating weight of the fragment connected with v'_1, v'_2. The proof is analogous for the fragment connected with vertices v'_3, v'_4. If within a gadget vertices v'_1, v'_3 are matched with a_{1,c_i}, a_{2,c_i}, then the weight of the edges within the gadget is equal to $-p_1$, which is equal to the alternating weight of the fragment connected with v_2, v_4. If within the gadget vertices v'_1, v'_4 are matched with a_{1,c_i}, a_{2,c_i}, then the weight of the edges within the gadget is either $-l_1/2 - l_2/2 - l_3/2 + l_4/2$ or $-p_1 - w(a_{1,c_i} v'_3) - p_2 - w(a_{2,c_i} v'_2)$. We check that since $l_2 + l_4 \geq l_5 + l_6$, then the weight of the edges within the gadget will always be equal to $-l_1/2 - l_2/2 - l_3/2 + l_4/2$. However the alternating weight of the fragment connected with v_2, v_3 is $-l_2$, so the difference in the weights is $-(l_2 + l_4)/2 + (l_1 + l_3)/2$. The weight of each edge in a bad square is $< \frac{2}{9} w(c_i)$, where $w(c_i)$ is the weight of the square, therefore the maximal difference between the weight of the pairs of edges is $1/9 w(c_i)$, which means that the matching lies on a square by an error at most $\frac{1}{18}$.   □

Clearly B in G' is *not* a good cycle cover of G (it is not even a cycle cover of G). Let us however point the analogies between B and a good cycle cover of G. Let us define for B a **quasi-alternating** multiset S_B. S_B will contain: (1) for each edge $e = (v_1, v_2)$ $|(Z_e \cap B) \setminus C|$ number of copies of e, where

$Z_e = \{(v_1, v_2), (v_1', v_2'), (v_1, v_2'), (v_1', v_2)\}$, (2) the set of edges $C \setminus B$, (3) for each gadget U_i it contains a fragment connected with v_1, v_2 iff v_1', v_2' are matched in B with vertices from the original graph G. For example in Figure 3 $S_B = C \oplus C'$. Another example is given in Figure 4.

The alternating weight of a multiset is defined in an analogous way so that the weight of the edge not in C is counted the number of times it occurs in the

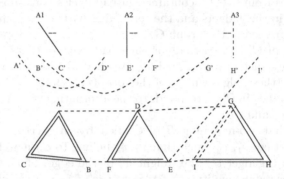

Fig. 3. The weight of the edges drawn with a solid line is 5, with a dashed line 3 and a minus near the edge shows that the edge has negative weight (-5 or -3). The weight of the remaining edges of G is 0. Cycle cover C of maximum weight consists of cycles ABC, DEF, GHI, a good cycle cover C' consists of cycle $ACBFEIHGD$ and a b matching B of maximum weight in G' consists of cycles ABC, GHI and path $G'DFEI'$ and edges $A1C', A2E', A3H', A'D', B'F'$. $C \oplus C'$ consists of two alternating cycles (AD, DF, FB, BA) and (DG, GI, IE, ED). Their alternating weight equals to correspondingly -4 and -2. $C \oplus B$ consists of edges $A1C', A'D', B'F', A2E'$ and $A3H', G'D, DE, EI'$. Let us notice that the alternating weights of these sets of edges are also -4 and -2 and that the weight of C' and B are the same.

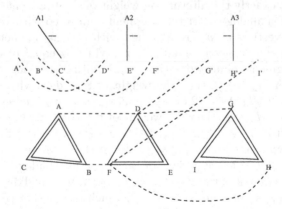

Fig. 4. The weight of the edges are represented in the same way as in Figure 3. A cycle cover C of maximum weight consists of cycles ABC, DEF, GHI. A b-matching B of maximum weight consists of cycles ABC, GHI, path $G'DEFH'$ and edges $A3I', A2E', A1C', A'D', B'F'$. $S_B = \{DA, AB, BF, FD, FD, DG, GH, HF\}$ (there is a mistake in the figure).

multiset and the weight of the edge in C is subtracted the number of times it occurs in the multiset. We have

Fact 2
$$w(B) = w(C) + w'(B \oplus C).$$

Next, we are going to bind good cycle covers with b-matchings in G'.

Lemma 2. *If C_1 is a good cycle cover, then there exists such a b-matching B in G' that $w'(S_B) \geq w'(C_1 \oplus C)$.*

Proof. By Fact 1 for each cycle c of length 3 or 4 there exists an alternating cycle K_c in $C \oplus C_1$ such that there are two different vertices v_1, v_2 on c such that the part of an alternating cycle K_c between vertices v_1, v_2 is a fragment (i.e. this part is on vertices solely from c).

For some cycles there are more such alternating cycles or there is more than one place of this kind on such an alternating cycle. Nevertheless for each cycle c we choose one such K_c and one subpath P_c on it. Next we build a b-matching B. Originally let all the edges from C belong to B. Each subpath P_c is encoded by the corresponding gadget and for the remaining edges of K_c we do as follows. If $e \in E \setminus C$, then we add e to B (or more precisely sometimes a corresponding edge between the copies of vertices) and if $e \in C$, then we remove it from B. For example in Figure 3 for cycles ABC and DEF we chose an alternating cycle (AD, DF, FB, BA) and for cycle GHI the other alternating cycle. We could also choose for cycle ABC the same alternating cycle (AD, DF, FB, BA), but for DEF and GHI the other one. Then b-matching B would consist of cycles ABC, GHI, path $A'FEDC'$ and edges $A1C', A2E', A3H', D'G', F'I'$. □

By Lemma 1 we have that $w'(B \oplus C) \geq w'(S_B) - \frac{1}{18}w(C)$ and therefore we get

Corollary 1. $w(B) \geq w(C') - \frac{1}{18}w(C)$. *Recall that C' denotes a good cycle cover of maximum weight.*

From $B \cup C$ we obtain a 4-regular graph H. We do it in the following way. At the beginning H consists of two copies of the cycle cover C (at this moment H is 4-regular). Next we compute S_B and apply it to H, that is we put $H := H \oplus (B \oplus S_B)$.

For example in Figure 4 we would get $H = \{AC, AC, CB, CB, AB, AD, BF, FE, FE, ED, ED, DG, FH, GI, GI, IH, IH, GH\}$ and in Figure 3 $H = \{AC, AC, CB, CB, AB, AD, BF, DF, DE, FE, FE, EI, DG, GI, IH, IH, HG, HG\}$.

3   Extracting a Heavy Tour

To 2-**path-color** graph G will mean to color its edges into two colors (each edge is colored into one color) so that the edges of the same color form a collection of vertex-disjoint paths.

To 2-**cycle-color** the graph will mean to color its edges into two colors so that the edges of each color form a collection of vertex-disjoint cycles. Since the graph can contain double edges, some of these cycles can be of length 2. To **well 2-cycle-color** the graph will mean to 2-cycle-color it so that each monochromatic cycle has length at least 5.

Since H is 4-regular, we can 2-cycle-color it. If we could well 2-cycle-color it, then we would put the edge of minimal weight from each monochroamtic in E', then E' would have weight at most $1/5w(H)$ and graph $H' = (V, E_H \setminus E')$ would be 2-path-colored. As one can easily check, however, there exist graphs that cannot be well 2-cycle-colored.

We can however restrict ourselves to considering graphs that (almost) do not contain triangles as we prove Lemma 3, which is the corollary of two lemmas from Section 4.

Lemma 3. *In Theorem 2 we can restrict ourselves to graphs H such that if a triangle T is a subgraph of H, then either (1) T contains two double edges or (2) T consists of single edges and each vertex of T is adjacent to a double edge.*

(If we can eliminate a triangle from a connected component having 5 vertices using lemmas from Section 4, then we do not do that but deal with such a component separately.) It would be nice to be able to restrict ourselves also to graphs that do not contain cycles of length 2 or 4. However, we have not been able to find an analogous way to that from lemmas in Section 4. Instead we will well 2-almost-cycle-color the graph, which we define as follows. To 2-*almost-cycle-color* the graph means to color the subset of its edges into two colors, so that the edges of each color form a collection of vertex-disjoint paths and cycles and the set of uncolored (called *blank*) edges is vertex-disjoint. (The set of blank edges can be empty.) To *well* 2-*almost-cycle-color* the graph means to 2-almost-cycle-color it so that each monochromatic cycle has length at least 5.

In Section 5, we give the algorithm for well 2-almost-cycle-coloring the graph. The key part of the algorithm is played by *disabling* cycles of length correspondingly 2, 3 and 4, which consists in such a colouring of a certain subset of the edges that whatever happens to the rest of the edges no monochromatic cycle of lengh 2, 3 or 4 will arise.

Once graph H gets well 2-almost-cycle-colored, we would like to find such a subset E' that $w(E') \leq 1/5w(H)$ and such that after the removal of E' from H, $H' = (V, E \setminus E')$ is 2-path-colored that is the edges that got colored in 2-almost-cycle-coloring keep their color and blank edges are colored into an appropriate color.

In Section 6 we will describe five phases of dealing with a well 2-almost-cycle-colored graph H: two red ones, two blue ones and one blank one. With the phases we will attach five disjoint subsets of edges $R_1, R_2, B_1, B_2, Blank$ such that in the i-th $(i = 1, 2)$ red phase we will obtain a graph $P_{R_i} = (V_t, E_t \setminus R_i)$, which after coloring the remaining blank edges red, will be 2-path-colored, analogously for the blue phases. In the blank subphase we will obtain a graph $P_{Bl} = (V_t, E_t \setminus Blank)$, which after coloring the remaining blank edges into an appropriate color will also be 2-path-colored. Thus each blank edge acts twice (i.e. in two phases) as a red edge, twice as a blue edge and once it is removed.

4   Eliminating Triangles

Definition 2. *Suppose we have a graph $J = (V_J, E_J)$ as in Theorem 2 (i.e. loopless, 4-regular, having at most 2 edges between a pair of vertices and such that each of its connected components has at least 5 vertices) and its subgraph S. We say that we can eliminate S from J iff there exists graph $K = (V_K, E_K)$ that does not contain S, has at least one vertex less than J and such that the solution from Theorem 2 for K can be transformed into a solution for J, which means that if we have a set $E'_K \subseteq E_K$ such that $w(E'_K) \leq 1/5w(K)$ a 2-path-coloring of graph $K' = (V_K, E_K \setminus E'_K)$, then we can find a set E'_J such that $w(E'_J) \leq 1/5w(J)$ and such that graph $J' = (V_J, E_J \setminus E'_J)$ can be 2-path-colored.*

First, we will eliminate triangles that contain exactly one double edge.

Lemma 4. *If a triangle T has exactly one double edge, then we can eliminate T.*

The proof is omitted due to space limits.

Next, we will eliminate triangles that do not contain any double edges.

Lemma 5. *If graph J does not contain triangles having exactly one double edge, but contains a triangle T, whose all edges are single and such that at least one vertex of T is not adjacent to a double edge, then we can eliminate T from J.*

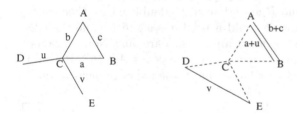

Proof. Suppose that a triangle T is on vertices A, B, C and let a, b, c denote the weights of the appropriate edges. First assume that C is not adjacent to a double edge and $c \geq \min\{a, b\}$. Thus C is also connected by two edges with two other vertices D, E and u, v are the weights of these edges. Without loss of generality, suppose that $u \leq v$. We build K as follows. At the beginning, it is the same as J. Next, in K we remove C and all four edges adjacent to it. Now A, B, D, E are of degree 3. We connect D and E with an additional edge of weight u. (Since J does not contain triangles with exactly one double edge, in K D and E will be connected with at most two edges.) We connect A and B with an additional edge of weight $a + v$ and we change the weight of the edge that had weight c to $c + b$. Note that $w(K) = w(J)$.

Assume we have found the set of edges E'_K and have 2-path-colored K'. Except for edges $b + c, a + v, u$, E'_J will contain the same edges as E'_K and the 2-path-coloring of J' will differ from that of K' only on these edges.

If from K none of the edges $c + b, a + v, u$ was removed then, in K' vertices A and B are connected with a double edge and suppose that the edge DE is blue, then in J' we color CD and CE into blue and AC, CB into red. If the edge u was removed and edges $b + c, a + v$ not, then E is in K' adjacent to at most three edges and suppose that two of them are blue. Then, we color EC into red and AC, CB into blue and AB is left as a red edge. If the edges $a + v, b + c$ were removed and the edge u not and is blue, then in J' we color CD, CE into blue. If $a + v \leq b + c$ and the edge $a + v$ was removed but the edges $u, b + c$ not, then

- if $b + c$ is red and u blue, then if A has only one red edge incident on it $(b + c)$, color AC into red, DC into *blue*, else if A has two red edges incident on it but is not connected with D via a blue path that does not contain u, then color AC, CD into blue (notice that we do not create a blue cycle), else if A is connected with D via a blue path that does not contain u, color AC, EC into blue (notice that E is not connected with A via a blue path not containing u).
- if $b + c, u$ are blue, then if A has at most one red edge incident on it, color AC into red and DC into blue, else if A has two red edges incident on it, color AC into blue and either EC or DC into blue so as not to create a cycle (since at most one of the vertices D, E is connected via a blue path not containing u with B, it is always possible).

If $a + v > b + c$ and the edge $b + c$ was removed but the edges $a + v, u$ not, then if u is blue, color CD, CE into blue. Since $c \geq a$, it is all right.

If $c < \min\{a, b\}$ and both A and B are adjacent to double edges, then everything goes as above. The only trouble could arise if $a + v > b + c$ and the edge $b + c$ was removed but the edges $a + v, u$ not and $a + v, u$ are blue. Now we cannot only color CD, CE into blue and not take any of the edges AC, BC. However, since B is adjacent to a double edge, it has at most one red edge incident on it, so we can additionally color BC into red.

Let us notice that $w(E'_J) \leq w(E'_K)$.   □

5   Disabling Cycles of Length < 5

By Lemma 4 we can assume that, if graph H contains a triangle T, then either it has two double eges or it consists only of single edges and each vertex of it is adjacent to a double edge. We will give the algorithm for well 2-almost-cycle-coloring H. We will use colors: blue and red. In the algorithm once the edge gets colored, it will not change its color and we will preserve the following invariant.

Invariant 1. *If at some step of the algorithm exactly two of the edges incident on vertex v are colored, then they have different colors.*

We remind that disabling cycles of length less than 5 consists in such a coloring of the subset of the edges of H, that however the rest of blank edges are coloured, the graph will not contain a monochromatic cycle of length less than 5. We begin from cycles of length 2. Disabling such cycles is very easy, we consider each double edge and colour it into two different colours: red and blue. As the graph does not contain connected components having less than 5 vertices, no monochromatic cycle of length less than 5 will arise. Next we disable caps. *A cap* is a triangle that has exactly two double edges or a square that has exactly three double edges. Let (v_1, v_2) denote the only non-double edge of a given cap C. Then the non-double edges incident at v_1 and v_2 different from (v_1, v_2) are called the *ribbons* of cap C. A ribbon may belong to two different caps.

We can eliminate some caps from the graph in a way similar to that in which we eliminated most kinds of triangles in the previous section.

Lemma 6. *If a cap C with two ribbons r_1, r_2 is such that r_1, r_2 do not share a vertex and are not connected by a double edge, then we can eliminate C.*

We eliminate C by removing cap C together with ribbons from the graph and connecting the other end vertices of ribbons with an edge.

If a cap C with two ribbons r_1, r_2 is such that r_1, r_2 share a vertex or are connected by a double edge, then we disable it by coloring r_1, r_2 into different colors.

We will say that a square or triangle is *active* if some of its edges (possibly all) are blank and there exists such a well 2-almost-cycle-coloring of all blank edges in the graph that this square or triangle is monochromatic.

We say that edge e is *active* if it is colored and included in some active square. In disabling squares we will maintain the following property of active edges.

Invariant 2. *If edge $e = (v_1, v_2)$, say red, is active, then v_1 has either two or four coloured edges incident to it and the same for v_2 and either e is double or neither v_1 nor v_2 has an active blue edge incident at it.*

Consider the following algorithm.

    while there are active squares do.

    if there is an active square s with two active edges, then color the two blank edges of s into different colors.

    else if there is an active square $s = (v_1, v_2, v_3, v_4)$ with one active edge $e = (v_1, v_2)$ (double or say, to fix the attention red), then check if there is another active square $s' = (v_1, v_2, v_3', v_4')$ that contains e but no other edge of s. Notice that since H does not contain triangles, s' cannot contain v_3 or v_4. Color (v_2, v_3) and (v_1, v_4) into red and (v_3, v_4) into blue. If s' exists and (v_1, v_2) is double color (v_2, v_3') and (v_1, v_4') into blue and (v_3', v_4') into red (otherwise do nothing, as it is not needed).

    else if there is an active square s with all blank edges, then color the edges of s alernately into blue and red. Next if the red edge e_2 belongs to an active square s_1, color the edges of s_1 adjacent to e_2 into red and the remaining one into blue. Next if the blue edge of s: e_1 or e_3 belongs to an active square, color it analogously: two edges into blue and one into red. Notice that if e_1 or e_3 belongs to an active square then it is not adjacent on the blue edge of s_1, because the graph does not contain triangles. Next, do the same with the remaining colored edges of s, if they belong to an active square.

6 Partition

In this phase we will give the algorithm that finds five disjoint subsets of edges $R_1, R_2, B_1, B_2, Blank$ corresponding to the five subphases: two red ones, two blue ones and one blank one such that after removing the edges from each one of these and coloring the blank edges, depending on the subphase: red (in the red subphases) or blue (in the blue subphases), the graph will contain only blue or red paths, that is will be 2-path-colored.

After disabling cycles of length less than 5, we arbitrarily color the rest of the edges, as a result in the graph no two adjacent blank edges will be left. Thus if we have a blank edge e between vertices v_1 and v_2, then the remaining three edges of v_1 are coloured: two into blue and one into red and the remaining three edges of $v2$ are coloured: two into red and one into blue or vice versa. We will say that the blue edges of v_1 and the red edges of v_2 are the blue or red *heads* of the edge e. We will also say that a red edge of v_1 is the red tail of e and a blue edge of v_2 is a blue tail of e. Let us notice that if we would like to colour a given blank edge blue or red, then we have to remove one blue or correspondingly red head. From the point of view of a given blank edge e the situation presents itself as follows. In the red subphases it is coloured red and one of its heads is removed in one red subphase and the other head is removed in the second subphase, thus one of its heads must belong to R_1 and the other one to R_2 and analogously in the blue subphases. In the blank subphase e is simply removed. Therefore we can see that a blank edge and its four heads fall into five different sets $R_1, R_2, B_1, B_2, Blank$. We will call all the blank edges and their heads *charged (edges)*. In the red and blue subphases we must be careful not to create cycles that consist solely of charged edges. We are not allowed to create such cycles, because we cannot afford to remove any edge from this cycle, as all of them must belong to the sets attached to other subphases (for example, $R_2, Blank$ if we are now in the first red subphase).

Lemma 7. *If the (blue or red) cycle c consists only of charged edges, then for every blank edge belonging to c we have that its head (the one that belongs to c) is a tail of another blank edge belonging to c.*

Proof. Suppose that c contains k (originally) blank edges. First let us notice that $k > 1$. Since c consists only of charged edges, all the edges between two consecutive blank edges on c must be charged. There are exactly k disjoint nonempty subsets of edges connecting the k blank edges on the cycle. Let us fix one direction of movement along c, say clockwise. Then each blank edge is either followed or preceded by its head. Let us observe, that if c consists only of charged edges we must have that either each blank edge is followed by its head or each blank edge is preceded by its head, because otherwise one set of edges connecting certain two blank edges would not contain a charged edge. Since all the edges must be charged, the sets connecting the blank edges must contain one edge each.   □

Due to space limits the description of the algorithm is omitted.

Acknowledgements. The first and second author would like to thank Łukasz Kowalik for many helpful discussions.

References

1. Barvinok, A., Gimadi, E.K., Serdyukov, A.I.: The maximun traveling salesman problem. In: Gutin, G., Punnen, A. (eds.) The Traveling Salesman Problem and its variations, pp. 585–607. Kluwer, Dordrecht (2002)
2. Bhatia, R.: Private communication
3. Bläser, M., Ram, L.S., Sviridenko, M.: Improved Approximation Algorithms for Metric Maximum ATSP and Maximum 3-Cycle Cover Problems. In: Dehne, F., López-Ortiz, A., Sack, J.-R. (eds.) WADS 2005. LNCS, vol. 3608, pp. 350–359. Springer, Heidelberg (2005)
4. Chen, Z.-Z., Okamoto, Y., Wang, L.: Improved deterministic approximation algorithms for Max TSP. Information Processing Letters 95, 333–342 (2005)
5. Chen, Z.-Z., Nagoya, T.: Improved approximation algorithms for metric Max TSP. J.Comb. Optim. 13, 321–336 (2007)
6. Engebretsen, L.: An explicit lower bound for TSP with distances one and two. Algorithmica 35(4), 301–319 (2003)
7. Engebretsen, L., Karpinski, M.: Approximation hardness of TSP with bounded metrics. In: Orejas, F., Spirakis, P.G., van Leeuwen, J. (eds.) ICALP 2001. LNCS, vol. 2076, pp. 201–212. Springer, Heidelberg (2001)
8. Fisher, M.L., Nemhauser, G.L., Wolsey, L.A.: An analysis of approximation for finding a maximum weight Hamiltonian circuit. Oper. Res. 27, 799–809 (1979)
9. Hassin, R., Rubinstein, S.: Better Approximations for Max TSP. Information Processing Letters 75, 181–186 (2000)
10. Hassin, R., Rubinstein, S.: A 7/8-approximation algorithm for metric Max TSP. Information Processing Letters 81(5), 247–251 (2002)
11. Kaplan, H., Lewenstein, M., Shafrir, N., Sviridenko, M.: Approximation Algorithms for Asymmetric TSP by Decomposing Directed Regualar Multigraphs. J. ACM 52(4), 602–626 (2005)
12. Kosaraju, S.R., Park, J.K., Stein, C.: Long tours and short superstrings. In: Proc. 35th Annual Symposium on Foundations of Computer Science (FOCS), pp. 166–177 (1994)
13. Kostochka, A.V., Serdyukov, A.I.: Polynomial algorithms with the estimates $\frac{3}{4}$ and $\frac{5}{6}$ for the traveling salesman problem of he maximum (in Russian). Upravlyaemye Sistemy 26, 55–59 (1985)
14. Kowalik, Ł., Mucha, M.: 35/44-Approximation for Asymmetric Maximum TSP with Triangle Inequality. In: Dehne, F., Sack, J.-R., Zeh, N. (eds.) WADS 2007. LNCS, vol. 4619, pp. 589–600. Springer, Heidelberg (2007)
15. Kowalik, Ł., Mucha, M.: Deterministic 7/8-approximation for the metric maximum TSP. In: Goel, A., Jansen, K., Rolim, J.D.P., Rubinfeld, R. (eds.) APPROX and RANDOM 2008. LNCS, vol. 5171, pp. 132–145. Springer, Heidelberg (2008)
16. Papadimitriou, C.H., Yannakakis, M.: The traveling salesman problem with distances one and two. Mathematics of Operations Research 18(1), 1–11 (1993)
17. Schrijver, A.: Nonbipartite Matching and Covering. In: Combinatorial Optimization, vol. A, pp. 520–561. Springer, Heidelberg (2003)
18. Serdyukov, A.I.: An Algorithm with an Estimate for the Traveling Salesman Problem of Maximum (in Russian). Upravlyaemye Sistemy 25, 80–86 (1984)

Approximation Algorithms for Domatic Partitions of Unit Disk Graphs

Saurav Pandit, Sriram V. Pemmaraju, and Kasturi Varadarajan[*]

Department of Computer Science,
The University of Iowa,
Iowa City, IA 52242-1419, USA
{spandit,sriram,kvaradar}@cs.uiowa.edu

Abstract. We prove a new structural property regarding the "skyline" of uniform radius disks and use this to derive a number of new sequential and distributed approximation algorithms for well-known optimization problems on unit disk graphs (UDGs). Specifically, the paper presents new approximation algorithms for two problems: *domatic partition* and *weighted minimum dominating set (WMDS)* on UDGs, both of which are of significant interest to the distributed computing community because of applications to energy conservation in wireless networks. Using the aforementioned skyline property, we derive the first constant-factor approximation algorithm for the domatic partition problem on UDGs. Prior to our work, the best approximation factor for this problem was $O(\log n)$, obtained by simply using the approximation algorithm for general graphs. From the domatic partition algorithm, we derive a new and simpler constant-factor approximation for WMDS on UDGs. Because of "locality" properties that our algorithms possess, both algorithms have relatively simple *constant-round* distributed implementations in the $\mathcal{LOCAL}$ model, where there is no bound on the message size. In addition, we obtain $O(\log^2 n)$-round distributed implementations of these algorithms in the $\mathcal{CONGEST}$ model, where message sizes are bounded above by $O(\log n)$ bits per message.

1 Introduction

We prove a new structural property regarding the "skyline" of uniform radius disks and use this to derive a number of new sequential and distributed approximation algorithms for well-known optimization problems on unit disk graphs (UDGs). Using the aforementioned skyline property, we derive the first constant-factor approximation algorithm for the *domatic partition* problem on UDGs. A number of researchers [4,5,6,11,15] have used this or related problems as an abstraction for the problem of deriving efficient *sleep schedules* in wireless networks. Prior to our work, the best approximation factor for this problem was $O(\log n)$, obtained by simply using the approximation algorithm for domatic partition

---

[*] Part of the work by this author was done when visiting The Institute of Mathematical Sciences, Chennai. He was also supported by NSF CAREER award CCR 0237431.

I. Dinur et al. (Eds.): APPROX and RANDOM 2009, LNCS 5687, pp. 312–325, 2009.

on general graphs [8]. We also derive here a distributed version of the domatic partition algorithm that can be implemented in $O(\log^2 n)$ rounds of communication in the $\mathcal{CONGEST}$ model; in this model all message sizes are bounded by $O(\log n)$ bits. Subsequently, we use the domatic partition algorithm to obtain a new and simple constant-factor approximation algorithm for the *weighted minimum dominating set* (WMDS) problem on UDGs. Our result also shows that the standard LP-relaxation for WMDS has constant integrality gap. Unlike the minimum dominating set (MDS) problem on UDGs, WMDS on UDGs has proved quite hard and only recently Ambuhl et al. [2] and subsequently Huang et al. [9] have presented the first constant-factor approximation algorithms for the the WMDS problem on UDGs. These WMDS algorithms [9,2] have an easy constant-round distributed implementation in the $\mathcal{LOCAL}$ model of distributed computation [13], in which there is no bound on message sizes. However these algorithms seem to require a huge amount of information exchange and it is not clear if they can be implemented in a sublinear number of rounds, in a model such as the $\mathcal{CONGEST}$ model in which message sizes are bounded. Our new algorithm for WMDS attains a constant-factor approximation in $O(\log^2 n)$ rounds in the $\mathcal{CONGEST}$ model. In [13], a reduction from the *facility location* problem on UDGs to WMDS on UDGs is presented. Our new WMDS algorithm, along with this reduction implies an $O(\log^2 n)$-round, constant-factor approximation algorithm in the $\mathcal{CONGEST}$ model for facility location on UDGs.

Domatic Partition. A standard approach for reducing energy consumption in wireless networks is to keep only a small fraction of nodes active (for sensing, communicating, etc.) at any time and put the rest of the nodes to sleep, thereby conserving energy. The problem of maximizing the number of nodes that are asleep at any given time while maintaining sufficient activity in the network is usually modeled as the problem of finding a small *dominating set* in the network. Once a small dominating set is found, the nodes in the dominating set collectively act as "coordinators" for the network and the rest of the nodes go to sleep. To maximize the lifetime of the network it is critical that the role of coordinators be rotated among the nodes in the network, so that every node gets a chance to sleep. Moscibroda and Wattenhofer [11] have abstracted the problem of rotating the responsibility of being a coordinator as the *domatic partition problem*. Given a graph $G = (V, E)$, a *dominating set* $D \subseteq V$ of G is a vertex-subset such that each vertex is either in D or has a neighbor in D. A *domatic partition* is a partition $\mathcal{D} = \{D_1, D_2, \ldots, D_t\}$ of V such that each block D_i of $\mathcal{D}$ is a dominating set of G. The *domatic partition problem* seeks a domatic partition $\mathcal{D}$ of largest cardinality. To understand the motivation, suppose that $\mathcal{D} = \{D_1, D_2, \ldots, D_t\}$ is a domatic partition of G. Then a simple schedule for the nodes would be for the nodes in D_1 to be active for some fixed period of time T, during which the rest of the nodes are asleep, followed by a period of time T in which nodes in D_2 are active, while the rest of the nodes are asleep, and so on. Such a schedule would imply that in the long run, each node is active for roughly $1/t$ of the time. Therefore maximizing t leads to minimizing this fraction, thereby maximizing the fraction of time nodes are asleep. Thus far, to solve this problem, researchers in the wireless networks community have either used heuristics [6], the $O(\log n)$-approximation that works for general graphs [11], or have settled for a fractional solution obtained by solving the LP-relaxation of the problem [4]. This paper shows how to obtain a constant-factor approximation to domatic partition on UDGs. A UDG is specified by giving the coordinates of its vertices, which are points in the plane; there is an edge between every pair of points whose Eucildean distance is at most one. We also show how to implement a distributed version of this algorithm in polylogarithmic rounds of communication and with small messages, i.e., in the $\mathcal{CONGEST}$ model.

Our algorithm in fact shows that in a UDG, we can obtain a domatic partition with cardinality proportional to the minimum degree. This combinatorial geometric result can be viewed as solving a special case of the problem of decomposing multiple coverings [12,1] with unit disks. The general problem is as follows. We have a set $\mathcal{D}$ of unit disks in the plane, and a set P of points so that each point in P is contained in at least k disks from $\mathcal{D}$. Can we partition $\mathcal{D}$ into $\Omega(k)$ sets each of which covers P? Our result on domatic partitions gives an affirmative answer for the special case when P equals the set of centers of $\mathcal{D}$. The general problem however remains open and has eluded a solution for several years now. Even for showing that for a sufficiently large k we can partition $\mathcal{D}$ into two covers of P, there is only an old, unpublished manuscript. We refer the reader to Pach and Toth [12].

Weighted Minimum Dominating Set. The input to the *weighted minimum dominating set* (WMDS) problem consists of a vertex-weighted graph $G = (V, E)$, with each vertex v assigned a non-negative weight $w(v)$. The problem seeks a dominating set D of G of minimum total weight. On UDGs, the minimum dominating set problem (i.e., the unit-weight version of WMDS) is easy to approximate since any maximal independent set is a constant-factor approximation of the MDS. However, the WMDS problem on UDGs seems fundamentally more difficult relative to the MDS problem since an optimal solution to WMDS can be arbitrarily dense (see Figure 1). The WMDS problem on UDGs did not have a constant-factor approximation until recently. Ambuhl et al. [2] presented the first constant-factor approximation, which was improved by Huang et al. [9]. Both of these WMDS algorithms are inherently "local" in the sense that a constant-factor approximation to the problem can be obtained by separately solving subproblems induced by diameter-1 square cells and then simply "unioning" the solutions. Obtaining a constant-factor approximation for the subproblems is enough to guarantee a constant-factor approximation for the original problem. If messages sizes are allowed to be unbounded, then the above locality property immediately implies that the above WMDS algorithms can be implemented in the $\mathcal{LOCAL}$ model in a constant number of rounds. However, these algorithms are not "lightweight" and make use of techniques such as dynamic programming, and making polynomially many solution guesses, etc. It is not clear that these techniques can be efficiently implemented in a distributed model (such as the $\mathcal{CONGEST}$ model) that places a restriction on the size of messages. Our paper presents a radically different constant-factor approximation algorithm for

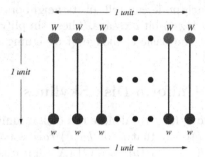

Fig. 1. Arrange nodes in the plane so that the induced UDG contains a clique, formed by the top nodes, a clique formed by the bottom nodes, and a perfect matching between the top and bottom nodes. Assuming that there are n nodes at the top (and therefore n nodes at the bottom) and picking the node weights w and W so that $w < W/n$, we see that the minimum weight dominating set of this UDG consists of all the bottom nodes, whereas a solution to MDS consists of one top node and one bottom node.

WMDS on UDGs, that is much simpler than the algorithms of Ambuhl et al. [2] and Huang et al. [9]. Our algorithm easily follows from the domatic partition algorithm and its simplicity makes it possible to implement it in a distributed setting in the $\mathcal{CONGEST}$ model in $O(\log^2 n)$ rounds.

The $\mathcal{LOCAL}$ and $\mathcal{CONGEST}$ models. As defined by Peleg [14], the $\mathcal{LOCAL}$ model is a message-passing model in which nodes run synchronously and each node is allowed to send a message of unbounded size in each round. The focus of this model is on the inherent "locality" of the problem. As mentioned earlier, both problems considered in this paper have a "locality" property that makes it easy to design constant-round, constant-factor distributed algorithms in the $\mathcal{LOCAL}$ model, provided that the corresponding subproblems admit constant-factor approximation algorithms. In this paper, we present distributed algorithms that run efficiently in the $\mathcal{CONGEST}$ model; this model differs from the $\mathcal{LOCAL}$ model only in that each node is only allowed to send a message of size $O(\log n)$ bits in each round. The $\mathcal{CONGEST}$ model is significantly more stringent than the $\mathcal{LOCAL}$ model, and "congestion" is an additional constraint that has to be dealt with. From a distributed computing point of view, this paper's major contribution is to show that the problems are not just local, but can be solved efficiently with limited amount of information exchange.

Since our problems and algorithms are motivated by the wireless networks setting, we assume that in each round each node v sends a single message via local broadcast that all neighbors of v can "hear" (i.e., receive) in that round. We also assume that each node has a unique ID and knows its coordinates in some globally consistent coordinate system. Furthermore, we assume that each "piece" of information that a node initially possesses can fit in $\log n$ bits. For example, each node v running our WMDS algorithm uses $\log n$ bits for ID_v, $\log n$ bits for each of its coordinates x_v and y_v, and $\log n$ bits for its weight w_v. This assumption basically means that each node can tell all its neighbors all about itself in one round, using a single $O(\log n)$-bit message. These simplifying assumptions serve to clarify and highlight the basic challenge of designing efficient algorithms in the $\mathcal{CONGEST}$ model.

2   Properties of Uniform Disk Skylines

For a point p in the plane, let $D(p)$ denote the disk of unit radius centered at p. It will be convenient to to say that p (or $D(p)$) *covers* any point $q \in D(p)$. For a set Q of points, let $D(Q) = \{D(p) \mid p \in Q\}$. A point a is *covered* by $D(Q)$ if a is covered by some disk in $D(Q)$. For a point q and a finite set Q of points, let $C_Q(q) = Q \cap D(q)$; this is the set of all points in Q covered by q.

Suppose that A is a set of points strictly below that x-axis and B a set of points strictly above the x-axis, such that each point in B covers at least one point in A. A point $p = (p.x, p.y)$ with $p.y \geq 0$ is said to belong to the *skyline* of $D(A)$ if p is covered by $D(A)$ and any $p' = (p'.x, p'.y)$ with $p'.x = p.x$ and $p'.y > p.y$ is not covered by $D(A)$. We say that a point $a \in A$ *contributes to the skyline* of $D(A)$ if $D(a)$ contains some point in the skyline of $D(A)$. Let $A' \subseteq A$ denote the set of points that contribute to the skyline of $D(A)$. See Figure 2 for an illustration of these definitions. It is easy to check that A' does not contain two points with the same x-coordinate. Thus, we think of A' as being ordered according to increasing x-coordinate. The set A' has two nice properties that are encapsulated in the lemmas below. The first of these is similar to observations used by Călinescu et al. [7]. These two lemmas will play a crucial role here.

Lemma 1. *For any $b \in B$, the set $C_{A'}(b)$ is a non-empty contiguous subset of the ordered set A'.*

Lemma 2. *Suppose that B has the property that for any two distinct elements b_1 and b_2 of B, $C_A(b_1)$ is not a subset of $C_A(b_2)$. If $C_{A'}(b_1) \cap C_{A'}(b_2)$ is non-empty for distinct elements b_1 and b_2 of B, then $C_{A'}(b_2)$ contains either the leftmost element or the rightmost element of $C_{A'}(b_1)$.*

We now prove Lemmas 1 and 2. Let a and a' be any two points on or below that x-axis such that $D(a)$ contains a point above the x-axis not in $D(a')$ and vice versa. We assume a is to the left of a'. If the boundaries of $D(a)$ and $D(a')$ intersect above the x-axis, let $\ell_{a,a'}$ be the vertical line through this intersection point. Otherwise, let $\ell_{a,a'}$ be any vertical line so that points above the x-axis within $D(a)$ lie to the left of the line and the points above the x-axis within $D(a')$ lie to the right of the line. See Figure 2 for an illustration. It is easy to observe that $D(a')$ does not contain any point on the skyline of $D(\{a, a'\})$ to the left of $\ell_{a,a'}$ and $D(a)$ does not contain any point on the skyline of $D(\{a, a'\})$ to the right of $\ell_{a,a'}$.

Lemma 3. *Let a_1, a_2, a_3 be three points below the x-axis such that $a_1.x \leq a_2.x \leq a_3.x$. Suppose there is a point b above the x-axis that is contained in $D(a_1)$ and $D(a_3)$ but not in $D(a_2)$. Then a_2 does not contribute to the skyline of $D(\{a_1, a_2, a_3\})$.*

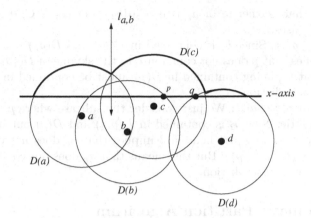

Fig. 2. Let $A = \{a, b, c, d\}$. The skyline of $D(A)$ is shown by darkened arcs above the x-axis. Point b does not contribute to the skyline and therefore $A' = \{a, c, d\}$. Disks $D(a)$ and $D(c)$ intersect above the x-axis and the line $\ell_{a,b}$ passes through this point of intersection. Disks $D(b)$ and $D(d)$ intersect below the x-axis and $\ell_{b,d}$ is any vertical line that intersects the x-axis between points p and q.

Proof. If $a_1.x = a_2.x$, then it must be that a_1 is above a_2, and in this case a_2 does not contribute to the skyline of $D(\{a_1, a_2\})$ and hence it does not contribute to the skyline of of $D(\{a_1, a_2, a_3\})$. The case where $a_2.x = a_3.x$ is handled similarly. So let us now look at the case where $a_1.x < a_2.x < a_3.x$. Let us consider the non-trivial situation where $D(a_2)$ contains some point above the x-axis not in $D(a_1)$ (resp. $D(a_3)$).

Since b is in $D(a_1)$ but not in $D(a_2)$, it is easy to see that b lies strictly to the left of ℓ_{a_1, a_2}. It follows that a_2 does not contain any point on the skyline of $D(\{a_1, a_2\})$ that is on or to the left of the vertical line through b. Similarly, we argue that a_2 does not contain any point on the skyline of $D(\{a_2, a_3\})$ that is on or to the right of the vertical line through b. The lemma follows. □

Proof of Lemma 1. It is clear that $C_{A'}(b)$ is non-empty. For consider the point p that is on the skyline of $D(A)$ and has the same x-coordinate as b. Let $a \in A'$ be such that $D(a)$ contains p. It is easy to see that $D(a)$ contains b as well.

To show that $C_{A'}(b)$ is a contiguous subsequence of A', suppose for a contradiction that $a_1, a_2, a_3 \in A$ are such that $a_1.x < a_2.x < a_3.x$ and $a_1, a_3 \in C_{A'}(b)$ but $a_2 \notin C_{A'}(b)$. By Lemma 3, we conclude that a_2 does not contribute to the skyline of $D(\{a_1, a_2, a_3\})$. Thus a_2 does not contribute to the skyline of $D(A)$, a contradiction. □

Proof of Lemma 2. The lemma follows easily when $C_{A'}(b_2)$ is not contained in $C_{A'}(b_1)$ because $C_{A'}(b_2)$ and $C_{A'}(b_1)$ are contiguous subsequences of A'.

Suppose $C_{A'}(b_2) \subseteq C_{A'}(b_1)$. By assumption, $C_A(b_2)$ has a point $p \in A$ that is not in $C_A(b_1)$. Let a_l and a_r be the leftmost and rightmost points of $C_{A'}(b_1)$. Consider the nontrivial case where a_l and a_r are distinct. Suppose for a contradiction that neither a_l or a_r is in $C_{A'}(b_2)$. Let $a_m \in C_{A'}(b_2) \cap C_{A'}(b_1)$. There are two cases.

$a_l.x \le p.x \le a_r.x$: Since b_1 is contained in $D(a_l)$ and $D(a_r)$ but not in $D(p)$, Lemma 3 implies that p does not contribute to the skyline of $D(\{a_l, p, a_r\})$. But this means that b_2, being contained in $D(p)$, must be contained in either $D(a_l)$ or $D(a_r)$, a contradiction.

$p.x > a_r.x$ or $p.x < a_l.x$: We just consider the subcase where $p.x > a_r.x$; the other is symmetric. Now b_2 is contained in $D(a_m)$ and $D(p)$ but not in $D(a_r)$. We have $a_m.x < a_r.x < p.x$. Lemma 3 implies that a_r does not contribute to the skyline of $D(\{a_m, a_r, p\})$. But this means that a_r does not contribute to the skyline of $D(A)$, a contradiction. $\qquad\square$

3   The Domatic Partition Algorithm

Here we address the domatic partition problem for a unit disk graph G induced by a set $P = \{p_1, \ldots, p_n\}$ of points in the plane. In such a graph, there is a vertex corresponding to each point p_i, and an edge corresponding to each pair (p_i, p_j) if and only if the Euclidean distance between p_i and p_j is at most 1.

It is convenient to treat the domatic partition as a special case of the *disk cover packing problem*, which we define below. An instance of the disk cover packing problem is an ordered pair (A, B) where A and B are finite subsets of the plane. We wish to find subsets $A_1, \ldots, A_\tau$ of A such that (1) $A_i \cap A_j = \emptyset$ for $i \ne j$, and (3) each $D(A_i)$ covers B, that is, each $b \in B$ is contained in some disk in $D(A_i)$. The *size* τ of the disk cover packing $\{A_1, \ldots, A_\tau\}$ is the quantity we wish to maximize. Clearly, the domatic partition problem for a unit disk graph induced by a set P of points in the plane is equivalent to the disk cover packing problem for the instance (P, P).

Define the *load* $L(A, B)$ of an instance (A, B) of the disk cover problem to be $\min_{b \in B} |C_A(b)|$. Evidently, $L(A, B)$ is an upper bound on the size of any disk cover packing for (A, B). For the domatic partition problem, on a unit disk graph G induced by a set P of points, the load $L(P, P)$ is just (one plus) the minimum degree of a vertex in G. Our constant-factor approximation is obtained by showing how to construct a domatic partition of size at least $L(P, P)/C$ for some constant C.

3.1   The First Reduction

We now reduce an instance (P, P) of the disk cover packing problem to instances (A^σ, B^σ) where (1) A^σ is a set of points lying inside a square σ of diameter 1, (2) B^σ is a set of points lying outside σ, and (3) $L(A^\sigma, B^\sigma) \ge L(P, P)/25$.

In this subsection, let L denote $L(P, P)$. Consider a square grid of side length $1/\sqrt{2}$, which subdivides the plane into square grid cells of diameter 1. By ensuring that the grid lines do not contain any point in P, each point in P will belong to a unique grid cell. Let us call a grid cell σ *heavy* if it contains at least $L/25$

points of P; we call σ *light* otherwise. Let A^σ denote the set of points in P that lie in grid cell σ. Let S denote the set of heavy grid cells. For each point $p \in P$ that lies in a light grid cell, there is a heavy cell σ such that $C_{A^\sigma}(p) \geq L/25$; pick one such cell σ and assign p to σ.[1]

For each $\sigma \in S$, we obtain an instance of the disk cover problem $(A^\sigma, A^\sigma \cup B^\sigma)$ where B^σ is the set of points assigned to σ. Notice that $|B^\sigma| \leq L$, $|A^\sigma| \geq L/25$, and $L(A^\sigma, A^\sigma \cup B^\sigma) \geq L/25$. Observe that if we have a disk cover packing of size t for each instance $(A^\sigma, A^\sigma \cup B^\sigma)$, then we can combine them in a straightforward way to obtain a disk cover packing of size t for instance (P, P).

Consider an instance $(A^\sigma, A^\sigma \cup B^\sigma)$. If $B^\sigma = \emptyset$, the set $\{\{a\} \mid a \in A^\sigma\}$ is a disk cover packing of size $|A^\sigma| = L(A^\sigma, A^\sigma \cup B^\sigma)$, since σ has diameter 1. If $B^\sigma \neq \emptyset$, it is easy to see that any disk cover packing for instance (A^σ, B^σ) is also a disk cover packing for instance $(A^\sigma, A^\sigma \cup B^\sigma)$, and $L(A^\sigma, B^\sigma) = L(A^\sigma, A^\sigma \cup B^\sigma)$.

3.2   The Second Reduction

We now consider an instance of (A, B) of the disk cover packing problem where A is a set of points inside a square σ and B is a set of points outside the square σ. In this subsection let L denote $L(A, B)$; we'll assume that $|B| \leq 25L$. We reduce this instance to four instances that are further specialized.

We first partition B into four sets B^e, B^n, B^w, B^s. A point $b \in B$ belongs to B^e (resp. B^n, B^w, B^s) if e is east (resp. north, west, south) of square σ; we break ties arbitrarily. We partition A into four sets A^e, A^n, A^w, A^s by independently throwing each $a \in A$ into one of the four sets uniformly at random.

Using the Chernoff bound and the union bound (observing that $|B| \leq 25L$), we can conclude that with high probability (probability smaller than 1 by a quantity exponentially small in L), we have

$$L(A^e, B^e), L(A^n, B^n), L(A^w, B^w), L(A^s, B^s) \geq L/8.$$

Furthermore, if we have disk cover packings of size at least t for each of the four instances (A^e, B^e), (A^n, B^n), (A^w, B^w), (A^s, B^s), then we can readily combine them to obtain a disk cover packing of size t for (A, B). Notice that for each instance (A^α, B^α), we have a line separating A^α and B^α.

3.3   The Separated Case

Consider an instance (A, B) of the disk cover packing problem where A is a set of points strictly below that x-axis and B a set of points strictly above the x-axis. In this section, we show that we can efficiently find a disk cover packing for this instance whose size is at least $L/4$, where now L denotes $L(A, B)$.

The DomPart *Algorithm.* Let E initially denote the set A. Let $j = 0$. We repeat the following steps as long as $D(E)$ covers B.

---

[1] We have not tried to optimize the constants being used. Here we use 25 as an upper bound on the maximum number of $\frac{1}{\sqrt{2}} \times \frac{1}{\sqrt{2}}$ grid cells that a unit disk can intersect.

1. $j \leftarrow j + 1$.
2. We repeatedly perform the following step on B till it is no longer applicable: if there exist distinct elements b_1 and b_2 in B so that $C_E(b_1) \subseteq C_E(b_2)$, we discard b_2 from B. Let B' denote the resulting B.
3. Let $A' \subseteq E$ be the points that contribute to the skyline of $D(E)$.
4. Compute a maximal (not necessarily maximum) subset $B'' \subseteq B'$ so that for any two distinct $b_1, b_2 \in B''$, $C_{A'}(b_1) \cap C_{A'}(b_2) = \emptyset$.
5. Construct A_j by adding to it the leftmost and rightmost points of $C_{A'}(b)$, for each $b \in B''$.
6. $E \leftarrow E \setminus A_j$.

Suppose that the algorithm terminates with $j = k$. It is clear that $A_1, \ldots, A_k$ are pairwise disjoint subsets of A. The first lemma below shows that each $D(A_j)$ covers B, and the second lemma shows that $k \geq L/4$.

Lemma 4. *For each* $1 \leq j \leq k$, $D(A_j)$ *covers* B.

Proof. The set $D(E)$ at the beginning of the j'th iteration covers B. By Lemma 1, the set $D(A')$ also covers B. Now for any $b \in B''$ we add the two endpoints of $C_{A'}(b)$ to A_j, so $D(A_j)$ covers b. Consider a $b' \in B' \setminus B''$. By the maximality of B'', there is a $b \in B''$ so that $C_{A'}(b) \cap C_{A'}(b') \neq \emptyset$. Since $C_{A'}(b)$ and $C_{A'}(b')$ are contiguous subsequences of A', $C_{A'}(b')$ must contain an endpoint of $C_{A'}(b)$ if $C_{A'}(b')$ is not a subset of $C_{A'}(b)$. On the other hand, if $C_{A'}(b') \subseteq C_{A'}(b)$, then Lemma 2 implies that $C_{A'}(b')$ contains an endpoint of $C_{A'}(b)$. We conclude that $D(A_j)$ covers b'.

Finally, consider a $b' \in B \setminus B'$. There is a $b \in B'$ such that $C_E(b) \subseteq C_E(b')$. Since A_j contains a point from $C_E(b)$ it also contains a point from $C_E(b')$.   □

Lemma 5. *At the end of the j-th iteration,* $|C_E(b)| \geq L - 4j$ *for each* $b \in B$.

Proof. Assume that the statement is true at the end of the $(j-1)$-th iteration. We let E^f (resp. E^s) denote the E at the end (resp. start) of the j-th iteration. Assume that $|C_{E^s}(b)| \geq L - 4(j-1)$ for each $b \in B$; we will argue that $|C_{E^f}(b)| \geq L - 4j$.

For any $b \in B''$, it is clear $|A_j \cap C_{E^s}(b)| = 2$, so $|C_{E^f}(b)| = |C_{E^s}(b)| - 2 > L - 4j$.

For any $b \in B' \setminus B''$, it is not hard to derive from Lemma 2 that $|A_j \cap C_{E^s}(b)| \leq 4$. So $|C_{E^f}(b)| \geq |C_{E^s}(b)| - 4 \geq L - 4j$.

For any $b \in B \setminus B'$, there is a $b' \in B'$ such that $C_{E^s}(b') \subseteq C_{E^s}(b)$. Thus $|C_{E^f}(b)| \geq |C_{E^f}(b')| \geq L - 4j$.   □

Corollary 1. *The number of covers k returned by the algorithm is at least $L/4$.*

Putting Sections 3.1, 3.2, and 3.3 together, we obtain:

Theorem 1. *Any unit disk graph admits a domatic partition whose cardinality is at least a constant fraction of the minimum degree, and hence at least a constant fraction of the optimal domatic partition. Such a domatic partition can be computed by a polynomial time sequential algorithm.*

3.4   Distributed Algorithm for Domatic Partition

Given that nodes are aware of their Euclidean coordinates, it is not too difficult to check that the reduction in Section 3.1 that decomposes the domatic partition problem into local subproblems can be implemented in a constant number of rounds in $\mathcal{CONGEST}$ model. The reduction in 3.2 can be implemented in $O(\log n)$ rounds in the $\mathcal{CONGEST}$ model to give the desired partition with high probability. In this section, we focus on obtaining a polylogarithmic round distributed implementation, DISTDOMPART, of the DOMPART algorithm in Section 3.3, thus obtaining a polylogarithmic round distributed algorithm for domatic partition in the $\mathcal{CONGEST}$ model. We first show that each iteration of the DOMPART algorithm can be executed in a constant number of communication rounds, yielding an $O(L(A,B))$-round algorithm. As in the DOMPART algorithm, let E denote A, initially. Let j be a variable local to each node, initially having the value 1.

1. Each node $a \in E$ broadcasts $(\text{ID}_a, (x_a, y_a))$. Since E is a clique, each node $a \in E$ now knows the IDs and coordinates of all other nodes in E.
2. Each node $b \in B$ broadcasts $(\text{ID}_b, (x_b, y_b))$.
3. Each node $a \in E$ determines, if it contributes to the skyline of $D(E)$; nodes that do make a contribution mark themselves as belonging to A'.
4. Each node $a \in A'$ computes $C_{A'}(b)$ for all $b \in N(a) \cap B$. Viewing each $C_{A'}(b)$ as an interval over the ordered set A', each node $a \in A'$ picks an inclusion-wise minimal interval I_a from $\{C_{A'}(b) \mid b \in N(a) \cap B\}$ and broadcasts $(\ell(I_a), r(I_a))$ (the IDs of the left and right endpoints of I_a).
5. Each node $a \in A'$ computes a maximal subset of disjoint intervals from $\{I_a \mid a \in A'\}$; if a is an endpoint of one of these chosen intervals, a marks itself as belonging to A_j and deletes itself from E. The remaining nodes in E increment the value of j.

It is worth noting that the property of each set $C_{A'}(b)$ being a contiguous subsequence of A' (Lemma 1) is quite critical in the above algorithm; it allows different nodes $a \in A'$ to exchange information about these sets using $O(\log n)$ size messages. The steps above correspond to one iteration of the DOMPART algorithm and can be repeated $L(A,B)/4$ times. Thus we have the following lemma.

Lemma 6. *There is a distributed algorithm, running in $O(L(A,B))$-rounds in the $\mathcal{CONGEST}$ model, that computes a disk cover packing of size $L(A,B)/4$ for an instance (A,B) where A and B are separated by a line.*

Let $n = |A \cup B|$. We now describe a distributed algorithm that computes an $\Omega(L(A,B))$-size domatic partition, while running in $O(\log^2 n)$ rounds in the $\mathcal{CONGEST}$ model. The key observation is that if A were partitioned into sets A^1 and A^2, we could run the above algorithm "in parallel" on the two sets. This motivates the idea of first partitioning A "equitably" and then run the DISTDOM-PART algorithm "in parallel" on each block of the partition. If $L(A,B) \leq c \cdot \log^2 n$ for some constant c, we can run the above algorithm as is. Otherwise, let T equal $\lceil L(A,B)/\log n \rceil$ and color each node $a \in A$ with a color $x \in \{1, 2, \ldots, T\}$ chosen uniformly at random. This partitions the set A into color classes $A^1, A^2, \ldots, A^T$ and the following lemma follows from an application of Chernoff bounds.

Lemma 7. *With probability at least* $1 - \frac{1}{n}$, $L(A^i, B) \geq \alpha \log n$, *for all* i, *and for some positive constant* α.

It is easy to check that the DISTDOMPART can be run independently and in parallel for each of the A^i's. Running $\alpha \log n / 4$ iterations of this algorithm yields, with probability at least $1 - \frac{1}{n}$, a disk cover packing of size at least

$$\frac{T \cdot \alpha \log n}{4} \geq \frac{\alpha}{4} \cdot L(A, B).$$

We therefore obtain:

Theorem 2. *There is a distributed algorithm that runs in* $O(\log^2 n)$ *rounds in the* $\mathcal{CONGEST}$ *model to produce a domatic partition that is a constant-factor approximation of the largest possible domatic partition in a given unit disk graph.*

The non-uniform case. Moscibroda and Wattenhofer [11] view the domatic partition problem as an instance of the problem of maximizing the lifetime of a wireless ad hoc network whose nodes are *uniform*. In the non-uniform version of the problem, each node i comes with an associated value $b(i) \in \mathbb{Z}^+$ that represents an upper bound on the number of dominating sets that i can belong to. The value $b(i)$ models the initial battery supply at node i and in the domatic partition problem we simply assumed that $b(i) = 1$ for all i. Using the algorithm described in the previous section, one can also obtain a constant-factor approximation for the problem of finding a maximum number of dominating sets such that each node i appears in at most $b(i)$ dominating sets.

4   Weighted Minimum Dominating Set

In this section we first present a new, sequential, constant-factor approximation algorithm for WMDS that uses the domatic partition algorithm presented in Section 3. Our analysis also shows that the integrality gap of the standard WMDS LP-relaxation is bounded above by a constant for UDGs. Subsequently, we show how this algorithm can be implemented in the distributed $\mathcal{CONGEST}$ model in $O(\log^2 n)$ rounds.

Let $G = (V, E)$ be a given UDG. Recall our assumption that we are given a geometric representation of G; so we assume that V is a set of points in the plane. For each $j \in V$, let $D(j)$ denote the unit disk centered at j. The WMDS LP-relaxation we use is

$$\min \sum_i w_i \cdot x_i \tag{1}$$

$$\sum_{i \in D(j)} x_i \geq 1 \quad \text{for all } j \in V$$

$$x_i \geq 0 \quad \text{for all } i \in V$$

Let $\{x_j^* \mid j \in V\}$ be an optimal solution to the above LP. For any j such that $x_j^* < 1/2n$, round x_j^* down to 0 and for any j such that $x_j^* \geq 1/2n$, round x_j^* up to the nearest $k/2n$ for positive integer k. Let this new "discretized" solution be denoted $\{\bar{x}_j \mid j \in V\}$. Two properties of this solution are worth noting.

Lemma 8. *(i)* $\sum_{i \in D(j)} \bar{x}_i \geq 1/2$ *and (ii)* $\sum_{i \in V} w_i \cdot \bar{x}_i \leq 2 \cdot OPT$, *where OPT is the weight of an optimal dominating set.*

Proof. (i) follows from the fact that the maximum decrease in $\sum_{i \in D(j)} x_i^*$ due to rounding down of x_i^*-values to 0 is less than $n \cdot \frac{1}{2n} = \frac{1}{2}$. (ii) follows from the fact that $\bar{x}_i \leq 2 \cdot x_i^*$, for all $i \in V$. Therefore,

$$\sum_{i \in V} w_i \cdot \bar{x}_i \leq 2 \cdot \sum_{i \in V} w_i \cdot x_i^* \leq 2 \cdot OPT. \qquad \square$$

Now construct a new set P of points, by making, for each $j \in V$, $2n \cdot \bar{x}_j$ copies of vertex j. Suppose that each point inherits the weight of the vertex that it is a copy of. By Lemma 8, the total weight of all the points in P, $w(P) = 2n \cdot \sum_{j \in V} w_j \cdot \bar{x}_j \leq 4n \cdot OPT$. The following lemma shows a lower bound on the load of the instance (P, V) of the disk cover packing problem.

Lemma 9. $L(P, V) \geq n$.

Proof. For all $j \in V$, the number of points $i \in D(j)$ is $2n \cdot \sum_{i \in D(j)} \bar{x}_i$. By Lemma 8, $\sum_{i \in D(j)} \bar{x}_i \geq 1/2$. The lemma follows. $\qquad \square$

We now subdivide the plane with a square grid of diameter 1, and call a grid cell σ *heavy* if the set $P^\sigma \subseteq P$ of points contained in σ has at least $n/25$ elements. Assign each point $j \in V$ to a heavy cell σ so that $|P^\sigma \cap D(j)| \geq n/25$. Let $V^\sigma \subseteq V$ be the points assigned to σ. Partition V^σ further into four sets V_e^σ, V_w^σ, V_n^σ, and V_s^σ that lie to the east, west, north, and south of σ respectively, plus an additional set consisting of the points in σ. Notice that $L(P^\sigma, V_\alpha^\sigma) \geq n/25$ for each $\alpha \in \{e, w, n, s\}$.

Hence, by executing the DomPart algorithm described in Section 3.3, we get a disk cover packing $(A_1, A_2, \ldots, A_t)$ for the instance $(P^\sigma, V_\alpha^\sigma)$ with $t \geq n/C$, for some constant C. Since $\sum_{i=1}^{t} w(A_i) = w(P^\sigma)$, there exists an A_i such that

$$w(A_i) \leq \frac{w(P^\sigma)}{t} \leq \frac{C \cdot w(P^\sigma)}{n}.$$

This A_i covers not only V_α^σ but also the points in V^σ that lie in σ. The union of such covers over all instances $(P^\sigma, V_\alpha^\sigma)$, where σ ranges over the heavy cells and α over $\{e, w, n, s\}$, is a cover (or dominating set) for V. Its weight is bounded by

$$\sum_\sigma \frac{4C \cdot w(P^\sigma)}{n} \leq \frac{4C \cdot w(P)}{n} \leq \frac{16Cn \cdot OPT}{n} \leq 16C \cdot OPT.$$

The algorithm implied by this analysis is the following.

The WMDS Algorithm

1. Solve the WMDS LP relaxation (1) to obtain a solution $\{x_i^* \mid i \in V\}$.
2. For each $i \in V$, set $\bar{x}_i = 0$ if $x_i^* < 1/2n$; otherwise, set $\bar{x}_i$ to $\min\{\frac{k}{2n} \mid k \in \mathbb{Z}^+, \frac{k}{2n} \geq x_i^*\}$.

3. Create a set of points P by making $2n \cdot \overline{x}_i$ copies of each $i \in V$.
4. Partition into disk cover packing instances $(P^\sigma, V_\alpha^\sigma)$ as above.
5. Computing a disk cover packing of $(P^\sigma, V_\alpha^\sigma)$ using the DOMPART algorithm of Section 3.3.
6. Output a disk cover in the packing that has smallest weight, and return the union over all instances $(P^\sigma, V_\alpha^\sigma)$.

A distributed implementation of this algorithm in $O(\log^2 n)$ rounds of communication in the $\mathcal{CONGEST}$ model can be obtained as follows. A constant-factor approximation to the WMDS LP relaxation can be obtained in $O(\log^2 n)$ rounds in the $\mathcal{CONGEST}$ model using the algorithms in [3,10]. The next three steps can be done locally at each node. The disk cover packing of $(P^\sigma, V_\alpha^\sigma)$ can be computed in $O(\log^2 n)$ rounds in the $\mathcal{CONGEST}$ model using the DISTDOM-PART algorithm in Section 3.4. Picking a cover of minimum weight from such a packing is a "local" task and takes a constant number of communication rounds in the $\mathcal{CONGEST}$ model.

Theorem 3. *There is a distributed algorithm that runs in $O(\log^2 n)$ rounds in the $\mathcal{CONGEST}$ model and produces a constant-factor approximation to the minimum-weight dominating set in a unit disk graph.*

References

1. Aloupis, G., Cardinal, J., Collette, S., Langerman, S., Orden, D., Ramos, P.: Decomposition of multiple coverings into more parts. In: Proceedings of SODA (2009)
2. Ambühl, C., Erlebach, T., Mihalák, M., Nunkesser, M.: Constant-factor approximation for minimum-weight (Connected) dominating sets in unit disk graphs. In: Díaz, J., Jansen, K., Rolim, J.D.P., Zwick, U. (eds.) APPROX-RANDOM 2006. LNCS, vol. 4110, pp. 3–14. Springer, Heidelberg (2006)
3. Bartal, Y., Byers, J.W., Raz, D.: Global optimization using local information with applications to flow control. In: FOCS 1997: Proceedings of the 38th Annual Symposium on Foundations of Computer Science, Washington, DC, USA, 1997, p. 303. IEEE Computer Society Press, Los Alamitos (1997)
4. Berman, P., Calinescu, G., Shah, C., Zelikovsky, A.: Power efficient monitoring management in sensor networks. In: Wireless Communications and Networking Conference (WCNC), vol. 4, pp. 21–24 (2004)
5. Calinescu, G., Kapoor, S., Olshevsky, A., Zelikovsky, A.: Network lifetime and power assignment in ad hoc wireless networks. In: Di Battista, G., Zwick, U. (eds.) ESA 2003. LNCS, vol. 2832, pp. 114–126. Springer, Heidelberg (2003)
6. Cardei, M., Thai, M.T., Li, Y., Wu, W.: Energy-efficient target coverage in wireless sensor networks. In: INFOCOM, pp. 1976–1984 (2005)
7. Călinescu, G., Măndoiu, I., Wan, P., Zelikovsky, A.: Selecting forwarding neighbors in wireless ad hoc networks. Mobile Networks and Applications 9, 101–111 (2004)
8. Feige, U., Halldorsson, M.M., Kortsarz, G., Srinivasan, A.: Approximating the domatic number. SIAM J. Comput. 32(1), 172–195 (2002)
9. Huang, Y., Gao, X., Zhang, Z., Wu, W.: A better constant-factor approximation for weighted dominating set in unit disk graph. Journal of Combinatorial Optimization (2008)

10. Kuhn, F., Moscibroda, T., Wattenhofer, R.: The price of being near-sighted. In: SODA 2006: Proceedings of the seventeenth annual ACM-SIAM symposium on Discrete algorithm, pp. 980–989. ACM Press, New York (2006)
11. Moscibroda, T., Wattenhofer, R.: Maximizing the lifetime of dominating sets. In: Proceedings of the 5th. IEEE International Workshop on Algorithms for Wireless, Mobile, Ad Hoc and Sensor Networks (2005)
12. Pach, J., Tóth, G.: Decomposition of multiple coverings into many parts. In: Proc. 23rd ACM Symp. on Computational Geometry, pp. 133–137 (2007)
13. Pandit, S., Pemmaraju, S.: Finding facilities fast. In: Proceedings of the 10th International Conference on Distributed Computing and Networks (January 2009)
14. Peleg, D.: Distributed computing: a locality-sensitive approach. Society for Industrial and Applied Mathematics (2000)
15. Pemmaraju, S.V., Pirwani, I.A.: Energy conservation via domatic partitions. In: MobiHoc, pp. 143–154 (2006)

On the Complexity of the Asymmetric VPN Problem

Thomas Rothvoß and Laura Sanità[*]

Institute of Mathematics, EPFL, Lausanne, Switzerland
{thomas.rothvoss,laura.sanita}@epfl.ch

Abstract. We give the first constant factor approximation algorithm for the asymmetric Virtual Private Network (VPN) problem with arbitrary concave costs. We even show the stronger result, that there is always a tree solution of cost at most $2 \cdot OPT$ and that a tree solution of (expected) cost at most $49.84 \cdot OPT$ can be determined in polynomial time.

For the case of linear cost we obtain a $(2 + \varepsilon \frac{\mathcal{R}}{\mathcal{S}})$-approximation algorithm for any fixed $\varepsilon > 0$, where $\mathcal{S}$ and $\mathcal{R}$ ($\mathcal{R} \geq \mathcal{S}$) denote the outgoing and ingoing demand, respectively.

Furthermore, we answer an outstanding open question about the complexity status of the so called *balanced* VPN problem by proving its **NP**-hardness.

1 Introduction

The asymmetric Virtual Private Network (VPN) problem is defined on a communication network represented as an undirected connected graph $G = (V, E)$ with cost vector $c : E \to \mathbb{Q}_+$, where c_e indicates the cost of installing one unit of capacity on edge e. Within this network, there is a set of terminal nodes that want to communicate with each other, but the amount of traffic between pairs of terminals is not known exactly. Instead, each vertex v has two thresholds $b_v^+, b_v^- \in \mathbb{N}_0$, representing the cumulative amount of traffic that v can send and receive, respectively. The bounds implicitly describe a set of *valid* traffic matrices which the network has to support. In particular, a traffic matrix specifies for each *ordered* pairs of vertices (u, v), a non-negative amount of traffic that u wishes to send to v. Such a set of traffic demands corresponds to a valid traffic matrix if and only if the total amount of traffic entering and leaving each terminal v does not exceed its bounds b_v^- and b_v^+, respectively.

A solution to an instance of the asymmetric VPN problem is given by a collection of paths $\mathcal{P}$ containing exactly one path for each ordered pair of terminals, and a capacity reservation $x : E \to \mathbb{Q}_+$. Such a solution $(\mathcal{P}, x)$ is *feasible* if every valid traffic matrix can be routed via the paths in $\mathcal{P}$ without exceeding the capacity reservation x. The aim is to find a feasible solution that minimizes the total cost of the installation.

---

[*] Supported by Swiss National Science Foundation within the project "Robust Network Design".

I. Dinur et al. (Eds.): APPROX and RANDOM 2009, LNCS 5687, pp. 326–338, 2009.
© Springer-Verlag Berlin Heidelberg 2009

A feasible solution is called a *tree solution* if the union of the selected paths induces a tree.

The VPN problem was introduced by Fingerhut et al. [1] and Gupta et al. [2], and it soon attracted a lot of attention in the network design community. In fact, the model is relevant for many practical applications where flexible communication scenarios are needed, e.g. to face phenomena like input data uncertainty, demands that are hard to forecast as well as traffic fluctuations, which are typical for instance in IP networks.

Such a high interest in the problem motivated several authors (see e.g. [3–10]) in the investigation of the model and its important variations. A recent survey on network design problems provided by Chekuri [11] reports a lot of interesting open questions concerning VPN models, some of them discussed below.

1.1 Related Work

The *asymmetric* VPN problem is **APX**-hard, even if we restrict to tree solutions [1, 2]. The current best approximation algorithm gives a ratio of 3.55 [4]. Still, the best known upper bound on the ratio between an optimal solution and an optimal tree solution is 4.74 [9].

A quite natural variant of this problem is the so-called *balanced* VPN problem, that is, when the following condition holds: $\sum_v b_v^+ = \sum_v b_v^-$. Italiano et al. [5] show that, differently from the asymmetric version, an optimal tree solution in this case can be found in polynomial time, and Eisenbrand et al. [4] obtain that an optimal tree solution is in fact a 2-approximate solution for the general case. Unfortunately, it has been recently shown that the cheapest solution does not always have a tree structure [12]. Nevertheless, the complexity of the balanced VPN problem is still an open question [5, 11].

Finally, an important variant of this problem is the *symmetric* VPN problem, where each vertex has one single integer bound b_v representing the total amount of traffic that v can exchange with the other nodes: in this case, a solution specifies an $u - v$ path for each *unordered* pair of nodes and a capacity reservation vector in such a way that every valid traffic matrix can be routed via the selected paths, where a valid traffic matrix now specifies an amount of flow that each unordered pair of nodes wishes to exchange, without exceeding the given threshold for each node. Both papers [1] and [2] show that an optimal tree solution can be computed in polynomial time. It has been conjectured in Erlebach et al. [3] and in Italiano et al. [5] that there always exists an optimal solution to the symmetric VPN problem that is a tree solution: this has become known as the *VPN tree routing conjecture*. The conjecture has first been proved for ring networks [6, 7], and was finally settled for general graphs by Goyal et al. [8].

Recently, Fiorini et al. [10] started the investigation of the symmetric VPN problem with concave costs. More precisely, the concave symmetric VPN problem is defined as the symmetric VPN problem, but the contribution of each edge to the total cost is proportional to some concave non-decreasing function of the capacity reservation. The motivation for studying this problem is due to the fact that buying capacity can often reflect an economy of scale principle: the more capacity is installed, the less is the per-unit reservation cost. They give a constant factor approximation algorithm for the problem, and show that also in this case there always exists an optimal solution that has a tree structure. An alternative subsequent proof of the latter result is also given by Goyal et al. [13]. The investigation of the concave asymmetric VPN problem has not been addressed so far.

The importance of tree solutions becomes more evident in the context of symmetric VPN and balanced VPN, where any tree solution has in fact a *central hub node*, as shown by [2] for the symmetric case and by [5] for the balanced case. More precisely, any tree solution in these cases has enough capacity such that *all* the terminal nodes could simultaneously route their traffic to some hub node r in network. Combining this with some simple observations, it follows that computing the cheapest tree solution reduces to computing the cheapest way to simultaneously send a given amount flow from the terminal nodes to some selected hub node r. In case of linear edge costs [2, 5], the latter min-cost flow problem becomes simply a shortest path tree problem. In case that the edge costs are proportional to a non-decreasing concave cost function [10], the latter min-cost flow problem is known as *Single Sink Buy-At-Bulk* (SSBB) problem (a formal definition is given in the next section). Differently, the above property does not hold for tree solutions of asymmetric VPN instances.

We point out that in the literature there is another possible definition of SSBB that does not compute costs according to a concave cost function, but instead deals with an input set of possible cable types that may be installed on the edges, each with different capacity and cost. For this latter version of the problem, the first constant approximation (roughly 2000) is due to Guha et al. [14], subsequently reduced to 216 by Talwar [15] and to 76.8 by Gupta et al. [16], with an algorithm based on random sampling. Refining their approach, the approximation was later reduced to 65.49 by Jothi and Raghavachari [17], and eventually to 24.92 by Grandoni and Italiano [18]. In this paper, according to the first definition, we however refer to SSBB as the problem of routing a given amount of flow from some terminal nodes to a hub node minimizing a concave cost function on the capacity installed on the edges. It is shown in [10] that the (expected) 24.92-approximation algorithm of Grandoni and Italiano [18] can be used to obtain a tree solution with the same approximation factor for our version of SSBB.

1.2   Our Contribution

We give the first constant factor approximation algorithm for the asymmetric VPN problem with arbitrary concave costs, showing that a tree solution of expected cost at most $49.84 \cdot OPT$ can be computed in polynomial time. Moreover, in case of linear cost, we show that for any fixed $\varepsilon > 0$ a $(2 + \varepsilon \frac{R}{S})$-approximate solution can be obtained in polynomial time, with $\mathcal{R} := \sum_v b_v^-$, $\mathcal{S} := \sum_v b_v^+$, and without loss of generality $\mathcal{R} \geq \mathcal{S}$.

The key-point of our approximation results is showing that there always exists a cheap solution with a *capacitated* central hub node, which in particular has a cost of at most twice the optimum. More precisely, there exists a 2-approximate solution with enough capacity such that any subset of terminals could simultaneously send their flow to a hub node r up to a cumulative amount of $\mathcal{S}$. Then, we show how to approximate such a centralized solution by using known results on SSBB. Based on this, we can then state that there exists a VPN *tree solution*, with cost at most $2 \cdot OPT$. This substantially improves the previous known upper bound of 4.74 on the ratio between an optimal solution and an optimal tree solution, which only applies in case of linear costs. We remark that our result holds considering any non-decreasing concave cost function.

The technique used to prove our results is substantially different from the previous approaches known in literature. In fact, approximation algorithms developed in the past mostly relate on computing bounds on the *global cost* of an optimal solution, e.g. showing that an approximate solution constructed out of several matchings or Steiner trees, has a total cost that is not that far from the optimum [4, 9, 16].

In contrast, we focus locally on the *capacity* installed on an edge, and we show that, given any feasible solution, we can obtain a new solution with a capacitated central hub node, such that, on average the capacity on an edge is at most doubled. This result is independent on the cost function. Still, we reinterpret the known fact that, given a set of paths, the minimal amount of capacity to install on an edge can be computed by solving a bipartite matching problem on some auxiliary graph. Using duality, we look instead at minimal *vertex covers* on such graphs, and this reinterpretation allows us to develop a very simple analysis for our statement.

Eventually, we answer the open question regarding the complexity status of the balanced VPN problem with linear costs. We prove that it is **NP**-hard even with unit thresholds on each node.

2   Description of the Problem

In this section we describe in detail the problem addressed in this paper, and other related problems that we will use to state our results.

(Concave/Linear) Virtual Private Network. An instance $\mathcal{I}$ of the *concave Virtual Private Network* (cVPN) problem consists of an undirected connected graph $G = (V, E)$ with edge costs $c : E \to \mathbb{Q}^+$, two non-negative integer vectors $b^+ \in \mathbb{Z}^V, b^- \in \mathbb{Z}^V$, as well as a concave non-decreasing function $f : \mathbb{Q}_+ \to \mathbb{Q}_+$.

A vertex v such that $b_v^+ + b_v^- > 0$ is referred to as a *terminal*: by duplicating nodes, we can assume without loss of generality that each terminal is either a *sender* s, with $b_s^+ > 0, b_s^- = 0$, or a *receiver* r, with $b_r^+ = 0, b_r^- > 0$. Let S and R be set of senders and receivers, respectively.

The vectors b^+ and b^- specify a set of valid traffic matrices that can be interpreted as follows. Let $K_{S,R}$ be the complete bipartite graph with nodes partitioned into senders and receivers: each valid traffic matrix corresponds to a *fractional b-matching* on $K_{S,R}$ and vice versa.

A solution to an instance of the problem is a pair $(\mathcal{P}, x)$, where $\mathcal{P}$ is a collection of paths $\mathcal{P} := \{P_{sr} \mid \forall r \in R, s \in S\}$, and $x \in \mathbb{Q}_+^E$ specifies the capacity to install on each edge of the network. A solution is *feasible* if the installed capacities suffice to route each valid traffic matrix via the selected paths $\mathcal{P}$. A feasible solution is *optimal* if it minimizes the emerging cost $\sum_{e \in E} c_e \cdot f(x_e)$.

If $f(x_e) = x_e$, that means we have *linear* costs on the edges, we term this problem just *Virtual Private Network* (VPN) problem. We call an instance of the problem *balanced* whenever $\mathcal{S} := \sum_{s \in S} b_s^+$ equals $\mathcal{R} := \sum_{r \in R} b_r^-$.

Given a collection of paths $\mathcal{P}$, the minimum amount of capacity x_e that has to be install on $e \in E$ to turn $(\mathcal{P}, x)$ into a feasible solution can be computed in polynomial time as follows (see [2, 4, 5] for details):

$$x_e = \text{maximal cardinality of a } b\text{-matching in } G_e = (S \cup R, E_e),$$
$$\text{with } (s, r) \in E_e \Leftrightarrow e \in P_{sr}$$

Notice that, since the graph G_e is bipartite, an optimum capacity reservation vector x will always be integer.

Single Sink Buy-At-Bulk. An instance of the *Single Sink Buy-At-Bulk* (SSBB) problem consists of an undirected connected graph $G = (V, E)$ with edge costs $c : E \to \mathbb{Q}_+$, a demand function $d : V \to \mathbb{N}$, a root $r \in V$ and a concave non-decreasing function $f : \mathbb{Q}_+ \to \mathbb{Q}_+$.

The aim is to find capacities $x_e \in \mathbb{Q}_+$ for the edges, sufficient to simultaneously route a demand of $d(v)$ from each node v to the root, such that the emerging cost $\sum_{e \in E} c_e \cdot f(x_e)$ is minimized.

Sometimes it is assumed that $d(v) \in \{0, 1\}$, and in this case the vertices $D = \{v \in V \mid d(v) = 1\}$ are called *clients*.

Single Sink Rent-or-Buy. An instance of the *Single Sink Rent-or-Buy* (SROB) problem consists of an undirected connected graph $G = (V, E)$ with edge costs $c : E \to \mathbb{Q}_+$, a demand function $d : V \to \mathbb{N}$, a root $r \in V$ and a parameter $M \geq 1$.

The aim is to find capacities $x_e \in \mathbb{Q}_+$ for the edges, sufficient to simultaneously route a demand of $d(v)$ from each node v to the root, such that the emerging cost $\sum_{e \in E} c_e \cdot \min\{x_e, M\}$ is minimized. Note that this problem is a special case of Single Sink Buy-At-Bulk.

Steiner Tree. An instance of the *Steiner tree* problem consists of an undirected connected graph $G = (V, E)$ with edge costs $c : E \to \mathbb{Q}_+$ and a set of terminals $K \subseteq V$.

The aim is to find the cheapest tree $T \subseteq E$ spanning the terminals.

3   Approximation Results

We now state the first constant factor approximation algorithm for CVPN, starting with some simplifying assumptions that we can make on a CVPN instance
without loss of generality.

First, by duplicating nodes, we may assume b^+, b^- to be 0/1 vectors, that
means, $b_s^+ = 1, b_s^- = 0$ for a sender s, and $b_r^+ = 0, b_r^- = 1$ for a receiver r. The
latter assumption is correct if we can guarantee that the paths in a solution
between copies of a terminal v and copies of a terminal u are all the same. Our
algorithm developed below can be easily adapted in such a way that it satisfies
the latter consistence property, and that it runs in polynomial time even if the
thresholds are not polynomially bounded. Note that, under these assumptions,
$\mathcal{S} = |S|$ and $\mathcal{R} = |R|$. Then by symmetry, suppose that $|R| \geq |S|$.

We propose the following algorithm.

Algorithm 1. CVPN algorithm

---

1. Choose a sender $s^* \in S$ uniformly at random as the *hub*
2. Compute a ρ_{SSBB}-approximate SSBB tree solution $(x_e)_{e \in E}$ for graph G with clients
   $S \cup R$, root s^* and cost function $c_e \cdot f(\min\{x_e, |S|\})$
3. Return $((P_{sr})_{s \in S, r \in R}, x')$ with path P_{sr} being the unique path in the tree defined
   by the support of x_e, and $x_e' = \min\{x_e, |S|\}$

Note that $f(\min\{x_e, |S|\})$ indeed is concave and non-decreasing in x_e. Let
$OPT := OPT_{\text{VPN}}(\mathcal{I})$ be the optimum cost for the CVPN instance $\mathcal{I}$.

Let us first argue, that the capacity reservation x_e' in fact suffices. Consider
an edge e, which is used by k paths in the SSBB solution. Then the capacity
reservation is $x_e' \geq \min\{k, |S|\}$. It is easy to see that this is sufficient for the
constructed CVPN solution. Clearly the cost of this solution is equal to the cost
of the SSBB-solution.

We will now show that indeed, there is a SSBB-solution of cost at most $2 \cdot OPT$
for the instance defined in Step (2) of the algorithm. As it was pointed out in
[10], any solution for SSBB can then be turned into a tree solution of at most
the same cost[1].

To prove this, we first define $\mathcal{I}'$ as a modified CVPN instance, which differs
from $\mathcal{I}$ in such a way that there is a *single* sender with *non-unit* threshold, and
in particular:

$$b_v^+(\mathcal{I}') = \begin{cases} |S| & \text{if } v = s^* \\ 0 & \text{otherwise} \end{cases} \quad \text{and} \quad b_v^-(\mathcal{I}') = \begin{cases} 1 & \text{if } v \in S \cup R \\ 0 & \text{otherwise} \end{cases}$$

Intuitively we reroute all flow through the hub s^*. We will now prove that this
new CVPN instance coincides with the SSBB problem, i.e. their optimum values
are identical.

---

[1] This is not true anymore, if the function f is not concave, but defined by a set of
cables. In that case one might loose a factor of 2 in the approximation.

Let OPT_{SSBB} be the cost of an optimum SSBB solution for the instance defined in Step (2) of the algorithm.

Lemma 1. $OPT_{\text{VPN}}(\mathcal{I}') = OPT_{\text{SSBB}}.$

Proof. Let P_{s^*v} be the paths in a CVPN solution for $\mathcal{I}'$. Consider an edge $e \in E$ and let $v_1, \ldots, v_k \in S \cup R$ be the nodes, such that $e \in P_{s^*v_i}$. If $k \leq |S|$ we can define a traffic matrix in which s^* sends 1 unit of flow to all v_i. If $k > |S|$, we may send 1 unit of flow from s^* to each node in $v_1, \ldots, v_{|S|}$. Anyway the needed capacity of e is $x_e = \min\{k, |S|\}$, which costs $c_e \cdot f(\min\{k, |S|\})$. This is the same amount, which an SSBB solution pays for capacity k on $e \in E$. Thus both problems are equal. $\qquad\square$

The critical point is to show that:

Lemma 2. $E[OPT_{\text{VPN}}(\mathcal{I}')] \leq 2 \cdot OPT_{\text{VPN}}(\mathcal{I}).$

Proof. Let $\mathcal{P} = \{P_{sr} \mid s \in S, r \in R\}$ be the set of paths in the optimum CVPN solution for $\mathcal{I}$ and x_e be the induced capacities. We need to construct a CVPN solution of $\mathcal{I}'$, consisting of s^*-v paths P'_{s^*v} for $v \in S \cup R$.

The solution is surprisingly simple: Choose a receiver $r^* \in R$ uniformly at random as a second hub. Take $P'_{s^*r} := P_{s^*r}$ as s^*-r path. Furthermore concatenate $P'_{s^*s} := P_{s^*r^*} + P_{r^*s}$ to obtain a s^*-s path. To be more precise we can shortcut the latter paths, such that they do not contain any edge twice.

We define a sufficient capacity reservation x'_e as follows: Install $|S|$ units of capacity on the path $P_{s^*r^*}$. Then for each sender $s \in S$ (receiver $r \in R$) install in a cumulative manner one unit of capacity on P_{sr^*} (on P_{s^*r}, respectively). Note that x'_e is a random variable, depending on the choice of s^* and r^*. We show that $E[x'_e] \leq 2x_e$. Once we have done this, the claim easily follows from Jensen's inequality and concavity of f:

$$E[OPT_{\text{VPN}}(\mathcal{I}')] \leq E[\sum_{e \in E} c_e f(x'_e)] \leq \sum_{e \in E} c_e f(E[x'_e]) \leq 2 \cdot OPT_{\text{VPN}}(\mathcal{I})$$

Now consider an edge $e \in E$. Since we want to bound the quantity $E[x'_e]$ in terms of the original capacity x_e, let us inspect, how this capacity is determined. Define the bipartite graph $G_e = (S \cup R, E_e)$ containing an edge $(s, r) \in E_e$ if and only if $e \in P_{sr}$. Then x_e must be the cardinality of a maximal matching in G_e. König's theorem (see e.g. [19, 20]) says that there is a vertex cover $C \subseteq S \cup R$ with $x_e = |C|$ (see Figure 1 for a visualization).

We now distinguish two cases and account their expected contribution to $E[x'_e]$.

1. <u>Case: $s^* \in S \cap C$ or $r^* \in R \cap C$.</u> We account the worst case of $|S|$ units of capacity. The expected contribution is then

$$\Pr[(s^* \in S \cap C) \vee (r^* \in R \cap C)] \cdot |S| \leq \frac{|S \cap C|}{|S|} \cdot |S| + \frac{|R \cap C|}{|R|} \cdot |S| \leq |C|$$

using $|R| \geq |S|$.

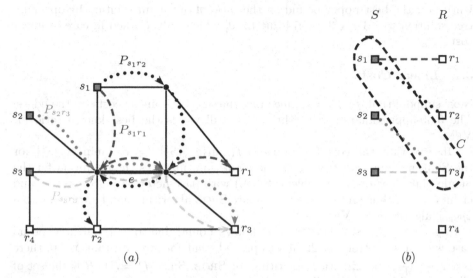

Fig. 1. Example of a CVPN instance in (a), where terminals are depicted as rectangles, senders are drawn solid. Only paths, crossing edge e are shown. In (b) the graph G_e with vertex cover C is visualized, implying that $x_e = 2$.

2. Case: $s^* \in S \backslash C, r^* \in R \backslash C$. We bound the probability of this case by 1. We know that edge (s^*, r^*) cannot exist in G_e since all edges need to be incident to C. Consequently e does not lie on the path $P_{s^*r^*}$. Thus we just have to install 1 unit of capacity for each sender s, such that $(s, r^*) \in E_e$. But only sender in $S \cap C$ may be adjacent to r^* in G_e, thus this number is at most $|S \cap C|$. A similar argument holds for the receivers. The expected contribution of this case is consequently upperbounded by $|S \cap C| + |R \cap C| = |C|$.

Combining the expected capacities for both cases we derive that $E[x'_e] \le 2|C| = 2x_e$, which implies the claim. $\qquad\square$

As a consequence, our algorithm yields a $2\rho_{\text{SSBB}}$-approximation. Using the expected 24.92-approximation of [18], we conclude

Theorem 1. *There is an expected 49.84-approximation algorithm for CVPN which even yields a tree solution.*

Using the derandomized SSBB algorithm of van Zuylen [21] with an approximation factor of 27.72 and the fact that all choices for $s^* \in S$ can be easily tried out, one obtains

Corollary 1. *There is a deterministic factor 55.44-approximation algorithm for CVPN, which even yields a tree solution.*

Corollary 2. *Given any CVPN solution of cost α, one can find deterministically and in polynomial time a tree solution of cost at most 2α.*

Until now the best upper bound on the ratio of optimum solution by optimum tree solution was $3 + \sqrt{3} \approx 4.74$ due to [9] which only worked in case of linear cost.

3.1   Linear Costs

Next suppose that $f(x_e) = x_e$, meaning that we have linear costs on the edges. The 3.55-approximation algorithm of [4] still yields the best known ratio for VPN.

Observe that the cost function $c_e \cdot f(\min\{x_e, |S|\}) = c_e \cdot \min\{x_e, |S|\}$ for the SSBB instance constructed in the algorithm, matches the definition for the Single Sink Rent-or-Buy problem (SROB) with parameter $M = |S|$, root s^* and clients $S \cup R$, thus any ρ_{SROB}-approximate SROB algorithm can be turned into a $2\rho_{\text{SROB}}$-algorithm for VPN.

In general $\rho_{\text{SROB}} \le 2.92$ is the best known bound due to [22], but in a special case we can do better. In [22] it was proved, that for any constant $\delta > 0$, there is a $1 + \delta\frac{|D|}{M}$-approximation algorithm for SROB. Since $D = S \cup R$ is the set of clients and $M = |S|$, this directly yields

Corollary 3. *For any fixed $\varepsilon > 0$, there is a polynomial time $(2 + \varepsilon\frac{R}{S})$-approximation algorithm for VPN.*

Recall that this result also holds in case of non-unit demands.

4   Hardness of Balanced VPN

We here consider the balanced VPN problem with linear costs. Recall that, while the asymmetric VPN is **NP**-hard even restricted to tree solutions, an optimal tree solution for this case can be computed in polynomial time as in the symmetric version [5]. So far, the complexity of the balance VPN was an open question [5, 11]: we now show that the problem is **NP**-hard even with unit thresholds on the nodes, by reduction from the Steiner Tree problem.

Given an instance $\mathcal{I}$ for Steiner Tree consisting of a graph $G = (V, E)$ with cost function $c : E \to \mathbb{Q}_+$, and set of $k + 1$ terminals $\{v_1, \ldots, v_k, v_{k+1}\}$, we construct an instance $\mathcal{I}'$ of the balanced VPN problem on a graph $G' = (V', E')$ as follows.

First, introduce two large numbers: $C := \sum_{e \in E} c_e + 1$, and $M \gg (k + 1)C$. To construct G' from G, add a vertex a_4 and make it adjacent to the vertices $v_1, v_2, \ldots, v_k$ by edges of cost C. Then, add a path $v_{k+1}, a_1, a_2, a_3, a_4$, where the first two edges of the path have cost M, while the last two edges have cost kM. Finally, add k vertices $w_1, w_2, \ldots, w_k$, each of them adjacent to a_2 with a zero cost edge, and add $2k - 1$ vertices $u_1, u_2, \ldots, u_{2k-1}$, each of them adjacent to a_3 with a zero cost edge. Figure 2 shows the resulting graph G'.

Define the set of senders as $S := \{a_1\} \cup \{u_1, u_2, \ldots, u_{2k-1}\}$ and the set of receivers as $R := \{v_1, v_2, \ldots, v_k\} \cup \{w_1, w_2, \ldots, w_k\}$. Note that indeed $|S| = |R|$.

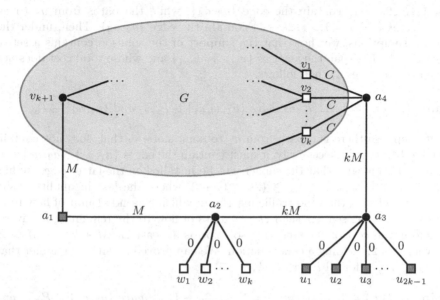

Fig. 2. VPN instance $\mathcal{I}'$. Edges are labeled with their cost. Terminals are depicted as rectangles, senders are drawn in gray.

Lemma 3. *There exists a solution to the Steiner tree instance $\mathcal{I}$ of cost at most C^* if and only if there exists a solution to the balanced VPN instance $\mathcal{I}'$ with cost at most $Z = 2k^2M + 2M + kC + C^*$.*

Proof. ($\Rightarrow$) The only if part is trivial. Suppose there exists a solution T to the Steiner tree instance $\mathcal{I}$ of cost C^*. We construct a solution to $\mathcal{I}'$ by defining the following paths:

- $P_{a_1 w_i} = \{a_1, a_2\} \cup \{a_2, w_i\}$, for $i = 1, \ldots, k$;
- $P_{a_1 v_i} = \{a_1, v_{k+1}\} \cup \{$the edges of the unique $(v_{k+1} - v_i)$-path induced by $T\}$, for $i = 1, \ldots, k$;
- $P_{u_j w_i} = \{u_j, a_3\} \cup \{a_3, a_2\} \cup \{a_2, w_i\}$, for $i = 1, \ldots, k$, $j = 1, \ldots, 2k - 1$;
- $P_{u_j v_i} = \{u_j, a_3\} \cup \{a_3, a_4\} \cup \{a_4, v_i\}$, for $i = 1, \ldots, k$, $j = 1, \ldots, 2k - 1$.

Finally, install the following amount of capacity on the edges of the graph: $x_e = k$ for $e = \{a_2, a_3\}$ and $e = \{a_3, a_4\}$, $x_e = 0$ for $e \in E \setminus T$ and $x_e = 1$ otherwise. It is easy to see that the resulting set of paths and the capacity vector x define a solution to $\mathcal{I}'$ of cost at most Z.

($\Leftarrow$) For the reverse direction, suppose we have a VPN solution $(\mathcal{P}, x)$ to $\mathcal{I}'$ with cost at most Z. Recall that we may assume x to be an integer vector. We now have to argue that in fact this solution must be of the same structure as suggested in the ($\Rightarrow$) part.

First, we show that the paths in $\mathcal{P}$ from a_1 to v_i ($i = 1, \ldots, k$) contain the edge $\{a_1, v_{k+1}\}$, while the paths from a_1 to w_i ($i = 1, \ldots, k$) contain the edge $\{a_1, a_2\}$. Similarly, we show that the paths from u_j to v_i ($i = 1, \ldots, k$ and

$j = 1, \ldots, 2k - 1$) contain the edge $\{a_3, a_4\}$, while the paths from u_j to w_i ($i = 1, \ldots, k$ and $j = 1, \ldots, 2k - 1$) contain the edge $\{a_2, a_3\}$. Then, under the above assumptions, we show that the support of the solution contains a set of edges $T \subseteq E$ that span the vertices $\{v_1, \ldots, v_{k+1}\}$ and whose total cost does not exceed C^*. The result then follows.

Claim 1. *For $i = 1, \ldots, k$, we have $\{a_1, v_{k+1}\} \in P_{a_1 v_i}$ and $\{a_1, a_2\} \in P_{a_1 w_i}$.*

Proof. Suppose there is a path from a_1 to some node v_i that does not contain the edge $\{a_1, v_{k+1}\}$. Necessarily, it must contain the edges $\{a_1, a_2\}$, $\{a_2, a_3\}$ and $\{a_3, a_4\}$. This means, that the capacity to be installed on the latter edges fulfills $x_{a_1 a_2} \geq 1$ and $x_{a_2 a_3} + x_{a_3 a_4} \geq (2k - 1) + 2$, where the last inequality easily follows considering the valid traffic matrix, in which a_1 sends 1 unit of flow to v_i and the remaining senders send $2k - 1$ units of flow to the remaining receivers. Therefore, the cost of the emerging solution is at least $2k^2 M + kM + M > Z$ for every $k \geq 2$, yielding a contradiction. We can prove in a similar manner that $\{a_1, a_2\} \in P_{a_1 w_i}$ for all $i = 1, \ldots, k$. $\qquad\square$

Claim 2. *For $i = 1, \ldots, k$ and $j = 1, \ldots, 2k - 1$, we have $\{a_3, a_4\} \in P_{u_j v_i}$ and $\{a_2, a_3\} \in P_{u_j w_i}$.*

Proof. First, we focus on the capacity installed on the edges $e = \{a_2, a_3\}$ and $e' = \{a_3, a_4\}$. Clearly, $x_e + x_{e'} \geq 2k - 1$. We now prove that in fact the inequality is strict.

Suppose it holds with equality. We inspect the bipartite graphs G_e and $G_{e'}$ (this time without a_1, since we already proved that it uses neither e nor e') and let $C_e, C_{e'}$ be the minimum vertex covers on $G_e, G_{e'}$, respectively. By hypothesis, $|C_e| + |C_{e'}| = 2k - 1$. That means that there is at least one node $r \in R$ that does not belong to any of the two covers. Now notice that G_e and $G_{e'}$ are complementary bipartite graphs, since the union of their edges gives the complete bipartite graph $K_{2k-1,2k}$. It follows than that $C_e \cup C_{e'} = S \setminus \{a_1\}$: otherwise, there would be a node $s \in S \setminus (\{a_1\} \cup C_e \cup C_{e'})$ with an incident edge (s, r), that is neither covered by C_e nor by $C_{e'}$, a contradiction.

As a conclusion all senders in C_e route to the $2k$ receivers on paths containing the edge e, while all senders in $C_{e'}$ route to the $2k$ receivers on paths containing the edge e'. Then it is easy to see, that the installed capacities satisfy $x_{v_{k+1} a_1} \geq k$ and $x_{a_1 a_2} \geq k$. Therefore, the cost of the emerging solution is at least $2k^2 M - kM + 2kM > Z$, for $k \geq 3$, a contradiction.

It follows that $x_e + x_{e'} \geq 2k$. Suppose now, there is a path from some u_j to some v_i that does not contain the edge $e = \{a_3, a_4\}$. Necessarily, it must contain the edges $\{a_1, a_2\}$ and $\{v_{k+1}, a_1\}$. Using the previous claim, it is easy to see that the installed capacities satisfy $x_{v_{k+1} a_1} \geq 2$ and similarly $x_{a_1 a_2} \geq 2$. Therefore, the cost of the emerging solution is at least $2k^2 M + 4M > Z$, again a contradiction.

We can prove in a similar manner that there is no path from some u_j to some w_i that does not contain the edge $e' = \{a_2, a_3\}$. $\qquad\square$

Putting all together, it follows that $x_{v_{k+1}a_1} \geq 1$, $x_{a_1a_2} \geq 1$, $x_{a_2a_3} \geq k$, $x_{a_3a_4} \geq k$, and the cost of the capacity installed on the latter edges is at least $2k^2M + 2kM$.

Now, consider the edges $\{a_4, v_i\}$, $i = 1, \ldots, k$: clearly, $\sum_{i=1,\ldots,k} x_{a_4v_i} \geq k$, since we can define a traffic matrix where k senders in $S \setminus \{a_1\}$ simultaneously send k units of flow to $v_1, \ldots, v_k$. It follows that $\sum_{i=1,\ldots,k} c_{a_4v_i} x_{a_4v_i} \geq k \cdot C$.

Finally, let T be the subset of edges of E that are in the support of the solution. Suppose that T does not span the nodes $v_1, \ldots, v_{k+1}$. Then there exists at least one node v_i such that the path from a_1 to v_i contains at least 2 edges with cost C. But in this case, we would have $\sum_{i=1,\ldots,k} x_{a_4v_i} \geq k + 1$ and the cost of the solution exceeds Z. We conclude that indeed T contains a Steiner tree and $c(T) \leq Z - (2k^2M + 2M + kC) = C^*$.   □

From the discussions above, it follows:

Theorem 2. *The balanced* VPN *problem is* **NP**-*hard.*

Note that the above reduction is not approximation preserving, i.e. in contrast to **NP**-hardness, the **APX**-hardness of Steiner tree [23] is not conveyed to balanced VPN. In other words, our reduction does not exclude the possible existence of a PTAS for balanced VPN.

References

1. Fingerhut, J., Suri, S., Turner, J.: Designing least-cost nonblocking broadband networks. Journal of Algorithms 24(2), 287–309 (1997)
2. Gupta, A., Kleinberg, J., Kumar, A., Rastogi, R., Yener, B.: Provisioning a virtual private network: A network design problem for multicommodity flow. In: Proc. of the 33rd ACM Symposium on Theory of Computing (STOC), pp. 389–398 (2001)
3. Erlebach, T., Rüegg, M.: Optimal bandwidth reservation in hose-model vpns with multi-path routing. In: Proc. of the 23rd Annual Joint Conference of the IEEE Computer and Communications Societies INFOCOM, vol. 4, pp. 2275–2282 (2004)
4. Eisenbrand, F., Grandoni, F., Oriolo, G., Skutella, M.: New approaches for virtual private network design. SIAM Journal on Computing 37(3), 706–721 (2007)
5. Italiano, G., Leonardi, S., Oriolo, G.: Design of trees in the hose model: the balanced case. Operation Research Letters 34(6), 601–606 (2006)
6. Hurkens, C., Keijsper, J., Stougie, L.: Virtual private network design: A proof of the tree routing conjecture on ring networks. SIAM Journal on Discrete Mathematics 21, 482–503 (2007)
7. Grandoni, F., Kaibel, V., Oriolo, G., Skutella, M.: A short proof of the vpn tree routing conjecture on ring networks. Operation Research Letters 36(3), 361–365 (2008)
8. Goyal, N., Olver, N., Shepherd, B.: The vpn conjecture is true. In: Proc. of the 40th annual ACM symposium on Theory of computing (STOC), pp. 443–450 (2008)
9. Eisenbrand, F., Grandoni, F.: An improved approximation algorithm for virtual private network design. In: Proc. of the 16th ACM-SIAM Symposium on Discrete Algorithms (SODA) (2005)
10. Fiorini, S., Oriolo, G., Sanità, L., Theis, D.: The vpn problem with concave costs (submitted manuscript) (2008)

11. Chekuri, C.: Routing and network design with robustness to changing or uncertain traffic demands. SIGACT News 38(3), 106–128 (2007)
12. Sanità, L.: Robust Network Design. PhD thesis, Università Sapienza di Roma (2008)
13. Goyal, N., Olver, N., Shepherd, B.: Personal communication (2008)
14. Guha, S., Meyerson, A., Munagala, K.: A constant factor approximation for the single sink edge installation problems. In: Proc. of the 33rd Annual ACM Symposium on Theory of Computing (STOC), New York, pp. 383–388 (2001)
15. Talwar, K.: The single-sink buy-at-bulk LP has constant integrality gap. In: Proc. of the 9th International Conference on Integer programming and combinatorial optimization (IPCO), pp. 475–486 (2002)
16. Gupta, A., Kumar, A., Roughgarden, T.: Simpler and better approximation algorithms for network design. In: Proc. of the 35th Annual ACM Symposium on Theory of Computing (STOC), New York, pp. 365–372 (2003)
17. Jothi, R., Raghavachari, B.: Improved approximation algorithms for the single-sink buy-at-bulk network design problems. In: Hagerup, T., Katajainen, J. (eds.) SWAT 2004. LNCS, vol. 3111, pp. 336–348. Springer, Heidelberg (2004)
18. Grandoni, F., Italiano, G.F.: Improved approximation for single-sink buy-at-bulk. In: Asano, T. (ed.) ISAAC 2006. LNCS, vol. 4288, pp. 111–120. Springer, Heidelberg (2006)
19. Cook, W., Cunningham, W., Pulleyblank, W., Schrijver, A.: Combinatorial optimization. John Wiley & Sons Inc., New York (1998)
20. Schrijver, A.: Combinatorial optimization. Polyhedra and efficiency. vol. A. Algorithms and Combinatorics, vol. 24. Springer, Berlin (2003)
21. van Zuylen, A.: Deterministic sampling algorithms for network design. In: Halperin, D., Mehlhorn, K. (eds.) Esa 2008. LNCS, vol. 5193, pp. 830–841. Springer, Heidelberg (2008)
22. Eisenbrand, F., Grandoni, F., Rothvoß, T., Schäfer, G.: Approximating connected facility location problems via random facility sampling and core detouring. In: Proc. of the 19th ACM-SIAM Symposium on Discrete Algorithms (SODA), pp. 1174–1183 (2008)
23. Chlebík, M., Chlebíková, J.: Approximation hardness of the steiner tree problem on graphs. In: Penttonen, M., Schmidt, E.M. (eds.) SWAT 2002. LNCS, vol. 2368, p. 170. Springer, Heidelberg (2002)

Deterministic Approximation Algorithms for the Nearest Codeword Problem

Noga Alon[1,*], Rina Panigrahy[2], and Sergey Yekhanin[3]

[1] Tel Aviv University, Institute for Advanced Study, Microsoft Israel
nogaa@tau.ac.il
[2] Microsoft Research Silicon Valley
rina@microsoft.com
[3] Microsoft Research Silicon Valley
yekhanin@microsoft.com

Abstract. The Nearest Codeword Problem (NCP) is a basic algorithmic question in the theory of error-correcting codes. Given a point $v \in \mathbb{F}_2^n$ and a linear space $L \subseteq \mathbb{F}_2^n$ of dimension k NCP asks to find a point $l \in L$ that minimizes the (Hamming) distance from v. It is well-known that the nearest codeword problem is NP-hard. Therefore approximation algorithms are of interest. The best efficient approximation algorithms for the NCP to date are due to Berman and Karpinski. They are a deterministic algorithm that achieves an approximation ratio of $O(k/c)$ for an arbitrary constant c, and a randomized algorithm that achieves an approximation ratio of $O(k/\log n)$.

In this paper we present new deterministic algorithms for approximating the NCP that improve substantially upon the earlier work. Specifically, we obtain:

- A polynomial time $O(n/\log n)$-approximation algorithm;
- An $n^{O(s)}$ time $O(k \log^{(s)} n/\log n)$-approximation algorithm, where $\log^{(s)} n$ stands for s iterations of log, e.g., $\log^{(2)} n = \log \log n$;
- An $n^{O(\log^* n)}$ time $O(k/\log n)$-approximation algorithm.

We also initiate a study of the following Remote Point Problem (RPP). Given a linear space $L \subseteq \mathbb{F}_2^n$ of dimension k RPP asks to find a point $v \in \mathbb{F}_2^n$ that is far from L. We say that an algorithm achieves a remoteness of r for the RPP if it always outputs a point v that is at least r-far from L. In this paper we present a deterministic polynomial time algorithm that achieves a remoteness of $\Omega(n \log k/k)$ for all $k \leq n/2$. We motivate the remote point problem by relating it to both the nearest codeword problem and the matrix rigidity approach to circuit lower bounds in computational complexity theory.

1 Introduction

The Nearest Codeword Problem (NCP) is a basic algorithmic question in the theory of error-correcting codes. Given a point $v \in \mathbb{F}_2^n$ and a linear space $L \subseteq \mathbb{F}_2^n$

* Research supported in part by an ERC advanced grant, by NSF grant CCF 0832797 and by the Ambrose Monell Foundation.

I. Dinur et al. (Eds.): APPROX and RANDOM 2009, LNCS 5687, pp. 339–351, 2009.

of dimension k NCP asks to find a point $l \in L$ that minimizes the (Hamming) distance from v. The nearest codeword problem is equivalent to the problem of finding a vector $x \in \mathbb{F}_2^k$ that minimizes the number of unsatisfied linear equations in the system $xG = v$, given a matrix $G \in \mathbb{F}_2^{k \times n}$ and a vector $v \in \mathbb{F}_2^n$. It is well-known that the NCP is NP-hard. Therefore approximation algorithms are of interest.

The best efficient approximation algorithms for the NCP to date are due to Berman and Karpinski [3]. They are a deterministic algorithm that achieves an approximation ratio of $O(k/c)$ for an arbitrary constant c, and a randomized algorithm that achieves an approximation ratio of $O(k/\log n)$.[1] There has been a substantial amount of work on hardness of approximation for the NCP [1, 2, 4]. The best result to date is due to Arora et al. [2]. It shows that one cannot approximate the NCP to within $2^{\log^{0.5-\epsilon} n}$, for any $\epsilon > 0$ unless NP is in $\mathrm{DTIME}\left(n^{poly(\log n)}\right)$. Alekhnovich [1] has made a conjecture that implies inapproximability of the NCP to within $n^{1-\epsilon}$, for every $\epsilon > 0$.

In this paper we develop new *deterministic* algorithms for approximating the NCP. Specifically, we obtain:

1. A polynomial time $O(n/\log n)$-approximation algorithm;
2. An $n^{O(s)}$ time $O(k \log^{(s)} n/\log n)$-approximation algorithm, where $\log^{(s)} n$ stands for s iterations of log, e.g., $\log^{(2)} n = \log \log n$;
3. An $n^{O(\log^* n)}$ time $O(k/\log n)$-approximation algorithm.

Our first algorithm matches the performance of the randomized algorithm of [3] for $k = \Omega(n)$. This is the regime that is of primary importance for the coding theory applications. Our second algorithm improves substantially upon the deterministic algorithm of [3], and nearly matches the randomized algorithm of [3] in terms of the approximation ratio. Finally, our third algorithm has the same approximation ratio as the randomized algorithm of [3] and a slightly super-polynomial running time. All our algorithms (as well as other known algorithms for the NCP in the literature) can be easily generalized to fields other than $\mathbb{F}_2$.

Remote Point Problem. In this work we also initiate a study of the following Remote Point Problem (RPP). Given a linear space $L \subseteq \mathbb{F}_2^n$ of dimension k RPP asks to find a point $v \in \mathbb{F}_2^n$ that is far from L. We say that an algorithm achieves a remoteness of r for the RPP if it always outputs a point v that is at least r-far from L. We present a deterministic polynomial time algorithm that achieves a remoteness of $\Omega(n \log k/k)$ for all $k \leq n/2$. Our algorithm for the remote point problem is closely related to our first approximation algorithm for the nearest codeword problem.

We motivate the remote point problem by relating it to the matrix rigidity approach to circuit lower bounds in computational complexity theory. The notion of matrix rigidity was introduced by Leslie Valiant in 1977 [10]. In what follows

---

[1] In fact, Berman and Karpinski [3] only claim that their randomized algorithm achieves a $O(k/\log k)$ approximation. However it is immediate from their analysis that they also get a $O(k/\log n)$ approximation.

we say that a set $A \subseteq \mathbb{F}_2^n$ is r-far from a linear space $L \subseteq \mathbb{F}_2^n$ if A contains a point that is r-far from L. (Observe, that this is quite different from the usual notion of distance between sets.) Valiant called a set $A \subseteq \mathbb{F}_2^n$ rigid if for some fixed $\epsilon > 0$, A is n^ϵ-far from every linear space $L \subseteq \mathbb{F}_2^n$, $\dim L = n/2$. Valiant showed that if a set $A \subseteq \mathbb{F}_2^n$ is rigid and $|A| = O(n)$; then the linear transformation from n bits to $|A|$ bits induced by a matrix whose rows are all elements of A can not be computed by a circuit of XOR gates that simultaneously has size $O(n)$ and depth $O(\log n)$.[2]

Valiant's work naturally led to the challenge of constructing a small explicit rigid set A, (since such a set yields an explicit linear map, for that we have a circuit lower bound). This challenge has triggered a long line of work. For references see [5, 7–9]. Unfortunately, after more than three decades of efforts, we are still nowhere close to constructing an explicit rigid set with the parameters needed to get implications in complexity theory. The smallest known explicit sets $A \subseteq \mathbb{F}_2^n$ (presented in the appendix) that are d-far from every linear space $L \subseteq \mathbb{F}_2^n$, $\dim L = n/2$ have size $2^{O(d)} n/d$.

In particular there are no known constructions of sets $A \subseteq \mathbb{F}_2^n$ of size $O(n)$ that are $\omega(1)$-far from linear spaces dimension $n/2$. Moreover if we restrict ourselves to sets A of size n; then we do not know how to construct an explicit set that is just 3-far from every linear space of dimension $n/2$, despite the fact that a random set A of cardinality n is $\Omega(n)$-far from every such space with an overwhelming probability.

In this paper we propose the remote point problem as an intermediate challenge that is less daunting than the challenge of designing a small rigid set, and yet could help us develop some insight into the structure of rigid sets. Recall that a rigid set is a set that is simultaneously n^ϵ-far from *every* linear space L, $\dim L = n/2$. Given the state of art with constructions of explicit rigid sets we find it natural to consider an easier algorithmic Remote Set Problem (RSP) where we are given a single linear space L, and our goal is to design an $O(n)$-sized set $A_L \subseteq \mathbb{F}_2^n$ that is n^ϵ-far from L. Clearly, if we knew how to construct explicit rigid sets, we could solve the RSP without even looking at the input. The remote point problem is a natural special case of the remote set problem. Here we are given a linear space $L \subseteq \mathbb{F}_2^n$ and need to find a single point that is far from L.

In this paper we present an algorithm that for every linear space $L \subseteq \mathbb{F}_2^n$, $\dim L = n/2$ generates a point that is $\Omega(\log n)$-far from L. (For spaces L of dimension $k < n/2$, our algorithm generates a point of distance at least $\Omega(n \log k/k)$ from L.) We are not aware of efficient algorithms to generate points (or $O(n)$-sized collections of points) further away from a given arbitrary linear space of dimension $n/2$.

---

[2] The original paper of Valiant [10] and the follow-up papers use a somewhat different language. Specifically, they talk about matrices A whose rank remains no less than $n/2$ even after every row is modified in less than n^ϵ coordinates; rather than about sets A that for every linear space $L \subseteq \mathbb{F}_2^n$, $\dim L = n/2$ contain a point $a \in A$ that is n^ϵ-far from L. However, it is not hard to verify that the two concepts above are equivalent.

The remote point problem can be viewed as a search variant of the covering radius problem: finding a point in space that is as far away as possible from a given code. The complexity of the covering radius problem has been studied in [6].

Organization. We present our first approximation algorithm for the NCP in section 2. We present our second and third algorithms in section 3. We present our algorithm for the remote point problem in section 4. We present a family of explicit subsets of $\mathbb{F}_2^n$ that are d-far from all linear spaces $L \subseteq \mathbb{F}_2^n$, $\dim L = n/2$ in the appendix.

2   An $O(n/\log n)$-Approximation Algorithm

We start with the formal statements of the NCP and of our main result.

Nearest Codeword Problem

- INSTANCE: A linear code $L = \{xG \mid x \in \mathbb{F}_2^k\}$ given by a generator matrix $G \in \mathbb{F}_2^{k \times n}$ and a vector $v \in \mathbb{F}_2^n$.
- SOLUTION: A codeword $l \in L$.
- OBJECTIVE FUNCTION (to be minimized): The Hamming distance $d(l, v)$.

Theorem 1. *Let $c \geq 1$ be an arbitrary constant. There exists a deterministic $n^{O(c)}$ time $\lceil n/c \log n \rceil$-approximation algorithm for the NCP.*

In order to proceed with the proof we need the following notation:

- For a positive integer d, let $B_d = \{x \in \mathbb{F}_2^n \mid d(0^n, x) \leq d\}$ denote a Hamming ball of radius d.
- For a collection of vectors $M \subseteq \mathbb{F}_2^n$, let $\mathrm{Span}(M)$ denote the smallest linear subspace of $\mathbb{F}_2^n$ containing M.
- For sets $A, B \subseteq \mathbb{F}_2^n$, we define $A + B = \{a + b \mid a \in A, b \in B\}$.

The next lemma is the core of our algorithm. It shows that a d-neighborhood of a linear space L can be covered by a (small) number of linear spaces M_S of larger dimension, in such a way that no linear space M_S contains points that are too far from L.

Lemma 1. *Let L be a linear space, and $d \leq t$ be positive integers. Let $B_1 \setminus \{0^n\} = \bigcup_{i=1}^{t} B_1^i$ be an arbitrary partition of the set of n unit vectors into t disjoint classes each of size $\lceil n/t \rceil$ or $\lfloor n/t \rfloor$. For every $S \subseteq [t]$ such that $|S| = d$ let $M_S = \mathrm{Span}\left(L \cup \left(\bigcup_{i \in S} B_1^i\right)\right)$. Then*

$$L + B_d \subseteq \bigcup_S M_S \subseteq L + B_{d\lceil n/t \rceil}, \tag{1}$$

where S runs over all subsets of $[t]$ of cardinality d.

Proof. We first show the left containment. Let v be an arbitrary vector in $L + B_d$. We have $v = l + e_{j_1} + \ldots + e_{j_{d'}}$, where $d' \leq d$, all e_{j_r} are unit vectors and $l \in L$. For every $r \in [d']$ let $i_r \in [t]$ be such that $j_r \in B_1^{i_r}$. Consider a set $S \subseteq [t]$ such that $|S| = d$ and $i_1, \ldots, i_{d'} \in S$. It is easy to see that $v \in M_S$.

We proceed to the right containment. Let $S = \{i_1, \ldots, i_d\}$ be an arbitrary subset of $[t]$ of cardinality d. Recall that the cardinality of every set $B_1^{i_r}$, $r \in [d]$ is at most $\lceil n/t \rceil$. Therefore every element $v \in M_S$ can be expressed as a sum $v = l + y$, where $l \in L$ and y is a sum of at most $d\lceil n/t \rceil$ unit vectors. Thus $v \in L + B_{d\lceil n/t \rceil}$.

We are now ready to proceed with the proof of the theorem.

Proof (of theorem 1). Observe that if the point v is more than $c \log n$-far from L; then any vector in L (for instance, the origin) is an $\lceil n/c \log n \rceil$-approximation for v. Let us assume that $d(v, L) \leq c \log n$ and set $t = \lceil c \log n \rceil$. Our algorithm iterates over values $d \in [0, \lceil c \log n \rceil]$. For each d we generate all linear spaces $M_S, S \subseteq [t], |S| = d$ as defined in lemma 1. We check whether v is contained in one of those spaces. Lemma 1 implies that after at most $d(v, L)$ iterations we get $v \in M_S$, for some $S = \{i_1, \ldots, i_d\}$. We expand v as a sum $v = l + y$ where $l \in L$ and y is a sum of at most $d\lceil n/c \log n \rceil$ unit vectors from $\bigcup_{r=1}^{d} B_1^{i_r}$. Obviously, $d(v, l) \leq d(v, L)\lceil n/c \log n \rceil$. We report l as our $\lceil n/c \log n \rceil$-approximation for v. The pseudo-code is below.

Set $t = \lceil c \log n \rceil$;
  For every $d \in [0, c \log n]$
    For every $S = \{i_1, \ldots, i_d\} \subseteq [t]$ such that $|S| = d$
      If $v \in M_S$ **Then**
        Begin
          Represent v as $v = l + y$,
             where $l \in L$ and y is a sum of unit vectors from $\bigcup_{r=1}^{d} B_1^{i_r}$;
          Output l;
          Terminate;
        End
  Output 0^n;

It is easy to see that the algorithm above runs in time $n^{O(c)}$. The first loop makes $O(c \log n)$ iterations. The second loop makes at most $2^{\lceil c \log n \rceil} = n^{O(c)}$ iterations. Finally, the internal computation runs in $n^{O(1)}$ time.

3 A Recursive $O(k \log^{(s)} n / \log n)$-Approximation Algorithm

The goal of this section is to prove the following

Theorem 2. *Let $s \geq 1$ be an integer and $c \geq 1$ be an arbitrary constant. There exists a deterministic $n^{O(cs)}$ time $\lceil k \log^{(s)} n / c \log n \rceil$-approximation algorithm for the NCP, where the constant inside the O-notation is absolute and $\log^{(s)} n$ denotes s iterations of the $\log$ function.*

Proof. Our proof goes by induction on s and combines ideas from our $O(n/\log n)$-approximation algorithm of section 2 with ideas from the deterministic approximation algorithm of Berman and Karpinski [3]. We start with some notation.

- Let $x^*G = l^* \in L$ denote some fixed optimal approximation of v by a vector in L.
- Let $E = \{i \in [n] \mid l_i^* \neq v_i\}$ be the set of coordinates where l^* differs from v.
- In what follows we slightly abuse the notation and use the letter G to denote the multi-set of columns of the generator matrix of L (as well as the generator matrix itself).
- We call a partition of the multi-set $G = \bigcup_i^h G_i$ into disjoint sets *regular* if for every $i \in [h]$, the vectors in G_i are linearly independent and:

$$\text{Span}(G_i) = \text{Span}\left(\bigcup_{j \geq i}^h G_j\right). \tag{2}$$

Again, in what follows we slightly abuse the notation and use symbols $G_i, i \in [h]$ to denote the sets of columns of the generator matrix, the corresponding subsets of $[n]$, and the sub-matrices of the generator matrix of L.
- We denote the restriction of a vector $u \in \mathbb{F}_2^n$ to coordinates in a set $S \subseteq [n]$, by $u|_S \in \mathbb{F}_2^{|S|}$.

The following claim (due to Berman and Karpinski [3]) constitutes the base case of the induction. We include the proof for the sake of completeness.

Base Case of the Induction: Let $c \geq 1$ be an arbitrary constant. There exists a deterministic $n^{O(c)}$ time $\lceil k/c \rceil$-approximation algorithm for the NCP.

Proof of the Base Case: We start with an informal description of the algorithm. Our goal is to "approximately" recover x^* from v (which is a "noisy" version of l^*). Recall that l^* and v differ in coordinates that belong to E. We assume that $|E| < n/\lceil k/c \rceil$ since otherwise any vector in the space L is a valid $\lceil k/c \rceil$-approximation for v. The algorithm has two phases. During the first phase we compute a regular partition of the multi-set G. Note that such a partition necessarily has at least $h \geq n/k$ classes. Therefore there is a class $G_i, i \in [h]$ such that

$$|G_i \cap E| \leq (n/\lceil k/c \rceil)/(n/k) \leq c.$$

During the second phase we iterate over all classes $G_i, i \in [h]$ of the regular partition, trying to "fix" the differences between $v|_{G_i}$ and $l^*|_{G_i}$ and thus "approximately" recover x^*. More specifically, for every $i \in [h]$ we solve the system $xG_i = u$ for x, for every u that differs from $v|_{G_i}$ in up to c coordinates. (In cases when the system $xG_i = u$ happens to be under-determined we take an arbitrary single solution.) This way every class in the regular partition gives us a number of candidate vectors x. In the end we select a single vector that yields the best approximation for v.

To see that the algorithm indeed produces a valid $\lceil k/c \rceil$-approximation for v, consider the smallest index i such that $|G_i \cap E| \leq c$. Note that one of the linear systems that we are going to solve while processing the i-th class of the regular partition is $xG_i = l^* |_{G_i}$. Let x be an arbitrary solution of the above system. Clearly,

$$d(xG, v) = \sum_{j=1}^{i-1} d\left(xG_j, v |_{G_j}\right) + \sum_{j=i}^{h} d\left(xG_j, v |_{G_j}\right). \tag{3}$$

However for every $j \leq i - 1$ we have

$$d\left(xG_j, v |_{G_j}\right) \leq k \leq c\lceil k/c \rceil \leq d\left(l^* |_{G_j}, v |_{G_j}\right) \lceil k/c \rceil, \tag{4}$$

by our choice of i. Also, $xG_i = l^* |_{G_i}$ and formula (2) yield

$$xG_j = l^* |_{G_j}, \tag{5}$$

for all $j \geq i$. Combining formulae (4), (5) and (3) we get $d(xG, v) \leq d(l^*, v)\lceil k/c \rceil$ and thus xG is a $\lceil k/c \rceil$-approximation for v. The pseudo-code of the algorithm is below:

Obtain a regular partition $G = \bigcup_{i \in h} G_i$;
Set $x_{\text{best}} = 0^k$;
For every $i \in [h]$
    For every vector y in $\mathbb{F}_2^{|G_i|}$ of Hamming weight at most c
      Begin
        Find an $x \in \mathbb{F}_2^k$ such that $xG_i = v |_{G_i} + y$;
        If $d(xG, v) < d(x_{\text{best}}G, v)$ **Then Set** $x_{\text{best}} = x$;
      End
Output $x_{\text{best}}G$;

It is easy to see that the algorithm above runs in time $n^{O(c)}$. The first loop makes $O(n)$ iterations. The second loop makes at most n^c iterations. Finally, obtaining a regular partition and the internal computation both run in $n^{O(1)}$ time.

We now proceed to the induction step.

Induction Step: Let $s \geq 1$ be an integer and $c \geq 1$ be an arbitrary constant. Suppose there exists a deterministic $n^{O(cs-c)}$ time $\lceil k \log^{(s-1)} n/c \log n \rceil$-approximation algorithm for the NCP; then there exists deterministic $n^{O(cs)}$ time $\lceil k \log^{(s)} n/c \log n \rceil$-approximation algorithm for the NCP.

Proof of the Induction Step: The high level idea behind our algorithm is to reduce the nearest codeword problem on an instance (G, v) to $n^{O(c)}$ (smaller) instances of the problem and to solve those instances using the algorithm from the induction hypothesis.

We start in a manner similar to the proof of the base case. Our goal is to "approximately" recover the vector x^* from v (which is a "noisy" version of l^*).

Recall that l^* and v differ in coordinates that belong to E. We assume that $|E| < n/\lceil k \log^{(s)} n / c \log n \rceil$ since otherwise any vector in the space L is a valid $\lceil k \log^{(s)} / c \log n \rceil$-approximation for v. Our algorithm has two phases. During the first phase we compute a regular partition of the multi-set G. Note that such a partition necessarily has at least $h \geq n/k$ classes. Therefore there is a class $G_i, i \in [h]$ such that

$$|G_i \cap E| \leq (n/\lceil k \log^{(s)} n / c \log n \rceil)/(n/k) \leq c \log n / \log^{(s)} n.$$

During the second phase we iterate over all classes $G_i, i \in [h]$ of the regular partition, trying to locate a large subset $W \subseteq G_i$ such that $l^*|_W = v|_W$. We use such a subset to restrict our optimization problem to $x \in \mathbb{F}_2^k$ that satisfy $xG|_W = v|_W$ and thus obtain a smaller instance of the NCP. More formally, during the second phase we:

1. Set

$$b = \left\lfloor \frac{c \log n}{\log^{(s)} n} \right\rfloor, \qquad t = \left\lceil \frac{2c \log n \log^{(s-1)} n}{\log^{(s)} n} \right\rceil. \qquad (6)$$

2. Set $x_{\text{best}} = 0^k$.
3. For every $i \in [h]$:
4. Set $G' = \bigcup_{j \geq i} G_j$.
   (a) If $k \geq t$ then
      i. Split the class G_i into a disjoint union of t sets $G_i = \bigcup_{r=1}^{t} G_i^r$, each of size $\lceil |G_i|/t \rceil$ or $\lfloor |G_i|/t \rfloor$.
      ii. For every $S \subseteq [t]$ such that $|S| = b$, set $W = \bigcup_{r \in [t] \setminus S} G_i^r$:
      iii. Consider an affine optimization problem of finding an $x \in \mathbb{F}_2^k$ that minimizes $d(xG', v|_{G'})$, subject to $xG|_W = v|_W$. Properties of the regular partition imply that here we are minimizing over an affine space L' of dimension $|G_i| - |W|$, in $\mathbb{F}_2^{|G'|}$.
      iv. Turn the problem above into a form of an NCP (in $\mathbb{F}_2^n$, padding both the target vector v and the matrix G' with zeros) and solve it approximately for x using the algorithm from the induction hypothesis. (Note that every affine optimization problem of minimizing $d(xJ + z, v)$ over x for $J \in \mathbb{F}_2^{k \times n}$ and $z, v \in \mathbb{F}_2^n$, can be easily turned into a form of an NCP, i.e., the problem of minimizing $d(xJ, v + z)$ over $x \in \mathbb{F}_2^k$.)
      v. If $d(xG, v) < d(x_{\text{best}}G, v)$ then set $x_{\text{best}} = x$.
   (b) Else
      i. For every vector y in $\mathbb{F}_2^{|G_i|}$ such that the Hamming weight of y is at most b :
      ii. Find an $x \in \mathbb{F}_2^k$ such that $xG_i = v|_{G_i} + y$;
      iii. If $d(xG, v) < d(x_{\text{best}}G, v)$ then set $x_{\text{best}} = x$.
5. Output $x_{\text{best}}G$.

We now argue that the algorithm above obtains a valid $\lceil k \log^{(s)} n / c \log n \rceil$-approximation for the NCP. We first consider (the easier) case when $k < t$. Our analysis is similar to the analysis of the base case of the induction. Let $i \in [h]$ be the smallest index such that $|G_i \cap E| \leq \lfloor c \log n / \log^{(s)} n \rfloor = b$. Note that one of the linear systems that we are going to solve while processing the i-th class of the regular partition is $x G_i = l^* |_{G_i}$. Let x be an arbitrary solution of the above system. We need to bound $d(xG, v)$ from above. Clearly,

$$d(xG, v) = \sum_{j=1}^{i-1} d\left(x G_j, v |_{G_j}\right) + d\left(x G', v |_{G'}\right). \tag{7}$$

However for every $j \leq i - 1$ we have

$$d\left(x G_j, v |_{G_j}\right) \leq k \leq \frac{c \log n}{\log^{(s)} n} \left\lceil k / \left(\frac{c \log n}{\log^{(s)} n}\right) \right\rceil \leq \tag{8}$$

$$d\left(l^* |_{G_j}, v |_{G_j}\right) \left\lceil \frac{k \log^{(s)} n}{c \log n} \right\rceil,$$

by our choice of i. Also, $x G_i = l^* |_{G_i}$ and formula (2) yield

$$x G' = l^* |_{G'}, \tag{9}$$

Combining formulae (8), (9) and (7) we get $d(xG, v) \leq d(l^*, v) \lceil k \log^{(s)} n / c \log n \rceil$.

We now proceed to the $k \geq t$ case. Again, let $i \in [h]$ be the smallest index such that $|G_i \cap E| \leq b$. Note that one of the sets $W \subseteq G_i$ considered when processing the class G_i will necessarily have an empty intersection with the set E. Let $x \in \mathbb{F}_2^k$ be an approximate solution of the corresponding problem of minimizing $d(x G', v |_{G'})$, subject to $x G |_W = v |_W$, produced by an algorithm from the induction hypothesis. We need to bound $d(xG, v)$ from above. Formulae (7) and (8) reduce our task to bounding $d(x G', v |_{G'})$. Observe that when minimizing $d(x G', v |_{G'})$, subject to $x G |_W = v |_W$, we are minimizing over an affine space of dimension k', where

$$k' \leq \lceil k/t \rceil b \leq \left\lceil \frac{k \log^{(s)} n}{2c \log n \log^{(s-1)} n} \right\rceil \frac{c \log n}{\log^{(s)} n}.$$

Note that $k \geq t$ implies

$$\left\lceil \frac{k \log^{(s)} n}{2c \log n \log^{(s-1)} n} \right\rceil \leq \frac{k \log^{(s)} n}{c \log n \log^{(s-1)} n}.$$

Therefore $k' \leq k / \log^{(s-1)} n$ and the approximation algorithm from the induction hypothesis yields a $\lceil k / c \log n \rceil$-approximate solution, i.e.,

$$d(x G', v |_{G'}) \leq d(l^* |_{G'}, v |_{G'}) \lceil k / c \log n \rceil. \tag{10}$$

Combining formulae (8), (10) and (7) we get $d(xG, v) \leq d(l^*, v) \lceil k \log^{(s)} n / c \log n \rceil$.

To estimate the running time note that the external loop of our algorithm makes $O(n)$ iterations and the internal loop makes at most $\binom{t}{b}$ iterations where each iteration involves a recursive $n^{O(cs-c)}$ time call if $k \geq t$. It is easy to see that

$$\binom{t}{b} \leq (et/b)^b \leq \left(\frac{4ec \log n \log^{(s-1)} n}{\log^{(s)} n} \frac{c \log^{(s)} n}{\log n} \right)^{c \log n / \log^{(s)} n} = n^{O(c)},$$

where the second inequality follows from $b \leq t/2$ and $t \leq 4c \log n \log^{(s-1)} n / \log^{(s)} n$. Combining the estimates above we conclude that the total running time of our algorithm is $n^{O(cs)}$.

Choosing $s = \lceil \log^* n \rceil$ in theorem 2 we obtain.

Theorem 3. *Let $c \geq 1$ be an arbitrary constant. There exists a deterministic $n^{O(c \log^* n)}$ time $\lceil k/c \log n \rceil$-approximation algorithm for the NCP.*

4   The Remote Point Problem

We start with a formal statement of the remote point problem.

Remote point problem

- INSTANCE: A linear code $L = \{xG \mid x \in \mathbb{F}_2^k\}$ given by a generator matrix $G \in \mathbb{F}_2^{k \times n}$.
- SOLUTION: A point $v \in \mathbb{F}_2^n$.
- OBJECTIVE FUNCTION (to be maximized): The Hamming distance $d(L, v)$ from the code L to a point v.

We start with an algorithm that generates $c \log n$-remote points for linear spaces of dimension $k \leq n/2$.

Theorem 4. *Let $c \geq 1$ be an arbitrary constant. There exists a deterministic $n^{O(c)}$ time algorithm that for a given linear space $L \subseteq \mathbb{F}_2^n, \dim L \leq n/2$ generates a point v such that $d(L, v) \geq c \log n$, provided n is large enough.*

Proof. At the first phase of our algorithm we set $d = \lceil c \log n \rceil$, $t = \lceil 4c \log n \rceil$ and use lemma 1 to obtain a family of $\binom{t}{d} = n^{O(c)}$ linear spaces M_S, $S \subseteq [t], |S| = d$ such that

$$L + B_{\lceil c \log n \rceil} \subseteq \bigcup_S M_S.$$

It is readily seen from the construction of lemma 1 that the dimension of every space M_S is at most $n/2 + n/3 = 5n/6$, provided n is large enough.

At the second phase of our algorithm we generate a point v that is not contained in the union $\bigcup_S M_S$, (and therefore is $\lceil c \log n \rceil$-remote from L.) We consider a potential function Φ that for every set $W \subseteq \mathbb{F}_2^n$ returns

$$\Phi(W) = \sum_S |W \cap M_S|,$$

where the sum is over all $S \subseteq [t], |S| = d$. We assume that n is large enough, so that

$$\Phi(\mathbb{F}_2^n) = \sum_S |M_S| = \binom{t}{d} |M_S| < 2^n.$$

We initially set $W = \mathbb{F}_2^n$ and iteratively reduce the size of W by a factor of two (cutting W with coordinate hyperplanes). At every iteration the value of $\Phi(W)$ gets reduced by a factor of two or more. Therefore after n iterations we arrive at a set W that contains a single point v such that $\Phi(\{v\}) = 0$. That point is $\lceil c \log n \rceil$-remote from L. For a set $W \subseteq \mathbb{F}_2^n$, $i \in [n]$, and $b \in \mathbb{F}_2$ let $W|_{x_i=b}$ denote the set $\{x \in W \mid x_i = b\}$. The pseudo-code of our algorithm is below:

Set $t = \lceil 4c \log n \rceil$ and $d = \lceil c \log n \rceil$;
Obtain $\binom{t}{d}$ linear spaces M_S as defined in lemma 1.
Set $W = \mathbb{F}_2^n$;
For every i in $[n]$
    If $\Phi(W|_{x_i=0}) \leq \Phi(W|_{x_i=1})$ **Set** $W = W|_{x_i=0}$; **Else Set** $W = W|_{x_i=1}$;
Output the single element of W;

Note that every evaluation of the potential function Φ in our algorithm takes $n^{O(c)}$ time, since all we need to do is compute the dimensions of $\binom{t}{d} = n^{O(c)}$ affine spaces $W \cap M_S$. The algorithm involves $2n$ such computations and therefore runs in $n^{O(c)}$ time.

Remark 1. It is easy to see that the algorithm of theorem 4 can be extended to generate points that are $c \log n$-far from a given linear space of dimension up to $(1 - \epsilon)n$ for any constant $\epsilon > 0$.

We now present our algorithm for the remote point problem in its full generality.

Theorem 5. *Let $c \geq 1$ be an arbitrary constant. There exists a deterministic $n^{O(c)}$ time algorithm that for a given linear space $L \subseteq \mathbb{F}_2^n, \dim L = k \leq n/2$ generates a point v such that $d(L, v) \geq \lfloor n/2k \rfloor \lceil 2c \log k \rceil$, provided n is large enough.*

Proof. We partition the multi-set of columns of the matrix G in $h = \lceil n/2k \rceil$ multi-sets G_i, $i \in [h]$ in such a way that every multi-set G_i, (with possibly a single exception) has size exactly $2k$. Next for all multi-sets G_i of size $2k$ we use the algorithm of theorem 4 to obtain a point v_i that is $2c \log k$-remote from the space $\{xG_i \mid x \in \mathbb{F}_2^k\} \subseteq \mathbb{F}_2^{2k}$. Finally, we concatenate all vectors v_i together (possibly padding the result with less than $2k$ zeros) to obtain a vector $v \in$ that is $\lfloor n/2k \rfloor \lceil 2c \log k \rceil$-remote from L.

5 Conclusion

In this paper we have given three new deterministic approximation algorithms for the nearest codeword problem. Our algorithms improve substantially upon the (previously best known) deterministic algorithm of [3]. Moreover, our algorithms approach (though do not match) the performance of the randomized algorithm of [3]. Obtaining a complete derandomization remains a challenging open problem.

We have also initiated a study of the remote point problem that asks to find a point far from a given linear space $L \subseteq \mathbb{F}_2^n$. We presented an algorithm that achieves a remoteness of $\Omega(n \log k/k)$ for linear spaces of dimension $k \leq n/2$. We consider further research on the remote point problem (and the related remote set problem) to be a promising approach to constructing explicit rigid matrices in the sense of Valiant [10].

Acknowledgement

Sergey Yekhanin would like to thank Venkat Guruswami for many helpful discussions regarding this work.

References

1. Alekhnovich, M.: More on average case vs. approximation complexity. In: Proc. of the 44th IEEE Symposium on Foundations of Computer Science (FOCS), pp. 298–307 (2003)
2. Arora, S., Babai, L., Stern, J., Sweedyk, Z.: Hardness of approximate optima in lattices, codes, and linear systems. Journal of Computer and System Sciences 54(2), 317–331 (1997)
3. Berman, P., Karpinski, M.: Approximating minimum unsatisfiability of linear equations. In: Proc. of ACM-SIAM Symposium on Discrete Algorithms (SODA), pp. 514–516 (2002)
4. Dumer, I., Miccancio, D., Sudan, M.: Hardness of approximating the minimum distance of a linear code. IEEE Transactions on Information Theory 49(1), 22–37 (2003)
5. Friedman, J.: A note on matrix rigidity. Combinatorica 13(2), 235–239 (1993)
6. Guruswami, V., Micciancio, D., Regev, O.: The complexity of the covering radius problem. Computational Complexity 14, 90–120 (2005)
7. Kashin, B., Razborov, A.: Improved lower bounds on the rigidity of Hadamard matrices. Mathematical Notes 63(4), 471–475 (1998)
8. Lokam, S.: Spectral methods for matrix rigidity with applications to size-depth trade-offs and communication complexity. Journal of Computer and System Sciences 63(3), 449–473 (2001)
9. Shokrollahi, M., Speilman, D., Stemann, V.: A remark on matrix rigidity. Information Processing Letters 64(6), 283–285 (1997)
10. Valiant, L.: Graph-theoretic arguments in low level complexity. In: Proc. of 6th Symposium on Mathematical Foundations of Computer Science (MFCS), pp. 162–176 (1977)

Appendix: Explicit Rigid Sets

The definition of a rigid set involves three parameters. Specifically, to get implications in complexity theory we want to obtain explicit subsets of $\mathbb{F}_2^n$ of *size*

$O(n)$ that for any linear space $L \subseteq \mathbb{F}_2^n$ of *dimension* $n/2$ contain a point at *distance* at least n^ϵ from L.

Given that we are currently very far from constructing explicit sets with the desired values of all three parameters it is natural to approach the problem by studying the trade-offs. Historically, the research on matrix rigidity [5, 7–9] has focused on the trade-off between the values of dimension and distance that can be obtained by explicit sets of size n.

In the next theorem we initiate a study of a trade-off between the values of *size* and *distance*, when the *dimension* is set to $n/2$.

Theorem 6. *For every* $0 \le d \le O(n)$ *there exists an explicit set* $A \subseteq \mathbb{F}_2^n$ *of size* $2^{O(d)}n/d$ *such that for any linear space* $L \subseteq \mathbb{F}_2^n$, $\dim L = n/2$ *one of the points of* A *is more than* d-*far from* L.

Proof. Observe that there exists a constant $c > 0$ such that for any linear space L of dimension $n/2$ there is a point in $\mathbb{F}_2^n$ that is more than cn-far from L.

To obtain the set A, split the coordinates into $n/d\lceil 1/c\rceil$ sets of size $d\lceil 1/c\rceil$ each, and in each set take all binary vectors with support on this set. A consists of all these vectors. Note that every vector in $\mathbb{F}_2^n$ is the sum of at most cn/d vectors of our set A, whose size is $2^{O(d)}n/d$.

Now suppose that L is a linear space of dimension $n/2$ and every vector in A is at most d-far from L. Then any vector of A is a sum of a vector of L and at most d unit vectors. Hence any vector in $\mathbb{F}_2^n$ is a sum of a vector of L and at most $d(cn/d)$ unit vectors, contradicting the fact that there exists a vector that are more than cn-far from L.

Strong Parallel Repetition Theorem for Free Projection Games

Boaz Barak[1,*], Anup Rao[2], Ran Raz[3], Ricky Rosen[4], and Ronen Shaltiel[5,**]

[1] Department of Computer Science, Princeton University
[2] Institute for Advanced Study
[3] Faculty of mathematics and computer science, Weizmann Institute
[4] Department of Computer Science, Tel-Aviv University
[5] Department of Computer Science, Haifa University

Abstract. The parallel repetition theorem states that for any two provers one round game with value at most $1 - \epsilon$ (for $\epsilon < 1/2$), the value of the game repeated n times in parallel is at most $(1 - \epsilon^3)^{\Omega(n/\log s)}$ where s is the size of the answers set [Raz98],[Hol07]. For *Projection Games* the bound on the value of the game repeated n times in parallel was improved to $(1 - \epsilon^2)^{\Omega(n)}$ [Rao08] and was shown to be tight [Raz08]. In this paper we show that if the questions are taken according to a product distribution then the value of the repeated game is at most $(1 - \epsilon^2)^{\Omega(n/\log s)}$ and if in addition the game is a *Projection Game* we obtain a *strong parallel repetition* theorem, i.e., a bound of $(1 - \epsilon)^{\Omega(n)}$.

1 Introduction

In a two provers one round game there are two *provers* and a *verifier*. The verifier selects randomly $(x, y) \in X \times Y$, a question for each prover, according to some distribution P_{XY} where X is the questions set of prover 1 and Y is the questions set of prover 2. Each prover knows only the question addressed to her, prover 1 knows only x and prover 2 knows only y. The provers cannot communicate during the transaction. The provers send their answers to the verifier, $a = a(x) \in A$ and $b = b(y) \in B$ where A is the answers set of the first prover and B is the answers set of the second prover. The verifier evaluates an acceptance predicate $V(x, y, a, b)$ and accepts or rejects based on the outcome of the predicate. The acceptance predicate as well as the distribution of the questions are known in advance to the provers. The provers answer the questions according to a strategy which is a pair of functions $f_a : X \to A$, $f_b : Y \to B$. The strategy of the provers is also called a protocol. If $P_{XY} = P_X \cdot P_Y$, that is P_{XY} is a product distribution, we say that the game is a *free game*.

The *value* of the game is the maximum of the probability that the verifier accepts, where the maximum is taken over all the provers strategies. More formally, the value of the game is:

---

[*] Supported by NSF grants CNS-0627526, CCF-0426582 and CCF-0832797, US-Israel BSF grant 2004288 and Packard and Sloan fellowships.
[**] This research was supported by BSF grant 2004329 and ISF grant 686/07.

I. Dinur et al. (Eds.): APPROX and RANDOM 2009, LNCS 5687, pp. 352–365, 2009.
© Springer-Verlag Berlin Heidelberg 2009

$$\max_{f_a,f_b} \mathbb{E}_{xy}\left[V\left(x,y,f_a(x),f_b(y)\right)\right]$$

where the expectation is taken with respect to the distribution P_{XY}.

Roughly speaking, the n-fold parallel repetition of a game G is a game in which the provers try to win simultaneously n copies of G and it is denoted by $G^{\otimes n}$. More precisely, the verifier sends n questions to each prover, $(x_1, x_2 \ldots, x_n)$ to prover 1 and $(y_1, y_2 \ldots, y_n)$ to prover 2 where for all i, (x_i, y_i) is distributed according to P_{XY} and is independent of the other questions. The provers generate n answers, $(a_1, a_2 \ldots, a_n)$ by prover 1 and $(b_1, b_2 \ldots, b_n)$ by prover 2. The verifier evaluates the acceptance predicate on each coordinate and accepts if and only if all the predicates accept, namely if and only if $V^{\otimes n} = \wedge_{i=1}^n V(x_i, y_i, a_i, b_i) = 1$. Note that the verifier treats each of the n games independently, but the provers may not; the answer of each question addressed to a prover may depend on all the questions addressed to that prover. There are examples of games where the value of the game repeated n times in parallel is strictly larger than the value of the original game to the power of n [For89], [FV02], [Raz08].

The Parallel Repetition Theorem. A series of papers deal with the nature of the value decrease of games repeated n times in parallel. The parallel repetition theorem of Raz [Raz98] states that for every game G with value at most $1 - \epsilon$ where $\epsilon < 1/2$, the value of $G^{\otimes n}$ is at most $(1 - \epsilon^{32})^{\Omega(n/\log s)}$ where s is the size of the answers support $s = |A \times B|$. In a recent elegant result, Holenstein [Hol07] improved the bound to $(1 - \epsilon^3)^{\Omega(n/\log s)}$ while simplifying the proof of [Raz98]. Subsequently, for the important special type of games known as *projection games*, Rao [Rao08] proved a bound of $(1-\epsilon^2)^{\Omega(n)}$ (for a special type of projection games known as *XOR games* such a bound was previously proven by Feige, Kindler and O'Donnell [FKO07]). Note that Rao's [Rao08] bound does not depend on the size of the answers set, s. In the general case, Feige and Verbitsky [FV02] showed that the dependency on s is tight (up to loglog factors).

Many researchers studied the problem of whether there exists a strong parallel repetition theorem in the general case or at least in some important special cases. Namely, is it the case that for a given game G of value $1 - \epsilon$, say, for $\epsilon < 1/2$, the value of $G^{\otimes n}$ is at most $(1 - \epsilon)^{\Omega(n/\log s)}$? This question was motivated by connections to hardness of approximation as well as connections to problems in geometry [FKO07], [SS07]. A recent result of Raz [Raz08] showed a counterexample for the general case, as well as for the case of projection games, unique games and XOR games. Raz [Raz08] showed that there is an example of a XOR game (thus also projection game and unique game) of value $1 - \epsilon$ such that for large enough n, the value of the game is at least $(1 - \epsilon^2)^{O(n)}$. For some extensions, generalization and applications see Barak, Hardt, Haviv, Rao, Regev and Steurer [BHH+08], Kindler, O'Donnell, Rao and Wigderson [KORW08] and Alon and Klartag [AK08].

Other related results: For the special case of unique games played on expander graphs Arora, Khot, Kolla, Steurer, Tulsiani and Vishnoi [AKK+08] proved an "almost" strong parallel repetition theorem (strong up to a polylogarithmic factor). For the special case of games where the roles of the two players are symmetric and the game is played on an expander graph that contains a self loop on every vertex, Safra and Schwartz [SS07] showed that $O(1/\epsilon)$ repetitions are sufficient to reduce the value of the game from $1 - \epsilon$ to some constant.

In this paper we prove a strong parallel repetition theorem for free projection games and we improve the known bound for every free game. More precisely:

1. For every **Free game** of value $\leq (1 - \epsilon)$ for $\epsilon < 1/2$, the value of $G^{\otimes n}$ is at most $(1 - \epsilon^2)^{\Omega(n/\log s)}$
2. For every **Free Projection game** of value $\leq (1 - \epsilon)$ for $\epsilon < 1/2$, the value of $G^{\otimes n}$ is at most $(1 - \epsilon)^{\Omega(n)}$

Techniques. The main technical contribution of this paper is the ability to work throughout the whole proof with relative entropy without the need to switch to ℓ_1 norm. In previous results [Raz98], [Hol07], [Rao08] a bound on the distance between a distribution "generated by the provers' strategies" and the original distribution was derived using the relative entropy between the two distributions. This bound was then used to obtain a bound on the ℓ_1 distance between those distributions. This was done using the fact that $\|P - Q\|_1 \leq O(\sqrt{D(P\|Q)})$ where $D(P\|Q)$ is the relative entropy between P and Q. Since the bound is quadratic, there is a loss when using the ℓ_1 norm instead of using directly the relative entropy. We show that for the special case of free games one can redo the whole proof using relative entropy, without switching to ℓ_1 norm. We bound the value of a game by using our Corollary 1 (that might be useful for other applications). We note that since we are only considering free games, the proof is simpler than the one for general games and we do not use much of the machinery used in previous results, e.g., [Raz98], [Hol07], [Rao08].

2 Preliminaries

2.1 Notations

General Notations. We denote an n-dimensional vector by a superscript n, e.g., $\phi^n = (\phi_1, \ldots, \phi_n)$ where ϕ_i is the ith coordinate. The function $\log(x)$ is the logarithm base 2 of x. We use the common notation $[n]$ to denote the set $\{1, \ldots, n\}$.

Random Variables and Sets. By slightly abusing notations, we will use capital letters to denote both sets and random variables distributed over these sets, and we will use lower case letters to denote values. For example, X, Y will denote sets as well as random variables distributed over these sets, and x, y will denote values in these sets that the random variables can take. Nevertheless, it will always be clear from the context whether we are referring to sets or random variables. For a random variable Z it will be convenient in some lemmas, such as Lemma 4, to think of $\Pr(Z)$ as a random variable.

Random Variables and their Distributions. For a random variable X, we denote by P_X the distribution of X. For an event U we use the notation $P_{X|U}$ to denote the distribution of $X|U$, that is, the distribution of X conditioned on the event U. If Z is an additional random variable that is fixed (e.g., inside an expression where an expectation over Z is taken), we denote by $P_{X|Z}$ the distribution of X conditioned on Z. In the same way, for two (or more) random variables X, Y, we denote their joint distribution by P_{XY}, and we use the same notations as above to denote conditional distributions. For example, for an event U, we write $P_{XY|U}$ to denote the distribution of X, Y conditioned on the event U, i.e., $P_{XY|U}(x, y) = \Pr(X = x, Y = y|U)$. For two (or more) random variables X, Y with distribution P_{XY}, we use the notation P_X to denote the marginal distribution of X.

The Game G. We denote a game by G and define X to be the set of questions to prover 1, Y to be the set of questions to prover 2 and P_{XY} to be the joint distribution according to which the verifier chooses a pair of questions to the provers. We denote by A the set of answers of prover 1 and by B the set of answers of prover 2. We denote the acceptance predicate by V. A game G with acceptance predicate V and questions distribution P_{XY} is denoted by $G(P_{XY}, V)$. As mentioned above, we also denote by X, Y, A, B random variables distributed over X, Y, A, B respectively. X, Y will be the questions addressed to the two provers, distributed over the question sets X and Y respectively. Fixing a strategy f_a, f_b for the game G, we can also think of the answers A and B as random variables distributed over the answer sets A and B respectively.

The Game G Repeated n Times. For the game G repeated n times in parallel, $G^{\otimes n} = G(P_{X^n Y^n}, V^{\otimes n})$, the random variable X_i denotes the question to prover 1 in coordinate i, and similarly, the random variable Y_i denotes the question to prover 2 in coordinate i. We denote by X^n the tuple $(X_1, \dots, X_n)$ and by Y^n the tuple $(Y_1, \dots, Y_n)$. Fixing a strategy f_a, f_b for $G^{\otimes n}$, the random variable A_i denotes the answer of prover 1 in coordinate i, and similarly, the random variable B_i denotes the answer of prover 2 in coordinate i. We denote by A^n the tuple $(A_1, \dots, A_n)$ and by B^n the tuple $(B_1, \dots, B_n)$. It will be convenient in some lemmas to denote $X^k = (X_{n-k+1}, \dots, X_n)$, i.e., the last k coordinates of X^n and in the same way, $Y^k = (Y_{n-k+1}, \dots, Y_n)$, $A^k = (A_{n-k+1}, \dots, A_n)$ and $B^k = (B_{n-k+1}, \dots, B_n)$. We also denote $X^{n-k} = (X_1, \dots, X_{n-k})$, i.e., the first $n - k$ coordinates of X^n, and similarly, $Y^{n-k} = (Y_1, \dots, Y_{n-k})$. For fixed $i \in [n-k]$, we denote $X^m = (X_1, \dots, X_{i-1}, X_{i+1}, \dots, X_{n-k})$, i.e., X^{n-k} without X_i, and similarly, $Y^m = (Y_1, \dots, Y_{i-1}, Y_{i+1}, \dots, Y_{n-k})$.

The Event W_i. For the game $G^{\otimes n} = G(P_{X^n Y^n}, V^{\otimes n})$ and a strategy

$$f_a : X^n \to A^n, f_b : Y^n \to B^n$$

we can consider the joint distribution:

$$P_{X^n, Y^n, A^n, B^n}(x^n, y^n, a^n, b^n) = \begin{cases} P_{X^n, Y^n}(x^n, y^n) & \text{if } a^n = f_a(x^n), b^n = f_b(y^n) \\ 0 & \text{otherwise} \end{cases}$$

We define the event W_i to be the event of winning the game in coordinate i, i.e., the event that the verifier accepts on coordinate i. Since the random variables A^n and B^n are functions of X^n and Y^n respectively, we can think of W_i as an event in the random variables X^n, Y^n.

2.2   Special Types of Games

Definition 1 (Free Games). *A game is* Free *if the distribution of the questions is a product distribution, i.e.,* $P_{XY} = P_X \times P_Y$

Definition 2 (Projection Games). *A* Projection *game is a game where for each pair of questions* x, y *there is a function* $f_{xy} : B \to A$ *such that* $V(x, y, a, b)$ *is satisfied if and only if* $f_{xy}(b) = a$.

2.3   Entropy and Relative Entropy

Definition 3 (Entropy). *For a probability distribution* ϕ *over a sample space* Ω *we define the entropy of* ϕ *to be*

$$H(\phi) = - \sum_{x \in \Omega} \phi(x) \log \phi(x) = -\mathbb{E}_{x \sim \phi} \log \phi(x) = \mathbb{E}_{x \sim \phi} \log \left(\frac{1}{\phi(x)} \right)$$

By applying Jensen's inequality on the concave function $\log(\cdot)$ one can derive the following fact:

Fact 1. *For every distribution* ϕ *over* Ω, $H(\phi) \leq \log(|\mathrm{supp}(\phi)|)$ *where*

$$\mathrm{supp}(\phi) = \{x \in \Omega | \phi(x) > 0\}$$

Definition 4 (Relative Entropy). *We define Relative Entropy, also called the Kullback-Leibler Divergence or simply* divergence. *Let* P *and* Q *be two probability distributions defined on the same sample space* Ω. *The relative entropy of* P *with respect to* Q *is:*

$$D(P\|Q) = \sum_{x \in \Omega} P(x) \log \frac{P(x)}{Q(x)}$$

where $0 \log \frac{0}{0}$ *is defined to be* 0 *and* $p \log \frac{p}{0}$ *where* $p \neq 0$ *is defined to be* ∞.

Vaguely speaking, we could think of the relative entropy as a way to measure the information we gained by learning that a random variable is distributed according to P when apriority we thought that it was distributed according to Q. This indicates how *far* Q is from P; if we don't gain much information then the two distributions are very *close* in some sense. Note that the relative entropy is not symmetric (and therefore is not a metric).

Fact 2. *Let* $\Phi^n = \Phi_1 \times \Phi_2 \times \cdots \times \Phi_n$ *and let* μ^n *be any distribution over the same sample space (not necessarily a product distribution) then* $\sum_{i=1}^n D(\mu_i \| \Phi_i) \leq$ $D(\mu^n \| \Phi^n)$ *thus* $\mathbb{E}_{i \in [n]} D(\mu_i \| \Phi_i) = \dfrac{1}{n} \sum_{i \in [n]} D(\mu_i \| \Phi_i) \leq \dfrac{D(\mu^n \| \Phi^n)}{n}$.

3   Our Results

We prove the following theorems:

Theorem 3 (Parallel Repetition For Free Games). *For every game* G *with value* $1 - \epsilon$ *where* $\epsilon < 1/2$ *and* $P_{XY} = P_X \times P_Y$ *(the questions are distributed according to some product distribution), the value of* $G^{\otimes n}$ *is at most* $(1 - \epsilon^2/9)^{n/(18 \log s + 3)}$

Theorem 4 (Strong Parallel Repetition For Free Projection Games). *For every projection game* G *with value* $1 - \epsilon$ *where* $\epsilon < 1/2$ *and* $P_{XY} = P_X \times P_Y$ *(the questions are distributed according to some product distribution), the value of* $G^{\otimes n}$ *is at most* $(1 - \epsilon/9)^{(n/33) - 1}$

3.1   Technical Lemma

Lemma 1. *For every* $0 \le p, q \le 1$ *define binary distributions* $P = (p, 1 - p)$ *and* $Q = (q, 1 - q)$, *over* $\{0, 1\}$, *if* $D(P\|Q) \le \delta$ *and* $p < \delta$ *then*

$$q \le 4\delta$$

Proof. If $\delta \ge \frac{1}{4}$ then the statement is obviously true. For the case that $\delta < \frac{1}{4}$, assume by way of contradiction that $q > 4\delta$. Since for $q > p$, $D(P\|Q)$ is decreasing in p and increasing in q,

$$D(P\|Q) = p \log \frac{p}{q} + (1 - p) \log \frac{1 - p}{1 - q}$$

$$> \delta \log\left(\frac{\delta}{4\delta}\right) + (1 - \delta) \log \frac{1 - \delta}{1 - 4\delta}$$

$$= -2\delta + (1 - \delta) \log \left(1 + \frac{3\delta}{1 - 4\delta}\right) \tag{1}$$

If $\delta \ge 1/7$ then $\log \left(1 + \frac{3\delta}{1 - 4\delta}\right) \ge 1$. Thus,

$$(1) \ge -2\delta + (1 - \delta) > \delta$$

where the last inequality follows since $\delta < 1/4$.

If $\delta < 1/7$ then $\frac{3\delta}{1 - 4\delta} < 1$. Using the inequality $\log_2(1 + x) \ge x$ for every $0 \le x \le 1$ we obtain,

$$(1) \ge -2\delta + (1 - \delta) \frac{3\delta}{1 - 4\delta} \ge -2\delta + 3\delta = \delta$$

where the last inequality follows since $\frac{1 - \delta}{1 - 4\delta} > 1$. Since we obtained a contradiction in both cases, the lemma holds.

Corollary 1. *For every probability distributions* P,Q *over the same sample space* Ω *and for every* $T \subseteq \Omega$, *if* $D(P\|Q) \le \delta$ *and* $P(T) \le \delta$ *then* $Q(T) \le 4\delta$

Proof. Denote $p = P(T)$ and $q = Q(T)$ and let $P' = (p, 1 - p)$, $Q' = (q, 1 - q)$. By the data processing inequality for mutual information $D(P\|Q) \ge D(P'\|Q')$ and the corollary follows.

3.2   Main Lemmas

We now state the main lemmas for general product distribution games.

Recall that for a coordinate i, W_i is the event of the provers winning the game played in this coordinate.

Lemma 2 (Main Lemma For General Free Games). *Let G be a free game with value $1 - \epsilon$. For any set T of k coordinates, $(T \subseteq [n]$ and $|T| = k)$, let W be the event of the provers winning the games in those k coordinates. If $\Pr(W) \geq 2^{-\epsilon(n-k)/9+k\log s}$ where s is the size of the answers set, then there is $i \notin T$ for which*

$$\Pr(W_i|W) \leq 1 - \frac{\epsilon}{9}$$

Lemma 3 (Main Lemma For Free Projection Games). *Let G be a free projection game with value $1 - \epsilon$. For any set T of k coordinates, $(T \subseteq [n]$ and $|T| = k)$, let W be the event of the provers winning the games in those k coordinates. If $\Pr(W) \geq 2^{-\epsilon(n-k)/144}$ and $n - k \geq (48/\epsilon)\log(8/\epsilon)$ then there is $i \notin T$ for which*

$$\Pr(W_i|W) \leq 1 - \frac{\epsilon}{9}$$

In the lemmas below we assume without loss of generality that the set T of k coordinates is the set of the last k coordinates. Recall that $P_{X^n Y^n} = P_{XY} \times \cdots \times P_{XY}$ n-times. Recall that $X^k = (X_{n-k+1}, \ldots, X_n)$, i.e., the last k coordinates of X^n and in the same way, $Y^k = (Y_{n-k+1}, \ldots, Y_n)$, $A^k = (A_{n-k+1}, \ldots, A_n)$ and $B^k = (B_{n-k+1}, \ldots, B_n)$. Recall that $X^{n-k} = (X_1, \ldots, X_{n-k})$, i.e., the first $n-k$ coordinates of X^n, and similarly, $Y^{n-k} = (Y_1, \ldots, Y_{n-k})$.

Lemma 4. *For any event[1] U, the following holds:*

$$\mathbb{E}_{X^k, Y^k, A^k|U} D\left(P_{X^{n-k}, Y^{n-k}|X^k, Y^k, A^k, U} \| P_{X^{n-k}, Y^{n-k}}\right)$$
$$\leq \log\left(\frac{1}{\Pr(U)}\right) + \mathbb{E}_{X^k, Y^k|U} H(P_{A^k|X^k, Y^k, U})$$

The proof is given in the full version of the paper.

We define W to be the event that the provers win all the games in the last k coordinates and define E to be

$$\left\{(a^k, x^k, y^k) \in A^k \times X^k \times Y^k \,\Big|\, \Pr(A^k = a^k|X^k = x^k, Y^k = y^k) \geq 2^{-\epsilon(n-k)/16})\right\}.$$

The event W' is defined as $W \wedge [(A^k, X^k, Y^k) \in E]$.

Proposition 1. *For W and W', the events defined above, the following holds:*

1. For general games and the event W

$$\mathbb{E}_{X^k, Y^k|W} H\left(P_{A^k|X^k, Y^k, W}\right) \leq k\log s \qquad [Raz98],[Hol07]$$

---

[1] We will use the lemma for events that depend only on X^k, Y^k, A^k, B^k, e.g., we will use it for the event W, see definition in Lemma 2.

2. *For projection games and the event W'*

$$\mathbb{E}_{X^k,Y^k|W'} \mathrm{H}\left(\mathrm{P}_{A^k|X^k,Y^k,W'}\right) \leq \epsilon(n-k)/16 \quad \text{[Rao08]}$$

Proof (For general games). We use the trivial bound on the size of the support, namely, for every x^k, y^k we can bound

$$|\mathrm{supp}(\mathrm{P}_{A^k|X^k=x^k,Y^k=y^k,W})| \leq |\mathrm{supp}(\mathrm{P}_{A^k})| \leq s^k$$

where s is the size of the answers set. Using Fact 1 we obtain:

$$\mathbb{E}_{X^k,Y^k|W} \mathrm{H}\left(\mathrm{P}_{A^k|X^k,Y^k,W}\right) \leq \mathbb{E}_{X^k,Y^k|W} \log(|\mathrm{supp}(\mathrm{P}_{A^k|X^k,Y^k,W})|) \leq \log s^k$$

Proof (For projection games). Using Fact 1 we can trivially bound:

$$\mathbb{E}_{X^k,Y^k|W'} \mathrm{H}\left(\mathrm{P}_{A^k|X^k,Y^k,W'}\right) \leq \mathbb{E}_{X^k,Y^k|W'} \log(|\mathrm{supp}(\mathrm{P}_{A^k|X^k,Y^k,W'})|) \quad (2)$$

Since for every x^k, y^k and $a^k \in \mathrm{supp}(\mathrm{P}_{A^k|X^k=x^k,Y^k=y^k,W'})$,

$$\Pr(A^k = a^k|X^k = x^k, Y^k = y^k) \geq 2^{-\epsilon(n-k)/16},$$

there are at most $2^{\epsilon(n-k)/16}$ such a^k. Hence,

$$(2) \leq \mathbb{E}_{X^k,Y^k|W'} \log\left(2^{\epsilon(n-k)/16}\right) = \epsilon(n-k)/16$$

Corollary 2. *For the events W, W' the following holds:*

1. *For general games and the event W*

$$\mathbb{E}_{i\in[n-k]}\mathbb{E}_{X^k,Y^k,A^k|W} \mathrm{D}\left(\mathrm{P}_{X_i,Y_i|X^k,Y^k,A^k,W} \| \mathrm{P}_{X_i,Y_i}\right)$$
$$\leq \frac{1}{n-k}\left(k\log s - \log(\Pr(W))\right)$$

2. *For projection games and the event W'*

$$\mathbb{E}_{i\in[n-k]}\mathbb{E}_{X^k,Y^k,A^k|W'} \mathrm{D}\left(\mathrm{P}_{X_i,Y_i|X^k,Y^k,A^k,W'} \| \mathrm{P}_{X_i,Y_i}\right)$$
$$\leq \frac{1}{n-k}\left(\epsilon(n-k)/16 - \log\left(\Pr(W) - 2^{-\epsilon(n-k)/16}\right)\right)$$

(for $z < 0$ we define $\log(z) = -\infty$.)

Proof. For the general case, fixing $U = W$ in Lemma 4 and using the bound on $\mathbb{E}_{X^k,Y^k|W} \mathrm{H}\left(\mathrm{P}_{A^k|X^k,Y^k,W}\right)$ from Proposition 1 we obtain:

$$\mathbb{E}_{X^k,Y^k,A^k|W} \mathrm{D}\left(\mathrm{P}_{X^{n-k},Y^{n-k}|X^k,Y^k,A^k,W} \| \mathrm{P}_{X^{n-k},Y^{n-k}}\right) \leq k\log s - \log(\Pr(W))$$

To complete the proof apply Fact 2.

For the projection game case, fix $U = W'$ in Lemma 4 and use the bound on $\mathbb{E}_{X^k,Y^k|W'} \mathrm{H}\left(\mathrm{P}_{A^k|X^k,Y^k,W'}\right)$ from Proposition 1 to obtain:

$$\mathbb{E}_{X^k,Y^k,A^k|W'} \mathrm{D}\left(\mathrm{P}_{X^{n-k},Y^{n-k}|X^k,Y^k,A^k,W'} \| \mathrm{P}_{X^{n-k},Y^{n-k}}\right)$$
$$\leq \epsilon(n-k)/16 - \log(\Pr(W'))$$

We bound $\Pr(W')$ in the following way:

$$\Pr(W') = \Pr(W \wedge [(A^k, X^k, Y^k) \in E]) = \Pr(W) - \Pr(W \wedge [(A^k, X^k, Y^k) \notin E])$$

We now bound the term $\Pr(W \wedge [(A^k, X^k, Y^k) \notin E])$. For every game G and strategy f_a, f_b, the probability of winning the game played with strategy f_a, f_b is

$$\mathbb{E}_{X,Y} \sum_{b \in B} \Pr(B = b|Y) \sum_{a \in A} \Pr(A = a|X) V(X, Y, a, b).$$

Recall that for every projection game G and every $x \in X, y \in Y, b \in B$ there is only one $a \in A$ for which $V(x, y, a, b) = 1$, this a is $f_{xy}(b)$ (recall that f_{xy} is the projection function, see Definition 2). Thus for every projection game G and strategy f_a, f_b, the probability of winning the game played according to f_a, f_b is:

$$\mathbb{E}_{XY} \sum_{(b, f_{XY}(b)) \in B \times A} \Pr(B = b|Y) \Pr(A = a|X).$$

For x^k, y^k we define $f_{x^k, y^k} : B^k \to A^k$ by $[f_{x^k, y^k}(b^k)]_i = f_{x_i, y_i}(b_i)$. We want to bound the probability of winning in the last k coordinates and that

$$(A^k, X^k, Y^k) \notin E.$$

Thus, for every x^k, y^k we want to sum $\Pr(B^k = b^k|Y^k = y^k) \Pr(A^k = a^k|X^k = x^k)$, only over $(b^k, f_{x^k, y^k}(b^k)) \in B^k \times A^k$ for which $(f_{x^k, y^k}(b^k), x^k, y^k) \notin E$. Thus

$$\Pr(W \wedge [(A^k, X^k, Y^k) \notin E])$$
$$= \mathbb{E}_{X^k, Y^k} \sum_{(b^k, f_{X^k, Y^k}(b^k)) \text{ s.t. } (f_{X^k, Y^k}(b^k), X^k, Y^k) \notin E} \Pr(B^k = b^k|Y^k) \cdot$$
$$\Pr(A^k = f_{X^k, Y^k}(b^k)|X^k, Y^k) < 2^{-\epsilon(n-k)/16} \tag{3}$$

where the last inequality follows since if $(a^k, x^k, y^k) \notin E$ then

$$\Pr(A^k = a^k|X^k = x^k) = \Pr(A^k = a^k|X^k = x^k, Y^k = y^k) < 2^{-\epsilon(n-k)/16}.$$

Thus $\Pr(W') > \Pr(W) - 2^{-\epsilon(n-k)/16}$. We now conclude that

$$\mathbb{E}_{X^k,Y^k,A^k|W'} \mathrm{D}\left(\mathrm{P}_{X^{n-k},Y^{n-k}|X^k,Y^k,A^k,W'} \| \mathrm{P}_{X^{n-k},Y^{n-k}}\right)$$
$$\leq \epsilon(n-k)/16 - \log\left(\Pr(W) - 2^{-\epsilon(n-k)/16}\right)$$

The corollary follows by using Fact 2.

Observation 5. *For any product distribution* $P_{\alpha,\beta} = P_\alpha \times P_\beta$ *and any event* τ *that is determined only by* α *(or only by* β*)* $P_{\alpha,\beta|\tau}$ *is a product distribution*

$$P_{\alpha,\beta|\tau} = P_{\alpha|\tau} \times P_{\beta|\tau} = P_{\alpha|\tau} \times P_\beta$$

(or $P_{\alpha,\beta|\tau} = P_\alpha \times P_{\beta|\tau}$*)*

Proposition 2. *For a free game* G, *an event* U *that is determined by* X^k, Y^k, A^k, B^k *and for every* x^k, y^k, a^k *the following holds:*

$$P_{X^{n-k}Y^{n-k}|X^k=x^k,Y^k=y^k,A^k=a^k,U} = P_{X^{n-k}|X^k=x^k,Y^k=y^k,A^k=a^k,U} \times$$
$$P_{Y^{n-k}|X^k=x^k,Y^k=y^k,A^k=a^k,U}$$

That is $P_{X^{n-k}Y^{n-k}|X^k=x^k,Y^k=y^k,A^k=a^k,U}$ *is a product distribution.*

The proof is given in the full version of the paper.

Corollary 3. *For a free game* G, *any event* U *that is determined by* X^k, Y^k, A^k, B^k *and for every* x^k, y^k, a^k, x, y *and every* $i \in [n-k]$ *the following holds:*

$$P_{X^{n-k}Y^{n-k}|X^k=x^k,Y^k=y^k,A^k=a^k,U,X_i=x,Y_i=y}$$
$$= P_{X^{n-k}|X^k=x^k,Y^k=y^k,A^k=a^k,U,X_i=x} \times P_{Y^{n-k}|X^k=x^k,Y^k=y^k,A^k=a^k,U,Y_i=y}$$

The proof is given in the full version of the paper.

Recall that for fixed $i \in [n-k]$, we denote

$$X^m = (X_1,\ldots,X_{i-1},X_{i+1},\ldots,X_{n-k}),$$

i.e., X^{n-k} without X_i, and similarly, $Y^m = (Y_1,\ldots,Y_{i-1},Y_{i+1},\ldots,Y_{n-k})$.

Proof (Of Lemma 2 and Lemma 3). For both $U = W$ and $U = W'$ and for every x^k, y^k, a^k and $i \in [n-k]$, we will use a strategy for the game $G(P_{X^n,Y^n}, V^{\otimes n})$ to obtain a strategy for the game $G(P_{X_iY_i|X^k=x^k,Y^k=y^k,A^k=a^k,U}, V)$. Fix any strategy, f_a, f_b, for the game $G(P_{X^nY^n}, V^{\otimes n})$, and apply the following to obtain a strategy for $G(P_{X_iY_i|X^k=x^k,Y^k=y^k,A^k=a^k,U}, V)$:

Algorithm 6. *Protocol for* $G(P_{X_iY_i|X^k=x^k,Y^k=y^k,A^k=a^k,U}, V)$ *for fixed* x^k, y^k, a^k, i

1. *When the game starts, prover 1 receives a question* x *and prover 2 receives a question* y *according to* $P_{X_iY_i|X^k=x^k,Y^k=y^k,A^k=a^k,U}$. *Define* $X_i = x, Y_i = y$ *(the provers will play this game in coordinate* i*).*
2. *Prover 1 randomly chooses*
   $x^m = (x_1,\ldots,x_{i-1},x_{i+1},\ldots,x_{n-k})$ *according to*

   $$P_{X^{n-k}|X^k=x^k,Y^k=y^k,A^k=a^k,U,X_i=x}$$

   and Prover 2 randomly chooses $y^m = (y_1,\ldots,y_{i-1},y_{i+1},\ldots,y_{n-k})$ *according to*

   $$P_{Y^{n-k}|X^k=x^k,Y^k=y^k,A^k=a^k,U,Y_i=y}$$

3. *Prover 1 answers* $[f_a(x^n)]_i$ *and prove 2 answers* $[f_b(y^n)]_i$.

Remark 1. Notice that in step 2, since both events $U = W$ and $U = W'$ are determined by X^k, Y^k, A^k, B^k, the joint distribution of x^m, y^m is

$$\mathrm{P}_{X^m,Y^m|X^k=x^k,Y^k=y^k,A^k=a^k,X_i=x,Y_i=y,U}$$

which follows from Corollary 3.

Remark 2. Notice that since Remark 1 holds, the probability of winning the game

$$G(\mathrm{P}_{X_iY_i|X^k=x^k,Y^k=y^k,A^k=a^k,U}, V)$$

is exactly

$$\Pr(W_i|X^k = x^k, Y^k = y^k, A^k = a^k, U).$$

Remark 3. Notice that this is a randomized algorithm. However, it is well known that since any randomized algorithm is a convex combination of deterministic algorithms, there is a deterministic algorithm that achieves the same value as the randomized algorithm. Namely, there is a deterministic protocol for which the probability of winning the game

$$G(\mathrm{P}_{X_iY_i|X^k=x^k,Y^k=y^k,A^k=a^k,U}, V)$$

is exactly

$$\Pr(W_i|X^k = x^k, Y^k = y^k, A^k = a^k, U).$$

Using this remark we will think of this algorithm as a deterministic algorithm.

Proof for General Games. In this version, due to space limitation, we omit this proof and only show the proof for Projection games.

Proof for Projection Games. From Corollary 2 we obtain:

$$\mathbb{E}_{i\in[n-k]}\mathbb{E}_{X^k,Y^k,A^k|W'}\mathrm{D}\left(\mathrm{P}_{X_i,Y_i|X^k,Y^k,A^k,W'}\|\mathrm{P}_{X_i,Y_i}\right) \tag{4}$$

$$\leq \frac{1}{n-k}\left(\epsilon(n-k)/16 - \log\left(\Pr(W) - 2^{-\epsilon(n-k)/16}\right)\right) \tag{5}$$

By the assumption in the lemma, $\Pr(W) \geq 2^{-\epsilon(n-k)/144}$ thus,

$$\mathbb{E}_{i\in[n-k]}\mathbb{E}_{X^k,Y^k,A^k|W'}\mathrm{D}\left(\mathrm{P}_{X_i,Y_i|X^k,Y^k,A^k,W'}\|\mathrm{P}_{X_i,Y_i}\right)$$

$$\leq \epsilon/16 - \frac{1}{n-k}\log\left(2^{-\epsilon(n-k)/144} - 2^{-\epsilon(n-k)/16}\right)$$

$$= \epsilon/16 - \frac{1}{n-k}\log\left(2^{-\epsilon(n-k)/16}\left(2^{\epsilon(n-k)/18} - 1\right)\right)$$

$$= \epsilon/16 + \epsilon/16 - \frac{1}{n-k}\log\left(2^{\epsilon(n-k)/18} - 1\right)$$

$$\leq \epsilon/8 \tag{6}$$

where the last inequality is due to the bound on $n - k$. Assume by way of contradiction that for all $i \in [n - k]$, $\Pr(W_i|W') > 1 - \epsilon/8$. Notice that since

$$\Pr(W_i|W') = \mathbb{E}_{X^k,Y^k,A^k|W'} \Pr(W_i|X^k,Y^k,A^k,W'),$$

an equivalent assumption is that for all $i \in [n - k]$,

$$\mathbb{E}_{X^k,Y^k,A^k|W'} \Pr(\neg W_i|X^k,Y^k,A^k,W') < \epsilon/8.$$

By a simple averaging argument, there are x^k, y^k, a^k and $i \in [n - k]$ for which both equations hold:

$$D\left(P_{X_i,Y_i|X^k=x^k,Y^k=y^k,A^k=a^k,W'}\middle\|P_{X_i,Y_i}\right) \leq \epsilon/4 \tag{7}$$

$$\Pr(\neg W_i|X^k = x^k, Y^k = y^k, A^k = a^k, W') < \epsilon/4 \tag{8}$$

For the strategy f_a, f_b, and for x^k, y^k, a^k, i for which both Equation (7) and Equation (8) hold consider the protocol suggested in Algorithm 6. Recall that by Remark 3 there is a deterministic protocol for which the provers win on coordinate i with probability

$$\Pr(W_i|X^k = x^k, Y^k = y^k, A^k = a^k, W').$$

Denote this deterministic protocol by h_a, h_b. For h_a, h_b, denote by R the set of all questions on which the provers err when playing according to this protocol. By our assumption

$$P_{X_i,Y_i|X^k=x^k,Y^k=y^k,A^k=a^k,W'}(R) < \epsilon/4. \tag{9}$$

Combining Equation (9) with Equation (7), we can apply Corollary 1 to obtain $P_{X_i,Y_i}(R) < \epsilon$. The provers can play h_a, h_b as a strategy for $G(P_{X_i,Y_i}, V)$ and err only on questions in R. Since $P_{X_i,Y_i}(R) < \epsilon$, the value of $G(P_{X_i,Y_i}, V) > 1 - \epsilon$. Since $P_{X_i,Y_i} = P_{XY}$ the value of $G(P_{XY}, V) > 1 - \epsilon$ which is a contradiction.

We showed that there is $i \in [n - k]$ for which

$$\Pr(W_i|W') \leq 1 - \epsilon/8$$

but we need to show that there is $i \in [n - k]$ for which $\Pr(W_i|W) \leq 1 - \epsilon/9$. This is done in the following way: Since $W' \subseteq W$

$$\Pr(W_i|W) = \Pr(W_i|W')\Pr(W'|W) + \Pr(W_i|\neg W')\Pr(\neg W'|W)$$
$$\leq \Pr(W_i|W') + \Pr(\neg W'|W).$$

Thus for all $i \in [n - k]$,

$$\Pr(W_i|W) \leq \Pr(W_i|W') + \Pr((A^k, X^k, Y^k) \notin E|W).$$

Since $\Pr((A^k, X^k, Y^k) \notin E|W) = \Pr(W \wedge [(A^k, X^k, Y^k) \notin E])/\Pr(W)$ we can use the bound in Equation (3), $\Pr(W \wedge [(A^k, X^k, Y^k) \notin E]) < 2^{-\epsilon(n-k)/16}$ and obtain that

$$\Pr(W_i|W) \leq \Pr(W_i|W') + 2^{-\epsilon(n-k)/16}/\Pr(W).$$

Therefore:

$$\Pr(W_i|W) \leq 1 - \epsilon/8 + 2^{-\epsilon(n-k)/16}/2^{-\epsilon(n-k)/144}$$
$$\leq 1 - \epsilon/8 + 2^{-\epsilon(n-k)/18}$$
$$\leq 1 - \epsilon/9$$

where the last inequality follows from the bound on $n - k$.

Proof (Of Theorem 3). In this version, due to space limitation, we omit this proof and only show the proof for Projection games.

Proof (Of Theorem 4). We first prove the case of $n \geq (50/\epsilon)\log(8/\epsilon)$. We show by induction, for every $k \leq (n/33) - 1$ there is a set $T \subseteq [n]$ of k coordinates ($|T| = k$) for which $\Pr(W) \leq (1 - \epsilon/9)^k$ where the event W is winning on all the coordinates in T. For $k = 0$ the statement trivially holds. Assume by induction that there is a set T of size k for which $\Pr(W) \leq (1 - \epsilon/9)^k$. If $\Pr(W) \leq (1 - \epsilon/9)^{k+1}$ then we are done, else

$$\Pr(W) \geq (1 - \epsilon/9)^{k+1} \geq 2^{-\epsilon(k+1)/4.5}.$$

In order to use Lemma 3 we need to make sure that

$$\Pr(W) \geq 2^{-\epsilon(n-k)/144}$$

and that

$$n - k \geq (48/\epsilon)\log(8/\epsilon)$$

Since $k \leq (n/33) - 1$,

$$\text{if } \Pr(W) \geq 2^{-\epsilon(k+1)/4.5} \text{ then } \Pr(W) \geq 2^{-\epsilon(n-k)/144}$$

Since $k \leq (n/33) - 1$ then $n - k \geq 32n/33 + 1$. Since $n \geq 50/\epsilon\log(8/\epsilon)$ then

$$32n/33 + 1 \geq (48/\epsilon)\log(8/\epsilon) + 1.$$

Therefore,

$$n - k \geq (48/\epsilon)\log(8/\epsilon)$$

Now we can apply Lemma 3 to obtain that there is $i \notin T$ for which $\Pr(W_i|W) \leq 1 - \epsilon/9$. Therefore,

$$\Pr(W_i \wedge W) = \Pr(W) \cdot \Pr(W_i|W) \leq (1 - \epsilon/9)^k(1 - \epsilon/9) = (1 - \epsilon/9)^{k+1}$$

For $k = (n/33) - 1$ there is a set $T \subseteq [n], |T| = k$ for which:

$$\Pr(W_1 \wedge \ldots \wedge W_n) \leq \Pr(\bigwedge_{i \in T} W_i) \leq (1 - \epsilon/9)^{(n/33)-1}$$

For the case of $n < (50/\epsilon)\log(8/\epsilon)$, as suggested in [Rao08], it can be shown that if the theorem was false for small n it would not hold for big n. If there was a strategy with success probability greater than $(1 - \epsilon/9)^{(n/33)-1}$ then for the same game played on $m \cdot n$ coordinates the success probability was at least $(1 - \epsilon/9)^{m((n/33)-1)}$ and for large enough m, this yield a contradiction.

References

[AK08]   Alon, N., Klartag, B.: Economical toric spines via cheeger's inequality (manuscript) (2008)

[AKK+08]   Arora, S., Khot, S.A., Kolla, A., Steurer, D., Tulsiani, M., Vishnoi, N.K.: Unique games on expanding constraint graphs are easy: extended abstract. In: STOC. ACM, New York (2008),
            http://doi.acm.org/10.1145/1374376.1374380

[BHH+08]   Barak, B., Hardt, M., Haviv, I., Rao, A., Regev, O., Steurer, D.: Rounding parallel repetitions of unique games. In: FOCS, pp. 374–383 (2008)

[FKO07]   Feige, U., Kindler, G., O'Donnell, R.: Understanding parallel repetition requires understanding foams. In: IEEE Conference on Computational Complexity, pp. 179–192 (2007)

[For89]   Fortnow, L.J.: Complexity - theoretic aspects of interactive proof systems. Technical Report MIT-LCS//MIT/LCS/TR-447, Department of Mathematics, Massachusetts Institute of Technology (1989)

[FV02]   Feige, U., Verbitsky, O.: Error reduction by parallel repetition—a negative result. Combinatorica 22(4), 461–478 (2002)

[Hol07]   Holenstein, T.: Parallel repetition: simplifications and the no-signaling case. In: STOC (2007)

[KORW08]   Kindler, G., O'Donnell, R., Rao, A., Wigderson, A.: Spherical cubes and rounding in high dimensions. In: FOCS, pp. 189–198 (2008)

[Rao08]   Rao, A.: Parallel repetition in projection games and a concentration bound. In: STOC (2008)

[Raz98]   Raz, R.: A parallel repetition theorem. SIAM J. Comput. 27(3), 763–803 (1998)

[Raz08]   Raz, R.: A counterexample to strong parallel repetition. In: FOCS (2008)

[SS07]   Safra, S., Schwartz, O.: On Parallel-Repetition, Unique-Game and Max-Cut (manuscript) (2007)

Random Low Degree Polynomials are Hard to Approximate

Ido Ben-Eliezer[1], Rani Hod[1], and Shachar Lovett[2],[*]

[1] Tel Aviv University
{idobene,ranihod}@tau.ac.il
[2] Weizmann Institute of Science
shachar.lovett@weizmann.ac.il

Abstract. We study the problem of how well a typical multivariate polynomial can be approximated by lower degree polynomials over $\mathbb{F}_2$. We prove that, with very high probability, a random degree $d + 1$ polynomial has only an exponentially small correlation with all polynomials of degree d, for all degrees d up to $\Theta(n)$. That is, a random degree $d + 1$ polynomial does not admit a good approximation of lower degree. In order to prove this, we prove far tail estimates on the distribution of the bias of a random low degree polynomial. Recently, several results regarding the weight distribution of Reed–Muller codes were obtained. Our results can be interpreted as a new large deviation bound on the weight distribution of Reed–Muller codes.

1 Introduction

Two functions $f, g : \mathbb{F}_2^n \to \mathbb{F}_2$ are said to be ϵ-*correlated* if

$$\Pr[f(x) = g(x)] \geq \frac{1 + \epsilon}{2}.$$

A function $f : \mathbb{F}_2^n \to \mathbb{F}_2$ is said to be ϵ-*correlated* with a set of functions $F \subseteq \mathbb{F}_2^n \to \mathbb{F}_2$ if it is ϵ-correlated with at least one function $g \in F$.

We are interested in functions that have a low correlation with the set of degree d polynomials; namely, functions that cannot be approximated by any polynomial of total degree at most d. How *complex* must such a function be? We use the most natural measure for complexity in these settings, which is the degree of the function when considered as a polynomial.

A simple probabilistic argument shows that for any constant $\delta < 1$ and for $d < \delta n$, a random function has an exponentially small correlation with degree d polynomials. However, a random function is complex as, with high probability, its degree is at least $n - 2$. In this work, we study how well a random degree $d + 1$ polynomial can be approximated by any lower degree polynomial, and show that with very high probability a random polynomial of degree $d + 1$ cannot be approximated by polynomials of lower degree in a strong sense. Thus, if we want

---

[*] Research supported by ISF grant 1300/05.

I. Dinur et al. (Eds.): APPROX and RANDOM 2009, LNCS 5687, pp. 366–377, 2009.

to find functions that are uncorrelated with degree d polynomials, considering degree $d + 1$ polynomials is enough.

It is worth noting that naïve volume estimates are not sufficient to get a substantial bound on the correlation.

The study of the correlation of functions with the set of low degree polynomials is interesting from both coding theory and complexity theory points of view.

Complexity Theory. Approximation of functions by low degree polynomials is one of the major tools used in proving lower bounds for constant depth circuits. For example, Razborov and Smolensky [18,19] provided an explicit function MOD3 that cannot be computed by a constant depth circuit with a subexponential number of AND, OR and XOR gates. The proof combines two arguments:

1. Any constant depth circuit of subexponential size has a very high correlation (that is, $1 - o\,(1)$) with some polynomial of degree poly $\log n$;
2. Such a low degree polynomial has a correlation of at most 2/3 with MOD3. (In fact, this is true for any polynomial of degree at most $\epsilon\sqrt{n}$ for some constant ϵ.)

The best known constructions of explicit functions that cannot be approximated by low degree polynomials (see, e. g., [3,4,18,19,21]) divide into two ranges:

- For large degrees ($d < n^{O(1)}$), there exists a symmetric function with a correlation of at most $O\,(1/\sqrt{n})$ with degree $O\,(\sqrt{n})$ polynomials;
- For small degrees ($d < \log n$) there are explicit functions having a correlation of at most $\exp(-n/c^d)$ with degree d polynomials for some constants c (best known is $c = 2$.)

Certain applications, e. g., pseudorandom generator constructions via the Nisan–Wigderson construction [17], require a function having an exponentially small correlation with low degree polynomials. This is only known for degrees up to $\log n$, while for larger degrees the best known bound is polynomial in n. Finding explicit functions with a better correlation is an ongoing quest with limited success. For more details, see a survey by Viola [20].

Coding Theory. The Reed–Muller code $\mathcal{RM}\,(n,d)$ is the linear code of all polynomials (over $\mathbb{F}_2$) in n variables of total degree at most d. This family of codes is one of the most studied objects in coding theory (see, e.g., [16]). Nevertheless, determining the weight distribution of these codes (for $d \geq 3$) is a long standing open problem. Interpreted in this language, our main lemma gives a new tail estimate on the weight distribution of Reed–Muller codes.

1.1   Our Results

We show that, with very high probability, a random degree d polynomial has an exponentially small correlation with polynomials of lower degree. We prove this for degrees ranging from a constant up to $\delta_{max}n$, where $0 < \delta_{max} < 1$ is some constant. All results hold for large enough n.

We now state our main theorem.

Theorem 1. *There exist a constant $0 < \delta_{\max} < 1$ and constants $c, c' > 0$ such that the following holds. Let f be a random n-variate polynomial of degree $d + 1$ for $d \leq \delta_{\max} n$. The probability that f has a correlation of $2^{-cn/d}$ with polynomials of degree at most d is at most $2^{-c'\binom{n}{\leq d+1}}$, where $\binom{n}{\leq d} = \sum_{i=0}^{d} \binom{n}{i}$.*

The main theorem is an easy corollary of the following lemma, which is the main technical contribution of the paper.

We define the *bias* of a function $f : \mathbb{F}_2^n \to \mathbb{F}_2$ to be

$$bias\,(f) = \mathbb{E}_x \left[(-1)^{f(x)} \right] = \Pr\left[f\,(x) = 0 \right] - \Pr\left[f\,(x) = 1 \right].$$

Lemma 2. *Fix $\epsilon > 0$ and let f be a random degree d polynomial for $d \leq (1 - \epsilon)\,n$. Then,*

$$\Pr\left[|bias\,(f)| > 2^{-c_1 n/d} \right] \leq 2^{-c_2 \binom{n}{\leq d}},$$

where $0 < c_1, c_2 < 1$ are constants depending only on ϵ.

Note that Lemma 2 holds for degrees up to $(1 - \epsilon)\,n$, while we were only able to prove Theorem 1 for degrees up to $\delta_{\max} n$.

The following proposition shows that the estimate in Lemma 2 is somewhat tight for degrees up to $n/2$.

Proposition 3. *Fix $\epsilon > 0$ and let f be a random degree d polynomial for $d \leq (1/2 - \epsilon)\,n$. Then,*

$$\Pr\left[|bias\,(f)| > 2^{-c_1' n/d} \right] \geq 2^{-c_2' \binom{n}{\leq d}},$$

where $0 < c_1', c_2' < 1$ are constants depending only on ϵ.

As a part of the proof of Lemma 2, we give the following tight lower bound on the dimension of truncated Reed–Muller codes, which is of independent interest.

Lemma 4. *Let $x_1, \dots, x_R$ be $R = 2^r$ distinct points in $\mathbb{F}_2^n$. Consider the linear space of degree d polynomials restricted to these points; that is, the space*

$$\{(p\,(x_1), \dots, p\,(x_R)) : p \in \mathcal{RM}\,(n, d)\}\,.$$

The linear dimension of this space is at least $\binom{r}{\leq d}$.

We have recently learned that this lemma appeared earlier in [15, Theorem 1.5]. Our proof, on the other hand, is independent and has an algorithmic flavor.

1.2   Related Work

Reed–Muller codes' weight distribution is completely known for $d = 2$ (see, for example, [5]) and some partial results are known also for $d = 3$. In the general case, there are estimates (see, e.g., [13,14]) on the number of codewords with weight between w and $2.5w$, where $w = 2^{-d}$ is the minimal weight of the code. Kaufman and Lovett [12] proved bounds for larger weights, and following Gopalan et al. [10], they used it to prove new bounds for the *list-decoding* of Reed–Muller codes.

The case of multilinear polynomials was considered by Alon et al. [2], who proved a tail estimate similar to Lemma 2 and used it to prove bounds on the size of distributions that fool low degree polynomials. Namely, they prove that for any distribution $\mathcal{D}$ that fools degree d polynomials with error ϵ,

$$|support(\mathcal{D})| \geq \Omega \left(\frac{(n/2d)^d}{\epsilon^2 \log (1/\epsilon)} \right) .$$

Substituting our Lemma 2 for [2, Lemma 1] yields

$$|support(\mathcal{D})| \geq \Omega \left(\frac{\binom{n}{d}}{\epsilon^2 \log (1/\epsilon)} \right) ,$$

improving the lower bound for the case of polynomials over $\mathbb{F}_2^n$ by a factor of roughly $(2e)^d$.

The Gowers Norm is a measure related to the approximability of functions by low degree polynomials. It was introduced by Gowers [7] in his seminal work on a new proof for Szemerédi's Theorem. Using the Gowers Norm machinery, it is easy to prove that a random polynomial of degree $d < \log n$ has a small correlation with lower degree polynomials. However, this approach fails for degrees exceeding $\log n$. In constrast, note that our result holds for degrees up to $\delta_{\max} n$.

Green and Tao [8] study the structure of biased multivariate polynomials. They prove that if their degree is at most the size of the field, then they must have structure — they can be expressed as a function of a constant number of lower degree polynomials. Kaufman and Lovett [11] strengthen this structure theorem for polynomials of every constant degree, removing the field size restriction.

The rest of the paper is organized as follows. Our main result, Theorem 1, is proved in Section 2. The proof of the lower bound on the bias (Proposition 3) is omitted due to space constraints.

2   Proof of the Main Theorem

First we show that Theorem 1 follows directly from Lemma 2 by a simple counting argument.

Let f be a random degree $d + 1$ polynomial for $d \leq \delta_{\max} n$, where $\delta_{\max}$ will be determined later. For every polynomial g of degree at most d, $f - g$ is also a random degree $d + 1$ polynomial. By the union bound for all possible choices of g,

$$\Pr_f\left[\exists g \in \mathcal{RM}(n,d) : |bias\,(f-g)| \geq 2^{-c_1 n/d}\right] \leq 2^{\binom{n}{\leq d} - c_2\binom{n}{\leq d+1}}$$

Choosing $\delta_{\max}$ to be a small enough constant, we get that there is a constant $c' > 0$ such that $c_2\binom{n}{\leq d+1} - \binom{n}{\leq d} \geq c'\binom{n}{\leq d+1}$ for all $d \leq \delta_{\max} n$ (see, for example, [9, Exercise 1.14]).

We now move on to prove Lemma 2. The rest of this section is organized as follows. Lemma 2 is proved in Subsection 2.1, where the technical claims are postponed to Subsection 2.2. Lemma 4 is proved in Subsection 2.3.

2.1   Proof of Lemma 2

We need to prove that a random degree d polynomial has a very small bias with very high probability. Denote by $\mathcal{RM}\,(n,d)^{\perp}$ the dual code of $\mathcal{RM}\,(n,d)$. We start by correlating the moments of the bias of a random degree d polynomial to short words in $\mathcal{RM}\,(n,d)^{\perp}$.

Proposition 5. *Fix $t \in \mathbb{N}$ and let $p \in \mathcal{RM}\,(n,d)$ and $x_1,\ldots,x_t \in \mathbb{F}_2^n$ be chosen independently and equiprobably. Then,*

$$\mathbb{E}\left[bias(p)^t\right] = \Pr\left[e_{x_1} + \cdots + e_{x_t} \in \mathcal{RM}(n,d)^{\perp}\right],$$

where e_x for $x \in \mathbb{F}_2^n$ is the unit vector in $\mathbb{F}_2^{2^n}$, having 1 in position x and 0 elsewhere.

In favor of not interrupting the proof, we postpone the proof of Proposition 5 and other technical propositions to Subsection 2.2.

We proceed by introducing the following definitions. Fix d. For $x \in \mathbb{F}_2^n$ let $\mathrm{eval}_d(x)$ denote its d-*evaluation*; that is, a (row) vector in $\mathbb{F}_2^{\binom{n}{\leq d}}$ whose coordinates are the evaluation of all monomials of degree up to d at the point x. Formally,

$$\mathrm{eval}_d(x) = \left(\prod_{i \in I} x(i)\right)_{I \subset [n],|I| \leq d}.$$

For points $x_1,\ldots,x_t \in \mathbb{F}_2^n$ let $\mathcal{M}_d(x_1,\ldots,x_t)$ denote their d-*evaluation matrix*; this is a $t \times \binom{n}{\leq d}$ matrix whose ith row is the d-evaluation of x_i. We denote the rank of $\mathcal{M}_d(x_1,\ldots,x_t)$ by $\mathrm{rank}_d(x_1,\ldots,x_t)$. As this value is independent of the order of $x_1,\ldots,x_t$, we may refer without ambiguity to the d-*rank* of a set $S \subseteq \mathbb{F}_2^n$ by $\mathrm{rank}_d(S)$.

According to Proposition 5, in order to bound the moments of the bias of a random polynomial we need to study the probability that a random word of length about[1] t is in $\mathcal{RM}\,(n,d)^{\perp}$.

Let $A = \mathcal{M}_d(x_1,\ldots,x_t)$. Note that $e_{x_1} + \cdots + e_{x_t} \in \mathcal{RM}(n,d)^{\perp}$ if and only if

$$p\,(x_1) + \cdots + p\,(x_t) = 0 \tag{1}$$

---

[1] We say "about t" as $x_1,\ldots,x_t$ might not be distinct.

for any degree d polynomial p. Therefore, $e_{x_1} + \cdots + e_{x_t} \in \mathcal{RM}(n,d)^{\perp}$ if and only if the sum of the rows of A is zero. It is sufficient to satisfy (1) only on the monomial basis of the degree d polynomials; that is, verify that each column in A sums to zero.

We turn to bound the probability that the rows of A sum to the zero vector for random $x_1, \ldots, x_t \in \mathbb{F}_2^n$. For this we divide the n variables into two sets: V' of size $n' = \lceil n(1 - 1/d) \rceil$ and V'' of size $n'' = n - n'$. Let $\alpha = n''/n \approx 1/d$. Instead of requiring that *every* column of A sums to zero, we require this only for columns corresponding to monomials that contain exactly one variable from V'' (and thus up to $d - 1$ variables from V').

For $i = 1, \ldots, t$ denote by x_i' ($\in \mathbb{F}_2^{n'}$) the restriction of $x_i \in \mathbb{F}_2^n$ to the variables in V'. The following proposition bounds the probability that sum of A's rows is zero in terms of the $(d-1)$-rank of $x_1', \ldots, x_t'$.

Proposition 6

$$\Pr_{\{x_i\}} \left[e_{x_1} + \cdots + e_{x_t} \in \mathcal{RM}(n,d)^{\perp} \right] \leq \mathbb{E}_{\{x_i'\}} \left[2^{-\mathrm{rank}_{d-1}(x_1', \ldots, x_t')\alpha n} \right].$$

To finish the proof, we provide a (general) lower bound on d-ranks of random vectors.

Proposition 7. *For all fixed $\beta < 1$ and $\delta < 1$, there exist constants $c > 0$ and $\eta > 1$ such that if $x_1, \ldots, x_t \in \mathbb{F}_2^n$ are chosen uniformly and independently, where $t \geq \eta \binom{n}{\leq d}$ and $d \leq \delta n$, then*

$$\Pr \left[\mathrm{rank}_d(x_1, \ldots, x_t) < \beta \binom{n}{\leq d} \right] \leq 2^{-c \binom{n}{\leq d+1}}.$$

We now put it all together, in order to complete the proof of Lemma 2. According to Proposition 6, we have

$$\Pr_{\{x_i\}} \left[e_{x_1} + \cdots + e_{x_t} \in \mathcal{RM}(n,d)^{\perp} \right] \leq \mathbb{E}_{\{x_i'\}} \left[2^{-\mathrm{rank}_{d-1}(x_1', \ldots, x_t')\alpha n} \right].$$

Applying Proposition 7 for $d - 1$ and n' (instead of d and n in the proposition statement), and assuming $t \geq \eta \binom{n'}{\leq d-1}$, we get that

$$\Pr \left[\mathrm{rank}_{d-1}(x_1', \ldots, x_t') < \beta \binom{n'}{\leq d-1} \right] < 2^{-c \binom{n'}{\leq d}}.$$

Therefore,

$$\Pr_{\{x_i\}} \left[e_{x_1} + \cdots + e_{x_t} \in \mathcal{RM}(n,d)^{\perp} \right] \leq 2^{-\beta \binom{n'}{\leq d-1}\alpha n} + 2^{-c \binom{n'}{\leq d}}.$$

Recalling that $n' = \lceil n(1 - 1/d) \rceil$ and $\alpha = 1 - n'/n = 1/d + O(1/n)$, we get that for any constant β (and $c = c(\beta)$) there is a constant c' such that

$$\Pr_{\{x_i\}} \left[e_{x_1} + \cdots + e_{x_t} \in \mathcal{RM}(n,d)^{\perp} \right] \leq 2^{-c' \binom{n}{\leq d}}.$$

This is because $\binom{n'}{\leq d-1} = \Theta\left(\binom{n}{\leq d}d/n\right)$ and $\binom{n'}{\leq d} = \Theta\left(\binom{n}{\leq d}\right)$.

We thus proved that there is a constant c' such that

$$\mathbb{E}_{f \in \mathcal{RM}(n,d)}\left[bias(f)^t\right] \leq 2^{-c'\binom{n}{\leq d}},$$

for $t = \eta\binom{n'}{\leq d-1} = \Theta\left(\binom{n}{\leq d-1}\right)$. Hence, $tn/d \leq c''\binom{n}{\leq d}$ for some constant c''.

For small enough $c_1 > 0$ such that $c_2 = c' - c''c_1 > 0$, by Markov inequality,

$$\Pr\left[|bias(f)| \geq 2^{-c_1 n/d}\right] \leq 2^{tc_1 n/d - c'\binom{n}{\leq d}} \leq 2^{(c''c_1 - c')\binom{n}{\leq d}} \leq 2^{-c_2\binom{n}{\leq d}}.$$

2.2   Proofs of Technical Propositions

Proof (of Proposition 5). Write p as

$$p(x) = \sum_{I \subset [n], |I| \leq d} \alpha_I \prod_{i \in I} x(i),$$

where $x(i)$ denotes the ith coordinate of $x \in \mathbb{F}_2^n$. As p was chosen uniformly, all α_I are uniform and independent over $\mathbb{F}_2$. Therefore,

$$\mathbb{E}_p\left[(bias(p))^t\right] = \mathbb{E}_p\left[\prod_{j=1}^t bias(p)\right]$$

$$= \mathbb{E}_{\{\alpha_I\}}\left[\prod_{j=1}^t \mathbb{E}_{x_j}\left[(-1)^{\sum_I \alpha_I \prod_{i \in I} x_j(i)}\right]\right]$$

$$= \mathbb{E}_{\{x_j\}}\left[\prod_I \mathbb{E}_{\alpha_I}\left[(-1)^{\alpha_I\left(\sum_{j=1}^t \prod_{i \in I} x_j(i)\right)}\right]\right]$$

$$= \mathbb{E}_{\{x_j\}}\left[\prod_I 1_{\left\{\sum_{j=1}^t \prod_{i \in I} x_j(i) = 0\right\}}\right]$$

$$= \Pr_{\{x_j\}}\left[\forall I \sum_{j=1}^t \prod_{i \in I} x_j(i) = 0\right]$$

$$= \Pr_{\{x_j\}}\left[e_{x_1} + \cdots + e_{x_t} \in \mathcal{RM}(n,d)^\perp\right]. \qquad \square$$

Proof (of Proposition 6). Let $A' = \mathcal{M}_{d-1}(x_1', \ldots, x_t')$ be the $t \times \binom{n'}{\leq d-1}$ submatrix of A corresponding to monomials of degree at most $d - 1$ in variables from V'. Let $\mathcal{E}$ be the event in which every column of A corresponding to a monomial that contains exactly one variable from V'' sums to zero. It is easy to see that this event is equivalent to the event that every column of A' is orthogonal to the set of vectors $\{(x_1(i), \ldots, x_t(i)) : i \in V''\}$.

Fix the variables in V'; this determines A'. As the variables in V'' are independent of those in V', the probability of $\mathcal{E}$ (given A') is

$$\left(2^{-\operatorname{rank}(A')}\right)|V''| = 2^{-\operatorname{rank}(A')\alpha n} = 2^{-\operatorname{rank}_{d-1}(x'_1,\dots,x'_t)\alpha n}.$$

This holds for every assignment for variables of V', hence the result follows. □

Proof (of Proposition 7). Let $B = M_d(x_1,\dots,x_t)$ be the $t \times \binom{n}{\leq d}$ d-evaluation matrix of the random $x_1,\dots,x_t \in \mathbb{F}_2^n$. We need to bound the probability that $\operatorname{rank}(B) < \beta\binom{n}{\leq d}$.

Fix some $b \leq \beta\binom{n}{\leq d}$, and let us consider the event that the first b rows of B span the entire row span of B. Denote by V the linear space spanned by the first b rows of B. Since all rows of B are d-evaluations of some points in $\mathbb{F}_2^n$, we need to study the maximum number of d-evaluations contained in a linear subspace of dimension b.

Assume there are at least 2^r distinct d-evaluations in V. By Lemma 4, $\dim(V) \geq \binom{r}{\leq d}$. Assume further that $\operatorname{rank}(B) < \beta\binom{n}{\leq d}$; we get that

$$\beta\binom{n}{\leq d} > \operatorname{rank}(B) \geq \dim(V) \geq \binom{r}{\leq d}.$$

By Proposition 8, $r \leq n(1 - \gamma/d)$, where γ is a constant depending only on β. In other words, out of the 2^n d-evaluations of all points in $\mathbb{F}_2^n$, at most $2^{n(1-\gamma/d)}$ fall in V and hence the probability that a random d-evaluation is in V is at most $2^{-\gamma n/d}$.

Assume the number of rows t is at least $\eta\binom{n}{\leq d}$ for some $\eta > 1$. The probability that all the remaining rows of B are in V is at most

$$\left(2^{-\gamma n/d}\right)^{t-b} \leq 2^{-(\eta-\beta)\binom{n}{\leq d}\gamma n/d} \leq 2^{-\gamma\rho(\eta-\beta)\binom{n}{\leq d+1}},$$

where the last inequality follows from the fact that there exists a constant $\rho > 0$ such that $(n/d)\binom{n}{\leq d} \geq \rho\binom{n}{\leq d+1}$ for all n and d.

Choosing η large enough (as a function of β), we get that when we union bound over all possible ways to choose at most $\beta\binom{n}{\leq d}$ rows out of $t \geq \eta\binom{n}{\leq d}$, the probability that any of them spans the rows of B is at most $2^{-c\binom{n}{\leq d+1}}$, where c depends only on β. □

Proposition 8. *For any $\beta, \delta < 1$, there is a constant $\gamma = \gamma(\beta,\delta)$ such that if $1 \leq d \leq \delta n$ and $r \geq d$ satisfy $\beta\binom{n}{\leq d} \geq \binom{r}{\leq d}$ then $r \leq n(1 - \gamma/d)$.*

Proof. We bound

$$\frac{1}{\beta} \leq \frac{\binom{n}{\leq d}}{\binom{r}{\leq d}} \leq \max_{0 \leq i \leq d} \frac{\binom{n}{i}}{\binom{r}{i}} = \frac{\binom{n}{d}}{\binom{r}{d}} \leq \left(\frac{n-d}{r-d}\right)^d = \left(1 + \frac{n-r}{r-d}\right)^d.$$

Assuming for the sake of contradiction that $r > n(1-\gamma/d)$ and taking logarithms, we get

$$\ln\frac{1}{\beta} \leq d\ln\left(1 + \frac{n-r}{r-d}\right) \leq \frac{d(n-r)}{r-d} < \frac{\gamma n}{r-d} < \frac{\gamma}{r/n-\delta} < \frac{\gamma}{1-\delta+\gamma/d}.$$

This can be made false by picking, e.g., $\gamma = (1-\delta)\ln(1/\beta)$. □

2.3   Proof of Lemma 4

Restating the lemma in terms of d-evaluations, we need to show that for every subset $S \subseteq \mathbb{F}_2^n$ of size $R = 2^r$, $\mathrm{rank}_d(S) \geq \binom{r}{\leq d}$. Let $S = \{x_1, \ldots, x_{2^r}\}$ be the set of points. We simplify S by applying a sequence of transformations that do not increase its d-rank until we arrive to the linear space $\mathbb{F}_2^r \times \{0\}^{n-r}$.

We now define our basic non-linear transformation Π, mapping the set S to a set $\Pi(S)$ of equal size and not greater d-rank. Informally, Π tries to set the first bit of each element in S to zero, unless this results in an element already in S (and in this case Π keeps the element unchanged). The operator Π was used in other contexts of extremal combinatorics, and is usually referred to as the *compressing* or *shifting* operator (see, e.g., [1,6].)

For $y = (y_1, \ldots, y_{n-1}) \in \mathbb{F}_2^{n-1}$, denote by $0y$ and $1y$ the elements $(0, y_1, \ldots, y_{n-1})$ and $(1, y_1, \ldots, y_{n-1})$ in $\mathbb{F}_2^n$, respectively. Extend this notation to sets by writing $0T = \{0y : y \in T\}$, $1T = \{1y : y \in T\}$ for a set $T \subseteq \mathbb{F}_2^{n-1}$.

We define the following three sets in $\mathbb{F}_2^{n-1}$.

$$T_* = \{y \in \mathbb{F}_2^{n-1} : 0y \in S \text{ and } 1y \in S\},$$
$$T_0 = \{y \in \mathbb{F}_2^{n-1} : 0y \in S \text{ and } 1y \notin S\},$$
$$T_1 = \{y \in \mathbb{F}_2^{n-1} : 0y \notin S \text{ and } 1y \in S\}.$$

Writing S as

$$S = 0T_* \cup 1T_* \cup 0T_0 \cup 1T_1,$$

we define $\Pi(S)$ to be

$$\Pi(S) = 0T_* \cup 1T_* \cup 0T_0 \cup 0T_1;$$

namely, we set to zero the first bit of all the elements in $1T_1$. It is easy to see that $|\Pi(S)| = |S|$ as $\Pi(S)$ introduces no collisions.

Proposition 9. $\mathrm{rank}_d(\Pi(S)) \leq \mathrm{rank}_d(S)$.

Proof. It will be easier to prove this using an alternative definition for $\mathrm{rank}_d(S)$.

Let $(x_1, \ldots, x_{2^r})$ be some ordering of S. For a degree d polynomial $p \in \mathcal{RM}(n, d)$, let $v_p \in \mathbb{F}_2^{2^r}$ be the evaluation of p on the points of S

$$v_p = (p(x_1), p(x_2), \ldots, p(x_{2^r})).$$

Consider the linear space of vectors v_p for all $p \in \mathcal{RM}(n, d)$. The dimension of this space is exactly $\mathrm{rank}_d(S)$, as the monomials used in the definition of d-rank form a basis for the space of polynomials.

But now, instead of the dimension, consider the co-dimension. We call a point x_i, $1 \leq i \leq 2^r$, *dependent* if there are coefficients $\alpha_1, \ldots, \alpha_{i-1} \in \mathbb{F}_2$ such that for all degree d polynomials

$$p(x_i) = \sum_{j=1}^{i-1} \alpha_j p(x_j).$$

We thus expressed $\text{rank}_d(S)$ as the number of independent points in S, which is the same as the difference between $|S| = 2^r$ and the number of dependent points in S. To prove that $\text{rank}_d(\Pi(S)) \leq \text{rank}_d(S)$, it suffices to show that Π maps dependent points in S to dependent images in $\Pi(S)$. Let us consider an ordering of S in which the elements of $1T_1$ come last. Since all other points in S are mapped to themselves by Π, it is clear that dependent points in S appearing before $1T_1$ are also dependent in $\Pi(S)$. It remains to prove the proposition for points in $1T_1$.

Let $t_1 = |T_1|$ and let $y_1, \ldots, y_{t_1}$ be some ordering of T_1. Assume $1y_i \in S$ is dependent and we will show that $0y_i \in \Pi(S)$ is also dependent. By definition, there exist coefficients $\alpha_y, \beta_y, \gamma_y, \delta_y$ such that, for any degree d polynomial,

$$p(1y_i) = \sum_{y \in T_*} \alpha_y p(0y) + \sum_{y \in T_*} \beta_y p(1y) + \sum_{y \in T_0} \gamma_y p(0y) + \sum_{y_j \in T_1 : j < i} \delta_{y_j} p(1y_j).$$

Each polynomial $p \in \mathcal{RM}(n, d)$ can be uniquely decomposed as

$$p(x_1, \ldots, x_n) = x_1 p'(x_2, \ldots, x_n) + p''(x_2, \ldots, x_n),$$

where $p' \in \mathcal{RM}(n - 1, d - 1)$ and $p'' \in \mathcal{RM}(n - 1, d)$. Moreover, for every $y \in \mathbb{F}_2^{n-1}$, we have that $p(0y) = p''(y)$ and $p(1y) = p'(y) + p''(y)$. Since p' and p'' are independent, we can decompose the dependency of $p(1y_i)$ into its p' and p'' components as follows.

$$p'(y_i) = \sum_{y \in T_*} \beta_y p''(y) + \sum_{y_j \in T_1 : j < i} \delta_{y_j} p'(y_j), \tag{2}$$

$$p''(y_i) = \sum_{y \in T_*} (\alpha_y + \beta_y) p''(y) + \sum_{y \in T_0} \gamma_y p''(y) + \sum_{y_j \in T_1 : j < i} \delta_{y_j} p''(y_j). \tag{3}$$

We now move to consider $\Pi(S)$. Every $1y_i$ for $y_i \in T_1$ is mapped to $0y_i$, so we should only consider the p'' component for T_1's elements. Also, by the definition of T_* and T_0, for each $y \in T_* \cup T_0$, $0y \in S \cap \Pi(S)$. By (3), for any $p \in \mathcal{RM}(n, d)$,

$$p(0y_i) = \sum_{y \in T_*} (\alpha_y + \beta_y) p(0y) + \sum_{y \in T_0} \gamma_y p(0y) + \sum_{y_j \in T_1 : j < i} \delta_{y_j} p(0y_j),$$

that is, $0y_i$ is also dependent in $\Pi(S)$.

Therefore, we have established that $\text{rank}_d(\Pi(S)) \leq \text{rank}_d(S)$. $\square$

We now combine our basic Π with invertible linear transformations to define a wider class of simplifying transformations. For any $u, v \in \mathbb{F}_2^n$ whose inner product is $\langle u, v \rangle = 1$, we define the mapping $\Pi_{u,v}$ as follows. Informally, $\Pi_{u,v}$ tries to add v to elements x of S for which $\langle u, x \rangle = 1$, unless this results in an element already in S. In other words, if both x and $x + v$ are in S, then $\Pi_{u,v}(S)$ maps them both to themselves. Otherwise, if just one of them is in S, it maps it to x if $\langle u, x \rangle = 0$, and to $x + v$ if $\langle u, x + v \rangle = 0$. This is well defined as $\langle u, v \rangle = 1$. Note that $\Pi_{e_1, e_1} \equiv \Pi$.

Formally, let A be an $n \times n$ invertible matrix such that $e_1^T A = u$ and $A^{-1} e_1 = v$. We can construct such invertible A since $\langle u, v \rangle = 1$ by setting the first row of A to be u and the remaining rows of A to be a basis for the $(n-1)$-dimensional space normal to v. Define $\Pi_{u,v} = A^{-1} \Pi A$.

Observe that invertible affine transformations do not change the d-rank of a set, as they act as permutations on the set of degree d polynomials. Combining this with Proposition 9, we get that $\Pi_{u,v}$ maintains the size of S without increasing the d-rank.

We now use a sequence of $\Pi_{u,v}$ applications to transform the set S into the linear space $V = \mathbb{F}_2^r \times \{0\}^{n-r}$ spanned by the first r unit vectors $e_1, \ldots, e_r$. We say that $x \in S$ is *good* if $x \in V$, and is *bad* otherwise. If all the elements of S are good then $S = V$ since all the elements of S are distinct. Otherwise, let $x \in S$ be some bad element and let $x' \in V \setminus S$. Since $x \notin V$, there must be some index $r < i \le n$ such that $x_i = 1$; set $u = e_i$ and $v = x + x'$.

We show that applying $\Pi_{u,v}$ maps x to x' and does not affect any good elements, thus increasing the number of good elements. First see that $\langle u, v \rangle = v_i = x_i + x_i' = 1 + 0 = 1$ since $x' \in V$ so $\Pi_{u,v}$ is well defined. See also that as $\langle u, x \rangle = x_i = 1$ and $x + v \notin S$, $\Pi_{u,v}$ will add v to x, transforming it to $x' \in V$. Also, any good element y is unchanged by $\Pi_{u,v}$ since $\langle u, y \rangle = y_i = 0$. In total, the number of good elements increased by at least one.

We repeat this until all elements are good, that is, until S is transformed to V, establishing that $\operatorname{rank}_d(S) \ge \operatorname{rank}_d(V)$. To finish the proof, observe that the restriction of polynomials in $\mathcal{RM}(n, d)$ to points in a linear space of dimension r is exactly $\mathcal{RM}(r, d)$. Since $|\mathcal{RM}(r, d)| = \binom{r}{\le d}$ (see [16]), we get that for any set S of size 2^r,

$$\operatorname{rank}_d(S) \ge \binom{r}{\le d},$$

as required.

Acknowledgements. We would like to thank Simon Litsyn and Michael Krivelevich for helpful discussions. We thank Noga Alon for pointing out an alternative proof of Lemma 4. The third author would like to thank his advisor, Omer Reingold, for helpful discussions.

References

1. Alon, N.: On the density of sets of vectors. Discrete Math. 46, 199–202 (1983)
2. Alon, N., Ben-Eliezer, I., Krivelevich, M.: Small sample spaces cannot fool low degree polynomials. In: Goel, A., Jansen, K., Rolim, J.D.P., Rubinfeld, R. (eds.) APPROX and RANDOM 2008. LNCS, vol. 5171, pp. 266–275. Springer, Heidelberg (2008)
3. Ben-Sasson, E., Kopparty, S.: Affine dispersers from subspace polynomials. In: proceedings of the 41st Annual ACM Symposium on Theory of Computing (STOC), pp. 65–74 (2009)
4. Babai, L., Nisan, N., Szegedy, M.: Multiparty protocols, pseudorandom generators for logspace and time-space trade-offs. J. of Computer and System Sciences 45(2), 204–232 (1992)

5. Cohen, G., Honkala, I., Litsyn, S., Lobstein, A.: Covering Codes. North Holland, Amsterdam (1997)
6. Frankl, P.: On the trace of finite sets. J. of Combinatorial Theory, Series A 34(1), 41–45 (1983)
7. Gowers, W.T.: A new proof of Szemerédi's theorem. Geometric and Functional Analysis 11(3), 465–588 (2001)
8. Green, B., Tao, T.: The distribution of polynomials over finite fields, with applications to the Gowers Norm (submitted) (2007)
9. Jukna, S.: Extremal Combinatorics with Applications in Computer Science. Springer, Heidelberg (2001)
10. Gopalan, P., Klivans, A., Zuckerman, D.: List-decoding Reed–Muller codes over small fields. In: Proceedings of the 40th Annual ACM Symposium on Theory of Computing (STOC), pp. 265–274 (2008)
11. Kaufman, T., Lovett, S.: Average case to worst case reduction for polynomials. In: Proceedings of the 49th Annual IEEE Symposium on Foundations of Computer Science (FOCS), pp. 166–175 (2008)
12. Kaufman, T., Lovett, S.: List size vs. decoding radius for Reed–Muller codes (submitted)
13. Kasami, T., Tokura, N.: On the weight structure of Reed–Muller codes. IEEE Transactions on Information Theory 16(6), 752–759 (1970)
14. Kasami, T., Tokura, N., Azumi, S.: On the weight enumeration of weights less than 2.5d of Reed–Muller codes 30(4), 380–395 (1976)
15. Keevash, P., Sudakov, B.: Set systems with restricted cross-intersections and the minimum rank of inclusion matrices. SIAM J. of Discrete Mathematics 18(4), 713–727 (2005)
16. MacWilliams, J., Sloane, N.: The Theory of Error Correcting Codes. North Holland, Amsterdam (1977)
17. Nisan, N., Wigderson, A.: Hardness vs. randomness. J. of Computer and System Sciences 49(22), 149–167 (1994)
18. Razborov, A.: Lower bounds on the size of bounded depth circuits over a complete basis with logical addition. Math. Notes 41(4), 333–338 (1987); Translated from Matematicheskie Zametki 41(4), 598–607 (1987)
19. Smolensky, R.: Algebraic methods in the theory of lower bounds for Boolean circuit complexity. In: Proceedings of the 19th Annual ACM Symposium on the Theory of Computation (STOC), pp. 77–82 (1987)
20. Viola, E.: Correlation bounds for polynomials over {0,1}. SIGACT News 40(1), 27–44 (2009)
21. Viola, E., Wigderson, A.: Norms, XOR lemmas, and lower bounds for $GF(2)$ polynomials and multiparty protocols. In: Proceedings of the 22nd IEEE Conference on Computational Complexity (CCC), pp. 141–154 (2007)

Composition of Semi-LTCs by Two-Wise Tensor Products*

Eli Ben-Sasson and Michael Viderman

Computer Science Department
Technion — Israel Institute of Technology
Haifa, 32000, Israel
{eli,viderman}@cs.technion.ac.il

Abstract. We continue the study of the local testability of error correcting codes constructed by taking the two-wise tensor product of a "base-code" with itself. We show that if the base-code is any locally testable code (LTC) or any expander code, then the code obtained by taking the *repeated* two-wise tensor product of the base-code with itself is locally testable. This extends the results of Dinur et al. in [11] in two ways. First, we answer a question posed in that paper by expanding the class of allowed base-codes to include all locally testable code, and not just so-called *uniform LTCs* whose associated tester queries all codeword entries with equal probability. Second, we show that *repeating* the two-wise tensor operation a constant number of times still results in a locally testable code, improving upon previous results which only worked when the tensor product was applied *once*.

To obtain our results we define a new tester for the tensor product of LTCs. Our tester uses the distribution of the tester associated with the base-code to sample rows and columns of the product code. This construction differs from previously studied testers for tensor product codes which sampled rows and columns *uniformly*.

1 Introduction

Locally testable codes (LTCs) are error correcting codes for which distinguishing, when given oracle access to a purported word w, between the case that w is a codeword and the case that it is very far from all codewords, can be accomplished by a randomized algorithm, called a *tester*, which reads a sublinear amount of information from w. Such codes are of interest in computer science due to their numerous connections to probabilistically checkable proofs (PCPs) and property testing. (See the surveys [12, 17] for more information.) By now several different constructions of LTCs are known including codes based on low-degree polynomials over finite fields [1, 7], constructions based on PCPs of proximity/assignment

---

* Research supported in part by a European Community International Reintegration Grant, an Alon Fellowship, and grants by the Israeli Science Foundation (grant number 679/06) and by the US-Israel Binational Science Foundation (grant number 2006104).

I. Dinur et al. (Eds.): APPROX and RANDOM 2009, LNCS 5687, pp. 378–391, 2009.

testers [2, 10], sparse random linear codes [14] and affine invariant codes [15]. Our work studies a different family of LTC constructions, namely, *tensor codes*. Given two linear error correcting codes $C \subseteq \mathbf{F}^n, R \subseteq \mathbf{F}^m$ over a finite field $\mathbf{F}$, we define their *tensor product* to be the subspace $R \otimes C \subseteq \mathbf{F}^{n \times m}$ consisting of $n \times m$ matrices M with entries in $\mathbf{F}$ having the property that every row of M is a codeword of R and every column is a codeword of C. If $C = R$ we use C^2 to denote $C \otimes C$ and for $i > 2$ define $C^i = C \otimes C^{i-1}$.

Ben-Sasson and Sudan suggested in [4] to use tensor product codes as a means to construct LTCs combinatorially. They showed that taking the *three-wise tensor* C^3 of any code $C \subseteq \mathbf{F}^n$ with sufficiently large distance results in a *robust* locally testable code. By *robust* we informally mean that the tester associated with C^3 has the property that given any word w that is far from C^3, the *local view* selected by the tester will be far, on average, from being consistent with a local view of a codeword of C^3. More formally, denoting by $w|_I$ the projection of w onto the set of queries $I \subset \{1, \ldots, n\}^3$ picked by the tester, and denoting by $C^3|_I = \{c|_I \mid c \in C^3\}$ the set of views that are *consistent* with C^3, the *robustness* of C^3 means that, on average, $w|_I$ will be far in Hamming distance from all elements of $C^3|_I$. This robustness allowed them to apply *composition* and prove that the *repeated* three-wise tensor product of C, namely, the code C^{3^t}, is locally testable. They also raised the question of whether the repeated *two-wise* tensor product of C also leads to robust LTCs.

There is a surprising difference between two- and three-wise tensor products. For two-wise products, large distance is not sufficient to guarantee robustness (whereas for three-wise products it is). This phenomena was discovered by Valiant who constructed in [18] a pair of codes R, C with large distance whose tensor product is not robust. (See [8, 13] for generalizations of this result.) Nevertheless, in another surprising turn of events, Dinur et al. [11] showed that if C is any so-called *smooth* code, on top of having sufficiently large distance, then C^2 is robust. The family of smooth codes includes low density parity check (LDPC) codes based on expander graphs with very good expansion properties, even though these codes are not necessarily locally testable [3], and *uniform* LTCs which are LTCs whose associated tester is *equally* likely to query any codeword symbol. (These results were generalized in our earlier work [6] to *weakly smooth* codes which include also unique-neighbor expander codes and locally correctable codes.)

One issue that has remained open in all previous works on two-wise tensor product codes is under what conditions can one compose such codes and apply *repeated* two-wise products. To see the problem consider C^2 where C is an expander code, which is smooth (as well as weakly smooth). The work of Dinur et al. showed that C^2 is robust, however, there is no reason to believe C^2 is smooth or weakly smooth. So one cannot argue that C^4 is a robust LTC and we cannot apply composition.[1] In terms of LTC constructions, this means that, using

---

[1] Close inspection of [6, 11] reveals that repeated products can result in robust LTCs if the base code is a *strong uniform LTC*, but it was not clear how to obtain similar results for expander codes or for nonuniform LTCs.

previous techniques, the smallest query complexity we could get in a two-wise tensor based construction would be at least $\Omega(\sqrt{n})$, where n is a blocklength of the constructed code. This contrasts once again with the case of three-wise tensors which can be composed again and again provided the base code C has (very) large distance, thus resulting in LTCs with polynomial rate and polylogarithmic query complexity.

Our main result uses a new family of testers for repeated two-wise tensor product codes that allows us to construct LTCs with query complexity n^ϵ for any $\epsilon > 0$ based on repeated two-wise tensors, where n is a blocklength of the constructed code. This result holds even for LTCs that are *nonuniform*, i.e., whose associated tester may sample some codeword bits more often than others (some LTCs, most notably those of [2, 5, 9], are indeed nonuniform). This result also answers a question raised in [11, Section 2.2], namely, the question of constructing robust testers for two-wise tensors of a nonuniform LTC with some other code.

Our proof follows by defining a new tester for two-wise tensor codes which differs from previous constructions and the key difference is that our tester also uses the distribution associated with the base code C to sample rows and columns of C^2. (Previous testers used only the uniform distribution to sample rows and columns of C^2.)

We end by pointing out that our result does not require the base code to have very large distance, hence it holds even over fields of small cardinality. This contrasts with previous works on iterative combinatorial constructions of LTCs due to Ben-Sasson and Sudan [4] and Meir [16] which required very large base-code distance implying large field size. Moreover, in [4] the required base-code distance (and thus a field cardinality) depends on the number of repeated tensor products that should be applied. In our work, the repeated tensor product can be applied any constant number of times even over binary field and the initial requirements about the base-codes are independent on the number of times that repeated tensor products should be applied.

Organization of the rest of the paper. After presenting the necessary definitions in the next section we state our main results in Section 3. In Section 4 we describe the notion of a *semi LTC* which is crucial for our proofs. This is followed by the definition of our suggested tester for two-wise tensor codes in Section 5 and we conclude in Section 6 by our main technical lemma.

2   Preliminary Definitions

The definitions appearing here are pretty much standard in the literature on tensor-based LTCs.

Throughout this paper $\mathbf{F}$ is a finite field, $[n]$ denotes the set $\{1, \ldots, n\}$ and $\mathbf{F}^n$ denotes $\mathbf{F}^{[n]}$. All codes discussed in this paper will be a linear. Let $C \subseteq \mathbf{F}^n$ be a linear code over $\mathbf{F}$.

For $w \in \mathbf{F}^n$ let $supp(w) = \{i | w_i \neq 0\}$, $|w| = |supp(w)|$ and $\mathrm{wt}(w) = \frac{|w|}{n}$. We define the *distance* between two words $x, y \in \mathbf{F}^n$ to be $\Delta(x, y) = |\{i \mid x_i \neq$

$y_i\}|$ and the relative distance to be $\delta(x, y) = \frac{\Delta(x,y)}{n}$. The relative distance of a code is denoted $\delta(C)$ and defined to be the minimal value of $\delta(x, y)$ for two distinct codewords $x, y \in C$. The distance of C is defined similarly as $\Delta(C) = \min_{x \neq y \in C}\{\Delta(x, y)\}$. For $x \in \mathbf{F}^n$ and $C \subseteq \mathbf{F}^n$, let $\delta(x, C) = \delta_C(x) = \min_{y \in C}\{\delta(x, y)\}$ denote the relative distance of x from the code C. If $\delta(x, C) \geq \epsilon$ we say that x is ϵ-far from C and otherwise x is ϵ-close to C. We let $\dim(C)$ denote the dimension of C. The vector inner product between u_1 and u_2 is denoted by $\langle u_1, u_2 \rangle$. We let $C^\perp = \{u \in \mathbf{F}^n \mid \forall c \in C : \langle u, c \rangle = 0\}$ be the dual code of C and $C_t^\perp = \{u \in C^\perp \mid |u| = t\}$. In a similar way we define $C_{<t}^\perp = \{u \in C^\perp \mid |u| < t\}$ and $C_{\leq t}^\perp = \{u \in C^\perp \mid |u| \leq t\}$.

For $w \in \mathbf{F}^n$ and $S = \{j_1, j_2, ..., j_m \mid j_k \in [n]\}$ we let $w|_S = (w_{j_1}, w_{j_2}, ..., w_{j_m})$ be the projection of w onto the subset S. Similarly, we let $C|_S = \{c|_S \mid c \in C\}$ denote the projection of the code C onto S.

2.1 Linear Testers as Distributions

A *standard q-tester* for a $[n, k, d]_\mathbf{F}$ code is a randomized algorithm with oracle access to a string w of length n over $\mathbf{F}$. The randomized algorithm makes q queries to the oracle and outputs accept or reject. For linear codes we can assume without loss of generality that testers are non-adaptive and have perfect completeness (see [3, Theorem 2]). We define a generalized tester which does not make queries, rather it returns a "view" which can be considered as a code by itself. Therefore we put forward a more general definition of tester (Definition 2).

Note that given a code $C \subseteq F^n$, the subset $I \subseteq [n]$ uniquely defines $C|_I$. The linearity of C implies that $C|_I$ is a linear subspace of $\mathbf{F}^I$.

Definition 1 ((Test of C)). *A q-test is a set of coordinates $I \subseteq [n]$ s.t. $|I| \leq q$.*

Definition 2 ((Tester of C)). *A q-tester T is a distribution $\mathbf{D}$ over q-tests, i.e., over subsets $I \subseteq [n]$ s.t. $|I| \leq q$.*

Although the tester does not output accept , reject, the way a standard tester does, it can be converted to output accept , reject as follows. Whenever the task is to test whether w is in C and a test I is selected by the tester, the tester can output accept if $w|_I \in C|_I$ and otherwise output reject.

Definition 3 ((LTCs)). *A code $C \subseteq F^n$ is a (q, ϵ, δ) LTC if it has a q-tester $\mathbf{D}$ such that $\forall w \in \mathbf{F}^n$, if $\delta(w, C) \geq \delta$ we have $\Pr_{I \sim \mathbf{D}}[w|_I \notin C|_I] \geq \epsilon$.*

Definition 4 ((Strong LTCs)). *A code $C \subseteq F^n$ is a (q, ϵ) strong LTC if it has a q-tester $\mathbf{D}$ such that $\forall w \in \mathbf{F}^n$, we have $\Pr_{I \sim \mathbf{D}}[w|_I \notin C|_I] \geq \epsilon \cdot \delta(w, C)$.*

2.2 Odd Expanders

Next we give standard definitions of codes based on expander graphs.

Definition 5 ((Neighbors)). *Let $G = (V, E)$ be a graph. For $S \subseteq V$, let*

- *$N(S)$ be the set of neighbors of S.*
- *$N^1(S)$ be the set of unique neighbors of S, i.e. vertices with exactly one neighbor in S.*
- *$N^{odd}(S)$ be the set of neighbors of S with an odd number of neighbors in S.*

Notice that $N^1(S) \subseteq N^{odd}(S)$.

Definition 6 ((Expansion)). *Let $c, d \in N$ and let $\gamma, \delta \in (0, 1)$.*

Define a (c, d)-bounded (γ, δ)-expander to be a bipartite graph (L, R, E) with vertex sets L, R such that all vertices in L have degree $\leq c$, and all vertices in R have degree $\leq d$;

- *G is called a (c, d, γ, δ)-expander if $\forall S \subseteq L$ s.t. $|S| \leq \delta n$ we have $|N(S)| > \gamma \cdot c|S|$*
- *G is called an (c, d, γ, δ)-odd expander if $\forall S \subseteq L$ s.t. $|S| \leq \delta n$ we have $|N^{odd}(S)| > \gamma \cdot c|S|$*

We say that a code C is an (c, d, γ, δ)-*odd expander* code if it has a parity check graph (see [11, Section 2.3]) that is an odd (c, d)-bounded (γ, δ)-expander.

We notice that the definition of an odd expander generalizes the definition of a unique neighbor expander, which was already shown in [6] to result in a robustly testable tensor code (see Definition 9).

2.3   Tensor Product Codes

For $x \in \mathbf{F}^I$ and $y \in \mathbf{F}^J$ we let $x \otimes y$ denote the tensor product of x and y (i.e. the matrix $M_{(i,j)} = x_i \cdot y_j$ where $(i, j) \in I \times J$). Let $R \subseteq \mathbf{F}^I$ and $C \subseteq \mathbf{F}^J$ be linear codes. We define the tensor product code $R \otimes C$ to be the linear space spanned by words $r \otimes c \in \mathbf{F}^{I \times J}$ for $r \in R$ and $c \in C$. Some immediate facts:

- The code $R \otimes C$ consists of all $I \times J$ matrices over $\mathbf{F}$ whose rows belong to R and whose columns belong to C.
- $\dim(R \otimes C) = \dim(R) \cdot \dim(C)$
- $\delta(R \otimes C) = \delta(R) \cdot \delta(C)$

We let $C^{2^0} = C$ and $C^{2^t} = C^{2^{t-1}} \otimes C^{2^{t-1}}$ for $t > 0$.

2.4   Robust Locally Testable Codes

Throughout this paper let $C \subseteq \mathbf{F}^n$ and $R \subseteq \mathbf{F}^m$ be linear codes over $\mathbf{F}$.

Definition 7 ((Test View)). *Let $w \in \mathbf{F}^n$ (think of the task of testing whether $w \in C$). Let I be a test of C and let $w|_I$ denote the projection of w to I. We call $w|_I$ the view of the test. If $w|_I \in C|_I$ we say that this view is consistent with C, or when C is clear from the context we simply sat $w|_I$ is consistent.*

When considering a tensor code $R \otimes C \subseteq \mathbf{F}^m \otimes \mathbf{F}^n$, the coordinate set of a test is some $I \subseteq [n] \times [m]$.

Definition 8 ((Local distance)). *Let C be a code and $w|_I$ be view on coordinate set I obtained from word w. The local distance of w from C with respect to I (also denoted the I-distance of w from C) is $\min_{c \in C}\{\Delta\left(w|_I, c|_I\right)\}$ and similarly the relative local distance of w from C with respect to I (relative I-distance of w from C) is $\min_{c \in C}\{\delta(w|_I, c|_I)\}$. When I is clear from context we omit reference to it.*

Informally, robustness implies that if a word is far from the code then, on average, a test's view is far from any consistent view that can be accepted on the same coordinate set I. We are ready now to provide a general definition of robustness.

Definition 9 ((Robustness)). *Given a tester (i.e. a distribution) $\mathbf{D}$ for the code $C \subseteq F^n$, we let*

$$\rho^{\mathbf{D}}(w) = \mathop{\mathbf{E}}_{I \sim \mathbf{D}}[\delta(w|_I, C|_I)]$$

be the expected relative local distance of input w. The robustness of the tester is defined as

$$\rho^{\mathbf{D}} = \min_{w \in F^n \setminus C} \frac{\rho^{\mathbf{D}}(w)}{\delta_C(w)}.$$

Let $\{C_n\}_n$ be a family of codes where C_n is of blocklength n and $\mathbf{D_n}$ is a tester for C_n. A family of codes $\{C_n\}_n$ is robustly testable with respect to testers $\{\mathbf{D_n}\}_n$ if there exists a constant $\alpha > 0$ such that for all n we have $\rho^{\mathbf{D_n}} \geq \alpha$.

Notice that the robustness of a tester is well defined since if $w \in F^n \setminus C$ we have $\delta_C(w) \neq 0$ and thus the denominator of the fraction $\frac{\rho^{\mathbf{D}}(w)}{\delta_C(w)}$ is nonzero.

3   Main Results

Our first main result says that codes obtained by the tensor product of a LTC code and some other code is robust with respect to our tester.

Theorem 10 ((Robust Tensor of LTCs)). *Let $R \subseteq \mathbf{F}^m$ be a code s.t. $\delta(R) = \delta_R$. Let $C \subseteq \mathbf{F}^n$ be a (q, ϵ, ρ) LTC s.t. $\delta(C) = \delta_C$ and $\rho \leq \frac{\delta_C}{4}$. Let T be our suggested tester for the code $R \otimes C$. Then,*

$$\rho^T \geq \min\left\{\frac{\rho \delta_C \delta_R}{12}, \frac{\epsilon \cdot \delta_R}{32q^2}\right\}.$$

Next theorem shows that the tensor product of an odd expander code and some other code is robust with respect to our tester. Note that even random expander will be odd-expander with high probability although it is not locally testable (see [3]).

Theorem 11 ((Robust Tensor of Expanders)). *Let $R \subseteq \mathbf{F}^m$ be a code s.t. $\delta(R) = \delta_R$. Let $C \subseteq \mathbf{F}^n$ be a (c, d, γ, δ)-odd expander code. Let T be our suggested tester for the code $R \otimes C$. Then,*

$$\rho^T \geq \frac{\gamma \delta \delta_R}{128d^2}.$$

The proofs of Theorem 10 and Theorem 11 are omitted due to space limitations. Informally, LTCs and odd expander codes are semi LTCs (see Definition 14) and thus by Lemma 18 result in robust tensor products.

3.1 Main Corollaries

We show that taking the repeated two-wise tensor product of either a strong LTC, or an expander code, results in a robust LTC. Recall that $C^{2^0} = C$ and $C^{2^t} = C^{2^{t-1}} \otimes C^{2^{t-1}}$ for $t > 0$.

Corollary 12. *Let $t > 0$ be an integer. Let $C \subseteq \mathbf{F}^n$ be a (q, ϵ) strong LTC s.t. $\delta(C) = \delta_C$. Then C^{2^t} is a (q, ϵ') strong LTC, where*

$$\epsilon' = \left(\frac{\epsilon}{48q^2} \right)^{2t} \left(\frac{\delta_C}{4} \right)^{4 \cdot 2^t}$$

Corollary 13. *Let $t > 0$ be an integer. Let $C \subseteq \mathbf{F}^n$ be a (c, d, γ, δ)-odd expander code s.t. $\delta(C) = \delta_C$. Then C^{2^t} is (n, ϵ') strong LTC, where*

$$\epsilon' = \frac{\gamma^t \cdot (\delta \delta_C)^{2^{t+1}}}{(96d^2)^t \cdot 8^{t^2}}.$$

4 Semi LTCs

We define semi LTCs and strong semi LTCs. It can be verified that LTCs, strong LTCs and odd-expanders are semi LTCs (strong semi LTCs). Note that nonuniform LTCs and odd-neighbor expanders were not known to be smooth (see [11]) or weakly smooth (see [6]) and thus did not facilitate a composition via robust two-vise tensor product.

Definition 14 ((Semi LTCs)). *Let $0 < \rho < 1$. We say that code C with $\delta(C) = \delta_C$ is (q, ϵ, ρ)-semi LTC (sLTC) if there exists q-tester $\mathbf{D}$ s.t. $\forall w \in \mathbf{F}^n$ if $\rho \delta_C / 3 \leq \mathrm{wt}(w) \leq \rho \delta_C$ then $\Pr_{I \sim \mathbf{D}} [w|_I \notin C|_I] \geq \epsilon$.*

Definition 15 ((Strong semi LTCs)). *Let $0 < \rho < 1$. We say that code C with $\delta(C) = \delta_C$ is (q, ϵ, ρ)-strong sLTC if there exists q-tester $\mathbf{D}$ s.t. $\forall w \in \mathbf{F}^n$ if $\mathrm{wt}(w) \leq \rho \delta_C$ then $\Pr_{I \sim \mathbf{D}} [w|_I \notin C|_I] \geq \epsilon \cdot \mathrm{wt}(w)$.*

In Proposition 16 (the proof is omitted) we show that strong sLTC property is preserved after tensor operation. Recall that in the previous works [6, 11] it was not known whether smooth (or weakly smooth) property is preserved after tensor operation, and thus previous works did not achieve composition via two-wise tensor codes.

Proposition 16. *Let $t > 0$ be an integer. Let C be a $[n, k, d]$ code, (q, ϵ, ρ) strong sLTC. Then C^{2^t} is a $[n^{2^t}, k^{2^t}, d^{2^t}]$ code, $(q, (\frac{3}{8})^t \epsilon, \frac{\rho^{2^t}}{4^t})$ strong sLTC.*

5   A New Tester for Two-Wise Tensor Product Codes

We present here our new tester which is used to prove the main theorems stated in the previous subsection. Our starting point is the *uniform row/column tester* used in all previous works on two-wise tensor codes [6, 8, 11, 13, 18].

We describe this tester for $R \otimes C \subseteq \mathbf{F}^m \otimes \mathbf{F}^n$. For $i \in [n]$ and $j \in [m]$ let the i-row $= \{i\} \times [m]$ and j-column $= [n] \times \{j\}$.

Uniform Row/Column Tester
 − With probability $\frac{1}{2}$ pick $i \in_U [n]$ and choose i-row.
 − With probability $\frac{1}{2}$ pick $j \in_U [m]$ and choose j-column.

The distribution over the tests of this tester is uniform over rows and columns and does not depend on the structure of the base-codes R, C.

Our suggested tester is a combination of the Uniform Row/Column Tester and the $\mathbf{D_C}$-distribution Tester which depends on the structure of the base code. Our tester for a code $R \otimes C$ picks views that will be $M|_S$ where S is either a row ($\{i\} \times [m]$) or a column ($[n] \times \{i\}$) or a *rectangle* $supp(u) \times \{i\}$ for small weight $u \in C^{\perp}$.

To define our suggested tester we assume that the code C has some distribution $\mathbf{D_C}$ over $C^{\perp}_{\leq q}$. The main place where we use our suggested tester is Main Lemma 18 where we assume that the code C is (q, ϵ, ρ)-sLTC (see Definition 14) and thus has a "corresponding" distribution $\mathbf{D_C}$ over $C^{\perp}_{\leq q}$.

Our Suggested Tester
 − with probability $\frac{1}{2}$ invoke Uniform Row/Column Tester described above
 − with probability $\frac{1}{2}$ invoke $\mathbf{D_C}$-distribution Testerdefined next

$\mathbf{D_C}$-*distribution Tester*

 − pick $u \in_{\mathbf{D_C}} C^{\perp}_{\leq q}$
   • with probability $\frac{1}{2}$ pick $i \in_U supp(u)$ and choose i-row
   • with probability $\frac{1}{2}$ pick $j \in_U [m]$ and choose $supp(u) \times \{j\}$

Finally we define Rectangle Tester which will be used only in the proof of Main Lemma 18. We start with the definition of rectangle.

Definition 17 ((Rectangle)). *Let $S_{rows} \subset [n]$, $T_{cols} \subset [m]$. We call $S_{rows} \times T_{cols}$ a rectangle coordinate set or simply a rectangle. For $M \in \mathbf{F}^m \otimes \mathbf{F}^n$ we call $M|_{(S_{rows} \times T_{cols})}$ a rectangle view.*

The rectangles we will use are of the form $supp(u) \times [n]$ for $u \in C^{\perp}$.
Now, we define Rectangle Tester which picks rectangles as views.

Rectangle Tester

- pick $u \in_{\mathbf{D_C}} C^{\perp}_{\leq q}$

- choose $Rect = supp(u) \times [m]$.

Notice that $\mathbf{D_C}$-distribution Tester is actually an invocation of Uniform Row/Column Tester on the view chosen by Rectangle Tester, so the view of our suggested tester will be either row, column or support of dual word of weight at most q.

For every word $M \in \mathbf{F}^n \times \mathbf{F}^m$ we let

$$\rho_{rect}(M) = \mathop{\mathbf{E}}_{u \in_{\mathbf{D_C}} C^{\perp}_{\leq q}} \left[\delta \left(M|_{supp(u) \times [m]}, (R \otimes C)|_{supp(u) \times [m]} \right) \right]$$

be the expected relative local distance of input M obtained by the Rectangle Tester.

Similarly, let $\rho_{row/col}(M)$ be the expected relative local distance of input M obtained by the Uniform Row/Column Tester. Let $\delta_R(M)$ be a relative distance of a typical row of M from R and $\delta_C(M)$ be a relative distance of a typical column of M from C. Then we have $\rho_{row/col}(M) = \frac{\delta_R(M) + \delta_C(M)}{2}$ since with probability $\frac{1}{2}$ the Uniform Row/Column Tester picks a random row and with probability $\frac{1}{2}$ the Uniform Row/Column Tester picks a random column.

Similarly, we let $\rho_{\mathbf{D_C}}(M)$ be the expected relative local distance of input M obtained by the $\mathbf{D_C}$-distribution Tester.

We let $\rho(M)$ the expected relative local distance of input M obtained by our suggested tester. Then we have

$$\rho(M) = \frac{\rho_{row/col}(M) + \rho_{\mathbf{D_C}}(M)}{2} \tag{1}$$

since our suggested tester invokes the Uniform Row/Column tester with probability $\frac{1}{2}$ and with probability $\frac{1}{2}$ our suggested tester invokes the $\mathbf{D_C}$-distribution Tester. From Equation 1 we have

$$\rho(M) \geq \frac{1}{2}\rho_{\mathbf{D_C}}(M), \text{ and} \tag{2}$$

$$\rho(M) \geq \frac{1}{2}\rho_{row/col}(M) \tag{3}$$

6   Semi LTCs Results in Robust Tensor

The following is the main technical lemma used to show that the tensor product of a sLTC with another code is robust with respect to our tester.

Lemma 18 ((Main Lemma)). *Let $R \subseteq \mathbf{F}^m$ be a code s.t. $\delta(R) = \delta_R$ and $C \subseteq \mathbf{F}^n$ be a $(q, \epsilon, \rho \le \frac{3}{4})$ sLTC s.t. $\delta(C) = \delta_C$. Let T be our new tester defined in Section 5 for the code $R \otimes C$. Then*

$$\rho^T \ge \min \left\{ \frac{\rho \delta_C \delta_R}{36}, \frac{\epsilon \cdot \delta_R}{32q^2} \right\}.$$

Notice that the distribution of the tester is over rows, columns and dual words of weight at most q.

Proof. We have $\frac{1}{36}\rho\delta_C\delta_R < \frac{1}{16}$ because $\rho, \delta_C, \delta_R \le 1$, so it is sufficient to show that for all $M \in (\mathbf{F}^m \otimes \mathbf{F}^n) \setminus (R \otimes C)$ we have

$$\frac{\rho(M)}{\delta_{R \otimes C}(M)} \ge \min \left\{ \frac{\rho \delta_C \delta_R}{36}, \frac{\epsilon \cdot \delta_R}{32q^2}, \frac{1}{16} \right\}$$

Fix $M \in (\mathbf{F}^m \otimes \mathbf{F}^n) \setminus (R \otimes C)$ and denote $\delta_{R \otimes C}(M)$ by $\delta(M)$. If $\frac{\rho(M)}{\delta(M)} \ge \frac{1}{16}$ or $\rho(M) \ge \frac{\rho\delta_C\delta_R}{36}$ we are done since $\delta(M) \le 1$ and hence $\frac{\rho(M)}{\delta(M)} \ge \min \left\{ \frac{\rho\delta_C\delta_R}{36}, \frac{1}{16} \right\}$. Thus in what follows we assume that

$$\rho(M) < \frac{\rho\delta_C\delta_R}{36}, \text{ and} \tag{4}$$

$$\delta(M) > 16\rho(M) \tag{5}$$

We prove that $\rho(M) \ge \frac{\epsilon \cdot \delta_R}{32q^2}$. Let $\delta^{row}(M) = \delta_{R \otimes \mathbf{F}^n}(M)$ denote the distance of M from the space of matrices whose rows are codewords of R, and define $\delta^{col}(M) = \delta_{\mathbf{F}^m \otimes C}(M)$ similarly. For row $i \in [n]$, let $r^{(i)} \in R$ denote the codeword of R closest to the i-th row of M. For column $j \in [m]$, let $c^{(j)} \in C$ denote the codeword of C closest to the j-th column of M. Let M_R denote the $n \times m$ matrix whose i-th row is $r^{(i)}$, and let M_C denote the matrix whose j-th column is $c^{(j)}$. Let $E = M_R - M_C$.

In what follows the matrices M_R, M_C and (especially) E will be the central objects of attention. We refer to E as the error matrix. We use the error matrix E for the analysis of robustness, note that the tester does not obtain a view of E but only of M and of course it is possible that some constraints that are unsatisfied on M are satisfied on E and vice versa.

Note that $\delta(M, M_R) = \delta^{row}(M)$ and $\delta(M, M_C) = \delta^{col}(M)$ thus $\rho_{row/col}(M) = \frac{\delta^{row}(M) + \delta^{col}(M)}{2}$ because the Uniform Row/Column Tester picks with probability $\frac{1}{2}$ a random row and with probability $\frac{1}{2}$ a random column. Let $\mathrm{wt}(E)$ be the relative weight of E, so

$$\mathrm{wt}(E) = \delta(M_R, M_C) \le \delta(M, M_R) + \delta(M, M_C) = 2\rho_{row/col}(M) \le 4\rho(M) \tag{6}$$

By Equations 4 and 6 it follows that

$$\mathrm{wt}(E) < 4 \cdot \frac{1}{36}\rho\delta_C\delta_R = \frac{1}{9}\rho\delta_C\delta_R \tag{7}$$

We want to prove that $\rho(M) \geq \frac{\epsilon \cdot \delta_R}{32q^2}$. It is sufficient to show that $\rho_{\mathbf{D_C}}(M) \geq \frac{\epsilon \cdot \delta_R}{16q^2}$ and then from Equation 2 we conclude $\rho(M) \geq \frac{\epsilon \cdot \delta_R}{32q^2}$.

To show that $\rho_{\mathbf{D_C}}(M) \geq \frac{\epsilon \cdot \delta_R}{16q^2}$ we prove Proposition 19 in Section 6.1, Proposition 20 in Section 6.2 and Proposition 21 in Section 6.3. The Main Lemma follows from Proposition 21 by the previous discussions.

Proposition 19. *Let $u \in C_{\leq q}^{\perp}$ and $S = supp(u) \times [m]$ be a rectangle coordinate set. If $u^T \cdot E \neq 0$ then $\Delta(M|_S, (R \otimes C)|_S) \geq \frac{\delta_R m}{2}$.*

Proposition 20. $\Pr_{u \in_{\mathbf{D_C}} C_{\leq q}^{\perp}} [u^T \cdot E \neq 0] \geq \frac{\epsilon}{2}$.

Proposition 19 and Proposition 20 will be used to conclude.

Proposition 21. $\rho_{\mathbf{D_C}}(M) \geq \frac{\epsilon \cdot \delta_R}{16q^2}$.

6.1   Proof of Proposition 19

Proposition 19 is the central observation in the Main Lemma. Recall that we use the error matrix E for the analysis of robustness and that the tester does not obtain a view of E but only of M.

We start from a simple claim (the proof is omitted) that will be crucial in the proof of Proposition 19. Recall that $R \subseteq \mathbf{F}^m$ is a linear code s.t. $\delta(R) = \delta_R$.

Claim 22. *Let $w \in \mathbf{F}^m$. If $c_1 \in R$ is the closest codeword of R to w then $\forall c_2 \in R \setminus \{c_1\}$ we have $\Delta(w, c_2) \geq \frac{\delta_R m}{2}$.*

We notice that $u|_{supp(u)} \cdot (E|_S) \neq 0$ if and only if $u \cdot (E) \neq 0$. Let $\hat{M}|_S$ be the consistent view that is closest to $M|_S$. There are two cases: either $\hat{M}|_S \neq M_R|_S$ or $\hat{M}|_S = M_R|_S$.

Case 1: $\hat{M}|_S \neq M_R|_S$ so, at least one row i of $\hat{M}|_S$ is not equal to row i of $M_R|_S$, but row i of $\hat{M}|_S$ is a codeword of R because $\hat{M}|_S$ is a consistent view and the row i of $M_R|_S$ is a codeword of R by definition of M_R. Row i of M_R is the closest codeword of R to row i of M, thus according to Claim 22 row i of M is at least $\frac{\delta_R m}{2}$ far from row i of $\hat{M}|_S$. So, $\Delta\left(\hat{M}|_S, M|_S\right) \geq \frac{\delta_R m}{2}$.

Case 2: $\hat{M}|_S = M_R|_S$ and thus $M_R|_S$ is the consistent view. We argue that it is impossible and show that $M_R|_S$ will not satisfy constraint u (or formally $u|_{supp(u)}$).

This is true since $0 \neq u^T \cdot E = u^T \cdot (M_R - M_C) = u^T \cdot M_R - u^T \cdot M_C$, every column of M_C satisfies u and so $u^T \cdot M_C = 0$ and thus $0 \neq u^T \cdot E = u^T \cdot M_R = u^T|_{supp(u)} \cdot M_R|_S = u^T|_{supp(u)} \cdot \hat{M}|_S$. Contradiction.

6.2   Proof of Proposition 20

We start from auxiliary Proposition 23 (the proof is omitted) that will be used later in the proof of Proposition 20.

Proposition 23. *There exists a rectangle $Rect = A \times B$ s.t. $A \subseteq [n]$, $\delta_C n/2 \leq |A|$ and $B \subseteq [m]$, $\frac{2}{3}\delta_R m \leq |B|$, and all rows and columns of $E|_{Rect}$ are non-zero and every column c of E indexed by a member of B has $\text{wt}(c) < \frac{1}{3}\rho\delta_C$.*

Recall that C is a (q, ϵ, ρ) sLTC and thus it has a distribution $\mathbf{D_C}$ over $C_{\leq q}^{\perp}$ such that

Claim 24. *For all $w \in \mathbf{F}^n$ if $(\rho/3)\delta(C) \leq \text{wt}(w) \leq \rho\delta(C)$ then*

$$\Pr_{u \in \mathbf{D_C} C_{\leq q}^{\perp}} [\langle u, w \rangle \neq 0] \geq \frac{\epsilon}{2}$$

Proof (of Proposition 20). We say that the column E_j of E is a light column if $0 < \text{wt}(E_j) \leq \frac{1}{3}\rho\delta_C$. Recall that Rectangle Tester obtains views $M|_{supp(u) \times [n]}$ for $u \in C_{\leq q}^{\perp}$. By Proposition 23 there exists non-zero rectangle $A \times B$ of E, namely $E|_{A \times B}$, s.t. for every column of E E_i indexed by a member of B it holds that E_i is a light column. We argue that

$$\Pr_{u \in \mathbf{D_C} C_{\leq q}^{\perp}} [u \cdot E \neq 0] \geq \epsilon.$$

It is sufficient to show that there exists a linear combination of columns of E, call it E_{res}, such that

$$\Pr_{u \in \mathbf{D_C} C_{\leq q}^{\perp}} [\langle u, E_{res} \rangle \neq 0] \geq \epsilon$$

because if $\langle u, E_{res} \rangle \neq 0$ then $u^T \cdot E \neq 0$ since for at least one column of E (E_j) we have $\langle u, E_j \rangle \neq 0$.

Let $LightCols = \{E_1, ..., E_k\}$ be a set of all columns of E indexed by B, note they all are light columns.

It holds that $|\bigcup_{E_j \in LightCols}(supp(E_j))| \geq \delta_C n/2$ because by Proposition 23 every row of $E|_{A \times B}$ is non-zero and $|A| \geq \delta_C n/2$. Throw them (E_j) one by one from $LightCols$ reducing their total support $(\bigcup_{E_j \in LightCols}(supp(E_j)))$, finally obtain set $(LightCols')$ of total support between $(\frac{2}{3})\rho\delta_C n$ and $\rho\delta_C n$, i.e. $\frac{2}{3}\rho\delta_C n \leq |\bigcup_{E_j \in LightCols}(supp(E_j))| \leq \rho\delta_C n$. There exists a linear combination (over $\mathbf{F}$) of $\{E_j \in LightCols'\}$, call it E_{res}, s.t. $\text{wt}(E_{res}) \geq (\frac{1}{3})\rho\delta_C$. Moreover, $\text{wt}(E_{res}) \leq \rho\delta_C$ because $|\bigcup_{E_j \in LightCols}(supp(E_j))| \leq \rho\delta_C$.

By Claim 24 it holds that

$$\Pr_{u \in \mathbf{D_C}} [\langle u, E_{res} \rangle \neq 0] \geq \frac{\epsilon}{2}.$$

As we said if $\langle u, E_{res} \rangle \neq 0$ then $u^T \cdot E \neq 0$ and so

$$\Pr_{u \in \mathbf{D_C}} [u^T \cdot E \neq 0] \geq \frac{\epsilon}{2}.$$

6.3   Proof of Proposition 21

We proceed as follows. We first show in Proposition 25 that $\rho_{rect}(M) \geq \frac{\epsilon \cdot \delta_R}{4q}$. We then show in Proposition 26 (the proof is omitted) that if $\rho_{rect}(M) \geq \alpha$ then $\rho_{\mathbf{D_C}}(M) \geq \frac{\alpha}{4q}$. Finally we conclude that $\rho_{\mathbf{D_C}}(M) \geq \frac{\epsilon \cdot \delta_R}{16q^2}$.

Proposition 25. $\rho_{rect}(M) \geq \frac{\epsilon \delta_R}{4q}$.

Proof. By Proposition 20 we have $\displaystyle \Pr_{u \in \mathbf{D_C} C^{\perp}_{\leq q}} \left[u^T \cdot E \neq 0 \right] \geq \frac{\epsilon}{2}$. By Proposition 19 whenever $u^T \cdot E \neq 0$ it holds that $\Delta \left(M|_{supp(u) \times [m]}, (R \otimes C)|_{supp(u) \times [m]} \right) \geq \delta_R m/2$. Thus, the expected distance of the view chosen by Rectangle Tester from a consistent view is at least $\frac{\epsilon}{2} \cdot \frac{\delta_R m}{2}$, i.e.

$$\mathop{\mathbf{E}}_{u \in \mathbf{D_C} C^{\perp}_{\leq q}} \left[\Delta \left(M|_{supp(u) \times [m]}, (R \otimes C)|_{supp(u) \times [m]} \right) \right] \geq \frac{\epsilon}{2} \cdot \frac{\delta_R m}{2}.$$

For any $u \in C^{\perp}_{\leq q}$ we have $|supp(u) \times [m]| \leq qm$. So,

$$\rho_{rect}(M) = \mathop{\mathbf{E}}_{u \in \mathbf{D_C} C^{\perp}_{\leq q}} \left[\delta \left(M|_{supp(u) \times [m]}, (R \otimes C)|_{supp(u) \times [m]} \right) \right] \geq \frac{\frac{\epsilon}{2} \cdot \frac{\delta_R m}{2}}{qm} = \frac{\epsilon \delta_R}{4q}.$$

Proposition 26. *If $\rho_{rect}(M) \geq \alpha$ then $\rho_{\mathbf{D_C}}(M) \geq \frac{\alpha}{4q}$.*

Proposition 25 and Proposition 26 imply $\rho_{\mathbf{D_C}}(M) \geq \epsilon \cdot \frac{\delta_R n}{16q^2}$.

Acknowledgements. We thank Prahladh Harsha for helpful discussions.

References

1. Arora, S., Lund, C., Mutwani, R., Sudan, M., Szegedy, M.: Proof verification and Intractability of Approximation Problems. Journal of ACM 45(3), 501–555 (1998); Preliminary version in FOCS, pp. 14–23 (1992)
2. Ben-Sasson, E., Goldreich, O., Harsha, P., Sudan, M., Vadhan, S.: Robust PCPs of proximity, Shorter PCPs and Applications to Coding. SIAM Journal of Computing 36(4), 889–974 (2006); Preliminary version in STOC 2004, pp. 120–134
3. Ben-Sasson, E., Harsha, P., Raskhodnikova, S.: Some 3CNF Properties are Hard to Test. SIAM Journal on Computing 35(1), 1–21 (2005); Preliminary version appeared in STOC 2003
4. Ben-Sasson, E., Sudan, M.: Robust locally testable codes and products of codes. In: Jansen, K., Khanna, S., Rolim, J.D.P., Ron, D. (eds.) APPROX-RANDOM 2004. LNCS, vol. 3122, pp. 286–297. Springer, Heidelberg (2004) (See ECCC TR04-046, 2004)
5. Ben-Sasson, E., Sudan, M.: Short PCPs with Poly-log Rate and Query Complexity. In: STOC 2005, pp. 266–275 (2005); Full version can be obtained from Eli Ben-Sasson's homepage at http://www.cs.technion.ac.il/~eli/

6. Ben-Sasson, E., Viderman, M.: Tensor products of weakly smooth codes are robust. In: Goel, A., Jansen, K., Rolim, J.D.P., Rubinfeld, R. (eds.) APPROX-RANDOM 2008. LNCS, vol. 5171, pp. 290–302. Springer, Heidelberg (2008)
7. Blum, M., Luby, M., Rubinfeld, R.: Self Testing/Correcting with applications to Numerical Problems. Journal of Computer and System Science 47(3), 549–595 (1993)
8. Copersmith, D., Rudra, A.: On the robust testability of tensor products of codes, ECCC TR05-104 2005 (2005)
9. Dinur, I.: The PCP Theorem by gap amplification. Journal of ACM 54(3) (2007); Preliminary version in STOC 2006, pp. 241–250
10. Dinur, I., Reingold, O.: Assignment testers: Towards combinatorial proofs of the PCP theorem. SIAM Journal of Computing 36(4), 975–1024 (2006); Preliminary version in FOCS 2004, pp. 155–164
11. Dinur, I., Sudan, M., Wigderson, A.: Robust local testability of tensor products of LDPC codes. In: Díaz, J., Jansen, K., Rolim, J.D.P., Zwick, U. (eds.) APPROX-RANDOM 2006. LNCS, vol. 4110, pp. 304–315. Springer, Heidelberg (2006)
12. Goldreich, O.: Short locally testable codes and proofs (survey), ECCC TR05-014 (2005)
13. Goldreich, O., Meir, O.: The tensor product of two good codes is not necessarily robustly testable. In: ECCC TR 2007 (2007)
14. Kaufman, T., Sudan, M.: Sparse random linear codes are locally decodable and testable. In: FOCS 2007 (2007)
15. Kaufman, T., Sudan, M.: Algebraic Property Testing: The role of Invariance, ECCC Technical Report, TR07-111 (2007)
16. Meir, O.: Combinatorial Construction of Locally Testable Codes. M.Sc. Thesis, Weizmann Institute of Science (2007)
17. Trevisan, L.: Some Applications of Coding Theory in Computational Complexity, Survey Paper. Quaderni di Matematica 13, 347–424 (2004)
18. Valiant, P.: The tensor product of two codes is not necessarily robustly testable. In: Chekuri, C., Jansen, K., Rolim, J.D.P., Trevisan, L. (eds.) APPROX-RANDOM 2005. LNCS, vol. 3624, pp. 472–481. Springer, Heidelberg (2005)

On the Security of Goldreich's One-Way Function

Andrej Bogdanov[1] and Youming Qiao[2]

[1] Dept. of Computer Science and Engineering, The Chinese University of Hong Kong
andrejb@cse.cuhk.edu.hk
[2] Institute for Theoretical Computer Science, Tsinghua University
jimmyqiao86@gmail.com

Abstract. Goldreich (ECCC 2000) suggested a simple construction of a candidate one-way function $f : \{0,1\}^n \to \{0,1\}^m$ where each bit of output is a fixed predicate P of a constant number d of (random) input bits. We investigate the security of this construction in the regime $m = Dn$, where $D(d)$ is a sufficiently large constant. We prove that for any predicate P that correlates with either one or two of its variables, f can be inverted with high probability.

We also prove an amplification claim regarding Goldreich's construction. Suppose we are given an assignment $x' \in \{0,1\}^n$ that has correlation $\epsilon > 0$ with the hidden assignment $x \in \{0,1\}^n$. Then, given access to x', it is possible to invert f on x with high probability, provided $D = D(d, \varepsilon)$ is sufficiently large.

1 Introduction

In a short note in 2000, Oded Goldreich [Gol00] proposed a very simple construction of a conjectured one-way function:

1. Choose a bipartite graph G with n vertices on the left, m vertices on the right, and regular right-degree d.
2. Choose a predicate $P : \{0,1\}^d \to \{0,1\}$.
3. Let $f = f_{G,P}$ be the function from $\{0,1\}^n$ to $\{0,1\}^m$ defined by

$$f(x)_i = \text{the } i\text{th bit of } f(x) = P(x_{\Gamma(i,1)}, \dots, x_{\Gamma(i,d)})$$

where $\Gamma_{(i,j)}$ is the jth neighbor of right vertex i of G.

Goldreich conjectured that when $m = n$ and d is constant, for "most" graphs G and predicates P, the resulting function is one-way.[1]

In this work we investigate Goldreich's construction in the setting where the graph G is random, d is constant, and $m = Dn$ for a sufficiently large constant

---

[1] More precisely, with constant probability over the choice of G and P (say 2/3), the corresponding family of functions as $n \to \infty$ is one-way. Goldreich also suggests specific choices of P and G.

I. Dinur et al. (Eds.): APPROX and RANDOM 2009, LNCS 5687, pp. 392–405, 2009.

$D = D(d)$. We show that for this setting of parameters, Goldreich's construction is not secure for most predicates P. In fact, our conclusion holds for every predicate P that exhibits a correlation with either one of its variables or a pair of its variables.

We also show that if we are given a "hint" x' – any assignment that has nontrivial correlation with the actual input x to the one-way function – it is possible to invert f on x, as long as D is a sufficiently large constant. However, D depends not only on d but also on the correlation between x and x'.

While our theorem does not rule out the security of Goldreich's construction when $m = n$, it indicates some possible difficulties in using this construction, as it reveals its sensitivity on the output length. It indicates that when the ratio m/n is a sufficiently large constant, the construction can be broken for a large class of predicates. It is also easy to see that when m/n is smaller than $1/(d-1)$ the function can also be inverted for every predicate P, as with high probability the "constraint hypergraph" splits into components of size $O(\log n)$ [SS85].

On the other hand, for certain choices of the predicate P to which our theorem does not apply, it has been conjectured that the function f is not only one-way but also a pseudorandom generator [MST03].[2]

1.1 Goldreich's Function and Cryptography in NC⁰

Goldreich's proposal for a one-way function has several features that were absent from all known earlier proposals: (1) It is extremely simple to implement, and (2) it is very fast to compute, especially in parallel. On the other hand, the conjectured security of Goldreich's function is not known to relate to any standard assumptions in cryptography, such as hardness of factoring or hardness of finding short vectors in lattices.

This paradigm of "NC⁰ cryptographic constructions" where every bit of the output depends only on a constant number of input bits has since been extended to other cryptographic primitives, in particular pseudorandom generators. Remarkably, Applebaum, Ishai, and Kushilevitz [AIK04] showed that a pseudorandom generator (and in particular a one-way function) in NC⁰ can be obtained assuming the hardness of the discrete logarithm problem; however, the stretch of this pseudorandom generator is only constant. In a different work [AIK06], the same authors gave a different construction of a pseudorandom generator with small linear stretch using the less standard assumption that certain random linear codes are hard to decode.

These constructions give evidence that cryptography in NC⁰ may be possible. However, the constructions are rather complicated and the parameters they yield are of little practical value. For example, it is not known whether it is possible to have a pseudorandom generator that stretches n bits of input into, say, $10n$ bits of output under comparable assumptions.

For this reason, we believe it is interesting to investigate the power and limitations of simple constructions such as the one of Goldreich, which may be more

---

[2] Actually [MST03] considers a slightly different function; see below.

useful in practice. A step in this direction was made by Mossel, Shpilka, and Trevisan [MST03]. They conjectured that the function $f : \{0,1\}^n \times \{0,1\}^n \to \{0,1\}^m$ where

$$f(x,y)_i = x_{\Gamma(i,1)} + x_{\Gamma(i,2)} + x_{\Gamma(i,3)} + y_{\Delta(i,1)} \cdot y_{\Delta(i,2)}$$

is a pseudorandom generator with high probability, where Γ and Δ are incidence lists of random (n,m) bipartite graphs of right-degree 3 and 2 respectively. As partial evidence towards their conjecture, Mossel et al. proved that f is pseudorandom against linear functions for, say, $m = n^{1.1}$. It is not difficult to see by the Linial-Nisan conjecture [LN90], which was recently proved [Bra09], f is also pseudorandom against constant-depth circuits.

Very recently, Cook, Etesami, Miller, and Trevisan [CEMT09] showed that a restricted class of algorithms called "myopic algorithms" take exponential time to invert Goldreich's construction. The kinds of algorithms used in this work are not myopic.

1.2 Our Results

We state our main results. They refer to the standard notion of "correlation" among strings and functions which is formally defined in Section 2.

Theorem 1. *Let K be a sufficiently large constant and $D > 2^{Kd}$. Suppose $P : \{0,1\}^d \to \{0,1\}$ is a predicate that has nonzero correlation with one of its inputs or a pair of its inputs. Consider the function $f_{G,P} : \{0,1\}^n \to \{0,1\}^m$, where $m = Dn$. Then, with high probability over G, $f_{G,P}$ is invertible on a $1 - 2^{-2^{-\Omega(d)}n}$-fraction of inputs as a one-way function.*

Theorem 2. *Let K be a sufficiently large constant and $D > (1/\varepsilon)^{Kd}$. Let $P : \{0,1\}^d \to \{0,1\}$ be any non-constant predicate. Then there is an algorithm A such that with high probability over G, with the following holds. Consider the function $f_{G,P} : \{0,1\}^n \to \{0,1\}^m$, where $m = Dn$. For a $1 - 2^{-\varepsilon^2 2^{-\Omega(d)}n}$ fraction of assignments x and any assignment x' that has correlation ε (in absolute value) with x, on input $G, P, f(x)$ and x', A outputs an inverse for $f_{G,P}(x)$. The running time of A is polynomial in n and $1/\varepsilon^d$.*

1.3 Our Approach

The problem of inverting Goldreich's function is somewhat analogous to the problem of reconstructing assignments to random 3SAT formulas in the planted 3SAT model. We exploit this analogy and show that several of the tools developed for planted 3SAT can be applied to our setting as well.

The proofs of Theorems 1 and 2 consist of two stages. In the first stage, we almost invert f in the sense that we find an assignment z that matches the hidden assignment x on a 99% fraction of positions. In the second stage we turn z into a true inverse for $f(x)$. The second stage is common to the proofs of both theorems.

To give some intuition about the first stage in Theorem 1, suppose for instance that P is the majority predicate. Then we try to guess a the value of the bit x_i by looking at all constraints where x_i appears and taking the majority of these values. Since x_i has positive correlation with the majority predicate, we expect this process to result in a good guess for most x_i that appear in a sufficiently large number of clauses. In fact, if f has about $n \log n$ bits of output, this reconstructs the assignment completely; if $m = Dn$ for a sufficiently large constant D, a large constant fraction of the bits of x is recovered. The same idea applies to any predicate with correlates to one of its variables.

For predicates correlating with a pair of their variables, we will argue that the output of f contains certain noisy information about the correlation between the pairs. In particular, it gives information as to whether the pair of variables have the same or different values. More precisely, it is possible to construct a graph G whose vertices correspond to variables of i and an edge between i and j appears independently, but with probability depending on the event $x_i = x_j$. The clusters in this graph correspond to variables taking the same value. Using known methods for clustering random graphs [Coj06] we can recover most of the values of x.

The first stage in the proof of Theorem 2 is based on the observation that if we start with some assignment x' that correlates with the input x to f, then the output bits of $f(x)$ give information about the values of various variables x_i, for an arbitrary predicate P. We prove this in Section 4.

For the second stage, we extend an algorithm of Flaxman [Fla03] (similar ones have also been given in [Vil07, KV06]) for reconstructing planted assignments of random 3CNF formulas. The planted 3SAT model can be viewed as a variant of our model where the predicate P corresponds to one of the eight predicates $z_1 \vee z_2 \vee z_3, \ldots, \overline{z_1} \vee \overline{z_2} \vee \overline{z_3}$. This algorithm starts from an almost correct assignment, then unsets a small number of the variables in this assignment according to some condition ("small support size"), so that with high probability all (but a constant number of) the remaining set variables are correct. Then the value of the unset variables can be inferred in polynomial time. We show that the notion of "small support size" can be generalized to arbitrary non-constant predicates, and this type of algorithm can be used to invert f. While we directly follow previous approaches, our proofs include a few technical simplifications.

2   Preliminaries

Some definitions. Let X, Y be random variables over $\{0, 1\}$. The *correlation* between X and Y is the value $\mathrm{E}[(-1)^{X+Y}]$. The correlation between a predicate $P : \{0, 1\}^d \to \{0, 1\}$ and a subset $(x_i)_{i \in S}$ of its inputs is the correlation between the random variables $P(X_1, \ldots, X_d)$ and $\sum_{i \in S} X_i$, where the sum is taken modulo 2, and $X_1, \ldots, X_n$ are uniformly distributed. We say P correlates with $(x_i)_{i \in S}$ if the above correlation is nonzero. The correlation between a pair of assignments $x, y \in \{0, 1\}^n$ is the correlation between the ith bit of x and y, where $i \in [n]$ is random.

We say a Bernoulli random variable $X \sim \{0,1\}$ is ε-*biased towards* 0 (resp. 1) if the probability of $X = 0$ is at most $1/2 - \varepsilon$ (resp. $1/2 + \varepsilon$).

We say an assignment $x \in \{0,1\}^n$ is ε-*balanced* if its correlation with the all zero assignment is at most ε in absolute value.

By analogy with the random 3SAT problem, we will refer to the input $x \in \{0,1\}^n$ on which we are interested the function $f_{G,P}(x)$ as the *planted assignment*. We will call an assignment $x' \in \{0,1\}^n$ d-*correct* if it is at hamming distance at most d from the planted assignment.

On the random graph model. In Goldreich's definition [Gol00], The bipartite graph G in the function $f_{G,P}$ is chosen from the following random graph model $\mathcal{G} = \{\mathcal{G}_{n,m}\}$: (1) Each graph G in $\mathcal{G}_n$ has n left vertices and $m = m(n)$ right vertices; (2) each right vertex v of G has d neighbors on the left, labeled by $\Gamma_1(v), \dots, \Gamma_d(v)$; (3) The neighbors of each right vertex are uniformly distributed (repetitions allowed) and independent of the neighbors of all other vertices.

The literature on planted 3SAT usually considers a different model where each of the clauses is included in the formula independently with probability $p = p(n)$. Our results can be extended in the corresponding model for G, but such a model is less natural for one-way functions.

3   Obtaining an Almost Correct Assignment

In this section, we show that for predicates correlating with one or a pair of inputs, we can get an assignment that agrees with the planted one on almost all variables.

3.1   For Predicates Correlating with One Input

When the predicate $P(z_1, \dots, z_k)$ correlates with one of its inputs, say z_1, then every output bit of $f_{G,P}(x)$ gives an indication about what the corresponding input bit should be. If we think of this indication as a vote, and take a majority of all the votes, we set most of the input bits correctly. The following proposition formalizes this idea.

Algorithm Majority Voting
INPUTS: A predicate $P(z_1, \dots, z_d)$ that correlates with z_k; the graph G; the value $f_{G,P}(x)$
ALGORITHM.

1. For every input variable i, calculate the majority among the values $f_{G,P}(x)_j$ where i occurs as the kth variable.
2. Set x'_i to equal this value if the correlation between P and z_k is positive, and the complement of this value otherwise.
3. Output the assignment x'.

Proposition 1. *Suppose $D > 4^d$ and P is a predicate that correlates with its kth variable. For a $1 - 2^{-\Omega(n/d^2 4^d)}$ fraction of $x \in \{0,1\}^n$ and with probability $1 - 2^{-\Omega(4^d n)}$ over the choice of G, the assignment x' produced by algorithm Majority Voting agrees with x on a $(1 - 2^{-\Omega(D/4^d)})n$ fraction of variables.*

Proof. Without loss of generality assume $k = 1$, and assume the correlation between P and z_1 is positive. Since this correlation is a multiple of 2^{-d}, it must then be at least 2^{-d}.

Now fix any input x that is $1/2d2^d$-balanced. We think of the constraint graph G as being chosen in the following manner: First, for each constraint in G the first variable i_1 is chosen uniformly at random. Then, for every i, among the constraints where i is the first variable, the other variables $i_2, \ldots, i_d$ are chosen at random. Let N_i denote the number of constraints with i as the first variable.

Now consider the random experiment where one samples $x_{i_2}, \ldots, x_{i_d}$ at random and outputs the value $b = P(x_i, x_{i_2}, \ldots, x_{i_d})$. If $x_{i_2}, \ldots, x_{i_d}$ were uniformly distributed in $\{0,1\}$, then b is a Bernoulli random variable whose output is at least 2^{-d}-biased towards x_i. However, $x_{i_2}, \ldots, x_{i_d}$ might not be uniformly distributed but only $1/2d2^d$-balanced. Since the statistical difference between the distributions $(x_{i_2}, \ldots, x_{i_d})$ when the samples are uniform and when they are uniformly balanced is at most $(d-1)/2d2^d \leq 2^{-(d+1)}$, it follows that b is at least $2^{-(d+1)}$-biased towards x_i.

Fix some i such that $N_i \geq D/2$. By Chernoff bounds, over the random choice of G, the value x_i' agrees with x_i with probability at least $1 - 2^{-\Omega(4^{-d}D)}$. By another Chernoff bound, the number of is among those N_i such that $N_i \geq D/2$ where x_i and x_i' disagree is at most $2^{-\Omega(4^{-d}D)}n$ with probability $2^{-\Omega(4^{-d}Dn)}$. Applying Lemma 4 with $\varepsilon = 4^d/D$ we obtain the theorem.   □

3.2   For Predicates Correlating with a Pair of Inputs

We illustrate the inversion of $f_{G,P}(x)$ for a predicate that correlates with a pair of its inputs by looking at the "all equal" predicate. Specifically, let $AE(z_1, z_2, z_3)$ be the predicate "$z_1 = z_2 = z_3$". Then AE does not correlate with any of its variables, but it correlates with the pair (z_1, z_2).

In this example, every constraint $(x_{i_1}, x_{i_2}, x_{i_3})$ where AE evaluates to 1 tells us that $x_{i_1} = x_{i_2}$. Now construct a graph H whose vertices are variables of x and such a constraint gives rise to an edge (i_1, i_2). Then the connected components in this graph indicate collections of variables x_i that must have the same value. When x is roughly balanced, because G is random, the induced subgraphs on the sets $\{i : x_i = 0\}$ and $\{i : x_i = 1\}$ are random graphs with constant average degree. Therefore with high probability, each of these subgraphs will have a giant connected component, giving two large sets of variables of x that must have the same value. By guessing the value of the variables within each set we obtain an assignment x' that agrees with x almost everywhere.

Now consider the majority predicate $MAJ(z_1, z_2, z_3)$. This predicate also correlates with its first pair of variables. Fix an almost balanced assignment x. Now suppose we see a constraint such that $MAJ(x_{i_1}, x_{i_2}, x_{i_3}) = 1$. While we cannot say with certainty that $x_{i_1} = x_{i_2}$, this constraint gives an indication that x_{i_1} and x_{i_2} are more likely to be different than equal. So we can hope to recover a large portion of the assignment x by looking for a large cut in the graph H.

For a general predicate that correlates with a pair of its variables, we can reconstruct a large portion of the assignment x by using a spectral partitioning algorithm on H. This idea was used by Flaxman [Fla03] in a related context. Coja-Oghlan [Coj06] proved a general "partitioning theorem" which, in particular, gives the following algorithm.

Theorem 3 (Theorem 1 of [Coj06], special case). *There is a polynomial-time algorithm* Partition *with the following property. Let C_0 be a sufficiently large constant. Let (S_0, S_1) be a partition of $[n]$ such that $|S_0|, |S_1| \geq n/3$. Fix probabilities $p_{00}, p_{11}, p_{01} \in [C_0/n, D/n]$. Suppose the graph H' is a random graph where each edge (i, j), where $i \in S_a, j \in S_b$ $(a \leq b)$ is included independently at random with probability p_{ab}. Assume that*

$$n(|p_{00} - p_{01}| + |p_{11} - p_{01}|) \geq C_0 \max(\sqrt{np_{00} \log(np_{00})}, \sqrt{np_{11} \log(np_{11})}) \ , \quad (1)$$

then with high probability Partition(H') *outputs a partition (S_0', S_1') of $[n]$ such that (S_0, S_1) and (S_0', S_1') differ on at most $(1 - O(D^{-10}))n$ vertices of H'.*

Condition (1) is a non-degeneracy condition which requires there to be a noticeable difference in edge densities. Otherwise, the information about the original partition is lost.

Algorithm Pairwise
INPUTS: A predicate $P(z_1, \ldots, z_d)$ that correlates with (z_k, z_r); the graph G; the value $f_{G,P}(x)$
ALGORITHM

1. Choose b such that $\Pr_z[z_k \neq z_r \mid P(z) = b] \neq \Pr_z[z_k = z_r \mid P(z) = b]$.
2. Construct the graph H on vertex set $[n]$ with edges (i_k, i_r) iff there is a constraint in G such that $P(x_{i_1}, \ldots, x_{i_d}) = b$. Let m_H denote the number of edges of H.
3. Sample M from the binomial distribution with $\binom{n}{2}$ samples, each with probability $m_H/2$. Let H' be the subgraph consisting of the first M distinct edges of G. (If there are not enough such edges, fail.)
4. Run Partition(H'). Call the partition output by the algorithm (S_0', S_1').
5. Output the pair of assignments $x', \bar{x}'$, where $x_i' = a$ iff $i \in S_a'$, and $\bar{x}'$ is the complementary assignment.

For step 1, it follows that such a choice of b is always possible by the assumption that P correlates with (z_k, z_r). Step 3 is a technical trick that allows us to pass from our random graph model, where the number of edges is fixed, to the model where each edge is sampled independently at random with probability $m_H/2$. We believe this step is not necessary, but since the algorithm *Partition* is analyzed in the latter model we include it for accuracy.

Proposition 2. *Fix a sufficiently large constant C. Suppose $D > Cd16^d$ and P is a predicate that correlates with (z_k, z_r). For a $1 - 2^{-\Omega(d4^d)}$ fraction of $x \in \{0,1\}^n$ and with high probability over the choice of G, one of the two assignments produces by algorithm Pairwise agrees with x on a $(1 - \Omega(D^{-10}))n$ fraction of variables.*

Proof. Without loss of generality assume $b = 1$, $k = 1$ and $r = 2$. Let $p_{\neq} = \Pr_z[z_1 \neq z_2 \mid P(z) = 1]$, $p_= = \Pr_z[z_1 = z_2 \mid P(z) = 1]$. The fact that P is correlated with (z_k, z_r) implies that $|p_= - p_{\neq}| \geq 4^{-d}$.

Let us first fix a balanced input x. Let S_0 and S_1 denote the 0 and 1 variables of x. Let m_H be the number of 1-outputs of $f_{G,P}(x)$. Conditioned on $P(x_{i_1}, \ldots, x_{i_d}) = 1$, we can think of $i_1, \ldots, i_d$ as chosen by the following process. First, we determine where in the partition (S_0, S_1) the indices i_1 and i_2 belong. Then we randomly sample i_1 and i_2 from the corresponding set in the partition. Then we choose $i_3, \ldots, i_d$. This process induces the following random graph H: For each of m_H edges, first randomly choose where in the partition the edge belongs. We put the edge in (S_0, S_0) and (S_1, S_1) with probability $p_=/2$ and in (S_0, S_1) with probability $p_{\neq}$. Then randomly choose an edge on that side of the partition.

Disregarding the possibility that step 3 fails, the graph H' is then a random graph with edge densities $p_{00}, p_{11} = p_= m_H/n(n-1)$, and $p_{01} = p_{\neq} m_H/n(n-1)$. By Chernoff bounds, $m_H > m/2^d$ with high probability. Then for $D > C_1 d16^d$ condition (1) will be satisfied and with high probability over the choice of G, the algorithm will return the correct partition.

To complete the proof we need to analyze the effect that the imbalance of x and the step 3 failure have on this ideal scenario. We now assume that x is $1/2d4^d$-balanced. It can be checked (similarly to the proof of Proposition 1) that this affects the probabilities p_{00}, p_{01}, p_{11} by at most $2^{-(2d+1)} m_H/n(n+1)$, so condition (1) will still be satisfied. By Chernoff bounds, step 3 succeeds with high probability. $\square$

4   Amplifying Assignments

In this section we give the proof of Theorem 2. As discussed, the proof goes in two stages. First, we find an assignmnent w that agrees with x on most inputs. Then we use Theorem 4 to invert f. We focus on the first stage.

The idea of the algorithm is to use the assignment x' in order to get empirical evidence about the values of each variable x_i in the hidden assignment. First, since the predicate $P(z)$ is nontrivial, it must depend on at least one of its variables, say z_1. To obtain evidence about the value of x_i, let's consider all constraints in which x_i appears as the first variable. Since G is random, we expect the number of such constraints to be fairly large. Moreover, the other variables appearing in the constraints are also random.

Now let us fix a pair of assignments x and x' with correlation ε, a variable i, and a value $b \in \{0,1\}$, and look at the probability distribution D_b generated by the following process:[3]

---

[3] It is easy to see that D_b does not depend on i.

1. Choose a random G.
2. Choose a random constraint j of $f_{G,P}$ where i appears as the first variable. Call the other variables $i_2, \ldots, i_d$.
3. Output $(x'_{i_2}, \ldots, x'_{i_d}, f(b, x_{i_2}, \ldots, x_{i_d})_j)$.

Our main observation (see Lemma 1 below) is that the distributions D_0 and D_1 are statistically far apart. Therefore we can determine the value $b = f(x)$ with good confidence by observing enough samples from one of these two distributions. But observing the values $f(x)_j$ in those constraints j where i appears as the first variable amounts exactly to sampling from this process. This suggests the following algorithm for computing w:

Algorithm Amplify. On input $P, G, f(x), \varepsilon$, an assignment x' that ε-correlates with x,

1. Compute the distributions D_0 and D_1 (see below).
2. For every i, compute the empirical distribution $\hat{D}_i$ defined as follows:
   (a) Choose a random constraint $(i, i_2, \ldots, i_d)$ of f where i is the first variable.
   (b) Output $(x'_{i_2}, \ldots, x'_{i_d}, f(b, x_{i_2}, \ldots, x_{i_d})_j)$.
3. Set $w_i = b$ if $\hat{D}_b$ is closer to D_b than to D_{1-b} in statistical distance.

Proposition 3. *Let G be random right regular bipartite graph with n left vertices and $2^{\varepsilon^{Dd}} n$ right vertices, where D is a sufficiently large constant. With high probability over the choice of G, for a $1 - 2^{-\Omega(\varepsilon^2 n)}$ fraction of assignments x and every assignment x' that has correlation ε with x, algorithm Amplify outputs assignments $w_1, \ldots, w_n$ so that at least one of them agrees with x in a $1 - \varepsilon$ fraction of places.*

As discussed above, the proof of this theorem consists of two steps. First, we show that the distributions D_0 and D_1 are statistically far apart. Then, we show that with high probability over G, for most i the distribution $\hat{D}_i$ is statistically close to D_{x_i}.

Lemma 1. *Let x and x' be two assignments such that x is $\varepsilon/2$-balanced and x' has correlation ε with x. Then the statistical distance between D_0 and D_1 is at least $\varepsilon^{-O(d)}$.*

We observe that the distance can be as small as $\varepsilon^{-\Omega(d)}$, for example if P is the XOR predicate on d variables, x is any balanced assignment, and x' is an assignment that equals 1 on a $1 - \varepsilon$ fraction of inputs and 0 on the other inputs.

Proof. We begin by giving alternate descriptions of the distributions D_b. To do this, we define a distribution $\mathcal{F}$ over $\{0,1\}^2$ as follows: First, choose $i \in [n]$ at random, then output the pair (x_i, x'_i). Let (a, a') denote a pair sampled from $\mathcal{F}$. It is not difficult to see that

$$\min(\Pr[a' = 0], \Pr[a' = 1]) \geq \varepsilon/2 \tag{2}$$

for if this were not the case, it would violate the assumptions on x and x'.

The distribution D_b can now be described as follows:

1. Uniformly and independently sample pairs $(a_i, a_i') \sim \mathcal{F}$ for $i = 2, \ldots, n$.
2. Output $(a_2', \ldots, a_d', P(b, a_2, \ldots, a_d))$.

Intuitively, this corresponds to the process of first sampling input bits from x', then evaluating P at a "noisy" version of x'. If there was no noise, it is easy to see that D_0 and D_1 must be far apart, as they have to differ for at least one setting of $a_2', \ldots, a_d'$, and by (2) this happens with probability at least $(\varepsilon/2)^{d-1}$.

To argue the general case, note that the statistical distance between D_0 and D_1 is bounded below by the quantity

$$\mathrm{sd}(D_0, D_1)$$

$$= \sum_{(a_2', \ldots, a_d') \in \{0,1\}^{d-1}} 2 \cdot \mathcal{F}^{d-1}(a_2', \ldots, a_d')$$

$$\cdot \left| \mathrm{E}_{\mathcal{F}^{d-1}}[P(0, a_2, \ldots, a_d) - P(1, a_2, \ldots, a_d) \mid a_2', \ldots, a_d'] \right|$$

$$\geq 2 \cdot (\varepsilon/2)^{d-1}$$

$$\cdot \max_{(a_2', \ldots, a_d')} \left| \mathrm{E}_{\mathcal{F}^{d-1}}[P(0, a_2, \ldots, a_d) - P(1, a_2, \ldots, a_d) \mid a_2', \ldots, a_d'] \right|$$

$$\geq 2 \cdot (\varepsilon/2)^{d-1}$$

$$\cdot \mathrm{E}_{(a_2', \ldots, a_d')} \left[\mathrm{E}_{\mathcal{F}^{d-1}}[P(0, a_2, \ldots, a_d) - P(1, a_2, \ldots, a_d) \mid a_2', \ldots, a_d']^2 \right]^{1/2}$$

where $\mathcal{F}^{d-1}(a_2', \ldots, a_d')$ denotes the probability of sampling $a_2', \ldots, a_d'$ in $d - 1$ independent copies of $\mathcal{F}$, the expectation $\mathrm{E}_{\mathcal{F}^{d-1}}$ is taken over independent choices of $a_2, \ldots, a_d$ where each a_i is sampled from the distribution $\mathcal{F}$ conditioned on a_i', and the expectation $\mathrm{E}_{(a_2', \ldots, a_d')}$ refers to a uniformly random choice of $(a_2', \ldots, a_d') \sim \{0,1\}^{d-1}$.

To lower bound the last quantity, we consider the linear operator T_{d-1} on the space $\mathbf{R}^{\{0,1\}^{d-1}}$ defined by

$$(T_{d-1}g)(a_2', \ldots, a_d') = \mathrm{E}_{\mathcal{F}^{d-1}}[g(a_2, \ldots, a_d) \mid a_2', \ldots, a_d'].$$

Let T_{d-1}^{-1} denote its inverse (whose existence will be argued) and $\|\cdot\|_2$ denote the ℓ_2 operator norm. Recall that for any linear operator T,

$$\|T\|_2 = \max_g \|Tg\|_2 / \|g\|_2 = \max|\sigma|$$

where the maximum ranges over the singular values σ of T. Applying this definition to the operator T_{d-1}^{-1}, we have that

$$\|T_{d-1}^{-1}\|_2 \cdot \mathrm{E}_{(a_2', \ldots, a_d')} \left[\mathrm{E}_{\mathcal{F}^{d-1}}[P(0, a_2, \ldots, a_d) - P(1, a_2, \ldots, a_d) \mid a_2', \ldots, a_d']^2 \right]^{1/2}$$

$$\geq \mathrm{E}_{(a_2, \ldots, a_d)} \left[(P(0, a_2, \ldots, a_d) - P(1, a_2, \ldots, a_d))^2 \right]^{1/2} \geq 2^{-d+1}$$

We are left with the task of upper bounding the quantity $\|T_{d-1}^{-1}\|_2$. It is bounded by the largest (in absolute value) singular value of the operator T_{d-1}^{-1}, which is the inverse of the smallest singular value of $T_{d-1} = T_1^{\otimes(d-1)}$. Putting everything together, we obtain that

$$\mathrm{sd}(D_0, D_1) \geq 2 \cdot (\varepsilon/4)^{d-1} \cdot |\sigma|^{d-1}$$

where σ is the smaller singular value of the operator T_1. A calculation of the singular values of T_1 (which we omit) shows that $|\sigma| = \Omega(\varepsilon)$, so $\mathrm{sd}(D_0, D_1) = \varepsilon^{-O(d)}$. $\qquad\qquad\square$

We now prove that the distributions $\hat{D}_i$ are mostly close to the distributions D_{x_i}. We will need the following crude bound on the number of samples needed in order to approximate a distribution with bounded support by its empirical average. It easily follows from Chernoff bounds.

Lemma 2. *Suppose $\mathcal{D}$ is a distribution on a set of size S and $\hat{\mathcal{D}}$ is the empirical average of N independent samples of $\mathcal{D}$, where $N \geq 3S^2/\gamma^2 \log(S/\delta)$. Then*

$$\Pr[\mathrm{sd}(\mathcal{D}, \hat{\mathcal{D}}) < \gamma] > 1 - \delta.$$

Lemma 3. *Fix any constants $\gamma, \varepsilon > 0$. Suppose G is a random graph with n left vertices and Dn right vertices, where $D \geq 24d2^d \log(3/\varepsilon)/\gamma^2$. With probability $1 - 2^{-\Omega(\varepsilon^2 n)}$ over the choice of G, for a $1 - 2^{\Omega(\varepsilon^2 n)}$ fraction of assignments x, for at least a $1 - \varepsilon$ fraction of i, for every assignment x' that has correlation ε with x, we have that $\mathrm{sd}(\hat{D}_i, D_{x_i}) < \gamma$.*

Proof. Fix an $\varepsilon/2$-balanced assignment x. We will show that

$$\Pr_G\left[|\{i : \mathrm{sd}(\hat{D}_i, D_{x_i}) \geq \gamma\}| > \varepsilon n\right] = 2^{-\Omega(\varepsilon^2 n)}.$$

Since at most $2^{-O(\varepsilon^2 n)}$ assignments x are not balanced, it follows that

$$\Pr_{x,G}\left[|\{i : \mathrm{sd}(\hat{D}_i, D_{x_i}) \geq \gamma\}| > \varepsilon n\right] < 2^{-\Omega(\varepsilon^2 n)}$$

from where the lemma follows by Markov's inequality.

We think of the constraint graph G as being chosen in the following manner: First, for each constraint in G the first variable i_1 is chosen uniformly at random. Then, for every i, among the constraints where i is the first variable, the other variables $i_2, \ldots, i_d$ are chosen at random. Let N_i denote the number of constraints with i as the first variable. Observe that conditioned on the choices of N_i, the events

$$\mathrm{sd}(\hat{D}_i, D_{x_i}) \geq \gamma$$

are independent of one another. Let E_i be an indicator variable for this event. Moreover, the distribution $\hat{D}_i$ is an empirical average of N_i samples from D_{x_i}, so by Lemma 2 we have that as long as $N_i \geq D/2$, $\Pr_G[E_i = 1 \mid N_i] \leq \varepsilon/3$.

Let I denote the set of those i such that $N_i < D/2$. Then

$$\Pr_G[\sum_{i \in [n]} E_i \geq \varepsilon n] \leq \Pr_G[\sum_{i \in [n]} E_i \geq \varepsilon n \mid |I| < \varepsilon n/3] + \Pr_G[|I| \geq \varepsilon n/3]$$

$$\leq \Pr_G[\sum_{i \notin I} E_i \geq 2\varepsilon n/3 \mid |I| < \varepsilon n/3] + \Pr_G[|I| \geq \varepsilon n/3]$$

$$\leq 2^{-\Omega(\varepsilon^2 n)} + \Pr_G[|I| \geq \varepsilon n/3] \qquad \text{(by the Chernoff bound)}$$

$$\leq 2^{-\Omega(\varepsilon^2 n)} \qquad \text{(by Lemma 4)} \qquad\qquad \square$$

To finish the proof of proposition 3, we argue that algorithm *Amplify* outputs the correct answer with high probability. First, observe that the algorithm needs to know the correlation between x and x'; we try all possible n values for this correlation. (In fact, it is sufficient to try $O(1/\varepsilon)$ approximate values.) Then proposition 3 follows by combining Lemma 1 and Lemma 3 with $\gamma = \varepsilon^{-Dd}$ for a sufficiently large constant D.

5   From Almost Correct to Correct

In this section, we show that if we start with an almost correct assignment, $f_{G,P}(x)$ can be inverted for any nontrivial predicate P, provided that the constraint to variable ratio $m/n = D$ is a sufficiently large constant (depending on d). Our proofs are an adaptation of known algorithms for planted random 3SAT [Fla03, KV06].

Proposition 4. *Let K be a sufficiently large constant and P be an arbitrary nonconstant predicate. Suppose $D > Kd^6 4^d$. There exists a polynomial-time algorithm such that for a $1 - 2^{-\Omega(d^2 4^d)}$ fraction of $x \in \{0,1\}^n$ and with high probability over the choice of G, on input G, P, $f_{G,P}(x)$, and $x' \in \{0,1\}^n$ that has correlation $1 - 1/Kd2^d D$ with x, outputs an inverse for $f_{G,P}(x)$.*

Together with propositions 1 and 2, we have proved theorem 1. With proposition 3, we have proved theorem 2.

The algorithm has three stages. In the first stage, the objective is to come up with an assignment that matches most "core" variables of x. Roughly speaking, the *core* of G with respect to the assignment x is the set of those variables that occur regularly in G, in the sense that their presence in various types of constraints of G occurs within a small error of the expectation. The core will comprise most of the variables of x. In the second stage, some of the variables are unassigned. At the end of this stage, all assigned variables are assigned as in x, and all core variables are assigned. In the third stage, an assignment for the remaining variables is found by brute force. (The final assignment may not be x, as there are likely to be many possible inverses for $f_{G,P}(x)$.)

Due to space constraints we defer the proof of proposition 4 to the full version of the paper.

Acknowledgment

We thank Eyal Rozenman for useful discussions at the initial stages of this work. Both of the authors' work was supported in part by the National Natural Science Foundation of China Grant 60553001, and the National Basic Research Program of China Grant 2007CB807900,2007CB807901.

References

[AIK04]   Applebaum, B., Ishai, Y., Kushilevitz, E.: Cryptography in NC^0. In: Proceedings of the 45th Annual Symposium on Foundations of Computer Science, pp. 166–175 (2004)

[AIK06]   Applebaum, B., Ishai, Y., Kushilevitz, E.: On pseudorandom generators with linear stretch in NC^0. In: Díaz, J., Jansen, K., Rolim, J.D.P., Zwick, U. (eds.) APPROX 2006 and RANDOM 2006. LNCS, vol. 4110, pp. 260–271. Springer, Heidelberg (2006)

[Bra09]   Braverman, M.: Polylogarithmic independence fools AC^0. Technical Report TR09-011, Electronic Colloquium on Computational Complexity (ECCC) (2009)

[CEMT09]   Cook, J., Etesami, O., Miller, R., Trevisan, L.: Goldreich's one-way function candidate and myopic backtracking algorithms. In: Proceedings of the 6th Theory of Cryptography Conference (TCC), pp. 521–538 (2009)

[Coj06]   Coja-Oghlan, A.: An adaptive spectral heuristic for partitioning random graphs. In: Bugliesi, M., Preneel, B., Sassone, V., Wegener, I. (eds.) ICALP 2006. LNCS, vol. 4051, pp. 691–702. Springer, Heidelberg (2006)

[Fla03]   Flaxman, A.: A spectral technique for random satisfiable 3CNF formulas. In: SODA 2003: Proceedings of the fourteenth annual ACM-SIAM symposium on Discrete algorithms, Baltimore, Maryland, pp. 357–363 (2003)

[Gol00]   Goldreich, O.: Candidate one-way functions based on expander graphs. Technical Report TR00-090, Electronic Colloquium on Computational Complexity (ECCC) (2000)

[KV06]   Krivelevich, M., Vilenchik, D.: Solving random satisfiable 3CNF formulas in expected polynomial time. In: SODA 2006: Proceedings of the seventeenth annual ACM-SIAM symposium on discrete algorithms, pp. 454–463. ACM Press, New York (2006)

[LN90]   Linial, N., Nisan, N.: Approximate inclusion-exclusion. Combinatorica 10(4), 349–365 (1990)

[MST03]   Mossel, E., Shpilka, A., Trevisan, L.: On ϵ-biased generators in NC^0. In: Proceedings of the 44th Annual Symposium on Foundations of Computer Science, pp. 136–145 (2003)

[SS85]   Schmidt, J.P., Shamir, E.: Component structure in the evolution of random hypergraphs. Combinatorica 5(1), 81–94 (1985)

[Vil07]   Vilenchik, D.: It's all about the support: a new perspective on the satisfiability problem. Journal on Satisfiability, Boolean Modeling, and Computation 3, 125–139 (2007)

Appendix: A Sampling Lemma

Lemma 4. *Fix $\varepsilon < 1/2$ and suppose $D > 2\log(1/\varepsilon)$. Let $N_1, \ldots, N_n$ be random variables taking values in the set $\{0, \ldots, Dn\}$ sampled uniformly conditioned on $N_1 + \cdots + N_n = Dn$. Then with probability $2^{-\Omega(\varepsilon Dn)}$, fewer than εn of the variables take value less than $D/2$.*

Proof. Let I denote the set of those i such that $N_i < D/2$. By a union bound, the probability of $|I| \geq \varepsilon n$ is at most $\binom{n}{\varepsilon n}$ times the probability that $N_1, \ldots, N_{\varepsilon n} < D/2$. We argue that for every i,

$$\Pr[N_i < D/2 \mid N_1, \ldots, N_{i-1} < D/2] = 2^{-\Omega(D)}$$

from where the claim follows. To show this, observe that conditioned on $N = N_1 + \cdots + N_{i-1}$, N_i is a sum of $(Dn - N)$ independent Bernoulli random variables with probability $1/(n-i)$ each. If $N_1, \ldots, N_{i-1} < D/2$, then the conditional expectation of N_i is at least D. By Chernoff bounds, the conditional probability that $N_i < D/2$ is then at most $2^{-\Omega(D)}$.   □

Random Tensors and Planted Cliques

S. Charles Brubaker and Santosh S. Vempala[*]

Georgia Institute of Technology
Atlanta, GA 30332
{brubaker,vempala}@cc.gatech.edu

Abstract. The r-parity tensor of a graph is a generalization of the adjacency matrix, where the tensor's entries denote the parity of the number of edges in subgraphs induced by r distinct vertices. For $r = 2$, it is the adjacency matrix with 1's for edges and -1's for nonedges. It is well-known that the 2-norm of the adjacency matrix of a random graph is $O(\sqrt{n})$. Here we show that the 2-norm of the r-parity tensor is at most $f(r)\sqrt{n}\log^{O(r)} n$, answering a question of Frieze and Kannan [1] who proved this for $r = 3$. As a consequence, we get a tight connection between the planted clique problem and the problem of finding a vector that approximates the 2-norm of the r-parity tensor of a random graph. Our proof method is based on an inductive application of concentration of measure.

1 Introduction

It is well-known that a random graph $G(n, 1/2)$ almost surely has a clique of size $(2 + o(1)) \log_2 n$ and a simple greedy algorithm finds a clique of size $(1 + o(1)) \log_2 n$. Finding a clique of size even $(1+\epsilon) \log_2 n$ for some $\epsilon > 0$ in a random graph is a long-standing open problem posed by Karp in 1976 [2] in his classic paper on probabilistic analysis of algorithms.

In the early nineties, a very interesting variant of this question was formulated by Jerrum [3] and by Kucera [4]. Suppose that a clique of size p is planted in a random graph, i.e., a random graph is chosen and all the edges within a subset of p vertices are added to it. Then for what value of p can the planted clique be found efficiently? It is not hard to see that $p > c\sqrt{n \log n}$ suffices since then the vertices of the clique will have larger degrees than the rest of the graph, with high probability [4]. This was improved by Alon et al [5] to $p = \Omega(\sqrt{n})$ using a spectral approach. This was refined by McSherry [6] and considered by Feige and Krauthgamer in the more general semi-random model [7]. For $p \geq 10\sqrt{n}$, the following simple algorithm works: form a matrix with 1's for edges and -1's for nonedges; find the largest eigenvector of this matrix and read off the top p entries in magnitude; return the set of vertices that have degree at least $3p/4$ within this subset.

---

[*] Supported in part by NSF award CCF-0721503 and a Raytheon fellowship.

I. Dinur et al. (Eds.): APPROX and RANDOM 2009, LNCS 5687, pp. 406–419, 2009.

The reason this works is the following: the top eigenvector of a symmetric matrix A can be written as

$$\max_{x:\|x\|=1} x^T A x = \max_{x:\|x\|=1} \sum_{ij} A_{ij} x_i x_j$$

maximizing a quadratic polynomial over the unit sphere. The maximum value is the spectral norm or 2-norm of the matrix. For a random matrix with $1, -1$ entries, the spectral norm (largest eigenvalue) is $O(\sqrt{n})$. In fact, as shown by Füredi and Komlós [8],[9], a random matrix with i.i.d. entries of variance at most 1 has the same bound on the spectral norm. On the other hand, after planting a clique of size $\sqrt{n}$ times a sufficient constant factor, the indicator vector of the clique (normalized) achieves a higher norm. Thus the top eigenvector points in the direction of the clique (or very close to it).

Given the numerous applications of eigenvectors (principal components), a well-motivated and natural generalization of this optimization problem to an r-dimensional tensor is the following: given a symmetric tensor A with entries $A_{k_1 k_2 \ldots k_r}$, find

$$\|A\|_2 = \max_{x:\|x\|=1} A(x, \ldots, x),$$

where

$$A(x^{(1)}, \ldots, x^{(r)}) = \sum_{i_1 i_2 \ldots i_r} A_{i_1 i_2 \ldots i_r} x_{i_1}^{(1)} x_{i_2}^{(2)} \ldots x_{i_r}^{(r)}.$$

The maximum value is the spectral norm or 2-norm of the tensor. The complexity of this problem is open for any $r > 2$, assuming the entries with repeated indices are zeros.

A beautiful application of this problem was given recently by Frieze and Kannan [1]. They defined the following tensor associated with an undirected graph $G = (V, E)$:

$$A_{ijk} = E_{ij} E_{jk} E_{ki}$$

where E_{ij} is 1 is $ij \in E$ and -1 otherwise, i.e., A_{ijk} is the parity of the number of edges between i, j, k present in G. They proved that for the random graph $G_{n,1/2}$, the 2-norm of the random tensor A is $\tilde{O}(\sqrt{n})$, i.e.,

$$\sup_{x:\|x\|=1} \sum_{i,j,k} A_{ijk} x_i x_j x_k \leq C \sqrt{n} \log^c n$$

where c, C are absolute constants. This implied that if such a maximizing vector x could be found (or approximated), then we could find planted cliques of size as small as $n^{1/3}$ times polylogarithmic factors in polynomial time, improving substantially on the long-standing threshold of $\Omega(\sqrt{n})$.

Frieze and Kannan ask the natural question of whether this connection can be further strengthened by going to r-dimensional tensors for $r > 3$. The tensor itself has a nice generalization. For a given graph $G = (V, E)$ the r-parity tensor is defined as follows. Entries with repeated indices are set to zero; any other entry is the parity of the number of edges in the subgraph induced by the subset of vertices corresponding to the entry, i.e.,

$$A_{k_1,\dots,k_r} = \prod_{1 \le i < j \le r} E_{k_i k_j}.$$

Frieze and Kannan's proof for $r = 3$ is combinatorial (as is the proof by Füredi and Komlós for $r = 2$), based on counting the number of subgraphs of a certain type. It is not clear how to extend this proof.

Here we prove a nearly optimal bound on the spectral norm of this random tensor for any r. This substantially strengthens the connection between the planted clique problem and the tensor norm problem. Our proof is based on a concentration of measure approach. In fact, we first reprove the result for $r = 3$ using this approach and then generalize it to tensors of arbitrary dimension. We show that the norm of the subgraph parity tensor of a random graph is at most $f(r)\tilde{O}(\sqrt{n})$ whp. More precisely, our main theorem is the following.

Theorem 1. *There is a constant C_1 such that with probability at least $1 - n^{-1}$ the norm of the r-dimensional subgraph parity tensor $A : [n]^r \to \{-1, 1\}$ for the random graph $G_{n,1/2}$ is bounded by*

$$\|A\|_2 \le C_1^r r^{(5r-1)/2} \sqrt{n} \log^{(3r-1)/2} n.$$

The main challenge to the proof is the fact that the entries of the tensor A are not independent. Bounding the norm of the tensor where every entry is independently 1 or -1 with probability $1/2$ is substantially easier via a combination of an ϵ-net and a Hoeffding bound. In more detail, we approximate the unit ball with a finite (exponential) set of vectors. For each vector x in the discretization, the Hoeffding inequality gives an exponential tail bound on $A(x, \dots, x)$. A union bound over all points in the discretization then completes the proof. For the parity tensor, however, the Hoeffding bound does not apply as the entries are not independent. Moreover, all the $\binom{n}{r}$ entries of the tensor are fixed by just the $\binom{n}{2}$ edges of the graph. In spite of this heavy interdependence, it turns out that $A(x, \dots, x)$ does concentrate. Our proof is inductive and bounds the norms of vectors encountered in a certain decomposition of the tensor polynomial. It is not clear whether the bound of Theorem 1 is optimal, though a lower bound of $\|A\|_2 = \Omega(\max\{\sqrt{n}, (2 \log n)^{r/2}\})$ is trivial.

Using Theorem 1, we can show that if the norm problem can be solved for tensors of dimension r, one can find planted cliques of size as low as $Cn^{1/r}\text{poly}(r, \log n)$. While the norm of the parity tensor for a random graph remains bounded, when a clique of size p is planted, the norm becomes at least $p^{r/2}$ (using the indicator vector of the clique). Therefore, p only needs to be a little larger than $n^{1/r}$ in order for the the clique to become the dominant term in the maximization of $A(x, \dots, x)$. More precisely, we have the following theorem.

Theorem 2. *Let G be random graph $G_{n,1/2}$ with a planted clique of size p, and let A be the r-parity tensor for G. For $\alpha \leq 1$, let $T(n,r)$ be the time to compute a vector x such that $A(x, \ldots, x) \geq \alpha^r \|A\|_2$ whp. Then, for p such that*

$$n \geq p > C_0 \alpha^{-2} r^5 n^{1/r} \log^3 n,$$

the planted clique can be recovered with high probability in time $T(n,r) + \text{poly}(n)$, where C_0 is a fixed constant.

On one hand, this highlights the benefits of finding an efficient (approximation) algorithm for the tensor problem. On the other, given the lack of progress on the clique problem, this is perhaps evidence of the hardness of the tensor maximization problem even for a natural class of random tensors. For example, if finding a clique of size $\tilde{O}(n^{1/2-\epsilon})$ is hard, then by setting $\alpha = n^{1/2r+\epsilon/2-1/4}$ we see that even a certain polynomial approximation to the norm of the parity tensor is hard to achieve.

Corollary 1. *Let G be random graph $G_{n,1/2}$ with a planted clique of size p, and let A be the r-parity tensor for G. Let $\epsilon > 0$ be a small constant and let $T(n,r)$ be the time to compute a vector x such that $A(x, \ldots, x) \geq n^{1/2+r\epsilon/2-r/4} \|A\|_2$. Then, for*

$$p \geq C_0 r^5 n^{\frac{1}{2}-\epsilon} \log^3 n,$$

the planted clique can be recovered with high probability in time $T(n,r) + \text{poly}(n)$, where C_0 is a fixed constant.

1.1 Overview of Analysis

The majority of the paper is concerned with proving Theorem 1. In Section 2.1, we first reduce the problem of bounding $A(\cdot)$ over the unit ball to bounding it over a discrete set of vectors that have the same value in every non-zero coordinate. In Section 2.2, we further reduce the problem to bounding the norm of an off-diagonal block of A, using a method of Frieze and Kannan. This enables us to assume that if $(k_1, \ldots, k_r)$ is a valid index, then the random variables E_{k_i,k_j} used to compute $A_{i_1,\ldots,i_r}$ are independent. In Section 2.3, we give a large deviation inequality (Lemma 3) that allows us to bound norms of vectors encountered in a certain decomposition of the tensor polynomial. This inequality gives us a considerably sharper bound than the Hoeffding or McDiarmid inequalities in our context. We then apply this lemma to bound $\|A\|_2$ for $r = 3$ as a warm-up and then give the proof for general r in Section 3.

In Section 4 we prove Theorem 2. The key idea is that any vector x that comes close to maximizing $A(\cdot)$ must have an indicator decomposition where the support of one of the vectors has a large intersection with the clique (Lemma 7). This intersection is large enough that the clique can be recovered.

For some lemmas, only a summary of the proof is given. The details of these proofs are available in full version of the paper.[1]

---

[1] `arXiv:0905.2381 [cs.DS]`.

2   Preliminaries

2.1   Discretization

The analysis of $A(x, \ldots, x)$ is greatly simplified when x is proportional to some indicator vector. Fortunately, analyzing these vectors is sufficient, as any vector can be approximated as a linear combination of relatively few indicator vectors.

For any vector x, we define $x^{(+)}$ to be vector such that $x_i^{(+)} = x_i$ if $x_i > 0$ and $x_i^{(+)} = 0$ otherwise. Similarly, let $x_i^{(-)} = x_i$ if $x_i < 0$ and $x_i^{(-)} = 0$ otherwise. For a set $S \subseteq [n]$, let χ^S be the indicator vector for S, where the ith entry is 1 if $i \in S$ and 0 otherwise.

Definition 1 (Indicator Decomposition). *For a unit vector x, define the sets $S_1, \ldots$ and $T_1, \ldots$ through the recurrences*

$$S_j = \left\{ i \in [n] : (x^{(+)} - \sum_{k=1}^{j-1} 2^{-k} \chi^{S_k})_i > 2^{-j} \right\}.$$

and

$$T_j = \left\{ i \in [n] : (x^{(-)} - \sum_{k=1}^{j-1} 2^{-k} \chi^{S_k})_i < -2^{-j} \right\}.$$

Let $y_0(x) = 0$. For $j \geq 1$, let $y^{(j)}(x) = 2^{-j} \chi^{S_j}$ and let $y^{(-j)}(x) = -2^{-j} \chi^{T_j}$. We call the set $\{y^{(j)}(x)\}_{-\infty}^{\infty}$ the indicator decomposition of x.

Clearly, $\|y^{(i)}(x)\| \leq \max\{\|x^{(+)}\|, \|x^{(-)}\|\} \leq 1$, and

$$\left\| x - \sum_{j=-N}^{N} y^{(j)}(x) \right\| \leq \sqrt{n} 2^{-N}. \tag{1}$$

We use this decomposition to prove the following theorem.

Lemma 1. *Let*

$$U = \{ k|S|^{-1/2} \chi^S : S \subseteq [n], k \in \{-1, 1\} \}.$$

For any tensor A over $[n]^r$ where $\|A\|_\infty \leq 1$

$$\max_{x^{(1)}, \ldots, x^{(r)} \in B(0,1)} A(x^{(1)}, \ldots x^{(r)}) \leq (2\lceil r \log n \rceil)^r \max_{x^{(1)}, \ldots, x^{(r)} \in U} A(x^{(1)}, \ldots, x^{(r)})$$

Proof. Consider a fixed set of vectors $x^{(1)}, \ldots, x^{(r)}$ and let $N = \lceil r \log_2 n \rceil$. For each i, let

$$\hat{x^{(i)}} = \sum_{j=-N}^{N} y^{(j)}(x^{(i)}).$$

We first show that replacing $x^{(i)}$ with $\hat{x}^{(i)}$ gives a good approximation to $A(x^{(1)}, \ldots, x^{(r)})$. Letting ϵ be the maximum difference between an $x^{(i)}$ and its approximation, we have from (1) that

$$\max_{i \in [r]} \|x^{(i)} - \hat{x}^{(i)}\| = \epsilon \leq \frac{n^{r/2}}{2r}$$

Because of the multilinear form of $A(\cdot)$ we have

$$|A(x^{(1)}, \ldots, x^{(r)}) - A(\hat{x}^{(1)}, \ldots, \hat{x}^{(r)})| \leq \sum_{i=1}^{r} \epsilon^i r^i \|A\| \leq \frac{\epsilon r}{1 - \epsilon r} \|A\| \leq 1.$$

Next, we bound $A(\hat{x}^{(1)}, \ldots, \hat{x}^{(r)})$. For convenience, let $Y^{(i)} = \cup_{j=-N}^{N} y^{(j)}(x^{(i)})$. Then using the multlinear form of $A(\cdot)$ and bounding the sum by its maximum term, we have

$$A(\hat{x}^{(1)}, \ldots, \hat{x}^{(r)}) \leq (2N)^r \max_{v^{(1)} \in Y^{(1)}, \ldots, v^{(r)} \in Y^{(r)}} A(v^{(1)}, \ldots, v^{(r)})$$

$$\leq (2N)^r \max_{v^{(1)}, \ldots, v^{(r)} \in U} A(v^{(1)}, \ldots, v^{(r)}).$$

2.2   Sufficiency of Off-Diagonal Blocks

Analysis of $A(x^{(1)}, \ldots, x^{(r)})$ is complicated by the fact that all terms with repeated indices are zero. Off-diagonal blocks of A are easier to analyze because no such terms exist. Thankfully, as Frieze and Kannan [1] have shown, analyzing these off-diagonal blocks suffices. Here we generalize their proof to $r > 3$.

For a collection $\{V_1, V_2, \ldots, V_r\}$ of subsets of $[n]$, we define

$$A|_{V_1 \times \ldots \times V_r}(x^{(1)}, \ldots, x^{(r)}) = \sum_{k_1 \in V_1, \ldots, k_r \in V_r} A_{k_1 \ldots k_r} x_{i_1}^{(1)} x_{i_2}^{(2)} \ldots x_{i_r}^{(r)}$$

Lemma 2. *Let P be the class of partitions of $[n]$ into r equally sized sets $V_1, \ldots, V_r$ (assume wlog that r divides n). Let $V = V_1 \times \ldots \times V_r$. Let A be a random tensor over $[n]^r$ where each entry is in $[-1, 1]$ and let $R \subseteq B(0, 1)$. If for every fixed $(V_1, \ldots V_r) \in P$, it holds that*

$$\Pr[\max_{x^{(1)}, \ldots, x^{(r)} \in R} A|_V(x^{(1)}, \ldots, x^{(r)}) \geq f(n)] \leq \delta,$$

then

$$\Pr[\max_{x^{(1)}, \ldots, x^{(r)} \in R} A(x^{(1)}, \ldots, x^{(r)}) \geq 2r^r f(n)] \leq \frac{\delta n^{r/2}}{f(n)}.$$

2.3   A Concentration Bound

The following concentration bound is a key tool in our proof of Theorem 1. We apply it for $t = \tilde{O}(N)$.

Lemma 3. *Let* $\{u^{(i)}\}_{i=1}^N$ *and* $\{v^{(i)}\}_{i=1}^N$ *be collections of vectors of dimension* N' *where each entry of* $u^{(i)}$ *is 1 or* -1 *with probability* $1/2$ *and* $\|v^{(i)}\|_2 \le 1$. *Then for any* $t \ge 1$,

$$\Pr[\sum_{i=1}^N (u^{(i)} \cdot v^{(i)})^2 \ge t] \le e^{-t/18}(4\sqrt{e\pi})^N.$$

We note that this lemma is stronger than what a naive application of standard theorems would yield for $t = \tilde{O}(N)$. For instance one might treat each $(u^{(i)} \cdot v^{(i)})^2$ as an independent random variable and apply a Hoeffding bound. The quantity $(u^{(i)} \cdot v^{(i)})^2$ can vary by as much as N', however, so the bound would be roughly $\exp(-ct^2/NN'^2)$ for some constant c. Similarly, treating each $u_j^{(i)}$ as an independent random variable and applying McDiarmid's inequality, we find that every $u_j^{(i)}$ can affect the sum by as much as 1 (simultaneously). For instance suppose that every $v_j^{(i)} = 1/\sqrt{N'}$ and every $u_j^{(i)} = 1$. Then flipping $u_j^{(i)}$ would have an effect of $|N' - ((N'-2)/\sqrt{N'})^2| \approx 4$, so the bound would be roughly $\exp(-ct^2/NN')$ for some constant c. The proof, which uses an epsilon-net argument, can be found in the full version of the paper.

3   A Bound on the Norm of the Parity Tensor

In this section, we prove Theorem 1. As a warm-up we consider the case where $r = 3$.

3.1   Warm-Up: Third Order Tensors

For $r = 3$ the tensor A is defined as follows:

$$A_{k_1 k_2 k_3} = E_{k_1 k_2} E_{k_2 k_3} E_{k_1 k_3}.$$

Theorem 3. *There is a constant* C_1 *such that with probability* $1 - n^{-1}$

$$\|A\| \le C_1 \sqrt{n} \log^4 n.$$

Proof. Let V_1, V_2, V_3 be a partition of the n vertices and let $V = V_1 \times V_2 \times V_3$. The bulk of the proof consists of the following lemma.

Lemma 4. *There is some constant* C_3 *such that*

$$\max_{x^{(1)}, x^{(2)}, x^{(3)} \in U} A|_V(x^{(1)}, x^{(2)}, x^{(3)}) \le C_3 \sqrt{n} \log n$$

with probability $1 - n^{-7}$.

If this bound holds, then Lemma 1 then implies that there is some C_2 such that

$$\max_{x^{(1)}, x^{(2)}, x^{(3)} \in B(0,1)} A|_V(x^{(1)}, x^{(2)}, x^{(3)}) \le C_2 \sqrt{n} \log^4 n.$$

And finally, Lemma 2 implies that for some constant C_1

$$\max_{x^{(1)}, x^{(2)}, x^{(3)} \in B(0,1)} A(x^{(1)}, x^{(2)}, x^{(3)}) \leq C_1 \sqrt{n} \log^4 n$$

with probability $1 - n^{-1}$.

Proof (Proof of Lemma 4). Define

$$U_k = \{x \in U : |\mathrm{supp}(x)| = k\} \tag{2}$$

and consider a fixed $n \geq n_1 \geq n_2 \geq n_3 \geq 1$. We will show that

$$\max_{(x^{(1)}, x^{(2)}, x^{(3)}) \in U_{n_1} \times U_{n_2} \times U_{n_3}} A|_V(x^{(1)}, x^{(2)}, x^{(3)}) \leq C_3 \sqrt{n} \log n$$

with probability n^{-10} for some constant C_3. Taking a union bound over the n^3 choices of n_1, n_2, n_3 then proves the lemma.

We bound the cubic form as

$$\max_{(x^{(1)}, x^{(2)}, x^{(3)}) \in U_{n_1} \times U_{n_2} \times U_{n_3}} A|_V(x^{(1)}, x^{(2)}, x^{(3)})$$

$$= \max_{(x^{(1)}, x^{(2)}, x^{(3)}) \in U_{n_1} \times U_{n_2} \times U_{n_3}} \sum_{k_1 \in V_1, k_2 \in V_2, k_3 \in V_3} A_{k_1 k_2 k_3} x_{k_1}^{(1)} x_{k_2}^{(2)} x_{k_3}^{(3)}$$

$$\leq \max_{(x^{(2)}, x^{(3)}) \in U_{n_2} \times U_{n_3}} \sqrt{\sum_{k_1 \in V_1} \left(\sum_{k_2 \in V_2, k_3 \in V_3} A_{k_1 k_2 k_3} x_{k_2}^{(2)} x_{k_3}^{(3)} \right)^2}$$

$$= \max_{(x^{(2)}, x^{(3)}) \in U_{n_2} \times U_{n_3}} \sqrt{\sum_{k_1 \in V_1} \left(\sum_{k_2 \in V_2} E_{k_1 k_2} x_{k_2}^{(2)} \sum_{k_3 \in V_3} E_{k_2 k_3} x_{k_3}^{(3)} E_{k_1 k_3} \right)^2}.$$

Note that each of the inner sums (over k_2 and k_3) are the dot product of a random $-1, 1$ vector (the $E_{k_1 k_2}$ and $E_{k_2 k_3}$ terms) and another vector. Our strategy will be to bound the norm of this other vector and apply Lemma 3.

To this end, we define the $-1, 1$ vectors $u_{k_3}^{(k_2)} = E_{k_2 k_3}$ and $u_{k_2}^{(k_1)} = E_{k_1 k_2}$, and the general vectors

$$v^{(k_1 k_2)}(x^{(3)})_{k_3} = x_{k_3}^{(3)} E_{k_1 k_3}$$

and

$$v^{(k_1)}(x^{(2)}, x^{(3)})_{k_2} = x_{k_2}^{(2)}(u^{(k_2)} \cdot v^{(k_1 k_2)}(x^{(3)})).$$

Thus, for each k_1,

$$\sum_{k_2 \in V_2} E_{k_1 k_2} x_{k_2}^{(2)} \sum_{k_3 \in V_3} E_{k_2 k_3} x_{k_3}^{(3)} E_{k_1 k_3}$$

$$= \sum_{k_2 \in V_2} E_{k_1 k_2} x_{k_2}^{(2)}(u^{(k_2)} \cdot v^{(k_1 k_2)}(x^{(3)}))$$

$$= u^{(k_1)} \cdot v^{(k_1)}(x^{(2)}, x^{(3)}). \tag{3}$$

Clearly, the u's play the role of the random vectors and we will bound the norms of the v's in the application of Lemma 3.

To apply Lemma 3 with k_1 being the index i and $u_{k_2}^{k_1} = E_{k_1 k_2}$ as above, we need a bound for every $k_1 \in V_1$ on the norm of $v^{(k_1)}(x^{(2)}, x^{(3)})$. We argue

$$\sum_{k_2} \left(x_{k_2}^{(2)} \sum_{k_3 \in V_3} E_{k_2 k_3} x_{k_3}^{(3)} E_{k_1 k_3} \right)^2$$

$$\leq \max_{k_1 \in V_1} \max_{x^{(2)} \in U_{n_2}} \max_{x^{(3)} \in U_{n_3}} \frac{1}{n_2} \sum_{k_2 \in \text{supp}(x^{(x_2)})} \left(\sum_{k_3} E_{k_2 k_3} x_{k_3}^{(3)} E_{k_1 k_3} \right)^2$$

$$\equiv F_1^2$$

Here we used the fact that $\|x^{(2)}\|_\infty \leq n_2^{-1/2}$. Note that by the definition above F_1 is a function of the random variables $\{E_{ij}\}$ only.

To bound F_1, we observe that we can apply Lemma 3 to the expression being maximized above, i.e.,

$$\sum_{k_2} \left(\sum_{k_3} E_{k_2 k_3} \left(x_{k_3}^{(3)} E_{k_1 k_3} \right) \right)^2$$

over the index k_2, with $u_{k_3}^{k_2} = E_{k_2 k_3}$. Now we need a bound, for every k_2 and k_1 on the norm of the vector $v^{(k_1 k_2)}(x^{(3)})$. We argue

$$\sum_{k_3} \left(x_{k_3}^{(3)} E_{k_1 k_3} \right)^2 \leq \|x^{(3)}\|_\infty^2 \sum_{k_3} E_{k_1 k_3}^2 \leq 1.$$

Applying Lemma 3 for a fixed $k_1, x^{(2)}$ and $x^{(3)}$ implies

$$\frac{1}{n_2} \sum_{k_2 \in \text{supp}(x^{(2)})} \left(\sum_{k_3} E_{k_2 k_3} x_{k_3}^{(3)} E_{k_1 k_3} \right)^2 > C_3 \log n$$

with probability at most

$$\exp\left(-\frac{C_3 n_2 \log n}{18}\right)(4\sqrt{e\pi})^{n_2}.$$

Taking a union bound over the $|V_1| \leq n$ choices of k_1, and the at most $n^{n_2} n^{n_3}$ choices for $x^{(2)}$ and $x^{(3)}$, we show that

$$\Pr[F_1^2 > C_3 \log n] \leq \exp\left(-\frac{C_3 n_2 \log n}{18}\right)(4\sqrt{e\pi})^{n_2} n n^{n_2} n^{n_3}.$$

This probability is at most $n^{-10}/2$ for a large enough constant C_3.

Thus, for a fixed $x^{(2)}$ and $x^{(3)}$, we can apply Lemma 3 to (3) with $F_1^2 = C_3 \log n$ to get:

$$\sum_{k_1 \in V_1} \left(\sum_{k_2 \in V_2} E_{k_1 k_2} \left(x_{k_2}^{(2)} \sum_{k_3 \in V_3} E_{k_2 k_3} x_{k_3}^{(3)} E_{k_1 k_3} \right) \right)^2 > F_1^2 C_3 n \log n$$

with probability at most $\exp(-C_3 n \log n/18)(4\sqrt{e\pi})^n$. Taking a union bound over the at most $n^{n_2} n^{n_3}$ choices for $x^{(2)}$ and $x^{(3)}$, the bound holds with probability

$$\exp(-C_3 n \log n/18)(4\sqrt{e\pi})^n n^{n_2} n^{n_3} \le n^{-10}/2$$

for large enough constant C_3.

Thus, we can bound the squared norm:

$$\max_{(x^{(1)},x^{(2)},x^{(3)}) \in U_{n_1} \times U_{n_2} \times U_{n_3}} A|_V(x^{(1)}, x^{(2)}, x^{(3)})^2$$

$$\le \sum_{k_1 \in V_1} \left(\sum_{k_2 \in V_2} E_{k_1 k_2} \left(x_{k_2}^{(2)} \sum_{k_3 \in V_3} E_{k_2 k_3} x_{k_3}^{(3)} E_{k_1 k_3} \right) \right)^2$$

$$\le C_3^2 n_1 \log^2 n$$

with probability $1 - n^{-10}$.

3.2 Higher Order Tensors

Let the random tensor A be defined as follows.

$$A_{k_1,\ldots,k_r} = \prod_{1 \le i < j \le r} E_{k_i k_j}$$

where E is an $n \times n$ matrix where each off-diagonal entry is -1 or 1 with probability $1/2$ and every diagonal entry is 1.

For most of this section, we will consider only a single off-diagonal cube of A. That is, we index over $V_1 \times \ldots \times V_r$ where V_i are an equal partition of $[n]$. We denote this block by $A|_V$. When k_i is used as an index, it is implied that $k_i \in V_i$.

The bulk of the proof consists of the following lemma.

Lemma 5. *There is some constant C_3 such that*

$$\max_{x^{(1)},\ldots x^{(r)} \in U} A|_V(x^{(1)}, \ldots, x^{(r)})^2 \le n(C_3 r \log n)^{r-1}$$

with probability $1 - n^{-9r}$.

The key idea is that the concentration inequality of Lemma 3 can be applied repeatedly to collections of u's and v's in a way analogous to (3). Each sum over $k_r, \ldots, k_2$ contributes a $C_3 r \log n$ factor and the final sum over k_1 contributes the factor of n.

If the bound holds, then Lemma 1 implies that there is some C_2 such that

$$\max_{x^{(1)},x^{(2)},x^{(3)} \in B(0,1)} A|_V(x^{(1)}, x^{(2)}, x^{(3)})^2 \le C_2^r r^{2r+r-1} n \log^{2r+(r-1)} n.$$

And finally, Lemma 2 implies that

$$\max_{x^{(1)},x^{(2)},x^{(3)} \in B(0,1)} A(x^{(1)}, x^{(2)}, x^{(3)}) \le C_1^r r^{2r+2r+(r-1)} n \log^{2r+r-1} n$$

$$= C_1^r r^{5r-1} n \log^{3r-1} n.$$

with probability $1 - n^{-1}$ for some constant C_1. For the complete proof, please see the full version of the paper.

4   Finding Planted Cliques

We now turn to Theorem 2 and to the problem of finding a planted clique in a random graph. A random graph with a planted clique is constructed by taking a random graph and then adding every edge between vertices in some subset P to form the planted clique. We denote this graph as $G_{n,1/2} \cup K_p$. Letting A be the rth order subgraph parity tensor, we show that a vector $x \in B(0,1)$ that approximates the maximum of $A(\cdot)$ over the unit ball can be used to reveal the clique, using a modification of the algorithm proposed by Frieze and Kannan [1].

This implies an interesting connection between the tensor problem and the planted clique problem. For symmetric second order tensors (i.e. matrices), maximizing $A(\cdot)$ is equivalent to finding the top eigenvector and can be done in polynomial time. For higher order tensors, however, the complexity of maximizing this function is open if elements with repeated indices are zero. For random tensors, the hardness is also open. Given the reduction presented in this section, a hardness result for the planted clique problem would imply a similar hardness result for the tensor problem.

Given an x that approximates the maximum of $A(\cdot)$ over the unit ball, the algorithm for finding the planted clique is given in Alg. 1. The key ideas of using the top eigenvector of subgraph and of randomly choosing a set of vertices to "seed" the clique (steps 2a-2d) come from Frieze-Kannan [1]. The major difference in the algorithms is the use of the indicator decomposition. Frieze and Kannan sort the indices so that $x_1 \geq \ldots x_n$ and select one set S of the form $S = [j]$ where $\|A|_{S \times S}\|$ exceeds some threshold. They run steps (2a-2d) only on this set. By contrast Alg. 1 runs these steps on every $S = \text{supp}(y^{(j)}(x))$ where $j = -\lceil r \log n \rceil, \ldots \lceil r \log n \rceil$.

The algorithm succeeds with high probability when a subset S is found such that $|S \cap P| \geq C\sqrt{|S| \log n}$, where C is an appropriate constant.

Lemma 6 (Frieze-Kannan). *There is a constant C_5 such that if $S \subseteq [n]$ satisfies $|S \cap P| \geq C_5\sqrt{|S| \log n}$, then with high probability steps 2a)-2d) of Alg. 1 find a set P' equal to P.*

To find such an subset S from a vector x, Frieze and Kannan require that $\sum_{i \in P} x_i \geq C \log n$. Using the indicator decomposition, as in the Alg. 1, however, reduces this to $\sum_{i \in P} x_i \geq C\sqrt{\log n}$. Even more importantly, using the indicator decomposition means that only one element of the decomposition needs to point in the direction of the clique. The vector x could point in a very different direction and the algorithm would still succeed. We exploit this fact in our proof of Theorem 2. The relevant claim is the following.

Lemma 7. *Let B' be a set of vectors $x \in B(0,1)$ such that*

$$|\text{supp}(y^{(j)}(x)) \cap P| < C_5\sqrt{|\text{supp}(y^{(j)}(x))| \log n}$$

for every $j \in \{-\lceil r \log n \rceil, \ldots, \lceil r \log n \rceil\}$. Then, there is a constant C_1' such that with high probability

$$\sup_{x \in B'} A(x, \ldots, x) \leq C_1'^r r^{5r/2} \sqrt{n} \log^{3r/2} n.$$

Algorithm 1. An Algorithm for Recovering the Clique

Input:
1) Graph G.
2) Integer $p = |P|$.
3) Unit vector x.

Output: A clique of size p or FAILURE.

1. Calculate $y^{-\lceil r \log n \rceil}(x), \ldots, y^{\lceil r \log n \rceil}(x)$ as defined in the indicator decomposition.
2. For each such $y^{(j)}(x)$, let $S = \mathrm{supp}(y^{(j)}(x))$ and try the following:
   (a) Find v, the top eigenvector of the $1, -1$ adjacency matrix $A|_{S \times S}$.
   (b) Order the vertices (coordinates) such that $v_1 \geq \ldots \geq v_{|S|}$. (Assuming dot-prod is $\sqrt{1/2}$ below)
   (c) For $\ell = 1$ to $|S|$, repeat up to $n^{30} \log n$ times:
       i. Select $10 \log n$ vertices Q_1 at random from $[\ell]$.
       ii. Find Q_2, the set of common neighbors of Q_1 in G.
       iii. If the set of vertices with degree at least $7p/8$, say P' has cardinality p and forms a clique in G, then return P'.
   (d) Return FAILURE.

Proof. By the same argument used in the discretization, we have that for any $x \in B'$

$$A(x, \ldots, x) \leq (2\lceil r \log n \rceil)^r \max_{x^{(1)} \in Y^{(1)}(x), \ldots x^{(r)} \in Y^{(r)}(x)} A(x^{(1)}, \ldots, x^{(r)})$$

$$\leq (2\lceil r \log n \rceil)^r \max_{x^{(1)}, \ldots x^{(r)} \in U'} A(x^{(1)}, \ldots, x^{(r)}), \tag{4}$$

where

$$U' = \{|S|^{-1/2} \chi^S : S \subseteq [n], |S \cap P| < C_5 \sqrt{|S| \log n}\}.$$

Consider an off-diagonal block $V_1 \times \ldots \times V_r$. For each $i \in 1 \ldots r$, let $P_i = V_i \cap P$ and let $R_i = V_i \setminus P$. Then, breaking the polynomial $A|_V(\cdot)$ up as a sum of 2^r terms, each corresponding to a choice of $S_1 \in \{P_1, R_1\}, \ldots, S_r \in \{P_r, R_r\}$ gives

$$\max_{x^{(1)}, \ldots, x^{(r)} \in U'} A|_V(x^{(1)}, \ldots, x^{(r)})$$

$$\leq 2^r \max_{x^{(1)}, \ldots, x^{(r)} \in U'} \sum_{S_1 \in \{P_1, R_1\}, \ldots, S_r \in \{P_r, R_r\}} A|_{S_1 \times \ldots \times S_r}(x^{(1)}, \ldots, x^{(r)}). \tag{5}$$

By symmetry, without loss of generality we may consider the case where $S_i = R_i$ for $i = 1 \ldots r - \ell$ and $S_i = P_i$ for $i = r - \ell + 1 \ldots r$ for some ℓ. Let $\tilde{V} = R_1 \times \ldots \times R_{r-\ell} \times P_{r-\ell+1} \times \ldots \times P_r$. Then,

$$\max_{x^{(1)}, \ldots, x^{(r)} \in U'} A|_{\tilde{V}}(x^{(1)}, \ldots, x^{(r)})$$

$$= \sum_{k_1 \in R_1} \ldots \sum_{k_{r-\ell} \in R_{r-\ell}} \prod_{i=1 \ldots r - \ell} x_{k_i}^{(i)} \prod_{i,j : i, j \leq r - \ell} E_{k_i k_j} B^{(k_1, \ldots, k_{r-\ell})},$$

where

$$B^{(k_1,\dots,k_{r-\ell})}\big(x^{(r-\ell+1)},\dots,x^r\big)$$

$$= \sum_{k_{r-\ell+1}\in P_{r-\ell+1}} \cdots \sum_{k_r\in P_r} \prod_{i=r-\ell+1\dots r} x_{k_i}^{(i)} \prod_{i,j:i,r-\ell+1<j} E_{k_i k_j}.$$

By the assumption that every $x^{(i)} \in U'$, this value is at most $(C_5 \log n)^{\ell/2}$. Thus,

$$\max_{x^{(1)},\dots,x^{(r)}\in U'} A|_{\tilde{V}}\big(x^{(1)},\dots,x^{(r)}\big)$$

$$\leq \sum_{k_1\in R_1} \cdots \sum_{k_{r-\ell}\in R_{r-\ell}} \prod_{i=1\dots r-\ell} x_{k_i}^{(i)} \prod_{i,j:i,j\leq r-\ell} E_{k_i k_j}(C_5 \log n)^{\ell/2}.$$

Note that every edge $E_{k_i k_j}$ above is random, so the polynomial may be bounded according to Lemma 5. Altogether,

$$\max_{x^{(1)},\dots,x^{(r)}\in U'} A|_{\tilde{V}}\big(x^{(1)},\dots,x^{(r)}\big) \leq (\max\{C_5,C_3\} \log n)^{r/2}.$$

Combining (4), (5), and applying Lemma 2 completes the proof with C_1' chosen large enough.

Proof (Proof of Theorem 2). The clique is found by finding a vector x such that $A(x,\dots,x) \geq \alpha^r |P|^{r/2}$ and then running Alg. 1 on this vector. Algorithm 1 clearly runs in polynomial time, so the theorem holds if the algorithm succeeds with high probability.

By Lemma 6 the algorithm does succeed with high probability when $x \notin B'$, i.e. when some $S \in \{\operatorname{supp}(y-\lceil r \log n\rceil(x)),\dots,\operatorname{supp}(y-\lceil r \log n\rceil(x)\}$ satisfies $|S \cap P| \geq C_5\sqrt{|S|\log n}$.

We claim $x \notin B'$ with high probability. Otherwise, for some $x \in B'$,

$$A(x,\dots,x) \geq \alpha^r p^{r/2} > C_0^r r^{5r/2}\sqrt{n}\log^{3r/2} n.$$

This is a low probability event by Lemma 7 if $C_0 \geq C_1'$.

References

1. Frieze, A., Kannan, R.: A new approach to the planted clique problem. In: Proc. of FST & TCS (2008)
2. Karp, R.: The probabilistic analysis of some combinatorial search algorithms. In: Algorithms and Complexity: New Directions and Recent Results, pp. 1–19. Academic Press, London (1976)
3. Jerrum, M.: Large cliques elude the metropolis process. Random Structures and Algorithms 3(4), 347–360 (1992)
4. Kucera, L.: Expected complexity of graph partitioning problems. Discrete Applied Mathematics 57, 193–212 (1995)

5. Alon, N., Krivelevich, M., Sudakov, B.: Finding a large hidden clique in a random graph. Random Structures and Algorithms 13, 457–466 (1998)
6. McSherry, F.: Spectral partitioning of random graphs. In: Proc. of FOCS, pp. 529–537 (2001)
7. Feige, U., Krauthgamer, R.: Finding and certifying a large hidden clique in a semi-random graph. Random Structures and Algorithms 16(2), 195–208 (2000)
8. Füredi, Z., Komlós, J.: The eigenvalues of random symmetric matrices. Combinatorica 1(3), 233–241 (1981)
9. Vu, V.H.: Spectral norm of random matrices. In: Proc. of STOC, pp. 423–430 (2005)

Sampling s-Concave Functions: The Limit of Convexity Based Isoperimetry

Karthekeyan Chandrasekaran[1], Amit Deshpande[2], and Santosh Vempala[1]

[1] School of Computer Science, Georgia Institute of Technology
karthe@cc.gatech.edu, vempala@cc.gatech.edu
[2] Microsoft Research India
amitdesh@microsoft.com

Abstract. Efficient sampling, integration and optimization algorithms for logconcave functions [BV04, KV06, LV06a] rely on the good isoperimetry of these functions. We extend this to show that $-1/(n-1)$-concave functions have good isoperimetry, and moreover, using a characterization of functions based on their values along every line, we prove that this is *the largest class* of functions with good isoperimetry in the spectrum from concave to quasi-concave. We give an efficient sampling algorithm based on a random walk for $-1/(n-1)$-concave probability densities satisfying a smoothness criterion, which includes heavy-tailed densities such as the Cauchy density. In addition, the mixing time of this random walk for Cauchy density matches the corresponding best known bounds for logconcave densities.

1 Introduction

Given a function $f : \mathbb{R}^n \to \mathbb{R}_+$, accessible by querying the function value at any point $x \in \mathbb{R}^n$, and an error parameter $\epsilon > 0$, three fundamental problems are: (i) Integration: estimate $\int f$ to within $1 \pm \epsilon$, (ii) Maximization: find x that approximately maximizes f, i.e., $f(x) \geq (1 - \epsilon) \max f$, and (iii) Sampling: generate x from density π with $d_{tv}(\pi, \pi_f) \leq \epsilon$ where d_{tv} is the total variation distance and π_f is the density proportional to f. The complexity of an algorithm is measured by the number of queries for the function values.

The most general class of functions for which these problems are known to have compexity polynomial in the dimension, is the class of logconcave functions. A function $f : \mathbb{R}^n \to \mathbb{R}_+$ is logconcave if its logarithm is concave on its support, i.e., for any two points $x, y \in \mathbb{R}^n$ and any $\lambda \in (0, 1)$,

$$f(\lambda x + (1 - \lambda)y) \geq f(x)^\lambda f(y)^{1-\lambda}. \tag{1}$$

Logconcave functions generalize indicator functions of convex bodies (and hence the problems subsume convex optimization and volume computation) as well as Gaussians. Following the polynomial time algorithm of Dyer, Frieze and Kannan [DFK91] for estimating the volume of a convex body, a long line of work

I. Dinur et al. (Eds.): APPROX and RANDOM 2009, LNCS 5687, pp. 420–433, 2009.

[AK91, Lov90, DF91, LS92, LS93, KLS97, LV07, LV06c, LV06b] culminated in the results that both sampling and integration have polynomial complexity for any logconcave density. Integration is done by a reduction to sampling and sampling also provides an alternative to the Ellipsoid method for optimization [BV04, KV06, LV06a]. Sampling itself is achieved by a random walk whose stationary distribution has density proportional to the given function. The key question is thus the rate of convergence of the walk, which depends (among other things) on the isoperimetry of the target function.

Informally, a function has good isoperimetry if one cannot remove a set of small measure from its domain and partition it into two disjoint sets of large measure. Logconcave functions satisfy the following isoperimetric inequality:

Theorem 1. *[DF91, LS93] Let $f : \mathbb{R}^n \to \mathbb{R}_+$ be a logconcave function with a convex support K of diameter D, $\int_{\mathbb{R}^n} f < \infty$, and S_1, S_2, S_3 be any partition of K into three measurable sets. Then, for a distribution π_f with density proportional to f,*

$$\pi_f(S_3) \geq \frac{2d(S_1, S_2)}{D} \min\{\pi_f(S_1), \pi_f(S_2)\},$$

where $d(S_1, S_2)$ refers to the minimum distance between any two points in S_1 and S_2.

Although the class of logconcave functions is fairly large, it does not capture all the functions with good isoperimetry. The definition of logconcavity says that, for every line segment in the domain, the value at its midpoint is at least the geometric mean of the values at its endpoints. This is a generalization of concavity where, for every line segment in the domain, the value at its midpoint is at least the arithmetic mean of the values at its endpoints. This motivates the following question: What condition should a function satisfy along every line segment to have good isoperimetry?

In this paper, using a characterization of functions based on generalized means, we present a class of functions with good isoperimetry that is the largest under this particular characterization. We also give an efficient algorithm to sample from these functions; a well-known example among these is the Cauchy density (which is not logconcave and is heavy-tailed).

To motivate and state our results, we begin with a discussion of one-dimensional conditions.

1.1 From Concave to Quasi-concave

Definition 1. *(s-concavity of probability density) A function $f : \mathbb{R}^n \to \mathbb{R}_+$ is said to be s-concave, for $-\infty \leq s \leq 1$, if*

$$f(\lambda x + (1 - \lambda)y) \geq (\lambda f(x)^s + (1 - \lambda)f(y)^s)^{1/s},$$

for all $\lambda \in [0, 1], \forall x, y \in \mathbb{R}^n$.

The following are some special cases: A function $f : \mathbb{R}^n \to \mathbb{R}_+$ is said to be

$$
\begin{cases}
\textit{concave if,} & f\left(\lambda x + (1-\lambda)y\right) \geq \lambda f(x) + (1-\lambda)f(y) \\[2ex]
\textit{logconcave if,} & f\left(\lambda x + (1-\lambda)y\right) \geq f(x)^\lambda f(y)^{1-\lambda} \\[2ex]
\textit{harmonic-concave if,} & f\left(\lambda x + (1-\lambda)y\right) \geq \left(\dfrac{\lambda}{f(x)} + \dfrac{(1-\lambda)}{f(y)}\right)^{-1} \\[2ex]
\textit{quasi-concave if,} & f\left(\lambda x + (1-\lambda)y\right) \geq \min\{f(x), f(y)\}
\end{cases}
$$

for all $\lambda \in [0,1], \forall x, y \in \mathbb{R}^n$.

These conditions are progressively weaker, restricting the function value at a convex combination of x and y to be at least the arithmetic average, geometric average, harmonic average and minimum, respectively. Note that s_1-concave functions are also s_2-concave if $s_1 > s_2$. It is thus easy to verify that:

concave $\subsetneq$ s-concave $(s > 0)$ $\subsetneq$ logconcave $\subsetneq$ s-concave $(s < 0)$ $\subsetneq$ quasi-concave.

Relaxing beyond quasi-concave would violate unimodality, i.e., there could be two distinct local maxima, which appears problematic for all of the fundamental problems. Also, it is well-known that quasi-concave functions have poor isoperimetry.

There is a different characterization of probability measures based on a generalization of the Brunn-Minkowski inequality. The Brunn-Minkowski inequality states that the Euclidean volume (or Lebesgue measure) μ satisfies

$$\mu\left(\lambda A + (1-\lambda)B\right)^{1/n} \geq \lambda \mu(A)^{1/n} + (1-\lambda)\mu(B)^{1/n},$$

for $\lambda \in [0,1]$ and compact subsets $A, B \subseteq \mathbb{R}^n$, where $\lambda A + (1-\lambda)B = \{\lambda a + (1-\lambda)b : a \in A, b \in B\}$ is the Minkowski sum.

Definition 2. *(κ-concavity of probability measure) A probability measure μ over $\mathbb{R}^n$ is κ-concave if*

$$\mu(\lambda A + (1-\lambda)B)^\kappa \geq \lambda \mu(A)^\kappa + (1-\lambda)\mu(B)^\kappa,$$

$\forall A, B \subseteq \mathbb{R}^n, \forall \lambda \in [0,1]$.

Note that the Euclidean volume (or Lebesgue measure) is quasi-concave according to Definition 1 but $1/n$-concave according to Definition 2. Borell [Bor74, Bor75] showed an equivalence between these two definitions as follows.

Lemma 1. *An absolutely continuous probability measure μ on a convex set $K \subseteq \mathbb{R}^n$ is κ-concave, for $-\infty < \kappa \leq 1/n$, if and only if there is a density function $p : \mathbb{R}^n \to \mathbb{R}_+$, which is s-concave for $s = \dfrac{\kappa}{1 - \kappa n}$.*

Thus, if the density function is s-concave for $s \in [-1/n, 0]$, then the corresponding probability measure is κ-concave for $\kappa = \frac{s}{1+ns}$. Bobkov [Bob07] proves the following isoperimetric inequality for κ-concave probability measures for $-\infty < \kappa \le 1$.

Theorem 2. *Given a κ-concave probability measure μ, for any measurable subset $A \subseteq \mathbb{R}^n$,*

$$\mu(\delta A) \ge \frac{c(\kappa)}{m} \min\{\mu(A), 1 - \mu(A)\}^{1-\kappa}$$

where m is the μ-median of the Euclidean norm $x \mapsto \|x\|$, for some constant $c(\kappa)$ depending on κ.

Therefore, by Lemma 1, we get an isoperimetric inequality for any s-concave function $f : \mathbb{R}^n \to \mathbb{R}_+$, for $s \in [-1/n, 0]$, as

$$\pi_f(\delta A) \ge \frac{c(s)}{m} \min\{\pi_f(A), 1 - \pi_f(A)\}^{1 - \frac{s}{1+ns}},$$

for any measurable set $A \subseteq \mathbb{R}^n$.

In comparison, we prove a stronger isoperimetric inequality for the class of $-1/(n-1)$-concave functions (which subsumes $-1/n$-concave functions) and we remove the dependence on s in the inequality completely.

1.2   The Cauchy Density

The generalized Cauchy probability density $f : \mathbb{R}^n \to \mathbb{R}_+$ parameterized by a positive definite matrix $A \in \mathbb{R}^{n \times n}$ and a vector $m \in \mathbb{R}^n$, is given by

$$f(x) \propto \frac{\det(A)^{-1}}{\left(1 + \|A(x - m)\|^2\right)^{(n+1)/2}}.$$

For simplicity, we assume $m = \bar{0}$ using a translation. It is easy to sample this distribution in full space (by an affine transformation it becomes spherically symmetric and therefore a one-dimensional problem) [Joh87]. We consider the problem of sampling according to the Cauchy density restricted to a convex set. This is reminiscent of the work of Kannan and Li who considered the problem of sampling a Gaussian distribution restricted to a convex set [KL96].

1.3   Our Results

Our first result establishes good isoperimetry for $-1/(n-1)$-concave functions in $\mathbb{R}^n$.

Theorem 3. *Let $f : \mathbb{R}^n \to \mathbb{R}_+$ be a $-1/(n-1)$-concave function with a convex support $K \subseteq \mathbb{R}^n$ of diameter D, and let $\mathbb{R}^n = S_1 \cup S_2 \cup S_3$ be a measurable partition of $\mathbb{R}^n$ into three non-empty subsets. Then*

$$\pi_f(S_3) \ge \frac{d(S_1, S_2)}{D} \min\{\pi_f(S_1), \pi_f(S_2)\}.$$

It is worth noting that the isoperimetric coefficient above is only smaller by a factor of 2 when compared to that of logconcave functions (Theorem 1).

Next, we prove that beyond the class of $-1/(n-1)$-concave functions, there exist functions with exponentially small isoperimetric coefficient.

Theorem 4. *For any $\epsilon > 0$, there exists a $-1/(n-1-\epsilon)$-concave function $f :$ $\mathbb{R}^n \to \mathbb{R}_+$ with a convex support K of finite diameter and a partition $\mathbb{R}^n = S \cup T$ such that*

$$\frac{\pi_f(\partial S)}{\min\left\{\pi_f(S), \pi_f(T)\right\}} \leq Cn(1+\epsilon)^{-\epsilon n}$$

for some constant $C > 0$.

Theorems 3 and 4 can be summarized by the following figure.

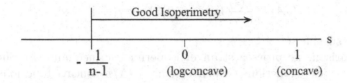

Fig. 1. Limit of isoperimetry for s-concave functions

We prove that the ball walk with a Metropolis filter can be used to sample efficiently according to $-1/(n-1)$-concave densities which satisfy a certain Lipschitz condition. In each step, the ball walk picks a new point y, uniformly at random from a small ball around the current point x, and moves to y with probability $\min\{1, f(y)/f(x)\}$. A distribution σ_0 is said to be an H-warm start ($H > 0$) for the distribution π_f if for all $S \subseteq \mathbb{R}^n$, $\sigma_0(S) \leq H\pi_f(S)$. Let σ_m denote the distribution after m steps of the ball walk with a Metropolis filter.

Definition 3. *We call a function $f : \mathbb{R}^n \to \mathbb{R}_+$ to be (α, δ)-smooth if*

$$\max\left\{\frac{f(x)}{f(y)}, \frac{f(y)}{f(x)}\right\} \leq \alpha,$$

for all x, y in the support of f with $\|x - y\| \leq \delta$.

Theorem 5. *Let $f : \mathbb{R}^n \to \mathbb{R}_+$ be proportional to an s-concave (α, δ)-smooth function, restricted to a convex body $K \subseteq \mathbb{R}^n$ of diameter D, where $s \geq -1/(n-1)$. Let K contain a ball of radius δ and σ_0 be an H-warm start. Then, after*

$$m \geq \left(\frac{CnD^2}{\delta^2} \log \frac{2H}{\epsilon}\right) \cdot \max\left\{\frac{nH^2}{\epsilon^2}, \frac{(\alpha^{-s}-1)^2}{s^2}\right\}$$

steps of the ball walk with radius $r \leq \min\left\{\frac{\epsilon\delta}{16H\sqrt{n}}, \frac{|2s\delta|}{\alpha^{-s}-1}\right\}$, *we have that*

$$d_{tv}(\sigma_m, \pi_f) \leq \epsilon,$$

for some absolute constant C, where $d_{tv}(\cdot, \cdot)$ *is the total variation distance.*

Applying the above theorem directly to sample according to the Cauchy density, we get a mixing time of $O\left(\left(\frac{n^3 H^2}{\epsilon^2}\log\frac{2H}{\epsilon}\right)\cdot\max\left\{\frac{H^2}{\epsilon^2}, n\right\}\right)$ using parameters $\delta = 1$, $\alpha = e^{\frac{n+1}{2}}$ and, $D = \frac{8\sqrt{2n}H}{\epsilon}$ (one can prove that the probability measure outside the ball of radius D around the origin is at most $\epsilon/2H$ for the chosen value of D). Using a more careful analysis (comparison of 1-step distributions), this bound can be improved to match the current best bounds for sampling logconcave functions.

Theorem 6. *Consider the Cauchy probability density f defined in Section 1.2, restricted to a convex set $K \subseteq \mathbb{R}^n$ containing a ball of radius $\|A^{-1}\|_2$ and let σ_0 be an H-warm starting distribution. Then after*

$$m \geq O\left(\frac{n^3 H^4}{\epsilon^4}\log\frac{2H}{\epsilon}\right)$$

steps with ball-walk radius $r = \epsilon/8\sqrt{n}$, we have

$$d_{tv}(\sigma_m, \pi_f) \leq \epsilon,$$

where $d_{tv}(.,.)$ is the total variation distance.

The proof of this theorem departs from its earlier counterparts in a significant way. In addition to isoperimetry, and the closeness of one-step distributions of nearby points, we have to prove that most of the measure is contained in a ball of not-too-large radius. For logconcave densities, this large-ball probability decays exponentially with the radius. For the Cauchy density it only decays linearly (Proposition 3).

All missing proofs are available in the full version of the paper[1].

2 Preliminaries

Let $r\mathbb{B}_x$ denote a ball of radius r around point x. One step of the ball walk at a point x defines a probability distribution P_x over $\mathbb{R}^n$ as follows.

$$P_x(S) = \int_{S\cap r\mathbb{B}_x}\min\left\{1, \frac{f(y)}{f(x)}\right\}dy.$$

For every measurable set $S \subseteq \mathbb{R}^n$ the ergodic flow from S is defined as

$$\Phi(S) = \int_S P_x(\mathbb{R}^n \setminus S)f(x)dx,$$

---

[1] http://arxiv.org/abs/0906.2448

and the measure of S according to π_f is defined as $\pi_f(S) = \int_S f(x)dx / \int_{\mathbb{R}^n} f(x)dx$. The s-conductance ϕ_s of the Markov chain defined by ball walk is

$$\phi_s = \inf_{s \leq \pi_f(S) \leq 1/2} \frac{\Phi(S)}{\pi_f(S) - s}.$$

To compare two distributions Q_1, Q_2 we use the total variation distance between Q_1 and Q_2, defined by $d_{tv}(Q_1, Q_2) = \sup_A |Q_1(A) - Q_2(A)|$. When we refer to the distance between two sets, we mean the minimum distance between any two points in the two sets. That is, for any two subsets $S_1, S_2 \subseteq \mathbb{R}^n$, $d(S_1, S_2) := \min\{|u - v| : u \in S_1, v \in S_2\}$. Next we quote a lemma from [LS93] which relates the s-conductance to the mixing time.

Lemma 2. *Let $0 < s \leq 1/2$ and $H_s = \sup_{\pi_f(S) \leq s} |\sigma_0(S) - \pi_f(S)|$. Then for every measurable $S \subseteq \mathbb{R}^n$ and every $m \geq 0$,*

$$|\sigma_m(S) - \pi_f(S)| \leq H_s + \frac{H_s}{s}\left(1 - \frac{\phi_s^2}{2}\right)^m.$$

Finally, the following localization lemma [LS93, KLS95] is a useful tool in the proofs of isoperimetric inequalities.

Lemma 3. *Let $g : \mathbb{R}^n \to \mathbb{R}$ and $h : \mathbb{R}^n \to \mathbb{R}$ be two lower semi-continuous integrable functions such that*

$$\int_{\mathbb{R}^n} g(x)dx > 0 \quad and \quad \int_{\mathbb{R}^n} h(x)dx > 0.$$

Then there exist two points $a, b \in \mathbb{R}^n$ and a linear function $l : [0,1] \to \mathbb{R}_+$ such that

$$\int_0^1 g((1-t)a + tb)l(t)^{n-1}dt > 0 \quad and \quad \int_0^1 h((1-t)a + tb)l(t)^{n-1}dt > 0.$$

3   Isoperimetry

Here we prove an isoperimetric inequality for functions satisfying a certain unimodality criterion. We further show that $-1/(n-1)$-concave functions satisfy this unimodality criterion and hence have good isoperimetry.

We begin with a simple lemma that will be used in the proof of the isoperimetric inequality.

Lemma 4. *Let $p : [0,1] \to \mathbb{R}_+$ be a unimodal function, and let $0 \leq \alpha < \beta \leq 1$. Then*

$$\int_\alpha^\beta p(t)dt \geq |\alpha - \beta| \min\left\{\int_0^\alpha p(t)dt, \int_\beta^1 p(t)dt\right\}.$$

Now we are ready to prove an isoperimetric inequality for functions satisfying a certain unimodality criterion.

Theorem 7. *Let $f : \mathbb{R}^n \to \mathbb{R}_+$ be a function whose support has diameter D, and f satisfies the following unimodality criterion: For any affine line $L \subseteq \mathbb{R}^n$ and any linear function $l : K \cap L \to \mathbb{R}_+$, $h(x) = f(x)l(x)^{n-1}$ is unimodal. Let $\mathbb{R}^n = S_1 \cup S_2 \cup S_3$ be a partition of $\mathbb{R}^n$ into three non-empty subsets. Then*

$$\pi_f(S_3) \geq \frac{d(S_1, S_2)}{D} \min \{\pi_f(S_1), \pi_f(S_2)\}.$$

Proof. Suppose not. Define $g : \mathbb{R}^n \to \mathbb{R}$ and $h : \mathbb{R}^n \to \mathbb{R}$ as follows.

$$g(x) = \begin{cases} \dfrac{d(S_1, S_2)}{D} f(x) & \text{if } x \in S_1 \\ 0 & \text{if } x \in S_2 \\ -f(x) & \text{if } x \in S_3 \end{cases} \quad \text{and} \quad h(x) = \begin{cases} 0 & \text{if } x \in S_1 \\ \dfrac{d(S_1, S_2)}{D} f(x) & \text{if } x \in S_2 \\ -f(x) & \text{if } x \in S_3. \end{cases}$$

Thus

$$\int_{\mathbb{R}^n} g(x) dx > 0 \quad \text{and} \quad \int_{\mathbb{R}^n} h(x) dx > 0,$$

Lemma 3 implies that there exist two points $a, b \in \mathbb{R}^n$ and a linear function $l : [0, 1] \to \mathbb{R}_+$ such that

$$\int_0^1 g((1 - t)a + tb)l(t)^{n-1} dt > 0 \quad \text{and} \quad \int_0^1 h((1 - t)a + tb)l(t)^{n-1} dt > 0. \quad (2)$$

Moreover, w.l.o.g. we can assume that the points a and b are within the support of f, and hence $\|a - b\| \leq D$. We may also assume that $a \in S_1$ and $b \in S_2$. Consider a partition of the interval $[0, 1] = Z_1 \cup Z_2 \cup Z_3$, where

$$Z_i = \{z \in [0, 1] : (1 - z)a + zb \in S_i\}.$$

For $z_1 \in Z_1$ and $z_2 \in Z_2$, we have

$$d(S_1, S_2) \leq d((1 - z_1)a + z_1b, (1 - z_2)a + z_2b) \leq |z_1 - z_2| \cdot \|a - b\| \leq |z_1 - z_2| D,$$

and therefore $d(S_1, S_2) \leq d(Z_1, Z_2)D$. Now we can rewrite Equation (2) as

$$\int_{Z_3} f((1 - t)a + tb)l(t)^{n-1} dt < \frac{d(S_1, S_2)}{D} \int_{Z_1} f((1 - t)a + tb)l(t)^{n-1} dt$$

$$\leq d(Z_1, Z_2) \int_{Z_1} f((1 - t)a + tb)l(t)^{n-1} dt$$

and similarly

$$\int_{Z_3} f((1 - t)a + tb)l(t)^{n-1} dt \leq d(Z_1, Z_2) \int_{Z_2} f((1 - t)a + tb)l(t)^{n-1} dt$$

Define $p : [0, 1] \to \mathbb{R}_+$ as $p(t) = f((1 - t)a + tb)l(t)^{n-1}$. From the unimodality assumption in our theorem, we know that p is unimodal. Rewriting the above equations, we have

$$\int_{Z_3} p(t) dt < d(Z_1, Z_2) \int_{Z_1} p(t) dt \quad \text{and} \quad \int_{Z_3} p(t) dt < d(Z_1, Z_2) \int_{Z_2} p(t) dt. \quad (3)$$

Now suppose Z_3 is a union of disjoint intervals, i.e., $Z_3 = \bigcup_i (\alpha_i, \beta_i)$, $0 \le \alpha_1 < \beta_1 < \alpha_2 < \beta_2 < \cdots \le 1$. By Lemma 4 we have

$$\int_{\alpha_i}^{\beta_i} p(t)dt \ge |\alpha_i - \beta_i| \cdot \min\left\{ \int_0^{\alpha_i} p(t)dt, \int_{\beta_i}^1 p(t)dt \right\}.$$

Therefore, adding these up we get

$$\int_{Z_3} p(t)dt = \sum_i \int_{\alpha_i}^{\beta_i} p(t)dt$$

$$\ge |\alpha_i - \beta_i| \cdot \sum_i \min\left\{ \int_0^{\alpha_i} p(t)dt, \int_{\beta_i}^1 p(t)dt \right\}$$

$$\ge d(Z_1, Z_2) \cdot \min\left\{ \int_{Z_1} p(t)dt, \int_{Z_2} p(t)dt \right\}.$$

The last inequality follows from the fact that either every interval in Z_1 or every interval in Z_2 is accounted for in the summation. Indeed, suppose some interval in Z_2 is not accounted for in the summation. Then, that interval has to be either the first or the last interval in $[0, 1]$ in which case all intervals in Z_1 are accounted for. But this is a contradiction to Inequality (3). This completes the proof of Theorem 7.

3.1 Isoperimetry of $-1/(n-1)$-Concave Functions

We show that $-1/(n-1)$-concave functions satisfy the unimodality criterion used in the proof of Theorem 7. Therefore, as a corollary, we get an isoperimetric inequality for $-1/(n-1)$-concave functions.

Proposition 1. *Let $f : \mathbb{R}^n \to \mathbb{R}_+$ be a smooth $-1/(n-1)$-concave function and $l : [0, 1] \to \mathbb{R}_+$ be a linear function. Now let $a, b \in \mathbb{R}^n$ and define $h : [0, 1] \to \mathbb{R}_+$ as $h(t) = f((1 - t)a + tb)l(t)^{n-1}$. Then h is a unimodal function.*

We get Theorem 3 as a corollary of Theorem 7 and Proposition 1.

3.2 Lower Bound for Isoperimetry

In this section, we show that $-1/(n - 1)$-concave functions are the limit of isoperimetry by showing a $-1/(n-1-\epsilon)$-concave function with poor isoperimetry for $0 < \epsilon \le 1$.

Proof (Proof of Theorem 4). The proof is based on the following construction. Consider $K \subseteq \mathbb{R}^n$ defined as follows.

$$K = \left\{ x \ : \ 0 \le x_1 < \frac{1}{1 + \delta} \text{ and } x_2^2 + x_3^2 + \ldots + x_n^2 \le (1 - x_1)^2 \right\},$$

where $\delta > 0$. K is a parallel section of a cone symmetric around the X_1-axis and is therefore convex. Now we define a function $f : \mathbb{R}^n \to \mathbb{R}_+$ whose support is K.

$$f(x) = \begin{cases} \dfrac{C}{(1 - (1 + \delta)x_1)^{n-1-\epsilon}} & \text{if } x \in K, \\ 0 & \text{if } x \notin K, \end{cases}$$

where C is the appropriate constant so as to make $\pi_f(K) = 1$. By definition, f is a $-1/(n - 1 - \epsilon)$-concave function.

Define a partition $\mathbb{R}^n = S \cup T$ as $S = \{x \in K \; : \; 0 \le x_1 \le t\}$ and $T = \mathbb{R}^n \setminus S$. It can be shown that the theorem holds for a suitable choice of t.

4 Sampling s-Concave Functions

Throughout this section, let $f : \mathbb{R}^n \to \mathbb{R}_+$ be an s-concave (α, δ)-smooth function given by an oracle such that $s \ge -1/(n - 1)$. Let K be the convex body over which we want to sample points according to f. We also assume that K contains a ball of radius δ and is contained in a ball of radius D. We state a technical lemma related to the smoothness and the concavity of the function.

Lemma 5. *Suppose $f : \mathbb{R}^n \to \mathbb{R}$ is a s-concave (α, δ)-smooth function. For any constant c such that $1 < c < \alpha$, if $\|x - z\| \le \frac{|cs\delta|}{\alpha^{-s}-1}$, then $\frac{f(x)}{f(z)} \le c$.*

The above lemma states that every s-concave (α, δ)-smooth function, is also $(c, \frac{|cs\delta|}{(\alpha^{-s}-1)})$-smooth for any constant c such that $1 < c < \alpha$. In particular, if $\alpha > 2$, this suggests that we may use the smoothness parameters to be $(\alpha' = 2, \delta' = \frac{|2s\delta|}{\alpha^{-s}-1})$ and if $\alpha \le 2$, then we may use $(\alpha' = 2, \delta' = \delta)$ as the parameters. Thus, the function can be assumed to be $(2, \min\{\delta, \frac{|2s\delta|}{\alpha^{-s}-1}\})$-smooth.

In order to sample, we need to show that K contains points of good local conductance. For this, define

$$K_r = \left\{ x \in K \; : \; \frac{\text{vol}\,(r\mathbb{B}_x \cap K)}{\text{vol}\,(r\mathbb{B}_x)} \ge \frac{3}{4} \right\}.$$

The idea is that, for appropriately chosen r, the log-lipschitz-like constraint will enforce that the points in K_r have good local conductance. Further, we have that the measure in K_r is close to the measure of f in K based on the radius r.

Lemma 6. *For any $r > 0$, the set K_r is convex and*

$$\pi_f(K_r) \ge 1 - \frac{4r\sqrt{n}}{\delta}.$$

4.1 Coupling

In order to prove conductance, we need to prove that when two points are geometrically close, then their one-step distributions overlap. We will need the following technical lemma about spherical caps to prove this.

Lemma 7. *Let H be a halfspace in $\mathbb{R}^n$ and B_x be a ball whose center is at a distance at most $tr/\sqrt{n}$ from H. Then*

$$e^{-\frac{t^2}{4}} > \frac{2\,\mathrm{vol}\,(H \cap r\mathbb{B})}{\mathrm{vol}\,(r\mathbb{B})} > 1 - t$$

Lemma 8. *For $r \leq \min\{\delta, \frac{|2s\delta|}{\alpha^{-s}-1}\}$, if $u, v \in K_r$, $\|u - v\| < r/16\sqrt{n}$, then*

$$d_{tv}(P_u, P_v) \leq 1 - \frac{7}{16}$$

Proof. We may assume that $f(v) \geq f(u)$. Then,

$$d_{tv}(P_u, P_v) \leq 1 - \frac{1}{\mathrm{vol}\,(r\mathbb{B})} \int_{r\mathbb{B}_v \cap r\mathbb{B}_u \cap K} \min\left\{1, \frac{f(y)}{f(v)}\right\} dy$$

Let us lower bound the second term in the right hand side.

$$
\begin{aligned}
\int_{r\mathbb{B}_v \cap r\mathbb{B}_u \cap K} \min\left\{1, \frac{f(y)}{f(v)}\right\} dy &\geq \int_{r\mathbb{B}_v \cap r\mathbb{B}_u \cap K} \min\left\{1, \frac{f(y)}{f(v)}\right\} dy \\
&\geq \left(\frac{1}{2}\right) \mathrm{vol}\,(r\mathbb{B}_v \cap r\mathbb{B}_u \cap K) \qquad \text{(By Lemma 5)} \\
&\geq \left(\frac{1}{2}\right) (\mathrm{vol}\,(r\mathbb{B}_v) - \mathrm{vol}\,(r\mathbb{B}_v \setminus r\mathbb{B}_u) - \mathrm{vol}\,(r\mathbb{B}_v \setminus K)) \\
&\geq \left(\frac{1}{2}\right) \left(\mathrm{vol}\,(r\mathbb{B}_v) - \frac{1}{16}\mathrm{vol}\,(r\mathbb{B}) - \frac{1}{16}\mathrm{vol}\,(r\mathbb{B})\right) \\
&\geq \left(\frac{7}{16}\right) \mathrm{vol}\,(r\mathbb{B})
\end{aligned}
$$

where the bound on $\mathrm{vol}\,(r\mathbb{B}_v \setminus r\mathbb{B}_u)$ is derived from Lemma 7 and $\mathrm{vol}\,(r\mathbb{B}_v \setminus K)$ is bounded using the fact that $v \in K_r$. Hence,

$$d_{tv}(P_u, P_v) \leq 1 - \frac{7}{16}$$

4.2   Conductance and Mixing Time

Consider the ball walk with metropolis filter using the s-concave (α, δ)-smooth density function oracle with ball steps of radius r.

Lemma 9. *Let $S \subseteq \mathbb{R}^n$ be such that $\pi_f(S) \geq \epsilon_1$ and $\pi_f(\mathbb{R}^n \setminus S) \geq \epsilon_1$. Then, for ball walk radius $r \leq \min\left\{\frac{\epsilon_1\delta}{8\sqrt{n}}, \frac{|2s\delta|}{\alpha^{-s}-1}\right\}$, we have that*

$$\Phi(S) \geq \frac{r}{2^9\sqrt{n}D} \min\{\pi_f(S) - \epsilon_1, \pi_f(\mathbb{R}^n \setminus S) - \epsilon_1\}$$

Using the above lemma, we prove Theorem 5.

Proof (Proof of Theorem 5). On setting $\epsilon_1 = \epsilon/2H$ in Lemma 9, we have that for ball-walk radius $r = \min\{\frac{\epsilon\delta}{16H\sqrt{n}}, \frac{|2s\delta|}{(\alpha^{-s}-1)}\}$,

$$\phi_{\epsilon_1} \geq \frac{r}{2^9\sqrt{n}D}.$$

By definition $H_s \leq H \cdot s$ and hence by Lemma 2,

$$|\sigma_m(S) - \pi_f(S)| \leq H \cdot s + H \cdot exp\left\{-\frac{mr^2}{2^{19}nD^2}\right\}$$

which gives us that beyond

$$m \geq \frac{2^{19}nD^2}{r^2} \log \frac{2H}{\epsilon}$$

steps, $|\sigma_m(S) - \pi_f(S)| \leq \epsilon$. Substituting for r, we get the theorem.

4.3   Sampling the Cauchy Density

In this section, we prove certain properties of the Cauchy density along with the crucial coupling lemma leading to Theorem 6. Without loss of generality, we may assume that the distribution given by the oracle is,

$$f(x) \propto \begin{cases} 1/(1 + ||x||^2)^{\frac{n+1}{2}} & \text{if } x \in K, \\ 0 & \text{otherwise.} \end{cases} \tag{4}$$

This is because, either we are explicitly given the matrix A of a general Cauchy density, or we can compute it using the function f at a small number of points and apply a linear transformation. Further, note that by the hypothesis of Theorem 6, we may assume that K contains a unit ball.

Proposition 2. *The Cauchy density function is $-1/(n-1)$-concave.*

Proposition 3 says that we can find a ball of radius $O(\sqrt{n}/\epsilon_1)$ outside which the Cauchy density has at most ϵ_1 probability mass.

Proposition 3

$$\Pr\left(||x|| \geq \frac{2\sqrt{2n}}{\epsilon_1}\right) \leq \epsilon_1.$$

Proposition 4 shows the smoothness property of the Cauchy density. This is the crucial ingredient used in the stronger coupling lemma. Define K_r as before. Then,

Proposition 4. *For $x \in K_r$, let*

$$C_x = \{y \in r\mathbb{B}_x : |x \cdot (x - y)| \leq \frac{4r||x||}{\sqrt{n}}\}$$

and $y \in C_x$. Then,

$$\frac{f(x)}{f(y)} \geq 1 - 4r\sqrt{n}$$

Finally, we have the following coupling lemma.

Lemma 10. *For $r \leq 1/\sqrt{n}$, if $u, v \in K_r$, $\|u - v\| < r/16\sqrt{n}$, then*

$$d_{tv}(P_u, P_v) < \frac{1}{2}.$$

The proof of conductance and mixing bound follow the proof of mixing bound for s-concave functions closely. Comparing the above coupling lemma with that of s-concave functions (Lemma 8), we observe that the improvement is obtained due to the constraint on the radius of the ball walk in the coupling lemma. In the case of Cauchy, a slightly relaxed radius suffices for points close to each other to have a considerable overlap in their one-step distribution.

4.4   Discussion

There are two aspects of our algorithm and analysis that merit improvement. The first is the dependence on the diameter, which could perhaps be made logarithmic by applying an appropriate affine transformation as in the case of logconcave densities. The second is eliminating the dependence on the smoothness parameter entirely, by allowing for sharp changes locally and considering a smoother version of the original function. Both these aspects seem to be tied closely to proving a tail bound on a 1-dimensional marginal of an s-concave function.

References

[AK91]   Applegate, D., Kannan, R.: Sampling and integration of near log-concave functions. In: STOC 1991: Proceedings of the twenty-third annual ACM symposium on Theory of computing, pp. 156–163. ACM, New York (1991)

[Bob07]   Bobkov, S.G.: Large deviations and isoperimetry over convex probability measures with heavy tails. Electronic Journal of Probability 12, 1072–1100 (2007)

[Bor74]   Borell, C.: Convex measures on locally convex spaces. Ark. Math. 12, 239–252 (1974)

[Bor75]   Borell, C.: Convex set functions in d-space. Periodica Mathematica Hungarica 6, 111–136 (1975)

[BV04]   Bertsimas, D., Vempala, S.: Solving convex programs by random walks. J. ACM 51(4), 540–556 (2004)

[DF91]   Dyer, M.E., Frieze, A.M.: Computing the volume of a convex body: a case where randomness provably helps. In: Proc. of AMS Symposium on Probabilistic Combinatorics and Its Applications, pp. 123–170 (1991)

[DFK91]   Dyer, M.E., Frieze, A.M., Kannan, R.: A random polynomial-time algorithm for approximating the volume of convex bodies. J. ACM 38(1), 1–17 (1991)

[Joh87]   Johnson, M.E.: Multivariate statistical simulation. Wiley, Chichester (1987)

[KL96]   Kannan, R., Li, G.: Sampling according to the multivariate normal density. In: FOCS 1996: Proceedings of the 37th Annual Symposium on Foundations of Computer Science, Washington, DC, USA, p. 204. IEEE Computer Society Press, Los Alamitos (1996)

[KLS95]   Kannan, R., Lovász, L., Simonovits, M.: Isoperimetric problems for convex bodies and a localization lemma. J. Discr. Comput. Geom. 13, 541–559 (1995)

[KLS97]   Kannan, R., Lovász, L., Simonovits, M.: Random walks and an $O^*(n^5)$ volume algorithm for convex bodies. Random Structures and Algorithms 11, 1–50 (1997)

[KV06]   Kalai, A., Vempala, S.: Simulated annealing for convex optimization. Math. Oper. Res. 31(2), 253–266 (2006)

[Lov90]   Lovász, L.: How to compute the volume? Jber. d. Dt. Math.-Verein, Jubiläumstagung 1990, 138–151 (1990)

[LS92]   Lovász, L., Simonovits, M.: On the randomized complexity of volume and diameter. In: Proc. 33rd IEEE Annual Symp. on Found. of Comp. Sci., pp. 482–491 (1992)

[LS93]   Lovász, L., Simonovits, M.: Random walks in a convex body and an improved volume algorithm. Random Structures and Alg 4, 359–412 (1993)

[LV06a]   Lovász, L., Vempala, S.: Fast algorithms for logconcave functions: Sampling, rounding, integration and optimization. In: FOCS 2006: Proceedings of the 47th Annual IEEE Symposium on Foundations of Computer Science, Washington, DC, USA, pp. 57–68. IEEE Computer Society Press, Los Alamitos (2006)

[LV06b]   Lovász, L., Vempala, S.: Hit-and-run from a corner. SIAM J. Computing 35, 985–1005 (2006)

[LV06c]   Lovász, L., Vempala, S.: Simulated annealing in convex bodies and an $O^*(n^4)$ volume algorithm. J. Comput. Syst. Sci. 72(2), 392–417 (2006)

[LV07]   Lovász, L., Vempala, S.: The geometry of logconcave functions and sampling algorithms. Random Struct. Algorithms 30(3), 307–358 (2007)

Average-Case Analyses of Vickrey Costs

Prasad Chebolu[1], Alan Frieze[2,*], Páll Melsted[3], and Gregory B. Sorkin[4]

[1] University of Liverpool, Liverpool L69 3BX U.K.
P.Chebolu@liverpool.ac.uk
[2] Carnegie Mellon University, Pittsburgh PA 15213
alan@random.math.cmu.edu
[3] Carnegie Mellon University, Pittsburgh PA 15213
pmelsted@cmu.edu
[4] IBM Research, Yorktown Heights NY 10598
sorkin@watson.ibm.com

Abstract. We explore the average-case "Vickrey" cost of structures in three random settings: the Vickrey cost of a shortest path in a complete graph or digraph with random edge weights; the Vickrey cost of a minimum spanning tree (MST) in a complete graph with random edge weights; and the Vickrey cost of a perfect matching in a complete bipartite graph with random edge weights. In each case, in the large-size limit, the Vickrey cost is precisely 2 times the (non-Vickrey) minimum cost, but this is the result of case-specific calculations, with no general reason found for it to be true.

Separately, we consider the problem of sparsifying a complete graph with random edge weights so that all-pairs shortest paths are preserved approximately. The problem of sparsifying a given graph so that for every pair of vertices, the length of the shortest path in the sparsified graph is within some multiplicative factor and/or additive constant of the original distance has received substantial study in theoretical computer science. For the complete digraph $\vec{K}_n$ with random edge weights, we show that **whp** $\Theta(n \ln n)$ edges are necessary and sufficient for a spanning subgraph to give good all-pairs shortest paths approximations.

Keywords: Average-case analysis, VCG auction, random graph, shortest path, minimum spanning tree, MST, Random Assignment Problem.

1 Introduction

"Algorithmic mechanism design", recognized by a Nobel Prize for Vickrey [Vic61], is even more important with today's "ad auctions" and other electronic commerce. The canonical "Vickrey-Clarke-Groves" (VCG) auction mechanism [Vic61, Cla71, Gro73] has benefits of "truthfulness" and "social welfare maximization", but may result in arbitrarily large overpayments. We are interested in whether VCG overpayments are reasonably small in an average-case setting. In this introduction we first recapitulate the VCG auction mechanism and introduce a small amount of notation, then state our average-case results.

---

* Research supported in part by NSF Grant DMS6721878.

I. Dinur et al. (Eds.): APPROX and RANDOM 2009, LNCS 5687, pp. 434–447, 2009.

1.1   The VCG Auction Mechanism

Suppose that in a graph, each edge is provided by an independent, selfish agent who incurs a cost for supplying it (or for allowing us to drive over it, transmit data over it, or whatever). This "private" cost, the price point at which the agent is neutral between selling the edge or not, is known only to herself. We wish to buy some structure, for example a path between two particular points, or a spanning tree, as cheaply as possible. An obvious "mechanism" to do this is to ask each agent the cost of her edge, find the cheapest structure, and pay each agent accordingly. The problem with this and many other mechanisms is that agents have an incentive to lie: by inflating her claimed cost, an agent will get more money (up to the point where she prices herself out of competition).

The Vickrey-Clarke-Groves (VCG) auction is a cleverly designed "truthful" mechanism: assuming that the agents act without collusion, in a VCG auction it is in each agent's best interest to name her true cost (if her edge is used, she will get paid at least this, but typically more). Under the same assumption, a VCG auction also maximizes "social welfare": the structure selected is the one that is genuinely cheapest (and so the least possible resource is consumed in road maintenance, data-server support, or whatever).

In a VCG auction, an "auctioneer" first finds a cheapest structure S^*, according to the edge costs $c(e)$ declared by the agents. (This might be a cheapest path, for example; VCG was first explicitly applied to the shortest-path problem in [NR99, NR01].) For each edge $e \in S^*$ in this structure, the auctioneer pays the corresponding agent not the stated cost $c(e)$ of the edge, but a measure of the benefit it provided, namely the difference between what a cheapest structure would have cost if the edge were not present or had infinite cost, call it $c(S_e^\infty)$, and what the cheapest structure would have cost if the edge were free, call it $c(S_e^0)$. It is clear that neither of these terms depends on $c(e)$. An agent whose edge is not used, $e \notin S^*$, is paid nothing.

We can now confirm three important properties of the auction. Define an "incentive cost" for edge e as

$$c^+(e) = c(S_e^\infty) - c(S^*). \tag{1}$$

Then for any edge, used or not, $c^+(e) \geq 0$, and assuming for convenience of discussion that there is a unique cheapest structure S^*, e is used iff $c(S^*) < c(S_e^\infty)$, i.e., iff $c^+(e) > 0$. If edge e is used, the payment for it is

$$c(S_e^\infty) - c(S_e^0) = [c(S_e^\infty) - c(S^*)] + [c(S^*) - c(S_e^0)] = c^+(e) + c(e),$$

so an edge is used iff the payment would exceed the agent's stated cost.

Second, this shows that the auction is truthful. Since by manipulating her price $c(e)$ an agent cannot influence what she would be paid, but only whether or not she will, her best strategy is to get paid iff the payment exceeds her true cost, which she can achieve by setting $c(e)$ to be the true cost. Setting $c(e)$ lower may result in her being paid less than cost; setting it higher may cause her to lose out on a profitable sale.

Finally, assuming that every agent does state her true cost, the structure selected is one that is genuinely cheapest, and social welfare is maximized.

We will later take advantage of an observation based on the fact that the incentive cost (1) is 0 for edges not used, namely that

$$\text{VCG} = \sum_{e \in S^*} [c(e) + (c(S_e^\infty) - c(S^*))]$$

$$= c(S^*) + \sum_{e \in E} (c(S_e^\infty) - c(S^*)) \tag{2}$$

$$= \sum_{e \in E} c(S_e^\infty) - (|E| - 1)c(S^*). \tag{3}$$

1.2   Average-Case Analysis

Naturally, a VCG auction pays more than the cost of the cheapest structure, and unfortunately the overpayment can be arbitrarily large. In [AT02, AT07] it is shown that any truthful mechanism has bad worst-case s–t path overpayment.

A worst-case analysis of VCG costs may be overly pessimistic. One alternative is to take real-world measurements, and a small step in this direction is included in [FPSS02]. Another alternative, and the one we adopt here, is to compare the VCG cost with the minimum cost in an average-case setting. This was done for shortest paths in certain graphs in [MPS03, CR04, KN05, FGS06]. Bad VCG costs can occur even in the average case. For example, a shortest path between two random vertices on an n-cycle has expected length $n/4$ and the alternative path expected length $3n/4$, making each incentive cost $\Omega(n)$, for a Vickrey cost of $\Omega(n^2)$ where the minimum cost is $O(n)$. The precise calculation is an easy exercise, and the result holds equally if the edge weights are random. However, we find that the expected VCG cost is only twice the minimum in three settings, in each of which the expected minimum cost is a classical result in the analysis of random structures.

We use either i.i.d. uniform $[0, 1]$ or exponential(1) edge weights, whichever is more convenient. It is a standard observation that since only low-cost edges are used, all that matters *asympotically* for i.i.d. weights is the distribution's density near 0; all distributions with constant density near 0 are equivalent. (The exact doubling we observe for MST, and the exact formulas for Random Assignment, depend on using the uniform and exponential distributions respectively.)

Shortest Paths. We first consider shortest paths in the complete graph K_n, or complete digraph $\vec{K}_n$, with i.i.d. exponential(1) edge weights, where exponential(1) denotes the exponential distribution with mean 1. (We use the terms edge weight, cost, or length interchangeably, and a shortest path is a cheapest path.) Janson [Jan99] has shown that **whp** the distance between two vertices, say 1 and n, in this model is $(1 + o(1)) \log n/n$. We prove that the asymptotic expected Vickrey cost is twice as large.

Theorem 1. *Suppose that the edges of the complete graph K_n (respectively, digraph $\vec{K}_n$) have i.i.d. exponential mean-1 edge weights. Let $\mathbf{E}(\mathrm{SP})$ be the expected cost of a shortest path from 1 to n. Then*

$$\mathbf{E}(\mathrm{VCG}) \sim 2\mathbf{E}(\mathrm{SP}).$$

In a small digression, we consider the problem of sparsifying the random edge-weighted digraph so that **whp** shortest path distances are (approximately) preserved. Janson [Jan99] also showed that the weighted diameter in this model is $(3 + o(1)) \log n/n$. It follows that **whp** the subgraph consisting of the $4n \log n$ cheapest edges contains the shortest path between each pair of vertices. If we only keep Dn edges (so D is the average in+out degree), with $D = D(n) = O(\log n)$, how good an approximation can we find in the all-pairs shortest path problem?

Theorem 2. *In a complete digraph $\vec{K}_n$ with i.i.d. exponential(1) edge weights, **whp**, for every edge subset of size at most Dn defining a sub-digraph H, some pair of vertices s, t has $d_H(s,t)/d_{\vec{K}_n}(s,t) \geq \frac{\log n}{4D}$, where $d_H(s,t)$ and $d_{\vec{K}_n}(s,t)$ denote shortest distance in H and $\vec{K}_n$ respectively.*

Minimum Spanning Tree. We next consider a minimum spanning tree of K_n with uniform $[0, 1]$ edge weights. It was shown by Frieze [Fri85] that the expected cost $\mathbf{E}(\mathrm{MST})$ of a minimum spanning tree on K_n satisfies $\lim_{n \to \infty} \mathbf{E}(\mathrm{MST}) = \zeta(3) = \sum_{i=1}^{\infty} i^{-3}$. Even though there is no nice expression for the exact expectation for finite n, we prove that the expected VCG cost is *exactly* (not just asymptotically) twice as large.

Theorem 3. *Suppose that the edges of the complete graph K_n have i.i.d. uniform $[0, 1]$ edge weights. Let $\mathbf{E}(\mathrm{MST})$ be the expected cost of a minimum spanning tree. Then*

$$\mathbf{E}(\mathrm{VCG}) = 2\mathbf{E}(\mathrm{MST}).$$

Assignment. Finally, we consider the VCG cost of a perfect matching in a complete bipartite graph with random edge weights, known as the "random assignment problem". When the edge weights are i.i.d. exponential(1) random variables, Mézard and Parisi [MP85, MP86, MP87] gave a sophisticated mathematical physics argument, using the "replica method" (related to the "cavity method"), that the minimum cost AP satisfies $\lim_{n \to \infty} \mathbf{E}(\mathrm{AP}) = \zeta(2) = \pi^2/6$. Aldous [Ald92, Ald01] made this mathematically rigorous through reasoning about a "Poisson weighted infinite tree". For finite values of n, Parisi [Par98] conjectured the expected cost to be $\sum_{i=1}^{n} i^{-2}$, Coppersmith and Sorkin [CS99] extended the conjecture to cheapest cardinality-k assignments in $K_{m,n}$, and these results were proved simultaneously, by different methods, by Linusson and Wästlund [LW04] and Nair, Prabhakar and Sharma [NPS05]. A beautiful short proof was later found by Wästlund [Wäs].

As in the previous cases, we find that the expected VCG cost is twice the minimum cost asymptotically (but not for finite n as it is for MST).

Theorem 4. *Suppose that the edges of the complete bipartite graph $K_{n,n}$ have i.i.d. exponential mean-1 edge weights. Let $\mathbf{E}(\mathrm{AP})$ be the expected cost of a minimum weight perfect matching. Then*

$$\mathbf{E}(\mathrm{VCG}) = \mathbf{E}(\mathrm{AP}) + n \left(\frac{1}{n-1} + \sum_{l=1}^{n-1} \frac{1}{l}\frac{n-l}{n} - \sum_{l=2}^{n-1} \frac{1}{l(l-1)} \sum_{i=0}^{l-1} \frac{n-i}{n} \prod_{j=i+1}^{l} \frac{(n-j)j}{(n-j+1)j-1} \right)$$

$$\sim 2\mathbf{E}(\mathrm{AP}).$$

In the remainder of the paper we prove Theorems 1—4. We conjecture that similar results hold for spanning arborescence in $\vec{K}_n$, perfect matching in K_n, symmetric and asymmetric TSP, etc., the key question being *why*.

2   Shortest Paths

The cost model for this section will be that each edge e of the complete graph K_n or digraph $\vec{K}_n$ is given an independent cost X_e where X_e is exponential with mean 1, i.e., $X_e \sim \mathrm{exponential}(1)$ are i.i.d. random variables. We will compute the expected Vickrey cost of a shortest path between two random vertices A and B, or equivalently between vertex 1 and a random vertex B. We start by computing the expected cost $\mathbf{E}(\mathrm{SP})$ of this path. We follow the analysis of Dijkstra's algorithm due to Janson [Jan99]. Janson actually considered the symmetric (undirected) case, but there is no essential difference in the analysis of the two. We begin with the directed (asymmetric) case.

Asymmetric (Directed) Model. In the complete digraph $\vec{K}_n$, if D_k is the distance to the kth closest vertex from vertex 1 then $D_1 = 0$ and for $k > 1$ we have

$$\mathbf{E}(D_{k+1}) = \mathbf{E}(D_k) + \frac{1}{k(n-k)}. \tag{4}$$

Explanation. Growing the shortest path tree T iteratively by Dijkstra's algorithm, let T_k be the set of vertices after k rounds (including the first, trivial round), so $|T_k| = k$. Writing $d(v)$ for the distance to vertex v, and D_k for the distance to the last vertex added, we claim that for all $v \in T_k$ and $w \notin T_k$, the edge length $C(v, w)$ is exponential mean-1 conditioned on being at least $D_k - d(v)$; by the memoryless property of the exponential, this means that $d_v + c(v, w) = D_k + X_{v,w}$. Furthermore, the set of these exponential variables is independent. Both assertions are easily checked inductively. It follows from (4) that

$$\mathbf{E}(D_{k+1}) = \sum_{i=1}^{k} \frac{1}{i(n-i)}. \tag{5}$$

Vertex B is added to T at a random stage of this process, so the shortest path to B has expectation

$$\mathbf{E}(\mathrm{SP}) = \frac{1}{n-1} \sum_{k=1}^{n-1} \sum_{i=1}^{k} \frac{1}{i(n-i)} = \frac{H_{n-1}}{n-1} \sim \frac{\ln n}{n}, \tag{6}$$

where H_n is the nth harmonic number.

From (2), the expected VCG cost satisfies

$$\mathbf{E}(\text{VCG} - c(S^*)) = \sum_{e \in E} (c(S_e^\infty) - c(S^*)) = n(n-1)\mathbf{E}(c(S_e^\infty) - c(S^*)), \qquad (7)$$

the final expectation taken over random edges e as well as random values of the other parameters (in this case, random edge weights and a random terminal B). We now compute this quantity.

Let $D_k^{(r,s)}$ denote the distance to the kth closest vertex v_k from vertex 1 when $e = (r,s)$ is excluded. In place of (5), with expectation taken over random edge weights but a fixed missing edge (r,s), we have

$$\mathbf{E}(D_{k+1}^{(r,s)}) = \sum_{i=1}^{k} \frac{1}{i(n-i)-\theta(r,s,i)},$$

where $\theta(r,s,i)$ is the indicator for $r \in T_i$, $s \notin T_i$ (in this event we are finding the cheapest of $i(n-i)-1$ edges rather than $i(n-i)$). Subtracting (5),

$$\mathbf{E}(D_{k+1}^{(r,s)} - D_{k+1}) = \sum_{i=1}^{k} \left(\frac{1}{i(n-i)-\theta(r,s,i)} - \frac{1}{i(n-i)} \right)$$

$$= \sum_{i=1}^{k} \frac{\theta(r,s,i)}{i(n-i)(i(n-i)-1)}. \qquad (8)$$

To compute (7), we must now compute the expectation of (8) over a random terminal and a random edge (r,s). Since we may compute the Dijkstra shortest-path tree first and then randomly choose which point B is the terminal, the step k at which B is discovered is uniformly random from 1 to $n-1$. However, this is not true of the appearance times of r and s: for example, if r is the first vertex then s cannot be the second.

For a random edge (r,s), compare the evolution of the Dijkstra process with (r,s) missing (the VCG process) with that on the complete graph (the original process). Let $i(r)$ and $i(s)$ denote the (random) steps at which vertices r and s are discovered (e.g., $i(r) = 1$ if r is vertex 1). For a given set of edge weights, the two processes evolve identically until vertex r is discovered. At that point, s (like any other vertex but r) remains undiscovered with probability $(n-r)/(n-1)$. Then, conditional on s still not being discovered at the start of step $j \geq i(r)$, the probability of it being discovered in that step is $(j-1)/(j(n-j)-1)$: there are $j-1$ edges from T_j into s, but j edges from T_j into each of the other undiscovered vertices. Thus, s remains undiscovered with probability $j(n-j-1)/(j(n-j)-1)$. It follows that, conditioned on $i(r) = \rho$, for a given $i \geq \rho$ the probability that s has not been discovered by the start of round i is

$$\mathbf{E}(\theta(r,s,i) \mid i(r) = \rho) = \frac{n-\rho}{n-1} \prod_{j=\rho}^{i-1} \frac{j(n-j-1)}{j(n-j)-1} = \frac{n-i}{n-1} \prod_{j=\rho}^{i-1} \left(1 + \frac{1}{j(n-j)-1} \right). \qquad (9)$$

The final product satisfies

$$1 \le \prod_{j=\rho}^{i-1} \left(1 + \frac{1}{j(n-j)-1}\right) \le \prod_{j=1}^{n-1} \exp\left(\frac{1}{j(n-j)-1}\right) \le \exp\left(\sum_{j=1}^{n-1} \frac{2}{n}(\frac{1}{j} + \frac{1}{n-j})\right)$$

$$= \exp\left(\frac{4H_{n-1}}{n}\right) = \exp(o(1)) = 1 + o(1).$$

Thus,

$$\mathbf{E}(\theta(r,s,i) \mid i(r) = \rho) \sim \frac{n-i}{n-1},$$

asymptotically the same that would be obtained were s uniformly distributed. As previously noted, r's discovery time $i(r)$ is uniformly random over $\{1, \ldots, n\}$, so we compute (7) via

$$\mathbf{E}(c(S_e^\infty) - c(S^*)) = \frac{1}{n-1} \sum_{k=1}^{n-1} \frac{1}{n} \sum_{\rho=1}^{n} \sum_{i=\rho}^{k} \frac{\mathbf{E}(\theta(r,s,i)|i(r)=\rho)}{i(n-i)(i(n-i)-1)} \tag{10}$$

$$\sim \frac{1}{n^3} \sum_{i=1}^{n-1} \frac{1}{i} = \frac{H_{n-1}}{n^3} \sim \frac{\ln n}{n^3}.$$

Returning to (7),

$$\mathbf{E}(\text{VCG} - c(S^*)) = n(n-1)\mathbf{E}(c(S_e^\infty) - c(S^*)) \sim \frac{\ln n}{n}. \tag{11}$$

Since (6) established that $\mathbf{E}(c(S^*)) \sim \frac{\ln n}{n}$, this proves the asymmetric case of Theorem 1.

Symmetric (Undirected) Model. Our analysis extends easily to the case where we have a complete graph K_n, as opposed to a complete digraph, still using i.i.d. exponential(1) edge weights.

3 All-Pairs Shortest Paths

We give an outline proof of Theorem 2, which is somewhat peripheral to the main theme of the paper. Let an edge $e = (s, t)$ be *bad* if (i) it is "short", with cost $c(e) \le 2D/n$, and (ii) $d_{K_n \backslash e}(s, t) \le \frac{\log n}{2n}$. For a random edge e, $\mathbf{P}(e \text{ is bad}) = \mathbf{P}(\text{i})\mathbf{P}(\text{ii})$ because these events are independent, depending respectively on the length of e and the lengths of the other edges. The theorem's conclusion is null unless $D = O(\log n)$, thus $D = o(n)$, in which case $\mathbf{P}(\text{i}) = 1 - \exp(-2D/n) \sim \frac{2D}{n}$. For (ii) we use that $d_H(s, t) \ge d_{\vec{K}_n}(s, t)$; we have seen in (6) that $\mathbf{E}(d_{\vec{K}_n}(s, t)) \sim \frac{\log n}{n}$ and Janson [Jan99] has shown that $\mathbf{Var}(d_{\vec{K}_n}(s, t)) = O(n^{-2})$. (These results are shown for undirected graphs but follow similarly for directed graphs.) By Cheychev's inequality, then, $\mathbf{P}(\text{ii}) =$

$O(1/\log^2 n)$. Multiplying $\mathbf{P}$(i) and $\mathbf{P}$(ii), the number of bad edges Z satisfies $\mathbf{E}(Z) = O(Dn/\log^2 n)$, and by Markov's inequality, $Z \leq Dn/\log n$ **whp**. From $\mathbf{P}$(i) $\sim 2D/n$ it is immediate that the (binomially distributed) number of such short edges is at least $\frac{3}{2}Dn$ **whp**. Assuming that all the above high-probability events hold, for any way of selecting Dn edges, there will be at least $Dn/2$ short edges not selected, and since $Z \leq Dn/\log n$, at least $Dn/3$ of these are not bad. These edges are short yet are not bad, thus must violate (ii), so that $d_{\vec{K}_n}(s,t) \leq c(e) \leq \frac{2D}{n}$ while $d_H(s,t) \geq d_{\vec{K}_n \backslash e}(s,t) > \frac{\log n}{2n}$. This completes the proof of Theorem 2, since for these edges $d_H(s,t)/d_{\vec{K}_n}(s,t) \geq \frac{\frac{1}{2}\log n/n}{2D/n} = \frac{\log n}{4D}$.

In fact it shows more: for a large number of edges ($\Theta(n)$ of them), not only is the approximation ratio poor, but the additive gap is large as well: of order $\Omega(\log n/n)$, the same order as a typical distance.

4   Minimum Spanning Tree

The cost model for this section will be that each edge of the complete graph K_n is given an independent cost X_e where X_e is uniform $[0,1]$ for all $e \in E(K_n)$. We use the integral formula of Avram and Bertsimas [AB92]: For a connected graph $G = (V, E)$ with uniform $[0,1]$ edge weights X_e, $e \in E$, let $\mathrm{MST}_G = \mathrm{MST}(G, X)$ denote the length of the minimum spanning tree with these edge weights. Then

$$\mathbf{E}(\mathrm{MST}_G) = \int_{p=0}^{1} \mathbf{E}(\kappa(G_p))dp - 1 \qquad (12)$$

where G_p is the random subgraph of G obtained by including each edge independently with probability p and $\kappa(G_p)$ is the number of components.

For $1 \leq s \leq n$ let $C_{s,m}$ denote the number of connected graphs of order s and size m. Using

$$\kappa(G) = \sum_{s=1}^{n} \binom{n}{s} \mathbf{1}(\text{the } s \text{ vertices induce a connected subgraph of } G)$$

$$= \sum_{s=1}^{n} \binom{n}{s} \sum_{m=s-1}^{\binom{s}{2}} \sum_{i=1}^{C_{s,m}} \mathbf{1}(\text{the } i\text{th } s, m \text{ graph is a component of } G), \quad (13)$$

$$\mathbf{E}(\mathrm{MST}_{K_n}) = \sum_{s=1}^{n} \sum_{m=s-1}^{\binom{s}{2}} \binom{n}{s} C_{s,m} \int_{p=0}^{1} p^m (1-p)^{\binom{s}{2}-m+s(n-s)} dp - 1$$

$$= \sum_{s=1}^{n} \sum_{m=s-1}^{\binom{s}{2}} C'_{s,m} - 1 \qquad (14)$$

where, by $\int_{p=0}^{1} p^m (1-p)^k dp = \frac{m!k!}{(m+k+1)!}$,

$$C'_{s,m} = \binom{n}{s} C_{s,m} \frac{m! \left(\binom{s}{2} - m + s(n-s)\right)!}{\left(\binom{s}{2} + s(n-s) + 1\right)!}. \qquad (15)$$

Going back to (12) and reasoning similarly as for (13),

$$\sum_{e\in K_n} \mathbf{E}(\mathrm{MST}_{K_n\setminus e}) = \sum_{e\in K_n}\left[\sum_{s=1}^{n}\sum_{m=s-1}^{\binom{s}{2}-\mathbf{1}(s=n)}\int_0^1 \mathbf{E}\big(\#(s\text{-vertex }m\text{-edge}\right.$$
$$\left.\text{components of }(K_n\setminus e)_p)\big)\,dp - 1\right]$$

$$= \sum_{s=1}^{n}\sum_{m=s-1}^{\binom{s}{2}-\mathbf{1}(s=n)} (A^0_{s,m} + A^2_{s,m} + A^1_{s,m}) - \binom{n}{2} \qquad (16)$$

where $A^0_{s,m}$ is the integral over p of the sum over e of the expected number of s-vertex m-edge components of a random probability-p subgraph of $K_n\setminus e$ containing *neither* endpoint of e, $A^2_{s,m}$ is the like sum for components containing *both* endpoints, and $A^1_{s,m}$ that for exactly *one* endpoint. For an edge e of K_s, let $\widehat{C}_{s,m}$ denote the number of m-edge spanning connected subgraphs of K_s containing e. We have

$$\binom{s}{2}\widehat{C}_{s,m} = mC_{s,m} \qquad (17)$$

since both sides of this equation count the number of pairs (f,H) where H is an m-edge spanning subgraph of $[s]$ and f is an edge of H. (Throughout, we take $\binom{n}{k} = 0$ if $n < k$. Above, if $s = 1$ the left side is 0, and so is the right, since $s = 1$ implies $m = 0$.) To calculate $A^0_{s,m}$ we select s vertices, whereupon edge e must have both endpoints outside these s, while on these s vertices any subgraph $C_{s,m}$ is acceptable, and we integrate the probability that the chosen subgraph is induced and isolated in $G = K_n\setminus e$:

$$A^0_{s,m} = \binom{n}{s}\binom{n-s}{2}C_{s,m}\int_{p=0}^1 p^m(1-p)^{\binom{s}{2}-m+s(n-s)}\,dp = \binom{n-s}{2}C'_{s,m},$$

where the last line uses (15). The formula correctly evaluates to 0 for $s > n-2$, and thus can safely be applied for all pairs s, m in the sum (16).

In calculating $A^2_{s,m}$, both endpoints of e are within the s vertices selected, but the graph under consideration is $G = K_n\setminus e$ and a subgraph of G cannot use e, so by (15) the number of valid subgraphs is $C_{s,m} - \widehat{C}_{s,m}$, and the probability that a subgraph is induced is adjusted to reflect that the size of G is $\binom{s}{2} - 1$:

$$A^2_{s,m} = \binom{n}{s}\binom{s}{2}(C_{s,m} - \widehat{C}_{s,m})\int_{p=0}^1 p^m(1-p)^{\binom{s}{2}-1-m+s(n-s)}\,dp$$
$$= \left(\binom{s}{2} - m\right)\frac{\binom{s}{2} + s(n-s) + 1}{\binom{s}{2} - m + s(n-s)}C'_{s,m}.$$

The formula properly evaluates to 0 for $s = 1$ (where $m = 0$) and for $m = \binom{s}{2}$. It cannot be applied when $s = n$, $m = \binom{n}{2}$, where it is 0/0, but anyway this pair is excluded from the sum (16).

Calculating $A^1_{s,m}$ is similar but now the missing edge is among the $s(n-s)$ cross edges, any of the $C_{s,m}$ subgraphs on the s vertices is acceptable, and the probability that a subgraph is isolated is adjusted to reflect that the number of cross edges is $s(n-s)-1$:

$$A^1_{s,m} = \binom{n}{s} s(n-s) C_{s,m} \int_{p=0}^{1} p^m (1-p)^{\binom{s}{2}-m+s(n-s)-1} dp$$

$$= s(n-s) \frac{\binom{s}{2} + s(n-s) + 1}{\binom{s}{2} - m + s(n-s)} C'_{s,m}.$$

This formula properly evaluates to 0 for $s = n$, except that like the formula for A^2 it is invalid for the pair $s = n$, $m = \binom{n}{2}$ excluded from (16).

Now observe that

$$A^2_{s,m} + A^1_{s,m} = \left(\binom{s}{2} - m + s(n-s) \right) \frac{\binom{s}{2} + s(n-s) + 1}{\binom{s}{2} - m + s(n-s)} C'_{s,m}$$

$$= \left(\binom{s}{2} + s(n-s) + 1 \right) C'_{s,m}.$$

The terms being canceled are 0 only in the case $m = \binom{s}{2}$ and $s = n$ excluded from (16). Thus

$$A^0_{s,m} + A^1_{s,m} + A^2_{s,m} = \left(\binom{n-s}{2} + \binom{s}{2} + s(n-s) + 1 \right) C'_{s,m} = \left(\binom{n}{2} + 1 \right) C'_{s,m}.$$

Substituting this into (16), and writing $N = \binom{n}{2}$ and $E = \mathbf{E}(\mathrm{MST}_{K_n})$, we have

$$\sum_{e \in K_n} \mathbf{E}(\mathrm{MST}_{K_n \setminus e}) = \sum_{s=1}^{n} \sum_{m=s-1}^{\binom{s}{2}-1(s=n)} (N+1) C'_{s,m} - N.$$

For $s = n$ we extend the sum to include $\binom{n}{2} = N$, subtracting out $(N+1)C'_{n,N}$ to correct:

$$= (N+1) \left(\sum_{s=1}^{n} \sum_{m=s-1}^{\binom{s}{2}} C'_{s,m} - C'_{n,N} \right) - N.$$

From (14) the double sum is $E+1$, while $C_{n,N} = 1$ and thus $C'_{n,N} = 1/(N+1)$, so this is

$$= (N+1) \left(E + 1 - \frac{1}{N+1} \right) - N = (N+1)E.$$

Going back to (3) we see that

$$\mathbf{E}(\mathrm{VCG}) = (N+1)E - (N-1)E = 2E$$

and this completes the proof of Theorem 3.

5   Assignment Problem

Let the weight of edge (i,j) be denoted $x_{i,j}$. Let X be the $n \times n$ matrix with entries $x_{i,j}$. Let Y_j, $j = 1, \ldots, n$, be the $(n-1) \times (n-1)$ matrices obtained from X by deleting row n and column j. Let T_k, $k = 0, 1, \ldots, n-1$, denote the minimum assignment costs of Y_j, $j = 1, \ldots, n$, **sorted into increasing order**. Nair, Prabhakar and Sharma [NPS05] proved that, for a matrix of i.i.d. exponential(1) variables, the increments $T_k - T_{k-1}$ are independent, and

$$T_k - T_{k-1} \sim \exp(k(n-k)). \tag{18}$$

A minimum assignment of a (random) $n \times n$ matrix is given by a value in the "missing" n'th row and the minimum assignment in the complementary submatrix. If $x_{n,\pi(n)}$ belongs to the minimum assignment, its Vickrey bonus is the cost difference between this assignment and the smallest assignment using a different element in row n. We are thus interested in gap between the smallest and second-smallest values of

$$x_{n,\sigma(j)} + T_j \tag{19}$$

where $x_{n,j} \sim$ exponential(1) are independent random variables and $\sigma(j)$ is a random permutation. This is the incentive cost of the nth row; the total Vickrey incentive cost is the sum of similar values for all rows, and thus the total expected Vickrey incentive cost is n times the expectation of the difference of the minimum and second-minimum values of (19).

We explicitly compute the expected Vickrey incentive cost. Going back to (18) and (19) we now fix the values of $T_i = t_i$ for $i = 0, \ldots, n-1$ and define $t_n = \infty$. Let $Y_j = x_{n,\sigma(j)} + t_j$ and $Y_{(1)} < Y_{(2)}$ be the two smallest values of the Y_j. We want to evaluate the following integral

$$\int_0^\infty x\mathbf{P}(Y_{(2)} = x)dx. \tag{20}$$

Assume $Y_{(1)} = Y_i$ and $Y_{(2)} = Y_j = x$. Then we must have $Y_i < x$, $Y_j = x$ and $Y_k > x$ for $k \in \{0, \ldots, n-1\} \setminus \{i,j\}$. Since $x > Y_i \geq t_i$ and $x = Y_j \geq t_j$ we must have $x \geq t_{\max\{i,j\}}$. We break up the integral (20) into integrals over $[t_l, t_{l+1}]$ where $\max\{i,j\} \leq l \leq n-1$. The integral (20) is then obtained by summing over all possible pairs i,j where $i \neq j$ giving

$$\sum_{i \neq j} \sum_{l=\max\{i,j\}}^n \int_{t_l}^{t_{l+1}} x\mathbf{P}(Y_i < x, Y_j = x, Y_k \geq x, k \in \{0, \ldots, n-1\} \setminus \{i,j\})dx$$

$$= \sum_{i \neq j} \sum_{l=\max\{i,j\}}^n \int_{t_l}^{t_{l+1}} x(1 - e^{-(x-t_i)})e^{-(x-t_j)} \prod_{k \in \{0,\ldots,n-1\} \setminus \{i,j\}} \mathbf{P}(Y_k \geq x)dx$$

$$= \sum_{i \neq j} \sum_{l=\max\{i,j\}}^n \int_{t_l}^{t_{l+1}} x(1 - e^{-(x-t_i)})e^{-(x-t_j)} \prod_{k \in 0,\ldots,l \setminus \{i,j\}} e^{-(x-t_k)}dx$$

$$= \sum_{i \neq j} \sum_{l = \max\{i,j\}}^{n} \int_{t_l}^{t_{l+1}} x \left(e^{-(lx - t_0 - \ldots - t_{i-1} - t_{i+1} - \ldots t_l)} - e^{-((l+1)x - t_0 - \ldots t_l)} \right) dx.$$

$$(21)$$

Setting $s_l = t_0 + \ldots + t_l$, the innermost integral in (21) can be evaluated as

$$\int_{t_l}^{t_{l+1}} x \left(e^{-(lx - s_l + t_i)} - e^{-((l+1)x - s_l)} \right) dx =$$

$$\left[-\frac{x}{l} e^{-(lx - s_l + t_i)} - \frac{1}{l^2} e^{-(lx - s_l + t_i)} + \frac{x}{l+1} e^{-((l+1)x - s_l)} + \frac{1}{(l+1)^2} e^{-((l+1)x - s_l)} \right]_{t_l}^{t_{l+1}}.$$

Note that our double summation in (21) can be split into four parts of the form

$$\sum_{i \neq j} \sum_{l = \max\{i,j\}}^{n} b_l a_i = \sum_{l=1}^{n-1} b_l \left(\sum_{i \neq j, i, j \leq l} a_i \right) = \sum_{l=1}^{n-1} b_l \left(\sum_{i=0}^{l} l a_i \right) = \sum_{l=1}^{n-1} l b_l \left(\sum_{i=0}^{l} a_i \right)$$

where $a_i = e^{-t_i}$ for the first two terms of (21) and $a_i = 1$ for the last two. We now evaluate each part of (21) separately, call them $I_1, \ldots, I_4$.

$$I_1 = \sum_{i \neq j} \sum_{l = \max\{i,j\}}^{n} \left[-\frac{x}{l} e^{-(lx - s_l + t_i)} \right]_{t_l}^{t_{l+1}} = t_1 + \sum_{l=1}^{n-1} t_l e^{-(lt_l - s_{l-1})}$$

where by abuse of notation we have used the fact that $t_n e^{-((n-1)t_n - s_{n-1})} = 0$. We have also used the identity, under the assumption $u_n = 0$,

$$\sum_{l=1}^{n-1} (u_l - u_{l+1}) \sum_{i=0}^{l} v_i = u_1 v_0 + \sum_{l=1}^{n-1} u_l v_l.$$

In this version we must omit the calculations of the remaining terms, but

$$I_2 = 1 + \sum_{l=1}^{n-1} \frac{1}{l} e^{-(lt_l - s_{l-1})}, \quad I_3 = -\sum_{l=1}^{n-1} t_l e^{-(lt_l - s_{l-1})}, \quad I_4 = -\sum_{l=1}^{n-1} \frac{1}{l(l+1)} e^{-(lt_l - s_{l-1})}.$$

Notice that

$$I_1 + I_3 = t_1, \quad I_2 + I_4 = 1 + \sum_{l=1}^{n-1} \frac{1}{l+1} e^{-(lt_l - s_{l-1})} - \sum_{l=2}^{n-1} \frac{1}{l(l-1)} e^{-(lt_l - s_l)} \left(\sum_{i=0}^{l-1} e^{t_l - t_i} \right).$$

The minimum Y_j has expectation $t_0 + 1 - \sum_{l=1}^{n-1} \frac{1}{l(l+1)} e^{-(lt_l - s_{l-1})}$ [NPS05], so the expected difference $D_n = I_1 + I_3 + I_2 + I_4 - \min Y_j$ is given by

$$D_n = (t_1 - t_0) + \sum_{l=1}^{n-1} \frac{1}{l} e^{-(lt_l - s_{l-1})} - \sum_{l=2}^{n-1} \frac{1}{l(l-1)} e^{-(lt_l - s_l)} \left(\sum_{i=0}^{l-1} e^{t_l - t_i} \right). \quad (22)$$

Taking the expectation over the T_i's we get from [NPS05] and (18) that

$$\mathbf{E}[e^{-(lT_l-S_{l-1})}] = \prod_{j=1}^{l} \mathbf{E}[e^{-j(T_j-T_{j-1})}] = \prod_{j=1}^{l} \frac{n-j}{n-j+1} = \frac{n-l}{n}.$$

Similarly we get for $i = 0, \ldots, l-1$

$$\mathbf{E}[e^{(T_l-T_i)}e^{-(lT_l-S_{l-1})}] = \mathbf{E}[e^{((T_l-T_{l-1})+\ldots+(T_{i+1}-T_i))-\sum_{j=1}^{l} j(T_j-T_{j-1})}]$$

$$= \frac{n-i}{n} \prod_{j=i+1}^{l} \frac{(n-j)j}{(n-j+1)j-1}.$$

Plugging this into (22) we get

$$D_n = \frac{1}{n-1} + \sum_{l=1}^{n-1} \frac{1}{l}\frac{n-l}{n} - \sum_{l=2}^{n-1} \frac{1}{l(l-1)} \sum_{i=0}^{l-1} \frac{n-i}{n} \prod_{j=i+1}^{l} \frac{(n-j)j}{(n-j+1)j-1}.$$

One can show that $D_n \sim \frac{\pi^2}{6n}$, completing the proof of Theorem 4.

References

[AB92]   Avram, F., Bertsimas, D.: The minimum spanning tree constant in geo-
           metrical probability and under the independent model: a unified approach.
           Annals of Applied Probability 2, 113–130 (1992)
[Ald92]   Aldous, D.: Asymptotics in the random assignment problem. Pr. Th. Re-
           lated Fields 93, 507–534 (1992)
[Ald01]   Aldous, D.J.: The $\zeta(2)$ limit in the random assignment problem. Random
           Structures Algorithms 18(4), 381–418 (2001)
[AT02]   Archer, A., Tardos, É.: Frugal path mechanisms. In: Proceedings of the
           Thirteenth Annual ACM-SIAM Symposium on Discrete Algorithms, San
           Francisco, California, January 06-08, 2002, pp. 991–999 (2002)
[AT07]   Archer, A., Tardos, É.: Frugal path mechanisms. ACM Trans. Algo-
           rithms 3(1), Art. 3, 22 (2007); MRMR2301829 (2008b:68013)
[Cla71]   Clarke, E.H.: Multipart pricing of public goods. Public Choice 8, 17–33
           (1971)
[CR04]   Czumaj, A., Ronen, A.: On the expected payment of mechanisms for task
           allocation (extended abstract). In: Proceedings of the Fifth ACM Confer-
           ence on Electronic Commerce, pp. 252–253 (2004)
[CS99]   Coppersmith, D., Sorkin, G.B.: Constructive bounds and exact expec-
           tations for the random assignment problem. Random Structures Algo-
           rithms 15(2), 113–144 (1999); MR2001j:05096
[Elk05]   Elkind, E.: True costs of cheap labor are hard to measure: edge deletion and
           VCG payments in graphs. In: Proceedings of the Sixth ACM Conference
           on Electronic Commerce, pp. 108–117 (2005)
[ESS04]   Elkind, E., Sahai, A., Steiglitz, K.: Frugality in path auctions. In: Pro-
           ceedings of the Fifteenth Annual ACM-SIAM Symposium on Discrete Al-
           gorithms, pp. 694–702 (2004)

[FGS06]   Flaxman, A., Gamarnik, D., Sorkin, G.B.: First-passage percolation on a width-2 strip and the path cost in a VCG auction. In: Spirakis, P.G., Mavronicolas, M., Kontogiannis, S.C. (eds.) WINE 2006. LNCS, vol. 4286, pp. 99–111. Springer, Heidelberg (2006)

[FPSS02]   Feigenbaum, J., Papadimitriou, C., Sami, R., Shenker, S.: A BGP-based mechanism for lowest-cost routing. In: PODC 2002: Proceedings of the Twenty-First Annual Symposium on Principles of Distributed Computing, pp. 173–182. ACM Press, New York (2002)

[Fri85]   Frieze, A.: On the value of a random minimum spanning tree problem. Discrete Applied Mathematics 10, 47–56 (1985)

[Gro73]   Groves, T.: Incentives in teams. Econometrica 41(4), 617–631 (1973)

[Jan99]   Janson, S.: One, two, three $\log n/n$ for paths in a complete graph with random weights. Combinatorics, Probability and Computing 8, 347–361 (1999)

[KN05]   Karger, D., Nikolova, E.: Brief announcement: on the expected overpayment of VCG mechanisms in large networks. In: PODC 2005: Proceedings of the 24th Annual ACM Symposium on Principles of Distributed Computing, pp. 126–126. ACM Press, New York (2005)

[LW04]   Linusson, S., Wästlund, J.: A proof of Parisi's conjecture on the random assignment problem. Probability Theory and Related Fields 128, 419–440 (2004)

[MP85]   Mézard, M., Parisi, G.: Replicas and optimization. J. Physique Lettres 46, 771–778 (1985)

[MP86]   Mézard, M., Parisi, G.: Mean-field equations for the matching and the travelling salesman problems. Europhys. Lett. 2, 913–918 (1986)

[MP87]   Mézard, M., Parisi, G.: On the solution of the random link matching problems. J. Physique Lettres 48, 1451–1459 (1987)

[MPS03]   Mihail, M., Papadimitriou, C., Saberi, A.: On certain connectivity properties of the Internet topology. In: FOCS 2003: Proceedings of the 44th Annual IEEE Symposium on Foundations of Computer Science, Washington, DC, USA, p. 28. IEEE Computer Society Press, Los Alamitos (2003)

[NPS05]   Nair, C., Prabhakar, B., Sharma, M.: Proofs of the Parisi and Coppersmith-Sorkin random assignment conjectures. Random Structures Algorithms 27(4), 413–444 (2005) MR MR2178256 (2006e:90050)

[NR99]   Nisan, N., Ronen, A.: Algorithmic mechanism design (extended abstract). In: Proceedings of the Thirty-First Annual ACM Symposium on Theory of Computing, Atlanta, Georgia, United States, pp. 129–140 (1999)

[NR01]   Nisan, N., Ronen, A.: Algorithmic mechanism design. Games Econom. Behav. 35(1-2), 166–196 (2001); Economics and artificial intelligence MR MR1822468 (2002a:68146)

[Par98]   Parisi, G.: A conjecture on random bipartite matching, Physics e-Print archive (January 1998), http://xxx.lanl.gov/ps/cond-mat/9801176

[Vic61]   Vickrey, W.: Counterspeculation, auctions and competitive sealed tenders. Journal of Finance 16, 8–37 (1961)

[Wäs]   Wästlund, J.: An easy proof of the zeta(2) limit in the random assignment problem. Electronic Journal of Probability (to appear)

A Hypergraph Dictatorship Test with Perfect Completeness

Victor Chen*

Massachusetts Institute of Technology
Computer Science and Artificial Intelligence Laboratory
victor@csail.mit.edu

Abstract. A hypergraph dictatorship test is first introduced by Samorodnitsky and Trevisan and serves as a key component in their unique games based PCP construction. Such a test has oracle access to a collection of functions and determines whether all the functions are the same dictatorship, or all their low degree influences are $o(1)$. Their test makes $q \geq 3$ queries, has amortized query complexity $1 + O\left(\frac{\log q}{q}\right)$, but has an inherent loss of perfect completeness. In this paper we give an (adaptive) hypergraph dictatorship test that achieves both perfect completeness and amortized query complexity $1 + O\left(\frac{\log q}{q}\right)$.

Keywords: Property testing, Gowers norm, Fourier analysis, PCP.

1 Introduction

Linearity and dictatorship testing have been studied in the past decade both for their combinatorial interest and connection to complexity theory. These tests distinguish functions which are linear/dictator from those which are far from being a linear/dictator function. The tests do so by making queries to a function at certain points and receiving the function's values at these points. The parameters of interest are the number of queries a test makes and the completeness and soundness of a test.

In this paper we shall work with boolean functions of the form $f : \{0,1\}^n \rightarrow \{-1, 1\}$. We say a function f is *linear* if $f = (-1)^{\sum_{i \in S} x_i}$ for some subset $S \subseteq [n]$. A *dictator* function is simply a linear function where $|S| = 1$, i.e., $f(x) = (-1)^{x_i}$ for some i. A dictator function is often called a *long code*, and it is first used in [1] for the constructions of probabilistic checkable proofs (PCPs), see e.g., [2,3]. Since then, it has become standard to design a PCP system as the composition of two verifiers, an outer verifier and an inner verifier. In such case, a PCP system expects the proof to be written in such a way so that the outer verifier, typically based on the verifier obtained from Raz's Parallel Repetition Theorem [4], selects some tables of the proof according to some distribution and then passes the control to the inner verifier. The inner verifier, with oracle access to these tables, makes queries into these tables and ensures that the tables are the encoding of some error-correcting codes and satisfy some joint constraint.

* Research supported in part by an NSF graduate fellowship, NSF Award CCR-0514915, the National Natural Science Foundation of China Grant 60553001, and the National Basic Research Program of China Grant 2007CB807900,2007CB807901.

I. Dinur et al. (Eds.): APPROX and RANDOM 2009, LNCS 5687, pp. 448–461, 2009.

The long code encoding is usually employed in these proof constructions, and the inner verifier simply tests whether a collection of tables (functions) are long codes satisfying some constraints. Following this paradigm, constructing a PCP with certain parameters reduces to the problem of designing a long code test with similar parameters.

One question of interest is the tradeoff between the soundness and query complexity of a tester. If a tester queries the functions at every single value, then trivially the verifier can determine all the functions. One would like to construct a dictatorship test that has the lowest possible soundness while making as few queries as possible. One way to measure this tradeoff between the soundness s and the number of queries q is *amortized query complexity*, defined as $\frac{q}{\log s^{-1}}$. This investigation, initiated in [5], has since spurred a long sequence of works [6,7,8,9]. All the testers from these works run many iterations of a single dictatorship test by reusing queries from previous iterations. The techniques used are Fourier analytic, and the best amortized query complexity from this sequence of works has the form $1 + O\left(\frac{1}{\sqrt{q}}\right)$.

The next breakthrough occurs when Samorodnitsky [10] introduces the notion of a *relaxed* linearity test along with new ideas from additive combinatorics. In property testing, the goal is to distinguish objects that are very structured from those that are pseudorandom. In the case of linearity/dictatorship testing, the structured objects are the linear/dictator functions, and functions that are far from being linear/dictator are interpreted as pseudorandom. The recent paradigm in additive combinatorics is to find the right framework of structure and pseudorandomness and analyze combinatorial objects by dividing them into structured and pseudorandom components, see e.g. [11] for a survey. One success is the notion of Gowers norm [12], which has been fruitful in attacking many problems in additive combinatorics and computer science. In [10], the notion of pseudorandomness for linearity testing is relaxed; instead of designating the functions that are far from being linear as pseudorandom, the functions having small low degree Gowers norm are considered to be pseudorandom. By doing so, an optimal tradeoff between soundness and query complexity is obtained for the problem of relaxed linearity testing. (Here the tradeoff is stronger than the tradeoff for the traditional problem of linearity testing.)

In a similar fashion, in the PCP literature since [13], the pseudorandom objects in dictatorship tests are not functions that are far from being a dictator. The pseudorandom functions are typically defined to be either functions that are far from all "juntas" or functions whose "low-degree influences" are $o(1)$. Both considerations of a dictatorship test are sufficient to compose the test in a PCP construction. In [14], building on the analysis of the relaxed linearity test in [10], Samorodnitsky and Trevisan construct a dictatorship test (taking the view that functions with arbitrary small "low-degree influences are pseudorandom) with amortized query complexity $1 + O\left(\frac{\log q}{q}\right)$. Furthermore, the test is used as the inner verifier in a conditional PCP construction (based on unique games [15]) with the same parameters. However, their dictatorship test suffers from an inherent loss of perfect completeness. Ideally one would like testers with one-sided errors. One, for aesthetic reasons, testers should always accept valid inputs. Two, for some hardness of approximation applications, in particular coloring problems (see e.g. [16] or [17]), it is important to construct PCP systems with one-sided errors.

In this paper, we prove the following theorem:

Theorem 1 (main theorem). *For every $q \geq 3$, there exists an (adaptive) dictatorship test that makes q queries, has completeness 1, and soundness $\frac{O(q^3)}{2^q}$; in particular it has amortized query complexity $1 + O\left(\frac{\log q}{q}\right)$.*

Our tester is a variant of the one given in [14]. Our tester is adaptive in the sense that it makes its queries in two stages. It first makes roughly $\log q$ nonadaptive queries into the function. Based on the values of these queries, the tester then selects the rest of the query points nonadaptively. Our analysis is based on techniques developed in [8,14,16,18].

1.1 Future Direction

Unfortunately, the adaptivity of our test is a drawback. The correspondence between PCP constructions and hardness of approximation needs the test to be fully nonadaptive. However, a more pressing issue is that our hypergraph dictatorship test does not immediately imply a new PCP characterization of NP. The reason is that a dictatorship test without "consistency checks" is most easily composed with the unique label cover defined in [15] as the outer verifier in a PCP reduction. As the conjectured NP-hardness of the unique label cover cannot have perfect completeness, the obvious approach in combining our test with the unique games-based outer verifier does not imply a new PCP result. However, there are variants of the unique label cover (e.g., Khot's d to 1 Conjecture) [15] that do have conjectured perfect completeness, and these variants are used to derive hardness of coloring problems in [17]. We hope that our result combined with similar techniques used in [17] may obtain a new conditional PCP construction and will motivate more progress on constraint satisfaction problems with bounded projection.

1.2 Related Works

The problem of linearity testing was first introduced in [19]. The framework of property testing was formally set up in [20]. The PCP Theorems were first proved in [2,3]; dictatorship tests first appeared in the PCP context in [1], and many dictatorship tests and variants appeared throughout the PCP literature. Dictatorship test was also considered as a standalone property testing in [21]. As mentioned, designing testers and PCPs focusing on amortized query complexity was first investigated in [5], and a long sequence of works [6,7,8,9] followed. The first tester/PCP system focusing on this tradeoff while obtaining perfect completeness was achieved in [16].

The orthogonal question of designing testers or PCPs with as few queries as possible was also considered. In a highly influential paper [13], Håstad constructed a PCP system making only three queries. Many variants also followed. In particular PCP systems with perfect completeness making three queries were also achieved in [18,22]. Similar to our approach, O'Donnell and Wu [23] designed an optimal three bit dictatorship test with perfect completeness, and later the same authors constructed a conditional PCP system [24].

2   Preliminaries

We fix some notation and provide the necessary background in this section. We let $[n]$ denote the set $\{1, 2, \ldots, n\}$. For a vector $v \in \{0, 1\}^n$, we write $|v| = \sum_{i \in [n]} v_i$. We let

$\wedge$ denote the boolean AND, where $a \wedge b = 1$ iff $a = b = 1$. For vectors $v, w \in \{0,1\}^n$, we write $v \wedge w$ to denote the vector obtained by applying AND to v and w component-wise. We abuse notation and sometimes interpret a vector $v \in \{0,1\}^n$ as a subset $v \subseteq [n]$ where $i \in v$ iff $v_i = 1$. For a boolean function $f : \{0,1\}^n \to \{0,1\}$, we make the convenient notational change from $\{0,1\}$ to $\{-1,1\}$ and write $f : \{0,1\}^n \to \{-1,1\}$.

2.1 Fourier Analysis

Definition 1 (Fourier transform). *For a real-valued function* $f : \{0,1\}^n \to \mathbb{R}$, *we define its Fourier transform* $\hat{f} : \{0,1\}^n \to \mathbb{R}$ *to be* $\hat{f}(\alpha) = \mathbb{E}_{x \in \{0,1\}^n} f(x) \chi_\alpha(x)$, *where* $\chi_\alpha(x) = (-1)^{\sum_{i \in [n]} \alpha_i x_i}$. *We say* $\hat{f}(\alpha)$ *is the* Fourier coefficient *of* f *at* α, *and the* characters *of* $\{0,1\}^n$ *are the functions* $\{\chi_\alpha\}_{\alpha \in \{0,1\}^n}$.

It is easy to see that for $\alpha, \beta \in \{0,1\}^n$, $\mathbb{E} \chi_\alpha \cdot \chi_\beta$ is 1 if $\alpha = \beta$ and 0 otherwise. Since there are 2^n characters, they form an orthonormal basis for functions on $\{0,1\}^n$, and we have the Fourier inversion formula $f(x) = \sum_{\alpha \in \{0,1\}^n} \hat{f}(\alpha) \chi_\alpha(x)$ and Parseval's Identity $\sum_{\alpha \in \{0,1\}^n} \hat{f}(\alpha)^2 = \mathbb{E}_x[f(x)^2]$.

2.2 Influence of Variables

For a boolean function $f : \{0,1\}^n \to \{-1,1\}$, the *influence* of the i-variable, $I_i(f)$, is defined to be $\Pr_{x \in \{0,1\}^n}[f(x) \neq f(x + e_i)]$, where e_i is a vector in $\{0,1\}^n$ with 1 on the i-th coordinate 0 everywhere else. This corresponds to our intuitive notion of influence: how likely the outcome of f changes when the i-th variable on a random input is flipped. For the rest of this paper, it will be convenient to work with the Fourier analytic definition of $I_i(f)$ instead, and we leave it to the readers to verify that the two definitions are equivalent when f is a boolean function.

Definition 2. *Let* $f : \{0,1\}^n \to \mathbb{R}$. *We define the influence of the i-th variable of f to be*

$$I_i(f) = \sum_{\alpha \in \{0,1\}^n:\, \alpha_i = 1} \hat{f}(\alpha)^2.$$

We shall need the following technical lemma, which is Lemma 4 from [14], and it gives an upper bound on the influence of a product of functions.

Lemma 1 (from [14]). *Let* $f_1, \dots, f_k : \{0,1\}^n \to [-1,1]$ *be a collection of k bounded real-valued functions, and define* $f(x) = \prod_{i=1}^k f_i(x)$ *to be the product of these k functions. Then for each* $i \in [n]$,

$$I_i(f) \leq k \cdot \sum_{j=1}^k I_i(f_j).$$

When $\{f_i\}$ are boolean functions, it is easy to see that $I_i(f) < \sum_{j=1}^k I_i(f_j)$ by the union bound.

We now define the notion of low-degree influence.

Definition 3. *Let w be an integer between 0 and n. We define the w-th degree influence of the i-th variable of a function $f : \{0,1\}^n \to \mathbb{R}$ to be*

$$\mathrm{I}_i^{\leq w}(f) = \sum_{\alpha \in \{0,1\}^n:\ \alpha_i=1,\ |\alpha| \leq w} \hat{f}(\alpha)^2.$$

2.3 Gowers Norm

In [12], Gowers uses analytic techniques to give a new proof of Szemerédi's Theorem [25] and in particular, initiates the study of a new norm of a function as a measure of pseudorandomness. Subsequently this norm is termed the *Gowers uniformity norm* and has been intensively studied and applied in additive combinatorics, see e.g. [11] for a survey. The use of the Gowers norm in computer science is initiated in [10,14].

Definition 4. *Let $f : \{0,1\}^n \to \mathbb{R}$. We define the d-th dimension Gowers uniformity norm of f to be*

$$\|f\|_{U_d} = \left(\mathop{\mathbb{E}}_{x,\ x_1,\ldots,x_d} \left[\prod_{S \subseteq [d]} f\left(x + \sum_{i \in S} x_i \right) \right] \right)^{1/2^d}.$$

For a collection of 2^d functions $f_S : \{0,1\}^n \to \mathbb{R}, S \subset [d]$, we define the d-th dimension Gowers inner product of $\{f_S\}_{S \subseteq d}$ to be

$$\langle \{f_S\}_{S \subseteq [d]} \rangle_{U_d} = \mathop{\mathbb{E}}_{x,\ x_1,\ldots,x_d} \left[\prod_{S \subseteq [d]} f_S\left(x + \sum_{i \in S} x_i \right) \right].$$

When f is a boolean function, one can interpret the Gowers norm as simply the expected number of "affine parallelepipeds" of dimension d.

For the analysis of hypergraph-based dictatorship test, we shall encounter the following expression.

Definition 5. *Let $\{f_S\}_{S \subseteq [d]}$ be a collection of functions where $f_S : \{0,1\}^n \to \mathbb{R}$. We define the d-th dimension Gowers linear inner product of $\{f_S\}$ to be*

$$\langle \{f_S\}_{S \subseteq [d]} \rangle_{LU_d} = \mathop{\mathbb{E}}_{x_1,\ldots,x_d} \left[\prod_{S \subseteq [d]} f_S\left(\sum_{i \in S} x_i \right) \right].$$

This definition is a variant of the Gowers inner product and is in fact upper bounded by the square root of the Gowers inner product as shown in [14]. Furthermore they showed that if a collection of functions has large Gowers inner product, then two functions must share an influential variable. Thus, one can infer the weaker statement that large linear Gowers inner product implies two functions have an influential variable.

Lemma 2 (from [14]). *Let $\{f_S\}_{S \subseteq [d]}$ be a collection of bounded functions of the form $f_S : \{0,1\}^n \to [-1,1]$. Suppose $\langle \{f_S\}_{S \subseteq [d]} \rangle_{LU_d} \geq \epsilon$ and $\mathbb{E} f_{[d]} = 0$. Then there exists some variable i, some subsets $S \neq T \subseteq [d]$ such that the influences of the i-th variable in both f_S and f_T are at least $\frac{\epsilon^4}{2^{O(d)}}$.*

3 Dictatorship Test

Definition 6 (dictatorship). *For $i \in [n]$, the i-th dictator is the function $f(x) = (-1)^{x_i}$.*

In the PCP literature, the i–th dictator is also known as the long code encoding of i, $\langle(-1)^{x_i}\rangle_{x\in\{0,1\}^n}$, which is simply the evaluation of the i-th dictator function at all points.

Now let us define a t-function dictatorship test. Suppose we are given oracle access to a collection of boolean functions $f_1, \ldots, f_t$. We want to make as few queries as possible into these functions to decide if all the functions are the same dictatorship, or no two functions have some common structure. More precisely, we have the following definition:

Definition 7. *We say that a test $T = T^{f_1,\ldots,f_t}$ is a t–function dictatorship test with completeness c and soundness s if T is given oracle access to a family of t functions $f_1, \ldots, f_t : \{0,1\}^n \to \{-1,1\}$, such that*

- *if there exists some variable $i \in [n]$ such that for all $a \in [t]$, $f_a(x) = (-1)^{x_i}$, then T accepts with probability at least c, and*
- *for every $\epsilon > 0$, there exist a positive constant $\tau > 0$ and a fixed positive integer w such that if T accepts with probability at least $s + \epsilon$, then there exist two functions f_a, f_b where $a, b \in [t], a \neq b$ and some variable $i \in [n]$ such that $I_i^{\leq w}(f_a), I_i^{\leq w}(f_b) \geq \tau$.*

A q-function dictatorship test making q queries, with soundness $\frac{q+1}{2^q}$ was proved in [14], but the test suffers from imperfect completeness. We obtain a $(q - O(\log q))$–dictatorship test that makes q queries, has completeness 1, soundness $\frac{O(q^3)}{2^q}$, and in particular has amortized query complexity $1 + O\left(\frac{\log q}{q}\right)$, the same as the test in [14]. By a simple change of variable, we can more precisely state the following:

Theorem 2 (main theorem restated). *For infinitely many t, there exists an adaptive t-function dictatorship test that makes $t + \log(t + 1)$ queries, has completeness 1, and soundness $\frac{(t+1)^2}{2^t}$.*

Our test is adaptive and selects queries in two passes. During the first pass, it picks an arbitrary subset of $\log(t + 1)$ functions out of the t functions. For each function selected, our test picks a random entry y and queries the function at entry y. Then based on the values of these $\log(t+1)$ queries, during the second pass, the test selects t positions nonadaptively, one from each function, then queries all t positions at once. The adaptivity is necessary in our analysis, and it is unclear if one can prove an analogous result with only one pass.

3.1 Folding

As introduced by Bellare, Goldreich, and Sudan [1], we shall assume that the functions are "folded" as only half of the entries of a function are accessed. We require our dictatorship test to make queries in a special manner. Suppose the test wants to query f at the point $x \in \{0,1\}^n$. If $x_1 = 1$, then the test queries $f(x)$ as usual. If $x_1 = 0$, then the test queries f at the point $\mathbf{1} + x = (1, 1 + x_2, \ldots, 1 + x_n)$ and negates the value it receives. It is instructive to note that folding ensures $f(\mathbf{1} + x) = -f(x)$ and $\mathbb{E}\, f = 0$.

3.2   Basic Test

For ease of exposition, we first consider the following simplistic scenario. Suppose we have oracle access to just one boolean function. Furthermore we ignore the tradeoff between soundness and query complexity. We simply want a dictatorship test that has completeness 1 and soundness $\frac{1}{2}$. There are many such tests in the literature; however, we need a suitable one which our hypergraph dictatorship test can base on. Our basic test below is a close variant of the one proposed by Guruswami, Lewin, Sudan, and Trevisan [18].

BASIC TEST T: with oracle access to f,

1. Pick x_i, x_j, y, z uniformly at random from $\{0,1\}^n$.
2. Query $f(y)$.
3. Let $v = \frac{1 - f(y)}{2}$. Accept iff

$$f(x_i)f(x_j) = f(x_i + x_j + (v\mathbf{1} + y) \wedge z).$$

Lemma 3. *The test T is a dictatorship test with completeness 1.*

Proof. Suppose f is the ℓ-th dictator, i.e., $f(x) = (-1)^{x_\ell}$. First note that

$$v + y_\ell = \frac{1 - (-1)^{y_\ell}}{2} + y_\ell,$$

which evaluates to 0. Thus by linearity of f

$$
\begin{aligned}
f(x_i + x_j + (v\mathbf{1} + y) \wedge z) &= f(x_i)f(x_j)f((v\mathbf{1} + y) \wedge z) \\
&= f(x_i)f(x_j)(-1)^{(v+y_\ell)\wedge z_\ell} \\
&= f(x_i)f(x_j)
\end{aligned}
$$

and the test always accepts. $\qquad\qquad\qquad\qquad\qquad\qquad\qquad\qquad\qquad\qquad\square$

To analyze the soundness of the test T, we first need to derive a Fourier analytic expression for the acceptance probability of T. Its proof is standard and omitted due to space limitation. Readers may find a proof in the full version of this paper [26].

Proposition 1. *Let p be the acceptance probability of T. Then*

$$p = \frac{1}{2} + \frac{1}{2} \sum_{\alpha \in \{0,1\}^n} \widehat{f}(\alpha)^3\, 2^{-|\alpha|} \left(1 + \sum_{\beta \subseteq \alpha} \widehat{f}(\beta) \right).$$

For sanity check, let us interpret the expression for p. Suppose $f = \chi_\alpha$ for some $\alpha \neq 0 \in \{0, 1\}^n$, i.e., $\widehat{f}(\alpha) = 1$ and all other Fourier coefficients of f are 0. Then clearly $p = \frac{1}{2} + 2^{-|\alpha|}$, which equals 1 whenever f is a dictator function as we have just shown. If $|\alpha|$ is large, then T accepts with probability close to $\frac{1}{2}$. We now analyze the soundness of the test.

Lemma 4. *The test T is a dictatorship test with soundness $\frac{1}{2}$.*

Proof. Suppose the test T passes with probability at least $\frac{1}{2} + \epsilon$, for some $\epsilon > 0$. By applying Proposition 1, Cauchy-Scharz Inequality, and Parseval's Identity, respectively, we obtain

$$\epsilon \leq \frac{1}{2} \sum_{\alpha \in \{0,1\}^n} \widehat{f}(\alpha)^3 \, 2^{-|\alpha|} \left(1 + \sum_{\beta \subseteq \alpha} \widehat{f}(\beta) \right)$$

$$\leq \frac{1}{2} \sum_{\alpha \in \{0,1\}^n} \widehat{f}(\alpha)^3 \, 2^{-|\alpha|} \left(1 + \left(\sum_{\beta \subseteq \alpha} \widehat{f}(\beta)^2 \right)^{\frac{1}{2}} \cdot 2^{\frac{|\alpha|}{2}} \right)$$

$$\leq \sum_{\alpha \in \{0,1\}^n} \widehat{f}(\alpha)^3 \, 2^{-\frac{|\alpha|}{2}}.$$

Pick the least positive integer w such that $2^{-\frac{w}{2}} \leq \frac{\epsilon}{2}$. Then by Parseval's again,

$$\frac{\epsilon}{2} \leq \sum_{\alpha \in \{0,1\}^n : |\alpha| \leq w} \widehat{f}(\alpha)^3$$

$$\leq \max_{\alpha \in \{0,1\}^n : |\alpha| \leq w} \left| \widehat{f}(\alpha) \right|.$$

So there exists some $\beta \in \{0, 1\}^n, |\beta| \leq w$ such that $\frac{\epsilon}{2} \leq \left| \widehat{f}(\beta) \right|$. With f being folded, $\beta \neq 0$. Thus, there exists an $i \in [n]$ such that $\beta_i = 1$ and

$$\frac{\epsilon^2}{4} \leq \widehat{f}(\beta)^2 \leq \sum_{\alpha \in \{0,1\}^n : \alpha_i = 1, |\alpha| \leq w} \widehat{f}(\alpha)^2.$$

$\square$

3.3 Hypergraph Dictatorship Test

We prove the main theorem in this section. The basis of our hypergraph dictatorship test will be very similar to the test in the previous section. We remark that we did not choose to present the exact same basic test for hopefully a clearer exposition.

We now address the tradeoff between query complexity and soundness. If we simply repeat the basic test a number of iterations independently, the error is reduced, but the query complexity increases. In other words, the amortized query complexity does not change if we simply run the basic test for many independent iterations. Following Trevisan [5], all the dictatorship tests that save query complexity do so by reusing queries made in previous iterations of the basic test. To illustrate this idea, suppose test T queries f at the points $x_1 + h_1$, $x_2 + h_2$, $x_1 + x_2 + h_{1,2}$ to make a decision. For the second iteration, we let T query f at the points $x_3 + h_3$ and $x_1 + x_3 + h_{1,3}$ and reuse the value $f(x_1 + h_1)$ queried during the first run of T. T then uses the three values to make a second decision. In total T makes five queries to run two iterations.

We may think of the first run of T as parametrized by the points x_1 and x_2 and the second run of T by x_1 and x_3. In general, we may have k points $x_1, \ldots, x_k$ and a graph on $[k]$ vertices, such that each edge e of the graph corresponds to an iteration of T parametrized by the points $\{x_i\}_{i \in e}$. We shall use a complete hypergraph on k vertices to save on query complexity, and we will argue that the soundness of the algorithm decreases exponentially with respect to the number of iterations.

Formally, consider a hypergraph $H = ([k], E)$. Let $\{f_a\}_{a \in [k] \cup E}$ be a collection of boolean functions of the form $f_a : \{0, 1\}^n \to \{-1, 1\}$. We assume all the functions are folded, and so in particular, $\mathbb{E} f_a = 0$. Consider the following test:

HYPERGRAPH H-TEST: with oracle access to $\{f_a\}_{a \in [k] \cup E}$,

1. Pick $x_1, \ldots, x_k, y_1, \ldots, y_k$, and $\{z_a\}_{a \in [k] \cup E}$ independently and uniformly at random from $\{0, 1\}^n$.
2. For each $i \in [k]$, query $f_i(y_i)$.
3. Let $v_i = \frac{1 - f_i(y_i)}{2}$.
   Accept iff for every $e \in E$,

   $$\prod_{i \in e} [f_i(x_i + (v_i \mathbf{1} + y_i) \wedge z_i)] = f_e \left(\sum_{i \in e} x_i + (\Sigma_{i \in e}(v_i \mathbf{1} + y_i)) \wedge z_e \right).$$

We make a few remarks regarding the design of H-Test. The hypergraph test by Samorodnitsky and Trevisan [14] accepts iff for every $e \in E$, $\prod_{i \in e} f_i(x_i + \eta_i)$ equals $f_e(\sum_{i \in e} x_i + \eta_e)$, where the bits in each vector η_a are chosen independently to be 1 with some small constant, say 0.01. The noise vectors η_a rule out the possibility that linear functions with large support can be accepted. To obtain a test with perfect completeness, we use ideas from [18,21,16] to simulate the effect of the noise perturbation.

Note that for y, z chosen uniformly at random from $\{0, 1\}^n$, the vector $y \wedge z$ is a $\frac{1}{4}$-noisy vector. As observed by Parnas, Ron, and Samorodnitsky [21], the test $f(y \wedge z) = f(y) \wedge f(z)$ distinguishes between dictators and linear functions with large support. One can also combine linearity and dictatorship testing into a single test of the form $f(x_1 + x_2 + y \wedge z)(f(y) \wedge f(z)) = f(x_1)f(x_2)$ as Håstad and Khot demonstrated [16]. However, iterating this test is too costly for us. In fact, Håstad and Khot also consider an adaptive variant that reads $k^2 + 2k$ bits to obtain a soundness of 2^{-k^2}, the same parameters as in [7], while achieving perfect completeness as well. Without adaptivity, the test in [16] reads $k^2 + 4k$ bits. While both the nonadaptive and adaptive tests in [16] have the same amortized query complexity, extending the nonadaptive test by Hstad and Khot to the hypergraph setting does not work for us. So to achieve the same amortized query complexity as the hypergraph test in [14], we also exploit adaptivity in our test.

Theorem 3 (main theorem restated). *For infinitely many t, there exists an adaptive t-function dictatorship test with $t + \log(t + 1)$ queries, completeness 1, and soundness $\frac{(t+1)^2}{2^t}$.*

Proof. Take a complete hypergraph on k vertices, where $k = \log(t+1)$. The statement follows by applying Lemmas 5 and 6. □

Lemma 5. *The H-Test is a $(k + |E|)$-function dictatorship test that makes $|E| + 2k$ queries and has completeness 1.*

Due to space limitation we omit the easy proof of Lemma 5. Readers can find the proof in the full version of this paper [26].

Lemma 6. *The H-Test has soundness $2^{k-|E|}$.*

Before proving Lemma 6 we first prove a proposition relating the Fourier transform of a function perturbed by noise to the function's Fourier transform itself.

Proposition 2. *Let $f : \{0,1\}^n \to \{-1,1\}$. Define $g : \{0,1\}^{2n} \to [-1,1]$ to be*

$$g(x; y) = \mathop{\mathbb{E}}_{z \in \{0,1\}^n} f(c' + x + (c + y) \wedge z),$$

where c, c' are some fixed vectors in $\{0,1\}^n$. Then

$$\widehat{g}(\alpha; \beta)^2 = \widehat{f}(\alpha)^2 \, 1_{\{\beta \subseteq \alpha\}} 4^{-|\alpha|}.$$

Proof. This is a straightforward Fourier analytic calculation. By definition,

$$\widehat{g}(\alpha; \beta)^2 = \left(\mathop{\mathbb{E}}_{x,y,z \in \{0,1\}^n} f(c' + x + (c + y) \wedge z) \chi_\alpha(x) \chi_\beta(y) \right)^2.$$

By averaging over x it is easy to see that

$$\widehat{g}(\alpha; \beta)^2 = \widehat{f}(\alpha)^2 \left(\mathop{\mathbb{E}}_{y,z \in \{0,1\}^n} \chi_\alpha((c + y) \wedge z) \chi_\beta(y) \right)^2.$$

Since the bits of y are chosen independently and uniformly at random, if $\beta \backslash \alpha$ is nonempty, the above expression is zero. So we can write

$$\widehat{g}(\alpha; \beta)^2 = \widehat{f}(\alpha)^2 \, 1_{\{\beta \subseteq \alpha\}} \left(\prod_{i \in \alpha \backslash \beta} \mathop{\mathbb{E}}_{y_i, z_i} (-1)^{(c_i + y_i) \wedge z_i} \cdot \prod_{i \in \beta} \mathop{\mathbb{E}}_{y_i, z_i} (-1)^{(c_i + y_i) \wedge z_i + y_i} \right)^2.$$

It is easy to see that the term $\mathbb{E}_{y_i, z_i} (-1)^{(c_i + y_i) \wedge z_i}$ evaluates to $\frac{1}{2}$ and the term $\mathbb{E}_{y_i, z_i} (-1)^{(c_i + y_i) \wedge z_i + y_i}$ evaluates to $(-1)^{c_i} \frac{1}{2}$. Thus

$$\widehat{g}(\alpha; \beta)^2 = \widehat{f}(\alpha)^2 \, 1_{\{\beta \subseteq \alpha\}} \, 4^{-|\alpha|}$$

as claimed. □

Now we prove Lemma 6.

Proof. Let p be the acceptance probability of H-test. Suppose that $2^{k-|E|} + \epsilon \leq p$. We want to show that there are two functions f_a and f_b such that for some $i \in [n]$, some fixed positive integer w, some constant $\epsilon' > 0$, it is the case that $I_{\bar{i}}^{\leq w}(f_a), I_{\bar{i}}^{\leq w}(f_b) \geq \epsilon'$. As usual we first arithmetize p. We write

$$p = \sum_{v \in \{0,1\}^k} \mathop{\mathbb{E}}_{\{x_i\}, \{y_i\}, \{z_a\}} \prod_{i \in [k]} \frac{1 + (-1)^{v_i} f_i(y_i)}{2} \prod_{e \in E} \frac{1 + \mathrm{Acc}(\{x_i, y_i, v_i, z_i\}_{i \in e}, z_e)}{2},$$

where

$$\mathrm{Acc}(\{x_i, y_i, v_i, z_i\}_{i \in e}, z_e) = \prod_{i \in e} [f_i(x_i + (v_i\mathbf{1} + y_i) \wedge z_i)]$$

$$\cdot f_e \left(\sum_{i \in e} x_i + (\Sigma_{i \in e}(v_i\mathbf{1} + y_i)) \wedge z_e \right).$$

For each $i \in [k]$, f_i is folded, so $(-1)^{v_i} f_i(y_i) = f_i(v_i\mathbf{1} + y_i)$. Since the vectors $\{y_i\}_{i \in [k]}$ are uniformly and independently chosen from $\{0,1\}^n$, for a fixed $v \in \{0,1\}^k$, the vectors $\{v_i\mathbf{1} + y_i\}_{i \in [k]}$ are also uniformly and independently chosen from $\{0,1\}^n$. So we can simplify the expression for p and write

$$p = \mathop{\mathbb{E}}_{\{x_i\}, \{y_i\}, \{z_a\}} \left[\prod_{i \in [k]} (1 + f_i(y_i)) \prod_{e \in E} \frac{1 + (\mathrm{Acc}\{x_i, y_i, \mathbf{0}, z_i\}_{i \in e}, z_e)}{2} \right].$$

Instead of writing $\mathrm{Acc}(\{x_i, y_i, \mathbf{0}, z_i\}_{i \in e}, z_e)$, for convenience we shall write $\mathrm{Acc}(e)$ to be a notational shorthand. Observe that since $1 + f_i(y_i)$ is either 0 or 2, we may write

$$p \leq 2^k \mathop{\mathbb{E}}_{\{x_i\}, \{y_i\}, \{z_a\}} \left[\prod_{e \in E} \frac{1 + \mathrm{Acc}(e)}{2} \right].$$

Note that the product of sums $\prod_{e \in E} \frac{1 + \mathrm{Acc}(e)}{2}$ expands into a sum of products of the form

$$2^{-|E|} \left(1 + \sum_{\emptyset \neq E' \subseteq E} \prod_{e \in E'} \mathrm{Acc}(e) \right),$$

so we have

$$\frac{\epsilon}{2^k} \leq \mathop{\mathbb{E}}_{\{x_i\}, \{y_i\}, \{z_a\}} \left[2^{-|E|} \sum_{\emptyset \neq E' \subseteq E} \prod_{e \in E'} \mathrm{Acc}(e) \right].$$

By averaging, there must exist some nonempty subset $E' \subseteq E$ such that

$$\frac{\epsilon}{2^k} \leq \mathop{\mathbb{E}}_{\{x_i\}, \{y_i\}, \{z_a\}} \left[\prod_{e \in E'} \mathrm{Acc}(e) \right].$$

Let Odd consists of the vertices in $[k]$ with odd degree in E'. Expanding out the definition of $\mathrm{Acc}(e)$, we can conclude

$$\frac{\epsilon}{2^k} \leq \underset{\{x_i\},\{y_i\},\{z_a\}}{\mathbb{E}} \left[\prod_{i \in \mathrm{Odd}} f_i(x_i + y_i \wedge z_i) \cdot \prod_{e \in E'} f_e \left(\sum_{i \in e} x_i + \left(\sum_{i \in e} y_i \right) \wedge z_e \right) \right].$$

We now define a family of functions that represent the "noisy versions" of f_a. For $a \in [k] \cup E$, define $g'_a : \{0,1\}^{2n} \to [-1,1]$ to be

$$g'_a(x; y) = \underset{z \in \{0,1\}^n}{\mathbb{E}} f_a(x + y \wedge z).$$

Thus we have

$$\frac{\epsilon}{2^k} \leq \underset{\{x_i\},\{y_i\}}{\mathbb{E}} \left[\prod_{i \in \mathrm{Odd}} g'_i(x_i; y_i) \cdot \prod_{e \in E'} g'_e \left(\sum_{i \in e} x_i; \sum_{i \in e} y_i \right) \right].$$

Following the approach in [8,14], we are going to reduce the analysis of the iterated test to one hyperedge. Let d be the maximum size of an edge in E', and without loss of generality, let $(1, 2, \ldots, d)$ be a maximal edge in E'. Now, fix the values of $x_{d+1}, \ldots, x_k$ and $y_{d+1}, \ldots, y_k$ so that the following inequality holds:

$$\frac{\epsilon}{2^k} \leq \underset{x_1,y_1,\ldots,x_d,y_d}{\mathbb{E}} \left[\prod_{i \in \mathrm{Odd}} g'_i(x_i; y_i) \cdot \prod_{e \in E'} g'_e \left(\sum_{i \in e} x_i; \sum_{i \in e} y_i \right) \right]. \qquad (1)$$

We group the edges in E' based on their intersection with $(1, \ldots, d)$. We rewrite Inequality 1 as

$$\frac{\epsilon}{2^k} \leq \underset{(x_1,y_1),\ldots,(x_d,y_d)\in\{0,1\}^{2n}}{\mathbb{E}} \left[\prod_{S \subseteq [d]} \prod_{a \in \mathrm{Odd} \cup E':a\cap[d]=S} g_a \left(\sum_{i \in S} x_i; \sum_{i \in S} y_i \right) \right], \qquad (2)$$

where for each $a \in [k] \cup E$, $g_a(x; y) = g'_a(c'_a + x; c_a + y)$, with $c'_a = \sum_{i \in a \setminus [d]} x_i$ and $c_a = \sum_{i \in a \setminus [d]} y_i$ fixed vectors in $\{0,1\}^n$.

By grouping the edges based on their intersection with $[d]$, we can rewrite Inequality 2 as

$$\frac{\epsilon}{2^k} \leq \underset{(x_1,y_1),\ldots,(x_d,y_d)\in\{0,1\}^{2n}}{\mathbb{E}} \left[\prod_{S \subseteq [d]} G_S \left(\sum_{i \in S} (x_i; y_i) \right) \right]$$

$$= \langle \{G_S\}_{S \subseteq [d]} \rangle_{LU_d},$$

where G_S is simply the product of all the functions g_a such that $a \in \mathrm{Odd} \cup E'$ and $a \cap [d] = S$.

Since $(1, \ldots, d)$ is maximal, all the other edges in E' do not contain $(1, \ldots, d)$ as a subset. Thus $G_{[d]} = g_{[d]}$ and $\mathbb{E}\, G_{[d]} = 0$. By Lemma 2, the linear Gowers inner product of a family of functions $\{G_S\}$ being positive implies that two functions from the family must share a variable with positive influence. More precisely, there exist $S \neq T \subseteq [d]$, $i \in [2n]$, $\tau > 0$, such that $I_i(G_S), I_i(G_T) \geq \tau$, where $\tau = \frac{\epsilon^4}{2^{O(d)}}$.

Note that $G_\emptyset$ is the product of all the functions g'_a that are indexed by vertices or edges outside of $[d]$. So $G_\emptyset$ is a constant function, and all of its variables clearly have influence 0. Thus neither S nor T is empty. Since G_S and G_T are products of at most 2^k functions, by Lemma 1 there must exist some $a \neq b \in [d] \cup E'$ such that $I_i(g_a), I_i(g_b) \geq \frac{\tau}{2^{2k}}$. Recall that we have defined $g_a(x; y)$ to be $\mathbb{E}_z\, f_a(c'_a + x + (c_a + y) \wedge z)$. Thus we can apply Proposition 2 to obtain

$$
\begin{aligned}
I_i(g_a) &= \sum_{(\alpha,\beta) \in \{0,1\}^{2n}; i \in (\alpha,\beta)} \widehat{g}_a(\alpha;\beta)^2 \\
&= \sum_{\alpha \in \{0,1\}^n; i \in \alpha} \sum_{\beta \subseteq \alpha} \widehat{f}_a(\alpha)^2\, 4^{-|\alpha|} \\
&= \sum_{\alpha \in \{0,1\}^n; i \in \alpha} \widehat{f}_a(\alpha)^2\, 2^{-|\alpha|}.
\end{aligned}
$$

Let w be the least positive integer such that $2^{-w} \leq \frac{\tau}{2^{2k+1}}$. Then it is easy to see that $I_i^{\leq w}(f_a) \geq \frac{\tau}{2^{2k+1}}$. Similarly, $I_i^{\leq w}(f_b) \geq \frac{\tau}{2^{2k+1}}$ as well. Hence this completes the proof.   □

Acknowledgments. I am grateful to Alex Samorodnitsky for many useful discussions and his help with the Gowers norm. I also thank Madhu Sudan for his advice and support and Swastik Kopparty for an encouraging discussion during the initial stage of this research.

References

1. Bellare, M., Goldreich, O., Sudan, M.: Free bits, PCPs, and nonapproximability—towards tight results. SIAM Journal on Computing 27(3), 804–915 (1998)
2. Arora, S., Safra, S.: Probabilistic checking of proofs: a new characterization of NP. J. ACM 45(1), 70–122 (1998)
3. Arora, S., Lund, C., Motwani, R., Sudan, M., Szegedy, M.: Proof verification and the hardness of approximation problems. J. ACM 45(3), 501–555 (1998)
4. Raz, R.: A parallel repetition theorem. SIAM J. Comput. 27(3), 763–803 (1998)
5. Trevisan, L.: Recycling queries in PCPs and in linearity tests (extended abstract). In: STOC 1998: Proceedings of the thirtieth annual ACM symposium on Theory of computing, pp. 299–308. ACM, New York (1998)
6. Sudan, M., Trevisan, L.: Probabilistically checkable proofs with low amortized query complexity. In: FOCS 1998: Proceedings of the 39th Annual Symposium on Foundations of Computer Science, Washington, DC, USA, p. 18. IEEE Computer Society Press, Los Alamitos (1998)
7. Samorodnitsky, A., Trevisan, L.: A PCP characterization of NP with optimal amortized query complexity. In: STOC 2000: Proceedings of the thirty-second annual ACM symposium on Theory of computing, Portland, Oregon, United States, pp. 191–199. ACM, New York (2000)
8. Håstad, J., Wigderson, A.: Simple analysis of graph tests for linearity and PCP. Random Struct. Algorithms 22(2), 139–160 (2003)

9. Engebretsen, L., Holmerin, J.: More efficient queries in PCPs for NP and improved approximation hardness of maximum CSP. In: Diekert, V., Durand, B. (eds.) STACS 2005. LNCS, vol. 3404, pp. 194–205. Springer, Heidelberg (2005)

10. Samorodnitsky, A.: Low-degree tests at large distances. In: STOC 2007: Proceedings of the thirty-ninth annual ACM symposium on Theory of computing, San Diego, California, USA, pp. 506–515. ACM, New York (2007)

11. Tao, T.: Structure and randomness in combinatorics. In: FOCS 2007: Proceedings of the forty-eighth annual ACM symposium on Foundations of computer science, pp. 3–15. ACM, New York (2007)

12. Gowers, T.: A new proof of Szemerédi's theorem. Geom. Funct. Anal. 11(3), 465–588 (2001)

13. Håstad, J.: Some optimal inapproximability results. J. of ACM 48(4), 798–859 (2001)

14. Samorodnitsky, A., Trevisan, L.: Gowers uniformity, influence of variables, and PCPs. In: STOC 2006: Proceedings of the thirty-eighth annual ACM symposium on Theory of computing, pp. 11–20. ACM, New York (2006)

15. Khot, S.: On the power of unique 2-prover 1-round games. In: STOC 2002: Proceedings of the thiry-fourth annual ACM symposium on Theory of computing, Montreal, Quebec, Canada, pp. 767–775. ACM, New York (2002)

16. Håstad, J., Khot, S.: Query efficient PCPs with perfect completeness. Theory of Computing 1(7), 119–148 (2005)

17. Dinur, I., Mossel, E., Regev, O.: Conditional hardness for approximate coloring. In: STOC 2006: Proceedings of the thirty-eighth annual ACM symposium on Theory of computing, pp. 344–353. ACM, New York (2006)

18. Guruswami, V., Lewin, D., Sudan, M., Trevisan, L.: A tight characterization of NP with 3 query PCPs. In: FOCS 1998: Proceedings of the 39th Annual Symposium on Foundations of Computer Science, Washington, DC, USA, p. 8. IEEE Computer Society Press, Los Alamitos (1998)

19. Blum, M., Luby, M., Rubinfeld, R.: Self-testing/correcting with applications to numerical problems. Journal of Computer and System Sciences 47(3), 549–595 (1993)

20. Rubinfeld, R., Sudan, M.: Robust characterizations of polynomials with applications to program testing. SIAM Journal on Computing 25(2), 252–271 (1996)

21. Parnas, M., Ron, D., Samorodnitsky, A.: Testing basic boolean formulae. SIAM Journal on Discrete Mathematics 16(1), 20–46 (2002)

22. Khot, S., Saket, R.: A 3-query non-adaptive PCP with perfect completeness. In: CCC 2006: Proceedings of the 21st Annual IEEE Conference on Computational Complexity, Washington, DC, USA, pp. 159–169. IEEE Computer Society Press, Los Alamitos (2006)

23. O'Donnell, R., Wu, Y.: 3-bit dictator testing: 1 vs. 5/8. In: SODA 2009: Proceedings of the Nineteenth Annual ACM -SIAM Symposium on Discrete Algorithms, Philadelphia, PA, USA, pp. 365–373. Society for Industrial and Applied Mathematics (2009)

24. O'Donnell, R., Wu, Y.: Conditional hardness for satisfiable-3CSPs. In: STOC 2009: Proceedings of the forty-first annual ACM symposium on Theory of computing. ACM, New York (to appear, 2009)

25. Szemerédi, E.: On sets of integers containing no k elements in arithmetic progression. Acta Arith. 27, 199–245 (1975)

26. Chen, V.: A hypergraph dictatorship test with perfect completeness (2008) arXiv/0811.1922

Extractors Using Hardness Amplification

Anindya De[*] and Luca Trevisan[**]

Computer Science Division
University of California, Berkeley, CA, USA
{anindya, luca}@cs.berkeley.edu

Abstract. Zimand [24] presented simple constructions of locally computable strong extractors whose analysis relies on the *direct product theorem* for one-way functions and on the *Blum-Micali-Yao* generator. For N-bit sources of entropy γN, his extractor has seed $O(\log^2 N)$ and extracts $N^{\gamma/3}$ random bits.

We show that his construction can be analyzed based solely on the direct product theorem for general functions. Using the direct product theorem of Impagliazzo et al. [6], we show that Zimand's construction can extract $\tilde{\Omega}_\gamma(N^{1/3})$ random bits. (As in Zimand's construction, the seed length is $O(\log^2 N)$ bits.)

We also show that a simplified construction can be analyzed based solely on the XOR lemma. Using Levin's proof of the XOR lemma [8], we provide an alternative simpler construction of a locally computable extractor with seed length $O(\log^2 N)$ and output length $\tilde{\Omega}_\gamma(N^{1/3})$.

Finally, we show that the *derandomized direct product theorem* of Impagliazzo and Wigderson [7] can be used to derive a locally computable extractor construction with $O(\log N)$ seed length and $\tilde{\Omega}(N^{1/5})$ output length. Zimand describes a construction with $O(\log N)$ seed length and $O(2^{\sqrt{\log N}})$ output length.

Keywords: Extractors, Direct product theorems, Hardness amplification.

1 Introduction

Randomness extractors, defined by Nisan and Zuckerman [25,13] are a fundamental primitive with several applications in pseudorandomness and derandomization. A function $Ext : \{0,1\}^N \times \{0,1\}^t \to \{0,1\}^m$ is a (K, ϵ)-extractor if, for every random variable X of min-entropy at least K, the distribution $Ext(X, U_t)$ has statistical distance at most ϵ from the uniform distribution over $\{0,1\}^m$.[1]

---

[*] Supported by the "Berkeley Fellowship for Graduate Study".

[**] This material is based upon work supported by the National Science Foundation under grant No. CCF-0729137 and by the US-Israel BSF grant 2006060.

[1] We use U_n to denote the uniform distribution over $\{0,1\}^n$, and recall that a distribution X is said to have min-entropy at least K if for every a we have $\mathbb{P}[X = a] \leq 2^{-K}$. Two random variables Y, Z ranging over the same universe $\{0,1\}^m$ have distance at most ϵ in statistical distance if for every statistical test $T : \{0,1\}^m \to \{0,1\}$ we have

$$|\mathbb{P}[T(Y) = 1] - \mathbb{P}[T(Z) = 1]| \leq \epsilon$$

Besides their original applications to extract randomness from weak random sources and as primitives inside pseudorandom generators for space bounded computation, extractors have found several other applications. As surveyed in [12,16] extractors are related to hashing and error-correcting codes, and have applications to pseudorandomness and hardness of approximation.

Extractors have also found several applications in cryptography, for example in unconditionally secure cryptographic constructions in the bounded-storage model [10,1,9]. For such applications, it is particularly desirable to have *locally computable* extractors, in which a bit of the output can be computed by only looking at the seed and at $poly \log n$ bits of the input. (The weaker notion of *online* extractors [2], however, is sufficient.)

The starting point of our paper is Zimand's [24] simple construction of a locally computable extractor based on the Blum-Micali-Yao pseudorandom generator, and his analysis via the reconstruction approach of [20]. The extractor is neither optimal in terms of the output length nor the seed length. For e.g., both Lu [9] and Vadhan [21] achieve an optimal seed length of $\Theta(\log n)$ for inverse polynomial error while extracting almost all the entropy of the source. In fact, [21] does better than [9] by extracting all but an arbitrarily small constant factor of the min-entropy while the latter has to lose an arbitrarily small polynomial factor. However, both these constructions are complicated in the sense that while Vadhan uses tools like samplers and extractors [15,26] from pseudorandomness machinery, Lu uses the extractor from [20] along with error-correcting codes based on expander graphs. In contrast, the extractor construction in Zimand [24] is extremely simple, only the analysis is non-trivial.

The idea of the reconstruction approach to the analysis of extractors is the following. Suppose we want to prove that $Ext : \{0,1\}^N \times \{0,1\}^t \to \{0,1\}^m$ is a (K, ϵ) extractor. Then, towards a contradiction, we suppose there is a test $T : \{0,1\}^m \to \{0,1\}$ and a random variable X of min entropy at least K such that

$$|\mathbb{P}[T(Ext(X, U_t)) = 1] - \mathbb{P}[T(U_m) = 1]| > \epsilon$$

In particular, there is a probability at least $\epsilon/2$ when sampling from X of selecting a bad x such that

$$|\mathbb{P}[T(Ext(x, U_t)) = 1] - \mathbb{P}[T(U_m) = 1]| > \frac{\epsilon}{2}$$

At this point, one uses properties of the construction to show that if x is bad as above, x can be *reconstructed* given T and a r-bit string of "advice." This means that there can be at most 2^r bad strings x, and if X has min-entropy K then the probability of sampling a bad x is at most $2^r/2^K$, which is a contradiction if $2^K > 2^{r+1}/\epsilon$.

In Zimand's extractor construction, one thinks of a sample from X as specifying a cyclic permutation $p : \{0,1\}^n \to \{0,1\}^n$ (where n is roughly $\log N$), then let $\bar{p}$ be a permutation obtained from p via a *hardness amplification* procedure, so that the ability to invert $\bar{p}$ on a small α fraction of inputs implies the ability of invert p on a large $1 - \delta$ fraction of inputs. Then the output of the extractor, for seed z, is $BMY(\overline{p}, z)$, the Blum-Micali-Yao generator applied to permutation $\overline{p}$ with seed z. If a test T distinguishes the output of the extractor from the uniform distribution, then there is an algorithm that, using T, can invert $\overline{p}$ on a noticeable fraction of inputs, and hence p on nearly all inputs. The proof is completed by presenting a counting argument showing an upper bound on the number of permutations that can be easily inverted on nearly all inputs.

Zimand's extractor uses a seed of length $O(\log^2 N)$ and, for a source of entropy γN, the output length is $N^{\gamma/3}$ bits.

We show that, by using only direct product theorems and XOR lemmas, we can improve the output length to roughly $N^{1/3}$. This is true both for Zimand's original construction[2], as well as for a streamlined version we describe below. The streamlined version is essentially the same construction as the locally computable extractor of Dziembowski and Maurer [4]. Our analysis via Levin's XOR lemma is rather different from the one in [4] which is based on information-theoretic arguments. It should be noted that using information theoretic arguments, Dziembowski and Maurer manage to get an output length of $N^{1-o(1)}$. However, at a conceptual level, we show that the same style of analysis can be used both for the extractor in [4] and [24][3].

Using the *derandomized* direct product theorem of Impagliazzo and Wigderson [7], we give a construction in which the seed length reduces to $O(\log N)$, but the output length reduces to $N^{1/5}$.

Our Constructions

Consider the following approach. View the sample from the weak random source as a boolean function $f : [N] \to \{0,1\}$, and suppose that the extractor simply outputs the sequence

$$f(x), f(x+1), \ldots, f(x+m-1)$$

where $x \in [N]$ is determined by the seed, and sums are computed $\mod N$. Then, by standard arguments, if T is a test that distinguishes the output of the extractor from the uniform distribution with distinguishing probability ϵ, then there is a predictor P, derived from T, and $i \leq m$ such that

$$\mathbb{P}[P(x, f(x-1), \ldots, f(x-i)) = f(x)] \geq \frac{1}{2} + \frac{\epsilon}{m} \tag{1}$$

Note that if the right-hand side of (1) were $1 - \delta$ for some small δ, instead of $1/2 + \epsilon/m$, then we could easily deduce that f can be described using about $m + \delta N + H(\delta) \cdot N$ bits (where $H()$ is the entropy function), and so we would be done.

---

[2] We actually do not show an improved analysis for this specific construction by Zimand but rather for the second construction in the same paper which achieves exactly the same parameters. Our improved analysis works equally well for both the constructions but is slightly notationally cumbersome for the first one.

[3] The fact that [4] gets a better output length suggests that neither the original analysis of [24] nor our improved analysis is tight.

To complete the argument, given the function $f : [N] \to \{0,1\}$ that we sample from the random source, we define the function $\overline{f} : [N]^k \to \{0,1\}$ as

$$\overline{f}(x_1, \ldots, x_k) := \bigoplus_{i=1}^{k} f(x_i)$$

where $k \approx \log N$, and our extractor outputs

$$\overline{f}(\overline{x}), \overline{f}(\overline{x}+1), \ldots, \overline{f}(\overline{x}+\mathbf{m}-1)$$

where $\overline{x} = (x_1, \ldots, x_k) \in [N]^k$ is selected by the seed of the extractor, $\mathbf{j}$ is the vector $(j, \ldots, j)$, and sums are coordinate-wise, and $\mathrm{mod} N$.

If T is a test that has distinguishing probability ϵ for our extractor, then there is a predictor P based on T such that

$$\mathbb{P}[P(\overline{x}, \overline{f}(\overline{x}-\mathbf{1}), \ldots, \overline{f}(\overline{x}-\mathbf{i})) = \overline{f}(\overline{x})] \geq \frac{1}{2} + \frac{\epsilon}{m} \tag{2}$$

from which we can use the proof of the XOR lemma to argue that, using P and some advice, we can construct a predictor P' such that

$$\mathbb{P}[P'(x, f(x-1), \ldots, f(x-i)) = f(x)] \geq 1 - \delta \tag{3}$$

and now we are done. Notice that we cannot use standard XOR lemmas as a black box in order to go from (2) to (3), because the standard theory deals with a predictor that is only given x, rather than $x, f(x-1), \ldots, f(x-i)$. The proofs, however, can easily be modified at the cost of extra non-uniformity. To adapt, for example, Levin's proof of the XOR Lemma, we see that, in order to predict $f(x)$, it is enough to evaluate P at $O(m^2/\epsilon^2)$ points $\overline{x}$, each of them containing x in a certain coordinate and fixed values everywhere else. For each such point, $F(\overline{x}-\mathbf{1}), \ldots, F(\overline{x}-\mathbf{i})$ can be specified using $i \cdot (k-1) \leq mk$ bits of advice. Overall, we need m^3k/ϵ^2 bits of advice, which is why we can only afford the output length m to be the cubed root of the entropy. The seed length is $k \log N$, which is $O(\log^2 N)$.

This type of analysis is robust to various changes to the construction. For example, we can view a sample from the weak random source as a function $f : \{0,1\}^n \to \{0,1\}^n$, define

$$\overline{f}(x_1, \ldots, x_k) := f(x_1), \ldots, f(x_k)$$

View the seed as specifying an input $\overline{x}$ for $\overline{f}()$ and a boolean vector r of the same length, and define the output of the extractor as

$$\langle \overline{f}(\overline{x}), r \rangle, \langle \overline{f}(\overline{x}+1), r \rangle, \cdots, \langle \overline{f}(\overline{x}+\mathbf{m}-1), r \rangle \tag{4}$$

Then using appropriate versions of Goldreich-Levin and of the direct product lemma of Impagliazzo et al. [6], we can show that the construction is an extractor provided that m is about $N^{1/3}$ [4]. Construction (4) is precisely the second construction by Zimand [24].

By applying the *derandomized* direct product theorem of Impagliazzo and Wigderson [7], we are able to reduce the seed length to $O(\log N)$, but our reconstruction step requires more non-uniformity, and so the output length of the resulting construction is only about $N^{1/5}$.

Organization of the Paper. In section 2, we present some notations which shall be used throughout the paper and an overview of the techniques recurrent in the proofs of all the three constructions. Section 3 presents the first of our constructions. Its proof of correctness is self contained. Improved analysis of the construction by Zimand [24] as well as the description and proof of the *derandomized* extractor are deferred to the full version of the paper.

2   Preliminaries and Overview of Proofs

Notations and Definitions

The following notations are used throughout the paper. A tuple $(y_1, y_2, \ldots, y_k)$ is denoted by $\otimes_{i=1}^{k} y_i$. The concatenation of two strings x and y is denoted by $x \circ y$. If x and y are tuples, then $x \circ y$ represents the bigger tuple formed by concatenating x and y. The uniform distribution on $\{0,1\}^n$ is denoted by U_n. For $z_1, \ldots, z_k \in \{0,1\}$, $\oplus_{i=1}^{k} z_i$ denotes the XOR of $z_1, \ldots, z_k$. Statistical distance between two distributions D_1 and D_2 is denoted by $||D_1 - D_2||$.

Next, we define extractors as well as a stronger variant called strong extractors.

Definition 1. *[15,25] $Ext : \{0,1\}^N \times \{0,1\}^t \to \{0,1\}^m$ is said to be a (K, ϵ) extractor if for every random variable X with min-entropy at least K, the statistical distance between output of the extractor and the uniform distribution is at most ϵ i.e. $||Ext(X, U_t) - U_m|| \le \epsilon$. Ext is said to be a strong extractor if the seed can be included with the output and the distribution still remains close to uniform i.e. $||U_t \circ Ext(X, U_t) - U_{t+m}|| \le \epsilon$. Here both the U_t refer to the same sampling of the uniform distribution.*

In the above definition, t is referred to as seed length, m as the output length and ϵ as the error of the extractor.

General Paradigm of Construction. All the three extractors can be described in the following general model. Let $Ext : \{0,1\}^N \times \{0,1\}^t \to \{0,1\}^m$ be the extractor (terminology is the same as Definition 1) with X representing the weak random source and $\bar{y}$ the seed. X is treated as truth table of a function $X : \{0,1\}^n \to \{0,1\}^l$ ($l = 1$ in the first and the third constructions and $l = n$ in

---

[4] Even using the 'concatenation lemma' of Goldreich et al. [5] which is a much more non-uniform version of the direct product theorem, we get $m = N^{\frac{1}{10}}$ for which is better than Zimand's analysis for entropy rates < 0.3.

the second construction). This implies that n is logarithmic in the input length N and more precisely $N = l2^n$. Further, we associate a cyclic group of size 2^n with $\{0,1\}^n$ (This can be any ordering of the elements in $\{0,1\}^n$ except that the addition in the group should be efficiently computable). To make it easier to remind us that X is treated as truth table of a function, the corresponding function shall henceforth be called f. The seed $\overline{y}$ is divided into two chunks i.e. $\overline{y} = \overline{x} \circ \overline{z}$. $\overline{x}$ is called the input chunk and $\overline{z}$ is called the encoding chunk. Also, let k be a parameter of the construction such that $|\overline{x}| = g(n,k)$ and $|\overline{z}| = h(n,k)$ and hence $t = g(n,k) + h(n,k)$. Ext is specified by two functions namely $Exp : \{0,1\}^{g(n,k)} \to (\{0,1\}^n)^k$ and $Com : (\{0,1\}^l)^k \times \{0,1\}^{h(n,k)} \to \{0,1\}$. Ext computes the output as follows

- On input $(X, \overline{y}) \equiv (f, \overline{x} \circ \overline{z})$, Ext first computes $Exp(\overline{x}) = (x_1, x_2, x_3, \ldots, x_k)$ which gives k candidate inputs for the function f.
- Subsequently, the ith bit of the output is computed by combining the evaluation of f at shifts of $(x_1, \ldots, x_k)$ using Com. More precisely, the ith bit is given by $Com(\otimes_{j=1}^k f(x_j + i - 1), \overline{z})$.

Our constructions differ from each other in the definition of the functions Exp and Com. It can be easily seen that as long as Exp and Com are efficiently computable i.e. both of them are computable in $poly(n,k)$ time and $k = O(n)$, the extractors shall be locally computable. This is true for all our constructions.

Proofs in the Reconstruction Paradigm. We now show the steps (following the reconstruction paradigm) which are used in the proof of correctness of all the constructions. We first note that proving $Ext : \{0,1\}^N \times \{0,1\}^t \to \{0,1\}^m$ is a $(\gamma N, 2\epsilon)$ strong extractor is equivalent to proving that for every boolean function $T : \{0,1\}^{m+t} \to \{0,1\}$ and random variable X of min-entropy at least γN

$$\left| Pr_{f \in X, \overline{y} \in U_t}[T(y, Ext(f, \overline{y})) = 1] - Pr_{u \in U_{t+m}}[T(u) = 1] \right| \le 2\epsilon \qquad (5)$$

We had earlier noted the following fact which we formally state below.

Observation 1. *In order to prove equation (5), it suffices to prove that for any* $T : \{0,1\}^{m+t} \to \{0,1\}$, *there are at most* $\epsilon 2^{\gamma N}$ *functions* f *such that*

$$\left| Pr_{\overline{y} \in U_t}[T(y, Ext(f, \overline{y})) = 1] - Pr_{u \in U_{t+m}}[T(u) = 1] \right| > \epsilon \qquad (6)$$

In order to bound the number of functions which satisfy (6), we use the reconstruction approach in [20][5] (and more generally used in the context of pseudorandom generators in [3,14]). In particular, we show that given any f which satisfies (6), we can get a circuit C_f (not necessarily small) which predicts value of f by querying f at some related points. More precisely, we show that for some

---

[5] This particular instance of reconstruction paradigm was used in context of extractors by Zimand [24] and earlier in context of pseudorandom generators by Blum, Micali and Yao [3,23].

$m > i \geq 0$, using c bits of advice, we can construct C_f which satisfies (7) for some $s \leq \frac{1}{2}$.

$$Pr_{x \in U_n}[C_f(x, \otimes_{j=1}^{i} f(x - j)) = f(x)] \geq 1 - s \qquad (7)$$

The next lemma shows how such a circuit C_f can be used to bound the number of functions f satisfying (6).

Lemma 1. *If for every f satisfying (6), using c bits of advice, we can get a circuit C_f satisfying (7) for some $s \leq \frac{1}{2}$, then there are at most $2^{c+2^n(sl+H(s))+ml}$ functions satisfying (6).*

Proof. Let the set BAD consist of points $x \in \{0,1\}^n$ such that $C_f(x, \otimes_{j=1}^{i} f(x - j)) \neq f(x)$. Since the size of the set BAD is at most $s2^n$, to fully specify the set, we require at most $\log_2 S$ bits where $S = \sum_{i=0}^{s2^n} \binom{2^n}{i}$. Further, to specify the value of f on the set BAD, we require at most $sl2^n$ bits. We now note that if we are given the value of f on any consecutive i points (say $[0, \ldots, i-1]$), which requires at most il bits, then using the circuit C_f, the set BAD and the value of f on points in BAD, one can fully specify f. We also use the following standard fact. (Log is taken base 2 unless mentioned otherwise)

Fact 2. *For $s \leq \frac{1}{2}, \sum_{i=0}^{s2^n} \binom{2^n}{i} \leq 2^{H(s)2^n}$ where $H(s) = -s \log s - (1-s) \log(1-s)$.*

Hence, we see that if we are given that f satisfies (6), then using T and $c + 2^n(s + H(s)) + il$ bits of advice, we can exactly specify f. Hence for any particular T, (using $i < m$) we get that there are at most $2^{c+2^n(sl+H(s))+ml}$ functions satisfying (6).

In light of lemma 1, given f satisfying (6), we should use T to construct a circuit C_f satisfying (7) with as minimum advice and as small s as possible. We first use the standard hybrid argument and Yao's distinguisher versus predictor argument to get a circuit which is a 'next-element' predictor. In particular, we create a circuit which predicts a particular position in the output of the extractor with some advantage over a random guess when given as input the value of the random seed as well as all the bits in the output preceding the bit to be predicted. The argument is by now standard and can be found in several places including [20,19,17]. We do not redo the argument here but simply state the final result.

Lemma 2. *Let f be any function satisfying (6) and $Ext(f, \overline{y})_i$ be the ith bit of the output. Then using $m + \log m + 3$ bits of advice, we can get a circuit T_2 such that for some $0 \leq i < m$, f satisfies (8).*

$$Pr_{\overline{y} \in U_t}[T_2(\overline{y}, \otimes_{j=1}^{m-i-1} Ext(f, \overline{y})_j) = Ext(f, \overline{y})_{m-i}] > \frac{1}{2} + \frac{\epsilon}{m} \qquad (8)$$

The proof of correctness of all our constructions start from the above equation and use more advice to finally get a circuit C_f satisfying (7). We now describe one of our constructions and its proof of correctness (Refer to the full version for the other two constructions).

3   Extractor from XOR Lemma

Description of the Construction. $Ext : \{0,1\}^{2^n} \times \{0,1\}^{kn} \to \{0,1\}^m$ is defined as follows. On input $(f, \bar{y})$, the seed $\bar{y}$ is partitioned into k chunks of length n - call it $(x_1, x_2, x_3, \ldots, x_k)$. The source f is treated as truth table of a function from $\{0,1\}^n$ to $\{0,1\}$. Then the ith bit of the output is given by the bitwise XOR of $f(x_1 + i - 1), \ldots, f(x_k + i - 1)$ i.e. $Ext(f, \bar{y})_i = \oplus_{i=1}^k f(x_j + i - 1)$. In terminology of the last section, $N = 2^n$, $g(k, n) = kn$ and $h(k, n) = 0$. Note that there is no encoding chunk in the seed and the entire seed is the input chunk. Further, the function Exp simply partitions a string of length kn into k chunks of length n while the function Com computes a bitwise XOR of its first input (the second input is the empty string).

Difference from Construction in [4]. As we have mentioned before, the construction in [4] is very similar though we have some minor simplifications. The extractor in [4] $Ext' : (\{0,1\}^{N+m-1})^k \times \{0,1\}^{k \log N} \to \{0,1\}^m$ can be described as follows. The weak source is treated as truth table of k functions $f_1, \ldots, f_k$ such that for each $j \in [k]$, $f_j : [N+m-1] \to \{0,1\}$. The seed is divided into k chunks $l_1, \ldots, l_k$ such that each l_j can be treated as an element in $[N]$. The ith bit of the output is computed as $\oplus_{j=1}^k f_j(l_j + i - 1)$. Thus, we avoid a minor complication of not having to divide the source into chunks. Our proof can be modified to work in this case as well at the cost of making it more cumbersome while conceptually remaining the same. However, the main difference is that we come up with an entirely different proof from the one in [4].

Main Theorem and Proof of Correctness

Theorem 3. *The function* $Ext : \{0,1\}^{2^n} \times \{0,1\}^{kn} \to \{0,1\}^m$ *is a* $(\gamma 2^n, 2\epsilon)$ *strong extractor for a (constant)* $\gamma > 0$, $\epsilon \geq 2^{-\frac{n}{7}}$, $m = \frac{\epsilon^{\frac{2}{3}} 2^{\frac{n}{3}}}{n^2}$ *and seed length* $kn = O\left(\frac{n \log \frac{m}{\epsilon}}{\gamma^2}\right)$.

Before proving Theorem 3, we see an immediate corollary of the above theorem with parameters of interest.

Corollary 1. *The function* Ext *as defined above is a* $(\gamma 2^n, 2\epsilon)$ *strong extractor for a (constant)* $\gamma > 0$, $2\epsilon = 2^{-n^{\frac{1}{4}}}$, $m = 2^{\frac{n}{3} - \sqrt{n}}$ *and seed length* $kn = O\left(\frac{n^2}{\gamma^2}\right)$.

In order to prove Theorem 3, we first state the following main technical lemma of this section and then see how Theorem 3 follows from it. Subsequently, we prove the lemma.

Lemma 3. *Let* $T : \{0,1\}^{m+kn} \to \{0,1\}$ *and* $f : \{0,1\}^n \to \{0,1\}$ *such that (6) holds. Also, let* $1 > \delta > 0$ *be such that* $\delta^k \leq \frac{\epsilon}{m}$ *and* $m \geq nk$. *Then with at most* $\frac{6nk^2 m^3}{\epsilon^2}$ *bits of advice, we can get a circuit* C_f *such that*

$$Pr_{x_1 \in U_n}[C_f(x_1, \otimes_{j=1}^i f(x_1 - j)) = f(x_1)] \geq \frac{1 + \delta}{2}$$

Before we formally prove Theorem 3 using Lemma 3, it is useful to mention that an application of δ is meaningful when it is close to 1 rather than 0. As can be seen from Lemma 3, we construct a circuit C_f which has correlation δ with f and hence we would like $1 - \delta$ to be small. This is different from the terminology used in Section 1 where we want to construct a circuit C_f which computes f with probability $1 - \delta$ and hence we would like δ to be close to 0.

Proof (of Theorem 3). In light of Observation 1, we note that it is sufficient to prove that for any statistical test $T : \{0,1\}^{m+kn} \to \{0,1\}$, the number of functions f satisfying (6) is at most $\epsilon 2^{\gamma N}$. Let δ be such that $\frac{1-\delta}{2} = \min\{10^{-3}, \frac{\gamma^2}{4}\}$. Also putting $k = \frac{C \log \frac{m}{\epsilon}}{\gamma^2} = O\left(\frac{n}{\gamma^2}\right)$ for some appropriate constant C clearly satisfies $\delta^k \leq \frac{\epsilon}{m}$. Further, $m = 2^{\Omega(n)}$ while $nk = O\left(\frac{n^2}{\gamma^2}\right)$. So, clearly $m \geq nk$ for constant γ and sufficiently large n. With this, we satisfy the conditions for applying lemma 3 and hence with $\frac{6nk^2m^3}{\epsilon^2}$ bits of advice, we can get a circuit C_f satsifying (7) with $s = \frac{1-\delta}{2}$. Using lemma 1, we can say that for any test T, the total number of functions satisfying (6) is at most $2^{\frac{6nk^2m^3}{\epsilon^2} + (\frac{1-\delta}{2} + H(\frac{1-\delta}{2}))2^n + m}$. We now use the following fact

Fact 4. *For any $0 \leq \alpha \leq 10^{-3}$, $\alpha + H(\alpha) \leq \sqrt{\alpha}$*

Putting everything together now, we get that the total number of functions satisfying (6) is at most (we consider the case when $\gamma > 0$ is a constant and n is large enough integer).

$$2^{\frac{6nk^2m^3}{\epsilon^2} + (\frac{1-\delta}{2} + H(\frac{1-\delta}{2}))2^n + m} \leq 2^{O(\frac{2^n}{n^3\gamma^4})}2^{\frac{\gamma}{2}2^n}2^{2^{\frac{n}{3}}} \leq 2^{-\frac{n}{7}}2^{\gamma 2^n} \leq \epsilon 2^{\gamma 2^n}$$

Proof (of Lemma 3). Using lemma 2, we get that for any f such that (6) holds, using $m + \log m + 3$ bits of advice, we can get a circuit T_2 such that

$$Pr[T_2(\overline{x}, \oplus_{j=1}^{k}f(x_j), \ldots, \oplus_{j=1}^{k}f(x_j+m-i-2)) = \oplus_{j=1}^{k}f(x_j+m-i-1)] > \frac{1}{2} + \frac{\epsilon}{m}$$

In the above, $x_1, x_2, \ldots, x_k$ are independent random variables drawn from U_n and $\overline{x}$ is the concatenation of $x_1, \ldots, x_k$. Unless otherwise stated, in this section, any variable picked randomly is picked from the uniform distribution (The domain shall be evident from the context). We now introduce some changes in the notation so as to make it more convenient. First of all, we note that $m - i - 1$ can be replaced by i as i runs from 0 to $m - 1$. Further, we can assume that the first k arguments in the input are changed from x_j to $x_j + i$ for all $1 \leq j \leq k$ and hence we get a circuit C such that

$$Pr[C(\overline{x}, \oplus_{j=1}^{k}f(x_j - i), \ldots, \oplus_{j=1}^{k}f(x_j - 1)) = \oplus_{j=1}^{k}f(x_j)] > \frac{1}{2} + \frac{\epsilon}{m}$$

In this proof, we closely follow the proof of XOR lemma due to Levin [8] as presented in [5]. As is done there, for convenience, we change the range of f

from $\{0,1\}$ to $\{-1,1\}$ i.e. $f(x)$ now changes to $(-1)^{f(x)}$. With this notational change, parity changes to product and prediction changes to correlation i.e.

$$\mathbb{E}[\prod_{j=1}^{k} f(x_j)C(\overline{x}, \prod_{j=1}^{k} f(x_j - i), \ldots, \prod_{j=1}^{k} f(x_j - 1))] > \frac{2\epsilon}{m}$$

In order to simplify the notation further, we make one more change. For any tuple $(x_1, x_2, \ldots, x_t) = \overline{x}$, $\prod_{j=1}^{t} f(x_j - s)$ is denoted by $\overline{f}(\overline{x} - s)$. Using the notation introduced earlier for denoting tuples, we get

$$\mathbb{E}_{\overline{x}}[\overline{f}(\overline{x})C(\overline{x}, \otimes_{j=1}^{i} \overline{f}(\overline{x} - j))] > \frac{2\epsilon}{m}$$

Let δ and η be such that $\delta^k \le \frac{\epsilon}{m}$ and $\eta = \frac{\epsilon}{km}$. Then the above equation can be rewritten as

$$\mathbb{E}_{\overline{x}}[\overline{f}(\overline{x})C(\overline{x}, \otimes_{j=1}^{i} \overline{f}(\overline{x} - j))] > \delta^k + k\eta \qquad (9)$$

Further, we can write $\overline{x}$ as $x_1 \circ \overline{y_1}$ where $x_1 \in \{0,1\}^n$ and $\overline{y_1} \in (\{0,1\}^n)^{k-1}$ and then the above can be rewritten as

$$\mathbb{E}_{x_1 \in U_n}[f(x_1)\Gamma(x_1, \otimes_{j=1}^{i} f(x_1 - j))] > \delta^k + k\eta \qquad (10)$$

where $\Gamma(x_1, \otimes_{j=1}^{i} f(x_1 - j)) = \mathbb{E}_{\overline{y_1} \in U_{(k-1)n}} \overline{f}(\overline{y_1})C(x_1 \circ \overline{y_1}, \otimes_{j=1}^{i} f(x_1 - j)\overline{f}(\overline{y_1} - j))$. At this stage, there are the following two possibilities.

1. $\forall x_1, |\Gamma(x_1, \otimes_{j=1}^{i} f(x_1 - j))| \le \delta^{k-1} + (k-1)\eta$.
2. $\exists x_1$ such that $|\Gamma(x_1, \otimes_{j=1}^{i} f(x_1 - j))| > \delta^{k-1} + (k-1)\eta$.

The following lemma shows how to construct the circuit in (7) in the first case. The second case follows by an inductive argument.

Lemma 4. *If for all x_1, $|\Gamma(x_1, \otimes_{j=1}^{i} f(x_1 - j))| \le \delta^{k-1} + (k-1)\eta$, then with $\frac{4nm}{\eta^2} + \log\left(\frac{4n}{\eta^2}\right) + 1$ bits of advice, we can get a circuit $C_f : \{0,1\}^n \times \{0,1\}^i \to \{-1,1\}$ such that*

$$\mathbb{E}_{x_1}[f(x_1)C_f(x_1, \otimes_{j=1}^{i} f(x_1 - j))] > \delta \qquad (11)$$

Proof. Let $\Gamma_1(x_1, \otimes_{j=1}^{i} f(x_1-j)) = \frac{\Gamma(x_1, \otimes_{j=1}^{i} f(x_1-j))}{\delta^{k-1}+(k-1)\eta} \in [-1,1]$. We note that (10) says that $\Gamma_1(x_1, \otimes_{j=1}^{i} f(x_1 - j))$ has high correlation with $f(x_1)$ and hence if we could compute Γ_1, then we could compute $f(x_1)$ with high probability . Since computing Γ_1 looks unlikely (without using 2^n bits of advice), we will approximate Γ_1 and still manage to compute f with high probability. In particular, we define a circuit C_1 such that for every x_1, C_1 approximates $\Gamma(x_1, \otimes_{j=1}^{i} f(x_1-j))$ within an additive error of η when given input x_1 and $\otimes_{j=1}^{i} f(x_1 - j)$. To do this, C_1 picks up $q = \frac{2n}{\eta^2}$ elements independently at random from $(\{0,1\}^n)^{(k-1)}$. Call these elements $\overline{w}_1, \ldots, \overline{w}_q$. C_1 then takes $\otimes_{j=0}^{i} \overline{f}(\overline{w}_l - j)$ for $l \in [q]$ as advice. Subsequently, it computes the function Γ_2 which is defined as follows. (Note that Γ_2 depends upon $\overline{w}_i$'s and the corresponding advice though $\overline{w}_i$'s are not explicitly included in the argument)

$$\Gamma_2(x_1, \otimes_{j=1}^i f(x_1 - j)) = \mathbb{E}_{l \in [q]} \overline{f}(\overline{w}_l) C(x_1 \circ \overline{w}_l, \otimes_{j=1}^i f(x_1 - j) \overline{f}(\overline{w}_l - j))$$

By Chernoff bound, we can say the following is true for all x_1. (The probability is over the random choices of $\overline{w}_l$ for $l \in [q]$)

$$Pr[|\Gamma_2(x_1, \otimes_{j=1}^i f(x_1 - j)) - \Gamma(x_1, \otimes_{j=1}^i f(x_1 - j))| > \eta] < 2^{-n}$$

We would like our estimate of $\Gamma(x_1, \otimes_{j=1}^i f(x_1 - j))$ to have absolute value bounded by $\delta^{k-1} + (k-1)\eta$. Hence, we define Γ_3 as follows.

1. If $|\Gamma_2(x_1, \otimes_{j=1}^i f(x_1 - j))| \le \delta^{k-1} + (k-1)\eta$ then Γ_3 is the same as Γ_2 i.e.
$\Gamma_3(x_1, \otimes_{j=1}^i f(x_1 - j)) = \Gamma_2(x_1, \otimes_{j=1}^i f(x_1 - j))$
2. If not, then Γ_3 has absolute value $\delta^{k-1} + (k-1)\eta$ with sign same as Γ_2 i.e.
$\Gamma_3(x_1, \otimes_{j=1}^i f(x_1 - j)) = \frac{|(\Gamma_2(x_1,\otimes_{j=1}^i f(x_1-j)))|}{(\Gamma_2(x_1,\otimes_{j=1}^i f(x_1-j)))}(\delta^{k-1} + (k-1)\eta)$

The final output of $C_1(x_1, \otimes_{j=1}^i f(x_1 - j))$ is $\Gamma_3(x_1, \otimes_{j=1}^i f(x_1 - j))$. Since Γ_3 is definitely at least as good a approximation of Γ as Γ_2 is, we can say the following (the probability is again over the random choices of $\overline{w}_l$ for $l \in [q]$ and as before $\overline{w}_l$ is not explicitly included in the argument).

$$Pr[|\Gamma_3(x_1, \otimes_{j=1}^i f(x_1 - j)) - \Gamma(x_1, \otimes_{j=1}^i f(x_1 - j))| > \eta] < 2^{-n}$$

By a simple union bound, we can see that there exists a q-tuple $\otimes_{l=1}^q \overline{w}_l$ is such that for all x_1, $|\Gamma_3(x_1, \otimes_{j=1}^i f(x_1 - j)) - \Gamma(x_1, \otimes_{j=1}^i f(x_1 - j))| \le \eta$. Hence with $qn(k-1) \le \frac{2n^2 k}{\eta^2}$ bits of advice, we can get such a tuple $\otimes_{l=1}^q \overline{w}_l$. Further, the advice required for getting $\otimes_{j=0}^i \overline{f}(\overline{w}_l - j)$ for each $l \in [q]$ is $(i+1)q \le \frac{2nm}{\eta^2}$ bits. So, we hardwire these 'good' values of $\overline{w}_l$ and $\otimes_{j=0}^i \overline{f}(\overline{w}_l - j)$ into C_1 (i.e. instead of taking random choices, it now works with these hardwired values) and we can say that

$$\mathbb{E}_{x_1}[f(x_1)C_1(x_1, \otimes_{j=1}^i f(x_1 - j))] \ge \mathbb{E}_{x_1}[f(x_1)\Gamma(x_1, \otimes_{j=1}^i f(x_1 - j))] - \eta \quad (12)$$

The above claim uses that the range of f is $[-1, 1]$. This can now be combined with (10) to give the following

$$\mathbb{E}_{x_1}[f(x_1)C_1(x_1, \otimes_{j=1}^i f(x_1 - j))] > \delta^k + (k-1)\eta \quad (13)$$

We now define $C_2(x_1, \otimes_{j=1}^i f(x_1 - j)) = \frac{C_1(x_1, \otimes_{j=1}^i f(x_1-j))}{\delta^{k-1}+(k-1)\eta}$. Note that the output of C_2 is in $[-1, 1]$ and hence by (13), we can say (using $\delta \le 1$)

$$\mathbb{E}_{x_1}[f(x_1)C_2(x_1, \otimes_{j=1}^i f(x_1 - j))] > \frac{\delta^k + (k-1)\eta}{\delta^{k-1} + (k-1)\eta} \ge \delta \quad (14)$$

C_2 is almost the circuit C_f we require except its output is in $[-1, 1]$ rather than $\{-1, 1\}$. To rectify this, we define a randomized circuit C_3 which computes $r = C_2(x_1, \otimes_{j=1}^i f(x_1 - j))$ and then outputs 1 with probability $\frac{1+r}{2}$ and -1 with probability $\frac{1-r}{2}$ otherwise. Clearly this randomized circuit C_3 has the same correlation with $f(x_1)$ as C_2 does. To fix the randomness of the circuit C_3 and to get C_f, we observe that the output of C_2 can only be in multiples of $\frac{\eta^2}{2n(\delta^{k-1}+(k-1)\eta)}$. Since the output is in the interval $[-1, 1]$, it suffices to pick a random string $\lceil \log \frac{4n(\delta^{k-1}+(k-1)\eta)}{\eta^2} \rceil$ bits long (rather than a random number in $[-1, 1]$). Hence by fixing this randomness using $\lceil \log \frac{4n}{\eta^2} \rceil \leq \log \frac{4n}{\eta^2} + 1$ bits of advice, we get a circuit C_f which satisfies $(11)^6$. Clearly, the total amount of advice required is at most $\frac{2n(m+nk)}{\eta^2} + \log\left(\frac{4n}{\eta^2}\right) + 1$ bits. Using $m \geq nk$, we get the bound on the advice stated in the lemma.

Hence, in the first case, we get a circuit C_f such that its expected correlation with f is greater than δ. Changing the $\{-1, 1\}$ notation to $\{0, 1\}$ notation, we get that

$$Pr_{x_1 \in U_n}[C_f(x_1, \otimes_{j=1}^i f(x_1 - j)) = f(x_1)] > \frac{1+\delta}{2}$$

Therefore, we have a circuit C_f satisfying the claim in the lemma. Now, we handle the second case. Let x_1 be such that $|\Gamma(x_1, \otimes_{j=1}^i f(x_1 - j))| > \delta^{k-1} + (k-1)\eta$. We take x_1, $\otimes_{j=1}^i f(x_1 - j)$ and the sign of $\Gamma(x_1, \otimes_{j=1}^i f(x_1 - j))$ (call it α) as advice (and this is at most $n + m$ bits) and define the circuit C^0 as follows.

$$C^0(\overline{y_1}, \otimes_{j=1}^i \overline{f}(\overline{y_1} - j)) = (-1)^\alpha C(x_1 \circ \overline{y_1}, \otimes_{j=1}^i f(x_1 - j)\overline{f}(\overline{y_1} - j))$$

By definition and the previous assumptions, we get the following

$$\mathbb{E}_{\overline{y_1} \in U_{(k-1)n}} \overline{f}(\overline{y_1})C^0(\overline{y_1}, \otimes_{j=1}^i \overline{f}(\overline{y_1} - j)) > \delta^{k-1} + (k-1)\eta$$

Note that the above equation is same as (10) except circuit C has been replaced by C^0 and the input has changed from a k-tuple in $\{0,1\}^n$ to a $k-1$-tuple. Hence, this can be handled in an inductive way and the induction can go for at most $k-1$ steps. Further, each descent step in the induction can require at most $n + m$ bits of advice. In the step where we apply Lemma 4, we require at most $\frac{4nm}{\eta^2} + \log\left(\frac{4n}{\eta^2}\right) + 1$ bits of advice^7. So, from T_2, with at most $(k-1)(m+n) + \frac{4nk^2m^3}{\epsilon^2} + \log\left(\frac{4nk^2m^2}{\epsilon^2}\right) + 1$ bits of advice, we can get a circuit $C_f : \{0,1\}^n \times \{0,1\}^i$ such that

$$Pr_{x_1 \in U_n}[C_f(x_1, \otimes_{j=1}^i f(x_1 - j)) = f(x_1)] \geq \frac{1+\delta}{2}$$

---

^{6} We remove the factor $\log(\delta^{k-1} + (k-1)\eta)$ in calculating the advice because $(\delta^{k-1} + (k-1)\eta)$ is at most 1 and hence what we are calculating is an upper bound on the advice.

^{7} Note that η does not change for every step and is the same $\eta = \frac{\epsilon}{km}$ that it was set to in the beginning. The only extra condition we need for applying Lemma 4 is that $m \geq kn$ which shall definitely continue to hold as k decreases.

Finally accounting for the advice to use Lemma 2, we get that the total amount of advice required to get C_f from the circuit T in the hypothesis is $(k-1)(m+n) + \frac{4nk^2m^3}{\epsilon^2} + \log\left(\frac{4nk^2m^2}{\epsilon^2}\right) + 2 + m + \log m + 3 \leq \frac{6nk^2m^3}{\epsilon^2}$.

4 Conclusion

All the three extractor constructions described in this paper apply to sources of constant entropy rate, which could be pushed to entropy about $N/poly(\log N)$. A result of Viola [22] implies that it is impossible to extract from sources of entropy $N^{.99}$ if the extractor is such that each bit of the output can be computed by looking only at $N^{o(1)}$ bits of the input and seed length is $N^{o(1)}$. Since our construction is such that every bit of the output can be computed by looking at only $poly \log N$ bits of the input, significant improvements in the entropy rate can only come from rather different constructions.

It remains an interesting open question to improve the output length, and match the performance of other constructions which do not use complexity-theoretic tools in the analysis. Perhaps it is possible to use advice in a much more efficient way than we do.

Acknowledgments. The authors would like to thank Madhur Tulsiani for his useful comments on an earlier draft. The first author would like to also thank him for some very enjoyable discussions throughout the course of the work.

References

1. Aumann, Y., Ding, Y.Z., Rabin, M.O.: Everlasting security in the bounded storage model. IEEE Transactions on Information Theory 48(6), 1668–1680 (2002)
2. Bar-Yossef, Z., Reingold, O., Shaltiel, R., Trevisan, L.: Streaming computation of combinatorial objects. In: Proceedings of the 17th IEEE Conference on Computational Complexity, pp. 165–174 (2002)
3. Blum, M., Micali, S.: How to generate cryptographically strong sequences of pseudorandom bits. SIAM Journal on Computing 13(4), 850–864 (1984); Preliminary version in Proc. of FOCS 1982
4. Dziembowski, S., Maurer, U.: Optimal randomizer efficiency in the bounded-storage model. Journal of Cryptology 17(1), 5–26 (2004); Preliminary version in Proc. of STOC 2002
5. Goldreich, O., Nisan, N., Wigderson, A.: On Yao's XOR lemma. Technical Report TR95-50, Electronic Colloquium on Computational Complexity (1995)
6. Impagliazzo, R., Jaiswal, R., Kabanets, V., Wigderson, A.: Direct product theorems: Simplified, Optimized and Derandomized. In: Proceedings of the 40th ACM Symposium on Theory of Computing, pp. 579–588 (2008)
7. Impagliazzo, R., Wigderson, A.: $P = BPP$ unless E has sub-exponential circuits. In: Proceedings of the 29th ACM Symposium on Theory of Computing, pp. 220–229 (1997)
8. Levin, L.: One-way functions and pseudorandom generators. Combinatorica 7(4), 357–363 (1987)

9. Lu, C.-J.: Encryption against storage-bounded adversaries from on-line strong extractors. Journal of Cryptology 17(1), 27–42 (2004)
10. Maurer, U.M.: Conditionally-perfect secrecy and a provably-secure randomized cipher. J. Cryptology 5(1), 53–66 (1992)
11. Nisan, N.: Extracting randomness: How and why. In: Proceedings of the 11th IEEE Conference on Computational Complexity, pp. 44–58 (1996)
12. Nisan, N., Ta-Shma, A.: Extrating randomness: A survey and new constructions. Journal of Computer and System Sciences (1998) (to appear); Preliminary versions in [11, 18]
13. Nisan, N., Zuckerman, D.: More deterministic simulation in Logspace. In: Proceedings of the 25th ACM Symposium on Theory of Computing, pp. 235–244 (1993)
14. Nisan, N., Wigderson, A.: Hardness vs randomness. Journal of Computer and System Sciences 49, 149–167 (1994); Preliminary version in Proc. of FOCS 1988
15. Nisan, N., Zuckerman, D.: Randomness is linear in space. Journal of Computer and System Sciences 52(1), 43–52 (1996); Preliminary version in Proc. of STOC 1993
16. Shaltiel, R.: Recent developments in extractors. Bulletin of the European Association for Theoretical Computer Science 77, 67–95 (2002)
17. Shaltiel, R., Umans, C.: Simple extractors for all min-entropies and a new pseudorandom generator. Journal of the ACM 52(2), 172–216 (2005)
18. Ta-Shma, A.: On extracting randomness from weak random sources. In: Proceedings of the 28th ACM Symposium on Theory of Computing, pp. 276–285 (1996)
19. Ta-Shma, A., Zuckerman, D., Safra, S.: Extractors from Reed-Muller codes. Technical Report TR01-036, Electronic Colloquium on Computational Complexity (2001)
20. Trevisan, L.: Extractors and pseudorandom generators. Journal of the ACM 48(4), 860–879 (2001)
21. Vadhan, S.P.: Constructing locally computable extractors and cryptosystems in the bounded-storage model. Journal of Cryptology 17(1), 43–77 (2004)
22. Viola, E.: The complexity of constructing pseudorandom generators from hard functions. Computational Complexity 13(3-4), 147–188 (2004)
23. Yao, A.C.: Theory and applications of trapdoor functions. In: Proceedings of the 23th IEEE Symposium on Foundations of Computer Science, pp. 80–91 (1982)
24. Zimand, M.: Simple Extractors via Constructions of Cryptographic Pseudo-random Generators. In: Caires, L., Italiano, G.F., Monteiro, L., Palamidessi, C., Yung, M. (eds.) ICALP 2005. LNCS, vol. 3580, pp. 115–127. Springer, Heidelberg (2005)
25. Zuckerman, D.: General weak random sources. In: Proceedings of the 31st IEEE Symposium on Foundations of Computer Science, pp. 534–543 (1990)
26. Zuckerman, D.: Simulating BPP using a general weak random source. Algorithmica 16(4/5), 367–391 (1996)

How Well Do Random Walks Parallelize?

Klim Efremenko and Omer Reingold*

Department of Computer Science and Applied Mathematics,
The Weizmann Institute of Science, Rehovot, 76100 Israel
klimefrem@gmail.com, omer.reingold@weizmann.ac.il

Abstract. A random walk on a graph is a process that explores the graph in a random way: at each step the walk is at a vertex of the graph, and at each step it moves to a uniformly selected neighbor of this vertex. Random walks are extremely useful in computer science and in other fields. A very natural problem that was recently raised by Alon, Avin, Koucky, Kozma, Lotker, and Tuttle (though it was implicit in several previous papers) is to analyze the behavior of k independent walks in comparison with the behavior of a single walk. In particular, Alon et al. showed that in various settings (e.g., for expander graphs), k random walks cover the graph (i.e., visit all its nodes), $\Omega(k)$-times faster (in expectation) than a single walk. In other words, in such cases k random walks efficiently "parallelize" a single random walk. Alon et al. also demonstrated that, depending on the specific setting, this "speedup" can vary from logarithmic to exponential in k.

In this paper we initiate a more systematic study of multiple random walks. We give lower and upper bounds both on the cover time *and on the hitting time* (the time it takes to hit one specific node) of multiple random walks. Our study revolves over three alternatives for the starting vertices of the random walks: the worst starting vertices (those who maximize the hitting/cover time), the best starting vertices, and starting vertices selected from the stationary distribution. Among our results, we show that the speedup when starting the walks at the worst vertices cannot be too large - the hitting time cannot improve by more than an $O(k)$ factor and the cover time cannot improve by more than $\min\{k \log n, k^2\}$ (where n is the number of vertices). These results should be contrasted with the fact that there was no previously known upper-bound on the speedup and that the speedup can even be *exponential* in k for random starting vertices. Some of these results were independently obtained by Elsässer and Sauerwald (ICALP 2009). We further show that for k that is not too large (as a function of various parameters of the graph), the speedup in cover time is $O(k)$ *even for walks that start from the best vertices* (those that minimize the cover time). As a rather surprising corollary of our theorems, we obtain a new bound which relates the cover time C and the mixing time mix of a graph. Specifically, we show that $C = O(m\sqrt{\text{mix}} \log^2 n)$ (where m is the number of edges).

Keywords: Markov Chains, Random Walks.

---

* Research supported by grant 1300/05 from the Israel Science Foundation.

I. Dinur et al. (Eds.): APPROX and RANDOM 2009, LNCS 5687, pp. 476–489, 2009.

1 Introduction

A random walk on a graph is a process of exploring the graph in a random way. A simple random walk starts at some node of a graph and at each step moves to a random neighbor. Random walks are fundamental in computer science. They are the basis of MCMC (Markov-Chain Monte-Carlo) algorithms, and have additional important applications such as randomness-efficient sampling (via random walks on expanders) [AKS87], and space-efficient graph connectivity algorithms [AKL+79]. Random walks became a common notion in many fields, such as computational physics, computational biology, economics, electrical engineering, social networks, and machine learning.

Assume that we have some network (e.g. a communication or a social network), and some node u sends a message. Assume that at each step this message is sent to a random neighbor of the last recipient. The message will travel through the network as a random walk on a graph. The expected time until the message will arrive to some other node v is called the hitting time $h(u, v)$. The expected time until the message will visit all the nodes is called the cover time C_u. The hitting time and the cover time of a random walk are thoroughly studied parameters (see surveys [AF99, LWP, Lov96]).

In this paper we consider the following natural question: What happens if we take multiple random walks instead of a single walk? Assume that instead of one copy, k copies of the same message were sent. How long would it take for one of these copies to reach some node v? How long would it take until each node receives at least one of the k copies? What are the speedups in the hitting and cover times of multiple walks compared with a single walk?

Multiple random walks were studied in a series of papers [BKRU89, Fei97, BF93] on time-space tradeoffs for solving undirected s-t connectivity. These papers considered upper bounds for the cover time of multiple random walks, each paper giving a different answer for different distributions of the starting vertices of the random walks. In randomized parallel algorithms, multiple random walks are a very natural way of exploring a graph since they can be easily distributed between different processes. For example, multiple random walks were used in [HZ96, KNP99] for designing efficient parallel algorithms for finding the connected components of an undirected graph.

Multiple random walks were suggested as a topic of independent interest by Alon, Avin, Koucky, Kozma, Lotker, and Tuttle [AAK+07]. Alon et al. [AAK+07] studied lower bounds on the relation between the cover time of a simple random walk and of multiple random walks when the walks start from the same node. The paper proves that if the number of random walks k is small enough (i.e., asymptotically less than $\frac{C}{h_{\max}}$, where C and $h_{\max}$ are the maximal cover time and hitting time respectively) then the relation between the cover time of a single random walk and of multiple random walks is at least $k - o(k)$. In such a case, we can argue that multiple random walks "parallelize" a single walk efficiently (as they don't increase the total amount of work by much). [AAK+07] also showed that there are graphs with logarithmic speedup (e.g., the cycle), and there are graphs with an exponential speedup for specific starting point (e.g., the so called barbell graph; we will shortly discuss a related example). [AAK+07] leaves open the question of upper bounds for the speedup.

The goal of this paper is to systematically study multiple random walks. In addition to the cover time of multiple random walks we will also discuss the hitting time, proving both lower and upper bounds on the speedup. We will extend the discussion to the case where not all the walks start from the same node.

Before getting into the details of our results, let us consider an example which illustrates how multiple random walks behave differently according to the choice of their starting vertices. Consider a graph G which is composed of two cliques of size n connected by a single edge (see Figure 1).

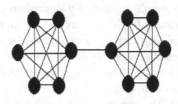

Fig. 1. Two cliques graph - how the speedup changes according to the starting vertices

While the cover time of a single random walk will not depend on the starting vertex and is $\Theta(n^2)$, the cover time of multiple random walks will be very different for different starting vertices of the random walks. When the walks start from the worst vertices (all walks start from the same clique) the cover time is $\Theta(\frac{n^2}{k})$. Even for $k = 2$, when the random walks start from the best vertices (one walk starts at one clique and the other from another clique) the cover time is $\Theta(n \log n)$. When the starting vertices of k random walks are drawn independently from the stationary distribution, then the probability that all starting vertices will fall into the same clique is 2^{-k}. Therefore, for $k \leq \log n - \log \log n$, the cover time in this case is $\Theta(2^{-k}n^2)$. When considering the hitting times, we get the same behavior for the worst starting vertices and for randomly-chosen starting vertices. The case of the best starting vertices is uninteresting when discussing the hitting time as the hitting time in such a case is zero (even for a single walk).

As we can see from the aforementioned example, both the cover and the hitting times heavily depend on the starting vertices. Therefore, we study these three scenarios separately: (1) The case when the random walks start from the nodes which maximize the cover/hitting time (worst starting vertices). (2) The case when the random walks start from the nodes which minimize the cover time (best starting vertices). (3) The case when the starting vertices are drawn independently according to the stationary distribution (random starting vertices).

Our Contribution

In this paper we systematically study multiple random walks and their speedup both in terms of the cover time and in terms of the hitting time. We give various lower and upper bounds for different ways of choosing the starting vertices. Our main bounds on the speedup of multiple random walks are summarized in Table 1.

Table 1. Summary of main bounds on the speedup. Notation: n - number of vertices; k - the number of walks; C - maximal cover time; $h_{\max}$ - maximal hitting time; mix - mixing time.

	Worst case	Average Case	Best Case
Hitting time Upper bounds	$O(k)$ *for any k*, Theorem 4	$k + o(k)$ *for $k \log n = o(\frac{h_{\max}}{\text{mix}})$* Theorem 20	Not applicable
Hitting time Lower bounds	$\Omega(k)$ *for $k \log n = O(\frac{h_{\max}}{\text{mix}})$* Theorem 8	k *for any k* Theorems 6	Not applicable
Cover time Upper bounds	$O(\min\{k^2, k \log n\})$ Theorems 12 & 13	$k + o(k)$ *for $k \log k = o(\frac{C}{\text{mix}})$* Theorem 19	$k + o(k)$ *for $k = o(\frac{C}{h_{\max}})$* Theorem 15
Cover time Lower Bounds	$(\frac{k}{\log n})(1 - o(1))$ for $k \log n = o(\frac{h_{\max}}{\text{mix}})$ Theorem 14 $k - o(k)$ for $k = o(\frac{C}{h_{\max}})$ Theorem 5 in [AAK$^+$07]	$\Longrightarrow$	$\Longrightarrow$

Upper bounds on the speedup. [AAK$^+$07] left open the question of upper bounding the speedup of multiple random walks. In this work we show that the answer depends on how the starting vertices are selected. In Theorem 4 we show that for the worst starting vertices, the speedup on *hitting time* is at most $O(k)$. In Section 4, we use this theorem to show that the speedup on the *cover time* is at most $O(\min(k^2, k \log n))$. As we can see from the example above, the speedup for the best or even for random starting vertices may be very large (e.g., exponential in k). Still, we are able to show in Section 4 that even in these cases, if the number of walks is small enough then the speedup will be at most $k + o(k)$. In Theorem 15 (see also Corollary 17) we show that for $k \ll \frac{C}{h_{\max}}$ the speedup for the best starting vertices is at most $k + o(k)$. This result is interesting for graphs with a large gap between the cover time and the hitting time. For random starting vertices, Theorem 19 (see also Corollary 21) shows that if $k \log k \ll \frac{C}{\text{mix}}$, then the speedup is at most $k + o(k)$. The mixing time, mix, of a graph is the number of steps a random walk has to make until its position is distributed almost according to the stationary distribution.

Lower bounds on the speedup. In Theorem 6 we show that the speedup for the hitting times is at least k when all the starting vertices are drawn from the stationary distribution. This theorem also allows us to prove lower bounds for the case of worst starting vertices for graphs with small mixing time. Using this theorem we prove in Theorem 8 that when the number of walks is less than $\tilde{O}(\frac{h_{\max}}{\text{mix}})$ the speedup for the hitting times is at least $\Omega(k)$. We get similar results for the cover time (Theorem 14). Namely, we show that the speedup for the cover time is at least $(\frac{k}{\log n})(1 + o(1))$, when k is less than $\tilde{o}(\frac{h_{\max}}{\text{mix}})$. This result improves the lower bound of $\Omega(\frac{k}{\log n \cdot \text{mix}})$ from [AAK$^+$07].

A new relation between the cover time and the mixing time. Finally, our study of multiple random walks gives a rather surprising implication on the study of a single random walk. Our results, together with the results of [BKRU89] about multiple random walks, imply a new relation between the cover time and the mixing time of a graph. Specifically, we prove that $C = O(m\sqrt{\text{mix}}\log^2 n)$. The best previous result we are aware of is due to Broder and Karlin [BK88]. In [BK88] it was proven that $C = O(\frac{m\log n}{1-\lambda(G)})$, where $\lambda(G)$ is the second eigenvalue of the normalized adjacency matrix. A known relation between $\lambda(G)$ and mix is that $\Omega(\frac{1}{1-\lambda(G)}) \leq \text{mix} \leq O(\frac{\log n}{1-\lambda(G)})$ (cf. [Sin92], Proposition 1). Therefore a corollary of [BK88] is that $C = O(\text{mix}m\log n)$. Our result improves this bound whenever mix $= \omega(\log^2 n)$.

Our new relation also has an application in electrical engineering. View a graph G as an electrical network with unit resistors as edges. Let R_{st} be the effective resistance between nodes s and t. Then it was shown in [CRRS89] that for any two nodes s and t it holds that $mR_{st} \leq C$. Therefore, together with our result it implies that $R_{st} = O(\sqrt{\text{mix}}\log^2 n)$. The best previous upper bound on the electrical resistance in terms of the mixing time was also obtained by Chandra et al. [CRRS89] and was $R_{st} \leq \frac{2}{1-\lambda(G)} = O(\text{mix})$.

Related Work. Independently of our work, Elsässer and Sauerwald [ES09] recently studied multiple random walks. Their most related results are upper bounds and lower bounds on the speed-up of cover time for worst case starting points. In fact, [ES09] gives an upper bound of $O(k\log n)$ on the speed-up of any graph (similarly to our Theorem 12) and a lower bound of $\Omega(\frac{k}{\log n})$ under some conditions on mixing time (similarly to our Theorem 14). Under some mild conditions, they are also able to prove an upper bound of $O(k)$. Another recent work on multiple random walks is due to [CCR09]. This work studies multiple random walks in random graphs, and among other result show that for random d-regular graph the speed-up is $O(k)$.

2   Notation

We will use standart definitions of the hitting time, the cover time and the mixing time. We briefly review the notation that will be used throughout the paper: The mixing time of a graph G is denoted mix. Let $\varsigma(u, v)$ be the time it takes for a random walk that starts at u to reach v i.e. $\varsigma(u, v) = \min\{t \mid X_u(t) = v\}$. Note that $\varsigma(u, v)$ is a random variable. Let the hitting time $h(u, v) = \mathbf{E}(\varsigma(u, v))$ be the expected time for the random walk to traverse from u to v. Let $h_{\max} = \max_{u,v \in V} h(u, v)$ and $h_{\min} = \min_{u,v \in V} h(u, v)$ be the maximal and minimal hitting times. Similarly let τ_u be the time for the simple random walk to visit all the nodes of the graph. Let $C_u = \mathbf{E}(\tau_u)$ be the cover time for a simple walk starting at u. The cover time $C = \max_u(C_u)$ is the maximal (over the starting vertex u) expected time it takes for a single walk to cover the graph. It will be convenient for us to define the following parameter of a graph: $H(G) = \frac{C}{h_{\max}}$.

The following theorem provides fundamental bounds on the cover time in terms of the hitting time (for more details see [LWP] Chapter 11 or [Mat88]):

Theorem 1   (cf. [Mat88]). *For every graph G with n vertices*

$$h_{min} \cdot \log n \leq C \leq h_{max} \cdot \log n.$$

Note that there also exists a trivial bound of $h_{max} \leq C$. It will be convenient for us to define the following parameter of a graph: $H(G) = \frac{C}{h_{max}}$. Note that $1 \leq H(G) \leq \log n$. Also note that there exist graphs where $H(G) = O(1)$ (for example the cycle), and there exist graphs with $H(G) = \Omega(\log n)$ (for example the complete graph).

For k parallel independent random walks we have the following notation: $\varsigma(\{u_1, u_2, \ldots u_k\}, v) = \min_{i=1}^{k} \varsigma(u_i, v)$ is the random variable corresponding to the hitting time of k random walks, where some of the u_i's may be equal. Let $h(\{u_1, u_2, \ldots u_k\}, v) = \mathbf{E}(\varsigma(\{u_1, u_2, \ldots u_k\}, v))$ be the hitting time of k random walks starting at vertices u_i. If all the walks start at the same vertex u we will write it as $h^k(u, v)$. Let $h_{\max}^k = \max_{u_i, v} h(\{u_1, u_2, \ldots u_k\}, v)$ be the maximal hitting time of k random walks. Similarly, for the cover time we define $\tau_{u_1, u_2, \ldots u_k} = \min\{t \mid \bigcup_{i=1}^{k}\{X_{u_i}(1), X_{u_i}(2), \ldots X_{u_i}(t)\} = V\}$ and define $C_{u_1, u_2, \ldots u_k} = \mathbf{E}\tau_{u_1, u_2, \ldots u_k}$ to be the expected cover time. Let $C^k = \max_{u_1, u_2, \ldots u_k} C_{u_1, u_2, \ldots u_k}$.

The proof of Theorem 1 (see [LWP] Chapter 11) easily extends to multiple walks implying the following theorem:

Theorem 2. *For every (strongly connected) graph G with n vertices, and for every k*

$$\frac{C_k}{h_{\max}^k} \leq \log n.$$

3 Hitting Time of Multiple Random Walks

In this section we study the behavior of the hitting time of k random walks. The first question we will consider is: what are the starting vertices of multiple random walks which maximize the hitting time? Later, we will give a lower bound on the maximal hitting time of multiple random walks. We will prove that $\frac{h_{\max}}{h_{\max}^k} = O(k)$. Then we will consider the case where the walks' starting vertices are chosen independently according to the stationary distribution. Note that in this setting the ratio between hitting times is not upper bounded by $O(k)$; in fact it may even be exponential in k. We will prove that in this setting the ratio between the hitting time of the single walk and the hitting time of k walks is at least k. Next we will use this theorem in order to prove that for graphs with small mixing time the ratio $\frac{h_{\max}}{h_{\max}^k} = \Omega(k)$. Finally we consider the evaluation of hitting times.

3.1 Worst to Start in a Single Vertex

Let us prove that the maximal hitting time is achieved when all the walks start from the same node.

Theorem 3. *For every graph $G = (V, E)$, for every $v \in V$ it holds that*

$$\max_{u_1, u_2, \ldots u_k} h(\{u_1, u_2, \ldots u_k\}, v) = \max_{u} h^k(u, v).$$

The proof of the theorem (which employs a generalization of Hölder's Inequality) is deferred to the full version.

3.2   Upper Bound on the Speedup of the Hitting Time of Multiple Random Walks

We will now prove that the ratio between the hitting time of a single random walk and the hitting time of k random walks is at most $O(k)$.

Theorem 4. *For any graph G it holds that* $h_{\max} \leq 4k h_{\max}^k$.

Loosely, the theorem is proved by deducing a bound of $\frac{1}{2k}$ on the probability that a single walk will hit the target vertex in $2h_{\max}^k$ steps. The formal proof is deferred to the full version. By a slightly more complicated argument we can replace the constant 4 in Theorem 4 by $e + o(1)$. However it seems plausible that the right constant is 1.

Open Problem 5. *Prove or disprove that for any graph G it holds that* $h_{\max} \leq k h_{\max}^k$.

3.3   Lower Bounds on the Speedup of the Hitting Time of Multiple Random Walks

In this section, we consider the case where the starting vertices of the random walks are selected according to the stationary distribution. Theorem 4 shows that for worst-case starting vertices the ratio between the hitting times of a single walk and multiple walks is at most $O(k)$. But as we will soon show, when the starting vertices of all walks are drawn independently from the stationary distribution then, loosely speaking, this ratio becomes at least k. Note that in some graphs the ratio of hitting times, when the starting vertices are selected according to the stationary distribution, may even become exponential in k. Indeed, such an example is given in Figure 1 and is discussed in the introduction (the discussion there is for the cover time but the analysis for the hitting time is very similar)

The next theorem gives a lower bound on the ratio between hitting times for random starting vertices.

Theorem 6. *Let $G(V, E)$ be a (connected) undirected graph. Let X be a random walk on G. Let $u, u_1, \ldots u_k \in V$ be independently chosen according to the stationary distribution of G. Then:*

$$E_u(h(u, v)) \geq k \left(E_{u_i} h(\{u_1, u_2, \ldots u_k\}, v) - 1 \right).$$

Remark 7. *In this theorem we assume continues model of random walk.*

As we will later see (in Corollary 22), when $k \log k = o(h(u, v)/\text{mix})$ then the speedup is at most $k + o(k)$ in the scenario of random starting vertices. Thus when $k \log k = o(h(u, v)/\text{mix})$ the speedup is k up to lower order terms.

The proof of the theorem is deferred to the full version.

Lower bound on the speedup for worst starting vertices. The lower bound on the speedup for walks starting at the stationary distribution translates into a lower bound that also applies to walks starting at the worst vertices: First let the walks converge to the stationary distribution and then apply the previous lower bound. The bounds that we obtain are especially meaningful when the mixing time of the graph is sufficiently smaller than the hitting time.

Theorem 8. *Let $G(V, E)$ be a (connected) undirected graph. Then*

$$h_{\max}^k \leq \frac{h_{\max}}{k} + O(\text{mix}(\log n + \log k)).$$

As a corollary we get:

Corollary 9. *Let $G(V, E)$ be a (connected) undirected graph such that $k\text{mix}(\log n + \log k) = o(h_{\max})$. Then:*

$$\frac{h_{\max}}{h_{\max}^k} \geq k(1 - o(1)).$$

3.4 Calculating the Hitting Time of Multiple Random Walks

We would like to address a question which is somewhat orthogonal to the main part of this paper. Namely, we would like to discuss how the hitting time of multiple walks can be calculated. Let us observe that multiple random walks on graph G can be presented as a single random walk on another graph G^k.

Definition 10. *Let $G = (V, E)$ be some graph. Then the graph $G^k = (V', E')$ is defined as follows: The vertices of G^k are k-tuples of vertices of G i.e.*

$$V' = \underbrace{V \oplus V \ldots \oplus V}_{k \text{ times}} = V^k.$$

For every k edges of G, (u_i, v_i) for $i = 1, \ldots, k$ we have an edge between $u' = (u_1, u_2, \ldots u_k)$ and $v' = (v_1, v_2 \ldots v_k)$ in G^k.

One can view k random walks on G as a single random walk on G^k where the first coordinate of G^k corresponds to the first random walk, the second coordinate corresponds to the second random walk, and so on.

Let $A \subset V^k$ be the set of all nodes of G^k which contain the node $v \in V$. Assume that we have k random walks beginning at $u_1, u_2, \ldots u_k$. Then the time it will take to hit v is equal to the time for a single random walk on G^k beginning at node $(u_1, u_2, \ldots u_k)$ to hit the set A. Thus instead of analyzing multiple random walks we can study a single random walk on G^k. There is a polynomial time algorithm for calculating hitting times of a single random walk (cf. [Lov96]). This gives us an algorithm, which is polynomial in n^k, for calculating $h(\{u_1, u_2, \ldots u_k\}, v)$. A natural question is whether there exist more efficient algorithms.

Open Problem 11. *Find a more efficient algorithm for calculating $h(\{u_1, u_2, \ldots u_k\}, v)$.*

4 Cover Time of Multiple Random Walks

Let us turn our attention from the hitting time to the cover time. As in the case of the hitting time, the cover time heavily depends on the starting vertices of the random walks. The graph given by Figure 1 and discussed in the introduction gives an example where the speedup in cover time of k random walks is linear in k for worst-case starting vertices, it is exponential in k for random starting vertices, and even for $k = 2$ it is $\Omega(n/\log n)$ for the best starting vertices.

Theorem 1 gives a relation between hitting times and cover times. Thus, our results on hitting times from the previous section also give us results on the cover times. In Subsection 4.1 we will give these results and will analyze the speedup, $\frac{C}{C_k}$, for worst starting vertices. We show that it is bounded by $\min\{k^2, k\log n\}$ for any k. We will also show that for k such that $k\log n\mathrm{mix} = O(h_{\max})$ the speedup is $\Omega(\frac{k}{\log n})$.

We will show in Subsection 4.2 that when k random walks begin from the best starting vertices for $k = o(H(G))$ the speedup is roughly k and is therefore essentially equal to the speedup for the worst case. In Subsection 4.3 we will show that when the starting vertices are drawn from the stationary distribution for k such that $\mathrm{mix}k\log(k) = o(C)$, the speedup is at most k.

4.1   The Worst Starting Vertices

As a simple corollary of Theorem 4 we obtain the following relation:

Theorem 12. *The speedup $\frac{C}{C^k}$ is at most $4kH(G) \le 4k\log n$*

Proof. Recall that $C^k \ge h_{\max}^k$ so $\frac{C}{C^k} \le \frac{C}{h_{\max}^k} = \frac{h_{\max}}{h_{\max}^k}H(G)$. From Theorem 4 it follows that $\frac{h_{\max}}{h_{\max}^k}H(G) \le 4kH(G)$. And finally from Theorem 1 we have that $4kH(G) \le 4k\log n$.

From this theorem it follows that for $k = \Omega(H(G))$ the speedup is $O(k^2)$. Theorem 15 implies that if $k < 0.01H(G)$ then the speedup $\frac{C}{C^k}$ is at most $2k$. Therefore, we can conclude a bound for every k:

Theorem 13. *For every (strongly connected) graph G and every k, it holds that $\frac{C}{C^k} = O(k^2)$.*

From Theorem 8 we can also deduce a lower bound on the speedup for rapidly-mixing graphs:

Theorem 14. *Let $G(V, E)$ be an undirected graph and let k be such that $k(\log n)\mathrm{mix} = o(h_{\max})$ then*

$$\frac{C}{C_k} \ge \frac{k}{\log n}(1 - o(1)).$$

Proof. From Theorem 2 it follows that $\frac{C}{C_k} \ge \frac{h_{\max}}{h_{\max}^k\log n}$. Since $k\log n\mathrm{mix} = o(h_{\max})$, Theorem 8 implies that $\frac{h_{\max}}{h_{\max}^k} = k(1 - o(1))$. Thus: $\frac{C}{C_k} \ge \frac{k}{\log n}(1 + o(1))$.

4.2   The Best Starting Vertices

As we discussed earlier, multiple random walks can be dramatically more efficient than a single random walk if their starting vertices are the best nodes (rather than the worst nodes). In fact, we have seen an example where taking two walks instead of one reduces the cover time by a factor of $\Omega(n/\log n)$. In this section we show that in graphs where the cover time is significantly larger than the hitting time, a few random walks cannot give such a dramatic speedup in the cover time, *even when starting at the best nodes*: If $k = o(H(G))$ (recall that $H(G) = \frac{C}{h_{max}}$), then the speedup $\frac{C}{C_{u_1,u_2,...u_k}}$ (where $u_1, u_2, ... u_k$ are best possible) is not much bigger than k. Note that in the case where $k = o(H(G))$ it has been shown in [AAK+07] that the speedup $\frac{C}{C_{u_1,u_2,...u_k}}$ is at least $k - o(k)$, even if $u_1, u_2, ... u_k$'s are worst possible. Combining the two results we get that the speedup is roughly k regardless of where the k walks start.

We want to show that the cover time of a single random walk is not much larger than k times the cover time of k random walks. For that we will let the single walk simulate k random walks (starting from vertices u_i) as follows: The single walk runs until it hits u_1, then it simulates the first random walk. Then it runs until it hits u_2 and simulates the second random walk and so on until hitting u_k and simulating the k'th random walk. The expected time to hit any vertex from any other vertex is bounded by h_{max}. Thus intuitively the above argument should imply the following bound: $C \leq kC_{u_1,u_2,...u_k} + kh_{max}$. Unfortunately, we do not know how to formally prove such a strong bound. The difficulty is that the above argument only shows how a single walk can simulate k walks for t steps, where t is fixed ahead of time. However, what we really need is for the single walk to simulate k walks until the walks cover the graph. In other words, t is not fixed ahead of time but rather a random variable which depends on the k walks. Nevertheless, we are still able to prove the following bound which is weaker by at most a constant factor:

Theorem 15. *For every graph G and for **any** k nodes $u_1, u_2, ... u_k$ in G, it holds that:*

$$C \leq kC_{u_1,u_2,...u_k} + O(kh_{max}) + O\left(\sqrt{kC_{u_1,u_2,...u_k} h_{max}}\right).$$

The proof will appear in the full version.

In [AAK+07] the following theorem was proved:

Theorem 16 (Theorem 5 from [AAK+07]). *Let G be a strongly connected graph and $k = o(H(G))$ then $\frac{C}{C_k} \geq k - o(k)$.*

In the case where $k = o(H(G))$ then $O(kh_{max}) + O(\sqrt{C_{u_1,u_2,...u_k} kh_{max}}) = o(C)$ and therefore $C \leq kC_{u_1,u_2,...u_k} + o(C)$. As a corollary we get:

Corollary 17. *Let G be a strongly connected graph and $k = o(H(G))$ then for any starting vertices $u_1, u_2, ... u_k$ it holds that: $\frac{C}{C_{u_1,u_2,...u_k}} = k \pm o(k)$*

It seems plausible that the speedup is at most k for *any* starting vertices, also when k is significantly larger than $H(G)$. When $k \geq e^{H(G)}$ we can give an example where $kC_{u_1,u_2,...u_k} << C$. Consider a graph G which is composed of a clique of size n and t vertices where each vertex is connected by one edge to some node of a clique. We will assume that $n >> t$. The maximal hitting time for this graph is $O(n^2)$. The cover time of this graph is $O(n^2 \log t)$ and $H(G) = \log t$. If $k = t$ then when k multiple random walks start from the t vertices which are not in the clique, then $C_{u_1,u_2,...u_k} = \frac{n \log n}{k} + O(1)$. Therefore, a natural open problem is the following:

Open Problem 18. *Prove or disprove that for some constant $\alpha > 0$, for any graph G, if $k \leq e^{\alpha H(G)}$ then $C \leq O(k)C_{u_1,u_2,...u_k}$.*

4.3 Random Starting Vertices

Finally we consider the cover time of k walks that start from vertices drawn from the stationary distribution. In this case, Theorem 6 loosely states that the ratio between the *hitting times* is at least k. Now let us show an upper bound on the ratio between the *cover time* of a single random walk and multiple random walks.

The intuition for the bound is quite similar to the intuition behind the proof of Theorem 15 (nevertheless, the proofs are quite a bit different). We will simulate k random walks by a single walk. The single random walk will first run $\ln(k)$mix steps, getting to a vertex that is distributed almost according to the stationary distribution. The walk then simulates the first of the k random walks. Next, the walk takes $\ln(k)$mix steps again and simulates the second random walk and so on until simulating the kth random walk. Since the start vertex of the k simulated walks are *jointly* distributed almost as if they were independently sampled from the stationary distribution it seems that we should obtain the following upper bound: $C \leq k\mathbf{E}_{u_i}C_{u_1,u_2,\ldots u_k} + k\ln(k)$mix, where $u_1, u_2, \ldots u_k$ are independently drawn from the stationary distribution. But as before we can not make this intuition formal, mainly because we do not know ahead of time how long the k random walks will take until they cover the graph. We will instead prove the following bound which again may be weaker by at most a constant factor:

Theorem 19. *Let $G = (V, E)$ be any (strongly connected) graph. Let $u_1, u_2, \ldots u_k$ be drawn from the stationary distribution of G. Then:*

$$C \leq k\mathbf{E}_{u_i}C_{u_1,u_2,\ldots u_k} + O(k\ln(k)\text{mix}) + O\left(k\sqrt{EC_{u_1,u_2,\ldots u_k}\text{mix}}\right).$$

Under some restrictions, the mixing time cannot be much larger than the maximal hitting time and often will be much smaller. In such cases, Theorem 19 may be more informative than Theorem 15 in the sense that it implies a bound of roughly k on the speedup as long as $k = \tilde{O}(\frac{C}{\text{mix}})$ (rather than $k = O(\frac{C}{h_{\text{max}}})$ as implied by Theorem 15). On the other hand, the starting vertices in Theorem 19 are according to the stationary distribution rather than arbitrary starting vertices as in Theorem 15.

The proof of Theorem 19 will appear in the full version. We note that the proof also works if we consider the hitting times (rather than the cover times), implying the following theorem:

Theorem 20. *Let $G = (V, E)$ be any (strongly connected) graph. Let u, v be any nodes of the graph and let $u_1, u_2, \ldots u_k$ be drawn from the stationary distribution of G. Then:*

$$h(u, v) \leq k\mathbf{E}_{u_i}h(\{u_1, u_2, \ldots u_k\}, v) + O(k\ln(k)\text{mix}) + O\left(k\sqrt{\mathbf{E}_{u_i}h(\{u_1, u_2, \ldots u_k\}, v)\text{mix}}\right).$$

As a corollary of Theorems 19 it follows that if $k\log k$mix is negligible relative to the cover time then the speedup of the cover time is at most k

Corollary 21. *Let $G = (V, E)$ be any (strongly connected) graph. Let $u_1, u_2, \ldots u_k$ be drawn from the stationary distribution of G. Then if $k\log(k) = o(C/\text{mix})$ then*

$$\frac{C}{\mathbf{E}_{u_i}C_{u_1,u_2,\ldots u_k}} \leq k + o(k).$$

Similarly, from Theorem 20 we obtain the following corollary:

Corollary 22. *Let* $G = (V, E)$ *be any (strongly connected) graph. Let* $u_1, u_2, \ldots u_k$ *be drawn from the stationary distribution of G and u, v any nodes. Then if $k \log(k) = o(h(u, v)/\mathrm{mix})$ then*

$$\frac{h(u, v)}{\boldsymbol{E}_{u_i} h(\{u_1, u_2, \ldots u_k\}, v)} \le k + o(k).$$

5 A New Relation between Cover and Mixing Time

In this section we will show how we can use the results proven above in order to prove a new upper bound on the cover time in terms of mixing time. In order to do this we will need the following bound from [BKRU89].

Theorem 23 (cf. [BKRU89] Theorem 1). *Let G be a connected undirected graph with n vertices and m edges. Let $u_1, u_2, \ldots u_k$ be drawn from the stationary distribution of G. Then:*

$$\boldsymbol{E}_{u_i}(C_{u_1, u_2, \ldots u_k}) \le O(\frac{m^2 \log^3 n}{k^2}).$$

As a rather intriguing corollary of Theorem 23 and Theorem 19 we get the following bound on the cover time.

Theorem 24. *Let G be a connected undirected graph with n vertices and m edges. Then:*

$$C \le O(m\sqrt{\mathrm{mix}} \log^2 n).$$

Proof. From Theorem 19 it follows that:

$$C(G) \le k\boldsymbol{E}_{u_i} C_{u_1, u_2, \ldots u_k}(G) + O(k \ln(k)\mathrm{mix}) + O\left(k\sqrt{\boldsymbol{E}C_{u_1, u_2, \ldots u_k}\mathrm{mix}}\right).$$

Thus from Theorem 23 we get the following bound on $C(G)$:

$$C(G) \le O(\frac{m^2 \log^3 n}{k}) + O(k \ln(k)\mathrm{mix}) + O(m \log^{1.5} n\sqrt{\mathrm{mix}}).$$

As long as k is at most polynomial in n it follows that $\log k = O(\log n)$. Thus:

$$C(G) \le O(\frac{m^2 \log^3 n}{k}) + O(k \ln(n)\mathrm{mix}) + o(m \log^2 n\sqrt{\mathrm{mix}}).$$

Setting $k = \frac{m \log n}{\sqrt{\mathrm{mix}}}$ implies the theorem.

6   Future Research

This paper systematically studies the behavior of multiple random walks. While we have given various upper and lower bounds for the speedup of multiple random walks, there is still much more that we do not know on this topic, with a few examples being Open Problems 5, 11 and 18. In this section, we will discuss a few additional directions for further research.

Our knowledge on the hitting time of multiple random walks is more complete than our knowledge on their cover time. Indeed, analyzing the hitting time seems easier than analyzing the cover time. Designing new tools for analyzing the cover time of multiple random walks is an important challenge. For example, we have proved that the maximal hitting time of multiple random walks is obtained when all the walks start from the same vertex (see Theorem 4), but we don't know if the same is also true for the cover times:

Open Problem 25. *Prove or disprove that for any graph G*

$$\max_{u_1,u_2,\ldots u_k} C^k_{u_1,u_2,\ldots u_k} = \max_u C^k_{u,u,\ldots u}.$$

We have proved that in the case of worst starting vertices the speedup of the hitting time is at most $4k$, and we raised the question of whether the correct constant is one (see Open Problem 5). It seems however, that for the cover time the speedup may be larger than k (though it is still possible that it is $O(k)$). Consider a walk on a "weighted" path $a - b - c$ with self loops such that the probability of staying in place is $1 - \frac{1}{x}$. In other words, consider a Markov chain $X(t)$ with the following transition probabilities:

$$\Pr[X(t) = b|X(t-1) = a] = \Pr[X(t) = b|X(t-1) = c] = \frac{1}{x}$$
$$\Pr[X(t) = c|X(t-1) = b] = \Pr[X(t) = a|X(t-1) = b] = \frac{1}{2x}$$

Calculating the cover times gives the following: The worst starting vertex of a single random walk is b and the cover time is $5x+o(x)$. The worst starting vertices of 2 random walks is when both walks start at a and the cover time in such a case is $2.25x + o(x)$. Thus, in this case the speedup for 2 walks is 2.222. It is an interesting question to find stronger examples (where the speedup is larger than k), and of course it would be interesting to find a matching upper bound on the speedup.

A technical issue that comes up in our analysis is that in order to understand the behavior of multiple random walks it may be helpful to understand the behavior of short random walks. For example, what kind of bound can be obtained on $\Pr[\varsigma(u, v) \geq h_{\max}/2]$ (for an undirected and connected graph).

Finally, it will be interesting to explore additional applications of multiple random walks, either in computer science or in other fields.

Acknowledgements

We thank Chen Avin, Uriel Feige, Gady Kozma, Zvi Lotker and Ofer Zeitouni for useful discussions. In particular, we are grateful to Gady for pointing out a flaw in one of our earlier arguments. We thank Rivka Efremenko for reviewing drafts of this paper.

References

[AAK+07]   Alon, N., Avin, C., Koucky, M., Kozma, G., Lotker, Z., Tuttle, M.R.: Many Random Walks Are Faster Than One. ArXiv e-prints 705 (May 2007), http://arxiv.org/abs/0705.0467

[AF99]   Aldous, D., Fill, J.: Reversible Markov Chains and Random Walks on Graphs (1999), http://www.stat.berkeley.edu/aldous/RWG/book.html

[AKL+79]   Aleliunas, R., Karp, R.M., Lipton, R.J., Lovász, L., Rackoff, C.: Random walks, universal traversal sequences, and the complexity of maze problems. In: 20th Annual Symposium on Foundations of Computer Science (FOCS), San Juan, Puerto Rico, October 29-31, 1979, pp. 218–223. IEEE Computer Society Press, Los Alamitos (1979)

[AKS87]   Ajtai, M., Komlós, J., Szemerédi, E.: Deterministic simulation in LOGSPACE. In: Proceedings of the Nineteenth Annual ACM Symposium on Theory of Computing (STOC), New York City, May 25-27, 1987, pp. 132–140 (1987)

[BF93]   Barnes, G., Feige, U.: Short random walks on graphs. In: STOC, pp. 728–737 (1993)

[BK88]   Broder, A.Z., Karlin, A.R.: Bounds on the cover time. Symposium on Foundations of Computer Science, 479–487 (1988)

[BKRU89]   Broder, A.Z., Karlin, A.R., Raghavan, P., Upfal, E.: Trading space for time in undirected s-t connectivity. In: STOC, pp. 543–549 (1989)

[CCR09]   Frieze, A., Cooper, C., Radzik, T.: Multiple random walks in random regular graphs. In: ICALP (2009)

[CRRS89]   Chandra, A.K., Raghavan, P., Ruzzo, W.L., Smolensky, R.: The electrical resistance of a graph captures its commute and cover times. In: STOC 1989: Proceedings of the twenty-first annual ACM symposium on Theory of computing, pp. 574–586. ACM, New York (1989)

[ES09]   Elsässer, R., Sauerwald, T.: Tight bounds for the cover time of multiple random walks. In: ICALP (2009)

[Fei97]   Feige, U.: A spectrum of time-space trade-offs for undirected s-t connectivity. J. Comput. Syst. Sci. 54(2), 305–316 (1997)

[HZ96]   Halperin, S., Zwick, U.: An optimal randomised logarithmic time connectivity algorithm for the erew pram. J. Comput. Syst. Sci. 53(3), 395–416 (1996)

[KNP99]   Karger, D.R., Nisan, N., Parnas, M.: Fast connected components algorithms for the EREW PRAM. SIAM J. Comput. 28(3), 1021–1034 (1999)

[Lov96]   Lovász, L.: Random walks on graphs: A survey. Combinatorics, Paul Erdos is Eighty 2, 353–398 (1996)

[LWP]   Levin, D.A., Wilmer, E., Peres, Y.: Markov Chains and Mixing Times. Oxford University Press, Oxford

[Mat88]   Matthews, P.: Covering problems for markov chains. The Annals of Probability 16(3), 1215–1228 (1988)

[Sin92]   Sinclair, A.: Improved bounds for mixing rates of markov chains and multicommodity flow. Combinatorics, Probability and Computing 1, 351–370 (1992)

An Analysis of Random-Walk Cuckoo Hashing

Alan Frieze[1,*], Páll Melsted[1], and Michael Mitzenmacher[2,**]

[1] Department of Mathematical Sciences
Carnegie Mellon University
Pittsburgh PA 15213
U.S.A.
[2] School of Engineering and Applied Sciences
Harvard University
Cambridge MA 02138

Abstract. In this paper, we provide a polylogarithmic bound that holds with high probability on the insertion time for cuckoo hashing under the random-walk insertion method. Cuckoo hashing provides a useful methodology for building practical, high-performance hash tables. The essential idea of cuckoo hashing is to combine the power of schemes that allow multiple hash locations for an item with the power to dynamically change the location of an item among its possible locations. Previous work on the case where the number of choices is larger than two has required a breadth-first search analysis, which is both inefficient in practice and currently has only a polynomial high probability upper bound on the insertion time. Here we significantly advance the state of the art by proving a polylogarithmic bound on the more efficient random-walk method, where items repeatedly kick out random blocking items until a free location for an item is found.

Keywords: Cuckoo Hashing, Random Walk Algorithm.

1 Introduction

Cuckoo hashing [12] provides a useful methodology for building practical, high-performance hash tables by combining the power of schemes that allow multiple hash locations for an item (e.g., [1,2,3,13]) with the power to dynamically change the location of an item among its possible locations. Briefly (more detail is given in Section 2), each of n items x has d possible locations $h_1(x), h_2(x), \ldots, h_d(x)$, where d is typically a small constant and the h_i are hash functions, typically assumed to behave as independent fully random hash functions. (See [11] for some justification of this assumption.) We assume each location can hold only one item. When an item x is inserted into the table, it can be placed immediately if one of its d locations is currently empty. If not, one of the items in its d locations must be

---

* Supported in part by NSF Grant CCF-0502793.
** Supported in part by NSF Grant CNS-0721491 and grants from Cisco Systems Inc., Yahoo!, and Google.

I. Dinur et al. (Eds.): APPROX and RANDOM 2009, LNCS 5687, pp. 490–503, 2009.

displaced and moved to another of its d choices to make room for x. This item in turn may need to displace another item out of one its d locations. Inserting an item may require a sequence of moves, each maintaining the invariant that each item remains in one of its d potential locations, until no further evictions are needed. Further variations of cuckoo hashing, including possible implementation designs, are considered in for example [5], [6], [7], [8], [9].

It is often helpful to place cuckoo hashing in a graph theoretic setting, with each item corresponding to a node on one side of a bipartite graph, each bucket corresponding to a node on the other side of a bipartite graph, and an edge between an item x and a bucket b if b is one of the d buckets where x can be placed. In this case, an assignment of items to buckets forms a matching and a sequence of moves that allows a new item to be placed corresponds to a type of augmenting path in this graph. We call this the cuckoo graph (and define it more formally in Section 2).

The case of $d = 2$ choices is notably different than for other values of d. When $d = 2$, after the first choice of an item to kick out has been made, there are no further choices as one walks through the cuckoo graph to find an augmenting path. Alternatively, in this case one can think of the cuckoo graph in an alternative form, where the only nodes are buckets and items correspond to edges between the buckets, each item connecting the two buckets corresponding to it. Because of these special features of the $d = 2$ case, its analysis appears much simpler, and the theory for the case where there are $d = 2$ bucket choices for each item is well understood at this point [4,10,12].

The case where $d > 2$ remains less well understood, although values of d larger than 2 rate to be important for practical applications. The key question is if when inserting a new item x all $d > 2$ buckets for x are already full, what should one do? A natural approach in practice is to pick one of the d buckets randomly, replace the item y at that bucket with x, and then try to place y in one of its other $d - 1$ bucket choices [6]. If all of the buckets for y are full, choose one of the other $d - 1$ buckets (other than the one that now contains x, to avoid the obvious cycle) randomly, replace the item there with y, and continue in the same fashion. At each step (after the first), place the item if possible, and if not randomly exchange the item with one of $d - 1$ choices. We refer to this as the random-walk insertion method for cuckoo hashing.

There is a clear intuition for how this random walk on the buckets should perform. If a fraction f of the items are adjacent to at least one empty bucket in the corresponding graph, then we might expect that each time we place one item and consider another, we should have approximately a probability f of choosing an item adjacent to an empty bucket. With this intuition, assuming the load of the hash table is some constant less than 1, the time to place an item would be at most $O(\log n)$ with high probability[1].

Unfortunately, it is not clear that this intuition should hold true; the intuition assumes independence among steps when the assumption is not necessarily

---

[1] An event $\mathcal{E}_n$ occurs *with high probability* if $\mathsf{P}(\mathcal{E}_n) = 1 - O(1/n^\alpha)$ for some constant $\alpha > 0$, see also discussion on page 494.

warranted. Bad substructures might arise where a walk could be trapped for a large number of steps before an empty bucket is found. Indeed, analyzing the random-walk approach has remained open, and is arguably the most significant open question for cuckoo hashing today.

Because the random-walk approach has escaped analysis, thus far the best analysis for the case of $d > 2$ is due to Fotakis et al. [6], and their algorithm uses a breadth-first search approach. Essentially, if the d choices for the initial item x are filled, one considers the other choices of the d items in those buckets, and if all those buckets are filled, one considers the other choices of the items in those buckets, and so on. They prove a constant expected time bound for an insertion for a suitably sized table and constant number of choices, but to obtain a high probability bound under their analysis requires potentially expanding a logarithmic number of levels in the breadth-first search, yielding only a polynomial bound on the time to find an empty bucket with high probability. It was believed this should be avoidable by analyzing the random-walk insertion method. Further, in practice, the breadth-first search would not be the choice for most implementations because of its increased complexity and memory needs over the random-walk approach.

In this paper, we demonstrate that, with high probability, for sufficiently large d the cuckoo graph has certain structural properties that yield that on the insertion of any item, the time required by the random-walk insertion method is polylogarithmic in n. The required properties and the intuition behind them are given in subsequent sections. Besides providing an analysis for the random-walk insertion method, our result can be seen as an improvement over [6] in that the bound holds for every possible starting point for the insertion (with high probability). The breadth-first search of [6] gives constant expected time, implying polylogarithmic time with probability $1 - o(1)$. However when inserting $\Omega(n)$ element into the hash table, the breadth-first search algorithm cannot guarantee a sub-polynomial running time for the insertion of each element. This renders the breadth-first search algorithm unsuitable for many applications that rely on guarantees for individual insertions and not just expected or amortized time complexities.

While the results of [6] provide a starting point for our work, we require further deconstruction of the cuckoo graph to obtain our bound on the performance of the random-walk approach.

Simulations in [6] (using the random-walk insertion scheme), indicate that constant expected insertion time is possible. While our guarantees do not match the running time observed in simulations, they give the first clear step forward on this problem for some time.

2 Definitions and Results

We begin with the relevant definitions, followed by a statement of and explanation of our main result.

Let $h_1, \ldots h_d$ be independent fully random hash functions $h_i : [n] \to [m]$ where $m = (1 + \varepsilon)n$. The necessary number of choices d will depend on ϵ, which gives the amount of extra space in the table. We let the *cuckoo graph* G be a bipartite graph with a vertex set $L \cup R$ and an edge set $\bigcup_{x \in L} \{(x, h_1(x)), \ldots (x, h_d(x))\}$, where $L = [n]$ and $R = [m]$. We refer to the left set L of the bipartite graph as *items* and the right set R as *buckets*.

An assignment of the items to the buckets is a left-perfect matching M of G such that every item $x \in L$ is incident to a matching edge. The vertices $F \subseteq R$ not incident to the matching M are called *free* vertices. For a vertex v the *distance* to a free vertex is the shortest M-alternating path from v to a free vertex.

We present the algorithm for insertion as Algorithm 1 below. The algorithm augments the current matching M with an augmenting path P. An item is assigned to a free neighbor if one exists; otherwise, a random neighbor is chosen to displace from its bucket, and this is repeated until an augmenting path is found. In practice, one generally sets an upper bound on the number of moves allowed to the algorithm, and a failure occurs if there remains an unassigned item after that number of moves. Such failure can be handled by additional means, such as stashes [8].

Algorithm 1. Insert-node

1: **procedure** INSERT-NODE(G,M,u)
2:     $P \leftarrow ()$
3:     $v \leftarrow u$
4:     $i \leftarrow d + 1$
5:     **loop**
6:         **if** $h_j(v)$ is not covered by M for some $j \in \{1, \ldots, d\}$ **then**
7:             $P \leftarrow P \oplus (v, h_j(v))$
8:             **return** Augment(M,P)
9:         **else**
10:            Let $j \in_R \{1, \ldots, d\} \setminus \{i\}$ and w be such that $(h_j(v), w) \in M$
11:            $P \leftarrow P \oplus (v, h_j(v)) \oplus (h_j(v), w)$
12:            $v \leftarrow w$
13:            $i \leftarrow j$
14:        **end if**
15:    **end loop**
16: **end procedure**

We note that our analysis that follows also holds when the table experiences deletions. This is because our result is based on the structure of the underlying graph G, and not on the history that led to the specific current matching. The statement of the main result is that given that G satisfies certain conditions, which it will with high probability, the insertion time is polylogarithmic with high probability. It is important to note that we have two distinct probability spaces, one for the hash functions which induce the graph G, and another for the randomness employed by the algorithm. For the probability space of hash function, we say that an event $\mathcal{E}_n$ occurs *with high probability* if $\mathsf{P}(\mathcal{E}_n) = 1 - O(n^{-2d})$.

For the probability space of randomness used by the algorithm we use the regular definition of *with high probability*.

Theorem 1. *Conditioned on an event of probability $1 - O(n^{4-2d})$ regarding the structure of the cuckoo graph G, the expected time for insertion into a cuckoo hash-table using Algorithm 1 is $O\left(\log^{1+\gamma_0+2\gamma_1} n\right)$ where $\gamma_0 = \frac{d+\log d}{(d-1)\log(d/3)}$ and $\gamma_1 = \frac{d+\log d}{(d-1)\log(d-1)}$, assuming $d \geq 8$ and if $\varepsilon \leq \frac{1}{6}$, $d \geq 4+2\varepsilon - 2(1+\varepsilon)\log\left(\frac{\varepsilon}{1+\varepsilon}\right)$. Furthermore, the insertion time is $O\left(\log^{2+\gamma_0+2\gamma_1} n\right)$ with high probability.*

The algorithm will fail if the graph does not have a left-perfect matching, which happens with probability $O(n^{4-2d})$ [6]. We show that all necessary structural properties of G hold with probability $1 - O(n^{-2d})$, so that the probability of failure is dominated by the probability that G has no left-perfect matching.

At a high level, our argument breaks down into a series of steps. First, we show that the cuckoo graph expands suitably so that most vertices are within $O(\log\log n)$ distance from a free vertex. Calling the free vertices F and this set of vertices near to the free vertices S, we note that if reach a vertex in S, then the probability of reaching F from there over the next $O(\log\log n)$ steps in the random walk process is inverse polylogarithmic in n, so we have a reasonable chance of getting to a free vertex and finishing. We next show that if the cuckoo graph has an expansion property, then from any starting vertex, we are likely to reach a vertex of S though the random walk in only $O(\log n)$ steps. This second part is the key insight into this result; instead of trying to follow the intuition to reach a free vertex in $O(\log n)$ steps, we aim for the simpler goal of reaching a vertex S close to F and then complete the argument.

As a byproduct of our Lemma 2 (below), we get an improved bound on the expected running time for the breadth-first variation on cuckoo hashing from [6], $\left(\frac{1}{\varepsilon}\right)^{O(1)}$ instead of $\left(\frac{1}{\varepsilon}\right)^{O(\log d)}$.

Theorem 2. *The breadth-first search insertion procedure given in [6] runs in*
$$O\left(\max\left\{d^4\left(\frac{1}{6\varepsilon}\right)^{\frac{1}{1-\frac{\log 6}{\log d}}}, d^5\right\}\right)$$
expected time, provided $d \geq 8$ and if $\varepsilon \leq \frac{1}{6}$ if $d \geq 4+2\varepsilon - 2(1+\varepsilon)\log\left(\frac{\varepsilon}{1+\varepsilon}\right)$.

We prove the necessary lemmas below. The first lemma shows that large subsets of R have large neighborhoods in L. Using this we can show in our second lemma that the number of R-vertices at distance k from F shrinks geometrically with k. This shows that the number of vertices at distance $\Omega(\log\log n)$ from F is sufficiently small. The second lemma shows that at least half the vertices of L are at a constant distance from F. The next two lemmas will be used to show that successive levels in a breath first search expand very fast i.e. at a rate close to $d - 1$. The first of these lemmas deals with the first few levels of the process by showing that small connected subgraphs cannot have to many "extra" edges. The second will account for subsequent levels through expansion.

3 Expansion and Related Graph Structure

We first show that large subsets of R have corresponding large neighborhoods in L.

Lemma 1. *If* $1/2 \leq \beta \leq 1 - \frac{2d^3 \log n}{n}$ *and* $\alpha = d - 1 - \frac{d + \log d}{1 - \log(1 - \beta)} > 0$ *then with high probability every subset* $Y \subseteq R$ *of size* $|Y| = (\beta + \varepsilon)n$, *has a G-neighborhood* $X \subseteq L$ *of size at least* $n \left(1 - \frac{1-\beta}{\alpha}\right)$.

Proof. We show by a union bound that with high probability, there does not exist a pair X, Y such that $|Y| = (\beta + \varepsilon)n$, $|X| < n - \frac{(1+\varepsilon)n - |Y|}{\alpha}$ and X is the neighborhood of Y in G.

Let $S = L \setminus X$ and $T = R \setminus Y$. Then $|S| \geq \frac{(1+\varepsilon)n - |Y|}{\alpha} = \frac{n(1-\beta)}{\alpha}$ and $|T| = (1 + \varepsilon)n - (\beta + \varepsilon)n = (1 - \beta)n$. Each vertex in L has all of its edges in T with probability $\left(\frac{1-\beta}{1+\varepsilon}\right)^d$ independently of other vertices. Thus for any T the size of S is a binomially distributed random variable, $\mathrm{Bin}(n, \left(\frac{1-\beta}{1+\varepsilon}\right)^d)$. Thus the probability of the existence of a pair X, Y is at most

$$\binom{(1+\varepsilon)n}{(\beta+\varepsilon)n} \mathsf{P}\left(|S| \geq \frac{(1-\beta)n}{\alpha}\right)$$

$$= \binom{(1+\varepsilon)n}{(1-\beta)n} \mathsf{P}\left(\mathrm{Bin}\left(n, \left(\frac{1-\beta}{1+\varepsilon}\right)^d\right) \geq \frac{(1-\beta)n}{\alpha}\right)$$

$$\leq \left(e\frac{1+\varepsilon}{1-\beta}\right)^{(1-\beta)n} \left(\frac{\left(\frac{1-\beta}{1+\varepsilon}\right)^d}{e^{\frac{1-\beta}{\alpha}}}\right)^{\frac{1-\beta}{\alpha}n}$$

$$= \left(\frac{\alpha e^{1+\alpha}(1+\varepsilon)^{\alpha-d}}{(1-\beta)^{\alpha-d+1}}\right)^{\frac{1-\beta}{\alpha}n} \tag{1}$$

where we have used the inequality $\mathsf{P}\left(\mathrm{Bin}(n, p) \geq \rho p n\right) \leq \left(\frac{e}{\rho}\right)^{\rho p n}$. (While there are tighter bounds, this is sufficient for our purposes.)

Taking logarithms, dropping the $1 + \varepsilon$ factor, and letting $D = d + \log d$ and $L = \log(1 - \beta)$ gives

$$\frac{\log((1))}{\frac{1-\beta}{\alpha}n} \leq \log\left(\frac{d - 1 - D/(1-L)}{d}\right) + D - \frac{D}{1-L} + \frac{DL}{1-L}$$

$$= \log\left(\frac{d - 1 - D/(1-L)}{d}\right) \leq \log\left(\frac{d-1}{d}\right)$$

Then we can upper bound the expression in (1) by

$$\left(\frac{d-1}{d}\right)^{\frac{1-\beta}{\alpha}n} \leq \exp\left(-\frac{1}{d}\frac{2d^3 \log n}{d}n\right) \leq n^{-2d}$$

The following Lemma corresponds to Lemma 8 in [6]. We give an improved bound on an important parameter k^* which gives an improvement for the running time of the breadth-first search algorithm.

Lemma 2. *Assume* $d \geq 8$ *and furthermore if* $\varepsilon \leq \frac{1}{6}$ *we assume* $d \geq 4 + 2\varepsilon - 2(1+\varepsilon) \log\left(\frac{\varepsilon}{1+\varepsilon}\right)$. *Then the number of vertices in* L *at distance at most* k^* *from* F *is at least* $\frac{n}{2}$, *where* $k^* = 4 + \frac{\log\left(\frac{1}{6\varepsilon}\right)}{\log\left(\frac{d}{6}\right)}$ *if* $\varepsilon \leq \frac{1}{6}$ *and* $k^* = 5$ *if* $\varepsilon \geq \frac{1}{6}$.

Proof. Omitted.   □

We can now give a proof of Theorem 2:

Proof. We follow the proof of Theorem 1 in [6]. The breadth-first search insertion procedure takes time $O(|T_v|)$ where T_v is a BFS tree rooted at the newly inserted vertex v, which is grown until a free vertex is found.

The expected size of T_v is bounded above by d^{k^*} which is at most d^5 for $\varepsilon \geq \frac{1}{6}$ and

$$d^{4+\log\left(\frac{1}{6\varepsilon}\right)/\log\left(\frac{d}{6}\right)} = d^4 \left(\frac{1}{6\varepsilon}\right)^{\frac{\log d}{\log d - \log 6}} = d^4 \left(\frac{1}{6\varepsilon}\right)^{\frac{1}{1-\frac{\log 6}{\log d}}}$$

if $\varepsilon \leq \frac{1}{6}$.   □

Let $k^* = \max\{4 + \frac{\log\left(\frac{1}{6\varepsilon}\right)}{\log\left(\frac{d}{6}\right)}, 5\}$ and let Y_k be the vertices in R at distance at most $k^* + k$ from F and let $|Y_k| = (\beta_k + \varepsilon)n$. We note that Lemma 2 guarantees that with high probability at most $\frac{n}{2}$ vertices in R are at distance more than k^* from F and so with high probability $\beta_k \geq 1/2$ for $k \geq 0$.

We now move to showing that for sufficiently large k of size $O(\log \log n)$, a large fraction of the vertices are within distance k of the free vertices F with high probability.

Lemma 3. *Suppose that* $d \geq 8$ *and if* $\varepsilon \leq \frac{1}{6}$ *assume* $d \geq 4 + 2\varepsilon - 2(1 + \varepsilon) \log\left(\frac{\varepsilon}{1+\varepsilon}\right)$. *Then with high probability* $1 - \beta_k = O\left(\frac{\log^{\gamma_0} n}{(d-1)^k}\right)$ *for* k *such that* $1/2 \geq 1 - \beta_k \geq 2d^3 \log n/n$.

The reader should not be put off by the fact that the lemma only has content for $k = \Omega(\log \log n)$. It is only needed for these values of k.

Proof. We must show that

$$1 - \beta_k = O\left(\frac{\log^{\gamma_0} n}{(d-1)^k}\right) \quad \text{whenever } 1 - \beta_k \geq 2d^3 \log n/n. \tag{2}$$

Assume that the high probability event in Lemma 1 occurs and $1 - \beta_k \geq \log n/n$ and the G-neighborhood X_k of Y_k in L has size at least $n - \frac{(1-\beta_k)n}{\alpha_k}$ where $\alpha_k = d - 1 - \frac{d + \log d}{1 - \log(1 - \beta_k)}$. Note that for $\beta_k \geq \frac{3}{4}$ and $d \geq 8$ this implies

$$\alpha_k \geq \frac{d}{3} \quad \text{and} \quad \frac{(d + \log d)/(d-1)}{1 - \log(1 - \beta_k)} \leq 0.9. \tag{3}$$

First assume that $\beta_0 \geq 3/4$, we will deal with the case of $\beta_0 \geq \frac{1}{2}$ later. Note now that $Y_{k+1} = F \cup M(X_k)$ where $M(X_k) = \{y : (x,y) \in M \text{ for some } x \in X_k\}$. Thus $|Y_{k+1}| = (\beta_{k+1} + \varepsilon)n \geq \varepsilon n + n - \frac{(1-\beta_k)n}{\alpha_k}$. This implies that

$$1 - \beta_{k+1} \leq \frac{1 - \beta_k}{\alpha_k} \tag{4}$$

$$= \frac{1 - \beta_k}{d - 1}\left(1 - \frac{(d + \log d)/(d-1)}{1 - \log(1 - \beta_k)}\right)^{-1}$$

$$\leq \frac{1 - \beta_k}{d - 1}\exp\left(h\left(\frac{(d + \log d)/(d-1)}{1 - \log(1 - \beta_k)}\right)\right) \tag{5}$$

$$\leq \frac{1 - \beta_k}{d - 1}\exp\left(h\left(\frac{(d + \log d)/(d-1)}{1 - \log(1 - \beta_0) + k\log(d/3)}\right)\right). \tag{6}$$

In (5) we let $h(x) = x + 3x^2$ and note that $(1 - x)^{-1} \leq \exp(h(x))$ for $x \in [0, .9]$. For (6) we have assumed that $1 - \beta_k \leq 3^k(1 - \beta_0)/d^k$, which follows from (3) and (4) provided $\beta_k \geq 3/4$.

For $\beta \in [\frac{1}{2}, \frac{3}{4}]$ note that α_k is increasing in d and β. Also starting with $\beta_0 = \frac{1}{2}$ and using $d = 8$ we see numerically that $1 - \beta_3 \leq \frac{\frac{1}{2}}{\alpha_0\alpha_1\alpha_2} \leq \frac{1}{4}$. Thus after at most 3 steps we can assume $\beta \geq 3/4$. To simplify matters we will assume $\beta_0 \geq 3/4$, since doing this will only "shift" the indices by at most 3 and distort the equations by a $O(1)$ factor.

Using inequality (6) repeatedly gives

$$1 - \beta_{k+1} \leq \frac{1 - \beta_0}{(d-1)^{k+1}} \times$$

$$\exp\left(\frac{d + \log d}{(d-1)\log(d/3)}\sum_{j=0}^{k}\frac{1}{j + \frac{1-\log(1-\beta_0)}{\log(d/3)}} + O\left(\sum_{j=0}^{k}\frac{1}{\left(j + \frac{1-\log(1-\beta_0)}{\log(d/3)}\right)^2}\right)\right)$$

$$\leq \frac{1 - \beta_0}{(d-1)^{k+1}}\exp\left(\frac{d + \log d}{(d-1)\log(d/3)}\log\left(\frac{1 - \log(1 - \beta_k)}{1 - \log(1 - \beta_0)}\right) + O(1)\right) \tag{7}$$

$$\leq O\left(\frac{\log^{\gamma_0} n}{(d-1)^{k+1}}\right).$$

Note that (7) is obtained as follows:

$$\sum_{j=0}^{k}\frac{1}{j + \frac{1-\log(1-\beta_0)}{\log(d/3)}} = \log\left(\frac{k + \zeta}{\zeta}\right) + O(1) \leq \log\left(\frac{1 - \log(1 - \beta_k)}{1 - \log(1 - \beta_0)}\right) + O(1) \tag{8}$$

where $\zeta = \frac{1-\log(1-\beta_0)}{\log(d/3)}$. Now $1 - \beta_k \leq 3^k(1 - \beta_0)/d^k$ implies that $k \leq \log((1 - \beta_0)/(1-\beta_k))/\log(d/3)$. Substituting this upper bound for k into the middle term of (8) yields the right hand side.

We now require some additional structural lemmas regarding the graph in order to show that a breadth first search on the graph expands suitably.

Lemma 4. Whp G *does not contain a connected subgraph H on $2k+1$ vertices with $2k + 3d$ edges, where $k + 1$ vertices come from L and k vertices come from R for $k \leq \frac{1}{6} \log_d n$.*

Proof. We put an upper bound on the probability of the existence of such a subgraph using the union bound. Let the vertices of H be fixed, any such graph H can be constructed by taking a bipartite spanning tree on the $k + 1$ and k vertices and adding j edges. Thus the probability of such a subgraph is at most

$$
\binom{n}{k+1} \binom{(1+\varepsilon)n}{k} k^{(k+1)-1}(k+1)^{k-1} (k(k+1))^j \left(\frac{j}{(1+\varepsilon)n} \right)^{2k+j}
$$

$$
\leq \left(\frac{en}{k+1} \right)^{k+1} \left(\frac{e(1+\varepsilon)n}{k} \right)^k k^k (k+1)^{k-1} \left(\frac{jk(k+1)}{(1+\varepsilon)n} \right)^j j^{2k} \left(\frac{1}{(1+\varepsilon)n} \right)^{2k}
$$

$$
\leq n \, (ej)^{2k} \left(\frac{jk(k+1)}{n} \right)^j \tag{9}
$$

For $j = 3d$ and $k \geq \frac{1}{6} \log_d n$ we have $\frac{3dk(k+1)}{n} \leq n^{-1+\frac{1}{4}}$ and $(3ed)^{2k} \leq \exp\left(\log(d^3) \frac{\log n}{3 \log d} \right) = n$ and (9) is at most $n^2 n^{-3d+2} = O(n^{-2d})$.

Lemma 5. Whp *there do not exist $S \subseteq L$, $T \subseteq R$ such that $N(S) \subseteq T$, $2d^2 \log n \leq s = |S| \leq n/d$, $t = |T| \leq (d - 1 - \theta_s)s$ and $\theta_s = \frac{d + \log d}{\log(n/((d-1)s))} \geq \frac{d + \log d}{\log(n/t)}$.*

Proof. The expected number of pairs S, T satisfying (i),(ii) can be bounded by

$$
\sum_{s=2d^2 \log n}^{n/d} \binom{n}{s} \binom{(1+\varepsilon)n}{t} \left(\frac{t}{(1+\varepsilon)n} \right)^{ds}
$$

$$
\leq \sum_{s=2d^2 \log n}^{n/d} \left(\frac{ne}{s} \right)^s \left(\frac{(1+\varepsilon)ne}{t} \right)^t \left(\frac{t}{(1+\varepsilon)n} \right)^{ds}
$$

$$
\leq \sum_{s=2d^2 \log n}^{n/d} \left(\frac{ne}{s} \right)^s e^{(d-1-\theta_s)s} \left(\frac{t}{(1+\varepsilon)n} \right)^{ds-(d-1-\theta_s)s}
$$

$$
\leq \sum_{s=2d^2 \log n}^{n/d} \left(\frac{t}{s} \frac{e^{d-\theta_s}}{(1+\varepsilon)^{1+\theta_s}} \left(\frac{t}{n} \right)^{\theta_s} \right)^s
$$

$$
\leq \sum_{s=2d^2 \log n}^{n/d} \left((d-1)e^{d-\theta_s} \left(\frac{t}{n} \right)^{\theta_s} \right)^s
$$

$$
\leq \sum_{s=2d^2 \log n}^{n/d} \left(\frac{d-1}{d} \right)^s = O\left(n^{-\frac{2d^2 \log n}{d}} \right) = O\left(n^{-2d} \right)
$$

4   Random Walks

Suppose now that we are in the process of adding u to the hash table. For our analysis, we consider exploring a subgraph of G using breadth-first search, starting with the root $u \in L$ and proceeding until we reach F. We emphasize that this is not the behavior of our algorithm; we merely need to establish some properties of the graph structure, and the natural way to do that is by considering a breadth-first search from u.

Let $L_1 = \{u\}$. Let the R-neighbors of x be $w_1, w_2, \ldots, w_d$ and suppose that none of them are in F. Let $R_1 = \{w_1, w_2, \ldots, w_d\}$. Let $L_2 = \{v_1, v_2, \ldots, v_d\}$ where v_i is matched with w_i in M, for $i = 1, 2, \ldots, d$. In general, suppose we have constructed L_k for some k. R_k consists of the R-neighbors of L_k that are not in $R_{\leq k-1} = R_1 \cup \cdots \cup R_{k-1}$ and L_{k+1} consists of the M-neighbors of R_i. An edge (x, y) from L_k to R is *wasted* if either (i) $y \in R_j, j < k$ or if there exists $x' \in L_k$, $x' < x$ such that the edge $(x', y) \in G$. We let

$$k_0 = \lfloor \log_{d-1}(n) - 1 \rfloor$$

and $\rho_k = |R_k|, \lambda_k = |L_k|$ for $1 \leq k \leq k_0$. Assume for the moment that

$$|R_k| \cap F = \emptyset \text{ for } 1 \leq k \leq k_0. \tag{10}$$

Lemma 6. *Assume that* (10) *holds. Then*

$$\rho_{k_0} = \Omega \left(\frac{n}{\log^{\gamma_1} n} \right). \tag{11}$$

Proof. We can assume that $1 - \beta_{k_0} \geq 2d^3 \log n / n$. If $1 \leq k \leq k_1 = \left\lceil \frac{\log_d n}{6} \right\rceil$ then Lemma 4 implies that we generate at most $3d$ wasted edge in the construction of $L_j, R_j, 1 \leq j \leq k$. If we consider the full BFS path tree, where vertices can be repeated, then each internal vertex of the tree L has $d - 1$ children. For every wasted edge we cut off a subtree of the full BFS tree, what remains when all the wasted edges have been cut is the regular BFS tree. Clearly the worst case is when all the subtrees cut off are close to the root, thus $3d$ wasted edges can at most stunt the growth of the tree for 4 levels ($d - 2$ edges cut at the 3 lowest levels and 6 edges cut off at the 4-th level). This means that

$$\rho_k > (d-1)^{k-5} \text{ for } 1 \leq k \leq k_1. \tag{12}$$

In particular $\rho_{k_1} = \Omega \left((d-1)^{\frac{\log_d n}{6}} \right) = \Omega \left(2d^2 \log n \right)$ so Lemma 5 applies to the BFS tree at this stage. In general Lemma 5 implies that for $j \geq k_1$

$$\rho_1 + \rho_2 + \cdots + \rho_j \geq (d - 1 - \theta_s)s \tag{13}$$

where

$$s = \lambda_1 + \lambda_2 + \cdots + \lambda_j = 1 + \rho_1 + \rho_2 + \cdots + \rho_{j-1}. \tag{14}$$

This follows from the fact that $\lambda_1 = 1$ and (10) implies $\lambda_j = \rho_{j-1}$ for $j \geq 2$.

Now $\lambda_j \leq (d-1)\lambda_{j-1}$ for $j \geq 3$ and so s in (14) satisfies $s \leq 1 + d + d(d-1) + \cdots + d(d-1)^{j-2} < (d-1)^{j-1}$. Thus θ_s in (13) satisfies

$$\theta_s \leq \phi_j = \frac{d + \log d}{\log n - j \log(d-1)}.$$

Thus, by (13) and (14) we have (after dropping a term)

$$\rho_j \geq (d - 2 - \phi_j)(\rho_1 + \rho_2 + \cdots + \rho_{j-1}). \tag{15}$$

An induction then shows that for $\ell \geq 1$,

$$\rho_{k_1+\ell} \geq (\rho_1 + \cdots + \rho_{k_1})(d - 2 - \phi_{k_1+\ell}) \prod_{k=1}^{\ell-1} (d - 1 - \phi_{k_1+k}). \tag{16}$$

Indeed the case $\ell = 1$ follows directly from (15). Then, by induction,

$$\rho_{k_1+\ell+1} \geq$$

$$(\rho_1 + \cdots + \rho_{k_1})(d - 2 - \phi_{k_1+\ell+1}) \left(1 + \sum_{k=1}^{\ell} (d - 2 - \phi_{k_1+k}) \prod_{i=1}^{k-1} (d - 1 - \phi_{k_1+i}) \right)$$

$$= (\rho_1 + \cdots + \rho_{k_1})(d - 2 - \phi_{k_1+\ell+1}) \prod_{k=1}^{\ell} (d - 1 - \phi_{k_1+k}). \tag{17}$$

To check (17) we can use induction. Assume that

$$1 + \sum_{k=2}^{\ell+1} (d - 2 - \phi_{k_1+k}) \prod_{i=2}^{k} (d - 1 - \phi_{k_1+i}) = \prod_{k=2}^{\ell+1} (d - 1 - \phi_{k_1+k})$$

and then multiply both sides by $d - 1 - \phi_{k_1+1}$.

We deduce from (12) and (16) that provided $k_1 + \ell \leq k_0$ (which implies $\frac{\phi_{k_1+\ell}}{d-1} \leq \frac{1}{2}$),

$$\rho_{k_1+\ell} \geq ((d-1)^{k_1-5} - 1)(d - 2 - \phi_{k_1+\ell}) \prod_{k=1}^{\ell-1} (d - 1 - \phi_{k_1+k})$$

$$\geq \frac{1}{2}(d-1)^{k_1+\ell-4} \exp\left\{ -\frac{1}{d-1} \sum_{k=1}^{\ell} \phi_{k_1+k} - \frac{1}{(d-1)^2} \sum_{k=1}^{\ell} \phi_{k_1+k}^2 \right\} \tag{18}$$

Note next that

$$\sum_{k=1}^{\ell} \phi_{k_1+k} = \frac{d + \log d}{\log(d-1)} \left(\log\left(\frac{\log n - k_1 \log(d-1)}{\log n - (k_1 + \ell) \log(d-1)} \right) + O(1) \right)$$

$$\leq \frac{d + \log d}{\log(d-1)} (\log\log n + O(1))$$

and $\sum_{k=1}^{\ell} \phi_{k_1+k}^2 = O(1)$. Thus, putting $\ell = k_0 - k_1$ we get

$$\rho_{k_0} = \Omega \left(\frac{(d-1)^{k_0}}{(\log n)^{(d+\log d)/((d-1)\log(d-1))}} \right)$$

and the lemma follows.

5 Proof of Theorem 1

Let S denote the set of vertices $v \in R$ at distance at most $\Delta = k^* + (\gamma_0 + \gamma_1) \log_{d-1} \log n + 2K$ from F, where K is a large constant and k^* is given in Lemma 2. Then by Lemma 3

$$|R \setminus S| \le \frac{n}{(d-1)^K \log^{\gamma_1}(n)}.$$

We have used $(d-1)^K$ to "soak up" the hidden constant in the statement of Lemma 3 and the requirement $1 - \beta_k \ge 2d^2 \log n/n$ in Lemma 3 does not cause problems. If it fails then at most $O(\log n)$ vertices are at distance greater than Δ from F.

If K is sufficiently large then Lemma 6 implies that

$$|R \setminus S| \le \rho_{k_0}/2. \tag{19}$$

Every vertex $v \in S$ has a path of length $l \le \Delta$ to a free vertex and the probability that the random walk follows this path is $\left(\frac{1}{d-1}\right)^l \ge \left(\frac{1}{d-1}\right)^\Delta$, which is a lower bound on the probability the algorithm finds a free vertex within Δ steps, starting from $v \in S$. We now split the random walk into rounds, and each round into two phases.

The first phase starts when the round starts and ends when the random walk reaches a vertex of S or after k_0 steps(possibly the first phase is empty). Then, the second phase starts and ends either when the random walk reaches a free vertex or after Δ steps, finishing this round. The length of the first phase is at most k_0 and in the second phase takes at most Δ steps.

Claim. Starting from a vertex $v \notin S$ the expected number of rounds until the random walk is in S is at most $O(\log^{\gamma_1} n)$. Indeed the probability that a random walk of length k_0 passes through S is at least $\frac{\rho_{k_0} - |R \setminus S|}{(d-1)^{k_0}} = \Omega(\log^{-\gamma_1} n)$.

By Claim 5 we have a $\Omega(\log^{-\gamma_1} n)$ chance of reaching S at the end of the first phase. When we start the second phase we have at least a $\left(\frac{1}{d-1}\right)^\Delta$ probability of reaching a free vertex, thus ending the random walk. Then the number of rounds until we reach a free vertex is dominated by a geometric distribution with parameter $\Omega\left(\left(\frac{1}{d-1}\right)^\Delta \log^{-\gamma_1} n\right)$ and thus the expected number of rounds is $O((d-1)^{\Delta} \log^{\gamma_1} n)$. Since both Lemma 3 and Claim 5 apply regardless of the starting vertex, this shows that the expected number of steps until we reach a free vertex is at most

$$O\left(k_0 \log^{\gamma_1} n(d-1)^{\Delta}\right) = O\left((\log n)(\log^{\gamma_1} n)(d-1)^{(\gamma_0+\gamma_1)\log_{d-1}\log n+O(1)}\right)$$
$$= O\left(\log^{1+\gamma_0+2\gamma_1} n\right).$$

There is still the matter of Assumption (10). This is easily dealt with. If we find $v \in R_k \cap F$ then we are of course delighted. So, we could just add a dummy tree extending $2(k_0 - k)$ levels from v where each vertex in the last level is in F. The conclusion of Claim 5 will remain unchanged. This completes the proof of Theorem 1.

6   Conclusion

We have demonstrated that for sufficiently large d with high probability the graph structure of the resulting cuckoo graph is such that, regardless of the staring vertex, the random-walk insertion method will reach a free vertex in polylogarithmic time with high probability. Obvious directions for improvement include reducing the value of d for which this type of result holds, and reducing the exponent in the time bound.

References

1. Azar, Y., Broder, A., Karlin, A., Upfal, E.: Balanced Allocations. SIAM Journal on Computing 29(1), 180–200 (1999)
2. Broder, A., Karlin, A.: Multilevel Adaptive Hashing. In: Proceedings of the 1st ACM-SIAM Symposium on Discrete Algorithms (SODA), pp. 43–53 (1990)
3. Broder, A., Mitzenmacher, M.: Using Multiple Hash Functions to Improve IP Lookups. In: Proceedings of the 20th IEEE International Conference on Computer Communications (INFOCOM), pp. 1454–1463 (2001)
4. Devroye, L., Morin, P.: Cuckoo Hashing: Further Analysis. Information Processing Letters 86(4), 215–219 (2003)
5. Dietzfelbinger, M., Weidling, C.: Balanced Allocation and Dictionaries with Tightly Packed Constant Size Bins. Theoretical Computer Science 380(1-2), 47–68 (2007)
6. Fotakis, D., Pagh, R., Sanders, P., Spirakis, P.: Space Efficient Hash Tables With Worst Case Constant Access Time. Theory of Computing Systems 38(2), 229–248 (2005)
7. Kirsch, A., Mitzenmacher, M.: Using a Queue to De-amortize Cuckoo Hashing in Hardware. In: Proceedings of the Forty-Fifth Annual Allerton Conference on Communication, Control, and Computing (2007)
8. Kirsch, A., Mitzenmacher, M., Wieder, U.: More Robust Hashing: Cuckoo Hashing with a Stash. In: Proceedings of the 16th Annual European Symposium on Algorithms, pp. 611–622 (2008)

9. Kirsch, A., Mitzenmacher, M.: The Power of One Move: Hashing Schemes for Hardware. In: Proceedings of the 27th IEEE International Conference on Computer Communications (INFOCOM), pp. 565–573 (2008)
10. Kutzelnigg, R.: Bipartite Random Graphs and Cuckoo Hashing. In: Proceedings of the Fourth Colloquium on Mathematics and Computer Science (2006)
11. Mitzenmacher, M., Vadhan, S.: Why Simple Hash Functions Work: Exploiting the Entropy in a Data Stream. In: Proceedings of the Nineteenth Annual ACM-SIAM Symposium on Discrete Algorithms (SODA), pp. 746–755 (2008)
12. Pagh, R., Rodler, F.: Cuckoo Hashing. Journal of Algorithms 51(2), 122–144 (2004)
13. Vöcking, B.: How Asymmetry Helps Load Balancing. Journal of the ACM 50(4), 568–589 (2003)

Hierarchy Theorems for Property Testing

Oded Goldreich[1], Michael Krivelevich[2], Ilan Newman[3], and Eyal Rozenberg[4]

[1] Faculty of Math. and Computer Science, Weizmann Institute, Rehovot, Israel
[2] School of Mathematical Sciences, Tel Aviv University, Tel Aviv 69978, Israel
[3] Department of Computer Science, Haifa University, Haifa, Israel
[4] Department of Computer Science, Technion, Haifa, Israel

Abstract. Referring to the query complexity of property testing, we prove the existence of a rich hierarchy of corresponding complexity classes. That is, for any relevant function q, we prove the existence of properties that have testing complexity $\Theta(q)$. Such results are proven in three standard domains often considered in property testing: generic functions, adjacency predicates describing (dense) graphs, and incidence functions describing bounded-degree graphs. While in two cases the proofs are quite straightforward, the techniques employed in the case of the dense graph model seem significantly more involved. Specifically, problems that arise and are treated in the latter case include (1) the preservation of distances between graph under a blow-up operation, and (2) the construction of monotone graph properties that have local structure.

Keywords: Property Testing, Graph Properties, Monotone Graph Properties, Graph Blow-up, One-Sided vs Two-Sided Error, Adaptivity vs Non-adaptivity.

1   Introduction

In the last decade, the area of property testing has attracted much attention (see the surveys of [F, R], which are already somewhat out-of-date). Loosely speaking, property testing typically refers to sub-linear time probabilistic algorithms for deciding whether a given object has a predetermined property or is far from any object having this property. Such algorithms, called testers, obtain local views of the object by making adequate queries; that is, the object is seen as a function and the testers get oracle access to this function (and thus may be expected to work in time that is sub-linear in the length of the object).

Following most work in the area, we focus on the query complexity of property testing, where the query complexity is measured as a function of the size of the object as well as the desired proximity (parameter). Interestingly, many natural properties can be tested in complexity that only depends on the proximity parameter; examples include linearity testing [BLR], and testing various graph properties in two natural models (e.g., [GGR, AFNS] and [GR1, BSS], respectively). On the other hand, properties for which testing requires essentially maximal query complexity were proved to exist too; see [GGR] for artificial

I. Dinur et al. (Eds.): APPROX and RANDOM 2009, LNCS 5687, pp. 504–519, 2009.

examples in two models and [BHR, BOT] for natural examples in other models. In between these two extremes, there exist natural properties for which the query complexity of testing is logarithmic (e.g., monotonicity [EKK+, GGL+]), a square root (e.g., bipartiteness in the bounded-degree model [GR1, GR2]), and possibly other constant powers (see [FM, PRR]).

One natural problem that arises is whether there exist properties of arbitrary query complexity. We answer this question affirmative, proving the existence of a rich hierarchy of query complexity classes. Such hierarchy theorems are easiest to state and prove in the generic case (treated in Section 2): Loosely speaking, for every sub-linear function q, *there exists a property of functions over $[n]$ that is testable using $q(n)$ queries but is not testable using $o(q(n))$ queries.*

Similar hierarchy theorems are proved also for two standard models of testing graph properties: the adjacency representation model (of [GGR]) and the incidence representation model (of [GR1]). For the incidence representation model (a.k.a the bounded-degree graph model), we show (in Section 3) that, for every sub-linear function q, *there exists a property of bounded-degree N-vertex graphs that is testable using $q(N)$ queries but is not testable using $o(q(N))$ queries.* Furthermore, one such property corresponds to the set of N-vertex graphs that are 3-colorable and consist of connected components of size at most $q(N)$.

The bulk of this paper is devoted to hierarchy theorems for the adjacency representation model (a.k.a the dense graph model), where complexity is measured in terms of the number of vertices rather than the number of all vertex pairs. Our main results for the adjacency matrix model are:

1. For every sub-quadratic function q, *there exists a graph property Π that is testable in q queries, but is not testable in $o(q)$ queries.* Furthermore, for "nice" functions q, it is the case that Π is in $\mathcal{P}$ and the tester can be implemented in poly(q)-time. (See Section 4.)
2. For every sub-quadratic function q, there exists a *monotone* graph property Π that is testable in $O(q)$ queries, but is not testable in $o(q)$ queries. (See Section 5.)

The adjacency representation model is further studied in Sections 6 and 7.

Organization of this version. Due to space limitations, several proofs have been either omitted or trimmed. Full proofs can be found in our technical report [GKNR].

Conventions. For sake of simplicity, we state all results while referring to query complexity as a function of the input size; that is, we consider a fixed (constant) value of the proximity parameter, denoted ϵ. In such cases, we sometimes use the term ϵ-testing, which refers to testing when the proximity parameter is fixed to ϵ. All our lower bounds hold for any sufficiently small value of the proximity parameter, whereas the upper bounds hide a (polynomial) dependence on (the reciprocal of) this parameter. In general, bounds that have no dependence on the proximity parameter refer to some (sufficiently small but) fixed value of this parameter.

A remotely related prior work. In contrast to the foregoing conventions, we mention here a result that refers to graph properties that are testable in (query) complexity that only depends on the proximity parameter. This result, due to [AS], establishes a (very sparse) hierarchy of such properties. Specifically, [AS, Thm. 4] asserts that for every function q there exists a function Q and a graph property that is ϵ-testable in $Q(\epsilon)$ queries but is *not* ϵ-testable in $q(\epsilon)$ queries.[1]

2   Properties of Generic Functions

In the generic function model, the tester is given oracle access to a function over $[n]$, and distance between such functions is defined as the fraction of (the number of) arguments on which these functions differ. In addition to the input oracle, the tester is explicitly given two parameters: a size parameter, denoted n, and a proximity parameter, denoted ϵ.

Definition 1. *Let $\Pi = \bigcup_{n \in \mathbb{N}} \Pi_n$, where Π_n contains functions defined over the domain $[n] \overset{\text{def}}{=} \{1, ..., n\}$. A* tester for a property *Π is a probabilistic oracle machine T that satisfies the following two conditions:*

1. *The tester accepts each $f \in \Pi$ with probability at least $2/3$; that is, for every $n \in \mathbb{N}$ and $f \in \Pi_n$ (and every $\epsilon > 0$), it holds that $\Pr[T^f(n, \epsilon) = 1] \geq 2/3$.*
2. *Given $\epsilon > 0$ and oracle access to any f that is ϵ-far from Π, the tester rejects with probability at least $2/3$; that is, for every $\epsilon > 0$ and $n \in \mathbb{N}$, if $f : [n] \to \{0, 1\}^*$ is ϵ-far from Π_n, then $\Pr[T^f(n, \epsilon) = 0] \geq 2/3$.*

We say that the tester has one-sided error *if it accepts each $f \in \Pi$ with probability 1 (i.e., for every $f \in \Pi$ and every $\epsilon > 0$, it holds that $\Pr[T^f(n, \epsilon) = 1] = 1$).*

Definition 1 does not specify the query complexity of the tester, and indeed an oracle machine that queries the entire domain of the function qualifies as a tester (with zero error probability...). Needless to say, we are interested in testers that have significantly lower query complexity. Recall that [GGR] asserts that in some cases such testers do not exist; that is, there exist properties that require linear query complexity. Building on this result, we show:

Theorem 2. *For every $q : \mathbb{N} \to \mathbb{N}$ that is at most linear, there exists a property Π of Boolean functions that is testable* (with one-sided error) *in $q + O(1)$ queries, but is not testable in $o(q)$ queries* (even when allowing two-sided error).

We start with an arbitrary property Π' of Boolean functions for which testing is known to require a linear number of queries (even when allowing two-sided error). The existence of such properties was first proved in [GGR]. Given

---

[1] We note that while Q depends only on q, the dependence proved in [AS, Thm. 4] is quite weak (i.e., Q is lower bounded by a non-constant number of compositions of q), and thus the hierarchy obtained by setting $q_i = Q_{i-1}$ for $i = 1, 2, ...$ is very sparse.

$\Pi' = \bigcup_{m \in \mathbb{N}} \Pi'_m$, we define $\Pi = \bigcup_{n \in \mathbb{N}} \Pi_n$ such that Π_n consists of "duplicated versions" of the functions in $\Pi'_{q(n)}$. Specifically, for every $f' \in \Pi'_{q(n)}$, we define $f(i) = f'(i \bmod q(n))$ and add f to Π_n, where $i \bmod m$ is (non-standardly) defined as the smallest positive integer that is congruent to i modulo m, The proof that Π satisfies the conditions of Theorem 2 appears in our technical report [GKNR].

Comment. Needless to say, Boolean functions over $[n]$ may be viewed as n-bit long binary strings. Thus, Theorem 2 means that, for every sub-linear q, there are properties of binary strings for which the query complexity of testing is $\Theta(q)$. Given this perspective, it is natural to comment that such properties exist also in $\mathcal{P}$. This comment is proved by starting with the hard-to-test property asserted in Theorem 7 of our technical report [GKNR] (or alternatively with the one in [LNS], which is in $\mathcal{L}$).

3   Graph Properties in the Bounded-Degree Model

The bounded-degree model refers to a fixed (constant) degree bound, denoted $d \geq 2$. An N-vertex graph $G = ([N], E)$ (of maximum degree d) is represented in this model by a function $g : [N] \times [d] \rightarrow \{0, 1, ..., N\}$ such that $g(v, i) = u \in [N]$ if u is the ith neighbor of v and $g(v, i) = 0$ if v has less than i neighbors.[2] Distance between graphs is measured in terms of their aforementioned representation; that is, as the fraction of (the number of) different array entries (over dN). Graph properties are properties that are invariant under renaming of the vertices (i.e., they are actually properties of the underlying unlabeled graphs).

Recall that [BOT] proved that, in this model, testing 3-Colorability requires a linear number of queries (even when allowing two-sided error). Building on this result, we show:

Theorem 3. *In the bounded-degree graph model, for every $q : \mathbb{N} \rightarrow \mathbb{N}$ that is at most linear, there exists a graph property Π that is testable* (with one-sided error) *in $O(q)$ queries, but is not testable in $o(q)$ queries* (even when allowing two-sided error). *Furthermore, this property is the set of N-vertex graphs of maximum degree d that are 3-colorable and consist of connected components of size at most $q(N)$.*

We start with an arbitrary property Π' for which testing is known to require a linear number of queries (even when allowing two-sided error). We further assume that Π' is downward monotone (i.e., if $G' \in \Pi'$ then any subgraph of G' is in Π'). Indeed, by [BOT], 3-Colorability is such a property. Given $\Pi' = \bigcup_{n \in \mathbb{N}} \Pi'_n$, we define $\Pi = \bigcup_{N \in \mathbb{N}} \Pi_N$ such that each graph in Π_N consists of connected components that are each in Π' and have size at most $q(N)$; that is, each connected component in any $G \in \Pi_N$ is in Π'_n for some $n \leq q(N)$ (i.e., n denotes this component's size). The proof that Π satisfies the conditions of Theorem 3 appears in our technical report [GKNR].

---

[2] For simplicity, we assume here that the neighbors of v appear in arbitrary order in the sequence $g(v, 1), ..., g(v, \deg(v))$, where $\deg(v) \stackrel{\text{def}}{=} |\{i : g(v, i) \neq 0\}|$.

Comment. The construction used in the proof of Theorem 3 is slightly different from the one used in the proof of Theorem 2: In the proof of Theorem 3 each object in Π_N corresponds to a sequence of (possibly different) objects in Π'_n, whereas in the proof of Theorem 2 each object in Π_N corresponds to multiples copies of a single object in Π'_n. While Theorem 2 can be proved using a construction that is analogous to one used in the proof of Theorem 3, the current proof of Theorem 2 provides a better starting point for the proof of the following Theorem 4.

4   Graph Properties in the Adjacency Matrix Model

In the adjacency matrix model, an N-vertex graph $G = ([N], E)$ is represented by the Boolean function $g : [N] \times [N] \to \{0,1\}$ such that $g(u,v) = 1$ if and only if u and v are adjacent in G (i.e., $\{u, v\} \in E$). Distance between graphs is measured in terms of their aforementioned representation; that is, as the fraction of (the number of) different matrix entries (over N^2). In this model, we state complexities in terms of the number of vertices (i.e., N) rather than in terms of the size of the representation (i.e., N^2). Again, we focus on graph properties (i.e., properties of labeled graphs that are invariant under renaming of the vertices).

Recall that [GGR] proved that, in this model, there exist graph properties for which testing requires a quadratic (in the number of vertices) query complexity (even when allowing two-sided error). It was further shown that such properties are in $\mathcal{NP}$. Slightly modifying these properties, we show that they can be placed in $\mathcal{P}$; see Appendix A of our technical report [GKNR]. Building on this result, we show:

Theorem 4. *In the adjacency matrix model, for every $q : \mathbb{N} \to \mathbb{N}$ that is at most quadratic, there exists a graph property Π that is testable in q queries, but is not testable in $o(q)$ queries.[3] Furthermore, if $N \mapsto q(N)$ is computable in $\mathrm{poly}(\log N)$-time, then Π is in $\mathcal{P}$, and the tester is relatively efficient in the sense that its running time is polynomial in the total length of its queries.*

We stress that, unlike in the previous results, the positive part of Theorem 4 refers to a two-sided error tester. This is fair enough, since the negative side also refers to two-sided error testers. Still, one may seek a stronger separation in which the positive side is established via a one-sided error tester. Such a separation is presented in Theorem 6 (except that the positive side is established via a tester that is not relatively efficient).

Outline of the proof of Theorem 4. The basic idea of the proof is to implement the strategy used in the proof of Theorem 2. The problem, of course, is that we need to obtain graph properties (rather than properties of generic Boolean functions). Thus, the trivial "blow-up" (of Theorem 2) that took place on the truth-table (or function) level has to be replaced by a blow-up on the vertex level. Specifically,

---

[3] Both the upper and lower bounds refer to two-sided error testers.

starting from a graph property Π' that requires quadratic query complexity, we consider the graph property Π consisting of N-vertex graphs that are obtained by a $(N/\sqrt{q(N)})$-factor blow-up of $\sqrt{q(N)}$-vertex graphs in Π', where G is a t-factor blow-up of G' if the vertex set of G can be partitioned into (equal size) sets that correspond to the vertices of G' such that the edges between these sets represent the edges of G'; that is, if $\{i,j\}$ is an edge in G', then there is a complete bipartite between the ith set and the jth set, and otherwise there are no edges between this pair of sets.[4]

Note that the notion of "graph blow-up" does not offer an easy identification of the underlying partition; that is, given a graph G that is as a t-factor blow-up of some graph G', it is not necessary easy to determine a t-way partition of the vertex set of G such that the edges between these sets represent the edges of G'. Things may become even harder if G is merely close to a t-factor blow-up of some graph G'. We resolve these as well as other difficulties by augmenting the graphs of the starting property Π'.

The proof of Theorem 4 is organized accordingly: In Section 4.1, we construct Π based on Π' by first augmenting the graphs and then applying graph blow-up. In Section 4.2 we lower-bound the query complexity of Π based on the query complexity of Π', while coping with the non-trivial question of *how does the blow-up operation affect distances between graphs*. In Section 4.3 we upper-bound the query complexity of Π, while using the aforementioned augmentations in order to obtain a tight result (rather than an upper bound that is off by a polylogarithmic factor).

4.1 The Blow-Up Property Π

Our starting point is any graph property $\Pi' = \bigcup_{n\in\mathbb{N}} \Pi'_n$ for which testing requires quadratic query complexity. Furthermore, we assume that Π' is in $\mathcal{P}$. Such a graph property is presented in Theorem 7 of our technical report [GKNR] (which builds on [GGR]).

The notion of graphs that have "vastly different vertex neighborhoods" is central to our analysis. Specifically, for a real number $\alpha > 0$, we say that a graph $G = (V, E)$ is α-dispersed if the neighbor sets of any two vertices differ on at least $\alpha \cdot |V|$ elements (i.e., for every $u \neq v \in V$, the symmetric difference between the sets $\{w : \{u,w\} \in E\}$ and $\{w : \{v,w\} \in E\}$ has size at least $\alpha\cdot|V|$). We say that a set of graphs is dispersed if there exists a constant $\alpha > 0$ such that every graph in the set is α-dispersed.[5]

The augmentation. We first augment the graphs in Π' such that the vertices in the resulting graphs are dispersed, while the augmentation amount to adding a linear number of vertices. The fact that these resulting graphs are dispersed will be useful for establishing both the lower and upper bounds. The augmentation

---

[4] In particular, there are no edges inside any set.
[5] Our notion of dispersibility has nothing to do with the notion of dispersers, which in turn is a weakening of the notion of (randomness) extractors (see, e.g., [S]).

is performed in two steps. First, setting $n' = 2^{\lceil \log_2(2n+1) \rceil} \in [2n+1, 4n]$, we augment each graph $G' = ([n], E')$ by $n' - n$ isolated vertices, yielding an n'-vertex graph $H' = ([n'], E')$ in which every vertex has degree at most $n - 1$. Next, we augment each resulting graph H' by a clique of n' vertices and connect the vertices of H' and the clique vertices by a bipartite graph that corresponds to a Hadamard matrix; that is, the ith vertex of H' is connected to the jth vertex of the clique if and only if the inner product modulo 2 of $i - 1$ and $j - 1$ (in $(\log_2 n')$-bit long binary notation) equals 1. We denote the resulting set of (unlabeled) graphs by Π'' (and sometimes refer to Π'' as the set of all labeled graphs obtained from these unlabeled graphs).

We first note that Π'' is indeed dispersed (i.e., the resulting $2n'$-vertex graphs have vertex neighborhoods that differ on at least $n \geq n'/4$ vertices). Next note that testing Π'' requires a quadratic number of queries, because testing Π' can be reduced to testing Π'' (i.e., ϵ-testing membership in Π'_n reduces to ϵ'-testing membership in $\Pi''_{2n'}$, where $n' \leq 4n$ and $\epsilon' = \epsilon/64$). Finally, note that Π'' is also in $\mathcal{P}$, because it is easy to distinguish the original graph from the vertices added to it, since the clique vertices have degree at least $n' - 1$ whereas the vertices of G' have degree at most $(n-1) + (n'/2) < n' - 1$ (and isolated vertices of H' have neighbors only in the clique).[6]

Applying graph blow-up. Next, we apply an (adequate factor) graph blow-up to the augmented set of graphs Π''. Actually, for simplicity of notation we assume, without loss of generality, that $\Pi' = \bigcup_{n \in \mathbb{N}} \Pi'_n$ itself is dispersed, and apply graph blow-up to Π' itself (rather than to Π''). Given a desired complexity bound $q : \mathbb{N} \to \mathbb{N}$, we first set $n = \sqrt{q(N)}$, and next apply to each graph in Π'_n an N/n-factor blow-up, thus obtaining a set of N-vertex graphs denoted Π_N. (Indeed, we assume for simplicity that both $n = \sqrt{q(N)}$ and N/n are integers.) Recall that G is a t-factor blow-up of G' if the vertex set of G can be partitioned into t (equal size) sets, called clouds, such that the edges between these clouds represent the edges of G'; that is, if $\{i, j\}$ is an edge in G', then there is complete bipartite between the ith cloud and the jth cloud, and otherwise there are no edges between this pair of clouds. This yields a graph property $\Pi = \bigcup_{N \in \mathbb{N}} \Pi_N$.

Let us first note that Π is in $\mathcal{P}$. This fact follows from the hypothesis that Π' is dispersed: Specifically, given any graph N-vertex graph G, we can cluster its vertices according to their neighborhood, and check whether the number of clusters equals $n = \sqrt{q(N)}$. (Note that if $G \in \Pi_N$, then we obtain exactly n (equal sized) clusters, which correspond to the n clouds that are formed in the N/n-factor blow-up that yields G.) Next, we check that each cluster has size N/n and that the edges between these clusters correspond to the blow-up of some n-vertex G'. Finally, we check whether G' is in Π'_n (relying on the fact that $\Pi' \in \mathcal{P}$). Proving that the query complexity of testing Π indeed equals $\Theta(q)$ is undertaken in the next two sections.

---

[6] Once this is done, we can verify that the original graph is in Π (using $\Pi \in \mathcal{P}$), and that the additional edges correspond to a Hadamard matrix.

4.2   Lower-Bounding the Query Complexity of Testing Π

In this section we prove that the query complexity of testing Π is $\Omega(q)$. The basic idea is reducing testing Π' to testing Π; that is, given a graph G' that we need to test for membership in Π'_n, we test its N/n-factor blow-up for membership in Π_N, where N is chosen such that $n = \sqrt{q(N)}$. This approach relies on the assumption that the N/n-factor blow-up of any n-vertex graph that is far from Π'_n results in a graph that is far from Π_N. (Needless to say, the N/n-factor blow-up of any graph in Π'_n results in a graph that is in Π_N.)

As shown by Arie Matsliah (see Appendix B of our technical report [GKNR]), the aforementioned assumption does *not* hold in the strict sense of the word (i.e., it is not true that the blow-up of any graph that is ϵ-far from Π' results in a graph that is ϵ-far from Π). However, for our purposes it suffices to prove a relaxed version of the aforementioned assumption that only asserts that *for any $\epsilon' > 0$ there exists an $\epsilon > 0$ such that the blow-up of any graph that is ϵ'-far from Π' results in a graph that is ϵ-far from Π.* Below we prove this assertion for $\epsilon = \Omega(\epsilon')$ and rely on the fact that Π' is dispersed. In Appendix B of our technical report [GKNR], we present a more complicated proof that holds for arbitrary Π' (which need not be dispersed), but with $\epsilon = \Omega(\epsilon')^2$.

Claim 4.1. *There exists a universal constant $c > 0$ such that the following holds for every n, ϵ', α and (unlabeled) n-vertex graphs G'_1, G'_2. If G'_1 is α-dispersed and ϵ'-far from G'_2, then for any t the (unlabeled) t-factor blow-up of G'_1 is $c\alpha \cdot \epsilon'$-far from the (unlabeled) t-factor blow-up of G'_2.*

Using Claim 4.1 we infer that *if G' is ϵ'-far from Π' then its blow-up is $\Omega(\epsilon')$-far from Π.* This inference relies on the fact that Π' is dispersed (and on Claim 4.1 when applied to $G'_2 = G'$ and every $G'_1 \in \Pi'$).

Proof. Let G_1 (resp., G_2) denote the (unlabeled) t-factor blow-up of G'_1 (resp., G'_2), and consider a bijection π of the vertices of $G_1 = ([t \cdot n], E_1)$ to the vertices of $G_2 = ([t \cdot n], E_2)$ that minimizes the size of the set (of violations)

$$\{(u, v) \in [t \cdot n]^2 : \{u, v\} \in E_1 \text{ iff } \{\pi(u), \pi(v)\} \notin E_2\}. \tag{1}$$

(Note that Eq. (1) refers to ordered pairs, whereas the distance between graphs refers to unordered pairs.) Clearly, if π were to map to each cloud of G_2 only vertices that belong to a single cloud of G_1 (equiv., for every u, v that belong to the same cloud of G_1 it holds that $\pi(u), \pi(v)$ belong to the same cloud of G_2), then G_2 would be ϵ'-far from G_1 (since the fraction of violations under such a mapping equals the fraction of violations in the corresponding mapping of G'_1 to G'_2). The problem, however, is that it is not clear that π behaves in such a nice manner (and so violations under π do not directly translate to violations in mappings of G'_1 to G'_2). Still, we show that things cannot be extremely bad. Specifically, we call a cloud of G_2 good if at least $(t/2) + 1$ of its vertices are mapped to it (by π) from a single cloud of G_1.

Letting 2ϵ denote the fraction of violations in Eq. (1) (i.e., the size of this set divided by $(tn)^2$), we first show that at least $(1 - (6\epsilon/\alpha)) \cdot n$ of the clouds of G_2 are good. Assume, towards the contradiction, that G_2 contains more that $(6\epsilon/\alpha) \cdot n$ clouds that are not good. Considering any such a (non-good) cloud, we observe that it must contain at least $t/3$ disjoint pairs of vertices that originate in different clouds of G_1 (i.e., for each such pair (v, v') it holds that $\pi^{-1}(v)$ and $\pi^{-1}(v')$ belong to different clouds of G_1).[7] Recall that the edges in G_2 respect the cloud structure of G_2 (which in turn respects the edge relation of G'_2). But vertices that originate in different clouds of G_1 differ on at least $\alpha \cdot tn$ edges in G_1. Thus, every pair (v, v') (in this cloud of G_2) such that $\pi^{-1}(v)$ and $\pi^{-1}(v')$ belong to different clouds of G_1 contributes at least $\alpha \cdot tn$ violations to Eq. (1).[8] It follows that the set in Eq. (1) has size greater than

$$\frac{6\epsilon n}{\alpha} \cdot \frac{t}{3} \cdot \alpha tn = 2\epsilon \cdot (tn)^2$$

in contradiction to our hypothesis regarding π. Having established that at least $(1 - (6\epsilon/\alpha)) \cdot n$ of the clouds of G_2 are good and recalling that a good cloud of G_2 contains a strict majority of vertices that originates from a single cloud of G_1, we consider the following bijection π' of the vertices of G_1 to the vertices of G_2: For each good cloud g of G_2 that contains a strict majority of vertices from cloud i of G_1, we map all vertices of the ith cloud of G_1 to cloud g of G_2, and map all other vertices of G_1 arbitrarily. The number of violations under π' is upper-bounded by four times the number of violations occuring under π between good clouds of G_2 (i.e., at most $4 \cdot 2\epsilon \cdot (tn)^2$) plus at most $(6\epsilon/\alpha) \cdot tn \cdot tn$ violations created with the remaining $(6\epsilon/\alpha) \cdot n$ clouds. This holds, in particular, for a bijection π' that maps to each remaining cloud of G_2 vertices originating in a single cloud of G_1. This π', which maps complete clouds of G_1 to clouds of G_2, yields a mapping of G'_1 to G'_2 that has at most $(8\epsilon + (6\epsilon/\alpha)) \cdot n^2$ violations. Recalling that G'_1 is ϵ'-far from G'_2, we conclude that $8\epsilon + (6\epsilon/\alpha) \geq 2\epsilon'$, and the claim follows (with $c = 1/7$).   □

Recall that Claim 4.1 implies that if G' is ϵ'-far from Π', then its blow-up is $\Omega(\epsilon')$-far from Π. Using this fact, we conclude that ϵ'-testing of Π' reduces to $\Omega(\epsilon')$-testing of Π. Thus, a quadratic lower bound on the query complexity of ϵ'-testing Π'_n yields an $\Omega(n^2)$ lower bound on the query complexity of $\Omega(\epsilon')$-testing Π_N, where $n = \sqrt{q(N)}$. Thus, we obtain an $\Omega(q)$ lower bound on the query complexity of testing Π, for some constant value of the proximity parameter.

---

[7] This pairing is obtained by first clustering the vertices of the cloud of G_2 according to their origin in G_1. By the hypothesis, each cluster has size at most $t/2$. Next, observe that taking the union of some of these clusters yields a set containing between $t/3$ and $2t/3$ vertices. Finally, we pair vertices of this set with the remaining vertices. (A better bound of $\lfloor t/2 \rfloor$ can be obtained by using the fact that a t-vertex graph of minimum degree $t/2$ contains a Hamiltonian cycle.)

[8] For each such pair (v, v'), there exists at least $\alpha \cdot tn$ vertices u such that exactly one of the (unordered) pairs $\{\pi^{-1}(u), \pi^{-1}(v)\}$ and $\{\pi^{-1}(u), \pi^{-1}(v')\}$ is an edge in G_1. Recalling that for every u, the pair $\{u, v\}$ is an edge in G_2 if and only if $\{u, v'\}$ is an edge in G_2, it follows that for at least $\alpha \cdot tn$ vertices u either $(\pi^{-1}(u), \pi^{-1}(v))$ or $(\pi^{-1}(u), \pi^{-1}(v'))$ is a violation.

4.3   An Optimal Tester for Property Π

In this section we prove that the query complexity of testing Π is at most q (and that this can be met by a relatively efficient tester). We start by describing this (alleged) tester.

Algorithm 4.2. *On input N and proximity parameter ϵ, and when given oracle access to a graph $G = ([N], E)$, the algorithm proceeds as follows:*

1. *Setting $\epsilon' \stackrel{\text{def}}{=} \epsilon/3$ and computing $n \leftarrow \sqrt{q(N)}$.*
2. *Finding n representative vertices; that is, vertices that reside in different alleged clouds, which corresponds to the n vertices of the original graph. This is done by first selecting $s \stackrel{\text{def}}{=} O(\log n)$ random vertices, hereafter called the* signature vertices, *which will be used as a basis for clustering vertices (according to their neighbors in the set of signature vertices). Next, we select $s' \stackrel{\text{def}}{=} O(\epsilon^{-2} \cdot n \log n)$ random vertices, probe all edges between these new vertices and the signature vertices, and cluster these s' vertices accordingly (i.e., two vertices are placed in the same cluster if and only if they neighbor the same signature vertices). If the number of clusters is different from n, then we reject. Furthermore, if the number of vertices that reside in each cluster is not $(1 \pm \epsilon') \cdot s'/n$, then we also reject. Otherwise, we select (arbitrarily) a vertex from each cluster, and proceed to the next step.*
3. *Note that the signature vertices (selected in Step 2) induce a clustering of all the vertices of G. Referring to this clustering, we check that the edges between the clusters are consistent with the edges between the representatives. Specifically, we select uniformly $O(1/\epsilon)$ vertex pairs, cluster the vertices in each pair according to the signature vertices, and check that their edge relation agrees with that of their corresponding representatives. That is, for each pair (u, v), we first find the cluster to which each vertex belongs (by making s adequate queries per each vertex), determine the corresponding representatives, denoted (r_u, r_v), and check (by two queries) whether $\{u, v\} \in E$ iff $\{r_u, r_v\} \in E$. (Needless to say, if one of the newly selected vertices does not reside in any of the n existing clusters, then we reject.)*
4. *Finally, using $\binom{n}{2} < q(N)/2$ queries, we determine the subgraph of G induced by the n representatives. We accept if and only if this induced subgraph is in Π'.*

Note that, for constant value of ϵ, the query complexity is dominated by Step 4, and is thus upper-bounded by $q(N)$. Furthermore, in this case, the above algorithm can be implemented in time $\text{poly}(n \cdot \log N) = \text{poly}(q(N) \cdot \log N)$. We comment that the Algorithm 4.2 is adaptive, and that a straightforward non-adaptive implementation of it has query complexity $O(n \log n)^2 = \widetilde{O}(q(N))$.

Remark 4.3. *In fact, a (non-adaptive) tester of query complexity $\widetilde{O}(q(N))$ can be obtained by a simpler algorithm that selects a random set of s' vertices and accepts if and only if the induced subgraph is ϵ'-close to being a $(s'/n$-factor) blow-up of some graph in Π'_n. Specifically, we can cluster these s' vertices by* *using them also in the role of the signature vertices. Furthermore, these vertices (or part of them) can also be designated for use in Step 3. We note that the analysis of this simpler algorithm does not rely on the hypothesis that Π' is dispersed.*

We now turn to analyzing the performance of Algorithm 4.2. We note that the proof that this algorithm accepts, with very high probability, any graph in Π_N relies on the hypothesis that Π' is dispersed.[9]

We first verify that any graph in Π_N is accepted with very high probability. Suppose that $G \in \Pi_N$ is a N/n-factor blow-up of $G' \in \Pi'_n$. Relying on the fact that Π' is dispersed we note that, for every pair of vertices in $G' \in \Pi'_n$, with constant probability a random vertex has a different edge relation to the members of this pair. Therefore, with very high (constant) probability, a random set of $s = O(\log n)$ vertices yields n different neighborhood patterns for the n vertices of G'. It follows that, with the same high probability, the s signature vertices selected in Step 2 induced n (equal sized) clusters on the vertices of G, where each cluster contains the cloud of N/n vertices (of G) that replaces a single vertex of G'. Thus, with very high (constant) probability, the sample of $s' = O(\epsilon^{-2} \cdot n \log n)$ additional vertices selected in Step 2 hits each of these clusters (equiv., clouds) and furthermore has $(1 \pm \epsilon') \cdot s'/n$ hits in each cluster. We conclude that, with very high (constant) probability, Algorithm 4.2 does not reject G in Step 2. Finally, assuming that Step 2 does not reject (and we did obtain representatives from each cloud of G), Algorithm 4.2 never rejects $G \in \Pi$ in Steps 3 and 4.

We now turn to the case that G is ϵ-far from Π_N, where we need to show that G is rejected with high constant probability (say, with probability $2/3$). We will actually prove that if G is accepted with sufficiently high constant probability (say, with probability $1/3$), then it is ϵ-close to Π_N. We call a set of s vertices **good** if (when used as the set of signature vertices) it induces a clustering of the vertices of G such that n of these clusters are each of size $(1 \pm 2\epsilon') \cdot N/n$. Note that good s-vertex sets must exist, because otherwise Algorithm 4.2 rejects in Step 2 with probability at least $1 - \exp(\Omega(\epsilon^2/n) \cdot s') > 2/3$. Fixing any good s-vertex set S, we call a sequence of n vertices $R = (r_1, ..., r_n)$ **well-representing** if (1) the subgraph of G induced by R is in Π'_n, and (2) at most ϵ' fraction of the vertex pairs of G have edge relation that is inconsistent with the corresponding vertices in R (i.e., at most ϵ' fraction of the vertex pairs in G violate the condition by which $\{u, v\} \in E$ if and only if $\{r_i, r_j\} \in E$, where u resides in the ith cluster (w.r.t S) and v resides in the jth cluster). Now, note that there must exist a good s-vertex set S that has a well-representing n-vertex sequence $R = (r_1, ..., r_n)$, because otherwise Algorithm 4.2 rejects with probability at least $2/3$ (i.e., if a ρ fraction of the s-vertex sets are good (but have no corresponding n-sequence that is well-representing), then Step 2 rejects with probability at least $(1 - \rho) \cdot 0.9$ and either Step 3 or Step 4 reject with probability $\rho \cdot \min((1 - (1 - \epsilon')^{\Omega(1/\epsilon)}), 1))$.

---

[9] In contrast, the proof that Algorithm 4.2 rejects, with very high probability, any graph that is ϵ-far from Π_N does not rely on this hypothesis.

Fixing any good s-vertex set S and any corresponding $R = (r_1, ..., r_n)$ that is well-representing, we consider the clustering induced by S, denoted $(C_1, ..., C_n, X)$, where X denotes the set of (untypical) vertices that do not belong to the n first clusters. Recall that, for every $i \in [n]$, it holds that $r_i \in C_i$ and $|C_i| = (1 \pm 2\epsilon') \cdot N/n$. Furthermore, denoting by $i(v)$ the index of the cluster to which vertex $v \in [N] \setminus X$ belongs, it holds that the number of pairs $\{u, v\}$ (from $[N] \setminus X$) that violate the condition $\{u, v\} \in E$ iff $\{r_{i(u)}, r_{i(v)}\} \in E$ is at most $\epsilon' \cdot \binom{N}{2}$. Now, observe that by modifying at most $\epsilon' \cdot \binom{N}{2}$ edges in G we can eliminate all the aforementioned violations, which means that we obtain n sets with edge relations that fit some graph in Π'_n (indeed the graph obtained as the subgraph of G induced by R, which was not modified). Recall that these sets are each of size $(1 \pm 2\epsilon') \cdot N/n$, and so we may need to move $2\epsilon' N$ vertices in order to obtain sets of size N/n. This movement may create up to $2\epsilon' N \cdot (N - 1)$ new violations, which can be eliminated by modifying at most $2\epsilon' \cdot \binom{N}{2}$ additional edges in G. Using $\epsilon = 3\epsilon'$, we conclude that G is ϵ-close to Π_N.

5   Revisiting the Adj. Matrix Model: Monotonicity

In continuation to Section 4, which provides a hierarchy theorem for generic graph properties (in the adjacency matrix model), we present in this section a hierarchy theorem for *monotone* graph properties (in the same model). We say that a graph property Π is **monotone** if adding edges to any graph that resides in Π yields a graph that also resides in Π. (That is, we actually refer to upward monotonicity, and an identical result for downward monotonicity follows by considering the complement graphs.)[10]

Theorem 5. *In the adjacency matrix model, for every* $q : \mathbb{N} \to \mathbb{N}$ *that is at most quadratic, there exists a* monotone *graph property* Π *that is testable in* $O(q)$ *queries, but is not testable in* $o(q)$ *queries.*

Note that Theorem 5 refers to two-sided error testing (just like Theorem 4). Theorems 4 and 5 are incomparable: the former provides graph properties that are in $\mathcal{P}$ (and the upper bound is established via relatively efficient testers), whereas the latter provides graph properties that are monotone.

Outline of the proof of Theorem 5. Starting with the proof of Theorem 4, one may want to apply a monotone closure to the graph property Π (presented in the proof of Theorem 4).[11] Under suitable tuning of parameters, this allows to retain the proof of the lower bound, but the problem is that the tester presented for the upper bound fails. The point is that this tester relies on the structure of graphs obtained via blow-up, whereas this structure is not maintained by

---

[10] We stress that these notions of monotonicity are different from the notion of monotonicity considered in [AS], where a graph property Π is called monotone if any subgraph of a graph in Π is also in Π.

[11] Indeed, this is the approach used in the proof of [GT, Thm. 1].

the monotone closure. One possible solution, which assumes that all graphs in Π have approximately the same number of edges, is to augment the monotone closure of Π with all graphs that have significantly more edges, where the corresponding threshold (on the number of edges) is denoted T. Intuitively, this way, we can afford accepting any graph that has more than T edges, and handle graphs with fewer edges by relying on the fact that in this case the blow-up structure is essentially maintained (because only few edges are added). Unfortunately, implementing this idea is not straightforward: On one hand, we should set the threshold high enough so that the lower bound proof still holds, whereas on the other hand such a setting may destroy the local structure of a constant fraction of the graph's vertices. The solution to this problem is to use an underlying property Π' that supports "error correction" (i.e., allows recovering the original structure even when a constant fraction of it is destroyed as above). (The actual proof of Theorem 5 is given in our technical report [GKNR].)

6   Revisiting the Adj. Matrix Model: One-Sided Error

In continuation to Section 4, which provides a hierarchy theorem for two-sided error testing of graph properties (in the adjacency matrix model), we present in this section a hierarchy theorem that refers to one-sided error testing. Actually, the lower bounds will hold also with respect to two-sided error, but the upper bounds will be established using a tester of one-sided error.

Theorem 6. *In the adjacency matrix model, for every* $q : \mathbb{N} \to \mathbb{N}$ *that is at most quadratic, there exists a graph property Π that is testable with one-sided error in $O(q)$ queries, but is not testable in $o(q)$ queries even when allowing two-sided error. Furthermore, Π is in $\mathcal{P}$.*

Theorems 4 and 6 are incomparable: in the former the upper bound is established via relatively efficient testers (of two-sided error), whereas in the latter the upper bound is established via one-sided error testers (which are not relatively efficient). (Unlike Theorem 5, both Theorems 4 and 6 do not provide monotone properties.)

Outline of the proof of Theorem 6. Starting with the proof of Theorem 4, we observe that the source of the two-sided error of the tester is in the need to approximate set sizes. This is unavoidable when considering graph properties that are blow-ups of some other graph properties, where blow-up is defined by replacing vertices of the original graph by *equal-size* clouds. The natural solution is to consider a *generalized* notion of blow-up in which each vertex is replaced by a (non-empty) cloud of arbitrary size. That is, G is a (generalized) blow-up of $G' = ([n], E')$ if the vertex set of G can be partitioned into n non-empty sets (of arbitrary sizes) that correspond to the n vertices of G' such that the edges between these sets represent the edges of G'; that is, if $\{i, j\}$ is an edge in G' (i.e., $\{i, j\} \in E'$), then there is a complete bipartite between the ith set and the

j^{th} set, and otherwise (i.e., $\{i, j\} \notin E'$) there are no edges between this pair of sets.

The actual proof of Theorem 6 is given in our technical report [GKNR]. Among other things, this proof copes with the non-trivial question of how does the *generalized* (rather than the standard) blow-up operation affect distances between graphs.

7 Concluding Comments

Theorems 4, 5 and 6 (and their proofs) raise several natural open problems, listed next. We stress that all questions refer to the adjacency matrix graph model considered in Sections 4–6.

1. *Preservation of distance between graphs under blow-up*: Recall that the proof of Theorem 4 relies on the preservation of distances between graphs under the blow-up operation. The partial results (regarding this matter) obtained in this work suffice for the proof of Theorem 4, but the problem seems natural and of independent interest.

   Recall that Claim 4.1 asserts that in some cases the distance between two unlabeled graphs is preserved *up to a constant factor* by any blow-up (i.e., "linear preservation"), whereas Theorem 8 of our technical report [GKNR] asserts a quadratic preservation for any pair of graphs. Also recall that it is not true that the distance between any two unlabeled graphs is *perfectly preserved* by any blow-up (see beginning of Appendix B in our technical report [GKNR]).

   In earlier versions of this work we raised the natural question of whether the distance between any two unlabeled graphs is preserved up to a constant factor by any blow-up. This question has been recently resolved by Oleg Pikhurko, who showed that *the distance is indeed preserved up to a factor of three* [P, Sec. 4]. Note that Arie Matsliah's counterexample to perfect preservation (presented in Appendix B of our technical report [GKNR]) shows that the said constant factor cannot be smaller than $6/5$. Indeed, determining the true constant factor remains an open problem.

2. *Combining the features of all three hierarchy theorems*: Theorems 4, 5 and 6 provide incomparable hierarchy theorems, each having an additional feature that the others lack. Specifically, Theorem 4 refers to properties in $\mathcal{P}$ (and testing, in the positive part, is relatively efficient), Theorem 5 refers to monotone properties, and Theorem 6 provides one-sided testing (in the positive part). Is it possible to have a single hierarchy theorem that enjoys all three additional feature? Intermediate goals include the following:

   (a) *Hierarchy of monotone graph properties in $\mathcal{P}$*: Recall that Theorem 4 is proved by using non-monotone graph properties (which are in $\mathcal{P}$), while Theorem 5 refers to monotone graph properties that are not likely to be in $\mathcal{P}$. Can one combine the good aspects of both results?

(b) *Hard-to-test monotone graph property in* $\mathcal{P}$: Indeed, before addressing Problem 2a, one should ask whether a result analogous to Theorem 7 of our technical report [GKNR] holds for a monotone graph property? Recall that [GT, Thm. 1] provides a monotone graph property in $\mathcal{NP}$ that is hard-to-test.

(c) *One-sided versus two-sided error testers*: Recall that the positive part of Theorem 6 refers to testing with one-sided error, but these testers are not relatively efficient. In contrast, the positive part of Theorem 4 provides relatively efficient testers, but these testers have two-sided error. Can one combine the good aspects of both results?

Acknowledgments

We are grateful to Ronitt Rubinfeld for asking about the existence of hierarchy theorems for the adjacency matrix model. Ronitt raised this question during a discussion that took place at the Dagstuhl 2008 workshop on sub-linear algorithms. We are also grateful to Arie Matsliah and Yoav Tzur for helpful discussions. In particular, we thank Arie Matsliah for providing us with a proof that the blow-up operation does not preserve distances in a perfect manner. Oded Goldreich was partially supported by the Israel Science Foundation (grant No. 1041/08). Michael Krivelevich was partially supported by a USA-Israel BSF Grant, by a grant from the Israel Science Foundation, and by Pazy Memorial Award.

References

[ABI]   Alon, N., Babai, L., Itai, A.: A fast and Simple Randomized Algorithm for the Maximal Independent Set Problem. J. of Algorithms 7, 567–583 (1986)

[AFKS]   Alon, N., Fischer, E., Krivelevich, M., Szegedy, M.: Efficient Testing of Large Graphs. Combinatorica 20, 451–476 (2000)

[AFNS]   Alon, N., Fischer, E., Newman, I., Shapira, A.: A Combinatorial Characterization of the Testable Graph Properties: It's All About Regularity. In: 38th STOC, pp. 251–260 (2006)

[AGHP]   Alon, N., Goldreich, O., Hastad, J., Peralta, R.: Simple constructions of almost k-wise independent random variables. Journal of Random structures and Algorithms 3(3), 289–304 (1992)

[AS]   Alon, N., Shapira, A.: Every Monotone Graph Property is Testable. SIAM Journal on Computing 38, 505–522 (2008)

[BSS]   Benjamini, I., Schramm, O., Shapira, A.: Every Minor-Closed Property of Sparse Graphs is Testable. In: 40th STOC, pp. 393–402 (2008)

[BLR]   Blum, M., Luby, M., Rubinfeld, R.: Self-Testing/Correcting with Applications to Numerical Problems. JCSS 47(3), 549–595 (1993)

[BHR]   Ben-Sasson, E., Harsha, P., Raskhodnikova, S.: 3CNF Properties Are Hard to Test. SIAM Journal on Computing 35(1), 1–21 (2005)

[BOT]   Bogdanov, A., Obata, K., Trevisan, L.: A lower bound for testing 3-colorability in bounded-degree graphs. In: 43rd FOCS, pp. 93–102 (2002)

[EKK+]   Ergun, F., Kannan, S., Kumar, S.R., Rubinfeld, R., Viswanathan, M.: Spot-checkers. JCSS 60(3), 717–751 (2000)

[F]   Fischer, E.: The art of uninformed decisions: A primer to property testing. Bulletin of the European Association for Theoretical Computer Science 75, 97–126 (2001)

[FM]   Fischer, E., Matsliah, A.: Testing Graph Isomorphism. In: 17th SODA, pp. 299–308 (2006)

[GGL+]   Goldreich, O., Goldwasser, S., Lehman, E., Ron, D., Samorodnitsky, A.: Testing Monotonicity. Combinatorica 20(3), 301–337 (2000)

[GGR]   Goldreich, O., Goldwasser, S., Ron, D.: Property testing and its connection to learning and approximation. Journal of the ACM, 653–750 (July 1998)

[GKNR]   Goldreich, O., Krivelevich, M., Newman, I., Rozenberg, E.: Hierarchy Theorems for Property Testing. ECCC, TR08-097 (2008)

[GR1]   Goldreich, O., Ron, D.: Property Testing in Bounded Degree Graphs. Algorithmica 32(2), 302–343 (2002)

[GR2]   Goldreich, O., Ron, D.: A Sublinear Bipartiteness Tester for Bounded Degree Graphs. Combinatorica 19(3), 335–373 (1999)

[GT]   Goldreich, O., Trevisan, L.: Three theorems regarding testing graph properties. Random Structures and Algorithms 23(1), 23–57 (2003)

[LNS]   Lachish, O., Newman, I., Shapira, A.: Space Complexity vs. Query Complexity. Computational Complexity 17, 70–93 (2008)

[NN]   Naor, J., Naor, M.: Small-bias Probability Spaces: Efficient Constructions and Applications. SIAM J. on Computing 22, 838–856 (1993)

[PRR]   Parnas, M., Ron, D., Rubinfeld, R.: Testing Membership in Parenthesis Laguages. Random Structures and Algorithms 22(1), 98–138 (2003)

[P]   Pikhurko, O.: An Analytic Approach to Stability (2009), http://arxiv.org/abs/0812.0214

[R]   Ron, D.: Property testing. In: Rajasekaran, S., Pardalos, P.M., Reif, J.H., Rolim, J.D.P. (eds.) Handbook on Randomization, vol. II, pp. 597–649 (2001)

[RS]   Rubinfeld, R., Sudan, M.: Robust characterization of polynomials with applications to program testing. SIAM Journal on Computing 25(2), 252–271 (1996)

[S]   Shaltiel, R.: Recent Developments in Explicit Constructions of Extractors. In: Current Trends in Theoretical Computer Science: The Challenge of the New Century. Algorithms and Complexity, vol. 1, pp. 67–95. World scientific, Singapore (2004); Preliminary version in Bulletin of the EATCS, vol. 77, pp. 67–95 (2002)

Algorithmic Aspects of Property Testing in the Dense Graphs Model

Oded Goldreich[1],[*] and Dana Ron[2],[**]

[1] Department of Computer Science, Weizmann Institute of Science, Rehovot, Israel
oded.goldreich@weizmann.ac.il
[2] Department of Electrical Engineering-Systems, Tel Aviv University, Tel Aviv Israel
danar@eng.tau.ac.il

Abstract. In this paper we consider two basic questions regarding the query complexity of testing graph properties in the adjacency matrix model. The first question refers to the relation between adaptive and non-adaptive testers, whereas the second question refers to testability within complexity that is inversely proportional to the proximity parameter, denoted ϵ. The study of these questions reveals the importance of algorithmic design (also) in this model. The highlights of our study are:

- A gap between the complexity of adaptive and non-adaptive testers. Specifically, there exists a (natural) graph property that can be tested using $\widetilde{O}(\epsilon^{-1})$ adaptive queries, but cannot be tested using $o(\epsilon^{-3/2})$ non-adaptive queries.
- In contrast, there exist natural graph properties that can be tested using $\widetilde{O}(\epsilon^{-1})$ non-adaptive queries, whereas $\Omega(\epsilon^{-1})$ queries are required even in the adaptive case.

We mention that the properties used in the foregoing conflicting results have a similar flavor, although they are of course different.

1   Introduction

In the last decade, the area of property testing has attracted much attention (see the surveys of [9,17], which are already out-of-date). Loosely speaking, property testing typically refers to sub-linear time probabilistic algorithms for deciding whether a given object has a predetermined property or is far from any object having this property. Such algorithms, called testers, obtain bits of the object by making adequate queries, which means that the object is seen as a function and the testers get oracle access to this function (and thus may be expected to work in time that is sub-linear in the length of the description of this object).

---

[*] Partially supported by the Israel Science Foundation (grants No. 460/05 and 1041/08).
[**] Partially supported by the Israel Science Foundation (grants No. 89/05 and 246/08).

I. Dinur et al. (Eds.): APPROX and RANDOM 2009, LNCS 5687, pp. 520–533, 2009.

Much of the aforementioned work (see, e.g., [11,2,4]) was devoted to the study of testing graph properties in the adjacency matrix model, which is also the setting of the current work. In this model, introduced in [11], graphs are viewed as (symmetric) Boolean functions over a domain consisting of all possible vertex-pairs (i.e., an N-vertex graph $G = ([N], E)$ is represented by the function $g :$ $[N] \times [N] \to \{0, 1\}$ such that $\{u, v\} \in E$ if and only if $g(u, v) = 1$). Consequently, an N-vertex graph represented by the function $g : [N] \times [N] \to \{0, 1\}$ is said to be ϵ-far from some predetermined graph property more than $\epsilon \cdot N^2$ entries of g must be modified in order to yield a representation of a graph that has this property. We refer to ϵ as the proximity parameter, and the complexity of testing is stated in terms of ϵ and the number of vertices in the graph (i.e., N).

Interestingly, many natural graph properties can be tested within query complexity that depends only on the proximity parameter; see [11], which presents testers with query complexity poly($1/\epsilon$), and [4], which characterizes the class of properties that are testable within query complexity that depends only on the proximity parameter (where this dependence may be an arbitrary function of ϵ). However, a common phenomenon in all the aforementioned works is that they utilize quite naive algorithms and their focus is on the (often quite sophisticated) analysis of these algorithms. This phenomenon is no coincidence: As shown in [2,15], when ignoring a quadratic blow-up in the query complexity, property testing (in this model) reduces to sheer combinatorics. Specifically, without loss of generality, the tester may just inspect a random induced subgraph (of adequate size) of the input graph.

In this paper we demonstrate that a more refined study of property testing (in this model) reveals the importance of algorithmic design (also in this model). This is demonstrated both by studying the advantage of adaptive testers over non-adaptive ones as well as by studying the class of properties that can be tested within complexity that is inversely proportional to the proximity parameter.

1.1   Two Related Studies

We start by reviewing the two related studies conducted in the current work.

Adaptivity vs. Non-adaptivity. A tester is called non-adaptive if it determines all its queries independently of the answers obtained for previous queries, and otherwise it is called adaptive. Indeed, by [2,15], the benefit of adaptivity (or, equivalently, the cost of non-adaptivity) is polynomially bounded: Specifically, any (possibly adaptive) tester (for any graph property) of query complexity $q(N, \epsilon)$ can be transformed into a non-adaptive tester of query complexity $O(q(N, \epsilon)^2)$. But is this quadratic gap an artifact of the known proofs (of [2,15]) or does it reflect something inherent?

A recent work by [16] suggests that the latter case may hold: For every $\epsilon > 0$, they showed that the set of N-vertex bipartite graphs of maximum degree $O(\epsilon N)$ is ϵ-testable (i.e., testable with respect to proximity parameter ϵ) by $\widetilde{O}(\epsilon^{-3/2})$ queries, while (by [7]) a non-adaptive tester for this set must use $\Omega(\epsilon^{-2})$ queries. Thus, there exists a case where non-adaptivity has the cost of increasing the query complexity; specifically, for any $c < 4/3$, the query complexity of the non-adaptive tester is greater than a c-power of the query complexity of the adaptive tester (i.e., $\widetilde{O}(\epsilon^{-3/2})^c = o(\epsilon^{-2})$). We stress that the result of [16] does not refer to property testing in the "proper" sense; that is, the complexity is not analyzed with respect to a varying value of the proximity parameter for a fixed property. It is rather the case that, for every value of the proximity parameter, a different property (which depends on this parameter) is considered and the (upper- and lower-) bounds refer to this combination (of a property tailored for a fixed value of the proximity parameter). Thus, *the work of [16] leaves open the question of whether there exists a single graph property such that adaptivity is beneficial for any value of the proximity parameter* (as long as $\epsilon > N^{-\Omega(1)}$). That is, the question is whether adaptivity is beneficial for the standard asymptotic-complexity formulation of property testing.

Complexity inversely proportional to the proximity parameter. As shown in [11], many natural graph properties can be tested within query complexity that is polynomial in the reciprocal of the proximity parameter (and independent of the size of the graph). We ask whether a linear complexity is possible at all, and if so which properties can be tested within query complexity that is linear (or almost linear) in the reciprocal of the proximity parameter.[1]

The first question is easy to answer (even when avoiding trivial properties).[2] Note that the property of being a clique (equiv., an independent set) can be tested by $O(1/\epsilon)$ queries, even when these queries are non-adaptive (e.g., make $O(1/\epsilon)$ random queries and accept if and only if all return 1). Still, we ask whether "more interesting"[3] graph theoretical properties can also be tested within similar complexity (either only adaptively or also non-adaptively).

1.2 Our Results

We address the foregoing questions by studying a sequence of natural graph properties. The first property in the sequence, called clique collection and denoted $\mathcal{CC}$, is the set of graphs such that each graph consists of a collection of isolated cliques. Testing this property corresponds to the following natural clustering problem: can a set of possibly related elements be partitioned into "perfect clusters" (i.e., two elements are in the same cluster if and only if they are related)? For this property (i.e., $\mathcal{CC}$), we prove a gap between adaptive and non-adaptive query complexity, where the adaptive query complexity is almost linear in the reciprocal of the proximity parameter. That is:

---

[1] Note that $\Omega(1/\epsilon)$ queries are required for testing any of the graph properties considered in the current work; for a more general statement see our technical report [13].

[2] A graph property Π is trivial for testing if for every $\epsilon > 0$ there exists $N_0 > 0$ such that for every $N \geq N_0$ either all N-vertex graphs belong to Π or all of them are ϵ-far from Π.

[3] A more articulated reservation towards the foregoing properties may refer to the fact that these graph properties contain a single N-vertex graph (per each N) and are represented by monochromatic functions.

Theorem 1. (the query complexity of clique collection):

1. *There exists an adaptive tester of query complexity $\widetilde{O}(\epsilon^{-1})$ for $\mathcal{CC}$. Furthermore, this tester runs in time $\widetilde{O}(\epsilon^{-1})$.[4]*
2. *Any non-adaptive tester for $\mathcal{CC}$ must have query complexity $\Omega(\epsilon^{-4/3})$.*
3. *There exists a non-adaptive tester of query complexity $O(\epsilon^{-4/3})$ for $\mathcal{CC}$. Furthermore, this tester runs in time $O(\epsilon^{-4/3})$.*

Note that the complexity gap (between Parts 1 and 2) of Theorem 1 matches the gap established by [16] (for "non-proper" testing). A larger gap is established for a property of graphs, called bi-clique collection and denoted $\mathcal{BCC}$, where a graph is in $\mathcal{BCC}$ if it consists of a collection of isolated bi-cliques (i.e., complete bipartite graphs). We note that bi-cliques may be viewed as the bipartite analogues of cliques (w.r.t. general graphs), and indeed they arise naturally in (clustering) applications that are modeled by bipartite graphs over two types of elements.

Theorem 2. (the query complexity of bi-clique collection):

1. *There exists an adaptive tester of query complexity $\widetilde{O}(\epsilon^{-1})$ for $\mathcal{BCC}$. Furthermore, this tester runs in time $\widetilde{O}(\epsilon^{-1})$.*
2. *Any non-adaptive tester for $\mathcal{BCC}$ must have query complexity $\Omega(\epsilon^{-3/2})$. Furthermore, this holds even if the input graph is promised to be bipartite.*

The furthermore clause (in Part 2 of Theorem 2) holds also for the model studied in [3], where the bi-partition of the graph is given.

Theorem 2 asserts that the gap between the query complexity of adaptive and non-adaptive testers may be a power of $1.5 - o(1)$. Recall that the results of [2,15] assert that the gap may not be larger than quadratic. We conjecture that this upper-bound can be matched.

Conjecture 3. (an almost-quadratic complexity gap): *For every positive integer $t \geq 5$, there exists a graph property Π for which the following holds:*

1. *There exists an adaptive tester of query complexity $\widetilde{O}(\epsilon^{-1})$ for Π.*
2. *Any non-adaptive tester for Π must have query complexity $\Omega(\epsilon^{-2+(2/t)})$.*

Furthermore, Π consists of graphs that are each a collection of "super-cycles" of length t, where a super-cycle is a set of t independent sets arranged on a cycle such that each pair of adjacent independent sets is connected by a complete bipartite graph.

We were able to prove Part 2 of Conjecture 3, but failed to provide a full analysis of an algorithm that we designed for Part 1. However, we were able to prove a *promise problem version of Conjecture 3*; specifically, this promise problem refers to inputs promised to reside in a set $\Pi' \supset \Pi$ and the tester is required to distinguish graphs in Π from graphs that are ϵ-far from Π. For further details see our technical report [13].

---

[4] We refer to a model in which elementary operations regarding pairs of vertices are charged at unit cost.

In contrast to the foregoing results that aim at identifying properties with a substantial gap between the query complexity of adaptive versus non-adaptive testing, we also study cases in which no such gap exists. Since query complexity that is linear in the reciprocal of the proximity parameter is minimal for many natural properties (and, in fact, for any property that is "non-trivial for testing" (see Footnote 2)), we focus on non-adaptive testers that (approximately) meet this bound. Among the results obtained in this direction, we highlight the following one.

Theorem 4. (the query complexity of collections of $O(1)$ cliques): *For every positive integer c, there exists a non-adaptive tester of query complexity $\widetilde{O}(\epsilon^{-1})$ for the set of graphs such that each graph consists of a collection of up to c cliques. Furthermore, this tester runs in time $\widetilde{O}(\epsilon^{-1})$.*

Discussion. The foregoing results demonstrate that a finer look at property testing of graphs in the adjacency matrix model reveals the role of algorithm design in this model. In particular, in some cases (see, e.g., Theorems 1 and 2), carefully designed adaptive algorithms outperform any non-adaptive algorithm. Indeed, this conclusion stands in contrast to [15, Thm. 2], which suggests that a less fine view (which ignores polynomial blow-ups)[5] deems algorithm design irrelevant to this model. We also note that, in some cases (see, e.g., Theorem 4 and Part 3 of Theorem 1), carefully designed non-adaptive algorithms outperform canonical ones.

As discussed previously, one of the goals of this work was to study the relation between adaptive and non-adaptive testers in the adjacency matrix model. Our results demonstrate that, in this model, the relation between the adaptive and non-adaptive query-complexities is not fixed, but rather varies with the computational problem at hand. In some cases (e.g., Theorem 4) the complexities are essentially equal (indeed, as in the case of sampling [8]). In other cases (e.g., Theorem 1), these complexities are related by a fixed power (e.g., 4/3) that is strictly between 1 and 2. And, yet, in other cases the non-adaptive complexity is quadratic in the adaptive complexity, which is the maximum gap possible (by [2,15]). Furthermore, for any $t \geq 4$, there exists a promise problem for which the aforementioned complexities are related by a power of $2 - (2/t)$.

Needless to say, the fundamental relation between adaptive and non-adaptive algorithms was studied in a variety of models, and the current work studies it in a specific natural model (i.e., of property testing in the adjacency matrix representation). In particular, this relation has been studied in the context of property testing in other domains. Specifically, in the setting of testing the satisfiability of linear constraints, it was shown that adaptivity offers absolutely no gain [6]. A similar result holds for testing monotonicity of sequences of positive integers [10]. In contrast, an exponential gap between the adaptive and non-adaptive complexities may exist in the context of testing other properties of functions [10]. Lastly, we mention that an even more dramatic gap exists in the setting of testing graph properties in the bounded-degree model (of [12]); see [18].

---

[5] Recall that [15, Thm. 2] asserts that canonical testers, which merely select a random subset of vertices and rule according to the induced subgraph, have query-complexity that is at most quadratic in the query-complexity of the best tester. We note that [15, Thm. 2] also ignores the time-complexity of the testers.

1.3   Open Problems

In addition to the resolution of Conjecture 3, our study raises many other open problems; the most evident ones are listed next.

1. What is the non-adaptive query complexity of $\mathcal{BCC}$? Note that Theorem 2 only establishes a lower-bound of $\Omega(\epsilon^{-3/2})$. We conjecture that an efficient non-adaptive algorithm of query complexity $\widetilde{O}(\epsilon^{-3/2})$ can be devised.
2. For which constants $c \in [1, 2]$ does there exist a property that has adaptive query complexity of $q(\epsilon)$ and non-adaptive query complexity of $\widetilde{\Theta}(q(\epsilon)^c)$? Note that Theorem 1 shows that 4/3 is such a constant, and the same holds for the constant 1 (see, e.g., Theorem 4). We conjecture that, for any $t \geq 2$, it holds that the constant $2 - (2/t)$ also satisfies the foregoing requirement. It may be the case that these constants are the only ones that satisfy this requirement.
3. Characterize the class of graph properties for which the query complexity of non-adaptive testers is almost linear in the query complexity of adaptive testers.
4. Characterize the class of graph properties for which the query complexity of non-adaptive testers is almost quadratic in the query complexity of adaptive testers.
5. Characterize the class of graph properties for which the query complexity of adaptive (resp., non-adaptive) testers is almost linear in the reciprocal of the proximity parameter.

Finally, we recall the well-known open problem (partially addressed in [5]) of providing a characterization of the class of graph properties that are testable within query complexity that is polynomial in the reciprocal of the proximity parameter.

1.4   Organization

Due to space limitations, this version only contains the proofs of the first two items of Theorem 1, and the proofs of all other results can be found in our technical report [13]. Specifically, in Section 2 we present an adaptive tester of almost-linear (i.e., $\widetilde{O}(\epsilon^{-1})$) query complexity for Clique Collection, and in Section 3 we contrast it with a (tight) $\Omega(\epsilon^{-4/3})$ lower-bound on the query complexity of non-adaptive testers.

2   The Adaptive Query Complexity of $\mathcal{CC}$

In this section we study the (adaptive) query complexity of clique collection, presenting an almost optimal (adaptive) tester for this property. Loosely speaking, the tester starts by finding a few random neighbors of a few randomly selected start vertices, and then examines the existence of edges among the neighbors of each start vertex as well as among these neighbors and the non-neighbors of each start vertex.

We highlight the fact that adaptivity is used in order to perform queries that refer only to pairs of neighbors of the same start vertex. To demonstrate the importance of this fact, consider the case that the N-vertex graph is partitioned into $O(1/\epsilon)$ connected components each having $O(\epsilon N)$ vertices. Suppose that we wish to tell whether the connected component that contains the vertex v is indeed a clique. Using adaptive queries we may first find two neighbors of v, by selecting $t \stackrel{\text{def}}{=} O(1/\epsilon)$ random vertices and checking whether each such vertex is adjacent to v, and then check whether these *two* neighbors are adjacent. In contrast, intuitively, a non-adaptive procedure cannot avoid making all $\binom{t}{2}$ possible queries.

The foregoing adaptive procedure is tailored to the case that the N-vertex graph is partitioned into $O(1/\epsilon)$ ("strongly connected") components, each having $O(\epsilon N)$ vertices. In such a case, it suffices to check that a constant fraction of these components are in fact cliques (or rather close to being so) and that there are no edges (or rather relatively few edges) from these cliques to the rest of the graph. However, if the components (and potential cliques) are larger, then we should check more of them. Fortunately, due to their larger size, finding neighbors requires less queries, and the total number of queries remains invariant. These considerations lead us to the following algorithm.

Algorithm 1. (adaptive tester for $\mathcal{CC}$): *On input N and ϵ and oracle access to a graph $G = ([N], E)$, set $t_1 = \Theta(1)$ and $t_2 = \Theta(\log^3(1/\epsilon))$, and proceed in $\ell \stackrel{\text{def}}{=} \log_2(1/\epsilon) + 2$ iterations as follows: For $i = 1, \dots, \ell$, select uniformly $t_1 \cdot 2^i$ start vertices and for each selected vertex $v \in [N]$ perform the following sub-test, denoted* sub-test$_i(v)$:

1. *Select at random a sample, S, of $t_2/(2^i \epsilon)$ vertices.*
2. *Determine $\Gamma_S(v) = S \cap \Gamma(v)$, by making the queries (v, w) for each $w \in S$.*
3. *If $|\Gamma_S(v)| \leq \sqrt{t_2/2^i \epsilon}$ then check that for every $u, w \in \Gamma_S(v)$ it holds that $(u, w) \in E$. Otherwise (i.e., $|\Gamma_S(v)| > \sqrt{t_2/2^i \epsilon}$), select a sample of $t_2/(2^i \epsilon)$ pairs in $\Gamma_S(v) \times \Gamma_S(v)$ and check that each selected pair is in E.*
4. *Select a sample of $t_2/(2^i \epsilon)$ pairs in $\Gamma_S(v) \times (S \setminus \Gamma_S(v))$ and check that each selected pair is* not *in E.*

The sub-test (i.e., sub-test$_i(v)$*) accepts if and only if all checks were positive (i.e., no edges were missed in Step 3 and no edges were detected in Step 4). The tester itself accepts if and only if all $\sum_{i=1}^{\ell} t_1 \cdot 2^i$ invocations of the sub-test accepted.*

The query complexity of this algorithm is $\sum_{i=1}^{\ell} t_1 2^i \cdot O(t_2/2^i \epsilon) = O(\ell \cdot t_1 t_2/\epsilon) = \tilde{O}(1/\epsilon)$, and evidently it is efficient. Clearly, this algorithm accepts (with probability 1) any graph that is in $\mathcal{CC}$. It remains to analyze its behavior on graphs that are ϵ-far from $\mathcal{CC}$.

Lemma 1. *If $G = ([N], E)$ is ϵ-far from $\mathcal{CC}$, then on input N, ϵ and oracle access to G, Algorithm 1 rejects with probability at least $2/3$.*

Part 1 of Theorem 1 follows.

Proof. We shall prove the contrapositive statement; that is, that if Algorithm 1 accepts with probability at least $1/3$ then the graph is ϵ-close to $\mathcal{CC}$. The proof evolves around the following notion of i-good start vertices (for $i \in [\ell]$). We first show that if Algorithm 1 accepts with probability at least $1/3$ then the number of "important" vertices that are not i-good is relatively small, and next show how to use the i-good vertices in order to construct a partition of the vertices that demonstrates that the graph is ϵ-close to $\mathcal{CC}$. The following definition refers to a parameter γ_2, which will be set to $\Theta(1/t_2)$.

Definition 1. *A vertex v is i-good if the following two conditions hold.*

1. *The number of missing edges in the subgraph induced by $\Gamma(v)$ is at most $\gamma_2 \cdot 2^i \epsilon \cdot |\Gamma(v)| \cdot N$.*

2. *For every positive integer $j \le j_0 \stackrel{\text{def}}{=} \log_2(|\Gamma(v)|/(\gamma_2 \cdot 2^i \epsilon N))$, the number of vertices in $\Gamma(v)$ that have at least $\gamma_2 \cdot 2^{i+j} \epsilon \cdot N$ edges going out of $\Gamma(v)$ is at most $2^{-j} \cdot |\Gamma(v)|$.*

Note that Condition 1 holds vacuously whenever $|\Gamma(v)| < \gamma_2 \cdot 2^i \cdot \epsilon \cdot N$. However, when $|\Gamma(v)| \gg \gamma_2 \cdot 2^i \epsilon \cdot N$, Condition 1 implies that at least 99% of the vertices in $\Gamma(v)$ have at least $0.99 \cdot |\Gamma(v)|$ neighbors in $\Gamma(v)$. Condition 2 implies that, when ignoring at most $2^{-j_0} \cdot |\Gamma(v)| < \gamma_2 \cdot 2^i \epsilon \cdot N$ vertices (in $\Gamma(v)$), the number of edges going out of $\Gamma(v)$ is at most $\sum_{j=1}^{j_0} 2^{-(j-1)} |\Gamma(v)| \cdot \gamma_2 2^{i+j} \epsilon N$, which is less than $4\ell \cdot \gamma_2 2^i \epsilon \cdot |\Gamma(v)| \cdot N$, since $j_0 \le \log_2(1/\gamma_2 2^i \epsilon) \le \log_2(1/\gamma_2 \epsilon) < 2\log_2(1/\epsilon)$.

Claim 2. *If v has degree at least $\gamma_2 \cdot 2^i \epsilon \cdot N$ and is not i-good, then the probability that sub-test$_i(v)$ accepts is less than 5%.*

The proof can be found in our technical report [13].

Claim 3. *If Algorithm 1 accepts with probability at least $1/3$, then for every $i \in [\ell]$ the number of vertices of degree at least $\gamma_2 \cdot 2^i \epsilon \cdot N$ that are not i-good is at most $\gamma_1 \cdot 2^{-i} \cdot N$, where $\gamma_1 \stackrel{\text{def}}{=} \Theta(1/t_1)$.*

Claim 3 follows by combining Claim 2 with the fact that Algorithm 1 invokes sub-test$_i$ on $t_1 \cdot 2^i$ random vertices (and using $(1 - \gamma_1 \cdot 2^{-i})^{t_1 \cdot 2^i} + 0.05 < 1/3$). Next, using the conclusion of Claim 3, we turn to construct a partition $(C_1, \dots, C_t)$ of $[N]$ such that the following holds: the total number of missing edges (in G) within the C_i's is at most $\epsilon \cdot N^2/2$ and the total number of (superfluous) edges between the C_i's is at most $\epsilon \cdot N^2/2$. The partition is constructed in iterations. *We start with a motivating discussion.*

Note that any i-good vertex, v, yields a set of vertices (i.e., $\Gamma(v)$) that is "close" to being a clique, where "closeness" has a stricter meaning when i is smaller. Specifically, by Condition 1, the number of missing edges between pairs of vertices in this set is at most $\gamma_2 \cdot 2^i \epsilon \cdot |\Gamma(v)| \cdot N$. But we should also care about how this set "interacts" with the rest of the graph, which is where Condition 2 comes into play. Letting C_v contain only the vertices in $\Gamma(v)$ that have less than $|\Gamma(v)|$ neighbors outside of $\Gamma(v)$, we upper-bound the number of edges going out of C_v as follows: We first note that these edges are either edges between C_v and $\Gamma(v) \setminus C_v$ or edges between C_v and $[N] \setminus \Gamma(v)$. The number of edges of the first type is upper-bounded by $|C_v| \cdot |\Gamma(v) \setminus C_v|$, which (by using Condition 2 and $j_0 = \log_2(|\Gamma(v)|/(\gamma_2 \cdot 2^i \epsilon N))$) is upper-bounded by $|C_v| \cdot 2^{-j_0}|\Gamma(v)| = |C_v| \cdot \gamma_2 2^i \epsilon N \leq \gamma_2 2^i \epsilon \cdot |\Gamma(v)| \cdot N$. The number of edges of the second type is upper-bounded by

$$\sum_{j=1}^{j_0} 2^{-(j-1)}|\Gamma(v)| \cdot \gamma_2 \cdot 2^{i+j}\epsilon \cdot N \;=\; 2j_0 \cdot \gamma_2 2^i \epsilon \cdot |\Gamma(v)| \cdot N, \tag{1}$$

by assigning each vertex $u \in C_v$ the smallest $j \in [j_0]$ such that $|\Gamma(u) \setminus \Gamma(v)| < \gamma_2 \cdot 2^{i+j}\epsilon \cdot N$, and using $\gamma_2 2^{i+j_0}\epsilon \cdot N = |\Gamma(v)|$. Thus, the total number of these edges is upper-bounded by $(2j_0 + 1) \cdot \gamma_2 2^i \epsilon \cdot |\Gamma(v)| \cdot N$, which is upper-bounded by $3\ell \cdot \gamma_2 2^i \epsilon \cdot |\Gamma(v)| \cdot N$ (since $j_0 \leq \log_2(1/(\gamma_2 \cdot 2^i \epsilon)) \leq \log_2(1/\gamma_2 \epsilon) = (1+o(1)) \cdot \ell$).

The foregoing paragraph identifies a single (good) clique, while we wish to identify all cliques. Starting with $i = 1$, the basic idea is to identify new cliques by using i-good vertices that are not covered by previously identified cliques. If we are lucky and the entire graph is covered this way then we halt. But it may indeed be the case that some vertices are left uncovered and that they are not i-good. At this point we invoke Claim 3 and conclude that these vertices either have low degree (i.e., have degree at most $\gamma_2 \cdot 2^i \epsilon \cdot N$) or are relatively few in number (i.e., their number is at most $\gamma_1 \cdot 2^{-i} \cdot N$). Ignoring (for a moment) the vertices of low degree, we deal with the remaining vertices by invoking the same reasoning with respect to an incremented value of i (i.e., $i \leftarrow i + 1$). The key observation is that the number of violations, caused by cliques identified in each iteration i, is upper-bounded by the product of the number of vertices covered in that iteration (which is linearly related to 2^{-i}) and the "density" of violations caused by each identified clique (which is linearly related to $2^i \epsilon$). Thus, intuitively, each iteration contributes $O(\ell \gamma_2 \epsilon \cdot N^2)$ violations, and after the last iteration (i.e., $i = \ell$) we are left with at most $\gamma_1 \cdot 2^{-i} \cdot N < \gamma_1 \epsilon N$ vertices, which we can afford to identify as a single clique (or alternatively as isolated vertices).

Two problems, which were ignored by the foregoing description, arise from the fact that vertices that are identified as belonging to the clique C_v (of some i-good vertex v) may belong either to previously identified cliques or to the set of vertices cast aside as having low degree. Our solution is to use only i-good vertices for which the majority of neighbors do not belong to these two categories (i.e., vertices v such that most of $\Gamma(v)$ belongs neither to previously identified cliques nor have low degree). This leads to the following description.

The partition reconstruction procedure. The iterative procedure is initiated with $C = L_0 = \emptyset$, $R_0 = [N]$ and $i = 1$, where C denotes the set of vertices "covered" (by cliques) so far, R_{i-1} denotes the set of "remaining" vertices after iteration $i - 1$ and L_{i-1} denotes the set of vertices cast aside (as having "low degree") in iteration $i - 1$. The procedure refers to a parameter $\beta = \Theta(1/\ell) \gg \gamma_2$, which determines the "low degree" threshold (for each iteration). The ith iteration proceeds as follows, where $i = 1, \ldots, \ell$ and F_i is initialized to $\emptyset$.

1. Pick an arbitrary vertex $v \in R_{i-1} \setminus C$ that satisfies the following three conditions
   (a) v is i-good.
   (b) v has sufficiently high degree; that is, $|\Gamma(v)| \geq \beta \cdot 2^i \epsilon \cdot N$.
   (c) v has relatively few neighbors in C; that is, $|\Gamma(v) \cap C| \leq |\Gamma(v)|/4$.
   If no such vertex exists, define $L_i = \{v \in R_{i-1} \setminus C : |\Gamma(v)| < \beta \cdot 2^i \epsilon \cdot N\}$ and $R_i = R_{i-1} \setminus (L_i \cup C)$. If $i < \ell$ then proceed to the next iteration, and otherwise terminate.
2. For vertex v as selected in Step 1, let $C_v = \{u \in \Gamma(v) : |\Gamma(u) \setminus \Gamma(v)| < |\Gamma(v)|\}$. Form a new clique with the vertex set $C'_v \leftarrow C_v \setminus C$, and update $F_i \leftarrow F_i \cup \{v\}$ and $C \leftarrow C \cup C'_v$.

Note that by Condition 1c, for every $v \in F_i$, it holds that $|C'_v| \geq |C_v| - (|\Gamma(v)|/4)$, whereas by i-goodness[6] (and $j_0 = \log_2(|\Gamma(v)|/(\gamma_2 \cdot 2^i \epsilon N)) \geq \log_2(\beta/\gamma_2) = \omega(1)$) we have $|C_v| > (1 - o(1)) \cdot |\Gamma(v)|$. Thus, quality guarantees that are quantified in terms of $|\Gamma(v)|$ translate well to similar guarantees in terms of $|C'_v|$. This fact, combined with the fact that C_v cannot contain many low degree vertices (i.e., vertices cast aside (in prior iterations) as having low degree), plays an important role in the following analysis.

Claim 4. *Referring to the partition reconstruction procedure, for every $i \in [\ell]$, the following holds.*

1. *The number of missing edges inside the cliques formed in iteration i is at most $8\gamma_2 \epsilon \cdot N^2$; that is,*

$$\left| \bigcup_{v \in F_i} \{(u,w) \in C'_v \times C'_v : (u,w) \notin E\} \right| \leq 8\gamma_2 \epsilon \cdot N^2.$$

2. *The number of ("superfluous") edges between cliques formed in iteration i and either R_i or other cliques formed in the same iteration is $24\ell \cdot \gamma_2 \epsilon \cdot N^2$; actually,*

$$\left| \bigcup_{v \in F_i} \{(u,w) \in C'_v \times (R_{i-1} \setminus C'_v) : (u,w) \in E\} \right| \leq 24\ell \cdot \gamma_2 \epsilon \cdot N^2.$$

3. $|R_i| \leq 2^{-i} \cdot N$ *and* $|L_i| \leq 2^{-(i-1)} \cdot N$.

Thus, the total number of violations caused by the cliques that are formed by the foregoing procedure is upper-bounded by $(24 + o(1))\ell^2 \cdot \gamma_2 \epsilon \cdot N^2 = o(\epsilon N^2)$. (We mention that the setting $\gamma_2 - o(\ell^2)$ is used for establishing Item 3.)

---

[6] Every $v \in F_i$ is i-good and thus satisfies $|C_v| > (1 - 2^{-j_0}) \cdot |\Gamma(v)|$.

Proof: We prove all items simultaneously, by induction from $i = 0$ to $i = \ell$. Needless to say, all items hold vacuously for $i = 0$, and thus we focus on the induction step.

Starting with Item 1, we note that every $v \in F_i$ is i-good and thus the number of edges missing in $C'_v \times C'_v \subseteq \Gamma(v) \times \Gamma(v)$ is at most $\gamma_2 2^i \epsilon \cdot |\Gamma(v)| \cdot N < 2\gamma_2 2^i \epsilon \cdot |C'_v| \cdot N$, where the inequality follows from $|C'_v| > |\Gamma(v)|/2$ (which follows by combining $|C'_v| \geq |C_v| - (\Gamma(v)|/4)$ and $|C_v| \geq (1 - 2^{-j_0}) \cdot |\Gamma(v)|$, where $j_0 = \log_2(|\Gamma(v)|/(\gamma_2 \cdot 2^i \epsilon N)) > 2$). Recall that the i-goodness of v (combined with $|\Gamma(v)| \geq \beta \cdot 2^i \epsilon \cdot N$) implies that $\Gamma(v)$ contains at least $0.99 \cdot |\Gamma(v)|$ vertices of degree exceeding $0.99 \cdot |\Gamma(v)|$. This implies that $|\Gamma(v) \cap (\bigcup_{j \in [i-1]} L_j)| < |C_v|/4$, because $|\Gamma(v)| \geq \beta 2^i \epsilon \cdot N$ whereas every vertex in $\bigcup_{j \in [i-1]} L_j$ has degree at most $\beta 2^{i-1} \epsilon \cdot N$. Observing that $C'_v = (C'_v \cap R_{i-1}) \cup (C'_v \cap \bigcup_{j \in [i-1]} L_j)$, it follows that $|\bigcup_{v \in F_i} C'_v \cap R_{i-1}| > |\bigcup_{v \in F_i} C'_v|/2$, and thus $\sum_{v \in F_i} |C'_v| \leq 2|R_{i-1}|$. Combining all these bounds, we obtain

$$\left| \bigcup_{v \in F_i} \{(u, w) \in C'_v \times C'_v : (u, w) \notin E\} \right| = \sum_{v \in F_i} |\{(u, w) \in C'_v \times C'_v : (u, w) \notin E\}|$$

$$\leq 2\gamma_2 2^i \epsilon \cdot \sum_{v \in F_i} |C'_v| \cdot N$$

$$\leq 2\gamma_2 2^i \epsilon \cdot 2|R_{i-1}| \cdot N.$$

Using the induction hypothesis regarding R_{i-1} (i.e., $|R_{i-1}| < 2^{-(i-1)} \cdot N$), Item 1 follows.

Item 2 is proved in a similar fashion. Here we use the fact[7] that i-goodness of v (which follows from $v \in F_i$) implies that the number of edges in $C'_v \times (R_{i-1} \setminus C'_v) \subseteq C_v \times ([N] \setminus C_v)$ is at most $3\ell \cdot \gamma_2 2^i \epsilon \cdot |\Gamma(v)| \cdot N$, which is upper-bounded by $6\ell \cdot \gamma_2 2^i \epsilon \cdot |C'_v| \cdot N$. Using again $\sum_{v \in F_i} |C'_v| < 2|R_{i-1}|$ and $|R_{i-1}| < 2^{-(i-1)} \cdot N$, we establish Item 2.

Turning to Item 3, we first note that $L_i \subseteq R_{i-1}$ and thus $|L_i| \leq |R_{i-1}| \leq 2^{-(i-1)} \cdot N$. As for R_i, it may contain only vertices that are neither in L_i nor in $\bigcup_{v \in F_i} C'_v$. It follows that for every $v \in R_i$ either v is not i-good (although it has degree at least $\beta \cdot 2^i \epsilon \cdot N$) or it has at least $|\Gamma(v)|/4$ neighbors in previously identified cliques (which implies $|\Gamma(v) \cap (\bigcup_{w \in \bigcup_{j \in [i]} F_j} C'_w)| \geq |\Gamma(v)|/4$). By Claim 3, the number of vertices of the first type is at most $\gamma_1 2^{-i} \cdot N$. As for vertices of the second type, each such vertex v (in R_i) requires at least $|\Gamma(v)|/4 \geq \beta \cdot 2^i \epsilon \cdot N/4$ edges from $C' \stackrel{\text{def}}{=} \bigcup_{w \in \bigcup_{j \in [i]} F_j} C'_w$ to it (because C' is the set of vertices covered by previously identified cliques at the time iteration i is

---

[7] This fact was established in the motivating discussion that precedes the description of the procedure (see Eq. (1) and its vicinity). Specifically, recall that the number of edges in $C_v \times ([N] \setminus C_v)$ is upper-bounded by the sum of $|C_v \times (\Gamma(v) \setminus C_v)|$ and the number of edges in $C_v \times ([N] \setminus \Gamma(v))$. Using Condition 2 of i-goodness, we upper-bound both $|\Gamma(v) \setminus C_v|$ and the number of edges of the second type, and the fact follows.

completed). By Item 2, the total number of edges going out from C' to R_i is at most $i \cdot 24\ell \cdot \gamma_2\epsilon \cdot N^2 \leq 24\ell^2 \cdot \gamma_2\epsilon \cdot N^2$. On the other hand, as noted above, each vertex of the second type has at least $\beta \cdot 2^i\epsilon \cdot N/4$ edges incident to vertices in C'. Hence, the number of vertices of the second type is upper-bounded by

$$\frac{24\ell^2 \cdot \gamma_2\epsilon \cdot N^2}{\beta \cdot 2^i\epsilon \cdot N} = \frac{24\ell^2 \cdot \gamma_2}{\beta} \cdot 2^{-i}N, \tag{2}$$

Thus, $|R_i| \leq (\gamma_1 + 24\ell^2\gamma_2\beta^{-1}) \cdot 2^{-i} \cdot N$. By the foregoing setting of γ_1, γ_2 and β (e.g., $\gamma_1 = 1/2$ and $\gamma_2 = \beta/(48\ell^2)$), it follows that $|R_i| \leq 2^{-i} \cdot N$. ∎

Completing the reconstruction and its analysis. The foregoing construction leaves "unassigned" the vertices in R_ℓ as well as some of the vertices in $L_1, \ldots, L_\ell$. (Note that some vertices in $\bigcup_{i=1}^{\ell-1} L_i$ may be placed in cliques constructed in later iterations, but there is no guarantee that this actually happens.) We now assign each of these remaining vertices to a singleton clique (i.e., an isolated vertex). The number of violation caused by this assignment equals the number of edges with both endpoints in $R' \stackrel{\text{def}}{=} R_\ell \cup \bigcup_{i=1}^\ell L_i$, because edges with a single endpoint in R' were already accounted for in Item 2 of Claim 4. Nevertheless, we upper-bound the number of violations by the total number of edges adjacent at R', which in turn is upper-bounded by

$$\sum_{v \in R_\ell \cup \bigcup_{i \in [\ell]} L_i} |\Gamma(v)| \leq |R_\ell| \cdot N + \sum_{i=1}^\ell \sum_{v \in L_i} |\Gamma(v)|$$

$$\leq \frac{\epsilon N}{4} \cdot N + \sum_{i=1}^\ell 2^{-(i-1)}N \cdot \beta 2^i\epsilon N$$

$$= \frac{\epsilon}{4} \cdot N^2 + 2\ell \cdot \beta \cdot \epsilon N^2.$$

By the foregoing setting of β (i.e., $\beta \leq 1/8\ell$), it follows that the number of these edges is smaller than $\epsilon N^2/2$. Combining this with the bounds on the number of violating edges (or non-edges) as provided by Claim 4, the lemma follows.

3 The Non-adaptive Query Complexity of $\mathcal{CC}$

In this section we establish Part 2 of Theorem 1. Specifically, for every value of $\epsilon > 0$, we consider two different sets of graphs, one consisting of graphs in $\mathcal{CC}$ and the other consisting of graphs that are ϵ-far from $\mathcal{CC}$, and show that a non-adaptive algorithm of query complexity $o(\epsilon^{-4/3})$ cannot distinguish between graphs selected at random in these sets.

The first set, denoted $\mathcal{CC}_\epsilon$, contains all N-vertex graphs such that each graph consists of $(3\epsilon)^{-1}$ cliques, and each clique has size $3\epsilon \cdot N$. It will be instructive to partition these $(3\epsilon)^{-1}$ cliques into $(6\epsilon)^{-1}$ pairs (each consisting of two cliques). The second set, denoted $\mathcal{BCC}_\epsilon$, contains all N-vertex graphs such that each graph consists of $(6\epsilon)^{-1}$ bi-cliques, and each bi-clique has $3\epsilon \cdot N$ vertices on each side.

Indeed, $\mathcal{CC}_\epsilon \subseteq \mathcal{CC}$, whereas each graph in $\mathcal{BCC}_\epsilon$ is ϵ-far from $\mathcal{CC}$ (because each of the bi-cliques must be turned into a collection of cliques).

In order to motivate the claim that a non-adaptive algorithm of query complexity $o(\epsilon^{-4/3})$ cannot distinguish between graphs selected at random in these sets, consider the (seemingly best such) algorithm that selects $o(\epsilon^{-2/3})$ vertices and inspects the induced subgraph. Consider the partition of a graph in $\mathcal{CC}_\epsilon$ into $(6\epsilon)^{-1}$ pairs of cliques, and correspondingly the partition of a graph in $\mathcal{BCC}_\epsilon$ into $(6\epsilon)^{-1}$ bi-cliques. Then, the probability that a sample of $o(\epsilon^{-2/3})$ vertices contains at least three vertices that reside in the same part (of $6\epsilon \cdot N$ vertices) is $o(\epsilon^{-2/3})^3 \cdot (6\epsilon)^2 = o(1)$. On the other hand, if this event does not occur, then the answers obtained from both graphs are indistinguishable (because in each case a random pair of vertices residing in the same part is connected by an edge with probability $1/2$). As is outlined next, this intuition extends to an arbitrary non-adaptive algorithm.

Specifically, by an averaging argument, it suffices to consider deterministic algorithms, which are fully specified by the sequence of queries that they make and their decision on each corresponding sequence of answers. Recall that these (fixed) queries are elements of $[N] \times [N]$. We shall show that, for every sequence of $o(\epsilon^{-4/3})$ queries, the answers provided by a randomly selected element of $\mathcal{CC}_\epsilon$ are statistically close to the answers provided by a randomly selected element of $\mathcal{BCC}_\epsilon$. We shall use the following notation: For an N-vertex graph G and a query (u, v), we denote the corresponding answer by $\mathrm{ans}_G(u, v)$; that is, $\mathrm{ans}_G(u, v) = 1$ if $\{u, v\}$ is an edge in G and $\mathrm{ans}_G(u, v) = 0$ otherwise.

Lemma 5. *Let G_1 and G_2 be random N-vertex graphs uniformly distributed in $\mathcal{CC}_\epsilon$ and $\mathcal{BCC}_\epsilon$, respectively. Then, for every sequence $(v_1, v_2), \ldots, (v_{2q-1}, v_{2q}) \in [N] \times [N]$, where the v_i's are not necessarily distinct, it holds that the statistical difference between $\mathrm{ans}_{G_1}(v_1, v_2), \ldots, \mathrm{ans}_{G_1}(v_{2q-1}, v_{2q})$ and $\mathrm{ans}_{G_2}(v_1, v_2), \ldots, \mathrm{ans}_{G_2}(v_{2q-1}, v_{2q})$ is $O(q^{3/2}\epsilon^2)$.*

The proof of Lemma 5 appears in our technical report [13], and Part 2 of Theorem 1 follows.

Tightness of the lower bound: We mention that the above lower bound is tight (indeed, as asserted in Part 3 of Theorem 1). This fact is proved, in our technical report [13], by presenting a non-adaptive algorithm that is *not canonical*. Recall that a canonical algorithm operates by selecting a random set of vertices and inspecting the induced subgraph. We mention that our algorithm improves over the $\widetilde{O}(\epsilon^{-2})$ bound of [5, Thm. 2] (which is obtained by a canonical algorithm).

Acknowledgments

We are grateful to Lidor Avigad for comments regarding a previous version of this work.

References

1. Alon, N.: On the number of subgraphs of prescribed type of graphs with a given number of edges. Israel J. Math. 38, 116–130 (1981)
2. Alon, N., Fischer, E., Krivelevich, M., Szegedy, M.: Efficient Testing of Large Graphs. Combinatorica 20, 451–476 (2000)
3. Alon, N., Fischer, E., Newman, I.: Testing of bipartite graph properties. SIAM Journal on Computing 37, 959–976 (2007)
4. Alon, N., Fischer, E., Newman, I., Shapira, A.: A Combinatorial Characterization of the Testable Graph Properties: It's All About Regularity. In: 38th STOC, pp. 251–260 (2006)
5. Alon, N., Shapira, A.: A Characterization of Easily Testable Induced Subgraphs. Combinatorics Probability and Computing 15, 791–805 (2006)
6. Ben-Sasson, E., Harsha, P., Raskhodnikova, S.: 3CNF properties are hard to test. SIAM Journal on Computing 35(1), 1–21 (2005)
7. Bogdanov, A., Trevisan, L.: Lower Bounds for Testing Bipartiteness in Dense Graphs. In: IEEE Conference on Computational Complexity, pp. 75–81 (2004)
8. Canetti, R., Even, G., Goldreich, O.: Lower Bounds for Sampling Algorithms for Estimating the Average. IPL 53, 17–25 (1995)
9. Fischer, E.: The art of uninformed decisions: A primer to property testing. Bulletin of the European Association for Theoretical Computer Science 75, 97–126 (2001)
10. Fischer, E.: On the strength of comparisons in property testing. Inform. and Comput. 189(1), 107–116 (2004)
11. Goldreich, O., Goldwasser, S., Ron, D.: Property testing and its connection to learning and approximation. Journal of the ACM, 653–750 (July 1998)
12. Goldreich, O., Ron, D.: Property Testing in Bounded Degree Graphs. Algorithmica 32(2), 302–343 (2002)
13. Goldreich, O., Ron, D.: Algorithmic Aspects of Property Testing in the Dense Graphs Model. ECCC, TR08-039 (2008)
14. Goldreich, O., Ron, D.: On Proximity Oblivious Testing. ECCC, TR08-041 (2008); Extended abstract in the proceedings of the 41st STOC (2009)
15. Goldreich, O., Trevisan, L.: Three theorems regarding testing graph properties. Random Structures and Algorithms 23(1), 23–57 (2003)
16. Gonen, M., Ron, D.: On the Benefit of Adaptivity in Property Testing of Dense Graphs. In: Charikar, M., Jansen, K., Reingold, O., Rolim, J.D.P. (eds.) RANDOM 2007 and APPROX 2007. LNCS, vol. 4627, pp. 525–539. Springer, Heidelberg (2007); To appear in Algorithmica (special issue of RANDOM and APPROX 2007)
17. Ron, D.: Property testing. In: Rajasekaran, S., Pardalos, P.M., Reif, J.H., Rolim, J.D.P. (eds.) Handbook on Randomization, vol. II, pp. 597–649 (2001)
18. Raskhodnikova, S., Smith, A.: A note on adaptivity in testing properties of bounded-degree graphs. ECCC, TR06-089 (2006)
19. Rubinfeld, R., Sudan, M.: Robust characterization of polynomials with applications to program testing. SIAM Journal on Computing 25(2), 252–271 (1996)

Succinct Representation of Codes with Applications to Testing

Elena Grigorescu[1,*], Tali Kaufman[1,**], and Madhu Sudan[2,***]

[1] MIT, Cambridge, MA, USA
{elena_g,kaufmant}@mit.edu
[2] Microsoft Research, Cambridge, MA, USA
madhu@microsoft.com

Abstract. Motivated by questions in property testing, we search for linear error-correcting codes that have the "single local orbit" property: they are specified by a single local constraint and its translations under the symmetry group of the code. We show that the dual of every "sparse" binary code whose coordinates are indexed by elements of $\mathbb{F}_{2^n}$ for prime n, and whose symmetry group includes the group of non-singular affine transformations of $\mathbb{F}_{2^n}$, has the single local orbit property. (A code is *sparse* if it contains polynomially many codewords in its block length.) In particular this class includes the dual-BCH codes for whose duals (BCH codes) simple bases were not known. Our result gives the first short ($O(n)$-bit, as opposed to $\exp(n)$-bit) description of a low-weight basis for BCH codes. If $2^n - 1$ is a Mersenne prime, then we get that every sparse *cyclic* code also has the single local orbit.

Keywords: Locally testable codes, affine/cyclic invariance, single orbit.

1 Introduction

Motivated by questions about the local testability of some well-known error-correcting codes, in this paper we examine their "invariance" properties. Invariances of codes are a well-studied concept (see, for instance, [16, Chapters 7, 8.5, and 13.9]) and yet we reveal some new properties of BCH codes. In the process we also find broad classes of sparse codes that are locally testable. We describe our problems and results in detail below.

A code $C \subseteq \mathbb{F}_2^N$ is said to be locally testable if membership of a word $w \in \mathbb{F}_2^N$ in the code C can be checked probabilitistically by a few probes into w. The famed "linearity test" of Blum, Luby and Rubinfeld [2] may be considered the first result to show that some code is locally testable. Locally testable codes were formally defined by Rubinfeld and Sudan [17]. The first substantial study of locally testable codes was conducted by Goldreich and Sudan [9], where the

---

* Funded in part by NSF grant CCR-0829672.
** Funded in part by NSF grant CCR-0829672 and NSF-0729011.
*** Research conducted when this author was at MIT CSAIL. Research was funded in part by NSF award CCR-0829672.

I. Dinur et al. (Eds.): APPROX and RANDOM 2009, LNCS 5687, pp. 534–547, 2009.
© Springer-Verlag Berlin Heidelberg 2009

principal focus was the construction of locally testable codes of high rate. Local testing of codes is effectively equivalent to property testing [17,8] with the difference being that the emphasis here is when C is an error-correcting code, i.e., elements of C are pairwise far from each other.

A wide variety of "classical" codes are by now known to be locally testable, including Hadamard codes [2], Reed-Muller codes of various parameters [17,1,13,10], dual-BCH codes [11,14], turning attention to the question: What broad characteristics of codes are necessary, or sufficient, for codes to be locally testable. One characteristic explored in the recent work of Kaufman and Sudan [15] is the "invariance group" of the code, which we describe next.

Let $[N]$ denote the set of integers $\{1, \ldots, N\}$. A code $C \subseteq \mathbb{F}_2^N$ is said to be invariant under a permutation $\pi : [N] \to [N]$ if for every $a = \langle a_1, \ldots, a_N \rangle \in C$, it is the case that $a \circ \pi = \langle a_{\pi(1)}, \ldots, a_{\pi(N)} \rangle$ is also in C. The set of permutations under which any code C is invariant forms a group under composition and we refer to it as the invariant group. [15] suggested that the invariant group of a code may play an important role in its testability. They supported their suggestion by showing that if the invariant group is an "affine group", then a "linear" code whose "dual" has the "single local orbit" property is locally testable. We explain these terms (in a restricted setting) below.

Let $N = 2^n$ and let $C \subseteq \mathbb{F}_2^N$ be a code. In this case we can associate the coordinate set $[N]$ of the code C with the field $\mathbb{F}_{2^n}$. Now consider the permutations $\pi : \mathbb{F}_{2^n} \to \mathbb{F}_{2^n}$ of the form $\pi(x) = \alpha x + \beta$ where $\alpha \in \mathbb{F}_{2^n} - \{0\}$ and $\beta \in \mathbb{F}_{2^n}$. This set is closed under composition and we refer to this as the *affine group*. If C is invariant under every π in the affine group, then we say that C is *affine-invariant*. We say that C is *linear* if it is a vector subspace of $\mathbb{F}_2^N$. The *dual* of C, denoted $C^\perp$, is the null space of C as a vector space.

We now define the final term above, namely, the "single local orbit property". Let G be a group of permutations mapping $[N]$ to $[N]$. For $b \in \mathbb{F}_2^N$, let its *weight*, denoted $\mathrm{wt}(b)$, be the number of non-zero elements of b. A code C is said to have the *k-single orbit property* under G if there exists an element $b \in \mathbb{F}_2^N$ of weight at most k such that $C = \mathrm{Span}(\{b \circ \pi | \pi \in G\})$, where $\mathrm{Span}(S) = \{\sum_i c_i b_i | c_i \in \mathbb{F}_2, b_i \in S\}$. Two groups are of special interest to us in this work. The first is the affine group on $\mathbb{F}_{2^n}$. A second group of interest to us is the "cyclic group" on $\mathbb{F}_{2^n}^* = \mathbb{F}_{2^n} - \{0\}$ given by the permutations $\pi_a(x) = ax$ for $a \in \mathbb{F}_{2^n}^*$. (Note that if ω is a multiplicative generator of $\mathbb{F}_{2^n}^*$ and the coordinates of C are ordered $\langle \omega, \omega^2, \ldots, \omega^{2^n-1} = 1 \rangle$ then each π_a is simply a cyclic permutation.)

The invariance groups of codes are well-studied objects. In particular codes that are invariant under cyclic permutations, known as cyclic codes, are widely studied and include many common algebraic codes (under appropriate ordering of the coordinates and with some slight modifications, see [18] or [16].) The fact that many codes are also affine-invariant is also explicitly noted and used in the literature [16].

Conditions under which codes have the single-orbit property under any given group, seem to be less well-studied. This is somewhat surprising given that the single-orbit property implies very succinct (nearly explicit) descriptions (of size

$k \log N$ as opposed to $\omega(N))^1$ of bases for codes (that have the k-single orbit property under some standard group.) Even for such commonly studied codes such as the BCH codes such explicit descriptions of bases were not known prior to this work. In retrospect, the single orbit property was being exploited in previous results in algebraic property testing [2,17,1,13,10] though this fact was not explicit until the work of [15].

In this work we explore the single orbit property under the affine group for codes on the coordinate set $\mathbb{F}_{2^n}$, as also the single orbit property under the cyclic group for codes over $\mathbb{F}_{2^n}^*$. We show that the dual of every "sparse" affine-invariant code (i.e., codes with at most polynomially many codewords in N) has the k-single orbit property under the affine group for some constant k, provided $N = 2^n$ for prime n (see Theorem 1.) When $N - 1$ is also prime, it turns out that the duals of sparse codes have the k-single orbit property under the cyclic group for some constant k yielding an even stronger condition on the basis (see Theorem 2.) Both theorems shed new light on well-studied codes including BCH codes. The actual families considered here are broader, but the BCH codes are typical in these collections. Lemma 1 explicitly characterizes the entire family of codes investigated in this paper.

In particular the first theorem has immediate implications for testing and shows that every sparse affine invariant code is locally testable. This merits comparison with the results of [14] who show that sparse high-distance codes are locally testable. While syntactically the results seem orthogonal (ours require affine-invariance whereas theirs required high-distance) it turns out (as we show in this paper) that all the codes we consider do have high-distance. Yet for the codes we consider our results are more constructive in that they not only prove the "existence" of a local test, but give a much more "explicit" description of the tester: Our tester is described by a single low-weight word in the dual and tests that a random affine permutation of this word is orthogonal to the word being tested.[2]

Given a code of interest to us, we first study the algebraic structure of the given code by representing codewords as polynomials and studying the degree patterns among the support of these polynomials. We interpret the single orbit property in this language; and this focusses our attention on a collection of closely related codes. We then turn to recent results from additive number theory [4,3,6,5,7] and apply them to the dual of the given code, as well as the other related codes that arise from our algebraic study, to lower bound their distance. In turn, using the MacWilliams identities (as in prior work [14]) this translates to some information

---

[1] One way to represent a sparse code C whose dual $C^\perp$ has a basis among the weight k codewords is to give $\Omega(N)$ codewords that generate $C^\perp$. This requires space $\Omega(kN \log N)$ bits. Alternately, if C is sparse and has N^t codewords, one can give $t \log N$ codewords that generate it; this requires $tN \log N = \Omega(N \log N)$ bits.

[2] In contrast the tester of [14] was less "explicit". It merely proved the existence of many low weight codewords in the dual of the code being tested and proved that the test which picked one of these low-weight codewords uniformly at random and tested orthogonality of the given word to this dual codeword was a sound test.

on the weight-distribution of the given code and the related ones. Some simple counting then yields that the given code must have the single-orbit property.

We believe that our techniques are of interest, beyond just the theorems they yield. In particular we feel that techniques to assert the single-orbit property are quite limited in the literature. Indeed in all previous results [2,17,1,13,10] this property was "evident" for the code: The local constraint whose orbit generated a basis for all constraints was explicitly known, and the algebra needed to prove this fact was simple. Our results are the first to consider the setting where the basis is not explicitly known (even after our work) and manages to bring in non-algebraic tools to handle such cases. We believe that the approach is potentially interesting in broader settings.

2   Definitions and Main Results

We recall some basic notation. $[N]$ denotes the set $\{1, \ldots, N\}$. $\mathbb{F}_q$ denotes the finite field with q elements and $\mathbb{F}_q^*$ will denote the non-zero elements of this field. We will consider codes contained in the vector space $\mathbb{F}_2^N$. For a word $a = \langle a_1, \ldots, a_N \rangle \in \mathbb{F}_2^N$ its support is the set $\mathrm{Supp}(a) = \{i | a_i \neq 0\}$ and its weight is the quantity $\mathrm{wt}(a) = |\mathrm{Supp}(a)|$. For $a = \langle a_i \rangle_i$, and $b = \langle b_i \rangle_i \in \mathbb{F}_2^N$ define the *relative* distance between a, b as $\delta(a, b) = \frac{1}{N} |\{i \mid a_i \neq b_i\}|$. Note $\delta(a, b) = \frac{\mathrm{wt}(a-b)}{N}$. A binary code $\mathcal{C}$ is a subset of $\mathbb{F}_2^N$. The (relative) distance of $\mathcal{C}$ is $\delta(\mathcal{C}) = \min_{a,b \in \mathcal{C}; a \neq b} \{\delta(a, b)\}$. For a set of vectors $S = \{v_1, \ldots, v_k\} \subseteq \mathbb{F}_2^N$, let $\mathrm{Span}(S) = \{\sum_{i=1}^k \alpha_i v_i | \alpha_1, \ldots, \alpha_k \in \mathbb{F}_2\}$ denote the linear span of S. $\mathcal{C}$ is a *linear* code if its codewords form a vector space in $\{0, 1\}^N$ over $\mathbb{F}_2$, i.e., if $\mathrm{Span}(\mathcal{C}) = \mathcal{C}$. For $a, b \in \mathbb{F}_2^N$, let $a \cdot b = \sum_i a_i b_i$ denote the inner product of a and b. The *dual* of $\mathcal{C}$ is the code $\mathcal{C}^\perp = \{b \in \mathbb{F}_2^N \mid b \cdot a = 0, \forall a \in \mathcal{C}\}$. We will alternate between viewing $a \in \mathbb{F}_2^N$ as a vector $a = \langle a_1, \ldots, a_N \rangle$ and as a function $a : D \to \mathbb{F}_2$ where D will be some appropriate domain of size N. Two particular domains of interest to us will be $\mathbb{F}_{2^n}$ and $\mathbb{F}_{2^n}^*$.

2.1   Invariance and the Single Local Orbit Property

Let $a \in \mathbb{F}_2^N$ be viewed as a function $a : D \to \mathbb{F}_2$ for some domain D of size N. Let $\pi : D \to D$ be a permutation of D. The π-rotation of a is the function $a \circ \pi : D \to \mathbb{F}_2$ given by $a \circ \pi(i) = a(\pi(i))$ for every $i \in D$.

Let D be a set of size N and let $\mathbb{F}_2^N$ denote the set of functions from $D \to \mathbb{F}_2$. A code $\mathcal{C} \subseteq \mathbb{F}_2^N$ is said to be *invariant* under a permutation $\pi : D \to D$ if for every $a \in \mathcal{C}$, it is the case that $a \circ \pi \in \mathcal{C}$. The set of permutations under which a code $\mathcal{C}$ is invariant forms a group under composition and we refer to it as the invariant group of a code.

We will be interested in studying codes that are invariant under some well-studied groups (i.e., whose invariant groups contain some well-studied groups.) Two groups of interest to us are the affine group over $\mathbb{F}_{2^n}$ and the cyclic group over $\mathbb{F}_{2^n}^*$. In what follows we let $N = 2^n$ and view $\mathbb{F}_2^N$ as the set of functions from $\mathbb{F}_{2^n}$ to $\mathbb{F}_2$ and $\mathbb{F}_2^{N-1}$ as the set of functions from $\mathbb{F}_{2^n}^*$ to $\mathbb{F}_2$.

Definition 1 (Affine invariance). *A function* $\pi : \mathbb{F}_{2^n} \to \mathbb{F}_{2^n}$ *is an* affine permutation *if there exist* $\alpha \in \mathbb{F}_{2^n}^*$ *and* $\beta \in \mathbb{F}_{2^n}$ *such that* $\pi(x) = \alpha x + \beta$*. The* affine group *over* $\mathbb{F}_{2^n}$ *consists of all the affine permutations over* $\mathbb{F}_{2^n}$*. A code* $\mathcal{C} \subseteq \mathbb{F}_2^N$ *is said to be* affine invariant *if the invariant group of* $\mathcal{C}$ *contains the* affine group.

Definition 2 (Cyclic invariance). *A function* $\pi : \mathbb{F}_{2^n}^* \to \mathbb{F}_{2^n}^*$ *is a* cyclic per- mutation *if it is of the form* $\pi(x) = \alpha x$ *for* $\alpha \in \mathbb{F}_{2^n}^*$*.* [3] *The* cyclic group *over* $\mathbb{F}_{2^n}^*$ *consists of all the cyclic permutations over* $\mathbb{F}_{2^n}^*$*. A code* $\mathcal{C} \subseteq \mathbb{F}_2^{N-1}$ *is said to be* cyclic invariant *(or simply* cyclic*) if the invariant group of* $\mathcal{C}$ *contains the* cyclic group.

Many well-known families of codes (with minor variations) are known to be affine- invariant and/or cyclic. In particular BCH codes are cyclic and Reed-Muller codes are affine-invariant. Furthermore under a simple "extension" operation BCH codes become affine-invariant, and vice versa under a simple puncturing operation, Reed-Muller codes become cyclic. We elaborate on these later.

In this paper our aim is to show that certain families of affine-invariant and cyclic codes have a simple description, that we call a "single-orbit description". We define this term next.

Definition 3 (k-single orbit code). *Let* $\mathbb{F}_2^N$ *be the collection of functions from* D *to* $\mathbb{F}_2$ *for some domain* D*. Let* G *be a group of permutations from* D *to* D*. A linear code* $\mathcal{C} \subseteq \mathbb{F}_2^N$ *is said to have the* k*-single orbit property under the group* G *if there exists* $a \in \mathcal{C}$ *with* $\mathrm{wt}(a) \leq k$ *such that* $\mathcal{C} = \mathrm{Span}(\{a \circ \pi | \pi \in G\})$*.*

In particular the k-single orbit property under the affine group has implications to testing that we discuss in Section 2.3.

2.2   Main Results

Our main results show that, under certain conditions, duals of "sparse" codes have the single orbit property for small k. By "sparse" we mean that the code has only polynomially many codewords in the length of the codewords.

Our first result considers affine-invariant codes.

Theorem 1 (Single orbit property in affine-invariant codes). *For every* $t > 0$ *there exists a* $k = k(t)$ *such that for every prime* n *the following holds: Let* $N = 2^n$ *and* $\mathcal{C} \subseteq \mathbb{F}_2^N$ *be a linear affine-invariant code containing at most* N^t *codewords. Then* $\mathcal{C}^\perp$ *has the* k*-single orbit property under the affine group.*

Next we present our main theorem for cyclic codes.

Theorem 2 (Single orbit property in cyclic codes). *For every* t *there exists a* k *such that the following holds: Let* n *be such that* $2^n - 1$ *is prime. Let* $\mathcal{C} \subseteq \mathbb{F}_2^{N-1}$ *be a linear, cyclic invariant, code with at most* N^t *codewords. Then* $\mathcal{C}^\perp$ *has the* k*-single orbit property under the cyclic group.*

---

[3] Note that this is a permutation of $\mathbb{F}_{2^n}^*$ if the elements of $\mathbb{F}_{2^n}^*$ are enumerated as $\langle \omega, \omega^2, \ldots, \omega^{N-1} \rangle$ where ω is a multiplicative generator of $\mathbb{F}_{2^n}^*$.

We remark that it is not known if there are infinitely many n such that $2^n - 1$ is prime. Of course if there are only finitely many such primes then our theorem becomes "trivial". Nevertheless, as things stand, the question of whether the number of such primes is infinite or not is unresolved (and indeed there are conjectures suggesting there are infinitely many such primes), and so unconditional result should remain interesting.

2.3 Implications to Property Testing

It follows from the work of [15] that codes with a single local orbit under the affine symmetry group are locally testable. We recall some basic definitions below and summarize the implication of our main theorem to testability.

Definition 4 (Locally testable code [9]). *A code $C \subseteq \mathbb{F}_2^N$ is (k, α)-locally testable if there exists a probabilistic algorithm T called the tester that, given oracle access to a vector $v \in \mathbb{F}_2^N$ makes at most k queries to the oracle for v and accepts $v \in C$ with probability 1, while rejecting $v \notin C$ with probability at least $\alpha \cdot \delta(v, C)$. C is said to be locally testable if there exist $k < \infty$ and $\alpha > 0$ such that C is (k, α)-locally testable.*

We note that the above definition corresponds to the strong definition of local testability ([9, Definition 2.2].) We now state the result of [15] on the testability of affine-invariant codes with the single local orbit property.

Theorem 3 ([15]). *If $C \subseteq \mathbb{F}_2^N$ is linear and has the k-single orbit property under the affine group, then C is $(k, \Omega(1/k^2))$-locally testable.*

We note that in [15] the single-orbit property under the affine group is described as the "strong formal characterization."

Our main theorem, Theorem 1, when combined with the above theorem, immediately yields the following implication for sparse affine invariant codes.

Corollary 1. *For every constant t there exists a constant k such that if n is prime, $N = 2^n$ and $C \subseteq \mathbb{F}_2^N$ is a linear, affine-invariant code with at most N^t codewords, then C is $(k, \Omega(1/k^2))$-locally testable.*

2.4 Implications to BCH Codes

In addition to the implications for the testability of sparse affine-invariant codes, our results also give new structural insight into the classical BCH codes. Even though these codes have been around a long time, and used often in the CS literature, some very basic questions about them are little understood. We describe the codes, the unanswered questions about them, and the implications of our work in this context below.

We start by defining the BCH codes and the extended-BCH codes. The former are classical cyclic codes, and the latter are affine-invariant. Let Trace : $\mathbb{F}_{2^n} \to \mathbb{F}_2$ be the function $\text{Trace}(x) = x + x^2 + \ldots + x^{2^{n-1}}$. We define the BCH codes by defining their dual.

Definition 5. *For every pair of integers n and t, the (binary) dual-BCH code with parameters n and t, denoted* $\mathrm{BCH}(n,t)^{\perp} \subseteq \mathbb{F}_2^{N-1}$ *consists of the evaluations of traces of polynomials of degree 2t over* $\mathbb{F}_{2^n}^*$. *I.e.,*

$$\mathrm{BCH}(n,t)^{\perp} = \{\langle \mathrm{Trace}(f(\alpha)) \rangle_{\alpha \in \mathbb{F}_{2^n}^*} | f \in \mathbb{F}_{2^n}[x], \deg(f) \leq 2t\}$$

The BCH code $\mathrm{BCH}(n,t)$ *is simply the dual of* $\mathrm{BCH}(n,t)^{\perp}$. *The extended dual-BCH code* $\mathrm{eBCH}(n,t)^{\perp} \subseteq \mathbb{F}_2^N$ *is simply the evaluation of the same functions over all of* $\mathbb{F}_{2^n}$, *and* $\mathrm{eBCH}(n,t)$ *is its dual.*

(We note that the more common definition of BCH codes is as the subfield subcodes of Reed Solomon codes, with $\mathrm{BCH}(n,t)$ being the subfield subcodes of RS codes of degree $N - 2t - 1$. But it is a folklore fact that the two definitions are equivalent.)

Even though the BCH codes are very classical codes, much is unknown about them. For instance, while it is easy to see (by a counting argument) that the BCH code $\mathrm{BCH}(n,t)$ must have codewords of weight $2t + 1$, such words are not known "explicitly," leading to the first question: "What is an explicit low-weight codeword of $BCH(n,t)$?" Till recently it was not known that the set of codes of low weight even generate the BCH code, and this was answered affirmatively only recently by Kaufman and Litsyn [12] who showed that words of weight $2t + 1$ and $2t + 2$ certainly include a basis for the BCH code. This proof remains "non-explicit" and the most "succinct" description of this basis is via $O(Nt)$ field elements of $\mathbb{F}_{2^n}$. This leads to the second, harder question: "What is an explicit basis of $BCH(n,t)$?"

Our result manages to make progress on the second question without making progress on the first, by showing that the affine orbit (or in some cases the cyclic orbit) of a single low-weight codeword gives a basis for the BCH code. While this single codeword is still not explicit, the rest of the basis is explicit given the codeword! We state these implications formally below.

Corollary 2. *For every t there exists a k such that for all prime n,* $\mathrm{eBCH}(n,t)$ *has the k-single orbit property under the affine group.*

The above follows from Theorem 1 using the observation that $\mathrm{eBCH}(n,t)^{\perp}$ is sparse (has $N^{O(t)}$ codewords) and affine invariant.

Corollary 3. *For every t there exists a k such that for all n such that* $2^n - 1$ *is prime,* $\mathrm{BCH}(n,t)$ *has the k-single orbit property under the cyclic group.*

The above follows from Theorem 2 using the observation that $\mathrm{BCH}(n,t)^{\perp}$ is sparse (has $N^{O(t)}$ codewords) and cyclic invariant.

We remark that questions of this nature are relevant not only to coding theory, but also to computing. For instance a recurring question in CS is to find explicit balls of small radius in tightly packed codes that contain many codewords. In such problems, the goal is to find an explicit vector (not in the code) along with explicit description of a large set of nearby codewords. Our study, in contrast, attempts to find an explicit description of a large set of codewords near the zero vector (a codeword.)

Finally, we point out that the need for various parameters (n and $2^n - 1$) being prime is a consequence of the application of some recent results in additive number theory that we use to show that certain codes have very high distance. We do not believe such assumptions ought to be necessary; however we do not see any immediate path to resolving the "stronger" number-theoretic questions that would arise by allowing n to be non-prime.

3 Overview of Techniques

Our main theorems are proved essentially by implementing the following plan:

1. We first show that every codeword in the codes we consider are expressible as the Traces of *sparse* polynomials. In the affine-invariant case we also show that these polynomials have somewhat low-degree, i.e., at most $N^{1-\epsilon}$. This part follows standard literature in coding theory (and similar steps were employed already in [15].)

2. We then apply the recent results in additive number theory to conclude that these codes have very high distance. This already suffices to show that the affine-invariant codes are testable by [14]. However the tests given there are "non-explicit" and we need to work further to get an "explicit" test for these codes, or to show the single-orbit condition.

3. The final, and the novel part of this work, is to show by a counting argument, that there exists one (in fact many) low-weight codewords in the dual of the codes we consider such that their orbit spans the dual.

We elaborate on these steps in detail below, laying out precise statements we will prove.

We start with some notation. Recall $N = 2^n$ and n is prime. Also, we view elements $c \in \mathbb{F}_2^N$ as functions $c : \mathbb{F}_N \to \mathbb{F}_2$. Let $\{\mathbb{F}_N \to \mathbb{F}_2\}$ denote the set of all such functions. Similarly we view elements $c \in \mathbb{F}_2^{N-1}$ as functions $\mathbb{F}_N^* \to \mathbb{F}_2$ and let $\{\mathbb{F}_N^* \to \mathbb{F}_2\}$ denote the set of all such functions.

For $d \in \{1, \ldots, N-2\}$, let $\text{orb}(d) = \{d, 2d(\bmod N-1), 4d(\bmod N-1), \ldots, 2^{n-1} d(\bmod N - 1)\}$ By the primality of n, we have that $|\text{orb}(d)| = n$ for every d. Let min-orb(d) denote the smallest integer in $\text{orb}(d)$, and let $\mathcal{D} = \{\text{min-orb}(d) \mid d \in \{1, \ldots, N-2\}\} \cup \{N-1\}$. Note that $|\mathcal{D}| = 1 + (N-2)/n$.

For $D \subseteq \mathcal{D}$ let

$$P_{N,D} = \{\alpha_0 + \sum_{d \in D} \alpha_d x^d \mid \alpha_d \in \mathbb{F}_N, \alpha_0, \alpha_{N-1} \in \{0,1\}\},$$

and $P_{N-1,D} = \{\sum_{d \in D} \alpha_d x^d \mid \alpha_d \in \mathbb{F}_N, \alpha_{N-1} \in \{0,1\}\}.$

The first step in our analysis of codes invariant over the affine group (resp. cyclic group) is that such codes can be associated uniquely with a set $D \subseteq \mathcal{D}$ so that every codeword in our code is the evaluation of the trace of a polynomial from the associated family $P_{N,D}$ over $\mathbb{F}_N$ (resp. $P_{N-1,D}$ over $\mathbb{F}_N^*$.)

Lemma 1. *For every cyclic-invariant code $C \subseteq \{\mathbb{F}_N^* \to \mathbb{F}_2\}$ there exists a set $D \subseteq \mathcal{D}$ such that $c \in C$ if and only if there exists a polynomial $p \in P_{N-1,D}$ such that $c(x) = \mathrm{Trace}(p(x))$ for every $x \in \mathbb{F}_N^*$. Furthermore $|D| \leq t$ iff $|C| \leq N^t$.*

Similarly, for every affine-invariant code $C \subseteq \{\mathbb{F}_N \to \mathbb{F}_2\}$ of cardinality N^t, there exists a set $D \subseteq \mathcal{D}$ such that $c \in C$ if and only if there exists a polynomial $p \in P_{N,D}$ such that $c(x) = \mathrm{Trace}(p(x))$ for every $x \in \mathbb{F}_N$. Furthermore, $|C| \leq 2N^t$ iff $|D| \leq t$ and $D \subseteq \{1, \ldots, N^{1-1/t}\}$.

Thus in both cases codes are represented by collections of t-sparse polynomials. And in the affine-invariant case, these are also somewhat low-degree polynomials. In what follows we use $C_N(D)$ to denote the code $\{\mathrm{Trace}(p(x)) | p \in P_{N,D}\}$ and $C_{N-1}(D)$ to denote the code $\{\mathrm{Trace}(p(x)) | p \in P_{N-1,D}\}$.

We next use a (small variant of a) theorem due to Bourgain [3] to conclude that the codes $C_N(D)$ and $C_{N-1}(D)$ have very high distance (under the given conditions on D.)

Theorem 4 ([3]). *For every $\epsilon > 0$ and $r < \infty$, there is a $\delta > 0$ such that for every prime n the following holds: Let $N = 2^n$ and $\mathbb{F} = \mathbb{F}_N$ and let $f(x) = \sum_{i=1}^{r} a_i x^{k_i} \in \mathbb{F}[x]$ with $a_i \in \mathbb{F}$, satisfy (1) $1 \leq k_i \leq N-1$, (2) $\gcd(k_i, N-1) < N^{1-\epsilon}$ for every $1 \leq i \leq r$, and (3) $\gcd(k_i - k_j, N-1) < N^{1-\epsilon}$ for every $1 \leq i \neq j \leq r$. Then*

$$\left| \sum_{x \in \mathbb{F}} (-1)^{\mathrm{Trace}(f(x))} \right| < N^{1-\delta}.$$

We note that strictly speaking, [3, Theorem 7], only considers the case where N is prime, and considers the sum of any character from $\mathbb{F}$ to the complexes (not just $(-1)^{\mathrm{Trace}(\cdot)}$.) We note that the proof extends to cases where $N = 2^n$ where n is prime as well. We comment on the places where the proof in [3] (and related papers) have to be changed to get the result in our case, in Appendix A.

In our language the above theorem implies that codes represented by sparse polynomials of somewhat low-degree have large distance. Furthermore if the polynomials are sparse, and $N - 1$ is prime, then also the codes have large distance. We thus get the following implication.

Lemma 2. *For every t there exists a δ such that the following holds for every $N = 2^n$ for prime n. Let $\mathcal{D} = \mathcal{D}(N)$ and let $D \subseteq \mathcal{D}$ be of size at most t. Then the code $C = C_N(D)$ satisfies $\frac{1}{2} - N^{-\delta} \leq \delta(C) \leq \frac{1}{2} + N^{-\delta}$.*

Similarly for every t there exists a δ such that the following holds for for every $N = 2^n$ such that $N - 1$ is prime. Let $\mathcal{D} = \mathcal{D}(N)$ and let $D \subseteq \mathcal{D}$ be of size at most t. Then the $C = C_{N-1}(D)$ satisfies $\frac{1}{2} - N^{-\delta} \leq \delta(C) \leq \frac{1}{2} + N^{-\delta}$.

We remark that such use of results from number theory in coding theory is also common. For example, the distance of the sparse dual-BCH codes is inferred by using the "Weil bound" on exponential sums in a similar manner.

We now move to the crucial part of the paper where we attempt to use counting style arguments to claim that the codes we are considering have the single orbit property for small k. Here our plan is as follows.

We first use a result from [14] to show that for any specific code $\mathcal{C}$ we consider and for every sufficiently large k, its dual has roughly $\binom{N}{k}/|\mathcal{C}|$ codewords of weight k (this bound is tight to within $1 \pm \Theta(1/N^c)$ factor, for large enough k, where k is independent of N and depends only on t, c and the δ of Lemma 2.) Specifically they show:

Theorem 5 ([14] Lemma 3.5). *For every* $c, t < \infty$ *and* $\delta > 0$ *there exists a* k_0 *such that for every* $k \geq k_0$ *and for every code* $\mathcal{C} \subseteq \mathbb{F}_2^N$ *with at most* N^t *codewords satisfying* $\frac{1}{2} - N^{-\delta} \leq \delta(\mathcal{C}) \leq \frac{1}{2} + N^{-\delta}$ *it is the case the* $\mathcal{C}^\perp$ *has* $\binom{N}{k}/|\mathcal{C}| \cdot (1 \pm \Theta(N^{-c})$ *codewords of weight* k.

Thus for any code $\mathcal{C} = \mathcal{C}(D)$ under consideration, this allows us to conclude that $\mathcal{C}^\perp$ has many codewords of weight k (for sufficiently large, but constant k.) What remains to be shown is that the orbit of one of these, under the appropriate group (affine or cyclic) contains a basis for the whole code $\mathcal{C}^\perp$. To do so, we consider any codeword x of weight k in the dual whose orbit under the group does *not* contain a basis for $\mathcal{C}^\perp$ (i.e., Span($\{x \circ \pi | \pi\}$) $\neq \mathcal{C}^\perp$.) We show that for every such word x there is a set $D' \subseteq D$ of size $|D'| = |D| + 1$ such that $x \in \mathcal{C}(D')^\perp$. The size of $\mathcal{C}(D')$ is roughly a factor of N larger than the size of $\mathcal{C}$ and thus $\mathcal{C}(D')^\perp$ is smaller than $\mathcal{C}^\perp$ by a factor of roughly N. We argue further that this code $\mathcal{C}(D')$ also satisfies the same invariant structure as $\mathcal{C}$ and so one can apply Lemma 2 and Theorem 5 to it and thereby conclude that the number of weight k codewords in $\mathcal{C}(D')^\perp$ are also smaller than the number weight k codewords in $\mathcal{C}^\perp$ by a factor of approximately N. Finally we notice that the number of sets D' is $o(N)$ and so the set $\cup_{D'} \mathcal{C}(D')^\perp$ can not include all possible weight k codewords in $\mathcal{C}^\perp$, yielding the k-single orbit property for $\mathcal{C}$. This leads to the proofs of Theorem 1 and 2 - see Section 5.

4 Representing Sparse Invariant Codes by Sparse Polynomials

In this section we study representations of affine-invariant and cyclic-invariant codes by polynomials. That leads to the proof of Lemma 1, which we defer to the full version, along with the other missing proofs of this section. (We will be using the definitions of the sets $\mathcal{D}$, $P_{N,D}$, and $P_{N-1,D}$ as defined in Section 3 heavily throughout this section.)

We start by recalling some standard properties of the Trace function. Recall that $\text{Trace}(x) = x + x^2 + x^4 + \cdots + x^{2^{n-1}}$. The Trace function is linear, i.e. $\text{Trace}(\alpha + \beta) = \text{Trace}(\alpha) + \text{Trace}(\beta) \ \forall \alpha, \beta \in \mathbb{F}_N$. Recall that every function from $\mathbb{F}_N$ to $\mathbb{F}_N$ and hence every function from $\mathbb{F}_N$ to $\mathbb{F}_2$ is the evaluation of polynomial from $\mathbb{F}_N[x]$. More useful to us is the fact that every function from $\mathbb{F}_N$ to $\mathbb{F}_2$ can also be expressed as the trace of a polynomial from $\mathbb{F}_N[x]$, however this representation is not unique. E.g., $\text{Trace}(x^d) = \text{Trace}(x^{2d}) = \text{Trace}(x^{2^i \cdot d})$. However if we restrict to the setting of polynomials from $P_{N,\mathcal{D}}$ then this representation is unique, as shown below.

Lemma 3. *For every word* $w : \mathbb{F}_N \to \mathbb{F}_2$ *(respectively* $w : \mathbb{F}_N^* \to \mathbb{F}_2$*) there is a unique polynomial* $p \in P_{N,\mathcal{D}}$ *(respectively* $p \in P_{N-1,\mathcal{D}}$*) such that* $w(x) = \text{Trace}(p(x))$.

Lemma 4. *Suppose* $\mathcal{C} \subseteq \{\mathbb{F}_N \to \mathbb{F}_2\}$ *is an affine invariant code containing the word* $w = \text{Trace}(p(x))$ *for some* $p \in P_{N,\mathcal{D}}$. *Then, for every monomial* x^e *in the support of* p, *the function* $\text{Trace}(x^e)$ *is in* $\mathcal{C}$. *Furthermore, if* $e \notin \{0, N-1\}$ *then for every* $\beta \in \mathbb{F}_N$, $\text{Trace}(\beta x^e) \in \mathcal{C}$.

Similarly if $\mathcal{C} \subseteq \{\mathbb{F}_N^* \to \mathbb{F}_2\}$ *is cyclic invariant code containing the word* $w = \text{Trace}(p(x))$ *for* $p \in P_{N-1,\mathcal{D}}$. *Then, for every monomial* x^e *in the support of* p, *the function* $\text{Trace}(x^e)$ *is in* $\mathcal{C}$. *If* $e \neq N-1$ *then for every* $\beta \in \mathbb{F}_N$, $\text{Trace}(\beta x^e) \in \mathcal{C}$.

We now use Lemma 4 to characterize cyclic invariant families, while also working towards the characterization of affine invariant families.

Lemma 5. *For every affine invariant code* $\mathcal{C} \subseteq \{\mathbb{F}_N \to \mathbb{F}_2\}$ *there exists a (unique) set* $D \subseteq \mathcal{D}$ *such that* $\mathcal{C} = \{\text{Trace}(p) | p \in P_{N,D}\}$.

For every cyclic invariant family $\mathcal{C} \subseteq \{\mathbb{F}_N^* \to \mathbb{F}_2\}$ *there exists a (unique) set* $D \subseteq \mathcal{D}$ *such that* $\mathcal{C} = \{\text{Trace}(p) | p \in P_{N-1,D}\}$.

Lemma 5 essentially suffices to yield Lemma 1 for the cyclic case (though we still need to verify that $|D|$ is small as claimed.) For the affine case we need to work a little harder to bound the size of the integers in D. To do so we note that affine-invariant properties have further constraints on the set D.

For non-negative integers d and e we say e is in the *shadow* of d (denoted $e \prec d$) if in the binary representations $d = \sum_i d_i 2^i$ and $e = \sum_i e_i 2^i$ with $d_i, e_i \in \{0, 1\}$, it is the case that $e_i \leq d_i$ for every i. We note that affine-invariant codes are characterized by codes with a "shadow-closure" property described below.

Lemma 6. *If* $\mathcal{C}$ *is an affine-invariant code,* $\text{Trace}(x^d) \in \mathcal{C}$ *and* $e \prec d$ *then* $\text{Trace}(x^e) \in \mathcal{C}$.

5   Proofs of Main Theorems

5.1   Analysis of the Cyclic Case

Proof (of Theorem 2). Let $\delta = \delta(t)$ and $\delta' = \delta'(t+1)$ be as given by Lemma 2 for the cyclic invariant case (so codes of length $N-1$ have distance roughly $1/2 - N^{-\delta}$.) Let $c = 2$ and let $k_0 = k_0(c, t, \delta)$ and $k_0' = k_0(c, t+1, \delta')$ be as given by Theorem 5. We prove the theorem for $k = \max\{k_0, k_0'\}$.

Fix N so that $N-1$ is prime and let $\mathcal{C} \subseteq \{\mathbb{F}_N^* \to \mathbb{F}_2\}$ be a cyclic code of cardinality at most N^t. Let $D \subseteq \mathcal{D}$ be as given by Lemma 1, so that $\mathcal{C} = \{\text{Trace}(p) | p \in P_{N-1,D}\}$. For $d \in \mathcal{D} - D$, let $\mathcal{C}(d) = \{\text{Trace}(p) | p \in P_{N-1,D\cup\{d\}}\}$. Our analysis below will show that (1) Every codeword in $w \in \mathcal{C}^\perp - \cup_{d \in \mathcal{D} - D}(\mathcal{C}(d)^\perp)$ generates the code $\mathcal{C}^\perp$ by its cyclic shifts, i.e., $\mathcal{C}^\perp = \text{Span}\{w(\alpha x) | \alpha \in \mathbb{F}_N^*\}$, and (2) There is a codeword of weight k in $\mathcal{C}^\perp - \cup_{d \in \mathcal{D} - D}(\mathcal{C}(d)^\perp)$. Putting the two together we get the proof of the theorem.

We start with the first part. Consider any codeword $w \in \mathcal{C}^\perp$. We claim that if $\text{Span}\{w(\alpha x)\} \neq \mathcal{C}^\perp$, then there must exist an element $d \in \mathcal{D} - D$ such that $w \in \mathcal{C}(d)^\perp$. To see this, first note that $\text{Span}\{w(\alpha x)\}$ is a code invariant under the cyclic group, and is contained in $\mathcal{C}^\perp$. Thus if $\text{Span}\{w(\alpha x)\} \neq \mathcal{C}^\perp$ then it must be strictly contained in $\mathcal{C}^\perp$ and so $(\text{Span}\{w(\alpha x)\})^\perp$ must be a strict superset of $\mathcal{C}$. Using Lemma 1 there must exist a set D' such that $(\text{Span}\{w(\alpha x)\})^\perp = P_{N-1,D'}$. Furthermore D' must be a strict superset of D and so there must exist an element $d \in D' - D$. We claim that $w \in \mathcal{C}(d)^\perp$. This is so since $\mathcal{C}(d) \subseteq (\text{Span}\{w(\alpha x)\})^\perp$ and so $w \in (\text{Span}\{w(\alpha x)\}) \subseteq \mathcal{C}(d)^\perp$. This concludes the proof of the first claim.

It remains to show that there is a codeword of weight k in $\mathcal{C}^\perp - \cup_{d \in \mathcal{D}-D}(\mathcal{C}(d)^\perp)$. For this we employ simple counting arguments. We first note that, using Lemma 2, that $\mathcal{C}$ is a code satisfying $\frac{1}{2} - N^{-\delta} \leq \delta(\mathcal{C}) \leq \frac{1}{2} + N^{-\delta}$. Hence we can apply Theorem 5 to conclude that $\mathcal{C}^\perp$ has at least $\binom{N}{k}/(|\mathcal{C}|) \cdot (1 - O(1/N^2))$ codewords of weight k. On the other hand, for every fixed $d \in \mathcal{D}-D$, we have (by Lemma 2 again) $\frac{1}{2} - N^{-\delta'} \leq \delta(\mathcal{C}(d)) \leq \frac{1}{2} + N^{-\delta'}$. Again applying Theorem 5 we have $\mathcal{C}(d)^\perp$ has at most $\binom{N}{k}/(|\mathcal{C}(d)|)(1 + O(1/N^2))$ codewords of weight k. In case $d = N - 1$, then $|\mathcal{C}(d)| = 2 \cdot |\mathcal{C}|$. In case $d \neq N - 1$ then $|\mathcal{C}(d)| = N \cdot |\mathcal{C}|$. Thus we can bound the total number of codewords of weight k in $\cup_{d \in \mathcal{D}-D}\mathcal{C}(d)^\perp$ from above by

$$\frac{\binom{N}{k}}{2 \cdot |\mathcal{C}|}(1 + O(\frac{1}{N^2})) + |\mathcal{D}| \cdot \frac{\binom{N}{k}}{N \cdot |\mathcal{C}|}(1 + O(\frac{1}{N^2})) \leq \frac{1}{2|\mathcal{C}|} \cdot \binom{N}{k}(1 + \frac{1}{\log_2 N} + O(\frac{1}{N^2})),$$

where above we use the fact that $|\mathcal{D}| \leq N/\log_2 N$. For sufficiently large N (i.e., when $1/\log_2 N + O(1/N^2) \leq 1/2$) we have that this quantity is strictly smaller than $\binom{N}{k}/(|\mathcal{C}|) \cdot (1 - O(1/N^2))$, which was our lower bound on the number of codewords of weight k in $\mathcal{C}^\perp$. We conclude that there is a codeword of weight k in $\mathcal{C}^\perp - \cup_{d \in \mathcal{D}-D}(\mathcal{C}(d)^\perp)$ as claimed. This concludes the proof of the theorem.

5.2 Analysis of the Affine-Invariant Case

Proof (of Theorem 1). The proof is similar to the proof of Theorem 2 with the main difference being that we need to argue that the polynomials associated with functions in $\mathcal{C}$ and $\mathcal{C}(d)$ are of somewhat low-degree (to be able to conclude that they have high-distance.) Details below.

Given t, let δ be from Lemma 2 and let k be large enough for application of Theorem 5. Fix $N = 2^n$ for prime n and and let $\mathcal{C}$ be an affine-invariant code of cardinality N^t. Let $D \subseteq \mathcal{D}$ be a set of cardinality at most t and consisting of integers smaller that $N^{1-1/t}$ such that $\mathcal{C} = \{\text{Trace}(p)|p \in P_{N,D}\}$ (as given by Lemma 1.) For $d \in \mathcal{D} - D$, let $\mathcal{C}(d) = \{\text{Trace}(p)|p \in P_{N,D\cup\{d\}}\}$. Let $\mathcal{D}' = (\mathcal{D} - D) \cap \{1, \ldots, \lfloor N^{1-1/t}\rfloor\}$. Similar to the proof of Theorem 2 we argue that if there is a weight k codeword w in $\mathcal{C}^\perp$ that is not in some $\mathcal{C}(d)^\perp$, but now only for every $d \in \mathcal{D}'$, then $\{\text{Span}(w(\alpha x + \beta)|\alpha \in \mathbb{F}_N^*, \beta \in \mathbb{F}_N\} = \mathcal{C}^\perp$. The same counting argument as in the proof of Theorem 2 suffices to show that such a word does exist.

Consider $w \in \mathcal{C}^{\perp}$ and the code $\{\text{Span}(w(\alpha x + \beta))|\alpha \in \mathbb{F}_N^*, \beta \in \mathbb{F}_N\}$, which is affine invariant and so is given by $P_{N,E}$ for some shadow-closed set E. If $\{\text{Span}(w(\alpha x + \beta))\}^{\perp} \neq \mathcal{C}$ then E strictly contains D and so there must exist some element $d' \in E - D$. Now consider smallest binary weight element $d \prec d'$ such that $d \in E - D$. We claim that the binary weight of d must be at most $t + 1$ (since elements of D have binary weight at most t.) We then conclude that $w \in \{\text{Span}(w(\alpha x + \beta))\} \subseteq \mathcal{C}(d)^{\perp}$ yielding the claim.

The counting argument to show there is a codeword of weight k in $\mathcal{C}^{\perp} - (\cup_{d \in \mathcal{D}'} \mathcal{C}(d)^{\perp}$ is now same as in the proof of Theorem 2 except that we use the affine-invariant part of Lemma 2. This completes the proof of Theorem 1.

Acknowledgments

We would like to thank Oded Goldreich for valuable suggestions and anonymous reviewers for detecting several omissions and errors in prior versions of this paper. We thank Swastik Kopparty for helpful discussions.

References

1. Alon, N., Kaufman, T., Krivelevich, M., Litsyn, S., Ron, D.: Testing Reed-Muller codes. IEEE Transactions on Information Theory 51(11), 4032–4039 (2005)
2. Blum, M., Luby, M., Rubinfeld, R.: Self-testing/correcting with applications to numerical problems. Journal of Computer and System Sciences 47(3), 549–595 (1993)
3. Bourgain, J.: Mordell's exponential sum estimate revisited. J. Amer. Math. Soc. 18(2), 477–499 (2005) (electronic)
4. Bourgain, J.: Some arithmetical applications of the sum-product theorems in finite fields. In: Geometric aspects of functional analysis. Lecture Notes in Math., vol. 1910, pp. 99–116. Springer, Berlin (2007)
5. Bourgain, J., Katz, N., Tao, T.: A sum-product estimate in finite fields, and applications. Geom. Funct. Anal. 14(1), 27–57 (2004)
6. Bourgain, J., Chang, M.-C.: A Gauss sum estimate in arbitrary finite fields. C. R. Math. Acad. Sci. Paris 342(9), 643–646 (2006)
7. Bourgain, J., Konyagin, S.V.: Estimates for the number of sums and products and for exponential sums over subgroups in fields of prime order. C. R. Math. Acad. Sci. Paris 337(2), 75–80 (2003)
8. Goldreich, O., Goldwasser, S., Ron, D.: Property testing and its connection to learning and approximation. JACM 45(4), 653–750 (1998)
9. Goldreich, O., Sudan, M.: Locally testable codes and PCPs of almost-linear length. J. ACM 53(4), 558–655 (2002); Preliminary version in FOCS 2002
10. Jutla, C.S., Patthak, A.C., Rudra, A., Zuckerman, D.: Testing low-degree polynomials over prime fields. In: FOCS 2004, pp. 423–432. IEEE Computer Society Press, Los Alamitos (2004)
11. Kaufman, T., Litsyn, S.: Almost orthogonal linear codes are locally testable. In: FOCS, pp. 317–326. IEEE Computer Society Press, Los Alamitos (2005)
12. Kaufman, T., Litsyn, S.: Long extended BCH codes are spanned by minimum weight words. In: Fossorier, M.P.C., Imai, H., Lin, S., Poli, A. (eds.) AAECC 2006. LNCS, vol. 3857, pp. 285–294. Springer, Heidelberg (2006)

13. Kaufman, T., Ron, D.: Testing polynomials over general fields. SIAM J. Comput. 36(3), 779–802 (2006)
14. Kaufman, T., Sudan, M.: Sparse random linear codes are locally decodable and testable. In: FOCS, pp. 590–600. IEEE Computer Society Press, Los Alamitos (2007)
15. Kaufman, T., Sudan, M.: Algebraic property testing: the role of invariance. In: Ladner, R.E., Dwork, C. (eds.) STOC, pp. 403–412. ACM Press, New York (2008)
16. MacWilliams, F.J., Sloane, N.J.A.: The Theory of Error-Correcting Codes. Elsevier/North-Holland, Amsterdam (1981)
17. Rubinfeld, R., Sudan, M.: Robust characterizations of polynomials with applications to program testing. SIAM Journal on Computing 25(2), 252–271 (1996)
18. van Lint, J.H.: Introduction to Coding Theory, 3rd edn. Graduate Texts in Mathematics, vol. 86. Springer, Berlin (1999)

A   On Using Results from Additive Number Theory

As pointed out earlier Theorem 7 of [3] only considers the analog of Theorem 4 where the field $\mathbb{F}$ is of prime cardinality N, and shows that for any additive character χ, $|\sum_{x \in \mathbb{F}} \chi(f(x))| \leq N^{1-\delta}$. Here we mention the modifications necessary to extend the proof to the case where $\mathbb{F}_N$ is of cardinality 2^n with n being prime.

In [3] the proof reduces to the two cases $r = 1$ and $r = 2$. The case $r = 1$ in the prime case was obtained in [7]. In our case, where $N = 2^n$, the $r = 1$ case was shown in [6]. For $r = 2$ the proof in the prime case applied the sum-product theorem from [5] and uses Proposition 1 of [4]. We note that Proposition 1 of [4] works also when the field is not of prime cardinality. As argued in [5], the sum-product statement might weaken for more general fields only when the field $\mathbb{F}_N$ contains somewhat large subfields. However, when n is prime $\mathbb{F}_{2^n}$ contains only the constant size base field $\mathbb{F}_2$. We conclude that when $\mathbb{F} = \mathbb{F}_{2^n}$ (n prime) it remains true that if a set $A \subset \mathbb{F}_N$ has size $1 < |A| < N^{1-\epsilon}$ for some given ϵ then $|A + A| + |A \cdot A| > C|A|^{1+\delta}$, for some $\delta = \delta(\epsilon)$. The key ingredient of the proof in [4] is an additional sum-product theorem in the additive/multiplicative group $\mathbb{F}_N \times \mathbb{F}_N$ with N prime, where addition and multiplication are defined coordinate-wise. The equivalent formulation for our case $\mathbb{F}_{2^n} \times \mathbb{F}_{2^n}$ follows exactly as in [4], and so does the rest of the proof.

Efficient Quantum Tensor Product Expanders and k-Designs

Aram W. Harrow[1] and Richard A. Low[2]

[1] Department of Mathematics, University of Bristol, Bristol, U.K.
[2] Department of Computer Science, University of Bristol, Bristol, U.K.
low@cs.bris.ac.uk

Abstract. Quantum expanders are a quantum analogue of expanders, and k-tensor product expanders are a generalisation to graphs that randomise k correlated walkers. Here we give an efficient construction of constant-degree, constant-gap quantum k-tensor product expanders. The key ingredients are an efficient classical tensor product expander and the quantum Fourier transform. Our construction works whenever $k = O(n/\log n)$, where n is the number of qubits. An immediate corollary of this result is an efficient construction of an approximate unitary k-design, which is a quantum analogue of an approximate k-wise independent function, on n qubits for any $k = O(n/\log n)$. Previously, no efficient constructions were known for $k > 2$, while state designs, of which unitary designs are a generalisation, were constructed efficiently in [1].

1 Introduction

Randomness is an important resource in both classical and quantum computing. However, obtaining random bits is often expensive, and so it is often desirable to minimise their use. For example, in classical computing, expanders and k-wise independent functions have been developed for this purpose and have found wide application. In this paper, we explore quantum analogues of these two tools.

In quantum computing, operations are unitary gates and randomness is often used in the form of random unitary operations. Random unitaries have algorithmic uses (e.g. [2]) and cryptographic applications (e.g. [3,4]). For information-theoretic applications, it is often convenient to use unitary matrices drawn from the uniform distribution on the unitary group (also known as the Haar measure, and described below in more detail). However, an n-qubit unitary is defined by 4^n real parameters, and so cannot even be approximated efficiently using a subexponential amount of time or randomness. Instead, we will seek to construct efficient pseudo-random ensembles of unitaries which resemble the Haar measure for certain applications. For example, a k-design (often referred to as a t-design, or a (k, k)-design) is a distribution on unitaries which matches the first k moments of the Haar distribution. This is the quantum analogue of k-wise independent functions. k-designs have found cryptographic uses (e.g. [5]) as well as physical applications [6], for which designs for large k are crucial.

I. Dinur et al. (Eds.): APPROX and RANDOM 2009, LNCS 5687, pp. 548–561, 2009.

Below, we will give an efficient construction of a k-design on n qubits for any k up to $O(n/\log(n))$. We will do this by first finding an efficient construction of a quantum 'k-copy tensor product expander' (defined later), which can then be iterated to produce a k-design. We will therefore need to understand some of the theory of expanders before presenting our construction.

Classical expander graphs have the property that a marker executing a random walk on the graph will have a distribution close to the stationary distribution after a small number of steps. We consider a generalisation of this, known as a k-tensor product expander (TPE) and due to [7], to graphs that randomise k different markers carrying out correlated random walks on the same graph. This is a stronger requirement than for a normal ($k = 1$) expander because the correlations between walkers (unless they start at the same position) must be broken. We then generalise quantum expanders in the same way, so that the unitaries act on k copies of the system. We give an efficient construction of a quantum k-TPE which uses an efficient classical k-TPE as its main ingredient. We then give as a key application the first efficient construction of a unitary k-design for any k.

While randomised constructions yield k-designs (by a modification of Theorem 5 of [5]) and k-TPEs (when the dimension is polynomially larger than k [7]) with near-optimal parameters, these approaches are not efficient. State k-designs, meaning ensembles of quantum states matching the first k moments of the uniform distribution on pure states, have been efficiently constructed in [1], but their approach does not appear to generalise to (unitary) k-designs. Previous efficient constructions of k-designs were known only for $k = 1, 2$, and no efficient constant-degree, constant-gap quantum k-TPEs were previously known, except for the $k = 1$ case corresponding to quantum expanders [8,3,9,10].

In Section 1.1, we will define quantum expanders and other key terms. Then in Section 1.2 we will describe our main result which will be proved in Section 2.

1.1 Quantum Expanders

If $\mathcal{S}_N$ denotes the symmetric group on N objects and $\pi \in \mathcal{S}_N$, then define

$$B(\pi) := \sum_{i=1}^{N} |\pi(i)\rangle\langle i| \tag{1}$$

to be the matrix that permutes the basis states $|1\rangle, \ldots, |N\rangle$ according to π.

We will only consider D-regular expander graphs here. We can think of a random walk on such a graph as selecting one of D permutations of the vertices randomly at each step. We construct the permutations as follows. Label the vertices from 1 to N. Then label each edge from 1 to D so that each edge label appears exactly once on the incoming and outgoing edges of each vertex. This gives a set of D permutations. Choosing one of these permutations at random (for some fixed probability distribution) then defines a random walk on the graph.

We now define a classical k-TPE:

Definition 1 ([7]). *Let ν be a probability distribution on $\mathcal{S}_N$ with support on $\leq D$ permutations. Then ν is an (N, D, λ, k) classical k-copy tensor product expander (TPE) if*

$$\left\| \mathbb{E}_{\pi \sim \nu} \left[B(\pi)^{\otimes k} \right] - \mathbb{E}_{\pi \sim \mathcal{S}_N} \left[B(\pi)^{\otimes k} \right] \right\|_\infty = \left\| \sum_{\pi \in \mathcal{S}_N} \left(\nu(\pi) - \frac{1}{N!} \right) B(\pi)^{\otimes k} \right\|_\infty \leq \lambda.$$

(2)

with $\lambda < 1$. Here $\mathbb{E}_{\pi \sim \nu}$ means the expectation over π drawn according to ν and $\mathbb{E}_{\pi \sim \mathcal{S}_N}$ means the expectation over π drawn uniformly from $\mathcal{S}_N$.

Here, as in the rest of the paper, the norms we use are Schatten p-norms. Setting $k = 1$ recovers the usual spectral definition of an expander. Note that a (N, D, λ, k) TPE is also a (N, D, λ, k') TPE for any $k' \leq k$. The largest meaningful value of k is $k = N$, corresponding to the case when ν describes a Cayley graph expander on $\mathcal{S}_N$.

The degree of the map is $D = |\operatorname{supp}\nu|$ and the gap is $1 - \lambda$. Ideally, the degree should be small and gap large. To be useful, these should normally be independent of N and possibly k. We say that a TPE construction is efficient if it can be implemented in poly log N steps. There are known constructions of efficient classical TPEs. The construction of Hoory and Brodsky [11] provides an expander with $D = \operatorname{poly} \log N$ and $\lambda = 1 - 1/\operatorname{poly}(k, \log N)$ with efficient running time. An efficient TPE construction is also known, due to Kassabov [12], which has constant degree and gap (independent of N and k).

Similarly, we define a quantum k-TPE. First we introduce the notation

$$U^{\otimes k,k} = U^{\otimes k} \otimes (U^*)^{\otimes k}.$$

The distribution on the unitary group that we use is the Haar measure. This distribution is the unique unitarily invariant distribution i.e. the only measure dU where $\int f(U)dU = \int f(UV)dU$ for all functions f and unitaries V. Now we define

Definition 2 ([7]). *Let ν be a distribution on $\mathcal{U}(N)$, the group of $N \times N$ unitary matrices, with $D = |\operatorname{supp}\nu|$. Then ν is an (N, D, λ, k) quantum k-copy tensor product expander if*

$$\left\| \mathbb{E}_{U \sim \nu} \left[U^{\otimes k,k} \right] - \mathbb{E}_{U \sim \mathcal{U}(N)} \left[U^{\otimes k,k} \right] \right\|_\infty \leq \lambda \qquad (3)$$

with $\lambda < 1$. Here $\mathbb{E}_{U \sim \mathcal{U}(N)}$ means the expectation over U drawn from the Haar measure.

Again, normally we want D and λ to be constants and setting $k = 1$ recovers the usual definition of a quantum expander. Note that an equivalent statement of the above definition is that, for all ρ,

$$\left\| \mathbb{E}_{U \sim \nu} \left[U^{\otimes k} \rho (U^\dagger)^{\otimes k} \right] - \mathbb{E}_{U \sim \mathcal{U}(N)} \left[U^{\otimes k} \rho (U^\dagger)^{\otimes k} \right] \right\|_2 \leq \lambda \left\| \rho \right\|_2 \qquad (4)$$

A natural application of this is to make an efficient unitary k-design. A unitary k-design is the same as a quantum k-TPE except is close in the 1-norm rather than the ∞-norm:

Definition 3. *Let ν be a distribution on $\mathcal{U}(N)$ with $D = |\operatorname{supp}\nu|$. Say that ν is an ϵ-approximate unitary k-design if*

$$\left\|\mathbb{E}_{U\sim\nu}[U^{\otimes k,k}] - \mathbb{E}_{U\sim U(N)}[U^{\otimes k,k}]\right\|_1 \leq \epsilon. \tag{5}$$

As for TPEs, we say that a unitary design is efficient if a $\operatorname{poly}\log(N)$-time algorithm exists to sample U from ν and to implement U.

Other definitions of approximate designs are possible; for example we can use the diamond norm [13] between the superoperators $\hat{\mathcal{E}}^k_{\mathcal{U}(N)}$ and $\hat{\mathcal{E}}^k_\nu$ where

$$\hat{\mathcal{E}}^k_{\mathcal{U}(N)}(\rho) = \mathbb{E}_{U\sim\mathcal{U}(N)}[U^{\otimes k}\rho(U^\dagger)^{\otimes k}] \tag{6}$$

and

$$\hat{\mathcal{E}}^k_\nu(\rho) = \mathbb{E}_{U\sim\nu}[U^{\otimes k}\rho(U^\dagger)^{\otimes k}] \tag{7}$$

We can then, following [14], define an ϵ-approximate k-design as a set of unitaries $\mathcal{U}$ with

$$\|\hat{\mathcal{E}}^k_{\mathcal{U}(N)} - \hat{\mathcal{E}}^k_\nu\|_\diamond \leq \epsilon. \tag{8}$$

While these norms are in general incomparable, our results work efficiently for both definitions and indeed for any norms that are related by a factor that is polynomial in dimension.

We can make an ϵ-approximate unitary k-design from a quantum k-TPE with $O(k\log N)$ overhead:

Theorem 1. *If $\mathcal{U}$ is an (N, D, λ, k) quantum k-TPE then iterating the map $m = \frac{1}{\log 1/\lambda}\log\frac{N^{2k}}{\epsilon}$ times gives an ϵ-approximate unitary k-design with D^m unitaries.*

Proof. Iterating the TPE m times gives

$$\left\|\mathbb{E}_{U\sim\nu}[U^{\otimes k,k}] - \mathbb{E}_{U\sim\mathcal{U}(N)}[U^{\otimes k,k}]\right\|_\infty \leq \lambda^m$$

This implies that

$$\left\|\mathbb{E}_{U\sim\nu}[U^{\otimes k,k}] - \mathbb{E}_{U\sim\mathcal{U}(N)}[U^{\otimes k,k}]\right\|_1 \leq N^{2k}\lambda^m$$

We take m such that $N^{2k}\lambda^m = \epsilon$ to give the result.

We omit the analogous claim for Eqn. 8, as it, and the proof, are essentially the same.

Corollary 1. *A construction of an efficient quantum (N, D, λ, k)-TPE yields an efficient approximate unitary k-design, provided $\lambda = 1 - 1/\operatorname{poly}\log N$. Further, if D and λ are constants, the number of unitaries in the design is $N^{(O(k))}$.*

Our approach to construct an efficient quantum k-TPE will be to take an efficient classical $2k$-TPE and mix it with a quantum Fourier transform. The degree is thus only larger than the degree of the classical expander by one. Since the quantum Fourier transform on $\mathbb{C}^N$ requires $\operatorname{poly}\log(N)$ time, it follows that if the classical expander is efficient then the quantum expander is as well. The main technical difficulty is to show for suitable values of k that the gap of the quantum TPE is not too much worse than the gap of the classical TPE.

A similar approach to ours was first used in [7] to construct a quantum expander (i.e. a 1-TPE) by mixing a classical 2-TPE with a phase. However, regardless of the set of phases chosen, this approach will not yield quantum k-TPEs from classical $2k$-TPEs for any $k \geq 2$.

1.2   Main Result

Let $\omega = e^{2\pi i/N}$ and define the N-dimensional Fourier transform to be $\mathcal{F} = \frac{1}{\sqrt{N}} \sum_{m=1}^{N} \sum_{n=1}^{N} \omega^{mn} |m\rangle\langle n|$. Define $\delta_{\mathcal{F}}$ to be the distribution on $\mathcal{U}(N)$ consisting of a point mass on $\mathcal{F}$. Our main result is that mixing $\delta_{\mathcal{F}}$ with a classical $2k$-TPE yields a quantum k-TPE for appropriately chosen k and N.

Theorem 2. *Let ν_C be a classical $(N, D, 1 - \epsilon_C, 2k)$-TPE, and for $0 < p < 1$, define $\nu_Q = p\nu_C + (1 - p)\delta_{\mathcal{F}}$. Suppose that*

$$\epsilon_A := 1 - 2(2k)^{4k}/\sqrt{N} > 0. \tag{9}$$

Then ν_Q is a quantum $(N, D + 1, 1 - \epsilon_Q, k)$-TPE where

$$\epsilon_Q \geq \frac{\epsilon_A}{12} \min(p\epsilon_C, 1 - p) > 0 \tag{10}$$

The bound in Eqn. 10 is optimised when $p = 1/(1 + \epsilon_C)$, in which case we have

$$\epsilon_Q \geq \frac{\epsilon_A \epsilon_C}{24}. \tag{11}$$

This means that any constant-degree, constant-gap classical $2k$-TPE gives a quantum k-TPE with constant degree and gap. If the the classical TPE is efficient then the quantum TPE is as well. Using Corollary 1, we obtain approximate unitary k-designs with polynomial-size circuits.

Unfortunately the construction does not work for all dimensions; we require that $N = \Omega((2k)^{8k})$, so that ϵ_A is lower-bounded by a positive constant. However, in applications normally k is fixed. An interesting open problem is to find a construction that works for all dimensions, in particular a $k = \infty$ expander. (Most work on $k = \infty$ TPEs so far has focused on the $N = 2$ case [15].) We suspect our construction may work for k as large as cN for a small constant c. On the other hand, if $2k > N$ then the gap in our construction drops to zero.

2   Proof of Theorem 2

2.1   Proof Overview

First, we introduce some notation. Define $\mathcal{E}_{\mathcal{S}_N}^{2k} = \mathbb{E}_{\pi \sim \mathcal{S}_N}[B(\pi)^{\otimes 2k}]$ and $\mathcal{E}_{\mathcal{U}(N)}^{k} = \mathbb{E}_{U \sim \mathcal{U}(N)}[U^{\otimes k,k}]$. These are both projectors onto spaces which we label $V_{\mathcal{S}_N}$ and $V_{\mathcal{U}(N)}$ respectively. Since $V_{\mathcal{U}(N)} \subset V_{\mathcal{S}_N}$, it follows that $\mathcal{E}_{\mathcal{S}_N}^{2k} - \mathcal{E}_{\mathcal{U}(N)}^{k}$ is a projector onto the space $V_0 := V_{\mathcal{S}_N} \cap V_{\mathcal{U}(N)}^\perp$. We also define $\mathcal{E}_{\nu_C}^{2k} = \mathbb{E}_{\pi \sim \nu_C}[B(\pi)^{\otimes 2k}]$ and $\mathcal{E}_{\nu_Q}^k = \mathbb{E}_{U \sim \nu_Q}[U^{\otimes k,k}]$.

The idea of our proof is to consider $\mathcal{E}_{\nu_C}^{2k}$ a proxy for $\mathcal{E}_{\mathcal{S}_N}^{2k}$; if λ_C is small enough then this is a reasonable approximation. Then we can restrict our attention to vectors in V_0, which we would like to show all shrink substantially under the action of our expander. This in turn can be reduced to showing that $\mathcal{F}^{\otimes k,k}$ maps any vector in V_0 to a vector that has $\Omega(1)$ amplitude in $V_{\mathcal{S}_N}^\perp$. This last step is the most technically involved step of the paper, and involves careful examination of the different vectors making up $V_{\mathcal{S}_N}$.

Thus, our proof reduces to two key Lemmas. The first allows us to substitute $\mathcal{E}_{\nu_C}^{2k}$ for $\mathcal{E}_{\mathcal{S}_N}^{2k}$ while keeping the gap constant.

Lemma 1 ([7] Lemma 1). *Let Π be a projector and let X and Y be operators such that $\|X\|_\infty \leq 1$, $\|Y\|_\infty \leq 1$, $\Pi X = X\Pi = \Pi$, $\|(I-\Pi)X(I-\Pi)\|_\infty \leq 1-\epsilon_C$ and $\|\Pi Y \Pi\|_\infty \leq 1 - \epsilon_A$. Assume $0 < \epsilon_C, \epsilon_A < 1$. Then for any $0 < p < 1$, $\|pX + (1-p)Y\|_\infty < 1$. Specifically,*

$$\|pX + (1-p)Y\|_\infty \leq 1 - \frac{\epsilon_A}{12} \min(p\epsilon_C, 1-p). \tag{12}$$

We will restrict to $V_{\mathcal{U}(N)}^\perp$, or equivalently, subtract the projector $\mathcal{E}_{\mathcal{U}(N)}^k$ from each operator. Thus we have $X = \mathcal{E}_{\nu_C}^{2k} - \mathcal{E}_{\mathcal{U}(N)}^k$, $\Pi = \mathcal{E}_{\mathcal{S}_N}^{2k} - \mathcal{E}_{\mathcal{U}(N)}^k$ and $Y = \mathcal{F}^{\otimes k,k} - \mathcal{E}_{\mathcal{U}(N)}^k$. According to Definition 1, we have the bound

$$\|(I-\Pi)X(I-\Pi)\|_\infty = \|\mathcal{E}_{\nu_C}^{2k} - \mathcal{E}_{\mathcal{S}_N}^{2k}\|_\infty \leq 1 - \epsilon_C. \tag{13}$$

It will remain only to bound $\lambda_A := 1 - \epsilon_A = \|(\mathcal{E}_{\mathcal{S}_N}^{2k} - \mathcal{E}_{\mathcal{U}(N)}^k)\mathcal{F}^{\otimes k,k}(\mathcal{E}_{\mathcal{S}_N}^{2k} - \mathcal{E}_{\mathcal{U}(N)}^k)\|_\infty$.

Lemma 2. *For $N \geq (2k)^2$,*

$$\lambda_A = \|(\mathcal{E}_{\mathcal{S}_N}^{2k} - \mathcal{E}_{\mathcal{U}(N)}^k)\mathcal{F}^{\otimes k,k}(\mathcal{E}_{\mathcal{S}_N}^{2k} - \mathcal{E}_{\mathcal{U}(N)}^k)\|_\infty \leq 2(2k)^{4k}/\sqrt{N}. \tag{14}$$

Combining Eqn. 13, Lemma 2 and Lemma 1 now completes the proof of Theorem 2.

2.2 Action of a Classical $2k$-TPE

We start by analysing the action of a classical $2k$-TPE. (We consider $2k$-TPEs rather than general k-TPEs since our quantum expander construction only uses these.) The fixed points are states which are unchanged when acted on by $2k$ copies of any permutation matrix. Since the same permutation is applied to all copies, any equal indices will remain equal and any unequal indices will remain unequal. This allows us to identify the fixed points of the classical expander: they are the sums over all states with the same equality and difference constraints. For example, for $k = 1$ (corresponding to a 2-TPE), the fixed points are $\sum_{n_1} |n_1, n_1\rangle$ and $\sum_{n_1 \neq n_2} |n_1, n_2\rangle$ (all off-diagonal entries equal to 1). In general, there is a fixed point for each partition of the set $\{1, 2, \dots, 2k\}$ into at most N non-empty parts. If $N \geq 2k$, which is the only case we consider, the $2k^{\text{th}}$ Bell number β_{2k} gives the number of such partitions (see e.g. [16]).

We now write down some more notation to further analyse this. If Π is a partition of $\{1, \ldots, 2k\}$, then we write $\Pi \vdash 2k$. We will see that $\mathcal{E}_{S_N}^{2k}$ projects onto a space spanned by vectors labelled by partitions. For a partition Π, say that $(i, j) \in \Pi$ if and only if elements i and j are in the same block. Now we can write down the fixed points of the classical expander. Let

$$I_\Pi = \{(n_1, \ldots, n_{2k}) : n_i = n_j \text{ iff } (i, j) \in \Pi\}. \tag{15}$$

This is a set of tuples where indices in the same block of Π are equal and indices in different blocks are not equal. The corresponding state is

$$|I_\Pi\rangle = \frac{1}{\sqrt{|I_\Pi|}} \sum_{\mathbf{n} \in I_\Pi} |\mathbf{n}\rangle \tag{16}$$

where $\mathbf{n} = (n_1, \ldots, n_{2k})$ and $|\Pi|$ is the number of blocks in Π. Note that the $\{I_\Pi\}_{\Pi \vdash 2k}$ form a partition $\{1, \ldots, N\}^{2k}$ and thus the $\{|I_\Pi\rangle\}_{\Pi \vdash 2k}$ form an orthonormal basis for V_{S_N}. This is because, when applying the same permutation to all indices, indices that are the same remain the same and indices that differ remain different. This implies that

$$\mathcal{E}_{S_N}^{2k} = \sum_{\Pi \vdash 2k} |I_\Pi\rangle\langle I_\Pi|. \tag{17}$$

To evaluate the normalisation, use $|I_\Pi| = (N)_{|\Pi|}$ where $(N)_n$ is the falling factorial $N(N-1)\ldots(N-n+1)$. We will later find it useful to bound $(N)_n$ with

$$\left(1 - \frac{n^2}{2N}\right) N^n \leq (N)_n \leq N^n. \tag{18}$$

We will also make use of the refinement partial order:

Definition 4. *The refinement partial order $\leq$ on partitions $\Pi, \Pi' \in \text{Par}(2k, N)$ is given by*

$$\Pi \leq \Pi' \text{ iff } (i, j) \in \Pi \Rightarrow (i, j) \in \Pi'. \tag{19}$$

For example, $\{\{1, 2\}, \{3\}, \{4\}\} \leq \{\{1, 2, 4\}, \{3\}\}$. Note that $\Pi \leq \Pi'$ implies that $|\Pi| \geq |\Pi'|$.

Turning Inequality Constraints into Equality Constraints. In the analysis, it will be easier to consider just equality constraints rather than both inequality and equality constraints as in I_Π. Therefore we make analogous definitions:

$$E_\Pi = \{(n_1, \ldots, n_{2k}) : n_i = n_j \forall (i, j) \in \Pi\} \tag{20}$$

and

$$|E_\Pi\rangle = \frac{1}{\sqrt{|E_\Pi|}} \sum_{\mathbf{n} \in E_\Pi} |\mathbf{n}\rangle. \tag{21}$$

Then $|E_\Pi| = N^{|\Pi|}$. For E_Π, indices in the same block are equal, as with I_Π, but indices in different blocks need not be different.

We will need relationships between I_Π and E_Π. First, observe that E_Π can be written as the union of some I_Π sets:

$$E_\Pi = \bigcup_{\Pi' \geq \Pi} I_{\Pi'}. \tag{22}$$

To see this, note that for $\mathbf{n} \in E_\Pi$, we have $n_i = n_j \forall (i,j) \in \Pi$, but we may also have an arbitrary number of additional equalities between n_i's in different blocks. The (unique) partition Π' corresponding to these equalities has the property that Π is a refinement of Π'; that is, $\Pi' \geq \Pi$. Thus for any $\mathbf{n} \in E_\Pi$ there exists a unique $\Pi' \geq \Pi$ such that $\mathbf{n} \in I_{\Pi'}$. Conversely, whenever $\Pi' \geq \Pi$, we also have $I_{\Pi'} \subseteq E_{\Pi'} \subseteq E_\Pi$ because each inclusion is achieved only be relaxing constraints.

Using Eqn. 22, we can obtain a useful identity involving sums over partitions:

$$N^{|\Pi|} = |E_\Pi| = \sum_{\Pi' \geq \Pi} |I_{\Pi'}| = \sum_{\Pi' \geq \Pi} N_{(|\Pi'|)}. \tag{23}$$

Additionally, since both sides in Eqn. 23 are degree $|\Pi|$ polynomials and are equal on $\geq |\Pi| + 1$ points (we can choose any N in Eqn. 23 with $N \geq 2k$), it implies that $x^{|\Pi|} = \sum_{\Pi' \geq \Pi} x_{(\Pi')}$ as an identity on formal polynomials in x.

The analogue of Eqn. 22 for the states $|E_\Pi\rangle$ and $|I_\Pi\rangle$ is similar but has to account for normalisation factors. Thus we have

$$\sqrt{|E_\Pi|}|E_\Pi\rangle = \sum_{\Pi' \geq \Pi} \sqrt{|I_{\Pi'}|}|I_{\Pi'}\rangle. \tag{24}$$

We would also like to invert this relation, and write $|I_\Pi\rangle$ as a sum over various $|E_{\Pi'}\rangle$. Doing so will require introducing some more notation. Define $\zeta(\Pi, \Pi')$ to be 1 if $\Pi \leq \Pi'$ and 0 if $\Pi \not\leq \Pi'$. This can be thought of as a matrix that, with respect to the refinement ordering, has ones on the diagonal and is upper-triangular. Thus it is also invertible. Define $\mu(\Pi, \Pi')$ to be the matrix inverse of ζ, meaning that for all Π_1, Π_2, we have

$$\sum_{\Pi' \vdash 2k} \zeta(\Pi_1, \Pi')\mu(\Pi', \Pi_2) = \sum_{\Pi' \vdash 2k} \mu(\Pi_1, \Pi')\zeta(\Pi', \Pi_2) = \delta_{\Pi_1, \Pi_2},$$

where $\delta_{\Pi_1, \Pi_2} = 1$ if $\Pi_1 = \Pi_2$ and $= 0$ otherwise. Thus, if we rewrite Eqn. 24 as

$$\sqrt{|E_\Pi|}|E_\Pi\rangle = \sum_{\Pi' \vdash 2k} \zeta(\Pi, \Pi')\sqrt{|I_{\Pi'}|}|I_{\Pi'}\rangle, \tag{25}$$

then we can use μ to express $|I_\Pi\rangle$ in terms of the $|E_\Pi\rangle$ as

$$\sqrt{|I_\Pi|}|I_\Pi\rangle = \sum_{\Pi' \vdash 2k} \mu(\Pi, \Pi')\sqrt{|E_{\Pi'}|}|E_{\Pi'}\rangle. \tag{26}$$

This approach is a generalisation of inclusion-exclusion known as Möbius inversion, and the function μ is called the Möbius function (see Chapter 3 of [16] for more background). For the case of the refinement partial order, the Möbius function is known:

Lemma 3 ([17], Section 7).

$$\mu(\Pi, \Pi') = (-1)^{|\Pi|-|\Pi'|} \prod_{i=1}^{|\Pi'|} (b_i - 1)!$$

where b_i is the number of blocks of Π in the ith block of Π'.

We can use this to evaluate sums involving the Möbius function for the refinement order.

Lemma 4.

$$\sum_{\Pi' \geq \Pi} |\mu(\Pi, \Pi')| \, x^{|\Pi'|} = x^{(|\Pi|)} \tag{27}$$

where x is arbitrary and $x^{(n)}$ is the rising factorial $x(x+1)\cdots(x+n-1)$.

Proof. Start with $|\mu(\Pi, \Pi')| = (-1)^{|\Pi|-|\Pi'|}\mu(\Pi, \Pi')$ to obtain

$$\sum_{\Pi' \geq \Pi} |\mu(\Pi, \Pi')| x^{|\Pi'|} = (-1)^{|\Pi|} \sum_{\Pi' \geq \Pi} \mu(\Pi, \Pi')(-x)^{|\Pi'|}$$

$$= (-1)^{|\Pi|} \sum_{\Pi' \geq \Pi} \mu(\Pi, \Pi') \sum_{\Pi'' \geq \Pi'} \zeta(\Pi', \Pi'')(-x)_{(|\Pi''|)}$$

using Eqn. 23. Then use Möbius inversion and $(-x)_{(n)} = (-1)^n x^{(n)}$ to prove the result.

We will mostly be interested in the special case $x = 1$:

Corollary 2.

$$\sum_{\Pi' \geq \Pi} |\mu(\Pi, \Pi')| = |\Pi|! \tag{28}$$

Using $|\mu(\Pi, \Pi')| \geq 1$ and the fact that $\Pi \geq \{\{1\}, \ldots, \{n\}\}$ for all $\Pi \vdash n$, we obtain a bound on the total number of partitions.

Corollary 3. *The Bell numbers β_n satisfy $\beta_n \leq n!$.*

2.3 Fixed Points of a Quantum Expander

We now turn to $V_{\mathcal{U}(N)}$, the space fixed by the quantum expander. By Schur-Weyl duality (see e.g. [18]), the only operators on $(\mathbb{C}^N)^{\otimes k}$ to commute with all $U^{\otimes k}$ are linear combinations of subsystem permutations

$$S(\pi) = \sum_{n_1=1}^{N} \cdots \sum_{n_k=1}^{N} |n_{\pi^{-1}(1)}, \ldots n_{\pi^{-1}(k)}\rangle\langle n_1, \ldots, n_k| \tag{29}$$

for $\pi \in \mathcal{S}_k$. The equivalent statement for $V_{\mathcal{U}(N)}$ is that the only states invariant under all $U^{\otimes k,k}$ are of the form

$$\frac{1}{\sqrt{N^k}} \sum_{n_1,\ldots,n_k \in [N]} |n_1, \ldots, n_k, n_{\pi(1)}, \ldots, n_{\pi(k)}\rangle, \tag{30}$$

for some permutation $\pi \in \mathcal{S}_k$. Since $\mathcal{E}^k_{\mathcal{U}(N)} = \mathbb{E}[U^{\otimes k,k}]$ projects onto the set of states that is invariant under all $U^{\otimes k,k}$, it follows that $V_{\mathcal{U}(N)}$ is equal to the span of the states in Eqn. 30.

Now we relate these states to our previous notation.

Definition 5. *For $\pi \in \mathcal{S}_k$, define the partition corresponding to π by*

$$P(\pi) = \{\{1, k + \pi(1)\}, \{2, k + \pi(2)\}, \ldots, \{k, k + \pi(k)\}\}.$$

Then the state in Eqn. 30 is simply $|E_{P(\pi)}\rangle$, and so

$$V_{\mathcal{U}(N)} = \mathrm{span}\{|E_{P(\pi)}\rangle : \pi \in \mathcal{S}_k\}. \tag{31}$$

Note that the classical expander has many more fixed points than just the desired $|E_{P(\pi)}\rangle$. The main task in constructing a quantum expander from a classical one is to modify the classical expander to decay the fixed points that should not be fixed by the quantum expander.

2.4 Fourier Transform in the Matrix Element Basis

Since we make use of the Fourier transform, we will need to know how it acts on a matrix element. We find

$$\mathcal{F}^{\otimes k,k}|\mathbf{m}\rangle = \frac{1}{N^k} \sum_{\mathbf{n}} \omega^{\mathbf{m}.\mathbf{n}}|\mathbf{n}\rangle$$

where

$$\mathbf{m}.\mathbf{n} = m_1 n_1 + \ldots + m_k n_k - m_{k+1} n_{k+1} - \ldots - m_{2k} n_{2k} \tag{32}$$

We will also find it convenient to estimate the matrix elements $\langle E_{\Pi_1}|\mathcal{F}^{\otimes k,k}|E_{\Pi_2}\rangle$. The properties we require are proven in the following lemmas.

Lemma 5. *Choose any $\Pi_1, \Pi_2 \vdash 2k$. Let $\mathbf{m} \in \Pi_1$ and $\mathbf{n} \in \Pi_2$. Call the free indices of $\mathbf{m}$ $\tilde{m}_i$ for $1 \leq i \leq |\Pi_1|$. Then let $\mathbf{m}.\mathbf{n} = \sum_{i=1}^{|\Pi_1|} \sum_{j=1}^{2k} \tilde{m}_i A_{i,j} n_j$ where $A_{i,j}$ is a $|\Pi_1| \times 2k$ matrix with entries in $\{0, 1, -1\}$ which depends on Π_1 (but not Π_2). Then*

$$\langle E_{\Pi_1}|\mathcal{F}^{\otimes k,k}|E_{\Pi_2}\rangle = N^{-k + \frac{|\Pi_1| - |\Pi_2|}{2}} \sum_{\mathbf{n} \in E_{\Pi_2}} \mathbb{I}\left(\sum_j A_{i,j} n_j \equiv 0 \bmod N \, \forall i\right) \tag{33}$$

where $\mathbb{I}$ is the indicator function.

Proof. Simply perform the $\mathbf{m}$ sum in

$$\langle E_{\Pi_1}|\mathcal{F}^{\otimes k,k}|E_{\Pi_2}\rangle = N^{-\left(k + \frac{|\Pi_1| + |\Pi_2|}{2}\right)} \sum_{\mathbf{m} \in E_{\Pi_1}} \sum_{\mathbf{n} \in E_{\Pi_2}} \omega^{\mathbf{m}.\mathbf{n}} \tag{34}$$

Lemma 6. $\langle E_{\Pi_1}|\mathcal{F}^{\otimes k,k}|E_{\Pi_2}\rangle$ *is real and positive.*

Proof. Since all entries in the sum in Eqn. 33 are nonnegative and at least one ($\mathbf{n}=0$) is strictly positive, Lemma 5 implies the result.

Lemma 7. *If $\Pi_1' \leq \Pi_1$ and $\Pi_2' \leq \Pi_2$ then*

$$\sqrt{|E_{\Pi_1}| \cdot |E_{\Pi_2}|}\langle E_{\Pi_1}|\mathcal{F}^{\otimes k,k}|E_{\Pi_2}\rangle \leq \sqrt{|E_{\Pi_1'}| \cdot |E_{\Pi_2'}|}\langle E_{\Pi_1'}|\mathcal{F}^{\otimes k,k}|E_{\Pi_2'}\rangle \tag{35}$$

Proof. We prove first the special case when $\Pi_1' = \Pi_1$, but $\Pi_2' \leq \Pi_2$ is arbitrary. Recall that $\Pi_2' \leq \Pi_2$ implies that $E_{\Pi_2} \subseteq E_{\Pi_2'}$. Now the LHS of Eqn. 35 equals

$$N^{-k} \sum_{\mathbf{m}\in E_{\Pi_1}, \mathbf{n}\in E_{\Pi_2}} \exp\left(\frac{2\pi i}{N}\mathbf{m}.\mathbf{n}\right)$$

$$= N^{|\Pi_1|-k} \sum_{\mathbf{n}\in E_{\Pi_2}} \mathbb{I}\left(\sum_j A_{i,j}n_j \equiv 0 \bmod N \,\forall i\right)$$

$$= N^{|\Pi_1|-k} \sum_{\mathbf{n}\in E_{\Pi_2'}} \mathbb{I}(\mathbf{n}\in E_{\Pi_2})\,\mathbb{I}\left(\sum_j A_{i,j}n_j \equiv 0 \bmod N \,\forall i\right)$$

$$\leq N^{|\Pi_1|-k} \sum_{\mathbf{n}\in E_{\Pi_2'}} \mathbb{I}\left(\sum_j A_{i,j}n_j \equiv 0 \bmod N \,\forall i\right)$$

$$= \sqrt{|E_{\Pi_1}|\,|E_{\Pi_2'}|}\langle E_{\Pi_1}|\mathcal{F}^{\otimes k,k}|E_{\Pi_2'}\rangle,$$

as desired. To prove Eqn. 35 we repeat this argument, interchanging the roles of Π_1 and Π_2 and use the fact that $\langle E_{\Pi_1}|\mathcal{F}^{\otimes k,k}|E_{\Pi_2}\rangle$ is symmetric in Π_1 and Π_2.

Lemma 8.

$$\langle E_{\Pi_1}|\mathcal{F}^{\otimes k,k}|E_{\Pi_2}\rangle \leq N^{-\frac{1}{2}|2k-(|\Pi_1|+|\Pi_2|)|} \tag{36}$$

Proof. Here, there are two cases to consider. The simpler case is when $|\Pi_1| + |\Pi_2| \leq 2k$. Here we simply apply the inequality

$$\sum_{\mathbf{m}\in E_{\Pi_1}, \mathbf{n}\in E_{\Pi_2}} \exp\left(\frac{2\pi i}{N}\mathbf{m}.\mathbf{n}\right) \leq |E_{\Pi_1}|\,|E_{\Pi_2}| = N^{|\Pi_1|+|\Pi_2|}$$

to Eqn. 34, and conclude that $\langle E_{\Pi_1}|\mathcal{F}^{\otimes k,k}|E_{\Pi_2}\rangle \leq N^{\frac{|\Pi_1|+|\Pi_2|}{2}-k}$.

Next, we would like to prove that

$$\langle E_{\Pi_1}|\mathcal{F}^{\otimes k,k}|E_{\Pi_2}\rangle \leq N^{k-\frac{|\Pi_1|+|\Pi_2|}{2}}. \tag{37}$$

Here we use Lemma 7 with $\Pi_1' = \Pi_1$ and $\Pi_2' = \{\{1\},\{2\},\ldots,\{2k\}\}$, the maximally refined partition. Note that $|E_{\Pi_2'}| = N^{2k}$ and $\mathcal{F}^{\otimes k,k}|E_{\Pi_2'}\rangle = |0\rangle$. Thus

$$\langle E_{\Pi_1}|\mathcal{F}^{\otimes k,k}|E_{\Pi_2}\rangle \leq N^{k-\frac{|\Pi_2|}{2}}\langle E_{\Pi_1}|\mathcal{F}^{\otimes k,k}|E_{\Pi_2'}\rangle = N^{k-\frac{|\Pi_2|}{2}}\langle E_{\Pi_1}|0\rangle = N^{k-\frac{|\Pi_1|+|\Pi_2|}{2}},$$

establishing Eqn. 37.

Lemma 9.
If $\Pi_1 = \Pi_2 = P(\pi)$ then $\langle E_{\Pi_1}|\mathcal{F}^{\otimes k,k}|E_{\Pi_2}\rangle = 1$. If, for any Π_1, Π_2 with $|\Pi_1| + |\Pi_2| = 2k$, either condition isn't met (i.e. either $\Pi_1 \neq \Pi_2$ or there does not exist $\pi \in \mathcal{S}_k$ such that $P(\pi) = \Pi_1 = \Pi_2$) then

$$\langle E_{\Pi_1}|\mathcal{F}^{\otimes k,k}|E_{\Pi_2}\rangle \leq \frac{2k}{N} \tag{38}$$

for $N > k$.

Proof. In Lemma 10, we introduce the $\Pi_1 \times \Pi_2$ matrix $\tilde{A}$ with the property that

$$\mathbf{m}.\mathbf{n} = \sum_{i=1}^{|\Pi_1|} \sum_{j=1}^{|\Pi_2|} \tilde{m}_i \tilde{A}_{i,j} \tilde{n}_j \tag{39}$$

for all $\mathbf{m} \in \Pi_1$ and $\mathbf{n} \in \Pi_2$ where $\tilde{m}_j$ and $\tilde{n}_j$ are the free indices of $\mathbf{m}$ and $\mathbf{n}$. This is similar to the matrix A introduced in Lemma 5 except only the free indices of $\mathbf{n}$ are considered.

For $\Pi_1 = \Pi_2 = P(\pi)$, Lemma 10 implies that $\tilde{A} = 0$, or equivalently $\mathbf{m}.\mathbf{n} = 0$ for all $\mathbf{m}, \mathbf{n} \in P(\pi)$. Using $|\Pi_1| + |\Pi_2| = 2k$, $\langle E_{\Pi_1}|\mathcal{F}^{\otimes k,k}|E_{\Pi_2}\rangle = 1$.

Otherwise we have $(\Pi_1, \Pi_2) \notin \{(P(\pi), P(\pi)) : \pi \in \mathcal{S}_k\}$ with $|\Pi_1| + |\Pi_2| = 2k$. For all these, Lemma 10 implies that $\tilde{A}$ is nonzero (for $N > k$, no entries in $\tilde{A}$ can be $> N$ or $< -N$ so $\tilde{A} \equiv 0 \bmod N$ is equivalent to $\tilde{A} = 0$). Fix an i for which the ith row of $\tilde{A}$ is nonzero. We wish to count the number of $(\tilde{n}_1, \ldots, \tilde{n}_{|\Pi_2|})$ such that $\sum_j \tilde{A}_{i,j} \tilde{n}_j \equiv 0 \bmod N$. Assume that each $\tilde{A}_{i,j}$ divides N and is nonnegative; if not, we can replace $\tilde{A}_{i,j}$ with $\text{GCD}(|\tilde{A}_{i,j}|, N)$ by a suitable change of variable for $\tilde{n}_j$.

Now choose an arbitrary j such that $\tilde{A}_{i,j} \neq 0$. For any values of $\tilde{n}_1, \ldots, \tilde{n}_{j-1}$, $\tilde{n}_{j+1}, \ldots, \tilde{n}_{|\Pi_2|}$, there are $|\tilde{A}_{i,j}| \leq 2k$ choices of $\tilde{n}_j$ such that $\sum_j \tilde{A}_{i,j} \tilde{n}_j \equiv 0 \bmod N$. Thus, there are $\leq 2kN^{|\Pi_2|-1}$ choices of $\tilde{n}$ such that $\sum_j \tilde{A}_{i,j} \tilde{n}_j \equiv 0 \bmod N$. Substituting this into Eqn. 33 (which we can trivially modify to apply for $\tilde{A}$ rather than just A), we find that

$$\langle E_{\Pi_1}|\mathcal{F}^{\otimes k,k}|E_{\Pi_2}\rangle \leq \frac{2k}{N} N^{-k+\frac{|\Pi_1|+|\Pi_2|}{2}} = \frac{2k}{N},$$

thus establishing Eqn. 38.

Lemma 10. *Let $\tilde{A}$ be the matrix such that $\mathbf{m}.\mathbf{n} = \sum_{i=1}^{|\Pi_1|} \sum_{j=1}^{|\Pi_2|} \tilde{m}_i \tilde{A}_{i,j} \tilde{n}_j$ for all $\mathbf{m} \in \Pi_1$ and $\mathbf{n} \in \Pi_2$ where $\tilde{m}_j$ and $\tilde{n}_j$ are the free indices of $\mathbf{m}$ and $\mathbf{n}$. Then $\tilde{A} = 0$ if and only if $\Pi_1 = \Pi_2 \geq P(\pi)$ for some $\pi \in \mathcal{S}_k$.*

The proof of this Lemma, as well as of Lemma 2, are omitted from this extended abstract, and can be found in [19].

Instead, we outline the arguments behind these proofs. To prove Lemma 10, we first show that if $\Pi_1 = \Pi_2 \geq P(\pi)$ for some π, then a direct calculation yields $\tilde{A} = 0$. The converse is more involved, and requires looking at the intersections of each block from Π_1 with each block from Π_2, and proving that there exists a permutation $\pi \in \mathcal{S}_k$ such that each $\{i, k + \pi(i)\}$ is contained in a single such intersection.

For Lemma 2, we would like to show that, for any unit vector $|\psi\rangle \in V_0$, $|\langle\psi|\mathcal{F}^{\otimes k,k}|\psi\rangle|^2 \leq 2(2k)^{4k}/\sqrt{N}$. Our strategy will be to calculate the matrix elements of $\mathcal{F}^{\otimes k,k}$ in the $|I_\Pi\rangle$ and $|E_\pi\rangle$ bases. While the $|I_\Pi\rangle$ states are orthonormal, we will see that the $\langle E_{\Pi_1}|\mathcal{F}^{\otimes k,k}|E_{\Pi_2}\rangle$ matrix elements are easier to calculate. We then use Möbius functions to express $|I_\Pi\rangle$ in terms of $|E_\Pi\rangle$. The calculations are broken into a number of cases, with a leading-order contribution of $k!$, and then several other cases each contributing $k^{O(k)}/\sqrt{N}$. As the $k!$ terms correspond to the fixed subspace $V_{\mathcal{U}(N)}$, we are left with the $k! + 1^{\text{st}}$ largest eigenvalue being $\leq k^{O(k)}/\sqrt{N}$.

3 Conclusions

We have shown how efficient quantum tensor product expanders can be constructed from efficient classical tensor product expanders. This immediately yields an efficient construction of unitary k-designs for any k. Unfortunately our results do not work for all dimensions; we require the dimension N to be $\Omega((2k)^{8k})$. While tighter analysis of our construction could likely improve this, our construction does not work for $N < 2k$. Constructions of expanders for all dimensions remains an open problem.

Acknowledgments

We are grateful for funding from the Army Research Office under grant W9111NF-05-1-0294, the European Commission under Marie Curie grants ASTQIT (FP6-022194) and QAP (IST-2005-15848), and the U.K. Engineering and Physical Science Research Council through "QIP IRC." RL would like to thank Markus Grassl and Andreas Winter for helpful discussions. RL is also extremely grateful to Andreas Winter and the rest of the Centre for Quantum Technologies, National University of Singapore, where part of this research was carried out, for their kind hospitality. We would also like to thank an anonymous referee for suggesting a shorter and tighter proof of Lemma 9.

References

1. Ambainis, A., Emerson, E.: Quantum t-designs: t-wise Independence in the Quantum World. Computational Complexity 2007 (2007) arXiv:quant-ph/0701126v2
2. Sen, P.: Random measurement bases, quantum state distinction and applications to the hidden subgroup problem. In: Complexity 2006, pp. 274–287 (2005) arXiv:quant-ph/0512085
3. Ambainis, A., Smith, A.: Small Pseudo-Random Families of Matrices: Derandomizing Approximate Quantum Encryption. In: Jansen, K., Khanna, S., Rolim, J.D.P., Ron, D. (eds.) APPROX-RANDOM 2004. LNCS, vol. 3122, pp. 249–260. Springer, Heidelberg (2004)

4. Hayden, P., Leung, D., Shor, P.W., Winter, A.: Randomizing Quantum States: Constructions and Applications. Communications in Mathematical Physics 250, 371–391 (2004) arXiv:quant-ph/0307104
5. Ambainis, A., Bouda, J., Winter, A.: Tamper-resistant encryption of quantum information (2008) arXiv:0808.0353
6. Low, R.A.: Large Deviation Bounds for k-designs (2009) arXiv:0903.5236
7. Hastings, M.B., Harrow, A.W.: Classical and Quantum Tensor Product Expanders (2008) arXiv:0804.0011
8. Ben-Aroya, A., Ta-Shma, A.: Quantum expanders and the quantum entropy difference problem (2007) arXiv:quant-ph/0702129
9. Harrow, A.W.: Quantum expanders from any classical Cayley graph expander. Q. Inf. Comp. 8(8/9), 715–721 (2008)
10. Gross, D., Eisert, J.: Quantum Margulis Expanders. Q. Inf. Comp. 8(8/9), 722–733 (2008)
11. Hoory, S., Brodsky, A.: Simple Permutations Mix Even Better (2004) arXiv:math/0411098
12. Kassabov, M.: Symmetric Groups and Expanders (2005) arXiv:math/0503204
13. Kitaev, A.Y., Shen, A.H., Vyalyi, M.N.: Classical and Quantum Computation. American Mathematical Society, Boston (2002)
14. Harrow, A.W., Low, R.A.: Random quantum circuits are approximate 2-designs (2008) arXiv:0802.1919
15. Bourgain, J., Gamburd, A.: New results on expanders. C. R. Acad. Sci. Paris, Ser. I 342, 717–721 (2006)
16. Stanley, R.: Enumerative Combinatorics. Cambridge University Press, Cambridge (1986)
17. Rota, G.C.: On the foundations of combinatorial theory I. Theory of Möbius Functions. Probability Theory and Related Fields 2(4), 340–368 (1964)
18. Goodman, R., Wallach, N.: Representations and Invariants of the Classical Groups. Cambridge University Press, Cambridge (1998)
19. Harrow, A.W., Low, R.A.: Efficient quantum tensor product expanders and k-designs (2008) arXiv:0811.2597

Hellinger Strikes Back: A Note on the Multi-party Information Complexity of AND

T.S. Jayram

IBM Almaden Research Center
650 Harry Rd
San Jose, CA 95120, USA
jayram@almaden.ibm.com

Abstract. The AND problem on t bits is a promise decision problem where either at most one bit of the input is set to 1 (No instance) or all t bits are set to 1 (YES instance). In this note, I will give a new proof of an $\Omega(1/t)$ lower bound on the information complexity of AND in the number-in-hand model of communication. This was recently established by Gronemeier, STACS 2009. The proof exploits the information geometry of communication protocols via Hellinger distance in a novel manner and avoids the analytic approach inherent in previous work. As previously known, this bound implies an $\Omega(n/t)$ lower bound on the communication complexity of multiparty disjointness and consequently a $\Omega(n^{1-2/k})$ space lower bound on estimating the k-th frequency moment F_k.

1 Introduction

Welcome to the magical world of *Hellinger distance*![1] In this note, I will describe a short proof of an $\Omega(1/t)$ lower bound for the information complexity of the AND function in the number-in-hand model of communication. I should mention at the forefront that the result is not new (perhaps for the constants involved) as was shown by Gronemeier [Gro09] recently. My focus, however, is show the power and beauty of Hellinger distance when applied to communication protocols, and in particular, the light that it sheds on the *information geometry* inherent in the structure of communication protocols. To describe this problem and its motivation, I must first take you on a detor into the world of communication complexity and data streams.

1.1 Data Stream Space Complexity of Frequency Moments

Space… the Final Frontier.

Star Trek

A major influence on the foundations of massive data sets as well as a pioneer of novel techniques has been the frequency moments problem and its variants

---

[1] This is a metric between probability distributions μ and σ whose *square* is given by $\frac{1}{2}\sum_x \left(\sqrt{\mu(x)} - \sqrt{\sigma(x)}\right)^2$. A different viewpoint of this definition is given in Section 3.

I. Dinur et al. (Eds.): APPROX and RANDOM 2009, LNCS 5687, pp. 562–573, 2009.

[FM85, AMS99, Ind06, BJK$^+$02, CCFC04, IW05]. In the k-th frequency moment problem F_k, the goal is to estimate the sum of the k-th power of the frequencies of items in a stream, presented as a sequence of non-negative updates. This paper deals with the case $k \geq 2$. The best space upper bound for this problem is $O(n^{1-2/k})$ up to polylogarithmic factors [AMS99, IW05] (a better dependence on the polylogarithmic term is in [BGKS06, MW]). But what about space lower bounds? It is here that communication complexity enters the picture.

Communication complexity [Yao79], one of the crown jewels of complexity theory, measures the necessary amount of communication in a distributed setting. It is often the case that computation problems are too complex or the models are to fine-grained to be amenable to analysis. Communication models judiciously abstract away details of the original problem and perhaps even weaken some of the restrictions of the original model. Their power resides in their simplicity, in which one can hope to gain traction for solving difficult problems. The complexity theorist's life, being hard as it is, sees some glimmer of hope in such things!

In this note, I will consider the *number-in-hand multiparty communication model*. Loosely speaking (see Section 2 for formal details), the input is partitioned amongst several players, and their goal is to compute some function of the input by exchanging messages via a shared blackboard. The communication cost of a protocol is the maximum length of this shared communication over all inputs. In a randomized protocol the players also have private access to random coins. The protocol solves the communication problem if the answer equals the value of the function to some desired confidence.

In order to show tight bounds for space, Alon, Matias and Szegedy [AMS99] introduced a generalization of set-disjointness in the t-party communication model. Each of the t players is given a subset of $[n]$ with the following promise: either the sets are pairwise disjoint (No instance) or they have a unique common element but are otherwise disjoint (Yes instance). They proved a communication complexity lower bound of $\Omega(n/t^4)$, which implies a space lower bound of $\Omega(n^{1-5/k})$ for estimating the frequency moment F_k, and thus is non-trivial only when $k > 5$. They left open the problem of getting a $\Omega(n/t)$ communication complexity lower bound on t-party set-disjointness in order to close this gap.

1.2 Information Complexity...

Bar-Yossef, Jayram, Ravi Kumar and Sivakumar [BJKS04] tackled this problem in an *information complexity* paradigm, hoping to prove the result via a direct sum argument. Information theoretic arguments have been used in previous work [Abl96, SS02, BCKO93], but information complexity was given first-class status as a resource measure first by Chakrabarti, Shi, Wirth, and Yao [CSWY01] for two-party simultaneous protocols in the context of proving direct sum theorems. Briefly speaking, information complexity of a communication problem f characterizes how much information about the inputs the players must reveal in a correct protocol for f. Bar-Yossef *et al.* considered a powerful generalization of this measure to general communication protocols. In particular, they introduced *conditional information complexity* as a means to handle non-product distributions that are essential for proving tight lower bounds for multiparty set-disjointness.

By proving a direct-sum theorem, they reduced the problem to giving an $\Omega(1/t)$ information complexity bound on the multiparty AND problem: the t players each have a single bit with the promise that either at most one bit is set to 1 (NO instance) or all t bits are set to 1 (YES instance). Proving sub-constant lower bounds for information theoretic measures is somewhat unusual in that domain. By translating this problem to the domain of statistical divergences, especially using Hellinger distances, they obtained a non-optimal $\Omega(1/t^2)$ lower bound for general protocols. On the other hand, by using analytic techniques involving Rényi divergences, they obtained a near-$\Omega(1/t)$ optimal lower bound for the restricted one-way protocols. This created a gap between the two models even though it yielded near-optimal $\Omega(n^{1-2/k})$ space bounds for one-pass F_k estimation for all $k > 2$.

The situation was somewhat remedied by Chakrabarti, Khot and Sun [CKS03] who proved an $\Omega(1/t \log t)$ information complexity lower bound for AND. Recently, Gronemeier [Gro09] closed the gap to $\Omega(1/t)$. A common thread to both these papers is that they expand the information theory expressions directly in terms of analytic expressions (via Kullback-Liebler distance) since Rényi divergences seem to offer no advantage while dealing with general protocols. By using analytic techniques on the logarithm and associated functions, they manage to avoid the loss incurred by Bar-Yossef *et al.* in taking the Hellinger distance route.

1.3   ... to Hellinger Distance

> A thing of beauty is a joy for ever:
> Its loveliness increases; it will never
> Pass into nothingness; but still will
> keep
>
> J. Keats

This brings me to the main thrust of this paper—proving an optimal $\Omega(1/t)$ information complexity lower bound for AND using Hellinger distance. An immediate concern is whether Hellinger distance is too weak to yield such an optimal bound, as perhaps has been the impression created in previous work either explicitly or implicitly. It is true that Hellinger distance can be arbitrarily smaller than Kullback-Liebler distance. On the other hand, expressing the information complexity of AND as a distance measure results in the *Jensen-Shannon* distance. Although this measure can be expressed using Kullback-Leibler distances, the form is quite restricted. Indeed, both Hellinger and Jensen-Shannon distances are within small constants of each other, so the *apriori* loss in transitioning to Hellinger distance is not significant. The *real* weakness in the Bar-Yossef *et al.* approach to using Hellinger distance amounted to the following: the expressions to be bounded involved the square of Hellinger distance and therefore, since Hellinger distance is a metric, but not its square, only a weak form of triangle inequality could be used. Unfortunately, that loss was significant and only yielded a sub-optimal $\Omega(1/t^2)$ bound.

I will demonstrate in this paper that Hellinger distance exposes the rich *geometric* structure of communication protocols. Since Hellinger distance is just a scaled Euclidean distance, its square is not a metric. Nevertheless, it has been studied extensively in the theory of metric spaces [DL97] under the area of *negative-type distances*. I will show that a simple geometric negative-type inequality suffices to overcome the lossy triangle inequality of the previous approach, thereby yielding an optimal bound for AND.

The inductive argument used in the paper is perhaps more intuitive because it explicitly shows where the protocol must *create* distances in order for it to be a valid communication protocol. Growing enough of these distances results in a (squared) Hellinger distance between a YES and NO instance of AND. For a correct protocol this must be constant, yielding the desired lower bound. As further evidence to the power of this geometric structure, Jayram and Woodruff [JW09] have shown that estimating the product norm $\ell_2 \circ \ell_0$ requires communication $\Omega(\sqrt{n})$, and Andoni, Jayram, and Patrascu [AJP09] have shown improved lower bounds for the communication complexity of edit distance.

A central message promoted in this paper is that transcript distributions have a natural place in the Euclidean space; taking square-roots of probabilities puts them in the unit sphere of ℓ_2. Since pure states in a quantum system are naturally described this way, it would be interesting to explore the applicability of the techniques in the paper to quantum communication.

Section 2 contains the preliminaries including a review of information complexity notions. In Section 3, I will describe the key properties of Hellinger distance including the new ingredient needed in the proof, namely a negative-type inequality. These ingredients are combined in Section 4 in order to prove the main result.

2 Preliminaries

Suppose there are $t \geq 2$ players jointly holding an input $x = (x_1, x_2, \ldots, x_t) \in \mathcal{X}^t$, where player i has x_i, for $i \in [t]$. Their goal is to solve some communication problem $f(x_1, x_2, \ldots, x_t)$, defined on a *subset* of $\mathcal{X}^t$, by sending messages to each other. In this paper, the standard *blackboard* model will be used where the messages are all written on a shared medium. A *protocol* $\mathcal{P}$ on $\mathcal{X}^t$ specifies the rules for the players to write their messages on the blackboard when the inputs come from (all of) $\mathcal{X}^t$. The resulting sequence of messages is called the *transcript*. The maximum length of the transcript (in bits) over all inputs is the *communication cost* of the protocol $\mathcal{P}$. For technical reasons, it will be convenient not to require that the transcript also contain the answer. Instead, there is some referee who outputs an answer by looking only at the transcript and not the inputs. The protocol is allowed to be randomized in which each player, as well as the referee, has *private* access to an unlimited supply of random coins. The protocol solves the communication problem if the answer equals $f(x_1, x_2, \ldots, x_t)$ with probability at least $1 - \delta$. Throughout this paper, δ will be a small constant and such protocols will be called as *correct* protocols. Note that the protocol itself is legally defined for all inputs in $\mathcal{X}^t$ although no restriction is placed on the answer of the protocol outside the domain of f.

A family of sets $S_1, S_2, \ldots, S_t \subseteq [n]$ is called a *sunflower with kernel T* if for every $i \neq j, S_i \cap S_j = T$. (These are also known as delta-systems.) In other words, if an element belongs to any distinct pair of sets then it belongs to all of them, so in fact, the kernel equals $\bigcap_i S_i$. The *multi-party set-disjointness* communication problem, $\mathrm{DISJ}_{t,n}$, with t players on a universe of size n is a (promise) decision problem where the input $S_1, S_2, \ldots, S_t \subseteq [n]$ to the players is a sunflower whose kernel is either *empty* (No instance) or a *singleton* (Yes instance). A randomized private-coin communication protocol $\mathscr{P}$ that solves $\mathrm{DISJ}_{t,n}$ should accept Yes instances and reject No instances with error probability at most δ.

To describe $\mathrm{DISJ}_{t,n}$ as a valid Boolean formula over promise instances, encode the input of the players as bits as follows. Let $x = (x_{ij})$ denote a $t \times n$ array of bits. The i-th row of x is the characteristic vector of the set S_i. Then,

$$\mathrm{DISJ}_{t,n}(x) = \bigvee_{j=1}^{n} \bigwedge_{i=1}^{t} x_{ij}.$$

Define

$$\mathrm{AND}_t(u_1, u_2, \ldots, u_t) \triangleq \bigwedge_{i=1}^{t} u_i,$$

with the promise that either at most one input bit is set to 1 (No instance) or all input bits are set to 1 (Yes instance). Letting $x^j \in \{0,1\}^t$ denote the j-th column of x,

$$\mathrm{DISJ}_{t,n}(x^1, x^2, \ldots, x^n) \triangleq \mathrm{DISJ}_{t,n}(x) = \bigvee_{j=1}^{n} \mathrm{AND}_t(x^j).$$

This way of splitting the input highlights the fact that the set-disjointness problem is an Or of n instances of the AND_t problem on t bits. It therefore suggests a direct-sum argument for proving communication lower bounds for $\mathrm{DISJ}_{t,n}$.

I will now briefly review the *information complexity* paradigm for proving communication lower bounds via direct sum arguments, as developed in [BJKS04], for multi-party number-in-hand communication protocols. Information complexity of a communication problem f characterizes how much information about the inputs the players must reveal in a correct protocol for f. The underlying distribution on the inputs to the players can be *independent* across the players but in many cases the tight bounds are obtained by requiring dependent input distributions. This causes some complications which are overcome by introducing *conditional independence* on the inputs via auxiliary random variables. This is formalized below.

Notation. *Random variables will be denoted by upper case Roman or Greek letters, and the values they take by corresponding lower case letters. Probability distributions will be denoted by lower case Greek letters. A random variable X with distribution μ is denoted by $X \sim \mu$. If μ is the uniform distribution over a set $\mathscr{W}$, then this is also denoted as $X \in_R \mathscr{W}$.*

Definition 1. A distribution μ over $\mathscr{X}^t$ is *partitioned* by η if there exists a joint probability space $(X_1, X_2, \ldots, X_t, F)$ such that $(X_1, X_2, \ldots, X_t) \sim \mu$, $F \sim \eta$, and $X_1, X_2, \ldots, X_t$ are jointly independent conditioned on F i.e. $P(X_1, X_2, \ldots, X_t \mid F) = \prod_i P(X_i \mid F)$ $\square$

Definition 2 (Information Complexity). Let $\mathscr{P}$ be a t-party randomized private-coin protocol on the input domain $\mathscr{X}^t$ and let its random coins be denoted by the random variable R. Suppose μ is a distribution over $\mathscr{X}^t$ partitioned by η in some joint probability space where $X = (X_1, X_2, \ldots, X_t) \sim \mu$ and $F \sim \eta$. Extend this to a joint probability space over (X, F, R) such that (X, F) is independent of R. Now, let $\Pi = \Pi(X, R)$ be the random variable denoting the transcript of the protocol, where the randomness is *both* over the input distribution and the random coins of the protocol $\mathscr{P}$. The *(conditional) information cost* of $\mathscr{P}$ under (μ, η) is defined to be $I(X : \Pi \mid F)$, i.e., the (Shannon) conditional mutual information between X and Π conditioned on F.

The *information complexity* of a communication problem f, denoted by $\mathrm{IC}_\mu(f \mid \eta)$, is defined to be the minimum information cost of a correct protocol for f under (μ, η).   □

Since $I(X : \Pi \mid D) \le H(\Pi) \le |\Pi|$, it suffices to prove lower bounds on the information cost of a correct protocol.

For the problem $\mathrm{DISJ}_{t,n} = \bigvee \mathrm{AND}_t$, I will first define a distribution (v, ζ) for AND_t. Let $(U_1, U_2, \ldots, U_t) \sim v$ and $G \sim \zeta$ be such that

1. $G \in_R [t]$. G picks a player whose bit will vary while the rest are fixed to 0.
2. Conditioned on the event $G = i$, let $(U_1, U_2, \ldots, U_t) \in_R \{0, e_i\}$. Here, e_i is the standard basis vector with a 1 in the i-th position and 0 elsewhere.

The distribution for $\mathrm{DISJ}_{t,n}$ is defined by letting $\mu = v^n$ and $\eta = \zeta^n$. In other words, if $X = (X^1, X^2, \ldots, X^n)$ is the input and $F = (F^1, F^2, \ldots, F^n)$ is the auxiliary random variable, then independently for each $j \in [n]$, $F^j \sim \zeta$ and $X^j \sim v$.

Proposition 3 (Direct Sum for Information Complexity [BJKS04])

$$\mathrm{CC}(\mathrm{DISJ}_{t,n}) \ge \mathrm{IC}_\mu(\mathrm{DISJ}_{t,n} \mid \eta) \ge n \cdot \mathrm{IC}_v(\mathrm{AND}_t \mid \zeta). \qquad \square$$

Consequently, I will show an $\Omega(1/t)$ lower bound on the information complexity of AND.

3 Hellinger Distance

Notation. Let $\|\cdot\|$ denote the standard ℓ_2 norm and $\|\cdot\|_1$ denote the standard ℓ_1 norm.

Let u be an input to a protocol $\mathscr{P}$. Let $\pi(u)$ denote the probability distribution over the transcripts induced by $\mathscr{P}$ on input u, where the randomness is over the private coins of $\mathscr{P}$. Let $\pi(u)_\tau$ denote the probability that the transcript equals τ. Viewing $\pi(u)$ as an element of ℓ_1, note that $\|\pi(u)\|_1 = \sum_\tau \pi(u)_\tau = 1$.

The following switch in viewpoint is the perhaps most important notion in this paper. Consider the element $\psi(u) \in \ell_2$ obtained via the square-root map $\pi(u) \mapsto \psi(u) = \sqrt{\pi(u)}$. This means $\psi(u)_\tau = \sqrt{\pi(u)_\tau}$ for all τ. The central tenet is that $\psi(u)$ is an object that deserves real attention on its own right from the standpoint of information complexity. Now, $\|\psi(u)\| = \|\pi(u)\|_1 = 1$, and so $\psi(u) \in \mathbb{S}_+$, where $\mathbb{S}_+$ denotes the unit sphere in ℓ_2 restricted to the non-negative orthant. In analogy with quantum physics, call $\psi(u)$ the *transcript wave function* of u in $\mathscr{P}$.

Definition 4 (Hellinger Distance). The *Hellinger distance* between $\psi_1, \psi_2 \in \mathbb{S}_+$ is a scaled Euclidean distance defined as $h(\psi_1, \psi_2) \triangleq \frac{1}{\sqrt{2}} \|\psi_1 - \psi_2\|$.   □

Since $\|\psi_1 - \psi_2\|^2 \leq \|\psi_1\|^2 + \|\psi_2\|^2 = 2$, the scaling ensures that Hellinger distance is always between 0 and 1. To emphasize the geometric nature of Hellinger distance, I will almost exclusively use the norm notation to refer to Hellinger distance.

The following properties of Hellinger distance are well-known (see [BJKS04]):

Proposition 5 (Hellinger distance and communication protocols). *Let $\mathscr{P}$ be a randomized t-party private-coin protocol on the input domain $\mathscr{X}^t$. Let g be a decision problem defined on a subset of $\mathscr{X}^t$. Let $u, v \in \mathscr{X}^t$ be two distinct inputs whose transcript wave functions in $\mathscr{P}$ are denoted by $\psi(u)$ and $\psi(v)$, respectively.*

1. *Mutual information to Hellinger distance: Suppose $U \in_R \{u, v\}$. If Π denotes the transcript random variable, then*

$$I(U : \Pi) \geq \tfrac{1}{2} \|\psi(u) - \psi(v)\|^2.$$

2. *Soundness: If $\mathscr{P}$ is a correct protocol for g, and $g(u) \neq g(v)$, then*

$$\tfrac{1}{2} \|\psi(u) - \psi(v)\|^2 \geq 1 - 2\sqrt{\delta}.$$

3. *Cut-and-paste: Let u' and v' denote the inputs obtained by performing some cut-and-paste on u and v. In other words for each $1 \leq i \leq t$, either (a) $u_i' = u_i$ and $v_i' = v_i$ or (b) $u_i' = v_i$ and $v_i' = u_i$. Then*

$$\|\psi(u) - \psi(v)\| = \|\psi(u') - \psi(v')\|.$$

Consequently, suppose the inputs to $\mathscr{P}$ are such that each player holds a single bit, i.e., $\mathscr{X} = \{0, 1\}$. Identify the input $u \in \mathscr{X}^t$ with the subset $A = \{i \mid u_i = 1\} \subseteq [t]$. Similarly, identify v with $B \subseteq [t]$. Then

$$\|\psi(A) - \psi(B)\| = \|\psi(A \cup B) - \psi(A \cap B)\|.   \qquad \square$$

Property 1 in the above proposition is just a restatement of the fact that the Jensen-Shannon distance between $\psi(u)$ and $\psi(v)$ is bounded from below by their Hellinger distance. Property 2 follows by relating Hellinger to variational distance and then invoking the correctness of the protocol. Property 3 generalizes the rectangle property of deterministic communication protocols to randomized protocols. The corollary to this property, where each player holds a single bit, follows by letting $u_i' = u_i \vee v_i$ and $v_i' = u_i \wedge v_i$ for all i.

The next inequality is the new key ingredient that enables the tight lower bound for AND_t via Hellinger distance:

Proposition 6. *For any $v_0, v_1, v_2, \ldots, v_s \in \ell_2$,*

$$\sum_{i=1}^{s} \|v_0 - v_i\|^2 \geq \frac{1}{s} \sum_{1 \leq i < j \leq s} \|v_i - v_j\|^2$$

Proof. This is a special case of a general class of negative-type inequalities [DL97] satisfied by the square of the ℓ_2-distance: for any set of real numbers $b_0, b_1, \ldots, b_s$ such that $\sum_{i=0}^{s} b_i = 0$, it holds that

$$\sum_{\substack{0 \leq i \leq s \\ 0 \leq j \leq s}} b_i b_j \|v_i - v_j\|^2 \leq 0.$$

The above inequality is simple to derive and I will show this below for the sake of completeness. Setting $b_0 = s$ and $b_1 = b_2 = \cdots = b_s = -1$ yields the statement of the proposition.

Observe that:

$$\sum_{\substack{0 \leq i \leq s \\ 0 \leq j \leq s}} b_i b_j \|v_i - v_j\|^2 = \sum_{\substack{0 \leq i \leq s \\ 0 \leq j \leq s}} b_i b_j \left(\|v_i\|^2 + \|v_j\|^2 - 2\langle v_i, v_j \rangle \right)$$

$$= \left(\sum_{0 \leq i \leq s} b_i \|v_i\|^2 \sum_{0 \leq j \leq s} b_j \right) + \left(\sum_{0 \leq j \leq s} b_j \|v_j\|^2 \sum_{0 \leq i \leq s} b_i \right)$$

$$- 2 \left(\sum_{0 \leq i \leq s} b_i v_i \right) \cdot \left(\sum_{0 \leq j \leq s} b_j v_j \right)$$

$$= 0 + 0 - 2 \left\| \sum_{0 \leq i \leq s} b_i v_i \right\|^2$$

$$\leq 0,$$

proving the inequality.

4 Information Complexity of AND$_t$

> Beauty is the first test: there is no permanent place in the world for ugly mathematics.
>
> _____
>
> G.H. Hardy

The following is the main technical result of this note.

Theorem 7. *Let $\mathscr{P}$ be a t-party protocol on the input domain $\{0, 1\}^t$. Identify every subset of $[t]$ with its characteristic vector in $\{0, 1\}^t$. Let $\psi(A)$ denote the transcript wave function of input $A \subseteq [t]$ in $\mathscr{P}$. Suppose $A_1, A_2, \ldots, A_s$ are a pairwise disjoint collection of $s = 2^k$ subsets of $[t]$, where $k \geq 0$. Set $A \triangleq \bigcup_i A_i$. Then,*

$$\sum_{i=1}^{s} \|\psi(\emptyset) - \psi(A_i)\|^2 \geq \|\psi(\emptyset) - \psi(A)\|^2 \cdot \prod_{\ell=1}^{k} \left(1 - \frac{1}{2^\ell} \right)$$

Proof. By induction on k. The base case $k = 0$ (i.e., $s = 1$) follows trivially with equality. For the induction step, let $k \geq 1$ so that $s = 2^k$ is even. Now,

$$\sum_{i=1}^{s} \|\psi(\emptyset) - \psi(A_i)\|^2$$

$$\geq \frac{1}{s} \sum_{1 \leq i < j \leq s} \|\psi(A_i) - \psi(A_j)\|^2 \qquad \text{(Proposition 6)}$$

$$= \frac{1}{s} \sum_{1 \leq i < j \leq s} \|\psi(\emptyset) - \psi(A_i \cup A_j)\|^2 \qquad \text{(Proposition 5, cut-and-paste)} \quad (1)$$

Associate $\{(i, j) \mid 1 \leq i < j \leq s\}$ with the edges of the complete graph K_s. Since s is even, K_s can be decomposed into an edge-disjoint union of $s - 1$ perfect matchings, $\mathcal{M}_1, \mathcal{M}_2, \ldots, \mathcal{M}_{s-1}$, each having $s/2$ edges. Using this, rewrite the expression within the sum in (1) as follows:

$$\sum_{1 \leq i < j \leq s} \|\psi(\emptyset) - \psi(A_i \cup A_j)\|^2 = \sum_{p=1}^{s-1} \sum_{\{i,j\} \in \mathcal{M}_p} \|\psi(\emptyset) - \psi(A_i \cup A_j)\|^2 \quad (2)$$

Fix a p within the sum. The sets $A_i \cup A_j$, for $\{i, j\} \in \mathcal{M}_p$, are a pairwise disjoint collection of $s/2 = 2^{k-1}$ sets. By the induction hypothesis,

$$\sum_{\{i,j\} \in \mathcal{M}_p} \|\psi(\emptyset) - \psi(A_i \cup A_j)\|^2 \geq \|\psi(\emptyset) - \psi(A)\|^2 \cdot \prod_{\ell=1}^{k-1} \left(1 - \frac{1}{2^\ell}\right)$$

Substitute this bound in (2) for every p, and then combine it with (1) to get:

$$\sum_{i=1}^{s} \|\psi(\emptyset) - \psi(A_i)\|^2 \geq \frac{1}{s}(s - 1) \cdot \|\psi(\emptyset) - \psi(A)\|^2 \cdot \prod_{\ell=1}^{k-1} \left(1 - \frac{1}{2^\ell}\right)$$

$$= \left(1 - \frac{1}{2^k}\right) \cdot \|\psi(\emptyset) - \psi(A)\|^2 \cdot \prod_{\ell=1}^{k-1} \left(1 - \frac{1}{2^\ell}\right)$$

$$= \|\psi(\emptyset) - \psi(A)\|^2 \cdot \prod_{\ell=1}^{k} \left(1 - \frac{1}{2^\ell}\right),$$

proving the theorem.

Corollary 8. *The information complexity of* AND_t *is* $\Omega(1/t)$.

Proof. Let $U \sim \nu$ and $G \sim \zeta$. Let $\mathscr{P}$ be a correct protocol for AND_t whose information cost under (ν, ζ) equals C. If Π denotes the transcript, then

$$C = I(U : \Pi \mid G) = \frac{1}{t} \sum_{i=1}^{t} I(U : \Pi \mid G = i)$$

Conditioned on $G = i$, $U \in_R \{0, e_i\}$. Applying the Mutual-information-to-Hellinger-distance property in Proposition 5,

$$C \geq \frac{1}{t} \sum_{i=1}^{t} \frac{1}{2} \|\psi(0) - \psi(e_i)\|^2 = \frac{1}{t} \sum_{i=1}^{t} \frac{1}{2} \|\psi(0) - \psi(\{i\})\|^2$$

Suppose for the moment that $t = 2^k$ is a power of 2 with $k \geq 1$. Applying Theorem 7 with $s = t$ and $A_i = \{i\}$, for $1 \leq i \leq t$, to the RHS above,

$$C \geq \frac{1}{t} \cdot \left(\frac{1}{2} \|\psi(0) - \psi([t])\|^2 \right) \cdot \prod_{\ell=1}^{k} \left(1 - \frac{1}{2^\ell} \right) \tag{3}$$

Since $\text{AND}_t(0) \neq \text{AND}_t([t])$, the soundness property in Proposition 5 applied to $\mathscr{P}$ implies the following:

$$\frac{1}{2} \|\psi(0) - \psi([t])\|^2 \geq 1 - 2\sqrt{\delta} \tag{4}$$

For the product term in (3),

$$\prod_{\ell=1}^{k} \left(1 - \frac{1}{2^\ell} \right) \geq \prod_{\ell=1}^{\infty} \left(1 - \frac{1}{2^\ell} \right) = 0.288788\ldots\,^2 \tag{5}$$

Substituting the bounds in (4) and (5) into (3) shows that the information cost of $\mathscr{P}$ is $\Omega(1/t)$.

For arbitrary values of t, a minor modification yields the same asymptotic bound. Let t' be be the largest power of 2 which is at most t. Partition $[t]$ in some arbitrary manner into a collection of t' sets $A_1, A_2, \ldots, A_{t'}$ of sizes 1 and 2 that are pairwise disjoint. For each set $\{i, j\}$ of size 2, apply Theorem 7 with $k = 1$ to bound

$$\|\psi(0) - \psi(\{i\})\|^2 + \|\psi(0) - \psi(\{j\})\|^2 \geq \frac{1}{2} \|\psi(0) - \psi(\{i, j\})\|^2.$$

Thus,

$$\sum_{i=1}^{t} \|\psi(0) - \psi(\{i\})\|^2 \geq \frac{1}{2} \cdot \sum_{k=1}^{t'} \|\psi(0) - \psi(A_k)\|^2.$$

Then proceed with the same argument.

Remark. *The proof shows that the constant in $\Omega(1/t)$ is $c(1 - 2\sqrt{\delta})$ where c equals $0.288788\ldots$, the digital search tree constant, if t is a power of 2, and half that value otherwise.*

---

[2] This constant is known as the digital search tree constant (see Sloane's A048651 [Slo]). It also has connections to random binary matrices. Euler studied this in the context of generating functions for integer partitions, and gave methods to compute the infinite product that converge fairly rapidly. Thanks to Laurens Gunnarsen for discussions on this topic.

References

[Abl96]   Ablayev, F.: Lower bounds for one-way probabilistic communication complexity and their application to space complexity. Theoretical Computer Science 157(2), 139–159 (1996)

[AJP09]   Andoni, A., Jayram, T.S., Patrascu, M.: Non-embeddability and sketching complexity via information geometry (2009)

[AMS99]   Alon, N., Matias, Y., Szegedy, M.: The space complexity of approximating the frequency moments. Journal of Computer and System Sciences 58(1), 137–147 (1999)

[BCKO93] Bar-Yehuda, R., Chor, B., Kushilevitz, E., Orlitsky, A.: Privacy, additional information, and communication. IEEE Transactions on Information Theory 39(6), 1930–1943 (1993)

[BGKS06] Bhuvanagiri, L., Ganguly, S., Kesh, D., Saha, C.: Simpler algorithm for estimating frequency moments of data streams. In: Proceedings of the Seventeenth Annual ACM-SIAM Symposium on Discrete Algorithms, SODA 2006, Miami, Florida, USA, January 22-26, 2006, pp. 708–713. ACM Press, New York (2006)

[BJK+02] Bar-Yossef, Z., Jayram, T.S., Kumar, R., Sivakumar, D., Trevisan, L.: Counting distinct elements in a data stream. In: Rolim, J.D.P., Vadhan, S.P. (eds.) RANDOM 2002. LNCS, vol. 2483, pp. 1–10. Springer, Heidelberg (2002)

[BJKS04] Bar-Yossef, Z., Jayram, T.S., Kumar, R., Sivakumar, D.: An information statistics approach to data stream and communication complexity. J. Comput. Syst. Sci. 68(4), 702–732 (2004)

[CCFC04] Charikar, M., Chen, K., Farach-Colton, M.: Finding frequent items in data streams. Theor. Comput. Sci. 312(1), 3–15 (2004)

[CKS03]   Chakrabarti, A., Khot, S., Sun, X.: Near-optimal lower bounds on the multiparty communication complexity of set-disjointness. In: Proceedings of the 18th Annual IEEE Conference on Computational Complexity, pp. 107–117 (2003)

[CSWY01] Chakrabarti, A., Shi, Y., Wirth, A., Yao, A.C.-C.: Informational complexity and the direct sum problem for simultaneous message complexity. In: Proceedings of the 42nd IEEE Annual Symposium on Foundations of Computer Science (FOCS), pp. 270–278 (2001)

[DL97]   Deza, M., Laurent, M.: Geometry of Cuts and Metrics. Springer, Heidelberg (1997)

[FM85]   Flajolet, P., Martin, G.N.: Probabilistic counting algorithms for data base applications. J. Comput. Syst. Sci. 31(2), 182–209 (1985)

[Gro09]   Gronemeier, A.: Asymptotically optimal lower bounds on the nih-multi-party information complexity of the and-function and disjointness. In: Albers, S., Marion, J.-Y. (eds.) STACS, Schloss Dagstuhl - Leibniz-Zentrum fuer Informatik, Germany Internationales Begegnungs- und Forschungszentrum fuer Informatik (IBFI), Schloss Dagstuhl, Germany. Dagstuhl Seminar Proceedings, vol. 09001, pp. 505–516 (2009)

[Ind06]   Indyk, P.: Stable distributions, pseudorandom generators, embeddings, and data stream computation. J. ACM 53(3), 307–323 (2006)

[IW05]   Indyk, P., Woodruff, D.P.: Optimal approximations of the frequency moments of data streams. In: STOC, pp. 202–208 (2005)

[JW09]   Jayram, T.S., Woodruff, D.: The data stream space complexity of cascaded norms (submitted, 2009)

[MW]   Monemizadeh, M., Woodruff, D.: l_p-sampling with applications (manuscript)
[Slo]   Sloane, N.: The on-line encyclopedia of integer sequences!,
          http://www.research.att.com/~njas/sequences/A048651
[SS02]   Saks, M., Sun, X.: Space lower bounds for distance approximation in the data
          stream model. In: Proceedings of the 34th Annual ACM Symposium on Theory
          of Computing (STOC), pp. 360–369 (2002)
[Yao79]   Yao, A.C.-C.: Some complexity questions related to distributive computing. In:
          Proceedings of the 11th ACM Symposium on Theory of Computing (STOC), pp.
          209–213 (1979)

Pseudorandom Generators and Typically-Correct Derandomization

Jeff Kinne[1,*], Dieter van Melkebeek[1,**], and Ronen Shaltiel[2,***]

[1] Department of Computer Sciences, University of Wisconsin-Madison, USA
{jkinne,dieter}@cs.wisc.edu
[2] Department of Computer Science, University of Haifa, Israel
ronen@cs.haifa.ac.il

Abstract. The area of derandomization attempts to provide efficient deterministic simulations of randomized algorithms in various algorithmic settings. Goldreich and Wigderson introduced a notion of "typically-correct" deterministic simulations, which are allowed to err on few inputs. In this paper we further the study of typically-correct derandomization in two ways.

First, we develop a generic approach for constructing typically-correct derandomizations based on seed-extending pseudorandom generators, which are pseudorandom generators that reveal their seed. We use our approach to obtain both conditional and unconditional typically-correct derandomization results in various algorithmic settings. We show that our technique strictly generalizes an earlier approach by Shaltiel based on randomness extractors, and simplifies the proofs of some known results. We also demonstrate that our approach is applicable in algorithmic settings where earlier work did not apply. For example, we present a typically-correct polynomial-time simulation for every language in BPP based on a hardness assumption that is weaker than the ones used in earlier work.

Second, we investigate whether typically-correct derandomization of BPP implies circuit lower bounds. Extending the work of Kabanets and Impagliazzo for the zero-error case, we establish a positive answer for error rates in the range considered by Goldreich and Wigderson. In doing so, we provide a simpler proof of the zero-error result. Our proof scales better than the original one and does not rely on the result by Impagliazzo, Kabanets, and Wigderson that NEXP having polynomial-size circuits implies that NEXP coincides with EXP.

1 Introduction

Randomized Algorithms and Derandomization. One of the central topics in the theory of computing deals with the power of randomness – can randomized procedures be efficiently simulated by deterministic ones? In some settings exponential

---

* Partially supported by NSF award CCR-0728809 and by a Cisco Systems Distinguished Graduate Fellowship.
** Research done while visiting the University of Haifa and the Weizmann Institute of Science. Partially supported by NSF award CCR-0728809.
*** Partially supported by BSF grant 2004329 and ISF grant 686/07.

I. Dinur et al. (Eds.): APPROX and RANDOM 2009, LNCS 5687, pp. 574–587, 2009.

gaps have been established between randomized and deterministic complexity; in some settings efficient derandomizations[1] are known; in others the question remains wide open. The most famous open setting is that of time-bounded computations, i.e., whether BPP=P, or more modestly, whether BPP lies in deterministic subexponential time. A long line of research gives "hardness versus randomness tradeoffs" for this problem (see [12] for an introduction). These are *conditional results* that give derandomizations assuming a hardness assumption (typically circuit lower bounds of some kind), where the efficiency of the derandomization depends on the strength of the hardness assumption. The latter is used to construct an efficient *pseudorandom generator*, which is a deterministic procedure G that stretches a short "seed" s into a longer "pseudorandom string" $G(s)$ with the property that the uniform distribution on pseudorandom strings is computationally indistinguishable from the uniform distribution on all strings. G allows us to derandomize a randomized procedure $A(x, r)$ that takes an input x and a string r of "coin tosses" as follows: We run the pseudorandom generator on all seeds to produce all pseudorandom strings of length $|r|$; for each such pseudorandom string we run A using that pseudorandom string as "coin tosses", and output the majority vote of the answers of A. Note that this derandomization procedure takes time that is exponential in the seed length of the pseudorandom generator. For example, efficient pseudorandom generators with logarithmic seed length imply that BPP=P, whereas subpolynomial seed length only yields simulations of BPP in deterministic subexponential time.

Typically-Correct Derandomization. Weaker notions of derandomization have been studied, in which the deterministic simulation is allowed to err on some inputs. Impagliazzo and Wigderson were the first to consider derandomizations that succeed with high probability on any efficiently samplable distribution; related notions have subsequently been investigated in [4, 8, 17, 20]. Goldreich and Wigderson [3] introduced a weaker notion in which the deterministic simulation only needs to behave correctly on most inputs of any given length. We refer to such simulations as "typically-correct derandomizations". The hope is to construct typically-correct derandomizations that are more efficient than the best-known everywhere-correct derandomizations, or to construct them under weaker assumptions than the hypotheses needed for everywhere-correct derandomization.

Previous Work on Typically-Correct Derandomization. Goldreich and Wigderson [3] had the key idea to obtain typically-correct derandomizations by "extracting randomness from the input": extract $r = E(x)$ in a deterministic way such that $B(x) = A(x, E(x))$ behaves correctly on most inputs. If this approach works (as such) and E is efficient, the resulting typically-correct derandomization B has essentially the same complexity as the original randomized procedure A. In

---

[1] In this paper the term "derandomization" always refers to "full derandomization", i.e., obtaining equivalent deterministic procedures that do not involve randomness at all.

principle, the approach is limited to algorithms A that use no more than $|x|$ random bits; by combining it with pseudorandom generators one can try to handle algorithms that use a larger number of random bits. Goldreich and Wigderson managed to get the approach to work *unconditionally* for logspace algorithms for undirected connectivity, a problem which has been fully derandomized by now [15]. Under a *hardness assumption* that is not known to imply BPP=P, namely that there are functions that are *mildly* hard on average for small circuits with access to an oracle for satisfiability, they showed that BPP has polynomial-time typically-correct derandomizations that err on very few inputs, namely at most a subexponential number. Their construction uses Trevisan's extractor [19].

Zimand [24] showed *unconditional* typically-correct derandomizations with polynomial overhead for sublinear-time algorithms, which can be viewed as randomized decision trees that use a sublinear number of random bits. Zimand's approach relies on a notion of randomness extractors called "exposure-resilient extractors" introduced in [23].

Shaltiel [16] described a generic approach to obtain typically-correct derandomization results. Loosely speaking he showed how to construct a typically-correct derandomization for any randomized procedure that uses a sublinear amount of randomness when given an extractor with exponentially small error that extracts randomness from distributions that are "recognizable by the procedure." We elaborate on Shaltiel's approach in Section 4. Using this approach and "off the shelf" randomness extractors, Shaltiel managed to reproduce Zimand's result for decision trees as well as realize *unconditional* typically-correct derandomizations for 2-party communication protocols and streaming algorithms.

Shaltiel also combined his approach with pseudorandom generator constructions to handle procedures that require a polynomial number of random bits. He obtained typically-correct derandomizations with a polynomially small error rate for randomized algorithms computable by polynomial-sized constant-depth circuits, based on the known hardness of parity for such circuits. He also derived a *conditional* typically-correct derandomization result for BPP under a hardness hypothesis that is incomparable to the Goldreich-Wigderson hypothesis (and is also not known to imply BPP=P), namely that there are functions that are *very* hard on average for small circuits without access to an oracle for satisfiability. The resulting error rate is exponentially small. For both results Shaltiel applies the pseudorandom generators that follow from the hardness versus randomness tradeoffs twice: once to reduce the need for random bits to sublinear, and once to construct the required randomness extractor with exponentially small error. Whereas the first pseudorandom generator application can do with functions that are *mildly* hard on average, the second one requires functions that are *very* hard on average.

Our Approach. In this paper we develop an alternative generic approach for constructing typically-correct derandomizations. The approach builds on "seed-extending pseudorandom generators" rather than "extractors". A seed-extending pseudorandom generator is a generator G which outputs the seed as part of the pseudorandom string, i.e., $G(s) = (s, E(s))$ for some function E.[2] The well-known Nisan-Wigderson pseudorandom generator construction [14] can easily be made seed-extending. We show that whenever a seed-extending pseudorandom generator passes certain statistical tests defined by the randomized procedure $A(x, r)$, the deterministic procedure $B(x) = A(x, E(x))$ forms a typically-correct derandomization of A, where the error rate depends on the error probability of the original randomized algorithm and on the error of the pseudorandom generator.

Note that this approach differs from the typical use of pseudorandom generators in derandomization, where the pseudorandom generator G is run on every seed. As the latter induces a time overhead that is exponential in the seed length, one aims for pseudorandom generators that are computable in time exponential in the seed length. A polynomial-time simulation is achieved only in the case of logarithmic seed lengths. In contrast, we run G *only once*, namely with the input x of the randomized algorithm as the seed. We use the pseudorandom generator to *select* one "coin toss sequence" $r = E(x)$ on which we run the randomized algorithm. As opposed to the traditional derandomization setting, our approach benefits from pseudorandom generators that are computable in time less than exponential in the seed length. With a pseudorandom generator computable in time polynomial in the output length, we obtain nontrivial polynomial-time typically-correct derandomizations even when the seed length is just subpolynomial.

Our approach has the advantage of being more direct than the one of [16], in the sense that it derandomizes the algorithm A in "one shot". More importantly, it obviates the second use of pseudorandom generators in Shaltiel's approach and allows us to start from the *weaker assumption* that there are functions which are *mildly* hard on average for small circuits without access to an oracle for satisfiability.

While our assumption is weaker than both the one in [3] and the one in [16], the error rate of our typically-correct derandomizations is only polynomially small. We can decrease the error rate by strengthening the hardness assumption. Under the same hardness assumption as [16] our approach matches the exponentially small error rate in that paper.

We can similarly relax the hardness assumption in a host of other settings. In some cases this allows us to establish new *unconditional* typically-correct derandomizations, namely for models where functions that are *very* hard on average are not known but functions which are only *mildly* hard on average are known unconditionally.

We also determine the precise relationship between our approach and Shaltiel's. We show that in the range of exponentially small error rates, "extractors for recognizable distributions" are equivalent to seed-extending pseudorandom generators

---

[2] Borrowing from the similar notion of "strong extractors" in the extractor literature, such pseudorandom functions have been termed "strong" in earlier papers. In coding-theoretic terms, they could also be called "systematic". However, we find the term "seed-extending" more informative.

that pass the statistical tests we need. This means that all the aforementioned results of [16] can also be obtained using our new approach. Since we can also handle situations where [16] does not apply, our approach is more generic.

Typically-Correct Derandomization and Circuit Lower Bounds. Kabanets and Impagliazzo [9] showed that subexponential-time derandomizations of BPP imply circuit lower bounds that seem beyond the scope of current techniques. We ask whether subexponential-time typically-correct derandomizations imply such lower bounds. A main contribution of our paper is an affirmative answer in the case of the error rates considered by Goldreich and Wigderson. The case of higher error rates remains open.

Our result is a strengthening of [9] from the everywhere-correct setting to the typically-correct setting. In developing it, we also obtain a simpler proof for the everywhere-correct setting. Our proof scales better than the one in [9], yields the same lower bound for a smaller class, and does not rely on the result from [6] that NEXP having polynomial-size circuits implies that NEXP coincides with EXP.

Organization. We start Section 2 with the formal definitions of the notions used throughout the rest of the paper, and the key lemma that shows how seed-extending pseudorandom generators yield typically-correct derandomizations. In Section 3 we state and discuss both the conditional and unconditional results we obtain by applying our approach using the Nisan-Wigderson pseudorandom generator construction. In Section 4 we give a detailed comparison of our approach with Shaltiel's extractor-based approach. In Section 5 we describe our results on circuit lower bounds that follow from typically-correct and everywhere-correct derandomization of BPP. Due to space limitations all formal proofs are deferred to the full version of this paper.

2   Typically-Correct Derandomization and the PRG Approach

Notation and Concepts. We use the following terminology throughout the paper. We view a randomized algorithm as defined by a deterministic algorithm $A(x, r)$ where x denotes the input and r the string of "coin tosses". We typically restrict our attention to one input length n, in which case A becomes a function $A : \{0,1\}^n \times \{0,1\}^m \to \{0,1\}$ where m represents the number of random bits that A uses on inputs of length n. We say that $A : \{0,1\}^n \times \{0,1\}^m \to \{0,1\}$ *computes* a function $L : \{0,1\}^n \to \{0,1\}$ with error ρ if for every $x \in \{0,1\}^n$, $\Pr_{R \leftarrow U_m}[A(x, R) \neq L(x)] \leq \rho$, where U_m denotes the uniform distribution over $\{0,1\}^m$. We say that the randomized algorithm A *computes* a language L with error $\rho(\cdot)$, if for every input length n, the function A computes the function L with error $\rho(n)$.

Given a randomized algorithm A for L, our goal is to construct a deterministic algorithm B of complexity comparable to A that is typically correct for L. By the latter we mean that B and L agree on most inputs of any given length, or equivalently, that the relative Hamming distance between B and L at any given length is small.

Definition 1 (typically-correct behavior). *Let $L : \{0,1\}^n \to \{0,1\}$ be a function. We say that a function $B : \{0,1\}^n \to \{0,1\}$ is within distance δ of L if $\Pr_{X \leftarrow U_n}[B(X) \neq L(X)] \leq \delta$. We say that an algorithm B computes a language L to within $\delta(\cdot)$ if for every input length n, the function B is within distance $\delta(n)$ of the function L.*

In general, a function $G : \{0,1\}^n \to \{0,1\}^\ell$ is ϵ-pseudorandom for a test $T : \{0,1\}^\ell \to \{0,1\}$ if $|\Pr_{S \leftarrow U_n}[T(G(S)) = 1] - \Pr_{R \leftarrow U_\ell}[T(R) = 1]| \leq \epsilon$. In this paper we are dealing with tests $T(x,r)$ that receive two inputs, namely x of length n and r of length m, and with corresponding pseudorandom functions G of the form $G(x) = (x, E(x))$, where x is of length n and $E(x)$ of length m. We call such functions "seed-extending".

Definition 2 (seed-extending function). *A function $G : \{0,1\}^n \to \{0,1\}^{n+m}$ is seed-extending if it is of the form $G(x) = (x, E(x))$ for some function $E : \{0,1\}^n \to \{0,1\}^m$. We refer to the function E as the extending part of G.*

Note that a seed-extending function G with extending part E is ϵ-pseudorandom for a test $T : \{0,1\}^n \times \{0,1\}^m \to \{0,1\}$ if

$$\left| \Pr_{X \leftarrow U_n, R \leftarrow U_m}[T(X,R) = 1] - \Pr_{X \leftarrow U_n}[T(X, E(X)) = 1] \right| \leq \epsilon.$$

A seed-extending $\epsilon(\cdot)$-pseudorandom generator for a family of tests T is a deterministic algorithm G such that for every input length n, G is a seed-extending $\epsilon(n)$-pseudorandom function for the tests in T corresponding to input length n.

The Seed-Extending Pseudorandom Generator Approach. Our key observation is that good seed-extending pseudorandom generators G for certain simple tests based on the algorithm A yield good typically-correct derandomizations of the form $B(x) = A(x, E(x))$. The following lemma states the quantitative relationship.

Lemma 1. *Let $A : \{0,1\}^n \times \{0,1\}^m \to \{0,1\}$ and $L : \{0,1\}^n \to \{0,1\}$ be functions such that*

$$\Pr_{X \leftarrow U_n, R \leftarrow U_m}[A(X,R) \neq L(X)] \leq \rho. \tag{1}$$

Let $G : \{0,1\}^n \to \{0,1\}^{n+m}$ be a seed-extending function with extending part E, and let $B(x) = A(x, E(x))$.

1. *If G is ϵ-pseudorandom for tests of the form $T(x,r) = A(x,r) \oplus L(x)$, then B is within distance $\rho + \epsilon$ of L*
2. *If G is ϵ-pseudorandom for tests of the form $T_{r'}(x,r) = A(x,r) \oplus A(x,r')$ where $r' \in \{0,1\}^m$ is an arbitrary string, then B is within distance $3\rho + \epsilon$ of L.*

Note that if A computes L with error ρ then condition (1) of the lemma is met. The two parts of the lemma differ in the complexity of the tests and in the error bound. The complexity of the tests plays a critical role for the existence of pseudorandom generators. In the first item the tests use the language L as an oracle, which may result in too high a complexity. In the second item we reduce the complexity of the tests at the cost of introducing non-uniformity and increasing the error bound. The increase in the error bound is often not an issue as we can easily reduce ρ by slightly amplifying the original algorithm A before applying the lemma.

The Nisan-Wigderson Construction. Some of the constructions of pseudorandom generators in the literature are seed-extending or can be easily modified to become seed-extending. One such example is the Nisan-Wigderson construction [14], which builds a pseudorandom generator for a given class of randomized algorithms out of a language that is hard on average for a related class of algorithms. We use the following terminology for the latter.

Definition 3 (hardness on average). *A language L is $\delta(\cdot)$-hard for a class of algorithms $\mathcal{A}$ if no $A \in \mathcal{A}$ is within distance $\delta(n)$ of L for almost all input lengths n.*

We use the Nisan-Wigderson construction for all our results in the next section. Some of the results are conditioned on reasonable but unproven hypotheses regarding the existence of languages that are hard on average. Others are unconditional because languages of the required hardness have been proven to exist.

3   Applications

3.1   Conditional Results

The first setting we consider is that of BPP. We use a modest hardness assumption to show that any language in BPP has a polynomial-time deterministic algorithm that errs on a polynomially small fraction of the inputs.

Theorem 1. *Let L be a language in BPP that is computed by a randomized bounded-error polynomial-time algorithm A. For any positive constant c, there is a positive constant d depending on c and the running time of A such that the following holds. If there is a language H in P that is $\frac{1}{n^c}$-hard for circuits of size n^d, then there is a deterministic polynomial-time algorithm B that computes L to within $\frac{1}{n^c}$.*

Comparison to Previous Work. We now compare Theorem 1 to previous conditional derandomization results for BPP. We first consider everywhere-correct results. Plugging our assumption into the hardness versus randomness tradeoffs of [14] gives the incomparable result that BPP is in deterministic subexponential time, i.e., in time 2^{n^ϵ} for every positive constant ϵ. We remark that to obtain this result one can relax the assumption and allow the language H to be in deterministic linear-exponential time, i.e., E=DTIME($2^{O(n)}$).

We next compare Theorem 1 to previous conditional results on typically-correct derandomization of BPP [3, 16]. The assumption that we use is weaker than the assumptions that are used by previous work. More specifically, [3] needs H to be $\frac{1}{n^c}$-hard for circuits of size n^d with a SAT oracle, and [16] requires that H be $(\frac{1}{2} - \frac{1}{2^{n^{\Omega(1)}}})$-hard for circuits of size n^d.

Thus, the two aforementioned results do not yield any typically-correct derandomization when starting from the modest assumption that we use. Under their respective stronger assumptions, the other approaches do yield typically-correct algorithms that are closer to L. We remark that we can match the distance in [16] if we are allowed to assume the same hardness hypothesis.

Extensions to Other Algorithmic Settings. [11] observed that the proof of the Nisan-Wigderson generator [14] relativizes and used this fact to give hardness versus randomness tradeoff results in a number of different algorithmic settings. This approach also works within our typically-correct derandomization framework.

Some consequences are listed in the table below for the classes AM, BP.$\oplus$P and BP.L, where the latter refers to randomized algorithms that run in logarithmic space and are allowed two-way access to their random coins [13]. We could also state similar results for the other settings considered by [11]. For each of these complexity classes we need to assume a different hardness assumption, where the difference lies in the type of circuits and in the uniform class to consider. We remark that for BP.$\oplus$P we only need a worst-case hardness assumption as in this setting worst-case hardness is known to imply average-case hardness [2].

Setting	Hardness Assumption	Conclusion
AM=BP.NP	NP $\cap$ coNP $\frac{1}{n^c}$-hard for $\mathrm{SIZE}^{\mathrm{SAT}}(n^d)$	AM within $\frac{1}{n^c}$ of NP
BP.$\oplus$P	$\oplus$P $\not\subseteq \mathrm{SIZE}^{\oplus\mathrm{SAT}}(n^d)$	BP.$\oplus$P within $\frac{1}{n^c}$ of $\oplus$P
BP.L	L $\frac{1}{n^c}$-hard for BP-SIZE(n^d)	BP.L within $\frac{1}{n^c}$ of L

In the table, SIZE(s) refers to Boolean circuits of size s, $\mathrm{SIZE}^O(\cdot)$ refers to Boolean circuits that have access to oracle gates for the language O, and BP-SIZE(s) refers to branching programs of size s. A class of languages is $\delta(\cdot)$-hard for $\mathcal{A}$ if it contains a language that is $\delta(\cdot)$-hard for $\mathcal{A}$.

3.2 Unconditional Results

Constant Depth Circuits. Our techniques imply typically-correct derandomization results for randomized constant-depth polynomial-size circuits. This result uses the fact that the parity function is $(\frac{1}{2} - \frac{1}{2^{n^{\Omega(1)}}})$-hard on average for constant-depth circuits [5] and gives an alternative and simpler proof of a result of [16] in this setting.

Constant Depth Circuits with Few Symmetric Gates. In contrast to the approach of [16], our techniques also yield results in settings where the best-known lower bounds only yield moderate hardness on average. One such model is that of constant-depth circuits that are allowed a small number of arbitrary symmetric gates, i.e., gates that compute functions which only depend on the Hamming weight of the input, such as parity and majority. In this setting Viola [21] constructed a function that is $(\frac{1}{2} - \frac{1}{n^{\Omega(\log n)}})$-hard on average. Via the Nisan-Wigderson construction, this in turn translates into a pseudorandom generator with stretch that is quasi-polynomial and error that is polynomially small in the output length, resulting in an error rate that is only quasipolynomially small. Thus, the approach of [16] does not apply, but ours can exploit these weak pseudorandom generators and gives the following result for both log-space and polynomial-time uniformity.

Theorem 2. *Let L be a language and A a uniform randomized bounded-error circuit of constant depth and polynomial size that uses $o(\log^2 n)$ symmetric gates such that A computes L with error at most ρ. Then there is a uniform deterministic circuit B of constant depth and polynomial size that uses exactly the same symmetric gates as A in addition to a polynomial number of parity gates such that B computes L to within $3\rho + \frac{1}{n^{\Omega(\log n)}}$.*

Multi-Party Communication Complexity. [16] proves a typically-correct derandomization result for two-party communication protocols. The proof of [16] is tailored to the two-party case and does not extend to the general case of k-party communication in which the players have the inputs on their foreheads [1]. Using our approach we can handle $k > 2$ and show that every uniform randomized k-party communication protocol has a uniform deterministic k-party communication protocol of comparable communication complexity that is typically correct. The following statement holds for both log-space and poly-time uniformity, where we call a communication protocol uniform if whenever a player sends a message, that message can be efficiently computed as a function of the player's view.

Theorem 3. *Let L be a language and A a uniform randomized communication protocol that computes L with error at most ρ and uses k players, q bits of communication, and m bits of public randomness, with k, q, m, and $\log(1/\epsilon)$ functions computable within the uniformity bounds. Then there is a uniform deterministic communication protocol B that computes L to within $3\rho + \epsilon$ and uses k players and $O(2^k \cdot m \cdot (q + \log(m/\epsilon)))$ bits of communication.*

For $k = 2$, Theorem 3 yields a weaker result than that of [16] – which gives a deterministic protocol with communication complexity $O(q + m)$ rather than $O(q \cdot m)$ – although we can also obtain the stronger result using our approach, as explained in the next section.

4   Comparison with the Extractor-Based Approach

We have seen several settings in which seed-extending pseudorandom generators allow us to prove typically-correct derandomization results that do not follow from the extractor-based approach of [16]. We now show that the approach of [16] is essentially equivalent to having seed-extending pseudorandom generators with *exponentially small error*. This reaffirms our claim that our approach is more general since we additionally obtain meaningful results using pseudorandom generators with larger error.

Overview of the Extractor-Based Approach. [16] uses a notion of "extractors for recognizable distributions" explained below. For every function $f : \{0,1\}^n \to \{0,1\}$ one can associate the distribution U_f that is *recognized* by f, which is the uniform distribution over $f^{-1}(1) = \{x : f(x) = 1\}$. A function $E : \{0,1\}^n \to \{0,1\}^m$ is a (k, ϵ)-extractor for distributions recognizable by some collection of functions $f : \{0,1\}^n \to \{0,1\}$, if for every such function f with $|f^{-1}(1)| \geq 2^k$, the distribution $E(U_f)$ has statistical distance at most ϵ from the uniform distribution on m bit strings.

[16] shows the following general approach towards typically-correct derandomization. Let $A : \{0,1\}^n \times \{0,1\}^m \to \{0,1\}$ be a randomized algorithm that computes some function L with error ρ. Let $\Delta = 100m$ and let E be an $(n - \Delta, 2^{-\Delta})$-extractor for distributions recognizable by functions of the form $f_{r_1,r_2}(x) = A(x, r_1) \oplus A(x, r_2)$ where $r_1, r_2 \in \{0,1\}^m$ are arbitrary strings. Then setting $B(x) = A(x, E(x))$ gives an algorithm that is within $3\rho + 2^{-10m}$ of L.

Comparison. The above approach requires extractors with error that is exponentially small in m, and breaks down completely when the error is larger. We now observe that an extractor with exponentially small error yields a seed-extending pseudorandom generator with exponentially small error. It follows that the extractors used in [16] can be viewed as seed-extending pseudorandom generators with exponentially small error.

Theorem 4. *Let $T : \{0,1\}^n \times \{0,1\}^m \to \{0,1\}$ be a function. Let $\Delta = m + \log(1/\epsilon) + 1$ and let $E : \{0,1\}^n \to \{0,1\}^m$ be an $(n - \Delta, 2^{-\Delta})$-extractor for distributions recognizable by functions of the form $f_r(x) = T(x, r)$ where $r \in \{0,1\}^m$ is an arbitrary string. Then, $G(x) = (x, E(x))$ is ϵ-pseudorandom for T.*

We remark that in some algorithmic settings, namely decision trees and 2-party communication protocols, the approach of [16] yields typically-correct derandomizations that are more efficient than the ones that follow from applying our methodology directly based on known hardness results. Nevertheless, by Theorem 4 the extractors used in [16] give rise to seed-extending pseudorandom generators that yield typically-correct derandomizations matching the efficiency of the extractor-based approach.

We also observe that seed-extending pseudorandom generators with error that is exponentially small in m yield extractors for recognizable distributions. Thus, the approach of [16] is essentially equivalent to the special case of seed-extending pseudorandom generators with error that is exponentially small.

Theorem 5. *Let $f : \{0,1\}^n \to \{0,1\}$ be a function and let $E : \{0,1\}^n \to \{0,1\}^m$ be a function such that $G(x) = (x, E(x))$ is seed-extending ϵ-pseudorandom for test $T(x, r)$ of the form $T_z(x, r) = f(x) \wedge (r = z)$ where $z \in \{0,1\}^m$ is an arbitrary string. Assume that $\epsilon < 2^{-3m}$ and let $\Delta = (\log(1/\epsilon) - m)/2 > m$. Then E is an $(n - \Delta, 2^{-\Delta})$-extractor for the distribution recognizable by f.*

Note that seed-extending pseudorandom generators with error $\epsilon < 2^{-m}$ must have $m < n$ (as there are only 2^n seeds). This is why the approach of [16] cannot directly handle randomized algorithms with a superlinear number of random bits. In contrast, in Theorems 1 and 2 we are able to directly handle algorithms with a superlinear number of random bits using pseudorandom generators with larger error.

5   Circuit Lower Bounds

From Everywhere-Correct Derandomization. It is well-known that the existence of pseudorandom generators for polynomial-size circuits (which yields everywhere-correct derandomization of BPP) implies that EXP does not have polynomial-size circuits; this is the easy direction of the hardness versus randomness tradeoffs. Impagliazzo et al. [6] showed that everywhere-correct derandomization of promise-BPP into NSUBEXP implies that NEXP does not have polynomial-size circuits. Building on [6], Kabanets and Impagliazzo [9] showed that everywhere-correct derandomization of BPP into NSUBEXP implies that NEXP does not have Boolean circuits of polynomial size or that the permanent over $\mathbb{Z}$ does not have arithmetic circuits of polynomial size. As a byproduct of our investigations, we obtain a simpler proof of the latter result.

We use the following terminology for the statements of our lower bound results. We consider arithmetic circuits with internal nodes representing addition, subtraction, and multiplication, and leaves representing variables and the constants 0 and 1. ACZ denotes the language of all arithmetic circuits that compute the zero polynomial over $\mathbb{Z}$. Perm denotes the permanent of matrices over $\mathbb{Z}$, and 0-1-Perm its restriction to matrices with all entries in $\{0, 1\}$. We measure the size of circuits by the string length of their description, and assume that the description mechanism is such that the description of a circuit of size s can easily be padded into the description of an equivalent circuit of size s' for any $s' > s$.

Our approach yields the following general statement regarding everywhere-correct derandomization of the specific BPP-language ACZ.

Theorem 6. *Let $a(n)$, $s(n)$, and $t(n)$ be functions such that $a(n)$ and $s(n)$ are constructible, $a(n)$ and $t(n)$ are monotone, and $s(n) \geq n$. The following holds as long as for every constant c and sufficiently large n,*

$$t\left((s(n))^c \cdot a((s(n)^c))\right) \leq 2^n.$$

If ACZ $\in$ NTIME($t(n)$) then (i) NTIME(2^n) $\cap$ coNTIME(2^n) does not have Boolean circuits of size $s(n)$, or (ii) Perm does not have arithmetic circuits of size $a(n)$.

We point out that part (i) states a lower bound for NEXP $\cap$ coNEXP rather than just for NEXP, and Theorem 6 does so for the entire range of the parameters; the proof in [9] only gives such a lower bound in the case where all the parameters are polynomially bounded. More importantly, due to its dependence on the result from [6] that NEXP having polynomial-size circuits implies that NEXP coincides with EXP, the proof in [9] only works when $s(n)$ is polynomially bounded; our proof gives nontrivial results for $s(n)$ ranging between linear and linear-exponential.

From Typically-Correct Derandomization. We initiate the study of whether typically-correct derandomization of BPP implies circuit lower bounds. We show that it does in the case of typically-correct derandomizations that run in NSUBEXP and are of the quality considered by Goldreich and Wigderson [3].

Theorem 7. *If for every positive constant ϵ there exists a nondeterministic Turing machine which runs in time $2^{n^{\epsilon}}$ and correctly decides ACZ on all but at most $2^{n^{\epsilon}}$ of the inputs of length n for almost every n, then (i) NEXP does not have Boolean circuits of polynomial size, or (ii) Perm does not have arithmetic circuits of polynomial size.*

Note that Theorem 7 strengthens the main result of [9], which establishes the theorem in the special case where the nondeterministic machines decide ACZ correctly on all inputs. We start with a proof sketch of this weaker result using our new approach, and then show how to adapt it to the setting of typically-correct derandomization with error rates of the order considered in [3].

The proof consists of two parts. We first show that $P^{0\text{-}1\text{-Perm}[1]}$ does not have circuits of fixed polynomial size, where $P^{0\text{-}1\text{-Perm}[1]}$ denotes the class of languages that can be decided in polynomial-time with one query to an oracle for 0-1-Perm. This follows because PH does not have circuits of fixed polynomial size [10], PH is contained in $P^{\#P[1]}$ [18], and 0-1-Perm is complete for #P under reductions that make a single query [22].

In the second step we assume that

(α) ACZ has derandomizations N_{ϵ} of the form described in the statement of Theorem 7 but without any errors, and
(β) Perm has polynomial-size arithmetic circuits,

and show that these hypotheses imply that $P^{0\text{-}1\text{-Perm}[1]}$ is contained in NSUBEXP. The crux is the following single-valued nondeterministic algorithm to compute the permanent of a given 0-1-matrix M over $\mathbb{Z}$.

1. Guess a polynomial-sized candidate arithmetic circuit C for Perm on matrices of the same dimension as M.
2. Verify the correctness of C. Halt and reject if the test fails.
3. Use the circuit C to determine the permanent of M in deterministic polynomial time.

The circuit in step 1 exists by virtue of hypothesis (β). By the downward self-reducibility of Perm, the test in step 2 just has to check an arithmetic circuit identity based on C, which can be verified in nondeterministic subexponential time by virtue of (α).

All together, the hypotheses (α) and (β) imply that NSUBEXP does not have circuits of fixed polynomial size, and therefore neither does NE. Since NE has a complete language under linear-time reductions, the latter implies that NEXP does not have polynomial-size circuits.

Now suppose that our nondeterministic algorithms N_{ϵ} for ACZ can err on a small number of inputs of any given length ℓ. The test in step 2 above may no longer be sound nor complete. We can make the test sound if we are given the number $fp(\ell, \epsilon)$ of false positives of length ℓ, i.e., the number of inputs of length ℓ that are accepted by N_ϵ but do not belong to ACZ. This is because we can guess the list of those $fp(\ell, \epsilon)$ inputs of length ℓ, verify that they are accepted by N_ϵ but do not compute the zero polynomial, and then check that the given input of length ℓ does not appear on the list. We can make the test complete by increasing ℓ a bit and exploiting the paddability of ACZ. Since the number of errors of N_ϵ is relatively small, for any correct circuit C there has to be a pad that N_ϵ accepts. Our test can guess such a pad and check it. If N_ϵ makes no more than 2^{ℓ^ϵ} errors at length ℓ, we obtain simulations of $P^{0\text{-}1\text{-}Perm[1]}$ in NSUBEXP with subpolynomial advice. We conclude that the latter class does not have circuits of fixed polynomial size, which implies that NSUBEXP doesn't, from which we conclude as before that NEXP does not have circuits of polynomial size. This ends our proof sketch of Theorem 7.

Extensions. We observe a few variations of Theorems 6 and 7. First, the theorems also hold when we simultaneously replace ACZ by AFZ (the restriction of ACZ to arithmetic formulas), and "arithmetic circuits" by "arithmetic formulas". Second, we can play with the underlying i.o. and a.e. quantifiers. For example, in the case of Theorem 7 it suffices for the nondeterministic machines N_ϵ to correctly decide ACZ on all but at most 2^{n^ϵ} of the inputs of length n for *infinitely many* n. Related to the latter variation, we point out that by [7] EXP differs from BPP iff all of BPP has deterministic typically-correct derandomizations that run in subexponential time and err on no more than a polynomial fraction of the inputs of length n for infinitely many n. Thus, extending this i.o.-version of Theorem 7 to the setting with polynomial error rates would show that EXP$\neq$BPP implies circuit lower bounds.

Acknowledgments. We would like to thank Oded Goldreich for suggesting the term "typically-correct derandomization," and Matt Anderson, Salil Vadhan, and anonymous reviewers for helpful comments. The third author thanks Salil Vadhan for suggesting this research direction to him and for collaboration at an early stage of this research.

References

[1] Babai, L., Nisan, N., Szegedy, M.: Multiparty protocols, pseudorandom generators for logspace, and time-space trade-offs. JCSS 45(2), 204–232 (1992)

[2] Feigenbaum, J., Fortnow, L.: Random-self-reducibility of complete sets. SICOMP 22(5) (1993)

[3] Goldreich, O., Wigderson, A.: Derandomization that is rarely wrong from short advice that is typically good. In: Rolim, J.D.P., Vadhan, S.P. (eds.) RANDOM 2002. LNCS, vol. 2483, pp. 209–223. Springer, Heidelberg (2002)

[4] Gutfreund, D., Shaltiel, R., Ta-Shma, A.: Uniform hardness versus randomness tradeoffs for Arthur-Merlin games. Comput. Compl. 12(3–4), 85–130 (2003)

[5] Håstad, J.: Computational limitations of small-depth circuits. MIT Press, Cambridge (1987)

[6] Impagliazzo, R., Kabanets, V., Wigderson, A.: In search of an easy witness: exponential time vs. probabilistic polynomial time. JCSS 65(4), 672–694 (2002)

[7] Impagliazzo, R., Wigderson, A.: Randomness vs time: Derandomization under a uniform assumption. JCSS 63(4), 672–688 (2001)

[8] Kabanets, V.: Easiness assumptions and hardness tests: Trading time for zero error. JCSS 63(2), 236–252 (2001)

[9] Kabanets, V., Impagliazzo, R.: Derandomizing polynomial identity tests means proving circuit lower bounds. Comput. Compl. 13(1/2), 1–46 (2004)

[10] Kannan, R.: Circuit-size lower bounds and nonreducibility to sparse sets. Inf. Cont. 55(1), 40–56 (1982)

[11] Klivans, A.R., van Melkebeek, D.: Graph nonisomorphism has subexponential size proofs unless the polynomial-time hierarchy collapses. SICOMP 31(5), 1501–1526 (2002)

[12] Miltersen, P.B.: Derandomizing complexity classes. In: Handbook of Randomized Computing, pp. 843–941. Kluwer Academic Publishers, Dordrecht (2001)

[13] Nisan, N.: On read-once vs. multiple access to randomness in logspace. Theor. Comp. Sci. 107(1), 135–144 (1993)

[14] Nisan, N., Wigderson, A.: Hardness vs. randomness. JCSS 49(2), 149–167 (1994)

[15] Reingold, O.: Undirected connectivity in log-space. JACM 55(4) (2008)

[16] Shaltiel, R.: Weak derandomization of weak algorithms: explicit versions of Yao's lemma. In: Proc. Conf. Comput. Compl. (2009)

[17] Shaltiel, R., Umans, C.: Low-end uniform hardness vs. randomness tradeoffs for AM. In: Proc. of the ACM Symp. Theory of Comp., pp. 430–439 (2007)

[18] Toda, S.: PP is as hard as the polynomial-time hierarchy. SICOMP 20(5), 865–877 (1991)

[19] Trevisan, L.: Extractors and pseudorandom generators. JACM 48(4), 860–879 (2001)

[20] Trevisan, L., Vadhan, S.P.: Pseudorandomness and average-case complexity via uniform reductions. Comput. Compl. 16(4), 331–364 (2007)

[21] Viola, E.: Pseudorandom bits for constant-depth circuits with few arbitrary symmetric gates. SICOMP 36(5), 1387–1403 (2006)

[22] Zanko, V.: #P-completeness via many-one reductions. Intl. J. Found. Comp. Sci. 2(1), 77–82 (1991)

[23] Zimand, M.: Exposure-resilient extractors. In: Proc. Conf. Comput. Compl., pp. 61–72 (2006)

[24] Zimand, M.: Exposure-resilient extractors and the derandomization of probabilistic sublinear time. Comput. Compl. 17(2), 220–253 (2008)

Baum's Algorithm Learns Intersections of Halfspaces with Respect to Log-Concave Distributions

Adam R. Klivans[1,*], Philip M. Long[2], and Alex K. Tang[3]

[1] UT-Austin
klivans@cs.utexas.edu
[2] Google
plong@google.com
[3] UT-Austin
tang@cs.utexas.edu

Abstract. In 1990, E. Baum gave an elegant polynomial-time algorithm for learning the intersection of two origin-centered halfspaces with respect to any symmetric distribution (i.e., any $\mathcal{D}$ such that $\mathcal{D}(E) = \mathcal{D}(-E)$) [3]. Here we prove that his algorithm also succeeds with respect to any mean zero distribution $\mathcal{D}$ with a log-concave density (a broad class of distributions that need not be symmetric). As far as we are aware, prior to this work, it was not known how to efficiently learn any class of intersections of halfspaces with respect to log-concave distributions.

The key to our proof is a "Brunn-Minkowski" inequality for log-concave densities that may be of independent interest.

1 Introduction

A function $f : \mathbb{R}^n \to \mathbb{R}$ is called a linear threshold function or *halfspace* if $f(x) = \text{sgn}(w \cdot x)$ for some vector $w \in \mathbb{R}^n$. Algorithms for learning halfspaces from labeled examples are some of the most important tools in machine learning.

While there exist several efficient algorithms for learning halfspaces in a variety of settings, the natural generalization of the problem — learning the intersection of two or more halfspaces (e.g., the concept class of functions of the form $h = f \wedge g$ where f and g are halfspaces) — has remained one of the great challenges in computational learning theory.

In fact, there are no nontrivial algorithms known for the problem of PAC learning the intersection of just two halfspaces with respect to an arbitrary distribution. As such, several researchers have made progress on restricted versions of the problem. Baum provided a simple and elegant algorithm for learning the intersection of two origin-centered halfspaces with respect to any symmetric distribution on $\mathbb{R}^n$ [3]. Blum and Kannan [4] and Vempala [16] designed polynomial-time algorithms for learning the intersection of any constant number

---

* Klivans and Tang supported by NSF CAREER Award CCF-643829, an NSF TF Grant CCF-728536, and a Texas Advanced Research Program Award.

I. Dinur et al. (Eds.): APPROX and RANDOM 2009, LNCS 5687, pp. 588–600, 2009.

of halfspaces with respect to the uniform distribution on the unit sphere in $\mathbb{R}^n$. Arriaga and Vempala [2] and Klivans and Servedio [13] designed algorithms for learning a constant number of halfspaces given an assumption that the support of the positive and negative regions in feature space are separated by a margin. The best bounds grow with the margin γ like $(1/\gamma)^{O(\log(1/\gamma))}$.

1.1 Log-Concave Densities

In this paper, we significantly expand the classes of distributions for which we can learn intersections of two halfspaces: we prove that Baum's algorithm succeeds with respect to any mean zero, log-concave probability distribution. We hope that this is a first step towards finding efficient algorithms that can handle intersections of many more halfspaces with respect to a broad class of probability distributions.

A distribution $\mathcal{D}$ is *log-concave* if it has a density f such that $\log f(\cdot)$ is a concave function. Log-concave distributions are a powerful class that capture a range of interesting scenarios: it is known, for example, that the uniform distribution over any convex set is log-concave (if the convex set is centered at the origin, then the corresponding density has mean zero). Hence, Vempala's result mentioned above works for a very special case of log-concave distributions (it is not clear whether his algorithm works for a more general class of distributions). Additionally, interest in log-concave densities among machine learning researchers has been growing of late [1, 7, 9, 10, 14].

There has also been some recent work on learning intersections of halfspaces with respect to the Gaussian distribution on $\mathbb{R}^n$, another special case of a log-concave density. Klivans et al. have shown how to learn (even in the agnostic setting) the intersection of a constant number of halfspaces to any constant error parameter in polynomial-time with respect to any Gaussian distribution on $\mathbb{R}^n$ [12]. Again, it is unclear how to extend their result to log-concave distributions.

1.2 Our Approach: Re-analyzing Baum's Algorithm

In this paper, we prove that Baum's algorithm from 1990 succeeds when the underlying probability distribution is not necessarily symmetric, but is log-concave.

Baum's algorithm works roughly as follows. Suppose the unknown target concept C is the intersection of the halfspace H_u defined by $u \cdot x \geq 0$ and the halfspace H_v defined by $v \cdot x \geq 0$. Note that if $x \in C$ then $(u \cdot x)(v \cdot x) \geq 0$, so that

$$\sum_{ij} u_i v_j x_i x_j \geq 0. \tag{1}$$

If we replace the original features $x_1, \ldots, x_n$ with all products $x_i x_j$ of pairs of features, this becomes a linear inequality. The trouble is that $(u \cdot x)(v \cdot x)$ is also positive if $x \in -C$, i.e., both $u \cdot x \leq 0$ and $v \cdot x \leq 0$. The idea behind Baum's algorithm is to eliminate all the negative examples in $-C$ by identifying a region N in the complement of C (the "negative" region) that, with high probability,

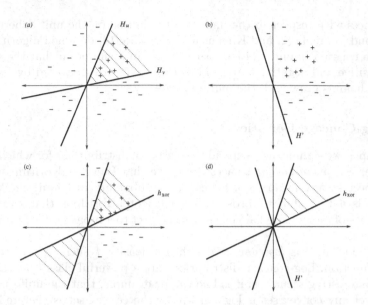

Fig. 1. Baum's algorithm for learning intersections of two halfspaces. (a) The input data, which is labeled using an intersection of two halfspaces. (b) The first step is to find a halfspace containing all the positive examples, and thus, with high probability, almost none of the reflection of the target concept through the origin. (c) The next step is to find a quadratic threshold function consistent with the remaining examples. (d) Finally, Baum's algorithm outputs the intersection of the halfspace found in step b and the classifier found in step c.

includes almost all of $-C$. Then, Baum finds a halfspace in an expanded feature space that is consistent with rest of the examples. (See Figure 1).

To compute N, Baum finds a halfspace H' containing a large set of positive examples in C, and then sets $N = -H'$. Here is where he uses the assumption that the distribution is symmetric: he reasons that if H' contains a lot of positive examples, then H' contains most of the measure of C, and, since the distribution is symmetric, $-H'$ contains most of the measure of $-C$. Then, if he draws more examples and excludes those in $-H'$, he is unlikely to obtain any examples in $-C$, and for each example x that remains, (1) will hold only if and only if $x \in C$. The output hypothesis classifies an example falling in N negatively, and uses the halfspace in the expanded feature space to classify the remaining examples.

We extend Baum's analysis by showing that, if the distribution is centered and log-concave, then the probability of the region in $-C$ that fails to be excluded by $-H'$ is not much larger than the probability of that part of C that is not covered by H'. Thus, if H' is trained with somewhat more examples, the algorithm can still ensure that $-H'$ fails to cover a small part of $-C$.

Thus, we arrive at the following natural problem from convex geometry: given a cone K whose apex is at the origin in $\mathbb{R}^n$, how does $\Pr(K)$ relate to $\Pr(-K)$ for distributions whose density is log-concave? Were the distribution uniform over a convex set centered at the origin, we could use the Brunn-Minkowski theory to argue that $\Pr(K)$ is always within a factor of n times $\Pr(-K)$ (see the discussion after the proof of Lemma 6). Instead, we are working with a mean zero log-concave distribution, and we do not know of an analog of the Brunn-Minkowski inequality for log-concave densities. Instead, we make use of the fact that the cones we are interested in are very simple and can be described by the intersection of just three halfspaces, and show that $\Pr(K)$ is within a *constant* factor of $\Pr(-K)$. Proving this makes use of tools for analyzing log-concave densities provided by Lovász and Vempala [14].

2 Preliminaries

2.1 VC Theory and Sample Complexity

We shall assume the reader is familiar with basic notions in computational learning theory such as Valiant's PAC model of learning and VC-dimension (see Kearns & Vazirani for an in-depth treatment [11]).

Theorem 1 ([6, 15]). *Let $\mathcal{C}$ be a class of concepts from the set X to $\{-1, 1\}$ whose VC dimension is d. Let $c \in \mathcal{C}$, and suppose*

$$M(\varepsilon, \delta, d) = O\left(\frac{d}{\varepsilon} \log \frac{1}{\varepsilon} + \frac{1}{\varepsilon} \log \frac{1}{\delta}\right)$$

examples $x_1, \ldots, x_M$ are drawn according to any probability distribution $\mathcal{D}$ over X. Then, with probability at least $1 - \delta$, any hypothesis $h \in \mathcal{C}$ that is consistent with c on $x_1, \ldots, x_M$ has error at most ε w.r.t. $\mathcal{D}$.

Lemma 1. *The class of origin-centered halfspaces over $\mathbb{R}^n$ has VC dimension n.*

Lemma 2. *Let $\mathcal{C}$ be a class of concepts from the set X to $\{-1, 1\}$. Let X' be a subset of X, and let $\mathcal{C}'$ be the class of concepts in $\mathcal{C}$ restricted to X'; in other words, let*

$$\mathcal{C}' := \{c|_{X'} \mid c \in \mathcal{C}\}.$$

Then, the VC dimension of $\mathcal{C}'$ is at most the VC dimension of $\mathcal{C}$.

2.2 Log-Concave Densities

Definition 1 (isotropic, log-concave). *A probability density function f over $\mathbb{R}^n$ is log-concave if $\log f(\cdot)$ is concave. It is isotropic if the covariance matrix of the associated probability distribution is the identity.*

We will use a number of facts that were either stated by Lovász and Vempala, or are easy consequences of their analysis.

Lemma 3 ([14]). *Any halfspace containing the origin has probability at least $1/e$ under a log-concave distribution with mean zero.*

Lemma 4 ([14]). *Suppose f is an isotropic, log-concave probability density function over $\mathbb{R}^n$. Then,*

(a) $f(0) \geq 2^{-7n}$.
(b) $f(0) \leq n(20n)^{n/2}$.
(c) $f(x) \geq 2^{-7n}2^{-9n\|x\|}$ *whenever* $0 \leq \|x\| \leq 1/9$.
(d) $f(x) \leq 2^{8n}n^{n/2}$ *for every* $x \in \mathbb{R}^n$.
(e) *For every line ℓ through the origin,* $\int_\ell f \leq (n-1)(20(n-1))^{(n-1)/2}$.

Proof. Parts a-d are immediate consequences of Theorem 5.14 of [14].

The proof of Part e is like the proof of an analogous lower bound in [14]. Change the basis of $\mathbb{R}^n$ so that ℓ is the x_n-axis, and let h be the marginal over the first $n-1$ variables. Then, by definition,

$$h(x_1, \ldots, x_{n-1}) = \int_\ell f(x_1, \ldots, x_{n-1}, t)\, dt,$$

so that $h(0) = \int_\ell f$. Applying the inequality of Part b gives Part e. $\quad\square$

3   Baum's Algorithm

Let H_u and H_v be the two origin-centered halfspaces whose intersection we are trying to learn. Baum's algorithm for learning $H_u \cap H_v$ is as follows:

1. First, define

$$m_1 := M(\varepsilon/2, \delta/4, n^2),$$
$$m_2 := M\big(\max\{\delta/(4e\kappa m_1), \varepsilon/2\}, \delta/4, n\big), \text{ and}$$
$$m_3 := \max\{2m_2/\varepsilon, (2/\varepsilon^2)\log(4/\delta)\},$$

   where κ is the constant that appears in Lemmas 6 and 7 below.
2. Draw m_3 examples. Let r denote the number of positive examples observed. If $r < m_2$, then output the hypothesis that labels every point as negative. Otherwise, continue to the next step.
3. Use linear programming to find an origin-centered halfspace H' that contains all r positive examples.
4. Draw examples until we find a set S of m_1 examples in H'. (Discard examples in $-H'$.) Then, use linear programming to find a weight vector $w \in \mathbb{R}^{n \times n}$ such that the hypothesis $h_{\mathrm{xor}} : \mathbb{R}^n \to \{-1, 1\}$ given by

$$h_{\mathrm{xor}}(x) := \mathrm{sgn}\left(\sum_{i=1}^n \sum_{j=1}^n w_{i,j} x_i x_j\right)$$

   is consistent with all examples in S.
5. Output the hypothesis $h : \mathbb{R}^n \to \{-1, 1\}$ given by

$$h(x) := \begin{cases} h_{\mathrm{xor}}(x) & \text{if } x \in H', \\ -1 & \text{otherwise.} \end{cases}$$

Outline of proof. In Theorem 2, we prove that Baum's algorithm learns $H_u \cap H_v$ in the PAC model, when the distribution on $\mathbb{R}^n$ is log-concave and has mean zero. Here we give an informal explanation of the proof. In step 3, the algorithm finds a halfspace H' that contains all but a small fraction of the positive examples. In other words, $\Pr(H_u \cap H_v \cap (-H'))$ is small. This implies that points in $-H'$ have a small chance of being positive, so we can just classify them as negative. To classify points in H', the algorithm learns a hypothesis h_{xor} in step 4. We must show that h_{xor} is a good hypothesis for points in H'. Under a log-concave distribution with mean zero, for *any* intersection of three halfspaces, its probability mass is at most a constant times the probability of its reflection about the origin; this is proved in Lemma 7. In particular,

$$\Pr\big((-H_u) \cap (-H_v) \cap H'\big) \leq \kappa \Pr\big(H_u \cap H_v \cap (-H')\big) \tag{2}$$

for some constant $\kappa > 0$. Therefore, since $\Pr\big(H_u \cap H_v \cap (-H')\big)$ is small, we can conclude that $\Pr\big((-H_u) \cap (-H_v) \cap H'\big)$ is also small. This implies that, with high probability, points in H' will *not* lie in $(-H_u) \cap (-H_v)$; thus, they must lie in $H_u \cap H_v$, $H_u \cap (-H_v)$, or $(-H_u) \cap H_v$. Such points are classified according to the symmetric difference $H_u \triangle H_v$ restricted to H'. (Strictly speaking, the points are classified according to the negation of the concept $H_u \triangle H_v$ restricted to H'; that is, we need to invert the labels so that positive examples are classified as negative and negative examples are classified as positive.) By Lemmas 1 and 2, together with the fact that h_{xor} can be interpreted as a halfspace over $\mathbb{R}^{n^2}$, the class of such concepts has VC dimension at most n^2. Hence, we can use the VC Theorem to conclude that the hypothesis h_{xor} has low error on points in H'.

Now, we describe the strategy for proving (2). In Lemma 7, we prove that $\Pr(-R) \leq \kappa \Pr(R)$, where R is the intersection of any three origin-centered halfspaces. This inequality holds when the probability distribution is log-concave and has mean zero. First, we prove in Lemma 6 that the inequality holds for the special case when the log-concave distribution not only has mean zero, but is also isotropic. Then, we use Lemma 6 to prove Lemma 7. We consider Lemma 7 to be a Brunn-Minkowski-type inequality for log-concave distributions (see the discussion after the proof of Lemma 6).

To prove Lemma 6, we will exploit the fact that, if R is defined by an intersection of three halfspaces, the probability of R is the same as the probability of R with respect to the marginal distribution over examples projected onto the subspace of $\mathbb{R}^n$ spanned by the normal vectors of the halfspaces bounding R — this is true, roughly speaking, because the dot products with those normal vectors are all that is needed to determine membership in R, and those dot products are not affected if we project onto the subspace spanned by those normal vectors. The same holds, of course, for $-R$.

Once we have projected onto a 3-dimensional subspace, we perform the analysis by proving upper and lower bounds on the probabilities of R and $-R$, and showing that they are within a constant factor of one another. We analyze the probability of R (respectively $-R$) by decomposing it into layers that are varying distances r from the origin. To analyze each layer, we will use upper and lower bounds on the density of points at a distance r. Since the sizes (even the shapes) of the regions at distance r are the same for R and $-R$, if the densities are close, then the probabilities must be close.

Lemma 5 provides the upper bound on the density in terms of the distance (the lower bound in Lemma 4c suffices for our purposes). We only need the bound in the case $n = 3$, but we go ahead and prove a bound for all n. Kalai, Klivans, Mansour, and Servedio prove a one-dimensional version in Lemma 6 of [9]. We adapt their proof to the n-dimensional case.

Lemma 5. *Let* $f : \mathbb{R}^n \to \mathbb{R}^+$ *be an isotropic, log-concave probability density function. Then,* $f(x) \le \beta_1 e^{-\beta_2 \|x\|}$ *for all* $x \in \mathbb{R}^n$, *where* $\beta_1 := 2^{8n} n^{n/2} e$ *and* $\beta_2 := \dfrac{2^{-7n}}{2(n-1)(20(n-1))^{(n-1)/2}}$.

Proof. We first observe that if $\|x\| \le 1/\beta_2$, then $\beta_1 e^{-\beta_2 \|x\|} \ge \beta_1 e^{-1} = 2^{8n} n^{n/2}$. By Lemma 4d, $f(x) \le \beta_1 e^{-\beta_2 \|x\|}$ if $\|x\| \le 1/\beta_2$. Now, assume there exists a point $v \in \mathbb{R}^n$ such that $\|v\| > 1/\beta_2$ and $f(v) > \beta_1 e^{-\beta_2 \|v\|}$. We shall show that this assumption leads to a contradiction. Let $[0, v]$ denote the line segment between the origin 0 and v. Every point $x \in [0, v]$ can be written as a convex combination of 0 and v as follows: $x = (1 - \|x\|/\|v\|)0 + (\|x\|/\|v\|)v$. Therefore, the log-concavity of f implies that

$$f(x) \ge f(0)^{1 - \|x\|/\|v\|} f(v)^{\|x\|/\|v\|}.$$

We assumed that $f(v) > \beta_1 e^{-\beta_2 \|v\|}$. So Lemma 4a implies

$$f(x) > \left(2^{-7n}\right)^{1 - \|x\|/\|v\|} \beta_1^{\|x\|/\|v\|} e^{-\beta_2 \|x\|}.$$

Because $2^{-7n} \le 1$ and $1 - \|x\|/\|v\| \le 1$, we know that $\left(2^{-7n}\right)^{1 - \|x\|/\|v\|} \ge 2^{-7n}$. Because $\beta_1 \ge 1$, we know that $\beta_1^{\|x\|/\|v\|} \ge 1$. We can therefore conclude that $f(x) > 2^{-7n} e^{-\beta_2 \|x\|}$. Integrating over the line ℓ through 0 and v, we get

$$\int_\ell f \ge \int_{[0,v]} f > \int_0^{\|v\|} 2^{-7n} e^{-\beta_2 r}\, dr = \frac{2^{-7n}}{\beta_2}\left(1 - e^{-\beta_2 \|v\|}\right).$$

We assumed that $\|v\| > 1/\beta_2$, so $1 - e^{-\beta_2 \|v\|} > 1 - e^{-1}$. Thus,

$$\int_\ell f > \frac{2^{-7n}}{\beta_2}(1 - e^{-1}) = 2(1 - e^{-1})(n - 1)(20(n - 1))^{(n-1)/2}.$$

Since $2(1 - e^{-1}) > 1$, we conclude that $\int_\ell f > (n - 1)(20(n - 1))^{(n-1)/2}$, but this contradicts Lemma 4e. $\square$

Now we are ready for Lemma 6, which handles the isotropic case.

Lemma 6. *Let* R *be the intersection of three origin-centered halfspaces in* $\mathbb{R}^n$. *Assume that the points in* $\mathbb{R}^n$ *are distributed according to an isotropic, log-concave probability distribution. Then,* $\Pr(-R) \le \kappa \Pr(R)$ *for some constant* $\kappa > 0$.

Proof. Let u_1, u_2, and u_3 be normals to the hyperplanes that bound the region R. Then,

$$R = \{x \in \mathbb{R}^n \mid u_1 \cdot x \geq 0 \text{ and } u_2 \cdot x \geq 0 \text{ and } u_3 \cdot x \geq 0\}.$$

Let U be the linear span of u_1, u_2, and u_3. Choose an orthonormal basis (e_1, e_2, e_3) for U and extend it to an orthonormal basis $(e_1, e_2, e_3, \ldots, e_n)$ for all of $\mathbb{R}^n$. Write the components of the vectors x, u_1, u_2, and u_3 in terms of this basis:

$$x = (x_1, x_2, x_3, x_4, \ldots, x_n),$$
$$u_1 = (u_{1,1}, u_{1,2}, u_{1,3}, 0, \ldots, 0),$$
$$u_2 = (u_{2,1}, u_{2,2}, u_{2,3}, 0, \ldots, 0),$$
$$u_3 = (u_{3,1}, u_{3,2}, u_{3,3}, 0, \ldots, 0).$$

Let $\text{proj}_U(x)$ denote the projection of x onto U; that is, let $\text{proj}_U(x) := (x_1, x_2, x_3)$. Likewise, let $\text{proj}_U(R)$ denote the projection of R onto U; that is, let $\text{proj}_U(R) := \{\text{proj}_U(x) \mid x \in R\}$. Observe that

$$x \in R \iff u_{j,1}x_1 + u_{j,2}x_2 + u_{j,3}x_3 \geq 0 \text{ for all } j \in \{1, 2, 3\}$$
$$\iff \text{proj}_U(x) \in \text{proj}_U(R). \tag{3}$$

Let f denote the probability density function of the isotropic, log-concave probability distribution on $\mathbb{R}^n$. Let g be the marginal probability density function with respect to (x_1, x_2, x_3); that is, define

$$g(x_1, x_2, x_3) := \int \cdots \int_{\mathbb{R}^{n-3}} f(x_1, x_2, x_3, x_4, \ldots, x_n) \, dx_4 \cdots dx_n.$$

Then, it follows from (3) that

$$\Pr(R) = \int \cdots \int_R f(x_1, x_2, x_3, x_4, \ldots, x_n) \, dx_1 \cdots dx_n$$
$$= \iiint_{\text{proj}_U(R)} g(x_1, x_2, x_3) \, dx_1 \, dx_2 \, dx_3.$$

Note that g is isotropic and log-concave, because the marginals of an isotropic, log-concave probability density function are isotropic and log-concave (see [14, Theorem 5.1, Lemma 5.2]). Thus, we can use Lemma 4c and Lemma 5 to bound g. The bounds don't depend on the dimension n, because g is a probability density function over $\mathbb{R}^3$. For brevity of notation, let $y := (x_1, x_2, x_3)$. By Lemma 4c, there exist constants κ_1 and κ_2 such that

$$g(y) \geq \kappa_1 e^{-\kappa_2 \|y\|} \quad \text{for } \|y\| \leq 1/9. \tag{4}$$

And by Lemma 5, there exist constants κ_3 and κ_4 such that

$$g(y) \leq \kappa_3 e^{-\kappa_4 \|y\|} \quad \text{for all } y \in \mathbb{R}^3. \tag{5}$$

Let $R' := \text{proj}_U(R) \cap B(0, 1/9)$, where $B(0, 1/9)$ denotes the origin-centered ball of radius $1/9$ in $\mathbb{R}^3$. Use (4) and (5) to derive the following lower and upper bounds:

$$\iiint\limits_{R'} \kappa_1 e^{-\kappa_2 \|y\|} \, dy_1 \, dy_2 \, dy_3 \leq \iiint\limits_{\text{proj}_U(R)} g(x_1, x_2, x_3) \, dx_1 \, dx_2 \, dx_3$$

$$\leq \iiint\limits_{\text{proj}_U(R)} \kappa_3 e^{-\kappa_4 \|y\|} \, dy_1 \, dy_2 \, dy_3. \tag{6}$$

Recall that

$$\Pr(R) = \iiint\limits_{\text{proj}_U(R)} g(x_1, x_2, x_3) \, dx_1 \, dx_2 \, dx_3.$$

Now, we transform the integrals in the lower and upper bounds in (6) to spherical coordinates. The transformation to spherical coordinates is given by $r := \sqrt{y_1^2 + y_2^2 + y_3^2}$, $\varphi := \arctan\left(\frac{y_2}{y_1}\right)$, $\vartheta := \arccos\left(\frac{y_3}{\sqrt{y_1^2+y_2^2+y_3^2}}\right)$. The determinant of the Jacobian of the above transformation is known to be $r^2 \sin\vartheta$ [5]. Thus (see [5]), inequality (6) becomes

$$\iiint\limits_{R'} \kappa_1 r^2 e^{-\kappa_2 r} \sin\vartheta \, dr \, d\varphi \, d\vartheta \leq \Pr(R) \leq \iiint\limits_{\text{proj}_U(R)} \kappa_3 r^2 e^{-\kappa_4 r} \sin\vartheta \, dr \, d\varphi \, d\vartheta.$$

Let A denote the surface area of the intersection of $\text{proj}_U(R)$ with the unit sphere S^2; that is, let

$$A := \iint\limits_{\text{proj}_U(R) \cap S^2} \sin\vartheta \, d\varphi \, d\vartheta.$$

Then, it follows that

$$A \int_0^{1/9} \kappa_1 r^2 e^{-\kappa_2 r} \, dr \leq \Pr(R) \leq A \int_0^{\infty} \kappa_3 r^2 e^{-\kappa_4 r} \, dr.$$

If we let

$$\kappa_5 := \int_0^{1/9} \kappa_1 r^2 e^{-\kappa_2 r} \, dr \quad \text{and} \quad \kappa_6 := \int_0^{\infty} \kappa_3 r^2 e^{-\kappa_4 r} \, dr,$$

then $\kappa_5 A \leq \Pr(R) \leq \kappa_6 A$. By symmetry, $\kappa_5 A \leq \Pr(-R) \leq \kappa_6 A$. Therefore, it follows that $\Pr(-R) \leq (\kappa_6/\kappa_5) \Pr(R)$. □

If the distribution were uniform over a convex set K whose centroid is at the origin, then the proof of Lemma 6 could be modified to show that the probabilities of R and $-R$ are within a factor of n without requiring that R is the intersection of three halfspaces; we would only need that R is a cone (closed under positive rescaling). This could be done by observing that the probability of R is proportional to the average distance of a ray contained in R to the boundary of K. Then we could apply the Brunn-Minkowski inequality (see [8, Lemma 29]) which states that for any direction x, the distance from the origin to the boundary of K in the direction of x is within a factor n of the distance to the boundary of K in the direction $-x$.

In Lemma 6, we assumed that the distribution is isotropic. The next lemma shows that this assumption can be removed (provided that the mean of the distribution is still zero). A key insight is that, under a linear transformation, the image of the intersection of three halfspaces is another intersection of three halfspaces. To prove the lemma, we use a particular linear transformation that brings the distribution into isotropic position. Then, we apply Lemma 6 to the transformed distribution and the image of the three-halfspace intersection.

Lemma 7. *Let R be the intersection of three origin-centered halfspaces in $\mathbb{R}^n$. Assume that the points in $\mathbb{R}^n$ are distributed according to a log-concave probability distribution with mean zero. Then, $\Pr(-R) \leq \kappa \Pr(R)$, where κ is the same constant that appears in Lemma 6.*

Proof. Let X be a random variable in $\mathbb{R}^n$ with a mean-zero, log-concave probability distribution. Let V denote the covariance matrix of X. Let W be a matrix square root of the inverse of V; that is, $W^2 = V^{-1}$. Then, the random variable $Y := WX$ is log-concave and isotropic. (Technically, if the rank of the covariance matrix V is less than n, then V would not be invertible. But, in that case, the probability distribution degenerates into a probability distribution over a lower-dimensional subspace. We just repeat the analysis on this subspace.) Let $W(R)$ and $W(-R)$ respectively denote the images of R and $-R$ under W. Notice that $W(-R) = -W(R)$. Also, notice that $X \in R \Leftrightarrow Y \in W(R)$ and that $X \in -R \Leftrightarrow Y \in W(-R) = -W(R)$. Let u_1, u_2, and u_3 be normals to the hyperplanes that bound R. Then,

$$\begin{aligned} W(R) &= \{Wx \mid x \in \mathbb{R}^n \text{ and } u_j^T x \geq 0 \text{ for all } j \in \{1,2,3\}\} \\ &= \{y \in \mathbb{R}^n \mid u_j^T W^{-1} y \geq 0 \text{ for all } j \in \{1,2,3\}\} \\ &= \{y \in \mathbb{R}^n \mid \left((W^{-1})^T u_j\right)^T y \geq 0 \text{ for all } j \in \{1,2,3\}\}. \end{aligned}$$

Therefore, $W(R)$ is the intersection of three origin-centered halfspaces, so we can apply Lemma 6 to obtain

$$\Pr(X \in -R) = \Pr\left(Y \in -W(R)\right) \leq \kappa \Pr\left(Y \in W(R)\right) = \kappa \Pr(X \in R). \qquad \square$$

Finally, we analyze Baum's algorithm using the probability bound given in Lemma 7.

Theorem 2. *In the PAC model, Baum's algorithm learns the intersection of two origin-centered halfspaces with respect to any mean zero, log-concave probability distribution in polynomial time.*

Proof. If the probability p of observing a positive example is less than ε, then the hypothesis that labels every example as negative has error less than ε; so the algorithm behaves correctly if it draws fewer than m_2 positive examples in this case. If $p \geq \varepsilon$, then by the Hoeffding bound,

$$\Pr(r < m_2) \leq \Pr\left(\frac{r}{m_3} < \frac{\varepsilon}{2}\right) \leq \Pr\left(\frac{r}{m_3} < p - \frac{\varepsilon}{2}\right) \leq e^{-m_3 \varepsilon^2 / 2} \leq \delta/4.$$

Thus, if $p \geq \varepsilon$, then the probability of failing to draw at least m_2 positive examples is at most $\delta/4$. For the rest of this proof, we shall assume that the algorithm succeeds in drawing at least m_2 positive examples.

Observe that the hypothesis output by the algorithm has error

$$\begin{aligned} \mathrm{err}(h) = \Pr(-H') \Pr(H_u \cap H_v \mid -H') \\ + \Pr(H') \Pr\big(h_{\mathrm{xor}}(x) \neq c(x) \mid x \in H'\big), \end{aligned} \tag{7}$$

where $c : \mathbb{R}^n \to \{-1, 1\}$ denotes the concept corresponding to $H_u \cap H_v$. First, we give a bound for

$$\begin{aligned} \Pr(-H') \Pr(H_u \cap H_v \mid -H') &= \Pr\big(H_u \cap H_v \cap (-H')\big) \\ &= \Pr(H_u \cap H_v) \Pr(-H' \mid H_u \cap H_v). \end{aligned}$$

Notice that $\Pr(-H' \mid H_u \cap H_v)$ is the error of the hypothesis corresponding to H' over the distribution conditioned on $H_u \cap H_v$. But the VC Theorem works for any distribution, so, since H' contains every one of $M\big(\max\{\delta/(4e\kappa m_1), \varepsilon/2\}, \delta/4, n\big)$ random positive examples, it follows from Lemma 1 that, with probability at least $1 - \delta/4$,

$$\Pr(-H' \mid H_u \cap H_v) \leq \max\left\{\frac{\delta}{4e\kappa m_1}, \frac{\varepsilon}{2}\right\}.$$

Since $\Pr(H_u \cap H_v) \leq 1$, it follows that

$$\Pr\big(H_u \cap H_v \cap (-H')\big) \leq \max\left\{\frac{\delta}{4e\kappa m_1}, \frac{\varepsilon}{2}\right\}.$$

Therefore, the left term in (7) is at most $\varepsilon/2$. All that remains is to bound the right term.

From Lemma 7, it follows that

$$\Pr\big((-H_u) \cap (-H_v) \cap H'\big) \leq \kappa \Pr\big(H_u \cap H_v \cap (-H')\big) \leq \frac{\delta}{4em_1}.$$

By Lemma 3, $\Pr(H') \geq 1/e$. Therefore,

$$\Pr\big((-H_u) \cap (-H_v) \mid H'\big) = \frac{\Pr\big((-H_u) \cap (-H_v) \cap H'\big)}{\Pr(H')} \leq \frac{\delta}{4m_1}.$$

Thus, each of the m_1 points in S has probability at most $\delta/4m_1$ of being in $(-H_u) \cap (-H_v)$, so with probability at least $1 - \delta/4$, none of the m_1 points are in $(-H_u) \cap (-H_v)$. Thus, each point in $x \in S$ lies in $H_u \cap H_v$, $H_u \cap (-H_v)$, or $(-H_u) \cap (H_v)$; if $x \in H_u \cap H_v$, then x is labeled as positive; if $x \in H_u \cap (-H_v)$ or $x \in (-H_u) \cap H_v$, then x is labeled as negative. In other words, the points in S are classified according to the negation of $H_u \triangle H_v$ restricted to the halfspace H'. Thus, the linear program executed in Step 4 successfully finds a classifier h_{xor} consistent with the examples in S. By Lemma 1 and Lemma 2, the class of symmetric differences of origin-centered halfspaces restricted to H' has VC dimension at most n^2. Therefore, the VC Theorem implies that, with probability at least $1 - \delta/4$,

$$\Pr\bigl(h_{\mathrm{xor}}(x) \neq c(x) \mid x \in H'\bigr) \leq \frac{\varepsilon}{2}.$$

Since $\Pr(H') \leq 1$, the right term in (7) is at most $\varepsilon/2$. Adding up the probabilities of the four ways in which the algorithm can fail, we conclude that the probability that $\mathrm{err}(h) > \varepsilon$ is at most $4(\delta/4) = \delta$. $\qquad\square$

References

1. Achlioptas, D., McSherry, F.: On spectral learning with mixtures of distributions. In: Auer, P., Meir, R. (eds.) COLT 2005. LNCS, vol. 3559, pp. 458–469. Springer, Heidelberg (2005)
2. Arriaga, R., Vempala, S.: An algorithmic theory of learning: Robust concepts and random projection. In: Proceedings of the 40th Annual Symposium on Foundations of Computer Science (FOCS), pp. 616–623 (1999)
3. Baum, E.: A polynomial time algorithm that learns two hidden unit nets. Neural Computation 2(4), 510–522 (1990)
4. Blum, A., Kannan, R.: Learning an intersection of a constant number of halfspaces under a uniform distribution. Journal of Computer and System Sciences 54(2), 371–380 (1997)
5. Blum, E.K., Lototsky, S.V.: Mathematics of Physics and Engineering. World Scientific, Singapore (2006)
6. Blumer, A., Ehrenfeucht, A., Haussler, D., Warmuth, M.K.: Learnability and the Vapnik-Chervonenkis dimension. JACM 36(4), 929–965 (1989)
7. Caramanis, C., Mannor, S.: An inequality for nearly log-concave distributions with applications to learning. IEEE Transactions on Information Theory 53(3), 1043–1057 (2007)
8. Dunagan, J.D.: A geometric theory of outliers and perturbation. PhD thesis, MIT (2002)
9. Kalai, A., Klivans, A., Mansour, Y., Servedio, R.: Agnostically learning halfspaces. In: Proceedings of the 46th IEEE Symposium on Foundations of Computer Science (FOCS), pp. 11–20 (2005)
10. Kannan, R., Salmasian, H., Vempala, S.: The spectral method for general mixture models. In: Proceedings of the Eighteenth Annual Conference on Learning Theory (COLT), pp. 444–457 (2005)
11. Kearns, M., Vazirani, U.: An introduction to computational learning theory. MIT Press, Cambridge (1994)

12. Klivans, A., O'Donnell, R., Servedio, R.: Learning geometric concepts via Gaussian surface area. In: Proc. 49th IEEE Symposium on Foundations of Computer Science (FOCS), pp. 541–550 (2008)
13. Klivans, A., Servedio, R.: Learning intersections of halfspaces with a margin. In: Proceedings of the 17th Annual Conference on Learning Theory, pp. 348–362 (2004)
14. Lovász, L., Vempala, S.: The geometry of logconcave functions and sampling algorithms. Random Structures and Algorithms 30(3), 307–358 (2007)
15. Vapnik, V.: Estimations of dependences based on statistical data. Springer, Heidelberg (1982)
16. Vempala, S.: A random sampling based algorithm for learning the intersection of halfspaces. In: Proc. 38th IEEE Symposium on Foundations of Computer Science (FOCS), pp. 508–513 (1997)

Tolerant Linearity Testing and Locally Testable Codes

Swastik Kopparty[1] and Shubhangi Saraf[2]

[1] CSAIL, MIT, Cambridge MA 02139, USA
swastik@mit.edu
[2] CSAIL, MIT, Cambridge MA 02139, USA
shibs@mit.edu

Abstract. We study tolerant linearity testing under general distributions. Given groups G and H, a distribution μ on G, and oracle access to a function $f : G \to H$, we consider the task of approximating the smallest μ-distance of f to a homomorphism $h : G \to H$, where the μ-distance between f and h is the probability that $f(x) \neq h(x)$ when x is drawn according to the distribution μ. This question is intimately connected to local testability of linear codes.

In this work, we give a general sufficient condition on the distribution μ for linearity to be tolerantly testable with a constant number of queries. Using this condition we show that linearity is tolerantly testable for several natural classes of distributions including low bias, symmetric and product distributions. This gives a new and simple proof of a result of Kaufman and Sudan which shows that sparse, unbiased linear codes over $\mathbb{Z}_2^n$ are locally testable.

1 Introduction

Let $\mathcal{C}$ be a class of functions from a finite set $\mathcal{D}$ to a finite set $\mathcal{R}$. In the task of *tolerant testing* for $\mathcal{C}$, we are given oracle access to a function $f : \mathcal{D} \to \mathcal{R}$, and we wish to determine using few queries to f, whether f is well approximable by functions in $\mathcal{C}$; equivalently, to distinguish between the case when f is *close* to some element of $\mathcal{C}$, and the case when f is *far* from all elements of $\mathcal{C}$. Tolerant property testing was introduced by Parnas, Ron and Rubinfeld in [PRR06] as a refinement of the problem of property testing [RS96], [GGR98] (where one wants to distinguish the case of f *in* $\mathcal{C}$ from the case when f is *far* from $\mathcal{C}$), and is now widely studied. The usual notion of closeness considered in the literature is via the distance measure $\Delta(f,g) = \Pr_{x \in \mathcal{D}}[f(x) \neq g(x)]$, where $x \in \mathcal{D}$ is picked according to the *uniform distribution* over $\mathcal{D}$.

We propose to study tolerant property testing under general distributions. Given a probability measure μ on $\mathcal{D}$, the μ-distance of f from g, where $f, g : \mathcal{D} \to \mathcal{R}$, is defined by

$$\Delta_\mu(f,g) = \Pr_{x \in \mu}[f(x) \neq g(x)].$$

I. Dinur et al. (Eds.): APPROX and RANDOM 2009, LNCS 5687, pp. 601–614, 2009.
© Springer-Verlag Berlin Heidelberg 2009

Then the measure of how well f can be approximated by elements of $\mathcal{C}$ is via the μ-distance

$$\Delta_\mu(f, \mathcal{C}) = \min_{g \in \mathcal{C}} \Delta_\mu(f, g).$$

The new goal in this context then becomes to approximate $\Delta_\mu(f, \mathcal{C})$ using only a few oracle calls to f. In this paper, we study a concrete instance of the above framework. We consider the original problem considered in the area of property testing, namely the classical problem of *linearity testing*.

The problem of linearity testing was introduced by Blum, Luby and Rubinfeld in [BLR93]. In this problem, we are given oracle access to a function $f : G \to H$, where G and H are abelian groups, and want to distinguish between the case that f is a *homomorphism* from G to H and the case that f is *far* from the class $\mathcal{C} = \mathsf{Hom}(G, H)$, of all homomorphisms from G to H. [BLR93] gave a simple 3-query test T that achieves this. In fact, this test also achieves the task of *tolerant linearity testing*; i.e., for any function $f : G \to H$, letting $\delta = \Pr[T^f \text{ rejects}]$, we have

$$C_1 \cdot \delta \leq \Delta_{U_G}(f, \mathsf{Hom}(G, H)) \leq C_2 \cdot \delta,$$

where C_1 and C_2 are absolute constants, and U_G is the uniform distribution on G. Hence the test of [BLR93], in addition to *testing* linearity, actually estimates how well f can be *approximated* by functions in $\mathcal{C} = \mathsf{Hom}(G, H)$.

Here we initiate the study of tolerant linearity testing over general probability distributions. Let μ be a probability distribution on an abelian group G. In the problem of *tolerant linearity testing under* μ, we wish to estimate the how well f may be approximated under μ by homomorphisms from G to H. For a given family $(G_n, H_n, \mu_n)_n$, we say *linearity is tolerantly testable* under $\mu = \mu_n$ with q queries, if there exists a q-query tester T_n and constants C_1, C_2 such that for any $f : G_n \to H_n$, letting $\delta = \Pr[T_n^f \text{ rejects}]$, we have

1. **Perfect completeness:** $\delta = 0$ if and only if $\Delta_\mu(f, \mathsf{Hom}(G_n, H_n)) = 0$.
2. **Distance approximation:** δ approximates $\Delta_\mu(f, \mathsf{Hom}(G_n, H_n))$:

$$C_1 \cdot \delta \leq \Delta_\mu(f, \mathsf{Hom}(G_n, H_n)) + o_n(1) \leq C_2 \cdot \delta, \tag{1}$$

We argue that this is indeed a natural definition under which to study linearity testing under general distributions. For one, this definition ensures that any "useful" queries made by the tester essentially have to be distributed according to μ. Without the "tolerant" aspect of the definition, a tester could potentially get access to "advice" by querying f at locations where μ has no support. For example, consider a scenario where f is given by a black box that runs in expected polynomial time under the distribution μ. In this setting, it is meaningful to ask how well f is approximated by linear functions under μ, although it is not as reasonable to expect access to evaluations of f at points not distributed according to μ. More importantly, the tolerant aspect of this definition give it a strong connection to locally testable codes (which we discuss shortly).

The problem of linearity testing (without the tolerant aspect) was studied in the setting of general distributions by [HK07]. They gave a simple 3-query test that, given oracle access to the function $f : G \to H$, distinguishes between $f \in \mathsf{Hom}(G, H)$ and f that are far from $\mathsf{Hom}(G, H)$. In fact this tester does not even require an explicit description of μ, it simply requires access to samples from $\mu!$[1] However, unlike the BLR test, the [HK07] test is intolerant: the queries it makes are not according to the distribution μ.

The main question is to determine for which μ is linearity tolerantly testable under μ. This seems to be a basic question worthy of further study. Furthermore, the notion of linearity being tolerantly testable under general distributions is intimately connected with the concept of locally testable linear codes [GS02], and we now elaborate on this connection.

Connection to locally testable codes: Let $C \subseteq \mathbb{Z}_2^N$ be a linear code (we restrict to binary codes in this discussion). C is called *locally testable* if there is a constant query tester, that given oracle access to any $r : [N] \to \mathbb{Z}_2$, distinguishes between the case that $r \in C$ and r being far from C (in Hamming distance).

Now let C be any linear code, and let $s_1, \ldots, s_N \in \mathbb{Z}_2^n$ be the columns of a generator matrix for C. Let μ be the uniform distribution over $\{s_1, \ldots, s_N\}$. Then, if linearity is tolerantly testable under μ, then C is locally testable. Indeed, given any $r : [N] \to \mathbb{Z}_2$, we may define the function $f : \mathbb{Z}_2^n \to \mathbb{Z}_2$ by $f(x) = r(j)$ if $x = s_j$, and $f(x) = 0$ otherwise. By the tolerant testability of linearity under μ, any useful query made by a tolerant linearity tester for μ must be to one of the s_j. The distance of f from linear under μ then translates directly into the Hamming distance of r from C, and the very same tester that tolerantly tests linearity under μ shows that C is locally testable.

A more general goal behind the study of linearity testing under general distributions is to develop a better understanding what makes a code locally testable. We also believe that the theory of property testing under nonuniform distributions for other classes of functions $\mathcal{C}$ will be a fruitful and enlightening pursuit.

1.1   Main Notions and Results

Our main contribution is to highlight a simple criterion, which we call *uniform-correlatability*, that lets us design and analyze tolerant linearity tests under a given distribution. Roughly speaking, a distribution μ over an abelian group G is uniformly-correlatable if one can "design" a distribution of small random matrices with entries from G with each entry of the matrix distributed according to μ, while all the row-sums and column-sums are nearly uniformly distributed. In this case, we show that linearity is tolerantly testable under μ with few queries (see Theorem 1). We complement this by demonstrating that many natural distributions satisfy this criterion (see Theorems 2, 3, 4).

Definition 1 (Uniformly-correlatable distribution). *Let μ be a probability distribution on an abelian group G. We say that μ is (ϵ, k)-uniformly-correlatable if there is a random variable $\mathbf{X} = (\mathbf{X}_{ij})_{i,j \in [k]}$ taking values in $G^{k \times k}$ such that:*

---

[1] Tolerant linearity testing, however, necessarily requires more information about μ.

1. For each $i, j \in [k]$, $X_{i,j}$ is distributed according to μ.
2. For $i \in [k]$, let Y_i be the random variable $\sum_{j\in[k]} X_{ij}$. For $j \in [k]$, Z_j be the random variable $\sum_{i\in[k]} X_{ij}$. Then the distribution of $((Y_i)_{i\in[k]}, (Z_j)_{j\in k})$ is ϵ-close to the uniform distribution over the set $\{((y_i)_{i\in[k]}, (z_j)_{j\in[k]}) \in G^{2k} \mid \sum_{i\in[k]} y_i = \sum_{j\in[k]} z_j\}$.

Our main result, given below, is that uniformly correlatable distributions are tolerantly testable.

Theorem 1 (Uniformly correlatable distributions are tolerantly testable). *Let μ be a distribution over G that is (ϵ, k)-uniformly-correlatable. Then there is a $4k$ query tester T such that for any $f : G \to H$, letting $\delta = \Pr[T^f \text{ rejects}]$, we have:*

1. **Perfect completeness:** $\delta = 0$ *if and only if* $\Delta_\mu(f, \mathsf{Hom}(G, H)) = 0$
2. **Distance approximation:**

$$\frac{\delta - 4\epsilon}{4k} \leq \Delta_\mu(f, \mathsf{Hom}(G, H)) \leq \frac{6k}{1 - 12\epsilon k} \cdot \delta.$$

Thus, for any μ which is (ϵ, k)-uniformly-correlatable for constant k and $\epsilon = o(1)$, we conclude that linearity is tolerantly testable under μ (in the sense of Equation (1)). We supplement the above theorem with following results, showing that some general classes of μ are all (ϵ, k)-uniformly-correlatable for suitable ϵ, k, and thus showing that linearity is tolerantly testable under all such μ.

Theorem 2 (Low-bias distributions are uniformly correlatable). *Let μ be a probability distribution on G such that for all nontrivial characters $\chi : G \to \mathbb{C}^\times$,*

$$\left| \mathop{\mathbf{E}}_{x \in \mu} [\chi(x)] \right| < |G|^{-\gamma}.$$

Then for $k = \Omega(1/\gamma)$, μ is $(|G|^{-\Omega(k\gamma)}, k)$-uniformly-correlatable.

Using the above theorem, we conclude that linearity is tolerantly testable under any low-bias distribution.

Corollary 1 (Low-bias distributions are tolerantly testable). *Let G be an abelian group, and let μ and γ be as in Theorem 2. Then there are constants $C_1 = C_1(\gamma)$, $C_2 = C_2(\gamma)$, and a $O(1/\gamma)$-query test T such that for any abelian group H and any function $f : G \to H$, letting $\delta = \Pr[T^f \text{ rejects}]$, we have*

$$C_1\delta \leq \Delta_\mu(f, \mathsf{Hom}(G, H)) + o(1/|G|) \leq C_2\delta.$$

For the special case of $G = \mathbb{Z}_2^n$, $H = \mathbb{Z}_2$, and μ being the uniform distribution over some set, via the connection described in Section 1, this gives a new proof of a result of Kaufman and Sudan [KS07], who proved that sparse, low-bias linear codes are locally testable (in particular, that sparse random linear codes are locally testable). Their proof used the machinery of Krawtchouk polynomials and nontrivial information about the distribution of their roots. The corollary above gives a new and simple proof of this fact[2], and generalizes it to arbitrary G and H.

In the next theorem, we show that product distributions over $\mathbb{Z}_2^n$ are uniformly correlatable, whenever the individual distributions are not too biased. We then use our main theorem to conclude that linearity is tolerantly testable under such distributions. The proof is by reducing to the $n = 1$ case, and is omitted in this version of the paper.

Theorem 3 (Product Distributions are uniformly correlatable). *Let $G = \mathbb{Z}_2^n$ and let $p_1, \ldots, p_n \in [\gamma, (1 - \gamma)]$. For each $i \in [n]$, let μ_i be the distribution over $\mathbb{Z}_2$ with $\mu_i(1) = p_i$, Let μ be the product distribution $\prod_{i=1}^{n} \mu_i$ on G. Then μ is $(0, O(1/\gamma))$-uniformly-correlatable.*

Corollary 2 (Product Distributions are tolerantly testable). *Let μ and γ be as in Theorem 3. Then there are constants $C_1 = C_1(\gamma)$, $C_2 = C_2(\gamma)$, and a $O(1/\gamma)$-query test T such that for any abelian group H and any function $f : G \to H$, letting $\delta = \Pr[T^f \text{ rejects}]$, we have*

$$C_1\delta \le \Delta_\mu(f, \mathrm{Hom}(G, H)) \le C_2\delta.$$

In the next theorem, we consider distributions on $\mathbb{Z}_2^n$ that are symmetric under permutations of the coordinates. For technical reasons, we only show correlatability for distributions supported on words of even Hamming weight, but this suffices to show testability for general symmetric distributions μ. The proof is by a volume growing argument, and is omitted in this version of the paper.

Theorem 4 (Symmetric distributions are uniformly correlatable). *Let $G' = \mathbb{Z}_2^n$ and let G be the subgroup of G' consisting of even weight words. Let μ be a distribution on G, symmetric under permutations of the coordinates of G', and supported on words whose weights lie in the interval $[\gamma n, (1 - \gamma)n]$. Then μ (viewed as a distribution on G) is $(0, O(1/\gamma))$-uniformly-correlatable.*

Corollary 3 (Symmetric distributions are tolerantly testable). *Let μ be a symmetric distribution on $G = \mathbb{Z}_2^n$ such that supported on words whose weights lie in the interval $[\gamma n, (1 - \gamma)n]$. Then there are constants $C_1 = C_1(\gamma)$, $C_2 = C_2(\gamma)$, and a $O(1/\gamma)$-query test T such that for any abelian group H and any function $f : G \to H$, letting $\delta = \Pr[T^f \text{ rejects}]$, we have*

$$C_1\delta \le \Delta_\mu(f, \mathrm{Hom}(G, H)) \le C_2\delta.$$

1.2   Other Related Work

Kiwi [Kiw03] considered puncturings of the Hadamard code and gave a sufficient condition for certain codes to be locally testable. There has been a large body

---

[2] The $o(1/|G|)$ term in this corollary, combined with the perfect completeness of the test, in fact shows that the corresponding code is "strongly" locally testable.

of work constructing short locally testable codes [GS02], [GR05], [BSSVW03], [BSGH+04], [BSS05], [Din06], [Mei08]. In our framework, these correspond to distributions μ over $\mathbb{Z}_2^n$ supported on sets of size poly(n) under which linearity is tolerantly testable. It would be interesting to obtain a better understanding of which μ with such small support/min-entropy are such that linearity is tolerantly testable under μ.

Property testing under nonuniform distributions has arisen naturally and studied in several other contexts (in addition to [HK07]). The problem of dictatorship testing under the p-biased distribution arose in the work of Dinur and Safra [DS02] on the inapproximability of VERTEX-COVER. Subsequently, the problem of junta-testing [KS03] was also considered under the p-biased distribution.

In [AKK+03], Alon et. al. gave a constant query test for low degree polynomials over $\mathbb{F}_2$ (this was later extended to larger fields by [KR04], [JPRZ04], [KS08]). A suitable application of this test shows that for any p of the form $\frac{a}{2^b}$ for constant b, there is a constant query test for low degree polynomials under the p-biased distribution. This lends optimism to the goal of understanding the more general question of testing low degree polynomials under general distributions.

Paper organization: In Section 2, we give an overview of our proofs. In Section 3, we prove that uniformly-correlatable distributions are tolerantly testable (Theorem 1). The proof of Theorem 2 appears in Section 4. Finally in Section 5, we discuss some problems and directions for further study.

2   Overview of Proofs

We first give some intuition for the uniform correlatability criterion. For T to be a tester for linearity under μ, it needs to satisfy the following minimum requirements: (1) each query made by the tester needs to be distributed essentially according to μ (so that the probability of rejection is upper bounded by the distance), and (2) the queries need to satisfy some linear relations (so that the tester has something to test). This already indicates that a tester will need to "design" a query distribution very carefully, so that both the above requirements are satisfied. This is where the uniformly-correlatable criterion comes in: given the uniformly-correlated distribution on matrices, it allows us to design other correlations quite flexibly, and in particular to produce queries distributed according to μ that satisfy linear relations.

The proof of Theorem 1 follows the rough outline of the original "self-correction" proof of the BLR linearity test (for linearity testing under the uniform distribution)[3]. It proceeds in 3 steps: we first define a "self-corrected" version of the function being tested and show that the function being tested is μ-close to that function. We then show that the self-corrected version is in fact self-corrected with

---

[3] Note that the uniform distribution is $(0, 1)$-uniformly-correlatable, and for this case, the test given by Theorem 1 essentially reduces to the BLR linearity test.

overwhelming probability. Finally we use the above two facts to show that the self-corrected function is in fact a homomorphism. We use the correlated matrix (given by uniform correlatability) to construct two tests: each of these tests helps with a particular step of the proof. In contrast, the BLR [BLR93] linearity test makes only one kind of test which miraculously suffices for all steps of the analysis.

We show Theorem 2 on uniform-correlatability of low-bias distributions by Fourier analysis, using a version of the Vazirani-XOR lemma. For Theorem 3, we use the closure property of uniform-correlatability under products, and it thus suffices to show that the p-biased distribution on $\mathbb{Z}_2$ is uniformly correlatable. We then exhibit such a correlation by a direct construction. Finally, to show Theorem 4, we use the closure property of uniform-correlatability under convex combinations. This reduces the question to showing that for any even $w \in [\gamma n, (1 - \gamma)n]$, the uniform distribution on vectors in $\mathbb{Z}_2^n$ of Hamming weight exactly w is uniformly correlatable (to get a uniform distribution on all words of even weight). This is a technically involved step, and is achieved by carefully analyzing the set of possible row-sums and column-sums of matrices with all entries being words of weight w. The correlatability of symmetric distributions supported on even weight words, along with an additional trick, then allows us to deduce that linearity is tolerantly testable under any symmetric distribution over words with weights in $[\gamma n, (1 - \gamma)n]$.

3   Uniformly Correlatable Distributions are Testable

Let μ be (ϵ, k)-uniformly-correlatable. Fix a distribution μ^{mat} over $G^{k \times k}$ witnessing this property. Without loss of generality, we may assume that all the rows and columns of μ^{mat} are identically distributed, and let μ^{row} be this distribution (indeed, we may take a random sample from μ^{mat}, randomly permute the rows and columns, and then transpose it with probability $\frac{1}{2}$: the distribution of the resulting matrix witnesses the correlatability property, and also has identical row and column distributions). We now define a few distributions related to it:

1. **Distribution $\mu^{\mathrm{mat}}_{(r,s)}$:** For $r, s \in G^k$ with $\sum_{i \in [k]} r_i = \sum_{j \in [k]} s_j$, $\mu^{\mathrm{mat}}_{(r,s)}$ is the distribution of samples X from μ^{mat} conditioned on $\sum_{j \in [k]} X_{ij} = r_i$ and $\sum_{i \in [k]} X_{ij} = s_j$.
2. **Distribution μ^{row}_r:** For $r \in G$, μ^{row}_r is the distribution of samples x from μ^{row} conditioned on $\sum_{i \in [k]} x_i = r$.
3. **Distribution μ^*_r:** This is the distribution of the random variables $(y, z) \in G^k \times G^k$ produced by the following random process. Let U_G be the uniform distribution on G. First sample r' from U_G. Then independently sample y from $\mu^{\mathrm{row}}_{r+r'}$ and z from $\mu^{\mathrm{row}}_{r'}$. In particular, $\sum_i y_i - \sum_j z_j = r$.

We may now describe the linearity test under μ.

Test T: With probability $1/2$, perform Test T1, and with probability $1/2$ perform Test T2.

- **Test T1:** Sample r from μ. Sample $(y, z) \in G^{2k}$ from μ_r^*. If $\sum_{i \in [k]} f(y_i) - \sum_{i \in [k]} f(z_i) = f(r)$, then accept, else reject.
- **Test T2:** Sample r from U_G. Independently sample (y, y') and (z, z') from μ_r^*. If $\sum_{i \in [k]} f(y_i) - \sum_{i \in [k]} f(y_i') = \sum_{i \in [k]} f(z_i) - \sum_{i \in [k]} f(z_i')$, then accept, else reject.

It is clear that this test has perfect completeness.

The following fact basic fact about distributions will be useful while analyzing the test.

Fact 5. *Let R, S be random variables, and let h be function such that the distribution of $h(R)$ is ϵ-close to the distribution of S. Consider the distribution of R' sampled as follows: first pick S, and then let R' be a sample of R conditioned on $h(R) = S$. Then the distribution of R' is ϵ-close to the distribution of R.*

We now prove that the Test T is indeed a tester for linearity under μ, hence completing the proof of Theorem 1.

Theorem 6. *Let $f : G \to H$ and let $\delta \overset{\text{def}}{=} \Pr[T^f \text{ rejects}]$. Then,*

$$\frac{\delta - 4\epsilon}{4k} \leq \Delta_\mu(f, \mathrm{Hom}(G, H)) \leq \frac{12k}{1 - 24\epsilon k} \cdot \delta.$$

Proof. Let $f : G \to H$ and let $\delta = \Pr[T^f \text{ rejects}]$. Notice that by Fact 5, each k-tuple of queries made by the test T is ϵ-close to the distribution μ^{row}. Therefore, the probability that no query is made to an element of G where f disagrees with its nearest homomorphism in $\mathrm{Hom}(G, H)$ is at most $4k \cdot \Delta_\mu(f, \mathrm{Hom}(G, H)) + 4\epsilon$. Thus $\delta \leq 4k \cdot \Delta_\mu(f, \mathrm{Hom}(G, H)) + 4\epsilon$, which is the first inequality.

We now show the second inequality. If $\delta \geq \frac{1}{12k} - 2\epsilon$, then the claim is trivial (since $\Delta_\mu(\cdot, \cdot) \leq 1$). Suppose $\delta \leq \frac{1}{12k} - 2\epsilon$. Let δ_1 be the probability that Test T1 rejects. Let δ_2 be the probability that Test T2 rejects. Then $\delta = \frac{1}{2}(\delta_1 + \delta_2)$.

For $x \in G$, define the "self-corrected" value $g(x)$ to be the most probable value of $\sum_{i \in [k]} f(y_i) - \sum_{i \in [k]} f(z_i)$, where $(y, z) \in \mu_x^*$.

Lemma 1 (g is close to f). $\Delta_\mu(f, g) < 2\delta_1$.

Proof. Let $B = \{x \in G : g(x) \neq f(x)\}$.

For any $x \in G$, define

$$p_x = \Pr_{(y,z) \in \mu_x^*} \left[\sum_{i \in [k]} f(y_i) - \sum_{j \in [k]} f(z_j) \neq f(x) \right].$$

Notice that for any $x \in B$, $p_x \geq 1/2$. By definition, $\delta_1 = \mathbf{E}_{x \in \mu}[p_x]$. Applying Markov's inequality, we conclude that

$$\Pr_{x \in \mu}[x \in B] \leq \Pr_{x \in \mu}[p_x \geq 1/2] \leq 2\delta_1.$$

We now show that g is in fact a homomorphism.

Lemma 2 (Majority votes of g are overwhelming majorities). *For all* $x \in G$,

$$\Pr_{(y,z)\in\mu_x^*}[\sum_{i\in[k]} f(y_i) - \sum_{j\in[k]} f(z_j) \neq g(x)] \leq 2\delta_2.$$

Proof. Let $x \in G$. Take two independent samples (y^1, z^1) and (y^2, z^2) from μ_x^*. We will show that

$$\Pr\left[\sum_{i\in[k]} f(y_i^1) - \sum_{j\in[k]} f(z_j^1) \neq \sum_{i\in[k]} f(y_i^2) - \sum_{j\in[k]} f(z_j^2)\right] \leq 2\delta_2. \tag{2}$$

The lemma follows immediately from this.

We now prove Equation 2. By definition, (y^1, z^1) was generated by picking $r^1 \in U_G$, and then picking $y^1 \in \mu_{r^1+x}^{\text{row}}$ and $z^1 \in \mu_{r^1}^{\text{row}}$. Similarly (y^2, z^2) was generated by picking $r^2 \in U_G$, and then picking $y^2 \in \mu_{r^2+x}^{\text{row}}$ and $z^2 \in \mu_{r^2}^{\text{row}}$. Observe that both (y^1, y^2) and (z^1, z^2) come from the distribution $\mu_{r^1-r^2}^*$ (albeit not independently). Let (w^1, w^2) be another sample from $\mu_{r^1-r^2}^*$ (independent of (y^1, y^2) and (z^1, z^2)).

We now rewrite and then bound the left hand side of Equation (2) by:

$$\Pr\left[\sum_{i\in[k]} f(y_i^1) - \sum_{j\in[k]} f(z_j^1) \neq \sum_{i\in[k]} f(y_i^2) - \sum_{j\in[k]} f(z_j^2)\right]$$

$$= \Pr\left[\sum_{i\in[k]} f(y_i^1) - \sum_{i\in[k]} f(y_i^2) \neq \sum_{j\in[k]} f(z_j^1) - \sum_{j\in[k]} f(z_j^2)\right]$$

$$\leq \Pr\left[\sum_{i\in[k]} f(y_i^1) - \sum_{i\in[k]} f(y_i^2) \neq \sum_{j\in[k]} f(w_j^1) - \sum_{j\in[k]} f(w_j^2)\right]$$

$$+ \Pr\left[\sum_{i\in[k]} f(z_i^1) - \sum_{i\in[k]} f(z_i^2) \neq \sum_{j\in[k]} f(w_j^1) - \sum_{j\in[k]} f(w_j^2)\right]$$

Finally, note that $r^1 - r^2$ is uniformly distributed over G. Since $(y_1, y_2), (w_1, w_2)$ are independent samples from $\mu_{r^1-r^2}^*$ (and similarly for $(z_1, z_2), (w_1, w_2)$), this implies that both the terms in the last expression above equal the rejection probability of Test T2 ($= \delta_2$). This completes the proof of the lemma.

Lemma 3 (g is linear). $g \in \text{Hom}(G, H)$.

Proof. Pick any $x, x' \in G^k$, and let $t = \sum_{i=1}^k x_i - \sum_{i=1}^k x_i'$. We will show that $\sum_{i=1}^k g(x_i) - \sum_{i=1}^k g(x_i') = g(t)$.

We now describe a random process. Pick $(\alpha, \beta) \in \mu_t^*$. Pick $r \in U_G$. Pick r^1, s^1, r^2, s^2 uniformly from G^k conditioned on $\sum_{i\in[k]} r_i^1 = \sum_{i\in[k]} s_i^1 = r$, $\sum_{i\in[k]} r_i^2 = r - \sum_{i\in[k]} \alpha_i$, and $\sum_{i\in[k]} s_i^2 = r - \sum_{i\in[k]} x_i$.

Now pick random matrices $A \in \mu^{\mathrm{mat}}_{(r^1, s^1)}$, $A' \in \mu^{\mathrm{mat}}_{(r^2, s^1 - \alpha)}$, $B \in \mu^{\mathrm{mat}}_{(r^1 - x, s^2)}$, and $B' \in \mu^{\mathrm{mat}}_{(r^2 - x, s^2 - \beta)}$.

$$
\begin{array}{cccc|cccc|c}
A_{11} & A_{12} & \cdots & A_{1k} & A'_{11} & A'_{12} & \cdots & A'_{1k} & \alpha_1 \\
A_{21} & A_{22} & \cdots & A_{2k} & A'_{21} & A'_{22} & \cdots & A'_{2k} & \alpha_2 \\
\vdots & \vdots & \ddots & \vdots & \vdots & \vdots & \ddots & \vdots & \vdots \\
A_{k1} & A_{k2} & \cdots & A_{kk} & A'_{k1} & A'_{k2} & \cdots & A'_{kk} & \alpha_k \\ \hline
B_{11} & B_{12} & \cdots & B_{1k} & B'_{11} & B'_{12} & \cdots & B'_{1k} & \beta_1 \\
B_{21} & B_{22} & \cdots & B_{2k} & B'_{21} & B'_{22} & \cdots & B'_{2k} & \beta_2 \\
\vdots & \vdots & \ddots & \vdots & \vdots & \vdots & \ddots & \vdots & \vdots \\
B_{k1} & B_{k2} & \cdots & B_{kk} & B'_{k1} & B'_{k2} & \cdots & B'_{kk} & \beta_k \\ \hline
x_1 & x_2 & \cdots & x_k & x'_1 & x'_2 & \cdots & x'_k & t
\end{array}
$$

Arrange these random variables in a matrix, as shown in the figure.

First notice that by Fact 5, the distribution of each α_i is ϵ-close to the distribution μ. Similarly, the distribution of each β_i is ϵ-close to the distribution μ.

Let us study the row distribution. Again by Fact 5, for each $i \in [k]$, the distribution of $(A_{i\bullet}, A'_{i\bullet})$ is 4ϵ-close to $\mu^*_{\alpha_i}$. Similarly, for each $i \in [k]$, the distribution of $(B_{i\bullet}, B'_{i\bullet})$ is 4ϵ-close to $\mu^*_{\beta_i}$.

Now consider the distribution of the columns. For each $j \in [k]$, the distribution of $(A_{\bullet j}, B_{\bullet j})$ is 4ϵ-close to $\mu^*_{x_j}$. Similarly, for each $j \in [k]$, the distribution of $(A'_{\bullet j}, B'_{\bullet j})$ is 4ϵ-close to $\mu^*_{x'_j}$.

Summarizing, each row distribution $(A_{i\bullet}, A'_{i\bullet}, \alpha_i)$ and $(B_{i\bullet}, B'_{i\bullet}, \beta_i)$ is 5ϵ-close to the distribution of the distribution of queries of Test T1. Thus by a union bound, with probability at least $1 - 2k \cdot (\delta_1 + 5\epsilon)$, both the following events occur:

- **Event 1:** For each $i \in [k]$, $\sum_{j \in [k]} f(A_{ij}) - \sum_{j \in [k]} f(A'_{ij}) = f(\alpha_i)$,
- **Event 2:** For each $i \in [k]$, $\sum_{j \in [k]} f(B_{ij}) - \sum_{j \in [k]} f(B'_{ij}) = f(\beta_i)$.

Each column distribution $(A_{\bullet j}, B_{\bullet j})$ is 4ϵ-close to $\mu^*_{x_j}$, and each column distribution $(A'_{\bullet j}, B'_{\bullet j})$ is 4ϵ-close to $\mu^*_{x'_j}$. Thus by Lemma 2 and a union bound, with probability at least $1 - 2k \cdot (2\delta_2 + 4\epsilon)$, both the following events occur.

- **Event 3:** For each $j \in [k]$, $\sum_{i \in [k]} f(A_{ij}) - \sum_{i \in [k]} f(B_{ij}) = g(x_j)$,
- **Event 4:** For each $j \in [k]$, $\sum_{i \in [k]} f(A'_{ij}) - \sum_{i \in [k]} f(B'_{ij}) = g(x'_j)$,

Finally, since (α, β) was picked from μ^*_t, Lemma 2 tells us that the following event occurs with probability at least $1 - 2\delta_2$:

- **Event 5:** $\sum_{j \in [k]} f(\alpha_j) - \sum_{j \in [k]} f(\beta_j) = g(t)$,

Thus Events 1,2,3, 4 and 5 all occur with probability at least $1 - (2k + 1) \cdot (\delta_1 + 2\delta_2 + 5\epsilon) > 0$, since we assumed that $\delta = \frac{\delta_1 + \delta_2}{2} < \frac{1}{12k} - 2\epsilon$. In this case, we see that

$$g(t) = \sum_{i=1}^{k} f(\alpha_i) - \sum_{i=1}^{k} f(\beta_i) \qquad\qquad\qquad \text{Event 5}$$

$$= \sum_{i=1}^{k} \left(\sum_{j=1}^{k} (f(A_{ij}) - f(A'_{ij})) \right) - \sum_{i=1}^{k} \left(\sum_{j=1}^{k} (f(B_{ij}) - f(B'_{ij})) \right) \qquad \text{Events 1 and 2}$$

$$= \sum_{j=1}^{k} \left(\sum_{i=1}^{k} (f(A_{ij}) - f(B_{ij})) \right) - \sum_{j=1}^{k} \left(\sum_{i=1}^{k} (f(A'_{ij}) - f(B'_{ij})) \right) \qquad \text{rearranging terms}$$

$$= \sum_{j=1}^{k} (g(x_j) - g(x'_j)) \qquad\qquad\qquad \text{Events 3 and 4.}$$

Hence, $\Pr[g(t) = \sum_{j=1}^{k} (g(x_j) - g(x'_j))] > 0$. However, this statement is a deterministic statement, and hence we conclude that $g(t) = \sum_{j=1}^{k} (g(x_j) - g(x'_j))$ Since this holds for every choice of x, x', g must be a homomorphism.

Thus, $\Delta_\mu(f, \mathsf{Hom}(G, H)) \leq \Delta_\mu(f, g) \leq 2\delta_1 \leq 4\delta \leq \frac{6k}{1 - 12\epsilon k} \cdot \delta$.

4   Low Bias Distributions are Uniformly Correlatable

In this section we prove Theorem 2.

Theorem 2. *Let μ be a probability distribution on G such that for all nontrivial characters $\chi : G \to \mathbb{C}^\times$,*

$$\left| \mathbf{E}_{x \in \mu} [\chi(x)] \right| < |G|^{-\gamma}.$$

Then for $k = \Omega(1/\gamma)$, μ is $(|G|^{-\Omega(k\gamma)}, k)$-uniformly-correlatable.

Proof. We begin with a lemma, which gives a simple criterion for checking that a distribution is close to uniform on a subgroup of G^t. It is an intermediate claim in the usual proof of the Vazirani XOR lemma [Gol95] which bounds the distance to uniform in terms of the maximum bias of the distribution. The full Vazirani XOR lemma turns out to be too weak for our purposes.

Lemma 4. *Let $S = \{(y_1, \ldots, y_t, z_1, \ldots, z_t) \in G^{2t} \mid \sum_{i \in [t]} y_i = \sum_{i \in [t]} z_i\}$. Let $(Y_1, \ldots, Y_t, Z_1, \ldots, Z_t)$ be an S-valued random variable. Suppose*

$$\sum_{\alpha_1, \ldots, \alpha_t, \beta_1, \ldots, \beta_t \in \hat{G} \text{ not all equal}} \left| \mathbf{E} \left[\prod_{i \in [t]} \chi_{\alpha_i}(Y_i) \cdot \prod_{j \in [t]} \overline{\chi_{\beta_j}(Z_j)} \right] \right|^2 \leq \lambda.$$

Then the distribution of $(Y_1, \ldots, Y_t, Z_1, \ldots, Z_t)$ is $\sqrt{\lambda}$-close to the uniform distribution over S.

We omit the proof of this lemma.

We can now prove the theorem. Let $\eta = |G|^{-\gamma}$. The distribution $\mathbf{X} = (\mathbf{X}_{ij})_{i,j\in[k]}$ is given by picking each $\mathbf{X}_{ij}$ independently from μ. For $i \in [k]$, let $\mathbf{Y}_i$ be the random variable $\sum_{j\in[k]} \mathbf{X}_{ij}$. For $j \in [k]$, $\mathbf{Z}_j$ be the random variable $\sum_{i\in[k]} \mathbf{X}_{ij}$. We wish to show that $(\mathbf{Y}_1,\ldots,\mathbf{Y}_k,\mathbf{Z}_1,\ldots,\mathbf{Z}_k)$ is $|G|^{-\Omega(k\gamma)}$-close to uniformly distributed on S.

In order to apply Lemma 4, we compute

$$\sum_{\substack{\alpha_1,\ldots,\alpha_k \\ \beta_1,\ldots,\beta_k \\ \text{n.a.e.}}} \left| \mathbf{E}\left[\prod_{i\in[k]} \chi_{\alpha_i}(\mathbf{Y}_i) \cdot \prod_{j\in[k]} \overline{\chi_{\beta_j}(\mathbf{Z}_j)} \right] \right|^2 = \sum_{\substack{\alpha_1,\ldots,\alpha_k \\ \beta_1,\ldots,\beta_k \\ \text{n.a.e.}}} \left(\mathbf{E}\left[\prod_{i,j} \chi_{\alpha_i-\beta_j}(\mathbf{X}_{ij}) \right] \right)^2$$

$$= \sum_{\alpha,\beta} \left(\prod_{i,j} \mathbf{E}[\chi_{\alpha_i-\beta_j}(\mathbf{X}_{ij})] \right)^2$$

(since the $\mathbf{X}_{ij}$ are independent)

$$\leq \sum_{\alpha,\beta} \left(\eta^{|\{(i,j)\in[k]^2 : \alpha_i \neq \beta_j\}|} \right)^2$$

Consider the term corresponding to $\alpha_1,\ldots,\alpha_k$ and $\beta_1,\ldots,\beta_k$. We classify the terms into 3 kinds, and separately bound the total contribution of terms of each kind.

Case A: the most frequently occurring element in $\alpha_1,\ldots,\alpha_k$ occurs at most $2k/3$ times. Then $|\{(i,j) \in [k]^2 : \alpha_i \neq \beta_j\}| \geq k \cdot k/3 = k^2/3$. Thus the sum of all terms in case A is at most $|G|^{2k} \cdot \eta^{2k^2/3}$.

Case B: the most frequently occurring element in $\alpha_1,\ldots,\alpha_k$ occurs at least $2k/3$ times, and that same element occurs in $\beta_1,\ldots,\beta_k$ at most $2k/3$ times. Then $|\{(i,j) \in [k]^2 : \alpha_i \neq \beta_j\}| \geq (k/3) \cdot (2k/3) = 2k^2/9$. Thus the sum of all terms in case B is at most $|G|^{2k} \cdot \eta^{2k^2/9}$.

Case C: Now suppose we are not in either of the above two cases. Suppose the most frequently occurring element in $\alpha_1,\ldots,\alpha_k$ occurs $a > 2k/3$ times, and that same element appears in $\beta_1,\ldots,\beta_k$ occurs $b > 2k/3$ times. Note that by the not all equal assumption, at most one of a,b can be equal to k. Then $|\{(i,j) \in [k]^2 : \alpha_i \neq \beta_j\}| \geq a \cdot (k-b) + b \cdot (k-a)$. Thus the total contribution of terms from Case C is at most (here we subtracted off the terms with $a = b = k$):

$$\sum_{a=2k/3}^{k} \sum_{b=2k/3}^{k} \binom{k}{a}\binom{k}{b} |G|^{k-a}|G|^{k-b}|G|\eta^{a(k-b)}\eta^{b(k-a)} - \binom{k}{k}\binom{k}{k}|G|.$$

This can be bounded from above by

$$|G| \cdot \left(\sum_{a=2k/3}^{k} \sum_{b=2k/3}^{k} \binom{k}{a}\binom{k}{b} |G|^{k-a}|G|^{k-b}\eta^{(k/2)(k-b)}\eta^{(k/2)(k-a)} - 1 \right),$$

which in turn may be upper bounded by

$$|G| \cdot \left((1 + |G|\eta^{(k/2)})^k (1 + |G|\eta^{(k/2)})^k - 1 \right)$$

$$\leq |G| \cdot (8k|G|\eta^{k/2}),$$

where the last inequality uses the fact that $k = \Omega(1/\gamma)$, and hence $\eta^{k/2} \ll (|G|k)^{-1}$. Summarizing, the sum of all the terms is at most $|G|^2 \eta^{\Omega(k)} + |G|^{2k} \eta^{\Omega(k^2)}$. For $k = \Omega(1/\gamma)$, this quantity is at most $|G|^{-\Omega(\gamma k)}$. Lemma 4 now implies the desired result.

5   Discussion, Problems and Directions

We believe that there are many fruitful and interesting questions waiting to be explored in tolerant property testing under nonuniform distributions in general and tolerant linearity testing under nonuniform distributions in particular.

As far as we know, every distribution of linear min-entropy with bias at most 0.9 (say) is uniformly $(o(1), O(1))$-uniformly correlatable, and hence tolerantly testable. In fact, we do not even know of a single μ of linear min-entropy under which linearity is not tolerantly testable.

Question 1. Let μ be a probability distribution on $\mathbb{Z}_2^n$ with min-entropy $\Omega(n)$. Find necessary and sufficient conditions on μ for linearity to be tolerantly testable under μ.

Via the connection between tolerant linearity testing and local testability of codes, we even venture the following conjecture.

Conjecture 1. Every linear code $C \subseteq \mathbb{Z}_2^N$ with $N^{O(1)}$ codewords is locally testable!

Acknowledgements

We are very grateful to Madhu Sudan and Tali Kaufman for encouragement and invaluable discussions.

References

[AKK+03]   Alon, N., Kaufman, T., Krivelevich, M., Litsyn, S., Ron, D.: Testing low-degree polynomials over GF(2). In: Arora, S., Jansen, K., Rolim, J.D.P., Sahai, A. (eds.) RANDOM 2003 and APPROX 2003. LNCS, vol. 2764, pp. 188–199. Springer, Heidelberg (2003)

[BLR93]   Blum, M., Luby, M., Rubinfeld, R.: Self-testing/correcting with applications to numerical problems. Journal of Computer and System Sciences 47(3), 549–595 (1993)

[BSGH+04]   Ben-Sasson, E., Goldreich, O., Harsha, P., Sudan, M., Vadhan, S.: Robust PCPs of proximity, shorter PCPs and applications to coding. In: Proceedings of the 36th Annual ACM Symposium on Theory of Computing, pp. 1–10. ACM Press, New York (2004)

[BSS05]   Ben-Sasson, E., Sudan, M.: Short PCPs with poly-log rate and query complexity. In: Proceedings of the 37th Annual ACM Symposium on Theory of Computing, pp. 266–275. ACM Press, New York (2005)

[BSSVW03]   Ben-Sasson, E., Sudan, M., Vadhan, S., Wigderson, A.: Randomness efficient low-degree tests and short PCPs via ε-biased sets. In: Proceedings of the 35th Annual ACM Symposium on Theory of Computing, pp. 612–621. ACM Press, New York (2003)

[Din06]   Dinur, I.: The PCP theorem by gap amplification. In: Proceedings of the 38th Annual ACM Symposium on Theory of Computing, pp. 241–250. ACM Press, New York (2006); Preliminary version appeared as an ECCC Technical Report TR05-046

[DS02]   Dinur, I., Safra, S.: The importance of being biased. In: Proceedings on 34th Annual ACM Symposium on Theory of Computing, Montreal, Quebec, Canada, May 19-21, 2002, pp. 33–42 (2002)

[GGR98]   Goldreich, O., Goldwasser, S., Ron, D.: Property testing and its connection to learning and approximation. JACM 45(4), 653–750 (1998)

[Gol95]   Goldreich, O.: Three xor-lemmas - an exposition. Electronic Colloquium on Computational Complexity (ECCC) 2(56) (1995)

[GR05]   Guruswami, V., Rudra, A.: Tolerant locally testable codes. In: Chekuri, C., Jansen, K., Rolim, J.D.P., Trevisan, L. (eds.) APPROX 2005 and RANDOM 2005. LNCS, vol. 3624, pp. 306–317. Springer, Heidelberg (2005)

[GS02]   Goldreich, O., Sudan, M.: Locally testable codes and PCPs of almost-linear length. In: Proceedings of the 43rd Annual IEEE Symposium on Foundations of Computer Science, Vancouver, Canada, November 16-19 (2002)

[HK07]   Halevy, S., Kushilevitz, E.: Distribution-free property-testing. SIAM J. Comput. 37(4), 1107–1138 (2007)

[JPRZ04]   Jutla, C.S., Patthak, A.C., Rudra, A., Zuckerman, D.: Testing low-degree polynomials over prime fields. In: FOCS 2004: Proceedings of the Forty-Fifth Annual IEEE Symposium on Foundations of Computer Science, pp. 423–432. IEEE Computer Society Press, Los Alamitos (2004)

[Kiw03]   Kiwi, M.A.: Algebraic testing and weight distributions of codes. Theor. Comput. Sci. 3(299), 81–106 (2003)

[KR04]   Kaufman, T., Ron, D.: Testing polynomials over general fields. In: Proceedings of the Forty-fifthth Annual Symposium on Foundations of Computer Science, pp. 413–422 (2004)

[KS03]   Kindler, G., Safra, S.: Noise-resistant boolean-functions are juntas, April 17 (2003)

[KS07]   Kaufman, T., Sudan, M.: Sparse random linear codes are locally decodable and testable. In: FOCS, pp. 590–600. IEEE Computer Society Press, Los Alamitos (2007)

[KS08]   Kaufman, T., Sudan, M.: Algebraic property testing: the role of invariance. In: Ladner, R.E., Dwork, C. (eds.) STOC, pp. 403–412. ACM Press, New York (2008)

[Mei08]   Meir, O.: Combinatorial construction of locally testable codes. In: Ladner, R.E., Dwork, C. (eds.) STOC, pp. 285–294. ACM Press, New York (2008)

[PRR06]   Parnas, M., Ron, D., Rubinfeld, R.: Tolerant property testing and distance approximation. J. Comput. Syst. Sci. 72(6), 1012–1042 (2006)

[RS96]   Rubinfeld, R., Sudan, M.: Robust characterizations of polynomials with applications to program testing. SIAM Journal on Computing 25(2), 252–271 (1996)

Pseudorandom Bit Generators
That Fool Modular Sums

Shachar Lovett[1,*], Omer Reingold[2,**], Luca Trevisan[3,***], and Salil Vadhan[4,†]

[1] Department of Computer Science, Weizmann Institute of Science,
Rehovot 76100, Israel
shachar.lovett@weizmann.ac.il
[2] Department of Computer Science, Weizmann Institute of Science,
Rehovot 76100, Israel
omer.reingold@weizmann.ac.il
[3] Computer Science Division, University of California, Berkeley, CA, USA
luca@cs.berkeley.edu
[4] School of Engineering and Applied Science, Harvard University,
Cambridge, MA 02138
salil@eecs.harvard.edu

Abstract. We consider the following problem: for given n, M, produce a sequence $X_1, X_2, \ldots, X_n$ of *bits* that fools every linear test modulo M. We present two constructions of generators for such sequences. For every constant prime power M, the first construction has seed length $O_M(\log(n/\epsilon))$, which is optimal up to the hidden constant. (A similar construction was independently discovered by Meka and Zuckerman [MZ]). The second construction works for every M, n, and has seed length $O(\log n + \log(M/\epsilon) \log(M \log(1/\epsilon)))$.

The problem we study is a generalization of the problem of constructing *small bias* distributions [NN], which are solutions to the $M = 2$ case. We note that even for the case $M = 3$ the best previously known constructions were generators fooling general bounded-space computations, and required $O(\log^2 n)$ seed length.

For our first construction, we show how to employ recently constructed generators for sequences of elements of $\mathbb{Z}_M$ that fool small-degree polynomials (modulo M). The most interesting technical component of our second construction is a variant of the derandomized graph squaring operation of [RV]. Our generalization handles a product of two distinct graphs with distinct bounds on their expansion. This is then used to produce pseudorandom-walks where each step is taken on a different regular directed graph (rather than pseudorandom walks on a single regular directed graph as in [RTV, RV]).

* Research supported by the Israel Science Foundation (grant 1300/05).
** Research supported by US-Israel BSF grant 2006060.
*** This material is based upon work supported by the National Science Foundation under grant No. CCF-0729137 and by the US-Israel BSF grant 2006060.
† Work done in part while visiting U.C. Berkeley, supported by the Miller Institute for Basic Research in Science and a Guggenheim Fellowship. Also supported by US-Israel BSF grant 2006060.

I. Dinur et al. (Eds.): APPROX and RANDOM 2009, LNCS 5687, pp. 615–630, 2009.

1   Introduction

Pseudorandomness is the theory of generating objects that "look random" despite being constructed using little or no randomness. A primary application of pseudorandomness is to address the question: *Are randomized algorithms more powerful than deterministic ones?* That is, how does randomization trade off with other computational resources? Can every randomized algorithm be converted into a deterministic one with only a polynomial slowdown (*i.e.*, does $\mathbf{BPP} = \mathbf{P}$) or with only a constant-factor increase in space (*i.e.*, does $\mathbf{RL} = \mathbf{L}$)? The study of both these questions has relied on pseudorandom bit generators that fool algorithms of limited computational powers. In particular, generators that fool space-bounded algorithms [AKS, BNS, Nis, INW] were highly instrumental in the study of the $\mathbf{RL}$ vs. $\mathbf{L}$ problem (e.g. used in the best known derandomization of $\mathbf{RL}$ [SZ]).

While the currently available space-bounded generators are extremely powerful tools, their seed length is still suboptimal. For example, if we want to fool a $\log n$-space algorithm then known generators require $\log^2 n$ truly random bits (the seed) in order to generate up to polynomially many pseudorandom bits. On the other hand, for several interesting special cases we do know generators with almost optimal seed length. The special case which serves as a motivation for our work is that of small-biased generators [NN]. These generators produce n bits $X_1, X_2, \ldots, X_n$ that fool all linear tests modulo 2. In other words, for each subset T of the bits, the sum $\Sigma_{i \in T} X_i \bmod 2$ is uniformly distributed up to bias ϵ. Explicit constructions of ϵ-biased generators are known with seed-length $O(\log(n/\epsilon))$, which is optimal up to the hidden constant [NN]. Even though linear tests may seem very limited, ϵ-biased generators have turned out to be very versatile and useful derandomization tools [NN, MNN, HPS, Nao, AM, AR, BSVW, BV, Lov, Vio].

Given the several applications of distributions that fool linear tests modulo 2, it is natural to consider the question of fooling modular sums for larger moduli. It turns out that the notion of small-biased generators can be generalized to larger fields. Such generators produce a sequence $X_1, X_2, \ldots, X_n$ of elements in a field $\mathbb{F}$ that fool every linear test over $\mathbb{F}$ [Kat, AIK+, RSW, EGL+, AM].[1] In this work, instead, we consider a different generalization of ϵ-biased generators where we insist on *bit*-generators. Namely we would like to generate a sequence $X_1, X_2, \ldots, X_n$ of bits that fool every linear test modulo a given number M. For every sequence $a_1, a_2, \ldots, a_n$ of integers in $\mathbb{Z}_M = \{0, 1, \ldots, M-1\}$ we want the sum $\sum_i a_i X_i \bmod M$ to have almost the same distribution (up to statistical distance at most ϵ) as in the case where the X_i's are uniform and independent random bits. (Note that this distribution may be far from the uniform distribution over $\mathbb{Z}_M$, particularly when only a few a_i's are nonzero.) It turns out that even for $M = 3$ and even if we limit all the a_i's to be either ones or zeros, the best

---

[1] More generally, an *ϵ-bias space* over a finite abelian group G is a distribution D on elements of G such that for every nontrivial character $\chi : G \to \mathbb{C}$, $|\mathbb{E}[\chi(D)]| \leq \epsilon$. The aforementioned results correspond to the special case $G = \mathbb{F}^n$, using the fact that the characters of $\mathbb{F}^n$ are in one-to-one correspondence with linear functions $\mathbb{F}^n \to \mathbb{F}$.

generators that were known prior to this work are generators that fool general space-bounded computations [Nis, INW], and required a seed of length $O(\log^2 n)$. Therefore, obtaining better pseudorandom bit generators that fool modular sums may be considered a necessary step towards improved space-bounded generators. In addition, we consider this notion to be a natural generalization of that of a small-bias generator, which is a central derandomization tool.

Our Results

We give two constructions of pseudorandom bit generators that fool modular sums. Similarly to [MST], each construction is actually comprised of two generators: one that fools summations $\sum_i a_i X_i$ in which only relatively few coefficients a_i are nonzero (the "low-weight" case) and one that fools summations $\sum_i a_i X_i$ in which many coefficients a_i are nonzero (the "high weight" case). The motivation is that fooling low-weight sums and fooling high-weight sums are tasks of a different nature. In the high-weight case, if R_i are truly random bits, then $\Sigma_i a_i R_i$ mod M is almost uniformly distributed in $\mathbb{Z}_M$ (at least when M is prime). Thus, in analyzing our generator, we just need to argue that $\Sigma_i a_i X_i$ mod M is close to uniform, where $X_1, \ldots, X_n$ is the output of the generator.

On the other hand, in the low-weight case the distribution may be far from uniform and therefore we may need to imitate the behavior of a random sequence of bits more closely.

Thus, in each construction, we shall present two generators: one that is pseudorandom against low-weight sums, and one that is pseudorandom against high-weight sums. We shall then combine them by evaluating them on independently chosen seeds and XORing the two resulting sequences.

Construction Based on Pseudorandom Generators for Polynomials

In our first construction, we handle the case of $M = 3$ and any other fixed prime modulus M (in fact, our construction works also for any fixed prime power). For these cases, our seed length is $O(\log(n/\epsilon))$ as in the case of ϵ-biased generators (but the hidden constant depends exponentially on M).

As mentioned above, for every fixed finite field $\mathbb{F}$, there are nearly-optimal known generators that construct a small-bias distribution $X_1, \ldots, X_n$ of *field elements*, while our goal is to generate *bits*. A natural approach to construct a bit generator would be to sample a sequence of field elements $X_1, \ldots, X_n$ from a small-bias distribution, and output a bit-sequence $g(X_1), \ldots, g(X_n)$ for an appropriate function $g : \mathbb{F} \to \{0, 1\}$. Unfortunately the pseudorandomness of $g(X_1), \ldots, g(X_n)$ against $\mathbb{F}$-linear tests does not seem to follow from the small-bias property of $X_1, \ldots, X_n$. Indeed, when $|\mathbb{F}|$ is odd, then g cannot be balanced, so at best we could hope is for $g(X_1), \ldots, g(X_n)$ to be indistinguishable by linear tests from a sequence of independent *biased* bits. But even this is not achievable in general, if we only assume the pseudorandomness of $X_1, \ldots, X_n$ against $\mathbb{F}$-linear tests(as per the definition of small-bias space).[2]

If, however, we start from a sequence of field elements $X_1, \ldots, X_n$ that fools *polynomials* over $\mathbb{F}$, then we can indeed show that $g(X_1), \ldots, g(X_n)$ is indistinguishable by linear tests from independent biased bits. The reason is that g can be chosen to be itself a polynomial (of degree $d = \Theta(|\mathbb{F}|)$), and thus any $\mathbb{F}$-linear test distinguisher on $g(X_1), \ldots, g(X_n)$ yields a degree d distinguisher on $X_1, \ldots, X_n$. Since we still only have indistinguishability from *biased* coins, we only apply this approach when the coefficient vector has sufficiently high weight so that both biased and unbiased random bits will yield a sum that is almost uniformly distributed over $|\mathbb{F}|$. Specifically, we need at least k non-zero coefficients a_i, where $k = O(M^2 \log 1/\epsilon)$. For fixed M, there are known constructions [BV, Lov, Vio] of pseudorandom generators that fool polynomials of degree d over $\mathbb{F} = \mathbb{Z}_M$, M prime, and which only require seed length $O_{M,d}(\log n/\epsilon)$.

In order to fool low-weight sums, we observe that a bit generator $X_1, \ldots, X_n$ which is ϵ-almost k-wise independent fools, by definition, every sum $\sum_i a_i X_i \mod M$ of weight at most k, and that such generators are known which require only seed length $O(\log n + k + \log 1/\epsilon)$.

A similar construction was independently discovered by Meka and Zuckerman [MZ].

Construction Based on the INW Generator

In our second construction, we give a pseudorandom bit generator that fools sums modulo *any* given M (not necessarily prime) with seed length $O(\log n + \log(M/\epsilon) \log(M \log(1/\epsilon)))$. In both the low-weight and high-weight cases, this generator relies on versions of the Impagliazzo–Nisan–Wigderson [INW] pseudorandom generator for space-bounded computation. Of course, modular sums are a special case of space-bounded computations, and thus we could directly apply the INW generator. But this would require seed length larger than $\log^2 n$. We obtain better bounds by more indirect use of the INW generator inside our construction.

The most interesting technical contribution underlying this construction is a new analysis of the derandomized graph squaring operation of [RV], which captures the effect of using the INW generator to derandomize random walks on graphs. Here we study the analogue of derandomized squaring for taking products of two distinct Cayley graphs over an abelian group (namely $\mathbb{Z}_M$). The advantage of the new analysis is that it handles graphs that have distinct bounds on their expansion, and works for bounding each eigenvalue separately. This is then used to produce pseudorandom walks where each step is taken on a different abelian Cayley graph (rather than pseudorandom walks on a single graph as in [RTV, RV]).

---

[2] Let $\mathbb{F} = \mathbb{Z}_3$, and $g : \mathbb{Z}_3 \to \{0, 1\}$ be any nonconstant function. Let a be the element of $\mathbb{Z}_3$ such that a is the unique preimage of $g(a)$. Let $(X_1, \ldots, X_n)$ be uniformly distributed over all elements of $\mathbb{Z}_3^n$ where the number of a's is divisible by 3. Then $\sum_i g(X_i) \mod 3$ is constant, but it can be shown that $(X_1, \ldots, X_n)$ is a $2^{-\Omega(n)}$-biased space.

For the purpose of this informal discussion we will assume that M is prime. (The idea for handling composite M's is to analyze each Fourier coefficient of the distribution of the sum separately. We defer further details to Section 2.1.)

Low-Weight Case. Let us first consider the case where the number of non-zero a_i's is at most $M' \cdot \log(1/\epsilon)$, for $M' = \text{poly}(M)$.[3] As before, we could use an almost k-wise independent distribution, but then our seed length would depend polynomially on M, while our goal is a polylogarithmic dependency.

First, we use a hash function to split the index set $[n] = \{1, 2, \ldots, n\}$ into $B = O(M')$ disjoint subsets T_j such that with high probability (say, $1 - \epsilon/10$) over the splitting, each set T_j contains at most $k = \log(1/\epsilon)$ indices i such that $a_i \neq 0$. We show that the selection of the hash function that determines the splitting can be done using $O(\log n + (\log M/\epsilon) \cdot \log(M \log 1/\epsilon))$ random bits.

Once we have this partition, it is sufficient to independently sample in each block from an ϵ/B-almost k-wise independent distribution, which requires $s = O(\log n + k + \log(B/\epsilon)) = O(\log n + \log(M/\epsilon))$ random bits per block. Then we argue that it is not necessary for the sampling in different blocks to be independent, and instead they can be sampled using a pseudorandom generator for space-bounded computation [Nis, INW]. (This relies on the fact the computation $\sum_i a_i X_i \bmod M$ can be performed in any order over the i's, in particular the order suggested by $\sum_j \sum_{i \in T_j} a_i \cdot X_i \bmod M$.) Using the INW generator, we can do all the sampling using $O(s + \log B \cdot (\log(B/\epsilon) + \log M)) = O(\log n + \log M \cdot \log(M/\epsilon))$ random bits.

High-Weight Case. We now discuss the generator that fools sums with more than $M' \cdot \log 1/\epsilon$ non-zero coefficients a_i, for $M' = \text{poly}(M)$. Here, we can think of the computation $\sum_i a_i X_i \bmod M$ as an n-step walk over $\mathbb{Z}_M$ that starts at 0. Unlike standard walks, *each step is taken on a different graph* (over the same set of vertices, namely $\mathbb{Z}_M$). Specifically, step i is taken on the (directed) Cayley graph where every node v has two outgoing edges. The first edge is labeled 0 and goes into v itself (*i.e.*, this edge is a self loop). The second edge is labeled 1 and goes into $v + a_i \bmod M$. Following the walk along the labels $X_1, X_2, \ldots, X_n$ arrives at the vertex $\sum_i a_i X_i \bmod M$. If the X_i's are uniform (*i.e.*, we are taking a random walk) then the end vertex will be almost uniformly distributed (because the number of steps is larger than $M^2 \cdot \log(1/\epsilon)$). What we are seeking is a pseudorandom walk that is generated using much fewer truly random bits but still converges to the uniform distribution (possibly slower, e.g. using $M' \cdot \log(1/\epsilon)$ steps).

Pseudorandom walk generators were constructed in [RTV, RV] for walks on a single regular and connected graph. In our case, we are walking not on a single graph but rather on a sequence of graphs, each of which is indeed regular. It turns out that the pseudorandom generators of [RTV, RV] still work for a

---

[3] In this preliminary version we did not try to optimize the various constants. In particular, in our analysis $M' = O(M^{24})$. We note that it can be made as small as $O(M^{2+\alpha})$ for any $\alpha > 0$.

sequence of graphs rather than a single graph. The more difficult aspect is that in our walk there is no uniform bound on the expansion of the graphs. Indeed, the graphs that correspond to $a_i = 0$ are not connected at all (they consist solely of self loops). In our setting, where the graphs are directed Cayley graphs for the abelian group $\mathbb{Z}_M$, we show how to generate pseudorandom walks on graphs with varying bounds on expansion.

We do this by a generalization of the derandomized graph product of [RV]. There, expanders are used to generate two steps on a degree-D graph using less than $2 \log D$ random bits, yet the (spectral) expansion of the resulting graph is almost as good as the square of the original graph. We analyze the analogous derandomization of two steps on two distinct (abelian Cayley) graphs for which we may have distinct bounds on their expansion. Moreover, to handle composite M, we show that the expansion can be analyzed in each eigenspace separately. (For example, for $\mathbb{Z}_6 = \mathbb{Z}_2 \times \mathbb{Z}_3$, a sequence of even coefficients a_i will yield a random walk that does not mix in the $\mathbb{Z}_2$ component, but may mix in the $\mathbb{Z}_3$ component, and our pseudorandom generator needs to preserve this property.)

To obtain our pseudorandom walk generator, we first randomly reorder the index set $[n]$ so that the nonzero coefficients are well-spread out, and then derandomize the walk by a recursive application of our aforementioned derandomized product. As discussed in [RV], the resulting pseudorandom walk generator is the same as the Impagliazzo–Nisan–Wigderson [INW] generator for space-bounded computation, with a different setting of parameters that enables a much smaller seed length than their analysis requires for general space-bounded algorithms.

Discussion

The natural open problem left by our work is to reduce the seed length further, ideally to $O(\log(nM/\epsilon))$, which can be shown to be possible via a nonconstructive probabilistic argument. For achieving such optimal parameters, the modular reduction is actually insignificant — it is equivalent to construct generators such that for every bounded coefficient vector $(a_1, \ldots, a_n) \in \mathbb{Z}^n$ where each $|a_i| \le M$, $\sum_i a_i X_i$ is statistically close to $\sum_i a_i R_i$ as distributions on $\mathbb{Z}$, where $(X_1, \ldots, X_n)$ is the output distribution of the generator, and $(R_1, \ldots, R_n)$ is the uniform distribution on $\{0, 1\}^n$. [4] As a result, such generators would also "fool" linear threshold functions (halfspaces) whose coefficients are polynomially bounded. Pseudorandom generators and related objects for threshold functions (with no bound on the coefficients) have recently been studied in [RS, DGJ+], with the latter achieving seed length $O((\log n) \cdot \log^2(1/\epsilon)/\epsilon^2)$.

2   Definitions and Tools

We denote by U_n the uniform distribution over $\{0, 1\}^n$. We fix an integer $M \ge 2$ for the rest of the paper. We will be interested in constructing pseudorandom bit

---

[4] Indeed, given any coefficient vector $(a_1, \ldots, a_n) \in \mathbb{Z}^n$, where each $|a_i| \le M$, we can apply the generator for modulus $M' = M \cdot n$ so that no modular reduction occurs.

generators that fool sums modulo M. We denote by $\mathbb{Z}_M$ the set $\{0, 1, \ldots, M-1\}$ with arithmetic modulo M. Due to space limitations, we defer many of the proofs to the full version of the paper.

Definition 1. *The statistical distance between two random variables X, Y taking values in $\mathbb{Z}_M$ is $\mathrm{dist}(X, Y) = \frac{1}{2} \sum_{i=0}^{M-1} |\Pr[X = i] - \Pr[Y = i]|$. The variables X and Y are said to be ϵ-close if their statistical distance is at most ϵ.*

Definition 2. *A random variable $X = (X_1, \ldots, X_n)$ taking values in $\{0,1\}^n$ is ϵ-pseudorandom against sums modulo M if for any $a_1, \ldots, a_n \in \mathbb{Z}_M$, the distribution of $a_1 X_1 + \cdots + a_n X_n$ modulo M, is ϵ-close (in statistical distance) to the distribution $a_1 R_1 + \cdots + a_n R_n$ modulo M, where $R_1, \ldots, R_n$ are uniform and independent random bits.*

Definition 3. *A function $G : \{0,1\}^r \to \{0,1\}^n$ is an ϵ-pseudorandom bit generator against sums modulo M if the distribution $G(U_r)$ is ϵ-pseudorandom against sums modulo M.*

Note that ϵ-biased generators is a special case of the definition of pseudorandom bit generators against sums modulo M, for $M = 2$.

Our goal is to build generators that fool sums modulo M, where M can be either prime or composite. Handling prime modulus is somewhat easier, and the approach in the following section allows handling both cases simultaneously. We will show that it is enough to construct pseudorandom generators which fools the bias of a sum modulo M, and under this approach, there is no major difference between primes and composites.

2.1 Small Bias Bit Generators

First we define the *bias* of a linear combination with coefficients $a_1, \ldots, a_n \in \mathbb{Z}_M$, given some distribution of $X = (X_1, \ldots, X_n) \in \{0,1\}^n$:

Definition 4. *Let $X = (X_1, \ldots, X_n)$ be a distribution over $\{0,1\}^n$, and $(a_1, \ldots, a_n) \in \mathbb{Z}_M^n$ a coefficient vector. We define the bias of $a_1, \ldots, a_n$ according to X to be*

$$\mathrm{bias}_X(a_1, .., a_n) = \mathbb{E}\left[\omega^{\sum a_i X_i}\right]$$

where $\omega = e^{2\pi i/M}$ is a primitive M-th root of unity.

Notice that the bias can in general be a complex number, of absolute value at most 1.

Definition 5. *We say a distribution $X = (X_1, \ldots, X_n)$ over n bits is ϵ-bit-biased against sums modulo M if for every coefficient vector $(a_1, \ldots, a_n) \in \mathbb{Z}_M^n$,*

$$|\mathrm{bias}_X(a_1, \ldots, a_n) - \mathrm{bias}_{U_n}(a_1, \ldots, a_n)| \le \epsilon$$

Let $G : \{0,1\}^r \to \{0,1\}^n$ be a bit generator. We shorthand $\mathrm{bias}_G(a_1, \ldots, a_n)$ for $\mathrm{bias}_{G(U^r)}(a_1, \ldots, a_n)$.

Definition 6. $G : \{0,1\}^r \to \{0,1\}^n$ *is an ϵ-bit-biased generator against sums modulo M if the distribution $G(U_r)$ is ϵ-bit-biased against sums modulo M. That is, for every coefficient vector $(a_1, \dots, a_n)$,*

$$|\mathrm{bias}_G(a_1, \dots, a_n) - \mathrm{bias}_{U_n}(a_1, \dots, a_n)| \leq \epsilon$$

The name "bit-biased" in the above definitions is meant to stress the difference from standard ϵ-biased generators modulo M. Here we compare the bias under the generator to the bias under uniformly selected bits (rather than uniformly selected elements in $\mathbb{Z}_M$).

We first reduce the problem of constructing pseudorandom modular generators to that of constructing ϵ-bit-biased modular generators.

Lemma 1. *Let $X = (X_1, \dots, X_n)$ be an ϵ-bit-biased distribution against sums modulo M. Then X is $(\epsilon\sqrt{M})$-pseudorandom against sums modulo M.*

From now on, we focus on constructing ϵ-bit-biased generators. We will need to differentiate two types of linear combinations, based on the number on non-zero terms in them.

Definition 7. *The weight of a coefficient vector $(a_1, \dots, a_n) \in \mathbb{Z}_M^n$ is the number of non-zero coefficients a_i.*

We will construct two generators: one fooling linear combination with small weights, and the other fooling linear combinations with large weight. Our final generator will be the be the bitwise-XOR of the two, where each is chosen independently. The following lemma shows this will result in an ϵ-bit-biased generator fooling all linear combinations.

Lemma 2. *Fix a weight threshold W. Let $X' = (X'_1, \dots, X'_n)$ be a distribution over $\{0,1\}^n$ such that for any vector coefficient $a_1, \dots, a_n$ of weight at most W,*

$$|\mathrm{bias}_{X'}(a_1, \dots, a_n) - \mathrm{bias}_{U_n}(a_1, \dots, a_n)| \leq \epsilon.$$

Let $X'' = (X''_1, \dots, X''_n)$ be a distribution over $\{0,1\}^n$ such that for any vector coefficient $a_1, \dots, a_n$ of weight at least W,

$$|\mathrm{bias}_{X'}(a_1, \dots, a_n) - \mathrm{bias}_{U_n}(a_1, \dots, a_n)| \leq \epsilon.$$

Let X be the bitwise-XOR of two independent copies of X' and X'', i.e.

$$X = X' \oplus X'' = (X'_1 \oplus X''_1, \dots, X'_n \oplus X''_n).$$

Then X is ϵ-bit-biased against sums modulo M.

Convergence of the Bias for Large Weights. The bias of a coefficient vector with respect to the uniform distribution can be large if there are only a few non-zero elements in the vector. However, when the weight is large, the bias is guaranteed to be small.

Lemma 3. *Let $(a_1, \ldots, a_n) \in \mathbb{Z}_M^n$ be a coefficient vector of weight w. Then*

$$|\mathrm{bias}_U(a_1, \ldots, a_n)| \leq \left(1 - \frac{1}{M^2}\right)^w$$

In particular, for $w \geq M^2 \log(1/\epsilon)$ the bias is at most $\epsilon/2$.

Notice that the above lemma holds for all coefficient vectors $(a_1, \ldots, a_n)$ and moduli M, even when M is composite and the coefficients are not relatively prime to M. For example, when $M = 6$ and $(a_1, \ldots, a_n) = (2, \ldots, 2)$. In such a case, $\sum_i a_i R_i \mod M$ does not converge to the uniform distribution on $\mathbb{Z}_M^n$, but the above lemma still says that the bias tends to zero.

A similar result holds if we consider the bias of a large weight coefficient vector under a skewed distribution.

Lemma 4. *Let $(a_1, \ldots, a_n) \in \mathbb{Z}_M^n$ be a coefficient vector of weight w. Let $Z_1, \ldots, Z_n \in \{0, 1\}$ be independently distributed with $\Pr[Z_i = 0] = (1 + \alpha)/2$. Then*

$$|\mathrm{bias}_{Z_1, \ldots, Z_n}(a_1, \ldots, a_n)| \leq \left(1 - \Omega\left(\frac{1 - \alpha^2}{M^2}\right)\right)^w$$

In particular, for $w \geq cM^2 \log(1/\epsilon)/(1 - \alpha^2)$ for a sufficiently large constant c, the bias is at most $\epsilon/2$.

2.2   Hashing

We use hashing as one of the ingredients in our construction. A family (multi-set) of functions $\mathcal{H} = \{h : [n] \to [k]\}$ is called a family of hash functions, if a randomly chosen function from the family behaves pseudorandomly under some specific meaning. We consider a hash function $H : [n] \to [k]$ to be a random variable depicting a randomly chosen function from the family. We say H can be generated *efficiently and explicitly* using s random bits, if a random function in the family can be sampled by a randomized polynomial-time algorithm using s random bits, and this function can be evaluated using a deterministic polynomial-time algorithm.

Fix $S \subset [n]$. We define the j-th bucket of H with respect to S, to be the set of elements of S mapped by H into j, i.e. $\{s \in S : H(s) = j\} = H^{-1}(j) \cap S$.

We will use the following three constructions of hash functions.

Lemma 5. *Assume k is a power of 2. There exists a hash function $H_1 : [n] \to [k]$ such that for every set $S \subset [n]$ of size at most $k \log(1/\epsilon)$, the probability that H_1 has a bucket $H_1^{-1}(j) \cap S$ with more than $100 \log(1/\epsilon)$ elements is at most $\epsilon/100$. Moreover, H_1 can be generated explicitly and efficiently using $O(\log n + \log(k/\epsilon) \log(k \log(1/\epsilon)))$ random bits.*

Lemma 6. *Assume k is a power of 2. There exists a hash function $H_2 : [n] \to [k]$ such that for every $S \subset [n]$ of size at least $100k^2$, the probability that H_2 has an empty bucket $H_2^{-1}(j) \cap S$ is at most $1/100$. Moreover, H_2 can be generated explicitly and efficiently using $O(\log n + \log^2 k)$ random bits.*

Lemma 7. *There exists a hash function* $H_3 : [n] \to [16\log(1/\epsilon)]$ *such that for every* $S \subset [n]$ *of size at least* $800k\log(1/\epsilon)$, *the probability that* H_3 *has at least* $\log(1/\epsilon)$ *buckets* $H_3^{-1}(j) \cap S$ *with at most* k *elements is at most* $\epsilon/100$. *Moreover,* H_3 *can be generated explicitly and efficiently using* $O(\log n + \log(1/\epsilon)\log(k\log(1/\epsilon)))$ *random bits.*

The constructions of the hashes in Lemmas 5, 6 and 7 are based on almost t-wise independence. A sequence of random variables $X_1, \ldots, X_n \in \{0, 1\}$ is said to be *t-wise independent* if any t random variables in it are independent. It is said to be δ-almost t-wise independent if any t random variables in it are δ-close in statistical distance to independent. Explicit constructions of δ-almost t-wise independent distributions are known, with nearly optimal seed length [NN, AGHP].

We identify a function $h : [n] \to [\ell]$, where ℓ is a power of 2, by a sequence of $n \log \ell$ bits. We construct the hash functions by choosing the sequence of bits according to an δ-almost t-wise independent distribution, where the values of δ and t differ in the three constructions. The main tool in our analysis is a tail bound on t-wise independent distributions, due to Bellare and Rompel [BR], extended to the case of δ-almost t-wise distributions. We defer further details to the full version of the paper.

2.3 Pseudorandom Generators for Small Space

An ingredient in our construction is the small-space pseudorandom generator of Impagliazzo, Nisan, and Wigderson [INW]. We first define branching programs, which form a non-uniform model of small-space computations.

Definition 8. *A* (read-once, oblivious) *branching program of length* n, *degree* d *and width* w *is a layered graph with* $n + 1$ *layers, where each layer contains at most* w *vertices. ¿From each vertex in the* i-th *layer* $(1 \leq i \leq n)$ *there are* d *outgoint edges, numbered* $0, 1, \ldots, d - 1$. *A vertex in the first layer is designated as the* start *vertex. Running the branching program on an input* $x_1, \ldots, x_n \in [d]$ *is done by following the path according to the inputs, starting at the start vertex. The* output *of the branching program is the vertex reached in the last layer.*

Definition 9. *A pseudorandom generator for branching programs of length* n, *degree* d *and width* w *with error* ϵ *is a function* $G : \{0, 1\}^r \to [d]^n$, *such that for every branching program of length* n, *degree* d *and width* w, *the statistical distance between the output of the branching program when run on uniform element in* $[d]^n$, *and the output when run on* $G(U_r)$, *is at most* ϵ.

Lemma 8. *[INW] There exists an explicit pseudorandom generators for branching programs of length* n, *degree* d, *width* w *with error* ϵ, *which uses* $r = O(\log d + (\log n)(\log(n/\epsilon) + \log w))$ *truly random bits.*

3   Construction Using PRG for Low-Degree Polynomials

We present in this section a simple construction for prime powers M, based on pseudorandom generators for low-degree polynomials. This construction is optimal for constant M, achieving a pseudorandom generator with seed length $O_M(\log(1/\epsilon))$ (where the constant depends exponentially on M).

Let $W = \Omega(M^3 \log 1/\epsilon)$. We will construct two generators: one for coefficient vectors of weight at most W, and one for coefficient vectors of weight at least W. Lemma 2 shows that the bitwise-XOR of the two generators is a pseudorandom generator for all coefficient vectors.

For small weights, we will use a distribution that is ϵ-almost W-wise independent. Such a distribution trivially fools coefficient vectors of weight at most W. It can be explicitly generated using $O(\log n + W + \log 1/\epsilon) = O_M(\log n/\epsilon)$ random bits [NN].

For large weights, let $(a_1, \ldots, a_n) \in \mathbb{Z}_M^n$ be a coefficient vector of weight at least W. Consider first the distribution of $a_1 R_1 + \ldots a_n R_n$ for independent and uniform bits $R_1, \ldots, R_n$. By Lemma 3, $|\mathrm{bias}_{U_n}(a_1, \ldots, a_n)| < \epsilon/2$.

Consider now $Z_i \in \{0,1\}$, where $\Pr[Z_i = 0] = c/M$ for some integer $1 \leq c \leq M - 1$. By Lemma 4,

$$|\mathrm{bias}_{Z_1, \ldots, Z_n \sim (c/M, 1-c/M)}(a_1, \ldots, a_n)| < \epsilon/4,$$

given that $W = \Omega(M^3 \log(1/\epsilon))$ with a large enough hidden constant.

The benefit of using this skewed distribution, is that it can be simulated by low-degree polynomials modulo M. Since we assume M is a prime power, there is a polynomial $g : \mathbb{Z}_M \to \mathbb{Z}_M$ that maps some c elements of $\mathbb{Z}_M$ to 0, and the rest to 1. For example, if $M = p^k$, the polynomial $g(x) = x^{(p-1)p^{k-1}}$ maps elements divisible by p to 0, and the rest to 1. The degree of this g is at most $M - 1$.

Let $Z_1, \ldots, Z_n \in \{0,1\}^n$ be generated by $g(Y_1), \ldots, g(Y_n)$, where $Y_1, \ldots, Y_n \in \mathbb{Z}_M$ are uniform and independent. We thus have:

$$|\mathrm{bias}_{Z_1, \ldots, Z_n \sim g(U_{Z_M})^n}(a_1, \ldots, a_n)| < \epsilon/4$$

Note that

$$\mathrm{bias}_{Z_1, \ldots, Z_n \sim g(U_{Z_M})^n}(a_1, \ldots, a_n) = \mathbb{E}_{Y_1, \ldots, Y_n \in \mathbb{Z}_M}[\omega^{a_1 g(Y_1) + \cdots + a_n g(Y_n)}],$$

and that $a_1 g(Y_1) + \cdots + a_n g(Y_n)$ is a polynomial of degree $\deg(g)$ in $Y_1, \ldots, Y_n$. Thus we can derandomize the choice of $Y_1, \ldots, Y_n$ using a a pseudorandom generator for low-degree polynomials [BV, Lov, Vio]. We note the results in these papers are stated for polynomials over prime finite fields, but they hold also for polynomials over $\mathbb{Z}_M$, using small-bias spaces for $\mathbb{Z}_M^n$ [Kat, AIK+, RSW, EGL+, AM] as a building block.

Lemma 9. *For every $M, n, d \in \mathbb{N}$, there is an explicit generator $G : \{0,1\}^r \to \mathbb{Z}_M^n$ such that for every polynomial $f : \mathbb{Z}_M^n \to \mathbb{Z}_M$ of degree at most d, the distribution of $f(\mathbb{Z}_M^n)$ and $f(G(U_r))$ are ϵ-close in statistical distance. The number of random bits required is $r = O(d2^d \log(M/\epsilon) + d \log(nM))$.*

We use the generator of Lemma 9 for error $\epsilon/4$ and degree $d = M - 1$. We thus get an explicit generator whose output distribution $(Y_1', \ldots, Y_n') \in \mathbb{Z}_M^n$, such that:

$$|\mathbb{E}_{(Y_1', \ldots, Y_n')}[\omega^{a_1 g(Y_1') + \ldots + a_n g(Y_n')}] - \mathbb{E}_{Y_1, \ldots, Y_n \in \mathbb{Z}_M^n}[\omega^{a_1 g(Y_1) + \ldots + a_n g(Y_n)}]| < \epsilon/4$$

Thus, if we define our generator G' to output $g(Y_1'), \ldots, g(Y_n')$, we have $Y_1', \ldots, Y_n'$ are the output of G, we get an explicit generator, such that $|\text{bias}_{G'}(a_1, \ldots, a_n)| < \epsilon/2$. Hence, we get that

$$|\text{bias}_{G'}(a_1, \ldots, a_n) - \text{bias}_G(a_1, \ldots, a_n)| < \epsilon$$

The randomness requirement of our generator comes directly from that of G, which is $O(M 2^{M-1} \log(M/\epsilon) + M \log(nM)) = O_M(\log(n/\epsilon))$ for constant M.

4   Construction Based on Pseudorandom Walk Generators

4.1   A Generator for Small Sums

We construct an ϵ-bit-biased generator for weights at most $W = 10^5 M^{24} \log(1/\epsilon)$. Let $(a_1, \ldots, a_n) \in \mathbb{Z}_M^n$ be a coefficient vector of weight at most W.

The construction has three stages:

1. Partitioning the set of indices $[n]$ into W buckets using the hash function H_1. Lemma 5 guarantees that with probability at least $1 - \epsilon/100$, each bucket contains at most $O(\log(1/\epsilon))$ non-zero coefficients.
2. For each bucket j, generate the X_i's for i's in the j'th bucket using an almost $O(\log(1/\epsilon))$-wise independent distribution.
3. Use the INW generator given by Lemma 8 to generate the W seeds for the $O(\log(1/\epsilon))$-wise independent distributions used for the different buckets.

Lemma 10. *The above construction is an ϵ-bit-biased generator against coefficient vectors of weight at most W, using $O(\log n + \log(M/\epsilon) \log(M \log(1/\epsilon)))$ random bits.*

4.2   A Generator for Large Sums

In this section we construct an ϵ-bit-biased distribution for coefficient vectors of weight at least $W = 10^5 M^{24} \log(1/\epsilon)$,

Recall that by Lemma 3, when the weight is large, the bias under the uniform distribution is small. Thus, to prove that a distribution is ϵ-bit-biased against large weight sums modulo M, it is enough to show that its bias is also small. We construct our ϵ-bit-biased generator in three steps:

- G_1: a generator that has bias at most $1 - 1/M^2$ on every coefficient vector which is not all zeros.
- G_2: a generator that has bias at most 0.91 on every coefficient vector of weight at least $100 M^{24}$.
- G_3: a generator that has bias at most $\epsilon/2$ on every coefficient vector of weight at least $10^5 M^{24} \log 1/\epsilon$.

The generator G_3 will be our ϵ-bit-biased generator for large weights. We will sketch the constructions of G_1, G_2 and G_3, deferring full details and proofs to the full version of the paper. The main ingredient in the construction will be a derandomized expander product, which we now define and analyze.

Derandomized Expander Products

Definition 10. *We say an undirected graph H is a $(2^r, 2^d, \lambda)$-expander if H has 2^r vertices, it is regular of degree 2^d and all eigenvalues but the first have absolute value at most λ. We will identify the vertices of H with $\{0,1\}^r$, and the edges exiting each vertex with $\{0,1\}^d$ in some arbitrary way.*

We will need explicit constructions of expanders, which can be obtained from various known constructions.

Lemma 11. *For some constant $Q = 2^q$, there exist an efficient sequence H_k of $(Q^k, Q, 1/100)$-expanders.*

Impagliazzo, Nisan, and Wigderson [INW] compose two pseudorandom generators using an expander as follows:

Definition 11. *Let $G', G'' : \{0,1\}^r \to \{0,1\}^t$ be two bit generators. Let H be a $(2^r, 2^d, \lambda)$-expander. We define $G' \otimes_H G'' : \{0,1\}^{r+d} \to \{0,1\}^{2t}$ to be the concatenation $(G'(x), G''(y))$, where x is a random vertex in H, and y is a random neighbor of x in H.*

Our main lemma relates the bias of $G' \otimes_H G''$ to the biases of G' and G'':

Lemma 12. *Let $G', G'' : \{0,1\}^r \to \{0,1\}^t$ be two bit generators and let H be a $(2^r, 2^d, \lambda)$-expander. Let $(a_1, \ldots, a_t), (b_1, \ldots, b_t)$ be two coefficient vectors. Then:*

$$|\mathrm{bias}_{(G' \otimes_H G'')(U_{r+d})}(a_1, \ldots, a_t, b_1, \ldots, b_t)|$$
$$\leq f_\lambda(|\mathrm{bias}_{G'(U_r)}(a_1, \ldots, a_t)|, |\mathrm{bias}_{G''(U_r)}(b_1, \ldots, b_t)|)$$

where $f_\lambda(x, y) = xy + \lambda \sqrt{1 - x^2} \sqrt{1 - y^2}$.

The bounds of [RV] imply that if $\max_{k \in \mathbb{Z}_M \setminus 0} |\mathrm{bias}_{G'(U_r)}(ka_1, \ldots, ka_t)| \leq x$ then $\max_{k \in \mathbb{Z}_M \setminus 0} |\mathrm{bias}_{(G' \otimes_H G')(U_{r+d})}(a_1, \ldots, a_t, a_1, \ldots, a_t)| \leq x^2 + \lambda \cdot (1 - x^2) = f_\lambda(x, x)$. If also $\max_{k \in \mathbb{Z}_M \setminus 0} |\mathrm{bias}_{G''(U_r)}(kb_1, \ldots, kb_t)| \leq y$, then [RV] proof can be extended to show $\max_{k \in \mathbb{Z}_M \setminus 0} |\mathrm{bias}_{(G' \otimes_H G')(U_{r+d})}(ka_1, \ldots, ka_t, kb_1, \ldots, kb_t)| \leq xy + \lambda \cdot (1 - xy)$, which is a worse than our bound $f(x, y)$ in case $x \neq y$ and does not suffice for our purposes. In addition, our result only requires a bound on the bias for the specific coefficient vectors $(a_1, \ldots, a_t), (b_1, \ldots, b_t)$ of interest, and not multiples of those coefficient vectors; this is crucial for our analysis when M is composite (cf., discussion after Lemma 3). On the other hand, the results of [RV] are more general in that they apply to generators G', G'' that correspond to random walks on any expander, not just Cayley graphs of $\mathbb{Z}_M$.

Construction of G_1. As in [INW, RV], we iterate the above product. Like [RV] we can use the constant-degree expander graphs $H_1, H_2, \ldots$ of Lemma 11 (as opposed to the expanders of degree poly(nw/ϵ) used by [INW] to prove Lemma 8). We define $G'_\ell : \{0,1\}^{\ell q} \to \{0,1\}^{2^{\ell-1}q}$ iteratively. $G'_1 : \{0,1\}^q \to \{0,1\}^q$ is the identity mapping, and $G'_\ell = G'_{\ell-1} \otimes_{H_{\ell-1}} G'_{\ell-1}$. We set $G_1 = G'_\ell$ for the minimal ℓ such that $2^{\ell-1}q \geq n$. We have:

Lemma 13. *Let* $(a_1, \ldots, a_n) \in \mathbb{Z}_M^n$ *be a coefficient vector, which is not all zeros. Then:*

$$\mathrm{bias}_{G_1}(a_1, \ldots, a_n) \leq 1 - \frac{1}{M^2}.$$

The seed-length of G_1 is $O(\log n)$.

Construction of G_2. We will construct G_2 based on G_1. Let $(a_1, \ldots, a_n)$ be a coefficient vector. Assume first a special case: Let $n = k2^s$, and partition the set of coefficients into 2^s consecutive parts, each of size k. Assume that each part contain at least one non-zero coefficient. By Lemma 13, applying G_1 to each part independently gives bias of at most $1 - 1/M^2$. We use this to analyze the bias of G_1 when applied in the special case:

Lemma 14. *Let $n = k2^s$. Let $a_1, \ldots, a_n$ be a coefficient vector such that for every $j \in [2^s]$, weight$(a_{jk+1}, a_{jk+2}, \ldots, a_{(j+1)k}) > 0$. Then:*

$$\mathrm{bias}_{G_1}(a_1, \ldots, a_n) \leq \min\left(1 - \left(\frac{9}{8}\right)^s \frac{1}{M^2}, 0.9\right).$$

In particular if $s \geq 12 \log M$, we have $\mathrm{bias}_{G_1}(a_1, \ldots, a_n) \leq 0.9$.

We now construct the generator G_2 in three steps:

- Obliviously partition the coefficients, using the hash function H_2. Re-order the coefficients according to the partition. This guarantees that with probability at least 0.99, the conditions of Lemma 14 hold.
- Use G_1 on the re-ordered coefficients.
- Return the pseudorandom bits back to the original order.

We have:

Lemma 15. *Let $(a_1, \ldots, a_n) \in \mathbb{Z}_M^n$ be a coefficient vector, of weight at least $100M^{24}$. Then:*

$$\mathrm{bias}_{G_2}(a_1, \ldots, a_n) \leq 0.91.$$

The seed length of G_2 is $O(\log n + \log^2 M)$.

Construction of G_3. We use G_2 to build our final ϵ-bit-biased generator G_3. The construction of G_3 has three parts:

- Use H_3 to partition the inputs to $O(\log(1/\epsilon))$ buckets, such that with probability $1 - \epsilon/100$, most buckets contain at least $100M^{24}$ non-zero coefficients.
- Use G_2 on each bucket.
- Combine the generators for the separate buckets using expander products, with expanders of growing degree as in [RV].

Lemma 16. *Let $(a_1, \ldots, a_n) \in \mathbb{Z}_M^n$ be a coefficient vector, of weight at least $10^5 M^{24} \log(1/\epsilon)$. Then:*

$$\mathrm{bias}_{G_3}(a_1, \ldots, a_n) \leq \epsilon/2.$$

The randomness required by G_3 is $O(\log n + \log(M/\epsilon) \log(M \log(1/\epsilon)))$.

Acknowledgments

We thank Emanuele Viola for drawing our attention to this problem. We thank Andrej Bogdanov for helpful discussions.

References

[AIK+]   Ajtai, M., Iwaniec, H., Komlós, J., Pintz, J., Szemerédi, E.: Construction of a thin set with small Fourier coefficients. Bull. London Math. Soc. 22(6), 583–590 (1990)

[AKS]   Ajtai, M., Komlós, J., Szemerédi, E.: Deterministic Simulation in LOGSPACE. In: Proceedings of the Nineteenth Annual ACM Symposium on Theory of Computing, New York City, pp. 132–140 (1987)

[AGHP]   Alon, N., Goldreich, O., Håstad, J., Peralta, R.: Simple constructions of almost k-wise independent random variables. Random Structures & Algorithms 3(3), 289–304 (1992)

[AM]   Alon, N., Mansour, Y.: ϵ-discrepancy sets and their application for interpolation of sparse polynomials. Information Processing Letters 54(6), 337–342 (1995)

[AR]   Alon, N., Roichman, Y.: Random Cayley graphs and expanders. Random Structures Algorithms 5(2), 271–284 (1994)

[BNS]   Babai, L., Nisan, N., Szegedy, M.: Multiparty protocols, pseudorandom generators for logspace, and time-space trade-offs. Journal of Computer and System Sciences, 204–232 (1989)

[BR]   Bellare, M., Rompel, J.: Randomness-Efficient Oblivious Sampling. In: 35th Annual Symposium on Foundations of Computer Science, Santa Fe, New Mexico, pp. 276–287. IEEE, Los Alamitos (1994)

[BSVW]   Ben-Sasson, E., Sudan, M., Vadhan, S., Wigderson, A.: Randomness-efficient low degree tests and short PCPs via epsilon-biased sets. In: Proceedings of the Thirty-Fifth Annual ACM Symposium on Theory of Computing, pp. 612–621. ACM, New York (2003) (electronic)

[BV]   Bogdanov, A., Viola, E.: Pseudorandom Bits for Polynomials. In: FOCS, pp. 41–51. IEEE Computer Society Press, Los Alamitos (2007)

[DGJ+]   Diakonikolas, I., Gopalan, P., Jaiswal, R., Servedio, R.A., Viola, E.: Bounded Independence Fools Halfspaces. CoRR abs/0902.3757 (2009)

[EGL+]   Even, G., Goldreich, O., Luby, M., Nisan, N., Veličković, B.: Efficient approximation of product distributions. Random Structures Algorithms 13(1), 1–16 (1998)

[HPS]   Håstad, J., Phillips, S., Safra, S.: A well-characterized approximation problem. Information Processing Letters 47(6), 301–305 (1993)

[INW]   Impagliazzo, R., Nisan, N., Wigderson, A.: Pseudorandomness for Network Algorithms. In: Proceedings of the Twenty-Sixth Annual ACM Symposium on the Theory of Computing, Montréal, Québec, Canada, pp. 356–364 (1994)

[Kat]   Katz, N.M.: An estimate for character sums. Journal of the American Mathematical Society 2(2), 197–200 (1989)

[Lov]   Lovett, S.: Unconditional pseudorandom generators for low degree polynomials. In: Ladner, R.E., Dwork, C. (eds.) STOC, pp. 557–562. ACM, New York (2008)

[MZ]   Meka, R., Zuckerman, D.: Small-Bias Spaces for Group Products. In: Dinur, I., et al. (eds.) APPROX and RANDOM 2009. LNCS, vol. 5687, Springer, Heidelberg (2009)

[MST]   Mossel, E., Shpilka, A., Trevisan, L.: On ϵ-biased generators in NC^0. Random Structures Algorithms 29(1), 56–81 (2006)

[MNN]   Motwani, R., Naor, J., Naor, M.: The probabilistic method yields deterministic parallel algorithms. Journal of Computer and System Sciences 49(3), 478–516 (1994)

[NN]   Naor, J., Naor, M.: Small-Bias Probability Spaces: Efficient Constructions and Applications. SIAM Journal on Computing 22(4), 838–856 (1993)

[Nao]   Naor, M.: Constructing Ramsey graphs from small probability spaces. Technical Report RJ 8810, IBM Research Report (1992)

[Nis]   Nisan, N.: Pseudorandom generators for space-bounded computation. Combinatorica 12(4), 449–461 (1992)

[RS]   Rabani, Y., Shpilka, A.: Explicit construction of a small epsilon-net for linear threshold functions. In: Mitzenmacher, M. (ed.) STOC, pp. 649–658. ACM, New York (2009)

[RSW]   Razborov, A., Szemerédi, E., Wigderson, A.: Constructing small sets that are uniform in arithmetic progressions. Combinatorics, Probability and Computing 2(4), 513–518 (1993)

[RTV]   Reingold, O., Trevisan, L., Vadhan, S.: Pseudorandom Walks In Regular Digraphs and the RL vs. L problem. In: Proceedings of the 38th Annual ACM Symposium on Theory of Computing, STOC 2006, May 21-23 (2006); Preliminary version on *ECCC* (February 2005)

[RV]   Rozenman, E., Vadhan, S.: Derandomized Squaring of Graphs. In: Chekuri, C., Jansen, K., Rolim, J.D.P., Trevisan, L. (eds.) APPROX and RANDOM 2005. LNCS, vol. 3624, pp. 436–447. Springer, Heidelberg (2005)

[SZ]   Saks, M., Zhou, S.: $BP_HSPACE(S) \subseteq DSPACE(S^{3/2})$. Journal of Computer and System Sciences 58, 376–403 (1999)

[Vio]   Viola, E.: The Sum of d Small-Bias Generators Fools Polynomials of Degree d. In: IEEE Conference on Computational Complexity, pp. 124–127. IEEE Computer Society, Los Alamitos (2008)

The Glauber Dynamics for Colourings of Bounded Degree Trees

Brendan Lucier[1], Michael Molloy[2], and Yuval Peres[3]

[1] Dept of Computer Science, University of Toronto
blucier@cs.toronto.edu
[2] Dept of Computer Science, University of Toronto
molloy@cs.toronto.edu
[3] Microsoft Research
peres@microsoft.com

Abstract. We study the Glauber dynamics Markov chain for k-colourings of trees with maximum degree Δ. For $k \geq 3$, we show that the mixing time on *every* tree is at most $n^{O(1+\Delta/(k \log \Delta))}$. This bound is tight up to the constant factor in the exponent, as evidenced by the complete tree. Our proof uses a weighted canonical paths analysis and a variation of the block dynamics that exploits the differing relaxation times of blocks.

1 Introduction

The Glauber dynamics is a Markov chain over configurations of spin systems on graphs, of which k-colourings is a special case. Such chains have generated a great deal of interest for a variety of reasons. For one thing, counting k-colourings is a fundamental #P-hard problem, and Markov chains that sample colourings can be used to obtain an FPRAS to approximately count them. For another, k-colourings are equivalent to the antiferromagnetic Potts model from statistical physics, and there is a large body of research into this and similar models.

The Glauber dynamics has received a very large part of this interest (see eg. [12]). It is particularly appealing because it is a natural and simple algorithm and it underlies more substantial procedures such as block dynamics and systematic scan (see [12,5]). It is also commonly used in practice, eg. in simulations, and is closely related to other important areas such as infinite-volume Gibbs distributions [2,10,14]. It is generally conjectured that the Glauber dynamics mixes in polynomial time on every graph of maximum degree Δ so long as $k \geq \Delta + 2$. Vigoda [19] has shown polynomial mixing time for $k \geq \frac{11}{6}\Delta$.

The focus of this paper will be the performance of the Glauber dynamics on trees. Of course, the task of sampling a k-colouring of a tree is not particularly difficult. Nevertheless, people have studied the Glauber dynamics on trees as a means of understanding its performance on more general graphs, and because

---

* This extended abstract presents two pieces of work. The first [13] proves the case $k \geq 4$ (amongst other things); it has been submitted to a journal. The second covers the case $k = 3$; a full version is in progress.

I. Dinur et al. (Eds.): APPROX and RANDOM 2009, LNCS 5687, pp. 631–645, 2009.
© Springer-Verlag Berlin Heidelberg 2009

the performance on trees is particularly relevant to related areas such as Gibbs distributions. Berger et al. [1] showed that the Glauber dynamics mixes in poly-nomial time on complete trees of maximum degree Δ, and Martinelli et al. [14] showed that this polynomial is $O(n \log n)$ so long as $k \geq \Delta + 2$.

Hayes, Vera and Vigoda [7] showed that it mixes in polytime for all planar graphs if $k \geq C\Delta/\log \Delta$ for a particular constant C. They remarked that this was best possible, up to the value of C: The chain takes superpolynomial time on every tree when $k = o(\Delta/\log n)$, and hence trees with $\Delta \geq n^\epsilon$ provide lower-bound examples for any constant ϵ. They asked whether such examples exist for smaller values of Δ; in particular, is the mixing time superpolynomial for the complete $(\Delta - 1)$-ary tree with $k = 3$ and $\Delta = O(1)$?

Proposition 2.5 of Berger et al. [1] shows that the mixing time is polynomial for every constant $k \geq 3$ and $\Delta \geq 2$ (in fact, it shows this for general particle systems on trees for which the Glauber dynamics is ergodic, of which proper colouring is a special case). Independently, Goldberg, Jerrum and Karpinski [6] and Lucier and Molloy [13] showed a lower bound of $n^{\Omega(1+\Delta/k \log \Delta)}$ on the mixing time for the case of the complete tree. Goldberg, Jerrum and Karpinski also give an upper bound of $n^{O(1+\Delta/\log \Delta)}$ for complete trees and constant Δ.

Our main result is an upper bound for *every* tree. Our bound is asymptotically tight, matching the lower bound up to a constant factor in the degree.

Theorem 1. *For $k \geq 3$, the Glauber dynamics on k-colourings of any tree with maximum degree Δ mixes in time at most $n^{O(1+\Delta/k \log \Delta)}$.*

Thus, for every $k \geq 3$ and $\Delta = O(1)$, we have polytime mixing on every tree. But if Δ grows with n, no matter how slowly, then on some trees (eg. complete trees) we require the $\Omega(\Delta/\log \Delta)$ colours for polytime mixing that Hayes, Vera and Vigoda noted were required at $\Delta = n^\epsilon$.

Let us describe the difficulties that occur when $k = o(\Delta/\log \Delta)$. If $k \geq \Delta + 2$ then no vertex will ever be *frozen*; i.e. there will always be at least one colour that it can switch to. (It also corresponds to the threshold for unique infinite-volume Gibbs distributions[10].) Much of the difficulty in showing rapid mixing for smaller values of k is in dealing with frozen variables. From this perspective, $k \geq C\Delta/\log \Delta$ for $C > 1$ is another natural threshold: if the neighbours of a vertex are assigned independently random colours then we expect that the vertex will not be frozen. But if $k < (1 - \epsilon)\Delta/\log \Delta$, then even in the steady state distribution most degree Δ vertices on a tree will be frozen.

If the children of a vertex u change colours enough times, then eventually u will become unfrozen and change colours. This allows vertices to unfreeze, level by level, much like in the level dynamics of [7]. This is a slow process: the number of times that the children of u have to change before u is unfrozen is (roughly) exponential in Δ/k. However, this value is manageable for $\Delta = O(1)$: the running time is a high degree polynomial rather than superpolynomial. For balanced trees, it is very helpful that there are only $O(\log n)$ levels. For taller trees, a more complicated approach is necessary.

The proofs of our main theorems use a variation of the well-known *block dynamics* which takes account of differing mixing times amongst the blocks. To the best of our knowledge, this is the first time that this variation has been used.

In order to apply the block dynamics, we need to analyze the mixing time of the Glauber dynamics on subtrees which have colours on their external boundaries fixed. This is equivalent to fixing the colours on some leaves of a tree. Markov chains on trees with fixed leaves are well-studied. When every leaf is fixed, Martinelli, Sinclair and Weitz [14] prove rapid mixing for $k \geq \Delta + 2$; at $k \leq \Delta + 1$ the chain might not be ergodic. In our setting, k may be much smaller and so we must bound the number of fixed leaves. Theorem 1 extends to show:

Theorem 2. *For any $k \geq 4$, the Glauber dynamics on k-colourings of any tree with maximum degree Δ and with the colours of any $b \leq k - 2$ leaves fixed mixes in time $n^{O(1+b+\Delta/k \log \Delta)}$.*

Due to space constraints, some proofs are omitted from this extended abstract and may be found in the full versions of the papers.

Remark 1. Our arguments can be extended to other instances of the Glauber dynamics, e.g. the Ising model. Details will appear in a full version of the paper.

2 Preliminaries

2.1 Graph Colourings

Let $G = (V, E)$ be a finite graph, and let $A = \{0, 1, \ldots, k - 1\}$ be a set of k colours. A *proper colouring* of G is an assignment of colours to vertices such that no two vertices connected by an edge are assigned the same colour. Define $\Omega \subset A^V$ to be the set of proper colourings of G. Given $\sigma \in \Omega$ and $x \in V$, we write $\sigma(x)$ to mean the colour of vertex x in σ. Given $S \subseteq V$, we write $\sigma(S)$ to refer to the assignment of colours to S in σ; that is, $\sigma(S)$ is σ restricted to S.

Given some $S \subseteq V$, Ω_S^σ is the set of proper colourings of G that are fixed to σ at all vertices not in S. We can think of Ω_S^σ as being equivalent to the set of proper colourings of S with boundary configuration σ. However, technically speaking, an element of Ω_S^σ will be viewed as a colouring of the entire graph G.

2.2 Glauber Dynamics

The *Glauber dynamics* for k-colourings of G is a Markov process over the space Ω of proper colourings. We make use of the continuous-time Metropolis version of the Glauber dynamics. (Standard methods, eg. [3,17], show that our theorems also hold for the heat-bath version.) Informally, the behaviour of this process is as follows: each vertex has a (rate 1) poisson clock. When the clock for vertex v rings, a colour a is chosen uniformly from A. The colour of v is set to a if a does not appear on any neighbour of v, otherwise the colouring remains unchanged.

More formally, recall that a continuous-time Markov process is defined by generator $\mathcal{L}$. We can think of $\mathcal{L}$ as a $|\Omega| \times |\Omega|$ matrix, whose non-diagonal entries represent the jump probabilities between colourings (and diagonal entries are such that all rows sum to 0). For $\sigma \neq \eta$, we will write $K[\sigma \to \eta]$ to denote the (σ, η) entry in this matrix. Under this framework, the jump probabilities for the Metropolis version of the Glauber dynamics are given by

$$K[\sigma \to \eta] = \begin{cases} \frac{1}{k} & \text{if } \sigma, \eta \text{ differ on exactly one vertex} \\ 0 & \text{otherwise} \end{cases}$$

Note that this process is symmetric and, for $k \geq 3$, ergodic on trees (see eg. [1]).

In many applications we are interested in the discrete analog of the Glauber dynamics. We then think of $K[\sigma \to \eta]$ as the probability of moving from colouring σ to colouring η, scaled by a factor of n. The mixing time for the discrete chain is precisely n times the mixing time for the corresponding continuous process (see eg. [1]), so our bounds on mixing time apply to the discrete setting.

We will additionally be interested in a variant of the Glauber dynamics, the 2-*path Glauber dynamics*, $\mathcal{L}_2$, that can also recolour pairs of adjacent vertices. That is, on each step of $\mathcal{L}_2$, a connected subgraph $S \subseteq T$ of size at most 2 is chosen uniformly at random. If the initial configuration is η, then the subgraph S is recoloured according to the uniform distribution on Ω_S^η.

2.3   Mixing Time

Given probability distributions π and μ over space Ω, the *total variation distance* between π and μ is defined as

$$\|\mu - \pi\|_{TV} = \frac{1}{2} \sum_{x \in \Omega} |\mu(x) - \pi(x)|.$$

Suppose $\mathcal{L}$ is the generator for an ergodic markov process over Ω. The *stationary distribution* for $\mathcal{L}$ is the unique measure π on Ω that satisfies $\pi\mathcal{L} = \pi$. It is well-known that the Glauber dynamics has uniform stationary distribution.

Given any $\sigma \in \Omega$, denote by μ_σ^t the measure on Ω given by running the process with generator $\mathcal{L}$ for time t starting from colouring σ. Then the *mixing time* of the process, $\mathcal{M}(\mathcal{L})$, is defined as

$$\mathcal{M}(\mathcal{L}) = \min \left\{ t : \sup_{\sigma \in \Omega} \|\mu_\sigma^t - \pi\|_{TV} \leq \frac{1}{4} \right\}.$$

We define the *spectral gap* of $\mathcal{L}$, $\text{Gap}(\mathcal{L})$, to be the second-largest eigenvalue of $-\mathcal{L}$. The *relaxation time* of $\mathcal{L}$, denoted $\tau(\mathcal{L})$, is defined as the inverse of the spectral gap. We will use the following standard bound (see eg. [17]):

$$M(\mathcal{L}) \leq \tau(\mathcal{L}) \log(|\Omega|) \leq (n \log k)\tau(\mathcal{L}) \qquad \text{since } |\Omega| \leq k^n. \qquad (1)$$

2.4   Colourings of Trees

Consider a (not necessarily complete) tree $G = (V, E)$ with maximum degree Δ. A subtree T of G is a connected induced subgraph of G. We shall write ∂T and

∂T to mean the exterior and interior boundaries of T. That is, $\partial T = \{x \in V \backslash T : N(x) \cap T \neq \emptyset\}$ and $\overline{\partial} T = \{x \in T : N(x) \cap \partial T \neq \emptyset\}$. Note that for each $x \in \partial T$ there is a unique $y \in \overline{\partial} T$ adjacent to x.

The following simple Lemma analyzes the ergodicity of the Glauber dynamics and 2-path Glauber dynamics on trees.

Lemma 1. *Let T be a subtree of G and suppose $k \geq \max\{3, |\partial T| + 2\}$. Then the Glauber dynamics is ergodic over Ω_T^σ for all $\sigma \in \Omega$. If additionally $k = 3$ and $|\partial T| \leq 2$, the 2-path Glauber dynamics is also ergodic over Ω_T^σ for all $\sigma \in \Omega$.*

3   Weighted Block Dynamics

In this section we present a generalization of the well-known block dynamics for local spin systems. We prove the result for the Glauber dynamics acting on a finite graph $G = (V, E)$. Our statement of the block dynamics actually applies to a more general setting, holding for all local update chains, including the 2-path Glauber dynamics defined above. We avoid a statement in full generality for succinctness. See [12] for a general treatment of local spin systems.

Suppose $D = \{V_1, \dots, V_r\}$ is a collection of subsets of V with $V = \cup_i V_i$. For each $1 \leq i \leq r$ and $\sigma \in \Omega$, let $\mathcal{L}_{V_i}^\sigma$ be the generator for the Glauber dynamics (or 2-path Glauber dynamics) restricted to V_i with boundary configuration σ. In other words, colours can change only for nodes in V_i.

Suppose that $\mathcal{L}_{V_i}^\sigma$ is ergodic for each i and σ. Let $\pi_{V_i}^\sigma$ denote the stationary distribution of $\mathcal{L}_{V_i}^\sigma$. For each i, define $g_i := \inf_{\sigma \in \Omega} \mathrm{Gap}(\mathcal{L}_{V_i}^\sigma)$, the minimum spectral gap for $\mathcal{L}_{V_i}^\sigma$ over all choices of boundary configurations. The *block dynamics* is a continuous-time Markov process with generator $\mathcal{L}_D$ defined by

$$K_D[\sigma \to \eta] = \begin{cases} \pi_{V_i}^\sigma[\eta] & \text{if there exists } i \text{ such that } \eta \in \Omega_{V_i}^\sigma \\ 0 & \text{otherwise.} \end{cases}$$

Note that $K_D[\sigma \to \eta] > 0$ precisely when η and σ differ only within a single block V_i. Informally, we think of the weighted block dynamics as having a poisson clock of rate 1 for each block V_i. When clock i rings, the colouring of V_i is replaced randomly according to $\pi_{V_i}^\sigma$, where σ is the previous colouring.

Using $\tau_{V_i} = 1/g_i$ to denote the maximum relaxation time of $\mathcal{L}_{V_i}^\sigma$ over all choices of boundary configurations, Proposition 3.4 of Martinelli [12] is:

Proposition 1. $\tau(\mathcal{L}_V) \leq \tau(\mathcal{L}_D) \times (\max_{1 \leq i \leq r} \tau_{V_i}) \times \sup_{x \in V} |\{i : x \in V_i\}|$.

We are now ready to define the *weighted block dynamics* corresponding to D. This is a continuous-time Markov process whose generator $\mathcal{L}_D^*$ is given by

$$K_D^*[\sigma \to \eta] = \begin{cases} g_i \pi_{V_i}^\sigma[\eta] & \text{for all } \eta, i \text{ such that } \eta \in \Omega_{V_i}^\sigma \\ 0 & \text{otherwise.} \end{cases}$$

The weighted block dynamics is similar to the block dynamics, but the transition probabilities for block V_i are scaled by a factor of g_i. The main result for this section is the following variant of Proposition 1:

Proposition 2. $\tau(\mathcal{L}_V) \le \tau(\mathcal{L}_D^*) \times \sup_{x \in V} |\{i : x \in V_i\}|.$

The proof of Proposition 2 is a simple modification to the proof of Proposition 1 [12]. It is worth noting the difference between Proposition 2 and the original block dynamics, Proposition 1. In the original version, the block dynamics Markov process can be thought of as having a poisson clock of rate g for each block, where g is the minimum over all g_i. In other words, each block is chosen with the same rate, that being the worst case over all blocks. On the other hand, in the modified version each block is chosen with the rate corresponding to that block. The original version yields a simpler Markov process, but a looser bound on the gap of the original process. In particular, applying the original block dynamics to our main result yields a mixing time of $n^{O(1+\Delta/k)}$, while the modified block dynamics tightens the bound to $n^{O(1+\Delta/k \log \Delta)}$ (see Remark 4).

We next show that the weighted block dynamics is equivalent to a related process. Informally, we wish to "collapse" each block to its set of internal boundary nodes. We will assign colours to these boundary nodes according to the probability such a boundary configuration would occur in the block dynamics. More formally, suppose $D = \{V_1, \ldots, V_m\}$ is a set of blocks of vertices of T. Let $B = \cup_{i=1}^m \overline{\partial} V_i$. That is, B contains all internal boundary nodes for the blocks in D. Note $B \cap V_i = \overline{\partial} V_i$. We define a Markov process $\mathcal{L}_B$ on Ω_B, which simulates the behaviour of $\mathcal{L}_D$ restricted to the nodes in B. Given distribution π over Ω_T, $S \subseteq T$, and $\eta \in \Omega_S$, write $\pi_T[\eta' : \eta'(S) = \eta(S)]$ to denote $\sum_{\eta' : \eta'(S) = \eta(S)} \pi_T[\eta']$, the probability that the configuration of S agrees with η. Then $\mathcal{L}_B$ is defined by

$$K_B[\sigma \to \eta] = \begin{cases} g_i \pi_{V_i}^{\sigma}[\eta' : \eta'(\overline{\partial} V_i) = \eta(\overline{\partial} V_i)] & \text{if } \sigma \text{ and } \eta \text{ differ only on } \overline{\partial} V_i \\ 0 & \text{otherwise.} \end{cases} \tag{2}$$

In other words, η is chosen according to the probability that η is the configuration on B after a step of the block dynamics. Our claim is that the relaxation times of $\mathcal{L}_D^*$ and $\mathcal{L}_B$ are the same; this is similar to Claim 2.9 due to Berger et al [1].

Proposition 3. $\tau(\mathcal{L}_D^*) = \tau(\mathcal{L}_B).$

4 An Upper Bound for General Trees

We now begin our proof of Theorem 1. Our approach is to decompose a tree into smaller subtrees, apply the block dynamics to the resulting subgraphs, and then use induction to bound the mixing time of the entire tree. Implicitly, this yields an iterative decomposition of the tree into smaller and smaller subtrees. How should we decompose a tree? A first idea is to root the tree at a vertex v, then take each subtree rooted at a child of v as a block (and v itself as a block of size 1). A nice property of this decomposition is that each subtree has at most one boundary node, adjacent to its root. In this case there will be h levels of recursion in the induction, where h is the height of tree T, and we will obtain a bound of the form c^h, where $c = c(\Delta, k)$ is the mixing time for an instance of the block dynamics. This method works for complete trees (and indeed was used by Berger et al. [1]) since they have logarithmic height. However, the height of a general tree could be much greater, leading to a super-polynomial bound.

Instead, we will partition the tree in a manner that guarantees each block has size at most half the size of the tree. This ensures that our recursion halts after logarithmically many steps, and yields a polynomial mixing time. To obtain such a partition, we choose a central node x and conceptually split the tree by removing x, obtaining at most Δ subtrees plus $\{x\}$.

There are difficulties with the above approach that must be overcome. First, a subtree T may have multiple boundary nodes, which complicates the behaviour of the block dynamics. We therefore make our choice of x carefully, so that boundaries are of size at most 2. Second, for non-complete trees we might have blocks of vastly differing sizes, which makes a tight analysis of the block dynamics more difficult. We therefore use the weighted version of the block dynamics.

In this section we describe our choice of blocks for the block dynamics. We then show that the upper bound of Theorem 1 holds, given a bound on the relaxation time of the block dynamics. The details of analyzing the block dynamics are encapsulated in Lemma 3, which is proved in Section 4.1.

Let T be any tree with maximum degree Δ. Suppose $|T| = n$ and $|\partial T| \leq 2$ (that is, T has at most two external boundary nodes). Let σ be a boundary configuration for T. If $k \geq 4$, then let $\mathcal{L}$ denote the Glauber dynamics on T with k colours and boundary configuration σ. If $k = 3$, then take $\mathcal{L}$ to be the 2-path Glauber dynamics on T with boundary configuration σ. Either way, since $|\partial T| \leq 2$, Lemma 1 implies that $\mathcal{L}$ is ergodic. Let τ_T^σ denote the relaxation time for $\mathcal{L}$. We wish to consider the maximum relaxation time over all boundary configurations and trees of a certain size. To this end, we define

$$\tau_T := \max_{\sigma \in \Omega} \tau_T^\sigma \qquad \text{and} \qquad \tau_i(n) := \max_{T:|T| \leq n,\ |\partial T| \leq i} \tau_T.$$

We will prove Theorem 1 by showing the slightly stronger result that $\tau_2(n) = n^{O(1+\Delta/k \log \Delta)}$. We will show that, for some fixed constant c and some $2 \leq i \leq \Delta$,

$$\tau_2(n) \leq ci^2 \left(\frac{k-1}{k-2}\right)^{i+1} \tau_2(\lfloor n/i \rfloor). \tag{3}$$

First let us show how (3) implies Theorem 1 when $k \geq 4$. By induction on n, (3) implies that $\tau_2(n) \leq n^{d(1+\Delta/k \log \Delta)}$ for some constant d (since we can assume $k \leq 2\Delta$, as otherwise the result is known [7]). By (1), the mixing time of the Glauber dynamics satisfies $\mathcal{M}(\mathcal{L}) \leq (n \log k)\tau_G \leq (n \log k)\tau_2(n) = n^{O(1+\Delta/k \log \Delta)}$ as required. For $k = 3$, (3) implies that the 2-path Glauber dynamics mixes in time $n^{O(1+\Delta/k \log \Delta)}$. Theorem 1 then follows from Lemma 2 below.

Lemma 2. *Let $\mathcal{L}_1$ denote the Glauber dynamics with $k = 3$ colours, and $\mathcal{L}_2$ denote the 2-path Glauber dynamics again with $k = 3$ colours. For any T with $|\partial T| \leq 1$ and boundary configuration ξ, $\tau(\mathcal{L}_1^\xi) \leq n^{O(\Delta/\log \Delta)}\tau(\mathcal{L}_2^\xi)$.*

Proof (sketch). We wish to apply the comparison method of Diaconis and Saloff-Coste [3]. We note that this application is not immediate, since a step of $\mathcal{L}_2$

cannot always be simulated by a small number of steps of $\mathcal{L}_1$. We therefore consider an intermediate process, which performs a cyclic shift of all colours of a subtree of T in one step. Such a process can be used to simulate a step of $\mathcal{L}_2$. To compare with $\mathcal{L}_1$, we simulate a rotation step by changing the colours of nodes in a bottom-up fashion. If these changes are ordered carefully, one can simulate a rotation of colours in $O(n)$ steps of $\mathcal{L}_1$, where each step has a congestion of $n^{O(\Delta/\log \Delta)}$. The term $n^{O(\Delta/\log \Delta)}$ derives from a bound on the number of siblings of ancestors of a given node. Details are given in the full version.

We now turn to proving (3). The following Lemma will be our main tool.

Lemma 3. *Suppose $k \geq 3$ and let T be a subtree of a tree G with $|\partial T| \leq 2$ and let $\sigma \in \Omega$ be a boundary condition for T. Choose $v \in T$ and let $D_v = \{\{v\}, V_1, \ldots, V_t\}$ be a partition of T into disjoint connected subtrees, where $1 \leq t \leq \Delta$. Suppose $|\partial V_i| \leq 2$ for each V_i. Then there exists constant c such that*

$$\tau_T^\sigma \leq c \max_{1 \leq i \leq t} i^2 \left(\frac{k-1}{k-2}\right)^i \tau_{V_i}.$$

We prove Lemma 3 in Section 4.1. Let us show how it implies (3). We first consider trees with boundaries of size one, then size two.

Lemma 4. *For some $2 \leq i \leq \Delta$, we have $\tau_1(n) \leq c i^2 \left(\frac{k-1}{k-2}\right)^i \tau_2(\lfloor n/i \rfloor)$.*

Proof. Suppose $|\partial T| \leq 1$. It is well-known that we can find a vertex $x \in T$ such that if $D_x = \{\{x\}, V_1, \ldots, V_t\}$, we will have $|V_i| \leq \lfloor n/2 \rfloor$ for all $1 \leq i \leq t$ (see eg. [11]). We will choose our indices so that $|V_1| \geq |V_2| \geq \ldots \geq |V_t|$. Since $|\partial T| \leq 1$, we have $|\partial V_i| \leq 2$ for all i. By Lemma 3, $\tau_T \leq c i^2 \left(\frac{k-1}{k-2}\right)^i \tau_{V_i}$ for some $1 \leq i \leq t$.

If $i \geq 2$, we get $\tau_{V_i} \leq \tau_2(|V_i|) \leq \tau_2(\lfloor n/i \rfloor)$, since the V_i are given by increasing size. Thus $\tau_T \leq c i^2 \left(\frac{k-1}{k-2}\right)^i \tau_2(\lfloor n/i \rfloor)$ for some $2 \leq i \leq t$ as required. If $i = 1$, then we recall that $|V_1| \leq \lfloor n/2 \rfloor$ by our choice of x. Hence $\tau_T \leq c \left(\frac{k-1}{k-2}\right) \tau_{V_1} < c(2)^2 \left(\frac{k-1}{k-2}\right)^2 \tau_2(\lfloor n/2 \rfloor)$ as required.

Proposition 4. *For some $2 \leq i \leq \Delta$, $\tau_2(n) \leq c^2 i^2 \left(\frac{k-1}{k-2}\right)^{i+1} \tau_2(\lfloor n/i \rfloor)$.*

Proof. Let T be a subtree with $|T| = n$ and $|\partial T| = 2$, say $\partial T = \{z_1, z_2\}$. Choose x as in Lemma 4, with x separating T into subtrees of size at most $\lfloor n/2 \rfloor$. We will call the unique path in G from z_1 to z_2 the *boundary path* for T. Suppose x is on the boundary path for T. Let $D_x = \{\{x\}, V_1, \ldots, V_t\}$ be a partition into disjoint connected subtrees, indexed so that $|V_1| \geq \ldots \geq |V_t|$; note that $|\partial V_i| \leq 2$ for all i. We then apply Lemma 3 as in Lemma 4 and obtain the desired result.

Now suppose that x is not on the boundary path for T. Consider T to be rooted at some $r \in \overline{\partial T}$. Let y be the least ancestor of x that lies on the boundary path. Consider $D_y = \{\{y\}, V_1, \ldots, V_t\}$. Since x separates T into subtrees of size at most $\lfloor n/2 \rfloor$, in particular the subtree containing y must have size at most $\lfloor n/2 \rfloor$. This implies that the subtree separated by y that contains x must contain at least $\lfloor n/2 \rfloor$ nodes, and is therefore V_1, the largest subtree separated by y. Also, $|\bar{\partial} V_i| \leq 2$ for all i, since y is on the boundary path for T. Lemma 3 implies

$$\tau_T \leq ci^2 \left(\frac{k-1}{k-2}\right)^i \tau_{V_i}$$

for some i. If $i > 1$ then we obtain the desired result since $|V_i| \leq \lfloor n/i \rfloor$. If $i = 1$, then since $|V_1| < n$ and $|\bar{\partial} V_1| = 1$ (by our choice of y), Lemma 4 implies

$$\tau_T \leq c\left(\frac{k-1}{k-2}\right) \tau_1(|V_1|) \leq c\left(\frac{k-1}{k-2}\right) \tau_1(n)$$

$$\leq c^2 i^2 \left(\frac{k-1}{k-2}\right)^{i+1} \tau_2(\lfloor n/i \rfloor) \quad \text{for some } 2 \leq i \leq \Delta.$$

We have now derived (3), completing the proof of Theorem 1.

4.1 Proof of Lemma 3

We now proceed with the proof of Lemma 3, which bounds the relaxation time on a tree with respect to the relaxation times for subtrees. Our approach is to use a canonical paths argument to bound the behaviour of the block dynamics. Indeed, there is a simple canonical path to move between configurations σ and η: modify the configuration of each V_i to an intermediate state so that v is free to change colour to $\eta(v)$, make that change to v, then set the configuration of each V_i to $\eta(V_i)$. The block dynamics paired with this path yields a bound on the relaxation time. However, that bound is not tight enough to imply the mixing rate we desire: it only implies a mixing time of $n^{O(\Delta)}$. We therefore apply the following sequence of improvements to the above approach.

1. We explicitly describe an intermediate configuration for the neighbours of v, in order to balance congestion over all start and end configurations. This improves the bound on the mixing time to $n^{O(\log \Delta + \log k + \Delta/k)}$.
2. Our path shifts between 3 different intermediate configurations to maximize the dependency on the start and end configurations at each step. This improves our bound to $n^{O(\log \Delta + \Delta/k)}$.
3. We apply the weighted block dynamics, to differentiate between large and small subtrees. We always change configurations of blocks in order of subtree size. This improves our bound to $n^{O(\log \Delta + \Delta/k \log \Delta)}$. See Remark 4.
4. We apply weights to our canonical path to discount the congestion on smaller subtrees. The net effect is that the presence of many small subtrees does not influence the congestion of our paths. This improves our bound to $n^{O(1+\Delta/k \log \Delta)}$. See Remark 3.

The Block Dynamics. Recall the conditions of Lemma 3. Suppose $k \geq 3$ and let T be a tree with $|\partial T| \leq 2$ and let $\sigma \in \Omega$ be a boundary condition for T. Choose $v \in T$ and consider $D = \{\{v\}, V_1, \ldots, V_t\}$, where $1 \leq t \leq \Delta$. Suppose we choose v so that $|\partial V_i| \leq 2$ for each V_i. We will think of T as being rooted at v; then let u_i denote the root of V_i (ie. the neighbour of v in V_i). Due to space limitations, we prove Lemma 3 under the assumption that $u_i \notin \overline{\partial} T$ for all i. The (simple) extension to remove this assumption is discussed at the conclusion of the section; see Remark 2.

Let $\mathcal{L}_D^*$ be the generator for the weighted block dynamics corresponding to D and boundary configuration σ. Let τ_D^σ denote the relaxation time of $\mathcal{L}_D^*$. Since no vertex lies in more than one block, Proposition 2 implies $\tau_T^\sigma \leq \tau_D^\sigma$.

Next recall the definition of graph B and dynamics $\mathcal{L}_B$ from Proposition 3. In this context, we can view $\mathcal{L}_B$ as a version of $\mathcal{L}_D$ wherein each block is treated like a single vertex. That is, B is a star with internal node v; we will refer to $u_1, \ldots, u_t$ as the leaf nodes of B. When such a leaf node, say u_i, is chosen by the dynamics, its colour updates with probability corresponding to the probability of seeing that colour as the root of V_i in $\mathcal{L}_D$. By Proposition 3, $\tau(\mathcal{L}_D^\sigma) = \tau(\mathcal{L}_B^\sigma)$. It is therefore sufficient to bound $\tau(\mathcal{L}_B^\sigma)$. Note that this is true even for the special case of $k = 3$, as $\mathcal{L}_B^\sigma$ depends only on the ergodicity of $\mathcal{L}$ (the 2-path Glauber dynamics) and its stationary distribution, which is uniform. The following simple Lemma bounds the transition probabilities of $\mathcal{L}_B^\sigma$.

Lemma 5. *Choose* $S \subseteq T$ *with* $|\partial S| \leq 2$ *and boundary configuration* ξ, *and suppose* $x \in \overline{\partial} S$. *Choose* $c \in A$ *and suppose there exists some* $\eta \in \Omega_S^\xi$ *with* $\eta(x) = c$. *Then* $\pi_S^\xi[\omega : \omega(x) = c] \geq 1/k$.

Corollary 1. *Suppose* $\alpha, \omega \in \Omega_B^\sigma$, $K_B[\alpha \to \omega] > 0$, *and* $\alpha(u_i) \neq \omega(u_i)$. *Then* $K_B[\alpha \to \omega] \geq (k\tau_{V_i}^\sigma)^{-1}$.

Defininition of Intermediate Configurations. Choose two colourings $\alpha, \eta \in \Omega_B$. Our goal is to define a sequence of steps of $\mathcal{L}_B$ that begins in state α and ends in state η. If $\alpha(v) = \eta(v)$ this sequence is simple: the colours of nodes $u_1, \ldots, u_t$ are changed from α to η one at a time. If $\alpha(v) \neq \eta(v)$, our strategy is to first change the colours of $u_1, \ldots, u_t$ so that none have colour $\eta(v)$, then change the colour of v to $\eta(v)$, and finally set the colours of the u_i nodes to match η. The obvious way to do this requires two "passes" of changes over the leaf nodes, but this method generates too much congestion (defined below). We therefore introduce a more complex path that uses three passes. We now define the colours used in the intermediate configurations of this path.

If $\alpha(v) \neq \eta(v)$ then for each $1 \leq i \leq t$ we will define three colours, a_i, b_i, and c_i, that depend on α and η. The first two colours are easy to define:

$$a_i = \begin{cases} \alpha(u_i) & \text{if } \alpha(u_i) \neq \eta(v) \\ \alpha(v) & \text{otherwise} \end{cases} \qquad b_i = \begin{cases} \eta(u_i) & \text{if } \eta(u_i) \neq \alpha(v) \\ \eta(v) & \text{otherwise} \end{cases}$$

That is, $(a_1, \ldots, a_t)$ are the colours of the children of v in α, with occurrences of $\eta(v)$ replaced with $\alpha(v)$. Note that our assumption that u_i is not adjacent to the external boundary of T ensures that there exists a configuration in which u_i has colour a_i. We define b_i in the same way, but with the roles of α and η reversed.

The definition of colour c_i is more involved. These will be the colours to which we set the leaf nodes to allow v to change from $\alpha(v)$ to $\eta(v)$. We will apply a function f that will map the colours $(\alpha(u_1), \ldots, \alpha(u_t))$ to a vector of colours $(c_1, \ldots, c_t)$ such that for all i, $c_i \notin \{\alpha(v), \eta(v)\}$. We want f to satisfy the following balance property: for all $1 \leq i \leq t$, writing $\mathbf{x}$ for $(x_1, \ldots, x_t)$,

$$\#\{\mathbf{x} : (x_j = \alpha(u_j) \ \forall j > i) \wedge (f(\mathbf{x})_j = c_j \ \forall j \leq i)\} \leq \left\lceil \left(\frac{k-1}{k-2}\right)^i \right\rceil. \quad (4)$$

That is, for any $1 \leq i \leq t$, if we are given $c_1, \ldots, c_i$ and $\alpha(u_{i+1}), \ldots, \alpha(u_t)$, there are at most $\left\lceil \left(\frac{k-1}{k-2}\right)^i \right\rceil$ possibilities for $\alpha(u_1), \ldots, \alpha(u_t)$. Such an f is guaranteed to exist; see Lucier and Molloy [13] for a construction.

The Path Definition. Let Γ be the transition graph over Ω_G with $(\omega, \beta) \in \Gamma$ if and only if $K_B[\omega \to \beta] > 0$. We are now ready to define a path $\gamma(\alpha, \eta)$ of transitions of Γ. If $\alpha(v) = \eta(v)$, our path simply changes the colour of each u_i from $\alpha(u_i)$ to $\eta(u_i)$, one at a time. If $\alpha(v) \neq \eta(v)$, we use the following path:

1. For each u_i in increasing order: recolour from $\alpha(u_i)$ to b_i, then to c_i.
2. Recolour v from $\alpha(v)$ to $\eta(v)$.
3. For each u_i in decreasing order: recolour from c_i to $\eta(u_i)$, then to a_i.
4. For each u_i in increasing order: recolour from a_i to $\eta(u_i)$.

The reader is encouraged to verify that all steps are valid transitions according to $\mathcal{L}_B^\sigma$. The number of changes to the colour of each u_i seems excessive, but we define our path this way to maintain an important property: each change is from a colour derived from α to a colour derived from η, or vice-versa. This will be important in our analysis of the path congestion, defined below.

Analysis of Weighted Path Congestion. We will now define the weighted congestion of our choice of paths. For each $(\omega, \beta) \in \Gamma$, we will define a weight $w(\omega, \beta) > 0$. Set $w(\omega, \beta) = 1$ if ω and β differ on the colour of v, and set $w(\omega, \beta) = i^{-2}$ if ω and β differ on the colour of vertex u_i. We define the weight of a path by $w(\gamma(\alpha, \eta)) = \sum_{(\omega, \beta) \in \gamma(\alpha, \eta)} w(\omega, \beta)$. Then note that for all $\gamma(\alpha, \eta)$, $w(\gamma(\alpha, \eta)) \leq 1 + 5 \sum_{i=1}^t i^{-2} < 1 + 5 \left(\frac{\pi^2}{6}\right) < 10$. For each edge $(\omega, \beta) \in \Gamma$, define the weighted congestion of that edge, $\rho_w(\omega, \beta)$, as

$$\rho_w(\omega, \beta) := \frac{1}{w(\omega, \beta)} \left(\sum_{\gamma(\alpha, \eta) \ni (\omega, \beta)} \frac{\pi[\alpha]\pi[\eta]w(\gamma(\alpha, \eta))}{\pi[\omega]K_B[\omega \to \beta]} \right).$$

The weighted congestion for our set of paths is $\rho_w := \sup_{\omega, \beta} \rho_w(\omega, \beta)$. The weighted canonical paths bound is $\tau_D^\sigma \leq \rho_w$. We note that this bound and its proof are implicit in [4] (see their Remark on page 38). The standard canonical path bound sets $w(\omega, \beta) = 1$ for all $(\omega, \beta) \in \Gamma$. Our choice of a different weight function will allow us to tighten the bound we obtain on τ_D^σ (see Remark 3).

Our result follows by bounding $\rho_w(\omega, \beta)$. Uniformity of π implies

$$\rho_w(\omega, \beta) \leq 10 \left(\frac{1}{w(\omega, \beta)} \times |\{\gamma(\alpha, \eta) \ni (\omega, \beta)\}| \times \frac{1}{(k-1)^{t+1} K_B[\omega \to \beta]} \right). \quad (5)$$

We now consider cases depending on the nature of the transition (ω, β).

Case 1: ω and β differ on the colour of v. Note that $w(\omega, \beta) = 1$. Also, from the definition of $\mathcal{L}_B$, we have $K_B[\omega \to \beta] = \inf_{\sigma \in \Omega} \text{gap}(\mathcal{L}_{\{v\}}^\sigma) \pi_{\{v\}}^\omega [\phi : \phi(v) = \beta(v)]$. But note that $\text{gap}(\mathcal{L}_{\{v\}}^\sigma) = 1$ for all boundary conditions, and $\pi_{\{v\}}^\omega$ is the uniform distribution over a set of at most $k - 1$ colours. We conclude

$$K_B[\omega \to \beta] \geq \frac{1}{k-1}. \quad (6)$$

Consider the number of (α, η) such that $(\omega, \beta) \in \gamma(\alpha, \eta)$. This occurs precisely when $\alpha(v) = \omega(v)$, $\eta(v) = \beta(v)$, and $\alpha(u_i) = \omega(u_i)$ for all u_i.

Consider the possibilities for η. Configuration β determines $\eta(v)$, and there are $(k-1)^t$ choices for η given $\eta(v)$ (consider choosing the colours for $u_1, \ldots, u_t$, which cannot be $\eta(v)$). Now consider α: the colour $\alpha(v)$ is determined by ω, as are $(c_1, \ldots, c_t)$. Thus by (4) there are at most $\lceil (\frac{k-1}{k-2})^\Delta \rceil$ possibilities for $(\alpha(u_1), \ldots, \alpha(u_t))$, which determines α. Putting this together, the total number of colourings α and η that satisfy $(\omega, \beta) \in \gamma(\alpha, \eta)$ is at most $(k-1)^t \left\lceil \left(\frac{k-1}{k-2} \right)^t \right\rceil$.

Substituting this and (6) into (5), we conclude

$$\rho_w(\omega, \beta) \leq 10(1)(k-1)^t \left\lceil \left(\frac{k-1}{k-2} \right)^t \right\rceil \frac{k-1}{(k-1)^{t+1}} \leq 20 \left(\frac{k-1}{k-2} \right)^t.$$

Case 2: ω and β differ on the colour of u_i for some i. In this case, $w(\gamma(\alpha, \eta)) = i^{-2}$. Also, since there exists a colouring of V_i in which u_i has colour $\beta(u_i)$ (recalling our assumption that $u_i \notin \partial T$), Corollary 1 implies

$$K_B[\omega \to \beta] \geq (k\tau_{V_i})^{-1}. \quad (7)$$

How many paths in $\gamma(\alpha, \eta)$ use the transition (ω, β)? We consider subcases for α and η. We give only one subcase here; the remaining 5 cases (which are very similar) are omitted due to space constraints.

Subcase: $\alpha(v) \neq \eta(v)$ and (ω, β) is the first change to u_i in $\gamma(\alpha, \eta)$. That is, (ω, β) is the first change in Step 1 of the canonical path description. In this case we know $\alpha(v) = \omega(v)$, $\alpha(u_j) = \omega(u_j)$ for all $j \geq i$, $b_i = \beta(u_i)$, and $c_j = \beta(u_j)$ for all $j < i$. How many α, η satisfy these conditions?

There are at most $k - 1$ possibilities for $\eta(v)$, since $\eta(v) \neq \alpha(v) = \omega(v)$. Given $\eta(v)$, there are $k - 1$ possibilities for $\eta(u_j)$ for each $j \neq i$. Note that β determines

b_i, from which $\eta(v)$ determines $\eta(u_i)$. Thus the total number of possibilities for η is $(k-1)^t$. Next consider α. ω determines $\alpha(v)$ and also $\alpha(u_j)$ for all $j \geq i$. Also, β determines c_j for all $j < i$. By (4), the number of possibilities for $\alpha(u_1), \ldots, \alpha(u_t)$ is at most $\left\lceil \left(\frac{k-1}{k-2}\right)^{i-1} \right\rceil$. The total number of α and η is therefore at most $\left\lceil \left(\frac{k-1}{k-2}\right)^{i-1} \right\rceil (k-1)^t$. This completes the subcase.

Summing up over all subcases, we get that the total number of possibilities for α and η, given that (ω, β) is a change in the colouring of u_i, is at most $12 \left(\frac{k-1}{k-2}\right)^i (k-1)^t$. Substituting this and (7) into (5), we have

$$\rho_w(\omega, \beta) \leq 120 i^2 \left(\frac{k-1}{k-2}\right)^i (k-1)^t \left(\frac{\tau_{V_i} k}{(k-1)^{t+1}}\right) \leq 180 i^2 \left(\frac{k-1}{k-2}\right)^i \tau_{V_i}.$$

This concludes our case analysis. Cases 1 and 2 (and the fact that $\tau_{V_i} \geq 1$) imply $\rho_w \leq \max_{1 \leq i \leq t} 180 i^2 \left(\frac{k-1}{k-2}\right)^i \tau_{V_i}$. Applying the canonical paths bound and Proposition 2 we conclude $\tau_T^\sigma \leq \tau_D^\sigma \leq 180 \max_{1 \leq i \leq t} i^2 \left(\frac{k-1}{k-2}\right)^i \tau_{V_i}$ as required.

Remark 2. Recall that in the analysis above we assumed that no u_i was in $\overline{\partial} T$. We now sketch the method for removing this assumption; additional details appear in the full version of this paper. We used the assumption to guarantee that no leaf of B was adjacent to the boundary of T. We modify our selection of blocks to maintain this property: we replace block $\{v\}$ with a block $R \subseteq T$ that contains v and any neighbouring nodes in $\overline{\partial} T$. Our new set of blocks D will contain R and all subtrees separated by R. Then B will no longer be a star, but rather a tree or forest with few internal nodes. We then bound the relaxation time of $\mathcal{L}_B$ as before, extending our set of canonical paths in the natural way. The congestion analysis for this set of paths is similar to the original, and we obtain the same result up to a constant factor.

Remark 3. We note the effect of using the weighted canonical paths bound. If we had used the standard canonical paths bound, then we would replace the factor of i^2 in Lemma 3 by the maximum length of a path, which is $5\Delta + 1$. However, this would lead to a bound of $n^{O(\log \Delta + \Delta/k \log \Delta)}$ on the mixing time of the Glauber dynamics, which is weaker than $n^{O(1+\Delta/k \log \Delta)}$.

Remark 4. We also note the effect of using the weighted block dynamics. If we had applied Proposition 1 instead of Proposition 2, the bound in (7) would become $K_B[\omega \to \beta] \geq (k\tau)^{-1}$, where $\tau = \max_i \tau_{V_i}$. This would lead to a bound of $\tau_T^\sigma \leq ct^2 \left(\frac{k-1}{k-2}\right)^t \max_{1 \leq i \leq t} \tau_{V_i}$ for Lemma 3. With this modified Lemma, the bound in (3) would become $\tau_2(n) \leq ct^2 \left(\frac{k-1}{k-2}\right)^t \tau_2(\lceil n/2 \rceil)$, leading to a mixing time bound of $n^{O(1+\Delta/k)}$, which is weaker than $n^{O(1+\Delta/k \log \Delta)}$.

5 Open Problems

Our results raise questions about the Glauber dynamics on planar graphs of bounded degree. Hayes, Vera and Vigoda [7] noted that when $\Delta \geq n^\eta$ for any $\eta > 0$ then certain trees require $k \geq c\Delta/\log\Delta$ for polytime mixing, where c is an absolute constant. The same is true for any Δ that grows with n [13]. But for $\Delta = O(1)$, Theorem 1 shows that no trees require $k > 3$. Is there a constant K such that for every $k \geq K$ and constant Δ, the Glauber dynamics mixes in polytime on k-colourings of every planar graph with maximum degree Δ?

Another question is how far Theorem 2 can be extended. In other words, how many leaves can we fix and still guarantee polytime mixing? It is easy to fix the colours of $k-1$ neighbours of each of two adjacent vertices u, v so that the chain is not ergodic, so the answer lies between $k - 2$ and $2k - 2$.

Acknowledgements

We thank Nayantara Bhatnagar, Jian Ding, Thomas Hayes, Juan Vera and Eric Vigoda for some helpful discussions.

References

1. Berger, N., Kenyon, C., Mossel, E., Peres, Y.: Glauber dynamics on trees and hyperbolic graphs. Prob. Th. Related Fields 131, 311–340 (2005)
2. Brightwell, G., Winkler, P.: Random colorings of a Cayley tree. In: Bollobás, B. (ed.) Contemporary Combinatorics. Bolyai Society Mathematical Studies, vol. 10, pp. 247–276 (2002)
3. Diaconis, P., Saloff-Coste, L.: Comparison theorems for reversible Markov chains. Ann. Appl. Prob. 3, 696–730 (1993)
4. Diaconis, P., Stroock, D.: Geometric bounds for eigenvalues of Markov chains. Ann. Appl. Prob. 1, 36–61 (1991)
5. Dyer, M., Goldberg, L., Jerrum, M.: Systematic scan for sampling colorings. Ann. Appl. Prob. 18, 185–230 (2006)
6. Goldberg, L., Jerrum, M., Karpinski, M.: The mixing time of Glauber dynamics for colouring regular trees (2008) arXiv:0806.0921v1
7. Hayes, T., Vera, J., Vigoda, E.: Randomly coloring planar graphs with fewer colors than the maximum degree. In: Proceedings of STOC 2007 (2007)
8. Jerrum, M.: A very simple algorithm for estimating the number of k-colourings of a low-degree graph. Rand. Struct. Alg. 7, 157–165 (1995)
9. Jerrum, M., Sinclair, A.: Approximating the permanent. Siam. Jour. Comput. 18, 1149–1178 (1989)
10. Jonasson, J.: Uniqueness of uniform random colourings of regular trees. Stat. & Prob. Letters 57, 243–248 (2002)
11. Jordan, C.: Sur les assemblages de lignes. J. Reine Angew. Math. 70, 185–190 (1869)
12. Martinelli, F.: Lectures on Glauber dynamics for discrete spin models. Lecture Notes in Mathematics, vol. 1717 (2000)

13. Lucier, B., Molloy, M.: The Glauber dynamics for colourings of bounded degree trees. Submitted to journal (2008)
14. Martinelli, F., Sinclair, A., Weitz, D.: Fast mixing for independent sets, colorings and other models on trees. Rand. Struc. & Alg. 31, 134–172 (2007)
15. Molloy, M.: Very rapidly mixing Markov Chains for (2Δ)-colourings and for independent sets in a 4-regular graph. Rand. Struc. & Alg. 18, 101–115 (2001)
16. Mossel, E., Sly, A.: Gibbs Rapidly Samples Colourings of G(n,d/n) (2007), http://front.math.ucdavis.edu/0707.3241
17. Randall, D.: Mixing. In: Proceedings of FOCS 2003 (2003)
18. Sinclair, A.: Improved bounds for mixing rates of Markov chains and multicommodity flow. Combinatorics, Probability and Computing 1, 351–370 (1992)
19. Vigoda, E.: Improved bounds for sampling colorings. J. Math. Physics 41, 1555–1569 (2000)

Testing ±1-Weight Halfspaces

Kevin Matulef[1], Ryan O'Donnell[2], Ronitt Rubinfeld[3], and Rocco A. Servedio[4]

[1] MIT
matulef@mit.edu
[2] Carnegie Mellon University
odonnell@cs.cmu.edu
[3] Tel Aviv University and MIT
ronitt@theory.csail.mit.edu
[4] Columbia University
rocco@cs.columbia.edu

Abstract. We consider the problem of testing whether a Boolean function $f : \{-1,1\}^n \rightarrow \{-1,1\}$ is a ±1-*weight halfspace*, i.e. a function of the form $f(x) = \text{sgn}(w_1 x_1 + w_2 x_2 + \cdots + w_n x_n)$ where the weights w_i take values in $\{-1,1\}$. We show that the complexity of this problem is markedly different from the problem of testing whether f is a general halfspace with arbitrary weights. While the latter can be done with a number of queries that is independent of n [7], to distinguish whether f is a ±1-weight halfspace versus ϵ-far from all such halfspaces we prove that nonadaptive algorithms must make $\Omega(\log n)$ queries. We complement this lower bound with a sublinear upper bound showing that $O(\sqrt{n} \cdot \text{poly}(\frac{1}{\epsilon}))$ queries suffice.

1 Introduction

A fundamental class in machine learning and complexity is the class of halfspaces, or functions of the form $f(x) = (w_1 x_1 + w_2 x_2 + \cdots + w_n x_n - \theta)$. Halfspaces are a simple yet powerful class of functions, which for decades have played an important role in fields such as complexity theory, optimization, and machine learning (see e.g. [1, 5, 8, 9, 11, 12]).

Recently [7] brought attention to the problem of *testing* halfspaces. Given query access to a function $f : \{-1,1\}^n \rightarrow \{-1,1\}$, the goal of an ϵ-testing algorithm is to output YES if f is a halfspace and NO if it is ϵ-far (with respect to the uniform distribution over inputs) from all halfspaces. Unlike a learning algorithm for halfspaces, a testing algorithm is not required to output an approximation to f when it is close to a halfspace. Thus, the testing problem can be viewed as a relaxation of the proper learning problem (this is made formal in [4]). Correspondingly, [7] found that halfspaces can be tested more efficiently than they can be learned. In particular, while $\Omega(n/\epsilon)$ queries are required to learn halfspaces to accuracy ϵ (this follows from e.g. [6]), [7] show that ϵ-testing halfspaces only requires $\text{poly}(1/\epsilon)$ queries, *independent of the dimension* n.

In this work, we consider the problem of testing whether a function f belongs to a natural subclass of halfspaces, the class of ±1-*weight halfspaces*. These are functions of the form $f(x) = \text{sgn}(w_1 x_1 + w_2 x_2 + \cdots + w_n x_n)$ where the weights w_i all take

I. Dinur et al. (Eds.): APPROX and RANDOM 2009, LNCS 5687, pp. 646–657, 2009.

values in $\{-1, 1\}$. Included in this class is the majority function on n variables, and all 2^n "reorientations" of majority, where some variables x_i are replaced by $-x_i$. Alternatively, this can be viewed as the subclass of halfspaces where all variables have the same amount of influence on the outcome of the function, but some variables get a "positive" vote while others get a "negative" vote.

For the problem of testing ± 1-weight halfspaces, we prove two main results:

1. **Lower Bound.** We show that any nonadaptive testing algorithm which distinguishes ± 1-weight halfspaces from functions that are ϵ-far from ± 1-weight halfspaces must make at least $\Omega(\log n)$ many queries. By a standard transformation (see e.g. [3]), this also implies an $\Omega(\log \log n)$ lower bound for adaptive algorithms. Taken together with [7], this shows that testing this natural subclass of halfspaces is more query-intensive then testing the general class of all halfspaces.

2. **Upper Bound.** We give a nonadaptive algorithm making $O(\sqrt{n} \cdot \mathrm{poly}(1/\epsilon))$ many queries to f, which outputs (i) YES with probability at least $2/3$ if f is a ± 1-weight halfspace (ii) NO with probability at least $2/3$ if f is ϵ-far from any ± 1-weight halfspace.

    We note that it follows from [6] that *learning* the class of ± 1-weight halfspaces requires $\Omega(n/\epsilon)$ queries. Thus, while some dependence on n is necessary for testing, our upper bound shows testing ± 1-weight halfspaces can still be done more efficiently than learning.

Although we prove our results specifically for the case of halfspaces with all weights ± 1, we remark that similar results can be obtained using our methods for other similar subclasses of halfspaces such as $\{-1, 0, 1\}$-weight halfspaces (± 1-weight halfspaces where some variables are irrelevant).

Techniques. As is standard in property testing, our lower bound is proved using Yao's method. We define two distributions D_{YES} and D_{NO} over functions, where a draw from D_{YES} is a randomly chosen ± 1-weight halfspace and a draw from D_{NO} is a halfspace whose coefficients are drawn uniformly from $\{+1, -1, +\sqrt{3}, -\sqrt{3}\}$. We show that a random draw from D_{NO} is with high probability $\Omega(1)$-far from every ± 1-weight halfspace, but that any set of $o(\log n)$ query strings cannot distinguish between a draw from D_{YES} and a draw from D_{NO}.

Our upper bound is achieved by an algorithm which uniformly selects a small set of variables and checks, for each selected variable x_i, that the magnitude of the corresponding singleton Fourier coefficient $|\hat{f}(i)|$ is close to to the right value. We show that any function that passes this test with high probability must have its degree-1 Fourier coefficients very similar to those of some ± 1-weight halfspace, and that any function whose degree-1 Fourier coefficients have this property must be close to a ± 1-weight halfspace. At a high level this approach is similar to some of what is done in [7], but in the setting of the current paper this approach incurs a dependence on n because of the level of accuracy that is required to adequately estimate the Fourier coefficients.

2 Notation and Preliminaries

Throughout this paper, unless otherwise noted f will denote a Boolean function of the form $f : \{-1, 1\}^n \to \{-1, 1\}$. We say that two Boolean functions f and g are ϵ-*far* if $\Pr_x[f(x) \neq g(x)] > \epsilon$, where x is drawn from the uniform distribution on $\{-1, 1\}^n$.

We say that a function f is *unate* if it is monotone increasing or monotone decreasing as a function of variable x_i for each i.

Fourier analysis. We will make use of standard Fourier analysis of Boolean functions. The set of functions from the Boolean cube $\{-1, 1\}^n$ to $\mathbf{R}$ forms a 2^n-dimensional inner product space with inner product given by $\langle f, g \rangle = \mathbf{E}_x[f(x)g(x)]$. The set of functions $(\chi_S)_{S \subseteq [n]}$ defined by $\chi_S(x) = \prod_{i \in S} x_i$ forms a complete orthonormal basis for this space. Given a function $f : \{-1, 1\}^n \to \mathbf{R}$ we define its *Fourier coefficients* by $\hat{f}(S) = \mathbf{E}_x[f(x)\chi_S]$, and we have that $f(x) = \sum_S \hat{f}(S)\chi_S$. We will be particularly interested in f's *degree*-1 coefficients, i.e., $\hat{f}(S)$ for $|S| = 1$; for brevity we will write these as $\hat{f}(i)$ rather than $\hat{f}(\{i\})$. Finally, we have *Plancherel's identity* $\langle f, g \rangle = \sum_S \hat{f}(S)\hat{g}(S)$, which has as a special case *Parseval's identity*, $\mathbf{E}_x[f(x)^2] = \sum_S \hat{f}(S)^2$. It follows that for every $f : \{-1, 1\}^n \to \{-1, 1\}$ we have $\sum_S \hat{f}(S)^2 = 1$.

Probability bounds. To prove our lower bound we will require the Berry-Esseen theorem, a version of the Central Limit Theorem with error bounds (see e.g. [2]):

Theorem 1. *Let* $\ell(x) = c_1 x_1 + \cdots + c_n x_n$ *be a linear form over the random ± 1 bits* x_i. *Assume* $|c_i| \leq \tau$ *for all i and write* $\sigma = \sqrt{\sum c_i^2}$. *Write F for the c.d.f. of $\ell(x)/\sigma$; i.e.,* $F(t) = \Pr[\ell(x)/\sigma \leq t]$. *Then for all $t \in \mathbf{R}$,*

$$|F(t) - \Phi(t)| \leq O(\tau/\sigma) \cdot \frac{1}{1 + |t|^3},$$

where Φ denotes the c.d.f. of X, a standard Gaussian random variable. In particular, if $A \subseteq \mathbf{R}$ *is any interval then* $|\Pr[\ell(x)/\sigma \in A] - \Pr[X \in A]| \leq O(\tau/\sigma)$.

A special case of this theorem, with a sharper constant, is also useful (the following can be found in [10]):

Theorem 2. *Let $\ell(x)$ and τ be as defined in Theorem 1. Then for any $\lambda \geq \tau$ and any* $\theta \in \mathbf{R}$ *it holds that* $\Pr[|\ell(x) - \theta| \leq \lambda] \leq 6\lambda/\sigma$.

3 A $\Omega(\log n)$ Lower Bound for Testing ± 1-Weight Halfspaces

In this section we prove the following theorem:

Theorem 3. *There is a fixed constant $\epsilon > 0$ such that any nonadaptive ϵ-testing algorithm $\mathcal{A}$ for the class of all ± 1-weight halfspaces must make at least $(1/26) \log n$ many queries.*

To prove Theorem 3, we define two distributions D_{YES} and D_{NO} over functions. The "yes" distribution D_{YES} is uniform over all 2^n ± 1-weight halfspaces, i.e., a function f drawn from D_{YES} is $f(x) = \text{sgn}(r_1 x_1 + \cdots r_n x_n)$ where each r_i is independently and uniformly chosen to be ± 1. The "no" distribution D_{NO} is similarly a distribution over halfspaces of the form $f(x) = \text{sgn}(s_1 x_1 + \cdots s_n x_n)$, but each s_i is independently chosen to be $\pm \sqrt{1/2}$ or $\pm \sqrt{3/2}$ each with probability $1/4$.

To show that this approach yields a lower bound we must prove two things. First, we must show that a function drawn from D_{NO} is with high probability far from any ± 1-weight halfspace. This is formalized in the following lemma:

Lemma 1. *Let f be a random function drawn from D_{NO}. With probability at least $1 - o(1)$ we have that f is ϵ-far from any ± 1-weight halfspace, where $\epsilon > 0$ is some fixed constant independent of n.*

Next, we must show that no algorithm making $o(\log n)$ queries can distinguish D_{YES} and D_{NO}. This is formalized in the following lemma:

Lemma 2. *Fix any set $x^1, \ldots, x^q$ of q query strings from $\{-1,1\}^n$. Let $\widetilde{D}_{YES}$ be the distribution over $\{-1,1\}^q$ obtained by drawing a random f from D_{YES} and evaluating it on $x^1, \ldots, x^q$. Let $\widetilde{D}_{NO}$ be the distribution over $\{-1,1\}^q$ obtained by drawing a random f from D_{NO} and evaluating it on $x^1, \ldots, x^q$. If $q = (1/26) \log n$ then $\|\widetilde{D}_{YES} - \widetilde{D}_{NO}\|_1 = o(1)$.*

We prove Lemmas 1 and 2 in subsections 3.1 and 3.2 respectively. A standard argument using Yao's method (see e.g. Section 8 of [3]) implies that the lemmas taken together prove Theorem 3.

3.1   Proof of Lemma 1

Let f be drawn from D_{NO}, and let $s_1, \ldots, s_n$ denote the coefficients thus obtained. Let T_1 denote $\{i : |s_i| = \sqrt{1/2}\}$ and T_2 denote $\{i : |s_i| = \sqrt{3/2}\}$. We may assume that both $|T_1|$ and $|T_2|$ lie in the range $[n/2 - \sqrt{n \log n}, n/2 + \sqrt{n \log n}]$ since the probability that this fails to hold is $1 - o(1)$. It will be slightly more convenient for us to view f as $\text{sgn}(\sqrt{2}(s_1 x_1 + \cdots + s_n x_n))$, that is, such that all coefficients are of magnitude 1 or $\sqrt{3}$.

It is easy to see that the closest ± 1-weight halfspace to f must have the same sign pattern in its coefficients that f does. Thus we may assume without loss of generality that f's coefficients are all $+1$ or $+\sqrt{3}$, and it suffices to show that f is far from the majority function $\text{Maj}(x) = \text{sgn}(x_1 + \cdots + x_n)$.

Let Z be the set consisting of those $z \in \{-1,1\}^{T_1}$ (i.e. assignments to the variables in T_1) which satisfy $S_{T_1} = \sum_{i \in T_1} z_i \in [\sqrt{n/2}, 2\sqrt{n/2}]$. Since we are assuming that $|T_1| \approx n/2$, using Theorem 1, we have that $|Z|/2^{|T_1|} = C_1 \pm o(1)$ for constant $C_1 = \Phi(2) - \Phi(1) > 0$.

Now fix any $z \in Z$, so $\sum_{i \in T_1} z_i$ is some value $V_z \cdot \sqrt{n/2}$ where $V_z \in [1,2]$. There are $2^{n-|T_1|}$ extensions of z to a full input $z' \in \{-1,1\}^n$. Let $C_{\text{Maj}}(z)$ be the fraction of those extensions which have $\text{Maj}(z') = -1$; in other words, $C_{\text{Maj}}(z)$ is the fraction of strings in $\{-1,1\}^{T_2}$ which have $\sum_{i \in T_2} z_i < -V_z \sqrt{n/2}$. By Theorem 1, this fraction is $\Phi(-V_z) \pm o(1)$. Let $C_f(z)$ be the fraction of the $2^{n-|T_1|}$ extensions of z which have $f(z') = -1$. Since the variables in T_2 all have coefficient $\sqrt{3}$, $C_f(z)$ is the fraction of strings in $\{-1,1\}^{T_2}$ which have $\sum_{i \in T_2} z_i < -(V_z/\sqrt{3})\sqrt{n/2}$, which by Theorem 1 is $\Phi(-V_z/\sqrt{3}) \pm o(1)$.

There is some absolute constant $c > 0$ such that for all $z \in Z$, $|C_f(z) - C_{\mathrm{Maj}}(z)| \geq c$. Thus, for a constant fraction of all possible assignments to the variables in T_1, the functions Maj and f disagree on a constant fraction of all possible extensions of the assignment to all variables in $T_1 \cup T_2$. Consequently, we have that Maj and f disagree on a constant fraction of all assignments, and the lemma is proved. $\square$

3.2   Proof of Lemma 2

For $i = 1, \dots, n$ let $Y^i \in \{-1,1\}^q$ denote the vector of $(x_i^1, \dots, x_i^q)$, that is, the vector containing the values of the ith bits of each of the queries. Alternatively, if we view the n-bit strings $x^1, \dots, x^q$ as the rows of a $q \times n$ matrix, the strings $Y^1, \dots, Y^n$ are the columns. If $f(x) = \mathrm{sgn}(a_1 x_1 + \cdots + a_n x_n)$ is a halfspace, we write $\mathrm{sgn}(\sum_{i=1}^n a_i Y^i)$ to denote $(f(x^1), \dots, f(x^q))$, the vector of outputs of f on $x^1, \dots, x^q$; note that the value $\mathrm{sgn}(\sum_{i=1}^n a_i Y^i)$ is an element of $\{-1,1\}^q$.

Since the statistical distance between two distributions D_1, D_2 on a domain $\mathcal{D}$ of size N is bounded by $N \cdot \max_{x \in \mathcal{D}} |D_1(x) - D_2(x)|$, we have that the statistical distance $\|\widetilde{D}_{YES} - \widetilde{D}_{NO}\|_1$ is at most $2^q \cdot \max_{Q \in \{-1,1\}^q} |\Pr_r[\mathrm{sgn}(\sum_{i=1}^n r_i Y^i) = Q] - \Pr_s[\mathrm{sgn}(\sum_{i=1}^n s_i Y^i) = Q]|$. So let us fix an arbitrary $Q \in \{-1,1\}^q$; it suffices for us to bound

$$\left| \Pr_r[\mathrm{sgn}(\sum_{i=1}^n r_i Y^i) = Q] - \Pr_s[\mathrm{sgn}(\sum_{i=1}^n s_i Y^i) = Q] \right|. \tag{1}$$

Let InQ denote the indicator random variable for the quadrant Q, i.e. given $x \in \mathbf{R}^q$ the value of $\mathrm{InQ}(x)$ is 1 if x lies in the quadrant corresponding to Q and is 0 otherwise. We have

$$(1) = \left| \mathbf{E}_r[\mathrm{InQ}(\sum_{i=1}^n r_i Y^i)] - \mathbf{E}_s[\mathrm{InQ}(\sum_{i=1}^n s_i Y^i)] \right| \tag{2}$$

We then note that since the Y^i vectors are of length q, there are at most 2^q possibilities in $\{-1,1\}^q$ for their values which we denote by $\widetilde{Y}^1, \dots, \widetilde{Y}^{2^q}$. We lump together those vectors which are the same: for $i = 1, \dots, 2^q$ let c_i denote the number of times that $\widetilde{Y}^i$ occurs in $Y^1, \dots, Y^n$. We then have that $\sum_{i=1}^n r_i Y^i = \sum_{i=1}^{2^q} a_i \widetilde{Y}^i$ where each a_i is an independent random variable which is a sum of c_i independent ± 1 random variables (the r_j's for those j that have $Y^j = \widetilde{Y}^i$). Similarly, we have $\sum_{i=1}^n s_i Y^i = \sum_{i=1}^{2^q} b_i \widetilde{Y}^i$ where each b_i is an independent random variable which is a sum of c_i independent variables distributed as the s_j's (these are the s_j's for those j that have $Y^j = \widetilde{Y}^i$). We thus can re-express (2) as

$$\left| \mathbf{E}_a[\mathrm{InQ}(\sum_{i=1}^{2^q} a_i \widetilde{Y}^i)] - \mathbf{E}_b[\mathrm{InQ}(\sum_{i=1}^{2^q} b_i \widetilde{Y}^i)] \right|. \tag{3}$$

Let us define a sequence of random variables that hybridize between $\sum_{i=1}^{2^q} a_i \widetilde{Y}^i$ and $\sum_{i=1}^{2^q} b_i \widetilde{Y}^i$. For $1 \leq \ell \leq 2^q + 1$ define

$$Z_\ell := \sum_{i<\ell} b_i \widetilde{Y}^i + \sum_{i\geq\ell} a_i \widetilde{Y}^i, \qquad \text{so} \qquad Z_1 = \sum_{i=1}^{2^q} a_i \widetilde{Y}^i \quad \text{and} \quad Z_{2^q+1} = \sum_{i=1}^{2^q} b_i \widetilde{Y}^i.$$

$$(4)$$

As is typical in hybrid arguments, by telescoping (3), we have that (3) equals

$$\left| \mathbf{E}_{a,b}[\sum_{\ell=1}^{2^q} \mathrm{InQ}(Z_\ell) - \mathrm{InQ}(Z_{\ell+1})]\right| = \left| \sum_{\ell=1}^{2^q} \mathbf{E}_{a,b}[\mathrm{InQ}(Z_\ell) - \mathrm{InQ}(Z_{\ell+1})]\right|$$

$$= \left| \sum_{\ell=1}^{2^q} \mathbf{E}_{a,b}[\mathrm{InQ}(W_\ell + a_\ell \widetilde{Y}^\ell) - \mathrm{InQ}(W_\ell + b_\ell \widetilde{Y}^\ell)]\right| \qquad (5)$$

where $W_\ell := \sum_{i<\ell} b_i \widetilde{Y}^i + \sum_{i>\ell} a_i \widetilde{Y}^i$. The RHS of (5) is at most

$$2^q \cdot \max_{\ell=1,\dots,2^q} | \mathbf{E}_{a,b}[\mathrm{InQ}(W_\ell + a_\ell \widetilde{Y}^\ell) - \mathrm{InQ}(W_\ell + b_\ell \widetilde{Y}^\ell)]|.$$

So let us fix an arbitrary ℓ; we will bound

$$\left| \mathbf{E}_{a,b}[\mathrm{InQ}(W_\ell + a_\ell \widetilde{Y}^\ell) - \mathrm{InQ}(W_\ell + b_\ell \widetilde{Y}^\ell)]\right| \leq B \qquad (6)$$

(we will specify B later), and this gives that $\|\widetilde{D}_{YES} - \widetilde{D}_{NO}\|_1 \leq 4^q B$ by the arguments above. Before continuing further, it is useful to note that W_ℓ, a_ℓ, and b_ℓ are all independent from each other.

Bounding (6). Let $N := (n/2^q)^{1/3}$. Without loss of generality, we may assume that the the c_i's are in monotone increasing order, that is $c_1 \leq c_2 \leq \dots \leq c_{2^q}$. We consider two cases depending on the value of c_ℓ. If $c_\ell > N$ then we say that c_ℓ is *big*, and otherwise we say that c_ℓ is *small*. Note that each c_i is a nonnegative integer and $c_1 + \dots + c_{2^q} = n$, so at least one c_i must be big; in fact, we know that the largest value c_{2^q} is at least $n/2^q$.

If c_ℓ is big, we argue that a_ℓ and b_ℓ are distributed quite similarly, and thus for any possible outcome of W_ℓ the LHS of (6) must be small. If c_ℓ is small, we consider some $k \neq \ell$ for which c_k is very big (we just saw that $k = 2^q$ is such a k) and show that for any possible outcome of a_ℓ, b_ℓ and all the other contributors to W_ℓ, the contribution to W_ℓ from this c_k makes the LHS of (6) small (intuitively, the contribution of c_k is so large that it "swamps" the small difference that results from considering a_ℓ versus b_ℓ).

Case 1: Bounding (6) when c_ℓ is big, i.e. $c_\ell > N$. Fix any possible outcome for W_ℓ in (6). Note that the vector $\widetilde{Y}^\ell$ has all its coordinates ± 1 and thus it is "skew" to each of the axis-aligned hyperplanes defining quadrant Q. Since Q is convex, there is some interval A (possibly half-infinite) of the real line such that for all $t \in \mathbf{R}$ we have $\mathrm{InQ}(W_\ell + t\widetilde{Y}^\ell) = 1$ if and only if $t \in A$. It follows that

$$| \Pr_{a_\ell}[\mathrm{InQ}(W_\ell + a_\ell \widetilde{Y}^\ell) = 1] - \Pr_{b_\ell}[\mathrm{InQ}(W_\ell + b_\ell \widetilde{Y}^\ell) = 1]| = | \Pr[a_\ell \in A] - \Pr[b_\ell \in A]|.$$

$$(7)$$

Now observe that as in Theorem 1, a_ℓ and b_ℓ are each sums of c_ℓ many independent zero-mean random variables (the r_j's and s_j's respectively) with the same total variance $\sigma = \sqrt{c_\ell}$ and with each $|r_j|, |s_j| \leq O(1)$. Applying Theorem 1 to both a_ℓ and b_ℓ, we get that the RHS of (7) is at most $O(1/\sqrt{c_\ell}) = O(1/\sqrt{N})$. Averaging the LHS of (7) over the distribution of values for W_ℓ, it follows that if c_ℓ is big then the LHS of (6) is at most $O(1/\sqrt{N})$.

Case 2: Bounding (6) when c_ℓ is small, i.e. $c_\ell \leq N$. We first note that every possible outcome for a_ℓ, b_ℓ results in $|a_\ell - b_\ell| \leq O(N)$. Let $k = 2^q$ and recall that $c_k \geq n/2^q$. Fix any possible outcome for a_ℓ, b_ℓ and for all other a_j, b_j such that $j \neq k$ (so the only "unfixed" randomess at this point is the choice of a_k and b_k). Let W'_ℓ denote the contribution to W_ℓ from these $2^q - 2$ fixed a_j, b_j values, so W_ℓ equals $W'_\ell + a_k \widetilde{Y}^k$ (since $k > \ell$). (Note that under this supposition there is actually no dependence on b_k now; the only randomness left is the choice of a_k.)

We have

$$| \Pr_{a_k}[\mathrm{InQ}(W_\ell + a_\ell \widetilde{Y}^\ell) = 1] - \Pr_{a_k}[\mathrm{InQ}(W_\ell + b_\ell \widetilde{Y}^\ell) = 1]|$$

$$= | \Pr_{a_k}[\mathrm{InQ}(W'_\ell + a_\ell \widetilde{Y}^\ell + a_k \widetilde{Y}^k) = 1] - \Pr_{a_k}[\mathrm{InQ}(W'_\ell + b_\ell \widetilde{Y}^\ell + a_k \widetilde{Y}^k) = 1]| \quad (8)$$

The RHS of (8) is at most

$$\Pr_{a_k}[\text{the vector } W'_\ell + a_\ell \widetilde{Y}^\ell + a_k \widetilde{Y}^k \text{ has any coordinate of magnitude at most } |a_\ell - b_\ell|].$$
$$(9)$$

(If each coordinate of $W'_\ell + a_\ell \widetilde{Y}^\ell + a_k \widetilde{Y}^k$ has magnitude greater than $|a_\ell - b_\ell|$, then each corresponding coordinate of $W'_\ell + b_\ell \widetilde{Y}^\ell + a_k \widetilde{Y}^k$ must have the same sign, and so such an outcome affects each of the probabilities in (8) in the same way – either both points are in quadrant Q or both are not.) Since each coordinate of $\widetilde{Y}^k$ is of magnitude 1, by a union bound the probability (9) is at most q times

$$\max_{\text{all intervals } A \text{ of width } 2|a_\ell - b_\ell|} \Pr_{a_k}[a_k \in A]. \quad (10)$$

Now using the fact that $|a_\ell - b_\ell| = O(N)$, the fact that a_k is a sum of $c_k \geq n/2^q$ independent ± 1-valued variables, and Theorem 2, we have that (10) is at most $O(N)/\sqrt{n/2^q}$. So we have that (8) is at most $O(Nq\sqrt{2^q})/\sqrt{n}$. Averaging (8) over a suitable distribution of values for $a_1, b_1, \ldots, a_{k-1}, b_{k-1}, a_{k+1}, b_{k+1}, \ldots, a_{2^q}, b_{2^q}$, gives that the LHS of (6) is at most $O(Nq\sqrt{2^q})/\sqrt{n}$.

So we have seen that whether c_ℓ is big or small, the value of (6) is upper bounded by

$$\max\{O(1/\sqrt{N}), O(Nq\sqrt{2^q})/\sqrt{n}\}.$$

Recalling that $N = (n/2^q)^{1/3}$, this equals $O(q(2^q/n)^{1/6})$, and thus $\|\widetilde{D}_{YES} - \widetilde{D}_{NO}\|_1 \leq O(q2^{13q/6}/n^{1/6})$. Recalling that $q = (1/26)\log n$, this equals $O((\log n)/n^{1/12}) = o(1)$, and Lemma 2 is proved.

4   A Sublinear Algorithm for Testing ± 1-Weight Halfspaces

In this section we present the ± 1-**Weight Halfspace-Test** algorithm, and prove the following theorem:

Theorem 4. *For any $36/n < \epsilon < 1/2$ and any function $f : \{-1,1\}^n \to \{-1,1\}$,*

- *if f is a ±1-weight halfspace, then* ±1-**Weight Halfspace-Test**(f, ϵ) *passes with probability $\geq 2/3$,*
- *if f is ϵ-far from any ±1-weight halfspace, then* ±1-**Weight Halfspace-Test**(f, ϵ) *rejects with probability $\geq 2/3$.*

The query complexity of ±1-**Weight Halfspace-Test**(f, ϵ) *is $O(\sqrt{n}\frac{1}{\epsilon^6}\log\frac{1}{\epsilon})$. The algorithm is nonadaptive and has two-sided error.*

The main tool underlying our algorithm is the following theorem, which says that if most of f's degree-1 Fourier coefficients are almost as large as those of the majority function, then f must be close to the majority function. Here we adopt the shorthand Maj_n to denote the majority function on n variables, and $\hat{\mathsf{M}}_n$ to denote the value of the degree-1 Fourier coefficients of Maj_n.

Theorem 5. *Let $f : \{-1,1\}^n \to \{-1,1\}$ be any Boolean function and let $\epsilon > 36/n$. Suppose that there is a subset of $m \geq (1 - \epsilon)n$ variables i each of which satisfies $\hat{f}(i) \geq (1 - \epsilon)\hat{\mathsf{M}}_n$. Then $\Pr[f(x) \neq \mathrm{Maj}_n(x)] \leq 32\sqrt{\epsilon}$.*

In the following subsections we prove Theorem 5 and then present our testing algorithm.

4.1   Proof of Theorem 5

Recall the following well-known lemma, whose proof serves as a warmup for Theorem 5:

Lemma 3. *Every $f : \{-1,1\}^n \to \{-1,1\}$ satisfies $\sum_{i=1}^n |\hat{f}(i)| \leq n\hat{\mathsf{M}}_n$.*

Proof. Let $G(x) = \mathrm{sgn}(\hat{f}(1))x_1 + \cdots + \mathrm{sgn}(\hat{f}(n))x_n$ and let $g(x)$ be the ±1-weight halfspace $g(x) = \mathrm{sgn}(G(x))$. We have

$$\sum_{i=1}^n |\hat{f}(i)| = \mathbf{E}[fG] \leq \mathbf{E}[|G|] = \mathbf{E}[G(x)g(x)] = \sum_{i=1}^n \hat{\mathsf{M}}_n,$$

where the first equality is Plancherel (using the fact that G is linear), the inequality is because f is a ±1-valued function, the second equality is by definition of g and the third equality is Plancherel again, observing that each $\hat{g}(i)$ has magnitude $\hat{\mathsf{M}}_n$ and sign $\mathrm{sgn}(\hat{f}(i))$.   □

Proof of Theorem 5. For notational convenience, we assume that the variables whose Fourier coefficients are "almost right" are $x_1, x_2, ..., x_m$. Now define $G(x) = x_1 + x_2 + \cdots x_n$, so that $\mathrm{Maj}_n = \mathrm{sgn}(G)$. We are interested in the difference between the following two quantities:

$$\mathbf{E}[|G(x)|] = \mathbf{E}[G(x)\mathrm{Maj}_n(x)] = \sum_S \hat{G}(S)\hat{\mathrm{Maj}}_n(S) = \sum_{i=1}^n \hat{\mathrm{Maj}}_n(i) = n\hat{\mathsf{M}}_n,$$

$$\mathbf{E}[G(x)f(x)] = \sum_S \hat{G}(S)\hat{f}(S) = \sum_{i=1}^n \hat{f}(i) = \sum_{i=1}^m \hat{f}(i) + \sum_{i=m+1}^n \hat{f}(i).$$

The bottom quantity is broken into two summations. We can lower bound the first summation by $(1 - \epsilon)^2 n \hat{\mathsf{M}}_n \geq (1 - 2\epsilon) n \hat{\mathsf{M}}_n$. This is because the first summation contains at least $(1-\epsilon)n$ terms, each of which is at least $(1-\epsilon)\hat{\mathsf{M}}_n$. Given this, Lemma 3 implies that the second summation is at least $-2\epsilon n \hat{\mathsf{M}}_n$. Thus we have

$$\mathbf{E}[G(x)f(x)] \geq (1 - 4\epsilon) n \hat{\mathsf{M}}_n$$

and hence

$$\mathbf{E}[|G| - Gf] \leq 4\epsilon n \hat{\mathsf{M}}_n \leq 4\epsilon\sqrt{n} \tag{11}$$

where we used the fact (easily verified from Parseval's equality) that $\hat{\mathsf{M}}_n \leq \frac{1}{\sqrt{n}}$.

Let p denote the fraction of points such that $f \neq \text{sgn}(G)$, i.e. $f \neq \text{Maj}_n$. If $p \leq 32\sqrt{\epsilon}$ then we are done, so we assume $p > 32\sqrt{\epsilon}$ and obtain a contradiction. Since $\epsilon \geq 36/n$, we have $p \geq 192/\sqrt{n}$. Let k be such that $\sqrt{\epsilon} = (4k+2)/\sqrt{n}$, so in particular $k \geq 1$. It is well known (by Stirling's approximation) that each "layer" $\{x \in \{-1,1\}^n : x_1 + \cdots + x_n = \ell\}$ of the Boolean cube contains at most a $\frac{1}{\sqrt{n}}$ fraction of $\{-1,1\}^n$, and consequently at most a $\frac{2k+1}{\sqrt{n}}$ fraction of points have $|G(x)| \leq 2k$. It follows that at least a $p/2$ fraction of points satisfy both $|G(x)| > 2k$ and $f(x) \neq \text{Maj}_n(x)$. Since $|G(x)| - G(x)f(x)$ is at least $4k$ on each such point and $|G(x)| - G(x)f(x)$ is never negative, this implies that the LHS of (11) is at least

$$\frac{p}{2} \cdot 4k > (16\sqrt{\epsilon}) \cdot (4k) \geq (16\sqrt{\epsilon})(2k + 1) = (16\sqrt{\epsilon}) \cdot \frac{\sqrt{\epsilon n}}{2} = 8\epsilon\sqrt{n},$$

but this contradicts (11). This proves the theorem. $\square$

4.2   A Tester for ± 1-Weight Halfspaces

Intuitively, our algorithm works by choosing a handful of random indices $i \in [n]$, estimating the corresponding $|\hat{f}(i)|$ values (while checking unateness in these variables), and checking that each estimate is almost as large as $\hat{\mathsf{M}}_n$. The correctness of the algorithm is based on the fact that if f is unate and most $|\hat{f}(i)|$ are large, then some *reorientation* of f (that is, a replacement of some x_i by $-x_i$) will make most $\hat{f}(i)$ large. A simple application of Theorem 5 then implies that the reorientation is close to Maj_n, and therefore that f is close to a ± 1-weight halfspace.

We start with some preliminary lemmas which will assist us in estimating $|\hat{f}(i)|$ for functions that we expect to be unate.

Lemma 4

$$\hat{f}(i) = \Pr_x[f(x^{i-}) < f(x^{i+})] - \Pr_x[f(x^{i-}) > f(x^{i+})]$$

where x^{i-} and x^{i+} denote the bit-string x with the ith bit set to -1 or 1 respectively.

We refer to the first probability above as the *positive influence* of variable i and the second probability as the *negative influence* of i. Each variable in a monotone function has only positive influence. Each variable in a *unate* function has only positive influence or negative influence, but not both.

Proof. (of Lemma 4) First note that $\hat{f}(i) = \mathbf{E}_x[f(x)x_i]$, then

$$\mathbf{E}_x[f(x)x_i] = \Pr_x[f(x) = 1, x_i = 1] + \Pr_x[f(x) = -1, x_i = -1]$$
$$- \Pr_x[f(x) = -1, x_i = 1] - \Pr_x[f(x) = 1, x_i = -1].$$

Now group all x's into pairs (x^{i-}, x^{i+}) that differ in the ith bit. If the value of f is the same on both elements of a pair, then the total contribution of that pair to the expectation is zero. On the other hand, if $f(x^{i-}) < f(x^{i+})$, then x^{i-} and x^{i+} each add $\frac{1}{2^n}$ to the expectation, and if $f(x^{i-}) > f(x^{i+})$, then x^{i-} and x^{i+} each subtract $\frac{1}{2^n}$. This yields the desired result.   □

Lemma 5. *Let f be any Boolean function, $i \in [n]$, and let $|\hat{f}(i)| = p$. By drawing $m = \frac{3}{pe^2} \cdot \log \frac{2}{\delta}$ uniform random strings $x \in \{-1,1\}^n$, and querying f on the values $f(x^{i+})$ and $f(x^{i-})$, with probability $1 - \delta$ we either obtain an estimate of $|\hat{f}(i)|$ accurate to within a multiplicative factor of $(1 \pm \epsilon)$, or discover that f is not unate.*

The idea of the proof is that if neither the positive influence nor the negative influence is small, random sampling will discover that f is not unate. Otherwise, $|\hat{f}(i)|$ is well approximated by either the positive or negative influence, and a standard multiplicative form of the Chernoff bound shows that m samples suffice.

Proof. (of Lemma 5) Suppose first that both the positive influence and negative influence are at least $\frac{\epsilon p}{2}$. Then the probability that we do not observe any pair with positive influence is $\leq (1 - \frac{\epsilon p}{2})^m \leq e^{-\epsilon pm/2} = e^{-(3/2\epsilon)\log(2/\delta)} < \frac{\delta}{2}$, and similarly for the negative influence. Therefore, the probability that we observe at least some positive influence and some negative influence (and therefore discover that f is not unate) is at least $1 - 2\frac{\delta}{2} = 1 - \delta$.

Now consider the case when either the positive influence or the negative influence is less than $\frac{\epsilon p}{2}$. Without loss of generality, assume that the negative influence is less than $\frac{\epsilon p}{2}$. Then the positive influence is a good estimate of $|\hat{f}(i)|$. In particular, the probability that the estimate of the positive influence is not within $(1 \pm \frac{\epsilon}{2})p$ of the true value (and therefore the estimate of $|\hat{f}(i)|$ is not within $(1 \pm \epsilon)p$), is at most $< 2e^{-mp\epsilon^2/3} = 2e^{-\log \frac{2}{\delta}} = \delta$ by the multiplicative Chernoff bound. So in this case, the probability that the estimate we receive is accurate to within a multiplicative factor of $(1 \pm \epsilon)$ is at least $1 - \delta$. This concludes the proof.   □

Now we are ready to present the algorithm and prove its correctness.

±1-Weight Halfspace-Test (inputs are $\epsilon > 0$ and black-box access to $f : \{-1,1\}^n \to \{-1,1\}$)

1. Let $\epsilon' = (\frac{\epsilon}{32})^2$.
2. Choose $k = \frac{1}{\epsilon'} \ln 6 = O(\frac{1}{\epsilon'})$ many random indices $i \in \{1, ..., n\}$.
3. For each i, estimate $|\hat{f}(i)|$. Do this as in Lemma 5 by drawing $m = \frac{24 \log 12k}{\hat{M}_n \epsilon'^2} = O(\frac{\sqrt{n}}{\epsilon'^2} \log \frac{1}{\epsilon'})$ random x's and querying $f(x^{i+})$ and $f(x^{i-})$. If a violation of unateness is found, reject.
4. Pass if and only if each estimate is larger than $(1 - \frac{\epsilon'}{2})\hat{M}_n$.

Proof. (of Theorem 4) To prove that the test is correct, we need to show two things: first that it passes functions which are ± 1-weight halfspaces, and second that anything it passes with high probability must be ϵ-close to a ± 1-weight halfspace. To prove the first, note that if f is a ± 1-weight halfspace, the only possibility for rejection is if any of the estimates of $|\hat{f}(i)|$ is less than $(1 - \frac{\epsilon'}{2})\hat{M}_n$. But applying lemma 5 (with $p = \hat{M}_n$, $\epsilon = \frac{\epsilon'}{2}$, $\delta = \frac{1}{6k}$), the probability that a particular estimate is wrong is $< \frac{1}{6k}$, and therefore the probability that any estimate is wrong is $< \frac{1}{6}$. Thus the probability of success is $\geq \frac{5}{6}$.

The more difficult part is showing that any function which passes the test whp must be close to a ± 1-weight halfspace. To do this, note that if f passes the test whp then it must be the case that for all but an ϵ' fraction of variables, $|\hat{f}(i)| > (1 - \epsilon')\hat{M}_n$. If this is not the case, then Step 2 will choose a "bad" variable – one for which $|\hat{f}(i)| \leq (1 - \epsilon')\hat{M}_n$ – with probability at least $\frac{5}{6}$. Now we would like to show that for any bad variable i, the estimate of $|\hat{f}(i)|$ is likely to be less than $(1 - \frac{\epsilon'}{2})\hat{M}_n$. Without loss of generality, assume that $|\hat{f}(i)| = (1 - \epsilon')\hat{M}_n$ (if $|\hat{f}(i)|$ is less than that, then variable i will be even less likely to pass step 3). Then note that it suffices to estimate $|\hat{f}(i)|$ to within a multiplicative factor of $(1 + \frac{\epsilon'}{2})$ (since $(1 + \frac{\epsilon'}{2})(1 - \epsilon')\hat{M}_n < (1 - \frac{\epsilon'}{2})\hat{M}_n$). Again using Lemma 5 (this time with $p = (1 - \epsilon')\hat{M}_n$, $\epsilon = \frac{\epsilon'}{2}$, $\delta = \frac{1}{6k}$), we see that $\frac{12}{\hat{M}\epsilon'^2(1-\epsilon')} \log 12k < \frac{24}{\hat{M}\epsilon'^2} \log 12k$ samples suffice to achieve discover the variable is bad with probability $1 - \frac{1}{6k}$. The total probability of failure (the probability that we fail to choose a bad variable, or that we mis-estimate one when we do) is thus $< \frac{1}{6} + \frac{1}{6k} < \frac{1}{3}$.

The query complexity of the algorithm is $O(km) = O(\sqrt{n}\frac{1}{\epsilon'^3} \log \frac{1}{\epsilon'}) = O(\sqrt{n} \cdot \frac{1}{\epsilon^6} \log \frac{1}{\epsilon})$.   □

5   Conclusion

We have proven a lower bound showing that the complexity of testing ± 1-weight halfspaces is is at least $\Omega(\log n)$ and an upper bound showing that it is at most $O(\sqrt{n} \cdot \text{poly}(\frac{1}{\epsilon}))$. An open question is to close the gap between these bounds and determine the exact dependence on n. One goal is to use some type of binary search to get a poly $\log(n)$-query adaptive testing algorithm; another is to improve our lower bound to $n^{\Omega(1)}$ for nonadaptive algorithms.

References

[1] Block, H.: The Perceptron: a model for brain functioning. Reviews of Modern Physics 34, 123–135 (1962)
[2] Feller, W.: An introduction to probability theory and its applications. John Wiley & Sons, Chichester (1968)
[3] Fischer, E.: The art of uninformed decisions: A primer to property testing. Bulletin of the European Association for Theoretical Computer Science 75, 97–126 (2001)
[4] Goldreich, O., Goldwasser, S., Ron, D.: Property testing and its connection to learning and approximation. Journal of the ACM 45, 653–750 (1998)

[5] Hajnal, A., Maass, W., Pudlak, P., Szegedy, M., Turan, G.: Threshold circuits of bounded depth. Journal of Computer and System Sciences 46, 129–154 (1993)

[6] Kulkarni, S., Mitter, S., Tsitsiklis, J.: Active learning using arbitrary binary valued queries. Machine Learning 11, 23–35 (1993)

[7] Matulef, K., O'Donnell, R., Rubinfeld, R., Servedio, R.: Testing halfspaces. SIAM J. Comp. (to appear); Extended abstract in Proc. Symp. Discrete Algorithms (SODA), pp. 256–264 (2009); Full version, http://www.cs.cmu.edu/

[8] Minsky, M., Papert, S.: Perceptrons: an introduction to computational geometry. MIT Press, Cambridge (1968)

[9] Novikoff, A.: On convergence proofs on perceptrons. In: Proceedings of the Symposium on Mathematical Theory of Automata, vol. XII, pp. 615–622 (1962)

[10] Petrov, V.V.: Limit theorems of probability theory. Oxford Science Publications, Oxford (1995)

[11] Shawe-Taylor, J., Cristianini, N.: An introduction to support vector machines. Cambridge University Press, Cambridge (2000)

[12] Yao, A.: On ACC and threshold circuits. In: Proceedings of the Thirty-First Annual Symposium on Foundations of Computer Science, pp. 619–627 (1990)

Small-Bias Spaces for Group Products

Raghu Meka* and David Zuckerman**

Department of Computer Science, University of Texas at Austin
{raghu,diz}@cs.utexas.edu

Abstract. Small-bias, or ϵ-biased, spaces have found many applications in complexity theory, coding theory, and derandomization. We generalize the notion of small-bias spaces to the setting of group products. Besides being natural, our extension captures some of the difficulties in constructing pseudorandom generators for constant-width branching programs - a longstanding open problem. We provide an efficient deterministic construction of small-bias spaces for solvable groups. Our construction exploits the fact that solvable groups have nontrivial normal subgroups that are abelian and builds on the construction of Azar et al. [AMN98] for abelian groups. For arbitrary finite groups, we give an efficient deterministic construction achieving constant bias. We also construct pseudorandom generators fooling linear functions mod p for primes p.

1 Introduction

In this work we generalize the notion of small-bias spaces to the setting of group products. Small-bias, or ϵ-biased, spaces over $\mathbb{Z}_2$ have been very useful in constructions of various pseudorandom objects. In particular, they are used in the construction of almost k-wise independent spaces ([NN93]), which in turn have many applications such as universal sets ([LY94], [BEG+94]). An application of interest to us is that ϵ-biased spaces fool branching programs of width two. Can we generalize this observation to fool constant-width branching programs? Our extension of small-bias spaces to finite groups besides being interesting on its own, could be useful for constructing pseudorandom generators for small width branching programs. We address the problem of explicitly constructing such small-bias spaces over finite groups, and give an efficient deterministic construction for solvable groups and a partial solution to the problem for arbitrary finite groups.

Constructing pseudorandom generators for constant-width branching programs is a fundamental problem with many applications in circuit lower bounds and derandomization. The problem is largely open even for strongly restricted classes such as width three *read-once permutation branching programs*

* Supported in part by NSF Grant CCF-0634811 and THECB ARP Grant 003658-0113-2007.
** Supported in part by NSF Grant CCF-0634811.

I. Dinur et al. (Eds.): APPROX and RANDOM 2009, LNCS 5687, pp. 658–672, 2009.

(ROPBPs) - branching programs where no variable is read more than once and the edges between any two layers with the same label define a permutation.

Our notion of small-bias spaces is motivated by the following group-theoretic formulation of the problem of constructing pseudorandom generators for constant-width ROPBPs. Consider the edges between layers i and $i + 1$ of a width-w ROPBP. By relabeling the nodes if necessary, we may assume that the permutation corresponding to the edges labeled 0 is the identity permutation. Then the permutation corresponding to the label 1 is some permutation $g_i \in S_w$, where S_w denotes the permutation group on w elements. Under this description, if the variable read by the branching program at layer i is x_i, the computation performed by the ROPBP can be written as follows: on input $(b_1, \ldots, b_n) \in \{0, 1\}^n$, output $g_1^{b_1} g_2^{b_2} \ldots g_n^{b_n} \in S_w$.

Thus, pseudorandom generators for width w ROPBPs are equivalent to functions $P : \{0, 1\}^r \to \{0, 1\}^n$ that fool *products of group elements* in the sense that, for all $g_1, \ldots, g_n \in S_w$, the distributions of $g_1^{b_1} g_2^{b_2} \ldots g_n^{b_n}$ with $b \in_u \{0, 1\}^n$ and $g_1^{P(y)_1} g_2^{P(y)_2} \ldots g_n^{P(y)_n}$ with $y \in_u \{0, 1\}^r$ are ϵ-close in variation distance (for a multi-set S, $x \in_u S$ denotes a uniformly sampled element of S.).

In this work we consider a dual of the above problem. A PRG for ROPBPs outputs Boolean exponents that fool products of arbitrary group elements raised to these exponents. Our notion of ϵ-biased space outputs group elements that fool products of these elements raised to arbitrary exponents. For convenience we use the max norm instead of variation distance.

Definition 1.1. *A multi-set $S \subseteq G^n$ is an ϵ-biased space for products over G[1] if for all $b_1, \ldots, b_n \in \{0, 1\}$ not all zero, and every $h \in G$*

$$\left| \Pr_{g \in_u S} [g_1^{b_1} g_2^{b_2} \ldots g_n^{b_n} = h] - \frac{1}{|G|} \right| \leq \epsilon.$$

Remark. The definition can naturally be extended to non-binary exponents $b_1, \ldots, b_n \in [|G|]$ and arbitrary permutations $\pi : [n] \to [n]$, where we look at products of the form $g_{\pi(1)}^{b_1} g_{\pi(2)}^{b_2} \ldots g_{\pi(n)}^{b_n}$. In this abstract we only consider the definition above for simplicity. Our results extend straightforwardly to arbitrary permutations π as well as for non-binary powers $b_i \in [|G|]$, provided $\gcd(b_1, \ldots, b_n, |G|) = 1$.

When the group G is the additive group $\mathbb{Z}_2$, the definition above coincides with the usual notion of small-bias spaces over $\mathbb{Z}_2$ of Naor and Naor [NN93]. Besides being a natural generalization of ϵ-biased spaces over $\mathbb{Z}_2$, the definition above captures some of the difficulties involved in constructing PRGs for constant-width ROPBPs, as PRGs for constant-width ROPBP imply ϵ-biased spaces for finite groups.

Theorem 1.1. *Given a PRG $G : \{0, 1\}^r \to \{0, 1\}^n$ for width w ROPBP with error at most ϵ and running time $t(n, \epsilon)$, for every group $H \subseteq S_w$ there exists*

---

[1] By convention, ϵ-biased spaces with no explicit mention of a group will correspond to ϵ-biased spaces over $\mathbb{Z}_2$. For brevity, we will refer to *small-bias spaces for products over G* simply as *small-bias spaces over G*.

an algorithm with running time $O(t(n,\epsilon)2^r)$ that outputs a 2ϵ-biased set over H of size $poly(2^r, n, 1/\epsilon)$.

We defer the proof to Section 6.

Remark. Azar et al. [AMN98] characterize ϵ-biased spaces for abelian groups in terms of the characters of the group. One could generalize this definition to non-abelian groups using the irreducible characters or irreducible representations of the group. However, there does not seem to be any connection between such objects and pseudorandom generators for constant-width read-once branching programs, our original motivation. As far as we know, a notion of small-bias spaces for finite groups in terms of irreducible representations is incomparable to our notion of small-bias spaces and to pseudorandom generators for constant-width ROPBPs.

By the probabilistic method it can be shown that for any group G, ϵ-biased sets of size $O(|G|n/\epsilon^2)$ exist. The constructions of Naor and Naor [NN93], Alon et al. [AGHP92], Azar et al. [AMN98] give explicit polynomial size ϵ-biased spaces for abelian groups. However, the problem seems to be considerably harder for non-abelian groups and the techniques of [NN93, AGHP92, AMN98] fail when the group is non-abelian. We prove the following results for general finite groups.

Theorem 1.2. *Let G be a finite group. There exists a deterministic algorithm running in time $poly(n)$ that takes as input n and outputs a set S of size $poly(n)$ that is α-biased over G, where $\alpha < 1/|G|$ is a fixed constant depending on $|G|$.*

Theorem 1.3. *Let G be a finite solvable group. There exists a deterministic algorithm running in time $poly(n, 1/\epsilon)$ that takes as input n, ϵ and outputs a set S of size $poly(n, 1/\epsilon)$ that is ϵ-biased over G.*

Our constructions are based on the ϵ-biased spaces for abelian groups of Azar et al. The construction for solvable groups is recursive and uses the fact that every solvable group has a nontrivial normal subgroup that is abelian.

It is instructive to examine the dual objects - PRGs and ϵ-biases spaces - for simpler families of constant-width ROPBPs such as the class of linear functions modulo a prime p. For this case, our notion of ϵ-biased spaces with group $G = \mathbb{Z}_p$ corresponds to the usual notion of ϵ-biased spaces over $\mathbb{Z}_p$ except that our notion assumes the linear functions have $\{0, 1\}$ coefficients. As far as we know there were no previous efficient constructions of PRGs for the ROPBPs corresponding to the family of linear functions modulo a prime p.

Definition 1.2. *A function $G : \{0, 1\}^r \to \{0, 1\}^n$ is said to be an ϵ-pseudorandom bit generator (ϵ-PBG) for sums mod p, if for every $v \in \mathbb{F}_p^n$, $v \neq 0$ and all $a \in \mathbb{F}_p$*

$$|\mathsf{Pr}_{z \in_u \{0,1\}^r}[\langle v, G(z)\rangle = a] - \mathsf{Pr}_{x \in_u \{0,1\}^n}[\langle v, x\rangle = a]| \leq \epsilon, \tag{1}$$

where the inner product is taken over $\mathbb{F}_p$.

Note that the existence of an efficient ϵ-PBG $G : \{0, 1\}^r \to \{0, 1\}^n$ with $r = O(\log n + \log(1/\epsilon))$ does not follow from the known constructions of ϵ-biased spaces for $\mathbb{Z}_p$. (The main difference is that ϵ-biased spaces mod p, by definition, take values in $\mathbb{Z}_p$.) We present a construction of ϵ-PBG with $r = O(\log n + \log(1/\epsilon))$ based on pseudorandom generators for low-degree polynomials obtained in [Vio08, Lov08, BV07]. Recently, independent of our work, Lovett et al. [LRTV09] constructed ϵ-PBG with better dependence on the field size p which also works for composite moduli. However, for our intended application, the field size is always a constant and the construction below is optimal up to constant factors.

Theorem 1.4. *For all $\epsilon > 0$ and primes p, there exists an efficient ϵ-PBG for $\mathbb{F}_p$, $G : \{0,1\}^r \to \{0,1\}^n$, with seed length $r = O(\log n + \log(1/\epsilon))$.*

Observe that a pseudorandom generator for width 3 read-once branching programs gives both an ϵ-biased set over $\mathbb{F}_2$ and an ϵ-PBG for $\mathbb{F}_3$. Motivated in part by our construction of ϵ-biased space for the permutation group on three elements - $\mathbb{S}_3$, a solvable group - we conjecture that a weak converse of the above statement holds.

Conjecture 1.5. *Let $G_1 : \{0,1\}^r \to S$ generate uniform samples from a ϵ-biased set $S \subseteq \{0,1\}^n$. Let $G_2 : \{0,1\}^r \to \{0,1\}^n$ be a ϵ-PBG for $\mathbb{F}_3$. Then, the sum $G_1 \oplus G_2 : \{0,1\}^r \times \{0,1\}^r \to \{0,1\}^n$ defined by $(G_1 \oplus G_2)(z_1, z_2) = G_1(z_1) \oplus G_2(z_2)$ is pseudorandom with respect to width 3 read-once permutation branching programs.*

We also provide an example showing that the sum of two constant-bias spaces over $\mathbb{Z}_2$ does not fool linear functions mod 3; in particular, the sum of two constant-bias spaces does not fool width 3 ROPBPs. Reingold and Vadhan [RV06] had asked whether the sum of two $n^{-O(1)}$-biased spaces fools logspace. Although we do not resolve the question, we remark that previously there was no known example ruling out the possibility that the sum of two constant-biased spaces gives a *hitting set* for logspace computations.

Theorem 1.6. *There exists an absolute constant α, $0 < \alpha < 1/2$, such that for all $n > 0$, there exists an α-biased space $S \subseteq \{0,1\}^n$, such that the dimension of the span of $S \oplus S = \{x \oplus y : x, y \in S\}$ viewed as a subset of the vector space $\mathbb{F}_3^n$ is $o(n)$. In particular, there exists a linear function f mod 3 such that f is constant on $S \oplus S$.*

2 Previous Work and Preliminaries

We first present the notions of pseudorandom generators for small width branching programs and small-bias spaces over abelian groups.

Definition 2.1. *A branching program (BP) of width w and length t ((w,t)-BP) is a rooted, layered directed acyclic graph with $t+1$ layers and at most w nodes at each layer. The nodes (internal nodes) at a layer $j \leq t$ in the graph are labeled with a variable x_i and have two outgoing edges each, corresponding to the two possible values of x_i. The nodes at the last layer (leaf nodes) are labeled* *either* 0 *or* 1. *An instance* $x \in \{0,1\}^n$ *defines a unique directed path through the branching program starting at the root and following the outgoing arc from internal nodes labeled by the value of the variable at that node. The output of the branching program is the label of the leaf node reached by this path.*

A branching program is read-once *(ROBP) if no variable occurs more than once on any path from the root to a leaf. A branching program is a* permutation branching program *(PBP) if any two edges pointing to the same node have different labels.*

Definition 2.2. *A pseudorandom generator (PRG) for width* w *BPs with error* ϵ *is a function* $G : \{0,1\}^r \to \{0,1\}^n$ *such that for every* (w,t)-*BP* A *with* $t = poly(n)$, $A(U_n)$ *is* ϵ-*close to* $A(G(U_r))$, *where* U_k *denotes the uniform distribution on* $\{0,1\}^k$. *Pseudorandom generators for ROBP and read-once permutation branching programs (ROPBP) are defined similarly.*

Constructing pseudorandom generators for constant-width branching programs with seed length $r = O(\log n + \log(1/\epsilon))$ is a fundamental open problem. It is known that ϵ-biased spaces over $\mathbb{F}_2$ fool width two branching programs ([SZ], [BDVY08]). Generalizing this observation, Bogdanov et al. [BDVY08] show that PRGs for degree k polynomials over GF(2) fool width two branching programs that are allowed to read up to k bits at each internal node.

However, for width more than two little is known. In fact, by Barrington's theorem ([Bar86]) constructing pseudorandom generators for width five branching programs would imply lower bounds for the circuit class NC^1 - a longstanding open problem in complexity theory. In view of the above, focus on the problem has been restricted to the class of read-once branching programs. Most known PRGs for ROBPs are based on their relation to space-bounded computations; nonuniform logspace computations in particular are equivalent to polynomial-width ROBPs. Even for width three ROBPs, the best generators are the much more powerful generators for logspace machines of Nisan [Nis92] and Impagliazzo et al. [INW94] that achieve a seed-length of $O(\log^2 n)$.

Constructing pseudorandom generators for logspace-computations with logarithmic seed-length is an outstanding open problem with progress being relatively slow. The main nontrivial results are those of [AKS87], [NZ96], [RR99], [Rei08]. In particular, Nisan and Zuckerman [NZ96] give a generator with seed-length $O(\log n)$ for logspace machines that use polylogarithmic randomness and Reingold [Rei08] gives a logspace algorithm for undirected st-connectivity.

The notion of ϵ-biased spaces over $\mathbb{Z}_2$ was introduced by Naor and Naor [NN93] who also gave efficient constructions of such spaces of size $poly(n, 1/\epsilon)$. Subsequently, Alon et al. [AGHP92] and Azar et al. [AMN98] obtained efficient constructions that work for arbitrary abelian groups.

In our construction of ϵ-biased spaces over solvable groups we make use of the fact that for abelian groups we can construct polynomial size sets that are *strongly* ϵ-biased in the following sense.

Definition 2.3. *Let* N *be an abelian group. A set* $S \subseteq N^n$ *is strongly* ϵ-*biased in* N *if, for all non-empty* $I = \{i_1, \ldots, i_k\} \subseteq [n]$, *automorphisms* $\Phi_{i_1}, \ldots, \Phi_{i_k} : N \to N$, *and* $h \in N$,

$$\left| \mathrm{Pr}_{g \in_u S} \left[\Phi_{i_1}(g_{i_1}) \Phi_{i_2}(g_{i_2}) \cdots \Phi_{i_k}(g_{i_k}) = h \right] - \frac{1}{|N|} \right| \leq \epsilon.$$

To get intuition for the above definition, consider the case when N is the cyclic group $\{1, \omega, \omega^2, \dots, \omega^{p-1}\}$ with ω a p'th root of unity for p prime. Then, the automorphisms of N are functions of the form $\Phi_a : N \to N$ defined by $\Phi_a(\omega^x) = \omega^{ax \bmod p}$, for $a \not\equiv 0 \bmod p$. Thus strongly ϵ-biased spaces for N in this case correspond to pseudorandom sets for linear functions mod p. The explicit constructions of ϵ-biased spaces of Azar et al. are in fact strongly ϵ-biased.

Theorem 2.1 ([AMN98]). *For every $d > 0$, there exists a deterministic algorithm running in time $poly(n, 1/\epsilon)$ that takes as input n, ϵ and outputs a set S of size $poly(n, 1/\epsilon)$ that is strongly ϵ-biased in $\mathbb{Z}_d$.*

Proof. Follows from the fact that small-bias spaces of Azar et al. fool the irreducible characters of $\mathbb{Z}_d$. $\qquad\square$

As a corollary we obtain strongly ϵ-biased sets for all abelian groups.

Corollary 2.2. *For an abelian group N, there exists a deterministic algorithm running in time $poly(n, 1/\epsilon)$ that takes as input n, ϵ and outputs a set S of size $poly(n, 1/\epsilon)$ that is strongly ϵ-biased in N.*

Proof. Follows from Theorem 2.1 and the fact that abelian groups are isomorphic to direct products of cyclic groups. $\qquad\square$

3   Constant-Bias Spaces for Arbitrary Groups

We now give a construction that achieves constant bias and works for arbitrary finite groups. We use the efficient constructions of small-bias spaces for $\mathbb{Z}_{|G|}$ given by Azar et al.

Proof of Theorem 1.2. Let $S \subseteq [|G|]^n$ be a $1/(2|G|)$-biased space in $\mathbb{Z}_{|G|}$ of size $poly(n)$ as given by setting $\epsilon = 1/(2|G|)$ in Theorem 2.1. Consider the set

$$T = \{(g^{x_1}, g^{x_2}, \dots, g^{x_n}) : g \in G, (x_1, \dots, x_n) \in S\}. \tag{2}$$

We claim that the set T is α-biased for a constant $\alpha < 1/|G|$. We will use the following lemma.

Lemma 3.1. *For any l with $\gcd(l, |G|) = 1$, the random variable $X = g^l$, where g is uniform in G, is uniformly distributed in G.*

Proof of lemma. The lemma follows from the fact that the map $\phi : G \to G$, $\phi(x) = x^l$ is bijective. For, if $g_1^l = g_2^l$, then for a, b such that $al + b|G| = 1$, $g_1 = g_1^{al+b|G|} = g_2^{al+b|G|} = g_2$. $\qquad\square$

Fix a sequence $b_1, \ldots, b_n \in \{0, 1\}$ and let $I = \{i_1, \ldots, i_k\} = \{i : b_i \neq 0\} \subseteq [n]$. Let $Y(g, x) = g^{x_{i_1} + \cdots + x_{i_k}}$. Note that for a fixed $x = (x_1, \ldots, x_n)$, if $\gcd(x_{i_1} + \ldots + x_{i_k}, |G|) = 1$, then by Lemma 3.1 $Y(U_G, x)$ is uniformly distributed in G, where U_G is the uniform distribution over G. Further, since $x \in_u S$ and S is $1/2|G|$-biased, $Pr[\gcd(x_{i_1} + \ldots + x_{i_k}, |G|) = 1] \geq \phi(|G|)(1/|G| - 1/2|G|)$, where ϕ is the Euler function. Thus, for a fixed $h \in G$,

$$Pr_{g \in_u G, x \in_u S}[Y(g, x) = h] \geq \frac{\beta}{|G|},$$

where $\beta = \phi(|G|)/2|G|$. Therefore T is α-biased in G, where $\alpha = (1 - \beta)/|G| < 1/|G|$ and $|T| = |G||S| = poly(n)$.   □

For abelian groups G, given a set that achieves constant bias, we can combine several independent copies of the constant bias space to obtain a space with smaller bias. The construction of ϵ-biased spaces in [NN93], at a high level, takes this approach. However, for non-abelian groups it is not clear how to perform such *amplification*. In particular, we ask the following question:

Question 3.2. *Let T be α-biased over G. Define*

$$T^k = \{(g_{11}g_{21} \cdots g_{k1}, g_{12}g_{22} \cdots g_{k2}, \ldots, g_{1n}g_{2n} \cdots g_{kn}) : g_i = (g_{i1}, g_{i2}, \ldots, g_{in})$$
$$\in T, 1 \leq i \leq k\}.$$

Then, is T^k ϵ-biased over G for $\epsilon = \alpha^{\Omega(k)}$?

For abelian groups the answer to the above question is yes, but the answer is not clear for non-abelian groups. An answer to the question even for the specific constant-bias space of equation (2) would be very interesting.

4   Small-Bias Spaces for Solvable Groups

We now address the case of solvable groups and prove Theorem 1.3. Our construction of ϵ-biased spaces is recursive, by using the fact that every solvable group G has a nontrivial abelian subgroup, say N, that is normal in the group G. We use the known constructions of ϵ-biased spaces for the abelian group N and combine them with an ϵ-biased space for the factor group G/N which can be obtained recursively, since G/N is also solvable. We first present some preliminaries from group theory.

Let N be a nontrivial normal subgroup of G and let $H = G/N$ be the factor group of N in G. Without loss of generality assume that the factor group H is given by elements of G which are in distinct cosets of N in G. Note that the representatives of H may not form a subgroup in G. In case of ambiguity in group operations we will denote multiplication in H by $\circ$.

Lemma 4.1. *Every $g \in G$ can be written uniquely as $g = nh$, where $n \in N$ and $h \in H$.*

The following lemma gives us a way to separate a *mixed* product $n_1 h_1 n_2 h_2 \ldots$
$n_k h_k$ with $n_i \in N$, $h_i \in H$ into products of elements in N and H respectively.

Lemma 4.2. *Let* $g_1 = n_1 h_1, g_2 = n_2 h_2, \ldots, g_k = n_k h_k$, *with* $n_i \in N$ *and* $h_i \in$
H. *Let* $\boldsymbol{h} = (h_1, \ldots, h_k)$. *Then,*

$$g_1 g_2 \ldots g_k = (n_1 h_1)(n_2 h_2) \ldots (n_k h_k) = n_1 \, \Phi_{1,\boldsymbol{h}}(n_2) \, \Phi_{2,\boldsymbol{h}}(n_3) \, \ldots \, \Phi_{k-1,\boldsymbol{h}}(n_k) \, a_{\boldsymbol{h}}$$
$$(h_1 \circ h_2 \circ \cdots \circ h_k),$$

where $\Phi_{i,\boldsymbol{h}} : N \to N$ *is an automorphism for* $1 \le i \le k-1$, *and* $a_{\boldsymbol{h}} \in N$ *depends
only on* $h_1, \ldots, h_k$.

Proof. For $1 \le i \le k-1$, define $\Phi_{i,\boldsymbol{h}} : N \to N$ by $\Phi_{i,\boldsymbol{h}}(n) = (h_1 \ldots h_i) \, n \, (h_1 \ldots$
$h_i)^{-1}$. Since N is a normal subgroup of G, $\Phi_{i,\boldsymbol{h}}$ are automorphisms on N. Observe
that,

$$(n_1 h_1)(n_2 h_2) \ldots (n_k h_k) = n_1 \, \Phi_{1,\boldsymbol{h}}(n_2) \, \Phi_{2,\boldsymbol{h}}(n_3) \, \ldots \, \Phi_{k-1,\boldsymbol{h}}(n_k) \, (h_1 h_2 \ldots h_k).$$

Further, $h_1 h_2 \ldots h_k$ and $h_1 \circ h_2 \circ \cdots \circ h_k$ (as elements of G) lie in the same coset
of N. Thus, there exists $a_{\boldsymbol{h}} \in N$ depending only on $h_1, h_2, \ldots, h_k$ such that
$h_1 h_2 \ldots h_k = a_{\boldsymbol{h}} \, (h_1 \circ h_2 \circ \cdots \circ h_k)$. The lemma now follows. $\qquad\square$

Definition 4.1. *A group* G *is said to be solvable if there exist subgroups* $G =$
$N_0 \supset N_1 \supset N_2 \supset \ldots \supset N_r = (e)$ *such that each* N_i *is normal in* N_{i-1} *and*
N_{i-1}/N_i *is abelian.*

The following properties of solvable groups can be found, for instance, in
[Her75].

Lemma 4.3 ([Her75]). *Let* G *be a solvable group. Then,*

- *For a normal subgroup* $N \subseteq G$, *the factor group* G/N *is solvable.*
- G *contains a nontrivial abelian subgroup* N *which is normal in* G.

Proof of Theorem 1.3. Let G be a solvable group. Let N be a nontrivial
abelian subgroup of G that is also normal and let $H = G/N$. Such an N is
guaranteed to exist by Lemma 4.3. As before, we assume that the factor group
H is given by elements of G which are in distinct cosets of N in G, with group
operation of H denoted by $\circ$.

Lemma 4.4. *Let* $S \subseteq N^n$ *be strongly* ϵ-*biased in* N *and let* $T \subseteq H^n$ *be* ϵ-*biased
in* H. *Let*

$$S \times T = \{(n_1 h_1, \ldots, n_n h_n) : (n_1, \ldots n_n) \in S, (h_1, h_2, \ldots, h_n) \in T\} \subseteq G^n. \quad (3)$$

Then, the set $S \times T$ *is* ϵ-*biased in* G.

Given the above lemma, Theorem 1.3 follows from Corollary 2.2 and induction,
as H is a solvable group with $|H| < |G|$. We now calculate the exact dependence
of the size of the small-bias space on $n, \epsilon, |G|$.

For an abelian group N, and sufficiently large n, Azar et al. give a strongly ϵ-biased set of size $(cn/\epsilon)^{2\log|N|}$, where c is an absolute constant independent of $n, \epsilon, |N|$. Combining the above with Lemma 4.4, for sufficiently large n, we get an ϵ-biased space for G of size $(cn/\epsilon)^{2\log|G|}$. In general, using a similar estimate of Azar et al. we get a bound of $(n/\epsilon)^{O(\log|G|)}$. $\qquad\square$

Proof of Lemma 4.4. Fix $b_1, \ldots, b_n \in \{0, 1\}$ and let $I = \{i : b_i \neq 0\}$. Without loss of generality, let $I = \{1, \ldots, k\}$. For $\boldsymbol{n} = (n_1, \ldots, n_n) \in N^n$ and $\boldsymbol{h} = (h_1, \ldots, h_n) \in H^n$, let $X(\boldsymbol{n}, \boldsymbol{h}) = (n_1 h_1)(n_2 h_2) \ldots (n_k h_k)$. Using the notation of Lemma 4.2, let $X(\boldsymbol{n}, \boldsymbol{h}) = Y_{\boldsymbol{h}}(\boldsymbol{n})Z(\boldsymbol{h})$, where

$$Y_{\boldsymbol{h}}(\boldsymbol{n}) = n_1 \Phi_{1,\boldsymbol{h}}(n_2)\Phi_{2,\boldsymbol{h}}(n_3) \ldots \Phi_{k-1,\boldsymbol{h}}(n_k)a_{\boldsymbol{h}} \in N,$$
$$Z(\boldsymbol{h}) = (h_1 \circ h_2 \circ \cdots \circ h_k) \in H.$$

Let $g_0 = n_0 h_0 \in G$ with $n_0 \in N, h_0 \in H$. Then, for a fixed $\boldsymbol{h}$, since S is strongly ϵ-biased in N,

$$|\mathrm{Pr}_{\boldsymbol{n}\in_u S}[Y_{\boldsymbol{h}}(\boldsymbol{n}) = n_0] - \frac{1}{|N|}| \leq \epsilon. \tag{4}$$

Further, since T is ϵ-biased in H,

$$|\mathrm{Pr}_{\boldsymbol{h}\in_u T}[Z(\boldsymbol{h}) = h_0] - \frac{1}{|H|}| \leq \epsilon. \tag{5}$$

Therefore,

$$\mathrm{Pr}_{\boldsymbol{n}\in_u S, \boldsymbol{h}\in_u T}[X(\boldsymbol{n}, \boldsymbol{h}) = g_0] = \mathrm{Pr}_{\boldsymbol{n}\in_u S, \boldsymbol{h}\in_u T}[Y_{\boldsymbol{h}}(\boldsymbol{n}) = n_0 \wedge Z(\boldsymbol{h}) = h_0]$$

$$= \sum_{\boldsymbol{h}:Z(\boldsymbol{h})=h_0} \frac{1}{|T|} \mathrm{Pr}_{\boldsymbol{n}\in_u S}[Y_{\boldsymbol{h}}(\boldsymbol{n}) = n_0]$$

$$\geq \sum_{\boldsymbol{h}:Z(\boldsymbol{h})=h_0} \frac{1}{|T|} \left(\frac{1}{|N|} - \epsilon\right) \quad \text{from equation (4)}$$

$$\geq \left(\frac{1}{|N|} - \epsilon\right) \mathrm{Pr}_{\boldsymbol{h}\in_u T}[Z(\boldsymbol{h}) = h_0]$$

$$\geq \left(\frac{1}{|N|} - \epsilon\right) \left(\frac{1}{|H|} - \epsilon\right) \quad \text{from equation (5)}$$

$$\geq \frac{1}{|G|} - \beta,$$

where $\beta = \epsilon/|N| + \epsilon/|H| - \epsilon^2/|G|$. As the above argument is applicable for all non-empty $I \subseteq [n]$, and $\beta < \epsilon$ for $|N|, |H| \geq 2$, we get that $S \times T$ is ϵ-biased in G. $\qquad\square$

4.1   Width 3 Branching Programs

We now study the particular case of width 3 ROPBPs and present some motivation for our Conjecture 1.5. Let $\mathbb{S}_3$ be the symmetric group on three elements

$\{1, 2, 3\}$. Let $a \in \mathbb{S}_3$ be the transposition (12), $b \in \mathbb{S}_3$ be the cyclic-shift (123), and e be the identity permutation. Then, the group $N = \{1, b, b^2\} \cong \mathbb{Z}_3$ is an abelian subgroup of $\mathbb{S}_3$ that is also normal. The factor group $\mathbb{S}_3/N$ is isomorphic to the group $\{1, a\} \cong \mathbb{Z}_2$. Thus, for the special case of $\mathbb{S}_3$, the construction presented in the previous section becomes,

$$S \times T = \{(a^{x_1}b^{y_1}, a^{x_2}b^{y_2}, \ldots, a^{x_n}b^{y_n}) : (x_1, \ldots, x_n) \in S, (y_1, \ldots, y_n) \in T\},$$

where $S \subseteq \{0, 1\}^n$ is ϵ-biased over $\mathbb{Z}_2$ and $T \subseteq \{0, 1, 2\}^n$ is ϵ-biased over $\mathbb{Z}_3$. This provides some motivation for Conjecture 1.5. Also, note that any pseudorandom generator for width three permutation branching programs must be pseudorandom with respect to linear functions mod 2 and mod 3 - a property satisfied by the generator of (1.5).

5   Pseudorandom Bit Generators for Modular Sums

As the existence of ϵ-PBG as required in Conjecture 1.5 does not follow directly from known constructions of ϵ-biased spaces, we provide an efficient construction of ϵ-PBG of size $poly(n, 1/\epsilon)$ below. Our construction is based on the pseudorandom-generators for low-degree polynomials obtained by Viola [Vio08].

Proof of Theorem 1.4. Suppose that p is an odd prime; for $p = 2$, ϵ-PBG can be obtained from ϵ-biased spaces straightforwardly. For the rest of this section, the arithmetic is over $\mathbb{F}_p$ and let $\oplus$ denote addition mod 2. To motivate our construction, let $v \in \mathbb{F}_p^n, v \neq 0$. We consider two cases depending on the support size of v. Let C be a large enough constant depending on p such that for all $v \in \mathbb{F}_p$ with $|support(v)| \geq C \log(1/\epsilon)$

$$|\mathsf{Pr}_{x \in_u \{0,1\}^n} [\langle v, x \rangle = a] - \frac{1}{p}| \leq \frac{\epsilon}{3}, \tag{6}$$

$$|\mathsf{Pr}_{y \in_u \mathbb{F}_p^n} [\sum_i v_i y_i^{p-1} = a] - \frac{1}{p}| \leq \frac{\epsilon}{3}.$$

Case 1. $|support(v)| \leq C \log(1/\epsilon)$. This case can be handled by a generator $H_1 : \{0, 1\}^r \to \{0, 1\}^n$ that generates a ϵ-almost $C \log(1/\epsilon)$-wise independent distribution. Such generators with $r = O(\log n + \log(1/\epsilon))$ are given in [NN93].

Case 2. $|support(v)| > C \log(1/\epsilon)$. Let $H : \{0, 1\}^r \to \mathbb{F}_p^n$ be such that for every degree at most $p - 1$ polynomial $P : \mathbb{F}_p^n \to \mathbb{F}_p$, and $a \in \mathbb{F}_p$,

$$|\mathsf{Pr}_{z \in_u \{0,1\}^r} [P(H(z)) = a] - \mathsf{Pr}_{y \in_u \mathbb{F}_p^n} [P(y) = a]| \leq \frac{\epsilon}{3}. \tag{7}$$

Pseudorandom generators for low-degree polynomials as above with $r = O(\log n + \log(1/\epsilon))$ were given by Viola building on the works of Bogdanov and Viola [BV07], Lovett [Lov08]. Define $H_2 : \{0, 1\}^r :\to \{0, 1\}^n$ by $H_2(z) = (y_1^{p-1}, \ldots, y_n^{p-1})$, where $(y_1, \ldots, y_n) = H(z)$. We will show that H_2 satisfies the conditions of ϵ-PBG. For $z \in \{0,1\}^r$, we have $\langle v, H_2(z) \rangle = \sum_i v_i(H(z)_i)^{p-1} = P_v(H(z))$, where P_v is the degree $p-1$ polynomial $P_v(y) = \sum_i v_i y_i^{p-1}$. Since H fools degree $p-1$ polynomials over $\mathbb{F}_p^n$,

$$|\mathrm{Pr}_{z \in_u \{0,1\}^r} [\langle v, H_2(z) \rangle = a] - \mathrm{Pr}_{y \in_u \mathbb{F}_p^n} [P_v(y) = a]| =$$
$$|\mathrm{Pr}_{z \in_u \{0,1\}^r} [P_v(H(z)) = a] - \mathrm{Pr}_{y \in_u \mathbb{F}_p^n} [P_v(y) = a]| \leq \frac{\epsilon}{3}. \quad (8)$$

From equations (6), (8), for $v \in \mathbb{F}_p^n$ with $|support(v)| > C \log(1/\epsilon)$,

$$|\mathrm{Pr}_{z \in_u \{0,1\}^r} [\langle v, H_2(z) \rangle = a] - \frac{1}{p}| \leq \frac{2\epsilon}{3}. \quad (9)$$

We now combine the generators H_1, H_2 from the above cases to obtain $(H_1 \oplus H_2) : \{0,1\}^{2r} \to \{0,1\}^n$ by defining $(H_1 \oplus H_2)(z_1, z_2) = H_1(z_1) \oplus H_2(z_2)$. Observe that for $b, c \in \{0,1\}$, $b \oplus c = b + c + (p-2)bc \mod p$. Let $v \in \mathbb{F}_p^n, v \neq 0$. Then,

$$\langle v, (H_1 \oplus H_2)(z_1, z_2) \rangle = \sum_i v_i(H_1(z_1)_i \oplus H_2(z_2)_i)$$
$$= \sum_i v_i \left(H_1(z_1)_i + H_2(z_2)_i + (p-2)H_1(z_1)_i H_2(z_2)_i \right)$$
$$= \sum_i v_i H_1(z_1)_i + \sum_i v_i(1 + (p-2)H_1(z_1)_i)H_2(z_2)_i$$
$$= \langle v, H_1(z_1) \rangle + \langle v(z_1), H_2(z_2) \rangle,$$

where $v(z_1)$ is the vector defined by $v(z_1)_i = v_i(1 + (p-2)H_1(z_1)_i)$. Note that $|support(v)| = |support(v(z_1))|$. Fix $a \in \mathbb{F}_p$ and suppose $|support(v)| > C \log(1/\epsilon)$. Then, for a fixed z_1, by equation (9)

$$|\mathrm{Pr}_{z_2 \in_u \{0,1\}^r} [\langle v(z_1), H_2(z_2) \rangle = a] - \frac{1}{p}| \leq \frac{2\epsilon}{3}.$$

Since z_1, z_2 are chosen independently, from the above equation and equation (6)

$$|\mathrm{Pr}_{z_1, z_2 \in_u \{0,1\}^r} [\langle v, (H_1 \oplus H_2)(z_1, z_2) \rangle = a] - \mathrm{Pr}_{x \in_u \{0,1\}^n} [\langle v, x \rangle = a]| \leq \epsilon.$$

Proceeding similarly for the case $|support(v)| \leq C \log(1/\epsilon)$, it follows that the above inequality holds for all $v \in \mathbb{F}_p^n, v \neq 0$ and $a \in \mathbb{F}_p$. Hence $H_1 \oplus H_2$ is an ϵ-PBG for sums mod p. $\qquad \square$

6   Relation to Branching Programs

Here we prove that PRGs for constant-width ROPBP imply small-bias spaces for finite groups.

Proof of Theorem (1.1). Assume we are given a PRG fooling constant-width ROPBPs. We want to construct an ϵ-biased space for a group H. Since finite groups are isomorphic to subgroups of the permutation groups, we can assume H to be a subgroup of the symmetric group $\mathbb{S}_w$ on w elements. Let $G : \{0,1\}^r \to \{0,1\}^n$ fool width w ROPBPs of length n with error at most ϵ. Consider the following procedure for generating a sequence in H^n:

1. Generate ϵ-almost k-wise independent sequences $(g_1, \ldots, g_n)$, $(h_1, \ldots, h_n) \in H^n$ for $k = O(\log(1/\epsilon))$ to be chosen later. For $k = O(\log n)$, Naor and Naor [NN93], Alon et al. [AGHP92] give efficient constructions of almost k-wise independent sequences using $O(\log n + \log(1/\epsilon))$ bits of randomness.
2. Choose $y \in_u \{0,1\}^r$ and output the sequence $(g_1 h_1^{G(y)_1}, g_2 h_2^{G(y)_2}, \ldots, g_n h_n^{G(y)_n})$.

Note that the procedure uses $O(\log n + \log(1/\epsilon)) + r$ bits of randomness. We will show that the multi-set of sequences generated by the above procedure is a $O(\epsilon)$-biased space over H. We need the following lemmas.

Let $g_1, \ldots, g_l \in H$, for l to be chosen later. Call a sequence of group elements $(h_1, \ldots, h_l)$ complete if $\{g_1 h_1^{x_1} g_2 h_2^{x_2} \cdots g_l h_l^{x_l} : x_i \in \{0,1\}\} = H$.

Lemma 6.1. *There exists a constant c such that for $l = c|H|^2$, a sequence $(h_1, \ldots, h_l) \in_u H^l$ is complete with probability at least $1/2$.*

Proof. For $1 \leq i \leq l$, let $S_i = \{g_1 h_1^{x_1} \cdots g_i h_i^{x_i} : x_j \in \{0,1\}\}$ and let random variable $X_i = |S_i|$. Note that given $h_1, \ldots, h_i$ and a $g \in H$, $g \notin S_i$, $\Pr_{h_{i+1} \in_u H}[g \in S_{i+1} \mid h_1, \ldots, h_i] = X_i/|H| \geq 1/|H|$. Thus, if $X_i < |H|$, then $\Pr[X_{i+1} \geq X_i + 1 \mid X_i] \geq 1/|H|$. The lemma now follows. $\qquad\square$

Lemma 6.2. *For any group H and $0 < \epsilon < 1/2$, there exists $k = O(\log(1/\epsilon))$, such that for all $t > k$ and $g_1, \ldots, g_t \in H$ the following holds. For $(x_1, \ldots, x_t) \in_u \{0,1\}^t$ and $h_1, \ldots, h_t \in H$ chosen from an ϵ-almost k-wise independent distribution, the distribution of $g_1 h_1^{x_1} g_2 h_2^{x_2} \ldots g_t h_t^{x_t}$ is 4ϵ-close in variation distance to the uniform distribution on H.*

Proof. Let $l = c|H|^2$ be as in Lemma 6.1. Let $k = 4ml \log(1/\epsilon)$, for m to be chosen later, and partition $(h_1, \ldots, h_k)$ into $4m \log(1/\epsilon)$ blocks of length l each. Then, by Lemma 6.1 and Chernoff bounds, for $(h_1, \ldots, h_k) \in H^k$ chosen from an ϵ-almost k-wise independent distribution, with probability at least $1 - (\exp(-\Omega(m \log(1/\epsilon)) + \epsilon)$, $m \log(1/\epsilon)$ of the $4m \log(1/\epsilon)$ blocks will be complete for $g_1, \ldots, g_k$.

Note that for any complete sequence $(h_{i_1}, \ldots, h_{i_l})$ the distribution of $g_{i_1} h_{i_1}^{x_1} g_{i_2} h_{i_2}^{x_2} \ldots g_{i_l} h_{i_l}^{x_l}$ for $x \in_u \{0,1\}^l$ is at least $\alpha = (1 - 1/2^l)$-close in variation distance to the uniform distribution on H. Thus, with probability at least $1 - (\exp(-\Omega(m \log(1/\epsilon))) + \epsilon)$, the distribution of $g_1 h_1^{x_1} g_2 h_2^{x_2} \ldots g_t h_k^{x_k}$ with $x_i \in_u \{0,1\}$ is $(1 - 1/2^l)^{m \log(1/\epsilon)}$-close in variation distance to the uniform distribution on H. The lemma now follows by taking $m = O(2^l)$. $\qquad\square$

Let $k = O(\log(1/\epsilon))$ be such that the above lemma holds for H. Let $I = \{i_1, \ldots, i_t\} = \{i : b_i \neq 0\} \subseteq [n]$. We consider two cases.

(a) $|I| = t \leq k$: Since, $(g_1, \ldots, g_n)$ is chosen independently of $(h_1, \ldots, h_n)$ and is ϵ-almost k-wise independent, the distribution of $g_{i_1} h_{i_1}^{G(y)_{i_1}} g_{i_2} h_{i_2}^{G(y)_{i_2}} \ldots g_{i_t} h_{i_t}^{G(y)_{i_t}}$ is ϵ-close to the uniform distribution on H.

(b) $|I| > k$: By relabeling the nodes according to the g_i, we can construct a width w ROPBP of length at most n such that on input $x_1, \ldots, x_t, \ldots, x_n$ the output is $g_{i_1} h_{i_1}^{x_{i_1}} g_{i_2} h_{i_2}^{x_{i_2}} \ldots g_{i_t} h_{i_t}^{x_{i_t}} \in \mathbb{S}_w$. Since G fools ROPBPs of width w and length at most n, we have for every $\pi \in \mathbb{S}_w$,

$$\left| \Pr_{y \in_u \{0,1\}^r}[g_{i_1} h_{i_1}^{G(y)_{i_1}} g_{i_2} h_{i_2}^{G(y)_{i_2}} \ldots g_{i_t} h_{i_t}^{G(y)_{i_t}} = \pi] - \right.$$
$$\left. \Pr_{x \in_u \{0,1\}^n}[i_1 h_{i_1}^{x_{i_1}} g_{i_2} h_{i_2}^{x_{i_2}} \ldots g_{i_t} h_{i_t}^{x_{i_t}} = \pi] \right| \leq \epsilon.$$

Now, by lemma 6.2 when $x \in_u \{0,1\}^n$, the distribution of $g_{i_1} h_{i_1}^{x_1} g_{i_2} h_{i_2}^{x_2} \ldots g_{i_t} h_{i_t}^{x_t}$ is 4ϵ-close to the uniform distribution on H. Therefore, for every $\pi \in \mathbb{S}_w$,

$$\left| \Pr_{y \in_u \{0,1\}^r}[g_{i_1} h_{i_1}^{G(y)_{i_1}} g_{i_2} h_{i_2}^{G(y)_{i_2}} \ldots g_{i_t} h_{i_t}^{G(y)_{i_t}} = \pi] - \frac{1}{|H|} \right| \leq 5\epsilon.$$

It follows that the generator defined above is $O(\epsilon)$-biased over H. □

7   Sum of Constant-Bias Spaces Does Not Fool Width 3

We now show that the sum of two constant-bias spaces over $\mathbb{Z}_2$ does not fool width 3 branching programs and prove Theorem 1.6. We do this by constructing a constant-bias space S over $\mathbb{Z}_2$ such that $S \oplus S$ is contained in a subspace of dimension $o(n)$ in $\mathbb{F}_3^n$. To avoid confusion in the following let $+$ denote addition in $\mathbb{F}_3$ and $\oplus$ denote addition in $\mathbb{F}_2$. For a set $T \subseteq \mathbb{F}_3^n$, let $d_3(T)$ denote the dimension of span of T in $\mathbb{F}_3^n$ and let $T \odot T = \{x \odot y = (x_1 y_1, \ldots, x_n y_n) : x = (x_1, \ldots, x_n), y = (y_1, \ldots, y_n) \in T\}$. We'll use the following lemmas.

Lemma 7.1. For any $T \subseteq \mathbb{F}_3^n$, $d_3(T \odot T) \leq d_3(T)^2$.

Proof. If $u_1, \ldots, u_k \in \mathbb{F}_3^n$ span T, then the $\binom{k}{2} + k$ vectors $u_i \odot u_j$ span $T \odot T$. □

Lemma 7.2. Let $T \subseteq \{0,1\}^n$. Then $d_3(T \oplus T) \leq 2d_3(T) + d_3(T)^2$.

Proof. Observe that for $x, y \in \{0,1\}$, $x \oplus y = x + y + xy$. Therefore the dimension of span of $T_1 \oplus T_2$ is at most the dimension of span of $T + T + T \odot T$. The lemma now follows from Lemma 7.1. □

Proof of Theorem 1.6. Let $n = \binom{d}{5}$. We will denote vectors $x \in \mathbb{F}_3^n$, by $(x_I)_{I \in \mathcal{C}}$, where $\mathcal{C} = \binom{[d]}{5}$ is the collection of subsets of $[d]$ of size 5. Let $p : \mathbb{F}_3^5 \to \mathbb{F}_3$ be the degree two multi-variate polynomial defined by

$$p(y_1, y_2, y_3, y_4, y_5) = (y_1 + y_2 + y_3 + y_4 + y_5)^2.$$

Let $q : \mathbb{F}_2^5 \to \mathbb{F}_2$ be the degree five multi-variate polynomial defined by

$$q(y_1, y_2, y_3, y_4, y_5) = \bigoplus_i y_i \oplus \bigoplus_{i \neq j} y_i y_j \oplus \bigoplus_{i,j,k,l \text{ distinct}} y_i y_j y_k y_l \oplus y_1 y_2 y_3 y_4 y_5.$$

Our construction is based on the observation - which can be verified by direct computation - that evaluated over the set $\{0,1\}^5$ the polynomials p and q are identical. That is, for all $(y_1, \ldots, y_5) \in \{0,1\}^5$, $p(y_1, \ldots, y_5) = q(y_1, \ldots, y_5)$. Now, let

$$S = \{(p(y_{i_1}, y_{i_2}, y_{i_3}, y_{i_4}, y_{i_5}))_{\{i_1, i_2, i_3, i_4, i_5\} \in \mathcal{C}} : (y_1, \ldots, y_d) \in \{0,1\}^d\}.$$

Let $T = \{(y_{i_1} + y_{i_2} + y_{i_3} + y_{i_4} + y_{i_5})_{\{i_1, i_2, i_3, i_4, i_5\} \in \mathcal{C}} : (y_1, \ldots, y_d) \in \{0,1\}^d\}$. Now, $d_3(T) \leq d$ and $S \subseteq T \odot T$. Therefore, by Lemma 7.1 $d_3(S) \leq d^2$. However, the dimension of the span of S viewed as a subset of $\mathbb{F}_2^n$ is $n = \binom{d}{5}$. In fact, for any non-zero $\alpha \in \mathbb{F}_2^n$, $\{\langle \alpha, x \rangle : x \in S\} = \{q_\alpha(y) : y \in \mathbb{F}_2^d\}$, where $q_\alpha : \mathbb{F}_2^d \to \mathbb{F}_2$ is a non-constant polynomial of degree 5. For,

$$q_\alpha(y) = \sum_{I \in \mathcal{C}} \alpha_I \prod_{i \in I} y_i + R_\alpha(y),$$

where R_α is a degree at most four polynomial. Since $\alpha \neq 0$, q_α has degree five. Using the fact that the minimum distance of the Reed Muller code of degree 5 over $\mathbb{F}_2$ is $1/32$, we get that for $a \in \{0,1\}$,

$$\mathsf{Pr}_{x \in_u S}[\langle \alpha, x \rangle = a] = \mathsf{Pr}_{y \in_u \mathbb{F}_2^d}[q_\alpha(y) = a] \leq \frac{31}{32}.$$

Thus, S is ϵ-biased over $\mathbb{F}_2$ for $\epsilon = 15/32$. Further, from Lemma 7.2 $d_3(S \oplus S) \leq 2d^2 + d^4 = o(n)$. $\qquad \square$

Acknowledgements. We thank Russell Impagliazzo for useful discussions and the anonymous referees for helpful comments.

References

[AGHP92]   Alon, N., Goldreich, O., Håstad, J., Peralta, R.: Simple construction of almost k-wise independent random variables. Random Struct. Algorithms 3(3), 289–304 (1992)

[AKS87]   Ajtai, M., Komlós, J., Szemerédi, E.: Deterministic simulation in logspace. In: STOC, pp. 132–140 (1987)

[AMN98]   Azar, Y., Motwani, R., Naor, J.: Approximating probability distributions using small sample spaces. Combinatorica 18(2), 151–171 (1998)

[Bar86]   Barrington, D.A.M.: Bounded-width polynomial-size branching programs recognize exactly those languages in NC^1. In: STOC, pp. 1–5 (1986)

[BDVY08]   Bogdanov, A., Dvir, Z., Verbin, E., Yehudayoff, A.: Pseudorandomness for width 2 branching programs (manuscript, 2008)

[BEG⁺94]   Blum, M., Evans, W.S., Gemmell, P., Kannan, S., Naor, M.: Checking the correctness of memories. Algorithmica 12(2/3), 225–244 (1994)

[BV07]   Bogdanov, A., Viola, E.: Pseudorandom bits for polynomials. In: FOCS, pp. 41–51 (2007)

[Her75]   Herstein, I.: Topics in algebra. Wiley, Chichester (1975)

[INW94]   Impagliazzo, R., Nisan, N., Wigderson, A.: Pseudorandomness for network algorithms. In: STOC, pp. 356–364 (1994)

[Lov08]   Lovett, S.: Unconditional pseudorandom generators for low degree polynomials. In: STOC, pp. 557–562 (2008)

[LRTV09]   Lovett, S., Reingold, O., Trevisan, L., Vadhan, S.: Pseudorandom bit generators that fool modular sums. In: RANDOM (2009)

[LY94]   Lund, C., Yannakakis, M.: On the hardness of approximating minimization problems. J. ACM 41(5), 960–981 (1994)

[Nis92]   Nisan, N.: Pseudorandom generators for space-bounded computation. Combinatorica 12(4), 449–461 (1992)

[NN93]   Naor, J., Naor, M.: Small-bias probability spaces: Efficient constructions and applications. SIAM J. Comput. 22(4), 838–856 (1993)

[NZ96]   Nisan, N., Zuckerman, D.: Randomness is linear in space. J. Comput. Syst. Sci. 52(1), 43–52 (1996)

[Rei08]   Reingold, O.: Undirected connectivity in log-space. J. ACM 55(4) (2008)

[RR99]   Raz, R., Reingold, O.: On recycling the randomness of states in space bounded computation. In: STOC, pp. 159–168 (1999)

[RV06]   Reingold, O., Vadhan, S.: Personal communication (2006)

[SZ]   Saks, M., Zuckerman, D. (unpublished manuscript)

[Vio08]   Viola, E.: The sum of d small-bias generators fools polynomials of degree d. In: IEEE Conference on Computational Complexity, pp. 124–127 (2008)

Small Clique Detection and Approximate Nash Equilibria

Lorenz Minder and Dan Vilenchik

Computer Science Division, University of California, Berkeley, CA 94720-1776
lorenz@eecs.berkeley.edu, danny.vilenchik@gmail.com

Abstract. Recently, Hazan and Krauthgamer showed [12] that if, for a fixed small ε, an ε-best ε-approximate Nash equilibrium can be found in polynomial time in two-player games, then it is also possible to find a planted clique in $G_{n,1/2}$ of size $C \log n$, where C is a large fixed constant independent of ε. In this paper, we extend their result to show that if an ε-best ε-approximate equilibrium can be efficiently found for arbitrarily small $\varepsilon > 0$, then one can detect the presence of a planted clique of size $(2+\delta) \log n$ in $G_{n,1/2}$ in polynomial time for arbitrarily small $\delta > 0$. Our result is optimal in the sense that graphs in $G_{n,1/2}$ have cliques of size $(2 - o(1)) \log n$ with high probability.

1 Introduction

The computational complexity of finding a Nash equilibrium in a given game has been the focus of extensive research in recent years: The problem of finding a best Nash equilibrium (i.e., an equilibrium that maximizes the sum of the expected payoffs) in a two-player game has been shown to be NP-hard by Gilboa and Zemel [9] in 1989. The easier problem of computing an arbitrary equilibrium in a finite two-player game was shown to be PPAD-complete by Chen et al [4] and Daskalakis et al [6].

Given these results, it is unlikely that Nash equilibria can be computed in polynomial time. However, some positive results show that Nash equilibria can at least to some extent be approximated. The most recent result, following extensive work in the area, provides a polynomial time algorithm that computes a 0.3393-equilibrium [15]. Another algorithm due to Lipton et al computes an ε-equilibrium in quasi-polynomial time $N^{\log N/\varepsilon^2}$, where $N \times N$ is the dimension of the game matrix. The latter result also extends to the case of an ε-equilibrium that maximizes the sum of payoffs.

Having a quasi-polynomial time approximation algorithm probably means that finding ε-equilibria is not NP-hard. It is however still not known whether the problem has a polynomial time approximation scheme.

Recently, Hazan and Krauthgamer [12] showed that for sufficiently small yet constant ε the problem of computing an ε-equilibrium whose sum of payoffs is off by at most ε from the best Nash equilibrium (for short, we call this problem

I. Dinur et al. (Eds.): APPROX and RANDOM 2009, LNCS 5687, pp. 673–685, 2009.

ε-best ε-equilibrium) is at least as hard as finding a planted k-clique in the random graph $G_{n,1/2}$, where $k = c \log n$ [1], and $c \approx 10^6$ is a fixed large constant (by "hard" we mean the standard notion of polynomial – maybe randomized – reductions). The planted k-clique problem consists of finding a clique of size k that was planted into an otherwise random graph with density $1/2$. This problem is a well-known notoriously hard combinatorial problem.

 - Despite considerable efforts, the currently best known efficient algorithm to solve the planted clique problem [2] needs a clique size of $k = \Omega(\sqrt{n})$.
 - The planted k-clique problem is (for certain values of k) related to the assumption that refuting low-density 3CNF formulas is hard on the average. This fact was used by Feige [7] to derive constant-factor hardness of approximation for several well-known problems.

1.1   Our Contribution

In this paper, we strengthen the result of Hazan and Krauthgamer in the following sense: We show that if a polynomial time approximation scheme exists that finds for any $\varepsilon > 0$ an ε-best ε-equilibrium, then for any $\delta > 0$ there is a polynomial time algorithm that detects the presence of a planted clique of size $(2 + \delta) \log n$ with high probability (*whp* for short).

Note that random graphs contain a clique of size $(2 - o(1)) \log n$. Hence the $2 \log n$ threshold that we achieve is a natural boundary implied by the problem statement. See in this context also the work of Juels and Peinado [11] for the planted clique problem when $k < 2 \log n$.

More formally, our main result can be stated as follows.

Theorem 1. *There exists a positive constant ε_0 so that if there is a polynomial time algorithm that finds in a two-player game the ε-best ε-equilibrium, $0 < \varepsilon \le \varepsilon_0$, then there is a probabilistic polynomial time algorithm that distinguishes whp between two graphs: $G \in G_{n,1/2}$ and H, an arbitrary graph on n nodes with a clique of size $(2 + 28\varepsilon^{1/8}) \log n$.*

As explained in Section 3.3, our analysis gives $\varepsilon_0 = 32^{-8}$, although this estimate is somewhat loose. The probability in the statement is taken over the choices of the algorithm and the distribution $G_{n,1/2}$.

Our result in particular implies that finding an ε-best ε-equilibrium is at least as hard as distinguishing between $G_{n,1/2}$ and $G_{n,1/2,k}$ (i.e., $G_{n,1/2}$ with a planted k-clique) for $k = (2 + 28\varepsilon^{1/8}) \log n$.

Let us also briefly mention that our analysis implies that for every fixed $\delta > 0$, given an efficient algorithm for finding an ε-best ε-equilibrium, one can efficiently *find* a planted clique of size $(3 + \delta) \log n$ in $G_{n,1/2}$. For details, see section 4.

In the next section we describe our technical contribution.

---

[1] In this paper, log denotes the base-2 logarithm.

1.2   Techniques

The main idea of the reduction, as put out by [12], is to incorporate the graph with the planted clique into a game so that the ε-best ε-equilibrium reflects in some useful sense that clique.

More formally, let G be a simple graph with self-loops added, and let A be its adjacency matrix. Construct the following game matrices R and C, composed of four blocks. $C = R^T$ so let us just describe R.

$$R = \begin{pmatrix} A & -B^T \\ B & \mathbf{0} \end{pmatrix}$$

Here, $\mathbf{0}$ stands for the all-0 matrix. The matrix B is constructed as follows. The constants $t \leq 2, p$ and s, whose values depend on ε, are chosen as in the proof of Proposition 1. B is an $n^s \times n$ matrix, each entry of which is a scaled Bernoulli random variable $B_{i,j}$ which takes the value t with probability p and 0 otherwise.

The difference from the construction in [12] is our different choice of parameters for the matrix B. The heart of the analysis outlined in [12] lies in proving that if a clique of size $c_1 \log n$ is planted in $G_{n,1/2}$ then a graph of size $c_2 \log n$, $c_2 \leq c_1$, with edge-density greater than, say, 0.55 can be recovered using the above construction and the ε-best ε-equilibrium which one assumes can be found efficiently. Since this graph is denser than what one expects in $G_{n,1/2}$, and the constant c_2 is sufficiently large, it has *whp* many edges in common with the planted clique. This fact can then be used to recover the clique. In [12], c_1 was a large constant, and so was c_2, and the question of how tight the gap between them can be was not considered. This is however exactly our main question. Our new choice of parameters and a refined analysis (based on [12]) allows us to get essentially the best possible ratio between c_1 and c_2 (which would be 1), and essentially the best value for c_1 (which would be 2). The "price" we pay for those optimal constants is that we are unable to find the planted clique, but rather distinguish between a random graph and a random graph with a slightly larger clique planted in it.

1.3   Notations

Let R and C (for "row" and "column") be two $N \times N$-matrices with entries in $\mathbb{R}$. Let x, y be in $\mathbb{R}^N$, with non-negative entries, and such that $\sum_{i=1}^{N} x_i = \sum_{i=1}^{N} y_i = 1$; such a pair (x, y) is called a pair of *mixed strategies*. The (expected) *payoff* of the row player is $x^T R y$, and the one of the column player is $x^T C y$.

The strategies (x, y) is an ε-*equilibrium* if none of the players can increase his payoff by more than ε by changing his strategy. In other words, the pair (x, y) is an ε-equilibrium if for all strategies $\tilde{x}$ and $\tilde{y}$, we have

$$\tilde{x}^T R y \leq x^T R y + \varepsilon \qquad \text{and} \qquad x^T C \tilde{y} \leq x^T C y + \varepsilon.$$

(For the definition of approximation we use the following conventional assumptions: the value of the equilibrium lies in $[0, 1]$ and the approximation is additive). A 0-equilibrium is more succinctly called a *Nash equilibrium*.

A Nash equilibrium is *best* if it maximizes the average payoff of the players, i.e., if it maximizes its *value*

$$\frac{1}{2}x^T(R+C)y.$$

A pair of strategies (x, y) is an *ε-best ε-equilibrium* if it is an ε-equilibrium and its value is at least as large as the value of the best Nash equilibrium minus ε, i.e.,

$$\left(\max_{\tilde{x},\tilde{y}} \frac{1}{2}\tilde{x}^T(R+C)\tilde{y}\right) - \varepsilon \le \frac{1}{2}x^T(R+C)y,$$

where $(\tilde{x}, \tilde{y})$ runs over the Nash equilibria of the game.

2   Preliminaries: Properties of the Matrix B

In this section we describe several properties that the matrix B has *whp*, which play a crucial role in our proof. We remind the reader that the entries of B are scaled Bernoulli variables of some scale t and probability p, i.e.,

$$B_{i,j} = \begin{cases} 0 & \text{with probability } 1-p, \\ t & \text{with probability } p, \end{cases}$$

where t and p are parameters to be selected.

Proposition 1. *Fix small enough $\beta > 0$, and let $c_1 = 2+7\beta^{1/2}$, $c_2 = 2+6\beta^{1/2}$. There exists parameters t, p and s such that the matrix B of size $n^s \times n$, filled with independent and identically distributed scaled Bernoulli-variables of scale t and parameter p, enjoys the following properties whp:*

(i) *Fix a set $I \subseteq [1, n]$ of $c_1 \log n$ indices (independently of B). For every row i of B,*

$$\frac{1}{c_1 \log n} \sum_{j \in I} B_{i,j} \le 1.$$

(ii) *For every set $J \subseteq [1, n]$ of $c_2 \log n$ indices, there exists a row $i = i(J)$ in B so that $B_{i,j} \ge 1 + 9\beta$ for every $j \in J$.*

The proof uses the following variant of the Chernoff bound. If $X_1, \ldots, X_m$ are m independent Bernoulli variables of scale t and parameter p, then

$$\Pr\left(m^{-1}\sum_{i=1}^{m} X_i \ge tp(1+\delta)\right) \le e^{mp[\delta - (1+\delta)\ln(1+\delta)]}, \tag{1}$$

for any $\delta > 0$.

Proof. **The c_1-calculation.** Write $m = c_1 \log n$, and fix a row j. Using (1), we see that

$$\Pr\left(m^{-1}\sum_{i=1}^{m} A_{j,i} \geq 1\right) \leq \exp\left(\frac{m[1 - tp + \ln(tp)]}{t}\right).$$

Hence using the union bound over all the rows, the first property does not hold with probability at most

$$n^{s + c_1[1 - tp + \ln(tp)]/(t \ln 2)}.$$

So the first property holds with high probability if

$$s < \frac{c_1}{t \ln 2}[tp - 1 - \ln(tp)]. \tag{2}$$

The c_2-calculation. For the second property to hold, we need $t \geq 1 + 9\beta$.

Fix a set $I = \{i_1, \ldots, i_{c_2 \log n}\} \subset [n]$ of $c_2 \log n$ indices. Then for a fixed row j, the probability that $B_{j,i} \geq t$ for every $i \in I$ is $p^{c_2 \log n}$, so the probability that there is no good row for the indices I is

$$(1 - p^{c_2 \log n})^{n^s},$$

hence by the union bound, the probability that there is a set of indices with no good row is at most

$$\binom{n}{c_2 \log n}(1 - p^{c_2 \log n})^{n^s} \leq \exp\left(c_2 \log n \ln n + n^s \ln(1 - p^{c_2 \log n})\right),$$

which tends to 0 with $n \to \infty$ if

$$n^s \ln(1 - p^{c_2 \log n}) < -c_2 \log n \ln n.$$

i.e., if $-n^{s - c_2 \log(p^{-1})} < -c_2 \log n \ln n$. So, for the second property to hold, it suffices to require that

$$s > c_2 \log(p^{-1}), \tag{3}$$

in which case $n^s \ln(1 - p^{c_2 \log n})$ goes polynomially fast to $-\infty$.

Choice of p and t. We can now deduce a sufficient condition by combining (2) and (3), which gives

$$\frac{c_1}{t \ln 2}[tp - 1 - \ln(tp)] > c_2 \log(p^{-1}).$$

Plugging in the values for c_1 and c_2 we obtain the following condition on p and t.

$$\frac{2 + 7\beta^{1/2}}{2 + 6\beta^{1/2}} > \frac{t \ln(p^{-1})}{tp - 1 - \ln(tp)}.$$

Now, the limit of the right hand side as $p \to 0$ equals t, hence if we set $t = 1 + 9\beta$, we see that the right hand side is indeed smaller than the left hand side for sufficiently small (yet constant) p, provided that

$$(2 + 7\beta^{1/2}) - (1 + 9\beta)(2 + 6\beta^{1/2}) > 0.$$

The left hand side of the above is a polynomial in $\beta^{1/2}$ with no constant term and whose dominant term, the coefficient in $\beta^{1/2}$, is positive (in fact, equal to one). Hence this polynomial is positive for small positive values.

3   Proof of Theorem 1

Before giving the actual details let us outline the proof in general lines.

3.1   Proof Outline

For the graph H (with the clique of size $(2 + 28\varepsilon^{1/8}) \log n$) we show that whp the game has a Nash equilibrium of social welfare at least 1. Then, given an ε-best ε-equilibrium, we show that a 3ε-best 7ε-equilibrium can be efficiently calculated whose support lies entirely on A (this is very similar to [12]). Then we show how to efficiently extract a very-dense subgraph D from that strategy (here our density is much higher than [12], we need this higher density as we don't have slackness in the size of the planted clique). On the other hand, we prove that $G \in G_{n,1/2}$ whp does not contain such subgraph, causing the algorithm to fail at some point. The graph D then allows us to distinguish H from G.

3.2   Formal Proof

In this section, we assume that the matrix B satisfies the properties of Proposition 1, which is the case whp. We assume that c_1 and c_2 are chosen according to Proposition 1, that is $c_1 = 2 + 7\beta^{1/2}, c_2 = 2 + 6\beta^{1/2}$.

Proposition 2. *If A represents a graph H with a clique of size at least $c_1 \log n$, then every equilibrium that maximizes the utilities of the players has value at least 1.*

Proof. Let C be a clique of H of size $c_1 \log n$. Consider the following strategies for both players: each player puts probability $|C|^{-1}$ on every row (column) of that clique. The value of that strategy is clearly 1 for both players. The first property of Proposition 1 guarantees that none of the players has an incentive to defect, thus ensuring that the strategies we chose indeed constitute a Nash equilibrium.

Proposition 3. *If (x, y) is a δ-equilibrium of value at least $1 - \delta$ then every player has at least $1 - 2\delta$ of his probability mass on A.*

Proof. The sum of payoffs of both players from the entries outside A is 0. If one player has more than 2δ of his probability mass outside A, then the value of the game cannot exceed (observing that the maximal entry in A is 1)

$$\frac{1}{2}(1 + (1 - 2\delta)) = 1 - \delta.$$

This contradicts our assumption on the value of the game.

Proposition 4. *Given a δ-equilibrium of value $1-\delta$ one can efficiently compute a 7δ-equilibrium of value at least $1 - 3\delta$ whose support is entirely on A.*

Proof. Given a δ-equilibrium (x, y), define (x', y') as follows: take the probabilities outside A in both x and y and spread them arbitrarily over A. Let us consider the row player (the column player is symmetric). The maximal entry outside A has value at most 2 (since $t \leq 2$), hence the payoff of the row player from the entries outside A is (in absolute value) at most $2\delta \cdot 2 = 4\delta$. The gain of relocating 2δ-probability to A is at most $1 \cdot 2\delta$ (he does not gain from the upper-right part of the matrix). Thus (x', y') is a $\delta + 4\delta + 2\delta = 7\delta$-equilibrium. As for its new total value, the total probability mass relocated for both players is 4δ, thus gaining at most $0.5 \cdot 4\delta \cdot 1 = 2\delta$ (0.5 factor comes from the definition of game-value, and the 1 is the maximal entry in A. The game outside A is zero-sum, so is disregarded).

Proposition 5. *Let (x, y) be a 7δ-equilibrium played entirely on A. Suppose also that the matrix B is generated with parameter $\beta \in [\delta, 1/9]$. Then every subset of the rows Σ whose probability in x is at least $1 - \beta$ satisfies $|\Sigma| \geq c_2 \log n$. The same applies for y.*

Proof. We shall prove for the column player, the proof of the row player is symmetric. For contradiction, say there exists a set Σ of columns whose total probability is at least $1 - \beta$ but $|\Sigma| \leq c_2 \log n$ (recall: $c_2 = (2 + 6\beta^{1/2})$). By the second property of B in Proposition 1, there exists a row in B in which all corresponding entries have value at least $1 + 9\beta$. If the row player relocates all his probability mass to that row, his new payoff is at least $(1 + 9\beta)(1 - \beta) > 1 + 7\beta \geq 1 + 7\delta$ (the last inequality is true for our choice of β). His current payoff is at most 1 (as all entries in A are bounded by 1), and so he will defect, contradicting the 7δ-equilibrium.

For two sets of vertices (not necessarily disjoint), we let $e(V, W)$ be the number of edges connecting a vertex from V with a vertex from W. We use $e(V)$ as a shorthand for $e(V, V)$. It is easy to see that the maximal number of edges is

$$K(V, W) = |V| \cdot |W| - \binom{|V \cap W|}{2}. \tag{4}$$

The density of the two sets is defined to be

$$\rho(V, W) = \frac{e(V, W)}{K(V, W)}. \tag{5}$$

Proposition 6. *Assume we are given a 7δ-equilibrium of value $1-3\delta$ played entirely on A, and the matrix B was generated with $\beta = 16\delta^{1/4}$. One can efficiently find two sets of vertices S_1, S_2 that enjoy the following properties:*

- *$|S_1|, |S_2| \geq (2 + 6\beta^{1/2}) \log n$,*
- *$\rho(S_1, S_2) > 1 - \beta$.*

Proof. First observe that if the value of the game (played on A) is $1 - 3\delta$, then each player's payoff is at least $1 - 6\delta$ (as the maximum payoff on A is 1). Let $e_i \in \mathbb{R}^n$ be the unit vector whose entries are 0 except the ith which is 1. Consider the following set of columns:

$$\Gamma_t = \{i : x^T A e_i \geq t\}, \qquad \bar{\Gamma}_t = \{i : x^T A e_i < t\}. \tag{6}$$

Since the payoff of the column player is at least $1 - 6\delta$ (and in particular at least $1 - 6\delta^{1/2}$), $\Gamma_{1-6\delta^{1/2}} \neq \emptyset$. We now claim that the total probability mass of y on the columns in $\bar{\Gamma}_{1-7\delta^{1/2}}$ is at most $16\delta^{1/2}$. If not, by relocating $16\delta^{1/2}$-probability from $\bar{\Gamma}_{1-7\delta^{1/2}}$ to $\Gamma_{1-6\delta^{1/2}}$ the gain is at least $(7 - 6)\delta^{1/2} \cdot 16\delta^{1/2} = 16\delta > 7\delta$, which contradicts the 7δ-equilibrium. Thus, by Proposition 5,

$$|\Gamma_{1-7\delta^{1/2}}| \geq (2 + 6\beta^{1/2}) \log n$$

(we can use Proposition 5 since $1 - 16\delta^{1/2} \geq 1 - \beta = 1 - 16\delta^{1/4}$).

For a set T of vertices, let $U_T \in \mathbb{R}^n$ be the uniform distribution over T and 0 elsewhere. The condition of Γ_t implies that (this is just a simple averaging argument)

$$x^T A U_{\Gamma_{1-7\delta^{1/2}}} \geq 1 - 7\delta^{1/2}. \tag{7}$$

Now define a set of rows Σ according to:

$$\Sigma = \{j : e_j^T A U_{\Gamma_{1-7\delta^{1/2}}} \geq 1 - 8\delta^{1/4}\}. \tag{8}$$

We claim that $\sum_{j \in \Sigma} x_j \geq 1 - \delta^{1/4}$. If not,

$$x^T A U_{\Gamma_{1-7\delta^{1/2}}} \leq \left(1 - \delta^{1/4}\right) \cdot 1 + \delta^{1/4} \cdot (1 - 8\delta^{1/4}) = 1 - 8\delta^{1/2}.$$

This contradicts (7). Applying Proposition 5 once more yields $|\Sigma| \geq (2 + 6\beta^{1/2}) \log n$. Equation (8) implies (again, averaging argument):

$$U_\Sigma^T A U_{\Gamma_{1-7\delta^{1/2}}} \geq 1 - 8\delta^{1/4}. \tag{9}$$

Finally we show how this gives the subgraph of correct density. Set $S_1 = \Sigma, S_2 = \Gamma_{1-7\delta^{1/2}}$. They are both of the required size, denoted s_1, s_2 respectively. The number of edges is (by $d_S(v)$ we denote the degree of v in the set S):

$$e(S_1, S_2) = \left(\sum_{v \in S_1} d_{S_2}(v)\right) - e(S_1 \cap S_2).$$

Here, $\sum_{v \in S_1} d_{S_2}(v)$ is just the total number of one-entries in the sub-matrix of A corresponding to $S_1 \times S_2$, which, by Equation (9), is at least $(1 - 8\delta^{1/4})s_1 s_2$, and we subtract the number of edges in the intersection (since they were counted twice). Thus,

$$e(S_1, S_2) \geq (1 - 8\delta^{1/4})s_1 s_2 - e(S_1 \cap S_2).$$

Recalling the definition of the density $\rho(S_1, S_2)$, Equation (5), we get

$$\rho(S_1, S_2) = \frac{e(S_1, S_2)}{K(S_1, S_2)}$$

$$\geq \frac{(1 - 8\delta^{1/4})s_1 s_2 - e(S_1 \cap S_2)}{s_1 s_2 - \binom{|S_1 \cap S_2|}{2}}$$

$$\geq \frac{(1 - 8\delta^{1/4})s_1 s_2 - \binom{|S_1 \cap S_2|}{2}}{s_1 s_2 - \binom{|S_1 \cap S_2|}{2}}.$$

This in turn equals

$$1 - \frac{8\delta^{1/4}s_1 s_2}{s_1 s_2 - \binom{|S_1 \cap S_2|}{2}}.$$

Observing that $s_1 s_2 - \binom{|S_1 \cap S_2|}{2} \geq s_1 s_2 / 2$, we get

$$\rho(S_1, S_2) \geq 1 - 2 \cdot 8\delta^{1/4} = 1 - 16\delta^{1/4} = 1 - \beta.$$

Finally we present the property of $G_{n,1/2}$ that we require.

Proposition 7. *The following assertion holds whp for $G_{n,1/2}$. For no $0 \leq \alpha \leq 1/8$, there exist two sets of vertices S_1, S_2 of size at least $(2 + 6\alpha) \log n$ each and such that $e(S_1, S_2) \geq (1 - \alpha^2)K(S_1, S_2)$.*

Proof. The proof idea is as follows. The expected number of such sets S_1, S_2 is at most (summing over all possible sizes for S_1 and S_2 and intersection size)

$$\mu \leq \sum_{y,z \geq (2+6\alpha)\log n} \sum_{x=0}^{\min\{y,z\}} \binom{n}{x}\binom{n}{y-x}\binom{n}{z-x} 2^{-K} \sum_{i=0}^{\alpha^2 K} \binom{K}{i}$$

where $K = K(S_1, S_2)$. The first term accounts for choosing the intersection vertices, then completing each of S_1 and S_2. Next choose which edges are present and finally multiply by the probability for edges/non-edges. We need to show that $\mu = o(1)$, and then the claim follows from Markov's inequality.

Define

$$f(x, y, z) = \binom{n}{x}\binom{n}{y-x}\binom{n}{z-x} 2^{-K} \sum_{i=0}^{\alpha^2 K} \binom{K}{i}, \tag{10}$$

so that

$$\mu \leq \sum_{y,z \geq (2+6\alpha)\log n} \sum_{x=0}^{\min(y,z)} f(x, y, z). \tag{11}$$

Our first goal is to estimate the sum $\sum_{i=0}^{\alpha^2 K} \binom{K}{i}$. We start out with the standard estimate

$$\sum_{i=0}^{\rho K} \binom{K}{i} \leq 2^{Kh(\rho)} \qquad \text{for } 0 \leq \rho \leq 1/2,$$

where $h(\rho) = -\rho \log(\rho) - (1-\rho) \log(1-\rho)$ is the binary entropy function. In the range of interest $0 \leq \rho = \alpha^2 \leq 1/64$, we get

$$h(\rho) \leq \rho(-\log(\rho) + 64 \log(64/63)),$$

by bounding $\log(1 - \rho)$ by the appropriate linear function.

Now, studying the first and second derivatives of $-\alpha^2 \log(\alpha^2)$, we see that this function is increasing in the range $0 \leq \alpha \leq 1/8$ and reaches its maximal slope in this range at $\alpha = 1/8$. The maximal slope is less than 1.2. Therefore,

$$h(\alpha^2) \leq 1.2\alpha + 64 \log\left(\frac{64}{63}\right)\alpha^2 \leq \left(1.2 + 8\log\left(\frac{64}{63}\right)\right)\alpha \leq \frac{3}{2}\alpha,$$

and so

$$\sum_{i=0}^{\alpha^2 K} \binom{K}{i} \leq 2^{\frac{3}{2}K\alpha}.$$

Going back to (10), bounding $\binom{n}{t}$ by n^t and recalling that $K = yz - \binom{x}{2}$, we get

$$\log f(x, y, z) \leq (y + z - x) \log n - K\left(1 - \frac{3}{2}\alpha\right)$$

$$= (y + z)\log n - yz\left(1 - \frac{3}{2}\alpha\right) - x\log n + \left(1 - \frac{3}{2}\alpha\right)\frac{x^2}{2}.$$

The maximum of the function $x \mapsto -x\log n + (1 - \frac{3}{2}\alpha)x^2/2$ in the range $x \in [0, \min\{y, z\}]$ is reached at the boundary $x = \min\{y, z\}$, which, assuming wlog $y \geq z$, is at $x = z$. Thus,

$$\log f(x, y, z) \leq (y + z)\log n - yz\left(1 - \frac{3}{2}\alpha\right) - z\log n + \left(1 - \frac{3}{2}\alpha\right)\frac{z^2}{2}$$

$$= y\log n - \left(1 - \frac{3}{2}\alpha\right)\left(yz - \frac{z^2}{2}\right).$$

Observe that $yz - z^2/2 \geq yz/2$, and hence

$$\log f(x, y, z) \leq y\log n - \left(1 - \frac{3}{2}\alpha\right)\frac{yz}{2} \leq y\left(\log n - \frac{z}{2}\left(1 - \frac{3}{2}\alpha\right)\right).$$

Recall our choice of z: $z \geq (2 + 6\alpha)\log n$ (and the same goes for y). Since $(1 - \frac{3}{2}\alpha)(2 + 6\alpha) \geq 2$ for our values of α,

$$\log f(x, y, z) = -\Omega(\log^2 n).$$

Plugging this into (11) one obtains

$$\mu \le \sum_{y,z \ge (2+6\alpha)\log n} \sum_{x=0}^{\min\{y,z\}} 2^{-\Omega(\log^2 n)} \le n^3 \cdot n^{-\Omega(\log n)} = o(1).$$

The proposition follows by Markov's inequality.

3.3 The Distinguishing Algorithm

Let $\mathcal{A}$ be a polynomial time algorithm that finds the ε-best ε-equilibrium in a two player game. We shall show that there exists an algorithm $\mathcal{B}$ that runs in polynomial time and distinguishes *whp* between a graph randomly chosen from $G_{n,1/2}$ and an arbitrary graph with a clique of size $c_1 \log n$.

The algorithm $\mathcal{B}$ does the following on an input graph G, which is either a graph from $G_{n,1/2}$ or a graph containing a clique of size at least $(2+28\varepsilon^{1/8})\log n$.

1. If any of the below steps fails, return "G belongs to $G_{n,1/2}$".
2. Generate the game matrix with parameter $\beta = 16\varepsilon^{1/4}$ in the matrix B.
3. Run $\mathcal{A}$ to receive an ε-best ε-equilibrium of that game.
4. Calculate a 7ε-equilibrium of value at least $1 - 3\varepsilon$ whose support lies entirely on $A(G)$ (according to the procedure in the proof of Proposition 4).
5. Use this equilibrium to find two sets S_1 and S_2 satisfying $|S_1|, |S_2| \ge (2 + 6\beta^{1/2})\log n$ and $\rho(S_1, S_2) \ge 1 - \beta$ (use the procedure in the proof of Proposition 6).
6. If succeeded, return "G does not belong to $G_{n,1/2}$".

We shall analyze the algorithm for $\varepsilon \le \varepsilon_0$. ε_0 is determined by the constraint of Proposition 7. Specifically, for the algorithm to answer correctly on $G_{n,1/2}$, it suffices if step 5 fails. For this we want to choose β so that *whp* $G_{n,1/2}$ does not contain two sets S_1, S_2 of the prescribed size and density. This is achieved by plugging $\alpha = \beta^{1/2}$, that is $\alpha = 4\varepsilon^{1/8}$, in Proposition 7, which translates to $\varepsilon_0 = (4 \cdot 8)^{-8}$.

It remains to prove that the algorithm answers correctly when G has a clique of size $\ge c_1 \log n$. Assume that the matrix B satisfies Proposition 1, which is the case *whp*. Propositions 2 and 4 then guarantee that Step 4 succeeds, and Step 5 succeeds by Proposition 6. Thus again the correct answer is returned.

4 Finding a Clique of Size $3\log n$

Let $\beta = 16\varepsilon^{1/4}$ as in the proof of Theorem 1, and let $H \in G_{n,1/2,k}$ with $k \ge (3 + 14\beta^{1/2})\log n$. Let C be the vertices of the planted clique. Observe that in no place in Section 1 did we use the actual size of the planted clique, just the "separation" properties as given by Proposition 1. Proposition 1 can be restated easily with $c_1 \ge (3 + 14\beta^{1/2})$ and $c_2 = (3 + 13\beta^{1/2})\log n$. Therefore by the same arguments as in the Proof of Theorem 1, we are guaranteed to efficiently find two sets S_1, S_2 of size at least $c_2 \log n$ each and density $1 - \beta$. Our first goal is to show that $S_1 \cup S_2$ must intersect the planted clique on many vertices. Suppose by contradiction that the intersection size is no more than $(1 + \beta^{1/2}) \log n$ vertices. Define $S_1' = S_1 \setminus C$ and similarly $S_2' = S_2 \setminus C$. Clearly, S_1', S_2' still contain at least $(2 + 12\beta^{1/2}) \log n$ vertices, and all edges between S_1' and S_2' are random edges of $G_{n,1/2}$. Finally let us compute the density $\rho(S_1', S_2')$. Recall Equation (9) which guarantees that in $A[S_1 \times S_2]$ there are at most $(\beta/2)S_1 S_2$ zeros. As for $A[S_1' \times S_2']$, the fraction of zeros is at most

$$\frac{(\beta/2)|S_1||S_2|}{|S_1'||S_2'|} \leq \frac{(\beta/2)|S_1||S_2|}{(2/3)^2|S_1||S_2|} = \frac{9\beta}{8}.$$

In the inequality we use $|S_1'| \geq 2|S_1|/3, |S_2'| \geq 2|S_2|/3$. Now the same arguments that follow Equation (9) give $\rho(S_1', S_2') \geq 1 - 2 \cdot \frac{9\beta}{8} \geq 1 - 3\beta$. To conclude, we found two sets of size at least $(2 + 12\beta^{1/2}) \log n$ each, and density $1 - 3\beta$, involving only edges of $G_{n,1/2}$. This however contradicts Proposition 7 (when plugging $\alpha^2 = 3\beta$ in that proposition).

Let us now assume that $S_1 \cup S_2$ contains at least $(1 + \beta^{1/2}) \log n$ vertices from the planted clique, call this set I. Further assume w.l.o.g. that $|S_1 \cup S_2| = O(\log n)$ (we can always do this since in Equations (6) and (8), which define S_1, S_2, we can limit the set size). Thus one can find I in polynomial time (using exhaustive search). Finally, let us compute the probability that a vertex $x \notin C$ has full degree in I. Since the planted clique was chosen independently of the random graph, this probability is at most

$$2^{-|I|} = 2^{-(1+\beta^{1/2}) \log n} = n^{-(1+\beta^{1/2})}.$$

Using the union bound, *whp* no such vertex exists. Now apply the following greedy procedure on I: go over the vertices of G and add each vertex if its degree in I is full. By the latter argument, this algorithm succeeds *whp* in reconstructing the planted clique.

5   Discussion

In this work we explored the technique of [12] in the regime where the planted clique is close to the smallest possible, that is of size $(2+\delta) \log n$ for small $\delta > 0$. We showed that for the problem of distinguishing $G_{n,1/2}$ from $G_{n,1/2,k}$, where $k = (2 + \delta) \log n$, the reduction works for arbitrarily small $\delta > 0$, provided that ε-best ε-approximate Nash equilibria can be found for a corresponding small $\varepsilon(\delta) > 0$.

We also showed that the problem of *finding* a planted clique of size $(3 + \delta) \log n$ for small $\delta > 0$ can be reduced to finding an ε-best ε-approximate Nash equilibrium for a sufficiently small $\varepsilon > 0$. But since the maximal clique in $G_{n,1/2}$ is only of size $(2 - o(1)) \log n$, this is possibly not optimal, and the question whether one could achieve the optimal $2 \log n$ clique size barrier for finding the clique is still open.

References

1. Alon, N., Kahale, N.: A spectral technique for coloring random 3-colorable graphs. SIAM Journal on Computation 26(6), 1733–1748 (1997)
2. Alon, N., Krivelevich, M., Sudakov, B.: Finding a large hidden clique in a random graph. Random Structures and Algorithms 13(3-4), 457–466 (1998)
3. Alon, N., Spencer, J.: The Probabilistic Method. Wiley, Chichester (1992)
4. Chen, X., Deng, X., Teng, S.-H.: Computing Nash equilibria: Approximation and smoothed complexity. In: FOCS 2006: Proceedings of the 47th Annual IEEE Symposium on Foundations of Computer Science, USA, pp. 604–612 (2006)
5. Conitzer, V., Sandholm, T.: Complexity Results about Nash Equilibria. In: 18th International Joint Conference on Artificial Intelligence, pp. 765–771 (2003)
6. Daskalakis, C., Goldberg, P.W., Papadimitriou, C.H.: The complexity of computing a Nash equilibrium. In: STOC 2006: Proceedings of the 38th annual ACM Symposium on Theory of computing, USA, pp. 71–78 (2006)
7. Feige, U.: Relations between average case complexity and approximation complexity. In: STOC 2002: Proceedings of the 34th annual ACM Symposium on Theory of Computing, pp. 534–543 (2002)
8. Feige, U., Ofek, E.: Easily refutable subformulas of large random 3CNF formulas. Theory of Computing 3(1), 25–43 (2007)
9. Gilboa, I., Zemel, E.: Nash and correlated equilibria: Some complexity considerations. Games and Economic Behavior 1(1), 80–93 (1989)
10. Jerrum, M.: Large Cliques Elude the Metropolis Process. Random Structures and Algorithms 3(4), 347–359 (1992)
11. Juels, A., Peinado, M.: Hiding Cliques for Cryptographic Security. Designs, Codes and Cryptography 20(3), 269–280 (2000)
12. Hazan, E., Krauthgamer, R.: How hard is it to approximate the best Nash equilibrium? In: ACM-SIAM Symposium on Discrete Algorithms, USA, pp. 720–727 (2009)
13. Kučera, L.: Expected Complexity of Graph Partitioning Problems. Discrete Applied Mathematics 57(2-3), 193–212 (1995)
14. Lipton, R.J., Markakis, E., Mehta, A.: Playing large games using simple strategies. In: 4th ACM conference on Electronic commerce, USA, pp. 36–41 (2003)
15. Tsaknakis, H., Spirakis, P.G.: An Optimization Approach for Approximate Nash Equilibria. In: Deng, X., Graham, F.C. (eds.) WINE 2007. LNCS, vol. 4858, pp. 42–56. Springer, Heidelberg (2007)

Testing Computability by Width Two OBDDs

Dana Ron* and Gilad Tsur

School of Electrical Engineering, Tel Aviv University, Tel Aviv Israel
danar@eng.tau.ac.il,
giladt@post.tau.ac.il

Abstract. Property testing is concerned with deciding whether an object (e.g. a graph or a function) has a certain property or is "far" (for some definition of far) from every object with that property. In this paper we give lower and upper bounds for testing functions for the property of being computable by a read-once width-2 *Ordered Binary Decision Diagram* (OBDD), also known as a *branching program*, where the order of the variables is known. Width-2 OBDDs generalize two classes of functions that have been studied in the context of property testing - linear functions (over $GF(2)$) and monomials. In both these cases membership can be tested in time that is linear in $1/\epsilon$. Interestingly, unlike either of these classes, in which the query complexity of the testing algorithm does not depend on the number, n, of variables in the tested function, we show that (one-sided error) testing for computability by a width-2 OBDD requires $\Omega(\log(n))$ queries, and give an algorithm (with one-sided error) that tests for this property and performs $\tilde{O}(\log(n)/\epsilon)$ queries.

1 Introduction

Property testing is concerned with deciding whether an object (e.g. a graph or a function) has a certain property or is "far" (for some definition of far) from every object with that property [9, 18]. Typical property testing algorithms are randomized, and perform queries regarding local properties of the object (e.g., the value of a function f on the input x), returning a correct answer with high probability. Generally, the distance parameter, denoted ϵ, is given as an input to the property testing algorithm and effects its running time. Property testing algorithms are designed to run in time that is sublinear in the size of the tested object or to perform a number of queries that is sublinear in it. Indeed, many of them use a number of queries that is independent of the object's size.

OUR RESULTS. In this paper we give lower and upper bounds for testing functions for the property of being computable by a read-once width-2 *Ordered Binary Decision Diagram* (OBDD), also known as a read-once *Oblivious Branching Program* of width 2, where the order of the variables is known. Width-2 OB-DDs generalize two classes of functions that have been studied in the context of property testing - linear functions (over $GF(2)$) [18] and monomials [15]. In

---

* This work was supported by the Israel Science Foundation (grant number 246/08).

I. Dinur et al. (Eds.): APPROX and RANDOM 2009, LNCS 5687, pp. 686–699, 2009.

both these cases membership can be tested in time that is linear in $1/\epsilon$. Interestingly, unlike either of these classes, in which the query complexity of the testing algorithm does not depend on the number of variables in the tested function, we show that (one-sided error) testing for computability by a width-2 OBDD requires $\Omega(\log(n))$ queries, and we give an algorithm (with one-sided error) that performs $\tilde{O}(\log(n)/\epsilon)$ queries.[1] We note that it is open whether allowing two-sided error can decrease the complexity of the problem. Observe that the logarithmic dependence on n is still much lower than the linear dependence that is necessary for learning this family. We later shortly discuss the extensive research on OBDDs in the learning literature.

Function classes for which property testing algorithms have been designed are usually characterized as either algebraic (e.g. [1, 3, 10, 11, 18]) or non-algebraic (e.g., [5, 7, 15]), though some results can be viewed as belonging to both categories. We view the family of functions we study as falling naturally into the second category, since it is described by a type of computational device and not by a type of algebraic formula. As opposed to many algorithms for algebraic families, algorithms for non-algebraic families generally rely on the fact that the functions in the family are close to *juntas*, that is, functions that depend on a small number of variables. This is true, by definition, for singletons [15] and juntas [7], but also for monomials, monotone DNF with a bounded number of terms [15], general DNF, decision lists and many other function classes, studied in [5]. In contrast, our algorithm tests for membership in a class of functions in which the function may depend (significantly) on many variables.

TECHNIQUES. Variables in functions that are computable by width-2 OBDDs can be divided into two groups. Variables that the function is "linear" in, and all other variables. This distinction is made more precise in Section 2. Our algorithm attempts to locate the last (according to the fixed order) $O(\log(1/\epsilon))$ non-linear variables, and uses this structural information to determine whether a function is computable by a width-2 OBDD. This can be contrasted with the results in [5] that cover a wide-range of non-algebraic function families. There, the algorithms detect a small number of subsets of variables, each containing a relevant variable. If the tested function belongs to the class in question then it depends almost entirely on this small set of relevant variables. Our algorithm learns something of the *structure* of the OBDD computing the function that relates to a small set of variables, yet many variables can have non-negligible influence.

Since our algorithm rejects only when it finds evidence that the tested function is not computable by a width-2 OBDD, it immediately follows that it always accepts functions in this family. The core of the proof is in showing that if the function is ϵ-far from the family, then the algorithm rejects with high constant probability. More precisely, we prove the contrapositive statement. Since our algorithm works by verifying that various restrictions of the tested function are close to having certain properties, the difficulty is in proving that we can "glue" together these restrictions and obtain a single width-2 OBDD.

---

[1] Here the notation $\tilde{O}(T)$ represents an upper bound that is linear in T up to a polylogarithmic factor.

ADDITIONAL RELATED WORK. We note that the type of question we ask differs from that studied by Newman [14]. Newman shows that a property testing algorithm exists for any property decidable by a constant width branching program. In [14] the property is defined with regard to a particular branching program, and the algorithm tests membership in a language decidable by that program. In contrast, in our result, the language we test for membership in is one where every word is the truth table of a width-2 branching program.

OBDDs and, in particular, bounded width OBDDs have been studied in the machine learning context rather extensively. In particular it has been shown that width-2 OBDDs are PAC-learnable, while width-3 and wider OBDDs are as hard to learn as DNF formulas [6]. These results were strengthened in [2, 4]. When membership queries are allowed and the underlying distribution is uniform, width-2 branching programs with a single 0 sink are efficiently learnable [2]. When both membership and equivalence queries are allowed then there are several positive results for other restricted types of branching programs [2, 8, 12, 13, 16].

ORGANIZATION. In Section 2 we provide some basic definitions and present general claims regarding OBDDs and width-2 OBDDs in particular. In Section 3 we give a lower bound for one-sided error testing of computability by width-2 OBDDs. In Section 4 we give a one-sided error testing algorithm for computability by a width-2 OBDD. All missing details can be found in the full version of this paper [17].

2 Preliminaries

2.1 Basic Definitions

Definition 1. *The* distance *between a function f and a function g (both from domain X to range R), denoted $d(f, g)$, is defined as $\Pr_x[f(x) \neq g(x)]$ where x is drawn uniformly at random from X. When $d(f, g) > \epsilon$ we say that f is ϵ-far from g. This definition extends to the distance of a function f from a family of functions G (also denoted $d(f, G)$). Here we have $d(f, G) = \min_{g \in G}\{d(f, g)\}$. When f is ϵ-far from all $g \in G$ we say that f is ϵ-far from G. We may occasionally describe a function f as* ϵ-close *to g. This means that f is not ϵ-far from g.*

Definition 2. *A* property testing algorithm *T for property P is given oracle access to an object in the form of a function f from domain X to range R and a distance parameter $0 < \epsilon < 1$.*

1. *If $f \in P$ then T accepts with probability at least 2/3 (over its internal coin tosses);*
2. *If f is ϵ-far from P then T rejects with probability at least 2/3 (over its internal coin tosses).*

A property testing algorithm is said to be a one-sided error property testing algorithm *if it accepts with probability 1 when $f \in P$.*

Of course, if we wish to accept or reject *incorrectly* with probability at most δ we can repeat the application of a property testing algorithm $\Theta(\log(1/\delta))$ times and take a majority vote. Later in this work we will routinely "amplify" probability of success as required.

A property testing algorithm that we use as a basic building block in our algorithm is the linearity tester, proposed by Blum, Luby and Rubinfeld [3]. In [3] it is assumed that for a linear function f it holds that $f(0^n) = 0$. For our purposes linearity allows for a single coefficient that doesn't depend on any variable, and the BLR algorithm is easily adapted for such a case.

Definition 3. *We say that $f : \{0,1\}^n \to \{0,1\}$ is a* linear *function if there exist coefficients $b_0, b_1, \ldots, b_n \in \{0,1\}$ such that for $x = x_1, \ldots, x_n \in \{0,1\}^n$, $f(x) = b_0 + \Sigma_{i=1}^n b_i x_i$.*

Theorem 1. [3] *There exists a one-sided error testing algorithm for linearity. Its query complexity is $O(1/\epsilon)$.*

Definition 4. *A* Binary Decision Diagram *(BDD), also known as a* branching program, *is an acyclic directed graph with a single source where sinks are labeled 0 or 1, and other nodes are labeled by a Boolean variable from $X = \{x_1, \ldots, x_n\}$. The Boolean function associated with a BDD is computed on a particular assignment to the variables in X by returning the label of the sink reached when these assignment values are used to trace a route through the graph.*

There are different definitions in the literature for Ordered Binary Decision Diagrams. Our results hold for the definition of a *strict* fixed width read-once binary decision diagram:

Definition 5. *A* read-once *Ordered Binary Decision Diagram (OBDD) is a BDD where each path from the root to a sink must pass through all variables in a fixed order (where each variable appears once). The* width *of an OBDD is the maximum number of nodes in a level of the rooted graph.*

Branching programs can, of course, compute functions from other, non-binary, finite domains and to other finite ranges. Indeed, it may sometimes be more convenient to consider such programs, and we do this in Claim 1. As each path from the root to a sink must pass through all variables in a given order, x_i's level in the OBDD is well defined.

Throughout the following text we refer to OBDD's over a fixed order of the variables $x_1, \ldots, x_n$ that is known to us. Once the order of the variables has been fixed there exists an automaton corresponding to any OBDD, that is, the OBDD can be thought of as an automaton without loops and with accepting states corresponding to 1 sinks of the OBDD.

2.2   Properties of OBDDs

Our first claim follows directly from the definition of OBDDs.

Claim 1. *A function $f : \{0,1\}^n \to [k]$ (where $[k] = \{0, \ldots, k-1\}$) is computable by a width-k OBDD if and only if $f(x_1, \ldots, x_n) = g_n(f_{n-1}(x_1, \ldots, x_{n-1}), x_n)$ where f_{n-1} is a function computable by a width-k OBDD (over 0 variables if $n = 1$) and g_n is a function from $[k] \times \{0,1\}$ to $[k]$.*

Definition 6. *A set S is said to be an i-Prefix Equivalence Class (or just an i-equivalence class) for a function $f : \{0,1\}^n \to \{0,1\}$ (where $i \leq n$) when S is a maximal subset of $\{0,1\}^i$ such that for all $x, y \in S$ and for all $z \in \{0,1\}^{n-i}$ it holds that $f(xz) = f(yz)$.*

As a corollary of Claim 1 we have:

Corollary 1. *A function $f : \{0,1\}^n \to \{0,1\}$ is computable by a width k OBDD if and only if $\forall i \in [1,n]$ there are at most k distinct i-prefix equivalence classes for f.*

Definition 7. *A string $z \in \{0,1\}^{n-i}$ is a distinguishing assignment for two i-prefix equivalence classes S_1, S_2 over a function f if for $x \in S_1, y \in S_2$ it holds that $f(xz) \neq f(yz)$.*

In describing our algorithm we routinely restrict a function over some of its input variables. We introduce a notation for such restrictions:

Definition 8. *Let f be a function of n boolean variables, and let $w \in \{0,1\}^m$. We define $f_{i,j,w}$, where $j = i + m - 1$, to be the function f with the variables $x_i, \ldots, x_j$ restricted to the values $w_1, \ldots, w_m$ accordingly. This means that*

$$f_{i,j,w}(x_1, \ldots, x_n) \equiv f(x_1, \ldots, x_{i-1}, w_1, \ldots, w_m, x_{j+1}, \ldots, x_n).$$

As the values $x_i, \ldots, x_j$ have no influence on $f_{i,j,w}$ we view it, interchangeably, as a function of n variables or as a function of $n - m$ variables.

For a width-2 OBDD we can arbitrarily equate each of the (at most) 2 nodes in the i'th level of the OBDD with a value in $\{0,1\}$. We denote by $f_i : \{0,1\}^i \to \{0,1\}$ the function that is given the string $x_1, \ldots, x_i$ and returns the value of the node in the $i + 1$'th level reached from the source by traversing the OBDD according to them. Thus, for a distinguishing assignment w of length $n - i$ we have that either $f_i(x_1, \ldots, x_i) = f_{i+1,n,w}(x_1, \ldots, x_i)$ for all $x_1, \ldots, x_i$, or that $f_i(x_1, \ldots, x_i) = \neg f_{i+1,n,w}(x_1, \ldots, x_i)$ for all $x_1, \ldots, x_i$.

Merging Levels and Bit Influence

Definition 9. *An OBDD is said to be 0-merging (1-merging) on the i'th level if in level $i + 1$ there exists a node with two incoming edges marked by 0 (1). For $\sigma \in \{0,1\}$, if an OBDD isn't σ-merging on the i'th level we say it is σ-non-merging on the i'th level.*

Claim 2. *Let M be a width-2 OBDD and let f_{i-1} be the function that maps the variables $x_1, \ldots, x_{i-1}$ to a node on the i'th level. If layers i to n are $0,1$-non-merging, then M computes a linear function of $f_{i-1}(x_1, \ldots, x_{i-1})$ and of $x_i, \ldots, x_n$. That is, there exists a linear function $\hat{f}$ such that $f(x_1, \ldots, x_n) = \hat{f}_i(f_{i-1}(x_1, \ldots, x_{i-1}), x_i, \ldots, x_n).$*

Definition 10. *Let x^i be the string x with the i'th bit flipped. The bit influence of the i'th bit in the function f, denoted $\inf_f(i)$ or just $\inf(i)$ is $\Pr[f(x) \neq f(x^i)]$ when x is drawn from the uniform distribution.*

Claim 3. *For a width-2 OBDD M where layer i is 0-merging or 1-merging, for $j < i$ it holds that $\inf_{f_i}(j) \leq \frac{1}{2} \inf_{f_{i-1}}(j)$.*

Claim 4. *For a function f computable by a width-2 OBDD M, the influence of the bits $x_1, \ldots, x_{i-1}$ in f_i is no greater than their influence in f_{i-1}.*

Claims 3 and 4 will later allow us to find several merging levels in width-2 OBDDs and to disregard all the variables coming before these levels. In Section 4 we will use the following corollary of these claims to show the correctness of our algorithm.

Claim 5. *Let f be a function of the form*

$$f(x_1, \ldots, x_n) = f'(g(x_1, \ldots, x_{n-m}), x_{n-m+1}, \ldots, x_n)$$

where f' is computable by a width-2 OBDD M. If M has at least k merging levels, then f is 2^{-k}-close to the function $f'(0, x_{n-m+1}, \ldots, x_n)$, that is computable by a width-2 OBDD.

3 A Lower Bound

Theorem 2. *Any one-sided error tester for computability by a width-2 OBDD requires $\Omega(\log(n))$ queries.*

We prove Theorem 2 by noting that any one-sided error property testing algorithm that rejects a function f must have queried, for some $i \in [1, n]$, several strings indicating that f has at least 3 different i-equivalence classes. If this were not the case, then by Claim 1, f may be computable by a width-2 OBDD. We describe a family of $\Theta(n)$ functions, all $1/4$-far from any function computable by a width-2 OBDD, where each function is fully specified by a unique value i where the function has more than two equivalence classes. We note that any one-sided error property testing algorithm for width two OBDDs would, in fact, implicitly correspond to an exact learning algorithm for members of this family, and that due to the family size this would require at least $\Omega(\log(n))$ queries. The construction is as follows: We define the function γ_j, for $1 < j < (n-1)$ as $x_1 \oplus \cdots \oplus (x_j \wedge x_{j+1}) \oplus \cdots \oplus x_n$. Essentially, for all prefixes except for the j'th this function has two equivalence classes, but it has more than two equivalence classes for the prefix $x_1, \ldots, x_j$, and it has a constant distance from any function computable by a width-2 OBDD.

Claim 6. *Let f be a function computable by a width-2 OBDD. For $1 < j < (n-1)$, it holds that $d(\gamma_j, f) \geq 1/4$.*

We prove this claim using the fact that for all j it holds that x_1 has influence 1 in γ_j, and that all the variables have influence greater than or equal to $1/2$ (all the variables have influence 1 except x_j and x_{j+1}, that have influence $1/2$).

The following is a well known fact.

Claim 7. *A (possibly randomized) algorithm L that is given oracle access to a* boolean *function $f \in F$, performs queries on f and determines the identity of f with probability greater than $2/3$ must perform at least $\Omega(\log(|F|))$ queries.*

Theorem 2 follows directly from Claims 6 and 7.

4   The Testing Algorithm

Imagine that we are promised that a function f given to us as input is either computable by a width-2 OBDD that has no merging levels, or that it is far from any function computable by a width-2 OBDD. We could check which of the above is the case using BLR's linearity test on f, as a function computable by a width-2 OBDD that has no merging levels is a linear function.

Now, imagine we are promised that f is either far from any function computable by a width-2 OBDD or that it is computable by a width-2 OBDD that has exactly one merging level, in the i'th level, where i is known. We could check to see which of the cases above holds by going through the following procedure. First we'd like to see if f has at most two i-equivalence classes. We cannot know this exactly, but we are able to tell if f is close to a function with 1, 2, or more i-equivalence classes using an algorithm we will describe below. If we only find one i-equivalence class for f it remains to check if f is a linear function of $x_{i+1}, \ldots, x_n$. If it is then f is computable by a width-2 OBDD with one merging level, i (and we can accept). If f has more than two i-equivalence classes then it is clearly not computable by a width-2 OBDD (of any kind), and we can reject. Finally, if f has two i-equivalence classes we must check that the function f_{i-1} (the function that maps the variables $x_1, \ldots, x_{i-1}$ to $(i-1)$-equivalence classes) is linear, and that the function which maps the i-equivalence class and the variables $x_{i+1}, \ldots, x_n$ to $f(x_1, \ldots, x_n)$ is linear, as well.

As a final hypothetical scenario, consider the following promise: f is either far from every function computable by a width-2 OBDD, or can be computed using a width-2 OBDD with a single *unknown* merging level. If we could locate the merging level, we know we could tell which of the two cases holds, as done in the previous paragraph. We note that as a consequence of Claim 3 any function that is computable by a width-2 OBDD and isn't linear is far from linear, so we would like to check f and see what parts of it are linear. We can do this by performing a binary search for a linear section. Begin by restricting the first $n/2$ variables to 0, and checking if the function computed on all the rest of the variables is (close to) linear. If it is, repeat the process with fewer variables restricted. If it isn't, repeat the process with more variables restricted. If we are, indeed, given a function that is computable by a width-2 OBDD that has only a single merging level, this process will allow us to detect the merging level with high probability.

The property testing algorithm we suggest for computability by a width-2 OBDD is based on the observations made above and on those made in the previous sections. In particular, we note that, as a consequence of Claim 5, any function f computable by a width-2 OBDD is ϵ-close to a function g computable by a width-2 OBDD that has at most $O(\log(1/\epsilon))$ merging levels, where the merging levels of g are all merging levels of f. When our algorithm is given as input a function computable by a width-2 OBDD, it will (with high probability) locate the last $O(\log(1/\epsilon))$ merging levels (if such merging levels exist, of course). Locating these levels will be done using a binary search technique reminiscent of the one suggested above. We will restrict the function on some of its bits $(x_1, \ldots, x_j)$ and test whether the restricted function is linear, using a version of the BLR linearity test. For any function f computable by a width-2 OBDD our algorithm will find the structure of a function g that is close to it, and for every function that passes our test, we will show that it is likely to be close to a function computable by a width-2 OBDD.

A notion that is used repeatedly is that of a function f that can be computed by a width-2 OBDD that accepts as input the value of a function $g(x_1, \ldots, x_t)$ and the bits $x_{t+1}, \ldots, x_n$ (in that order) and outputs the value $f(x_1, \ldots, x_n)$. We define this formally:

Definition 11. *A function $f : \{0,1\}^n \to \{0,1\}$ is said to be a* W2-function *of $g : \{0,1\}^t \to \{0,1\}$ and of $x_{t+1}, \ldots, x_n$ if there exists a width-2 OBDD that accepts as input the value $g(x_1, \ldots, x_t)$ and the bits $x_{t+1}, \ldots, x_n$ (in that order) and outputs the value $f(x_1, \ldots, x_n)$.*

In Fig. 1 we present the testing algorithm for computability by a width-2 OBDD. In the algorithm we use the value ϵ', which intuitively stands for the amount of error we are willing to accumulate during each round of the algorithm. We set $\epsilon' = \epsilon/(4\log(1/\epsilon))$. The algorithm uses two sub-procedures, **Get-linear-level** and **Count-equiv-classes**, both described after the algorithm.

Theorem 3. *Algorithm* **Test-width-**2 *is a one-sided error testing algorithm for the property of being computable by a width-2 OBDD, that runs using $\tilde{O}(\log(n)/\epsilon)$ queries.*

We now proceed to discuss the probabilistic procedures used in **Test-width-**2. We later return to proving Theorem 3. We have already mentioned the BLR linearity test, one procedure that we will use as an internal building block in our own sub-procedures. We now turn to describe an additional building block - a procedure that is given access to a function f and a number i, and attempts to check whether f has $1, 2$ or more i-equivalence classes. Despite the fact it only counts up to 2, or perhaps up to "many", we dub this sub-procedure **Count-equiv-classes**. A precise description of the algorithm appears in Fig. 2. The straightforward approach to performing this task may be to take a set of prefixes of length i and compare each two (or all of them) on a set of suffixes, trying to find prefixes that belong to different equivalence classes. A simple analysis implies a procedure that performs $\Theta(1/\epsilon^2)$ queries. The approach we take is slightly

---

Test-width-2

Input: Oracle access to a function f; Precision parameter ϵ.

1. Let $f^0 = f$. Let $r = 1$ and $t^0 = n$.

   The variable t will represent the number of variables of f that we haven't re-stricted, and r will be the number of the current round. The indexes on t and f will indicate the round r and help us keep track of different values for the analysis.

2. While $t^{r-1} \neq 1$ and $r \leq \log(1/\epsilon) + 2$

   (a) **Locate linear section:** Run **Get-linear-level**$(f^{r-1}, \epsilon'/3, \frac{1}{6(\log(1/\epsilon)+2)})$.

   This locates the last index j such that f^{r-1} is $(\epsilon'/3)$-close to a linear function of $f^{r-1}_{j+1,t^{r-1},w}(x_1, \ldots, x_j)$ and of $x_{j+1}, \ldots, x_{t^{r-1}}$ for a distinguishing sequence w.

   (b) If **Get-linear-level** indicated the existence of more than 2 different i-equivalence classes on some level i, reject.

   (c) Otherwise, let j be the level returned by **Get-linear-level** and let w be the distinguishing sequence returned by it.

   (d) Let $g^r = f^{r-1}_{j+1,t^{r-1},w}$ and let $\tilde{t}^r = j$.

   (e) If $j \neq 1$

      i. Run **Count-equiv-classes**$(g^r, \tilde{t}^r - 1, \epsilon'/3, \frac{1}{12\log(1/2\epsilon)})$.

         This tells us whether the number of $(\tilde{t}^r - 1)$-equivalence classes in g^r is $1, 2$ or more with precision $\epsilon'/3$.

      ii. If a single equivalence class is found, accept.

      iii. If more than 2 equivalence classes are found, reject.

      iv. Let w' denote the distinguishing assignment (of size 1) between the 2 equivalence classes found (returned by **Count-equiv-classes**). Let $f^r = g^r_{j,j,w'}$ and let $t^r = j - 1$.

   (f) Else, let $f^r = g^r$ and let $t^r = \tilde{t}^r$.

   (g) $r = r + 1$.

3. return accept.

Fig. 1. Algorithm Test-width-2

different. We start with the arbitrary string 0^i, which belongs to some equiva-lence class. To identify a second equivalence class we simply test the equality of $f(x_1, \ldots, x_n)$ with $f_{1,i,0^i}(x_{i+1}, \ldots, x_n)$. If a second equivalence class is detected then we use a similar technique to try and find a third equivalence class (with a small adjustment). This approach leads to a $\Theta(1/\epsilon)$ algorithm.

The proof of the following claim can be found in [17].

Claim 8. *The algorithm* **Count-equiv-classes***, given oracle access to a function f acts as follows:*

1. *If f is ϵ-far from every function with one i-equivalence class, then with prob-ability at least $1 - \delta$* **Count-equiv-classes** *will return representatives of at least two equivalence classes.*

2. *If f is ϵ-far from every function with two i-equivalence classes, then with probability at least $1 - \delta$* **Count-equiv-classes** *will return representatives of three equivalence classes.*

Count-equiv-classes

Input: Oracle access to a function f; Integer value $0 < i \leq n$; Precision parameter ϵ; Confidence parameter δ.

1. Select $m = \Theta(\log(1/\delta)/\epsilon)$ strings $x^1, \ldots, x^m$ from $\{0,1\}^n$.
2. If $f(x) = f_{1,i,0^i}(x_{i+1}, \ldots, x_n)$ for all $x \in \{x^1, \ldots, x^m\}$ output that 1 equivalence class was found. Otherwise, let $y \in \{0,1\}^i$, $w \in \{0,1\}^{n-i}$ be such that $f(yw) \neq f(0^i w)$.
3. Select $m = \Theta(\log(1/\delta)/\epsilon)$ new strings $z^1, \ldots, z^m \in \{0,1\}^i$.
4. Define $g(z^j)$ as 0 if $f(0^i, w) = f(z_1^j, \ldots, z_i^j, w)$, and as 1 otherwise. Compute $g(z^j)$ for all j.
5. For all j, if $g(z^j) = 0$ and $f(z^j) \neq f(0^i, z_{i+1}^j, \ldots, z_n^j)$, then output representatives of 3 different i-equivalence classes $(0^i, y$ and $z_1^j, \ldots, z_i^j)$ and distinguishing assignments for them $(w$ and $z_{i+1}^j, \ldots, z_n^j)$. Do the same if $g(z^j) = 1$ and $f(z^j) \neq f(y, z_{i+1}^j, z_n^j)$.
6. Output the representatives of 2 equivalence classes $(0^i$ and $y)$ and a distinguishing assignment for them (w).

Fig. 2. Algorithm Count-equiv-classes

3. *In any case* **Count-equiv-classes** *does not indicate the existence of more than the number of i-equivalence classes of f.*
4. *Conditioned on Count-equiv-classes(f) returning the representatives of 2 different i-equivalence classes and a distinguishing assignment, with probability at least $1 - \delta$ it holds that f is ϵ-close to a function of $f_{i+1,n,w}(x_1, \ldots, x_i)$ (where w is the distinguishing assignment) and of the variables $x_{i+1}, \ldots, x_n$.*

The algorithm performs $O\left(\frac{\log(1/\delta)}{\epsilon}\right)$ queries.

An additional building block we use is the algorithm **Get-linear-level**, presented in Fig. 3. The general approach here is to perform a binary search for a

---

Get-linear-level

Input: Oracle access to a function f; Precision parameter ϵ; Confidence parameter δ.

1. Let $min = 1$ and $max = n$.
2. Let w be the empty string.
3. While $min < max$
   (a) Let $mid = \lfloor (max + min)/2 \rfloor$
   (b) Run **Test-level-linearity**$(f, mid, \epsilon, \delta' = \delta/\log(n))$. If **Test-level-linearity** finds 3 different mid-equivalence classes, reject.
   (c) If **Test-level-linearity** returns accept set $max = mid$ and set w to be the distinguishing sequence.
   (d) Otherwise, set $min = mid + 1$.
4. return mid and w.

Fig. 3. Algorithm Get-linear-level

"linear level" (as described in the claim below), at each point counting the equivalence classes using **Count-equiv-classes**, and testing each equivalence class for linearity using the BLR linearity test. The algorithm uses a sub-procedure called **Test-level-linearity** (described in [17]) that tests whether a particular level of the input behaves as expected. The proof of the next claim can be found in [17].

Claim 9. *When* **Get-linear-level** *is given oracle access to a function f (of n variables), a precision parameter ϵ and a confidence parameter δ it acts as follows.* **Get-linear-level** *rejects only if more than 2 different i-equivalence classes were located for some i. Otherwise, with probability greater or equal to $1 - \delta$ it returns a value $1 \le i \le n$ and a string w so that the following hold:*

1. *The function f is ϵ-close to a linear function of $f_{j+1,n,w}(x_1, \ldots, x_j)$ and of the variables $x_{j+1}, \ldots, x_n$.*
2. *If $j \ne 1$, then the function f is not a linear function of $f_{j,n,w}(x_1, \ldots, x_{j-1})$ and of the variables $x_j, \ldots, x_n$ for any² w.*

The algorithm performs $O\left(\frac{\log(n)\,\log(\log(n)/\delta)}{\epsilon}\right)$ queries.

Before proving Theorem 3 we prove a small claim that will assist us in the proof. In both the claim and the proof of the claim we describe a situation where none of the (probabilistic) sub-procedures used by **Test-width-2** fail. By "Procedures not failing" we mean, e.g., that if the BLR test accepts, then the function is indeed ϵ-close to a linear function.

Claim 10. *Assuming none of the sub-procedures used by it fail, at the end of the r'th round of* **Test-width-2**, *the function f^{r-1} is ϵ'-close to a W2 function of $f^r(x_1, \ldots, x_{t^r})$ and the variables $x_{t^r+1}, \ldots, x_{t^{r-1}}$.*

The relationship between f^{r-1} and f^r is demonstrated in Fig. 4.

Proof: By Claim 9 we have that at the end of Item 2d of **Test-width-2** in the r'th round, the function f^{r-1} is $(\epsilon'/3)$-close to a W2 function of $g^r(x_1, \ldots, x_j)$ and the variables $x_{j+1}, \ldots, x_{\tilde{t}^r}$. When $j = 1$ this suffices (as we set $f^r = g^r$). When $j > 1$ we have by Claim 8 that at Item 2(e)iv f^r is set to a function $\epsilon'/3$-close to a W2 function of g^r and of the variable $x_{\tilde{t}^r}$. This function is surely a W2 function and thus f^{r-1} is $2\epsilon'/3$-close to a W2 function of $f^r(x_1, \ldots, x_{t^r})$ and the variables $x_{t^r+1}, \ldots, x_{t^{r-1}}$, as required.   □

Proof of Theorem 3: We prove Theorem 3 in three stages. We first show the correctness of the algorithm assuming that none of the probabilistic procedures it performed erred in any way. We follow this by bounding the probability of error for the different probabilistic procedures, and finally, we analyze the query complexity, concluding the proof.

---

² Note that this is true with probability 1 due to the one-sided rejection criteria of **Test-level-linearity**, but the claim as is suffices.

CORRECTNESS ASSUMING THE SUCCESS OF SUB-TESTS involves proving the following:

1. **Completeness: Test-width-2**, given oracle access to a function computable by a width-2 OBDD, accepts.
2. **Soundness:Test-width-2**, given oracle access to a function ϵ-far from any function computable by a width-2 OBDD, rejects with probability at least $2/3$.

Proof of the completeness condition is straightforward: Rejection by **Test-width-2** occurs only when 3 different i-equivalence classes are detected. By Corollary 1 this never happens in a function computable by a width-2 OBDD. As **Test-width-2** always terminates either by accepting a function or rejecting it, the completeness condition holds.

We prove the soundness condition by proving the contrapositive - any function that passes the tester with probability greater than $1/3$ is ϵ-close to a function computable by a width-2 OBDD. To this end we assume that all the sub-procedures performed by **Test-width-2** succeed and show that in such a case any function passing the test is, indeed, ϵ-close to a function computable by a width-2 OBDD. We later prove that the cumulative probability of the "sub-procedures" failing is less than $2/3$, thus ensuring that **Test-width-2** is indeed a one-sided error property testing algorithm.

We define the function α^0 to be f. We next construct for every round r of the algorithm a function α^r that has the following properties:

1. The function α^r is close to the function α^{r-1}. In particular $d(\alpha^r, \alpha^{r-1}) \leq \epsilon' = \epsilon/(4\log(1/\epsilon))$.
2. The function α^r is a W2 function of $f^r(x_1, \ldots, x_{t^r})$ and of $x_{t^r+1}, \ldots, x_n$, and has at least $r-1$ merging levels. The W2 function that accepts as input the values $f^r(x_1, \ldots, x_{t^r})$ and $x_{t^r+1}, \ldots, x_n$ is denoted β^r.

We construct α^r based on α^{r-1} as follows: By Claim 10, at the end of the r'th round the function f^{r-1} is ϵ'-close to a W2 function, which we denote ψ^r, of $f^r(x_1, \ldots, x_{t^r})$ and the variables $x_{t^r+1}, \ldots, x_{t^{r-1}}$. Let

$$\alpha^r = \beta^{r-1}(\psi^r(f^r(x_1, \ldots, x_{t^r}), x_{t^r+1}, \ldots, x_{t^{r-1}}), x_{t^{r-1}+1}, x_n).$$

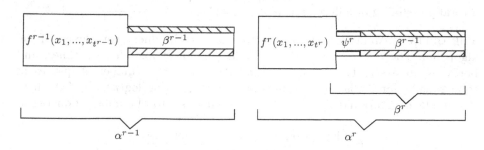

Fig. 4. An illustration for the construction of α^r and β^r

As we wish to view α^r as equivalent to $\beta^r(f^r(x_1,\ldots,x_{t^r}),x_{t^r+1})$, we define β^r accordingly (see Fig. 4). We have that $d(\alpha^r,\alpha^{r-1}) \leq \epsilon'$. We note that β^r is computable by a width-2 OBDD by a straightforward construction, and that unless $j = 1$ on the r'th round, the new width-2 OBDD constructed by this procedure (that computes β^r) has one more merging level than the one on the $r-1$'th round.

Denoting the last round of **Test-width-2** as s we now note that α^s is $(\epsilon/4)$-close to a function computable by a width-2 OBDD (assuming f passed the test). There are three ways the test can terminate successfully:

1. The test reaches the $(\log(1/\epsilon)+2)$'th round. In such a case α^s is ϵ' close to a W2 function (that accepts $f^s(x_1,\ldots,x_{t^s})$ and $x_{t^s+1},\ldots,x_n$ as input) with $\log(1/\epsilon) + 2$ merging levels in the OBDD computing it, and by Claim 5 is $(\epsilon/4)$-close to a function computable by a width-2 OBDD.
2. The test terminates because $t^s = 1$. In such a case, by Claim 10, at the end of the s'th round the function f^s is a function of 0 variables (a constant function), surely computable by a width-2 OBDD.
3. The test terminates because a single equivalence class was found in Item 2e. In such a case f^s is $(\epsilon'/3)$-close to a constant function, as above.

Let h be a function computable by a width-2 OBDD that's $(\epsilon/2)$-close to α^s, and let $W2$ be the set of functions computable by width 2 OBDDs. We have

$$d(f,W2) \leq d(f,h) \tag{1}$$
$$\leq d(f,\alpha^1) + d(\alpha^1,\alpha^2) + \cdots + d(\alpha^{s-1},\alpha^s) + \epsilon/2 \tag{2}$$
$$\leq \epsilon/2 + (\log(1/\epsilon) + 2)(\epsilon/(4\log(1/\epsilon))) \tag{3}$$
$$\leq \epsilon \tag{4}$$

as required.

THE PROBABILITY OF ANY SUB-TEST FAILING. The cumulative probability of any sub-procedure used by **Test-width-2** of failing during **Test-width-2**'s execution is less than $2/3$. This is due to the fact that in each of at most $\log(1/\epsilon)+2$ rounds the algorithm performs two probabilistic sub-procedures, each with a probability of failure of at most $\frac{1}{6(\log(1/\epsilon)+2)}$. Using a simple union bound we get a total probability of failure of at most $2(\log(1/\epsilon) + 2) \cdot \frac{1}{6(\log(1/\epsilon)+2)} = 1/3$.

THE QUERY COMPLEXITY. It remains to analyze the query complexity of the algorithm. The tester repeats the outer loop at most $\log(1/\epsilon) + 2$ times, and performs queries in two Items - 2a and 2e, where the number of queries in Item 2a is by far the larger and sums up to $\Theta(\log(n)(\log(\log(n)/\delta')/\epsilon''))$ where $\epsilon'' = \epsilon/(12\log(1/\epsilon))$ and $\delta' = \frac{1}{6(\log(1/\epsilon)+2)}$, giving us a total number of queries of

$$\Theta\left(\frac{\log(n)\log(\log(n)\log(1/\epsilon))\log(1/\epsilon)}{\epsilon}\right)$$

and the proof is complete. $\square$

References

1. Alon, N., Krivelevich, M., Kaufman, T., Litsyn, S., Ron, D.: Testing Reed-Muller codes. IEEE Transactions on Information Theory 51(11), 4032–4038 (2005)
2. Bergadano, F., Bshouty, N., Tamon, C., Varricchio, S.: On learning branching programs and small depth circuits. In: COLT 1997, pp. 150–161 (1997)
3. Blum, M., Luby, M., Rubinfeld, R.: Self-testing/correcting with applications to numerical problems. Journal of the ACM 47, 549–595 (1993)
4. Bshouty, N., Tamon, C., Wilson, D.: On learning width two branching programs. Information Processing Letters 65, 217–222 (1998)
5. Diakonikolas, I., Lee, H.K., Matulef, K., Onak, K., Rubinfeld, R., Servedio, R.A., Wan, A.: Testing for concise representations. In: FOCS 2007, pp. 549–557 (2007)
6. Ergün, F., Kumar, R.S., Rubinfeld, R.: On learning bounded-width branching programs. In: COLT 1995, pp. 361–368 (1995)
7. Fischer, E., Kindler, G., Ron, D., Safra, S., Samorodnitsky, S.: Testing juntas. Journal of Computer and System Sciences 68(4), 753–787 (2004)
8. Gavalda, R., Guijarro, D.: Learning ordered binary decision diagrams. In: Zeugmann, T., Shinohara, T., Jantke, K.P. (eds.) ALT 1995. LNCS, vol. 997, pp. 228–238. Springer, Heidelberg (1995)
9. Goldreich, O., Goldwasser, S., Ron, D.: Property testing and its connection to learning and approximation. Journal of the ACM 45(4), 653–750 (1998)
10. Jutla, C.S., Patthak, A.C., Rudra, A., Zuckerman, D.: Testing low-degree polynomials over prime fields. In: FOCS 2004 (2004)
11. Kearns, M., Ron, D.: Testing problems with sub-learning sample complexity. Journal of Computer and System Sciences 61(3), 428–456 (2000)
12. Nakamura, A.: Query learning of bounded-width OBDDs. Theoretical Computer Science 241, 83–114 (2000)
13. Nakamura, A.: An efficient query learning algorithm for OBDDs. Information and Computation 201, 178–198 (2005)
14. Newman, I.: Testing membership in languages that have small width branching programs. SIAM Journal on Computing 31(5), 1557–1570 (2002)
15. Parnas, M., Ron, D., Samorodnitsky, A.: Testing basic boolean formulae. SIAM Journal on Discrete Math 16(1), 20–46 (2002)
16. RagHavan, V., Wilkins, D.: Learning branching programs with queries. In: COLT 1993, pp. 27–36 (1993)
17. Ron, D., Tsur, G.: Testing computability by width two obdds (2009), http://www.eng.tau.ac.il/~danar
18. Rubinfeld, R., Sudan, M.: Robust characterization of polynomials with applications to program testing. SIAM Journal on Computing 25(2), 252–271 (1996)

Improved Polynomial Identity Testing for Read-Once Formulas

Amir Shpilka and Ilya Volkovich

Faculty of Computer Science, Technion, Haifa 32000, Israel
{shpilka,ilyav}@cs.technion.ac.il

Abstract. An *arithmetic read-once formula* (ROF for short) is a formula (a circuit whose underlying graph is a tree) in which the operations are $\{+, \times\}$ and such that every input variable labels at most one leaf. In this paper we study the problems of giving deterministic identity testing and reconstruction algorithms for ROFs. Our main result is an $n^{\mathcal{O}(k+\log n)}$ time deterministic algorithm for checking whether a black box holding the sum of k n-variate ROFs computes the zero polynomial. In other words, we provide a hitting set of size $n^{\mathcal{O}(k+\log n)}$ for the sum of k ROFs. This result greatly improves [27] where an $n^{\mathcal{O}(k^2+\sqrt{n})}$ algorithm was given for the problem.

Using our new results we obtain a deterministic reconstruction algorithms for read-once formulas that runs in time $n^{\mathcal{O}(\log n)}$.

In fact, our results also hold for the more general model of *preprocessed* read-once formulas that we define in this paper. In this model we are allowed to replace each variable x_i with a polynomial $T_i(x_i)$.

Our techniques are very close to the techniques in [27]. The main difference is that we obtain several tighter versions of the tools first used there. In particular we obtain a better version of the *hardness of representation* approach which was first used in [27]. This technique can be thought of as a very explicit way of transforming (mild) hardness of a very structured polynomial to an identity testing algorithm.

1 Introduction

In this paper we study the polynomial identity testing problem for several models based on read-once formulas. In the polynomial identity testing problem (PIT for short) we are given (either explicitly or via black-box access) an arithmetic circuit (or formula) and we have to decide whether the circuit computes the zero polynomial. Schwartz and Zippel [25, 28] gave a black-box randomized algorithm for the problem, that was later improved in several cases [2, 8, 16]. However, we are interested in the question of giving a deterministic algorithm to the problem.

In general, the PIT problem is believed to be very difficult and several results connecting deterministic algorithms for PIT and lower bounds for arithmetic circuits are known [1, 3, 10, 12]. However, for several special cases in which the underlying circuit comes from a restricted class of arithmetic circuits, efficient deterministic PIT algorithms were found. For example, efficient deterministic identity testing algorithms are known for depth-2 arithmetic circuits (that computes

I. Dinur et al. (Eds.): APPROX and RANDOM 2009, LNCS 5687, pp. 700–713, 2009.
© Springer-Verlag Berlin Heidelberg 2009

sparse polynomials) [5, 15, 17] and references within, for depth-3 arithmetic circuits with bounded top fan-in (also known as $\Sigma\Pi\Sigma(k)$ circuits) [4, 9, 13, 14, 24] and for non-commutative arithmetic formulas [20]. Interestingly, [3] showed that polynomial time deterministic black-box PIT algorithms for depth-4 arithmetic circuits imply exponential lower bounds on the size of general arithmetic circuits and a quasi-polynomial time algorithm for the general PIT problem. Indeed, efficient deterministic PIT algorithms are known only for a very restricted class of depth-4 circuits [4, 23] (and even those algorithms are non black-box).

In view of the difficulty in providing efficient deterministic PIT algorithms and the tight connection to lower bounds it is natural to study the PIT problem for models for which lower bounds are known. In particular, the recent results of [18, 19, 21, 22] on lower bounds for multilinear circuits and formulas suggest that giving efficient deterministic PIT algorithms for multilinear formulas may be possible. Unfortunately, except for the models of multilinear depth-2 and multilinear $\Sigma\Pi\Sigma(k)$ circuits no such algorithm is known. As a consequence the problem of PIT for read-once formulas, which can be thought of as the simplest form of multilinear formulas,[1] was considered in [27]. There, sub-exponential time deterministic PIT algorithms for (sums of) read-once arithmetic formulas in the black-box and non black-box models were given.

1.1   Our Results

In this work we improve and extend the results from [27] (obtained by the same authors as this paper). There are two aspects for this improvement. First, we give quasi-polynomial time identity testing algorithms, which greatly improve upon the previous sub-exponential time algorithms. Secondly, we consider the more general model of *preprocessed read-once formulas*. These are read-once formulas in which we substitute a univariate polynomial $T_i(x_i)$ for every variable x_i. Using our new results we obtain new identity-testing algorithms for multilinear depth-3 circuits and preprocessed multilinear depth-3 circuits, which are a restricted class of depth-4 circuits.

Our main result is the following Black-Box identity testing algorithm for the sum of k PROFs.

Theorem 1. *Given black-box access to $F = F_1 + \cdots + F_k$, where the F_i-s are preprocessed-read-once formulas in n variables, with individual degrees at most d, there is a deterministic algorithm that checks whether $F \equiv 0$. The running time of the algorithm is $(nd)^{\mathcal{O}(k+\log n)}$.*

We note that this result greatly improves the previous $n^{\mathcal{O}(k^2+\sqrt{n})}$ time algorithm of [27] that worked for ROFs. Using the techniques of [27] this result implies a deterministic reconstruction algorithm for a single PROF that runs in time $(nd)^{\mathcal{O}(\log n)}$. We also obtain a non black-box analog of the PIT result.

---

[1] Indeed, a read-once formula is a restricted form of multilinear formulas in which a variable can label at most one leaf of the formula (see Definition 3).

Theorem 2. *Given k preprocessed-read-once formulas in n variables, with individual degrees at most d, there is a deterministic algorithm that checks whether they sum to zero or not. The running time of the algorithm is $(nd)^{\mathcal{O}(k)}$.*

This result improves the $n^{\mathcal{O}(k^2)}$ time algorithm [27] (that worked for ROFs). Besides these two results we also improve the results of [27] of PIT for the sum of small depth PROFs and obtain an $(nd)^{\mathcal{O}(D+k)}$ time deterministic algorithm for checking whether a black-box holding the sum of k depth-D preprocessed read-once formulas on n-variables, with individual degrees at most d, computes the zero polynomial.

As a corollary of the above result we obtain an $n^{\mathcal{O}(k)}$ time PIT algorithm for multilinear $\Sigma\Pi\Sigma(k)$ circuits (a multilinear $\Sigma\Pi\Sigma(k)$ circuit can be considered as a sum of k ROFs of depth 2). We note that this result *does not* rely on bounds on the rank of zero $\Sigma\Pi\Sigma(k)$ circuits, which is the main tool for giving black-box PIT algorithms for depth-3 circuits. This result also extends to preprocessed multilinear depth-3 circuits.

1.2 Proof Technique

Most of our tools are taken from our previous paper [27]. The main difference is that we obtain tighter versions of the tools developed there combined with several basic observations.

The first step for obtaining a black-box identity testing algorithm is to start with a single ROF. We note that if the top gate of the formula is a $+$ then P can be written as a sum of two, variable-disjoint ROPs $P = P_1 + P_2$ and the two basic observation are that w.l.o.g P_1 depends on at most $n/2$ variables and that if P_1 depends on (a variable) x_i then so does P. Using these observations we show that for $\ell = \lceil \log_2 t \rceil + 1$ the function $G_\ell : \mathbb{F}^{2\ell} \to \mathbb{F}^n$, that we define next, is a hitting set generator for PROFs: Let $a_1, \ldots, a_n \in \mathbb{F}$ be n distinct[2] field elements. Let $u_1(w), \ldots, u_n(w)$ be univariate polynomials of degree $n-1$ such that $u_i(a_j) = \delta_{i,j}$. For every $i \in [n]$ let $G_k^i : \mathbb{F}^{2k} \to \mathbb{F}$ be defined as $G_k^i(y_1, \ldots, y_k, z_1, \ldots, z_k) \triangleq u_i(y_1) \cdot z_1 + \ldots + u_i(y_k) \cdot z_k$. Set $G_k(y_1, \ldots, y_k, z_1, \ldots, z_k) \triangleq (G_k^1, G_k^2, \ldots, G_k^n)$. In particular, by evaluating G_ℓ on $[n^2 d]^{2\ell}$ we get a hitting set for PROFs with individual degrees at most d, of size $(nd)^{\mathcal{O}(\log n)}$.

This simplifies and greatly improves the generator given in [27] that constructed a hitting set of size $n^{\mathcal{O}(\sqrt{n})}$ for ROFs. The extension from a PIT algorithm for a single PROF to an algorithm for the sum of k PROFs is via *hardness of representation* approach, first used in [27]. This technique enables us to transform a mild lower bound for a very structured polynomial into a PIT for sum of (preprocessed) ROFs. The idea behind this approach is the following: A *common justifying assignment* to a set $\{F_m\}_{m \in [k]}$ of PROFs is an assignment $\rho = (\rho_1, \ldots, \rho_n)$ such that (for every m) $F_m(\rho_1, \ldots, \rho_{i-1}, x_i, \rho_{i+1}, \ldots, \rho_n)$ depends on x_i if and only if $F_m(x_1, \ldots, x_n)$ depends on x_i (see Definition 1). Our hardness of representation theorem shows that if ρ is a justifying assignment for

---

[2] If $\mathbb{F}$ is too small then we move to an extension field.

$\{F_1, \ldots, F_k\}$, where $k = \mathcal{O}(n)$, then if $F_1 + \ldots + F_k = g(x_1, \ldots, x_n) \cdot \prod_{i=1}^n (x_i - \rho_i)$, for some polynomial g, then $g \equiv 0$. This form of the theorem generalizes and strengthens the corresponding theorem of [27] that only held for ROFs and that only showed that $\sqrt{n}$ ROFs are needed.

We apply this theorem in the following way. Assume that $F = F_1 + \ldots + F_k$ is a sum of k ROFs. Using the PIT algorithm for a single PROF we can find (a-la [27]) a common justifying assignment to all the F_m-s (actually we find a set, of polynomial size, of assignment that contains a common justifying assignment). Then, by the idea of [27] we evaluate F on all points of the form $\rho + v$, where $v \in [d+1]^n$ has weight at most $3k$. If all the evaluations are zero then we conclude that $F \equiv 0$. To get an intuitive feeling why this work, assume w.l.o.g. that $\rho = 0$. We thus evaluate F on all the points of weight at most $3k$ in $[d+1]^n$. Fix $x_n = 0$. Using induction we conclude that if F vanishes on all those assignments in which $x_n = 0$ then $F|_{x_n=0} \equiv 0$. In other words, x_n divides F. As this holds for every x_i we get that $x_1 \cdot \ldots \cdot x_n$ divides F. By the hardness of representation result we now get that $F \equiv 0$. Note, that the basis for the induction is when $n = 3k$ and then clearly evaluating a polynomial with individual degrees $\leq d$ on all the points in $[d+1]^{3k}$ tells us whether the polynomial is zero or not.

1.3 Comparison to Previous Works

Read-once arithmetic formulas were mostly studied in the context of computational learning theory. Various works considered the problem of reconstructing the unknown read-once formula using membership queries. A membership query to a ROF $f(\bar{x})$ is simply a query that asks for the value of $f(\bar{x})$ on a specific input. In [11] a deterministic learning algorithm for read-once arithmetic formulas that uses membership and *equivalence* queries was given. An equivalence query gives the oracle holding the unknown formula a certain hypothesis, $h(\bar{x})$, and the oracle answers "equal" if $f \equiv h$ or returns an input $\bar{\alpha}$ such that $f(\bar{\alpha}) \neq h(\bar{\alpha})$. In [7] a different approach was taken. They considered *randomized* learning algorithms that only use membership queries. It is not difficult to see that using randomness one does not need to use equivalence queries any more. The learning algorithm of [7] can reconstruct, with high probability, arithmetic read-once formulas that also use division gates (and not just $+, \times$ gates). This result was later generalized by [6] who gave a randomized reconstruction algorithm for read-once formulas that use additions, multiplications, divisions and exponentiations.

In [27] a sub-exponential time (i.e. $n^{\mathcal{O}(\sqrt{n})}$ time) PIT algorithm for black-box read-once formulas was given. Using this algorithm together with the learning methods of [7, 11] a deterministic sub-exponential time reconstruction algorithm for arithmetic ROFs was obtained. In addition, PIT algorithms for sum of ROFs both in the black-box and in the non black-box models were obtained in that work.

In this work we improve the PIT algorithms for sum of ROFs. In the black-box case we give a deterministic algorithm that runs in $n^{\mathcal{O}(k+\log n)}$ time. In the non black-box case our algorithm runs in time $n^{\mathcal{O}(k)}$. Moreover, both algorithms also work for the general model of preprocessed ROFs.

We are also able to apply our methods to depth-3 circuits, also known as $\Sigma\Pi\Sigma(k)$ circuits. This model was extensively studied in recent years [4, 9, 13, 14, 24, 26, 27] as it stands between the simpler depth-2 case and the depth-4 case that is as hard as the general case, for lower bounds and polynomial identity testing [3]. Prior to this work the best known black-box PIT algorithm for $\Sigma\Pi\Sigma(k, d)$ circuits had running time $n^{\mathcal{O}(k^3 \log d)}$ for the general case [13, 24] and $n^{\mathcal{O}(k^2)}$ in the multilinear case [27]. We improve the algorithm for the multilinear case and obtain an $n^{\mathcal{O}(k)}$ algorithm that also works in the preprocessed case.

1.4   Organization

In Section 3 we define the notion of ROF and state several of their important properties (mostly results proved in [27]). In Section 4 we consider the case of a single ROF. In Section 5 we extend the result from the previous section to sums of ROFs, proving Theorems 2 and 1. Due to space limitations we will not manage to give more proofs and, moreover, we only give proofs for the case of ROFs and not PROFs.

2   Preliminaries

For a positive integer n we denote $[n] = \{1, \ldots, n\}$. As usual, we define the *Hamming weight* of a vector $\bar{a} \in \mathbb{F}^n$ as: $\mathrm{w_H}(\bar{a}) \overset{\Delta}{=} |\{i \mid a_i \neq 0\}|$. For a polynomial $P(x_1, \ldots, x_n)$ a variable x_i and a field element α we denote with $P|_{x_i=\alpha}$ the polynomial resulting after substituting α to the variable x_i. The following definitions, taken from [27], are for a polynomial $P \in \mathbb{F}[x_1, \ldots, x_n]$ and an assignment $\bar{a} \in \mathbb{F}^n$. We say that P *depends* on x_i if there exist two inputs $\bar{a}$ and $\bar{b}$ that differ only on the i-th coordinate such that $P(\bar{a}) \neq P(\bar{b})$. We denote $\mathrm{var}(P) \overset{\Delta}{=} \{x_i \mid P \text{ depends on } x_i\}$. Clearly, P depends on x_i if x_i "appears" when P is listed as a sum of monomials. Given a subset $I \subseteq [n]$ we say that P is *defined on I* if $\mathrm{var}(P) \subseteq I$. E.g., the constant function is defined on every subset of $[n]$. Given a subset $I \subseteq [n]$ and an assignment $\bar{a} \in \mathbb{F}^n$ we define $P|_{x_I=\bar{a}_I}$ to be the polynomial resulting from substituting a_i to the variable x_i for every $i \in I$. In particular $P|_{x_I=\bar{a}_I}$ is defined on $[n] \setminus I$. It is clear that if $J \subseteq I \subseteq \mathrm{var}(P)$ are subsets of $\mathrm{var}(P)$, then for every assignment $\bar{a} \in \mathbb{F}^n$ it must be the case that $\mathrm{var}(P|_{x_I=\bar{a}_I}) \subseteq \mathrm{var}(P|_{x_J=\bar{a}_J}) \subseteq \mathrm{var}(P) \setminus J$. That is, by substituting a value to a variable of P we, obviously, eliminate the dependence of P on this variable, however we may also eliminate the dependence of P on other variables and thus lose more information than intended. For the purposes of reconstruction and identity testing we cannot allow losing any information as it would affect our final answer. We now define a lossless type of an assignment. Similar definitions were given in [11] and [7], but we repeat the definitions here to ease the reading of the paper (we also slightly change some of the definitions).

Definition 1 (Justifying assignment). *Given an assignment $\bar{a} \in \mathbb{F}^n$ we say that $\bar{a}$ is a justifying assignment of P if for each subset $I \subseteq \mathrm{var}(P)$ we have that* $\mathrm{var}(P|_{x_I=\bar{a}_I}) = \mathrm{var}(P) \setminus I$.

Note that $\bar{a} \in \mathbb{F}^n$ is a justifying assignment of P if and only if $\mathrm{var}(P|_{x_I = \bar{a}_I}) = \mathrm{var}(P) \setminus I$ for every subset I of size $|\mathrm{var}(P)| - 1$. Given an assignment $\bar{a} \in \mathbb{F}^n$ we say that $\bar{a}$ is a *weakly-justifying assignment* of P if $\mathrm{var}(P|_{x_I = \bar{a}_I}) = \mathrm{var}(P) \setminus I$ for every $|I| = 1$. Clearly, justification implies weak-justification, but not vice versa. We also define justification as a property of polynomials: A polynomial P is $\bar{a}$-*justified* if $\bar{a}$ is justifying assignment of P. We define the term weakly-$\bar{a}$-justified in a similar manner. By shifting we can convert any polynomial to a $\bar{0}$-justified form. The following lemma is similar to Lemma 2.3 of [27].

Lemma 1. *Let* $\bar{a} \in \mathbb{F}^n$ *and let* $f(\bar{x})$ *be a (weakly)* $\bar{a}$-justified *polynomial. Then* $f_{\bar{a}}(\bar{x}) \stackrel{\Delta}{=} f(\bar{x} + \bar{a})$ *is a (weakly)* $\bar{0}$-justified *polynomial. In addition,* $f_{\bar{a}} \equiv 0$ *if and only if* $f \equiv 0$.

2.1 Partial Derivatives

Discrete partial derivatives will play a important role in our proofs.

Definition 2. *Let* P *be an* n *variate polynomial over a field* $\mathbb{F}$. *We define the discrete partial derivative of* P *with respect to* x_i *as* $\frac{\partial P}{\partial x_i} = P|_{x_i=1} - P|_{x_i=0}$.

If P is a multilinear polynomial then this definition coincides with the "analytical" one when $\mathbb{F} = \mathbb{R}$ or $\mathbb{F} = \mathbb{C}$. The following lemma is easy to verify.

Lemma 2 (Lemma 2.5 of [27]). *The following properties hold for a multilinear polynomial* P: P *depends on* x_i *iff* $\frac{\partial P}{\partial x_i} \not\equiv 0$. $\frac{\partial P}{\partial x_i}$ *does not depend on* x_i *(in particular* $\frac{\partial^2 P}{\partial x_i^2} \equiv 0$). $\frac{\partial^2 P}{\partial x_i \partial x_j} = \frac{\partial}{\partial x_i}(\frac{\partial P}{\partial x_j}) = \frac{\partial^2 P}{\partial x_j \partial x_i}$. $\forall i \neq j$ $\frac{\partial P}{\partial x_i}|_{x_j=a} = \frac{\partial}{\partial x_i}(P|_{x_j=a})$. $\bar{a} \in \mathbb{F}^n$ *is a justifying assignment of* P *iff* $\forall i \in \mathrm{var}(P)$ *it holds* $\frac{\partial P}{\partial x_i}(\bar{a}) \neq 0$. *"Chain rule": Let* $Q(y, \bar{z})$ *be a multilinear polynomial such that* $P(\bar{x}, \bar{z}) \equiv Q(G(\bar{x}), \bar{z})$. *Then* $\frac{\partial P}{\partial x_i} = \frac{\partial Q}{\partial y} \cdot \frac{\partial G}{\partial x_i}$ *and in addition,* $\frac{\partial Q}{\partial y}$ *does not depend on* y.

Since the lemma trivially hold for multilinear polynomials, we will use it implicitly. However, notice these basic properties do not hold for general polynomials. For example, when $P(x) = x^2 - x$ we get that $\frac{\partial P}{\partial x} \equiv 0$.

3 Read-Once Formulas

In this section we discuss our computational model. Due to space limitations we only consider the model of read-once formulas and not the more general model of preprocessed-read-once formulas. Most of the definitions in this section are from [11]. The proofs of most properties can be found in Section 3 of [27].

Definition 3. *A read-once arithmetic formula (ROF for short) over a field* $\mathbb{F}$ *in the variables* $\bar{x} = (x_1, \ldots, x_n)$ *is a binary tree whose leafs are labeled with the input variables and whose internal nodes are labeled with the arithmetic operations* $\{+, \times\}$ *and with a pair of field elements* [3] $(\alpha, \beta) \in \mathbb{F}^2$. *Each input variable can label at most one leaf. The computation is performed in the following*

---

[3] This is a slightly more general model than the usual definition of read-once formulas.

way. A leaf labeled with the variable x_i and with (α, β) computes the polynomial $\alpha \cdot x_i + \beta$. If a node v is labelled with the operation op and with (α, β), and its children compute the polynomials f_{v_1} and f_{v_2} then the polynomial computed at v is $f_v = \alpha \cdot (f_{v_1} \text{ op } f_{v_2}) + \beta$. We say that a ROF (instance) f is non-degenerate if it depends on all the variables appearing in it.

A polynomial $P(\bar{x})$ is a *read-once polynomial* (ROP for short) if it can be computed by a read-once formula. A special class of ROFs that will play an important role in our proofs is the class of multiplicative ROFs.

Definition 4. *A ROF is called* multiplicative ROF *if it has no addition gates. A polynomial computed by a multiplicative ROF is called a* multiplicative ROP.

We shall later see (Lemma 6) that this notion is well defined. Note that from our definition of the ROF model, the polynomial $(5x_1 \cdot x_2 + 1) \cdot ((-x_3 + 2) \cdot (2x_4 - 1) + 5)$ has a multiplicative ROF. We now give some basic properties of ROPs that were mostly proved in [27]. As all this results were already proved in [27] we only list them here. Next is the structural lemma of ROPs.

Lemma 3 (Lemma 3.3 of [27]). *Every ROP $P(\bar{x})$ such that $|\text{var}(P)| \geq 2$ can be presented in exactly one of the following forms: $P(\bar{x}) = P_1(\bar{x}) + P_2(\bar{x})$ or $P(\bar{x}) = P_1(\bar{x}) \cdot P_2(\bar{x}) + c$, where P_1, P_2 are non-constant variable disjoint ROPs and c is a constant.*

The following is a simple lemma saying that we can remove any gate of a ROF and still obtain a ROF.

Lemma 4 (Lemma 3.4 of [27]). *Let $P(\bar{x})$ be a ROP and v a node in a ROF f computing P. We denote by $p_v(\bar{x})$ the polynomial that is computed by v. Then there exists a polynomial $Q(y, \bar{x})$ such that $Q(p_v(\bar{x}), \bar{x}) \equiv P(\bar{x})$ and, in addition, p_v and Q are variable-disjoint ROPs.*

The next two lemmas shows that ROPs are somewhat robust. Namely that a factor and partial derivative of a ROP are also ROPs.

Lemma 5 (Lemmas 3.5, 3.6 and 3.10 of [27]). *A partial derivative and a factor of a ROP is a ROP. Moreover, if the polynomial is weakly-$\bar{0}$-justified then so are its factors and partial derivatives.*

Lemma 6 (Proposition 3.8 of [27]). *A ROP P is a multiplicative ROP iff $\forall x_i \neq x_j \in \text{var}(P)$ we have that $\frac{\partial^2 P}{\partial x_i \partial x_j} \not\equiv 0$.*

The following is an extension of Lemma 3.9 of [27] that explains the structure of multiplicative ROFs. Recall that in our model a multiplicative ROF can compute more than a simple polynomial (due to the linear shifts at the gates).

Lemma 7. *Let $P(x_1, x_2, \ldots, x_n)$ be a multiplicative ROP with $|\text{var}(P)| \geq 2$. Then for every variable $x_i \in \text{var}(P)$ there exists another variable $x_j \in \text{var}(P)$ such that $\frac{\partial P}{\partial x_j} = (x_i - \alpha)h(\bar{x})$ for some $\alpha \in \mathbb{F}$ and ROP $h(\bar{x})$, such that $\text{var}(h) = \text{var}(P) \setminus \{x_i, x_j\}$ (in particular, $\frac{\partial P}{\partial x_j}|_{x_i = \alpha} \equiv 0$). If, in addition, P is weakly-$\bar{0}$-justified then so is $h(\bar{x})$. Moreover, $\alpha \neq 0$ and there exists **at most** one element $\beta \neq \alpha \in \mathbb{F}$ such that $P|_{x_i = \beta}$ is not weakly-$\bar{0}$-justified .*

4   Black-Box PIT for Read-Once Polynomials

In this section we prove Theorem 1 for a single ROF (i.e. the case $k = 1$). The main idea is to convert a ROP P, that has many variables, each with a "small" degree, to a polynomial P' with a smaller number of variables while maintaining a reasonable degree, such that $P' \equiv 0$ if and only if $P \equiv 0$. In fact, we will construct a generator for ROPs. In other words, we give a mapping $\mathcal{G} : \mathbb{F}^q \to \mathbb{F}^n$, for some "small" q, such that if $F \not\equiv 0$ then $F \circ \mathcal{G} \not\equiv 0$. We shall assume that $|\mathbb{F}| > n$ as we are allowed to use elements from an appropriate extension field. Throughout the entire section we fix a set $A = \{\alpha_1, \alpha_2, \ldots, \alpha_n\} \subseteq \mathbb{F}$ of n distinct elements.

Definition 5. *For every $i \in [n]$ let $u_i(w) : \mathbb{F} \to \mathbb{F}$ be the i-th Lagrange interpolation polynomial for the set A. Namely, each $u_i(w)$ is polynomial of degree $n - 1$ that satisfies $u_i(\alpha_j) = 1$ iff $j = i$ (and 0 otherwise). For every $i \in [n]$ and $k \geq 1$ we define $G_k^i : \mathbb{F}^{2k} \to \mathbb{F}$ as $G_k^i(y_1, \ldots, y_k, z_1, \ldots, z_k) \overset{\Delta}{=} u_i(y_1) \cdot z_1 + \ldots + u_i(y_k) \cdot z_k$. Finally, let $G_k : \mathbb{F}^{2k} \to \mathbb{F}^n$ be defined as $G_k(y_1, \ldots, y_k, z_1, \ldots, z_k) \overset{\Delta}{=} (G_k^1, G_k^2, \ldots, G_k^n)$.*

Denote with $\bar{e}_i \in \{0, 1\}^n$ the vector that has 1 in the i-th coordinate and 0 elsewhere. From the definition it is clear that $G_{k+1} = G_k + \sum_{i=1}^{n} u_i(y_{k+1}) \cdot z_{k+1} \cdot \bar{e}_i$. Hence, for every $k \geq 1$ and $\alpha_m \in A$ it holds that $G_{k+1}|_{y_{k+1} = \alpha_m} = G_k + z_{k+1} \cdot \bar{e}_m$.
Now we can construct a low-degree generator for ROPs.

Lemma 8. *Let $P \in \mathbb{F}[x_1, \ldots, x_n]$ be a non-zero ROP with $|\mathrm{var}(P)| \leq 2^t$, for some $t \geq 0$. Then $P(G_{t+1}) \not\equiv 0$. Moreover, if P is a non-constant polynomial then so is $P(G_{t+1})$.*

Proof. We prove the claim by induction on $|\mathrm{var}(P)|$. For $|\mathrm{var}(P)| = 0$ or 1 the claim is trivial. Now assume that $|\mathrm{var}(P)| \geq 2$ (which implies $t \geq 1$). By Lemma 3 we get that P can be written in exactly one of the two forms:

Case $P(\bar{x}) = P_1(\bar{x}) + P_2(\bar{x})$: Since P_1 and P_2 are variable disjoint we can assume w.l.o.g. that $|\mathrm{var}(P_1)| \leq |\mathrm{var}(P)|/2 \leq 2^{t-1}$ (in particular $|\mathrm{var}(P_1)| < |\mathrm{var}(P)|$). By the induction hypothesis we see that $P_1(G_t) \not\equiv 0$ is a non-constant polynomial. The next lemma shows that there is a variable x_m such that even after substituting all the other G_i-s, P_1 still depends on x_m.

Lemma 9. *Let $P \in \mathbb{F}[x_1, \ldots, x_n]$ be a polynomial and let $G = (G^1, \ldots, G^n) : \mathbb{F}^\ell \to \mathbb{F}^n$ satisfy that $P(G)$ is a non-constant polynomial. Then there exists $x_m \in \mathrm{var}(P)$ such that $P\left(G^1, \ldots, G^{m-1}, x_m, G^{m+1}, \ldots, G^n\right)$ (the polynomial resulting from substituting G^i for x_i for every $i \neq m$) depends on x_m.*

Let $x_m \in \mathrm{var}(P_1)$ be as promised by the lemma. As $x_m \notin \mathrm{var}(P_2)$ we obtain that $P\left(G_t^1, \ldots, G_t^{m-1}, x_m, G_t^{m+1}, \ldots, G_t^n\right)$ depends on x_m as well. Recall that $P(G_{t+1})|_{y_{t+1} = \alpha_m} = P\left(G_t^1, \ldots, G_t^{m-1}, G_t^m + z_{t+1}, G_t^{m+1}, \ldots, G_t^n\right)$. Thus, as z_{t+1} only appears in the m-th coordinate it follows that $P(G_{t+1})|_{y_{t+1} = \alpha_m}$

depends on z_{t+1}. As a conclusion we get that $P(G_{t+1})$ is a non-constant polynomial and in particular $P(G_{t+1}) \not\equiv 0$.

Case $P(\bar{x}) = P_1(\bar{x}) \cdot P_2(\bar{x}) + c$: As P_1 and P_2 are non-constant and variable-disjoint ROPs it holds that $1 \leq |\mathrm{var}(P_1)|, |\mathrm{var}(P_2)| < |\mathrm{var}(P)| \leq 2^t$. Hence, we can apply the induction hypothesis on both P_1 and P_2. As $P(G_{t+1}) = P_1(G_{t+1}) \cdot P_2(G_{t+1}) + c$ it follows that $P(G_{t+1})$ is a non-constant polynomial (since $P_1(G_{t+1})$ and $P_2(G_{t+1})$ are non-constant as well). □

Note that the $P(G_k) \not\equiv 0$ for the appropriate value of k regardless of the degree of P. We also note that the requirement that $|\mathbb{F}| > n$ is needed for the definition of G_k. The case $k = 1$ of Theorem 1 follows from the next theorem.

Theorem 3. *Let P be an n-variate ROP that depends on at most t variables.*[4] *Denote $\ell = \lceil \log_2 t \rceil + 1$. Let $W \subseteq \mathbb{F}$ be a set of size n^2. Let $\mathcal{H} = G_\ell\left(W^{2\ell}\right) \subseteq \mathbb{F}^n$ (that is, we take the image of $W^{2\ell}$ under G_ℓ). Then $P \equiv 0$ if and only if $P|_{\mathcal{H}} \equiv 0$.*

Proof. If $P \equiv 0$ then the claim is trivial. Assume that $P \not\equiv 0$. By Lemma 8 we get that $P(G_\ell) \not\equiv 0$. From the definition, G_ℓ^i depends on 2ℓ variables $\{y_j, z_j\}_{j \in [\ell]}$. The degrees of each y_j and each z_j in G_ℓ are $n-1$ and 1, respectively. Hence, the degrees of each y_j and each z_j in $P(G_\ell)$ are bounded by $(n-1)n$ and n, respectively. This immediately implies that $P \not\equiv 0$ if and only if $P|_{\mathcal{H}} \not\equiv 0$. Finally, we note that $|\mathcal{H}| \leq (n^2)^\ell = n^{\mathcal{O}(\log t)}$. □

In particular, since every ROP depends on at most n variables, we obtain a quasi-polynomial $n^{\mathcal{O}(\log n)}$ black-box PIT algorithm for ROPs. When the ROF is of small depth we obtain a faster algorithm using similar techniques. However, due to space limitation this proof is omitted from this version.

5   PIT for Sum of Read-Once Formulas

In this section we prove Theorems 1 and 2: we are given k ROPs $\{F_m\}_{m \in [k]}$ and we have to find whether they sum to zero. In other words, let $F = F_1 + \ldots + F_k$, then we have to check whether $F \equiv 0$. Our algorithm for the problem has two steps. First we find a *common* justifying assignment to $F_1, \ldots, F_k$ using the PIT algorithm for a single ROF from Section 4. Once we have a common justifying assignment we can assume w.l.o.g. that all the input formulas are $\bar{0}$-justified (see Lemma 1). In the second step we simply verify that F vanishes on a relatively small set of vectors, each of weight at most $3k$. Theorem 5 then guarantees that $F \equiv 0$. The main tool in the proof is Theorem 6 that shows that we cannot represent $\mathcal{P}_n \overset{\Delta}{=} \prod_{i=1}^n x_i$ as a sum of less than $\frac{1}{3}n$ $\bar{0}$-justified ROPs. We call this approach a *hardness of representation* approach as the proof is based on the fact that a simple polynomial cannot be represented (computed) by a sum of a "small" number of $\bar{0}$-justified ROPs.

---

[4] Clearly $t \leq n$ but we choose this more general statement.

We now show how to obtain a common justifying assignment from a PIT algorithm. For this we first note that our PIT for a single ROF actually gives a *generator*. Namely, a map $\mathcal{G} = (\mathcal{G}^1, \ldots, \mathcal{G}^n) : \mathbb{F}^q \to \mathbb{F}^n$ such that for every non-zero n-variate ROP P it holds that $P(\mathcal{G}) \not\equiv 0$. By Lemma 8 we get that for $\ell = \lceil \log_2 n \rceil + 1$ the mapping $G_\ell : \mathbb{F}^{2\ell} \to \mathbb{F}^n$ is a generator for ROPs. Recall that the individual degrees of the G_ℓ^i-s are bounded by $\delta = n - 1$. The following lemma is based on Lemmas 5.1 and 5.3 from [27]. The lemma summarizes the connection between PIT algorithms for ROPs and algorithms for finding a common justifying assignment, in both black-box and non black-box settings.

Lemma 10. *Let $\mathbb{F}$ be a field with $|\mathbb{F}| > n$ and let $\{P_m\}_{m \in [k]}$ be a set of ROPs. Then there is an algorithm that computes a common justifying assignment $\bar{a}$ for $\{P_m\}_{m \in [k]}$ in time $\mathcal{O}(n^4 k^2)$.*

Moreover, assume that $\mathcal{G} = (\mathcal{G}^1, \ldots, \mathcal{G}^n) : \mathbb{F}^q \to \mathbb{F}^n$ is a generator for ROPs, with individual degrees of each $\mathcal{G}^i$ bounded by δ and let $V \subseteq \mathbb{F}$ be an arbitrary subset of size $|V| = kn^2\delta + 1$. Then the set $\mathcal{J}_{\mathcal{G}}^k \triangleq \mathcal{G}(V^q)$ contains a common justifying assignment for $P_1, \ldots, P_k$. Clearly, $|\mathcal{J}_{\mathcal{G}}^k| \leq (kn^2\delta + 1)^q$.

We are now ready to prove Theorems 1 and 2. The next theorem shows that a common justifying assignment implies a PIT algorithm for sum of ROFs. First we need the following definition. For a set $W \subseteq \mathbb{F}$, such that $0 \in W$, and $k \leq n$ we define $\mathcal{A}_k^n(W)$ to be the set of all vectors in W^n with Hamming weight at most k, that is vectors that have **at most** k non-zero coordinates. Formally: $\mathcal{A}_k^n(W) \triangleq \{\bar{a} \in W^n \mid \mathrm{w}_H(\bar{a}) \leq k\}$. An immediate conclusion is that $|\mathcal{A}_k^n(W)| = (n \cdot (|W| - 1))^{\mathcal{O}(k)}$.

Theorem 4. *Let $W = \{0, 1\}$ and $\bar{a}$ be a common justifying assignment for the ROPs $F_1, F_2, \ldots, F_k$. Consider $F = \sum_{m=1}^k F_m$. Then $F \equiv 0$ iff $F|_{\mathcal{A}_{3k}^n(W)} \equiv 0$. Thus, if $\mathcal{G} = (\mathcal{G}^1, \ldots, \mathcal{G}^n) : \mathbb{F}^q \to \mathbb{F}^n$ is a generator as in Lemma 10 then $F \equiv 0$ iff for every $\bar{\gamma} \in \mathcal{J}_{\mathcal{G}}^k$ it holds that $F_{\bar{\gamma}}|_{\mathcal{A}_{3k}^n(W)} \equiv 0$.*

The theorem implies an $n^{\mathcal{O}(k)}$ time *non black-box* PIT algorithm as we can acquire a justifying assignment $\bar{a}$ in $\mathcal{O}(n^4 k^2)$ time given the F_m's (Lemma 10), and an $\mathcal{O}(|\mathcal{J}_{\mathcal{G}}^k| \cdot |\mathcal{A}_{3k}^n(W)|) = (kn^2\delta + 1)^q \cdot (n)^{\mathcal{O}(k)} = (n\delta)^{\mathcal{O}(k+q)}$ time *black-box* PIT algorithm. When δ is the bound on the individual degrees of the $\mathcal{G}^i$-s. Recalling Lemma 8 we get a black-box PIT algorithm for the sum of k ROPs of running time $n^{\mathcal{O}(k+2\ell)} = n^{\mathcal{O}(k+\log n)}$. The correctness of Theorem 4 follows from Lemma 1, Lemma 10 and Theorem 5.

Theorem 5. *Let $F(\bar{x}) = F_1(\bar{x}) + \ldots + F_k(\bar{x})$ be a sum of k $\bar{0}$-justified ROPs over $\mathbb{F}$. Set $W = \{0, 1\}$. Then $F \equiv 0$ if and only if $F|_{\mathcal{A}_{3k}^n(W)} \equiv 0$.*

The main in proving Theorem 5 is a hardness of representation theorem for ROPs. Such a theorem was first given in [27] and here we prove an improved version of it. Namely, we show that the polynomial $\mathcal{P}_n \triangleq x_1 \cdot x_2 \cdot \ldots \cdot x_n$ cannot be represented as a sum of $k \leq \frac{n}{3}$ $\bar{0}$-justified ROPs. We note that over a large field ($|\mathbb{F}| > n$) the polynomial $\mathcal{P}_n(\bar{x})$ can be represented as a sum of n 0-justified ROPs (by interpolation). Thus our theorem is nearly optimal.

Proof. We prove the Theorem by induction on n. Our base case is when $n \leq 3k$. In this case F is a multilinear polynomial in $n \leq 3k$ variables so clearly $F|_{\mathcal{A}_{3k}^n(W)} \equiv 0$ iff $F \equiv 0$. We now assume that $n > 3k \geq 4$. Let $\ell \in [n]$. Consider the restriction of the F_m's and F to the subspace $x_\ell = 0$. We now show that the required conditions hold for $F' \triangleq F|_{x_\ell=0}$ and $\left\{F_m' \triangleq F_m|_{x_\ell=0}\right\}_{m\in[k]}$ as well. Indeed, the $\{F_m'\}_{m\in[k]}$ are $\bar{0}$-justified ROPs. Moreover, $F'|_{\mathcal{A}_{3k}^{n-1}(W)} = F'|_{\mathcal{A}_{3k}^n(W)} \equiv 0$. From the induction hypothesis we conclude that $F|_{x_\ell=0} = F' \equiv 0$ and therefore x_ℓ is a factor of F. As this holds for every $\ell \in [n]$ we get that $\mathcal{P}_n(\bar{x})$ divides $F(\bar{x})$ or equivalently $F(\bar{x}) = c \cdot \mathcal{P}_n(\bar{x})$ for some $c \in \mathbb{F}$. It follows that $c \cdot \mathcal{P}_n(\bar{x})$ is a sum of k $\bar{0}$-justified ROPs. As $n > 3k$ we get by the next theorem (Theorem 6) that we must have that $c = 0$. Hence $F = c \cdot \mathcal{P}_n \equiv 0$. □

Theorem 6. $\mathcal{P}_n(\bar{x})$ *cannot be represented as sum of* $k \leq \frac{n}{3}$ *weakly-$\bar{0}$-justified ROPs.*

Proof. We shall use the following notation in the proof: For a non-empty subset $I = \{i_1, \ldots, i_{|I|}\} \subseteq [n]$, we define the iterated partial derivative with respect to I in the following way: $\partial_I P \triangleq \frac{\partial^{|I|} P}{\partial x_{i_1} \partial x_{i_2} \partial x_{i_3} \cdots \partial x_{i_{|I|}}}$.

Let $\{F_m(\bar{x})\}_{m\in[k]}$ be k weakly-$\bar{0}$-justified ROPs over $\mathbb{F}[x_1, \ldots, x_n]$. We prove the claim by induction on k. For $k = 0, 1$ the claim follows from the definition of $\bar{0}$-weak-justification. We now assume that $k \geq 2$ and that $n \geq 3k$. We shall assume for a contradiction that $\sum_{m=1}^{k} F_m = \mathcal{P}_n$. The idea of the proof is to eliminate a "large" number of ROPs at a cost of a "small" number of variables. More specifically, we find a small set of (indices of) input variables $J \subseteq [n-1]$ and a constant $\alpha \neq 0 \in \mathbb{F}$ such that after we take a partial derivative with respect to all of the variables in J and substitute $x_n = \alpha$ (that is we consider the ROPs $\{\partial_J F_m|_{x_n=\alpha}\}_{m\in[k]}$) we eliminate "many" polynomials such that the rest of the ROPs remain weakly-$\bar{0}$-justified. This way we get that a representation of polynomial $\partial_J \mathcal{P}_n|_{x_n=\alpha} = \alpha \cdot \mathcal{P}_{\hat{n}}$ (for a relatively large $\hat{n}$) as a sum of a "small" number of weakly-$\bar{0}$-justified ROPs. Then we use the induction hypothesis to reach a contradiction. We now proceed with the proof. There are two cases to consider.

Case 1: There exist $i \neq j \in [n]$ and $m \in [k]$ such that $\frac{\partial^2 F_m}{\partial x_i \partial x_j} \equiv 0$ (namely, F_m does not contain $x_i \cdot x_j$ in any of its monomials). Assume w.l.o.g. that $i = n - 1, j = n$ and $m = k$. By considering the partial derivatives with respect to $\{x_n, x_{n-1}\}$ we get that $\sum_{m=1}^{k-1} \frac{\partial^2 F_m}{\partial x_n \partial x_{n-1}} = \mathcal{P}_{n-2}$. It may be the case that more than one F_m vanishes when we take a partial derivative w.r.t. $\{x_n, x_{n-1}\}$, however they cannot all vanish simultaneously (as $\mathcal{P}_n$ contains $x_n \cdot x_{n-1}$). By Lemma 5 we have that the polynomials $\left\{\frac{\partial^2 F_m}{\partial x_n \partial x_{n-1}}\right\}$ are weakly-$\bar{0}$-justified

ROPs. Hence, we obtain a representation of $\mathcal{P}_{n-2}$ as a sum of $0 < \hat{k} \leq k - 1$ weakly-$\bar{0}$-justified ROPs such that $0 < 3\hat{k} \leq 3(k-1) = 3k - 3 < n - 2$ which contradicts the induction hypothesis.

Case 2: For every $i \neq j \in [n]$ and $m \in [k]$ we have that $\frac{\partial^2 F_m}{\partial x_i \partial x_j} \neq 0$. Thus, by Lemma 6 we get that the polynomials $\{F_m\}_{m \in [k]}$ are multiplicative ROPs. In addition, for every $m \in [k]$ we have that $\text{var}(F_m) = [n]$. In particular, $|\text{var}(F_m)| \geq 6$. Lemma 7 implies that $\forall m \in [k]$ there exist $j_m \in [n]$, $\alpha_m \neq 0 \in \mathbb{F}$ and a ROP $h_m(\bar{x})$ such that $\frac{\partial F_m}{\partial x_{j_m}} = (x_n - \alpha_m)h_m(\bar{x})$. Let $A = \{\alpha_m \mid m \in [k]\}$.

Clearly $0 \notin A$. For every $\alpha \in A$ we define: $E_\alpha \triangleq \{m \in [k] \mid \alpha_m = \alpha\}$, $B_\alpha \triangleq \{m \in [k] \mid \alpha_m \neq \alpha \wedge F_m|_{x_n = \alpha}$ is not weakly-$\bar{0}$-justified $\}$. Intuitively, E_α is set of the ROPs that can be eliminated by substituting $x_n = \alpha$ and B_α is set of ("bad") ROPs that will become non weakly-$\bar{0}$-justified upon the substitution and thus require a special treatment. From the definition of A we have that $|E_\alpha| \geq 1$ and $\sum_{\alpha \in A} |E_\alpha| = k$. More specifically, the E_α's form a partition of $[k]$. Similarly, Lemma 7 implies that for each $\alpha \neq \alpha' \in A$ the sets B_α and $B_{\alpha'}$ are disjoint (since for every ROP there exists at most one bad value β of x_n) and therefore $\sum_{\alpha \in A} |B_\alpha| \leq k$. Hence, there exists $\alpha_0 \in A$ such that $|B_{\alpha_0}| \leq |E_{\alpha_0}|$.

Now, let $I = E_{\alpha_0} \cup B_{\alpha_0}$ and $J = \{j_m \mid m \in I\}$. From the definition, $I \subseteq [k]$ and $J \subseteq [n]$. In addition, $1 \leq |J| \leq |I| \leq |E_{\alpha_0}| + |B_{\alpha_0}| \leq 2|E_{\alpha_0}|$ and $n \notin J$. Consider the following ROPs for every $m \in [k]$: $F'_m \triangleq \partial_J F_m$. Then the ROPs F'_m's have the following properties.

1. By Lemma 5 we get that every F'_m is a weakly-$\bar{0}$-justified ROP.
2. For every $m \in I$ we have that $F'_m = (x_n - \alpha_m)h'_m(\bar{x})$ for some ROP $h'_m(\bar{x})$. Indeed, as $j_m \in J$ we have that $F'_m = \partial_J F_m = \partial_{J \setminus \{j_m\}}(\frac{\partial F_m}{\partial x_{j_m}}) = \partial_{J \setminus \{j_m\}}((x_n - \alpha_m)h_m(\bar{x})) = (x_n - \alpha_m) \cdot \partial_{J \setminus \{j_m\}} h_m(\bar{x})$
3. For every $m \in I$ we have that $h'_m(\bar{x})$ is a weakly-$\bar{0}$-justified ROP (this follows from Lemma 5 and the previous two properties).

For $m \in [k]$ consider the following ROPs: $F''_m \triangleq \partial_J F_m|_{x_n = \alpha_0} = F'_m|_{x_n = \alpha_0}$. Based on the above we can conclude that:

- For every $m \in E_{\alpha_0}$ it holds that $F''_m = (\alpha_0 - \alpha_m)h'_m(\bar{x}) \equiv 0$ (by definition of E_{α_0} we have that $\alpha_m = \alpha_0$).
- For every $m \in B_{\alpha_0}$ we have that $F''_m = (\alpha_0 - \alpha_m)h'_m(\bar{x})$ is a non-zero weakly-$\bar{0}$-justified ROP. Notice that in contrary to F_m, the structure of F'_m guarantees that it remains weakly-$\bar{0}$-justified when substituting $x_n = \alpha_0$.
- For $m \in [k] \setminus I$ the definitions of E_{α_0} and B_{α_0} guarantee that $F_m|_{x_n = \alpha_0}$ is a weakly-$\bar{0}$-justified ROP. Lemma 5 implies that the same holds for $F''_m = \partial_J(F_m|_{x_n = \alpha_0})$ as well. Note that in this case it is also possible that $F''_m \equiv 0$.

Thus, $F''_m \equiv 0$ for $m \in E_{\alpha_0}$ and F''_m is a weakly-$\bar{0}$-justified ROP for $m \in [k] \setminus E_{\alpha_0}$. W.l.o.g. let us assume that $J = \{\hat{n} + 1, \hat{n} + 2, \ldots, n - 2, n - 1\}$ for some $\hat{n}$. We get that $\sum\limits_{m=1}^{k} F''_m = \partial_J \mathcal{P}_n|_{x_n=\alpha_0} = \alpha_0 \cdot \mathcal{P}_{\hat{n}}$. That is, we found a representation of $\alpha_0 \cdot \mathcal{P}_{\hat{n}}$ as a sum of weakly-$\bar{0}$-justified ROPs, where at least $|E_{\alpha_0}|$ of the ROPs are zeros. Notice that $2|E_{\alpha_0}| \geq |J| = (n-1) - \hat{n}$ and $|E_\alpha| \geq 1$. Therefore, we have found a representation of $\alpha_0 \cdot \mathcal{P}_{\hat{n}}$ as a sum of $0 \leq \hat{k} < k$ weakly-$\bar{0}$-justified ROPs such that $0 \leq 3\hat{k} \leq 3(k - |E_\alpha|) = 3k - 3|E_\alpha| \leq n - 3|E_\alpha| \leq \hat{n} + 1 - |E_\alpha| \leq \hat{n}$. By our induction hypothesis we get that $\alpha_0 = 0$, which is a contradiction (recall that $\alpha_0 \in A$ and $0 \notin A$). Hence, $\mathcal{P}_n$ cannot be represented as a sum of less than $\frac{n}{3}$ weakly-$\bar{0}$-justified ROPs. This completes the proof.   □

6   Conclusions

In this short version we gave a black-box identity testing algorithm for the sum of k ROFs of running time $n^{\mathcal{O}(k+\log n)}$. In the full version of this paper[5] we actually show that a similar result also holds for the more general model of preprocessed ROFs. It is an interesting question to obtain a polynomial time identity testing algorithm for this model, or to even just improve the running time to be of the form $f(k) \cdot n^{\mathcal{O}(1)}$ for some function f not depending on n. As shown in [27] a PIT algorithm for ROFs implies a reconstruction (learning) algorithm for them. Thus our results imply a quasi-polynomial time deterministic reconstruction algorithm for (preprocessed) ROFs. It will be interesting to get a reconstruction algorithm for sums of ROFs. Another question is obtaining a deterministic polynomial time reconstruction algorithm for a single ROF.

References

1. Agrawal, M.: Proving lower bounds via pseudo-random generators. In: Sarukkai, S., Sen, S. (eds.) FSTTCS 2005. LNCS, vol. 3821, pp. 92–105. Springer, Heidelberg (2005)
2. Agrawal, M., Biswas, S.: Primality and identity testing via chinese remaindering. JACM 50(4), 429–443 (2003)
3. Agrawal, M., Vinay, V.: Arithmetic circuits: A chasm at depth four. In: Proceedings of the 49th FOCS, pp. 67–75 (2008)
4. Arvind, V., Mukhopadhyay, P.: The monomial ideal membership problem and polynomial identity testing. In: Tokuyama, T. (ed.) ISAAC 2007. LNCS, vol. 4835, pp. 800–811. Springer, Heidelberg (2007)
5. Ben-Or, M., Tiwari, P.: A deterministic algorithm for sparse multivariate polynomial interpolation. In: Proceedings of the 20th STOC, pp. 301–309 (1988)
6. Bshouty, D., Bshouty, N.H.: On interpolating arithmetic read-once formulas with exponentiation. J. of Computer and System Sciences 56(1), 112–124 (1998)
7. Bshouty, N.H., Hancock, T.R., Hellerstein, L.: Learning arithmetic read-once formulas. SICOMP 24(4), 706–735 (1995)
8. Chen, Z., Kao, M.: Reducing randomness via irrational numbers. In: Proceedings of the 29th STOC, pp. 200–209 (1997)

---

[5] See http://www.cs.technion.ac.il/~shpilka/publications/PROF.pdf

9. Dvir, Z., Shpilka, A.: Locally decodable codes with 2 queries and polynomial identity testing for depth 3 circuits. SICOMP 36(5), 1404–1434 (2006)
10. Dvir, Z., Shpilka, A., Yehudayoff, A.: Hardness-randomness tradeoffs for bounded depth arithmetic circuits. In: Proceedings of the 40th STOC, pp. 741–748 (2008)
11. Hancock, T.R., Hellerstein, L.: Learning read-once formulas over fields and extended bases. In: Proceedings of the 4th COLT, pp. 326–336 (1991)
12. Kabanets, V., Impagliazzo, R.: Derandomizing polynomial identity tests means proving circuit lower bounds. Computational Complexity 13(1-2), 1–46 (2004)
13. Karnin, Z.S., Shpilka, A.: Deterministic black box polynomial identity testing of depth-3 arithmetic circuits with bounded top fan-in. In: Proceedings of the 23rd CCC, pp. 280–291 (2008)
14. Kayal, N., Saxena, N.: Polynomial identity testing for depth 3 circuits. Computational Complexity 16(2), 115–138 (2007)
15. Klivans, A., Spielman, D.: Randomness efficient identity testing of multivariate polynomials. In: Proceedings of the 33rd STOC, pp. 216–223 (2001)
16. Lewin, D., Vadhan, S.: Checking polynomial identities over any field: Towards a derandomization? In: Proceedings of the 30th STOC, pp. 428–437 (1998)
17. Lipton, R.J., Vishnoi, N.K.: Deterministic identity testing for multivariate polynomials. In: Proceedings of the 14th SODA, pp. 756–760 (2003)
18. Raz, R.: Multilinear $NC_1 \neq$ Multilinear NC_2. In: Proceedings of the 45th FOCS, pp. 344–351 (2004)
19. Raz, R.: Extractors with weak random seeds. In: Proceedings of the 37th STOC, pp. 11–20 (2005)
20. Raz, R., Shpilka, A.: Deterministic polynomial identity testing in non commutative models. Computational Complexity 14(1), 1–19 (2005)
21. Raz, R., Shpilka, A., Yehudayoff, A.: A lower bound for the size of syntactically multilinear arithmetic circuits. SICOMP 38(4), 1624–1647 (2008)
22. Raz, R., Yehudayoff, A.: Lower bounds and separations for constant depth multilinear circuits. In: 40th CCC, pp. 128–139 (2008)
23. Saxena, N.: Diagonal circuit identity testing and lower bounds. In: Aceto, L., Damgård, I., Goldberg, L.A., Halldórsson, M.M., Ingólfsdóttir, A., Walukiewicz, I. (eds.) ICALP 2008, Part I. LNCS, vol. 5125, pp. 60–71. Springer, Heidelberg (2008)
24. Saxena, N., Seshadhri, C.: An almost optimal rank bound for depth-3 identities. CoRR, abs/0811.3161 (2008)
25. Schwartz, J.T.: Fast probabilistic algorithms for verification of polynomial identities. JACM 27(4), 701–717 (1980)
26. Shpilka, A.: Interpolation of depth-3 arithmetic circuits with two multiplication gates. SICOMP 38(6), 2130–2161 (2009)
27. Shpilka, A., Volkovich, I.: Read-once polynomial identity testing. In: Proceedings of the 40th STOC, pp. 507–516 (2008)
28. Zippel, R.: Probabilistic algorithms for sparse polynomials. In: Symbolic and algebraic computation, pp. 216–226 (1979)

Smooth Analysis of the Condition Number and the Least Singular Value

Terence Tao and Van Vu

[1] Department of Mathematics, UCLA, Los Angeles CA 90095-1555
[2] Department of Mathematics, Rutgers, Piscataway, NJ 08854

Abstract. A few years ago, Spielman and Teng initiated the study of Smooth analysis of the condition number and the least singular value of a matrix. Let x be a complex variable with mean zero and bounded variance. Let N_n be the random matrix of sie n whose entries are iid copies of x and M a deterministic matrix of the same size. The goal of smooth analysis is to estimate the condition number and the least singular value of $M + N_n$.

Spielman and Teng considered the case when x is gaussian. We are going to study the general case when x is arbitrary. Our investigation reveals a new and interesting fact that, unlike the gaussian case, in the general case the "core" matrix M does play a role in the tail bounds for the least singular value of $M + N_n$. Consequently, our estimate involves the norm $\|M\|$ and it is a challenging question to determine the right magnitude of this involvement. When $\|M\|$ is relatively small, our estimate is nearly optimal and extends or refines several existing result.

1 Introduction

Let M be an $n \times n$ matrix and $s_1(M) \geq \cdots \geq s_n(M)$ its singular values. The condition number of A, as defined by numerical analysts, is

$$\kappa(M) := s_1(M)/s_n(M) = \|M\|\|M^{-1}\|.$$

This parameter is of fundamental importance in numerical linear algebra and related areas, such as linear programming. In particular, the value

$$L(M) := \log \kappa(M)$$

measures the (worst case) loss of precision the equation $Mx = b$ can exhibit [22,2].

The problem of understanding the typical behavior of $\kappa(M)$ and $L(M)$ when the matrix M is random has a long history. This was first raised by von Neumann and Goldstein in their study of numerical inversion of large matrices [32]. Several years later, the problem was restated in a survey of Smale [22] on the efficiency of algorithm of anaylsis. One of Smale's motivations was to understand the efficiency of the simplex algorithm in linear programming. The problem is also

I. Dinur et al. (Eds.): APPROX and RANDOM 2009, LNCS 5687, pp. 714–737, 2009.

at the core of Demmel's plan about the investigation of the probability that a numerical analysis problem is difficult [8].

To make the problem precise, the most critical issue is to choose a probability distribution for M. A convenient model has been random matrices with independent gaussian entries (either real of complex). An essential feature of this model is that here the joint distribution of the eigenvalues can be written down precisely

$$(Real\ Gaussian)\ c_1(n) \prod_{1 \le i < j \le n} |\lambda_i - \lambda_j| \exp(-\sum_{i=1}^{n} \lambda_i^2/2). \tag{1}$$

$$(Complex\ Gaussian)\ c_2(n) \prod_{1 \le i < j \le n} |\lambda_i - \lambda_j|^2 \exp(-\sum_{i=1}^{n} \lambda_i^2/2). \tag{2}$$

Here $c_1(n), c_2(n)$ are normalization factors whose explicit formulae can be seen in, for example, [16].

Most questions about the spectrum of these random matrices can then be answered by estimating a properly defined integral with respect to these measures. Many advanced techniques have been worked out to serve this purpose (see, for instance [16]). In particular, the condition number is well understood, thanks to works of Kostlan, Oceanu [22], Edelman [6] and many others (see Section 2).

The gaussian model, however, has serious shortcomings. As pointed out by many researchers (see, for example [3,24]), the gaussian model does not reflex the arbitrariness of the input. Let us consider, for example, a random matrix with independent real gaussian entries. By sharp concentration results, one can show that the fraction of entries with absolute values at most 1, is, with overwhelming probability, close to the absolute constant $\frac{1}{\sqrt{2\pi}} \int_{-1}^{1} \exp(-t^2/2)dt$. Many classes of matrices that occur in practice just simply do not possess this property. This problem persists even when one replaces gaussian by another fixed distribution, such as Bernoulli.

About 10 years ago, Spielman and Teng [24,25], motivated by Demmel's plan and the problem of understanding the efficiency of the simplex algorithm proposed a new, exciting distribution. Spielman and Teng observed that while the ideal input maybe a fixed matrix M, it is likely that the computer will work with a perturbation $M + N$, where N is a random matrix representing random noise. Thus, it raised the issue of studying the distribution of the condition number of $M + N$. This problem is at the heart of the so-called Spielman-Teng *smooth* analysis. (See [24,25] for a more detailed discussion and [3,4,5,26,9] for many related works on this topics.) Notice that the special case $M = 0$ corresponds to the setting considered in the previous paragraphs.

The Spielman-Teng model nicely addresses the problem about the arbitrariness of the inputs, as in this model every matrix generates a probability space of its own. In their papers, Spielman and Teng considered mostly gaussian noise (in some cases they also considered other continuous distributions such as uniform on $[-1, 1]$). However, in the digital world, randomness often *does not* have gaussian nature. To start with, all real-world data are finite. In fact, in many problems (particularly those in integer programming) all entries of the matrix are integers. The random errors made by the degital devices (for example, sometime a bit gets flipped) are obviously of discrete nature. In other problems, for example those in engineering, the data may contain measurements where it would be natural to assume gaussian errors. On the other hand, data are usually strongly truncated. For example, if an entry of our matrix represents the mass of an object, then we expect to see a number like 12.679 (say, tons), rather than 12.6792347043641259. Thus, instead of the gaussian distribution, we (and/or our computers) often work with a discrete distribution, whose support is relatively small and does not depend on the size of the matrix. (A good toy example is random Bernoulli matrix, whose entries takes values ± 1 with probability half.) This leads us to the following question

Question. (Smooth analysis of the condition number) Estimate the condition number of a random matrix $M_n := M + N_n$, where M is a fixed matrix of size n, and N_n a general random matrix ?

The goal of this paper is to investigate this question, where, as a generalization of Spielman-Teng model, we think of N_n as a matrix with independent random entries which (instead as being gaussian) have arbitrary distributions. Our main result will show that with high probability, M_n is well-conditioned. This result could be useful in further studies of smooth analysis in linear programming. The Spielman-Teng smooth analysis of the simplex algorithm [24,25] was done with gaussian noise. It is a natural and (from the practical point of view) important question to repeat this analysis with discrete noise (such as Bernoulli). This question was posed by Spielman to the authors few years ago. The paper [24] also contains a specific conjecture on the least singular value of random Bernoulli matrix.

In connection, we should mention here a recent series of papers by Burgisser, Cucker and Lotz [3,4,5], which discussed the smooth analysis of condition number under a somewhat different setting (they considered the notion of *conic* condition number and a different kind of randomness).

Before stating mathematical results, let us describe our notations. We use the usual asymptotic notation $X = O(Y)$ to denote the estimate $|X| \le CY$ for some constant $C > 0$ (independent of n); $X = \Omega(Y)$ to denote the estimate $X \ge cY$ for some $c > 0$ independent of n, and $X = \Theta(Y)$ to denote the estimates $X = O(Y)$ and $X = \Omega(Y)$ holding simultaneously. In some cases, we write $X \ll Y$ instead of $X = O(Y)$ and $X \gg Y$ instead of $X = \Omega(Y)$. Notations such as $X = O_{a,b}(Y)$ or $X \ll_{a,b} (Y)$ mean that the hidden constant in O or $\ll$ depend on previously defined constants a and b. We use $o(1)$ to denote any quantity that goes to zero as $n \to \infty$. $X = o(Y)$ means that $X/Y = o(1)$.

Recall that

$$\kappa(M) := s_1(M)/s_n(M) = \|M\| \|M^{-1}\|.$$

Since $\|M\|^2 \ge \sum_{ij} |m_{ij}|^2/n$ (where m_{ij} denote the entries of M) it is expected that $\|M\| = n^{\Omega(1)}$. Following the literature, we say that M is well-conditioned (or well-posed) if $\kappa(M) = n^{O(1)}$ or (equivalently) $L(M) = O(\log n)$.

By the triangle inequality,

$$\|M\| - \|N_n\| \le \|M + N_n\| \le \|M\| + \|N_n\|.$$

Under very general assumptions, the random matrix N_n satisfies $\|N_n\| = n^{O(1)}$ with overwhelming probability (see many estimates in Section 3). Thus, in order to guarantee that $\|M + N_n\|$ is well-conditioned (with high probability), it is natural to assume that

$$\|M\| = n^{O(1)}. \tag{3}$$

This is not only a natural, but fairly safe assumption to make (with respect to the applicability of our studies). Most large matrices in practice satisfy this assumption, as their entries are usually not too large compared to their sizes.

Our main result shows that under this assumption and a very general assumption on the entries of N_n, the matrix $M + N_n$ is well-conditioned, with high probability. This result extends and bridges several existing results in the literature (see next two sections).

Notice that under assumption (3), if we want to show that $M + N_n$ is typically well-conditioned, it suffices to show that

$$\|(M + N_n)^{-1}\| = s_n(M + N_n)^{-1} = n^{O(1)}$$

with high probability. Thus, we will formulate most results in a form of a tail bound for the least singular value of $M + N_n$. The typical form will be

$$\mathbf{P}(s_n(M + N_n) \le n^{-B}) \le n^{-A}$$

where A, B are positive constants and A increases with B. The relation between A and B is of importance and will be discussed in length.

2   Previous Results

Let us first discuss the gaussian case. Improving results of Kostlan and Oceanu [22], Edelman [6] computed the limiting distribution of $\sqrt{n}s_n(N_n)$ when N_n is gaussian. His result implies

Theorem 1. *There is a constant $C > 0$ such that the following holds. Let x be the real gaussian random variable with mean zero and variance one, let N_n be the random matrix whose entries are iid copies of x. Then for any constant $t > 0$*

$$\mathbf{P}(s_n(N_n) \le t) \le n^{1/2}t.$$

Concerning the more general model $M + N_n$, Sankar, Spielman and Teng proved [26]

Theorem 2. *There is a constant $C > 0$ such that the following holds. Let x be the real gaussian random variable with mean zero and variance one, let N_n be* *the random matrix whose entries are iid copies of* x, *and let* M *be an arbitrary fixed matrix. Let* $M_n := M + N_n$. *Then for any* $t > 0$

$$\mathbf{P}(s_n(M_n) \leq t) \leq Cn^{1/2}t.$$

Once we give up the gaussian assumption, the study of the least singular value s_n becomes much harder (in particular for discrete distributions such as Bernoulli, in which $x = \pm 1$ with equal probability $1/2$). For example, it is already non-trivial to prove that the least singular value of a random Bernoulli matrix is positive with probability $1 - o(1)$. This was first done by Komlós in 1967 [13], but good quantitative lower bounds were not available until recently. In a series of papers, Tao-Vu and Rudelson-Vershynin addressed this question [27,29,19,20] and proved a lower bound of the form $n^{-\Theta(1)}$ for s_n with high probability.

We say that x is *subgaussian* if there is a constant $B > 0$ such that

$$\mathbf{P}(|x| \geq t) \leq 2\exp(-t^2/B^2)$$

for all $t > 0$. The smallest B is called the *subgaussian moment* of x. The following is a corollary of a more general theorem by Rudelson and Vershynin [20, Theorem 1.2]

Theorem 3. *Let* x *be a subgaussian random variable with zero mean, variance one and subgaussian moment* B *and* A *be an arbitrary positive constant. Let* N_n *be the random matrix whose entries are iid copies of* x. *Then there is a positive constant* C *(depending on* B*) such that for any* $t \geq n^{-A}$ *we have*

$$\mathbf{P}(s_n(N_n) \leq t) \leq Cn^{1/2}t.$$

A similar result was obtained in [20] assuing only fourth moment control on x, but with a right-hand side which was $o(1)$ for $t = o(n^{-1/2})$.

We again turn to the general model $M + N_n$. In [17], the results of [20] were extended to this case assuming that the operator norm of $M + N_n$ was $O(n^{1/2})$, which largely restricts the range of applicability to the case when x has bounded fourth moment. A variant of this result also appears in [11], in which the operator norm of $M + N_n$ was allowed to be somewhat larger (in particular, treating the case in which x has bounded $2 + \eta$ moment for some $\eta > 0$) but the right-hand side was again of the form $o(1)$. In [29], the present authors proved

Theorem 4. *[29, Theorem 2.1] Let* x *be a random variable with non-zero variance. Then for any constants* $A, C > 0$ *there exists a constant* $B > 0$ *(depending on* A, C, x*) such that the following holds. Let* N_n *be the random matrix whose entries are iid copies of* x, *and let* M *be any deterministic* $n \times n$ *matrix with norm* $\|M\| \leq n^C$. *Then*

$$\mathbf{P}(s_n(M + N_n) \leq n^{-B}) \leq n^{-A}.$$

Notice that this theorem requires very little about the variable x. It does not need to be sub-gaussian nor even has bounded moments. All we ask is that the variance is bounded from zero, which basically means x is indeed "random". Thus, it guarantees the well-conditionness of $M + N_n$ in a very general setting.

The weakness of this theorem is that the dependence of B on A and C, while explicit, is too generous. The main result of this paper, Theorem 6, will improve this dependence significantly and provide a common extension of Theorem 4 and Theorem 3.

In a slightly different direction,

3   Main Result

As already pointed out, an important point is the relation between the constants A, B in a bound of the form

$$\mathbf{P}(s_n(M + N_n) \leq n^{-B}) \leq n^{-A}.$$

In Theorem 2, we have a simple (and optimal) relation $B = A + 1/2$. It is natural to conjecture that this relation holds for other, non-gaussian, models of random matrices. In fact, this conjecture was our starting point of this study. Quite surprisingly, it turns out not to be the case.

Theorem 5. *There are positive constants c_1 and c_2 such that the following holds. Let N_n be the $n \times n$ random Bernoulli matrix with n even. For any $L \geq n$, there is an $n \times n$ deterministic matrix M such that $\|M\| = L$ and*

$$\mathbf{P}(s_n(M + N_n) \leq c_1 \frac{n}{L}) \geq c_2 n^{-1/2}.$$

The assumption n is even is for convenience and can easily be removed by replacing the Bernoulli matrix by a random matrix whose entries take values $0, \pm 1$ with probability $1/3$ (say). Notice that if $L = n^D$ for some constant D then we have the lower bound

$$\mathbf{P}(s_n(M + N_n) \leq c_1 n^{-D+1}) \geq c_2 n^{-1/2},$$

which shows that one cannot expect Theorem 2 to hold in general and that the norm of M should play a role in tail bounds of the least singular value.

The main result of this paper is the following.

Theorem 6. *Let x be a random variable with mean zero and bounded second moment, and let $\gamma \geq 1/2$, $A \geq 0$ be constants. Then there is a constant c depending on x, γ, A such that the following holds. Let N_n be the random matrix of size n whose entries are iid copies of x, M be a deterministic matrix satisfying $\|M\| \leq n^\gamma$, and let $M_n := M + N_n$. Then*

$$\mathbf{P}(s_n(M_n) \leq n^{-(2A+1)\gamma}) \leq c\Big(n^{-A+o(1)} + \mathbf{P}(\|N_n\| \geq n^\gamma)\Big).$$

Note that this theorem only assumes bounded second moment on x. The assumption that the entries of N_n are iid is for convenience. A slightly weaker result would hold if one omit this assumption.

Corollary 1. *Let x be a random variable with mean zero and bounded second moment, and let $\gamma \geq 1/2$, $A \geq 0$ be constants. Then there is a constant c_2 depending on x, γ, A such that the following holds. Let N_n be the random matrix of size n whose entries are iid copies of x, M be a deterministic matrix satisfying $\|M\| \leq n^\gamma$, and let $M_n := M + N_n$. Then*

$$\mathbf{P}(\kappa(M_n) \geq 2n^{(2A+2)\gamma}) \leq c\left(n^{-A+o(1)} + \mathbf{P}(\|N_n\| \geq n^\gamma)\right).$$

Proof. Since $\kappa(M_n) = s_1(M_n)/s_n(M_n)$, it follows that if $\kappa(M_n) \geq n^{(2A+2)\gamma}$, then at least one of the two events $s_n(M_n) \leq n^{-(2A+1)\gamma}$ and $s_1(M_n) \geq 2n^\gamma$ holds. On the other hand,

$$s_1(M_n) \leq s_1(M) + s_1(N_n) = \|M\| + \|N_n\| \leq n^\gamma + \|N_n\|.$$

The claim follows.

In the rest of this section, we deduce a few corollaries and connect them with the existing results.

First, consider the special case when x is subgaussian. In this case, it is well-known that one can have a strong bound on $\mathbf{P}(\|N_n\| \geq n^\gamma)$ thanks to the following theorem (see [20] for references).

Theorem 7. *Let B be a positive constant. There are positive constants C_1, C_2 depending on B such that the following holds. Let x be a subgaussian random variable with zero mean, variance one and subgaussian moment B and N_n be the random matrix whose entries are iid copies of x. Then*

$$\mathbf{P}(\|N_n\| \geq C_1 n^{1/2}) \leq \exp(-C_2 n).$$

If one replaces the subgaussian condition by the weaker condition that x has forth moment bounded B, then one has a weaker conclusion that

$$\mathbf{E}(\|N_n\|) \leq C_1 n^{1/2}.$$

From Theorem 6 and Theorem 7 we see that

Corollary 2. *Let A and γ be arbitrary positive constants. Let x be a subgaussian random variable with zero mean and variance one and N_n be the random matrix whose entries are iid copies of x. Let M be a deterministic matrix such that $\|M\| \leq n^\gamma$ and set $M_n = M + N_n$. Then*

$$\mathbf{P}(s_n(M_n) \leq (n^{1/2} + \|M\|)^{-2A-1}) \leq n^{-A+o(1)}. \tag{4}$$

In the case $\|M_n\| = O(n^{1/2})$ (which of course includes the $M_n = 0$ special case), (4) implies

Corollary 3. *Let A be arbitrary positive constant. Let x be a subgaussian random variable with zero mean and variance one and N_n be the random matrix*

whose entries are iid copies of x. *Let* M *be a deterministic matrix such that* $\|M\| = O(n^{1/2})$ *and set* $M_n = M + N_n$. *Then*

$$P(s_n(M_n) \leq n^{-A-1/2}) \leq n^{-A+o(1)}. \tag{5}$$

Up to a loss of magnitude $n^{o(1)}$, this matches Theorem 3, which treated the base case $M = 0$.

If we assume bounded fourth moment instead of subgaussian, we can use the second half of Theorem 7 to deduce

Corollary 4. *Let* x *be a random variable with zero mean, variance one and bounded forth moment moment and* N_n *be the random matrix whose entries are iid copies of* x. *Let* M *be a deterministic matrix such that* $\|M\| = n^{O(1)}$ *and set* $M_n; = M + N_n$. *Then*

$$P(s_n(M_n) \leq (n^{1/2} + \|M\|)^{-1+o(1)}) = o(1). \tag{6}$$

In the case $\|M\| = O(n^{1/2})$, this implies that almost surely $s_n(M_n) \leq n^{-1/2+o(1)}$. For the special case $M = 0$, this matches (again up to the $o(1)$ term) Theorem [20, Theorem 1.1].

Let us now take a look at the influence of $\|M\|$ on the bound. Obviously, there is a gap between (4) and Theorem 5. On the other hand, by setting $A = 1/2$, $L = n^\gamma$ and assuming that $P(\|N_n\| \geq n^\gamma)$ is negligible (i.e., super-polynomially small in n), we can deduce from Theorem 6 that

$$P(s_n(M_n) \leq c_1 L^{-2}) \leq c_2 n^{-1/2+o(1)}.$$

This, together with Theorem 5, suggests that the influence of $\|M\|$ in $s_n(M_n)$ is of polynomial type.

In the next discussion, let us normalize and assume that x has variance one. One can deduce a bound on $\|N_n\|$ from the simple computation

$$\mathbf{E}\|N_n\|^2 \leq \mathbf{E}\, \mathrm{tr} N_n N_n^* = n^2.$$

By Chebyshev's inequality we thus have

$$P(\|N_n\| \geq n^{1+A/2}) \leq n^{-A}$$

for all $A \geq 0$.

Applying Theorem 6 we obtain

Corollary 5. *Let* x *be a random variable with mean zero and variance one and* N_n *be the random matrix whose entries are iid copies of* x. *Then for any constant* $A \geq 0$

$$P(s_n(N_n) \leq n^{-1-\frac{5}{2}A-A^2}) \leq n^{-A+o(1)}.$$

In particular, $s_n(N_n) \geq n^{-1-o(1)}$ *almost surely.*

It is clear that one can obtain better bounds for s_n, provided better estimates on $\|N_n\|$. The idea of using Chebyshev's inequality is very crude (we just like to give an example) and there are more sophisticated tools. One can, for instance, use higher moments. The expectation of a k-th moment can be expressed a sum of many terms, each correspond to a certain closed walk of length k on the complete graph of n vertices (see [12,33]). If the higher moments of N_n (while not bounded) do not increase too fast with n, then the main contribution in the expectation of the kth moment still come from terms which correspond to walks using each edge of the graph either 0 and 2 times. The expectation of such a term involves only the second moment of the entries in N_n. The reader may want to work this out as an exercise.

One can also use the following nice estimate of Seginer [23]

$$\mathbf{E}\|N_n\| = O\Big(\mathbf{E}\max_{1\le i\le n}\sqrt{\sum_{j=1}^{n} x_{ij}^2} + \mathbf{E}\max_{1\le j\le n}\sqrt{\sum_{i=1}^{n} x_{ij}^2}\Big).$$

Due to space limitation, we will give all proofs in the appendices. We will, however, discuss the main ideas in the next section.

4   Main Ideas for Proving Lower Bounds on the Least Singular Value

There are two recent sequences of papers that investigate the least singular values of random matrices, [19,20,21] and [27,28,29]. The main idea can be sketched (with some simplifications) as follows. Let d_i be the distance from the ith row vector of $M + N_n$ to the subspace spanned by the rest of the rows. Elementary linear algebra shows

$$\|(M + N_n)^{-1}\| = n^{O(1)}(\min_{1\le i\le n} d_i)^{-1}. \tag{7}$$

Ignoring various factor of $n^{O(1)}$ for a moment, our main ask with be to understand the distribution of d_i, for a given i. If $v = (v_1, \dots, v_n)$ is the normal vector of a hyperplane V, then the distance from a random vector $(a_1 + \xi_1, \dots, a_n + \xi_n)$ to V is

$$|\sum_{i=1}^{n} v_i(a_i + \xi_i)| = |S + \sum_{i=1}^{n} a_i v_i|,$$

where $S := \sum_{i=1}^{n} v_i \xi_i$.

Playing the key role now will be the relation between $\mathbf{P}(|S + \sum_{i=1}^{n} a_i v_i| \le \beta)$, for some given β, and the structure of the normal vector v.

If v is such that the probability in question is small, then we can be done using a relatively simple conditioning argument. The main task is to deal with the exceptional v when the probability in question is large. One tries to classify these vectors and show that there are not too many of them, and then use a direct counting argument. Here the main ingredient is the so-called *Inverse Littlewood-Offord theorems*, introduced in [27]. These theorems lead to a sufficiently strong characterization and consequently an efficient counting.

There are two types of Inverse theorems in the above mentioned sequences of papers. They have led to results of different strengths. The approach in [27,28,29] results in theorems which hold under very mild assumptions iid on the random entries (of N_n) and the "core" matrix M, but with non-optimal relation between the parameters (such as A and B in the exponent). The approach in [19,20,21] treats more restrictive models (in particular the matrix M is zero and the entries of N_n are sub-gaussian or at least have bounded 4th moment), but gives near optimal dependence between the parameters.

The results of this paper, as discussed, are sort of "best of two worlds" type. They provide strong quantitative bounds under very general assumptions. Not too surprisingly, the arguments that we will use combine ideas from both approaches. We are goping to rely on a counting lemma (Theorem C.8) from [29] which gives a sharp estimate for the number of exceptional normal vector v. Another important ingredient is Lemma F.1 from [20] which gives agood way to reduce the error term $n^{O(1)}$ in (7). These two ingredients and some additional technical ideas turn out to be sufficient to finish the job.

Finally, let us mention a very recent development in [30], where we managed to compute the asymptotic of the least singular value in the case M is zero. The approach there is very different and does not seem to extend to the case with general M.

5   Theorem 5: The Influence of M

Let M' be the $n - 1 \times n$ matrix obtaining by concatenating the matrix LI_{n-1} with an all L column, where L is a large number. The $n \times n$ matrix M is obtained from M' by adding to it a (first) all zero row; thus

$$M = \begin{pmatrix} 0 & 0 & \dots & 0 & 0 \\ L & 0 & \dots & 0 & L \\ 0 & L & \dots & 0 & L \\ \vdots & \vdots & \ddots & \vdots & \vdots \\ 0 & 0 & \dots & L & L \end{pmatrix}.$$

It is easy to see that

$$\|M\| = \Theta(L).$$

Now consider $M_n := M + N_n$ where the entries of N_n are iid Bernoulli random variables.

$$\mathbf{P}(s_n(M_n) \ll n^{1/4}L^{-1/2}) \gg n^{-1/2}.$$

Let M_n' be the (random) $(n-1) \times n$ matrix formed by the last $n-1$ rows of M_n. Let $v \in \mathbf{R}^n$ be a unit normal vector of the $n-1$ rows of M_n'. By replacing v with $-v$ if necessary we may write v in the form

$$v = \left(\frac{1}{\sqrt{n}} + a_1, \frac{1}{\sqrt{n}} + a_2 + \ldots, \frac{1}{\sqrt{n}} + a_{n-1}, \frac{-1}{\sqrt{n}} + a_n \right),$$

where $\frac{-1}{\sqrt{n}} + a_n \leq 0$.

Multiplying v with the first row of M_n', we have

$$0 = (L + \xi_1)(1 + a_1) + (L + \xi_n)(-1 + a_n)$$
$$= L(a_1 + a_n) + (\xi_1 - \xi_n) + \xi_1 a_1 + \xi_n a_n.$$

Since $|a_i| = O(1)$, it follows that $|a_1 + a_n| = O(\frac{1}{L})$. Repeating the argument with all other rows, we conclude that $|a_i + a_n| = O(\frac{1}{L})$ for all $1 \leq i \leq n-1$.

Since v has unit norm, we also have

$$1 = \|v\|^2 = \sum_{i=1}^{n-1} \left(\frac{1}{\sqrt{n}} + a_i \right)^2 + \left(\frac{-1}{\sqrt{n}} + a_n \right)^2,$$

which implies that

$$\frac{2}{\sqrt{n}}(a_1 + \cdots + a_{n-1} - a_n) + \sum_{i=1}^{n} a_i^2 = 0.$$

This, together with the fact that $|a_i + a_n| = O(\frac{1}{L})$ and all $1 \leq i \leq n-1$, yields

$$na_n^2 - 2na_n\left(\frac{1}{\sqrt{n}} + \frac{1}{L}\right) = O\left(\frac{\sqrt{n}}{L} + \frac{1}{L^2}\right).$$

Since $-\frac{1}{\sqrt{n}} + a_n \leq 0$ and $L \geq n$, it is easy to show from here that $|a_n| = O(\frac{1}{L})$. It follows that $|a_i| = O(\frac{1}{L})$ for all $1 \leq i \leq n$.

Now consider

$$\|M_n v\| = \left| \sum_{i=1}^{n-1} \left(\frac{1}{\sqrt{n}} + a_i\right)\xi_i + \left(-\frac{1}{\sqrt{n}} + a_n\right)\xi_n \right|.$$

Since n is even, with probability $\Theta(\frac{1}{\sqrt{n}})$, $\xi_1 + \cdots + \xi_{n-1} - \xi_n = 0$, and in this case

$$\|M_n v\| = \left| \sum_{i=1}^{n} a_i \xi_i \right| = O\left(\frac{n}{L}\right),$$

as desired.

6   Controlled Moment

It is convenient to establish some more quantitative control on x. We recall the following notion from [29].

Definition 1 (Controlled second moment). *Let $\kappa \geq 1$. A complex random variable x is said to have κ-controlled second moment if one has the upper bound*

$$\mathbf{E}|x|^2 \leq \kappa$$

(in particular, $|\mathbf{E}x| \leq \kappa^{1/2}$), and the lower bound

$$\mathbf{E}\mathrm{Re}(zx - w)^2 \mathbf{I}(|x| \leq \kappa) \geq \frac{1}{\kappa}\mathrm{Re}(z)^2 \qquad (8)$$

for all complex numbers z, w.

Example. The Bernoulli random variable ($\mathbf{P}(x = +1) = \mathbf{P}(x = -1) = 1/2$) has 1-controlled second moment. The condition (8) asserts in particular that x has variance at least $\frac{1}{\kappa}$, but also asserts that a significant portion of this variance occurs inside the event $|x| \leq \kappa$, and also contains some more technical phase information about the covariance matrix of $\mathrm{Re}(x)$ and $\mathrm{Im}(x)$.

The following lemma was established in [29]:

Lemma 1. *[29, Lemma 2.4] Let x be a complex random variable with finite non-zero variance. Then there exists a phase $e^{i\theta}$ and a $\kappa \geq 1$ such that $e^{i\theta}x$ has κ-controlled second moment.*

Since rotation by a phase does not affect the conclusion of Theorem 6, we conclude that we can assume without loss of generality that x is κ-controlled for some κ. This will allow us to invoke several estimates from [29] (e.g. Lemma 3 and Theorem 9 below).

Remark 1. The estimates we obtain for Theorem 6 will depend on κ but will not otherwise depend on the precise distribution of x. It is in fact quite likely that the results in this paper can be generalised to random matrices N_n whose entries are independent and are all κ-controlled for a single κ, but do not need to be identical. In order to simplify the exposition, however, we focus on the iid case.

7   Small Ball Bounds

In this section we give some bounds on the small ball probabilities $\mathbf{P}(|\xi_1 v_1 + \cdots + \xi_n v_n - z| \leq \varepsilon)$ under various assumptions on the random variables ξ_i and the coefficients v_i. As a consequence we shall be able to obtain good bounds on the probability that Av is small, where A is a random matrix and v is a fixed unit vector.

We first recall a standard bound (cf. [29, Lemmas 4.2, 4.3, 5.2]):

Lemma 2 (Fourier-analytic bound). *Let $\xi_1, \ldots, \xi_n$ be independent variables. Then we have the bound*

$$\mathbf{P}(|\xi_1 v_1 + \cdots + \xi_n v_n - z| \leq r) \ll r^2 \int_{w \in \mathbf{C}: |w| \leq 1/r} \exp\left(-\Theta\left(\sum_{j=1}^{n} \|w v_j\|_j^2\right)\right) dw$$

for any $r > 0$ and $z \in \mathbf{C}$, and any unit vector $v = (v_1, \ldots, v_n)$, where

$$\|z\|_j := (\mathbf{E}\|\mathrm{Re}(z(\xi_j - \xi_j'))\|_{\mathbf{R}/\mathbf{Z}}^2)^{1/2}, \tag{9}$$

ξ_j' is an independent copy of ξ_j, and $\|x\|_{\mathbf{R}/\mathbf{Z}}$ denotes the distance from x to the nearest integer.

Proof. By the Esséen concentration inequality (see e.g. [31, Lemma 7.17]), we have

$$\mathbf{P}(|\xi_1 v_1 + \cdots + \xi_n v_n - z| \leq r) \ll r^2 \int_{w \in \mathbf{C}: |w| \leq 1/r} |\mathbf{E}(e(\mathrm{Re}(w(\xi_1 v_1 + \cdots + \xi_n v_n))))| \, dw$$

for any $c > 0$, where $e(x) := e^{2\pi i x}$. We can write the right-hand side as

$$r^2 \int_{w \in \mathbf{C}: |w| \leq 1/r} \prod_{j=1}^{n} f_j(w v_j)^{1/2} \, dw$$

where

$$f_j(z) := |\mathbf{E}(e(\mathrm{Re}(\xi_j z)))|^2 = \mathbf{E} \cos(2\pi \mathrm{Re}(z(\xi_j - \xi_j'))).$$

Using the elementary bound $\cos(2\pi\theta) \leq 1 - \Theta(\|\theta\|_{\mathbf{R}/\mathbf{Z}}^2)$ we conclude

$$f_j(z) \leq 1 - \Theta(\|z\|_j^2) \leq \exp(-\Theta(\|z\|_j^2))$$

and the claim follows.

Next, we recall some properties of the norms $\|z\|_j$ in the case when ξ_j is κ-controlled.

Lemma 3. *Let $1 \leq j \leq n$, let ξ_j be a random variable, and let $\|\ \|_j$ be defined by (9).*

(i) *For any $w \in \mathbf{C}$, $0 \leq \|w\|_j \leq 1$ and $\| - w\|_j = \|w\|_j$.*
(ii) *For any $z, w \in \mathbf{C}$, $\|z + w\|_j \leq \|z\|_j + \|w\|_j$.*
(iii) *If ξ_j is κ-controlled for some fixed κ, then for any sufficiently small positive constants $c_0, c_1 > 0$ we have $\|z\|_j \geq c_1 \mathrm{Re}(z)$ whenever $|z| \leq c_0$.*

Proof. See [29, Lemma 5.3].

We now use these bounds to estimate small ball probabilities. We begin with a crude bound.

Corollary 6. *Let $\xi_1, \ldots, \xi_n$ be independent variables which are κ-controlled. Then there exists a constant $c > 0$ such that*

$$\mathbf{P}(|\xi_1 v_1 + \cdots + \xi_n v_n - z| \leq c) \leq 1 - c \tag{10}$$

for all $z \in \mathbf{C}$ and all unit vectors $(v_1, \ldots, v_n)$.

Proof. Let $c > 0$ be a small number to be chosen later. We divide into two cases, depending on whether all the v_i are bounded in magnitude by $\sqrt{c}$ or not.

Suppose first that $|v_i| \leq \sqrt{c}$ for all c. Then we apply Lemma 2 (with $r := c^{1/4}$) and bound the left-hand side of (10) by

$$\ll c^{1/2} \int_{w \in \mathbf{C}: |w| \leq c^{-1/4}} \exp\left(-\Theta\left(\sum_{j=1}^{n} \|wv_j\|_j^2\right)\right) dw.$$

By Lemma 3, if c is sufficiently small then we have $\|wv_j\|_j \geq c_1 \mathrm{Re}(wv_j)$, for some positive constant c_1. Writing each v_j in polar coordinates as $v_j = r_j e^{2\pi i \theta_j}$, we thus obtain an upper bound of

$$\ll c^{1/2} \int_{w \in \mathbf{C}: |w| \leq c^{-1/4}} \exp\left(-\Theta\left(\sum_{j=1}^{n} r_j^2 \mathrm{Re}(e^{2\pi i \theta_j} w)^2\right)\right) dw.$$

Since $\sum_{j=1}^{n} r_j^2 = 1$, we can use Hölder's inequality (or Jensen's inequality) and bound this from above by

$$\ll \sup_j c^{1/2} \int_{w \in \mathbf{C}: |w| \leq c^{-1/4}} \exp\left(-\Theta(\mathrm{Re}(e^{2\pi i \theta_j} w)^2)\right) dw$$

which by rotation invariance and scaling is equal to

$$\int_{w \in \mathbf{C}: |w| \leq 1} \exp\left(-\Theta(c^{-1/4} \mathrm{Re}(w)^2)\right) dw.$$

From the monotone convergence theorem (or direct computation) we see that this quantity is less than $1 - c$ if c is chosen sufficiently small. (If necessary, we allow c to depend on the hidden constant in Θ.)

Now suppose instead that $|v_1| > \sqrt{c}$ (say). Then by freezing all of the variables $\xi_2, \ldots, \xi_n$, we can bound the left-hand side of (10) by

$$\sup_w \mathbf{P}(|\xi_1 - w| \leq \sqrt{c}).$$

But by the definition of κ-control, one easily sees that this quantity is bounded by $1 - c$ if c is sufficiently small (compared to $1/\kappa$), and the claim follows.

As a consequence of this bound, we obtain

Theorem 8. *Let N_n be an $n \times n$ random matrix whose entries are independent random variables which are all κ-controlled for some constant $\kappa > 0$. Then there are positive constants c, c' such that the following holds. For any unit vector v and any deterministic matrix M,*

$$\mathbf{P}(\|(M + N_n)v\| \leq cn^{1/2}) \leq \exp(-c'n).$$

Proof. Let c be a sufficiently small constant, and let $X_1, \ldots, X_n$ denote the rows of $M + N_n$. If $\|(M + N_n)v\| \leq cn^{1/2}$, then we have $|\langle X_j, v \rangle| \leq c$ for at least $(1 - c)n$ rows. As the events $\mathbf{I}_j := |\langle X_j, v \rangle| \leq c$ are independent, we see from the Chernoff inequality (applied to the sum $\sum_j \mathbf{I}_j$ of indicator variables) that it suffices to show that

$$\mathbf{E}(I_j) = \mathbf{P}(|\langle X_j, v \rangle| \leq c) \leq 1 - 2c$$

(say) for all j. But this follows from Corollary 6 (after adjusting c slightly), noting that each X_j is a translate (by a row of M) of a vector whose entries are iid copies of x.

Now we obtain some statements of inverse Littlewood-Offord type.

Definition 2 (Compressible and incompressible vectors). *For any $a, b > 0$, let $\text{Comp}(a, b)$ be the set of unit vectors v such that there is a vector v' with at most an non-zero coordinates satisfying $\|v - v'\| \leq b$. We denote by $\text{Incomp}(a, b)$ the set of unit vectors which do not lie in $\text{Comp}(a, b)$.*

Definition 3 (Rich vectors). *For any $\varepsilon, \rho > 0$, let $S_{\varepsilon, \rho}$ be the set of unit vectors v satisfying*

$$\sup_{z \in \mathbf{C}} \mathbf{P}(|X \cdot v - z| \leq \varepsilon) \geq \rho,$$

where $X = (x_1, \ldots, x_n)$ is a vector whose coefficients are iid copies of x.

Lemma 4 (Very rich vectors are compressible). *For any $\varepsilon, \rho > 0$ we have*

$$S_{\varepsilon, \rho} \subset \text{Comp}\left(O(\frac{1}{n\rho^2}), O(\frac{\varepsilon}{\rho})\right).$$

Proof. We can assume $\rho \gg n^{-1/2}$ since the claim is trivial otherwise. Let $v \in S_{\varepsilon, \rho}$, thus

$$\mathbf{P}(|X \cdot v - z| \leq \varepsilon) \geq \rho$$

for some z. From Lemma 2 we conclude

$$\varepsilon^2 \int_{w \in \mathbf{C}: |w| \leq \varepsilon^{-1}} \exp(-\Theta(\sum_{j=1}^{n} \|wv_j\|_j^2)) \, dw \gg \rho. \tag{11}$$

Let $s > 0$ be a small constant (independent of n) to be chosen later, and let A denote the set of indices i for which $|v_i| \geq s\varepsilon$. Then from (11) we have

$$\varepsilon^2 \int_{w \in \mathbf{C}: |w| \leq \varepsilon^{-1}} \exp(-\Theta(\sum_{j \in A} \|wv_j\|_j^2)) \, dw \gg \rho.$$

Suppose A is non-empty. Applying Hölder's inequality, we conclude that

$$\varepsilon^2 \int_{w \in \mathbf{C}: |w| \leq \varepsilon^{-1}} \exp(-\Theta(|A| \|wv_j\|_j^2)) \, dw \gg \rho$$

for some $j \in A$. By the pigeonhole principle, this implies that

$$|\{w \in \mathbf{C} : |w| \leq \varepsilon^{-1}, |A| \| wv_j \|_j^2 \leq k\}| \gg k^{1/2} \varepsilon^{-2} \rho \tag{12}$$

for some integer $k \geq 1$.

If $|A| \ll k$, then the set in (12) has measure $\Theta(\varepsilon^{-2})$, which forces $|A| \ll \rho^{-2}$. Suppose instead that $k \leq s|A|$ for some small $s' > 0$. Since $|v_j| \geq s\epsilon$, we have $s'/|v_j| \leq s'/s\epsilon$. We will choose s' sufficiently small to make sure that this ratio is smaller than the constant c_0 in Lemma 3. By Lemma 3, we see that the intersection of the set in (12) with any ball of radius $s'/|v_j|$ has density at most $\sqrt{k/|A|}$, and so by covering arguments we can bound the left-hand side of (12) from above by $\ll k^{1/2}|A|^{-1/2}\varepsilon^{-2}$. Thus we have $|A| \ll \rho^{-2}$ in this case also. Thus we have shown in fact that $|A| \ll \rho^{-2}$ in all cases (the case when A is empty being trivial).

Now we consider the contribution of those j outside of A. From (11) and Lemma 3 we have

$$\varepsilon^2 \int_{w \in \mathbf{C} : |w| \leq \varepsilon^{-1}} \exp(-\Theta(\sum_{j \notin A} \mathrm{Re}(wv_j)^2))\, dw \gg \rho.$$

Suppose that A is not all of $\{1, \ldots, n\}$. Using polar coordinates $v_j = r_j e^{2\pi i\theta_j}$ as before, we see from Hölder's inequality that

$$\varepsilon^2 \int_{w \in \mathbf{C} : |w| \leq \varepsilon^{-1}} \exp(-\Theta(r^2 \mathrm{Re}(we^{2\pi i\theta_j})^2))\, dw \gg \rho$$

for some $j \notin A$, where $r^2 := \sum_{j \notin A} r_j^2$. After scaling and rotation invariance, we conclude

$$\int_{w \in \mathbf{C} : |w| \leq 1} \exp(-\Theta(\frac{r^2}{\varepsilon^2} \mathrm{Re}(w)^2))\, dw \gg \rho.$$

The left-hand side can be computed to be at most $O(\varepsilon/r)$. We conclude that $r \ll \varepsilon/\rho$. If we let v' be the restriction of v to A, we thus have $\|v - v'\| \ll \varepsilon/\rho$, and the claim $v \in \mathrm{Comp}(O(\frac{1}{n\rho^2}), O(\frac{\varepsilon}{\rho}))$ follows. (The case when $A = \{1, \ldots, n\}$ is of course trivial.)

Roughly speaking, Lemma 4 gives a complete characterization of vectors v such that

$$\sup_{z \in \mathbf{C}} \mathbf{P}(|X \cdot v - z| \leq \varepsilon) \geq \rho,$$

where $\rho > Cn^{-1/2}$, for some large constant C. The lemma shows that such a vector v can be approximated by a vector v' with at most $\frac{C'}{\rho^2}$ non-zero coordinates such that $\|v - v'\| \leq \frac{C''\epsilon}{\rho}$, where C', C'' are positive constants.

The dependence of parameters here are sharp, up to constant terms. Indeed, in the Bernoulli case, the vector $v = (1, \ldots, 1, 0, \ldots, 0)$ consisting of k 1s lies in $S_{0, \Theta(1/\sqrt{k})}$ and lies in $\mathrm{Comp}(a, 0)$ precisely when $an \geq k$ (cf. [7]). This shows that the $O(\frac{1}{n\rho^2})$ term on the right-hand side cannot be improved. On the other hand, in the Gaussian case, observe that if $\|v\| \le b$ then $X \cdot v$ will have magnitude $O(\varepsilon)$ with probability $O(\varepsilon/b)$, which shows that the term $O(\frac{\varepsilon}{\rho})$ cannot be improved.

Lemma 4 is only non-trivial in the case $\rho \ge Cn^{-1/2}$, for some large constant C. To handle the case of smaller ρ, we use the following more difficult entropy bound from [29].

Theorem 9 (Entropy of rich vectors). *For any ε, ρ, there is a finite set $S'_{\varepsilon,\rho}$ of size at most $n^{-(1/2-o(1))n}\rho^{-n} + \exp(o(n))$ such that for each $v \in S_{\varepsilon,\rho}$, there is $v' \in S'_{\varepsilon,\rho}$ such that $\|v - v'\|_\infty \le \varepsilon$.*

Proof. See [29, Theorem 3.2]. ∎

8   Proof of Theorem 6: Preliminary Reductions

We now begin the proof of Theorem 6. Let N_n, M, γ, A be as in that theorem. As remarked in Section 6, we may assume x to be κ-controlled for some κ. We allow all implied constants to depend on κ, γ, A. We may of course assume that n is large compared to these parameters. We may also assume that

$$\mathbf{P}(\|N_n\| \ge n^\gamma) \le \frac{1}{2} \qquad (13)$$

since the claim is trivial otherwise. By decreasing A if necessary, we may furthermore assume that

$$\mathbf{P}(\|N_n\| \ge n^\gamma) \le n^{-A+o(1)}. \qquad (14)$$

It will then suffice to show (assuming (13), (14)) that

$$\mathbf{P}(s_n(M_n) \le n^{-(2A+1)\gamma}) \ll n^{-A+\alpha+o(1)}$$

for any constant $\alpha > 0$ (with the implied constants now depending on α also), since the claim then follows by sending α to zero very slowly in n.

Fix α, and allow all implied constants to depend on α. By perturbing A and α slightly we may assume that A is not a half-integer; we can also take α to be small depending on A. For example, we can assume that

$$\alpha < \{2A\}/2 \qquad (15)$$

where $\{2A\}$ is the fractional part of $2A$.

Using the trivial bound $\|N_n\| \ge \sup_{1 \le i,j \le n} |x_{ij}|$, we conclude from (13), (14) that

$$\mathbf{P}(|x_{ij}| \ge n^\gamma \text{ for some } i, j) \le \min(\frac{1}{2}, n^{-A+o(1)}).$$

Since x_{ij} are iid copies of x, the n^2 events $|x_{ij}| \ge n^\gamma$ are independent with identical probability. It follows that

$$\mathbf{P}(|x| \ge n^\gamma) \le n^{-A-2+o(1)}. \qquad (16)$$

Let F be the event that $s_n(M_n) \leq n^{-(2A+1)\gamma}$, and let G be the event that $\|N_n\| \leq n^\gamma$. In view of (14), it suffices to show that

$$\mathbf{P}(F \wedge G) \leq n^{-A+\alpha+o(1)}.$$

Set

$$b := \beta n^{1/2-\gamma} \tag{17}$$

and

$$a := \frac{\beta}{\log n}, \tag{18}$$

where β is a small positive constant to be chosen later. We then introduce the following events:

- F_{Comp} is the event that $\|M_n v\| \leq n^{-(2A+1)\gamma}$ for some $v \in \text{Comp}(a, b)$.
- F_{Incomp} is the event that $\|M_n v\| \leq n^{-(2A+1)\gamma}$ for some $v \in \text{Incomp}(a, b)$.

Observe that if F holds, then at least one of F_{Comp} and F_{Incomp} holds. Theorem 6 then follows immediately from the following two lemmas.

Lemma 5 (Compressible vector bound). *If β is sufficiently small, then*

$$\mathbf{P}(F_{\text{Comp}} \wedge G) \leq \exp(-\Omega(n)).$$

Lemma 6 (Incompressible vector bound). *We have*

$$\mathbf{P}(F_{\text{Incomp}} \wedge G) \leq n^{-A+o(1)}.$$

In these lemmas we allow the implied constants to depend on β.

The proof of Lemma 5 is simple and will be presented in the next section. The proof of Lemma 6 is somewhat more involved and occupies the rest of the paper.

9   Treatment of Compressible Vectors

If $F_{\text{Comp}} \wedge G$ occurs, then by the definition of $\text{Comp}(a, b)$, there are unit vectors v, v' such that $\|M_n v\| \leq n^{-(2A+1)\gamma}$ and v' has support on at most an coordinates and $\|v - v'\| \leq b$.

By the triangle inequality and (17) we have

$$\|M_n v'\| \leq n^{-(2A+1)\gamma} + \|M_n\|\|v - v'\|$$
$$\leq n^{-(2A+1)\gamma} + n^\gamma b$$
$$\leq 2\beta n^{1/2}.$$

A set $\mathcal{N}$ of unit vectors in $\mathbf{C}^m$ is called a *δ-net* if for any unit vector v, there is a vector w in $\mathcal{N}$ such that $\|v - w\| \leq \delta$. It is well known that for any $0 < \delta < 1$, a δ-net of size $(C\delta^{-1})^m$ exists, for some constant C independent of δ and m.

Using this fact, we conclude that the set of unit vectors with at most an non-zero coordinates admits an b-net $\mathcal{N}$ of size at most

$$|\mathcal{N}| \leq \binom{n}{an}(Cb^{-1})^{an},$$

Thus, if $F_{\mathrm{Comp}} \wedge G$ occurs, then there is a unit vector $v'' \in \mathcal{N}$ such that

$$\|M_n v''\| \leq 2\beta n^{1/2} + \|M_n\| b = 3\beta n^{1/2}.$$

On the other hand, from Theorem 8 we see (for $\beta \leq c/3$) that for any fixed v'',

$$\mathbf{P}(\|M_n v''\| \leq 3\beta n^{1/2}) \leq \exp(-c'n),$$

where c and c' are the constants in Theorem 8.

By the union bound, we conclude

$$\mathbf{P}(F_{\mathrm{Comp}} \wedge G) \leq \binom{n}{an}(b^{-1})^{an}\exp(-c'n).$$

But from (17), (18) we see that the right-hand side can be made less than $\exp(-c'n/2)$, given that β is sufficiently small. This concludes the proof of Lemma 5.

10 Treatment of Incompressible Vectors

We now begin the proof of Lemma 6. We now fix β and allow all implied constants to depend on β.

Let X_k be the kth row vector of M_n, and let dist_k be the distance from X_k to the subspace spanned by $X_1, \ldots, X_{k-1}, X_{k+1}, \ldots, X_n$. We need the following, which is a slight extension of a lemma from [20].

Lemma 7. *For any $\varepsilon > 0$, and any event E, we have*

$$\mathbf{P}(\{\|Mv\| \leq \varepsilon b n^{-1/2} \text{ for some } v \in \mathrm{Incomp}(a,b)\} \wedge E) \leq \frac{1}{an}\sum_{k=1}^{n}\mathbf{P}(\{\mathrm{dist}_k \leq \varepsilon\} \wedge E).$$

Proof. See [20, Lemma 3.5]. The arbitrary event E was not present in that lemma, but one easily verifies that the proof works perfectly well with this event in place.

Applying this to our current situation with

$$\varepsilon := \frac{1}{\beta}n^{-2A\gamma}, \tag{19}$$

we obtain

$$\mathbf{P}(F_{\mathrm{Incomp}} \wedge G) \ll \frac{\log n}{n}\sum_{k=1}^{n}\mathbf{P}(\{\mathrm{dist}_k \leq \varepsilon\} \wedge G).$$

To prove Lemma 6, it therefore suffices (by symmetry) to show that

$$\mathbf{P}(\{\mathrm{dist}_n \leq \varepsilon\} \wedge G) \ll n^{-A+\alpha+o(1)}.$$

Notice that there is a unit vector X_n^* orthogonal to $X_1, \ldots, X_{n-1}$ such that

$$\mathrm{dist}_k = |X_n \cdot X_n^*|. \tag{20}$$

If there are many such X_n^*, choose one arbitrarily. However, note that we can choose X_n^* to depend only on $X_1, \ldots, X_{n-1}$ and thus be independent of X_n.

Let $\rho := n^{-A+\alpha}$. Let X be the random vector of length n whose coordinates are iid copies of x. From Definition 3 (and the observation that X_n has the same distribution as X after translating by a deterministic vector (namely the nth row of the deterministic matrix M), we have the conditional probability bound

$$\mathbf{P}(\mathrm{dist}_n \leq \varepsilon | X_n^* \notin S_{\varepsilon, \rho}) \leq \rho = n^{-A+\alpha}.$$

Thus it will suffice to establish the exponential bound

$$\mathbf{P}(\{X_n^* \in S_{\varepsilon, \rho}\} \wedge G) \leq \exp(-\Omega(n)).$$

Let

$$J := \lfloor 2A \rfloor \tag{21}$$

be the integer part of $2A$. Let $\alpha_1 > 0$ be a sufficiently small constant (independent of n and γ, but depending on α, A, J) to be chosen later. Set

$$\varepsilon_j := n^{(\gamma+\alpha_1)j}\varepsilon = \frac{1}{\beta}n^{(\gamma+\alpha_1)j}n^{-2A\gamma} \tag{22}$$

and

$$\rho_j := n^{(1/2-\alpha_1)j}\rho = n^{(1/2-\alpha_1)j}n^{-A+\alpha} \tag{23}$$

for all $0 \leq j \leq J$.

By the union bound, it will suffice to prove the following lemmas.

Lemma 8. *If α_1 is sufficiently small, then for any $0 \leq j < J$, we have*

$$\mathbf{P}(\{X_n^* \in S_{\varepsilon_j, \rho_j}\} \wedge \{X_n^* \notin S_{\varepsilon_{j+1}, \rho_{j+1}}\} \wedge G) \leq \exp(-\Omega(n)). \tag{24}$$

Lemma 9. *If α_1 is sufficiently small, then we have*

$$\mathbf{P}(X_n^* \in S_{\varepsilon_J, \rho_J}) \leq \exp(-\Omega(n)).$$

11   Proof of Lemma 8

Fix $0 \leq j < J$. Note that by (15), we have

$$\rho_j \leq n^{(J-1)/2}n^{-A+\alpha} \leq n^{-1/2-\{2A\}/2+\alpha} \leq n^{-1/2}.$$

We can then use Theorem 9 to conclude the existence of a set $\mathcal{N}$ of unit vectors such that every vector in S_{ε_j,ρ_j} lies within ε_j in l^∞ norm to a vector in $\mathcal{N}$, and with the cardinality bound

$$|\mathcal{N}| \le n^{-(1/2-o(1))n}\rho_j^{-n}. \tag{25}$$

Suppose that the event in Lemma 8 holds, then we can find $u \in \mathcal{N}$ such that $\|u - X_n^*\|_{l^\infty} \le \varepsilon_j$, and thus $\|u - X_n^*\| \le n^{1/2}\varepsilon_j$. On the other hand, since X_n^* is orthogonal to $X_1, \ldots, X_{n-1}$ and $\|M_n\| \ll n^\gamma$, we have

$$
\begin{aligned}
\Big(\sum_{i=1}^{n-1} |X_i \cdot u|^2\Big)^{1/2} &= \Big(\sum_{i=1}^{n-1} |X_i \cdot (u - X_n^*)|^2\Big)^{1/2} \\
&= \|M(u - X_n^*)\| \\
&\ll n^\gamma n^{1/2}\varepsilon_j \\
&\ll n^{1/2}n^{-\alpha_1}\varepsilon_{j+1}.
\end{aligned}
$$

On the other hand, from (24) and Definition 3 we have

$$\mathbf{P}(|X \cdot X_n^* - z| \le \varepsilon_{j+1}) \le \rho_{j+1} \tag{26}$$

for all $z \in \mathbf{C}$, where $X = (x_1, \ldots, x_n)$ consists of iid copies of x.

To conclude the proof, we will need the following lemma.

Lemma 10. *If w is any vector with $\|w\|_{l^\infty} \le 1$, then*

$$\mathbf{P}(|X \cdot w| \ge n^{\gamma+\alpha_1}) \ll n^{-A}.$$

Proof. Write $w = (w_1, \ldots, w_n)$ and $X = (x_1, \ldots, x_n)$. Observe from (14) that with probability $O(n^{-A-1}) = O(n^{-A})$, all the coefficients in X are going to be of magnitude at most n^γ. Thus it suffices to show that

$$\mathbf{P}(|w_1\tilde{x}_1 + \ldots + w_n\tilde{x}_n| \ge n^{\gamma+\alpha_1}) \ll n^{-A}$$

where $\tilde{x}_1, \ldots, \tilde{x}_n$ are iid with law equal to that of x conditioned to the event $|x| \ll n^\gamma$. As x has mean zero and bounded second moment, one verifies from (14) and Cauchy-Schwarz that the mean of the $\tilde{x}_i$ is $O(n^{-(A+2)/2})$. Thus if we let $x_i' := \tilde{x}_i - \mathbf{E}(\tilde{x}_i)$, we see that it suffices to show that

$$\mathbf{P}(|w_1x_1' + \ldots + w_nx_n'| \ge \frac{1}{2}n^{\gamma+\alpha_1}) \ll n^{-A}.$$

We conclude the proof by the moment method, using the following estimate

$$\mathbf{E}(|w_1x_1' + \ldots + w_nx_n'|^{2k}) \ll_k n^{2k\gamma}$$

for any integer $k \ge 0$. This is easily verified by a standard computation (using the hypothesis $\gamma \ge 1/2$), since all the x_i' have vanishing first moment, a second moment of $O(1)$, and a jth moment of $O_j(n^{(j-2)\gamma})$ for any $j > 2$. Now take k to be a constant sufficiently large compared to A/α_1.

We are now ready to finish the proof of Lemma 8. From lemma 10 and the bound $\|u - X_n^*\| \leq \varepsilon_j$ we see that

$$\mathbf{P}(|X \cdot (X_n^* - u)| \geq \varepsilon_{j+1}) \leq n^{-A} \leq \rho_{j+1};$$

combining this with (26) using the triangle inequality, we see that

$$\sup_{z \in \mathbf{C}} \mathbf{P}(|X \cdot u - z| \leq \varepsilon_{j+1}) \ll \rho_{j+1}. \tag{27}$$

We can therefore bound the left-hand side of (24) by

$$\sum_{u \in \mathcal{N}:(27) \text{ holds}} \mathbf{P}\left((\sum_{i=1}^{n-1} |X_i \cdot u|^2)^{1/2} \ll n^{1/2} n^{-\alpha_1} \varepsilon_{j+1}\right).$$

Now suppose that $u \in \mathcal{N}$ obeys (27). If we have $\sum_{i=1}^{n-1} |X_i \cdot u|^2)^{1/2} \ll n^{1/2} n^{-\alpha_1} \varepsilon_{j+1}$, then the event $|X_i \cdot u| \leq \varepsilon_{j+1}$ must hold for at least $n - O(n^{1-2\alpha_1})$ values of i. On the other hand, from (27) we see that each of these events $|X_i \cdot u| \leq \varepsilon_{j+1}$ only occurs with probability $O(\rho_{j+1})$. We can thus bound

$$\mathbf{P}(\sum_{i=1}^{n-1} |X_i \cdot u|^2)^{1/2} \ll n^{1/2} n^{-\alpha_1} \varepsilon_{j+1}) \leq \binom{n}{n - O(n^{1-2\alpha_1})} (O(\rho_{j+1}))^{n - O(n^{1-2\alpha_1})}$$

$$\ll n^{o(n)} \rho_{j+1}^n.$$

Applying (25), we can thus bound the left-hand side of (24) by

$$\ll n^{-(1/2 - o(1))n} \rho_j^{-n} \rho_{j+1}^n = n^{-(\alpha_1 - o(1))n}$$

and the claim follows.

12   Proof of Lemma 9

Suppose that X_n^* lies in $S_{\varepsilon_J, \rho_J}$. Then by Lemma 4, we have

$$X_n^* \subset \mathrm{Comp}(O(\frac{1}{n \rho_J^2}), O(\frac{\varepsilon_J}{\rho_J})).$$

Note from (23) and (21) that

$$\frac{1}{n \rho_J^2} = n^{2A - J - 1 + 2\alpha_1 J - 2\alpha} \leq n^{-\alpha_1}$$

if α_1 is sufficiently small. Thus, by arguing as in Section 9, the set $\mathrm{Comp}(O(\frac{1}{n\rho_J^2}), O(\frac{\varepsilon_J}{\rho_J}))$ has a $O(\frac{\varepsilon_J}{\rho_J})$-net $\mathcal{N}$ in l^2 of cardinality

$$|\mathcal{N}| \ll \binom{n}{\frac{1}{n\rho_J^2}} (O(\frac{\varepsilon_J}{\rho_J}))^{\frac{1}{n\rho_J^2}} = \exp(o(n)).$$

If we let $u \in \mathcal{N}$ be within $O(\frac{\varepsilon_J}{\rho_J})$ of X_n^*, then we have $|X_i \cdot u| \ll \frac{\varepsilon_J}{\rho_J}$ for all $1 \le i \le n - 1$. Thus we can bound

$$\mathbf{P}(X_n^* \in S_{\varepsilon_J, \rho_J}) \le \sum_{u \in \mathcal{N}} \mathbf{P}(|X_i \cdot u| \ll \frac{\varepsilon_J}{\rho_J} \text{ for all } 1 \le i \le n - 1).$$

Now observe from (22), (23), (21) and the hypothesis $\gamma \ge 1/2$ that

$$\frac{\varepsilon_J}{\rho_J} = n^{-\alpha + 2\alpha_1 J} n^{-(2A-J)(\gamma - 1/2)} \le n^{-\alpha/2}$$

(say) if α_1 is sufficiently small. Thus by Corollary 6 (or by a minor modification of Theorem 8) we see that

$$\mathbf{P}(|X_i \cdot u| \ll \frac{\varepsilon_J}{\rho_J} \text{ for all } 1 \le i \le n - 1) \ll \exp(-\Omega(n))$$

for each $u \in \mathcal{N}$, and the claim follows.

References

1. Bai, Z., Silverstein, J.: Spectral analysis of large dimensional random matrices. Science Press (2007)
2. Bau, D., Trefethen, L.N.: Numerical linear algebra. SIAM, Philadelphia (1997)
3. Burgisser, P., Cucker, F., Lotz, M.: The probability that a slightly perturbed numberical analysis problem is difficult. Math of Computation 77, 1559–1583 (2008)
4. Burgisser, P., Cucker, F., Lotz, M.: General formulas for the smooth analysis of condition numbers. C. R. Acad. Sc. Paris 343, 145–150 (2006)
5. Burgisser, P., Cucker, F., Lotz, M.: Smooth analysis of connic condition numbers. J. Math. Pure et Appl. 86, 293–309 (2006)
6. Edelman, A.: Eigenvalues and condition numbers of random matrices. SIAM J. Matrix Anal. Appl. 9(4), 543–560 (1988)
7. Erdös, P.: On a lemma of Littlewood and Offord. Bull. Amer. Math. Soc. 51, 898–902 (1945)
8. Demmel, J.: The probability that a numberical analysis problem is difficult. Math. Comp. 50, 449–480 (1988)
9. Dunagan, J., Spielman, D.A., Teng, S.H.: Smoothed Analysis of the Renegar's Condition Number for Linear Programming (preprint)
10. Golub, G., Van Loan, C.: Matrix computations, 3rd edn. Johns Hopkins Press (1996)
11. Götze, F., Tikhomirov, A.: The Circular Law for Random Matrices (preprint)
12. Füredi, Z., Komlós, J.: The eigenvalues of random symmetric matrices. Combinatorica 1(3), 233–241 (1981)
13. Komlós, J.: On the determinant of $(0, 1)$ matrices. Studia Sci. Math. Hungar. 2, 7–22 (1967)
14. Latala, R.: Some estimates of norms of random matrices. Proc. Amer. Math. Soc. 133, 1273–1282 (2005)
15. Litvak, A., Pajor, A., Rudelson, M., Tomczak-Jaegermann, N.: Smallest singular value of random matrices and geometry of random polytopes. Adv. Math. 195(2), 491–523 (2005)

16. Mehta, M.L.: Random Matrices and the Statistical Theory of Energy Levels. Academic Press, New York (1967)
17. Pan, G., Zhou, W.: Circular law, Extreme Singular values and Potential theory (preprint)
18. Pastur, L.A.: On the spectrum of random matrices. Teoret. Mat. Fiz. 10, 102–112 (1973)
19. Rudelson, M.: Invertibility of random matrices: Norm of the inverse. Annals of Mathematics (to appear)
20. Rudelson, M., Vershynin, R.: The Littlewood-Offord problem and the condition number of random matrices. Adv. Math. (to appear)
21. Rudelson, M., Vershynin, R.: The least singular value of a random rectangular matrix (submitted)
22. Smale, S.: On the efficiency of algorithms of analysis. Bullentin of the AMS (13), 87–121 (1985)
23. Seginer, Y.: The expected norm of random matrices. Combin. Probab. Comput. 9(2), 149–166 (2000)
24. Spielman, D.A., Teng, S.H.: Smoothed analysis of algorithms. In: Proceedings of the International Congress of Mathematicians, Beijing, vol. I, pp. 597–606. Higher Ed. Press, Beijing (2002)
25. Spielman, D.A., Teng, S.H.: Smoothed analysis of algorithms: why the simplex algorithm usually takes polynomial time. J. ACM 51(3), 385–463 (2004)
26. Sankar, A., Teng, S.H., Spielman, D.A.: Smoothed Analysis of the Condition Numbers and Growth Factors of Matrices. SIAM J. Matrix Anal. Appl. 28(2), 446–476 (2006)
27. Tao, T., Vu, V.: Inverse Littlewood-Offord theorems and the condition number of random discrete matrices. Annals of Mathematics (to appear)
28. Tao, T., Vu, V.: The condition number of a randomly perturbed matrix. In: STOC (2007)
29. Tao, T., Vu, V.: Random matrices: The circular law. Communications in Contemporary Mathematics 10, 261–307 (2008)
30. Tao, T., Vu, V.: Random matrices: The distribution of the smallest singular values, to appear in GAFA
31. Tao, T., Vu, V.: Additive Combinatorics. Cambridge University Press, Cambridge (2006)
32. von Neumann, J., Goldstein, H.: Numerical inverting matrices of high order. Bull. Amer. Math. Soc. 53, 1021–1099 (1947)
33. Vu, V.: Spectral norm of random matrices. Combinatorica 27(6), 721–736 (2007)
34. Wigner, P.: On the distribution of the roots of certain symmetric matrices. Annals of Math. 67, 325–327

Erratum: Resource Minimization Job Scheduling

Julia Chuzhoy[1] and Paolo Codenotti[2]

[1] Toyota Technological Institute, Chicago, IL 60637
Supported in part by NSF CAREER award CCF-0844872
cjulia@tti-c.org
[2] Department of Computer Science, University of Chicago, Chicago, IL 60637
paaloc@cs.uchicago.edu

I. Dinur et al. (Eds.): APPROX and RANDOM 2009, LNCS 5687, pp. 70–83, 2009.
Springer-Verlag Berlin Heidelberg 2009

DOI 10.1007/978-3-642-03685-9_54

A claim on page 75 of this paper, stating that $J = J^1 \cup \cdots \cup J^k$ was found to be incorrect, invalidating the main result of the paper. The best current approximation ratio for this problem therefore remains $O(\sqrt{log\ n/log\ log\ n})$. We thank Kirk Pruhs for pointing this error out to us.

The original online version for this chapter can be found at
http://dx.doi.org/10.1007/978-3-642-03685-9_6

Author Index